Antarctic Palaeoenvironments and Earth-Surface Processes

SCAR/GSL publishing agreement

This volume is published under an agreement between the Scientific Committee on Antarctic Research and the Geological Society of London and arises from the 11th International Symposium on Antarctic Earth Sciences (ISAES) held in Edinburgh, Scotland, 10–16 July 2011.

GSL is the publisher of choice for books related to SCAR's geoscience activities, and SCAR receives a fee for all books published under this agreement.

Books published under this agreement are subject to the Society's standard rigorous proposal and manuscript review procedures.

It is recommended that reference to all or part of this book should be made in one of the following ways:

Hambrey, M. J., Barker, P. F., Barrett, P. J., Bowman, V., Davies, B., Smellie, J. L. & Tranter, M. (eds) 2013. *Antarctic Palaeoenvironments and Earth-Surface Processes*. Geological Society, London, Special Publications, **381**.

Foley, D. J., Stump, E., van Soest, M., Whipple, K. X. & Hodges, K. V. 2013. Differential Movement across Byrd Glacier, Antarctica, as indicated by Apatite (U–Th)/He thermochronology and geomorphological analysis. *In:* Hambrey, M. J., Barker, P. F., Barrett, P. J., Bowman, V., Davies, B., Smellie, J. L. & Tranter, M. (eds) *Antarctic Palaeoenvironments and Earth-Surface Processes.* Geological Society, London, Special Publications, **381**, 39–46. First published online August 6, 2013, http://dx.doi.org/10.1144/SP381.25

GEOLOGICAL SOCIETY SPECIAL PUBLICATION NO. 381

Antarctic Palaeoenvironments and Earth-Surface Processes

EDITED BY

M. J. HAMBREY
Aberystwyth University, Wales

P. F. BARKER
British Antarctic Survey, UK

P. J. BARRETT
Victoria University of Wellington, New Zealand

V. BOWMAN
University of Leeds, UK

B. DAVIES
Aberystwyth University, Wales

J. L. SMELLIE
University of Leicester, UK

and

M. TRANTER
University of Bristol, UK

2013
Published by
The Geological Society
London

THE GEOLOGICAL SOCIETY

The Geological Society of London (GSL) was founded in 1807. It is the oldest national geological society in the world and the largest in Europe. It was incorporated under Royal Charter in 1825 and is Registered Charity 210161.

The Society is the UK national learned and professional society for geology with a worldwide Fellowship (FGS) of over 10 000. The Society has the power to confer Chartered status on suitably qualified Fellows, and about 2000 of the Fellowship carry the title (CGeol). Chartered Geologists may also obtain the equivalent European title, European Geologist (EurGeol). One fifth of the Society's fellowship resides outside the UK. To find out more about the Society, log on to www.geolsoc.org.uk.

The Geological Society Publishing House (Bath, UK) produces the Society's international journals and books, and acts as European distributor for selected publications of the American Association of Petroleum Geologists (AAPG), the Indonesian Petroleum Association (IPA), the Geological Society of America (GSA), the Society for Sedimentary Geology (SEPM) and the Geologists' Association (GA). Joint marketing agreements ensure that GSL Fellows may purchase these societies' publications at a discount. The Society's online bookshop (accessible from www.geolsoc.org.uk) offers secure book purchasing with your credit or debit card.

To find out about joining the Society and benefiting from substantial discounts on publications of GSL and other societies worldwide, consult www.geolsoc.org.uk, or contact the Fellowship Department at: The Geological Society, Burlington House, Piccadilly, London W1J 0BG: Tel. +44 (0)20 7434 9944; Fax +44 (0)20 7439 8975; E-mail: enquiries@geolsoc.org.uk.

For information about the Society's meetings, consult *Events* on www.geolsoc.org.uk. To find out more about the Society's Corporate Affiliates Scheme, write to enquiries@geolsoc.org.uk.

Published by The Geological Society from:
The Geological Society Publishing House, Unit 7, Brassmill Enterprise Centre, Brassmill Lane, Bath BA1 3JN, UK

The Lyell Collection: www.lyellcollection.org
Online bookshop: www.geolsoc.org.uk/bookshop
Orders: Tel. +44 (0)1225 445046, Fax +44 (0)1225 442836

British Library Cataloguing in Publication Data

A catalogue record for this book is available from the British Library.
ISBN 978-1-86239-363-9
ISSN 0305-8719

Distributors
For details of international agents and distributors see:
www.geolsoc.org.uk/agentsdistributors

Typeset by Techset Composition India (P) Ltd., Bangalore and Chennai, India.
Printed by Berforts Information Press Ltd, Oxford, UK

Contents

Introduction

A. Palaeozoic and Mesozoic evolution of the Antarctic Continent

B. Cenozoic glaciation and impacts

Peter Frank Barker (1939–2012)

Dr Peter Barker died on 25 June 2012, aged 73, in the midst of editing papers for this volume. During his research career at the University of Birmingham and the British Antarctic Survey he pioneered studies to unravel the tectonic and environmental history of the Southern Ocean. He took over the reins of the Antarctic Marine Group at Birmingham in 1965, when the plate tectonic revolution was still in its infancy, and was quick to recognize the opportunities the new paradigm presented. In addition to leading a series of research cruises on RRS *Shackleton* and RRS *Bransfield* that provided the foundations for understanding of the geological structure and tectonic evolution of the Scotia Sea region, Peter became deeply involved in scientific ocean drilling. During his career he was Co-Chief Scientist on four expeditions of the Deep Sea Drilling Project and Ocean Drilling Program that made fundamental contributions to our knowledge of the tectonic and environmental history of the South Atlantic region and the Southern Ocean. In 1986, Peter was appointed as Head of the British Antarctic Survey Geophysics Division, and moved to Cambridge. During his first few years in Cambridge, he played an important part in specifying the scientific facilities on a new polar research vessel, RRS *James Clark Ross*, on which he went on to lead several cruises. Peter retired in 1999 and held an E-fellowship with British Antarctic Survey until 2005, when he and his wife Jenny moved to rural Shropshire. He remained active as a sea-going scientist in retirement and his final research cruise was as an invited participant in a cruise to the Scotia Sea on a US research vessel in 2008. An international symposium on Scotia Arc geology was held in Spain in May 2013 in Peter's honour.

Robert D. Larter
British Antarctic Survey

Antarctic Palaeoenvironments and Earth-Surface Processes in context

MICHAEL J. HAMBREY* & BETHAN J. DAVIES

Centre for Glaciology and Climate Change Consortium of Wales, Institute of Geography and Earth Sciences, Aberystwyth University, Aberystwyth, Ceredigion SY23 3DB, UK

**Corresponding author (e-mail: mjh@aber.ac.uk)*

Abstract: The 26 papers in this volume comprise those presented at the International Symposium on Antarctic Earth Sciences in Edinburgh in July 2011, as well as several invited review papers, on Antarctic geology and geomorphology. These papers demonstrate a remarkable diversity of Earth science interests in the Antarctic. It is evident that during recent decades there has been a marked change in emphasis from pre-Cenozoic geology to topics related to the 34 million years of directly documented Antarctic glacial history. Increasing emphasis is being placed on the documentation and understanding of the palaeorecord of ice-sheet growth and decay. Furthermore, Antarctic Earth history is providing us with important information about potential future trends, as the impact of global warming is increasingly felt on the continent and its ocean.

Immediately before and during the XIth International Symposium on Antarctic Earth Sciences, held in Edinburgh, Scotland, in July 2011, the Scientific Steering Committee agreed to produce three complementary volumes:

(1) *Antarctica and Supercontinent Evolution*, edited by Simon Harley *et al.*, which covers the early, mainly crustal history of Antarctica (Harley *et al.* in press); and
(2) *Antarctic Palaeoenvironments and Earth-Surface Processes*, which covers Palaeozoic and subsequent geological and palaeoenvironmental history, as well as providing an understanding of modern processes operating in ice-free terrestrial areas (this volume).
(3) *Antarctic Earth Sciences*, a volume of keynote symposium presentations edited by Siegert & Bradwell (in press).

This volume aims to present state-of-the-art accounts in several key fields of study. The following contributions reflect the changed balance of Antarctic Earth science research since the previous international symposium in Santa Barbara, California, in 2007. There has been a strong change of focus from Palaeozoic and Mesozoic terrestrial stratigraphy and structural geology, towards the palaeoenvironmental record of the Cenozoic Era. Interest in the history of the Antarctic Ice Sheet and its potential impact on sea-level and climate has been driven by fears of recession of the enormous mass of ice on the continent. Nevertheless, in the review papers, the earlier geological history is well catered for. For ease of use, the papers in this volume have been grouped under three major themes.

Palaeozoic and Mesozoic evolution of Antarctica

In an invited review paper, David Elliot provides a comprehensive and concise account of the geological and tectonic history of one of the world's great ranges, the Transantarctic Mountains. Although much of them is hidden from view under the present ice sheet, there are plentiful exposures, especially where they flank the Ross Ice Shelf and Ross Sea. The mountains were initiated in Neoproterozoic time with rifting of the supercontinent of Rodinia, followed by intrusion and metamorphism during the Cambrian Ross Orogeny. Subsequent erosion of the orogeny produced a major erosion surface, upon which Devonian to Jurassic sediments accumulated in an intra-cratonic basin, while magmatic arcs developed on the surrounding continental margins. Gondwana break-up in Early Jurassic time was associated with the emplacement of the extensive Ferrar Large Igneous Province, events that marked the reorganization of plates and uplift of the Transantarctic Mountains from Cretaceous time onwards.

A novel approach to deriving the history of faulting, uplift and unroofing of the Transantarctic Mountains is presented by Daniel Foley, Edmund Stump and others. Using a combination of apatite

From: Hambrey, M. J., Barker, P. F., Barrett, P. J., Bowman, V., Davies, B., Smellie, J. L. & Tranter, M. (eds) 2013. *Antarctic Palaeoenvironments and Earth-Surface Processes*. Geological Society, London, Special Publications, **381**, 1–5. First published online September 24, 2013, http://dx.doi.org/10.1144/SP381.27

(U–Th)–He thermochronology and geomorphological analysis from satellite imagery, they show that the denudation rate of the mountains amounted to *c.* 0.04 mm per year from *c.* 140 to 40 Ma. They also infer differential movement of the order of 1 km across an inferred fault beneath one of Antarctica's largest outlet glaciers, the Byrd Glacier. The results support a model of relatively uniform cooling and unroofing of the region, with post-40 Ma fault-displacement that uplifted the south side of the glacier relative to the north.

In another tectonically focused paper, Adolfo Maestro and colleagues present the results of a study of intraplate stress fields under multiple remote stresses on the South Orkney micro-continent. Using a combination of fault, joint and tension gash population analyses in late Triassic to early Jurassic metamorphic rocks on Signy Island, they have determined compressive, strike-slip and extensional stress states. Overall, NNW–SSE horizontal compression was dominant, resulting from an inferred subduction-related tectonic regime, which then changed to north–south extension in Middle Jurassic to Early Cretaceous time. Complex additional stress changes took place during spreading of the Powell Basin from Eocene/Miocene time, and generated extensional structures across the micro-continent.

One of the most important stratigraphic units in Antarctica is the Beacon Supergroup, which is exposed throughout most of the length of the Transantarctic Mountains (separating the East and West Antarctic ice sheets) and in the Ellsworth Mountains in the middle of the West Antarctic Ice Sheet. These strata have a long history of investigation, starting with Scott's first expedition in 1901–1903. In an invited review, Margaret Bradshaw describes the Taylor Group, which is the lower division of the Beacon Supergroup and is of Devonian age. Resting on a major unconformity, the Kukri Erosion Surface, the Taylor Group is a cyclic series of coastal quartzose sandstones deposited in deserts, alluvial plain and shallow marine settings. Early Devonian marine faunas from the Ohio Range reveal links with other Southern Hemisphere continents, whereas mid to late Devonian terrestrial fish faunas from southern Victoria Land have global connections.

Two further papers concern palaeontological studies of specific Antarctic faunas. Firstly, Marcelo A. Reguero and colleagues describe Late Cretaceous terrestrial vertebrates from marine sediments of the James Ross Basin in the NW Weddell Sea. They focus on 10 occurrences of non-avian dinosaurs, notably on Snow Hill Island, and find that there is a Gondwanan 'hallmark' in the fauna that reflects climate-driven provinciality. Avian dinosaurs are also present in the region but are rare. The paper by Thomas Saucede *et al.* is one of the first to investigate quantitatively the palaeobiology of Antarctic echinoids. The paper explores this topic through four time intervals from Late Cretaceous time up to the present day, and provides palaeogeographical maps of the distribution of echinoids, all within the context of the evolution of Antarctic palaeogeography.

Cenozoic environments and glaciation

Cenozoic studies have gained precedence in Earth science research in the last two decades, stimulated by our desire to better understand the impact of the Antarctic cryosphere on global climate and sea-level. The Cenozoic Era embraces a pre-glacial phase, prior to *c.* 34 Ma (the Eocene/Oligocene boundary) followed by a phase of permanent and stable ice sheet-scale glaciations.

For the pre-glacial era, Piotr Jadwiszczak describes the taxonomic diversity and palaeoecology of Eocene Antarctic penguins. He has identified at least 10 species of Eocene penguins preserved in the La Meseta Formation of Seymour Island, NW Weddell Sea. The earliest reported bones of the giant penguin (genus *Anthropornis*) date from Late Eocene times (*c.* 53 Ma); a range of smaller penguin species is also described.

In a different discipline, the central Scotia Sea was the site for the development of seven heat-flow stations in order to determine magnetic anomaly-based ages of the basement, as reported by Peter Barker and co-authors. Although results are ambiguous, they affirm that basin extension began in Eocene time, supporting the concept of a relatively young back-arc origin for the central Scotia Sea, but do not affect prior interpretations concerning the onset of the Antarctic Circumpolar Current.

The glacial record of Antarctica (post *c.* 34 Ma) is represented first by two papers focusing on the terrestrial record. Duanne White and colleagues present evidence from the Prince Charles Mountains for palaeotopographic changes in the Lambert Glacier–Amery Ice Shelf system. These authors identify up to 1–2 km of glacial incision into a pre-glacial palaeosurface. This incision by major Cenozoic outlet glaciers has driven uplift of up to 50 m per million years since mid-Miocene time. This landscape evolution was sufficient to focus and influence Cenozoic ice-drainage patterns.

Brenda Hall and co-workers summarize the behaviour of the Antarctic Ice Sheet during the late Quaternary Period by investigating ice dynamics in the Ross Sea sector during the Last Glacial Maximum. In this invited review, the authors found ice-sheet thicknesses of up to 1000 m along the coastline of the central and southern Transantarctic Mountains and Marie Byrd Land. In that region, the Late Glacial Maximum occurred around 18 000 years ago, but as late as 7000–10 000 years ago further inland. Significant thinning took place between 13 000 and 14 600 years ago, and Hall and colleagues conclude that the Ross Sea sector was unlikely to be a major contributor to a sharp eustatic rise in sea-level around 14 000 years ago, referred to as Meltwater Pulse 1A.

Several papers deal with the glacial record in the offshore realm, combining geophysical and core-sampling of sea-floor sediment, and reflecting the major investment in marine geological research in state-of-the-art ice-breakers by a number of nations. A paper by Kari Strand and colleagues focuses on a well-known core from the Ocean Drilling Program's Leg 188 to Prydz Bay in East Antarctica. This region is second only to the western Ross Sea for studies of the long-term glaciation of Antarctica, and is important because it lies at the mouth of a major conduit of ice from the heart of the continent, the Lambert Graben. By examining the lithofacies in the core using a variety of techniques to define provenance and transport history, the authors conclude that the first evidence of glaciers dates from late Eocene time. This signal is not direct, but preserved in 110 m of alluvial plain sediments that are believed to be glaciofluvial. Full-scale glaciation is indicated by 20 m of overlying glaciomarine sediments, confirming the results from earlier cores in Prydz Bay and elsewhere in Antarctica.

Stephen Pekar and co-workers present a comprehensive investigation of multichannel seismic data, collected offshore from New Harbour in the western Ross Sea, in order to investigate the Cenozoic stratigraphy and tectonic history of a region of extensional basins bordering the Transantarctic Mountains since late Eocene time. The seismic data are tied in to previously drilled sites, so are well constrained by stratigraphy and age-determinations. The work will provide a basis for possible future continental shelf drilling in the region.

In an invited review, John Anderson and colleagues assess marine geological evidence for the behaviour of the West Antarctic, East Antarctic and Antarctic Peninsula ice sheets over the last 30 000 years. These ice sheets had largely receded by early Holocene time, completing their contribution to eustatic sea-level rise. Ice-sheet thicknesses, based on grounding depths, range from 640 to 1640 m on the inner continental shelf. Retreat around the continent was asynchronous and stepwise, with factors specific to particular drainage basins controlling ice-sheet behaviour. The resulting episodic sea-level rise is inferred to come from ice-stream collapse.

Part of the Antarctic continental shelf that has seen little attention until now has been the Amundsen Sea Embayment, which is the subject of a paper by Katharina Hochmuth and Karsten Gohl. Using a combination of geophysical and core-sampling techniques, they found that the embayment is characterized by large bathymetric depressions, formed by former ice streams of the West Antarctic Ice Sheet. By focusing on one of the largest linear depressions, the Abbot glacial trough, and analysing its glacial depositional and erosional history, they found that the delivery of sediment by grounded ice was influenced by several basement highs. The main Abbott glacier system was influenced by the extensions and convergence of several ice streams we see today, including Pine Island, Thwaites, Cosgrove and Abbott glaciers.

Continuing with the marine geological theme, Daniela Sprenk and co-workers used marine sediment cores from two deep-sea sites in the Scotia Sea to record bioproductivity changes in the Southern Ocean through the last glacial cycle. These cores, resolved on a decadal-scale, have a biogenic opal flux that records the influences of sea ice distribution and summer sea surface temperatures. South of the Polar Front, the lowest bioproductivity occurred during the Last Glacial Maximum, when the upwelling of mid-depth water was reduced and sea-ice growth intensified. Bioproductivity increased abruptly around 17 000 years ago, in concert with decreasing seasonal sea ice coverage.

Maryline Vautravers and co-workers provide data and interpretation of a marine sediment core dating back to 75 ka that was recovered from a depth of 2300 m off the west coast of the Antarctic Peninsula. The core is dominated by glaciomarine sediment, but there are sufficient planktonic foraminifera to provide a stable-isotope stratigraphy and relative palaeomagnetic intensity record. They show that, during Marine Isotope Stage 3, foraminiferal abundance peaks correlate with warming in Antarctica, as expressed in the oxygen-isotope record of an ice core from eastern Dronning Maud Land. Ice-rafted debris is most abundant in sediment from the last deglaciation (*c.* 19–11 ka), as sea-level rise destabilized the West Antarctic and Antarctic Peninsula ice sheets that had previously advanced over the shelf during the Late Glacial Maximum low-stand.

In another core analysis, Naresh Pant and colleagues examine the much longer U1359 core from Expedition 318 of the Integrated Ocean Drilling Program, to determine the glacial history of the East Antarctic Ice Sheet off the coast of Wilkes Land. They established a record of multiple sourcing of iceberg debris from the East Antarctic craton, the Ross Orogen and the Ferrar Large Igneous Province, the last two of which lie a considerable distance to the SE. From 60 m of core, these authors used a mix of heavy mineral analysis, X-ray diffraction and micro-beam techniques to characterize the sediment, for which they assigned a Pleistocene age.

Earth-surface processes

Glacial and periglacial processes in Antarctica have become increasingly recognized as important for interpreting the behaviour of former ice sheets and the periglacial zones beyond them. Several papers deal with these topics, emphasizing the processes occurring in cold-arid and semi-arid areas of Antarctica. Other topics covered in this section include investigations of snow-cover and atmosphere–ocean–solid Earth interaction.

Cold-based glaciers are an important part of the cryospheric system in Antarctica and, until recently, it was argued that they had little geomorphological significance. However, a new body of evidence suggests that cold-based glaciers can erode, transport and deposit debris, leading to a distinctive sediment–landform assemblage for cold-based glaciers. Cliff Atkins reviews the evidence for cold-based glacier activity in South Victoria Land, emphasizing the dual ability of these glaciers both to preserve landscapes and to erode and modify them. In this invited review, Atkins explores recent improvements in our understanding of cold-based glacier structure, process and interaction with various substrates. In the light of these advances in understanding, the landscape exposure record can be better interpreted, and previously unknown glacial events can be recognized.

South Victoria Land is also the scene for research by David Marchant and co-workers, who analyse geomorphological processes and landforms in the Dry Valleys. The Dry Valleys have three microclimates (a coastal thaw zone, an inland mixed zone, and an upland stable zone), resulting in considerably different landscapes, with unique landforms in each climatic zone. This invited review of landscape development has implications for our understanding of long-term climate change and ice-sheet stability.

Landscape evolution is increasingly seen as an important part of interpreting past glacial events, and Bethan Davies and co-workers present new modern and palaeoprocess–landform data from the northern Antarctic Peninsula. This paper investigates the behaviour of the Antarctic Peninsula Ice Sheet from the Last Glacial Maximum and through the Holocene, culminating with the modern periglacial environment. The authors emphasize the interplay between glacial, periglacial and paraglacial processes, and the importance in recognizing this when interpreting the palaeogeomorphological record.

A little-studied aspect of earth-surface processes in Antarctica is the development of soils. In a paper by Megan Balks, a wide range of soils in different topographic settings are considered. The regional coverage includes the Transantarctic Mountains (including the Dry Valleys), coastal areas of East Antarctica and the Antarctic Peninsula. The authors explore how soils develop under different regional climates and in different topographic settings. They argue that organisms, time and parent material are the dominant influences on soil properties only in localized situations. The authors conclude that Antarctic soils are not well categorized, and there is a need to introduce new categories.

Schirmacher Oasis and Larsemann Hills are relatively little-studied regions of Antarctic as far as Quaternary glacial history and contemporary processes are concerned, and they receive welcome attention in a paper by Rajesh Asthana and colleagues. These authors document a wide range of glacial and periglacial phenomena, and use scanning-electron microscopy of quartz grains and granulometry to pinpoint the transfer mechanisms of sediment, particularly through glaciers and the East Antarctic Ice Sheet.

Kevin Hall provides a thorough and thought-provoking review of periglacial processes and landforms around the Antarctic. This invited review takes into account a wide range of environments, from the relatively wet and mild sub-Antarctic oceanic islands through to the cold dry continent. These modern environments offer insights into conditions in the Northern Hemisphere during the last glacial, and also provide analogues for extraterrestrial periglacial environments, such as those on Mars. Periglacial processes and products can also be used to monitor long-term climate change. This review focuses on new research directions, including linkages between biotic components and abiotic responses, periglacial synergies and resulting landform development.

Measuring the response of the Earth's crust to ice sheet unloading is the subject of a paper by Stephanie Konfal and co-authors. Using the first values of crustal tilt resulting from the differential uplift of lacustrine strandlines obtained in the McMurdo Dry Valleys, they link these results with age-data to provide a history of solid earth deformation since deglaciation. They found that the magnitude of gradients increases exponentially with age, indicating an ongoing response to deglaciation since the Late Glacial Maximum. This tilt pattern contradicts that predicted by models of glacial isostatic adjustment for Antarctica.

Masaki Kanao and colleagues address a topic of rapidly developing interest – the physical interaction between atmosphere, ocean and the solid Earth (including the cryosphere). At Syowa Station, East Antarctica, they installed an infrasound sensor that measures microseism-baroms, thereby recording the impact of storms and sea-ice fracturing. These attributes are a useful proxy for characterizing ocean-wave climate, and the measurements here form part of a network of such studies in the southern high-latitudes.

Measurement of annual snow is also important for long-term monitoring. This is time-consuming and best done using remote imaging techniques. Carla Mora and co-workers investigated the possibility of using Envisat scenes for snow cover classification. However, they found a high degree of noise, which meant that high-resolution snow mapping was not possible.

Conclusions

The papers in this volume represent only a fraction of those presented at the International Symposium on Antarctic Earth Sciences in Edinburgh in July 2011. However, combined with several invited reviews, they demonstrate a remarkable diversity of Earth science interests in the Antarctic. Most noteworthy is that the last decade demonstrates a marked shift in emphasis from pre-Cenozoic geology to topics related to the 34 or so million years of Antarctic glacial history. This trend reflects the recognition, documentation and understanding of the palaeorecord of ice-sheet growth and decay. Furthermore, Antarctic Earth history is critically important for informing us about potential future trends, as the impact of global warming is increasingly felt on the continent and its surroundings.

We thank our co-editors, Peter Barrett, Vanessa Bowman and John Smellie, as well as John Howe, the Geological Society's Publication Committee representative, for their helpful comments and added insight that have helped improve this contribution. M. J .H. thanks these and the other editors, Peter Barker, Bethan Davies and Martin Tranter, for their care and patience in ensuring Geological Society standards were maintained. Below, we honour the contribution of the late Peter Barker, who sadly passed away during the early stages of editing the papers for this volume. His legacy in marine geology and geophysics has added substantially to our knowledge of Antarctica's tectonic and climatic history, especially in the Antarctic Peninsula–Scotia Arc region. We are also indebted to our 45 referees (listed below alphabetically by surname), whose efforts have helped improve all of the contributions substantially: John Anderson; Robert Anderson; Tatsuro Ando; Carlo Baroni; Michael Bentley; Manfred Boelter; Fernando Bohoyo; John Bradshaw; Judd Case; Alistair Crame; Ian Dalziel; Renato de Moraes; David Elliot; Jeffrey Evans; Warren Dickinson; Graham Eagles; Stephen Emslie; Sean Fitzsimons; Andrew Fountain; Neil Glasser; Melanie Grob; Mauro Guglielmin; Brenda Hall; Sidney Hemming; Stuart Henrys; Claus-Dieter Hillenbrand; Richard Kelly; Gerhard Kuhn; Malcolm Laird; Jennifer Latimer; Amy Leventer; Frank Lisker; Barrie McKelvey; Rich Mooi; Ana Navas; Didier Néraudeau; Peter Nienow; Philip O'Brien; Anya Reading; David Sugden; Colin Thorn; Rudolph Trouw; Slawek Tulaczyk; Alan Vaughan; Dan Zwartz.

References

HARLEY, S. L., FITZSIMONS, I. C. W. & ZHAO, Y. In press. *Antarctica and Supercontinent Evolution*. Geological Society, London, Special Publications, **383**.

SIEGERT, M. J. & BRADWELL, T. (eds) In press. *Antarctic Earth Sciences. Earth and Environmental Science*. Transactions of the Royal Society of Edinburgh, **104**.

The geological and tectonic evolution of the Transantarctic Mountains: a review

DAVID H. ELLIOT

Byrd Polar Research Center and School of Earth Sciences, The Ohio State University, Columbus, OH 43210, USA (e-mail: elliot.1@osu.edu)

Abstract: The development of the Transantarctic Mountains was initiated with the rifting of Rodinia and the formation of a late Neoproterozoic passive continental margin. In Cambrian time this rift setting evolved into an active margin with batholith emplacement into deformed and lightly metamorphosed upper Neoproterozoic–Cambrian strata, creating the Ross Orogen. Denudation and erosion of the Ross Orogen led to the formation of the pre-Devonian Kukri Erosion Surface on which Devonian quartzose sandstones accumulated in a continental setting. Palaeozoic magmatic arcs were intermittently active along the distal Panthalassan margin. Intracratonic basins developed in Permian time, one of which evolved into a foreland basin clearly related to a Permo-Triassic magmatic arc. The Palaeozoic–early Mesozoic arcs can be traced into both Australasia and South America. In Early Jurassic time the margin migrated outboard simultaneously with the advent of proximal silicic volcanism, emplacement of the Ferrar Large Igneous Province and Gondwana break-up. These events marked the onset of plate margin reorganization, and with it the early uplift of the Transantarctic Mountains. During Cretaceous and later time episodic uplift of the Transantarctic Mountains was accompanied by formation of a major crustal and lithospheric boundary marking the edge of the East Antarctic craton and the regions of crustal attenuation in the Ross (West Antarctic Rift System) and Weddell embayments.

The frontal scarp of the Transantarctic Mountains forms a geographic divide across the Antarctic interior, separating East and West Antarctica (Fig. 1). The geological record of the mountains themselves spans the Archean to Late Cenozoic, although in outcrop it is limited prior to the late Neoproterozoic and after Early Jurassic time. In the broadest outline, the Transantarctic Mountains comprise a pre-Devonian basement overlain by a Devonian–Jurassic sedimentary sequence, which in Early Jurassic time was overlain by extrusive basaltic rocks and intruded by sills. Younger rocks consist of Upper Cenozoic alkaline basaltic rocks and glacigenic strata.

This review, from the perspective of the central Transantarctic Mountains and south Victoria Land, will address primarily the mid-Palaeozoic and younger tectonic events, concentrating on the Ross Sea sector and on tectonic relations with the broader Gondwana margin. Through Cenozoic times the evolution of the Transantarctic Mountains has been intimately tied to development of a continental-scale rift system in West Antarctica, today the site of Earth's only marine ice sheet. The history of the mountain range is also significant for understanding the history of Antarctic glaciation and climate.

Archean to Neoproterozoic time: windows into the cratonic margin

The oldest rocks, restricted to the region of the upper Nimrod Glacier (Fig. 1), are Archean gneisses and schists intruded by 2.5 Ga granites (Goodge & Fanning 1999). These rocks were reworked by Palaeoproterozoic metamorphism and syn-tectonic magmatism at 1.7 Ga, and again during the Cambrian–earliest Ordovician Ross Orogeny (Goodge *et al.* 2001; Goodge & Fanning 2002). Glacial erratics in the Miller Range region indicate the presence also of Grenville-age (*c.* 1 Ga) crust in the hinterland (Goodge *et al.* 2010). Palaeoproterozoic to lower Cambrian igneous, metamorphic and sedimentary rocks crop out in the Shackleton Range (Fig. 1), forming three east–west structural terranes bounded by thrust fault systems (Buggisch & Kleinschmidt 2007), which detailed geochemical and isotopic data (Will *et al.* 2009, 2010) demonstrate have markedly different histories. Fossiliferous siliciclastic Ordovician strata, the Blaiklock Glacier Group, unconformably overlie older rocks in the Shackleton Range (Clarkson *et al.* 1995). The Miller and Shackleton ranges form part of the ring of Precambrian outcrops around East Antarctica that include pan-African age belts, Grenville age provinces, Proterozoic domains and Archean cratons (Fitzsimons 2003; Will *et al.* 2010; Boger 2011).

Late Neoproterozoic to early Palaeozoic (Cambrian) time: passive margin to subduction

Excluding the Miller and Shackleton ranges, basement rocks in the Transantarctic Mountains

From: Hambrey, M. J., Barker, P. F., Barrett, P. J., Bowman, V., Davies, B., Smellie, J. L. & Tranter, M. (eds) 2013. *Antarctic Palaeoenvironments and Earth-Surface Processes*. Geological Society, London, Special Publications, **381**, 7–35. First published online July 1, 2013, http://dx.doi.org/10.1144/SP381.14

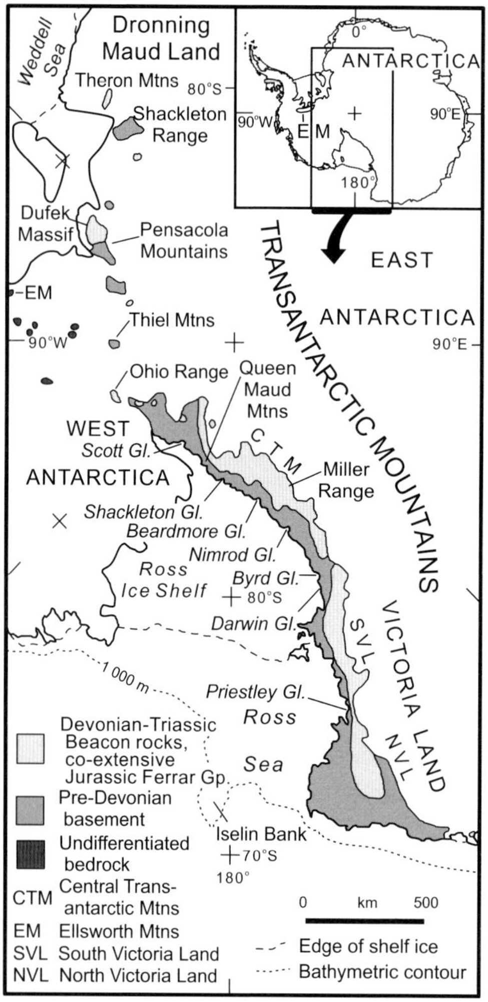

Fig. 1. Transantarctic Mountains region of Antarctica, showing the distribution of basement rocks (Proterozoic–early Palaeozoic) and regional cover beds (Devonian–Jurassic). Inset shows map area.

(Fig. 2) consist of Neoproterozoic strata and lower Cambrian to Ordovician(?) metasedimentary and metavolcanic rocks, all affected by deformation and granitoid emplacement associated with the Cambrian–earliest Ordovician Ross Orogeny (Rowell *et al.* 2001; Goodge *et al.* 2004*a*, *b*). Neoproterozoic siliciclastic strata crop out in the Pensacola Mountains (Hannah Ridge Formation) and the central Transantarctic Mountains (Beardmore Group). These rocks, which have yielded detrital zircon ages of latest Neoproterozoic (*c.* 556 Ma), form passive margin sequences deposited subsequent to the break-up of Rodinia. More widespread between the Pensacola Mountains and the Byrd Glacier are lower Cambrian to Ordovician siliciclastic and carbonate strata (Patuxent Formation in the Pensacola Mountains region and Byrd Group in the central Transantarctic Mountains). Cambrian silicic volcanic rocks are common in the Queen Maud and Thiel mountains (Pankhurst *et al.* 1988; Wareham *et al.* 2001). Deformation of these rocks was accompanied by intrusion of both pre- and syn-tectonic granitoids, the age of which ranges from about 560 Ma to, in isolated cases, 480 Ma (Borg *et al.* 1990; van Schmus *et al.* 1997; Rowell *et al.* 2001; Paulsen *et al.* 2007). There is evidence for early Cambrian deformation of the Hannah Ridge Formation, and a well-documented second episode of folding and low-grade metamorphism that is no younger than about 505 Ma. Deformation was followed by intrusion of post-tectonic granitoids. A basement discontinuity in the Byrd Glacier region (Stump *et al.* 2004; Carosi *et al.* 2007) marks a significant change in the metamorphic grade of basement rocks and a possible terrane boundary.

To the north, in south Victoria Land, the basement is dominated by medium- to high-grade metamorphic rocks of the Horney Formation (Borg *et al.* 1989) and the Koettlitz and Skelton groups (Goodge *et al.* 2004*a*; Cooper *et al.* 2011), and is intruded by a Neoproterozoic to Cambrian alkaline suite (Read *et al.* 2002) that pre-dates the more typical calc-alkaline granitoids of the Ross Orogen (Allibone & Wysoczanski 2002). The alkaline magma suite was emplaced between about 550 and 530 Ma, whereas the Ross granitoids are mainly younger than about 530 Ma, range to about 480 Ma, and again show both syn- and post-tectonic phases. The Skelton Group has been correlated with Neoproterozoic strata in the central Transantarctic Mountains (Goodge *et al.* 2004*a*; Cooper *et al.* 2011).

North Victoria Land, with a more complex Neoproterozoic to Ordovician history, comprises three fault-bounded lithotectonic terranes (Laird 1991). The Wilson terrane metamorphic rocks are intruded by Cambrian–lowest Ordovician granitoids with ages ranging from about 545 to 480 Ma (Bomparola *et al.* 2007; Giacomini *et al.* 2007; Federico *et al.* 2009) and form the continuation of the Ross Orogen in north Victoria Land. The narrow Cambrian-age allochthonous Bowers terrane, bounded by the Lanterman and Leap Year faults, lies outboard of the Wilson terrane (Bradshaw *et al.* 2009). It consists of a thick sequence (− 10 km) of fossiliferous marine siliciclastic and volcaniclastic strata together with mafic volcanic rocks, overlain by conglomeratic strata, and comprises the Bowers Supergroup. The Robertson Bay Group, outboard of the Bowers terrane, is a Cambrian(?) to Ordovician turbidite sequence, which,

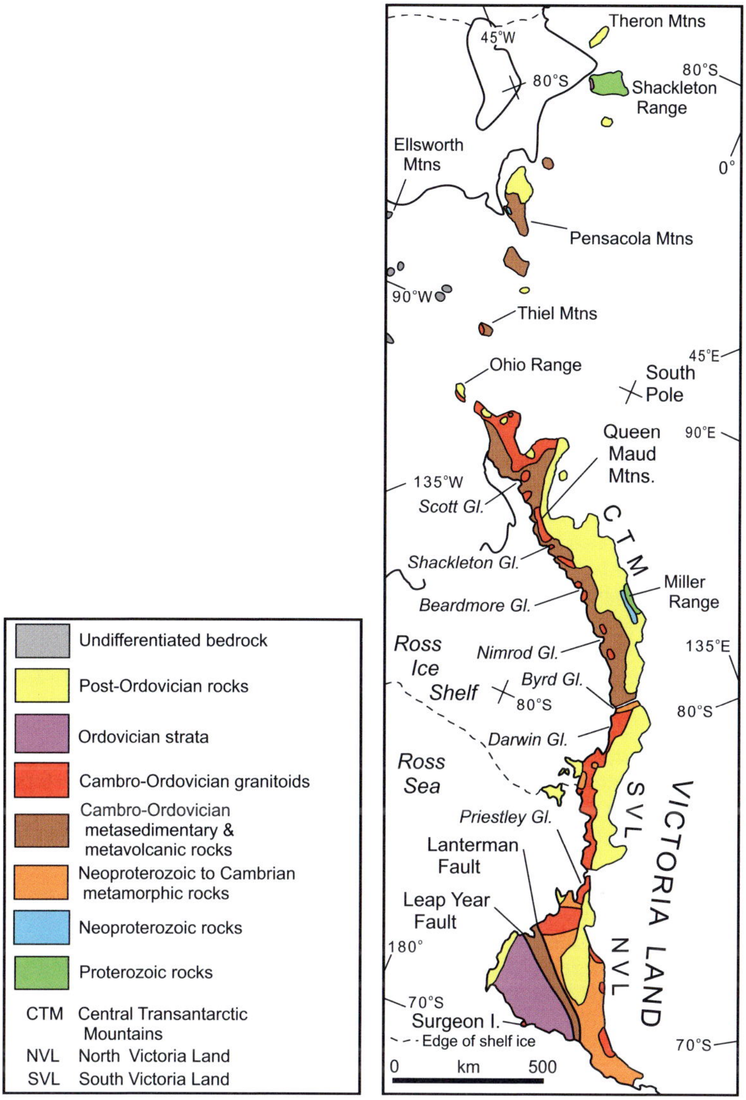

Fig. 2. Distribution of Neoproterozoic and Ross Orogen rocks along the Transantarctic Mountains.

similar to the Bowers, is strongly folded but only slightly metamorphosed, as indicated by illite crystallinity (Buggisch & Kleinschmidt 1991). A siliciclastic sequence of similar age and composition, the Swanson Formation (Bradshaw *et al.* 1983), crops out in Marie Byrd Land and is probably a correlative deep-sea fan sequence.

The nature of the north Victoria Land fault boundaries and the tectonic history remain controversial. Adjacent to the Lanterman Fault and forming the contact rocks of the Wilson terrane there are bodies of high pressure and ultra-high pressure rocks (Palmeri *et al.* 2003). The Leap Year Fault includes a zone of schists incorporating slices of both Bowers and Robertson Bay terrane rocks, and thus may represent a zone of imbricate thrusting. The contrasts in stratigraphic successions and metamorphic grade between the terranes have led to alternative tectonic hypotheses. In general, one hypothesis postulates major along-strike translation of the two outboard terranes and the other accretion and shortening of the terranes by major east-directed overthrusts, or some combination of both (Tessensohn & Henjes-Kunst 2005 and references therein), and with all events completed in early Palaeozoic time (Early Ordovician). More recent hypotheses are laid out by Bracciali *et al.* (2009) and Rocchi *et al.* (2011), who propose a more complex scenario of subduction, island arc formation in an extensional setting, and subsequent accretion and closure.

A Ross age syn-orogenic clastic sequence, the Douglas Conglomerate of the Byrd Group (Laird 1991) in the central Transantarctic Mountains, has been described, but no post-orogenic strata are known except for the Robertson Bay Group in north Victoria Land. Erosion of the Ross Orogen commenced in late Cambrian time and denudation continued through to the Late Silurian or Early Devonian, creating the widespread and distinctive Kukri Erosion Surface (Isbell 1999).

The Ross Orogen, including the Wilson terrane, reflects subduction of Panthalassan oceanic crust beneath the edge of the East Antarctic basement (Borg & DePaolo 1991), and clearly is correlative with the Delamerian Orogen of Australia (Finn *et al.* 1999; Glen 2005; Foden *et al.* 2006; Bradshaw 2007). The Bowers terrane has similarities to Cambrian basaltic volcanic rocks and siliciclastic sequences of the Delamerian Orogen, and also has been correlated with the Takaka terrane in New Zealand (Bradshaw *et al.* 2009). A single Ross-age granitoid is known from New Zealand (Allibone *et al.* 2009). Further afield, the Ross Orogen can be correlated with the Saldanian Orogen in South Africa (Scheppers & Armstrong 2002).

The Cambrian Surgeon Island granite (*c.* 512 Ma), which lies within the Robertson Bay terrane, but is not seen in outcrop to intrude those rocks (Fioretti *et al.* 2005), is possibly part of a terrane accreted in late Ross–Delamerian time (Cayley 2011). A rhyolite dredged from the Iselin Bank and having a late Proterozoic/early Cambrian age (Mortimer *et al.* 2011), and Cambrian orthogneiss and paragneisses in Marie Byrd Land (Pankhurst *et al.* 1998; Mukasa & Dalziel 2000), indicate that rocks with the same age as the Ross Orogen are present in West Antarctica. The Robertson Bay terrane, these isolated Ross-age rocks and the complexities of correlations between north Victoria Land, New Zealand and south-eastern Australia are discussed further in the next section

Ordovician to Late Triassic time: sedimentation and subduction

Devonian strata

The Kukri Erosion Surface is the defining contact in the Transantarctic Mountains, separating Ross-age granitoids and deformed rocks from essentially undeformed Devonian and younger rocks (Fig. 3). Although initially described as a peneplain (Gunn & Warren 1962), it is not a planar surface, but has topography as shown by local Devonian strata thickness variations of up to 700 m (Isbell 1999; Isbell *et al.* 2008). In south Victoria Land (Fig. 4), the surface is overlain by the 1500 m-thick quartzose strata of the Taylor Group (Bradshaw, this volume, in press). Sedimentary structures and trace fossils have been used to propose both marine and non-marine depositional environments, with only the Aztec Siltstone being clearly terrestrial. The age, based on palynomorphs, ranges from Early to Late Devonian (Kyle 1977). However, fish faunas from the Aztec Siltstone suggest an age no younger than late Middle–early Late Devonian (Young & Long 2005). Palaeocurrent directions are quite variable (Barrett & Kohn 1975), but the quartz-rich composition implies derivation from granitic rocks. The Devonian rocks assigned to the Taylor Group south of Byrd Glacier (Alexandra Formation and, in the Starshot Glacier region, the underlying Castle Crags Formation) have significantly lesser thicknesses and lack finer-grained beds except for the Castle Crags Formation. The age of the Castle Crags and Alexandra formations is unconstrained palaeontologically, and thus a Carboniferous age for the upper part of the Alexandra cannot be excluded. Beyond the most southerly outcrop of the Alexandra Formation at the Ramsey Glacier, Devonian rocks are present in the Ohio Range (Bradshaw 2013), where they form a 56 m-thick, fossiliferous, dominantly marine, siliciclastic sequence (Horlick Formation) with a palaeontological age of Emsian (Early Devonian; Boucot *et al.* 1963; Bradshaw & McCartan 1991). Siliciclastic strata with similar depositional environments and an inferred Devonian age are present in the Pensacola Mountains (Dover Sandstone; Cathcart & Schmidt 1977). The Crashsite Quartzite in the Ellsworth Mountains is Devonian in its upper part (Sporli 1992). However, at that time the mountains were located in Gondwana reconstructions between Africa and Antarctica (Watts & Bramall 1981; Grunow *et al.* 1987), and the geology is much more comparable with that of South Africa (Curtis & Storey 1996). Nevertheless, the Ohio Range and Ellsworth Mountains Devonian sequences carry the Malvinokafric fauna and hence can be linked to South Africa, the Falkland Islands and South America, each representing part of a widespread shallow marine environment. Further, there are links with Devonian marine faunas in New Zealand (Bradshaw & McCartan 1991).

The quartz-rich Devonian succession in the Transantarctic Mountains (Fig. 4), although principally derived from granitic sources located mainly in the East Antarctic, did include West Antarctic sources (Barrett & Kohn 1975; McCartan & Bradshaw 1987), and was deposited in a

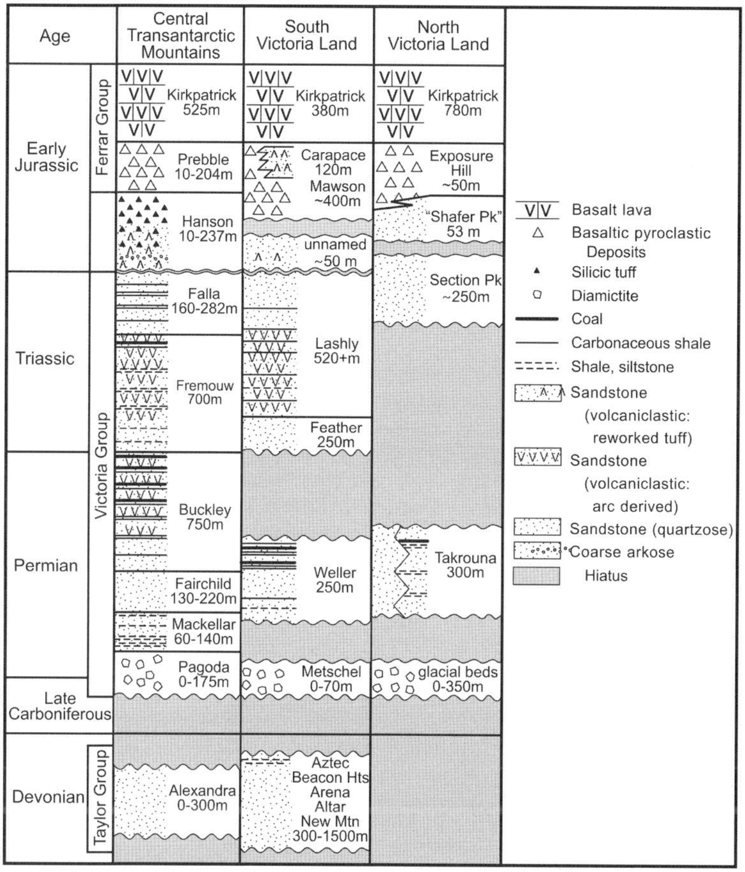

Fig. 3. Simplified stratigraphic columns for Devonian to Jurassic rocks overlying the Kukri Erosion Surface in the Transantarctic Mountains. *Sources*: Taylor and Victoria Groups in the central Transantarctic Mountains and south Victoria Land, Barrett (1991); Falla Formation, Elliot (1996); Hanson Formation, Elliot *et al.* (2007); Prebble Formation, Hanson & Elliot (1996); unnamed strata, south Victoria Land, Elliot & Grimes (2011); Mawson Formation, Ross *et al.* (2008); north Victoria Land glacial beds and Takrouna Formation, Collinson *et al.* (1986); Section Peak and 'Shafer Peak' formations, Schöner *et al.* (2007, 2011 respectively); Exposure Hill Formation, Viereck-Goette *et al.* (2007); Kirkpatrick Basalt, Elliot & Fleming (2008).

passively subsiding basin floored by Ross Orogen rocks. This setting is in marked contrast to the mid-Palaeozoic record of Marie Byrd Land, which is characterized by Devonian and Carboniferous plutons and metamorphic rocks (see the next section).

Early Palaeozoic (Ordovician to Early Carboniferous) plate margin

The Devonian Taylor Group and the late Devonian–Carboniferous magmatic belt of north Victoria Land and Marie Byrd Land need to be placed in the broader temporal and spatial context of the Pacific margin of Gondwana. This margin is better documented in the Lachlan Orogen of southeastern Australasia, which was widely developed outboard of the Delamerian Orogen that previously formed the Panthalassan margin (Gray & Foster 2004; Glen 2005), and in the Patagonian region of South America (Vaughan & Pankhurst 2008). Along the length of the margin from Australia/New Zealand to South America, there is a record of pre-Late Devonian events.

The Robertson Bay Group (north Victoria Land), the Swanson Group of Marie Byrd Land and the Greenland Group in New Zealand (Wandres & Bradshaw 2005) all have similarities to the Ordovician quartzose turbidite sequences in the Lachlan Orogen, and all represent erosion of continental sources in the Ross–Delamerian Orogen. All were

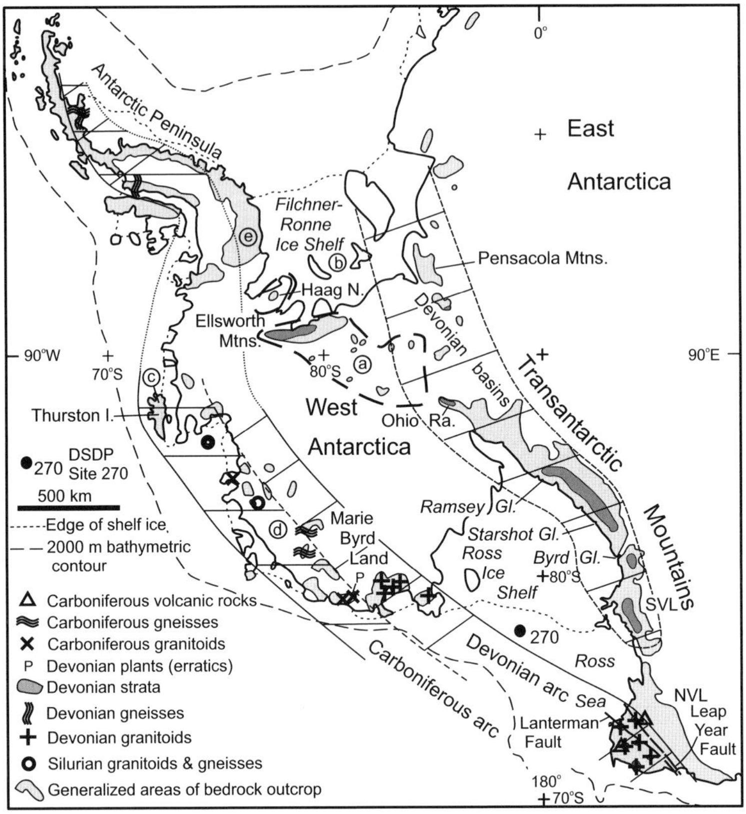

Fig. 4. Present-day distribution of Devonian sedimentary rocks and Silurian–Devonian–Carboniferous granitoids and gneisses in Marie Byrd Land and the Antarctic Peninsula. Note that during Palaeozoic time, (**a**) the Ellsworth–Whitmore Mountains block and Haag Nunataks (heavy dash outlines) were located between Africa and Antarctica in the proto-Weddell Sea region; (**b**) the Filchner-Ronne Ice Shelf region is highly extended; (**c**) Thurston Island was displaced relative to the Antarctic Peninsula and the Ellsworth–Whitmore Mountains block; (**d**) Marie Byrd Land, prior to crustal stretching in the Cretaceous and Cenozoic, was located closer to the Transantarctic Mountains; and (**e**) the base of the Antarctic Peninsula would also have been closer to the Transantarctic Mountains.

deformed in an early phase of the Lachlan Orogen (Gray & Foster 2004; Glen 2005) and its inferred extension into West Antarctica; it is also recorded in lower Palaeozoic schists on the Campbell Plateau (Adams 2008). There is a single (poorly constrained) Ordovician age (Rb–Sr) from orthogneisses at Clark Island, eastern Marie Byrd Land (Fig. 4; Pankhurst *et al.* 1998). Granite clasts in conglomerates from the Antarctic Peninsula, one from the Marguerite Bay area (Loske *et al.* 1997) and several from View Point (Pankhurst *et al.* 2003; Bradshaw *et al.* 2012) have yielded Late Ordovician zircon ages. Ordovician ages are also reported for the Cordillera Darwin Metamorphic Complex and from a granite clast in overlying Jurassic strata (Hervé *et al.* 2010*b*). In the Deseado Massif region of Patagonia, granitic cobbles in a Permian sequence also suggest Ordovician magmatism (Pankhurst *et al.* 2003), which is better expressed in the Famatinian belt farther north (Pankhurst *et al.* 2001).

Silurian magmatism is documented in the Lachlan Orogen (Gray & Foster 2004; Glen 2005), but again evidence of the orogen in Antarctica is fragmentary. A calc-silicate gneiss at Deep Sea Drilling Project Site 270 in the Ross Sea yielded a titanite metamorphic age of 437 Ma (Mortimer *et al.* 2011), and a Silurian gneiss crops out in Marie Byrd Land (Pankhurst *et al.* 1988). In the Antarctic Peninsula an orthogneiss in northern Palmer Land has a protolith Silurian age (Millar *et al.* 2002) and granite cobbles from Marguerite Bay give Silurian ages (Tangeman *et al.* 1996; Loske *et al.* 1997). Late Silurian magmatism is also recorded in the Deseado Massif, Patagonia (Pankhurst *et al.* 2003).

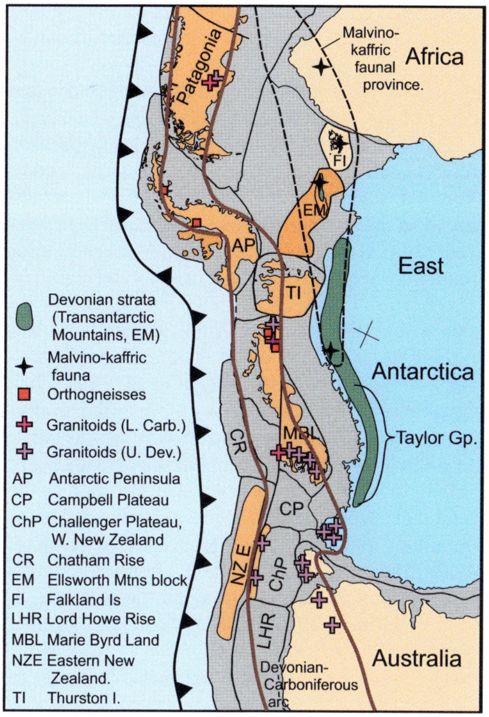

Fig. 5. Gondwana plate margin in mid-Palaeozoic time, showing the distribution of Devonian strata in the Transantarctic Mountains, and granitoids and gneisses in the Devonian to Early Carboniferous orogen in West Antarctica, the Antarctic Peninsula and adjacent regions. The arrangement of the cratonic regions and the size and shape of many of the continental blocks are taken from a reconstruction provided by the PLATES Project at the Institute of Geophysics at the University of Texas. Palaeomagnetic studies (Grunow *et al.* 1991; DiVenere *et al.* 1996; see also Dalziel & Lawver 2001) provide constraints on the disposition of some of the Antarctic continental blocks along the Gondwana plate margin. Palaeomagnetic data for New Zealand are limited to two Permian poles (Haston *et al.* 1989) and one Late Triassic–Early Jurassic pole (Grindley *et al.* 1980), both from southern South Island and east of the Alpine Fault. The possibility that the Haag Nunataks may form a separate entity distinct from the Ellsworth–Whitmore Mountains block is not considered here. The Agulhas Plateau and Maurice Ewing Bank, both of African geological affinities, are also not considered. Continuity of geological provinces, as discussed by Vaughan & Storey (2000) and others, is assumed. The possibility of strike-parallel movements within the plate margin belt (e.g. Adams *et al.* 1998) makes precise locations for the plate margin terranes tenuous. Likewise, the possibility of pre-Jurassic accretion of continental blocks, continental ribbons and island arc terranes (e.g. Haston *et al.* 1989; Mortimer 2004) introduces further uncertainty. The reconstruction shown in this figure, and in Figures 7 and 9, is therefore necessarily schematic. A detailed discussion of the constraints and uncertainties in the reconstruction of the Gondwana plate margin is beyond the scope of this paper.

The Ordovician through Middle Devonian record in West Antarctica and the Antarctic Peninsula is meagre at best, but it does link Australasia and South America, and suggests they are all part of one, episodically active, Panthalassan margin. The lack of granitoids and rarity of gneisses older than Late Devonian in north Victoria Land and Marie Byrd Land may reflect ice cover and tectonic history.

The Late Devonian–Carboniferous record of Marie Byrd Land (Fig. 5) is characterized by I-type Devonian plutons with ages of 375 ± 5 Ma, and Carboniferous plutonic (with both I- and S-type characteristics) and metamorphic rocks dated at about 330–340 Ma (Weaver *et al.* 1991; Pankhurst *et al.* 1998; Mukasa & Dalziel 2000). These suggest that a subduction-related magmatic arc bordered the earlier depositional system of the Taylor Group and its correlatives. This arc is expressed in north Victoria Land by the Devonian Admiralty Intrusives and the Carboniferous Gallipoli Porphyries and Salamander Granite complex (Henjes-Kuntz & Kreuzer 2003 and references therein), which were emplaced across all three north Victoria Land terranes and thus constrain the age of any movement on the Leap Year and Lanterman faults to be pre-Late Devonian. However, the granitoids have S- and I-type characteristics, and their spatial relations have been interpreted to suggest NE-directed subduction (Borg & DePaolo 1991), which would imply that the Robertson Bay, Bowers and the eastern part of the Wilson terranes are allochthonous and were amalgamated with the western Wilson terrane in post-Devonian time. This is incompatible with strong evidence for early Palaeozoic amalgamation of the north Victoria Land terranes.

There is no indication of volcanic detritus in the Taylor Group, nor is there deformation that might be interpreted as far-field effects of subduction. The former suggests that initiation of the Devonian magmatic arc in Marie Byrd Land and north Victoria Land post-dated deposition of the Taylor Group, and the latter that either the arc was far distant or the arc was allochthonous/para-autochthonous and accreted in post-Devonian time, which again is in conflict with evidence for early Palaeozoic amalgamation of north Victoria Land.

The Late Devonian–Early Carboniferous magmatic arc in Marie Byrd Land extended into the Western Province of New Zealand, with granitoids present in the Buller Terrane and forming parts of the Median Batholith (Mortimer 2004; Wandres &

Bradshaw 2005; Allibone *et al.* 2009). The arc continued into the southern part of the Lachlan Orogen in Tasmania and southeastern Australia (Gray & Foster 2004; Glen 2005). The Devonian–Carboniferous arc in the Antarctic Peninsula is represented only by orthogneisses (Millar *et al.* 2002). However, the active Gondwana margin continued through to Patagonia, where there is a fragmentary record of Devonian–Carboniferous magmatism (both S- and I-type) in the Deseado Massif and Cordillera Darwin, and also further north in the San Rafael block and the Sierras Pampeanas (Pankhurst *et al.* 2003; Hervé *et al.* 2010*b*).

The Lachlan Orogen, which persisted into the Carboniferous, has a well-documented complex history, involving subduction, rifting, accretion of arc terranes and incorporation of slices of continental crust and oceanic basalt. Gray & Foster (2004) have drawn analogies between the Lachlan Orogen and the present-day western Pacific. Along-strike variations in tectonic and magmatic events in orogenic belts are normal, and hence difficulties in correlation between Antarctica, New Zealand and Australia are not unexpected. The occurrence of Cambrian and Ordovician gneisses in Marie Byrd Land suggests an older crustal history in a belt far outboard of the Ross Orogen. Further, Nd model ages and U–Pb systematics of zircons from Marie Byrd Land granitoids (Pankhurst *et al.* 1998; Mukasa & Dalziel 2000), and Re–Os studies on Cenozoic spinel peridotite xenoliths (Handler *et al.* 2003), all suggest Proterozoic basement. Thus the possibility of autochthonous or para-autochthonous continental ribbons cannot be excluded. Accretion of continental blocks and arc terranes, often oblique, has been proposed for the Lachlan Orogen (Glen 2005; Cayley 2011) and for the upper Palaeozoic of Patagonia (Pankhurst *et al.* 2006). Rifting and later closure might explain the disjunct between the undeformed Lower to Middle Devonian quartzose Taylor Group and its correlative sequences, and the Gondwana margin, along which magmatic and compressional events were episodic throughout the early Palaeozoic, as well-documented in the Lachlan Orogen.

The Permo-Triassic Gondwana sequence

The Carboniferous time interval is represented in the Transantarctic Mountains by the Maya Erosion Surface, which is superimposed on Devonian beds but also merges with the older Kukri Erosion Surface where Devonian strata were stripped. Like the Kukri, the Maya surface has broad scale relief. Although subduction-related Carboniferous magmatism continued along the Panthalassan margin, most of the Transantarctic Mountains, as in the Devonian, must have been sufficiently distant to remain unaffected by any associated deformation and sedimentation. If any sediment had accumulated, whether derived from the craton or the inferred magmatic arc, it must have been removed by subsequent Late Carboniferous to Permian glaciation.

The Permian to Triassic Victoria Group, the Antarctic Gondwana sequence (Figs 3 & 6), occurs along the length of the Transantarctic Mountains and in at least three depositional basins. In both north and south Victoria Land, Victoria Group strata were deposited in intra-cratonic basins (Collinson *et al.* 1994). From the Byrd Glacier to the Ohio Range the Victoria Group accumulated in a cratonic basin that evolved in mid-Permian time into a foreland basin (Collinson *et al.* 1994). It is probable that by Late Triassic time the central Transantarctic Mountains basin was connected to the south Victoria Land basin but only peripherally, if at all, to north Victoria Land. The Permian Gondwana rocks from the Thiel Mountains to the Theron Mountains and beyond are geographically isolated platform sequences. Those in the Theron Mountains were derived mainly from a cratonic source to the SW (Brook 1972), that is, outboard of East Antarctica.

The north Victoria Land basin (Collinson *et al.* 1986) was probably narrow and possibly rift-related, and has only the eastern flank preserved. During Permian time a thick diamictite of probable glacial origin accumulated and was overlain by a fluvial sequence (Takrouna Formation) that changed to coal-bearing westwards away from the rift margin. Younger beds, the quartzose sandstones of the Section Peak Formation of Triassic–Jurassic age (Schöner *et al.* 2011), are geographically separated from the Permian rocks, overlie Wilson terrane granitic and metamorphic basement, and constitute a platform sequence. The Triassic *Dicroidium* flora has been found at one small isolated outcrop in north Victoria Land (Tessensohn & Mädler 1987), but its relationship to the Section Peak Formation is not known. In south Victoria Land, Victoria Group strata crop out between the David and Darwin glaciers, although it is only from Allan Hills southwards that correlations are clear. The thin glacial strata, rather than being related to a massive ice sheet, are now interpreted as glaciomarine and associated with two separate small ice sheets (Isbell 2010). These beds are overlain by Permian coal-bearing beds and then by Triassic fluvial strata that are in part carbonaceous. Clastic detritus was derived from flanking granitic sources to the east and west (Gondwana margin and East Antarctic basement, respectively) during Permian time. In Early Triassic time, there was an abrupt change in palaeocurrent direction to northerly, but it was not until Late Triassic time that the provenance changed with a significant influx of

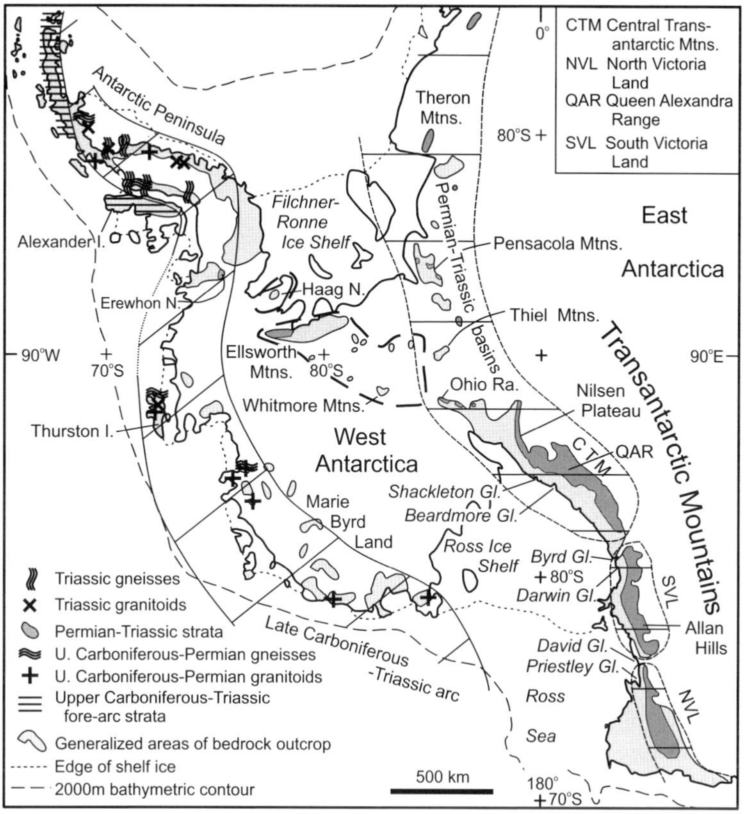

Fig. 6. Present-day distribution of Permo-Triassic Gondwana rocks in the Transantarctic Mountains and the Ellsworth Mountains, and Upper Carboniferous to Triassic granitoids and gneisses in West Antarctica and the Antarctic Peninsula. See caption for Figure 4 regarding the allochthonous Ellsworth–Whitmore Mountains block, etc.

silicic volcanic detritus. The shift in direction of palaeoflow and the incoming of volcanic detritus has suggested a palaeodrainage link between the basin south of Byrd Glacier and that in south Victoria Land, and to possible over-topping of the Ross High inferred for the Byrd Glacier region (Collinson *et al.* 1994).

The Permo-Triassic Gondwana sequence crops out from south of the Byrd Glacier to the Ohio Range and beyond, but is thicker and more complete in the Beardmore–Shackleton glacier region (central Transantarctic Mountains) than anywhere else (Fig. 3). The glacial strata, formerly thought of as terrestrial and deposited from a single large ice sheet, are now interpreted as glaciomarine, deposited in distinct basins, and to represent much smaller-scale ice sheets (Isbell *et al.* 2008). These re-interpretations of the Permian glacial beds lead to a significantly different view of Gondwana glaciation in Antarctica (Isbell *et al.* 2008; Isbell 2010). The glacial strata in the central Transantarctic Mountains are overlain by post-glacial shales and fine-sandstones (Mackellar Formation) deposited in a large post-glacial lake or inland sea (Miller & Isbell 2010). Towards the Ohio Range, equivalent strata were deposited in environments ranging from freshwater/brackish to marine (Collinson *et al.* 1994), and in an inland sea that extended to South Africa and South America. Mackellar beds are overlain by a deltaic to fluvial sequence (Fairchild Formation), which then passes up into fluvial coal-bearing strata of the Buckley Formation. Palaeoflow was generally southeastward along the mountain range until the middle Permian, when there was a swing to a northwesterly direction. Correlative Permian strata are exposed through to the Ohio Range, and carbonaceous *Glossopteris*-bearing Permian beds occur at isolated nunataks as far as the Theron Mountains. The Permian–Triassic boundary, with one documented exception in the Shackleton Glacier region, appears to be a local disconformity (Collinson *et al.* 2006 and references therein).

Triassic strata in the central Transantarctic Mountains crop out from the Queen Alexandra

Range to the Nilsen Plateau (Long *et al.* 2009), but are absent farther along the mountain range. The Triassic beds (Fremouw and Falla formations) are fluvial and in the upper part carbonaceous with thin coals. Palaeoflow was northwesterly. The approximate 180° reversal in uniform flow directions suggests that the extant record in the central Transantarctic Mountains mainly represents the axial part of the depositional basin (e.g. Barrett 1991). It is only in the Shackleton Glacier region that palaeoflow directly indicates derivation of detrital grains from the basin (arc) flank (Collinson *et al.* 1994), although north of the Beardmore Glacier there is some indication of flow from the craton during the Permian (Barrett *et al.* 1986).

Sedimentary petrology indicates a change in provenance in mid-Permian time from one dominated by basement igneous and metamorphic rocks to one with a significant volcanic input (Barrett *et al.* 1986). Palaeocurrent data from the Shackleton Glacier region demonstrate that the source of the intermediate to silicic detritus lay in West Antarctica. Detrital zircon studies (Elliot & Fanning 2008) on volcaniclastic upper Buckley sandstones revealed an abundance of primary igneous grains, and within the limitations of the method, suggest that magmatism may have started in the Early Permian. Although the overwhelming volcanic character of the upper Buckley in the Shackleton Glacier region is not repeated in the overlying Triassic strata, young zircon grains are present in those beds and show that magmatism continued through Late Triassic time (Elliot & Fanning 2011).

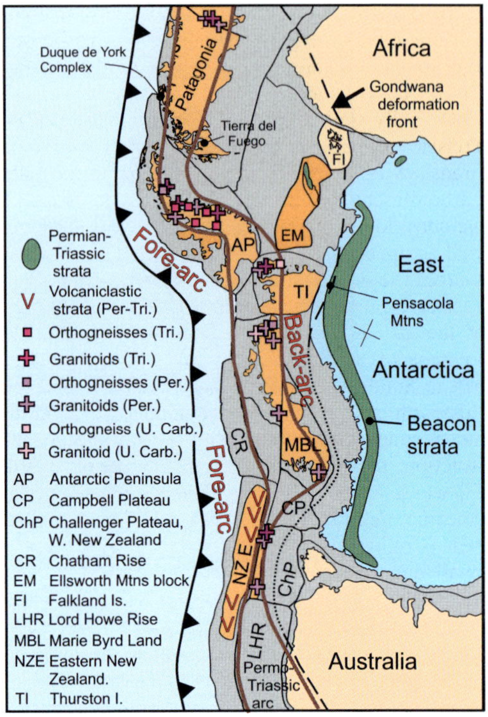

Fig. 7. Gondwana plate margin during the late Palaeozoic–early Mesozoic, showing the distribution of Permian-Triassic strata in the Transantarctic Mountains and Late Carboniferous to Triassic granitoids and gneisses in Marie Byrd Land, the Antarctic Peninsula and adjacent regions. See caption for Figure 5 for information on the reconstruction.

Late Carboniferous to Triassic plate margin

An upper Carboniferous orthogneiss (309 Ma) from Thurston Island and a granitoid from eastern Marie Byrd Land (Pankhurst *et al.* 1993) may represent an early phase of the younger Permo-Triassic arc of the Gondwana Panthalassan margin (Fig. 7). The Permo-Triassic magmatic arc is documented sparsely through West Antarctica by granitoids (Pankhurst *et al.* 1993, 1998; Mukasa & Dalziel 2000) and magmatism is supported by detrital zircon geochronology of Buckley, Fremouw and Falla sandstones (Elliot & Fanning 2008, 2011).

The arc and associated environments are documented in New Zealand by the Permian Brook Street terrane of oceanic volcanic rocks, Upper Permian to Triassic volcaniclastic strata of the Murihiku, Maitai and Caples volcanic arc-related terranes, by the Rakaia terrane, which differs in that it was derived from a continental magmatic arc (Mortimer 2004; Wandres & Bradshaw 2005), and by a Permian to Triassic intra-oceanic I-type batholith complex (Price *et al.* 2011). Farther afield, it is documented in the New England Orogen of eastern Australia where initiation of the arc, with both intrusive and extrusive activity, was in Late Carboniferous time and extended through the Permian Period (Glen 2005).

Magmatic activity in the Antarctic Peninsula is recorded by granitoids (many S-type) and orthogneisses, mainly of Triassic age (Vaughan & Storey 2000; Millar *et al.* 2002), and also by detrital zircons, dominantly Permian, in the fore-arc Trinity Peninsula Group (Barbeau *et al.* 2010; Bradshaw *et al.* 2012). In South America magmatic rocks are found in the coastal ranges of Chile (Hervé *et al.* 1988), and northern Patagonia (Pankhurst *et al.* 2006), where they are part of the Carboniferous–Triassic record of subduction and accretion. A Permian metamorphic event is also documented in basement rock from Tierra del Fuego, Patagonia (Hervé *et al.* 2010*a*). Detrital igneous zircons of Permian age occur in the Duque de York complex on the Pacific Coast (Sepúlveda *et al.* 2010) and in a quartzite from the Cordillera Darwin (Hervé *et al.* 2010*b*).

In the Antarctic Peninsula, fore-arc assemblages are represented by the Upper Carboniferous and younger accretionary complex of the LeMay Group of Alexander Island (Doubleday *et al.* 1994; Kelly *et al.* 2001) and the Carboniferous(?)–Triassic upper plate Trinity Peninsula Group (see Bradshaw *et al.* 2012). In southern Chile the Late Carboniferous to Permian Madre de Dios Terrane includes lower plate accretionary complexes as well as upper plate strata derived from continental sources (Sepúlveda *et al.* 2008, 2010).

Gondwanide deformation, located well inboard of the arc, is documented from the Sierra de la Ventana of northern Patagonia, to the Cape fold belt in South Africa, the Falkland Islands, and the Ellsworth and Pensacola mountains. Located in the Weddell Sea sector of a Gondwana reconstruction, the Permian Polarstar Formation of the Ellsworth Mountains is marked by folded strata, volcanic detritus in sandstones and sparse tuff beds (Collinson *et al.* 1992; Curtis 2001). Isolated small outcrops of dipping volcaniclastic strata in the southernmost Antarctic Peninsula (Erehwon Nunatak region) contain *Glossopteris* (Laudon 1991) and represent beds deposited proximal to the orogen. Deformation of Permian beds in Antarctica is displayed in the Ellsworth Mountains and to a significantly lesser extent in the Pensacola Mountains where the Permian diamictites of the Gale Mudstone form open folds, which suggests that the latter region was near the inboard limit of Gondwanide deformation. The timing of deformation in Antarctica is poorly constrained to a post-Permian age (Curtis 2001), although in South Africa, deformation is inferred to have been episodic between 278 and 230 Ma (Hälbich *et al.* 1983). How far towards the Ross Sea sector deformation continued is unknown, but Permian–Triassic deformation is well documented in the New England Orogen of eastern Australia (Glen 2005). Nevertheless, the major influx of volcanic detritus into the central Transantarctic Mountains Victoria Group basin in mid-Permian time in the Shackleton Glacier region and the evolution of the foreland basin suggest continuing tectonism and magmatism related to the plate margin (Elliot & Fanning 2011).

The extant part of the foreland basin in the central Transantarctic Mountains is the axial region and the thickness of the whole sequence at its margins suggests that the original basin was much wider, extending towards the orogen on one flank and to a forebulge on the basement flank. The axial drainage migrated across the foreland, reflecting the interplay between accommodation space, sediment input and tectonism (Collinson *et al.* 1994).

Structure contours on the Kukri Erosion Surface in the central Transantarctic Mountains suggest a gently southward plunging syncline (Barrett 1969). Although it might be interpreted in terms of the foreland basin, the syncline appears to have a relatively steep limb on the craton flank, opens out to the SE, and its axial trace is not coincident with the dominant flow directions particularly for Triassic strata. Thus the syncline is more likely to represent mainly Jurassic and/or younger warping associated with uplift of, and faulting within, the Transantarctic Mountains.

The central Transantarctic Mountains Permo-Triassic Victoria Group strata reflect sedimentation and tectonism associated with the Gondwanide orogen. The magmatic arc itself is documented in Marie Byrd Land by granitoids, by igneous zircons in the Victoria Group, and by igneous and metamorphic rocks in the Antarctic Peninsula. A fold-and-thrust belt in the Weddell Sea sector may have extended into the Ross Sea sector. The changes in zircon age provinces recorded in the Victoria Group sandstones may reflect thrusting events bringing rocks of differing ages to the surface and/or changing sediment sources (Elliot & Fanning 2008). The orogen played a major role in the development of the foreland basin, in contrast to the intra-cratonic basins of north and south Victoria Land, which were not directly influenced. The influx of volcanic detritus in the Triassic of south Victoria Land may simply record the integration of drainage with that of the central Transantarctic Mountains, which originated in a region of volcaniclastic sediment input.

Early Jurassic time: rifting

Lower Jurassic strata

The far-field effects of subduction along the Gondwana active margin, which had influenced the evolution of the Transantarctic Mountains during Permian and Triassic time, were brought to a close with the events leading up to the basaltic magmatism which formed the precursor to Gondwana break-up. The pre-basalt Lower Jurassic stratigraphic record, characterized by silicic volcaniclastic beds, is sparse and largely confined to those areas where the Ferrar extrusive rocks are still present (Fig. 8).

These rocks are exposed primarily in the central Transantarctic Mountains (Hanson Formation; Elliot 1996) and in north Victoria Land (Shafer Peak Formation; Schöner *et al.* 2007). In the central Transantarctic Mountains the sequence, disconformable on the Falla Formation, consists of tuffaceous sandstones and finer-grained beds and tuffs. Arkosic sandstones with coarse K-feldspar clasts are interbedded in the lower part of the sequence, indicate a proximal basement source

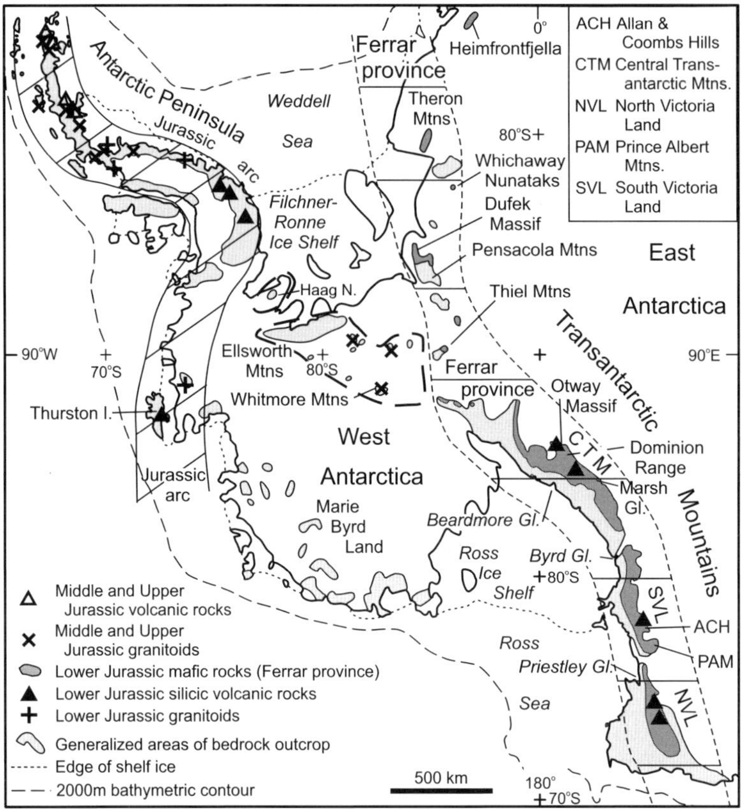

Fig. 8. Present-day distribution of Lower Jurassic silicic and basaltic rocks in the Transantarctic Mountains and Jurassic magmatic rocks in West Antarctica and the Antarctic Peninsula.

and suggest the possibility of deposition in a rift (Elliot & Larsen 1993).

In south Victoria Land there is only one location known so far where silicic shard-bearing sedimentary rocks overlie the Triassic Lashly Formation, and that is at Coombs Hills (Elliot & Grimes 2011). The section is no more than a few tens of metres thick, and the upper contact is missing. However, silicic tuff clasts in the overlying Mawson Formation basaltic pyroclastic rocks suggest a formerly more widespread occurrence of silicic strata, now removed by erosion.

By contrast, a more extensive record is present north of the Priestley Glacier in north Victoria Land (Schöner *et al.* 2007; Viereck-Goette *et al.* 2007). There, the siliciclastic Section Peak Formation passes up into a silicic shard-bearing sandstone sequence informally named the Shafer Peak Formation. This sequence consists of tuffaceous sandstones and siltstones, and is reported to be interbedded with Ferrar basaltic pyroclastic rocks of the Exposure Hill Formation (redefined by Viereck-Goette *et al.* 2007).

The silicic shard-bearing sequences are overlain by basaltic pyroclastic rocks and then flood lavas of the Ferrar Large Igneous Province (Elliot & Fleming 2008). The pyroclastic rocks are principally lahars, tuff breccias and tuffs. In the Queen Alexandra Range, central Transantarctic Mountains, they form stratigraphic sequences up to *c.* 200 m thick, whereas in the Prince Albert Mountains they are only a few tens of metres thick, although the basal and top contacts are not exposed. At Allan Hills (south Victoria Land) stratified pyroclastic rocks are cut by diatremes (Elliot *et al.* 2005; Ross *et al.* 2008). In contrast, in both north Victoria Land and at Coombs Hills (south Victoria Land), and probably also at Otway Massif (central Transantarctic Mountains), the pyroclastic rocks fill shallow-level phreatic vents or diatremes in which the vertical extent of the deposit may be several hundred metres (Elliot & Hanson 2001; White & McClintock 2001; Viereck-Goette *et al.* 2007). Thin sequences of pyroclastic rocks also occur, both as rafts in the vents and overlying the diatremes.

With the few exceptions where breccias and lavas are interbedded, the pyroclastic activity was abruptly replaced by effusion of massive lava flows that form sequences as much as 900 m thick (Elliot & Fleming 2008). Thin sedimentary interbeds occur low in the lava succession and beneath the capping lava flow. Many of the flows are more than 100 m thick and locally attain as much as 230 m at the Otway Massif. In the Prince Albert Mountains a massive pillow palagonite succession (*c*. 100 m thick) occurs at the base of the lava pile and together with the great thickness of many lava flows in the central Transantarctic Mountains and north Victoria Land suggest confining topography at the time of emplacement. In support of a rift setting is the occurrence of monoclinal structures in the central Transantarctic Mountains, one of which, west-facing, offsets the Hanson Formation and has pyroclastic breccia of the Prebble Formation injected into the fault plane (Elliot & Larsen 1993), and another of which displaces Kirkpatrick lavas. These data are interpreted to suggest that the Jurassic rocks were emplaced in a rift that was tectonically active through much of the Early Jurassic. A west-facing faulted monocline displaces Permian beds east of the Marsh Glacier (Barrett *et al.* 1970), a NW-facing normal fault offsets Buckley and Fremouw beds at Coalsack Bluff (Collinson & Elliot 1984), and an uninvestigated east-facing faulted monocline is also visible on the west face of the Dominion Range, offsetting Victoria Group beds (Elliot *et al.* 1974). Further, a small graben structure offsets Victoria Group beds near the head of the Beardmore Glacier (Barrett & Elliot 1973). These undated structures are attributed to Early Jurassic tectonism because they occur some distance back from the Cenozoic escarpment (discussed later).

Sills ranging in thickness up to 700 m, but more commonly 100–200 m, were emplaced into the underlying Beacon sequence throughout the Transantarctic Mountains, and locally into basement rocks (Elliot & Fleming 2004). In complete Beacon sections, the cumulative thickness of sills may be as much as 1500 m, suggesting that the sills are the dominant component of the Ferrar Large Igneous Province.

The source, or sources, for the Ferrar magmas, and their dispersal, remains uncertain. Ferrar magmas are characterized by enriched initial Sr and Nd isotope ratios and by trace element characteristics with a strong crustal imprint (Mensing *et al.* 1984; Hergt *et al.* 1989*a*; Fleming *et al.* 1995, 1997; Molzahn *et al.* 1996). Almost all show great geochemical coherence. The exceptions are the capping lavas together with some of the sills in the Theron Mountains and Whichaway Nunataks, which have a single similar but very distinct, coherent and evolved composition (Elliot *et al.* 1999; Leat *et al.* 2006). The geochemical characteristics of the latter have been interpreted to demonstrate derivation from a single source that was located in the region of Gondwana break-up between Antarctica and Africa and from which magmas were transported at depth throughout the Transantarctic Mountains (Elliot *et al.* 1999). This origin has been extended to the Ferrar as a whole, which includes southeastern Australasia and lava in South Africa of possible Ferrar composition (Elliot & Fleming 2000, 2004). In this scenario, the Ferrar magmas acquired their enriched crustal signature at the point of generation, probably a mantle plume (Storey 1995; Storey & Kyle 1997; Storey *et al.* 2001). Subsequent evolution of the majority of magmas was by fractionation with a small amount of superimposed crustal contamination (Fleming *et al.* 1995, 1997). Although sourcing of the magmas from the Dufek Massif and transport through supracrustal sills has been proposed (Storey & Kyle 1997; Ferris *et al.* 2003), it seems unlikely given the single unique composition of the capping lava flow(s), which are spread over 1600 km, the contrasting geochemical differences (e.g. MgO content) in the majority of the lavas from any one region compared with the next (central Transantarctic Mountains v. south Victoria Land v. north Victoria Land), and the presence of basement highs (Ross High, and between south Victoria Land and north Victoria Land; Elliot & Fleming 2008). Alternatively, rather than a plume origin, derivation of the magmas from a linear source in the subjacent heterogeneous lithospheric mantle, possibly enriched by subducted sediment during the Palaeozoic and Early Mesozoic, has been proposed (Kyle *et al.* 1983; Cox 1988; Hergt *et al.* 1989*b*, 1991; Molzahn *et al.* 1996). However, given the heterogeneous mantle along a length of 3000 km and the varied crustal provinces through which the magmas would then have ascended, the coherence of the geochemical characteristics over such a distance seems to make this distributed source a problematic origin. Magma mixing and crustal anatexis have also been proposed as important mechanisms for developing the Ferrar geochemistry (e.g. Mensing *et al.* 1984, 1991).

The rift setting proposed for the Ferrar rocks and the underlying volcaniclastic beds suggests that the setting might have been a series of linked rift basins (Elliot & Fleming 2008). The rift basins are now marked by phreatomagmatic deposits that preceded eruption of the lavas, and these sites formed the foci for subsequent major sill and lava emplacement. The magmas themselves were transported long distances at deep crustal levels or at the crust–mantle boundary prior to vertical migration into and through the upper crust and supracrustal rocks. Major dyke swarms for magma transport have not been recognized in outcrop, although they have been inferred from magnetic data near the Dufek Massif (Ferris *et al.* 2003) and in the central Transantarctic Mountains (Goodge & Finn 2010). Bounding faults for this proposed rift system have not been identified, although the Hanson Formation arkoses and quartzose pebbly beds indicate that at least one fault could not have been too far distant; given the local thickness of Victoria Group strata, that fault could have had a cumulative displacement of as much as 2 km. It is possible that the major discontinuity, the Marsh Fault, trending north–south along the Marsh Glacier (central Transantarctic Mountains) and juxtaposing the Miller Range Proterozoic rocks with the Victoria Group to the east (Goodge & Finn 2010), represents such a bounding fault. Faults, defining a small graben, also cut Victoria Group beds at the head of the Beardmore Glacier, and lie along-strike from the Marsh Fault, but the offsets are at least an order of magnitude less.

The Ferrar Large Igneous Province, as defined by its geochemical characteristics, extends into Tasmania and southeastern Australia (Hergt *et al.* 1991), New Zealand (Mortimer *et al.* 1995), and possibly into South Africa (Elliot & Fleming 2000). The precise temporal and source region relationships with the Karoo Large Igneous Province remain uncertain, and compared with the Ferrar province, Karoo rocks have a wide range of distinct magma compositions (Marsh *et al.* 1997).

The relationship between the silicic and basaltic magmatism remains unclear. Silicic magmatism was initiated with the onset of Hanson Formation deposition, and although a hiatus is inferred at the contact of the Falla and Hanson formations, the timing is not adequately constrained. Zircons from a tuff at Mt Kirkpatrick, about 100 m below the Hanson/Prebble contact, gave a super high resolution ion microprobe U–Pb age of about 194 Ma, and from a tuff 50 m below the contact yielded an age of about 186 Ma (Elliot & Fanning unpublished data). A super high resolution ion microprobe U–Pb zircon age of 182.7 ± 1.8 Ma on a silicic tuff clast in the basaltic Prebble Formation at the Otway Massif (Elliot *et al.* 2007) together with silicic shards in the upper interbed (palaeosol) at Mount Bumstead, central Transantarctic Mountains (Elliot *et al.* 1991), suggest that silicic volcanism continued into and during Ferrar time. At the type section of the Hanson Formation, the upper tuff unit includes accretionary lapilli and other evidence for proximal volcanism. Thus the older volcanism recorded in the Hanson Formation could have been distal plate margin related, whereas the youngest volcanism was proximal and possibly then anatectic and related to emplacement of Ferrar magmas, which have a best estimate U–Pb zircon and baddeleyite thermal ionization mass spectrometry age of 183.6 ± 1.8 Ma (Encarnación *et al.* 1996). The ages of the Ferrar rocks are under further investigation (Burgess *et al.* 2011).

Early Jurassic plate margin

The long-standing active margin of Gondwana continued into the Jurassic Period (Fig. 9). The record is clear in New Zealand, where it is documented by forearc volcaniclastic strata of the Murihiku and Pahau terranes (Adams *et al.* 2002; Mortimer 2004; Wandres & Bradshaw 2005), and a single Lower Jurassic granite from the remote Bounty Island (Wandres & Bradshaw 2005; Adams 2008), but it is very sparse in West Antarctica where there are but two igneous rocks of Early Jurassic age in the Thurston Island block (Pankhurst *et al.* 1993). In the Antarctic Peninsula the record is more substantial, with both intrusive (Leat *et al.* 1995, 2009; Vaughan & Storey 2000) and extrusive rocks (Pankhurst *et al.* 2000; Riley

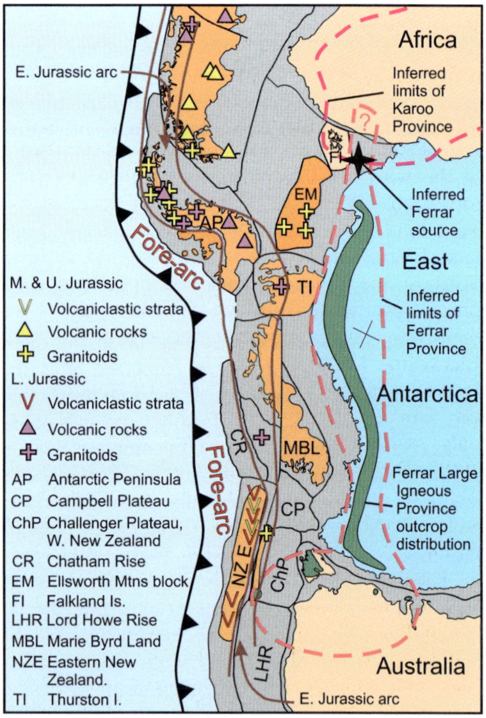

Fig. 9. Early Jurassic Gondwana reconstruction showing the distribution of magmatic rocks associated with the Karoo–Ferrar Large Igneous Province and the magmatic arc along the Gondwana margin. See caption for Figure 5 for information on the reconstruction.

et al. 2001; Hunter *et al.* 2006) demonstrating igneous activity in the Early Jurassic, although magmatism was much more extensive in later Jurassic time. The arc continued into southern South America, where it is recorded by I-type granites in the sub-cordilleran belt of Patagonia (Rapela *et al.* 2005), although the principal igneous activity of Early Jurassic age was eruption of the silicic Marifil Province in northern Patagonia (Pankhurst *et al.* 2000).

The setting for the Late Triassic to Early Jurassic magmatism in the Antarctic Peninsula is postulated to have been extensional (Storey *et al.* 1996; Vaughan & Storey 2000; Vaughan *et al.* 2013). However, the magmatic rocks emplaced during this time interval are interpreted to be a mixture of subduction-related and intra-plate granitoids (Millar *et al.* 2001). If the extensional setting of the Antarctic Peninsula sector of the Gondwana margin continued into West Antarctica, then the rift setting inferred for the Ferrar Large Igneous Province may have been a function of related back-arc extension, resulting from trench rollback, which thus created the deep crustal pathway for dispersal of Ferrar basaltic magmas.

Mid-Jurassic to mid-Cretaceous time: plate margin reorganization

This is a crucial time interval in the tectonic history of Antarctica, leading from an intact Gondwana to the now-isolated continent with the Ellsworth Mountains–Whitmore Mountains tectonic block, Thurston Island block, and others such as the Falkland Islands, moved from their postulated locations in the proto-Weddell sea region to their present positions. Based on palaeomagnetic data, there was major reorganization and rotation of the Ellsworth–Whitmore block (*c.* 90°) and the Falkland Islands (*c.* 150°) by 175 Ma, whereas the oldest Thurston Island block rotation is constrained only to be earlier than 125 Ma (Grunow *et al.* 1991). By that time the Ellsworth–Whitmore and Thurston Island blocks and the Antarctic Peninsula formed a single entity referred to as Weddellia. Subsequently Weddellia moved essentially as a single unit, rotating to its present position with respect to East Antarctica by about 110 Ma (Grunow *et al.* 1987, 1991). Displacements and rotations of the fragmented Gondwana margin have been attributed to the arrival of the Ferrar–Karoo mantle plume (Dalziel *et al.* 2000; Storey *et al.* 2001). However, the actual mechanisms and paths for these early movements remain unclear, although an analogy has been drawn with the Afar region (Dalziel & Lawver 2001) and with back-arc spreading and subduction rollback (Martin 2007). Further, aerogeophysical data for the head of the Weddell Sea have been interpreted to support a rift–rift–rift triple junction above a rising plume (Ferris *et al.* 2000).

Plate margin rearrangement must have occurred during a short time interval, for which there is evidence of extension in the Antarctic Peninsula region and continuing, although diminished, subduction (Vaughan & Storey 2000). Despite extensive marine geophysical surveys in the southern South Atlantic and Weddell Sea (König & Jokat 2006), the early stages of break-up are still obscure and it is only after anomaly M20 (167 Ma), the oldest identified anomaly which is located at the head of the Weddell Sea, that data allow interpretation of seafloor evolution. That anomaly determines by when the movement of the Ellsworth–Whitmore block and formation of the thinned crust of the Weddell embayment must have been effectively completed. Nevertheless, following those major movements, several phases of tectonism have been recognized for the Antarctic Peninsula, including possible accretion of suspect terranes (Vaughan & Storey 2000).

All the major translations and rotations must have occurred in a 15 m.y. interval following Karoo–Ferrar magmatism at about 183 Ma, and must have entailed movements along major shear zones and rotations between the various blocks. Further, there must have been stretching of continental crust and development of small oceanic basins (Dalziel & Lawver 2001). Later events (167–125 Ma) must have resulted from continuing Weddell Sea opening and plate margin development.

Cretaceous to Cenozoic time: rifting and uplift/denudation

The original thickness of the Ferrar lava sequence is uncertain and it is possible that a substantial section has been stripped. Studies of zeolites in basalt sequences have led to the recognition of zonations that must be related to temperature and thickness of the volcanic pile, fluid composition and exothermic hydration reactions (e.g. Walker 1960; Neuhoff *et al.* 2000). Zeolite zonation has been used to infer original thicknesses of volcanic piles. In Ferrar extrusive rocks, zeolite minerals filling amygdales, vugs and cavities between pillow basalts consist principally of stilbite and heulandite, with lesser scolecite and mesolite, plus apophyllite (Fleming *et al.* 1999). The absence of laumontite suggests that the original thickness of Kirkpatrick lavas could have been as much as 2500 m, although it was more likely to be some 1500 m or less (Neuhoff *et al.* 2000). Further, there is the possibility that the overburden included a now-eroded

Jurassic to Cretaceous sedimentary cover that accumulated within the postulated Ferrar rift.

Fission track studies throughout the Transantarctic Mountains (see Miller *et al.* 2010 and references therein) have shown several late Mesozoic episodes of denudation (Fig. 10), but the principal event was initiated in Eocene time. The oldest and only Jurassic (165–150 Ma) denudation is recorded in the Thiel Mountains (Fitzgerald & Baldwin 2007); elsewhere in the Transantarctic Mountains significant middle Cretaceous (ca. 125–90 Ma) denudation is recorded in the Scott Glacier region (Fitzgerald & Stump 1997), central Transantarctic Mountains (Fitzgerald 1994) and both south and north Victoria Land (Fitzgerald & Gleadow 1988; Fitzgerald 1992; Balestrieri *et al.* 1997; Lisker 2002; Lisker *et al.* 2006; Storti *et al.* 2008). In support of Cretaceous denudation are the Rb–Sr and $^{40}Ar/^{39}Ar$ age determinations on apophyllite in Ferrar basaltic rocks, which are interpreted to reflect uplift and reorganization of ground water systems with consequent mineral precipitation (Fleming *et al.* 1999; Molzahn *et al.* 1999). Further, in north Victoria Land, low-temperature alteration of Ferrar lavas in the Cretaceous is also recorded by Rb–Sr relationships, K–Ar dates and palaeomagnetic pole positions (summarized in Faure & Mensing 1993; Fleming *et al.* 1993). The Transantarctic Mountains are segmented geographically and topographically, and this is consistent with the fission-track data, which suggest a number of individual blocks, delineated by major outlet glaciers that have slightly different histories. Those outlet glaciers undoubtedly follow major faults, which are also suggested by offsets of the Kukri Erosion Surface. These offsets of a few hundred metres are small compared with the lateral persistence of this erosion surface over a distance of more than 2000 km, which implies a broadly similar uplift history along the length of the Transantarctic Mountains (Fitzgerald 1994, 2002). Apophyllite dates from replacement deposits in Permian sandstone provide further evidence of groundwater events, possibly related to uplift, at about 65 Ma (Elliot *et al.* 2004).

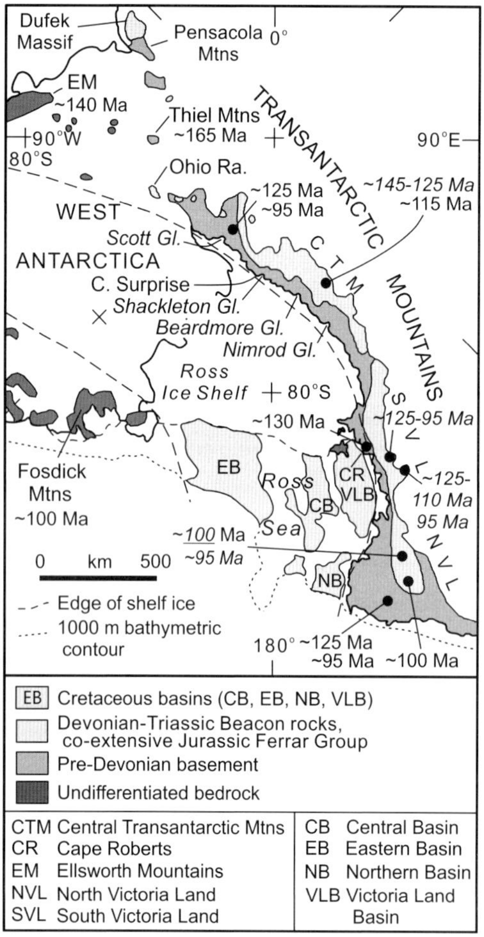

Fig. 10. Late Cretaceous extension and indicators in the Transantarctic Mountains of the onset of uplift/denudation during Late Jurassic and Cretaceous time. The West Antarctic Rift System is located between the long dashed lines; inferred Cretaceous basins from Decesari *et al.* (2007). Fission track ages are in upright lettering; in italics are Rb–Sr and $^{40}Ar/^{39}Ar$ apophyllite ages; underlined are ages for Rb–Sr ‘errorchrons’ for Kirkpatrick Basalt lavas. See text for references.

The Cenozoic uplift/denudation is associated with faulting along the Transantarctic Mountains front, which coincides with the rapid change in crustal thickness (40–20 km) from the East Antarctic basement to the Ross embayment (Lawrence *et al.* 2006), that is, the West Antarctic Rift System (e.g. Behrendt 1999). Faults along the front have been inferred or documented in south Victoria Land and the central Transantarctic Mountains (Beardmore and Shackleton glacier areas) from offset Beacon strata or basement rock thermochronology (Fitzgerald 1992, 1994; Miller *et al.* 2010). Only at one locality, Cape Surprise at the mouth of the Shackleton Glacier, is there outcrop of Permian Victoria Group strata that enables a direct estimate of the offset (*c.* 2.5 km) of the Kukri Erosion Surface along the mountain front (Miller *et al.* 2010). However, an even larger offset (>3 km) has been estimated from the Cape Roberts Project drill site number three (CRP-3), 10 km off Cape Roberts, south Victoria Land, which encountered Taylor Group (Devonian) quartz sandstone *c.* 1100 m below sea-level beneath Cenozoic glacimarine strata (Cape Roberts Science Team 2000, Barrett 2007). Comparable faults doubtless occur along other segments of the Transantarctic Mountains.

The uplift of the Transantarctic Mountains is clearly linked in some manner to the change in crustal thickness and the extensional regime of the West Antarctic Rift System. Investigations in Marie Byrd Land have shown that the extensional regime was initiated in Late Cretaceous time at about 105 Ma, and is documented by a core complex that formed while subduction along the margin was still active. This is supported by a fission track age of *c.* 100 Ma, indicating uplift/denudation at that time (Siddoway 2008). Correlated erosion surfaces in Marie Byrd Land and the South Island of New Zealand have been linked to this denudation in Late Cretaceous time (LeMasurier & Landis 1996). Much of the extension, up to 600 km, in the West Antarctic rift system is attributed to these Cretaceous events, during which four sedimentary basins were formed and filled by up to 4 km of sediment (Decesari *et al.* 2007). Compared with the Cenozoic, Cretaceous tectonism on the Transantarctic Mountains flank of the extensional rift system was limited, although widespread, and is documented in the denudation inferred from fission track data (Fig. 10). The separation of Campbell Plateau SE of New Zealand from Antarctica (Marie Byrd Land) started at about 84 Ma, and is a separate event that is notable for cutting across pre-existing extensional structures.

Based on fission track data, major denudation and uplift of the Transantarctic Mountains were initiated during the Eocene at about 50–55 Ma (Fig. 11), and there is a suggestion that the denudation event becomes younger southward from north Victoria Land to the central Transantarctic Mountains (Fitzgerald 2002; Miller *et al.* 2010). The onset in north Victoria Land is somewhat earlier than the opening of the Adare Trough, which was active from 43 to 26 Ma (Cande & Stock 2006). Adare Trough opening is linked to further extension in the Ross Sea basins, contributing to the increased thicknesses (up to −14 km in the Terror Rift) of strata in these basins (Trey *et al.* 1999). In a detailed analysis of seismic and core data from the Cape Roberts/McMurdo Sound region, five stages of tectonic activity and sedimentation have been recognized for the Victoria Land Basin (Fielding *et al.* 2006). The Ross Sea basins, located in rifts with thin crust, as little as 4 km in the Terror Rift (Davey & Brancolini 1995), surely extend south beneath the Ross Ice Shelf, and have been inferred from depth to basement and gravity anomalies (Decesari *et al.* 2007). Nevertheless, a 140 km-long seismic transect from the Ross Ice Shelf into the mountain front between the Nimrod and Beardmore glaciers revealed neither a large offset in thick sedimentary sequences beneath the western margin of the Ross Ice Shelf, nor a graben that might be the extension of the Victoria Land Basin/Terror Rift system, just where it might be expected given the known offsets in the frontal part of the range in the central Transantarctic

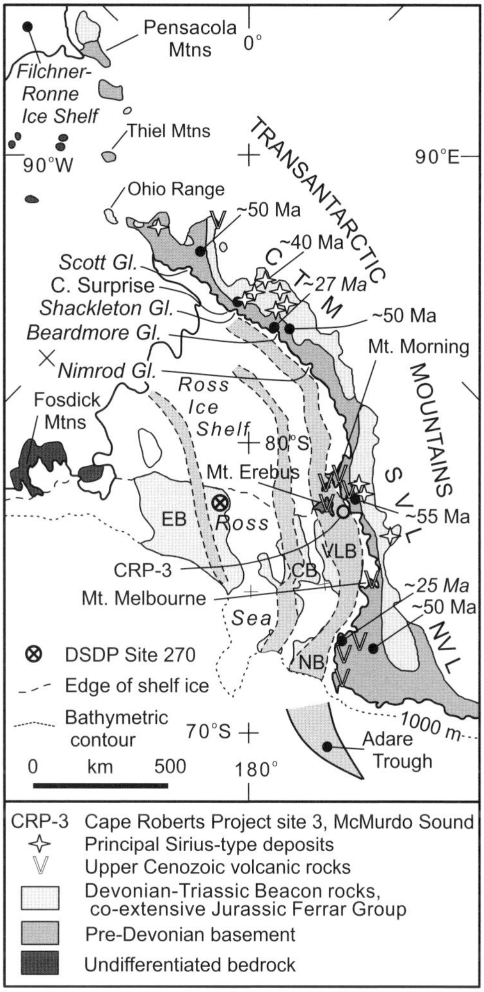

Fig. 11. Distribution of Cenozoic volcanic rocks, Sirius Group deposits in the Transantarctic Mountains, location of sedimentary basins and areas of extension in the Ross Sea, and interpreted basins (and extension) beneath the Ross Ice Shelf. Fission track ages are for onset of the principal episode of uplift/denudation in Eocene time, except for those in italics, which are for youngest ages given by fission track analysis. CB, Central Basin; EB, Eastern Basin; NB, Northern Basin; VLB, Victoria Land Basin. Light shading marks the extent of the inferred Cretaceous basins beneath the Ross Sea (see Fig. 10); heavy dashed lines with darker shading mark crust highly extended in Cenozoic time (Decesari *et al.* 2007) and the Cenozoic basins beneath the Ross Sea and by inference beneath the Ross Ice Shelf.

Mountains (ten Brink *et al.* 1993; Fitzgerald 1994). Another apparent anomaly is the relative youth of the oldest strata yet cored compared with the start of denudation of the Transantarctic Mountains: earliest Oligocene (34 Ma) for the Victoria Land Basin (Cape Roberts Project drill site number three, or CRP-3; Barrett 2007); and late Oligocene for the Eastern Basin (DSDP Site 270; Ford & Barrett 1975; Fig. 11). However, because sedimentation was occurring in a region of numerous faults, this anomaly may simply reflect where cores have been obtained.

The early Cenozoic phase of extension was followed by the onset of alkaline basaltic magmatism of the McMurdo Volcanic Group (LeMasurier & Thomson 1987), which was initiated by 48 Ma, the age of an alkali gabbro in north Victoria Land (Rocchi *et al.* 2003), although the age of the oldest extrusive rock, an ash from the Cape Roberts Project drill site number three (CRP-3), is much younger (24 Ma; Barrett 2007). Alkaline volcanism extended as far south as Mt Morning in south Victoria Land, but with two outliers near the head of the Scott Glacier (Stump *et al.* 1980). Active volcanism continues today at Mt Erebus and fumarolic activity is present at Mt Melbourne. On the other West Antarctic Rift System flank (Marie Byrd Land) alkaline volcanism (LeMasurier & Thomson 1987) started earlier (*c.* 36 Ma; Wilch & McIntosh 2000) and continues to the present in the fumarolic activity at Mt Berlin (76°S 136°W). The relationship, if any, between the Marie Byrd Land and McMurdo volcanic provinces, separated by several hundred kilometres, remains uncertain, although both tap deep mantle plume sources (Hole & LeMasurier 1994). The relationships between onset of rapid uplift (*c.* 55 Ma), the apparent onset of sedimentation (*c.* 34 Ma) and the onset of major alkaline volcanism (*c.* 24 Ma) remain uncertain.

The Transantarctic Mountains have the appearance of a classic rift shoulder, but the mechanism for uplift remains controversial. The range clearly marks a major geological boundary with contrasting geological provinces in West Antarctica. However, the topographic expression is much subdued in the segment from the Ohio Range through to the Weddell Sea, where the Ellsworth–Whitmore Mountains block abuts the Transantarctic Mountains (Storey *et al.* 1988), and in the Pensacola Mountains region where a significant geological boundary (crustal thickness change and/or major faulting) is recognized geophysically (Behrendt *et al.* 1974).

Discussion of the origin of the high topography of the Transantarctic Mountains, the mechanism of its formation and the relationship to the abrupt change in crustal thickness has been centred on the Ross Sea sector. More data exist for the Ross embayment than for the Weddell Sea sector, where there is a comparable vast province of thin crust beneath the Filchner–Ronne Ice Shelf. However, it should be noted that the latter region must differ in origin, dating from the initial break-up of Gondwana and probably for the most part completed by mid-Cretaceous time. Further, Stern *et al.* (2005) proposed that as much as half of the peak elevation of the Transantarctic Mountains in the Ross Sea sector may be from isostatic rebound, resulting from a combination of glacial incision and permanently frozen high peaks for the last 14 Ma, rather than large-scale tectonic processes.

An origin for the Transantarctic Mountains in the Ross Sea sector as simple rift shoulder uplift, resulting from extension in the West Antarctic Rift System, has been the long-standing interpretation (Davey & Brancolini 1987), and in this interpretation Marie Byrd Land has been considered the opposing rift shoulder. Uplift has been attributed to flexural uplift at the boundary between the contrasting regimes of the cold thick East Antarctic basement and the hot thin West Antarctic Rift System (e.g. Stern & ten Brink 1989). Lack of data to support the theoretical models led ten Brink *et al.* (1997) to propose that transtensional motion during the Eocene between East and West Antarctica, and resulting from plate re-organization in the SW Pacific Ocean, was the cause. This motion promoted a crustal break that enabled uplift of isostatically uncompensated crust and created the Transantarctic Mountains escarpment. There is accumulating evidence for transtensional motion in the north Victoria Land–western Ross Sea region (e.g. Wilson 1995; Storti *et al.* 2008), linking the Adare Trough, NW–SE structures in north and south Victoria Land, and the north–south oriented Victoria Land Basin.

A different approach has been through numerical modelling, which suggested the possibility that West Antarctica was a region of thick crust and high elevation, and subsequent extensional collapse has left the Transantarctic Mountains as a plateau margin remnant (Studinger *et al.* 2004; Bialas *et al.* 2007). This model formed the starting point for more recent numerical modelling of the relations between the Transantarctic Mountains and the West Antarctic Rift System (van Wijk *et al.* 2008). In support of the plateau, remnants of earlier drainage patterns scattered through the Transantarctic Mountains, with flow away from West Antarctica, have been advocated (Huerta 2007). The plateau, presumably coincident with the future West Antarctic Rift System, is postulated to have existed for some time until the mid-Cretaceous (*c.* 110 Ma), when subduction on the Pacific margin ceased and extension began with core complex development in Marie Byrd Land (Siddoway 2008). However, from break-up of Gondwana at *c.* 183 Ma until mid-Cretaceous time, the plate margin was being reorganized. This involved major rotations and displacements of various continental blocks including Marie Byrd Land, which according to palaeomagnetic data underwent major translation in mid-Cretaceous time (DiVenere *et al.* 1996). It is unclear how a high plateau could have been maintained under those circumstances. Further, it is unclear how plateau drainage would be distinguished from that on the gently sloping back side of a rift shoulder uplift. Cretaceous strata with a West Antarctic provenance on the East Antarctic flank of the Transantarctic Mountains would be an argument for the plateau model.

Fission track dating indicates uplift/denudation episodes as young as 24 Ma between north and south Victoria Land (Storti *et al.* 2008) and 20 Ma in the Beardmore Glacier region (Fitzgerald 1994). However, the amount of uplift in the last 14 Ma is relatively modest. In the Shackleton Glacier region it is estimated to be 290–790 m (Miller *et al.* 2010). In south Victoria Land geological evidence in the form of fjord deposits indicates a similar uplift of about 600 m in the last 10 myr (Webb & Wrenn 1982), and terrestrial volcanic cones show a maximum uplift of <300 m (permissibly 0 m) in the last 2.6 myr (Wilch *et al.* 1993). Further, the elevation of Miocene near-shore marine sediments off the south Victoria Land coast is within a few hundred metres of sea-level in the Cenozoic Investigations in the western Ross Sea (CIROS-1) and Cape Roberts Project (CRP) drill sites in McMurdo Sound, and likewise suggest relative Late Cenozoic tectonic stability of the rift margin (Fitzgerald 2002; Barrett 2007).

Regardless of the mechanism of formation, the uplifted Transantarctic Mountains have been considered one of the centres for initiation of Antarctic glaciation, having become a well-defined feature of the continent by 34 Ma (Wilson *et al.* 2012). Glacial deposits at high elevations along the Transantarctic Mountains, and known as the Sirius Group, appear to represent early ice sheets warmer and more dynamic than today, reaching the continental margin while the range was still rising (Barrett 2013). The deposits occur in a variety of topographic settings many of which are unrelated to present topography and drainage, and at elevations ranging from 1100 to more than 3500 m above sea-level. The oldest of these Sirius deposits could date back to the early Oligocene and might exist in the region between the heads of the Beardmore and Shackleton glaciers or at the head of the Scott Glacier. In a few localities they have yielded a wide array of fossils suggesting tundra conditions. Interbedded ash has allowed dating of what are considered to be the youngest of such high-elevation glacial deposits in south Victoria Land at 14 Ma, and dates the transition from warm to cold glaciation (Lewis *et al.* 2007, 2008).

The development of the present-day Transantarctic Mountains was a multi-stage process, with initial rifting associated with Gondwana fragmentation and the emplacement of the Ferrar province. A Late Jurassic to Early Cretaceous episode is recorded by fission track ages from the continental interior to south Victoria Land, but there is no obvious link either to the plate margin, which was active during that time, or to a rifting event that is otherwise documented. The mid-Cretaceous marks the onset of widespread extension, documented in Marie Byrd Land by the core complex, fission track and other isotopic data from the Transantarctic Mountains, and formation of the West Antarctic Rift System. The final Cenozoic stage is recorded by further extension in the West Antarctic Rift System, fission track data from the Transantarctic Mountains, and the alkaline basaltic provinces in Marie Byrd Land and Victoria Land. The alkaline basaltic phreatomagmatic volcanic edifices at the head of the Scott Glacier, dated at about 20 Ma, suggest that other basaltic centres may be scattered along the length of the range and either were eroded or are concealed by ice, and of course they directly indicate that ice existed on a substantially uplifted Transantarctic Mountains during early Miocene time.

Summary

The Transantarctic Mountains constitute one of Earth's great intra-continental mountain belts, extending for more than 3000 km across the continent. Their origins go back to the late Neoproterozoic break-up of Rodinia when a rifted margin formed along the site of the Transantarctic Mountains. This rifted margin has played a significant role in subsequent geological events, beginning with the Cambrian to earliest Ordovician Ross orogeny.

A period of major denudation was followed by widespread siliciclastic sedimentation in Early Devonian time, and then by another hiatus through Late Devonian and Carboniferous time, possibly in part a consequence of continental glaciation. Sedimentation resumed during the Permian and Triassic Periods in intra-cratonic basins, one of which in the central Transantarctic Mountains evolved into a foreland basin. From early Palaeozoic through Triassic time the Panthalassan margin had a complex history that included episodic magmatic activity, and possibly rifting, continental ribbon formation and its subsequent accretion, and island arc accretion. Regardless of the margin evolution, there was no clear impact on sedimentation until middle Permian time when the central part of the mountain range (central Transantarctic Mountains) received a major influx of volcanic detritus. The setting changed with the inception of early Jurassic silicic volcanism, probably distal initially but proximal later. Silicic magmatism associated with back-arc extension along the Panthalassan margin was followed by major rifting and emplacement of the Ferrar Large Igneous Province.

Subsequent evolution was dominated first by disruption and reorganization of the Gondwana plate margin, with the rotation and translation of the Ellsworth–Whitmore Mountains block and other continental fragments, and then by episodic Transantarctic Mountains uplift culminating in a widespread 55 Ma event and subsequent denudation of the range. The episodic uplift of the range is linked in some uncertain manner, across a major crustal discontinuity, to extension in the adjacent Ross embayment. That extension, initiated in late Cretaceous time, formed the West Antarctic Rift System and its associated continental shelf basins. The mismatch between the major early Cenozoic uplift/denudation and mid to late Cenozoic basin filling and volcanism has yet to be resolved.

Preparation of this manuscript has been supported by the United States National Science Foundation Grant ANT-0944662. The author also wishes to acknowledge significant research support over many years from the Office of Polar Programs, National Science Foundation. Reviews by J. Bradshaw, I. Dalziel and A. Vaughan, and comments by the editor, P. Barrett, are greatly appreciated and have significantly improved the manuscript. Byrd Polar Research Center contribution no C-1424.

References

Adams, C. J. 2008. Geochronology of Paleozoic Terranes at the Pacific margin of Zealandia. *Gondwana Research*, **13**, 250–258.

Adams, C. J., Barley, M. E., Fletcher, I. R. & Pickard, A. L. 1998. Evidence from U–Pb zircon and $^{40}Ar/^{39}Ar$ muscovite detrital ages in metasandstones for movement of the Torlesse suspect terrane around the eastern margin of Gondwanaland. *Terra Nova*, **10**, 183–189.

Adams, C. J., Barley, M. E., Maas, R. & Doyle, M. G. 2002. Provenance of Permian–Triassic volcaniclastic sedimentary Terranes in New Zealand: evidence from their radiogenic isotope characteristics and detrital mineral age patterns. *New Zealand Journal of Geology and Geophysics*, **45**, 221–242.

Allibone, A. & Wysoczanski, R. 2002. Initiation of magmatism during the Cambrian–Ordovician Ross orogeny in southern Victoria Land, Antarctica. *Geological Society of America Bulletin*, **114**, 1007–1018.

Allibone, A. H., Jongens, R. *et al.* 2009. Plutonic rocks of the Median Batholith in eastern and central Fjordland, New Zealand: field relations, geochemistry, correlation and nomenclature. *New Zealand Journal of Geology and Geophysics*, **52**, 101–148.

Balestrieri, M. L., Bigazzi, G. & Ghezzo, C. 1997. Uplift-denudation of the Transantarctic Mountains between the David and the Mariner glaciers, northern Victoria Land (Antarctica): constraints by apatite fission track analysis. *In*: Ricci, C. A. (ed.) *The Antarctic Region: Geological Evolution and Processes*. Terra Antarctica, Siena, 547–554.

Barbeau, D. L., Davis, J. T., Murray, K. E., Valencia, V., Gehrels, G. E., Zahid, K. M. & Gombosi, D. J. 2010. Detrital-zircon geochronology of the metasedimentary rocks of north-western Graham Land. *Antarctic Science*, **22**, 65–78.

Barrett, P. J. 1969. *Stratigraphy and Petrology of the Mainly Fluviatile Permian and Triassic Beacon Rocks, Beardmore Glacier Area, Antarctica*. Institute of Polar Studies Report **34**. The Ohio State University, Columbus, OH.

Barrett, P. J. 1991. The Devonian to Triassic Beacon Supergroup of the Transantarctic Mountains and correlatives in other parts of Antarctica. *In*: Tingey, R. J. (ed.) *The Geology of Antarctica*. Oxford Monographs on Geology and Geophysics. Oxford University Press, Oxford, **17**, 120–152.

Barrett, P. J. 2007. Cenozoic climate and sea level history from glacimarine strata off the Victoria Land Coast, Cape Roberts Project, Antarctica. *International Association of Sedimentologists Special Publication*, **39**, 259–287.

Barrett, P. J. 2013. Resolving views on Antarctic Neogene glacial history – the Sirius debate. *Earth and Environmental Science Transactions of the Royal Society of Edinburgh* (*Proceedings of the 11th ISAES*), **104**, 29–51.

Barrett, P. J. & Elliot, D. H. 1973. *Reconnaissance Geologic map of the Buckley Island Quadrangle, Transantarctic Mountains, Antarctica*. United States Geological Survey, Antarctica, 1:250 000 Map A-3.

Barrett, P. J. & Kohn, B. P. 1975. Changing sediment transport directions from Devonian to Triassic in the Beacon Super-Group of south Victoria Land, Antarctica. *In*: Campbell, K. S. W. (ed.) *Gondwana Geology*. Australian National University Press, Canberra, 15–35.

Barrett, P. J., Lindsay, J. F. & Gunner, J. 1970. *Reconnaissance Geologic Map of the Mount Rabot Quadrangle, Transantarctic Mountains, Antarctica*. United States Geological Survey, Antarctica, 1:250 000 Map No. 1.

Barrett, P. J., Elliot, D. H. & Lindsay, J. F. 1986. The Beacon Supergroup (Devonian–Triassic) and Ferrar Group (Jurassic) in the Beardmore Glacier area, Antarctica. *In*: Turner, M. D. & Splettstoesser, J. F. (eds) *Geology of the Central Transantarctic Mountains*. Antarctic Research Series, American Geophysical Union, Washington, DC, **36**, 339–428.

Behrendt, J. C. 1999. Crustal and lithospheric structure of the West Antarctic Rift System from geophysical investigations – a review. *Global and Planetary Change*, **23**, 25–44.

BEHRENDT, J. C., HENDERSON, J. R., MEISTER, L. J. & RAMBO, W. L. 1974. *Geophysical investigations of the Pensacola Mountains and adjacent glacierized areas of Antarctica*. US Geological Survey, Reston, VA, Professional Papers, **844**, 27.

BIALAS, R. W., BUCK, W. R., STUDINGER, M. & FITZGERALD, P. G. 2007. Plateau collapse model for the Transantarctic Mountains–West Antarctic rift system: insights from numerical experiments. *Geology*, **35**, 687–690.

BOGER, S. D. 2011. Antarctica before and after Gondwana. *Gondwana Research*, **19**, 335–371.

BOMPAROLA, R. M., GHEZZO, C., BELOUSOVA, E., GRIFFIN, W. L. & O'REILLY, S. Y. 2007. Resetting of the U–Pb zircon system in Cambro-Ordovician intrusives in the Deep Freeze Range, northern Victoria Land, Antarctica. *Journal of Petrology*, **48**, 327–364.

BORG, S. G. & DEPAOLO, D. J. 1991. A tectonic model of the Antarctic Gondwana margin with implication for southeastern Australia: isotopic and geochemical evidence. *Tectonophysics*, **196**, 339–358.

BORG, S. G., DEPAOLO, D. J., WENDLANDT, E. D. & DRAKE, T. G. 1989. Studies of granites and metamorphic rocks, Byrd Glacier area. *Antarctic Journal of the United States*, **24**, 19–21.

BORG, S. G., DEPAOLO, D. J. & SMITH, B. M. 1990. Isotopic structure and tectonics of the central Transantarctic Mountains. *Journal of Geophysical Research*, **95**, 6647–6667.

BOUCOT, A. J., CASTER, K. E., IVES, D. & TALENT, A. J. 1963. Relationships of a new Lower Devonian terebratuloid (Brachiopod) from Antarctica. *Bulletins of American Paleontology*, **46**, 81–151.

BRACCIALI, L., DI VINCENZO, G., ROCHI, S. & GHEZZO, C. 2009. The Tiger Gabbro from northern Victoria Land, Antarctica: the roots of an island arc within the Palaeozoic margin of Gondwana. *Journal of the Geological Society, London*, **166**, 711–724.

BRADSHAW, J. D. 2007. The Ross Orogen and Lachlan Fold Belt in Marie Byrd Land, Northern Victoria Land and New Zealand: implications for the tectonic setting of the Lachlan Fold Belt in Antarctica. *In*: COOPER, A. K., RAYMOND, C. R. ET AL. (eds) *Antarctica: A Keystone in a Changing World*. Online Proceedings of the 10th ISAES, USGS Open-File Report **2007-1047**, Short Research Paper 059, 4; doi: 10.3133/of2007-1047.srp059.

BRADSHAW, M. A. In press. The Taylor Group (Beacon Supergroup): the Devonian sediments of Antarctica. *In*: HAMBREY, M. J. ET AL. (eds) *Antarctic Palaeoenvironments and Earth-Surface Processes*. Geological Society, London, Special Publications, **381**, http://dx.doi.org/10.1144/SP381.23.

BRADSHAW, M. A. & MCCARTEN, L. 1991. Paleoecology and systematics of Lower Devonian bivalves from the Horlick Formation, Ohio Range, Antarctica. *Alcheringa*, **15**, 1–42.

BRADSHAW, J. D., ANDREWS, P. B. & FIELD, B. D. 1983. Swanson Formation and related rocks of Marie Byrd Land and a comparison with the Robertson Bay Group of northern Victoria Land. *In*: OLIVER, R. L., JAMES, P. R. & JAGO, J. B. (eds) *Antarctic Earth Science*. Australian Academy of Science, Canberra, 274–279.

BRADSHAW, J. D., GUTJHAR, M., WEAVER, S. D. & BASSETT, K. N. 2009. Cambrian intra-oceanic arc accretion to the austral Gondwana margin: constraints on the location of proto-New Zealand. *Australian Journal of Earth Sciences*, **56**, 587–594.

BRADSHAW, J. D., VAUGHAN, A. P. M., MILLAR, I. L., FLOWERDEW, M. J., TROUW, R. A. J., FANNING, C. M. & WHITEHOUSE, M. J. 2012. Permo-Carboniferous conglomerates in the Trinity Peninsula Group at View Point, Antarctic Peninsula: sedimentology, geochronology and isotope evidence for provenance and tectonic setting in Gondwana. *Geological Magazine*, **159**, 626–644, doi: 10.1017/S001675681100080X.

BROOK, D. 1972. Stratigraphy of the Theron Mountains. *British Antarctic Survey Bulletin*, **29**, 67–89.

BUGGISCH, W. & KLEINSCHMIDT, G. 1991. Recovery and recrystallization of quartz and 'crystallinity' of illite in the Bowers and Robertson bay terranes, northern Victoria Land, Antarctica. *In*: THOMSON, M. R. A., CRAME, J. A. & THOMSON, J. W. (eds) *Geological Evolution of Antarctica*. Cambridge University Press, Cambridge, 155–160.

BUGGISCH, W. & KLEINSCHMIDT, G. 2007. The Pan-African nappe tectonics in the Shackleton Range. *In*: COOPER, A. K., RAYMOND, C. R. ET AL. (eds) *Antarctica: A Keystone in a Changing World*. Online Proceedings of the 10th ISAES, USGS Open-File Report **2007-1047**, Short Research Paper 058, 4, doi: 10.3133/of2007-1047.srp058.

BURGESS, S. D., FLEMING, T. H., ELLIOT, D. H. & BOWRING, S. A. 2011. High-precision U–Pb zircon geochronology of the Ferrar large Igneous Province: can sill emplacement be linked with the Pliensbachian–Toarcian extinction event? *Geological Society of America, Abstracts with Programs*, **43**, 372.

CANDE, S. C. & STOCK, J. 2006. Constraints on timing of extension in the Northern Basin, Ross Sea. *In*: FUTTERER, D. K., DAMASKE, D., KLEINSCHMIDT, G., MILLER, H. & TESSENSOHN, F. (eds) *Antarctica: Contributions to Global Earth Sciences*. Springer, Berlin, 319–326.

CAPE ROBERTS SCIENCE TEAM 2000. Studies from the Cape Roberts Project, Ross Sea, Antarctica. Initial Report on CRP-3. *Terra Antarctica*, **7**, 1–209 with supplement.

CAROSI, R., GIACOMINI, F., TALARICO, F. & STUMP, E. 2007. Geology of the Byrd Glacier discontinuity (Ross Orogen): new survey data from the Britannia Range, Antarctica. *In*: COOPER, A. K., RAYMOND, C. R. ET AL. (eds) *Antarctica: A Keystone in a Changing World*. Online Proceedings of the 10th ISAES, U.S.G.S Open-File Report **2007-1047**, Short Research Paper 030, 6; doi: 10.3133/of2007-1047.srp030.

CATHCART, J. B. & SCHMIDT, D. L. 1977. *Middle Paleozoic sedimentary phosphate in the Pensacola Mountains, Antarctica*. US Geological Survey, Reston, VA, Professional Papers, **456E**, 1–18.

CAYLEY, R. A. 2011. Exotic block accretion to the eastern Gondwanaland margin in the late Cambrian–Tasmania, the Selwyn Block, and implications for the Cambrian–Silurian evolution of the Ross, Delamerian, and Lachlan orogens. *Gondwana Research*, **19**, 628–649.

CLARKSON, P. D., TESSENSOHN, F., THOMSON, J. W. ET AL. 1995. *Geological Map of the Shackleton Range, Antarctica*. BAS GEOMAP Series, Sheet 4, 1:250 000,

Geological map and supplementary text. British Antarctic Survey, Cambridge.

COLLINSON, J. M. & ELLIOT, D. H. 1984. Geology of Coalsack Bluff, Antarctica. *In*: TURNER, M. D. & SPLETTSTOESSER, J. F. (eds) *Geology of the Central Transantarctic Mountains*. Antarctic Research Series. American Geophysical Union, Washington, DC, **36**, 97–102.

COLLINSON, J. W., PENNINGTON, D. C. & KEMP, N. R. 1986. Stratigraphy and petrology of Permian and Triassic fluvial deposits in northern Victoria Land, Antarctica. *In*: STUMP, E. (ed.) *Geological Investigations in Northern Victoria Land*. Antarctic Research Series. American Geophysical Union, Washington DC, **46**, 211–242.

COLLINSON, J. W., VAVRA, C. L. & ZAWASKIE, J. M. 1992. Sedimentology of the Polarstar Formation (Permian), Ellsworth Mountains, West Antarctica. *In*: WEBERS, G. F., CRADDOCK, C. & SPLETTSTOESSER, J. F. (eds) *Geology and Paleontology of the Ellsworth Mountains, West Antarctica*. Geological Society of America, Boulder, CO, Memoirs, **170**, 63–79.

COLLINSON, J. W., ELLIOT, D. H., ISBELL, J. L. & MILLER, J. M. G. 1994. Permian-Triassic Transantarctic Basin. *In*: VEEVERS, J. J. & POWELL, C. McA. (eds) *Permian–Triassic Pangaean Basins and Foldbelts along the Panthalassan Margin of Gondwanaland*. Geological Society of America, Boulder, CO, Memoirs, **184**, 173–222.

COLLINSON, J. C., HAMMER, W. R., ASKIN, R. A. & ELLIOT, D. H. 2006. Permian–Triassic boundary in the central Transantarctic Mountains, Antarctica. *Bulletin of the Geological Society of America*, **118**, 747–763.

COOPER, A. F., MAAS, R., SCOTT, J. N. & BARBER, A. J. W. 2011. Dating of volcanism and sedimentation in the Skelton Group, Transantarctic Mountains: implications for the Rodinia-Gondwana transition in southern Victoria Land, Antarctica. *Geological Society of America Bulletin*, **123**, 681–702.

COX, K. G. 1988. The Karroo Province. *In*: MACDOUGALL, J. D. (ed.) *Continental Flood Basalts*. Kluwer, Boston, MA, 239–271.

CURTIS, M. L. 2001. Tectonic history of the Ellsworth Mountains, West Antarctica: reconciling an Antarctic enigma. *Geological Society of America Bulletin*, **113**, 939–958.

CURTIS, M. L. & STOREY, B. C. 1996. A review of geological constraints on the pre-break-up position of the Ellsworth Mountains within Gondwana: implications for Weddell Sea evolution. *In*: STOREY, B. C., KING, E. C. & LIVERMORE, R. A. (eds) *Weddell Sea Tectonics and Gondwana Break-up*. Geological Society, London, Special Publications, **108**, 11–30.

DALZIEL, I. W. D. & LAWVER, L. A. 2001. The lithospheric setting of the West Antarctic Ice Sheet. *In*: ALLEY, R. B. & BINDSCHADLER, R. A. (eds) *The West Antarctic Ice Sheet: Behavior and Environment*. Antarctic Research Series. American Geophysical Union, Washington, DC, **77**, 29–44.

DALZIEL, I. W. D., LAWVER, L. A. & MURPHY, J. B. 2000. Plumes, orogenesis, and supercontinent fragmentation. *Earth and Planetary Science Letters*, **178**, 1–11.

DAVEY, F. J. & BRANCOLINI, G. 1987. The Late Mesozoic and Cenozoic structural setting of the Ross Sea region. *In*: COOPER, A. K., BARKER, P. F. & BRANCOLINI, G. (eds) *Geology and Seismic Stratigraphy of the Antarctic Margin*. Antarctic Research Series. American Geophysical Union, Washington, DC, **68**, 167–182.

DAVEY, F. J. & BRANCOLINI, G. 1995. The late Mesozoic and Cenozoic structural setting of the Ross Sea region. *In*: COOPER, A. K., BARKER, P. F. & BRANCOLINI, G. (eds) *Geology and Seismic Stratigraphy of the Antarctic Margin*. Antarctic Research Series, American Geophysical Union, Washington, DC, **68**, 167–182.

DECESARI, R. C., WILSON, D. S., LUYENDYK, B. P. & FAULKNER, M. 2007. Cretaceous and Tertiary extension throughout the Ross Sea, Antarctica. *In*: COOPER, A. K., RAYMOND, C. R. *ET AL.* (eds) *Antarctica: A Keystone in a Changing World*. Online Proceedings of the 10th ISAES, USGS Open-File Report **2007-1047**, Short Research Paper 098, 4; doi: 10.3133/of2007-1047.srp098.

DIVENERE, V., KENT, D. V. & DALZIEL, I. W. D. 1996. Summary of paleomagnetic results from West Antarctica: implications for the tectonic evolution of the Pacific margin of Gondwana during the Mesozoic. *In*: STOREY, B. C., KING, E. C. & LIVERMORE, R. A. (eds) *Weddell Sea Tectonics and Gondwana Break-up*. Geological Society, London, Special Publications, **108**, 31–43.

DOUBLEDAY, P. A., LEAT, P. T., ALABASTER, A., TRANTER, T. R. & NELL, P. A. R. 1994. Allochthonous oceanic basalts within the Mesozoic accretionary complex of Alexander Island, Antarctica: remnants of proto-Pacific oceanic crust. *Journal of the Geological Society, London*, **152**, 65–78.

ELLIOT, D. H. 1996. The Hanson Formation: a new stratigraphical unit in the Transantarctic Mountains, Antarctica. *Antarctic Science*, **8**, 389–394.

ELLIOT, D. H. & FANNING, C. M. 2008. Detrital zircons from upper Permian and lower Triassic Victoria Group sandstones, Shackleton Glacier region, Antarctica: evidence for multiple sources along the Gondwana plate margin. *Gondwana Research*, **13**, 259–274.

ELLIOT, D. H. & FANNING, C. M. 2011. Age constraints on the Gondwana sequence in the central Transantarctic Mountains from SHRIMP zircon geochronology. *11th International Symposium on Antarctic Earth Sciences*, Abstracts, 481.

ELLIOT, D. H. & FLEMING, T. H. 2000. Weddell triple junction: the principal focus of Ferrar and Karoo magmatism during initial breakup of Gondwana. *Geology*, **28**, 539–542.

ELLIOT, D. H. & FLEMING, T. H. 2004. Occurrence and dispersal of magmas in the Jurassic Ferrar Large Igneous Province, Antarctica. *Gondwana Research*, **7**, 223–237.

ELLIOT, D. H. & FLEMING, T. H. 2008. Physical volcanology and geological relationships of the Ferrar large Igneous Province, Antarctica. *Journal of Volcanology and Geothermal Research*, **172**, 20–37.

ELLIOT, D. H. & GRIMES, C. G. 2011. Triassic and Jurassic strata at Coombs Hills, south Victoria Land: stratigraphy, petrology and cross-cutting breccia pipes. *Antarctic Science*, **23**, 268–280, doi: 10.1017/S0954102010000994.

ELLIOT, D. H. & HANSON, R. E. 2001. Origin of widespread, exceptionally thick basaltic phreatomagmatic tuff breccia in the Middle Jurassic Prebble and Mawson formations, Antarctica. *Journal of Volcanology and Geothermal Research*, **111**, 183–201.

ELLIOT, D. H. & LARSEN, D. 1993. Mesozoic volcanism in the Transantarctic Mountains: depositional environment and tectonic setting. *In*: FINDLAY, R. H., UNRUG, R., BANKS, M. R. & VEEVERS, J. J. (eds) *Gondwana 8: Assembly, Evolution and Dispersal*. A.A. Balkema, Rotterdam, 397–410.

ELLIOT, D. H., BARRETT, P. J. & MAYEWSKI, P. A. 1974. *Reconnaissance Geologic Map of the Plunket Point Quadrangle, Transantarctic Mountains, Antarctica*. United States Geological Survey, Antarctica, 1:250 000 Map A-4.

ELLIOT, D. H., BIGHAM, J. & JONES, F. S. 1991. Interbeds and weathering profiles in the Jurassic basalt sequence, Beardmore Glacier region, Antarctica. *In*: ULBRICH, H. & ROCHA CAMPOS, A. C. (eds) *Gondwana Seven Proceedings*. Instituto Geosciências, Universidade de São Paulo, São Paulo, 653–667.

ELLIOT, D. H., FLEMING, T. H., KYLE, P. R. & FOLAND, K. A. 1999. Long distance transport of magmas in the Jurassic Ferrar Large Igneous Province, Antarctica. *Earth and Planetary Science Letters*, **167**, 87–104.

ELLIOT, D. H., FOLAND, K. A. & FLEMING, T. H. 2004. Dating paleohydrologic events with authigenic apophyllite: an example from the Permian of the central Transantarctic Mountains, Antarctica. *Geological Society of America, Abstracts with Programs*, **36**, 473.

ELLIOT, D. H., FORTNER, E. H. & GRIMES, C. B. 2005. Mawson breccias intrude Beacon strata at Allan Hills, south Victoria Land: regional implications. *In*: FUTTERER, D. K., DAMASKE, D., KLEINSCHMIDT, G., MILLER, H. & TESSENSOHN, F. (eds) *Antarctica: Contributions to Global Earth Sciences*. Springer, Berlin, 289–296.

ELLIOT, D. H., FLEMING, T. F., FOLAND, K. A. & FANNING, C. M. 2007. Jurassic silicic volcanism in the Transantarctic Mountains: is it related to plate margin processes or to Ferrar magmatism? *In*: COOPER, A. K., RAYMOND, C. R. ET AL. (eds) *Antarctica: A Keystone in a Changing World*. Online Proceedings of the 10th ISAES, USGS Open-File Report **2007-1047**, Short Research Paper 051, 5; doi: 10.3133/of2007-1047.srp051.

ENCARNACIÓN, J., FLEMING, T. H., ELLIOT, D. H. & EALES, J. V. 1996. Synchronous emplacement of Ferrar and Karoo dolerites and the early breakup of Gondwana. *Geology*, **24**, 535–538.

FAURE, G. & MENSING, T. M. 1993. K–Ar dates and paleomagnetic evidence for Cretaceous alteration of Mesozoic basaltic lava flows, Mesa Range, northern Victoria Land, Antarctica. *Chemical Geology*, **109**, 305–315.

FEDERICO, L., CAPPONI, G., CRISPINI, L. & BRADSHAW, J. D. 2009. The Cambrian Ross Orogeny in northern Victoria Land (Antarctica) and New Zealand: a synthesis. *Gondwana Research*, **15**, 188–196.

FERRIS, J. K., VAUGHAN, A. P. M. & STOREY, B. C. 2000. Relics of a complex triple junction in the Weddell Sea embayment. *Earth and Planetary Science Letters*, **178**, 215–230.

FERRIS, J. K., STOREY, B. C., VAUGHAN, A. P. M., KYLE, P. R. & JONES, P. C. 2003. The Dufek and Forrestal intrusions, Antarctica: a centre for Ferrar Large Igneous Province dike emplacement? *Geophysical Research Letters*, **30**, 1348, doi: 10.1029/2002GL016719.

FIELDING, C. R., HENRYS, S. A. & WILSON, T. J. 2006. Rift history of the western Victoria Land Basin: a new perspective based on integration of cores with seismic reflection data. *In*: FUETTERER, D. K., DAMASKE, D., KLEINSCHMIDT, G., MILLER, H. & TESSENSOHN, F. (eds) *Antarctica: Contributions to Global Earth Sciences*. Springer, Berlin, 309–318.

FINN, C. A., MOORE, D., DAMASKE, D. & MACKEY, T. 1999. Aeromagnetic legacy of early Paleozoic subduction along the Pacific margin of Gondwana. *Geology*, **27**, 1087–1090.

FIORETTI, A., CAPPONI, G., BLACK, L. P., VARNE, R. & VISONÀ, D. 2005. Surgeon Island granite SHRIMP zircon ages: a clue for the Cambrian tectonic setting and evolution of the Paleopacific margin of Gondwana (northern Victoria Land, Antarctica). *Terra Nova*, **17**, 242–249.

FITZGERALD, P. G. 1992. The Transantarctic Mountains of southern Victoria Land: the application of apatite fission track analysis to a rift shoulder uplift. *Tectonics*, **11**, 634–662.

FITZGERALD, P. G. 1994. Thermochronologic constraints on post-Paleozoic tectonic evolution of the central Transantarctic Mountains, Antarctica. *Tectonics*, **13**, 634–662.

FITZGERALD, P. G. 2002. Tectonics and landscape evolution of the Antarctic Plate since the breakup of Gondwana, with an emphasis on the West Antarctic rift system and the Transantarctic Mountains. *In*: GAMBLE, J. A., SKINNER, D. N. B. & HENRYS, S. (eds) *Antarctica at the Close of a Millennium*. The Royal Society of New Zealand, Bulletin, Wellington, New Zealand, **35**, 453–469.

FITZGERALD, P. G. & BALDWIN, S. L. 2007. Thermochronologic constraints on Jurassic flank denudation in the Thiel Mountains, Antarctica. *In*: COOPER, A. K., RAYMOND, C. R. ET AL. (eds) *Antarctica: A Keystone in a Changing World*. Online Proceedings of the 10th ISAES, USGS Open-File Report **2007-1047**, Short Research Paper 044, 4; doi: 10.3133/of2007-1047.srp044.

FITZGERALD, P. G. & GLEADOW, A. J. W. 1988. Fission track geochronology, tectonics and structure of the Transantarctic Mountains in northern Victoria Land, Antarctica. *Isotope Geoscience*, **73**, 169–198.

FITZGERALD, P. G. & STUMP, E. 1997. Cretaceous and Cenozoic episodic denudation of the Transantarctic Mountains, Antarctica: new constraints from apatite fission track thermochronology in the Scott Glacier region. *Journal of Geophysical Research*, **102**, 7747–7765, doi: 10.1029/96JB03898.

FITZSIMONS, I. C. W. 2003. Proterozoic basement provinces of southern and southwestern Australia and their correlation with Antarctica. *In*: YOSHIDA, M. & WINDLEY, B. F. (eds) *Proterozoic East Gondwana: Supercontinet Assembly and Breakup*. Geological Society, London, Special Publications, **206**, 93–130.

FLEMING, T. H., ELLIOT, D. H., FOLAND, K. A., JONES, L. M. & BOWMAN, J. R. 1993. Mid-Cretaceous disturbance of Rb-Sr and K-Ar isotopic systems in the Kirkpatrick basalt, north Victoria Land, Antarctica. *In*: FINDLAY, R. H., UNRUG, R., BANKS, M. R. & VEEVERS, J. J. (eds) *Gondwana 8: Assembly, Evolution and Dispersal*. A.A. Balkema, Rotterdam, 411–424.

FLEMING, T. H., FOLAND, K. A. & ELLIOT, D. H. 1995. Isotopic and chemical constraints on the crustal evolution and source signature of Ferrar magmas, North Victoria Land, Antarctica. *Contributions to Mineralogy and Petrology*, **121**, 217–236.

FLEMING, T. H., HEIMANN, A., FOLAND, K. A. & ELLIOT, D. H. 1997. $^{40}Ar/^{39}Ar$ geochronology of Ferrar Dolerite sills from the Transantarctic Mountains, Antarctica: implications for the age and origin of the Ferrar magmatic province. *Geological Society of America Bulletin*, **109**, 533–546.

FLEMING, T. H., FOLAND, K. A. & ELLIOT, D. H. 1999. Apophyllite $^{40}Ar/^{39}Ar$ and Rb-Sr geochronology: potential utility and application to the timing of secondary mineralization of the Kirkpatrick Basalt, Antarctica. *Journal of Geophysical Research*, **104**, 20 081–20 095.

FODEN, J. D., ELBURG, M. A., DOUGHERTY-PAGE, J. & BURTT, A. 2006. The timing and duration of the Delamerian Orogeny: correlation with the Ross Orogen and implications for Gondwana assembly. *Journal of Geology*, **114**, 189–210.

FORD, A. B. & BARRETT, P. J. 1975. Basement rocks of the south-central Ross Sea, Site 270. *In*: HAYES, D. E. & FRAKES, L. A. (eds) *Initial Reports of the Deep Sea Drilling Project*, U.S. Government Printing Office, Washington, DC, **28**, 861–868.

GIACOMINI, F., TIEPOLO, M., DALLAI, L. & GHEZZO, C. 2007. On the onset and evolution of the Ross Orogeny magmatism in north Victoria Land, Antarctica. *Chemical Geology*, **240**, 103–128.

GLEN, R. A. 2005. The Tasmanides of eastern Australia. *In*: VAUGHAN, A. M. P., LEAT, P. T. & PANKHURST, R. J. (eds) *Terrane Processes at the Margin of Gondwana*. Geological Society, London, Special Publications, **246**, 23–93.

GOODGE, J. W. & FANNING, C. M. 1999. 2.5 b.y. of punctuated Earth history as recorded in a single rock. *Geology*, **27**, 1007–1010.

GOODGE, J. W. & FANNING, C. M. 2002. Precambrian history of the Nimrod Group, central Transantarctic Mountains. *In*: GAMBLE, J. A., SKINNER, D. N. B. & HENRYS, S. (eds) *Antarctica at the Close of a Millennium*. Royal Society of New Zealand, Bulletin, Wellington, New Zealand, **35**, 43–50.

GOODGE, J. W. & FINN, C. A. 2010. Glimpses of East Antarctica: aeromagnetic and satellite magnetic view from the central Transantarctic Mountains of West Antarctica. *Journal of Geophysical Research*, **115**, 22, doi: 10.1029/2009JB006890.

GOODGE, J. W., FANNING, C. M. & BENNET, V. C. 2001. U–Pb evidence of ~1.7 Ga crustal tectonism during the Nimrod Orogeny in the Transantarctic Mountains, Antarctica: implications for Proterozoic plate reconstructions. *Precambrian Research*, **112**, 261–288.

GOODGE, J. W., WILLIAMS, I. S. & MYROW, P. 2004*a*. Provenance of Neoproterozoic and lower Paleozoic siliciclastic rocks of the central Ross orogen, Antarctica: detrital record of rift-, passive-, and active-margin sedimentation. *Geological Society of America Bulletin*, **116**, 1253–1279.

GOODGE, J. W., MYROW, P., PHILIPS, D., FANNING, C. M. & WILLIAMS, I. S. 2004*b*. Siliciclastic record of rapid denudation in response to convergent-margin orogenesis, Ross orogen, Antarctica. *In*: BERNET, M. & SPIEGEL, C. (eds) *Detrital Thermochronology*. Geological Society of America, Boulder, CO, Special Papers, **378**, 105–126.

GOODGE, J. W., FANNING, C. M., BRECKE, D. M., LICHT, K. J. & PALMER, E. F. 2010. Continuation of the Laurentian Grenville province in western East Antarctica. *Journal of Geology*, **118**, 601–619.

GRAY, D. R. & FOSTER, D. A. 2004. Tectonic evolution of the Lachlan Orogen, southeast Australia: historical review, data synthesis and modern perspectives. *Australian Journal of Earth Sciences*, **51**, 772–817.

GRINDLEY, G. W., OLIVER, P. J. & SUKROO, J. C. 1980. Lower Mesozoic position of southern New Zealand determined from paleomagnetism of the Glenham porphyry, Murihiku terrane, Eastern Southland. *In*: CRESSWELL, M. M. & VELLA, P. (eds) *Gondwana Five*. A.A. Balkema, Rotteerdam, 319–326.

GRUNOW, A. M., KENT, D. V. & DALZIEL, I. W. D. 1987. Mesozoic evolution of West Antarctica and the Weddell Sea Basin: new paleomagnetic constraints. *Earth and Planetary Science Letters*, **86**, 16–26.

GRUNOW, A. M., KENT, D. V. & DALZIEL, I. W. D. 1991. New paleomagnetic data from Thurston Island: implications for the tectonics of West Antarctica and Weddell Sea opening. *Journal of Geophysical Research*, **96**, 17 935–17 954.

GUNN, B. M. & WARREN, G. 1962. Geology of Victoria Land between the Mawson and Mulock Glaciers, Antarctica. *Bulletin of the Geological Survey of New Zealand*, **71**, 157.

HÄLBICH, I. W., FITCH, F. J. & MILLER, J. A. 1983. Dating the cape orogeny. *In*: SÖHNGE, A. P. G. & HÄLBICH, I. W. (eds) *Geodynamics of the Cape Fold Belt*. Geological Society of South Africa, Special Publications, Johannesburg, South Africa, **12**, 75–100.

HANDLER, M. R., WYSOCZANSKI, R. J. & GAMBLE, J. A. 2003. Proterozoic lithosphere in Marie Byrd Land, West Antarctica: Re-Os systematic of spinel peridotite xenoliths. *Chemical Geology*, **23**, 131–145.

HANSON, R. H. & ELLIOT, D. H. 1996. Rift-related Jurassic basaltic phreatomagmatic volcanism in the central Transantarctic Mountains: precursory stage to flood-basalt effusion. *Bulletin of Volcanology*, **58**, 327–347.

HASTON, R. B., LUYENDYK, B. P., LANDIS, C. A. & COOMBS, D. S. 1989. Paleomagnetism and question of original location of the Permian Brook Street terrane, New Zealand. *Tectonics*, **8**, 791–801.

HENJES-KUNTZ, F. & KREUZER, H. 2003. Mid-Paleozoic igneous activity in northern Victoria Land, Antarctica: implications of new geochronological data. *Geologisches Jahrbuch*, **B85**, 271–302.

HERGT, J. M., CHAPPELL, B. W., FAURE, G. & MENSING, T. 1989*a*. The geochemistry of Jurassic dolerites from Portal Peak, Antarctica. *Contributions to Mineralogy and Petrology*, **102**, 298–305.

HERGT, J. M., CHAPPELL, B. W., MCCULLOCH, T. M., MCDOUGALL, I. & CHIVAS, A. R. 1989*b*. Geochemical and isotopic constraints on the origin of the Jurassic dolerites of Tasmania. *Journal of Petrology*, **30**, 341–383.

HERGT, J. M., PEATE, D. W. & HAWKESWORTH, C. J. 1991. The petrogenesis of Mesozoic Gondwana low-Ti flood basalts. *Earth and Planetary Science Letters*, **105**, 134–148.

HERVÉ, F., MUNIZAGA, F., PARADA, M. A., BROOK, M., PANKHURST, R. J., SNELLING, N. J. & DRAKE, R. 1988. Granitoids of the Coast Range of central Chile: geochronology and geologic setting. *Journal of South American Earth Sciences*, **1**, 185–194.

HERVÉ, F., CALDERÓN, M., FANNING, C. M., KRAUS, S. & PANKHURST, R. J. 2010*a*. SHRIMP chronology of the Magellanes basin basement, Tierra del Fuego: cambrian plutonism and Permian high-grade metamorphism. *Andean Geology*, **37**, 253–275.

HERVÉ, F., FANNING, C. M., PANKHURST, R. J., MPODOZIS, C., KLEPEIS, K., CALDERÓN, M. & THOMSON, S. N. 2010*b*. Detrital zircon SHRIMP U–Pb age study of the Cordillera Darwin Metamorphic Complex of Tierra del Fuego: sedimentary sources and implications for the evolution of the Pacific margin of Gondwana. *Journal of the Geological Society, London*, **167**, 555–568.

HOLE, M. J. & LE MASURIER, W. E. 1994. Tectonic controls on the geochemical composition of Cenozoic, mafic alkaline rocks from West Antarctica. *Contributions to Mineralogy and Petrology*, **117**, 187–202.

HUERTA, A. D. 2007. Byrd drainage system: evidence of a Mesozoic West Antarctic Plateau. *In*: COOPER, A. K., RAYMOND, C. R. ET AL. (eds) *Antarctica: A Keystone in a Changing World*. Online Proceedings of the 10th ISAES, USGS Open-File Report 2007-1047, Extended Abstract 091.

HUNTER, M. A., RILEY, T. R., CANTRILL, D. J., FLOWERDEW, M. A. & MILLAR, I. L. 2006. A new stratigraphy for the Latady Basin, Antarctic Peninsula: Part 1, Ellsworth Land Volcanic Group. *Geological Magazine*, **143**, 777–796.

ISBELL, J. L. 1999. The Kukri Erosion Surface; a reassessment of its relationship to the rocks of the Beacon Supergroup in the central Transantarctic Mountains, Antarctica. *Antarctic Science*, **11**, 228–238.

ISBELL, J. L. 2010. Environmental and paleogeographic implications of glaciotectonic deformation of glaciomarine deposits within Permian strata of the Metschel Tillite, southern Victoria Land, Antarctica. *In*: LÓPEZ-GAMUNDÍ, O. R. & BUATOIS, L. A. (eds) *Late Paleozoic Glacial Events and Postglacial Transgressions in Gondwana*. Geological Society of America, Boulder, CO, Special Papers, **468**, 81–100; doi: 10.1130/2010.2468(03).

ISBELL, J. L., KOCH, Z. J., SZABLEWSKI, G. M. & LENAKER, P. A. 2008. Permian glacigenic deposits in the Transantarctic Mountains, Antarctica. *In*: FIELDING, C. R., FRANK, T. D. & ISBELL, J. L. (eds) *Resolving the Late Paleozoic Ice Age in Time and Space*. Geological Society of America, Boulder, CO, Special Papers, **441**, 59–70, doi: 10.1130/2008.2441(04).

KELLY, S. R. A., DOUBLEDAY, P. A., BRUNTON, C. H. C., DICKENS, J. M., SEVASTOPULO, G. D. & TAYLOR, P. D. 2001. First Carboniferous and (?)Permian marine macrofaunas from Antarctica and their tectonic implications. *Journal of the Geological Society, London*, **158**, 219–232.

KÖNIG, M. & JOKAT, W. 2006. The Mesozoic breakup of the Weddell Sea. *Journal of Geophysical Research*, **111**, B12102, doi: 10.1029/2005JB004035.

KYLE, R. A. 1977. Devonian palynomorphs from the basal Beacon Supergroup of South Victoria Land (note). *New Zealand Journal of Geology and Geophysics*, **20**, 1147–1150.

KYLE, P. R., PANKHURST, R. J. & BOWMAN, J. R. 1983. Isotopic and chemical variations in Kirkpatrick Basalt group rocks from southern Victoria Land. *In*: OLIVER, R. L., JAMES, P. R. & JAGO, J. B. (eds) *Antarctic Earth Science*. Australian Academy of Sciences, Canberra, 234–237.

LAIRD, M. G. 1991. The Late Proterozoic-Middle Palaeozoic rocks of Antarctica. *In*: TINGEY, R. J. (ed.) *The Geology of Antarctica*. Oxford Monographs on Geology and Geophysics. Oxford University Press, Oxford, **17**, 74–119.

LAUDON, T. S. 1991. Petrology of sedimentary rocks from the English Coast, eastern Ellsworth land. *In*: THOMSON, M. R. A., CRAME, J. A. & THOMSON, J. W. (eds) *Geological evolution of Antarctica*. Cambridge University Press, Cambridge, 455–460.

LAWRENCE, J. F., WIENS, D. A., NYBLADE, A. A., ANANDAKRISHNAN, S., SHORE, P. J. & VOIGT, D. 2006. Crust and upper mantle structure of the Transantarctic Mountains and surrounding regions from receiver functions, surface waves, and gravity: implications for uplift models. *Geochemistry, Geophysics, Geosystems*, **7**, Q10011; doi: 10.1029/2006GC001282.

LEAT, P. T., SCARROW, J. H. & MILLAR, I. L. 1995. On the Antarctic Peninsula Batholith. *Geological Magazine*, **132**, 399–412.

LEAT, P. T., LUTTINEN, A. V., STOREY, B. C. & MILLAR, I. L. 2006. Sills of the Theron Mountains, Antarctica: evidence for long distance transport of mafic magmas during Gondwana break-up. *In*: HANSKI, E., MERTANEN, S., RÄMÖ, T. & VUOLLO, J. (eds) *Dyke Swarms – Time Markers of Crustal Evolution*. Taylor & Francis, London, 183–199.

LEAT, P. T., FLOWERDEW, M. J., RILEY, T. R., WHITEHOUSE, M. J., SCARROW, J. H. & MILLAR, I. L. 2009. Zircon U–Pb dating of Mesozoic volcanic and tectonic events in north-west Palmer Land and south-west Graham Land, Antarctica. *Antarctic Science*, **21**, 633–641.

LEMASURIER, W. E. & THOMSON, J. W. 1987. Volcanoes of the Antarctic Plate and Southern Ocean. *American Geophysical Union, Antarctic Research Series*, **48**, 487.

LEMASURIER, W. E. & LANDIS, C. A. 1996. Mantle–plume activity recorded by low-relief erosion surfaces in West Antarctica and New Zealand. *Geological Society of America Bulletin*, **108**, 1450–1466.

LEWIS, A. R., MARCHANT, D. R., ASHWORTH, A. C., HEMMING, S. R. & MACHLUS, M. L. 2007. Major Middle Miocene global climate change: evidence from East Antarctica and the Transantarctic Mountains. *Geological Society of America Bulletin*, **119**, 1449–1461.

LEWIS, A. R., MARCHANT, D. R., ASHWORTH, A. C. ET AL. 2008. Mid-Miocene cooling and the extinction of tundra in continental Antarctica. *Proceedings of the National Academy of Sciences of the United States of America*, **105**, 10676–10680.

LISKER, F. 2002. Review of fission track studies in northern Victoria Land, Antarctica: passive margin evolution versus uplift of the Transantarctic Mountains. *Tectonophysics*, **349**, 57–73.

LISKER, F., LÄUFER, A. L., OLESCH, M., ROSSETTI, F. & SCHÄFER, T. 2006. Transantarctic Basin: new insights from fission track and structural data from the USARP Mountains and adjacent areas (Northern Victoria Land, Antarctica). *Basin Research*, **18**, 497–520.

LONG, W. E., MCLELLAND, D., COLLINSON, J. W. & ELLIOT, D. H. 2009. Geology of the Nilsen Plateau, Queen Maud Mountains, Transantarctic Mountains. *Terra Antartica*, **15**, 239–253.

LOSKE, W., HERVÉ, F., MILLER, H. & PANKHURST, R. J. 1997. Rb–Sr and U–Pb studies of the pre-Andean and Andean magmatism in the Horseshoe Island area, Marguerite Bay (Antarctic Peninsula). *In*: RICCI, C. A. (ed.) *The Antarctic Region: Geological Evolution and Processes*. Terra Antartica, Siena, 353–360.

MARSH, J. S., HOOPER, P. R., REHACEK, J., DUNCAN, R. A. & DUNCAN, A. R. 1997. Stratigraphy and age of Karoo basalts of Lesotho and implications for correlations within the Karoo Igneous Province. *In*: MAHONEY, J. J. & COFFIN, M. F. (eds) *Large Igneous Provinces: Continental, Oceanic, and Planetary Flood Volcanism*. American Geophysical Union, Washington, DC, Geophysical Monographs, **100**, 247–272.

MARTIN, A. K. 2007. Gondwana breakup via double-saloon-door rifting and seafloor spreading in a backarc basin during subduction rollback. *Tectonophysics*, **445**, 245–272.

MCCARTAN, L. & BRADSHAW, M. A. 1987. Petrology and sedimentology of the Horlick Formation (Lower Devonian), Ohio Range, Transantarctic Mountains. *United States Geological Survey, Bulletins*, **1780**, 1–31.

MENSING, T. M., FAURE, G., JONES, L. M. & BOWMAN, J. R. 1984. Petrogenesis of the Kirkpatrick Basalt, Solo Nunatak, north Victoria Land, Antarctica. *Contributions to Mineralogy and Petrology*, **87**, 101–108.

MENSING, T. M., FAURE, G., JONES, L. M. & HOEFS, J. 1991. Stratigraphic correlation and magma evolution of the Kirkpatrick Basalt in the Mesa Range, northern Victoria Land, Antarctica. *In*: ULBRICH, H. & ROCHA CAMPOS, A. C. (eds) *Gondwana Seven Proceedings*. Instituto Geosciências, Universidade de São Paulo, São Paulo, 653–667.

MILLAR, I. L., WILLAN, R. C. R., WAREHAM, C. D. & BOYCE, A. J. 2001. The role of crustal and mantle sources in the genesis of granitoids of the Antarctic Peninsula and adjacent crustal blocks. *Journal of the Geological Society, London*, **158**, 855–867.

MILLAR, I. L., PANKHURST, R. J. & FANNING, C. M. 2002. Basement chronology of the Antarctic Peninsula: recurrent magmatism and anatexis in the Palaeozoic Gondwana margin. *Journal of the Geological Society, London*, **159**, 145–157.

MILLER, M. F. & ISBELL, J. L. 2010. Reconstruction of a high-latitude, post-glacial lake: Mackellar formation (Permian), Transantarctic Mountains. *In*: LÓPEZ-GAMUNDÍ, O. R. & BUATOIS, L. A. (eds) *Late Paleozoic Glacial Events and Postglacial Transgressions in Gondwana*. Geological Society of America, Boulder, CO, Special Papers, **468**, 193–207; doi: 10.1130/2010.2468(09).

MILLER, S. R., FITZGERALD, P. G. & BALDWIN, S. L. 2010. Cenozoic range-front faulting and development of the Transantarctic Mountains near Cape Surprise, Antarctica: thermochronologic and geomorphologic constraints. *Tectonics*, **29**, TC 1003, doi: 10.1029/2009TC 002457.

MOLZAHN, M., REISBERG, L. & WÖRNER, G. 1996. Os, Sr, Nd, Pb, O isotope and trace element data from the Ferrar flood basalts, Antarctica: evidence for an enriched subcontinental lithospheric source. *Earth and Planetary Science Letters*, **144**, 529–546.

MOLZAHN, M., WÖRNER, G., HENJES-KUNST, F. & ROCHOLL, A. 1999. Constraints on the Cretaceous thermal event in the Transantarctic Mountains from alteration processes in Ferrar flood basalts. *Global and Planetary Change*, **23**, 45–60.

MORTIMER, N. 2004. New Zealand's geological foundations. *Gondwana Research*, **7**, 261–272.

MORTIMER, N., PARKINSON, D., RAINE, J. I., ADAMS, C. J., GRAHAM, I. J., OLIVER, P. J. & PALMER, K. 1995. Ferrar magmatic province rocks discovered in New Zealand: implications for Mesozoic Gondwana geology. *Geology*, **23**, 185–188.

MORTIMER, N., PALIN, J. M., DUNLAP, W. J. & HAUFF, F. 2011. Extent of the Ross Orogen in Antarctica: new data from DSDP 270 and Iselin Bank. *Antarctic Science*, **23**, 297–306.

MUKASA, S. B. & DALZIEL, I. W. D. 2000. Marie Byrd Land, West Antarctica: evolution of Gondwana's Pacific margin constrained by zircon U–Pb geochronology and feldspar common-Pb isotopic compositions. *Geological Society of America Bulletin*, **112**, 611–627.

NEUHOFF, P. S., FRIDRIKSSON, T. & BIRD, D. K. 2000. Zeolite paragenesis in the North Atlantic igneous province: implications for geotectonics and groundwater quality of basaltic crust. *International Geology Review*, **42**, 15–44.

PALMERI, R., GHIRIBELLI, B., TALARICO, F. & RICCI, C. A. 2003. Ultra-high pressure metamorphism in felsic rocks: the garnet-phengite gneisses and quartzites of the Lanterman Range, Antarctica. *European Journal of Mineralogy*, **15**, 513–525.

PANKHURST, R. J., STOREY, B. C., MILLAR, I. L., MACDONALD, D. I. M. & VENNUM, W. R. 1988. Cambrian–Ordovician magmatism in the Thiel Mountains, Transantarctic Mountains, and implications for the Beardmore orogeny. *Geology*, **16**, 246–249.

PANKHURST, R. J., MILLAR, I. L., GRUNOW, A. M. & STOREY, B. C. 1993. The pre-Cenozoic magmatic history of the Thurston Island crustal block, West Antarctica. *Journal of Geophysical Research*, **98**, 11835–11850.

PANKHURST, R. J., WEAVER, S. D., BRADSHAW, J. D., STOREY, B. C. & IRELAND, T. R. 1998. Geochronolgy and geochemistry of pre-Jurassic superterranes in Marie Byrd Land, Antarctica. *Journal of Geophysical Research*, **103**, 2529–2547.

PANKHURST, R. J., RILEY, T. R., FANNING, C. M. & KELLEY, S. R. 2000. Episodic silicic volcanism along the proto-Pacific margin of Patagonia and the Antarctic

Peninsula: plume and subduction influences associated with the break-up of Gondwana. *Journal of Petrology*, **41**, 605–625.

PANKHURST, R. J., RAPELA, C. W. & FANNING, C. M. 2001. Age and origin of coeval TTG, I- and S-type granites in the Famatinian belt of NW Argentina. *Transactions of the Royal Society of Edinburgh, Earth Sciences*, **91**, 151–168.

PANKHURST, R. J., RAPELA, C. W., LOSKE, W. P., MARQUEZ, M. & FANNING, C. M. 2003. Chronological study of the pre-Permian basement rocks of southern Patagonia. *Journal of South American Earth Sciences*, **16**, 27–44.

PANKHURST, R. J., RAPELA, C. W., FANNING, C. M. & MARQUEZ, M. 2006. Gondwanide continental collision and the origin of Patagonia. *Earth Science Reviews*, **76**, 235–257.

PAULSEN, T. S., ENCARNACIÓN, J. & GRUNOW, A. M. 2007. New age constraints for a short pulse in Ross orogen deformation triggered by East-West Gondwana suturing. *Gondwana Research*, **12**, 417–427, doi: 10.1016/j.gr.2007.05.011.

PRICE, R., SPANDLER, C., ARCULUS, R. & REAY, A. 2011. The Longwood Igneous Complex, Southland, New Zealand: a Permo-Jurassic, intra-oceanic, subduction-related, I-type batholithic complex. *Lithos*, **126**, 1–21.

RAPELA, C. W., PANKHURST, R. J., FANNING, C. M. & HERVE, F. 2005. Pacific subduction coeval with the Karoo mantle plume: the Early Jurassic Subcordilleran belt of northwestern Patagonia. *In*: VAUGHAN, A. P. M., LEAT, P. T. & PANKHURST, R. J. (eds) *Terrane processes at the margins of Gondwana*. Geological Society, London, Special Publications, **246**, 217–239.

READ, S. E., COOPER, A. F. & WALKER, N. W. 2002. Geochemistry and U–Pb geochronology of the Neoproterozoic-Cambrian Koettlitz Glacier alkaline province. *In*: GAMBLE, J., SKINNER, D. N. B. & HENRYS, S. (eds) *Antarctica at the Close of a Millenium*. Royal Society of New Zealand, Bulletin, Wellington, New Zealand, **35**, 143–151.

RILEY, T. R., LEAT, P. T., PANKHURST, R. J. & HARRIS, C. 2001. Origins of large volume rhyolitic volcansim in the Antarctic Peninsula and Patagonia by crustal melting. *Journal of Petrology*, **42**, 1043–1065.

ROCCHI, S., STORTI, F. & DI VINCENZO, G. 2003. Intraplate strike-slip tectonics as an alternative to mantle plume activity for the Cenozoic rift magmatism in the Ross sea region, Antarctica. *In*: STORTI, F., HOLDSWORTH, R. E. & SALVINI, F. (eds) *Intraplate Strike-Slip Deformation Belts*. Geological Society, London, Special Publications, **210**, 145–158.

ROCCHI, S., BRACCIALI, L., DI VINCENZO, G., GEMELLI, M. & GHEZZO, C. 2011. Arc accretion to the early Paleozoic Antarctic margin of Gondwana in Victoria Land. *Gondwana Research*, **19**, 594–607.

ROSS, P.-S., WHITE, J. D. L. & MCCLINTOCK, M. K. 2008. Physical volcanology of mafic volcaniclastic deposits and lavas in the Coombs-Allan Hills area, Ferrar large igneous province, Antarctica. *Journal of Volcanology and Geothermal Research*, **172**, 38–60.

ROWELL, A. J., VAN SCHMUS, W. R., STOREY, B. C., FETTER, A. H. & EVANS, K. R. 2001. Latest Neoproterozoic to mid-Cambrian age for the main deformation phases of the Transantarctic Mountains: new stratigraphic and isotopic constraints from the Pensacola Mountains, Antarctica. *Journal of the Geological Society, London*, **158**, 295–308.

SCHEPPERS, R. & ARMSTRONG, R. 2002. New U–Pb SHRIMP zircon ages of the Cape Granite Suite: implications for the magmatic evolution of the Saldania Belt. *South African Journal of Geology*, **105**, 241–256.

SCHÖNER, R., VIERECK-GOETTE, L., SCHNEIDER, J. & BOMFLEUR, B. 2007. Triassic–Jurassic sediments and multiple volcanic events in North Victoria Land, Antarctica: a revised stratigraphy. *In*: COOPER, A. K., RAYMOND, C. R. ET AL. (eds) *Antarctica: A Keystone in a Changing World*. Online Proceedings of the 10th ISAES, USGS Open-File Report **2007-1047**, Short Research Paper 102, 5; doi: 10.3133/of2007-1047.srp102.

SCHÖNER, R., BOMFLEUR, B., SCHNEIDER, J. & VIERECK-GÖTTE, L. 2011. A systematic description of the Triassic to Lower Jurassic Section Peak Formation in North Victoria Land (Antarctica). *Polarforschung*, **80**, 71–87.

SEPÚLVEDA, F. A., HERVÉ, F., CALDERÓN, M. & LACASSIE, J. P. 2008. Petrological and geochemical characteristics of metamorphic and igneous units from the allochthonous Madre de Dios Terrane, Southern Chile. *Gondwana Research*, **13**, 238–249.

SEPÚLVEDA, F. A., PALMA-HELDT, S., HERVÉ, F. & FANNING, C. M. 2010. Permian depositional age of metaturbidites of the Duque de York Complex, southern Chile: U–Pb SHRIMP data and palynology. *Andean Geology*, **37**, 375–397.

SIDDOWAY, C. S. 2008. Tectonics of the West Antarctic Rift System: new light on the history and dynamics of distributed intracontinental extension. *In*: COOPER, A. K., RAYMOND, C. R. ET AL. (eds) *Antarctica: A Keystone in a Changing World*. National Academies Press, Washington, DC, 91–114.

SPORLI, K. B. 1992. Stratigraphy of the Crashsite Group, Ellsworth Mountains, West Antarctica. *In*: WEBERS, G. F., CRADDOCK, C. & SPLETTSTOESSER, J. F. (eds) *Geology and Paleontology of the Ellsworth Mountains, West Antarctica*. Geological Society of America, Boulder, CO, Memoirs, **170**, 21–35.

STERN, T. A. & TEN BRINK, U. S. 1989. Flexural uplift of the Transantarctic Mountains. *Journal of Geophysical Research*, **94**, 10 315–10 330.

STERN, T. A., BAXTER, A. K. & BARRETT, P. J. 2005. Isostatic rebound due to glacial erosion within the Transantarctic Mountains. *Geology*, **33**, 221–224.

STOREY, B. C. 1995. The role of mantle plumes in continental break-up: case histories from Gondwanaland. *Nature*, **377**, 301–308.

STOREY, B. C. & KYLE, P. R. 1997. An active mantle mechanism for Gondwana break-up. *Journal of African Earth Sciences*, **100**, 283–290.

STOREY, B. C., DALZIEL, I. W. D., GARRETT, S. W., GRUNOW, A. M., PANKHURST, R. J. & VENNUM, W. R. 1988. West Antarctica in Gondwanaland: crustal blocks, reconstruction and breakup processes. *Tectonophysics*, **155**, 381–390.

STOREY, B. C., VAUGHAN, A. P. M. & MILLAR, I. L. 1996. Geodynamic evolution of the Antarctic Peninsula during Mesozoic times and its bearing on Weddell Sea history. *In*: Storey, B. C., King, E. C. & Livermore, R. A. (eds) *Weddell Sea Tectonics and Gondwana Break-up*. Geological Society, London, Special Publications, **108**, 87–103.

Storey, B. C., Leat, P. T. & Ferris, J. K. 2001. The location of mantle plume centers during the initial stages of Gondwana break-up. *In*: Ernst, R. E. & Buchan, K. L. (eds) *Mantle Plumes: Their Identificaion through Time*. Geological Society of America, Boulder, CO, Memoirs, **352**, 71–80.

Storti, F., Balestrieri, M. L., Balsamo, F. & Rosetti, F. 2008. Structural and thermochronological constraints to the evolution of the West Antarctic Rift System in central Victoria Land. *Tectonics*, **27**, 1–21, doi: 10.1029/2006TC002066.

Studinger, M., Bell, R. E., Buck, W. R., Karner, G. D. & Blankenship, D. B. 2004. Sub-ice geology inland of the Transantarctic mountains in light of new aerogeophysical data. *Earth and Planetary Science Letters*, **220**, 391–408.

Stump, E., Sheridan, M. F., Borg, S. G. & Sutter, J. F. 1980. Early Miocene Subglacial Basalts, the East Antarctic ice sheet, and uplift of the Transantarctic Mountains. *Science*, **207**, 757–759.

Stump, E., Gootee, B. F., Talarico, F., van Schmus, W. R., Brand, P. K., Foland, K. A. & Fanning, C. M. 2004. Correlation of Byrd and Selbourne Groups, with implications for the Byrd Glacier discontinuity, central Transantarctic Mountains, Antarctica. *New Zealand Journal of Geology and Geophysics*, **47**, 157–171.

Tangeman, J. A., Mukasa, S. B. & Grunow, A. M. 1996. Zircon U–Pb geochronology of plutonic rocks from the Antarctic Peninsula: confirmation of the presence of unexposed Paleozoic crust. *Tectonics*, **15**, 1309–1324.

ten Brink, U. S., Bannister, S., Beaudoin, B. G. & Stern, T. A. 1993. Geophysical investigations of the tectonic boundary between East and West Antarctica. *Science*, **261**, 45–50.

ten Brink, U. S., Hackney, R. I., Bannister, S., Stern, T. A. & Makovsky, Y. 1997. Uplift of the Transantarctic Mountains and the bedrock beneath the East Antarctic ice sheet. *Journal of Geophysical Research*, **102**, 27 603–27 621.

Tessensohn, F. & Henjes-Kunst, F. 2005. Northern Victoria Land terranes, Antarctica: far-travelled or local products. *In*: Vaughan, A. M. P., Leat, P. T. & Pankhurst, R. J. (eds) *Terrane Processes at the Margin of Gondwana*. Geological Society, London, Special Publications, **246**, 275–292.

Tessensohn, F. & Mädler, K. 1987. Triassic plant fossils from North Victoria Land, Antarctica. *Geologisches Jahrbuch*, **B66**, 187–201.

Trey, H., Cooper, A. K., Pellis, G., della Vedova, B., Cochrane, G., Brancolini, G. & Makris, J. 1999. Transect across the West Antarctic rift system in the Ross Sea, Antarctica. *Tectonophysics*, **301**, 61–74.

van Schmus, W. R., McKenna, L. W., Gonzales, D. A. & Rowell, A. J. 1997. U/Pb geochronology of parts of the Pensacola, Harold Byrd, and Queen Maud Mountains, Antarctica. *In*: Ricci, C. A. (ed.) *The Antarctic Region: Geological Evolution and Processes*. Terra Antartica, Siena, 187–200.

van Wijk, J. W., Lawrence, J. F. & Driscoll, N. W. 2008. Formation of the Transantarctic Mountains related to extension of the West Antarctic rift system. *Tectonophysics*, **458**, 117–126.

Vaughan, A. P. M. & Pankhurst, R. J. 2008. Tectonic overview of the West Gondwana margin. *Gondwana Research*, **13**, 150–162.

Vaughan, A. P. M. & Storey, B. C. 2000. The eastern Palmer Land shear zone: a new terrane accretion model for the Mesozoic development of the Antarctic Peninsula. *Journal of the Geological Society, London*, **157**, 1243–1256.

Vaughan, A. P. M., Eagles, G. & Flowerdew, M. J. 2013. Evidence for a two-phase Palmer Land event from cross-cutting structural relationships and emplacement timing of the Lassiter Coast Intrusive Suite, Antarctic Peninsula: implications for mid-Cretaceous Southern Ocean plate configuration. *Tectonics*, **31**, doi: 10.1029/2011TC003006.

Viereck-Goette, L., Schöner, R., Bomfleur, B. & Schneider, J. 2007. Multiple shallow level sill intrusions coupled with hydromagmatic explosive eruptions marked the initial phase of Ferrar large igneous province magmatism in northern Victoria Land. *In*: Cooper, A. K., Raymond, C. R. *et al.* (eds) *Antarctica: A Keystone in a Changing World*. Online Proceedings of the 10th ISAES, USGS Open-File Report **2007-1047**, Short Research Paper 104, 5; doi: 10.3133/of2007-1047.srp104.

Walker, G. P. L. 1960. Zeolite zones and dyke distribution in relation to the structure of the basalts of eastern Iceland. *Journal of Geology*, **68**, 515–528.

Wandres, A. M. & Bradshaw, J. G. 2005. New Zealand tectonostratigraphy and implications from conglomeratic rocks for the configuration of the SW pacific margin of Gondwana. *In*: Vaughan, A. P. M., Leat, P. T. & Pankhurst, R. J. (eds) *Terrane Processes at the Margins of Gondwana*. Geological Society, London, Special Publications, **246**, 179–216.

Wareham, C. D., Stump, E., Storey, B. C., Millar, I. L. & Riley, T. R. 2001. Petrogenesis of the Cambrian Liv Group, a bimodal volcanic rock suite from the Ross orogen, Transantarctic Mountains, Antarctica. *Geological Society of America Bulletin*, **113**, 360–372.

Watts, D. R. & Bramall, A. M. 1981. Paleomagnetic evidence for a displaced terrain in western Antarctica. *Nature*, **293**, 638–641.

Weaver, S. D., Bradshaw, J. D. & Adams, C. J. 1991. Granitoids of the Ford Ranges, Marie Byrd Land, Antarctica. *In*: Thomson, M. R. A., Crame, J. A. & Thomson, J. W. (eds) *Geological Evolution of Antarctica*. Cambridge University Press, Cambridge, 345–351.

Webb, P. N. & Wrenn, J. H. 1982. Upper Cenozoic micropaleontology and biostratigraphy of eastern Taylor Valley, Antarctica. *In*: Craddock, C. (ed.) *Antarctic Geoscience*. University of Wisconsin Press, Madison, WI, 1117–1122.

White, J. D. L. & McClintock, M. K. 2001. Immense vent complex marks flood-basalt eruption in a wet, failed rift: Coombs Hills, Antarctica. *Geology*, **29**, 935–938.

Wilch, T. I. & McIntosh, W. C. 2000. Eocene and Oligocene volcanism at Mount Petras, Marie Byrd Land; implications for middle Cenozoic ice sheet reconstructions in West Antarctica. *Antarctic Science*, **12**, 477–491.

WILCH, T. M., DENTON, G. H., LUX, D. R. & MCINTOSH, W. C. 1993. Limited Pliocene glacier extent and surface uplift in middle Taylor Valley, Antarctica. *Geografiska Annaler*, **75**, 331–351.

WILL, T. M., ZEH, A., GERDES, A., FRIMMEL, H. E., MILLAR, I. L. & SCHMAEDICKE, E. 2009. Palaeoproterozoic to Palaeozoic magmatic and metamorphic events in the Shackleton Range, East Antarctica; constraints from zircon and monazite dating, and implications for the amalgamation of Gondwana. *Precambrian Research*, **172**, 25–45.

WILL, T. M., FRIMMEL, H. E., ZEH, A., LE ROUX, P. & SCHMAEDICKE, E. 2010. Geochemical and isotopic constraints on the tectonic and crustal evolution of the Shackleton Range, East Antarctica, and correlation with other Gondwana crustal segments. *Precambrian Research*, **180**, 85–112.

WILSON, T. J. 1995. Cenozoic transtension along the Transantarctic Mountains–West Antarctic rift boundary, southern Victoria Land, Antarctica. *Tectonics*, **14**, 531–545.

WILSON, D. S., JAMIESON, S. S. R., BARRETT, P. J., LEITCHENKOV, G., GOHL, K. & LARTER, R. D. 2012. Antarctic topography at the Eocene–Oligocene boundary. *Palaeogeography, Palaeoclimatology, Palaeoecology*, **335-336**, 24–34. doi: 10.1016/j.palaeo.2011.05.028.

YOUNG, G. C. & LONG, J. A. 2005. Phyllolepid placoderm fish remains from the Devonian Aztec Siltstone, southern Victoria Land, Antarctica. *Antarctic Science*, **17**, 387–408.

Differential Movement across Byrd Glacier, Antarctica, as indicated by Apatite (U–Th)/He thermochronology and geomorphological analysis

D. J. FOLEY, E. STUMP*, M. VAN SOEST, K. X. WHIPPLE & K. V. HODGES

School of Earth and Space Exploration, Arizona State University, Tempe, AZ 85287, USA

**Corresponding author (e-mail: ed.stump@asu.edu)*

Abstract: The objectives of this study were to assess possible differential movement across an inferred fault beneath Byrd Glacier, and to measure the timing of unroofing in this portion of the Transantarctic Mountains. Apatites separated from rock samples collected from known elevations at various locations north and south of Byrd Glacier were dated using single crystal (U–Th)/He analysis. Results indicate a denudation rate of *c.* 0.04 mm a^{-1} in the time range *c.* 140–40 Ma. Distinct age v. elevation plots from north and south of Byrd Glacier indicate an offset of *c.* 1 km across the glacier with south side up. A Landsat image of the Byrd Glacier area was overlain on an Aster Global Digital Elevation Model and spot elevations of the Kukri erosion surface to the north and south of Byrd Glacier were mapped. The difference in elevation of the erosion surface across Byrd Glacier also shows an offset of *c.* 1 km with south side up. Results support a model of relatively uniform cooling and unroofing of the region with later, post-40 Ma fault displacement that uplifted the south side of Byrd Glacier relative to the north.

Supplementary material: Sample and apatite (U–Th)/He data are listed at http://www.geolsoc.org.uk/SUP18671

Byrd Glacier, one of the major outlet glaciers crossing the Transantarctic Mountains (Fig. 1), marks a major discontinuity in the Neoproterozoic–early Palaeozoic Ross orogen, with plutonics and upper amphibolite-grade metamorphics to the north and lower greenschist-grade metamorphics and limestones to the south, indicating considerable differential unroofing of the orogen prior to deposition of the Beacon Supergroup on the Kukri erosion surface (Stump *et al.* 2005). Grindley & Laird (1969) mapped post-Beacon faults beneath Byrd Glacier and other outlet glaciers in the region, but cited no evidence other than the presence of the glaciers themselves.

This study asked the question, whether there are indications of displacement across the inferred Byrd fault, and if yes, what the amount of throw was. To assess possible differential movement across Byrd Glacier, we have used a two-pronged approach combining thermochronology, (U–Th)/He analysis of apatite and geomorphology, comparing elevations of the Kukri erosion surface on the north and south sides of Byrd Glacier.

Background

Our understanding of the history of uplift and denudation of the Transantarctic Mountains is based primarily on low-temperature thermochronological studies using apatite fission track analysis of samples collected from 'vertical profiles' (Gleadow & Fitzgerald 1987; Fitzgerald & Gleadow 1988; Fitzgerald 1992; Fitzgerald & Stump 1997; Lisker 2002). The Transantarctic Mountains are ideally suited for such studies because magmatism throughout the entirety of the range occurred in a brief period during the Jurassic (Fleming *et al.* 1997), totally annealing older fission tracks, except at two known localities where Palaeozoic apatite fission track ages survive (the western, inland margin of northern Victoria Land (Fitzgerald & Gleadow 1988) and the Miller Range at the inland margin of the central Transantarctic Mountains (Fitzgerald 1994)). When data are plotted on age v. elevation diagrams, they indicate that exhumation of the entire Transantarctic Mountains began in the Early Cenozoic, ranging between 45 and 55 Ma depending on the region. The central Transantarctic Mountains and the Scott Glacier area also have indica tions of pulses of exhumation in the Cretaceous (Stump & Fitzgerald 1992; Fitzgerald 1994).

One study utilizing (U–Th)/He analysis was conducted in the Ferrar Glacier area of southern Victoria Land where vertical profiles have been analysed on opposite sides of the glacier (Fitzgerald *et al.* 2006). Although the data are very scattered, probably because of a variety of complications

From: HAMBREY, M. J., BARKER, P. F., BARRETT, P. J., BOWMAN, V., DAVIES, B., SMELLIE, J. L. & TRANTER, M. (eds) 2013. *Antarctic Palaeoenvironments and Earth-Surface Processes*. Geological Society, London, Special Publications, **381**, 37–43. First published online August 6, 2013, http://dx.doi.org/10.1144/SP381.25

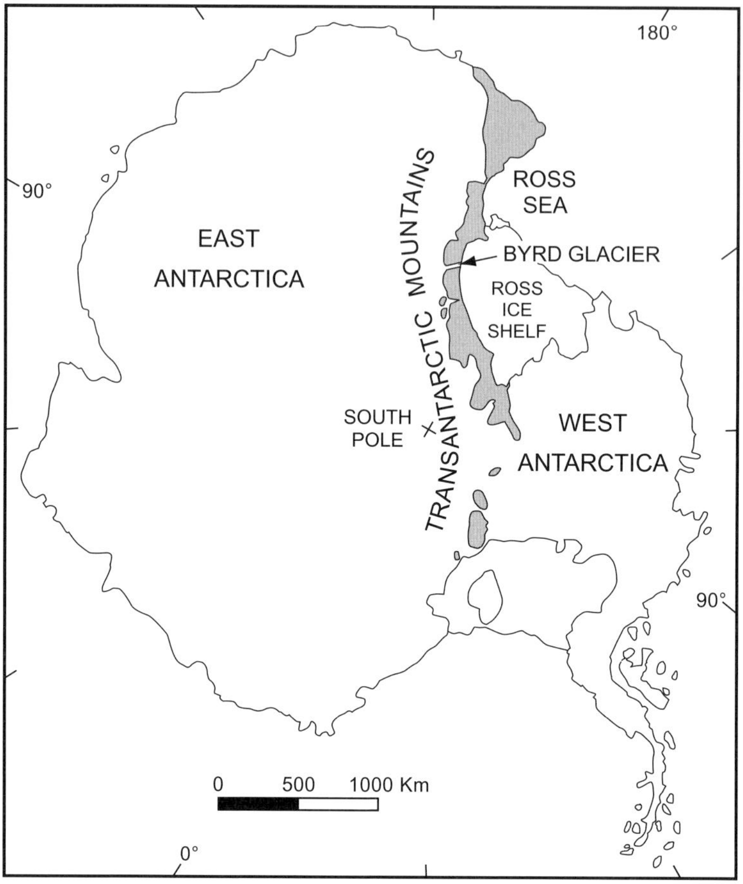

Fig. 1. Location map, Byrd Glacier.

associated with the technique, in general they indicate slow cooling in the Late Cretaceous and an increased cooling rate in the early Cenozoic.

(U–Th)/He thermochronology

Analytical procedures

The (U–Th)/He dating method is based on the production of ^{4}He nuclei (α particles) by uranium and thorium series as well as ^{147}Sm decay (Zeitler *et al.* 1987; Farley 2002). The helium closure temperature for apatite is *c.* 70 °C, making this method well suited for quantifying the cooling history of rocks as they pass through the upper 1–3 km of the crust (Reiners & Farley 2001; Ehlers & Farley 2003).

We chose eight samples of igneous and metamorphic rocks with known elevations from north and south of Byrd Glacier (Table 1, Fig. 2). The elevation of rock samples ESPR and JIF are estimated to be accurate within 10 m based on helicopter altimeter readings at the collecting sites. Elevations of the other rocks are based on mapping localities on the base map with a 200 m contour interval and are estimated to be accurate within 50 m.

In order to extract apatites for analysis, standard crushing, magnetic and heavy liquid separation

Table 1. *Sample and apatite (U/Th)/He data*

Rock	Side	Elevation (m)	Elevation error (m)	Apatite mean age (Ma)
ESPR	North	3450	10	142.9 ± 17.5
JIF	North	1800	10	102.1 ± 9.3
JIP	North	800	50	52.8 ± 18.0
JID	North	600	50	69.2 ± 18.2
JMG	North	200	50	50.5 ± 6.8
JHG	South	1750	50	54.2 ± 16.6
JHI	South	1700	50	53.9 ± 19.2
JMZ	South	1150	50	39.3 ± 14.5

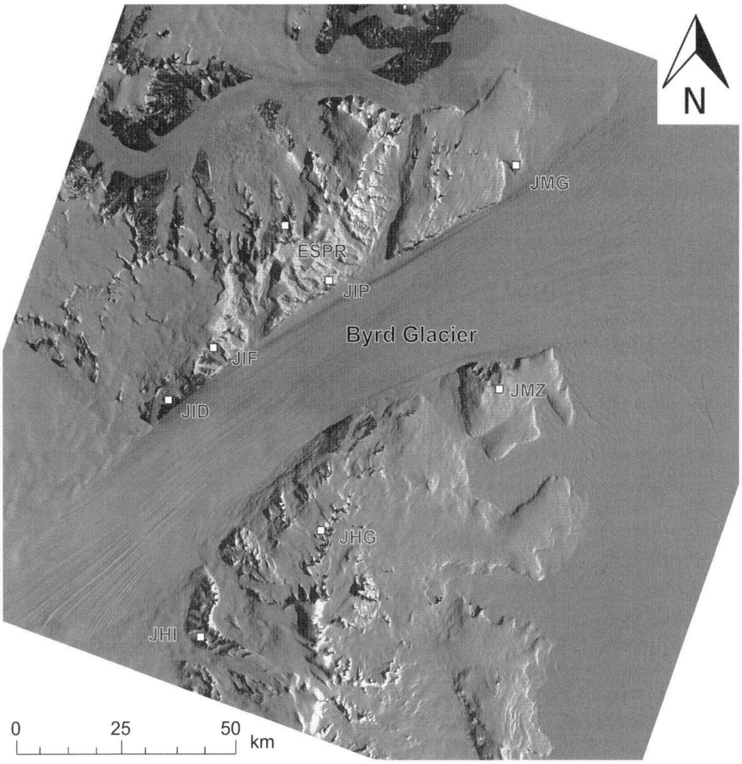

Fig. 2. Sample locations, Byrd Glacier area.

procedures were conducted on the samples. Separated grains were then hand-picked under a binocular microscope on the basis of morphology, size, clarity, euhedral crystal shape and lack of optically detectable inclusions. At least five apatite crystals from each sample were dated following procedures described in Schildgen *et al.* (2009) and updated in van Soest *et al.* (2011). These involved releasing helium by laser heating of apatite grains in an ASI Alphachron and spiking the gas with ^{3}He for ^{4}He/^{3}He analysis in the Noble Gas Geochemistry and Geochronology Laboratory at Arizona State University. Next, apatites were dissolved and solutions were analysed for uranium and thorium with an Thermo X series quadrupole inductively coupled plasma mass spectrometer in the W. M. Keck Foundation Laboratory for Environmental Geochemistry at Arizona State University. Ages were then calculated with an iterative process using blank corrected helium, thorium and uranium values.

Results

Mean (U–Th)/He ages are given in Table 1. The mean age for the suite of apatites from each rock sample is listed along with 2σ errors, which range from 39 ± 15 to 143 ± 18 Ma. The complete apatite (U–Th)/He dataset is available as Supplementary Material. To assess cooling histories and derive exhumation rates, all data were plotted in an age v. elevation diagram (Fig. 3) and a least squares regression through the data was computed in Matlab utilizing the equations of York (1969). The age variable was regressed using 2σ error uncertainties for each data point and a mean squared weighted deviation (MSWD) value was derived. From the slope of the derived regression equation for all apatite data, the exhumation rate between *c.* 140 and 40 Ma is 0.04 ± 0.01 (2σ) mm a^{-1} with a mean squared weighted deviation of 8.39.

To further assess cooling histories and possible differential movement north and south of Byrd Glacier, the apatite age v. elevation data from north and south of Byrd Glacier were regressed separately. The least squares regression of the data from north of Byrd Glacier indicates an exhumation rate of 0.035 ± 0.006 mm a^{-1} between *c.* 140 and 50 Ma with a mean squared weighted deviation of 0.93 (Fig. 4). Least squares regression of the data from south of Byrd Glacier indicates an

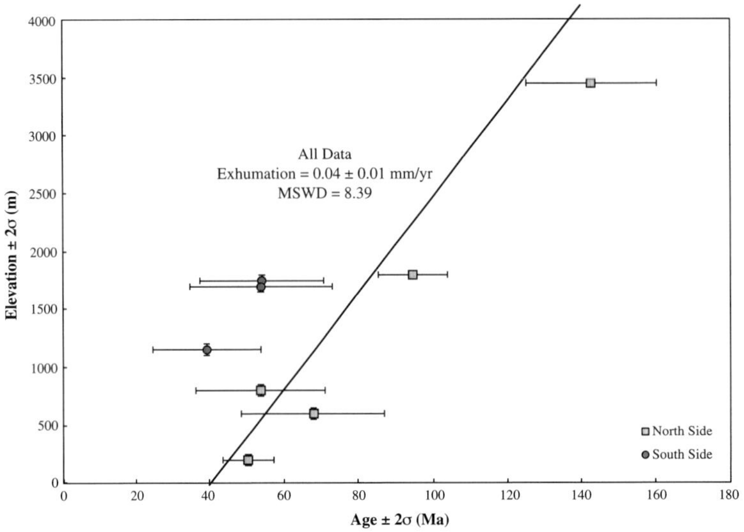

Fig. 3. Total apatite age v. elevation plot. Error bars are 2σ.

exhumation rate of 0.04 ± 0.05 mm a^{-1} between *c.* 55 and 40 Ma with a mean squared weighted deviation of 0.006 (Fig. 4). Separation of apatite age v. elevation data from north and south of Byrd Glacier produces exhumation rates similar to that derived from the total sample set, but separating the two sets of data leads to improved mean squared weighted deviation values.

When comparing apatite (U–Th)/He data across Byrd Glacier, Figure 4 shows that at similar elevations ages from the south are younger than ages from the north and at similar ages sample elevations from south of the glacier are higher than those from north of the glacier, indicating differential movement across Byrd Glacier. Two models could plausibly explain this inferred differential movement. In the first model, a relatively uniform exhumation across Byrd Glacier is followed by later fault displacement. The second model involves differential exhumation with the

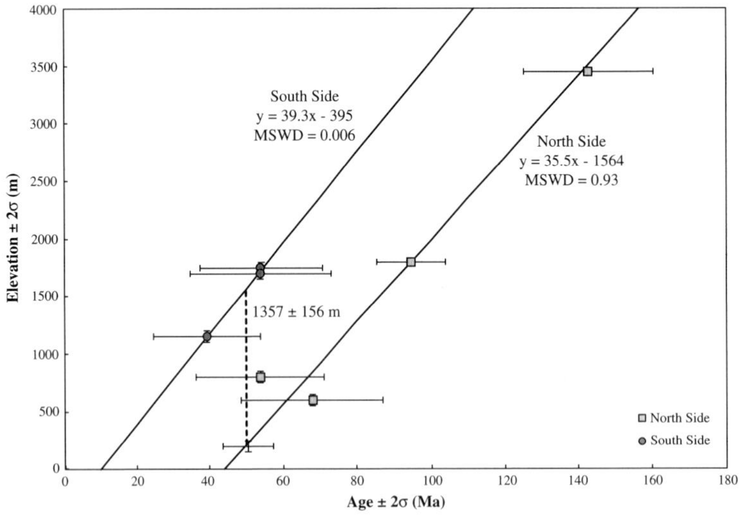

Fig. 4. Apatite age v. elevation plot with data separated between north (squares) and south (circles) samples. Bold lines are least squares regressions. Error bars are 2σ. Offset at the 50 Ma datum is plotted as a dashed line.

south side exhumed at a more rapid rate than the north side. The exhumation rates inferred from apatite data across Byrd Glacier are identical within error, 0.035 ± 0.006 mm a^{-1} to the north and 0.04 ± 0.05 mm a^{-1} to the south, suggesting that the differential movement was more probably attributable to an offset from fault displacement.

In order to quantify offset across Byrd Glacier, regression line equations of apatite age v. elevation data from north and south of Byrd Glacier were compared to calculate the difference in elevation (Fig. 4). The regression line equations of $y = 35.5x - 1564$ and $y = 39.3x - 395$, respectively, north and south of Byrd Glacier were utilized in calculations, with y representing elevation and x representing age. Offset was calculated by inputting an age value x as the predictor variable into each regression equation, resulting in a predicted elevation value for a given age. Then the north side elevation was subtracted from the south to determine the difference in elevation of the inferred rock at the same age across Byrd Glacier. The apatite data across Byrd Glacier is well constrained for comparison in the range of *c.* 40–60 Ma as rocks in this age range are represented in both north and south datasets. Therefore, an age value of 50 Ma was used for x in both north and south regression equations. The resulting elevation difference across Byrd Glacier is 1357 ± 156 (2σ) m.

Geomorphology

Analytical procedures

For assessing possible differential movement in this area from a geomorphological standpoint, the Kukri erosion surface was chosen as a datum, with the assumption that at the time of its formation the erosion surface had a relatively uniform elevation. The objective was to determine the current elevation of the Kukri erosion surface north and south of Byrd Glacier in order to quantify any displacement.

To map the elevation of the Kukri erosion surface across Byrd Glacier, imagery from the 30 m resolution Landsat Image Mosaic of Antarctica was draped over the 30 m resolution Aster Global Digital Elevation Model of the region. These were

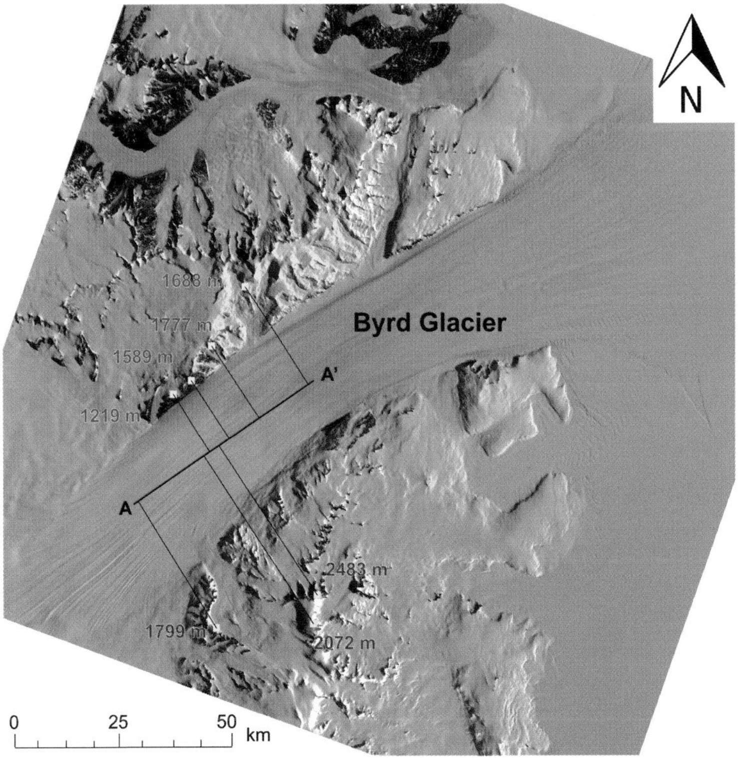

Fig. 5. Locations of Kukri erosion surface with corresponding elevations plotted on Byrd Glacier digital elevation model overlain with Landsat image. Note that distance values in Figure 6 and Table 2 are measured from the A end of the A–A′ transect, where the projection of the 1799 m elevation intersects the transect.

Table 2. *Kukri elevation and distance data*

N distance (m)	Elevation N (m)	S distance (m)	Elevation S (m)
20 677	1219	0	1799
27 404	1589	17 671	2072
33 679	1777	22 767	2483
45 087	1688		

N distance and S distance are measured from the A end of the A–A′ transect, where the projection of the 1799 m elevation intersects the transect.

processed in Arc Geographic Information System programs, Arc Map, Arc Scene and Arc View.

Based on field photographs, the Kukri erosion surface was located at seven points on Landsat imagery of the area, four points on the north side of Byrd Glacier and three points on the south side as shown in Figure 5. From these locations, elevations were extracted from the digital elevation model. Orthogonal projections of these points were then made onto a transect parallel to Byrd Glacier. The distances between points on transect A–A′ are measured from the southwesternmost point of the transect (Fig. 5).

Results

Elevation and distance data is presented in Table 2 and plotted in Figure 6, revealing that the observed points on the Kukri erosion surface are at higher elevations on the south side relative to the north. In addition, both north and south points show elevations higher in the east, indicating a slight dip toward the west. In order to accurately compare the elevation of the projected erosion surface across Byrd Glacier, the dips were calculated with a least-squares regression. The results are west dips of the Kukri erosion surface of 1.79° on the north side and 1.81° on the south side of Byrd Glacier.

As the dips of the erosion surface differ by only 0.02°, where the projections of the erosion surface overlap on the A–A′ transect, the elevation of the erosion surface across Byrd Glacier can be compared. However, this does involve the uncertainties of assuming the erosion surface has consistent strikes across Byrd Glacier and is not significantly deformed. The strikes could not be assessed in this study as locations of the Kukri erosion surface could not be extended farther north and south of the measured points owing to limited visible exposure and no available on-site data.

To measure the offset of the Kukri erosion surface, the elevation difference was calculated across Byrd Glacier by comparing the erosion surface elevations on either side at a point where the traces of the erosion surface overlap (Fig. 6). The erosion surface elevation data across Byrd Glacier overlap along the A–A′ transect between *c.* 20.7 and 22.7 km; therefore, a distance value of 21.7 km was chosen. The derived regression

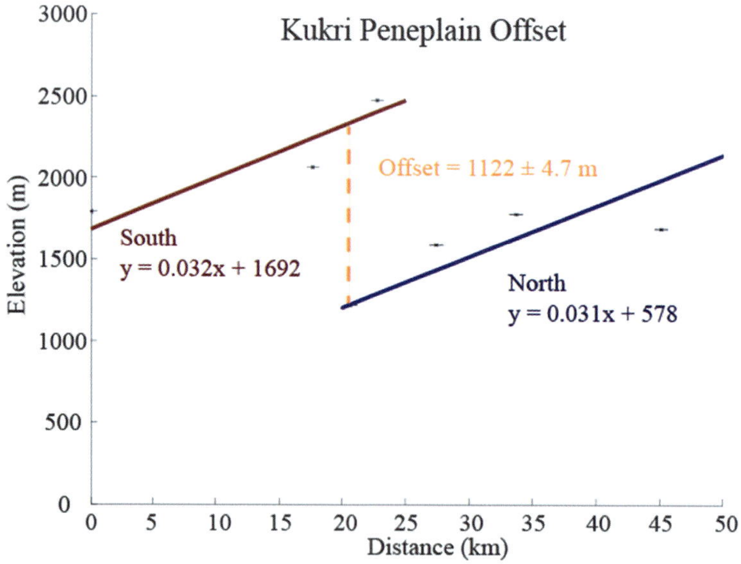

Fig. 6. Plot of projections of elevation data from Kukri erosion surface separated between north (blue) and south (red) points. Bold lines are least squares regressions of the points. Offset at the 21.7 km distance (measured from A on the A–A′ transect) is plotted in yellow.

equations from the Kukri erosion surface were $y = 0.0312x + 578$ (north side) and $y = 0.0316x + 1692$ (south side) with y representing elevation and x representing distance. Elevations were calculated based on an x value of 21.7 km as the predictor. At this location the calculated difference in elevation of the Kukri erosion surface across Byrd Glacier is 1122 ± 4.7 (2σ) m.

Whereas the offset in elevations of the Kukri erosion surface could be interpreted as due to horizontal displacement along a strike-slip fault, the thermochronologic data indicate that the dominant element of movement is vertical uplift.

Conclusions

Both thermochronologic and geomorphologic data indicate that the south side of Byrd Glacier was displaced upward relative to the north side, supporting the original suggestion by Grindley & Laird (1969) that a fault exists beneath Byrd Glacier. A comparison of offset as indicated by apatite (U–Th)/He thermochronologic and geomorphologic data demonstrates a close correspondence in the amount of vertical differential movement, 1357 ± 156 v. 1122 ± 4.7 m. The results suggest a model of relatively uniform exhumation followed by fault displacement. The thermochronological data suggest that the timing of the fault displacement occurred sometime after *c.* 40 Ma.

The authors would like to acknowledge B. Adams for his help with the MATLAB coding. NSF grant OPP 9909463 funded collection of the samples used in this study. F. Lisker and M. Laird are thanked for their helpful reviews.

References

EHLERS, T. A. & FARLEY, K. A. 2003. Apatite (U–Th)/He thermochronometry: methods and applications to problems in tectonic and surface processes. *Earth and Planetary Science Letters*, **206**, 1–14.

FARLEY, K. A. 2002. (U–Th)/He dating: techniques, calibrations, and applications. *Noble Gas Geochemistry, Reviews in Mineralogy and Geochemistry*, **47**, 819–843.

FITZGERALD, P. G. 1992. The Transantarctic Mountains of southern Victoria land: the application of apatite fission track analysis to a rift shoulder uplift. *Tectonics*, **11**, 634–662.

FITZGERALD, P. G. 1994. Thermochronologic constraints on post-Paleozoic tectonic evolution of the central Transantarctic Mountains, Antarctica. *Tectonics*, **13**, 818–836.

FITZGERALD, P. G. & GLEADOW, A. J. W. 1988. Fission-track geochronology, tectonics and structure of the Transantarctic Mountains in Northern Victoria land, Antarctica. *Chemical Geology: Isotope Geoscience*, **73**, 169–198.

FITZGERALD, P. G. & STUMP, E. 1997. Cretaceous and Cenozoic episodic denudation of the Transantarctic Mountains, Antarctica: new constraints from apatite fission track thermochronology in the Scott Glacier region. *Journal of Geophysical Research*, **102**, 7747–7765.

FITZGERALD, P. G., BALDWIN, S. L., WEBB, L. E. & O'SULLIVAN, P. B. 2006. Interpretation of (U–Th)/He single grain ages from slowly cooled crustal terranes: a case study from the Transantarctic Mountains of southern Victoria land. *Chemical Geology*, **225**, 91–120.

FLEMING, T. H., HEIMANN, A., FOLAND, K. A. & ELLIOT, D. H. 1997. $^{40}Ar/^{39}Ar$ geochronology of Ferrar Dolerite sills from the Transantarctic Mountains, Antarctica: implications for the age and origin of the Ferrar magmatic province. *Geological Society of America Bulletin*, **109**, 533–546.

GLEADOW, A. J. W. & FITZGERALD, P. G. 1987. Uplift history and structure of the Transantarctic Mountains: new evidence from fission track dating of basement apatites in the Dry Valleys area, southern Victoria land. *Earth and Planetary Science Letters*, **82**, 1–14.

GRINDLEY, G. W. & LAIRD, M. G. 1969. *Geology of the Shackleton coast*. American Geographical Society, Antarctic Map Folio Series, Folio 12, Sheet XIV.

LISKER, F. 2002. Review of fission studies in northern Victoria Land, Antarctica – passive margin evolution v. uplift of the Transantarctic Mountains. *Tectonophysics*, **349**, 57–73.

REINERS, W. & FARLEY, K. A. 2001. Influence of crystal size on apatite (U–Th)/He thermochronology: an example from the Bighorn Mountains, Wyoming. *Earth and Planetary Science Letters*, **188**, 413–420.

SCHILDGEN, T. F., EHLERS, T. A., WHIPP, D. M. JR, VAN SOEST, M. C., WHIPPLE, K. X. & HODGES, K. V. 2009. Quantifying canyon incision and Andean Plateau surface uplift, southwest Peru: a thermochronometer and numerical modeling approach. *Journal of Geophysical Research, Earth Surface*, **114**, F04014, doi: 10.1029/2009JF001305

STUMP, E. & FITZGERALD, P. G. 1992. Episodic uplift of the Transantarctic Mountains. *Geology*, **20**, 161–164.

STUMP, E., GOOTEE, B. F. & TALARICO, F. 2005. Tectonic model for the Byrd glacier discontinuity and surrounding regions of the Transantarctic Mountains during the Neoproterozoic-early Paleozoic. *In*: FUTTERER, D. K., DAMASKE, D., KLEINSCHMIDT, G., MILLER, H. & TESSENSOHN, F. (eds) *Antarctica: Contributions to Global Earth Sciences*. Springer, Berlin, 179–188.

VAN SOEST, M. C., HODGES, K. V. *ET AL.* 2011. (U–Th)/He dating of terrestrial impact structures: the Manicouagan example. *Geochemistry, Geophysics, Geosystems*, **12**, Q0AA16, doi: 10.1029/2010GC003465.

YORK, D. 1969. Least-squares fitting of a straight line. *Canadian Journal of Physics*, **44**, 1079–1086.

ZEITLER, P. K., HERCZIG, A. L., MCDOUGALL, I. & HONDA, M. 1987. U–Th–He dating of apatite: a potential thermochronometer. *Geochimica et Cosmochimica Acta*, **51**, 2865–2868.

Mesozoic to recent evolution of intraplate stress fields under multiple remote stresses: The case of Signy Island (South Orkney Microcontinent, Antarctica)

A. MAESTRO[1,2*], J. LÓPEZ-MARTÍNEZ[2] & F. BOHOYO[1]

[1]*Instituto Geológico y Minero de España, C/ Calera 1, 28760 Tres Cantos, Madrid, Spain*

[2]*Departamento de Geología y Geoquímica, Facultad de Ciencias, Universidad Autónoma de Madrid, 28049 Madrid, Spain*

**Corresponding author (e-mail: a.maestro@igme.es)*

Abstract: Palaeostresses inferred from approximately 3200 brittle mesostructures measured at 43 sites in Signy Island show a stress field characterized by compressional, strike-slip and extensional stress states. There is a dominant NW–SE horizontal compression direction as well as three (NNW–SSE, WNW–ESE and ENE–WSW) secondary σ_1 horizontal stress directions. Orientation of σ_3 shows a main mode trending NW–SE together with secondary north–south and NE–SW extensional stress directions. The NNW–SSE horizontal compression is consistent with the stretching–shearing direction defined from the fold-axis trend. This is owing to a subduction-related tectonic regime during early to middle Jurassic times. In middle Jurassic/early Cretaceous times, the tectonic regime changes to extensional with north–south extensional stress directions. The early processes of subduction of the northern sector of the Weddell Sea oceanic crust below the South Orkney Microcontinent, from the Paleocene until the middle Miocene, is related to the σ_1 direction WNW–ESE to NW–SE. The spreading of Powell Basin took place from the late Eocene to the early Miocene and is the cause of east–west and NE–SW extensional processes in the southern South Orkney Microcontinent. Finally, the NW–SE σ_3 direction is related to the current left-lateral strike-slip movement between the Scotia and Antarctic plates.

The South Orkney Microcontinent is the exposed part of a continental fragment on the southern limb of the Scotia Arc, between South America and the Antarctic Peninsula (Dalziel 1984; King & Barker 1988; Barker *et al.* 1991; King *et al.* 1994) (Fig. 1a). The northern margin of the South Orkney Microcontinent is the Antarctic–Scotia plate boundary with a sinistral strike-slip movement accomplished by an oblique plate convergence in the western part and a divergence component eastward toward Bruce Bank. The SE and SW South Orkney Microcontinent passive margins are connected to the Powell and the Jane oceanic basins, respectively. They formed after the rifting between the microcontinent and the Antarctic Peninsula in the Eocene/early Oligocene (Coren *et al.* 1997; Eagles & Livermore 2002) and in the early to middle Miocene, respectively (Bohoyo *et al.* 2002). The southern margin of the South Orkney Microcontinent is connected to the oceanic Weddell Basin (Barker *et al.* 1984, 1991; King & Barker 1988; Lawver *et al.* 1991).

The South Orkney archipelago, a unique emerged outcrop of the South Orkney Microcontinent, is composed of four major islands and a larger number of smaller islands. The principal islands are Coronation, Laurie, Powell and Signy (Fig. 1c). These islands, which have steep relief and are nearly 90% covered by glaciers, are composed of metapelites and metagreywackes of probable Triassic sedimentary age. Deformations related to an accretionary wedge setting, along with associated metamorphism from anchizone to the greenschist facies, are of early Jurassic age (176–200 Ma; Trouw *et al.* 1997).

Owing to the particular tectonic framework of the archipelago, the reconstruction of palaeostress fields in terms of either successive tectonic events or local stress perturbations from outcrop-scale structures is not easy when a variety of stress directions are recorded. The South Orkney Microcontinent geodynamic evolution is characterized by complex plate motions. As a result of this, a number of distinct remote tectonic forces with different tectonic genetic components may be merged into the regional stress field together with second- or third-order tectonic stress systems induced by local structures, anisotropies or heterogeneities (Heidbach *et al.* 2007). From this, actual complex stress fields, strongly varying in both space and time, develop. Each tectonic genetic component varies according to its own spatial gradients and temporal rates. The total stress tensor can be relatively stable only if one genetic component strongly

From: Hambrey, M. J., Barker, P. F., Barrett, P. J., Bowman, V., Davies, B., Smellie, J. L. & Tranter, M. (eds) 2013. *Antarctic Palaeoenvironments and Earth-Surface Processes*. Geological Society, London, Special Publications, **381**, 45–65. First published online April 19, 2013, http://dx.doi.org/10.1144/SP381.4

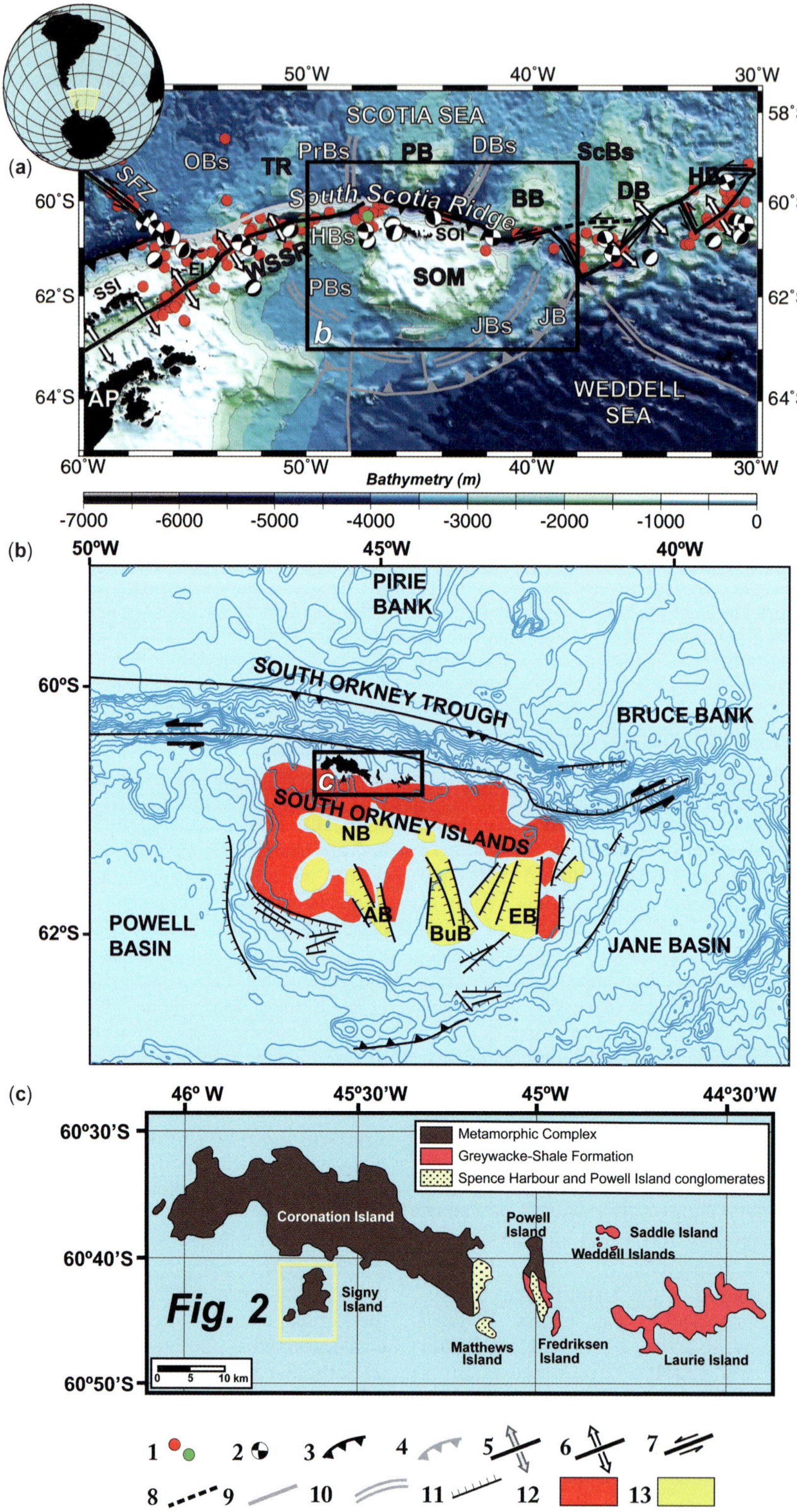
(a)
SCOTIA SEA
OBs
PrBs
PB
DBs
ScBs
TR
SFZ
South Scotia Ridge
BB
DB
HB
HBs
SOI
EI
WSSR
SOM
SSI
PBs
b
JBs
JB
AP
WEDDELL SEA
50°W
40°W
30°W
60°W
58°S
60°S
62°S
64°S
Bathymetry (m)
-7000
-6000
-5000
-4000
-3000
-2000
-1000
0
(b)
50ºW
45ºW
40ºW
60ºS
62ºS
PIRIE BANK
SOUTH ORKNEY TROUGH
BRUCE BANK
c
SOUTH ORKNEY ISLANDS
NB
AB
BuB
EB
POWELL BASIN
JANE BASIN
(c)
46º W
45º30'W
45ºW
44º30'W
60º30'S
60º40'S
60º50'S
Metamorphic Complex
Greywacke-Shale Formation
Spence Harbour and Powell Island conglomerates
Coronation Island
Powell Island
Saddle Island
Weddell Islands
Fig. 2
Signy Island
Matthews Island
Fredriksen Island
Laurie Island
0
5
10 km
1
2
3
4
5
6
7
8
9
10
11
12
13

prevails. Variability will be generally higher if several components with analogous but fluctuant magnitudes co-exist (Caputo 2005).

In this paper we present palaeostress indicators obtained from fault, joint and tension gash population analyses in the early late Triassic–early Jurassic metamorphic rocks of Signy Island (South Orkney archipelago). In the context of contributing to the knowledge of the tectonic evolution of the northern Antarctic Peninsula region and the South Orkney Microcontinent, the main objectives of this work are: (i) to characterize the Cenozoic to recent compressional and extensional stress fields in the South Orkney Microcontinent from brittle structure analysis data of Signy Island and the earthquake focal mechanisms; and (ii) to determine the stress evolution that may have been responsible for the orientation of the stress fields acting in this area during the Cenozoic period.

Geological setting

Signy Island forms part of the South Orkney archipelago, an emergent part of the platform of the South Orkney Microcontinent that represents the largest (about 70 000 km^2) continental fragment of the southern part of the Scotia Arc. The South Orkney Microcontinent was part of the former continental connection between South America and the Antarctic Peninsula prior to the Palaeogene continental break-up and played an important role in the configuration of the present Scotia Arc.

South Orkney Microcontinent margins and surrounding basins

Large amounts of geophysical data have been acquired in this region in the last 20 years by several countries. Additionally, the ODP Leg 113 has drilled three sites, with two of them located on the continental platform (Barker *et al.* 1988). The structural setting and morphology of the western and southern-to-southeastern margins of the South Orkney Microcontinent are clearly related to the evolution of Powell and Jane basins, respectively (King & Barker 1988; Maldonado *et al.* 1998; Busetti *et al.* 2000; Bohoyo *et al.* 2002; Fig. 1b). King & Barker (1988) have identified two main structural or tectonic domains and showed two orthogonal structural trends within the microcontinent (Fig. 1b): (1) an older east–west trend comprising a magmatic arc, a fore-arc-basin filled with 5 km of sediments, called Newton Basin, and an accretionary complex created when the South Orkney Microcontinent was at the Pacific active margin during the Mesozoic era; and (2) a younger structural set, trending approximately north–south, and comprising three basins, Airy, Bouguer and Eötvös from west to east, structured in a sequence of horsts and grabens filled by Eocene/Oligocene sediments (King & Barker 1988; Busetti *et al.* 2000), associated with the opening of Powell and Jane basins during the Cenozoic.

The western, southern and eastern margins of the South Orkney Microcontinent are clearly related to continental rifting and the subsequent development of the surrounding Powell and Jane basins (Fig. 1b). The deformational style of the western margin is characterized by both steep and gentle slopes as well as detached elevations in places (King & Barker 1988). However, the eastern margin is more irregular, with less steepness overall. The shelf break deepens from 600 to 1300 m, north to south, and the rotated fault blocks of the eastern margin of the South Orkney Microcontinent are observed further away westwards, up to 70 km.

The opening of Powell Basin was part of the regional response to the N55°W relative plate motion of South America away from Antarctica during the Eocene–Oligocene (Eagles & Livermore 2002) and opened in an ENE–WSW direction as

Fig. 1. (**a**) Regional tectonic framework and bathymetric contour map of the northern Antarctic Peninsula and South Orkney Microcontinent region. Bathymetric map derived from satellite and ship track data (Smith & Sandwell 1997). Earthquake focal mechanisms obtained from Harvard Seismology Centroid Moment Tensor Catalog (Dziewonski *et al.* 1981). Symbols of focal mechanisms: white, compression; black, extension. Earthquake epicentres obtained from NEIC (http://www.neic.cr.usgs.gov). Legend: 1, earthquake epicentre (shallow events in red and intermediate in green); 2, focal mechanism; 3, compressive or subduction zone; 4, inactive subduction zone; 5, transform fault; 6, extensional zone; 7, transcurrent fault and sense; 8, incipient transform fault; 9, Inactive transform fault; 10, Inactive spreading ridge; 11, Inactive extensional fault; 12, structural basement high; 13, basin. AB, Airy Basin; AP, Antarctic Peninsula; BB, Bruce Bank; BoB, Bourguer Basin; DBs, Dove Basin; DB, Discovery Bank; EB: Eötvös Basin; EI, Elephant Island; SFZ, Shackleton Fracture Zone; JBs, Jane Basin; JB, Jane Bank; OBs, Ona Basin; PB, Pirie Bank; PBs, Powell Basin; PrBs, Protector Basin; SOI, South Orkney Islands; SOM, South Orkney Microcontinent; SSI, South Shetland Islands; TR, Terror Rise. (**b**) Structural map of the South Orkney Microcontinent, obtained from interpretation of seismic data modified from King & Barker (1988) and Busetti *et al.* (2000). Two main domains with different structural trends are recognizable: the east–west trend following the transform plate boundary, affecting the structure of the basement high and the Newton Basin; and the north trend of Airy, Bouguer and Eötvös basins. (**c**) Geological map of South Orkney Islands modified from Trouw *et al.* (1997).

suggested by the basin margins and axis features (Fig. 1b). The reconstruction of the movement of the South Orkney Microcontinent during the development of the Scotia Arc indicates that it most likely did not rotate much during its separation from the Antarctic Peninsula (King & Barker 1988; Trouw *et al.* 1997). Extensional stress tensors, such as those that developed Powell Basin, have also been recognized on Seymour Island, in the northwestern Weddell Sea (Maestro *et al.* 2008). The timing of the opening of the basin was between the late Eocene and early Miocene periods according to a variety of authors (King & Barker 1988; Lawver *et al.* 1994; Coren *et al.* 1997; Rodríguez-Fernández *et al.* 1997; Eagles & Livermore 2002).

Jane Basin is a backarc basin related to subduction of the Weddell oceanic crust below the southern and southeastern borders of the South Orkney Microcontinent (Barker *et al.* 1984; Bohoyo *et al.* 2002; Fig. 1a, b). The multi-channel seismic profiles across the basin (Kavoun & Vinnikovskaya 1994; Bohoyo 2004; Lodolo *et al.* 2010), together with gravity models (Bohoyo 2004), show heterogeneous thinning by way of continental fragmentation processes of the continental crust at the southern and southeastern margins of the South Orkney Microcontinent. Bohoyo *et al.* (2002), based on the interpretation of sea-floor magnetic anomalies in Jane Basin, have established an opening period occurring between chron C5Dn (17.6 Ma) and C5ADn (14.4 Ma), after the collision and partial subduction of the Weddell spreading centre below the South Orkney Microcontinent (Barker *et al.* 1984; Bohoyo *et al.* 2002). Additionally, important reverse faults in the oceanic crust of the Weddell Sea and Jane Basin were detected in the multi-channel seismic profiles. These structures are younger than the age of development of Jane Basin, and they most likely mark the end of Jane Basin spreading owing to a change of the tectonic regime in the area. A thrusting of the South Orkney Microcontinent over the oceanic crust of Jane Basin and duplication of this oceanic crust are detected in the gravity modelling, which emphasizes the importance of these reverse faults (Bohoyo 2004).

The northern and northeastern margins of the South Orkney Microcontinent, with an ESE–WNW orientation, are both characterized by a convergent component that produces the subduction of the Scotia plate below the Antarctic plate (Fig. 1b), as proposed by Kavoun & Vinnikovskaya (1994), Klepeis & Lawver (1996), Lodolo *et al.* (1997, 2010), Maldonado *et al.* (1998), Busetti *et al.* (2000) and Bohoyo *et al.* (2007). In general, the northern border of the South Orkney Microcontinent is characterized by a linear depression associated with a gravity minimum that may be related to the thrusting or sinistral transcurrent displacement of the South Orkney Microcontinent in relation to the Scotia Sea. A deep (up to 6000 m), NW–SE-oriented trough (Bruce Deep) separates the South Orkney Microcontinent from Bruce Bank, suggesting a possible shifted continuation of the South Orkney trough. Towards the east, the convergence appears less evident, producing mainly compressional structures in the oceanic sediments with folds and reverse faults (Busetti *et al.* 2000; Lodolo *et al.* 2010). The crustal nature of the South Orkney Microcontinent was derived from gravity modelling along seismic profiles using geological constraints from seismic interpretation for the upper part of the model and the estimated rock densities according to the depths. The results provide a crustal thickness of approximately 25 km for both the South Orkney Microcontinent and Bruce Bank (King & Barker 1988; Busetti *et al.* 2000) and a progressive continental thinning towards the southern and eastern borders (Bohoyo 2004).

South Orkney Islands: Signy Island

The South Orkney Islands expose the northern margin of the South Orkney Microcontinent and show the early Jurassic tectonic events at the proto-Pacific margin that are not observed anywhere else along the Antarctic Peninsula. The rocks comprise predominantly Permian–Triassic sandstones and mudstones (metapelites and metagreywackes corresponding to the Greywacke Shale Formation/Scotia Metamorphic Complex) deposited by turbidity currents in a trench/slope environment (Smellie 1981; Dalziel 1982, 1984; Storey & Meneilly 1985; Trouw *et al.* 1997; Flowerdew *et al.* 2011; Fig. 1c).

The Scotia Metamorphic Complex on Signy Island was probably deposited after the late Permian period considering their prominent detrital zircon age population at *c.* 263 Ma (Flowerdew *et al.* 2011). Intense deformations and accompanying epidote amphibolite–facies metamorphism (*c.* 7 kbar at 520 °C on Signy Island) interleaved ocean floor sequences, including marble, pelite with metachert horizons, Triassic radioalarian chert and metabasaltic rocks with an ocean island chemical signature. Where oceanic elements are observed, they are mapped as a Scotia Metamorphic Complex mixed unit. The interleaved oceanic units, however, comprise only a small volume of this mixed unit, which is predominantly continent-derived material. Mineral geochronology from the metamorphic rocks indicates that deformation and metamorphism most likely first occurred at *c.* 190 Ma, and a major phase of cooling and exhumation occured at *c.* 175 Ma (Trouw *et al.* 1997; Flowerdew *et al.* 2011).

Matthews & Maling (1967) divided the lithological sequence of Signy Island into three series, the Marble, Amphibolite and Moe Island Series, based on the relative proportions of the three main rock types. Although the lithologies can be crudely grouped into the above units, Storey & Meneilly (1985) concluded that there is no justification for this stratigraphic relationship and proposed a lithological division that consists of a complex interlayered sequence of mica-schist, marble and amphibolite. These authors divided the structural history into five phases of deformation that are believed to form part of a single orogenic cycle. The most prominent structural features of Signy Island rocks are the strong NNW–SSE trending lineation of nearly all fold axes (Fig. 2a), bedding foliation (Fig. 2b), mineral lineations, quartz rods and augen lineations. This lineation defines the direction of shearing throughout the deformation history during ductile thrusting towards the NNW (Meneilly & Storey 1986).

The South Orkney Islands do not show an important set of north–south brittle deformation or faulting with the exception of the straits separating the three main islands (King & Barker 1988; Trouw *et al.* 1997). They are also considered to be of the same early Cretaceous–late Jurassic age as the subaerially exposed conglomerates (Elliot & Wells 1982), but may be related to the separation of the South Orkney Microcontinent and the Antarctic Peninsula during the development of the Scotia Arc (Trouw *et al.* 1997).

The palaeostress database

Palaeostress analysis was performed using brittle mesostructures, mainly striated fault planes and tension and shear joints, measured at 43 sites in late Triassic–early Jurassic metamorphic rocks (Fig. 2c). Five-hundred and seventeen striated faults (33 sites), approximately 2609 extension and shear joints (43 sites) and 69 quarzitic tension gashes (two sites) were measured in the field and analysed (Figs 3 & 4). Overall, fractures on the outcrop scale show an east–west orientation maximum.

Normal faults (approximately 240) dominate on the scale of exposure, although 125 reverse faults and 77 left-lateral and 73 right-lateral strike-slip faults were recognized (Fig. 4a–d). In both cases, the slip on the majority of fault planes varies from centimetres up to a few metres. Most strike-slip-fault planes are vertical, whereas normal and reverse faults dip between 20 and 80°. Directional analysis of faults shows a maximum striking east–west and a relative maxima striking NE–SW (Fig. 5a). The fault surfaces do not show pseudotachylites, fault breccias or gouges.

The analysis of fault data has been made using a sequence of three methods, each one providing a different approach to the stress determination problem, and whose joint usefulness has been broadly tested (Casas *et al.* 1992; Casas & Maestro 1996) and enables the complete definition of stress tensors (Fig. 6). These three methods are:

(1) The Right Dihedra method (Angelier & Mechler 1977) – this is a simple geometrical approach which provides an initial estimate of stress directions (Fig. 6a).

(2) The y–R diagram (Simón 1986) – this is a 2D approximation in which one of the principal stress axes is supposed to be vertical so that tensors may be represented only by two parameters: y (azimuth of the maximum horizontal stress, σ_y) and R (stress ratio in Bott's equation; Bott 1959), $R = (\sigma_z - \sigma_x)/(\sigma_y - \sigma_x)$. The y, R pairs satisfying one individual fault give rise to a curve. The 'knots' where these curves intersect show a preliminary spectrum of all possible solutions and their relative weight with respect to the whole fault population (especially useful in the case of polyphase tectonics or when a gradual tensor change occurs; Fig. 6b). The knots of a y–R diagram must be corroborated by other three-dimensional methods (e.g. Etchecopar's method) before we can conclude that the solutions determined represent true stress tensors showing a vertical axis (σ_1, σ_2 or σ_3, according to R value).

(3) Etchecopar's method (Etchecopar *et al.* 1981) – this is a numerical method that allows the exploration of the solutions suggested by the y–R and Right Dihedra diagrams. These solutions will eventually be confirmed and refined to obtain the final, complete solution: 3D orientations of the three principal stress axes and stress ratios (Fig. 6c1, c2, c3). Etchecopar's method is based on the minimization of the angles between real and theoretical striations, and it allows for the separation of different stress tensors by means of an adequate management of the percentage of data submitted to minimization. After a preliminary analysis using Etchecopar's method, we chose the solution that explained most of the faults with a good histogram of angular discrepancies and a good distribution of explained faults on the Mohr circle. In monophase populations, a satisfactory solution is normally obtained for 80–90% of faults, thus rejecting 10–20% of spurious data. In suspected polyphase populations, the initial required percentage should be lower (30–50%). The program chooses the best fitting faults representing this initial percentage, which are then discarded in order to find a second stress

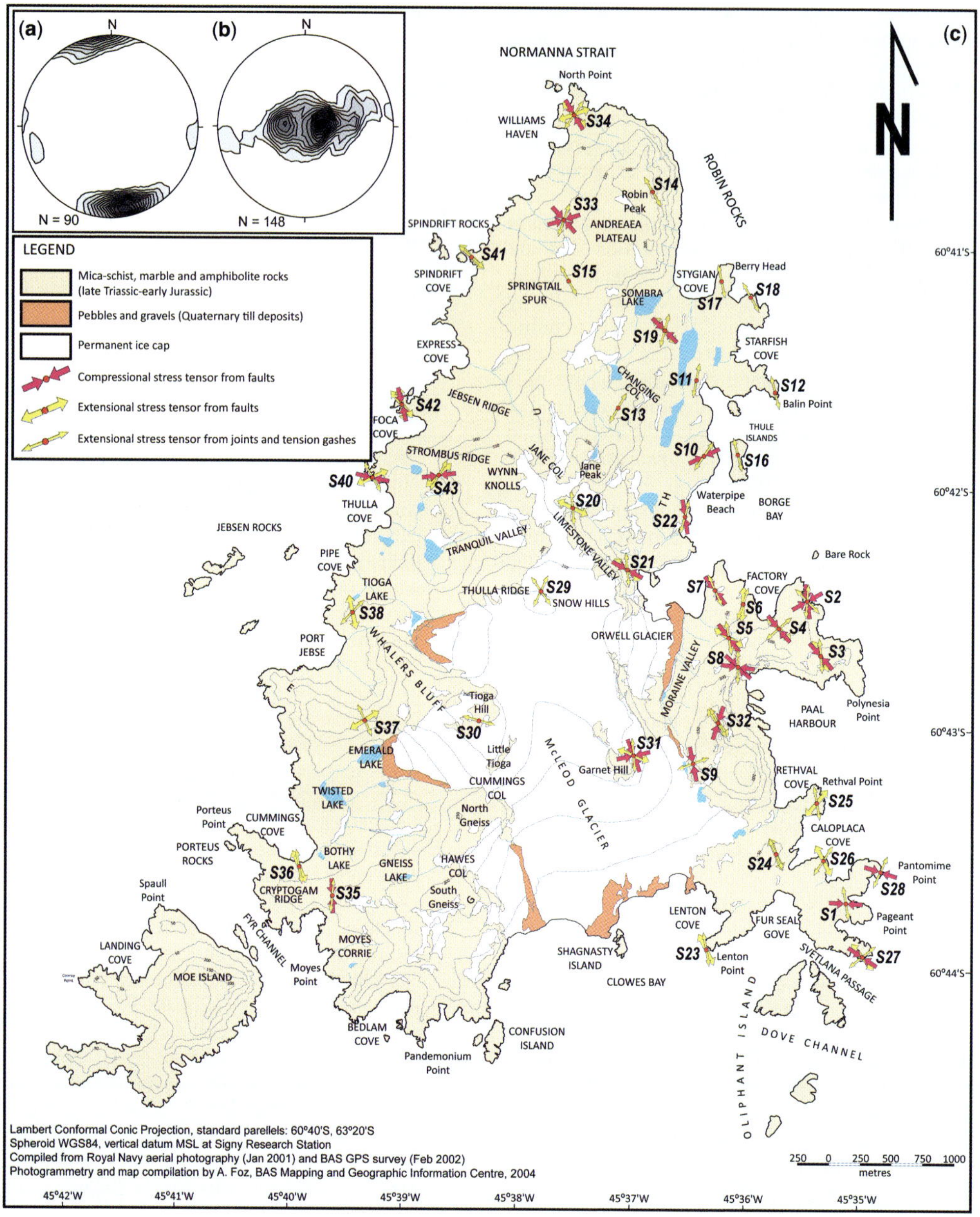

Fig. 2. Density stereoplots representing the fold axes (**a**) and the bedding foliation poles (**b**) obtained in the study sites (equal area projection, lower hemisphere; contour interval 2%; *N* is the number of measurements). (**c**) Geological sketch map of Signy Island (modified from Flowerdew *et al.* 2011) and location of study sites with the compressional and extensional stress tensors obtained from faults, joints and tension gashes.

solution. The reliability of the calculated solutions can be tested using random sub-sampling analysis (Arlegui & Simón 1998). The occurrence of different stress tensors with similar orientations at the same site could be related to local perturbations of the regional stress field.

In addition to faults, tension gashes and joints have also been used to establish palaeostress fields

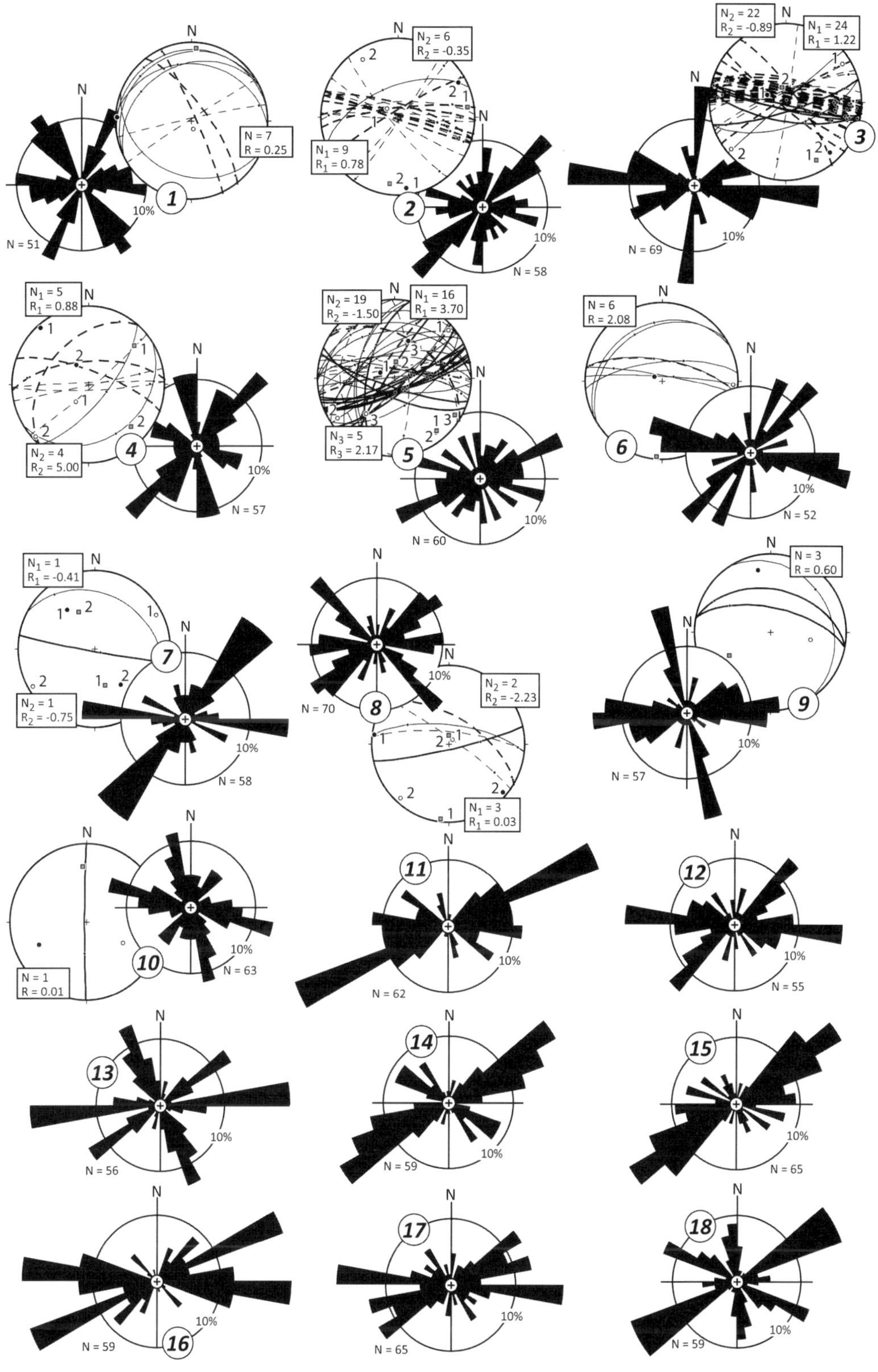

Fig. 3. *Continued.*

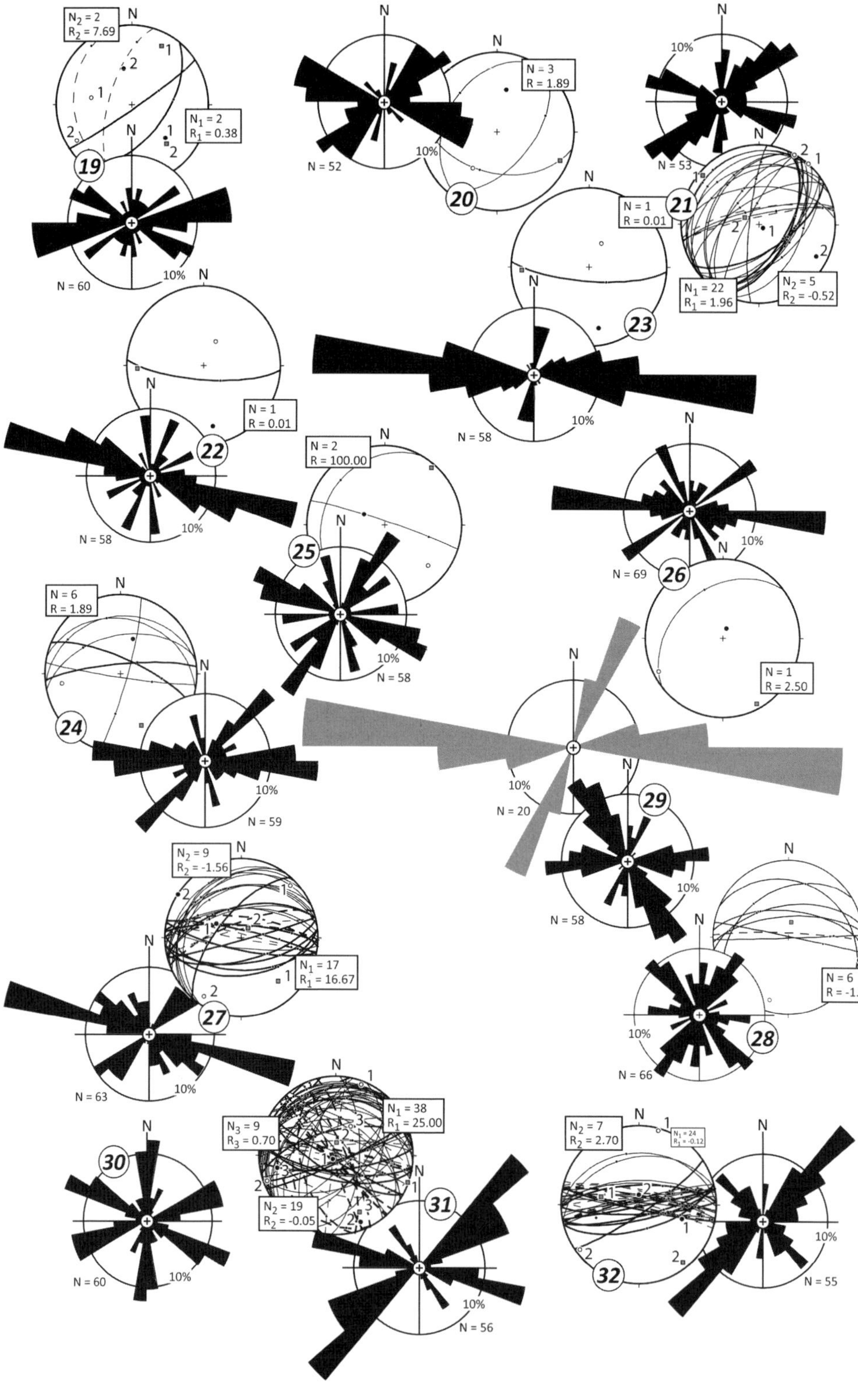

Fig. 3. *Continued.*

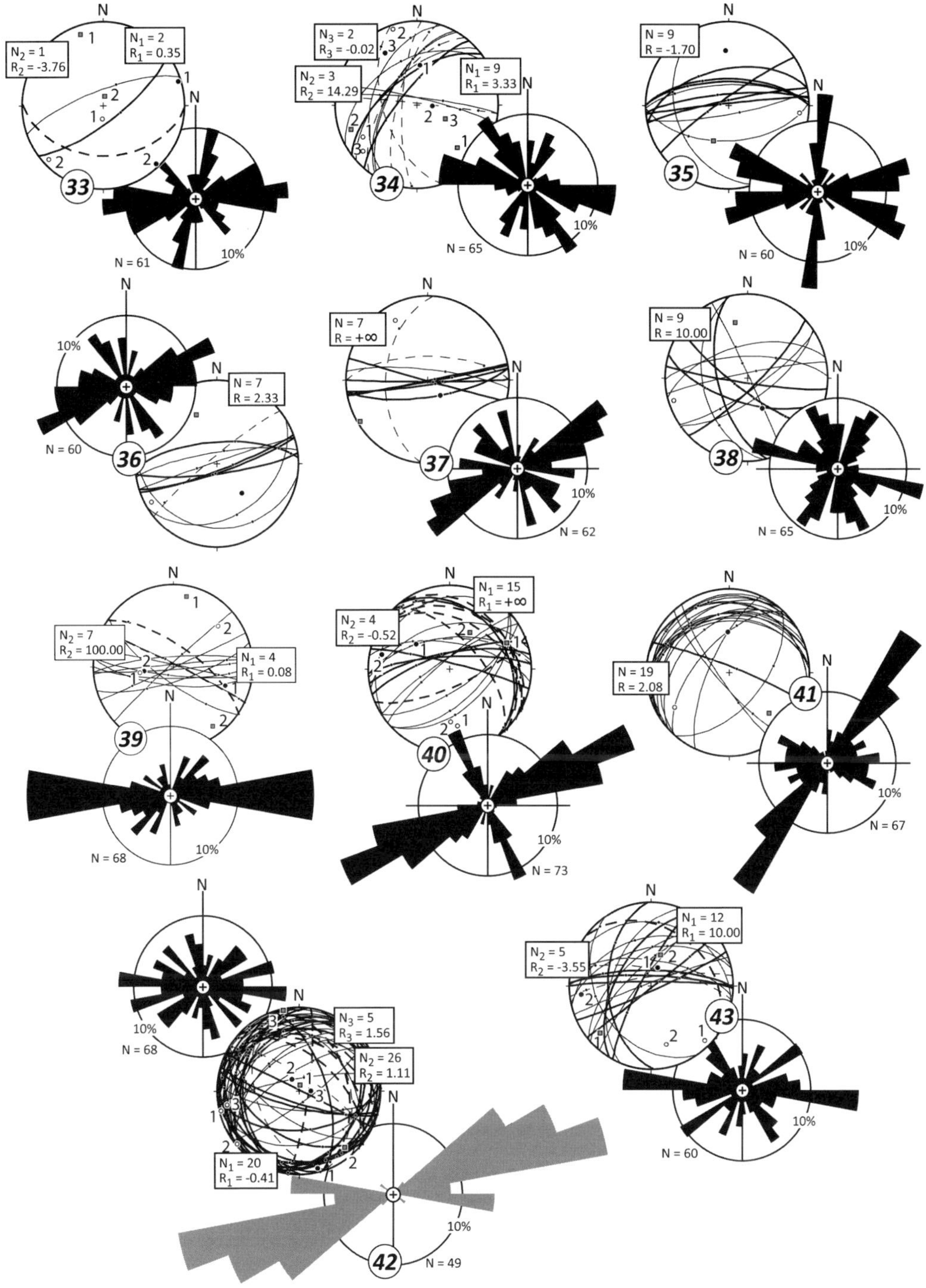

Fig. 3. Stereoplots and rose diagrams representing the sites data. Stereoplots include fault planes (thin line, normal fault; thick line, reverse fault, dashed thin line, right lateral strike-slip fault; dashed thick line, left lateral strike-slip fault), slickenside striations, stress axes (σ_1, white circle; σ_2, white square; and σ_3, black triangle), stress ratio (R) and number of data (N). Rose diagrams of orientation of joints (black roses) and tension gashes (grey roses) on the outcrop scale (outer circle represents 10% of the data set in the site, and N represents the number of data).

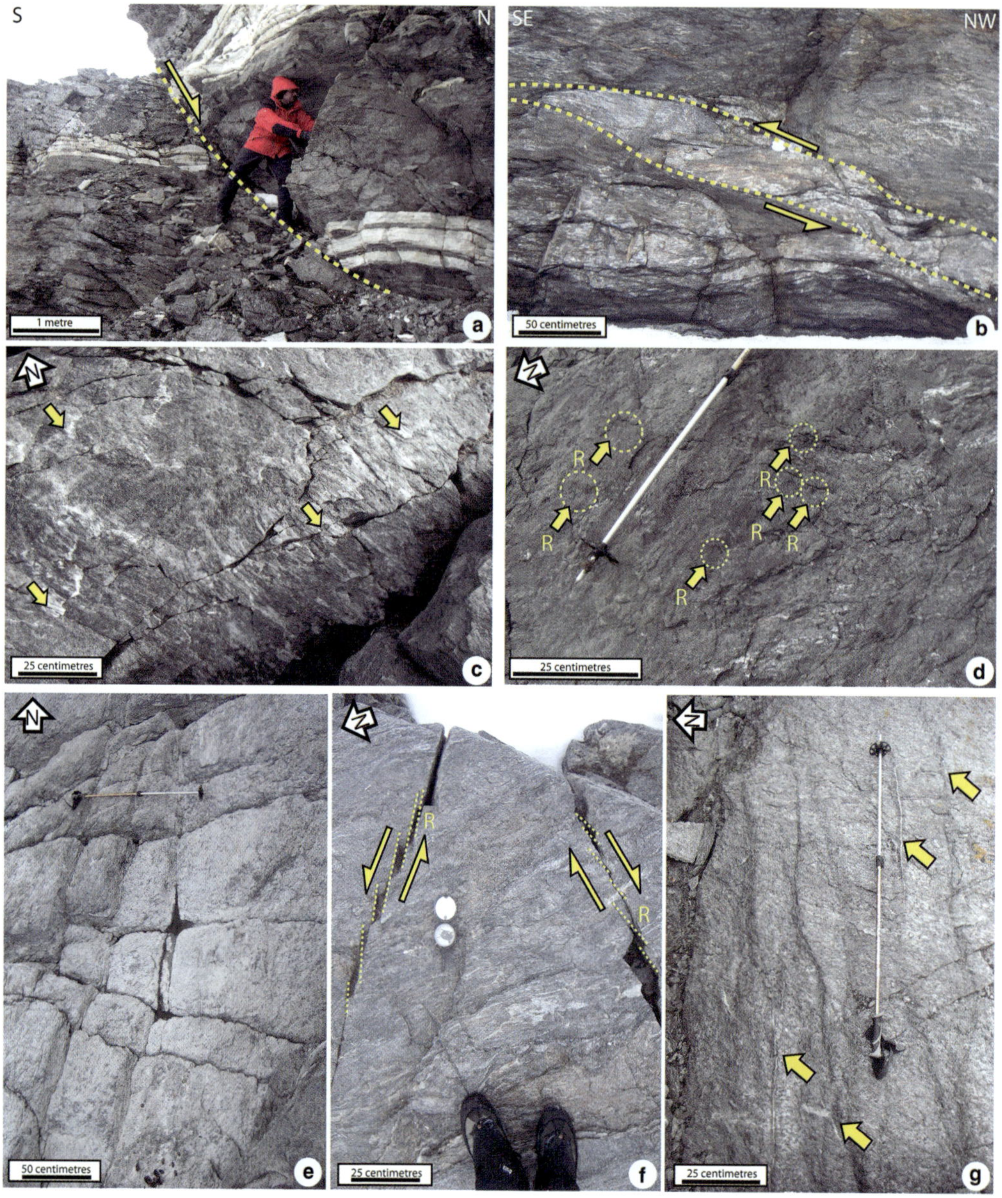

Fig. 4. Faults, joints and tension gashes are the main brittle structures analysed in this study to carry out the palaeostress reconstruction. (**a**) Stratigraphic criteria reinforce the sense of movement deduced by slickenside kinematic indicators. (**b**) Reverse fault determined by trains of sigmoid en echelon synthetic hybrid fractures. (**c**) Crystal fibres correspond to neoformed minerals growing congruously in shadow zones. (**d**) *R* fractures with no tectonic rupture of the tip of the dihedron formed by *R* and the main plain, so that the rupture surface determines incongruous fracture steps. (**e**) Orthogonal sets of tension joints. (**f**) Shear (left lateral and right lateral) joints showing en echelon Riedel minor fractures (*R*) with no observed offset. (**g**) Wedge-shaped tension gashes propagate perpendicularly to the direction of maximum extension.

during the Cenozoic. In this study, the joint and tension gashes population data have been processed by statistical methods that allow for the establishment of the preferential strikes affecting the metamorphic rocks outcropping in the area. A total of 2609 joints and 69 tension gashes have

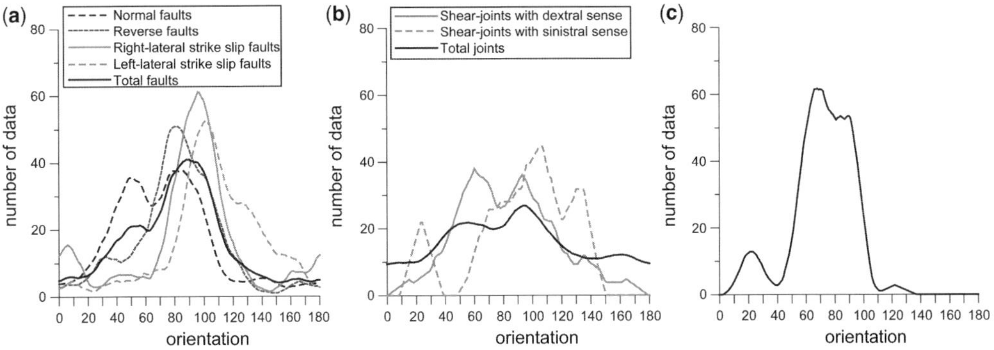

Fig. 5. Smoothed frequency curves of the orientation of faults (**a**), joints (**b**) and tension gashes (**c**).

been measured in two and 43 different sites, respectively (Figs 2 & 3). In all of them, the amount of data is enough to characterize the reliable fracture pattern at each site according to Arlegui & Simón (2001). Most joints and tension gash planes are vertical.

Several studies focus on the mechanical interpretation of joints as regional stress indicators (e.g. Engelder & Geiser 1980; Hancock & Engelder 1989). Many authors consider the existence of tension and shear joints (e.g. Hancock 1985), which provide a very accurate interpretation of the maximum horizontal (σ_y) and extensional (σ_3) stress trends, although no quantitative estimation of the stress ratio can be made. Other authors assume that every joint is a mode I fracture normal to σ_3 (e.g. Engelder & Geiser 1980; Pollard & Aydin 1988; Fig. 4e). However, this assumption cannot be considered owing to the abundant field evidence demonstrating the existence of shear and hybrid/shear tension joints. First, synchronous joints are arranged in conjugate pairs showing low dihedral angles (Hancock 1985; Bergerat *et al.* 1992). Additionally, joints oblique to σ_1 and σ_3 axes are independently determined from different structures (Bergerat *et al.* 1992). Finally, the existence of joint spectra as defined by Hancock (1985) and kinematic indicators on joint surfaces must also be considered. Quite often, joints arranged in low-dihedral conjugate pairs display en echelon microfractures randomly located within the parent joints. As they are not confined to the fringe zone, they cannot be interpreted as *f* joints (Hodgson 1961) related to fracture propagation and plume formation (Bahat 1991). On the contrary, they are shear Riedel fractures (Riedel 1929), that may be used as shear indicators (Engelder 1989; Arlegui & Simón 2001; Simón *et al.* 2006). In this study, several joint planes showing shear Riedel fractures have been observed (Fig. 4f) and reveal two relative maxima striking east–west and NE–SW (Figs 3 & 5b).

The tension gashes develop perpendicular to the least compressive stress axis (σ_3) (Ramsay & Huber 1983). These structures generally range from a few centimetres up to a metre long with wedge shapes and display a fibrous quartz crystal filling perpendicular to the block walls. The tension gashes observed show a separation between blocks of approximately 1 cm (Fig. 4g). The tension gash data show a maximum striking ENE–WSW and a relative maxima striking NE–SW (Figs 3b, c & 5c).

We conclude that, although our tectonic analysis is primarily based on palaeostress reconstructions using fault-slip data, the orientations of the other fractures, and especially of the joints, provide a reliable complementary record of palaeostress orientations (Figs 2 & 3, and Table 1).

Characterization of stress fields

The results of palaeostress analysis (by means of the three methods mentioned in 'the palaeostress database' section) are summarized in Table 1 and shown graphically in Figure 3 by means of 33 fault stereoplots, 43 joint and two tension gashes rose diagrams. Stress tensors are included, giving σ_1, σ_2, σ_3 and σ_y orientations and the values of R and Re. The number of faults used to define each stress state *v*. the number of faults in the fault population is also shown.

To show a graphical view of the stress tensor directions obtained in Signy Island, two types of representation were prepared: (1) the projection of the σ_1, σ_3 and σ_y axes of all the stress tensors obtained from fault slip data sets by Etchecopar's method, joints and tension gashes, respectively (Fig. 7a–c); and (2) a two-coordinate diagram that only allows representation of stress tensors, with one axis assumed to be vertical. When this condition is not accomplished, the axis that approaches a vertical alignment is considered as vertical for plotting

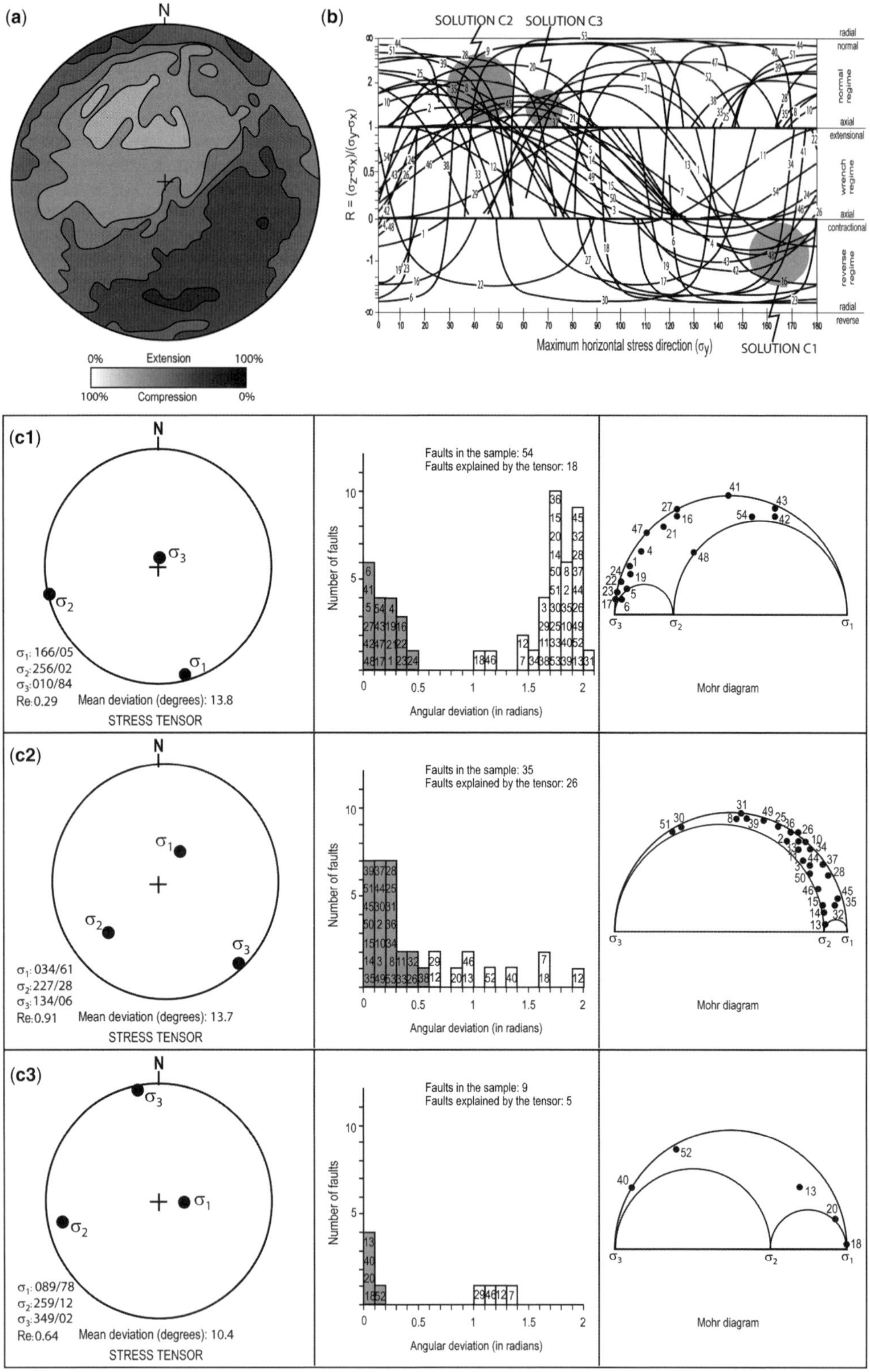
(a)
N
0% Extension 100%
100% Compression 0%
(b)
SOLUTION C2
SOLUTION C3
SOLUTION C1
R = (σz-σx)/(σy-σx)
Maximum horizontal stress direction (σy)
radial
normal
normal regime
axial
extensional
wrench regime
axial
contractional
reverse regime
radial
reverse
(c1)
σ1: 166/05
σ2: 256/02
σ3: 010/84
Re: 0.29
Mean deviation (degrees): 13.8
STRESS TENSOR
Faults in the sample: 54
Faults explained by the tensor: 18
Number of faults
Angular deviation (in radians)
Mohr diagram
(c2)
σ1: 034/61
σ2: 227/28
σ3: 134/06
Re: 0.91
Mean deviation (degrees): 13.7
STRESS TENSOR
Faults in the sample: 35
Faults explained by the tensor: 26
Number of faults
Angular deviation (in radians)
Mohr diagram
(c3)
σ1: 089/78
σ2: 259/12
σ3: 349/02
Re: 0.64
Mean deviation (degrees): 10.4
STRESS TENSOR
Faults in the sample: 9
Faults explained by the tensor: 5
Number of faults
Angular deviation (in radians)
Mohr diagram

the stress tensor. This representation is similar to the graphics of the y–R diagram (Simón 1986). This type of representation allows the orientation of the maximum horizontal stress axes and their associated stress regimes to be shown (Fig. 7d).

From the orientations of stress axes, it can be observed that there is a dominant NW–SE horizontal compression direction and three (NNW–SSE, WNW–ESE and ENE–WSW) secondary σ_1 horizontal stress directions (Fig. 7a). The orientation of σ_3 shows a main mode trending NNW–SSE, and NW–SE and NE–SW secondary σ_3 stress directions (Fig. 7b). Orientations of σ_y show a main mode trending ENE–WSW and two secondary modes with east–west and NW–SE directions (Fig. 7c). The NE–SW to east–west maximum stress direction shows an R relationship ranging from normal to wrench regimes (Fig. 7d). The NW–SE mode corresponds to reverse regimes (Fig. 7d).

The relative chronologies of stress tensors could be established at three sites. The chronology was determined from superposition of striae on a single fault plane. At site 31, the north–south σ_3 direction was prior to the NW–SE orientation of σ_3, and at sites 39 and 42 the NW–SE horizontal compression direction was prior to the NW–SE orientation of σ_3.

Geodynamic evolution of the intraplate stress fields within the South Orkney Microcontinent and relative chronology

Owing to the complexity of the structural setting of the area and the different tectonic regimes (transform, convergence and divergence) in the surroundings, the South Orkney Microcontinent has a complex geological history. The Mesozoic evolution of the South Orkney Microcontinent has been regarded as relatively straightforward by several authors. During this time the South Orkney Microcontinent had been part of the continental connection between South America and the Antarctic Peninsula (Dalziel & Elliott 1971; De Wit 1977; Barker 2001). Dalziel (1984) described north–south-trending lineations (D_1–D_3) in the South Orkney Microcontinent and attributed them to a subduction-related tectonic regime in early to middle Jurassic times (180–200 Ma). Barker & Dalziel 1983 suggest that the South Orkney Microcontinent can be rotated counter-clockwise to fit back into its original position occupying the present oceanic Powell Basin. Barker *et al.* (1991) interpreted the east–west trending structure of the islands, the presence of a major graben south of the islands called the Newton graben and the east–west-trending magnetic anomaly south of this graben (Harrington *et al.* 1972; King & Barker 1988) as an accretionary prism, a fore-arc spreading basin and a magmatic arc triplet associated with south-directed subduction (in the present orientation). Meneilly & Storey (1986) suggest that the sudbduction complex rocks of Signy Island have suffered intense shearing throughout their deformation history in a consistent NNW–SSE direction during the ductile thrusting toward the NNW, consistent with a NNW–SSE dominant fold-axis trend (Fig. 2a, b). Thus, the consistent stretching and shearing direction of Signy Island most likely indicates the thrusting direction in the South Orkney Islands block as a whole and is consistent with the NNW–SSE horizontal compression direction obtained from a brittle mesostructure analysis.

Trouw *et al.* (1997) established at Powell Island a later extension (middle Jurassic–early Cretaceous) in the north–south direction with respect to the present position of the islands that caused D_4 structures, with similar direction to D_1–D_3, but opposite shear sense to the subduction movement. They conclude this is possibly related to the development of the Newton graben, interpreted by King & Barker (1988) as a fore-arc basin. This extensional stage is related to retrograde metamorphic conditions owing to an uplift that favoured the development of brittle structures with an east–west trend (D_5) in Cenozoic times. Most of the fractures measured in Signy Island show a predominant east–west orientation.

The process of rifting and separation of the Antarctic Peninsula and South America began in the late Eocene (Meissner *et al.* 1988; Barker *et al.* 1991; Barker 2001). The oceanic crust of the Scotia Plate mainly formed in a NW–SE spreading direction. During this process, the dextral transcurrent fault systems of the South Scotia Ridge developed. Powell Basin, which exhibits a spreading axis oriented NNW–SSE along the centre of

Fig. 6. Example of standard stress analysis of a fault population (station 42). See location on map and equal-area plot of fault data in Figure 2. (**a**) Right Dihedra diagram (Schmidt net, lower hemisphere); isolines express the percentage of faults compatible with an extension axis. (**b**) y–R diagram; y is the azimuth of the maximum horizontal stress σ_y; $R = (\sigma_z - \sigma_x)/(\sigma_y - \sigma_x)$; the shaded ellipses indicate the zones showing the highest density of curve intersections, and therefore the optimum stress tensors. (**c**) The results of Etchecopar's method: stereoplot of the inferred stress axes, $R_e = (\sigma_2 - \sigma_3)/(\sigma_1 - \sigma_3)$; histogram of angular deviations between actual striation and theoretical shear component on each fault; Mohr diagram where the planes corresponding to the explained faults are plotted.

Table 1. *Summary of stress tensors and stress orientations obtained from brittle mesostructure population analysis*

Site	Foliation	n/N	σ_1	σ_2	σ_3	σ_y	Re	R	Structure
S1	025/48E 017/18W	7/9	273/2	165/81	004/8	273/2	0.25	0.25	Faults
S1	025/48E 017/18W	51	–	–	170/00	080/00	–	–	Tensional joints
S2	031/18E	9/18	173/9	306/76	082/10	173/9	0.78	0.78	Faults
S2	031/18E	6/18	061/11	328/15	187/71	61/11	0.26	−0.35	Faults
S2	031/18E	58	–	–	004/00	094/00	–	–	Shear joints
S2	031/18E	58	–	–	135/00	045/00	–	–	Tensional joints
S3	133/49W	24/48	291/69	057/12	151/16	057/12	0.82	1.22	Faults
S3	133/49W	22/48	138/14	230/7	345/74	138/14	0.47	−0.89	Faults
S3	133/49W	69	–	–	175/00	085/00	–	–	Shear joints
S4	066/18S	5/9	319/05	217/67	051/23	319/05	0.88	0.88	Faults
S4	066/18S	4/9	327/66	226/05	134/23	226/05	0.20	5.00	Faults
S4	066/18S	57	–	–	050/00	140/00	–	–	Shear joints
S5	028/34W 163/24E	16/41	291/74	049/08	141/14	049/08	0.27	3.70	Faults
S5	028/34W 163/24E	19/41	142/13	235/11	006/73	142/13	0.60	−1.50	Faults
S5	028/34W 163/24E	5/41	21/48	219/41	121/09	219/41	0.46	2.17	Faults
S5	028/34W 163/24E	60	–	–	170/00	080/00	–	–	Shear joints
S6	166/37E	6/7	301/81	093/07	184/04	093/07	0.48	2.08	Faults
S6	166/37E	52	–	–	020/00	110/00	–	–	Tensional joints
S7	018/38W 163/24E	1/2	324/38	062/09	163/50	324/38	0.29	−0.41	Faults
S7	018/38W 163/24E	1/2	143/42	239/07	336/47	143/42	0.43	−0.75	Faults
S7	018/38W 163/24E	58	–	–	155/00	65/00	–	–	Shear joints
S8	090/25N 178/41W	3/5	277/03	041/84	186/04	277/03	0.03	0.03	Faults
S8	090/25N 178/41W	2/5	131/06	222/07	359/81	131/06	0.69	−2.23	Faults
S8	090/25N 178/41W	70	–	–	180/00	090/00	–	–	Shear and tensional joints
S9	160/12W	3/3	348/20	100/45	241/38	348/20	0.60	0.60	Faults
S9	160/12W	57	–	–	070/00	160/00	–	–	Tensional joints
S10	009/15E	1/1	245/32	119/44	356/30	245/32	0.01	0.01	Faults
S10	009/15E	63	–	–	035/00	125/00	–	–	Shear joints
S11	150/20W	62	–	–	010/00	100/00	–	–	Shear joints
S12	170/44E 164/46W	55	–	–	165/00	075/00	–	–	Shear joints
S13	177/20W	56	–	–	030/00	120/00	–	–	Shear joints
S14	005/27E 175/30W	59	–	–	150/00	060/00	–	–	Tensional joints
S15	171/42E 162/31W	65	–	–	150/00	060/00	–	–	Tensional joints
S16	001/52E 179/44W	59	–	–	165/00	075/00	–	–	Shear joints
S17	125/33E 151/20W	65	–	–	168/00	078/00	–	–	Shear joints
S18	165/13E 184/62W	59	–	–	150/00	060/00	–	–	Tensional joints
S19	187/24W	2/4	134/39	279/45	28/18	134/39	0.38	0.38	Faults
S19	187/24W	2/4	347/52	238/15	138/33	238/15	0.13	7.69	Faults

S19	187/24W	60	–	–	015/00	105/00	–	–	Shear joints
S20	177/38W	3/3	13/45	214/43	114/11	214/43	0.53	1.89	Faults
S20	177/38W	52	–	–	175/00	085/00	–	–	Shear joints
S21	010/13W	22/29	130/84	040/00	310/6	040/00	0.51	1.96	Faults
S21	010/13W	5/29	118/16	028/01	295/74	118/16	0.34	−0.52	Faults
S21	010/13W	53	–	–	175/00	085/00	–	–	Shear joints
S22	004/37E 005/40W	1/1	171/23	028/62	267/15	171/23	0.01	0.01	Faults
S22	004/37E 005/40W	58	–	–	015/00	105/00	–	–	Tensional joints
S23	116/13N 160/19W	3/5	353/67	246/07	153/22	246/07	0.10	10.00	Faults
S23	116/13N 160/19W	58	–	–	170/00	080/00	–	–	Tensional joints
S24	085/35N 156/10W	6/7	021/51	261/22	157/30	261/22	0.53	1.89	Faults
S24	085/35N 156/10W	59	–	–	155/00	065/00	–	–	Shear joints
S25	035/17W	2/2	297/66	132/24	040/06	132/24	0.01	100.00	Faults
S25	035/17W	58	–	–	170/00	080/00	–	–	Shear and tensional joints
S26	063/13S	1/1	022/79	243/08	152/07	243/08	0.40	2.50	Faults
S26	063/13S	69	–	–	034/00	124/00	–	–	Shear joints
S27	087/10S 055/16N	17/30	298/60	044/09	139/28	044/09	0.06	16.67	Faults
S27	087/10S 055/16N	9/30	303/01	213/12	039/78	303/01	0.61	−1.56	Faults
S27	087/10S 055/16N	63	–	–	052/00	142/00	–	–	Shear joints
S28	002/10W 035/04E	6/8	107/01	197/16	012/74	107/01	0.55	−1.22	Faults
S28	002/10W 035/04E	66	–	–	025/00	115/00	–	–	Shear joints
S29	022/18E 119/13S	58	–	–	025/00	115/00	–	–	Shear joints
S29	022/18E 119/13S	20	–	–	145/00	055/00	–	–	Tension gashes
S30	177/11E	60	–	–	105/00	095/00	–	–	Shear joints
S31	069/16N	38/67	213/85	020/05	110/01	020/05	0.04	25.00	Faults
S31	069/16N	19/67	159/11	250/05	006/77	159/11	0.05	−0.05	Faults
S31	069/16N	9/67	258/23	028/56	157/23	258/23	0.70	0.70	Faults
S31	069/16N	56	–	–	175/00	085/00	–	–	Shear joints
S32	169/14W	24/32	108/41	015/03	281/49	108/41	0.11	−0.12	Faults
S32	169/14W	7/32	359/80	233/06	142/08	233/06	0.37	2.70	Faults
S32	169/14W	55	–	–	150/00	060/00	–	–	Tensional joints
S33	072/47S 174/32W	2/3	073/04	182/77	342/12	073/04	0.35	0.35	Faults
S33	072/47S 174/32W	1/3	137/05	226/08	011/81	137/05	0.79	−3.76	Faults
S33	072/47S 174/32W	61	–	–	020/00	110/00	–	–	Shear joints
S34	035/24W	9/14	005/51	240/25	136/29	240/25	0.30	3.33	Faults
S34	035/24W	3/14	094/74	342/06	250/15	342/06	0.07	14.29	Faults
S34	035/24W	2/14	328/28	230/15	115/58	328/28	0.02	−0.02	Faults
S34	035/24W	65	–	–	035/00	125/00	–	–	Shear joints
S35	157/37W	9/11	357/35	096/12	203/52	357/35	0.63	−1.70	Faults
S35	157/37W	60	–	–	005/00	095/00	–	–	Shear joints
S36	174/28W	7/11	138/51	240/09	337/37	240/09	0.43	2.33	Faults

(Continued)

Table 1. *Continued*

Site	Foliation	n/N	σ_1	σ_2	σ_3	σ_y	*Re*	*R*	Structure
S36	174/28W	60	–	–	170/00	080/00	–	–	Tensional joints
S37	140/22W	7/9	139/70	331/20	239/04	331/20	0.00	∞	Faults
S37	140/22W	62	–	–	155/00	065/00	–	–	Tensional joints
S38	024/10W 170/23E	9/13	153/56	253/07	347/33	253/07	0.10	10.00	Faults
S38	024/10W 170/23E	65	–	–	045/00	135/00	–	–	Shear joints
S39	167/70W	4/14	110/31	260/55	011/14	110/31	0.08	0.08	Faults
S39	167/70W	7/14	261/60	049/26	146/13	049/26	0.01	100.00	Faults
S39	167/70W	68	–	–	180/00	090/00	–	–	Tensional joints
S40	158/49W	15/22	306/48	172/32	066/24	172/32	0.00	∞	Faults
S40	158/49W	4/22	282/16	179/37	030/48	282/16	0.34	−0.52	Faults
S40	158/49W	73	–	–	150/00	060/00	–	–	Tensional joints
S41	002/16E	19/24	358/50	238/23	134/31	238/23	0.48	2.08	Faults
S41	002/16E	67	–	–	130/00	040/00	–	–	Tensional joints
S42	011/14E	18/55	166/05	256/02	010/84	166/05	0.29	−0.41	Faults
S42	011/14E	26/55	330/77	230/02	140/13	230/02	0.91	1.11	Faults
S42	011/14E	5/55	089/78	259/12	349/02	259/12	0.64	1.56	Faults
S42	011/14E	68	–	–	165/00	075/00	–	–	Shear joints
S42	011/14E	49	–	–	165/00	075/00	–	–	Tension gashes
S43	171/39W	12/21	021/71	135/08	228/17	135/08	0.10	10.00	Faults
S43	171/39W	5/21	263/15	165/28	018/58	263/15	0.78	−3.55	Faults
S43	171/39W	60	–	–	030/00	120/00	–	–	Shear joints

Sites are located as in Figure 2. n/N, number of data used for defining each stress state/number of faults in the fault population; σ_1, σ_2 and σ_3, values of principal stress axes; σ_y, maximum horizontal stress; *Re* stress ratio $= (\sigma_2 - \sigma_3)/(\sigma_1 - \sigma_3)$ (Etchecopar *et al.* 1981); *R* stress ratio $= (\sigma_z - \sigma_x)/(\sigma_y - \sigma_x)$ (Bott 1959).

the basin, opened from the late Eocene to the early Miocene (King & Barker 1988; Lawver *et al.* 1994; Coren *et al.* 1997; Rodríguez-Fernández *et al.* 1997; Eagles & Livermore 2002). This basin connects with Jane Basin to the SE, which is a backarc basin (Barker *et al.* 1984; Bohoyo *et al.* 2002). The subduction of the oceanic crust of the Weddell Sea toward the NNW below the South Orkney Microcontinent began in the Paleocene (Barker 2001; Bohoyo *et al.* 2002) and ended during the middle Miocene when Jane Basin was fully opened after the spreading axis subduction

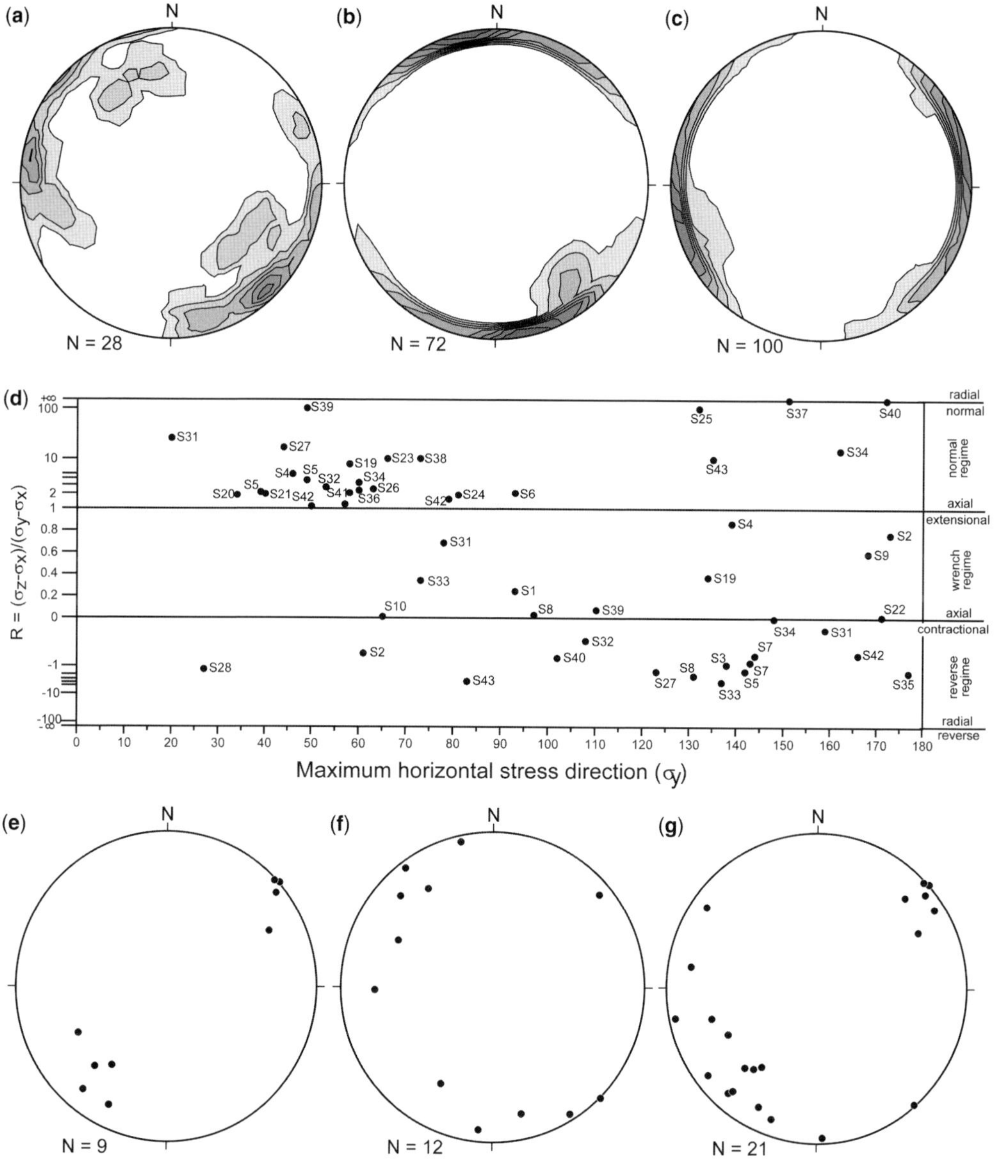

Fig. 7. Density stereoplots equal area projection, lower hemisphere, contour interval 2%, representing (**a**) compression (σ_1), (**b**) extension (σ_3), and (**c**) greatest horizontal stress (σ_y) axes obtained from the analysis of brittle mesostructures. (**d**) *y–R* diagram of the fault planes measured in this work. σ_y is the azimuth of the major horizontal axis; $R = (\sigma_z - \sigma_x)/(\sigma_y - \sigma_x)$, where σ_z is vertical and $\sigma_y > \sigma_x$ (Bott 1959). Density stereoplots equal area projection, lower hemisphere, contour interval 2%, representing (**e**) compression (σ_1), (**f**) extension (σ_3) and (**g**) greatest horizontal stress (σ_y) axes obtained from the analysis of earthquake focal mechanisms.

that definitively fixed the South Orkney Microcontinent to the Antarctic Plate (Bohoyo *et al.* 2002; Bohoyo 2004). The early processes of subduction of the northern sector of the oceanic crust of the Weddell Sea below the South Orkney Microcontinent could be related to the WNW–ESE to NW–SE σ_1 direction identified in the present work, prior to the perpendicular rifting processes related to the opening of Airy, Bouguer and Eötvös basins in the southern part of the South Orkney Microcontinent.

The rifting processes related to the opening of Powell and Jane basins caused the development of the southern Airy, Bouguer and Eötvös basins along an east–west trend as well as possibly other NE–SW extensional processes within the South Orkney Microcontinent. The eastward displacement of the South Orkney Microcontinent could be the origin of ENE–WSW compressive stages closer to the northern border and could be related to the opposition of other continental fragments in a proto South Scotia Ridge with a dextral transcurrent character (Bohoyo 2004; Bohoyo *et al.* 2007).

Finally, the extrapolation of present-day stress orientation data from earthquake focal mechanisms located on the South Scotia Ridge from the World Stress Map (www.world-stress-map.org), US Geological Survey National Earthquake Information Center (NEIC) and Harvard Seismology Centroid Moment Tensor Catalog (Dziewonski *et al.* 1981) from 1973 to the present, suggests NE–SW compression (Fig. 1a) related to the convergence between the South American and Antarctic plates accommodated by the left-lateral strike-slip movement of the Scotia Plate boundaries and a perpendicular SE–NW subhorizontal extension (Pelayo & Wiens 1989; Galindo-Zaldívar *et al.* 1996; González-Casado *et al.* 2000; Giner-Robles *et al.* 2003; Robertson *et al.* 2003; Bohoyo *et al.* 2007; Maestro *et al.* 2007). The stress tensors obtained from the analysis of earthquake focal mechanisms show a NE–SW subhorizontal compressional direction (Fig. 7e), a perpendicular NW–SE subhorizontal extensional trend (Fig. 7f) and a NE–SW orientation of the maximum horizontal stress (Fig. 7g). The fault type associated with the individual focal mechanism along the South Scotia Ridge varies from dip-slip to normal-oblique faults. In the study area, the recent tectonic regime is extensional with a predominant direction of σ_3 NW–SE.

Conclusions

The analysis of brittle mesostructures in Triassic–Jurassic metamorphic rocks enables the characterization of the Mesozoic–recent tectonic stress fields on Signy Island and its relationships within the South Orkney Microcontinent. From the orientations of stress axes, it can be observed that there is a dominant NW–SE horizontal compression direction and three (NNW–SSE, WNW–ESE and ENE–WSW) secondary σ_1 horizontal stress directions. The orientation of σ_3 shows a main mode trending NNW–SSE, and NW–SE and NE–SW extensional stress directions. The orientations of σ_y show a main mode trending ENE–WSW and two secondary modes with east–west and NW–SE directions. The NE–SW to east–west maximum stress direction shows an R relationship ranging from normal to wrench regimes. The NW–SE mode corresponds to reverse regimes.

The NNW–SSE horizontal compression direction obtained from a brittle mesostructure analysis is consistent with the stretching/shearing direction defined by fold-axis trends and most likely indicates the thrusting direction that the South Orkney Microcontinent suffered during early to middle Jurassic times. Subsequently, the tectonic regime changed to extensional with north–south orientation followed by uplift and erosion in middle Jurassic–early Cretaceous times. The process of rifting and separation between the Antarctic Peninsula and South America began during the Eocene period. The spreading of Powell Basin took place from the Late Eocene to early Miocene and was the origin of the translation of the South Orkney Microcontinent toward the ENE between the proto-South Scotia Ridge to the north and other transcurrent faults to the south. This displacement of the South Orkney Microcontinent, together with the prior rifting processes, is related to the development of the approximately north–south trend of Airy, Bouguer and Eötvös basins and other NE–SW extensional processes within of the microcontinent. This could be the origin of ENE–WSW compressive stages closer to the northern border and may be related to the movement of the proto south Scotia Ridge. In the same way, the first stages of the subduction of the northern sector of the oceanic crust of the Weddell Sea below the South Orkney Microcontinent, which began in the Paleocene, could be related to the WNW–ESE to NW–SE σ_1 direction. Finally, the NW–SE σ_3 direction obtained from brittle mesostructures population analysis is consistent with the current left-lateral strike-slip movement between the Scotia and Antarctic plates. The stress tensors obtained from the analysis of earthquake focal mechanisms show a NE–SW subhorizontal compressional direction and a perpendicular NW–SE subhorizontal extensional trend defined by dip-slip and normal-oblique faults. The chronology described is coherent with the relative relationship determined from the superposition of striations on a single fault plane measured in the studied sites.

The field work was carried out with the logistic support of the British Antarctic Survey. The authors express their gratitude to the director and staff of BAS and the Signy base personnel. We would like to thank Peter Barker, Renato de Moraes and Rudolph Trouw for their useful comments that improved the manuscript. This work was supported by the projects CGL2005-03256, CGL2007-28812E/ANT, CTM2011-13902E/ANT, CTM2011-13970E/ANT, CTM2011-26372 and CTM2011-30241-C02-02 of the Spanish National R&D Plan.

References

ANGELIER, J. & MECHLER, P. 1977. Sur une méthode graphique de recherche des contraintres principales également utilisable en tectonique et en séismologie: la méthode des diédres droits. *Bulletin de la Société Géologique de France*, **19**, 1309–1318.

ARLEGUI, L. E. & SIMÓN, J. L. 1998. Reliability of palaeostress analysis from fault striations in near multidirectional extension stress fields. Example from the Ebro Basin, Spain. *Journal of Structural Geology*, **20**, 827–840.

ARLEGUI, L. E. & SIMÓN, J. L. 2001. Geometry and distribution of regional joint sets in a non-homogeneous stress field: case study in the Ebro basin (Spain). *Journal of Structural Geology*, **23**, 297–313.

BAHAT, D. 1991. *Tectonofractography*. Springer, Berlin.

BARKER, P. F. 2001. Scotia Sea regional tectonic evolution: implications for mantle flow and palaeocirculation. *Earth-Science Reviews*, **55**, 1–39.

BARKER, P. F. & DALZIEL, I. W. D. 1983. Progress in geodynamics in the Scotia Arc region. *In*: CABRE, R. (ed.) *Geodynamics of the Eastern Pacific Region, Caribean and Scotia Arcs*. American Geophysical Union, Geodynamic Service, Washington, DC, 137–170.

BARKER, P. F., BARBER, P. G. & KING, E. C. 1984. An Early Miocene ridge crest-trench collision on the South Scotia Ridge near 36°W. *Tectonophysics*, **102**, 315–332.

BARKER, P. F., KENNETT, J. P. & SCIENTIFIC PARTY. 1988. Weddell Sea palaeoceanography: preliminary results of ODP leg 113. *Palaeogeography, Palaeoclimatology, Palaeoecology*, **67**, 75–102.

BARKER, P. F., DALZIEL, I. W. D. & STOREY, B. C. 1991. Tectonic development of the Scotia Arc region. *In*: TINGEY, R. J. (ed.) *Antarctic Geology*. Oxford University Press, Oxford, 215–248.

BERGERAT, F., BOUROZ, C. & ANGELIER, J. 1992. Paleostresses inferred from macrofractures, Colorado Plateau, western U.S.A. *Tectonophysics*, **206**, 219–243.

BOHOYO, F. 2004. *Fragmentación continental y desarrollo de cuencas oceánicas en el sector meridional del Arco de Scotia, Antártida*. PhD thesis, Granada University, Spain.

BOHOYO, F., GALINDO-ZALDÍVAR, J., MALDONADO, A., SCHRIDER, A. A. & SURIÑACH, E. 2002. Basin development subsequent to ridge-trench collision: the Jane Basin, Antarctica. *Marine Geophysical Researches*, **23**, 413–421.

BOHOYO, F., GALINDO-ZALDÍVAR, J., JABALOY, A., MALDONADO, A., RODRÍGUEZ-FERNÁNDEZ, J., SCHREIDER, A. A. & SURIÑACH, E. 2007. Extensional deformations and development of deep basins associated to the sinistral transcurrent fault zone of the Scotia-Antarctica plate boundary. *Geological Society, London, Special Publications*, **290**, 203–217, http://dx.doi.org/10.1144/SP290.6

BOTT, M. H. P. 1959. The mechanics of oblique slip faulting. *Geological Magazine*, **96**, 109–117.

BUSETTI, M., ZANOLLA, C. & MARCHETTI, A. 2000. Geological structure of the South Orkney Microcontinent. *Terra Antartica*, **8**, 1–8.

CAPUTO, R. 2005. Stress variability and brittle tectonic structures. *Earth-Science Reviews*, **70**, 103–127.

CASAS, A. M. & MAESTRO, A. 1996. Deflection of a compressional stress field by large-scale basement faults. A case study from the Tertiary Almazán basin (Spain). *Tectonophysics*, **255**, 135–156.

CASAS, A. M., SERÓN, F. J. & SIMÓN, J. L. 1992. Stress deflection in a tectonic compressional field: a model for the Northwestern Iberian Chain, Spain. *Journal of Geophysical Research*, **97**, 7183–7192.

COREN, F., CECCONE, G., LODOLO, E., ZANOLLA, C., ZITELLINI, N., BONAZZI, C. & CENTONZE, J. 1997. Morphology, seismic structure and tectonic development of the Powell Basin, Antarctica. *Journal of the Geological Society, London*, **154**, 849–862.

DALZIEL, I. W. D. 1982. The early (pre-middle Jurassic) history of the Scotia Arc Region: a review and progress report. *In*: CRADDOCK, C. (ed.) *Antarctic Geoscience*. University of Wisconsin Press, Madison, WI, 111–126.

DALZIEL, I. W. D. 1984. *Tectonic Evolution of a Forearc Terrane, Southern Scotia Ridge, Antarctica*. Geological Society of America, Boulder, CO, Special Papers, **220**, 32.

DALZIEL, I. W. D. & ELLIOT, D. H. 1971. Evolution of the Scotia arc. *Nature*, **233**, 246–252.

DE WIT, M. J. 1977. The evolution of the Scotia Arc as a key to the reconstruction of the southwestern Gondwanaland. *Tectonophysics*, **37**, 53–81.

DZIEWONSKI, A. M., CHOU, T. A. & WOODHOUSE, J. H. 1981. Determination of earthquake source parameters from waveform data for studies of global and regional seismicity. *Journal of Geophysical Research*, **86**, 2825–2852.

EAGLES, G. & LIVERMORE, R. A. 2002. Opening history of Powell Basin, Antarctic Peninsula. *Marine Geology*, **185**, 195–205.

ELLIOT, D. H. & WELLS, N. A. 1982. Mesozoic alluvial fans of the South Orkney Islands. *In*: CRADDOCK, C. (ed.) *Antarctic Geoscience*. University of Wisconsin Press, Madison, WI, 235–244.

ENGELDER, T. 1989. Analysis of pinnate joints in the Mount Desert Island granite: implications for postintrusion kinematics in the coastal volcanic belt, Maine. *Geology*, **17**, 564–567.

ENGELDER, T. & GEISER, P. 1980. On the use of regional joint sets as trajectories of paleostress fields during the development of the Appalachian Plateau, New York. *Journal of Geophysical Research*, **85**, 6319–6341.

ETCHECOPAR, A., VASSEUR, G. & DAIGNIERES, M. 1981. An inverse problem in microtectonics for the determination of stress tensors from fault population analysis. *Journal of Structural Geology*, **3**, 51–65.

Flowerdew, M. J., Riley, T. R. & Haselwimmer, C. E. 2011 *Geological Map of the South Orkney Islands (1:1 50 000 scale)*. BAS GEOMAP 2 Series, Sheet 3, British Antarctic Survey, Cambridge.

Galindo-Zaldívar, J., Jabaloy, A., Maldonado, A. & Sanz de Galdeano, C. 1996. Continental fragmentation along the South Scotia Ridge transcurrent plate boundary (Antarctic Peninsula). *Tectonophysics*, **259**, 275–301.

Giner-Robles, J. L., González-Casado, J. M., Gumiel, P., Martín-Velázquez, S. & García-Cuevas, C. 2003. A kinematic model of the Scotia plate (SW Atlantic Ocean). *Journal of South American Earth Sciences*, **16**, 179–191.

González-Casado, J. M., Giner, J. L. & López-Martínez, J. 2000. Bransfield Basin, Antarctic Peninsula: not a normal backarc basin. *Geology*, **28**, 1043–1046.

Hancock, P. L. 1985. Brittle microtectonics: principles and practice. *Journal of Structural Geology*, **7**, 437–457.

Hancock, P. L. & Engelder, T. 1989. Neotectonic joints. *Bulletin Geological Society of America*, **101**, 1197–1208.

Harrington, H. J., Barker, P. F. & Griffiths, D. H. 1972. Crustal structure of the South Orkney Islands are from seismic refraction and magnetic measurements. *In*: Adie, R. J. (ed.) *Antarctic Geology and Geophysics Symposium on Antarctic Geology and Solid Earth Geophysiscs*. 6–15 August 1970, Universitetsvolaget Oslo, 27–32.

Heidbach, O., Reinecker, J., Tingay, M., Müller, B., Sperner, B., Fuchs, K. & Wenzel, F. 2007. Plate boundary forces are not enough: second- and third-order stress patterns highlighted in the World Stress Map database. *Tectonics*, **26**, TC6014, doi: 10.1029/2007TC0022133, 2007

Hodgson, R. A. 1961. Classification of structures on joint surfaces. *American Journal of Science*, **259**, 493–502.

Kavoun, M. & Vinnikovskaya, O. 1994. Seismic stratigraphy and tectonics of the northwestern Weddell Sea (Antarctica) inferred from marine geophysical surveys. *Tectonophysics*, **240**, 299–341.

King, E. C. & Barker, P. F. 1988. The margins of the South Orkney microcontinent. *Journal of the Geological Society, London*, **145**, 317–331.

King, E. C., Leitchenkov, G., Galindo-Zaldívar, J. & Maldonado, A. 1994. Basement distribution in Powell Basin: understanding the tectonic controls on sedimentation. *Terra Antarctica*, **1**, 307–308.

Klepeis, K. A. & Lawver, L. A. 1996. Tectonics of the Antarctic-Scotia plate boundary near Elephant and Clarence Islands, West Antarctica. *Journal of Geophysical Research*, **101**, 20211–20231.

Lawver, L. A., Della Vedova, B. & Von Herzen, R. P. 1991. Heat flow in Jane Basin, Northern Weddell Sea. *Journal of Geophysical Research*, **96/B2**, 2019–2038.

Lawver, L. A., Williams, T. & Sloan, B. 1994. Seismic stratigraphy and heat flow of Powell Basin. *Terra Antarctica*, **1**, 309–310.

Lodolo, E., Coren, F., Schreider, A. A. & Ceccone, G. 1997. Geophysical evidence of a relict oceanic crust in the South-western Scotia Sea. *Marine Geophysical Research*, **19**, 439–450.

Lodolo, E., Civile, D., Vuan, A., Tassone, A. & Geletti, R. 2010. The Scotia–Antarctica plate boundary from 35°W to 45°W. *Earth and Planetary Science Letter*, **93**, 200–215.

Maestro, A., Somoza, L., Rey, J., Martínez-Frías, J. & López-Martínez, J. 2007. Active tectonics, fault patterns, and stress field of Deception Island: a response to oblique convergence between the Pacific and Antarctic plates. *Journal of South American Earth Sciences*, **23**, 256–268.

Maestro, A., López-Martínez, J., Bohoyo, F., Montes, M., Nozal, F., Santillana, S. & Marenssi, S. 2008. Geodynamic implications of Cenozoic to recent stress field, Seymour Island (Western Antarctica). *Antarctic Science*, **20**, 173–184.

Maldonado, A., Zitellini, N. *et al.* 1998. Small ocean basin development along the Scotia-Antarctica plate boundary and in the northern Weddell Sea. *Tectonophysics*, **296**, 371–402.

Matthews, D. H. & Maling, D. H. 1967. *The geology of the South Orkney Islands. I. Signy Island.* Scientific Reports Falkland Islands Dependencies Survey, **25**.

Meissner, R., Henriet, J. P. & Grape, T. 1988. Tectonic features northwest of the Antarctic Peninsula: new evidence from magnetic and seismic studies. *Servicio Científico INACH*, **38**, 89–105.

Meneilly, A. W. & Storey, B. 1986. Ductile thrusting within subduction complex rocks on Signy Island, South Orkney Islands. *Journal of Structural Geology*, **8**, 457–72.

Pelayo, A. & Wiens, D. 1989. Seismotectonics and relative plate motions in the Scotia Sea region. *Journal of Geophysical Research*, **96**, 7293–7320.

Pollard, D. D. & Aydin, A. 1988. Progress in understanding jointing over the past century. *Geological Society of America Bulletin*, **100**, 1181–1204.

Ramsay, J. G. & Huber, M. I. 1983. *The Techniques of Modern Structural Geology, Volume 2: Folds and fractures*. Academic Press, London.

Riedel, W. 1929. Zur mechanik geologischer brucherscheinungen. *Zentral-blatt fur Mineralogie, Geologie und Paleontologie, B*, 354–368.

Robertson, S. D., Wiens, D. A., Shore, P. J., Vera, E. & Dorman, L. M. 2003. Seismicity and tectonics of the South Shetland Islands and Bransfield Strait from a regional broadband seismograph deployment. *Journal of Geophysical Research Solid Earth*, **108**, 2156–2202, doi: 10.1029/2003JB002416

Rodríguez-Fernández, J., Balanya, J. C., Galindo-Zaldívar, J. & Maldonado, A. 1997. Tectonic evolution and growth patterns of a restricted ocean basin: the Powell Basin (northeastern Antarctic Peninsula). *Geodinamica Acta*, **10**, 159–174.

Simón, J. L. 1986. Análisis of gradual change in stress regime (example from the eastern Iberian Chain, Spain). *Tectonophysics*, **124**, 37–53.

Simón, J. L., Arlegui, L. E. & Pocoví, J. 2006. Fringe cracks and plumose structures in layered rocks: stepping senses and their implications for palaeostress interpretation. *Journal of Structural Geology*, **28**, 1103–1113.

SMELLIE, J. L. 1981. A complete arc-trench system recognized in Gondwana sequences of the Antarctic Peninsula region. *Geological Magazine*, **118**, 139–159.

SMITH, W. H. F. & SANDWELL, D. Y. 1997. Global seafloor topography from satellite altimetry and ship depth sounding. *Science*, **277**, 1957–1962.

STOREY, B. C. & MENEILLY, W. 1985. Petrogenesis of metamorphic rocks within a subduction-accretion terrane, Signy Island, South Orkney Islands. *Journal of Metamorphic Geology*, **3**, 21–42.

TROUW, R. A. J., PASSCHIER, C. W., SIMOES, L. S. A., ANDREIS, R. R. & VALERIANO, C. M. 1997. Mesozoic tectonic evolution of the South Orkney Microcontinent, Scotia arc, Antarctica. *Geological Magazine*, **134**, 383–401.

The Taylor Group (Beacon Supergroup): the Devonian sediments of Antarctica

MARGARET A. BRADSHAW

Department of Geological Sciences, University of Canterbury, Private Bag 4800, Christchurch 8140, New Zealand
(e-mail: margaret.bradshaw@canterbury.ac.nz)

Abstract: The Taylor Group, the lower division of the Beacon Supergroup, comprises mainly quartzose sandstones of Devonian age deposited after the development of the Kukri Erosion Surface across the Cambrian Ross orogen. Devonian sediments accumulated in a McMurdo Basin that now incorporates most of the Transantarctic Mountains, and a larger Ellsworth Basin that extends from West Antarctica into southern Africa. In southern Victoria Land (McMurdo Basin) the seven formations total around 1200 m in thickness. Sequence stratigraphy suggests five sedimentary cycles, the lower four of which may show shallow marine influence. Provenance studies indicate derivation solely from Ross Orogen sources. The Taylor Group in the central Transantarctic Mountains consists of two thin formations, only one of which is extensive. In the Ellsworth Basin, Taylor Group equivalents rest concordantly on thick piles of post-Ross Palaeozoic sediment. Shallow-water Devonian sediments in the Ohio Range contain a marine fauna linked with those of the Ellsworth Mountains, southern Africa, South America, New Zealand and Australia. A seaway probably existed along the Pacific edge of the East Antarctic craton, allowing these faunal links, and may also have communicated with the McMurdo Basin to drive the five sedimentary sequences.

The Beacon Supergroup, made up of Devonian, Permian and Triassic sediments, extends throughout the Transantarctic Mountains from the Pacific margin in northern Victoria Land to Dronning Maud Land on the eastern margin of the Weddell Sea (Fig. 1). Within East Antarctica, similar rocks occur at Horn Bluff and in the northern Prince Charles Mountains. In West Antarctica they are found in the allochthonous Ellsworth Mountains. Possible equivalents occur on the English Coast, and also in the Falkland Islands (Hunter & Lomas 2003). The Supergroup was deposited on the Kukri Erosion Surface and is divided by a regional unconformity into a lower Devonian Taylor Group (1200 m) and a more extensive Permian and Triassic Victoria Group (2400 m). Throughout the Transantarctic Mountains the Beacon Supergroup is cut by thick dolerite sills (Ferrar Group) of Early Jurassic age. In the Transantarctic Mountains the Beacon rocks dip gently *c.* 5° to the west and are cut by faults related to Jurassic sill intrusion, and also to block faulting resulting from Gondwana break-up and later uplift of the Transantarctic Mountains. In the Weddell Sea sector, rocks such as those in the Ellsworth and Pensacola Mountains were strongly deformed during the Permian–Triassic Gondwanian folding (Schmidt & Ford 1969; Curtis 2001; Curtis & Storey 2003).

History

Beacon Supergroup rocks were first observed by the early explorers when Scott's *Discovery* anchored at Hut Point (Ross Island) in 1901 and a 'band' of horizontally bedded pale rock was observed along the length of the distant 'Western Mountains' (Royal Society Range). Lieutenants Armitage and Skelton pioneered a route into the mountains up the Ferrar and Upper Taylor Glaciers (Fig. 2). They reported that the pale rocks were sandstones, easily accessible from a point 30 miles inland up towards the 'Inland Ice' (Polar Plateau) beyond Finger Mountain (Ferrar 1907). Armitage gave the name 'Beacon Heights' to two prominent peaks composed of these sandstones, which could be seen clearly from part way up the Ferrar Glacier. A party that included the geologist Ferrar accompanied Scott to the edge of the Polar Plateau and then returned separately to investigate rocks on either side of the Upper Taylor Glacier and middle Ferrar Glacier. Ferrar (1907) called these rocks the Beacon Sandstone after Beacon Heights, but did not provide a type section. Most of his attention was focussed on the western side of Tabular Mountain, Finger Mountain, the Inland Forts, Knobhead and Terra Cotta Mountain, as well as Beacon Heights (Fig. 2), which much later became the type area. During his trip, Ferrar discovered

From: Hambrey, M. J., Barker, P. F., Barrett, P. J., Bowman, V., Davies, B., Smellie, J. L. & Tranter, M. (eds) 2013. *Antarctic Palaeoenvironments and Earth-Surface Processes*. Geological Society, London, Special Publications, **381**, 67–97. First published online August 6, 2013, http://dx.doi.org/10.1144/SP381.23

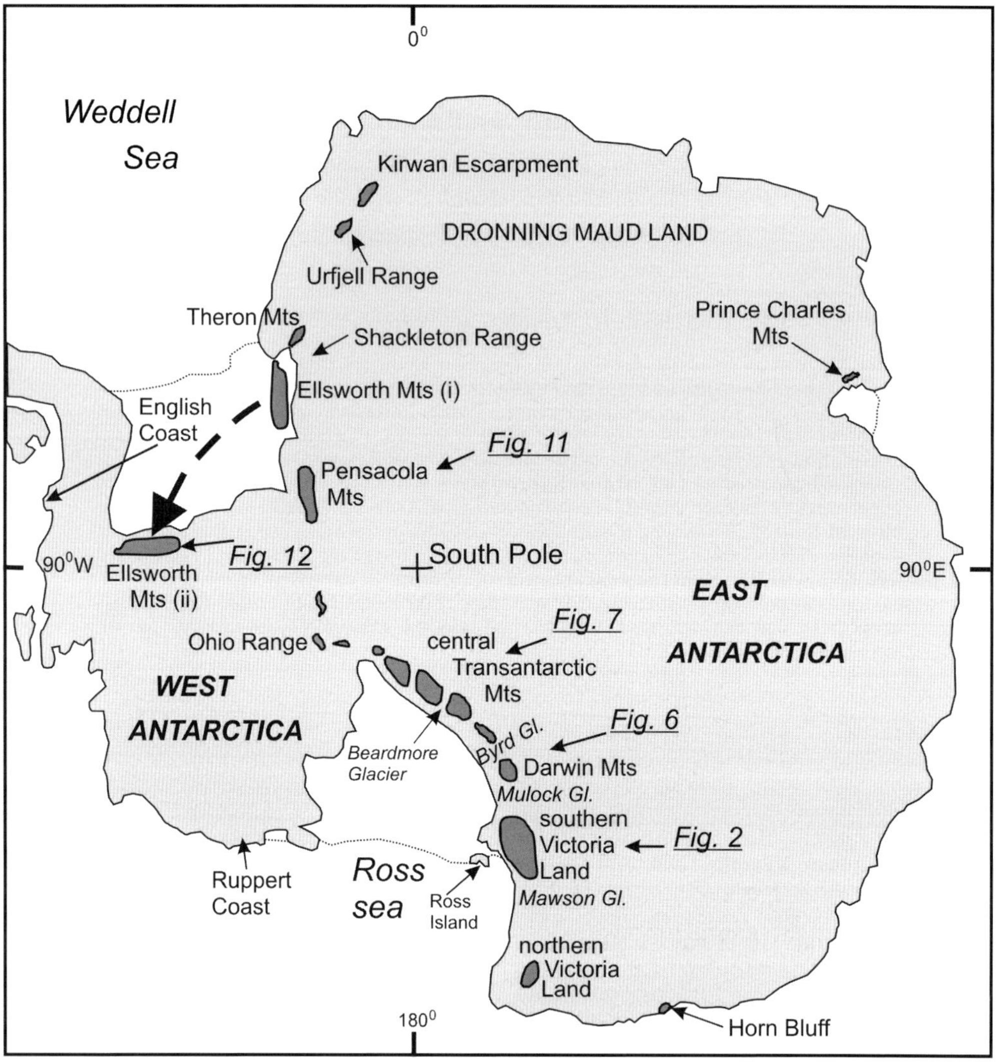

Fig. 1. Map of Antarctica showing the stylized locations of Beacon Supergroup rocks and their equivalents. The heavy dashed arrow indicates the movement path of the Ellsworth Mountains from their original position against East Antarctica (i) to their current position in West Antarctica (ii). The smaller arrows with underlined figure numbers mark the location of maps used in this paper. The Shackleton Range, where Beacon equivalents were thought to exist, is marked by an arrow.

'carbonaceous matter' and 'cylindrical rugosities' at several localities, which he saw as evidence of past life in Antarctica.

Further Beacon rocks were discovered in the central Transantarctic Mountains during Shackleton's British Antarctic Expedition of 1907–1909, when Wild discovered sandstones containing layers of coal up to 'eight feet thick' near the top of Mount Buckley at the head of the Beardmore Glacier (Fig. 3). During Scott's later *Terra Nova* Expedition (1910–1913), the Polar Party, on their return journey down the Beardmore Glacier, collected leaf impressions (*Glossopteris*) from the same deposits on Mount Buckley (Seward 1914; Edwards 1928). During the same expedition, a northern party discovered fossil fish material in blocks from a moraine on Mount Suess (Fig. 2). These were thought to have originated from Beacon rocks, and were later identified as Devonian in age (Woodward 1921).

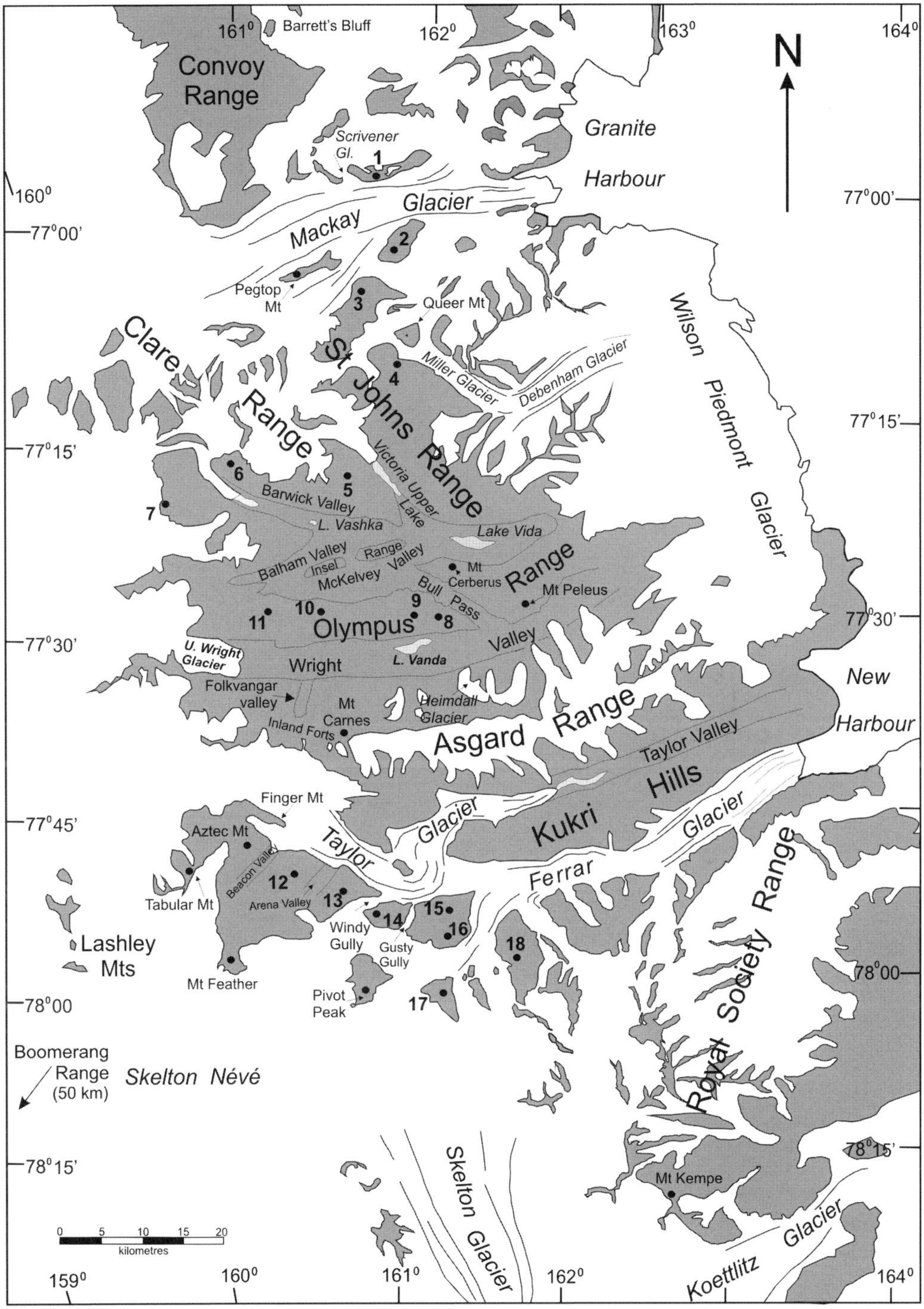

Fig. 2. Taylor Group localities in the McMurdo Dry Valleys region between the Skelton Névé and the Convoy Range. Numbers refer to the following localities: 1, Mount Allan Thomson; 2, Mount Suess; 3, Sperm Bluff; 4, Gargoyle Turrets; 5, Sponsors Peak; 6, The Fortress; 7, Mount Bastion; 8, Mount Jason; 9, Mount Hercules; 10, Mount Boreas; 11, Mount Electra; 12, Beacon Heights; 13, New Mountain; 14, Terra Cotta Mountain; 15, Knobhead; 16, Mount Handsley; 17, Rotunda; 18, Table Mountain.

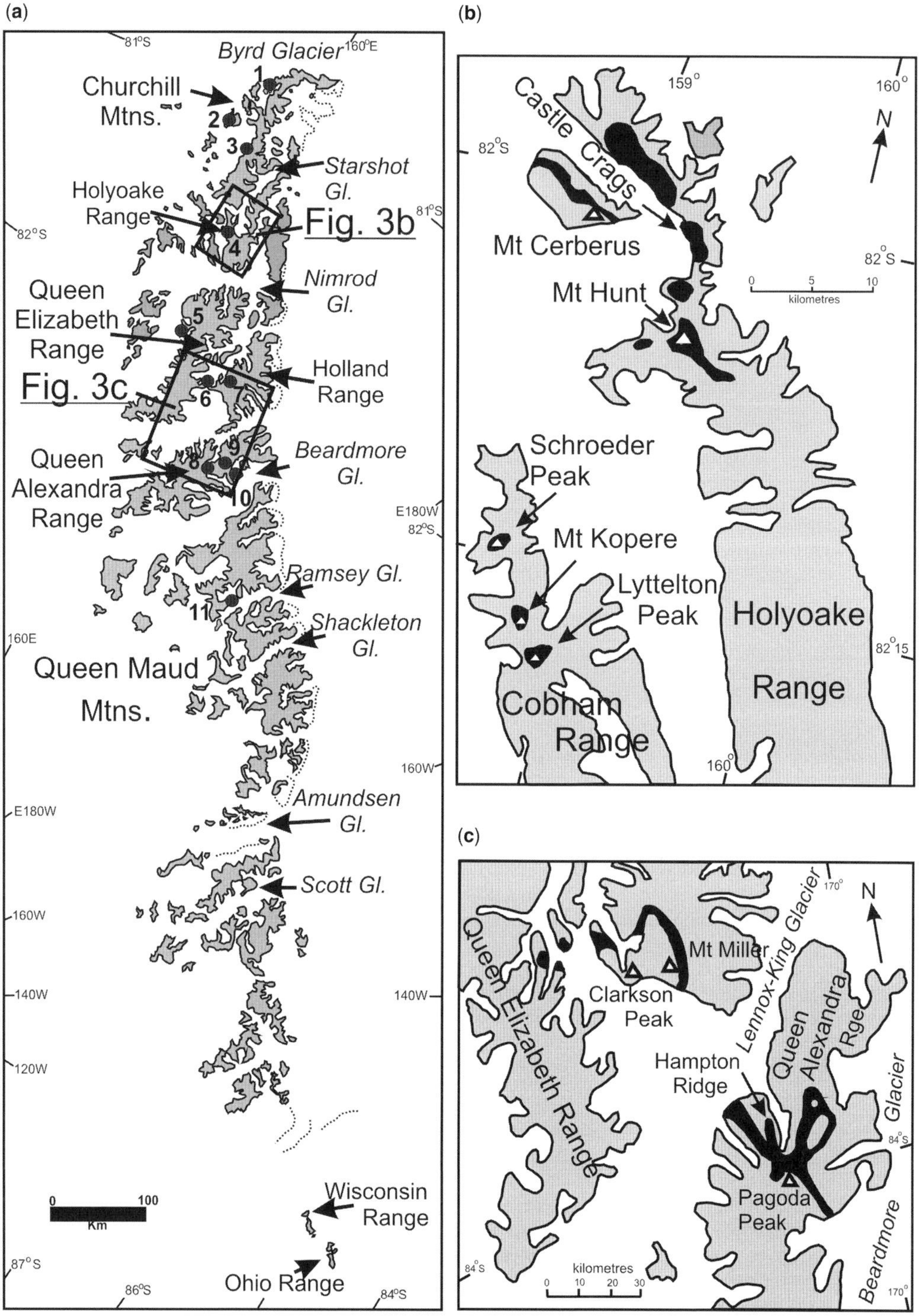

Fig. 3. Location map for the central Transantarctic region. (**a**) Relatively thin (up to 320 m) and scattered outcrops of Taylor Group rocks occur between the Queen Maud Mountains and the Byrd Glacier (at top). In the Devonian, a basement high between the Shackleton Glacier and the Wisconsin Range appears to have separated the McMurdo Basin from the Ohio–Ellsworth Basin, with the Ohio Range close to the margin. Numbers refer to localities: 1, Mount

The known outcrop of Beacon rocks was extended even further south to the Queen Maud Mountains (Fig. 3) by Byrd's Antarctic expedition of 1929 (Gould 1931) and into East Antarctica (Horn Bluff) by Mawson's 1911–1914 Australasian Antarctic Expedition (Mawson 1940). At that time, the Beacon was regarded as a single very thick, uniform body of rock because all outcrops looked much the same from a distance. No attempt was made to subdivide the rocks or to explore the age differences between parts of the succession.

Little more was achieved on Beacon rocks until the International Geophysical Year/Trans-Antarctic Expedition period. Significant exploratory work was accomplished on both the basement and overlying sedimentary rocks by Gunn & Warren (1962) between the Mawson and Mulock Glaciers of southern Victoria Land (Fig. 1) during the summer of 1957–1958. Constrained to travel on snow because they were using dog teams, they avoided the McMurdo Dry Valleys, but discovered new Beacon sections, chiefly along the western edge of the mountains, where, in the Boomerang Range, Lashley Mountains and Mount Feather (Fig. 2) they found *in situ* Devonian fossil fish (Gunn & Warren 1962). They established that the Beacon rocks rested on an erosion surface cut across crystalline basement and named this the Kukri Peneplain (now Kukri Erosion Surface). During the same season, two adventurous third year students from Victoria University, Peter Webb and Barrie McKelvey, started work in the McMurdo Dry Valleys. After a visit to the upper Taylor Valley, they subdivided the Beacon Formation into three informal members and named them, from bottom to top, A–C (McKelvey & Webb 1959).

Harrington (1958) attempted to clarify the range of informal names that had resulted from the work of the 1957–1958 Trans-Antarctic Expedition and other workers in the Ross Sea region, and suggested that the very thick sandy sequence be incorporated into a Beacon System, and the name Beacon Sandstone (Group) be retained only for the type rocks in the McMurdo Sound area. Because of conflicts with different codes of stratigraphic nomenclature, the use of 'Beacon System' was subsequently dropped, and 'Beacon Sandstone' was retained for all sandstones in the McMurdo area (Gunn & Warren 1962).

During the 1958–1960 field seasons, Beacon rocks along the south side of the upper Taylor Glacier were studied by four separate helicopter-supported parties, (Zeller *et al.* 1961; Harrington & Speden 1962; Angino & Owen 1962; Hamilton & Hayes 1963). Hamilton & Hayes (1963) attempted to establish a composite Beacon type section in the Taylor Glacier area based on the lower part of the New Mountain section of Zeller *et al.* (1961), a middle section that they themselves had measured on West Beacon Heights, and McKelvey & Webb's (1959) section on the west side of Beacon Dry Valley.

Geological mapping during 1959–1960 in the northern part of the McMurdo Dry Valleys (Barwick, Balham and McKelvey valleys, Fig. 2) by Allen (1962) emphasized the fact that the Beacon Group was inadequately defined, and Allen advised against incorporating such varied rocks under a single name. Allen proposed using the almost continuous succession on Mount Bastion and The Fortress at the head of the Barwick Valley (Fig. 2) as type localities for three formations comprising the Beacon sandstone (members A–C of McKelvey & Webb 1959), describing them as Webb Sandstone (lower), followed by the Fortress Sandstone (middle), overlain disconformably by Permian Bastion Coal Measures.

Mirsky (1964) recommended that the use of 'Beacon', either as a sandstone, formation or group, be discontinued, as there were several units of different ages, some separated by periods of erosion. Mirsky proposed that the clean quartz sandstones in the lower part of the succession be named the Razorback Formation, and the basal sediments below be called the Windy Gully Formation. The overlying carbonaceous sediments were named the Mount Bastion Formation. However, Mirsky's recommendations were not well received as the name Beacon, either as group or supergroup, was felt to be too useful to discard (see discussion, Mirsky 1964).

Although Allen's (1962) proposal for type sections in the Barwick Valley was never accepted, and the Beacon Heights–New Mountain massif

Fig. 3. (*Continued*) Egerton; 2, Wallabies Nunatak; 3, Mount Albert Markham; 4, Mount Hunt; 5, Mount Counts; 6, Turnabout Ridge; 7, Mount Miller; 8, Pagoda Peak; 9, Mount Elizabeth; 10, The Cloudmaker; 11, Head of Ramsey Glacier. The rectangles indicate the position of the detailed maps at right. (**b**) Map of Taylor Group outcrops, shown in black, in the Holyoake and Cobham Ranges north of the Nimrod Glacier. The type section for the Castle Crags Formation is on Castle Crags. In the Holyoake Range Taylor Group rocks rest on a basement of Cambrian Shackleton Limestone, while in the Cobham Range they rest on the Precambrian Goldie Formation (after Laird *et al.* 1971). (**c**) Map of Taylor Group outcrops, shown in black, in the Queen Elizabeth and Queen Alexandra Ranges. The type section of the Alexandra Formation is on Hampton Ridge. Devonian outcrops are limited and most of the Beacon rocks belong to the younger Victoria Group (after Barrett *et al.* 1986).

became the recognized type area, his identification of a disconformity below the Permian coal-bearing beds was an important discovery. A similar surface was noted by Webb (1963) on Aztec Mountain (Fig. 2), where Permian plant-bearing beds rest disconformably on Devonian Aztec Siltstone, while Grindley (1963) maintained that this same surface could be recognized in the Beardmore Glacier area. Harrington (1965) used this surface (Maya Erosion Surface) to subdivide the Beacon rocks into a lower Taylor Group (Devonian) and an upper Victoria Group (Permian–Jurassic at that time). However it was an Australian group led by Cliff McElroy, with extensive experience in the Sydney Basin, who investigated the Beacon Heights area under the auspices of the US Antarctic Research Program, and, through measuring and defining type sections for four additional formations, set the framework for future Beacon stratigraphy (McElroy 1969).

Subsequently Barrett (1971) formalized the term Beacon Supergroup to encompass both Taylor and Victoria groups, and an Antarctic-wide review of Barrett *et al.* (1972), along with a review of southern Victoria Land stratigraphy by McKelvey *et al.* (1972), firmly established this usage. The break in sedimentation between the two groups was later estimated to be *c.* 50 Ma in duration (Barrett 1981, p. 448). Cox *et al.* (2013) have provided a comprehensive 1:250 000 map, text and appendix with all stratigraphic units in southern Victoria Land.

Taylor Group

There are several earlier reviews of the Taylor Group (Boucot *et al.* 1963*b*, 1972; Barrett 1981, 1991; Bradshaw & Webers 1988; Bradshaw 1991). The Taylor Group has a more limited outcrop than the Victoria Group. It occurs principally in southern Victoria Land (Fig. 2) and the central Transantarctic Mountains (Fig. 3), with correlatives in the Ohio Range, Pensacola Mountains and Ellsworth Mountains region (Fig. 1). A small outcrop of quartzose sandstones on the English Coast of Ellsworth Land may be a displaced part of the Ellsworth basin (Laudon 1991). Highly siliceous sandstones and conglomerates (Urfjell Group) in the Urfjell Range of the Kirwan Escarpment (Dronning Maud Land; Fig. 1) were correlated with the Taylor Group of the Transantarctic Mountains (Wolmarans & Kent 1982), but were also linked to the Blaiklock Glacier Group, Shackleton Range (Fig. 1), which has since yielded an Early Ordovician age (Pankhurst *et al.* 1983; Laird 1991*b*). The overlying Amelang Plateau Formation, however, appears much more like parts of the Taylor Group and the 'problematica' figured and described by Wolmarans & Kent (1982, fig. 58) look much like *Heimdallia*, a distinctive and common burrow in parts of the Taylor Group. However, woody tissue from a thin carbonaceous shale was tentatively considered Carboniferous or Permian (Anderson in Aucamp *et al.* 1972). Confirmation is needed to eliminate the possibility that the sediments could be Devonian.

The Taylor Group is unknown in northern Victoria Land, where Late Devonian to Carboniferous subaerial volcanoes were active (Fioretti *et al.* 1997). Northern Victoria Land was probably a highland when the Taylor Group was deposited in southern Victoria Land.

Age of the Taylor Group

Body fossils and plant remains are extremely rare in the Taylor Group, and consequently the age is poorly constrained. What has been found suggests that the group is entirely Devonian in age. Poorly preserved palynomorphs from the Terra Cotta Siltstone near the base of the group in southern Victoria Land (Fig. 4) were tentatively included in *Emphanisporites*, and their coarse morphology, together with an abundance of small, simple and smooth-walled microfossils, suggests an Early Devonian age, possibly Emsian (Fig. 5), and probably post-Lochkovian (post-Gedinnian in Kyle 1977). Playford (1990, fig. 6.2) tentatively regarded Kyle's Terra Cotta Siltstone microflora as Pragian. The underlying Windy Gully Sandstone, the lowest formation in South Victoria Land, may therefore be Emsian, Pragian or Lochkovian, but it almost certainly is no older than Early Devonian.

An Emsian age has been more reliably applied to a thin sequence (Horlick Formation) that rests on basement in the Ohio Range (Fig. 1). This contains an abundant shelly fauna with Malvinokaffric elements (Boucot *et al.* 1963*a*; Doumani *et al.* 1965; Bradshaw & McCartan 1991). Kemp (1972) also records 16 species of palynomorphs and two chitonozoa from the same beds, which though carbonized and not well preserved, support the Emsian age of the shelly fauna. Psilophytes in thin carbonaceous beds are identified as having a *Hostinella* aspect (Rigby & Schopf 1969). A photograph from Schopf's collection was figured by Edwards (1990, fig. 8.2–5) as an indeterminate psilophyte axis, but was considered reminiscent of Pragian specimens in Australia and Canada.

A similar, but limited, Emsian shelly fauna is found at a single locality (Planck Point) in the Crashsite Group of the Ellsworth Mountains, demonstrating that at least part of this thick succession is Early Devonian in age and that there was faunal communication via a seaway with both the Ohio Range site and also with the contiguous

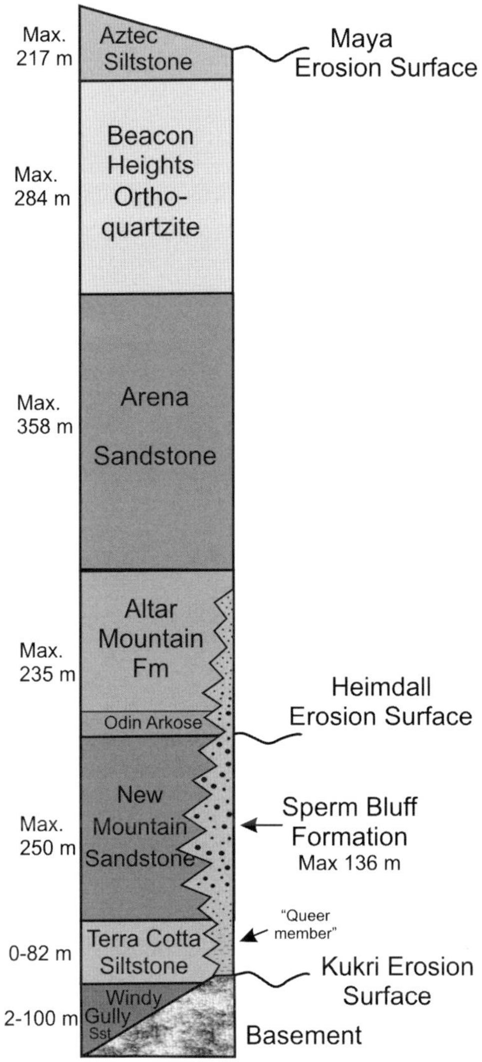

Fig. 4. Stratigraphic divisions of the Taylor Group in the McMurdo Dry Valleys region. Maximum thickness shown to left. Erosion surfaces indicated to right. Suggested stratigraphic relationship of the Sperm Bluff Formation that is found in the north is also shown (Savage *et al.* in press). Historically the Taylor Group was named by Harrington (1965) with stratigraphic names first defined as follows: Windy Gully Sandstone and Terra Cotta Siltstone by Zeller *et al.* (1961), New Mountain Sandstone by Hamilton & Hayes (1963), Altar Mountain Sandstone and Arena Sandstone by McElroy (1969), and the Beacon Heights Orthoquartzite and Aztec Siltstone by Webb (1963).

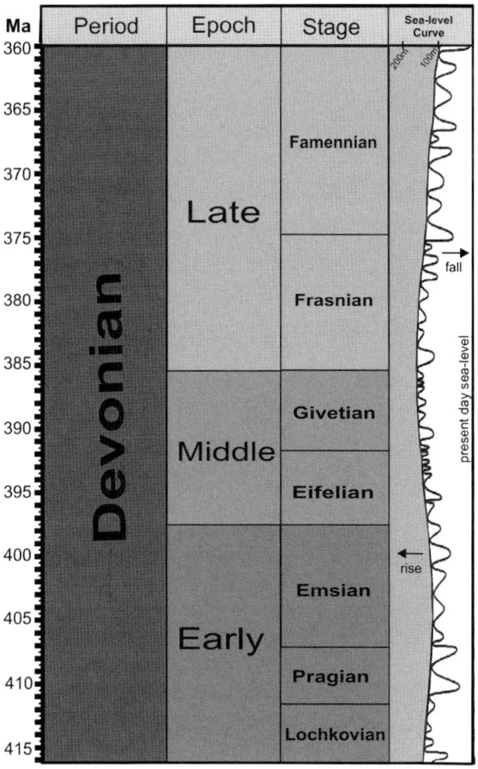

Fig. 5. The Devonian time scale, showing epochs and stages. The sea-level curve for the Devonian at right is taken from Haq & Schutter 2008. A fall in sea-level during the late Middle (Givetian) and early Late Devonian (Frasnian) coincides with the deposition of the Aztec Siltstone.

South Africa and Falkland Islands (Boucot *et al.* 1963*a*; Webers *et al.* 1992; Hunter & Lomas 2003). Based on the occurrence of a New Zealand brachiopod (*Tanerhynchia*) in the Ohio Range fauna, and an Ohio Range Malvinokaffric brachiopod (*Pleurothyrella*) in New Zealand, a coastal link must have existed between the Ohio Range and New Zealand via shelf seas marginal to East Antarctica (Bradshaw 1991, p. 196).

The occurrence of *Haplostigma irregularis* and *H. lineare* in sandstones of the Beacon Heights Orthoquartzite and Aztec Siltstones in southern Victoria Land (Plumstead 1962, 1964; McKelvey *et al.* 1972; McLoughlin & Long 1994) and at Okanogan Nunatak in the SW Pensacola Mountains (Schopf 1968), suggests a Middle to very early Late Devonian age, based on the common occurrence of this herbaceous lycophyte in the upper part of the Middle Devonian Bokkeveld Group and lower part of the Late Devonian Witteberg Group of South Africa, and also in the Falkland Islands (Edwards 1990). Spores and compressions of *Cyclostigma*-like stems in carbonaceous layers near the Okanogan Nunatak *Haplostigma* locality confirm a Middle Devonian age.

The firmest age control for the Taylor Group comes from numerous fish-fossil horizons in the Aztec Siltstone (Fig. 4) of southern Victoria Land at the top of the Group. The age of these beds was initially regarded as Late Devonian by Woodward (1921), modified to late Middle Devonian by White (1968) when more material was found, and now firmly established as late Middle Devonian (Givetian) to early Late Devonian (Frasnian) by Young (1988, 1993, 1996), Long (1989, 1995) and Turner & Young (1992). Isolated scales from the lower Aztec Siltstone, previously seen in morainic fragments at Mount Suess (Fig. 2), have been described as a new agnathan fish *Turinia antarctica* by Turner & Young (1992) and an Eifelian age (early Middle Devonian) suggested.

A Middle to Late Devonian age for the Aztec Siltstone is supported by a microflora collected on Aztec Mountain (Helby & McElroy 1969). The microflora is dominated by *Geminospora lemurata* and is strongly reminiscent of assemblages in Western Australia, whose age is considered either Frasnian (based on brachiopods) or Givetian (based on corals). Queensland floras that include *G. lemurata* are considered Givetian. Although Helby & McElroy (1969) suggest a probable Frasnian age in order to accord with the Late Devonian age at that time suggested for the fish fossils, a Givetian age for the microflora would be consistent with the slightly older age now proposed for the fish fauna.

Erratics containing *Haplostigma irregularis* and *Drepanophycus* of Middle to early Late Devonian age were found at Milan Rock near the Ruppert Coast of Marie Byrd Land (Fig. 1) (Mildenhall & Schopf 1980; Grindley & Mildenhall 1981). Edwards (1990) considered the fossils to be the best preserved Devonian plants in Antarctica. Their source, however, is still unknown. Based on lithology Grindley & Mildenhall (1981) correlated the specimens with the Wilkins Formation of Wilkins Nunatak 45 km to the NE. This appeared to confirm the Late Devonian to Carboniferous age proposed for the conformably overlying Ruppert Coast Volcanics. Subsequent extensive geochronological studies of Marie Byrd Land (Pankhurst *et al.* 1998; Mukasa & Dalziel 2000) have failed to find any Devonian volcanic rocks and it is more likely that the Ruppert Coast Volcanics are Cretaceous in age. Consequently the source of the Milan Rock erratics may be a Devonian Terrane distinct from that containing the Taylor Group.

The Kukri Erosion Surface

The active continental margin arc and accreted terranes of the Ross Orogen had developed along the Pacific margin of the Gondwana continent in the Cambrian (Stump *et al.* 2002; Goodge 2007; Federico *et al.* 2009), but by Early Devonian times had been deeply eroded to form the basement beneath the Taylor Group. The basal contact was described as the Kukri Peneplain by Gunn & Warren (1962), and said to have a 'very low relief at known localities except at Mt Suess' (Fig. 2), where they recorded a relief of 7 m with 2 m of weathering in the underlying granite. Zeller *et al.* (1961) observed an even stronger local relief of 30 m at Windy Gully (Fig. 6a). At this locality a zone of dark, highly altered granodiorite extends 6 m down from the Kukri Erosion Surface, and was initially thought by Zeller *et al.* (1961) to represent a pre-Devonian weathered horizon. However, a subsequent study of mineralogy and textures in both the granodiorite and the overlying Beacon Sandstone strongly suggests that the alteration is due to the hydrothermal effects of a nearby Ferrar Group dolerite sill (Craw & Findlay 1984).

During further exploration of the Transantarctic Mountains, the Kukri surface was found to be remarkably extensive, underlying the Taylor Group as far south as the Ohio Range (Figs 3 & 7), a distance of over 1500 km and usually with a rolling relief. Locally in the central Transantarctic Mountains (Fig. 3), and in particular where it was cut on Cambrian limestones, the Kukri surface displays a strong karstic relief of between 120 m near Mount Hunt in the Holyoake Range (Laird *et al.* 1971) and 700 m in the Churchill Mountains (Anderson 1979), and consequently, the more appropriate name of Kukri Erosion Surface was adopted (McKelvey *et al.* 1977). Substantial depressions on the Kukri Erosion Surface formed on Cambrian limestone became infilled with dark-coloured muddy sediments, possibly lacustrine or lagoonal in origin (Castle Crags Formation, Laird *et al.* 1971), before typical quartzose Taylor Group sedimentation commenced. On karstic surfaces elsewhere, valleys and caves show a thick fill (<15 m) of coarse lower Beacon sediment (Anderson 1979).

The deposition in southern Victoria Land of over 1 km of mainly sandy sediment of the Taylor Group required general subsidence of the region (Woolfe & Barrett 1995), although subsidence in the central Transantarctic Mountains may have been less. Isbell (1999) implied there was differential subsidence and that this pattern of subsidence continued during Permian glacial sedimentation with onlap onto basement highs and only limited erosion of Taylor Group rocks. However, the consideration of Permian sedimentation patterns cannot reliably be extended to the Taylor Group. Apart from the Churchill Mountains, where a karstic scenery developed before the Taylor Group was deposited, the single extensive formation, the

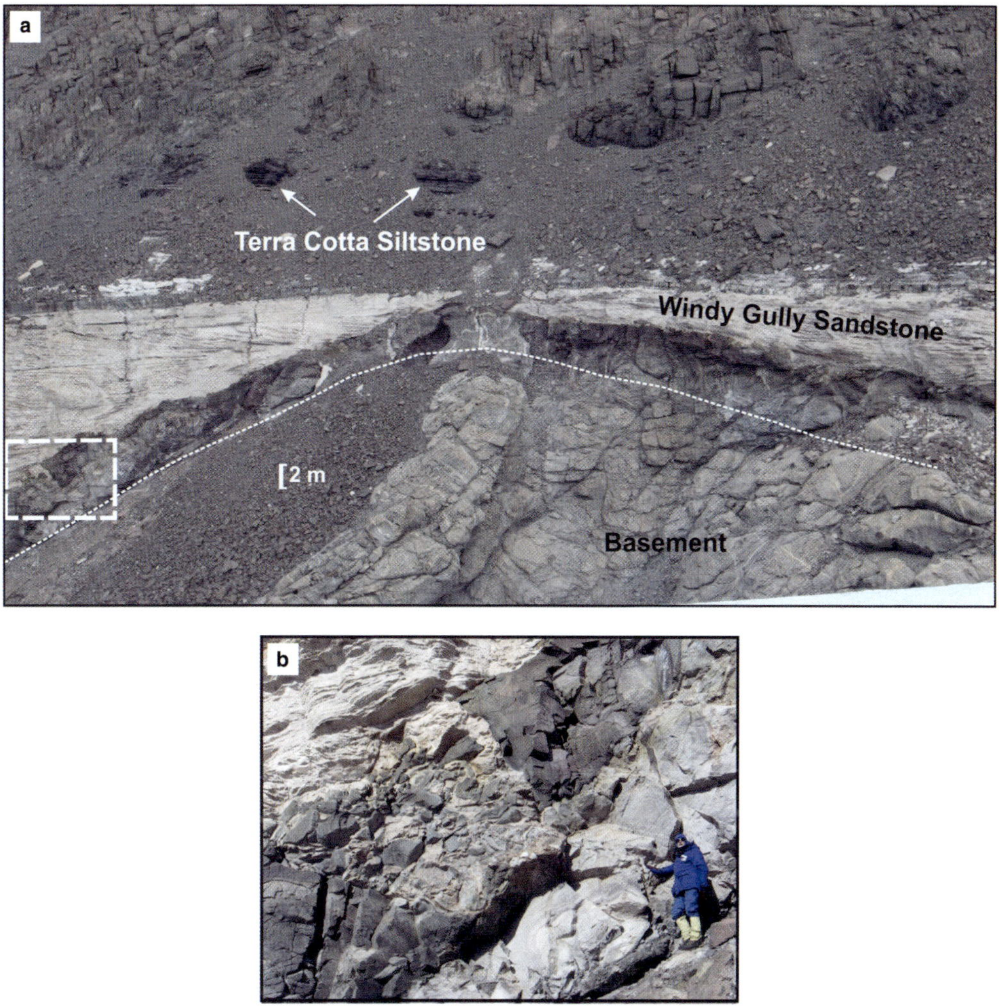

Fig. 6. Kukri Erosion Surface and basal sediments at the bottom of New Mountain immediately west of the Windy Gully Glacier. (**a**) A basement ridge on the Kukri Erosion Surface viewed from the Taylor Glacier. The outcrop is separated from the glacier by a large wind scoop. The Kukri surface (white dotted line) is mantled by a rubble of large clasts of Granite Harbour Intrusives (altered granodiorite), the detail of which is shown in (b) as indicated by the white dashed rectangle. The ridge and its rubble mantle became progressively buried by the onlapping cross-bedded Windy Gully Sandstone. Position of the Terra Cotta Siltstone indicated. The cliffs at the top of the photograph are part of a Jurassic dolerite sill. (**b**) Detail of the very large basement blocks within the basal layer of the Windy Gully Sandstone. The matrix of the boulder bed is sandstone. The figure on the right (1.8 m tall), is standing on the Kukri Erosion Surface.

Alexandra Sandstone, appears to be very uniform and consistent. There is no proof, however, that a greater thickness of sediment did not extend over the whole area to be later stripped down to basement during the Permian glaciation. This may be supported by the observation of Barrett *et al.* (1986) that 40% of the clasts in Permian glacial sediments in the Beardmore Glacier area were of Taylor Group origin, suggesting that a significant amount of Taylor Group rocks had been eroded. In the Ohio Range, where the Taylor Group equivalent (Horlick Formation) is only 56 m thick, there are also examples of Permian glacial erosion truncating the Devonian sediments down to basement (Aitchison *et al.* 1988). Blocks of Horlick Formation in the Buckeye Formation, with lithologies not observed in outcrop, also suggest that higher Devonian horizons had been eroded.

Beyond the Ohio Range, in the Pensacola and Ellsworth mountains, despite the presence of

Fig. 7. A horizontal Kukri Erosion Surface on Darling Ridge in the Ohio Range (dashed lines and white arrows). The prominent basal sandstone of the Horlick Formation is up to 2 m thick. The erosion surface truncates an undulating weathered profile in the underlying granitoid basement (black arrows).

Devonian rocks, a distinct erosion surface defining the base of them cannot be demonstrated.

Taylor Group depositional basins

The outcrop of the Taylor Group forms a long, narrow strip parallel to strike along the length of the block-faulted Transantarctic Mountains. The original form of sedimentary basins that developed on the Kukri Erosion Surface during the Devonian is consequently extremely difficult to determine. Barrett (1981) suggested two elongate basins: a land-locked McMurdo Basin that stretched from southern Victoria Land to near the Shackleton Glacier in the Queen Maud Range (Fig. 3), and a separate Ellsworth Basin stretching from the Ohio Range towards southern Africa. Barrett (1991, fig. 3.5b) later showed a seaway indirectly linking the two basins with a connection to the Pacific margin of Gondwana.

Dalziel & Elliot (1982), Elliot (1986) and Barrett (1991, fig 3.8a) suggest a Devonian volcanic arc in Marie Byrd Land, and a magmatic arc is supported by the presence of Devonian 'I'-type granites (Weaver *et al.* 1991). However, outcrops of Devonian volcanic rocks have not been confirmed (Pankhurst *et al.* 1998; Mukasa & Dalziel 2000). Late Devonian and Carboniferous volcanic rocks in northern Victoria Land (Findlay & Jordan 1984; Laird 1991*a*, *b*; Fioretti *et al.* 1997) may represent the western margin of the same magmatic belt. The age of these rocks, however, suggests eruption after the deposition of the Taylor Group.

However, the existence of an earlier volcanic arc near the Ellsworth Mountains is supported by the presence of fresh angular plagioclase and basaltic volcanic fragments of calc-alkaline affinity in the Early Devonian Wyatt Earp Formation of the Crashsite Group (Spörli 1992). This suggests that the arc formed earlier here in the Devonian and spread along the Pacific coast in the Late Devonian. There is no evidence of Devonian volcanic activity during the deposition of the Taylor Group sediments in southern Victoria Land. Volcanic clasts do occur (Sperm Bluff Formation), but these are of Cambro–Ordovician age based on U–Pb zircon ages (Wysoczanski *et al.* 2003; Savage *et al.* in press).

Within the McMurdo Basin (Fig. 8), the Taylor Group is far more diverse and thicker in southern Victoria Land than in the adjacent central Transantarctic Mountains (see Barrett 1981, fig. 2b), with seven formations in the McMurdo Dry Valleys

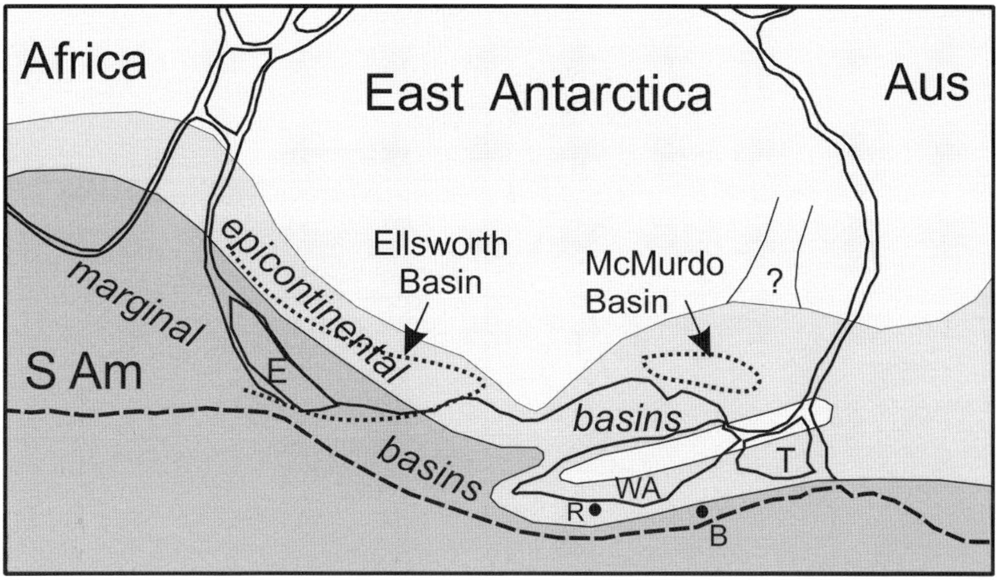

Fig. 8. Palaeogeographic map of the Pacific edge of Gondwana during the Early to Middle Devonian, showing the likely position of epicontinental and marginal basins. The heavy dashed line indicates the edge of the Gondwana continent. Light dashed lines indicate the position of the Devonian McMurdo Basin and the ?Ordovician to mid-Devonian Ellsworth Basin. West Antarctica (WA) was probably part of a linear continental ridge that continued into northern Victoria Land and western Tasmania, which may have acted as the eastern source of material in the northern part of the McMurdo basin. R marks the possible site of Reefton (NZ) marine faunas in an epicontinental basin that had faunal links with the Ohio Range along the coastline of Gondwana. B marks the possible site of Baton River marine faunas on the edge of a marginal basin. E indicates the original position of the Ellsworth block as part of the Ellsworth sedimentary basin. Size and shape of some continental blocks taken from a reconstruction provided by the PLATES Project at the Institute of Geophysics at the University of Texas.

compared with only two in the central Transantarctic Mountains, where the full thickness of the Taylor Group is unknown. In southern Victoria Land there is a definite traceable continuity of Taylor Group sediment between the McMurdo Dry Valleys region (Fig. 2) and the Darwin Mountains (Fig. 9a), confirming that they were parts of a single basin, but it may be significant that thick Taylor Group outcrops terminate abruptly on the north side of the Byrd Glacier (Britannia Range). Beyond the Byrd Fault (south of the Byrd Glacier), the Taylor Group continues but now rests unconformably on Shackleton Limestone rather than on crystalline basement (Anderson 1979). Coarse marble-rich basal conglomerates and cave deposits are overlain by more typical quartzose sandstones, similar to those in the Britannia Range and Darwin Mountains. The Byrd Fault has been recognized as a significant discontinuity and possible Cambrian terrane boundary where the Beardmore microcontinent (Borg *et al.* 1990) moved from the SSE, colliding with crust lying to the north of Byrd Glacier before the Devonian (Stump *et al.* 2006). Studies by Carosi *et al.* (2007) in the Britannia Range and eastern Cook Mountains north of the Byrd Glacier confirm differences in Ross Orogen basement rocks on either side of the Byrd Fault. They also noted that the age of the oldest Ross plutonic pulses in the north occurred at the same time as deposition of the Shackleton Limestone south of the Byrd fault, and that younger plutonic intrusions occurred on both sides, suggesting docking during the latest Cambrian to Ordovician. There are interesting differences between sections of the Taylor Group north and south of the proposed Byrd Fault. North of the Byrd Glacier there is a progressive decrease in grain-size and increase in sediment maturity up-section that is not seen to the south (Anderson 1979). Similarly, palaeocurrent directions are almost reversed between the two areas and there is a complete absence of limestone clasts in the sediment north of the Byrd Glacier, whereas they are common to the south. These data support Anderson's (1979) contention that the Beacon successions seen on either side of the glacier are closer

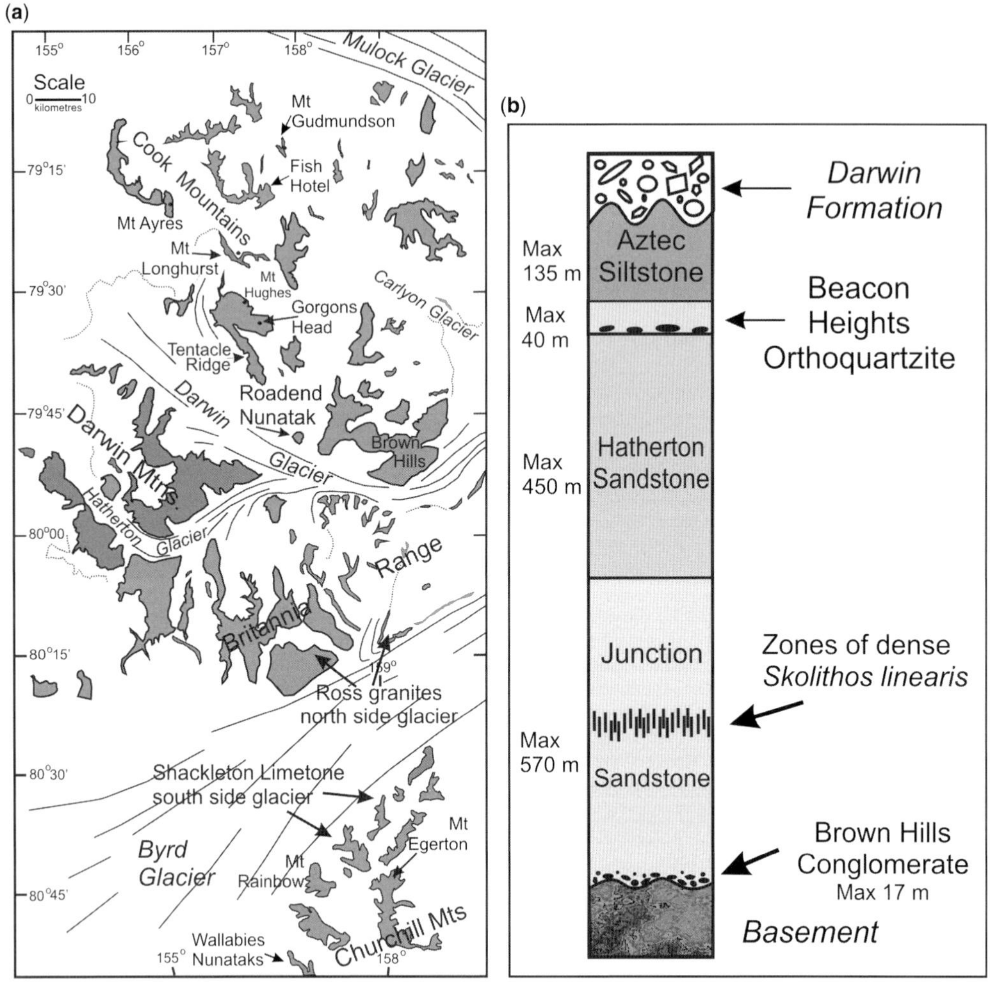

Fig. 9. (**a**) Location map for the Darwin and Cook Mountains and the most northerly part of the Churchill Mountains. The Byrd Glacier is thought to mark a fault that may have separated the main basin to the north from a more slowly subsiding slope leading to the central Transantarctic Mountains. This is emphasized by the contrast between a Kukri Erosion Surface cut across Ross granites north of the glacier and the same surface cut across Cambrian limestones to the south. There are also differences in the lithology and provenance of Taylor Group sediments on either side of the glacier (see text). (**b**) Stratigraphic divisions of the Taylor Group in the Darwin and Cook Mountains area. The Aztec Siltstone and Beacon Heights Orthoquartzite correlate with units in the Skelton Glacier to McMurdo Dry Valleys further north. The Hatherton Sandstone is correlated with the Arena Sandstone, and the thick Junction Sandstone is probably equivalent to all the combined lower formations in the north. The zone of dense *Skolithos linearis* between 300 and 380 m could correspond to similar zones on Table Mountain, Rotunda and Mount Handsley, and might equate with the Heimdall Erosion Surface further north.

now than at the time of their deposition, but the topic requires further investigation.

Southern Victoria Land (Convoy Range to Britannia Range)

For the northern part of this area, from the McMurdo Dry Valleys to the Convoy Range, a large amount of information has been presented in six 1:50 000 geological maps prepared by McElroy & Rose (1987), Woolfe *et al.* (1989), Allibone *et al.* (1991), Pocknall *et al.* (1994), Turnbull *et al.* (1994) and Isaac *et al.* (1996).

The Taylor Group in the McMurdo Dry Valleys region currently comprises seven formations (Fig. 4). Originally the Windy Gully Sandstone and Terra Cotta Siltstone were known only on New

Mountain and were given member status within the overlying New Mountain Sandstone. Subsequent work by Plume (1978) demonstrated their regional extent and raised them to formation status. This distinction has proved valuable in understanding the pattern of Taylor Group sedimentation.

The Darwin Glacier region, representing the southern part of the basin, was not visited until 1964, and although the Kukri Erosion Surface was found there, only two formations, the Brown Hills Conglomerate and Hatherton Sandstone, were recognized at two separate localities (Haskell *et al.* 1965). In a preliminary report, Barrett *et al.* (1971) later recorded a further 289 m of section below the Hatherton Sandstone, although no base was found. Subsequently this unit was named the Junction Sandstone by Askin *et al.* (1971).

Later fieldwork in the Cook Mountains between the Darwin and Mulock Glaciers (Bradshaw *et al.* 1990) discovered Junction Sandstone overlying the Kukri Erosion Surface on Tentacle Ridge (Fig. 9a), with two further formations above the Hatherton Sandstone (Beacon Heights Orthoquartzite and Aztec Siltstone; Fig. 9b), both of which have yielded fossils (Woolfe *et al.* 1990). However, the overall succession cannot be easily correlated with the McMurdo Dry Valleys area, except for the two youngest formations. In the following description of the McMurdo Dry Valleys formations, possible links to the Darwin succession will be suggested.

In southern Victoria Land, from the Mulock Glacier in the south to the Mawson Glacier in the north (Fig. 1), there is evidence of a progressive onlap of the Taylor Group onto basement across a low-lying but irregular erosion surface (Kukri Erosion Surface). This is shown by the distribution of the seven formations into which the Group is divided.

Windy Gully Sandstone

The oldest unit overlying the Kukri Erosion Surface, the Windy Gully Sandstone, comprises medium- to coarse-grained sandstones and rare thin mudstones. The most southerly outcrops are found on Mount Kempe, beyond which exposure is lost, and the most northerly outcrops are on Mount Jason (Fig. 2). Thickness varies with relief. This is well illustrated at the foot of New Mountain near Windy Gully (Fig. 2), where thickness varies from 30 to 2 m over a short distance as a basement ridge becomes buried (Fig. 6a). The basal conglomerate here is a mass of large granodiorite boulders that mantle the unconformity, some rounded and others subangular and weathered, and up to 3 m across (Fig. 6b). The nature of this deposit and the way it has been infiltrated and buried by clean cross-bedded, flaggy sandstones is similar to modern well-sorted shoreline sands that can be observed adjacent to coastal cliffs (O'Toole 2010). Elsewhere the basal conglomerate is thinner. The Windy Gully Sandstone contains occasional *Diplichnites gouldi* trackways on foresets (Bradshaw 1981). Cross beds are 0.5–2 m thick with current directions predominantly towards the west (Plume 1982). The formation also contains horizons of dense *Heimdallia chatwini* burrows that are interbedded in places within flaggy cross-sets (Bradshaw 1981). Glauconitic pellets were observed near the top of the formation (McElroy & Rose 1987) and Angino & Owen (1962) reported possible brachiopod fragments from the type section, both of which would suggest a marine environment. However, no further discoveries have been made. Thin green siltstones occur towards the top of the formation, many showing desiccation cracks (Barrett & Webb 1973; Plume 1982). Synaeresis cracks on asymmetrical ripples suggest deposition in tidal estuaries (O'Toole 2010).

Taken together, these data suggest that a combination of rocky coastline and estuaries occurred along the western margin of the uplifted source area.

Terra Cotta Siltstone

The Windy Gully Sandstone grades up into the Terra Cotta Siltstone with increasing silt. The formation is thickest in the south at Mount Kempe (82 m) (Plume 1982). The southern limit of the Terra Cotta Siltstone is unknown owing to lack of outcrop, but there is a definite northward thinning, and no outcrops are present north of Mount Jason or west of Mount Hercules in the Olympus Range, although a similar lithology occurs at Gargoyle Turrets at the base of one outcrop of the Sperm Bluff Formation (Savage *et al.* in press).

The Terra Cotta Siltstone is a series of interbedded sandstones (often cross-bedded) and siltstones (often ripple-laminated or showing lenticular bedding) that are black, maroon or green. Small calcareous nodules may be present, especially where silt beds are thick, and Plume (1982), while noting several features that could have formed in either a marine or non-marine environment, favoured non-marine on the basis of C^{13} values from the nodules. A contrary marine, estuarine or lagoonal origin is based on common synaeresis cracks (Bradshaw 1981, O'Toole 2010), mud-draped ripples (Savage *et al.* in press) and the presence of rare small arthropod traces (*Cruziana*, *Rusophycus*) and U-shaped burrows (Bradshaw 1981). Thin siltstones with desiccation cracks occur higher in the formation, and at several localities the top is marked by deep, sand-filled cracks and the suggestion of a poorly developed vein network

(Plume 1978), indicating a period of prolonged aerial exposure. Using a sequence stratigraphic approach, O'Toole (2010) regarded this surface at the top of the Terra Cotta Siltstone, previously called the '?Windy Gully Erosion Surface' (McKelvey *et al.* 1977, fig. 5) as the end of the first Taylor Group sedimentary cycle.

New Mountain Sandstone

The overlying New Mountain Sandstone, crops out from Rotunda in the south (Fig. 2), northwards to the Mackay Glacier (McKelvey *et al.* 1977; Woolfe *et al.* 1989). The formation has the same quartzose composition as the Windy Gully Sandstone and shows similar cross-bedding, but with some sets over 3 m thick. Coarse, feldspathic material occurs in the lower part of the formation (Gilmer 2008; O'Toole 2010), indicating the influx of fresh detritus into the basin. On Table Mountain Woolfe *et al.* (1989) reported subangular quartz pebbles 3 cm in diameter, rounded sandstone intraclasts and 1 cm diameter angular grains of orthoclase feldspar. To the south (Mount Handsley), rip-up clasts of mudstone occur in feldspathic sandstone (24–32% feldspar) close to the basal contact. Taken together these suggest the start of another sedimentary cycle (O'Toole 2010). Trace fossils are also identical to those in the Windy Gully Sandstone (*Heimdallia chatwini* (Fig. 10), *Diplichnites gouldi*, but not *Beaconites* of Wizevich (1997, fig. 9), which I believe is *Heimdallia*). Horizons of dense *Skolithos linearis* occur in the top of the New Mountain Sandstone on Table Mountain (Bradshaw 1981; Woolfe *et al.* 1989; Wizevich 1997) and on Mount Handsley (O'Toole 2010), and will be discussed later. Giant cross-beds at Table Mountain, up to 7 m high with troughs up to 60 m wide and traceable down dip for 150 m (Woolfe *et al.* 1989) may be aeolian dune deposits (Plume 1982; Wizevich 1997). A Pinstripe cross-bedded lithofacies seen further north in the Barwick valley is also regarded as aeolian (Gilmer 2008).

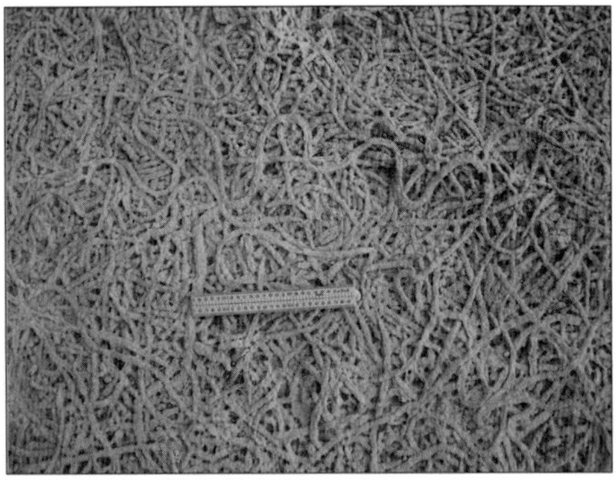

Fig. 10. Bedding plane view of planar bed in the New Mountain Sandstone showing dense concentrations of the burrow *Heimdallia chatwini*. The vertical meniscate structure of this burrow is not visible in this photograph owing to the effects of wind-driven sand blasting. The length of the scale is 25 cm. Mount Hercules, Olympus Range.

On Mount Jason, where the formation is especially well exposed above Terra Cotta Siltstone and Windy Gully Sandstone, the lowermost beds are pebbly, cross-bedded feldspathic sandstone (6–25% feldspar, Gilmer 2008). Higher beds show the usual alternation of *Heimdallia*-bioturbated planar beds (Fig. 10) and low-angle, flaggy cross-beds with *Diplichnites*. Towards the top of the formation, slump horizons appear, as well as 'pit and pile' structures, previously thought to be organic (Wizevich 1997, fig. 8) and now considered to be water escape features (Gilmer 2008). Larger sand volcano-like structures are also regarded as dewatering features, caused by an imbalance of sediment pore-pressure and weight of overlying water column. The highest beds contain fewer trace fossils, but desiccation cracks and adhesion ripples appear, and a large cross-set at the top of the section may be a dune. These features are interpreted as reflecting a gradual shallowing of the body of water in which the New Mountain Sandstone was deposited and possible emergence of the formation at this site (Gilmer 2008). Gilmer (2008) considered this water body to be marine.

West of Mount Jason the New Mountain Sandstone rests directly on granite basement, often with a thin, poorly sorted breccia at its base (Boreas Subgreywacke Member of McKelvey *et al.* 1977). There is a change in thickness from 150 m below Mount Electra to only 49 m a short distance away on Mount Boreas (McKelvey *et al.* 1977), reflecting the basin's progressive onlap across a rugged landscape. The New Mountain Sandstone appears to be absent over the Balham Basement High (Bradshaw 1981), where basement is covered by younger Altar Mountain Formation. New Mountain Sandstone is present in the Barwick Valley resting on basement (Fig. 11b) and at Sponsors Peak to the north (Fig. 2). At both of these localities the basal sediments are obviously infilling depressions in the landscape, with large scattered, angular to subrounded fragments of vein quartz, rhyolite and rare boulders of granite in the basal sediments. West of Sponsors Peak, New Mountain Sandstone is cross-bedded, flaggy and quartzose with an ichnofauna of *Heimdallia chatwini* and *Diplichnites gouldi* (personal observation).

Beacon sediments east of Bull Pass in the Olympus Range, and in the St Johns and Clare

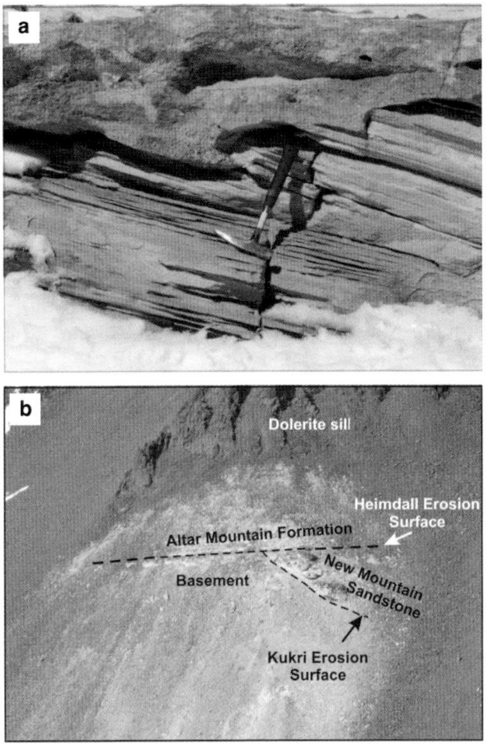

Fig. 11. (**a**) Heimdall Erosion Surface showing undercutting of cross-beds in the quartzose New Mountain Sandstone on Mount Jason. The undercutting indicates that the top of New Mountain Sandstone was already partially lithified when the Heimdall Surface was cut. The surface is overlain by the Odin Arkose, at this locality, less pebbly than usual, with up to 30% feldspar. West side of Mount Jason, Olympus Range. (**b**) View southwards across the Barwick Valley from near Vashka Crag, showing a level Heimdall Erosion Surface (with thin pebble conglomerate) truncating both basement rocks and a steeply sloping Kukri Erosion Surface overlain by New Mountain Sandstone containing scattered basement boulders up to 1 m in diameter. The outcrop is approximately 1 km across.

Ranges to the north (Fig. 2), have been mapped as undifferentiated Taylor Group because characteristic lithologies such as the Terra Cotta Siltstone and Altar Mountain Formation could not be identified (Turnbull *et al.* 1994). Outcrops are not extensive but significant occurrences of a rounded pebble, cobble and boulder fraction are present at some of these sites (Turnbull *et al.* 1994). At only one site (Sperm Bluff) do coarse conglomerates rest directly on the Kukri Erosion Surface. Elsewhere, finer beds of sandstone separate the conglomerate horizons from the Kukri Erosion Surface. Turnbull *et al.* (1994) incorporated the coarse beds in the Sperm Bluff Conglomerate (modified to Sperm Bluff Formation by Savage 2005), and correlated the formation with the Altar Mountain Formation Beacon Heights Orthoquartzite. The presence of thick conglomerates in the east, from Mount Suess in the north to Mount Peleus in the south (see Fig. 2), suggests an eastern provenance and proximity to the eastern side of the basin. Clasts include metaquartzite, acid volcanics and rare lithified Beacon sediment (Savage 2005). U–Pb zircon dating of the volcanic clasts (Wysoczanski *et al.* 2003; Savage 2005) and geochemistry suggest a relationship with Dry Valley 2 Suite (DV2) of the Ross age Granite Harbour Intrusives (Cox *et al.* 2000; Allibone & Wysoczanski 2002). Extrusive equivalents of the Ross Intrusives are not known in outcrop but may have become preserved in down-faulted areas prior to the development of the McMurdo Basin in the Early Devonian (Savage 2005).

Savage *et al.* (in press) in a study of the Sperm Bluff Formation as a whole, correlated it with New Mountain Sandstone, Terra Cotta Siltstone and Altar Mountain formations to the south (Fig. 4) and suggested marginal-marine to marine environments of deposition. The authors agreed with McKelvey *et al.* (1977) that the flaggy cross-bedded sandstones (10–25% feldspar, Savage 2005) overlying the Kukri Erosion Surface in the type section on Mount Suess was New Mountain Sandstone based on lithology and ichnology, and that the conformably overlying polymict conglomerates and sandstones were Altar Mountain Formation. The presence of locally dense *Heimdallia chatwini* burrows, rarer *Diplichnites* trackways and *Zoophycus* at Gargoyle Turrets, Sperm Bluff and Mount Suess, suggests the New Mountain Sandstone may have extended this far north and incorporated a significant amount of coarse material.

New Mountain Sandstone-like sediments at Mount Allan Thomson (Fig. 2) north of the Mackay Glacier are preserved between the basement and a Ferrar dolerite sill (Pocknall *et al.* 1994). Below the Kukri Erosion Surface, the Granite Harbour Intrusives are 'weathered' to a depth of 10 m, and the overlying Beacon sediments (20 m) appear to be infilling shallow depressions on the surface. Both the lithology and an ichnofauna of dense *Heimdallia chatwini* (Sykes & Pocknall 1991) suggest that this may be the most northerly outcrop of the New Mountain Sandstone.

The Altar Mountain Formation

In the northern part of the basin, from Sponsors Peak and Barwick Valley, south to the Beacon Heights region, the New Mountain Sandstone is truncated by the Heimdall Erosion Surface. Apart from small scale undercutting of foresets (Fig. 11a) and truncation of slump folds, all of which suggest partial lithification of the top of the New Mountain Sandstone, the Heimdall Erosion Surface is remarkably level (0–50 cm relief, Gilmer 2008), far more so than the Kukri Erosion Surface, and can be intermittently traced across *c.* 2300 km^2. On the south side of the Barwick valley, the Heimdall Erosion Surface truncates a steeply sloping Kukri Erosion Surface and overlying New Mountain Sandstone, and continues laterally across basement rocks (Fig. 11b). Wherever it occurs, the Heimdall Erosion Surface is covered by a basal conglomerate (lower Odin Arkose Member) of the Altar Mountain Formation (McKelvey *et al.* 1970). The conglomerate may be only a few pebbles thick but in the western Asgard Range forms a bed almost 1 m thick. Clasts include quartz, rhyolite, lithified grey quartzite and numerous clasts of what is probably New Mountain Sandstone. Clasts of the latter are often pink or brown and some have flat biscuit-like shapes. O'Toole (2010) records possible fluvial channelling in the Heimdall Erosion Surface in the western Asgard Range (Folkvanger Valley; Fig. 2) below the basal conglomerate. Gilmer (2008) believed that initially the Heimdall Erosion Surface was created by a drop in sea-level that exposed and eroded the top of the New Mountain Sandstone. Following a rise in sea-level, the sea transgressed across a shore platform, and both locally derived material and exotic clasts accumulated in the basal sediments. *Skolithos* occurs in the basal conglomerate, penetrating across the erosion surface down into the New Mountain Sandstone (Gilmer 2008). No change in provenance was detected across the non-conformity (Gilmer 2008; O'Toole 2010).

The conglomerate is succeeded by cross-bedded, feldspathic sandstones (11–29% feldspar, Gilmer 2008) that usually contain dense concentrations of *Skolithos linearis*. Higher up the succession, U-shaped burrows appear (*Diplocraterian*), as well as horizontal burrows showing arthropod appendage marks (?*Didymaulypnomos*; Bradshaw 1981) and surface *Helminthopsis*. The feldspathic sediments grade up into more quartzose and better sorted sandstones of the remaining Altar Mountain Formation, usually with a zone rich in *Beaconites* burrows near the top. The Altar Mountain Formation extends as far north as Mount Suess, where McKelvey *et al.* (1977) record 120 m of sediment above the level of polymict conglomerate, and below a Ferrar dolerite sill. The lower part of this unit is highly feldspathic, reminiscent of the Odin Arkose Member, with similar cross bedding, and is overlain by quartzose sandstones containing *Beaconites* (Savage *et al.* in press). The basal contact of the feldspathic unit is, unfortunately, obscured by debris.

The Heimdall Erosion Surface is difficult to recognize in the south. On Beacon Heights (Fig. 2), Barrett & Kohn (1971) record a surface cut on New Mountain Sandstone with a relief of about 30 cm, with lenses of subrounded pebbles in the overlying gritty sandstone. A non-erosive contact on Mount Handsley between *Skolithos*-rich quartz sandstone (New Mountain Sandstone) and a sudden appearance of coarse feldspar-rich sediment (5–20%, increasing to 40% in higher beds), also containing *Skolithos*, probably mark the equivalent position (O'Toole 2010). On Rotunda, a little further south, New Mountain Sandstone grades up abruptly into coarse feldspathic beds (10–15% feldspar) of the lower Altar Mountain Formation (O'Toole 2010). *Skolithos linearis* occurs both below and above this contact, as on Mount Handsley (Barrett *et al.* 1973; O'Toole 2010) and in the upper New Mountain Sandstone on Table Mountain (Bradshaw 1981; Woolfe *et al.* 1989).

The Heimdall Erosion Surface marks a flooding event that affected a huge area and was likely to have been slightly diachronous. It is difficult to see how such an even surface could be created in a fluvial system over such a large area, and a rise in sea-level is more probable. The Heimdall Erosion Surface marks the beginning of the third sedimentary sequence (O'Toole 2010).

Local instability of sediment in the Altar Mountain Formation is shown by four zones of large-scale intraformational folding in the lower Arena Valley caused by penecontemporaneous subaqueous gravitational slumping (McElroy & Rose 1987). *Skolithos linearis* is common, sometimes intimately associated with desiccation cracked mudstones (McElroy & Rose 1987) that become more common towards the top of the formation. Some of the cracks are infilled with pink, well-rounded quartz granules that could have been wind-blown (e.g. New Mountain). On Mount Handsley a blotchy cream and red colouration in the Altar Mountain Formation ('Handsley Beds') cross-cuts both stratification and trace fossils and is most likely the result of alteration due to a large Ferrar dolerite sill rather than indicating a primary red-bed.

In the Darwin Glacier region (Fig. 9a), the equivalents of the basal four formations in the McMurdo Dry Valleys are represented by the 570 m-thick Junction Sandstone, which is difficult to subdivide (Fig. 9b). A thin basal conglomerate on Tentacle Ridge thickens eastwards to 17 m on Roadend Nunatak (Brown Hills Conglomerate Member; Bradshaw *et al.* 1990; Bradshaw & Harmsen 2007). The Junction Sandstone is coarse, feldspathic and poorly sorted near its base, becoming more quartzose up-section, although still showing influxes of coarser material at intervals. Deposition was in broad, shallow channels and thin shales with desiccation cracks occur. Woolfe (1993) noted the presence of fining upward cycles. Zones of dense *Skolithos linearis* occur between 300 and 380 m above base (Bradshaw *et al.* 1990; Bradshaw & Harmsen 2007) that may correlate with *Skolithos*-rich horizons in the Table Mountain and Mount Handsley area. The Junction Sandstone trace fossils were mistakenly reported by Woolfe (1993) as part of a *Skolithos–Beaconites* association, when only one suspect *Beaconites* has ever been observed in these beds.

Arena Sandstone

The Arena Sandstone is a thick, mature, mainly fine- to medium-grained quartz sandstone with interbedded olive-green siltstones. The sandstones have a clay cement. In the New Mountain area Woolfe *et al.* (1989) record a sharp lower contact with a thin lag gravel, including well-rounded quartz pebbles up to 3 cm across, but further north in the Asgard Range Allibone *et al.* (1991) place the boundary at a marked break in slope that coincides with a change in bedding and grain size. This break is also close to the lowest occurrence of *Beaconites* in the Taylor Group (personal observation). The boundary may also mark the start of a fourth sedimentary sequence. The formation appears to thicken southwards to a maximum of 450 m on New Mountain, Knobhead and Pivot Peak (Woolfe *et al.* 1989). Hummocky cross-stratification was noted on Mount Carnes (Allibone *et al.* 1991, fig. 13), and intraformational slump folds occur in the upper Arena Valley (McElroy *et al.* 1987).

Trace fossils are less abundant than in older formations and seem to be mainly scattered *Beaconites antarcticus*, *B. barretti* (Bradshaw 1981) and *Skolithos* (McElroy & Rose 1987), but *Diplichnites gouldi*, *Diplocraterion*, *Helminthopsis* and *Didymaulyponomos* also occur (Allibone *et al.* 1991). Strange vertical structures identified as burrows by Allibone *et al.* (1991) are more likely to be water-escape conduits.

Woolfe *et al.* (1989) established the Pivot Coal Measures Member based on outcrops on Rotunda and Pivot Peak (Fig. 2) that were believed at the time to contain 50 cm-thick beds of coal. The name was later emended to Pivot Member when the coal was found to be bedding parallel titaniferous ironstone, formed as a secondary replacement related to Jurassic sill intrusion (Woolfe *et al.* 1995).

Arena Sandstone occurs at Pegtop Mountain (Fig. 2) on the south side of the Mackay Glacier (Turnbull *et al.* 1994) and continues north into the Convoy Range. In the Darwin/Cook Mountains area, the Hatherton Sandstone overlies the Junction Sandstone. It is the first unit that can be convincingly correlated with the McMurdo Dry Valleys succession and equates to the Arena Sandstone. There is little difference between outcrops in the Cook Mountains (Hatherton Sandstone) and those in the Boomerang Range across the Mulock Glacier (Arena Sandstone).

The Hatherton Sandstone is 450 m thick and has a gradational contact with the Junction Sandstone, placed at the first (40 cm) black shale horizon (Bradshaw & Harmsen 2007). Woolfe (1993) records a sharp contact and a thin lag gravel on Gorgons Head. The Hatherton Sandstone comprises cross-bedded, well-sorted, medium-grained mature sandstones with minor conglomerate and hummocky cross-stratification. The formation has a rich ichnofauna that includes *Diplichnites*, *Palaeohelcura*, *Beaconites antarcticus* and *B. barretti* (Bradshaw *et al.* 1990). The first three ichnospecies named occur on, or in, foresets, whereas the larger *B. barretti*, along with *Skolithos linearis*, penetrate down from bounding surfaces of the same cross-sets (Bradshaw *et al.* 1990; Bradshaw & Harmsen 2007). The environment is viewed as marginal to fully marine, in contrast to Woolfe (1993), who placed the trace fossils into a *Scoyenia* ichnofacies, although this ichnogenus is never present.

Beacon Heights Orthoquartzite

The Beacon Heights Orthoquartzite is a cliff-forming (Fig. 12) white medium-grained mature sandstone with well-rounded quartz grains and secondary silica overgrowths. The contact with the Arena Sandstone may be sharp, gradational (McElroy & Rose 1987) or irregular (Kohn & McPherson 1973). Sandstones are often cross-bedded and indicate a NE flow (Barrett & Kohn 1975) with channelling up to 2–3 m deep and *c.* 50 m wide (Allibone *et al.* 1991), and large ripples (0.3 × 2 m; McElroy & Rose 1987). Quartz pebbles also occur scattered or in local lenses (McElroy & Rose 1987). Trace fossils are rare and include *Skolithos* and *Beaconites*. Fossil lycopods *Haplostigma irregulare* (loose block near outcrop) and *H. lineare* (loose block in moraine at foot of mountain; Plumstead 1962) were found by Harrington & Speden (1962) on West Beacon Heights.

The Beacon Heights Orthoquartzite is the most widespread of all the Taylor Group formations, showing a continuing onlap northwards, and reaching almost to the Mawson Glacier. At Pegtop Mountain, Arena Sandstone is overlain by >240 m of Beacon Heights Orthoquartzite, which can be traced across the Mackay Glacier into the Convoy Range. Burgess *et al.* (1981) initially mapped it as a single formation, but Pocknall *et al.* (1994) divided the outcrop into Arena Sandstone (130 m)

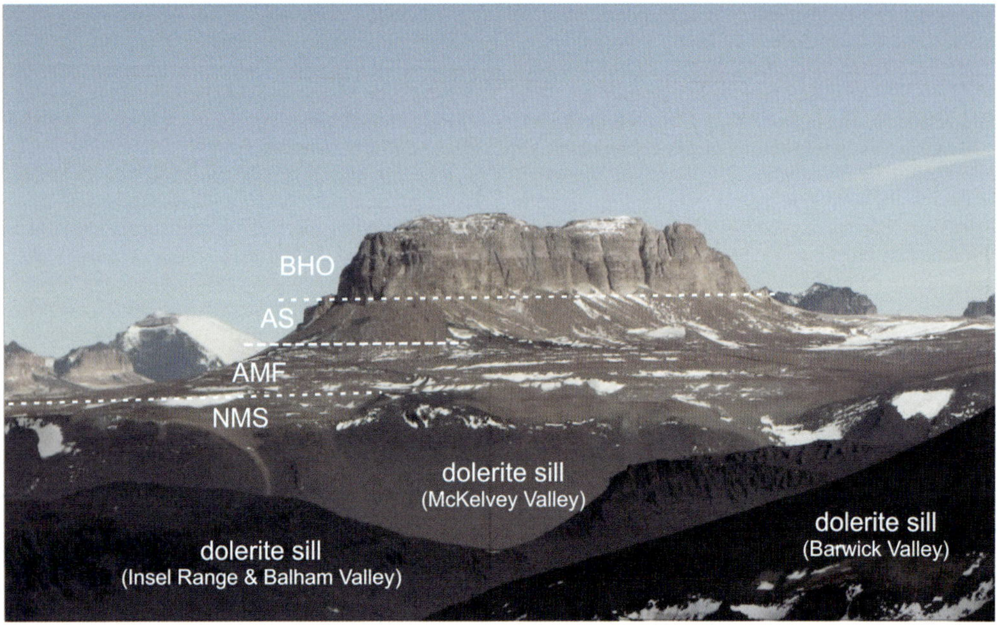

Fig. 12. Mount Boreas in the Olympus Range seen from the north, with the Balham and McKelvey valleys separated by the Insel Range in the middle distance. Formation boundaries in the Taylor Group are indicated by dashed lines. The cliffs on Mount Boreas (*c.* 300 m high) are of Beacon Heights Orthoquartzite (BHO). The overlying Aztec Siltstone has been removed by erosion. The lower slopes of the mountain are of Arena Sandstone (AS), and the gently sloping areas are underlain by Altar Mountain Formation (AMF) and New Mountain Sandstone (NMS). Rare outcrops on the SW and SE side of this mountain show both Windy Gully Sandstone and New Mountain Sandstone resting directly on granitic basement. On the southern side of the McKelvey Valley the New Mountain Sandstone has been intruded by a thick Jurassic dolerite sill that extends into the foreground (as labelled).

and Beacon Heights Orthoquartzite (70 m), with a gradational contact over a 20–30 m zone. In this region the Beacon Heights Orthoquartzite is dominated by multistacked channels, and Burgess *et al.* (1981) reported a coarsening-upward near the top of the formation to medium- and coarse-grained sandstones with vein quartz pebbles and large channels up to 10 × 40 m. Very large slumps (up to 100 × 50 m) in the upper part of the formation are described at several localities (Burgess *et al.* 1981). Formation thickness may have been greater prior to pre-Victoria Group erosion.

On Gorgons Head in the Cook Mountains (Fig. 2) in the south part of the basin, a cliff-forming, quartz-cemented sandstone, *c.* 30 m thick was discovered in 1990 overlying the Hatherton Sandstone and was correlated with the Beacon Heights Orthoquartzite (Bradshaw *et al.* 1990). The correlation was supported by the presence of *Haplostigma lineare* (originally described as *Protolepidodendron lineare* (Plumstead 1962) and *Malanzania* sp., both lycopods, and *Archaeosigillaria*, 15 m above the base of the formation (McLoughlin & Long 1994). The presence of these plants in the Beacon Heights Orthoquartzite is interesting as they were early colonists of the land. *Haplostigma* was a herbaceous semi-arboreal lycopod found in the littoral belt, forming stands along the edges of lagoons, deltas, swamps and estuaries (Scott *et al.* 1997). Woolfe (1993) included these beds in the top of the Hatherton Sandstone, but as they can be easily mapped in the northern Cook Mountains, they should remain as a separate unit. Where visible, the basal contact on burrowed Hatherton Sandstone is sharp but irregular, with rare lithic cobble/pebble lenses in the lowest metre of sandstone (Bradshaw & Harmsen 2007) possibly indicating the start of a fifth sedimentary cycle. The formation is noticeably thinner than in the McMurdo Dry Valleys region, but thickens to 40 m at Mount Gudmundson, increasing across the Mulock Glacier to 217 m in the Boomerang Range, where a thin conglomerate occurs at its base (personal observation).

Aztec Siltstone

The Beacon Heights Orthoquartzite passes up gradationally into the Aztec Siltstone, the base of which is placed at the first mudstone horizon (McElroy & Rose 1987). The Aztec Siltstone, a series of green or red siltstones interbedded with cross-bedded largely quartzose sandstones, was established by Webb (1963) with a type section on Aztec Mountain (56 m; Fig. 2), but thickest on Mount Ritchie (217 m) in the Boomerang Range. Thickness variations are most likely due to erosion before deposition of the Metschel Tillite or Weller Coal Measures (McKelvey *et al.* 1977). The most northerly exposures are on Mount Bastion (Fig. 2). The Aztec Siltstone contains ripple-marked or mud-cracked siltstones, vein networks and calcrete nodules, as well as root horizons in the sandstones, features consistent with subaerial and fluvial deposition (McPherson 1978, 1979). Rare plant remains occur (Cook Mountains) and include *Praeramunculus alternatiramus* and *Haplostigma lineare* (McLoughlin & Long 1994). Several of the sandstone beds show exceptionally large specimens of *Beaconites* in their upper surfaces (Cook Mountains), and a new ichnospecies may be appropriate. Conchostracans have been found in three sections but could not be dated (McPherson 1978).

The formation is best known for its prolific fish remains, first discovered as morainic fragments on Mount Suess (Debenham 1921). A Late Devonian age was suggested by Woodward (1921) based on eight fish taxa. *In situ* fish-bearing sediments, later discovered in the Boomerang Range, Mount Crean and Mount Feather by Gunn & Warren (1962), were described by White (1968), who suggested a late Middle Devonian age. Research by Young (1988, 1989, 1993, 1996), Young & Long (2005), Long (1989, 1995), Turner & Young (1992) and Long & Young (1995) now suggest a Middle to Late Devonian age for the Aztec Siltstone, supported by a microflora collected on Aztec Mountain (Helby & McElroy 1969).

New outcrops of Aztec Siltstone were discovered in the southern Cook Mountains in 1988 (Bradshaw *et al.* 1990). They contain well-developed palaeosols, desiccation cracks and rootlet horizons (mistakenly described as *Skolithos* by Woolfe 1993). A provisional analysis of fish material from a 15 cm-thick bed found near the top of the 100 m-thick unit can be correlated with the youngest biozone of Young (1988), indicating an early Late Devonian (Frasnian) age (Woolfe *et al.* 1990).

Further outcrops in the northern Cook Mountains (e.g. Mount Gudmundson and 'Fish Hotel' (name informal); Fig. 9a) produced a wealth of new fish material, mainly disarticulated plates in sandstone units, including new shark material (Long & Young 1995). The large size of some of the fishes, estimated to be 3 m long based on the size of cranial remains (Long 1989), indicates that they lived in substantial bodies of water (Young 1989).

These collections show that the Aztec Siltstone contains one of the most diverse fossil fish assemblages of Middle to Late Devonian age, showing close biogeographic affinity with eastern Australia (Young & Long 2005). Forty-five taxa have been described and nearly half of the fauna belong to the Placodermi (armoured fishes). Six 'biozones' have been determined in the Aztec Siltstone (Young 1988), but the lower ones are known only north of the Mulock Glacier, while the upper zones are well represented in the Cook Mountains to the south. This indicates that the entire Aztec Siltstone may be diachronous, becoming younger to the south (Young & Long 2005). Although the fish faunas are regarded as non-marine, the close affinity with Australia suggests that the species had a regional distribution and were capable of migration between drainage systems in marginal marine conditions. The Aztec Siltstone suggests that prograding non-marine sediments were building south, ending the fifth sedimentary cycle that had begun with the Beacon Heights Orthoquartzite.

The central Transantarctic Mountains

Compared with the thickness of the overlying Victoria Group, the Taylor Group is only modestly represented in the central Transantarctic Mountains and little new research has occurred. In the Churchill Mountains (Fig. 3a), the Taylor Group rests upon a highly irregular karstic topography developed on Cambrian Shackleton Limestone. Skinner (1964, 1965) recorded two separate instances on Mount Egerton (Fig. 3b) and Mount Albert Markham of limestone hillocks on the Kukri Erosion Surface (235 and 60 m high) buried by Taylor Group sediments. Subsequent work by Anderson (1979) reported a maximum relief of 700 m on the Kukri Erosion Surface in the Churchill Mountains, with Taylor Group sediments infilling palaeovalleys and burying individual hills up to 50 m high. Gradients on the pre-Devonian palaeokarst surface of 21° were reported. The Taylor Group comprises 710 m of quartzo-feldspathic sandstone with a thin (10 m) basal conglomerate. A strong northerly palaeo-flow is indicated. Anderson regarded these Taylor Group sediments as non-marine, and considered that they were significantly different to sediments north of the Byrd Glacier, where he believed the upper Hatherton Sandstone represented a marine transgression from the east (Anderson 1979). Wallabies Nunatak (Figs 3a & 9a), the most westerly of the Churchill Mountains, was visited by a US party who measured a minimum of 300 m of Taylor Group sediment below Permian glacial deposits, but provided no details (Isbell 1999).

Over the rest of the central Transantarctic Mountains, the Taylor Group appears to contain only two units. In the Holyoake Range (near Mount Hunt; Fig. 3b) north of the Nimrod Glacier, a 123 m-thick succession of shale with interbedded thin sandstones comprising the Castle Crags Formation overlies Shackleton Limestone on a level Kukri Erosion Surface (Laird *et al.* 1971). The basal sediments of the Castle Crags Formation include 5 m of bedded breccia with clasts of limestone, gneiss, granite and quartz in a clean sandy matrix (Laird *et al.* 1971). The bulk of the formation is of carbonaceous shale and siltstone, subdivided by two coarse sandstone units up to 6 m thick. There is a 30 cm-thick carbonaceous limestone near the top of the unit. The fineness of this unit indicates low energy, and its limited lateral extent led Laird *et al.* (1971) to suggest that the formation was deposited in a substantial depression on the Kukri Erosion Surface where a lake or lagoon had formed.

The Castle Crags Formation grades up over 3 m into the strongly cross-bedded, mature sandstones of the Alexandra Formation (Laird *et al.* 1971), which Grindley (1963) first described on Pagoda Peak in the Queen Alexandra Range (Fig. 3c), where *c.* 300 m of medium-grained quartzarenites rest upon a level Kukri Erosion Surface cut on basement metasediments. Other outcrops in the Queen Alexandra Range occur on Mount Elizabeth and the Cloudmaker (Fig. 3a, c). Well-rounded quartz pebbles occur sporadically, and are more abundant in the basal few metres. The Alexandra Formation also occurs in the Holland Range (near Mount Miller; Fig. 3c) and eastern Queen Elizabeth Range further north (Lindsay 1969; Barrett *et al.* 1986), where the formation thins westwards from 244 m on Turnabout Ridge (Fig. 3a) to zero on Solitary Peak 8 km to the NW. It is not clear if this is depositional thinning against the edge of a basin, or due to erosion prior to Permian glacial sedimentation. At Mount Counts in the western Queen Elizabeth Range (Fig. 3a), where the Permian Pagoda Formation rests directly on Cambrian limestone basement, Barrett *et al.* (1968) describe a 10 m diameter cave, 10 m below the erosion surface, that is filled with coarse green sandstone and large limestone blocks. Lindsay (1970) thought there had been two phases of cave fill: the first when there was collapse of the cave roof to produce numerous limestone blocks, and a second phase where surface conditions changed and fluvial sediment washed into the cave. The coarse sand fill is unlike the overlying Pagoda Formation, but does appear similar to the Alexandra Formation cropping out to the east. If so, this would suggest that the palaeokarst (and partially infilled cave) became buried by Alexandra Formation sandstone during the Devonian, but became exhumed by erosion before deposition of Permian glacial sediments.

At Mount Miller in the Holland Range (Fig. 3c), only the upper 100 m of the Alexandra Formation is exposed and Barrett *et al.* (1986) record 6 m of black shale at the bottom of the exposure, reminiscent of the Castle Crags Formation, with thin fine-grained limestone interbeds, and sparse sandstone layers with ripples and desiccation cracks. Most of the Alexandra Formation is well-sorted with well-rounded quartz grains and an authigenic quartz cement (Barrett *et al.* 1986, fig. 9). In the Holyoake Range Laird *et al.* (1971) observed two separate grain shape populations (rounded–subrounded and subangular) and the presence of minor grains of fresh microcline feldspar. They noted that currents were from the NW for the Alexandra Formation, but from the SE for the underlying Castle Crags Formation. In the Queen Alexandra and Holland Ranges, palaeocurrents are variable and have no clear pattern, leading Barrett *et al.* (1986) to initially think that this was consistent with coastal and near-shore current systems. Those in the Queen Elizabeth Range indicate more focussed currents flowing towards the east.

No Taylor Group fossils have been found in the central Transantarctic Mountains and trace fossils are relatively sparse and have not yet been identified (Laird *et al.* 1971; Barrett *et al.* 1986). A few small algal mats up to 30 mm across were found in calcite cemented sandstone near the base of Pagoda Peak (Fig. 3c) in the Queen Alexandra Range (Barrett *et al.* 1986). Strontium 87/86 ratios from these mats suggest a non-marine environment, while the textural and mineralogical maturity of the Alexandra Formation indicates extensive reworking under either shallow marine or continental conditions, with the variation in palaeocurrents supporting the former (Barrett *et al.* 1986).

The most southerly recorded outcrop of Taylor Group is a 65 m thick section at the head of Ramsey Glacier (Fig. 3a), but no details are reported (Isbell 1999). The central Queen Maud Land Mountains and Wisconsin Range (Fig. 3a) almost certainly formed a basement high during Taylor Group times, separating the central Transantarctic Mountains basin from the edge of the Ellsworth basin in the Ohio Range (Isbell 1999). Taylor Group rocks do not occur and were probably never deposited (Barrett 1965; Katz & Waterhouse 1970; Isbell 1999), so that Permian sediments rest directly on basement. The Butters Formation, described by Wade *et al.* (1965) as the possible equivalent of the Alexandra Formation on the Nilsen Plateau, has subsequently been recognized as a periglacial sequence of the Permian glacial Pagoda Formation (Isbell *et al.* 2001).

Ellsworth Basin

Understanding the development of this basin has been difficult owing to the uncertainty of relative positions of components during Palaeozoic times. It is now clear that the Ellsworth and Falkland Islands blocks are allochthonous. The Ellsworth block originated on the eastern side of the Weddell Sea (Fig. 1) and has migrated and rotated 90° towards eastern Marie Byrd Land (Curtis 2001). The Falklands Islands block is thought to have originated between the Ellsworth Block and the Antarctic Craton (Rapela *et al.* 2005).

Ohio Range

In the Early Devonian, the Ohio Range lay at the southern edge of the Ellsworth Basin, separated from the central Transantarctic Mountains Basin by a wide barrier of granitic basement in the southern Queen Maud Mountains and Wisconsin Range (Fig. 3a). The Ohio Range was first visited during International Geophysical Year (1958–1959) when Devonian marine fossils were found at the bottom of a flat-lying succession on Discovery Ridge. The Devonian unit was called the Horlick Formation and these were the first marine fossils of Devonian age to be found in Antarctica. Large collections were made in subsequent years (Boucot *et al.* 1963*b*; Doumani *et al.* 1965; Young 1986; Bradshaw & McCartan 1991) and an Early Devonian age (Emsian) suggested. The stratigraphy of the range was also clarified (Long 1965*a*, *b*).

Subsequent visits have shown that, of the 1200 m of Beacon rocks in the Ohio Range, only 56 m are Devonian, lying below Permian glacial and post-glacial deposits (McCartan *et al.* 1986). Nineteen detailed sections have been measured through the succession (McCartan & Bradshaw 1987; Bradshaw *et al.* 2002). The Devonian rocks rest on a wave-trimmed platform cut across deeply weathered granite (Fig. 7) and comprise mainly shallow marine subtidal inner shelf to shoreline sediments (Bradshaw & McCartan 1983; McCartan & Bradshaw 1987; Bradshaw *et al.* 2002). Brachiopod shells occur within 1 cm of the unconformity, confirming the marine transgression. Lycopod plant fossils and spores occur in thin carbonaceous interbeds close to the base of the succession. Trace fossils are common in higher energy sandstone beds where body fossils are few (Bradshaw 2002, 2010) and abundant marine fossils occur in thin tempestites (Bradshaw & McCartan 1991). The shelly fauna has Malvinokaffric affinities (Boucot *et al.* 1963*a*), pointing to strong links with South Africa and South America, but also to brachiopod genera shared with New Zealand, suggesting a peri-Gondwana seaway between South America and South Africa, New Zealand and Australia (Bradshaw & Webers 1988).

Sporadic, very thin phosphate-rich beds within the succession may have resulted from renewed marine transgressions across the basin. Rather deeper water may be indicated by the appearance of spiriferids at the top of the sequence (Bradshaw *et al.* 2002). One lithofacies identified as fluvial is found at a single locality, and suggests the mouth of a river draining an upland that lay towards Marie Byrd Land.

Pensacola Mountains

A 4000 m-thick clastic succession in the Neptune Range (Neptune Group and the overlying Dover Sandstone; Fig. 13a, b) is regarded as a possible equivalent of the Taylor Group (Schmidt *et al.* 1965; Schmidt & Ford 1969; Bradshaw & Webers 1988; Barrett 1991, table 3.1), in the upper part at least, although lithological differences with southern Victoria Land are marked. The age of the strata is poorly constrained but is younger than Upper Cambrian to Lower Ordovician intrusives that affect the underlying Cambrian rocks, and predates the Permian glacial Gale Mudstone (Schmidt & Ford 1969). The oldest unit of the Neptune Group is the Brown Ridge Conglomerate that rests with angular unconformity on Late Cambrian sediments and Ordovician intrusives. The conglomerate is poorly bedded and sorted, characteristically red in colour, with clast size varying from pebble to boulder. Clasts have been locally derived from basement sedimentary and volcanogenic rocks. Thickness is variable, and the formation reaches a maximum of 1000 m (Schmidt *et al.* 1965; Schmidt & Ford 1969; Laird 1991*a*). The remainder of the Neptune Group is mainly of cross-bedded quartz sandstones with well-rounded grains. The Elliott Sandstone conformably overlies the Brown Ridge Conglomerate and contains volcanic fragments, likely to have been derived from the Cambrian volcanics of the Gambacorta Formation, that become less common higher in the group (Williams 1969). The Elliott Formation is thickest in the south of the Neptune Range (1500 m), but thins north to zero *c.* 65 km away (Schmidt *et al.* 1965). The overlying Elbow Formation comprises interbedded red siltstones and grey cross-bedded sandstones, and appears to have an inverse relationship to the underlying Elliott Sandstone, being thickest in the north (330 m) and thinning and becoming coarser in the south (140 m) (Schmidt *et al.* 1965; Bradshaw & Webers 1988). Cross-bedding is abundant and ripple marks common (Laird 1991*a*, p. 100). Irregular tubes at a steep angle to bedding in mottled red and grey siltstone are probably animal burrows (Bradshaw & Webers 1988; Laird 1991*a*). Cathcart &

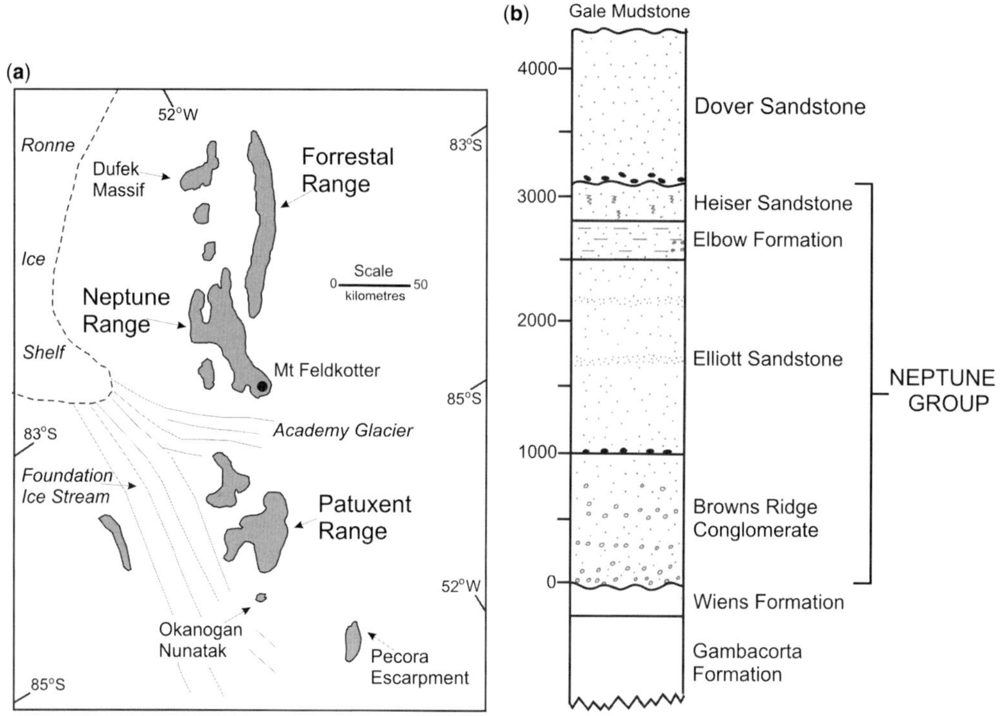

Fig. 13. (**a**) Location map for the Pensacola Mountains. (**b**) Stratigraphic column showing Devonian Dover Sandstone and underlying Neptune Group in the Pensacola Mountains.

Schmidt (1977) record primary phosphatic sandstones, and sandstones containing derived intraformational phosphatic pebbles in the Elbow Formation at Mount Feldkotter in the south of the range. The presence of phosphorite indicates temperate marine waters near an upwelling (Cathcart & Schmidt 1977). C^{13} and O^{18} ratios suggest that the marine basin may have been restricted (Cathcart & Schmidt 1977 p. E16).

The Elbow Formation is gradationally overlain by the Heiser Formation, which also thins southwards from 330 to 125 m (Cathcart & Schmidt 1977). The formation is noted for an abundance of *Skolithos* (Schmidt *et al.* 1965). Williams (1969 p. 1461) emphasized the similarity of the Heiser Sandstone to Taylor Group sediments in southern Victoria Land.

The Heiser Sandstone is disconformably overlain by the Dover Sandstone (1200 m), a tan to white, medium- to coarse-grained quartzose sandstone with extremely rare silt layers (Schmidt *et al.* 1965) and a 3–4 m-thick basal conglomerate of rounded pebbles of quartz and phosphate (Schmidt & Ford 1969). The overall lithology of this formation, including well-rounded quartz grains, shows similarities with the sandstone at Okanogan Nunatak in the south that contains *Haplostigma* and a microflora suggesting a Middle Devonian age. Barrett (1991) supported a shallow marine origin for the Heiser Sandstone, but believed that sedimentation reverted to non-marine for the Dover Sandstone. The phosphatic pebbles and cobbles in the Elbow Formation and in the base of the Dover Sandstone are reminiscent of ones in the Horlick Formation of the Ohio Range (Phosphatic Lithofacies; Bradshaw *et al.* 2002), occurring in layers that also contain fish material, phosphatized trilobites and brachiopods. The thin basal conglomerate in the Horlick Formation also contains black phosphatic pebbles. A marine origin for the Elbow Formation, Heiser Sandstone and at least the lower part of the Dover Sandstone should be considered, and the Dover disconformity regarded as representing a transgressive event.

Williams (1969) interpreted the Brown Ridge Conglomerate at the bottom of the Neptune Group as a coarse post-orogenic deposit derived from a rising fold-mountain belt (Ross Orogen), which was followed by alluvial fans (Elliott Sandstone). Restricted marine sedimentation occurred during the deposition of the Elbow Formation, with fully marine conditions for the Heiser and Dover

Sandstones. Laird (1991*a*) concludes that the Neptune Group probably ranges in age from Ordovician to Early Devonian.

The most characteristic Taylor Group lithology in the Pensacola Mountains crops out on a single nunatak (Okanogan Nunatak) at the SW end of the Patuxent Range (Schmidt *et al.* 1964; Fig. 13a). It is a well-sorted fine to coarse-grained quartz sandstone, red or white weathering, with a thickness of ≥180 m. Correlation is supported by the discovery of *Haplostigma* in a loose fragment of quartzite at this locality. In addition rare carbonaceous interbeds contain numerous spores and compressions of *Cyclostigma*-type stems that suggest a Middle Devonian age (Schopf 1968; Rigby & Schopf 1969).

Ellsworth Mountains

In the Ellsworth Mountains a very thick concordant succession of sediments (13 000 m) was deposited between the Early Cambrian and Permian with no readily detected unconformities (Webers *et al.* 1992). Part of this succession, the upper Crashsite Group, is Early Devonian in age and can be correlated with the Taylor Group. Unlike the Taylor Group of the Transantarctic Mountains, the Ellsworth sediments, including the overlying Permian glacial beds (Whiteout Conglomerate and Polarstar Formation), have been affected by Permian–Triassic folding.

The Crashsite Group (3000 m) comprises shallow water, mainly marine, quartzites and mudstones (Spörli 1992). The Group is made up of three formations, a lower Howard Nunatak Formation, followed by the Mount Liptack Formation with the Mount Wyatt Earp Formation as the youngest unit. The Crashsite Group overlies the Early to Late Cambrian Heritage Group, usually through a series of 'transition beds', and the contact was initially regarded as conformable (Spörli 1992), confirmed by Late Cambrian trilobite fossils in the 'transition beds' (Shergold & Webers 1992). The remainder of the Crashsite Group was regarded as Ordovician to Devonian.

Despite the concordant nature of the succession as a whole, and the presence of so-called 'transition beds', it is likely that there is one or more significant unconformities within it. Four pieces of evidence point to this:

(1) At Soholt Peaks (Fig. 14) the Heritage Group is overlain by 62 m of cobble to pebble conglomerate at the base of the Crashsite Group (Goldstrand *et al.* 1994).

(2) The report of clasts of cleaved, low-grade metamorphic clasts in the basal sandstone of the Crashsite Group (Yoshida 1982) suggests a pre-Crashsite Group tectonic event.

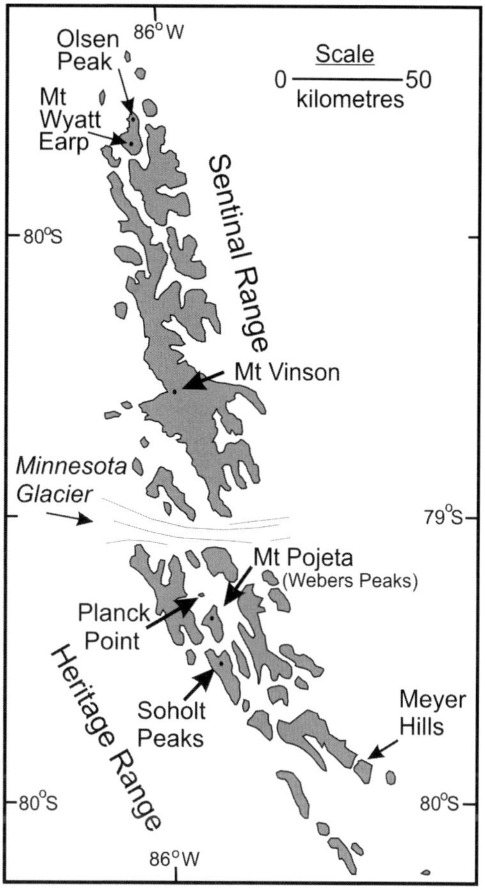

Fig. 14. Location map for the Ellsworth Mountains in their present position in West Antarctica. During the Palaeozoic, the Ellsworth crustal block lay close to the cratonic edge of East Antarctica between the Pensacola and Theron Mountains as part of the Ellsworth sedimentary basin.

(3) There is a local angular discordance of Crashsite rocks on the younger part of the Cambrian in the Webers Peaks range (Fig. 14; Duebendorfer & Rees 1998).

(4) The basal conglomerate and sandstone of the Crashsite Group truncates the 'transition beds' at Pojeta Peak (Webers Peaks; Fig. 14) and rests directly on Heritage Group limestone 2 km to the north, suggesting a low angle unconformity (Duebendorfer & Rees 1998).

The Mount Wyatt Earp Formation (≥300 m), the highest unit of the Crashsite Group, is well exposed but has yielded only a sparse, poorly preserved Devonian fauna in siltstones and mudstones from a single locality near Planck Point in the Heritage

Range (Fig. 14; Webers *et al.* 1992). The fossils include large brachiopods and molluscs, trilobites and a single fish spine *Machaeracanthus* cf. *kayseri*. The fauna shows Malvinokaffric affinities, and has been correlated with the Early Devonian (Emsian) Horlick Formation (Ohio Range) and the Falkland Islands (Webers *et al.* 1992).

Most of the formation is poorly sorted and laterally variable feldspathic arenites and quartzites. Heavy minerals suites are dominated by garnet and epidote, indicating a cratonic source (Spörli 1992). Fresh angular plagioclase and basaltic volcanic rock fragments are characteristic of this formation, indicating that the sediments are more immature than the two underlying formations of the Crashsite Group (Spörli 1992).

The contact between the Wyatt Earp Formation and the overlying glacial Whiteout Conglomerate is ambiguous, and the considered opinion has been that it was gradational, based on interbedded sandstones and diamictites (Meyer Hills, Heritage Range; Fig. 14), a dropstone 8 m below the Crashsite/Whiteout contact (near Olsen Peak, Sentinel Range; Fig. 14; Spörli 1992), and similar palaeocurrent directions (towards the north) in both formations (Matsch & Ojakangas 1992, p. 51). However, quartz sandstone clasts in the Whiteout Conglomerate resemble Crashsite sandstones, suggesting that the Crashsite Group could have been the source and that in places the contact may be erosional. A gradational contact of the Wyatt Earp Formation with the Whiteout Conglomerate would be surprising, unless the lower Whiteout could be firmly demonstrated to be Carboniferous or latest Devonian in age. The date given by Ojakangas & Matsch (1981) is Upper Carboniferous to Lower Permian, and it is more likely that the upper contact of the Crashsite Group is a disconformity.

Summary and conclusions

Taylor Group sediments or related older rocks were deposited in two basins along the slowly sinking continental edge of Gondwana (Fig. 8). The Ellsworth Basin was the first to form during the early or Middle Cambrian (Webers & Spörli 1983), probably as a shallow marginal, rift-related basin where subsidence kept pace with at least 13 000 m of predominantly shallow water sedimentation. Breaks in the sequence reflect tectonic activity beyond the region.

In the Ellsworth Mountains part of the basin, there was a dated change in the Late Cambrian from calcareous sedimentation (Heritage Group) to clastic sedimentation (Crashsite Group). Local disconformities and disconcordances followed by conglomerates suggest significant erosion at some sites, while cleaved clasts in local basal sandstones suggest a pre-Crashsite Group tectonic event that is believed to be related to the Ross Orogeny. The Mount Wyatt Earp Formation at the top of the Crashsite Group contains a sparse Early Devonian shelly marine fauna.

In the Pensacola Mountains part of the basin, the 3000 m-thick Neptune Group rests with angular unconformity on Late Cambrian sediments of the Wiens Formation and Lower Ordovician intrusives, affecting these sediments. The Group begins with the thick, coarse Brown Ridge Conglomerate which, like the cleaved clasts in the basal Crashsite Group, reflect the distant effects of the Ross Orogeny (Curtis *et al.* 2004). The Neptune Group shows a transition from alluvial fan to alluvial plain (Browns Ridge Conglomerate and Elliott Sandstone), and to restricted marine conditions (Elbow Formation), as suggested by phosphatic pebbles in the latter and abundant *Skolithos* in the overlying Heiser Sandstone. The Dover Sandstone, tenuously dated as Middle Devonian, overlies the Neptune Group with black phosphatic pebbles in the basal sediments. This may mark a marine transgression into the basin similar to that seen in the Ohio Range, where Emsian fossiliferous sediments (Horlick Formation) transgress onto Ross Orogen granitoids at the outer limit of the Ellsworth Basin. The Ohio Range fauna can be correlated with the Wyatt Earp Formation in the Ellsworth Mountains, as well as to faunas in South America, South Africa and New Zealand.

The McMurdo Basin was not initiated until the earliest Devonian on the folded and thickened margin of Gondwana following the Ross–Delamerian Orogeny. In the northern McMurdo Dry Valleys region of the basin, Taylor Group sediments contain clasts of quartzite and rhyolite of Ross–Delamerian age. The clasts indicate local uplift and erosion, and their deposition suggests complementary continental downwarping. The earliest Devonian age of the lower Taylor Group formations may reflect the distant effects of the Early Devonian Bindian Orogeny of Australia (VandenBerg *et al.* 2000). Within the southern Victoria Land part of the McMurdo basin, thick sediments accumulated between the Byrd Glacier and the Wright Dry Valley, with progressive northward onlap onto basement rocks. Palaeocurrents and clast provenance indicate a basement high to the east in what is now the Ross Sea. The lower four Taylor Group formations near the type area (Beacon Heights) cannot be recognized in the Darwin Glacier area, where they may equate with the thick Junction Sandstone. Horizons in the latter that are rich in *Skolithos* may correlate with *Skolithos*-rich sediments overlying the Heimdall Erosion Surface north of the Taylor Glacier. The environment of the Taylor Group is still uncertain. The initial flooding of the irregular Kukri Erosion Surface and the abrupt burial of very coarse, cliff-derived rubble by well-sorted sandstones (Windy Gully Sandstone) is easier to explain by an encroaching shoreline, and would be consistent with a lagoonal origin for the overlying Terra Cotta Siltstone, a coastal origin for the densely bioturbated horizons in both the Windy Gully Sandstone and New Mountain sandstones, as well as the large cross beds that may be coastal sand-dunes. The extent and flatness of the subsequent Heimdall Erosion Surface is also easier to explain by marine processes than by alluvial ones, and the fact that the surface dies out southwards suggests that the basin deepened in this direction. It is possible that the Taylor Group up to the top of the Beacon Orthoquartzite is marine in origin and represents up to five sedimentary cycles driven by sea-level changes. The cycles terminate with the alluvial plain sediments of the Aztec Siltstone, whose fish fossils suggest a diachronous younging southwards with progradation into the basin.

The McMurdo Basin appears to have continued south into the central Transantarctic Mountains, but beyond the Byrd Glacier there is little evidence that the basin subsided as much as in southern Victoria Land. The burial of the high karstic relief in the Churchill Mountains requires initial major subsidence, but further to the south the basal uniformity is less dramatic and slower subsidence is suggested by the thinner and muddy Castle Crags Formation (123 m). Elsewhere, the sandy Alexandra Formation rests directly on an even Kukri Erosion Surface and is nowhere more than 300 m thick. It is not clear whether this is a thin remnant left after pre-Permian erosion, or whether it is a thin succession resulting from less subsidence in this area.

The extent of the McMurdo Basin is impossible to deduce from the limited exposures available. It could be that this epicontinental basin extended deeper into Antarctica, and may have connected with comparable basins in Australia, as suggested by the similarity between horizons of dense *Heimdallia chatwini* in southern Victoria Land and the Tumblagooda Sandstone in Western Australia (Trewin 1993), the only other place where these burrows occur.

The author wishes to thank P. J. Barrett, B. C. McKelvey and D. H. Elliot for reviewing and improving this manuscript.

References

AITCHISON, J. C., BRADSHAW, M. A. & NEWMAN, J. 1988. Lithofacies and origin of the Buckeye formation: late Paleozoic glacial and glaciomarine sediments, Ohio Range, Transantarctic Mountains, Antarctica. *Palaeogeography, Palaeoclimatology, Palaeoecology*, **64**, 93–104.

ALLEN, A. D. 1962. Geological investigations in southern Victoria Land Antarctica. Part 7 – formations of the Beacon Group in the Victoria Valley region. *New Zealand Journal of Geology and Geophysics*, **5**, 278–294.

ALLIBONE, A. H. & WYSOCZANSKI, R. 2002. Initiation of magamatism during the Cambro–Ordovician Ross Orogeny in southern Victoria Land. *Geological Society of America Bulletin*, **114**, 1007–1018.

ALLIBONE, A. H., FORSYTH, P. J., SEWELL, R. J., TURNBULL, I. M. & BRADSHAW, M. A. 1991. *Geology of the Thundergut area, southern Victoria Land, Antarctica, scale 1:50 000*. New Zealand Geological Survey miscellaneous series map **15**.

ANDERSON, J. M. 1979. The geology of the Taylor Group, Beacon Supergroup, Byrd Glacier area, Antarctica. *New Zealand Antarctic Record*, **2**, 6–11.

ANGINO, E. E. & OWEN, D. E. 1962. Sedimentologic study of two members of the Beacon formation, Windy Gully, Victoria Land, Antarctica. *Transactions of the Kansas Academy of Science*, **65**, 61–69.

ASKIN, R. A., BARRETT, P. J., CURREEN, S. & YOUNG, G. 1971. Section E5 – Junction Spur. *In*: BARRETT, P. J. (ed.) *Stratigraphic sections of the Beacon Supergoup (Devonian and older (?) to Jurassic) in the Darwin Mountains, Antarctica*. Victoria University of Wellington, Wellington, Antarctic Data Series, **1**, 19–20.

AUCAMP, A. P. H., WOLMARANS, L. G. & NEETHLING, D. C. 1972. The Urfjell Group, a deformed (?) Early Palaeozoic sedimentary sequence, Kirwanveggan, western Dronning Maud Land. *In*: ADIE, R. A. (ed.) *Antarctic Geology and Geophysics*. Universitetsforlaget, Oslo, 557–562.

BARRETT, P. J. 1965. Geology of the area between the Axel Heiberg and Shackleton Glaciers, Queen Maud Range, Antarctica. *New Zealand Journal of Geology and Geophysics*, **8**, 344–370.

BARRETT, P. J. 1971. Stratigraphy and palaeogeography of the Beacon Supergrup in the Transantarctic Mountains, Antarctica. *Second Symposium on Gondwana Stratigraphy*, Capetown and Johannesburg, July 1970, 249–256.

BARRETT, P. J. 1981. History of the Ross Sea region during the deposition of the Beacon Supergroup 400–180 million years ago. *Journal of the Royal Society of New Zealand*, **11**, 447–458.

BARRETT, P. J. 1991. The Devonian to Jurassic Beacon Supergroup of the Transantarctic Mountains and correlatives in other parts of Antarctica. *In*: TINGEY, R. T. (ed.) *The Geology of Antarctica*. Clarendon Press, Oxford, 120–152.

BARRETT, P. J. & KOHN, B. P. 1971. Section BH – Beacon Heights. *In*: BARRETT, P. J. (ed.) *Stratigraphic Sections of the Beacon Supergroup (Devonian and older (?) to Jurassic) in South Victoria Land*. Victoria University of Wellington, Wellington, Antarctic Data Series, **2**, 67.

BARRETT, P. J. & KOHN, B. P. 1975. Changing sediment directions from Devonian to Triassic in the Beacon Super-Group of south Victoria Land. *In*: CAMPBELL,

K. S. W. (ed.) *Gondwana Geology*. Australian National University Press, Canberra, 15–35.

Barrett, P. J. & Webb, P. N. (eds) 1973. *Stratigraphic Sections of the Beacon Supergroup (Devonian and older (?) to Jurassic) in South Victoria Land*. Victoria University of Wellington, Wellington, Antarctic Data Series, **2**, 165.

Barrett, P. J., Elliot, D. H., Gunner, J. & Lindsay, J. F. 1968. Geology of the Beardmore Glacier area, Transantarctic Mountains. *US Antarctic Journal*, **3**, 102–106.

Barrett, P. J., Kohn, B. P., Askin, R. A. & McPherson, J. G. 1971. Preliminary report on Beacon Supergroup studies between the Hatherton and Mackay Glaciers, Antarctica. *New Zealand Journal of Geology and Geophysics*, **14**, 605–614.

Barrett, P. J., Grindley, G. W. & Webb, P. N. 1972. The Beacon Supergroup of East Antarctica. *In*: Adie, R. J. (ed.) *Antarctic Geology and Geophysics*. Universitetsforlaget, Oslo, 319–331.

Barrett, P. J., Kyle, R. A., Bamford, D. A., Bright, D. V. & Chinn, A. N. 1973. Section E1 – Rotunda. *In*: Barrett, P. J. & Webb, P. N. (eds) *Stratigraphic sections of the Beacon Supergroup (Devonian and older (?) to Jurassic) in south Victoria Land*. Victoria University of Wellington, Wellington, Antarctic Data Series, **3**, 90–96.

Barrett, P. J., Elliot, D. H. & Lindsay, J. F. 1986. The Beacon Supergroup (Devonian–Triassic) and Ferrar Group (Jurassic) in the Beardmore Glacier area, Antarctica. *In*: Turner, M. D. & Splettstoesser, J. (eds) *Geology of the Central Transantarctic Mountains*. American Geophysical Union, Washington, DC, Antarctic Research Series, **36**, 339–428.

Borg, S. G., DePaolo, D. J. & Smith, B. M. 1990. Isotopic structure and tectonics of the central Transantarctic Mountains. *Journal of Geophysical Research*, **95**, 6647–6667.

Boucot, A. J., Caster, K. E., Ives, D. & Talent, J. A. 1963*a*. Relationships of a new Lower Devonian terebratuloid (Brachiopoda) from Antarctica. *Bulletin of American Paleontology*, **46**, 81–151.

Boucot, A. J., Doumani, G. A., Johnson, J. G. & Webers, G. F. 1963*b*. The Devonian of Antarctica. *In*: Oswald, D. H. (ed.) *International Symposium on the Devonian System*. Alberta Society of Petroleum Geologists, Calgary, 639–648.

Bradshaw, M. A. 1981. Palaeoenvironmental interpretations and systematics of Devonian trace fossils from the Taylor Group (lower Beacon Supergroup), Antarctica. *New Zealand Journal of Geology and Geophysics*, **24**, 615–652.

Bradshaw, M. A. 1991. The Devonian Pacific margin of Antarctica. *In*: Thomson, M. R. A., Crame, J. A. & Thomson, J. W. (eds) *Geological Evolution of Antarctica*. Cambridge University Press, Cambridge, 193–197.

Bradshaw, M. A. 2002. A new ichnogenus *Catenarichnus* from the Devonian of the Ohio Range, Antarctica. *Antarctic Science*, **14**, 422–424.

Bradshaw, M. A. 2010. Devonian trace fossils of the Horlick formation, Ohio Range, Antarctica: systematic description and Palaeoenvironmental interpretation. *Ichnos*, **17**, 58–114.

Bradshaw, M. A. & Harmsen, F. J. 2007. *The paleoenvironmental significance of trace fossils in Devonian sediments (Taylor Group), Darwin Mountains to the Dry Valleys, southern Victoria Land*. US Geological Survey and the National Academies, Reston, VA, USGS OF-2007-1047, Extended Abstract, **133**, 1–5.

Bradshaw, M. A. & McCartan, L. 1983. The depositional environment of the Lower Devonian Horlick Formation, Ohio Range. *In*: Oliver, R. L., James, P. R. & Jago, J. B. (eds) *Antarctic Earth Science*. Australian Academy of Science, Canberra, 238–241.

Bradshaw, M. A. & McCartan, L. 1991. Palaeoecology and systematics of Lower Devonian bivalves from the Horlick Formation, Ohio Range, Antarctica. *Alcheringa*, **15**, 1–42.

Bradshaw, M. A. & Webers, G. 1988. The Devonian rocks of Antarctica. *In*: McMillan, N. J., Embry, A. F. & Glass, D. J. (eds) *Devonian of the World*. Canadian Society of Petroleum Geologists, Calgary, Memoirs, **14**, 783–795.

Bradshaw, M. A., Harmsen, F. J. & Kirkbride, M. P. 1990. Preliminary results of the 1988–1989 expedition to the Darwin Glacier area. *New Zealand Antarctic Record*, **10**, 28–48.

Bradshaw, M. A., Newman, J. & Aitchison, J. C. 2002. The sedimentary geology, palaeoenvironments and ichnocoenoses of the Lower Devonian Horlick formation, Ohio Range, Antarctica. *Antarctic Science*, **14**, 395–411.

Burgess, C. J., Palmer, A. & Anderson, J. M. 1981. The geology of the Fry Glacier area, south Victoria Land, Antarctica, with particular reference to the Taylor Group. *New Zealand Journal of Geology and Geophysics*, **24**, 373–388.

Carosi, R., Giacomini, F., Talarico, F. & Stump, E. 2007. *Geology of the Byrd Glacier Discontinuity (Ross Orogen): New Survey Data from the Britannia Range, Antarctica*. US Geological Survey and the National Academies, Reston, VA, USGS OF-2007-1047, Short Research Paper **030**, doi: 10.3133/of2007-1047.srp030

Cathcart, J. B. & Schmidt, D. L. 1977. *Middle Paleozoic Sedimentary Phosphate in the Pensacola Mountains, Antarctica*. US Geological Survey, Reston, VA, Professional Papers, **456-E**, E1–E18.

Cox, S. C., Parkinson, D. L., Allibone, A. H. & Cooper, A. F. 2000. Isotopic character of Cambro-Ordovician plutonism, southern Victoria Land, Antarctica. *New Zealand Journal of Geology and Geophysics*, **43**, 501–520.

Cox, S. C., Turnbull, I. M., Isaac, M. J., Townsend, D. & Smith Lyttle, B. (compilers). 2013. *Geology of the southern Victoria Land, Antarctica*. Institute of Geological and Nuclear Sciences 1:250 000 geological map **22**. GNS Science, Lower Hutt, New Zealand, in press.

Craw, D. & Findlay, R. H. 1984. Hydrothermal alteration of Lower Ordovician granitoids and Devonian Beacon Sandstone at Taylor Glacier, McMurdo Sound. *New Zealand Journal of Geology and Geophysics*, **27**, 465–475.

Curtis, M. L. 2001. Tectonic history of the Ellsworth Mountains, West Antarctica: reconciling a Gondwana enigma. *Geological Society of America Bulletin*, **113**, 939–958.

Curtis, M. L. & Storey, B. C. 2003. Early Palaeozoic near-surface deformation in the Neptune Range, Antarctica: implications for the Ross and Gondwanian orogenies. *Journal of the Geological Society, London*, **160**, 629–642.

Curtis, M. L., Millar, I. L., Storey, B. C. & Fanning, M. 2004. Structural and geochronological constraints of early Ross orogenic deformation in the Pensacola Mountains, Antarctica. *Geological Society of America Bulletin*, **116**, 619–636.

Dalziel, I. W. D. & Elliot, D. H. 1982. West Antarctica: problem child of Gondwanaland. *Tectonics*, **1**, 3–19.

Debenham, F. 1921. The sandstone, etc., of the McMurdo Sound, Terra Nova Bay, and Beardmore Glacier regions. British Antarctic (Terra Nova) Expedition, 1910. Natural History Report. *Geology*, **1**, 101–119.

Doumani, G. A., Boardman, R. S. *et al.* 1965. Lower Devonian fauna of the Horlick Formation, Ohio Range, Antarctica. *In*: Hadley, J. B. (ed.) *Geology and Paleontology of the Antarctic*. American Geophysical Union, Washington, DC, Antarctic Research Series, **6**, 241–281.

Duebendorfer, E. M. & Rees, M. N. 1998. Evidence for Cambrian deformation in the Ellsworth–Whitmore Mountain terrane, Antarctica: stratigraphic and tectonic implications. *Geology*, **26**, 55–58.

Edwards, W. N. 1928. The occurrence of *Glossopteris* in the Beacon Sandstone of Ferrar Glacier. *Geological Magazine*, **65**, 323–327.

Edwards, D. 1990. Silurian–Devonian paleobotany: problems, progress, and potential. *In*: Taylor, T. N. & Taylor, E. L. (eds) *Antarctic Paleobiology: Its role in the reconstruction of Gondwana*. Springer, Berlin, 89–101.

Elliot, D. H. 1986. Physical geography-geological evolution. *In*: Bonner, W. N. & Walton, D. W. H. (eds) *Key Environments – Antarctica*. Pergamon Press, Oxford, 39–61.

Federico, L., Crispini, L., Capponi, G. & Bradshaw, J. D. 2009. The Cambrian Ross Orogeny in northern Victoria Land (Antarctica) and New Zealand: a synthesis. *Gondwana Research*, **15**, 188–196.

Ferrar, H. T. 1907. Report on the field-geology of the region explored during the Discovery Antarctic Expedition, 1901–4. *In*: *National Antarctic Expedition 1901–1904, Natural History*. Trustees of the British Museum, London, **1**, Geology, 1–100.

Fioretti, A. M., Visona, D., Cavazzini, G. & Lombardo, B. 1997. Devonian magmatism: implications for the evolution of northern Victoria Land, and correlation with southeastern Australia and northeastern Tasmania. *In*: Ricci, C. A. (ed.) *The Antarctic Region: Geological Evolution and Processes*. Terra Antartica, Siena, 293–296.

Findlay, R. H. & Jordan, H. 1984. The volcanic rocks of Mount Black Prince and Lawrence Peaks, north Victoria Land, Antarctica. *Geologisches Jahrbuch*, **B60**, 143–151.

Gilmer, G. J. 2008. *Paleoenvironmental interpretations of the Lower Taylor Group, Olympus Range area, southern Victoria Land, Antarctica*. MSc thesis, University of Canterbury.

Goldstrand, P. M., Fitzgerald, P. G., Redfield, T. F., Stump, E. & Hobbs, C. 1994. Stratigraphic evidence for the Ross Orogeny in the Ellsworth Mountains, West Antarctica: implication for the evolution of the paleo-Pacific margin of Gondwana. *Geology*, **22**, 427–430.

Goodge, J. W. 2007. Metamorphism in the Ross Orogen and its bearing on Gondwana margin tectonics. *In*: Cloos, M., Carlson, W. D., Gilbert, M. C., Liou, J. G. & Sorensen, S. S. (eds) *Convergent Margin Tectonics and Associated Regions: A tribute to W. G. Ernst*. Geological Society of America, Boulder, CO, Special Papers, **419**, 185–203.

Gould, L. M. 1931. Some geographical results of the Byrd Antarctic Expedition. *The Geographical Review*, **21**, 177–200.

Grindley, G. W. 1963. The geology of the Queen Alexandra Range, Beardmore Glacier, Ross Dependency, Antarctica; with notes on the correlation of Gondwana sequences. *New Zealand Journal of Geology and Geophysics*, **6**, 307–347.

Grindley, G. W. & Mildenhall, D. C. 1981. Geological background to a Devonian plant fossil discovery, Ruppert Coast, Marie Byrd Land, West Antarctica. *In*: Cresswell, M. M. & Vella, P. (eds) *Gondwana Five*. Balkema, Rotterdam, 23–30.

Gunn, B. M. & Warren, G. 1962. Geology of Victoria Land between the Mawson and Mulock Glaciers, Ross Dependency, Antarctica. *New Zealand Geological Survey Bulletin*, **71**, 1–157.

Hamilton, W. & Hayes, P. T. 1963. *Type Section of the Beacon Sandstone of Antarctica*. US Geological Survey, Reston, VA, Professional Papers, **456-A**, A1–A18.

Haq, B. U. & Schutter, S. R. 2008. A chronology of Paleozoic sea-level changes. *Science*, **322**, 64–68.

Harrington, H. J. 1958. Nomenclature of rock units in the Ross Sea region, Antarctica. *Nature*, **182**, 290.

Harrington, H. J. 1965. The geology and morphology of Antarctica. *In*: van Mieghem, J., van Oye, P. & Schell, E. (eds) *Biogeography and Ecology in Antarctica*. Springer, The Hague, 20–37.

Harrington, H. J. & Speden, I. G. 1962. Section through the Beacon Sandstone at Beacon Height West, Antarctica. *New Zealand Journal of Geology and Geophysics*, **5**, 707–717.

Haskell, T. R., Kennett, P. & Prebble, W. M. 1965. Geology of the Brown Hills and Darwin Mountains, southern Victoria Land, Antarctica. *Transactions of the Royal Society of New Zealand Geology*, **2**, 231–248.

Helby, R. J. & McElroy, C. T. 1969. Microfloras from the Devonian and Triassic of the Beacon Group, Antarctica. *New Zealand Journal of Geology and Geophysics*, **12**, 376–82.

Hunter, M. A. & Lomas, S. A. 2003. Reconstructing the Siluro–Devonian coastline of Gondwana: insights from sedimentology of the Port Stephens Formation, Falkland Islands. *Journal of the Geological Society, London*, **160**, 459–476.

Isaac, M. J., Chinn, T. J., Edbrooke, S. W. & Forsythe, P. J. 1996. *Geology of the Olympic Range area, southern Victoria Land, Antarctica scale 1:50 000*. Institute of Geological and Nuclear Sciences geological map **20**.

ISBELL, J. L. 1999. The Kukri Erosion Surface; a reassessment of its relationship to rocks of the Beacon Supergroup in the central Transantarctic Mountains, Antarctica. *Antarctic Science*, **11**, 228–239.

ISBELL, J. L., MILLER, M. F., BABCOCK, L. E. & HASIOTIS, S. T. 2001. Ice-marginal environment and ecosystem prior to initial advance of the late Palaeozoic ice sheet in the Mount Butters area of the central Transantarctic Mountains, Antarctica. *Sedimentology*, **48**, 953–970.

KATZ, H. R. & WATERHOUSE, B. C. 1970. Geological reconnaissance of the Scott Glacier area, southeastern Queen Maud Range, Antarctica. *New Zealand Journal of Geology and Geophysics*, **13**, 1030–1037.

KEMP, E. M. 1972. Lower Devonian palynomorphs from the Horlick Formation, Ohio Range, Antarctica. *Palaeontographica Abt B*, **139**, 105–124.

KOHN, B. P. & MCPHERSON, J. G. 1973. Section BS – Mt Boreas. *In*: BARRETT, P. J. & WEBB, P. N. (eds) *Stratigraphic Sections of the Beacon Supergroup (Devonian and older (?) to Jurassic) in South Victoria Land.* Victoria University of Wellington, Wellington, Antarctic Data Series, **3**, 57–59.

KYLE, R. A. 1977. Devonian palynomorphs from the basal Beacon Supergroup of south Victoria Land, Antarctica. *New Zealand Journal of Geology and Geophysics*, **20**, 1147–1150.

LAIRD, M. G. 1991*a*. The Late Proterozoic–Middle Palaeozoic rocks of Antarctica. *In*: TINGEY, R.T. (ed.) *The Geology of Antarctica*. Clarendon Press, Oxford, 74–119.

LAIRD, M. G. 1991*b*. Lower to mid-Palaeozoic sedimentation and tectonic patterns on the palaeo-Pacific margin of Antarctica. *In*: THOMSON, M. R. A., CRAME, J. A. & THOMSON, J. W. (eds) *Geological Evolution of Antarctica*. Cambridge University Press, Cambridge, 177–185.

LAIRD, M. G., MANSERGH, G. D. & CHAPELL, J. M. A. 1971. Geology of the central Nimrod Glacier area, Antarctica. *New Zealand Journal of Geology and Geophysics*, **14**, 427–68.

LAUDON, T. S. 1991. Petrology of sedimentary rocks from the English Coast, eastern Ellsworth Land. *In*: THOMSON, M. R. A., CRAME, J. A. & THOMSON, J. W. (eds) *Geological Evolution of Antarctica*. Cambridge University Press, Cambridge, 455–460.

LINDSAY, J. F. 1969. *Stratigraphy and sedimentation of lower Beacon rocks in the central Transantarctic Mountains, Antarctica*. Institute of Polar Studies, Ohio State University, Report **33**, 1–58.

LINDSAY, J. F. 1970. Paleozoic cave deposit in the Central Transantarctic Mountains. *New Zealand Journal of Geology and Geophysics*, **13**, 1018–1049.

LONG, W. E. 1965*a*. Stratigraphy of the Ohio Range, Antarctica. *In*: HADLEY, J. B. (ed.) *Geology and Paleontology of the Antarctic*. American Geophysical Union, Washington, DC, Antarctic Research Series, **6**, 71–116.

LONG, W. E. 1965*b*. The Stratigraphy of the Horlick Mountains. *In*: ADIE, R. J. (ed.) *Antarctic Geology*. North-Holland, Amsterdam, 352–363.

LONG, J. A. 1989. Biogeographic and biostratigraphic significance of the Devonian fish fauna from the Aztec Siltstone, Antarctica. *New Zealand Antarctic Record*, **9**, 45–51.

LONG, J. A. 1995. A new groenlandaspidid arthrodire (Pices; Placodermi) from the Middle Devonian Aztec Siltstone, southern Victoria Land, Antarctica. *Records of the Western Australian Museum*, **17**, 35–41.

LONG, J. A. & YOUNG, G. C. 1995. Sharks from the Middle–Late Devonian Aztec Siltstone, southern Victoria Land, Antarctica. *Records of the Western Australian Museum*, **17**, 287–308.

MATSCH, C. L. & OJAKANGAS, R. W. 1992. Stratigraphy and sedimentology of the Whiteout Conglomerate; An upper Paleozoic glacigenic unit, Ellsworth Mountains, West Antarctica. *In*: WEBERS, G. F., CRADDOCK, C. & SPLETTSTOESSER, J. F. (eds) *Geology and Paleontology of the Ellsworth Mountains, West Antarctica.* Geological Society of America, Boulder, CO, Memoirs, **170**, 37–62.

MAWSON, D. 1940. *Sedimentary rocks*. Australasian Antarctic Expedition 1911–1914. Scientific Reports Series A, **4**, 349–367.

MCCARTAN, L. & BRADSHAW, M. A. 1987. Petrology and sedimentology of the Horlick Formation (Lower Devonian), Ohio Range, Transantarctic Mountains. *United States Geological Survey Bulletin*, **1780**, 1–31.

MCCARTAN, L., BRADSHAW, M. A. & KELLOGG, K. S. 1986. *Reconnaissance geologic map of the Ohio Range, Transantarctic Mountains, Antarctica, scale 1:50 000*. United States Geological Survey geologic map **A-13**.

MCELROY, C. T. 1969. Comparative lithostratigraphy of Gondwana sequences, eastern Australia and Antarctica. *In*: AMOS, A. H. (ed.) *Gondwana Stratigraphy*. IUGS Symposium, Buenos Aires. UNESCO, Paris, 441–466.

MCELROY, C. T. & ROSE, G. 1987. *Geology of the Beacon Heights area, southern Victoria Land, Antarctica. 1:50 000*. New Zealand Geological Survey miscellaneous series map **15**.

MCELROY, C. T., ROSE, G. & BRYAN, J. H. 1967. A unique occurrence of deformed sedimentary rocks of the Beacon Group, Antarctica. *Antarctic Journal of the United States*, **2**, 241–244.

MCKELVEY, B. C. & WEBB, P. N. 1959. Geological investigations in south Victoria Land, Antarctica; part 2 – geology of upper Taylor Glacier region. *New Zealand Journal of Geology and Geophysics*, **2**, 718–728.

MCKELVEY, B. C., WEBB, P. N., GORTON, M. P. & KOHN, B. P. 1970. Stratigraphy of the Beacon Supergroup between the Olympus and Boomerang Ranges, Victoria Land, Antarctica. *Nature*, **227**, 1126–1128.

MCKELVEY, B. C., WEBB, P. N., GORTON, M. P. & KOHN, B. P. 1972. Stratigraphy of the Beacon Supergroup between the Olympus and Boomerang Ranges, Victoria Land. *In*: ADIE, R. J. (ed.) *Antarctic Geology and Geophysics*. Universitetsforlaget, Oslo, 345–352.

MCKELVEY, B. C., WEBB, P. N. & KOHN, B. P. 1977. Stratigraphy of the Taylor Group and lower Victoria Groups (Beacon Supergroup) between the Mackay Glacier and Boomerang Range, Antarctica. *New Zealand Journal of Geology and Geophysics*, **20**, 813–863.

McLoughlin, S. & Long, J. A. 1994. New records of Devonian plants from southern Victoria Land, Antarctica. *Geological Magazine*, **131**, 81–90.

McPherson, J. G. 1978. Stratigraphy and sedimentology of the Upper Devonian Aztec Siltstone, southern Victoria Land, Antarctica. *New Zealand Journal of Geology and Geophysics*, **21**, 667–683.

McPherson, J. G. 1979. Calcrete (caliche) palaeosols in fluvial redbeds of the Aztec Siltstone (Upper Devonian), southern Victoria Land, Antarctica. *Sedimentary Geology*, **22**, 267–285.

Mildenhall, D. C. & Schopf, J. M. 1980. A Mid-Late Devonian flora from the Ruppert Coast, Marie Byrd Land, West Antarctica. *Journal of the Royal Society of New Zealand*, **10**, 271–285, doi. org/10.1080/03036758.1980.10415333

Mirsky, A. 1964. Reconsideration of the 'Beacon' as a stratigraphic name in Antarctica. *In*: Adie, R. J. (ed.) *Antarctic Geology*. North-Holland, Amsterdam, 364–378.

Mukasa, S. B. & Dalziel, I. W. D. 2000. Marie Byrd Land, West Antarctica: evolution of Gondwana's Pacific margin constrained by zircon U–Pb geochronology and feldspar common-Pb isotopic compositions. *Geological Society of America Bulletin*, **112**, 611–627.

Ojakangas, R. W. & Matsch, C. L. 1981. The Late Palaeozoic Whiteout Conglomerate; A glacial and glaciomarine sequence in the Ellsworth Mountains, West Antarctica. *In*: Hambrey, M. J. & Harland, W. B. (eds) *Earth's pre-Pleistocene Glacial Record*. Cambridge University Press, Cambridge, 241–244.

O'Toole, T. F. 2010. *The lower Taylor Group: Taylor and Wright Valleys, southern Victoria Land; Palaeoenvironmental interpretations and sequence stratigraphy*. MSc thesis, University of Canterbury.

Pankhurst, R. J., Marsh, P. D. & Clarkson, P. D. 1983. A geochronological investigation of the Shackleton Range. *In*: Oliver, R. L., James, P. R. & Jago, J. B. (eds) *Antarctic Earth Science*. Australian Academy of Science, Canberra, 176–182.

Pankhurst, R. J., Weaver, S. D., Bradshaw, J. D., Storey, B. C. & Ireland, T. R. 1998. Geochronology and geochemistry of pre-Jurassic superterranes in Marie Byrd Land, Antarctica. *Journal of Geophysical Research*, **103**, 2529–2547.

Playford, G. 1990. Proterozoic and Paleozoic palynology of Antarctica: a review. *In*: Taylor, T. N. & Taylor, E. L. (eds) *Antarctic Paleobiology: Its Role in the Reconstruction of Gondwana*. Springer, Berlin, 51–70.

Plume, R. W. 1978. A revision of the existing stratigraphy of the New Mountain Sandstone (Beacon Supergroup), South Victoria Land, Antarctica. *New Zealand Journal of Geology and Geophysics*, **21**, 167–173.

Plume, R. W. 1982. Sedimentology and paleocurrent analysis of the basal part of the Beacon Supergroup (Devonian [and older?] to Triassic) in south Victoria Land. *In*: Craddock, C. (ed.) *Antarctic Geoscience*. University of Wisconsin Press, Madison, WI, 571–580.

Plumstead, E. P. 1962. *Fossil floras of Antarctica*. Trans-Antarctic Expedition 1955–1958, Scientific Report 9 (Geology), 1–132.

Plumstead, E. P. 1964. Palaeobotany of Antarctica. *In*: Aidie, R. J. (ed.) *Antarctic Geology*. North-Holland, Amsterdam, 637–654.

Pocknall, D. T., Chinn, T. J., Sykes, R. & Skinner, D. N. B. 1994. *Geology of the Convoy Range area, southern Victoria Land scale 1:50 000*. Institute of Geological and Nuclear Sciences geological map **11**.

Rapela, C. W., Pankhurst, R. J., Fanning, C. M. & Hervé, F. 2005. Pacific subduction coeval with the Karoo mantle plume: the Early Jurassic subcordilleran belt of northwestern Patagonia. *In*: Vaughan, A. M. P., Leat, P. T. & Pankhurst, R. J. (eds) *Terrane Processes at the Margins of Gondwana*. Geological Society, London, 217–240.

Rigby, J. F. & Schopf, J. M. 1969. Stratigraphic implications of Antarctic paleobotanical studies. *In*: Amos, A. J. (ed.) *Gondwana Stratigraphy*. UNESCO, Paris, **2** (Earth Sciences), 91–106.

Savage, J. E. 2005. *Provenance analysis of the Sperm Bluff Formation, southern Victoria Land, Antarctica*. MSc thesis, University of Canterbury.

Savage, J. E., Bradshaw, M. A. & Bassett, K. N. In press. Marginal marine depositional setting and correlation of the Devonian Sperm Bluff Formation (Taylor Group), southern Victoria Land, Antarctica. Antarctic Science, doi:10.1017/S0954102013000205

Schmidt, D. L. & Ford, A. B. 1969. *Geology of the Pensacola and Thiel Mountains*. Antarctic Map Folio Series, Folio 12 Geology, Plate 5.

Schmidt, D. L., Dover, J. H., Ford, A. B. & Brown, R. D. 1964. Geology of the Patuxent Mountains. *In*: Adie, R. J. (ed.) *Antarctic Geology*. North-Holland, Amsterdam, 276–283.

Schmidt, D. L., Williams, P. L., Nelson, W. H. & Ege, J. R. 1965. *Upper Precambrian and Paleozoic Stratigraphy and Structure of the Neptune Range, Antarctica*. US Geological Survey, Reston, VA, Professional Papers, **525-D**, D112–D119.

Schopf, J. M. 1968. Studies in Antarctic Paleobotany. *Antarctic Journal*, **20**, 176.

Scott, L., Anderson, H. M. & Anderson, J. M. 1997. Vegetation history. *In*: Cowling, R. M., Richardson, D. M. & Pierce, S. M. (eds) *Vegetation of Southern Africa*. Cambridge University Press, Cambridge, 62–84.

Seward, A. C. 1914. *Antarctic fossil plants*. British Antarctica (Terra Nova) Expedition, Natural History Reports, Geology 1.

Shergold, J. H. & Webers, G. F. 1992. Late Dresbachian (Idamean) and other trilobite faunas from the Heritage Range, Ellsworth Mountains, West Antarctica. *In*: Webers, G. F., Craddock, C. & Splettstoesser, J. F. (eds) *Geology and Paleontology of the Ellsworth Mountains, West Antarctica*. Geological Society of America, Boulder, CO, Memoirs, **170**, 125–168.

Skinner, D. N. B. 1964. A summary of the Geology of the region between Byrd and Starshot Glaciers, south Victoria Land. *In*: Adie, R. J. (ed.) *Antarctic Geology*. North-Holland, Amsterdam, 284–292.

Skinner, D. N. B. 1965. Petrographic criteria of the rock units between the Byrd and Starshot Glaciers, south Victoria Land, Antarctica. *New Zealand Journal of Geology and Geophysics*, **8**, 292–303.

SPÖRLI, K. B. 1992. Stratigraphy of the Crashsite Group, Ellsworth Mountains, West Antarctica. *In*: WEBERS, G. F., CRADDOCK, C. & SPLETTSTOESSER, J. F. (eds) *Geology and Paleontology of the Ellsworth Mountains, West Antarctica*. Geological Society of America, Boulder, CO, Memoirs, **170**, 21–35.

STUMP, E., EDGERTON, D. G. & KORSCH, R. J. 2002. Geological relationships at Cotton Plateau, Nimrod Glacier area, bearing on the tectonic development of the Ross Orogen, Transantarctic Mountains, Antarctica. *Terra Antarctica*, **9**, 3–18.

STUMP, E., GOOTEE, B. F. & TALARICO, F. 2006. Tectonic models for the development of the Byrd Glacier discontinuity and the surrounding regions of the Transantarctic Mountains during the Neoproterozoic-Early Paleozoic. *In*: FUTTERER, D. K., DAMASKE, D., KLEINSCHMIDT, G., MILLER, H. & TESSENSOHN, F. (eds) *Antarctic Contributions to Global Earth Science*. Springer, Berlin, 181–194.

SYKES, R. & POCKNALL, D. T. 1991. The stratigraphy of Taylor Group (Devonian) in the Convoy Range, southern Victoria Land, Antarctica. *New Zealand Geological Survey Record*, **43**, 129–135.

TREWIN, N. H. 1993. *Mixed Aeolian Sandsheet and Fluvial Deposits in the Tumblagooda Sandstone, Western Australia*. Geological Society, London, Special Publications, **73**, 219–230.

TURNBULL, I. M., ALLIBONE, A. H., FORSYTH, P. J. & HERON, D. W. 1994. *Geology of the Bull Pass–St Johns Range area, southern Victoria Land, Antarctica scale 1:50 000*. Institute of Geological and Nuclear Sciences geological map **14**.

TURNER, S. & YOUNG, G. C. 1992. Thelodont scales from the Middle–Late Devonian Aztec Siltstone, southern Victoria Land, Antarctica. *Antarctic Science*, **4**, 89–105.

VANDENBERG, A. H. M., WILLMAN, C. E. *ET AL.* 2000. *The Tasman Fold belt in Victoria*. Geological Survey of Victoria, Melbourne, Special Publications.

WADE, F. A., YEATS, V. L., EVERETT, J. R., GREENLEE, D. W., LAPRADE, K. E. & SHENK, J. C. 1965. Geology of the central portion of the Queen Maud Range, Transantarctic Mountains. *Science*, **150**, 1808–1809.

WEAVER, S. D., BRADSHAW, J. D. & ADAMS, C. J. 1991. Granitoids of the Ford Ranges, Marie Byrd Land, Antarctica. *In*: THOMSON, M. R., CRAME, J. A. & THOMSON, J. W. (eds) *Geological Evolution of Antarctica*. Cambridge University Press, Cambridge, 345–351.

WEBB, P. N. 1963. Beacon Group of the Wright Valley and Taylor Glacier region. *New Zealand Journal of Geology and Geophysics*, **6**, 361–387.

WEBB, P. N. & GORTON, M. P. 1973. Section 12 – Mt Handsley. *In*: BARRETT, P. J. & WEBB, P. N. (eds) *Stratigraphic Sections of the Beacon Supergroup (Devonian and older (?) to Jurassic) in south Victoria Land*. Victoria University of Wellington, Wellington. Antarctic Data Series, **3**, 26–27.

WEBERS, G. F. & SPÖRLI, K. B. 1983. Palaeontological and stratigraphic investigations in the Ellsworth Mountains, West Antarctica. *In*: OLIVER, R. L., JAMES, P. R. & JAGO, J. B. (eds) *Antarctic Earth Science*. Australian Academy of Science, Canberra, 261–264.

WEBERS, G. F., GLENISTER, B., POJETA JR, J. & YOUNG, G. 1992. Devonian fossils from the Ellsworth Mountains, West Antarctica. *In*: WEBERS, G. F., CRADDOCK, C. & SPLETTSTOESSER, J. F. (eds) *Geology and Paleontology of the Ellsworth Mountains, West Antarctica*. Geological Society of America, Boulder, CO, Memoirs, **170**, 269–278.

WHITE, E. I. 1968. Devonian fishes of the Mawson–Mulock area, Victoria Land, Antarctica. Trans-Antarctic expedition 1955–1958, Scientific Reports. *Geology*, **16**, 1–26.

WILLIAMS, P. L. 1969. Petrology of Upper Cambrian and Paleozoic sandstones in the Pensacola Mountains, Antarctica. *Journal of Sedimentary Petrology*, **39**, 1455–1465.

WIZEVICH, M. C. 1997. Fluvial-eolian deposits in the Devonian New Mountain Sandstone, Table Mountain, southern Victoria Land, Antarctica: sedimentary architecture, genesis and stratigraphic evolution. *In*: RICCI, C. A. (ed.) *The Antarctic Region: Geological Evolution and Processes*. Terra Antarctica, Siena, 933–944.

WOLMARANS, L. G. & KENT, L. E. 1982. Geological investigations in western Dronning Maud Land, Antarctica – a synthesis. *South African Journal of Antarctic Research*, **14**, 1–93.

WOODWARD, A. S. 1921. *Fish-remains from the Upper Old Red Sandstone of Granite Harbour, Antarctica*. British Antarctic (Terra Nova) Expedition, 1910. Natural History Report, Geology **1**, 51–62.

WOOLFE, K. J. 1993. Devonian depositional environments in the Darwin Mountains: marine or non-marine? *Antarctic Science*, **5**, 211–220.

WOOLFE, K. J. & BARRETT, P. J. 1995. Constraining the Devonian to Triassic Tectonic evolution of the Ross Sea sector. *Terra Antarctica*, **2**, 7–21.

WOOLFE, K. J., KIRK, P. A. & SHERWOOD, A. M. 1989. *Geology of the Knobhead area, southern Victoria Land, Antarctica scale 1:50 000*. New Zealand Geological Survey miscellaneous series map **19**.

WOOLFE, K. J., LONG, J. A., BRADSHAW, M. A., HARMSEN, F. J. & KIRKBRIDE, M. P. 1990. Fish-bearing Aztec Siltstone (Devonian) in the Cook Mountains, Antarctica. *New Zealand Journal of Geology and Geophysics*, **33**, 511–514.

WOOLFE, K. J., ARNOTT, M. J. & BRADLEY, G. M. 1995. Jurassic titaniferous ironstone in a Devonian host: pivot peak coal measures expunged. *Antarctic Science*, **7**, 293–301.

WYSOCZANSKI, R. J., FORSYTH, P. J. & WOOLFE, K. J. 2003. Zircon dating and provenance of rhyolitic clasts in Beacon conglomerate, southern Victoria Land, Antarctica. *Terra Antarctica*, **10**, 67–80.

YOSHIDA, M. 1982. Superposed deformation and its implications to the geologic history of the Ellsworth Mountains, West Antarctica. *Tokyo National Institute of Polar Studies, Memoirs, Special Issue*, **21**, 120–171.

YOUNG, V. T. 1986. Early Devonian fish material from the Horlick Formation, Ohio Range Antarctica. *Alcheringa*, **10**, 35–44.

YOUNG, G. C. 1988. Antiarchs (placoderm fishes) from the Devonian Aztec Siltstone, southern Victoria Land. *Palaeontographica (A)*, **202**, 1–125.

YOUNG, G. C. 1989. The Aztec fish fauna of southern Victoria Land – evolutionary and biogeographic significance. *In*: CRAME, J. A. (ed.) *Origins and Evolution of the Antarctic biota*. Geological Society, London, Special Publications, **47**, 43–62.

YOUNG, G. C. 1993. Middle Palaeozoic macrovertebrate biostratigraphy of Eastern Gondwana. *In*: LONG, J. A. (ed.) *Palaeozoic Vertebrate Biostratigraphy and Biogeography*. Belhaven Press, London, 208–251.

YOUNG, G. C. 1996. Devonian (chart 4). *In*: YOUNG, G. C. & LAURIE, J. R. (eds) *An Australian Phanerozoic timescale*. Oxford University Press, Melbourne, 96–109.

YOUNG, G. C. & LONG, J. A. 2005. Phyllolepid placoderm fish remains from the Devonian Aztec Siltstone, southern Victoria Land, Antarctica. *Antarctic Science*, **17**, 387–408.

ZELLER, E. J., ANGINO, E. E. & TURNER, M. D. 1961. Basal sedimentary section at Windy Gully, Taylor Glacier, Victoria Land, Antarctica. *Geological Society of America Bulletin*, **72**, 781–786.

Late Cretaceous dinosaurs from the James Ross Basin, West Antarctica

MARCELO A. REGUERO[1,3,4*], CLAUDIA P. TAMBUSSI[1,4], RODOLFO A. CORIA[2,4] & SERGIO A. MARENSSI[3,4]

[1]*División Paleontología de Vertebrados, Museo de La Plata, Paseo del Bosque s/n, B1900FWA, La Plata, Argentina*

[2]*Universidad Nacional de Río Negro, Sede Alto Valle – Subsecretaría de Cultura de Neuquén – Museo Carmen Funes, Av. Córdoba 55 (8318), Plaza Huincul, Neuquén, Argentina*

[3]*Instituto Antártico Argentino, Cerrito 1248, C1010AAZ Ciudad Autónoma de Buenos Aires, Argentina*

[4]*Consejo Nacional de Investigaciones Científicas y Técnicas (CONICET), Rivadavia 1917, C1033AAJ Ciudad Autónoma de Buenos Aires, Argentina*

**Corresponding author (e-mail: regui@fcnym.unlp.edu.ar)*

Abstract: The fossil record of terrestrial vertebrates from the Late Cretaceous of Antarctica is currently composed of non-avian and avian dinosaurs from the marine sediments of the James Ross Basin, Antarctic Peninsula (West Antarctica). Although two dinosaurian formational assemblages (Late Campanian/Early Maastrichtian and Late Maastrichtian) are known, the record is still scattered, and evolutionary scenarios are tentative. Ten non-avian dinosaurs have been reported from Coniacian to Maastrichtian deposits, along with possible sauropod footprints of Early Maastrichtian age from Snow Hill Island. Five avian dinosaurs have been recorded or described exclusively from the Maastrichtian. The presence of an advanced titanosaur with characteristic procoelous mid-caudal vertebrae in Snow Hill Island Formation at Santa Marta Cove implies that the group achieved a global distribution by the Late Campanian. The Late Campanian/Early Maastrichtian non-avian dinosaur (ankylosaurs, ornithopods and dromaeosaurid theropods) clades probably attained a near-cosmopolitan distribution before the Late Cretaceous, and some aspects of this hallmark 'Gondwanan' fauna may therefore reflect climate-driven provinciality, not vicariant evolution driven by continental fragmentation. Antarctic Late Cretaceous avian dinosaurs are rare. They are restricted to the Maastrichtian and consist of a cariamid?, gaviids, a charadriiform and the basal Anseriformes *Vegavis*, and provide the first strong evidence for a basal radiation of birds known to exist in the Cretaceous.

Although the Late Cretaceous marine deposits of the James Ross Basin have been extensively surveyed by vertebrate palaeontologists, they have mostly recovered marine reptiles (del Valle *et al.* 1977; Gasparini *et al.* 1984; Chatterjee & Small 1989; Martin *et al.* 2002; Novas *et al.* 2002*a*; Martin 2006; Martin & Crame 2006; Reguero & Gasparini 2007; de la Fuente *et al.* 2010) and other marine vertebrate groups (Woodward 1908; Cione & Medina 1987; Grande & Chatterjee 1987; Richter & Ward 1990; Stahl & Chatterjee 1999, 2002; Kriwet 2003; Kriwet *et al.* 2006; Figs 1 & 2). In contrast, fewer terrestrial vertebrates have been discovered, and these are currently represented only by dinosaurs.

From 2008 to 2011, James Ross Basin was the scene of concentrated palaeontological and geological fieldwork. New dinosaur remains were found, notably in the Snow Hill Island Formation at Santa Marta Cove, James Ross Island (Cerda *et al.* 2012; Coria *et al.* 2013) (Fig. 1b). A significant amount of taxonomic and palaeobiological information has been gained from these fossils, which are important for understanding their phylogenetic relationships with other Gondwanan and Nearctic palaeofaunas, comprehending vertebrate biostratigraphy and revealing the role of the Antarctic Peninsula in the terrestrial vertebrate dispersal routes.

Late Cretaceous sedimentary rocks are exposed only around the northern part of West Antarctica, on the South Shetland Islands and in the area of James Ross Island, Weddell Sea, with dinosaur remains found only in the latter area. The sequence there is made up of marine clastic rocks deposited within a back-arc basin (Elliot 1988; Hathway

From: Hambrey, M. J., Barker, P. F., Barrett, P. J., Bowman, V., Davies, B., Smellie, J. L. & Tranter, M. (eds) 2013. *Antarctic Palaeoenvironments and Earth-Surface Processes*. Geological Society, London, Special Publications, **381**, 99–116. First published online July 24, 2013, http://dx.doi.org/10.1144/SP381.20

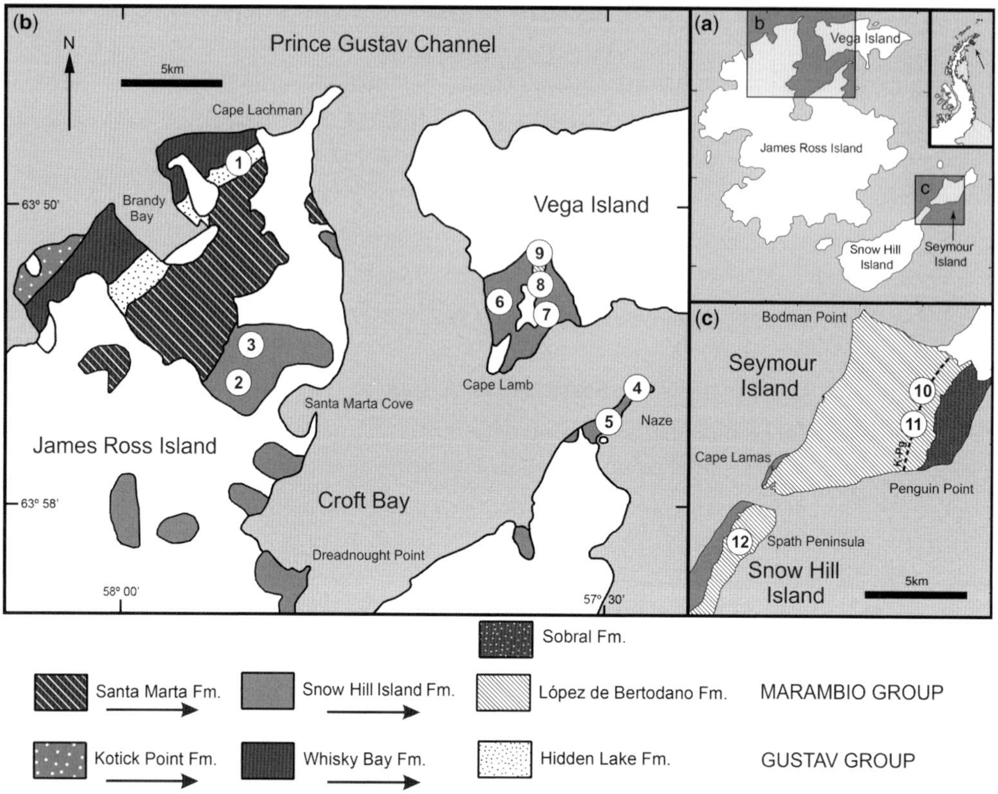

Fig. 1. Schematic geological maps of key dinosaur locations in James Ross Basin, north-eastern Antarctic Peninsula (see locality map on right). (**a**) Locality map with areas in B and C shaded. White areas are either the James Ross Island Volcanic Group or snow/ice cover. (**b**) NE James Ross and SW Vega islands. Key dinosaur locations: 1, Brandy Bay, Cape Lachman, James Ross Island, Hidden Lake Formation (Coniacian), megalasaur-like theropod site; 2, Santa Marta Cove, Snow Hill Island Formation (Late Campanian), *Antarctopelta* and ornithopod gen. et sp. nov. sites; 3, Santa Marta Cove, Snow Hill Island Formation (Late Campanian), titanosaur site; 4, the Naze, James Ross Island, Terrapin Hill, Cape Lamb Member (Early Maastrichtian), Snow Hill Island Formation, dromaeosaur site; 5, the Naze, Fortress Hill, James Ross Island, Cape Lamb Member (Early Maastrichtian), Snow Hill Island Formation, iguanodontid site; 6, Cape Lamb, Vega Island, Cape Lamb Member (Early Maastrichtian), Snow Hill Island Formation, charadriform site; 7, Cape Lamb South, Vega Island, Cape Lamb Member (Early Maastrichtian), Snow Hill Island Formation, hypsilophodontid site; 8, Sandwich Bluff, Vega Island, Sandwich Bluff Member (Late Maastrichtian), López de Bertodano Formation, *Vegavis, Polarornis* and cariamid? sites; 9, Sandwich Bluff, Vega Island, Sandwich Bluff Member (Late Maastrichtian), López de Bertodano Formation, 'Reptile horizon', hadrosaur site. (**c**) NE Snow Hill and SW Seymour islands. Key dinosaur locations: 10, Seymour Island, Unit Klb 9 (Late Maastrichtian), López de Bertodano Formation, hadrosaur site; 11, Seymour Island, Unit Klb 9 (Late Maastrichtian), López de Bertodano Formation, gaviiform site; 12, Spath Peninsula, Tessore Hill, Snow Hill Island, and López de Bertodano Formation (Early Maastrichtian), sauropod? footprint site. Position of the Cretaceous–Palaeogene boundary on Seymour Island indicated by dashed line.

2000) and represents the topmost beds of a regressive mega-sequence (Pirrie *et al.* 1991).

The Late Cretaceous non-avian and avian dinosaurs of the James Ross Basin, Antarctic Peninsula, are numerically scarce and some of them are poorly known or their studies are pending. It should be emphasized that all of these terrestrial taxa were deposited in shallow-marine settings and are thus found more or less where they once lived.

The aim of this paper is to review the current state of the stratigraphical record and biogeography of Antarctic Late Cretaceous terrestrial vertebrates.

Stratigraphy of James Ross Basin

The Late Cretaceous–Palaeogene strata of the James Ross Basin (Figs 1 & 2) comprise a thick sequence of marine sedimentary rocks divided into

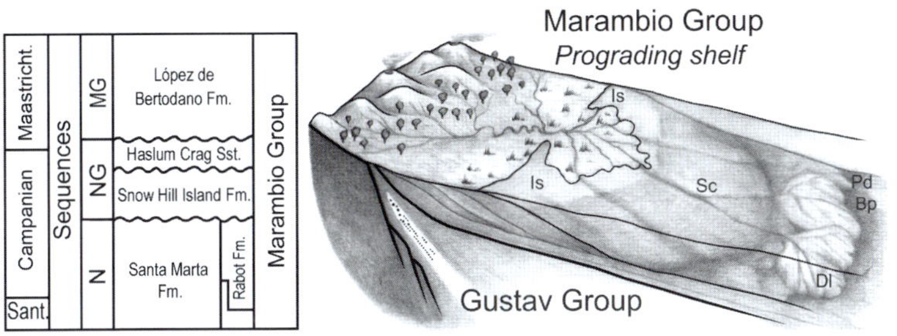

Fig. 2. Relationship between sequences and formations of the Marambio Group, a cartoon illustrating its prograding shelf setting (from Olivero 2012). Pd-Bp, prodelta–basin plain settings; Dl, depositional lobes; Sc, slope–channel complex; Is, inner shelf settings.

three lithostratigraphic groups: the basal Gustav Group (Aptian–Coniacian), the intermediate Marambio Group (Santonian–Danian), and the upper Seymour Island Group (Palaeogene; e.g. Rinaldi 1982; Crame *et al.* 1991; Riding & Crame 2002; Fig. 3).

The 2100 m-thick marine sediments of the Gustav Group are exposed along the NW coast of James Ross Island (Fig. 1b). The coarse-grained lower three units of the group (Lagrelius Point, Kotick Point and Whisky Bay formations) were deposited in a deep marine setting and represent proximal submarine-fan and slope-apron depositional systems (Ineson 1989; Buatois & López-Angrinman 1992). The fourth and youngest unit, Hidden Lake Formation, is composed of distinctive brown-weathering, volcaniclastic sediments, deposited on a fan delta, and readily traced along the length of NW James Ross Island (Whitham *et al.* 2006).

The finer-grained Marambio Group is exposed in James Ross, Vega, Humps, Snow Hill, Seymour and Cockburn islands, and has been recently summarized by Olivero (2012). The Group has a stratigraphic thickness of around 3000 m, with sediment supplied by a volcanic arc in the west deposited on a

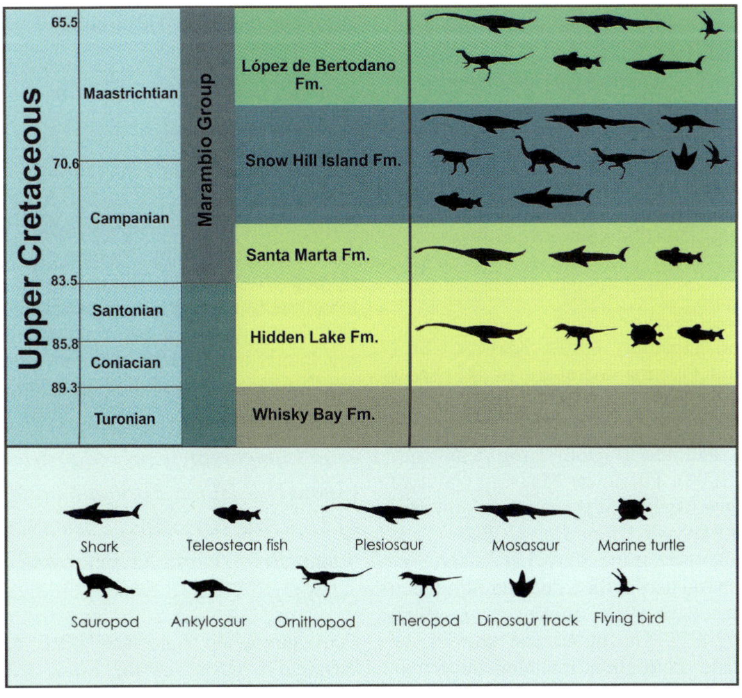

Fig. 3. Time scale, stratigraphy and vertebrate fossil record for the Late Cretaceous rocks in the James Ross Basin, Antarctic Peninsula, West Antarctica. Temporal and sedimentary units not to scale.

shelf prograding eastward into the Weddell Sea (Fig. 2, Olivero 2012). The section represents three major sedimentary cycles: the N (Santonian–early Campanian); NG (Late Campanian–early Maastrichtian); and MG (early Maastrichtian–Danian) sequences, the names of the sequences being derived from the most common kossmaticeratid ammonites that characterize each of them: N for *Natalites*; NG for *Neograhamites* and *Gunnarites*; and MG for *Maorites* and *Grossouvrites* (Olivero & Medina 2000).

Outcrops of the N Sequence (Santonian–early to basal Late Campanian) are exposed in Brandy Bay, James Ross Island (Fig. 1b) and comprise the *c.* 1000 m-thick Santa Marta Formation that includes the Alpha and Beta members following the original subdivision established in Olivero *et al.* (1986) and Scasso *et al.* (1991). These members are approximately equivalent to the Lachman Crags Member of Crame *et al.* (1991) and Pirrie *et al.* (1997). The Gamma Member (Olivero *et al.* 1986) and the Hamilton Point Member (Pirrie *et al.* 1997) are incorporated into the Snow Hill Island Formation instead of the Santa Marta Formation as originally defined (see Olivero & Medina 2000; Olivero *et al.* 2008). Notably, this member includes dinosaur bones (Olivero *et al.* 1986, 1991; Coria *et al.* 2008; Cerda *et al.* 2012), including a partial skeleton of the ankylosaurian dinosaur *Antarctopelta oliveroi* Salgado & Gasparini (2006). The Gamma Member is approximately equivalent to the Herbert Sound Member of Crame *et al.* (1991). Ammonites are relatively abundant in the upper third of the Gamma Member, characterized by the dominance of the kossmaticeratid genus *Gunnarites* (Olivero 2012).

Outcrops of the NG Sequence (Late Campanian–early Maastrichtian) are widely exposed in the James Ross Basin, from Snow Hill and Seymour islands to Santa Marta Cove area in James Ross Island (Fig. 1b, c). The sequence consists of *c.* 1000 m of fine-grained sediments comprising the Hamilton Point, Sanctuary Cliffs, Karlsen Cliffs, Cape Lamb and Gamma members of the Snow Hill Island Formation (Olivero *et al.* 1986; Crame *et al.* 1991, 1996; Pirrie *et al.* 1997; Olivero & Medina 2000) and the Haslum Crag Sandstone (Olivero *et al.* 2008). The lower Maastrichtian part of the Marambio Group at Seymour and Snow Hill islands is well exposed (Crame *et al.* 2004), and includes the upper part of the Snow Hill Island Formation and the basal part of the López de Bertodano Formation (Macellari 1988; Pirrie *et al.* 1997; Olivero *et al.* 2007). The uppermost part of the Snow Hill Island Formation includes the informal *Rotularia* Unit 1, and the basal López de Bertodano Formation, representing the *Rotularia* Units 2–5 of Macellari (1988). Absolute dating based on $^{87}Sr/^{86}Sr$ chronostratigraphy suggests an early Maastrichtian age for both stratigraphic units in these areas (Crame *et al.* 2004), thus supporting previous data based on ammonite biostratigraphy (cf. Olivero & Medina 2000; Crame *et al.* 2004).

The MG sequence (early Maastrichtian/Danian) consists mainly of the *c.* 1000 m-thick mudstone and silty, very fine-grained sandstone succession of the López de Bertodano Formation exposed in the ESE of the basin (Fig. 1c). Localized outcrops of a very distinctive coarse-grained volcaniclastic unit, the Sandwich Bluff Member, are located at Cape Lamb, Vega Island (Fig. 1b). The base of the MG Sequence is a marked unconformity that cuts a maximum of about 60–70 m of the underlying Haslum Crag Sandstone (Pirrie *et al.* 1997; Olivero *et al.* 2007, 2008). At the top of the MG Sequence, another marked unconformity also associated with a deeply incised erosional surface separates the Danian beds of the López de Bertodano Formation from the Paleocene Sobral Formation.

Dinosaurs from the James Ross Basin

Late Cretaceous dinosaurs found in James Ross Basin are summarized in Table 1 and discussed in stratigraphic order below. They come from Coniacian, Late Campanian, Early Maastrichtian and Late Maastrichtian strata on James Ross, Vega, Seymour and Snow Hill islands.

Coniacian (Hidden Lake Formation on James Ross Island)

A theropod fragment was collected from the Coniacian Hidden Lake Formation (Gustav Group) at Cape Lachman, James Ross Island (Fig. 1b, loc. 1), and described by Molnar *et al.* (1996) as a distal portion of tibia. This is the first Antarctic non-avian theropod with unquestionable affinities to taxa from other Gondwanan landmasses, suggesting faunal interchange between eastern and western Gondwana during the Late Cretaceous.

Late Campanian/Early Maastrichtian (Snow Hill Island Formation and basal part of López de Bertodano Formation on James Ross, Vega, Seymour and Snow Hill islands)

Four non-avian dinosaurs were exhumed from different horizons of the Snow Hill Island Formation at Santa Marta Cove, northeastern of the James Ross Island (Fig. 1b). The small-sized ankylosaur *Antarctopelta oliveroi* Salgado & Gasparini

Table 1. *Taxonomic list, stratigraphy, geographic locations, age and references for dinosaurs from the Late Cretaceous of the James Ross Basin, Antarctic Peninsula*

Taxon	Stratigraphy (Formation)	Geographic location	Age	References
Dinosauria (non-avian)				
Theropoda; gen. et sp. indet.	Hidden Lake Fm.	James Ross Island (Cape Lachman)	Coniacian	Molnar *et al.* (1996)
Ankylosauria: *Antarctopelta oliveroi*	GM, Snow Hill Island Fm.	James Ross Island (Santa Marta Cove)	Late Campanian	Salgado & Gasparini (2006)
Ornithopoda: gen. et sp. nov.	GM, Snow Hill Island Fm.	James Ross Island (Santa Marta Cove)	Late Campanian	Coria *et al.* (2008, 2013)
Ornithopoda: gen. et sp. indet.	GM, Snow Hill Island Fm.	James Ross Island (Santa Marta Cove)	Late Campanian	Coria *et al.* (2007)
Titanosauria: gen. et sp. indet.	GM, Snow Hill Island Fm.	James Ross Island (Santa Marta Cove)	Late Campanian	Cerda *et al.* (2012)
Iguanodontidae: gen. et sp. indet.	CLM, Snow Hill Island Fm.	Vega Island (Cape Lamb)	Early Maastrichtian	Hooker *et al.* (1991)
Dromaeosauridae: gen. et sp. indet.	CLM, Snow Hill Island Fm.	James Ross Island (the Naze)	Early Maastrichtian	Case *et al.* (2007)
Iguanodontidae?: gen. et sp. indet.	CLM, Snow Hill Island Fm.	James Ross Island (the Naze)	Early Maastrichtian	Cambiaso *et al.* (2002)
Hadrosauridae: gen. et sp. indet.	SBM, López de Bertodano Fm.	Vega Island (Cape Lamb)	Late Maastrichtian	Case *et al.* (2000)
?Hadrosauridae: gen. et sp. indet.	Klb9, López de Bertodano Fm.	Seymour Island	Late Maastrichtian	Rich *et al.* (1999)
Sauropod? footprints	López de Bertodano Fm.	Snow Hill Island (Tesore Hill)	Late Maastrichtian	Olivero *et al.* (2007)
Dinosauria (avian)				
Charadriiformes: gen. et sp. indet.	CLM, Snow Hill Island Fm.	Vega Island (Cape Lamb)	Early Maastrichtian	Case & Tambussi (1999)
Gaviiformes: *Polarornis gregorii*	Klb9?, López de Bertodano Fm.	Seymour Island	Late Maastrichtian	Chatterjee (1989)
Gaviiformes: *Polarornis* sp.	SBM, López de Bertodano Fm.	Vega Island (Cape Lamb)	Late Maastrichtian	Chatterjee *et al.* (2006)
Anseriformes: *Vegavis iaai*	SBM, López de Bertodano Fm.	Vega Island (Cape Lamb)	Late Maastrichtian	Clarke *et al.* (2005)
Cariamidae?: gen. et sp. indet.	SBM, López de Bertodano Fm.	Vega Island (Cape Lamb)	Late Maastrichtian	Case *et al.* (2006)

Abbreviations: CLM, Cape Lamb Member; GM, Gamma Member; SBM, Sandwich Bluff Member.

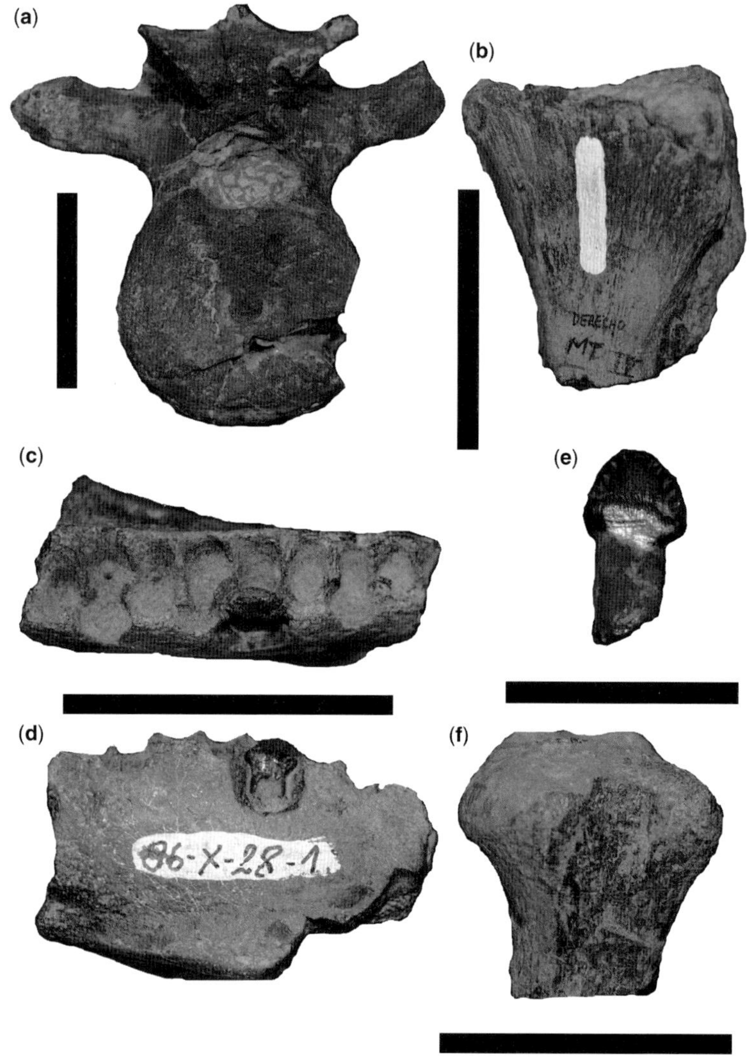

Fig. 4. *Antarctopelta oliveroi* (Salgado & Gasparini 2006), MLP 86-X-28-1, holotype. (**a**) Posterior cervical vertebra in anterior view (scale bar 50 mm); (**b**) right metatarsal IV? in lateral view (scale bar 50 mm); (**c**) left dentary in occlusal view; (**d**) left dentary in medial view (scale bar 50 mm); (**e**) tooth II in lingual view (scale bar 50 mm); (**f**) distal end of metapodial (scale bar 50 mm).

2006 was found in the Late Campanian Gamma Member of the Snow Hill Island Formation (Gasparini *et al.* 1987; Olivero 2012; Fig. 1b, loc. 2). The specimen, MLP 86-X-28-1, is represented by a partial skeleton consisting of a fragment of the left dentary with one tooth in place (Fig. 4), cervical, dorsal and sacral vertebrae, metacarpals and several scutes. The James Ross Island ankylosaur was collected from shallow marine deposits, associated with various marine invertebrates such as bivalves, gastropods and ammonites that indicate a late Early to basal Late Campanian age (Olivero *et al.* 1991; Olivero 2012). *Antarctopelta* is considered by Thompson *et al.* (2012) to be the most basal nodosaurid. In many features it appears intermediate between the ankylosaurid and nodosaurid conditions (Salgado & Gasparini 2006).

Also from the same area of the Santa Marta Cove, two ornithopods, an isolated ungueal phalanx and a semiarticulated and incomplete skeleton, have been reported by Coria *et al.* (2007, 2008). These specimens come from the same stratigraphical horizon and close to the site of *Antarctopelta* (Fig. 1b, loc. 2). The ornithopod skeleton

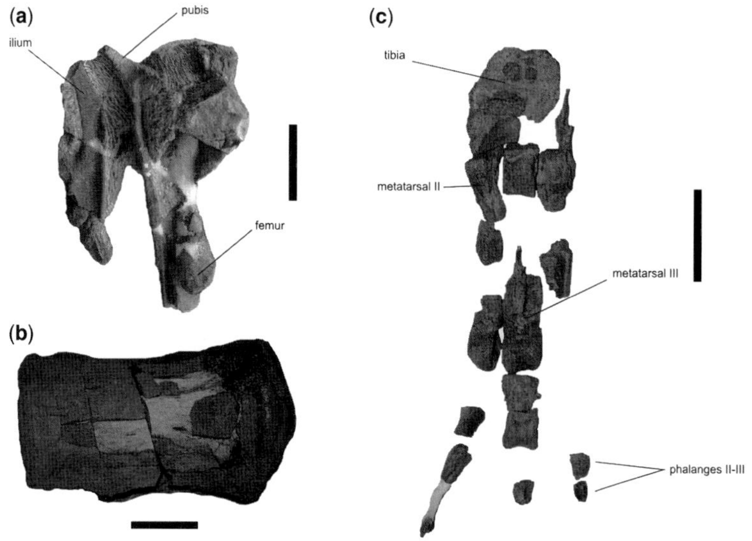

Fig. 5. (**a**) Ornithopod gen. et sp. nov., associated pubis, ilium and femur of the skeleton MLP 08-III-1-1. Scale bar: 50 mm; (**b**) Lithostrotian gen. et sp. indet., MLP 11-II-20-1, caudal vertebra centrum, right lateral view (scale bar 50 mm); (**c**) Dromaeosauridae gen. et sp. indet., partial left hindlimb, ankle and foot (scale bar 100 mm).

represents a new genus and new species of a small basal ornithopod (Coria *et al.* 2008, 2013; Fig. 5a). This new form, MLP 08-III-1-1, is currently under study and its phylogenetic relationships will be useful for insights about the palaeobiogeography of this group (Coria *et al.* 2013).

Recently, an incomplete vertebra of a sauropodomorph titanosaur was recovered in shallow marine shelf deposits referred to the Late Campanian Snow Hill Island Formation in the northern part of James Ross Island (Cerda *et al.* 2012; Fig. 1b, loc. 3). The size and morphology of the specimen (MLP 11-II-20-1) allowed its identification as a caudal vertebra of a lithostrotian titanosaur (Fig. 5b). This discovery increases the fossil dinosaur diversity of Antarctica and enhances our understanding of the global distribution of sauropod dinosaurs (Cerda *et al.* 2012).

The dinosaur assemblage from the Early Maastrichtian Cape Lamb Member of the Snow Hill Island Formation consists of four non-avian and one avian dinosaurs. Hooker *et al.* (1991) reported a partial skeleton of a small-bodied (4–5 m) ornithopod dinosaur from the Cape Lamb Member of the Snow Hill Island Formation (Crame *et al.* 2004; Fig. 1b, loc. 7). Numerous skeletal characteristics indicate that this specimen (BMNH BAS R.2450) is related to basal ornithopods but few postcranial features also resemble those of Dryomorpha (Iguanodontia) (Milner *et al.* 1992).

During the 2006–2007 field season J.A. Case and his field crew (including M.A.R.) collected vertebrate remains of a partial skeleton of a small-bodied theropod dinosaur (Case *et al.* 2007) from the lower portion of the Cape Lamb Member of the Snow Hill Island Formation (Crame *et al.* 2004) on the Naze, northern James Ross Island (Fig. 1b, loc. 4). The remains included a metatarsal II with a lateral expansion caudal to metatarsal III, a third metatarsal that is proximally narrow and distally wide, a metatarsal III with a distal end that is incipiently ginglymoidal and a second pedal digit with sickle-like ungual. Together these are diagnostic of a Dromaeosauridae theropod (Fig. 5c). The Dromaeosauridae is a family of highly derived small to mid-sized theropod dinosaurs characterized by the presence of a raptorial second pedal digit and a tail stiffened by the elongated prezygapophyses and chevrons of the medial to distal posterior vertebrae. The Antarctic dromaeosaur exhibits diagnostic characters of the Dromaeosauridae, that is Mt II, with a lateral expansion caudal to Mt III, and with Mt III proximally narrow and distally wide (Novas & Pol 2005; Case *et al.* 2007).

From similar-aged horizons and close to the dromaeosaur site on the Naze (Fig. 1a) portions of a hind leg and foot of an iguanodontid (MACN-19777) were recovered (Cambiaso *et al.* 2002; Novas *et al.* 2002*b*). The exact stratigraphic and geographic location of this iguanodontid is uncertain, although field evidence suggests that it came from Fortress Hill next to Terrapin Hill in the Naze (Fig. 1b, loc. 5), possibly the same

Fig. 6. Hadrosauriidae? MLP 96-I-6-2, distal end of metatarsal: (**a**) lateral view, (**b**) dorsal view, (**c**) ventral view (scale bar 50 mm).

lithostratigraphic unit as the dromaeosaur mentioned above.

Additional indications of non-avian dinosaurs from the James Ross Basin are aligned depressions interpreted as sauropod footprints by Olivero *et al.* (2007) in the basal López de Bertodano Formation, southwestern part of the Spath Peninsula, Tessore Hill, Snow Hill Island (Fig. 1c, loc. 12). These ichnites suggest a more extensive distribution of dinosaurs within the basin and southward. This circumstance could be reflecting opportunistic dispersal of members of the group at the end of the Cretaceous among southern Gondwanan areas via Antarctica, as was previously proposed (Sampson *et al.* 1998), during periods of more favourable climatic conditions in the region.

One avian dinosaur fragment was reported by Case & Tambussi (1999) from the Early Maastrichtian Cape Lamb Member of the Snow Hill Island Formation on Cape Lamb, Vega Island (Fig. 1b, loc. 6). These authors assigned the tarsometatarsus (MLP 98-I-10-25) to an undetermined charadriiform bird. The diaphysis of the MLP 98-I-10-25 is slender, being much longer than wide, and is eroded, with the trochleas broken and the proximal epiphysis absent (Fig. 7d). The preservation of the material does not permit its positive assignment. This specimen represents the oldest neognathous recorded in Antarctica.

Late Maastrichtian (López de Bertodano Formation on Vega and Seymour islands)

The Late Maastrichtian dinosaur assemblage from the James Ross Basin consists of two non-avian and four avian dinosaurs. Case *et al.* (2000) reported the presence of dental and pedal material of a hadrosaurine hadrosaur from the Late Maastrichtian Sandwich Bluff Member of the López de Bertodano Formation on Vega Island. The 'Reptile Horizon', named for the numerous mosasaur and plesiosaur specimens recovered from this stratigraphic level, is in the upper third of the Sandwich Bluff Member (Crame *et al.* 1991; Pirrie *et al.* 1991; Martin 2006; Fig. 1b, loc. 9), Unit C (Olivero *et al.* 1992) or K3 (Marenssi *et al.* 1992) of the López de Bertodano Formation. This member is composed of nearshore marine, fine-grained, ferruginous, loosely consolidated sandstones that are

Late Maastrichtian in age (approximately 66–68 Ma), based on correlations of ammonite and palynological taxa (Crame *et al.* 1991; Pirrie *et al.* 1991; Olivero & Medina 2000; Olivero 2012).

Rich *et al.* (1999) described a distal end of an ornistichian metatarsal (Fig. 6), MLP 96-I-6-2, which comes from the upper levels of the López de Bertodano Formation (Klb 9 *sensu* Macellari 1988) on Seymour Island (Fig. 1c, loc. 10). These authors attributed this metatarsal tentatively to a hadrosaurine dinosaur.

The four avian dinosaurs were all reported from horizons dated as Late Maastrichtian from Seymour and Vega islands. Chatterjee (1989, 1997) described the first Antarctic avian dinosaur, *Polarornis gregorii*, from the upper levels of the López de Bertodano Formation of Seymour Island (Klb9?, Late Maastrichtian). The material consists of incomplete and poorly preserved remains of one individual (a fragment of the bill and the adjacent distal part of the skull, parts of the otic region, four vertebrae, a small fragment of the sternum, a femur and a proximal tibiotarsus). Chatterjee (1997) assigned this species to the neornithine family Gaviidae (loons). Mayr (2004) pointed out that *Polarornis* differs from modern loons in some important features. He criticized the original description of the species because of the specimen's completeness and its

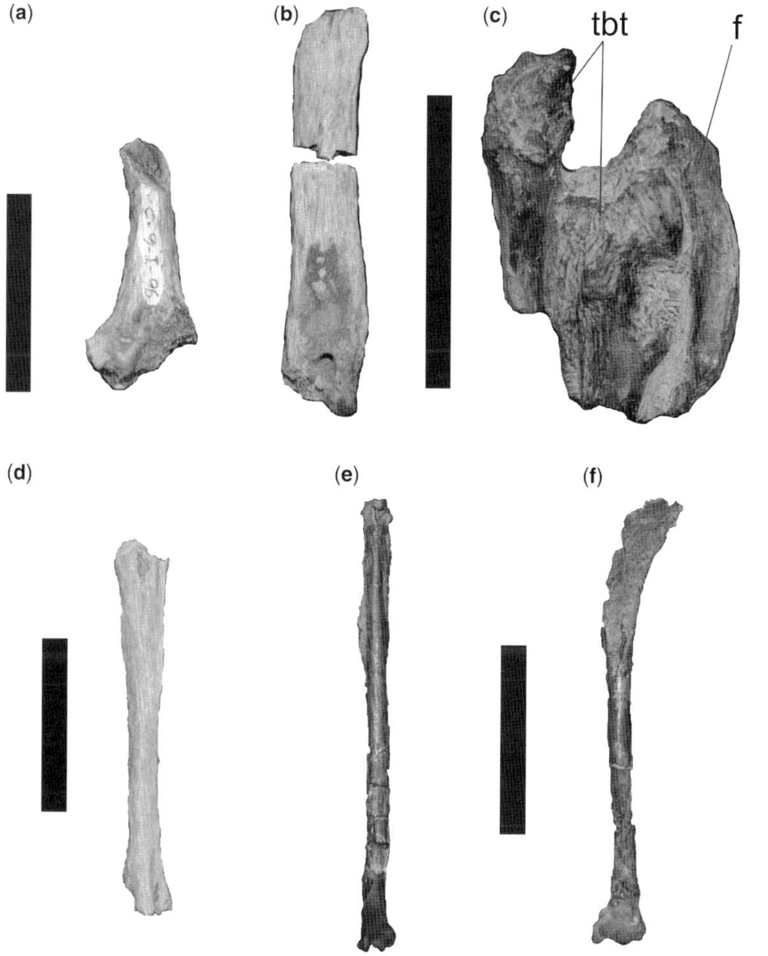

Fig. 7. (**a**) *Polarornis gregorii?* MLP 96-I-6-4, distal end of left femur (scale bar 50 mm); (**b**) *Polarornis gregorii?* MLP 96-I-6-3, distal end of left tibiotarsus (scale bar 50 mm); (**c**) *Polarornis gregorii* (Chatterjee 1989), MLP 96-I-6-2, concretion containing the two tibiotarsi and the right femur of the same individual (scale bar 50 mm); (**d**) Charadriiform gen. et sp. indet. MLP 98-I-10–25, fragment of left tarsometatarsus (scale bar 20 mm); (**e**) *Vegavis iaai* (Clarke *et al.* 2005), MLP 93-I-3-1, holotype, right humerus; (**f**) *Vegavis iaai* Clarke, Tambussi, Noriega, Erickson & Ketcham 2005, MLP 93-I-3-1, holotype, left tibiotarsus (scale bar 30 mm). *Abbreviations*: f, femur; tbt, tibiotarsus.

assignation to Gaviidae, and stated that '[*Polarornis*] may be a synonym of *Neogaeornis* – a possibility already raised by Olson (1992) but not discussed by Chatterjee (2002)' (Mayr 2004, p. 5). *Neogaeornis wetzeli*, from the Upper Cretaceous of Chile, based on a tarsometatarsus fragment, has been interpreted as the earliest fossil gaviiform (Lambrecht 1929; Olson 1992), resembling the highly derived bone of modern loons (Mayr 2004). Additional material (MLP 96-I-6-2/3) possibly belonging to the same taxon is figured here for the first time (Fig. 7a–c).

A second Late Maastrichtian bird, first referred to Presbyornithidae by Noriega & Tambussi (1995), was collected at Cape Lamb (locality VEG 9303, Fig. 1b, loc. 8), Vega Island from the Late Maastrichtian Sandwich Bluff Member of the López de Bertodano Formation. The specimen (MLP 93-I-3-1) consists of a disarticulated partial postcranium preserved in the two halves of a concretion (Fig. 8a, b), with a right humerus and left tibiotarsus out of the blocks (Fig. 7e, f). *Vegavis iaai* (Clarke *et al.* 2005, 2006) was subjected to phylogenetic analysis, which supported a basal position within the Anseriformes. Whereas this discovery pushes the neornithine record as far back as the Maastrichtian, this is still considerably later than is suggested by most molecular clock studies. *Vegavis* is assessed to be a basal anseriform close to Anatidae. It is important as the best support for a part of theradiation (crown clade) of birds in the Cretaceous, and directly relevant to the timing of this radiation.

Chatterjee *et al.* (2006) reported a well-preserved skeleton of a new gracile species of a fossil loon, *Polarornis*, from the Late Maastrichtian Sandwich Bluff Member of the López de Bertodano Formation on Vega Island (Fig. 1b, loc. 8). Although *Polarornis* is often claimed to be an ancestor of modern loons (Gaviiformes), its relationships are unclear (Tambussi & Acosta Hospitaleche 2007).

Case *et al.* (2006) reported a left femur recovered from the Sandwich Bluff area of the Cape Lamb on Vega Island (Fig. 1b, loc. 8). Its stratigraphic position places it near the base of the Sandwich Bluff Member and at an equivalent stratigraphic level to *Vegavis iaai*. It shows striking similarities to modern cursorial predatory birds of South America (Seriemas, Cariamidae) and of Africa (Secretarybirds, Sagittariidae).

Late Cretaceous Weddellian dinosaurs: palaeogeography and biogeography

The role of Antarctica in the palaeobiogeography of terrestrial vertebrates during the Cretaceous has remained essentially unknown. By the beginning of the Late Cretaceous (99.6 Ma), the major Gondwanan continental blocks (South America, Africa, Antarctica, Madagascar, the Indian subcontinent and Australia) were well into the process of break-up and dispersion. However, many issues concerning the timing and sequence of this process remain unresolved. The majority of proposed Cretaceous palaeogeographic scenarios (e.g. Smith *et al.* 1995; Roeser *et al.* 1996; Scotese 1997) detail a rapid dissociation of Gondwanan landmasses, with India–Madagascar separating completely from the Antarctica–Australia complex by the upper Barremian (Early Cretaceous, *c.* 125 Ma; Gradstein *et al.* 1995). According to these hypotheses, South America severed physical ties with Africa during the Aptian or Albian (prior to 100 Ma), but portions of these landmasses remained in fairly close proximity throughout much of the Cretaceous (Scotese & Golonka 1993). India and Madagascar were among the last southern landmasses to dissociate, separating at *c.* 88.0 Ma (Storey *et al.* 1995). The isolation of South America was finally completed in the Palaeogene, when it split apart from Antarctica. West Antarctica itself is a complex of microcontinental blocks (Dalziel & Elliot 1982) that have moved independently over the past 200 million years in the West Antarctic rift system, analogous to the Basin and Range Province of North America (Bialas *et al.* 2007). By the mid-Late Cretaceous (latter part of the Santonian, end-Cretaceous Normal Superchron, 83.5 Ma), South America, West Antarctica and Australia were the only Gondwanan continents still connected.

Relative plate motions suggest that there could not have been any land bridge between South America, Africa, Antarctica, India and Madagascar after 108–112 Ma. However, Sereno & Brusatte (2008), based on the phylogenetic relationship of theropod dinosaurs Abelisauroidea to Carnotaurinae and the finding of an abelisaurid fossil from the Cenomanian of Niger, proposed the existence of land bridges between South America and Africa and between India and East Antarctica at 97 Ma. Upchurch *et al.* (2002), based on a cladistic biogeographic study on all dinosaur groups, supported the hypothesis that many clades spread across Gondwana prior to Cretaceous continental fragmentation.

Dispersion of terrestrial vertebrates between South America and Australia–New Zealand across Late Cretaceous land bridges requires their presence during that interval in the Antarctic Peninsula. The knowledge of Late Cretaceous terrestrial vertebrates from West Antarctica is still limited and lacks any taxonomically identifiable small terrestrial and freshwater vertebrates (e.g. reptiles

Fig. 8. *Vegavis iaai* (Clarke *et al.* 2005), MLP 93-I-3-1, holotype. (**a**) The half of the concretion preserving most of the skeleton of the holotype; (**b**) the other half of the concretion with few bones. *Abbreviations*: c, coracoids; cv, cervical vertebra; f, femora; h, humerus; il, ilium; p, pubis; r, radius; tmt, tarsometatarsus; tv, thoracic vertebrae; s, sacrum; sc, scapula; u, ulna (scale bar 50 mm).

as turtles, lizards, snakes, crocodyliforms or even mammals). However, this knowledge has expanded substantially over the last decade and includes the discoveries of avian and non-avian dinosaurs discussed here. By the Late Maastrichtian to approximately Late Palaeocene, the northern tip of the Antarctic Peninsula and southernmost South America (Magellanic Region, Patagonia) were physically connected allowing the dispersal of plants and animals between these two areas (Marenssi *et al.* 1994; Shen 1995; Woodburne & Case 1996; Reguero *et al.* 1998, 2002; Reguero & Marenssi 2010). Woodburne & Case (1996) inferred that the initial marsupial dispersal, from South America into and through West Antarctica and onward to Australia, must have occurred during the Late Cretaceous (?Maastrichtian), but ended by 64 Ma, when the South Tasman Rise submerged

(McGowran 1991). The basal stocks of these clades, currently not represented by fossils in the Late Cretaceous of South America, West Antarctica and Australia, are supported by stratigraphically calibrated phylogenies revealing long ghost lineages that extended into the Late Cretaceous (e.g. Meredith *et al.* 2008).

The high-latitude Weddellian Biogeographical Province, conceived by Zinsmeister (1979, 1982) on the basis of marine molluscan, echinoderm and arthropod faunas, was defined as a cool temperate, shallow-water region that extended from southern South America (Magellanic Region in Chile, and Tierra del Fuego and Santa Cruz provinces in Argentina), along West Antarctica (including Antarctic Peninsula), to New Zealand, Tasmania and southeastern Australia. Later, Case (1988) expanded this concept to a 'biogeographic province' including terrestrial plants and mammals. This province existed from the Late Cretaceous through the Eocene when Australia, Antarctica and southernmost South America were in proximity (Woodburne & Zinsmeister 1984).

Non-avian dinosaurs from the Late Campanian Snow Hill Island Formation are represented by an ankylosaur, an indeterminate ornithopod, a new genus and species of ornithopod and a titanosaurian sauropod (Salgado & Gasparini 2006; Coria *et al.* 2007, 2008; Cerda *et al.* 2012). These Late Campanian non-avian dinosaurs are thus of interest to address a variety of questions, ranging from the dispersal patterns of major vertebrate clades to Gondwanan endemism. The ankylosaurian *Antarctopelta oliveroi* would have lived at about 60°S when the Antarctic Peninsula was in close proximity to southern South America. *Antarctopelta oliveroi* represents an Antarctic Late Cretaceous divergence of Ankylosauria and a rigorous phylogenetic analysis of this taxon would be necessary to test or support a hypothesis of vicariance. Salgado & Gasparini (2006) consider *A. oliveroi* to be an Ankylosauria *incertae sedis*, and to represent a different species from the ankylosaur found at Salitral Moreno in Río Negro province, Patagonia (Salgado & Coria 1996).

The Late Campanian Snow Hill Island Formation sauropod is the second sauropodomorph dinosaur recorded from Antarctica, and the first from West Antarctica. The first one is the basal sauropodomoph *Glacialisaurus hammeri* collected from the Beardmore Glacier region of the Central Transantarctic Mountains (Early Jurassic, Hanson Formation), East Antarctica (Hammer & Hickerson 1994; Smith & Pol 2007). The absence of sauropodomorph material between the Lower Jurassic and the Upper Cretaceous in Antarctica is more probably related to limited sampling than a genuine absence of members of this lineage during this time. The Hanson Formation sauropodomorph belongs to a geographically widespread clade that includes members from Africa, South America, Australia and Asia (Smith *et al.* 2007). Cerda *et al.* (2012) state that the presence of an advanced titanosaur with characteristic procoelous mid-caudal vertebrae in West Antarctica implies that the group achieved a global distribution by the Late Campanian. However, the presence of advanced titanosaurians in the Late Cretaceous of West Antarctica rests on fragmentary remains. The lack of an adequate sampling of the known fossil record is limiting our understanding of the evolution and biogeography of advanced titanosaurians in the Late Cretaceous of West Antarctica (Molnar & Wiffen 2007; Cerda *et al.* 2012).

Perhaps the most reliable fossil evidence for evaluating terrestrial connections among Late Cretaceous Gondwanan landmasses is derived from examining the stratigraphically calibrated phylogenetic relationships of theropods from the James Ross Basin with those of Australia and South America. It is interesting to note that some clades of theropods (e.g. abelisaurids and noasaurids), now extensively documented from the Late Cretaceous of South America, Madagascar and the Indian subcontinent, are unknown in Antarctica. With large body sizes, cursorial locomotion and diminutive forelimbs, these animals were almost certainly incapable of long-distance swimming or rafting across significant marine barriers.

Gondwanan dromaeosaurs are not North American immigrants into South America and Antarctica. Until recently, the dromaeosaur fossil record has been geographically restricted to the Northern Hemisphere (Norell & Makovicky 2004). However, recent discoveries documented from Turonian through Maastrichtian beds of Argentina and the reassessment of some theropod taxa in South America (Makovicky *et al.* 2005; Novas & Pol 2005) and Australia (Agnolin *et al.* 2010) demonstrated that a distinctive dromaeosauriid subfamily Unenlagiinae were present and highly diversified in the southern landmasses. This suggests that an important adaptive radiation took place in Gondwana during the Late Cretaceous (Gianechini & Apesteguía 2011). The record of dromaeosaurids in South America and Australia is still poor compared with that of North America and Asia, taking into account the lesser number of taxa and the fragmentary nature of the specimens. Nevertheless, the recent discoveries in Argentina, mainly of *Buitreraptor* and *Austroraptor*, have provided much information about the anatomy and phylogenetic relationships of this group of Gondwanan theropods. Thus, the dromaeosauriid Unenlagiinae taxon is restricted exclusively to Gondwana. Case *et al.* (2007), based on the plesiomorphic nature of the ankle and pes of the Naze dromaeosaur, support the hypothesis that dromaeosaurs were in fact the product of a cosmopolitan distribution, which has been separated by the vicariant break-up of Pangea, creating an endemic Gondwanan clade of dromaeosaurs.

The presence of hadrosaurid remains in the Late Maastrichtian of Antarctica is of particular interest for understanding the palaeobiogeographic dispersal events occurring during this time span, since this group of dinosaurs is considered to be of Laurasian origin (Horner *et al.* 2004). The dispersal of hadrosaurs into Antarctica by the Maastrichtian may coincide with and provide additional support for a predicted first wave of marsupials from South America to Antarctica (Woodburne & Case 1996; Goin *et al.* 1999).

Conclusions

Compared with those of most other continents, dinosaurs from Antarctica are still poorly known, and have thus been neglected in most discussions of Cretaceous biogeography. However, two important assemblages of non-avian dinosaurs from the Late Campanian and Early Maastrichtian Snow Hill Island Formation, one distributed in the northern part of the James Ross Island, and the second distributed in James Ross and Vega islands, provide much needed data on dinosaur distribution across Gondwana.

The main localities that have yielded Late Cretaceous Antarctic dinosaurs to date are situated within a relatively small area of approximately 350 km^2 at the NW corner of the James Ross Basin (Fig. 1b). This western margin of the basin is known to have been the approximate site of the palaeoshoreline during the Late Cretaceous, which progressed eastward up-section (Olivero 2012). Seven non-avian and four avian dinosaurs collected from a comparatively small geographical region suggest that the overall diversity on the Antarctic Peninsula was reasonably high.

One area of the James Ross Basin is particularly promising for future investigations. A well-exposed sequence of rocks of the Santa Marta Cove, James Ross Island spans the last 15 million years that dinosaurs existed, providing great potential for recovering Antarctic Late Cretaceous dinosaurs.

The Late Campanian dinosaur assemblage of Santa Marta Cove, James Ross Island, provides unique data on the composition of high-latitude Southern Hemisphere dinosaurs. The presence of an exclusive ankylosaur species, *Antarctopelta oliveroi*, in combination with other recent discoveries in the same area of clades previously thought to be restricted to Laurasia, ornithopod and titanosaurian species, provides evidence for cosmopolitan distribution of several dinosaur clades during the Late Cretaceous, and undermines previous suggestions that the dinosaur fauna of Antarctica was either largely endemic or predominantly Gondwanan in composition. Two of the three Santa Marta dinosaurs, *Antarctopelta oliveroi* and the new genus and new species of ornithopod, appear to show substantially greater morphological differences than the contemporaneous taxa beyond the Antarctic Peninsula region.

The majority of the non-avian dinosaurs from West Antarctica (megalosaur-like theropod, ankylosaur, dromaeosaurid, the ornithopods and the sauropod) are remnants of a cosmopolitan dinosaur fauna more typical of other areas (e.g. Campanian–Maastrichtian of Patagonia). Preliminary reports of both theropods and titanosaurs from the Late Cretaceous deposits in West Antarctica are suggestive in this regard. Furthermore, developing a more complete fossil record from other intervals and areas, for example Early and Late Maastrichtian of Vega Island, will provide information critical for reconstructing the biogeographical history of Late Cretaceous Gondwana. Thus, it is relevant to note that the recent discovery and/or re-appraisal of the non-avian dinosaur faunas of Australia and New Zealand have revealed typical Gondwanan elements, including the presence of lithostrotian titanosaur and theropods from the Early Cretaceous (Valanginian–Albian; Smith *et al.* 2008; Hocknull *et al.* 2009). This firmly establishes the presence of these clades in East Gondwana and South America (Patagonia) before the Late Cretaceous.

We support the suggestion that higher clades of dinosaurs present in the Late Campanian/Early Maastrichtian of the James Ross Basin (e.g. Ankylosauria, Ornithopoda, Theropoda) attained a global or near-global distribution before the Late Cretaceous, prior to fragmentation of the Pangean supercontinent, and some aspects of the hallmark 'Gondwanan' fauna of Gondwana may therefore reflect climate-driven provinciality. They do not provide useful data for appraising the existence of vicariance in dinosaur distributions. Instead, the relationships among lower-level taxa, representing Cretaceous evolutionary divergences (i.e. *Antarctopelta*), should be used to support hypotheses of vicariance. For this to be effective, greater sample sizes may be required for many clades (Benson *et al.* 2012).

Finally, the Late Cretaceous avian dinosaurs recorded in the James Ross Basin are rare, being restricted to one charadriform, a cariamid? and, with more certainty, a loon and a basal Anseriformes. The last two are flying birds whose life stories are linked mainly to water. If the assignment of the loon to *Neoageornis* is correct, and it seems that it is, then we have further clear evidence of affinities of the Late Cretaceous Antarctic fauna with that of southern Chile, although gaviids are clearly Holarctic. *Vegavis iaai* gives the first good evidence for a basal part of the extant avian radiation in the Cretaceous.

We especially acknowledge the Instituto Antártico Argentino and Fuerza Aérea Argentina, which provided logistical support for our participation in the Antarctic fieldwork. We also have benefited from collaborative effort in the field (prospecting and picking) of J. J. Moly, S. Santillana, S. Salgado, I. Cerda, A. Paulina Carabajal, T. Mors and J. Hängstrom. We thank P. Barrett, J. A. Crame and J. A. Case for their critical reviews and helpful suggestions that improved the manuscript considerably. M. Sosa drew Figures 1 and 2. Part of this study was funded by Instituto Antártico Argentino (PICTA 2004 and 2008), Agencia Nacional de Promoción Científica y Tecnológica (ANPCYT, PICT 0365 and PICTO 0093) and Consejo Nacional de Investigaciones Científicas y Técnicas (CONICET).

References

Agnolin, F. L., Ezcurra, M. D., Pais, D. F. & Salisbury, S. W. 2010. A reappraisal of the Cretaceous non-avian dinosaur faunas from Australia and New Zealand: evidence for their Gondwanan affinities. *Journal of Systematic Palaeontology*, **8**, 257–300.

Benson, R. B. J., Rich, T. H., Vickers-Rich, P. & Hall, M. 2012. Theropod fauna from Southern Australia indicates high polar diversity and climate-driven dinosaur provinciality. *PLoS ONE*, **7**, e37122, doi: 10.1371/journal.pone.0037122.

Bialas, R. W., Buick, W. R., Studinger, M. & Fitzgerald, P. G. 2007. Plateau collapse model for the Transantarctic Mountains–West Antarctic Rift System: insights from numerical experiments. *Geology*, **35**, 687–690.

Buatois, L. A. & López-Angriman, A. O. 1992. Trazas fósiles y sistemas deposicionales, Grupo Gustav, Cretácico de la isla James Ross. *In*: Rinaldi, C. A. (ed.) *Geología de la Isla James Ross, Antártida*. Instituto Antártico Argentino, Buenos Aires, 239–262.

Cambiaso, A., Novas, F., Lirio, J. M. & Núñez, H. 2002. Un nuevo dinosaurio del Cretácico Superior de la Isla James Ross, Península Antártica. *VIII Congreso Argentino de Paleontología y Bioestratigrafía*, Corrientes, Resúmenes, **61**.

Case, J. A. 1988. Paleogene floras from Seymour Island, Antarctic Peninsula. *In*: Feldmann, R. M. & Woodburne, M. O. (eds) *Geology and Paleontology of Seymour Island, Antarctic Peninsula*. Geological Society of America, Boulder, CO, Memoirs, **169**, 523–530.

Case, J. A. & Tambussi, C. P. 1999. Maestrichtian record of neornithine birds in Antarctica: comment on a Late Cretaceous radiation of modern birds. *Journal of Vertebrate Paleontology*, **19**, 37R.

Case, J. A., Martin, J. E., Chaney, D. S., Reguero, M., Marenssi, S. A., Santillana, S. M. & Woodburne, M. O. 2000. The first duck-billed dinosaur (Hadrosauridae) from Antarctica. *Journal of Vertebrate Paleontology*, **20**, 612–614, doi: 10.1671/0272-4634(2000)020[0612:TFDBDF]2.0.CO;2

Case, J. A., Reguero, M., Martin, J. E. & Cordes-Person, A. 2006. A cursorial bird from the Maastrichtian of Antarctica. *Journal of Vertebrate Paleontology*, **26**, 48A.

Case, J. A., Martin, J. E. & Reguero, M. A. 2007. *A dromaeosaur from the Maastrichtian of James Ross Island and the Late Cretaceous Antarctic dinosaur fauna*. US Geological Survey and The National Academies; USGS OF-2007-1047, Short Research Paper, 083, doi: 10.3133/of2007-1047.srp083

Cerda, I. A., Paulina Carabajal, A., Salgado, L., Coria, R. A., Reguero, M. A., Tambussi, C. P. & Moly, J. J. 2012. The first record of a sauropod dinosaur from Antarctica. *Naturwissenschaften*, **99**, 83–87, doi: 10.1007/s00114-011-0869-x

Chatterjee, S. 1989. The oldest Antarctic bird. *Journal of Vertebrate Paleontology*, **9**, 16A.

Chatterjee, S. 1997. *The Rise of Birds*. Johns Hopkins University Press, Baltimore, MD.

Chatterjee, S. 2002. The morphology and systematics of *Polarornis*, a Cretaceous loon (Aves: Gaviidae) from Antarctica. *In*: Zhou, Z. & Zhang, F. (eds) *Proceedings of the 5th Symposium of the Society of Avian Paleontology and Evolution, Beijing, 1–4 June 2000*. Science Press, Beijing, 125–155.

Chatterjee, S. & Small, B. J. 1989. New plesiosaurs from the Upper Cretaceous of Antarctica. *In*: Crame, J. A. (ed.) *Origin and Evolution of the Antarctic Biota*. Geological Society, London, Special Publications, **47**, 197–215.

Chatterjee, S., Martinioni, D., Novas, F., Mussel, F. & Templin, R. 2006. A new fossil loon from the Late Cretaceous of Antarctica and early radiation of foot-propelled diving birds. *Journal of Vertebrate Paleontology*, **26**, 49A.

Cione, A. L. & Medina, F. 1987. A record of *Notidanodon pectinatus* (Chondrichtyes, Hexanchiformes) in the Upper Cretaceous of Antarctic Peninsula. *Mesozoic Research*, **1**, 79–88.

Clarke, J. A., Tambussi, C. P., Noriega, J. I., Erickson, G. M. & Ketcham, R. A. 2005. Definitive fossil evidence for the extant avian radiation in the Cretaceous. *Nature*, **433**, 305–308.

Clarke, J. A., Tambussi, C. P., Noriega, J. I., Erickson, G. M. & Ketcham, R. A. 2006. Corrigendum to definitive fossil evidence for the extant avian radiation in the Cretaceous. *Nature*, **444**, 780.

Coria, R. A., Tambussi, C., Moly, J. J., Santillana, S. & Reguero, M. 2007. Nuevos restos de Dinosauria del Cretácico de las islas James Ross y Marambio, Península Antárctica. *VI Simposio Argentino y III Latinoamericano sobre Investigaciones Antárcticas, Dirección Nacional del Antárctico/Instituto Antárctico Argentino*, 10–14 September 2007, Instituto Antárctico Argentino, Buenos Aires.

Coria, R. A., Moly, J. J., Reguero, M. & Santillana, S. 2008. Nuevos restos de Ornithopoda (Dinosauria, Ornithischia) de la Fm. Santa Marta, Isla James Ross, Antártida. *Ameghiniana*, **45**, 25R.

Coria, R. A., Moly, J. J., Reguero, M., Santillana, S. & Marenssi, S. 2013. New ornithopod (Dinosauria;

Ornithischia) from the Late Cretaceous of James Ross Island, Antarctica. *Cretaceous Research*, **41**, 186–193.

CRAME, J. A., PIRRIE, D., RIDING, J. B. & THOMSON, M. R. A. 1991. Campanian–Maastrichtian (Cretaceous) stratigraphy of the James Ross Island area, Antarctica. *Journal of the Geological Society, London*, **148**, 1125–1140, doi: 10.1144/gsjgs.148.6.1125

CRAME, J. A., LOMAS, S. A., PIRRIE, D. & LUTHER, A. 1996. Late Cretaceous extinction patterns in Antarctica. *Journal of the Geological Society, London*, **153**, 503–506.

CRAME, J. A., FRANCIS, J. E., CANTRILL, D. J. & PIRRIE, D. 2004. Maastrichtian stratigraphy of Antarctica. *Cretaceous Research*, **25**, 411–423.

DALZIEL, I. W. D. & ELLIOT, D. H. 1982. West Antarctica: problem child of Gondwanaland. *Tectonics*, **1**, 3–19.

DE LA FUENTE, M., NOVAS, F. E., ISASI, M. P., LIRIO, J. M. & NUÑEZ, H. J. 2010. First Cretaceous turtle from Antarctica. *Journal of Vertebrate Paleontology*, **30**, 1275–1278.

DEL VALLE, R., MEDINA, F. & BRANDONI, Z. 1977. Nota preliminar sobre los hallazgos de reptiles fósiles marinos del suborden Plesiosauria en las islas James Ross y Vega, Antártida. *Contribuciones del Instituto Antártico Argentino*, **212**, 1–13.

ELLIOT, D. H. 1988. Tectonic setting and evolution of the James Ross Basin, northern Antarctic Peninsula. *In*: FELDMANN, R. M. & WOODBURNE, M. O. (eds) *Geology and Paleontology of Seymour Island, Antarctic Peninsula*. Geological Society of America, Boulder, CO, Memoirs, **169**, 541–555.

GASPARINI, Z., DEL VALLE, R. & GOÑI, R. 1984. Un elasmosáurido (Reptilia, Plesiosauria) del Cretácico Superior de la Antártida. *Contribuciones del Instituto Antártico Argentino*, **305**, 1–24.

GASPARINI, Z., OLIVERO, E., SCASSO, R. & RINALDI, C. 1987. Un ankylosaurio (Reptila, Ornithischia) campaniano en el continente antártico. *Anais IV Congresso Brasileiro de Paleontologia*, Rio de Janeiro, **1**, 131–141.

GIANECHINI, F. A. & APESTEGUÍA, S. 2011. Unenlagiinae revisited: dromaeosaurid theropods from South America. *Anais da Academia Brasileira de Ciências*, **83**, 163–195.

GOIN, F. J., CASE, J. A., WOODBURNE, M. O., VIZCAINO, S. F. & REGUERO, M. A. 1999. New discoveries of 'opossum-like' marsupials from Antarctica (Seymour Island, Middle Eocene). *Journal of Mammalian Evolution*, **6**, 335–365.

GRADSTEIN, F. M., AGTERBERG, F. P., OGG, J. G., HARDENBOL, J., VAN VEEN, P., THIERRY, J. & HUANG, Z. 1995. A Triassic, Jurassic, and Cretaceous timescale. *In*: BERGGREN, W. A., KENT, D. V., AUBRY, M. P. & HARDENBOL, J. (eds) *Geochronology Time Scales and Global Stratigraphic Correlation*. Society of Sedimentary Geology, Tulsa, OK, Special Publications, **54**, 95–126.

GRANDE, L. & CHATTERJEE, S. 1987. New Cretaceous fish fossils from Seymour Island, Antarctic Peninsula. *Palaeontology*, **30**, 829–837.

HAMMER, W. R. & HICKERSON, W. J. 1994. A crested theropod dinosaur from Antarctica. *Science*, **264**, 828–830.

HATHWAY, B. 2000. Continental rift to back-arc basin: Jurassic–Cretaceous stratigraphical and structural evolution of the Larsen Basin, Antarctic Peninsula. *Journal of the Geological Society, London*, **157**, 417–432.

HOCKNULL, S. A., WHITE, M. A., TISCHLER, R. R., COOK, A. G., CALLEJA, N. D., SLOAN, T. & ELLIOTT, D. A. 2009. New mid-Cretaceous (latest Albian) dinosaurs from Winton, Queensland, Australia. *PLoS ONE*, **4**, e6190, doi: 10.1371/journal.pone.0006190

HOOKER, J. J., MILNER, A. C. & SEQUEIRA, S. E. K. 1991. An ornithopod dinosaur from the Late Cretaceous of West Antarctica. *Antarctic Science*, **3**, 331–332.

HORNER, J. R., WEISHAMPEL, D. B. & FORSTER, C. A. 2004. Hadrosauridae. *In*: WEISHAMPEL, D. B., DODSON, P. & OSMÓLSKA, H. (eds) *The Dinosauria*, 2nd edn. University of California Press, Berkeley, CA, 438–463.

INESON, J. R. 1989. Coarse-grained submarine fan and slope apron deposits in a Cretaceous back-arc basin, Antarctica. *Cretaceous Research*, **7**, 141–159.

KRIWET, J. 2003. First record of Early Cretaceous shark (Chondrichthyes, Neoselachii) from Antarctica. *Antarctic Science*, **15**, 519–523.

KRIWET, K., LIRIO, J. M., NUÑEZ, H. J., PUCEAT, E. & LÉCUYER, C. 2006. Late Cretaceous Antarctic fish diversity. *In*: FRANCIS, J. E., PIRRIE, D. & CRAME, J. A. (eds) *Cretaceous–Tertiary High-Latitude Palaeonvironments, James Ross Basin, Antarctica*. Geological Society, London, Special Publications, **258**, 83–100.

LAMBRECHT, K. 1929. Mesozoische und tertiäre Vogelreste aus Siebenbürgen. *In*: CSIKI, E. (ed.) *X Congrés International de Zoologie*. Stephaneum, Budapest, 1262–1275.

MACELLARI, C. E. 1988. Stratigraphy, sedimentology, and paleoecology of Upper Cretaceous/Paleocene shelf-deltaic sediments of Seymour Island. *In*: FELDMANN, R. M. & WOODBURNE, M. O. (eds) *Geology and Paleontology of Seymour Island, Antarctic Peninsula*. Geological Society of America, Boulder, CO, Memoirs, **169**, 25–53.

MAKOVICKY, P. J., APESTEGUÍA, S. & AGNOLIN, F. L. 2005. The earliest dromaeosaurid from South America. *Nature*, **437**, 1007–1011.

MARENSSI, S. A., LIRIO, J. M., SANTILLANA, S. N., MARTINIONI, D. R. & PALAMARCZUK, S. 1992. The Upper Cretaceous of southern James Ross Island, Antarctica. *In*: RINALDI, C. A. (ed.) *Geología de la Isla James Ross, Antártida*. Instituto Antártico Argentino, Buenos Aires, 89–99.

MARENSSI, S. A., REGUERO, M. A., SANTILLANA, S. N. & VIZCAÍNO, S. F. 1994. Eocene land mammals from Seymour Island, Antarctica: palaeogeographical implications. *Antarctic Science*, **6**, 3–15.

MARTIN, J. E. 2006. Biostratigraphy of the Mosasauridae (Reptikia) from the Cretaceous of Antarctica. *In*: FRANCIS, J. E., PIRRIE, D. & CRAME, J. A. (eds) *Cretaceous–Tertiary High-Latitude Palaeonvironments, James Ross Basin, Antarctica*. Geological Society, London, Special Publications, **258**, 101–108.

MARTIN, J. E. & CRAME, J. A. 2006. Palaeobiological significance of high-latitude Late Cretaceous vertebrate fossils from James Ross Basin. Antarctica. *In*: FRANCIS, J. E., PIRRIE, D. & CRAME, J. A.

(eds) *Cretaceous–Tertiary High-Latitude Palaeoenvironments, James Ross Basin, Antarctica*. Geological Society, London, Special Publications, **258**, 109–124.

Martin, J. E., Bell, G., jr. *et al.* 2002. Late Cretaceous mosasaurs (Reptilia) from the Antarctic Peninsula. *In*: Gamble, J. A., Skinner, D. N. B. & Henrys, S. (eds) *Antarctica and the Close of the Millenium, 8th International Symposium on Antarctic Earth Sciences*. Bulletin of the Royal Society of New Zealand, **35**, 293–299.

Mayr, G. 2004. A partial skeleton of a new fossil loon (Aves, Gaviiformes) from the early Oligocene of Germany with preserved stomach content. *Journal of Ornithology*, **145**, 281–286, doi: 10.1007/s10336-004-0050-9

McGowran, B. 1991. Maastrichtian and early Cainozoic, southern Australia: planktonic foraminiferal biostratigraphy. *In*: Williams, M. A. J., de Decker, P. & Kershaw, A. P. (eds) *The Cainozoic in Australia: A Re-Appraisal of the Evidence*. Geological Society of Australia, Special Publications, **18**, 79–98.

Meredith, R. W., Westerman, M., Case, J. A. & Springer, M. S. 2008. A phylogeny and timescale for marsupial evolution based on sequences for five nuclear genes. *Journal of Mammalian Evolution*, **15**, 1–36.

Molnar, R. E. & Wiffen, J. 2007. A presumed titanosaurian vertebra from the late Cretaceous of North Island, New Zealand. *Arquivos do Museu Nacional, Rio de Janeiro*, **65**, 505–510.

Molnar, R. E., López-Angriman, A. & Gasparini, Z. 1996. An Antarctic Cretaceous theropod. *Memoirs of the Queensland Museum*, **39**, 669–674.

Milner, A. C., Hooker, J. J. & Sequeira, S. E. K. 1992. An ornithopod dinosaur from the upper Cretaceous of the Antarctic Peninsula. *Journal of Vertebrate Paleontology*, **12**, 44A.

Norell, M. A. & Makovicky, P. J. 2004. Dromaeosauridae. *In*: Weishampel, D. B., Dodson, P. & Osmólska, H. (eds) *The Dinosauria*. 2nd edn. University of California Press, Berkeley, CA, 196–209.

Noriega, J. I. & Tambussi, C. P. 1995. A Late Cretaceous Presbyornithidae (Aves: Anseriformes) from Vega Island, Antarctic Peninsula. *Paleobiogeographic implications. Ameghiniana*, **32**, 57–61.

Novas, F. E. & Pol, D. 2005. New evidence on deinonychosaurian dinosaurs from the Late Cretaceous of Patagonia. *Nature*, **433**, 858–861.

Novas, F., Fernández, M., Gasparini, Z. B., Lirio, J. M., Núñez, H. J. & Puerta, P. 2002*a*. *Lakumasaurus antarcticus* n. gen. et sp., a new mosasaur (Reptilia, Squamata) from the Upper Cretaceous of Antarctica. *Ameghiniana*, **39**, 245–249.

Novas, F. E., Cambiasso, A. V., Lirio, J. M. & Núñez, H. J. 2002*b*. Paleobiogeografía de los dinosaurios polares de Gondwana. *Ameghiniana*, **39**, 15R.

Olivero, E. B. 2012. Sedimentary cycles, ammonite diversity and palaeoenvironmental changes in the Upper Cretaceous Marambio Group, Antarctica. *Cretaceous Research*, **34**, 348–366, doi: 10.1016/j.cretres.2011.11.015

Olivero, E. B. & Medina, F. A. 2000. Patterns of Late Cretaceous ammonite biogeography in southern high latitudes: the Family Kossmaticeratidae in Antarctica. *Cretaceous Research*, **21**, 269–279.

Olivero, E. B., Scasso, R. A. & Rinaldi, C. A. 1986. Revision of the Marambio Group, James Ross Island, Antarctica. *Contribuciones del Instituto Antártico Argentino*, **331**, 1–28.

Olivero, E., Gasparini, Z., Rinaldi, C. & Scasso, R. 1991. First record of dinosaurs in Antarctica (Upper Cretaceous, James Ross Island): paleogeographical implications. *In*: Thomson, M. R. A., Crame, J. A. & Thomson, J. W. (eds) *Geological Evolution of Antarctica*. Cambridge University Press, Cambridge, 617–622.

Olivero, E. B., Martinioni, D. R. & Mussel, F. J. 1992. Upper Cretaceous sedimentology and biostratigraphy of western Cape Lamb (Vega Island, Antarctica). Implications on sedimentary cycles and evolution of the basin. *In*: Rinaldi, C. A. (ed.) *Geología de la Isla James Ross*. Instituto Antártico Argentino, Buenos Aires, 147–166.

Olivero, E. B., Ponce, J. J., Marsicano, C. A. & Martinioni, D. R. 2007. Depositional settings of the basal López de Bertodano Formation, Maastrichtian, Antarctica. *Revista de la Asociación Geológica Argentina*, **62**, 521–529.

Olivero, E. B., Ponce, J. J. & Martinioni, D. R. 2008. Sedimentology and architecture of sharp-based tidal sandstones in the Upper Marambio Group, Maastrichtian of Antarctica. *Sedimentary Geology*, **210**, 11–26.

Olson, S. 1992. *Neogaeornis wetzeli* Lambrecht, a Cretaceous loon from Chile (Aves: Gaviidae). *Journal of Vertebrate Paleontology*, **12**, 122–124.

Pirrie, D., Crame, J. A. & Riding, J. B. 1991. Late Cretaceous stratigraphy and sedimentology of Cape Lamb, Vega Island, Antarctica. *Cretaceous Research*, **12**, 227–258.

Pirrie, D., Crame, J. A., Lomas, S. A. & Riding, J. B. 1997. Late Cretaceous stratigraphy of the Admiralty Sound region, James Ross Basin, Antarctica. *Cretaceous Research*, **18**, 109–137.

Reguero, M. A. & Gasparini, Z. 2007. Late Cretaceous–Early Tertiary marine and terrestrial vertebrates from James Ross Basin, Antarctic Peninsula: a review. *In*: Rabassa, J. & Borla, M. L. (eds) *Antarctic Peninsula and Tierra del Fuego: 100 Years of Swedish–Argentine Scientific Cooperation at the end of the World*. Taylor & Francis, London, 55–76.

Reguero, M. A. & Marenssi, S. A. 2010. Paleogene climatic and biotic events in the terrestrial record of the Antarctic Peninsula: an overview. *In*: Madden, R., Carlini, A. A., Vucetich, M. G. & Kay, R. (eds) *The Paleontology of Gran Barranca: Evolution and Environmental Change through the Middle Cenozoic of Patagonia*. Cambridge University Press, Cambridge, 383–397.

Reguero, M. A., Vizcaíno, S. F., Goin, F. J., Marenssi, S. A. & Santillana, S. N. 1998. Eocene high-latitude terrestrial vertebrates from Antarctica as biogeographic evidence. *In*: Casadio, S. (ed.) *Paleógeno de América del Sur y de la Península Antártica*. Asociación Paleontológica Argentina, Publicación Especial, Buenos Aires, **5**, 185–198.

Reguero, M. A., Marenssi, S. A. & Santillana, S. N. 2002. Antarctic Peninsula and Patagonia Paleogene terrestrial environments: biotic and biogeographic relationships. *Palaeogeography, Palaeoclimatology, Palaeoecology*, **2776**, 1–22.

RICH, T., VICKERS-RICH, P., FERNÁNDEZ, M. & SANTILLANA, S. 1999. A probable hadrosaur from Seymour Island, Antarctica Peninsula. *In*: TOMIDA, Y., RICH, T. & VICKERS-RICH, P. (eds) *Proceedings of the Second Gondwana Dinosaur Symposium*. National Science Museum, Tokyo, 219–222.

RICHTER, M. & WARD, D. J. 1990. Fish remains from Santa Marta Formation (Late Cretaceous) of James Ross Island, Antarctica. *Antarctic Science*, **2**, 67–76.

RIDING, J. B. & CRAME, J. A. 2002. Aptian to Coniacian (Early–Late Cretaceous) palynostratigraphy of the Gustav Group, James Ross Basin, Antarctica. *Cretaceous Research*, **23**, 739–760.

RINALDI, C. A. 1982. The Upper Cretaceous in the James Ross Island Group. *In*: CRADDOCK, C. (ed.) *Antarctic Geoscience*. The University of Wisconsin Press, Madison, WI, 331–337.

ROESER, H. A., FRITSCH, J. & HINZ, K. 1996. The development of the crust off Dronning Maud Land, East Antarctica. *In*: STOREY, B. C., KING, E. C. & LIVERMORE, R. A. (eds) *Weddell Sea Tectonics and Gondwana Break-up*. Geological Society, London, Special Publications, **108**, 243–264.

SALGADO, L. & CORIA, R. 1996. First evidence of an armoured ornithischian dinosaur in the late Cretaceous of North Patagonia, Argentina. *Ameghiniana*, **33**, 367–371.

SALGADO, L. & GASPARINI, Z. 2006. Reappraisal of an ankylosaurian dinosaur from the Upper Cretaceous of James Ross Island (Antarctica). *Geodiversitas*, **28**, 119–135.

SAMPSON, S. D., WITMER, L. M., FORSTER, C. A., KRAUSE, D. W., CONNOR, P. O., DODSON, P. & RAVOAVY, F. 1998. Predatory dinosaur remains from Madagascar: implications for the Cretaceous biogeography of Gondwana. *Science*, **280**, 1048–1051.

SCASSO, R. A., OLIVERO, E. B. & BUATOIS, L. A. 1991. Lithofacies, biofacies and ichnoassemblage evolution of a shallow submarine volcaniclastic fan-shelf depositional system (Upper Cretaceous, James Ross Island, Antarctica). *Journal of South American Earth Sciences*, **4**, 239–260.

SCOTESE, C. R. 1997. *Paleogeographic Atlas*. PALEOMAP Progress Report 90-0497. Department of Geology, University of Texas at Arlington, Arlington, TX.

SCOTESE, C. R. & GOLONKA, J. 1993. *Paleogeographic Atlas*. PALEOMAP project. University of Texas at Arlington, Arlington, TX.

SERENO, P. C. & BRUSATTE, S. L. 2008. Basal abelisaurid and carcharodontosaurid theropods from the Lower Cretaceous Elrhaz Formation of Niger. *Acta Palaeontologica Polonica*, **53**, 15–46.

SHEN, Y. 1995. Subdivision and correlation of Cretaceous to Paleogene volcanosedimentary sequence from Fildes Peninsula, King George Island, Antarctica. *In*: SHEN, Y. (ed.) *Stratigraphy and Palaeontology of Fildes Peninsula, King George Island, Antarctica*. State Antarctic Committee/Science Press, Beijing, Monographs, **3**, 1–36.

SMITH, N. D. & POL, D. 2007. Anatomy of a basal sauropodomorph dinosaur from the Early Jurassic Hanson Formation of Antarctica. *Acta Palaeontologica Polonica*, **52**, 657–674.

SMITH, A. G., SMITH, D. G. & FUNNELL, B. M. 1995. *Atlas of Mesozoic and Cenozoic Coastlines*. Cambridge University Press, Cambridge.

SMITH, N. D., MAKOVICKY, P., HAMMER, W. R. & CURRIE, P. J. 2007. Osteology of *Cryolophosaurus ellioti* from the Early Jurassic of Antarctica and implications for early theropod evolution. *Zoological Journal of the Linnean Society*, **151**, 377–421.

SMITH, N. D., MAKOVICKY, P. J., AGNOLIN, F. L., EZCURRA, M. D., PAIS, D. F. & SALISBURY, S. W. 2008. A Megaraptor-like theropod (Dinosauria: Tetanurae) in Australia: support for faunal exchange across eastern and western Gondwana in the Mid-Cretaceous. *Proceedings of the Royal Society B: Biological Sciences*, **275**, 2085–2093.

STAHL, B. J. & CHATTERJEE, S. 1999. A Late Cretaceous chimaerid (Chondrichthyes, Holocephali) from Seymour Island, Antarctica. *Palaeontology*, **42**, 979–989.

STAHL, B. J. & CHATTERJEE, S. 2002. A Late Cretaceous callorhynchid (Chondrichthyes, Holocephali) from Seymour Island, Antarctica. *Journal of Vertebrate Paleontology*, **22**, 848–850.

STOREY, M., MAHONEY, J. J., SAUNDERS, A. D., DUNCAN, R. A., KELLEY, S. P. & COFFIN, M. F. 1995. Timing of hot-spot related volcanism and the break-up of Madagascar and India. *Science*, **267**, 852–855.

TAMBUSSI, C. P. & ACOSTA HOSPITALECHE, C. 2007. Antarctic birds (Aves) during the Cretaceous–Eocene times. *Revista de la Asociación Geológica Argentina*, **62**, 604–617.

THOMPSON, R. S., PARISH, J. C., MAIDMENT, S. C. R. & BARRETT, P. M. 2012. Phylogeny of the ankylosaurian dinosaurs (Ornithischia: Thyreophora). *Journal of Systematic Palaeontology*, **10**, 301–312.

UPCHURCH, P., HUNN, C. A. & NORMAN, D. B. 2002. An analysis of dinosaurian biogeography: evidence for the existence of vicariance and dispersal patterns caused by geological events. *Proceedings of the Royal Society of London B*, **269**, 613–621.

WHITHAM, A. G., INESON, J. R. & PIRRIE, D. 2006. Marine volcaniclastic of the Hidden Lake Formation (Coniacian) of James Ross Island, Antarctica: an enigmatic element in the history of a back-arc basin. *In*: FRANCIS, J. E., PIRRIE, D. & CRAME, J. A. (eds) *Cretaceous–Tertiary High-Latitude Palaeoenvironments, James Ross Basin, Antarctica*. Geological Society, London, Special Publications, **258**, 21–47.

WOODBURNE, M. O. & CASE, J. A. 1996. Dispersal, vicariance, and the post-Gondwana Late Cretaceous to early Tertiary biogeography from South America to Australia. *Journal of Mammalian Evolution*, **3**, 121–161.

WOODBURNE, M. O. & ZINSMEISTER, W. J. 1984. The first land mammal from Antarctica and its biogeographic implications. *Journal of Paleontology*, **54**, 913–948.

WOODWARD, A. S. 1908. On fossil fish-remains from Snow Hill and Seymour Islands. *Wissenschaftliche Ergebnisse der Schwedischen Südpolar-Expedition 1901–1903*, **3**, 1–4.

ZINSMEISTER, W. J. 1979. Biogeographic significance of the Late Mesozoic and Early Tertiary molluscan faunas of Seymour Island (Antarctic Peninsula) to the final breakup of Gondwanaland. *In*: GRAY, J. & BOUCOT, A. (eds) *Historical Biogeography, Plate Tectonics and the Changing Environment, Proceedings, 37th Annual Biological Colloquium and Selected Papers*, Oregon State University Press, Corvallis, OR, 349–355.

ZINSMEISTER, W. J. 1982. Late Cretaceous–Early Tertiary molluscan biogeography of southern Circum-Pacific. *Journal of Paleontology*, **56**, 84–102.

Palaeobiogeography of Austral echinoid faunas: a first quantitative approach

THOMAS SAUCEDE*, BENJAMIN PIERRAT, ARNAUD BRAYARD & BRUNO DAVID

UMR CNRS 6282 Biogéosciences, université de Bourgogne, 6, boulevard Gabriel, 21000 Dijon, France

**Corresponding author (e-mail: thomas.saucede@u-bourgogne.fr)*

Abstract: Few studies have been devoted to the palaeobiogeography of Antarctic echinoids, all of them analysing and discussing distribution patterns in a qualitative way. The present work aims at exploring the evolution of palaeobiogeographic relationships of Austral echinoid faunas through four time intervals, from the Maastrichtian to the present day, using a quantitative approach: the Bootstrapped Spanning Network procedure. Analyses were successfully performed and improve our knowledge of biogeographic relationships between the different Austral regions. Biogeographic maps were produced that can be easily and intuitively discussed. Our results mostly agree with palaeobiogeographic studies performed on other benthic invertebrates and are congruent with the palaeogeographic evolution of Antarctica. However, two main points markedly contrast with other works: there is no evidence of an Austral provincialism at the end of the Cretaceous and early Cenozoic, and echinoid data suggest isolation of southern Argentina from other Austral regions, including Antarctica, in the Early Miocene.

Earth history has had a deep influence on the evolution of the biosphere and consequently on the distribution of life through time. Regarding the Southern Ocean, the final break-up of Gondwana followed by the onset of the southern Pacific and Antarctic circumpolar surface currents during the Cenozoic have had a determinant role explaining the biogeography and diversity of modern Austral faunas (Zinsmeister 1979, 1981; Zinsmeister & Camacho 1980; Beu *et al.* 1997; Del Rio 2002; Pearse *et al.* 2009). Such events are considered to have promoted the two biogeographic processes that have long been invoked to explain the geographically discontinuous distribution patterns of taxa on Earth, namely (1) divergence by vicariance, an outcome of the fragmentation of Gondwana, and (2) dispersal, alternatively promoted or restricted in particular directions by southern surface currents. Both processes have probably worked in unison to shape the modern distribution of Austral faunas (Pearse *et al.* 2009).

Stressing the similarities among the gastropod faunas of Australasia, Antarctica and southern South America in the Late Cretaceous and early Palaeogene, Zinsmeister postulated the existence of a unique Austral faunistic province at that time, the so-called Weddellian province (Zinsmeister 1979, 1981; Zinsmeister & Camacho 1980). Then, the final break-up between Australia and Antarctica along with the northward drift of Australia and New Zealand to warmer conditions by the Late Eocene was associated with the appearance of distinctive faunas in these two latter regions. Faunal connections between Antarctica and southern South America persisted until the Drake Passage opened to depth about the Eocene–Oligocene boundary (Zinsmeister 1981; Beu *et al.* 1997; Scher & Martin 2006; Lawver *et al.* 2011). The onset of a southern Pacific surface current followed by the Antarctic circumpolar current in the Late Eocene was also considered to have influenced the evolution of Austral biogeography (Beu *et al.* 1997; Pearse *et al.* 2009). Hence, Beu *et al.* (1997) and Del Rio (2002) highlighted the existence of faunal similarities between New Zealand and southern South America owing to both eastward and westward dispersals from the Oligocene to the Holocene, while Pearse *et al.* (2009) stressed the role of the Antarctic circumpolar current in promoting speciation within brooding clades of the Southern Ocean.

Most palaeobiogeographic studies of Upper Cretaceous and Cenozoic Austral benthic invertebrates have been based on bivalve and gastropod faunas (Zinsmeister 1979; Zinsmeister & Camacho 1980; Crame 1996; Beu *et al.* 1997; Stilwell *et al.* 2004; Beu 2009), whereas few have considered fossil data across other benthic invertebrate groups such as brachiopods and echinoderms (Zinsmeister 1981). With only approximately 20 fossil species recorded in Antarctica so far, echinoids could appear at first sight as non-informative for palaeobiogeographic purposes. However, most fossils collected in the northern tip of the Antarctic Peninsula for more

From: Hambrey, M. J., Barker, P. F., Barrett, P. J., Bowman, V., Davies, B., Smellie, J. L. & Tranter, M. (eds) 2013. *Antarctic Palaeoenvironments and Earth-Surface Processes*. Geological Society, London, Special Publications, **381**, 117–127. First published online April 19, 2013, http://dx.doi.org/10.1144/SP381.6

than one century are Maastrichtian, Late Eocene and Early Miocene in age, namely three key time periods to document respectively the situation before the Gondwanan final break-up, the outcome of the northward continental drift of Australia and New Zealand, and the biogeographic consequences of the opening of the Drake Passage and setting up of the Antarctic circumpolar current. Fossil echinoids of the same time periods are recorded in southern Argentina, South Australia and New Zealand, so that the evolution of palaeobiogeographic relationships among these nowadays distinct parts of Gondwana can be investigated using echinoid data. Although palaeobiogeographic relationships among Austral fossil echinoid faunas were formerly discussed (Hotchkiss 1982; Néraudeau *et al.* 2000; Néraudeau & Mathey 2000), it was only in a qualitative way. The present work aims at exploring the palaeobiogeographic relationships through the three time periods mentioned above and comparing them with the present-day situation using the same quantitative approach.

Material and method

Fossil data and taxonomic issues

The three studied geological periods (the Maastrichtian, Late Eocene and Early Miocene) are known in the southern high latitudes from stratigraphic levels that crop out in Antarctica (Table 1), southern Argentina, Australia and New Zealand. The time resolution of the analysis was determined by the biostratigraphic resolution and incompleteness of the known Antarctic (Table 1) and Argentinian fossil record. Seven areas were determined where fossil deposits from the three studied time periods crop out: the western part of the Antarctic Peninsula (including The James Ross basin and King George island), southern Argentina (Patagonia), southeastern and southwestern Australia, Chatham Island, and North and South New Zealand. Present-day distribution data correspond to echinoid samples collected during oceanographic campaigns led in close nearshore marine areas (Fig. 1).

Analyses were performed at genus level to make up for systematic bias and low stratigraphic resolution. Material from GNS Science in Lower Hutt (New Zealand) and from the Museo Argentino de Ciencas Naturales in Buenos Aires (Argentina) was examined to check on systematic reliability and consistency, a recurrent issue in palaeobiogeographic studies (Brayard *et al.* 2007). This was completed with data from the literature (Table 1) and from the online database FRED (Fossil Record Electronic Database) for fossil echinoids from New Zealand. Present-day occurrence data come from the updated version of the Antarctic Echinoid Database (David *et al.* 2005), which integrates most records collected during oceanographic cruises led in the Southern Ocean until 2003. They were augmented by new records collected during cruises led in Antarctica since 2003 (ANTXXIII/8, BENTART, 2003, 2006 and JR230) and with northern data collected as far as latitude 35°S. Specimens housed in public collections were examined (National Institute of Water and Atmospheric Research in Wellington, Museo Argentino de Ciencas Naturales in Buenos Aires, Melbourne Museum in Melbourne, Australian Museum in Sydney). The dataset used in this work corresponds to species occurrences recorded off the land areas where fossil deposits crop out. It covers the parts of continental shelves that best fit with the location of fossil sampling areas (Fig. 1).

Palaeobiogeographic analyses

Regional similarities were analysed using the Bootstrapped Spanning Network (BSN) procedure, namely a non-hierarchical clustering method recently developed for palaeobiogeographic studies and efficient for identification of both biogeographic groupings and transitional areas (Brayard *et al.* 2007). This approach gives a simple, intuitively legible picture of the nested as well as gradational taxonomic similarity relationships, hence providing a good synthesis (and additional insights) between hierarchical clustering and ordination in reduced space results. Associated with each occurrence matrix, a dissimilarity matrix was here computed using the Bray & Curtis (1957) coefficient, which gives a double weight to shared presences and thus relative underweight to absence and unique occurrence as an indication of faunal differences (Brayard *et al.* 2007). Similarity relationships were displayed as a connected network supported by 10 000 bootstrap iterations giving confidence intervals for each connection between bioregions (see Brayard *et al.* 2007 for an exhaustive description of the procedure). Displayed on (palaeo)geographic maps, the BSN allows a robust appraisal of faunal exchange pathways, including faunal gradients or reticulated biogeographic structures. BSN were computed using the software BSN v. 1.0 (Brayard *et al.* 2007) and results were visualized using the program PAJEK v. 1.07 (Batagelj & Mrvar 2005).

A total of 845 fossil occurrences, which correspond to the different locations and geological time periods for which a genus is recorded (17 for the Antarctic Peninsula, 69 for southern Argentina, 316 for South Australia, 33 for Chatham island, 244 for North New Zealand and 166 for South New Zealand) and 3158 present-day occurrences

Table 1. *Fossil echinoids identified in Antarctica, from Late Cretaceous to Early Miocene*

Echinoid taxa	Synonym	Geologic Formation	Time period	Location	Literature consulted
Abatus kieri McKinney, McNamara & Wiedman, 1988	*?Schizaster antarcticus* Lambert, 1910	La Meseta	Late Eocene	Seymour Island	Lambert, 1910; McKinney, McNamara & Wiedman, 1988
Almucidaris durhami Blake & Zinsmeister, 1991		López de Bertodano	Maastrichtian	Seymour Island	Blake & Zinsmeister, 1991
Austrocidaris seymourensis Radwanska, 1996		La Meseta	Eocene	Seymour Island	Radwanska, 1996
Cyathocidaris erebus Lambert, 1910			Maastrichtian	Seymour Island	Lambert, 1910
Cyathocidaris nordenskjoeldi Lambert, 1910		López de Bertodano	Maastrichtian	Seymour Island	Lambert, 1910; Néraudeau, Crame & Kooser, 2000
Cyathocidaris patera Lambert, 1910		López de Bertodano	Maastrichtian	Seymour Island	Lambert, 1910; Néraudeau, Crame & Kooser, 2000
Giraliaster lorioli (Lambert, 1910)	*Holaster lorioli* Lambert, 1910	Snow Hill	Maastrichtian	Snow Hill Island	Lambert, 1910; Néraudeau, Crame & Kooser, 2000
Hemiaster (Bolbaster) vomer (Lambert, 1910)	*Vomeraster vomer* Lambert, 1910	Snow Hill	Maastrichtian	Snow Hill Island	Lambert, 1910; Néraudeau, Crame & Kooser, 2000
Huttonechinus antarctica Néraudeau, Crame & Kooser, 2000		Santa Marta	Late Cretaceous	James Ross Island	Néraudeau, Crame & Kooser, 2000
?Iheringiella sp.			Late Eocene	Black Island, McMurdo Sound	Hotchkiss & Fell, 1972
Micraster aff. regularis Arnaud, 1883		Santa Marta	Late Cretaceous	James Ross Island	Néraudeau, Crame & Kooser, 2000
Nordenskjoeldaster antarctica Lambert, 1910		Snow Hill	Maastrichtian	Snow Hill Island	Lambert, 1910; Bernasconi, 1959
Nordenskjoeldaster australis Néraudeau, Crame & Kooser, 2000		Santa Marta	Campanian	James Ross Island	Néraudeau, Crame & Kooser, 2000
?Notocidaris sp.		Cape Melville	Early Miocene	King George Island	Jesionek-Szymanska, 1987
?Schizaster sp.		Cape Melville	Early Miocene	King George Island	Jesionek-Szymanska, 1987
?Sterechinus sp.	*?Isechinus sp.* Gazdzicki & Wrona, 1982	Cape Melville	Early Miocene	King George Island	Jesionek-Szymanska, 1987
Stigmatopygus andersoni (Lambert, 1910)	*Cassidulus andersoni* Lambert, 1910	La Meseta	Eocene	Seymour Island	Lambert, 1910; McKinney, McNamara & Wiedman, 1988

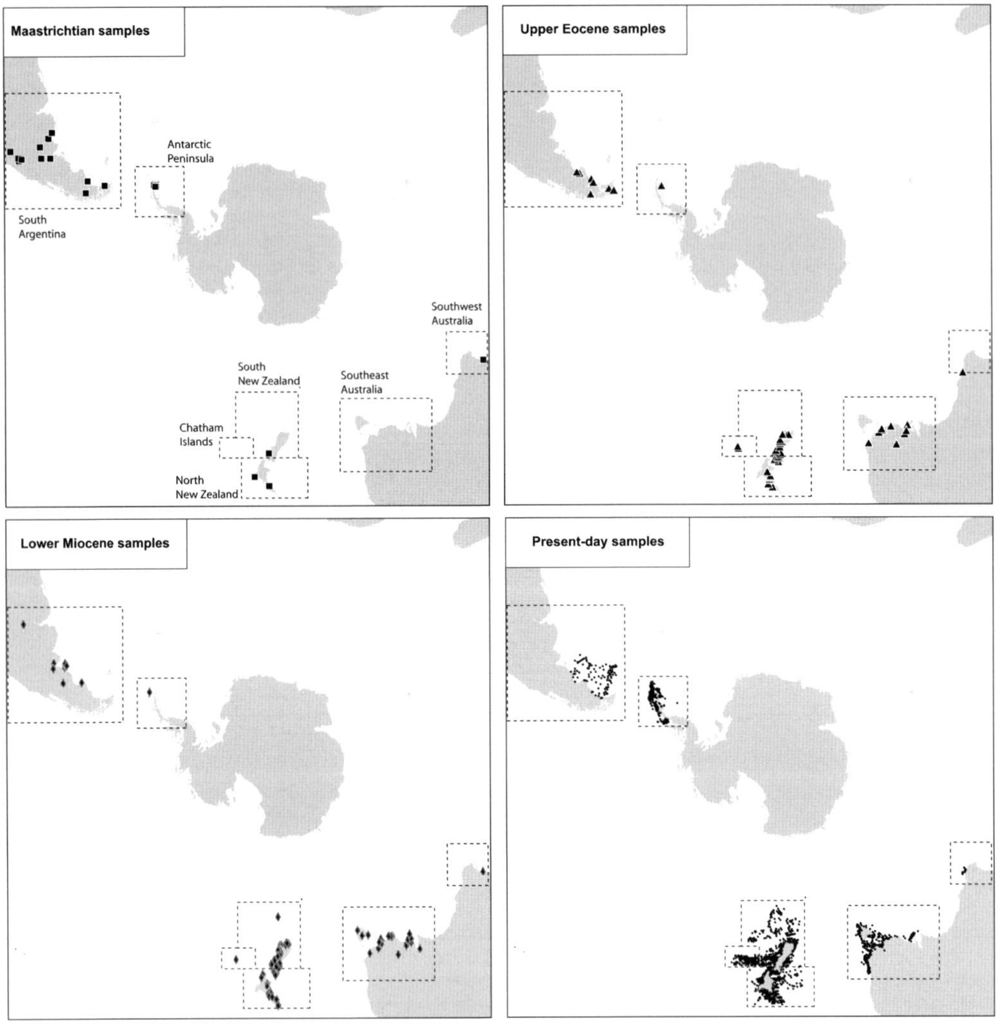

Fig. 1. Location of sampling sites for the Maastrichtian (squares), Late Eocene (triangles), Early Miocene (diamonds) and present day (solid circles).

(465 for the Antarctic Peninsula, 115 for southern Argentina, 704 for South Australia, 186 for Chatham island, 843 for North New Zealand and 845 for South New Zealand) were included within our final dataset and analysed (Fig. 1).

Fossil record and taxonomic richness

Genus richness values contrasted between both studied areas and time periods (Fig. 2). Overall, the total genus richness increases from 15 genera recorded in the Maastrichtian to 42 in the Late Eocene and 65 in the Early Miocene. The total present-day richness is represented by 165 genera. The incompleteness of our knowledge of the fossil record is obvious and could potentially bias the analyses. To assess this potential bias, the correlation significance between taxonomic dissimilarities and richness differences was addressed using a Mantel test. The test was performed between the matrix of taxonomic dissimilarities and the matrix of richness differences computed for each pairwise comparison between regions as the ratio of richness difference to total richness. The correlation value is 0.5 and significance was tested ($p < 0.01$; 5000 bootstrap iterations) using the software PAST v. 1. 92 (Hammer *et al.* 2001). Taxonomic dissimilarities are thus partly explained by richness differences. The relationship between faunal similarities and genus richness can be

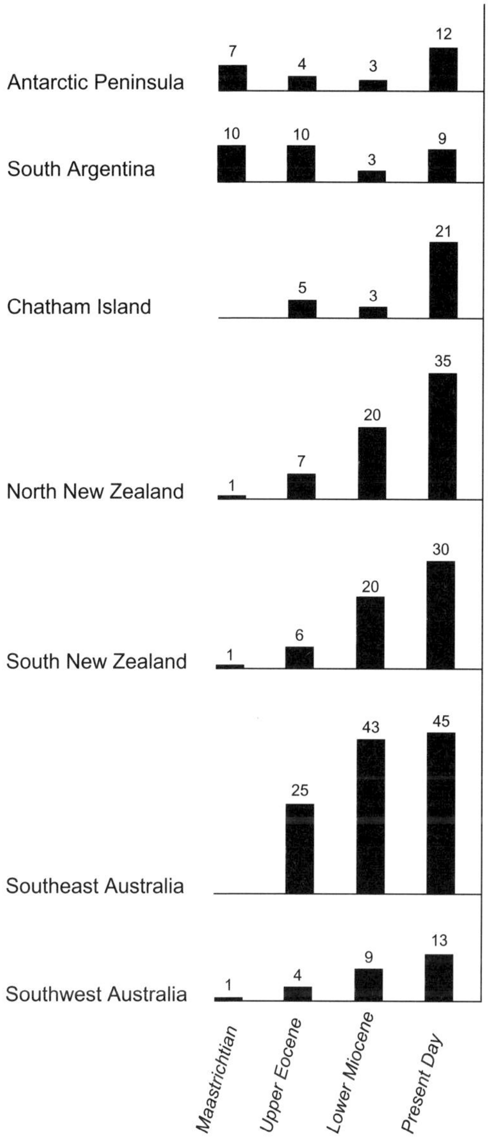

Fig. 2. Genus richness values for the seven studied areas and four time intervals.

interpreted either as biogeographically significant (e.g. provincialism and regional burst of diversity) or as a result of the uneven quality of data.

This stands out clearly when richness values are compared with the number of occurrences that increases from the Maastrichtian to the present day. On Chatham Island, in southern Argentina and the Antarctic Peninsula, richness values decrease from the Maastrichtian to the Early Miocene, then increase to the present day. Richness values and the number of fossil deposits investigated in the Antarctic Peninsula and Chatham Island vary in the same way; richness variation can therefore be interpreted as an outcome of collecting biases and should be interpreted with caution. However, in southern Argentina richness values do not seem to be correlated with the number of sampling sites, which are the most numerous for the Early Miocene (13) and Present Day (115), whereas richness values are the lowest for these time periods (Fig. 2). Therefore, the low richness values of the Early Miocene and present day in southern Argentina do not seem to be entirely determined by sampling biases and probably correspond to a distinctive biogeographic context. The low Argentinian richness sharply contrasts with the burst of echinoid richness recorded in the Late Eocene and Early Miocene of Australia and New Zealand (Fig. 2). These high richness values reflect the radiation of several regular and irregular families already documented in previous works (Philip & Foster 1971, 1977; Zinsmeister & Camacho 1980; Hotchkiss 1982) and obviously show up compared with the situation in the Maastrichtian, a period characterized by a poor fossil record in these regions.

Palaeobiogeographic relationships

BSN analyses were successfully performed and bootstrap support values stress robust, significant connections between regions despite the unevenness of fossil and present-day data. In Figure 3, faunal similarities among studied areas are shown as a connected network for each time period. Bold lines correspond to robust connections with a 100% confidence interval whereas weaker connections are depicted as dashed lines associated with their respective bootstrap support values.

Maastrichtian

Fossil faunas of the Maastrichtian remain insufficiently known and all computed faunal similarities are weak or not sufficiently contrasted. Thus, no robust network was computed nor displayed for this time period.

Despite the limited fossil record, especially in Australia and New Zealand, a qualitative survey of the echinoid faunas of the Late Cretaceous suggests the predominance of cosmopolitan genera with European (Boreal and Tethyan) affinities. Some authors also pointed out taxonomic similarities between faunas of West Antarctica and southern Argentina (Furque & Camacho 1949; Bernasconi 1954; Hotchkiss 1982). This is evidenced by the common occurrence of the genera *Cyathocidaris*, *Nordenskjoeldaster* and *Hemiaster* in both regions.

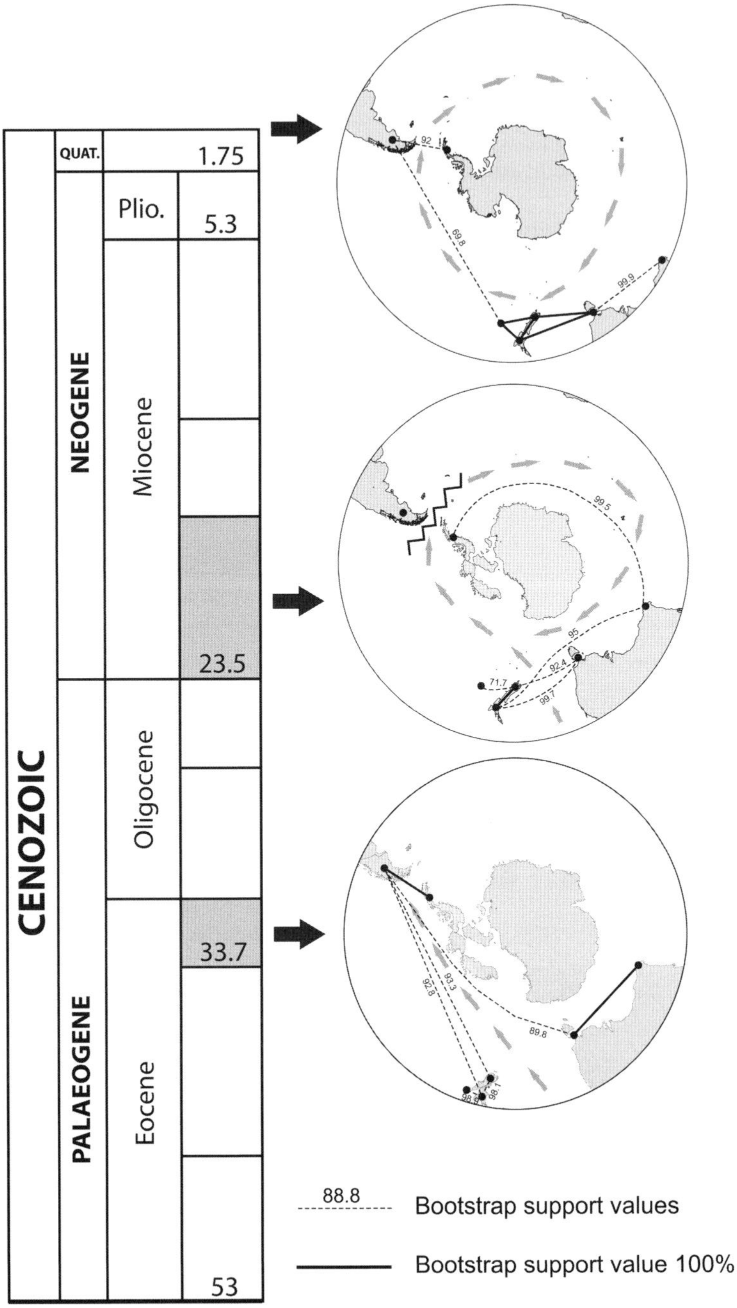

Fig. 3. Bootstrapped spanning networks showing echinoid faunal similarities for the Late Eocene, Early Miocene and present day (ages in millions of years ago). Connection robustness is given by bootstrap support values. Zigzag line symbolizes the biogeographic isolation of southern Argentina from Austral areas in the Early Miocene. Grey arrows depict the position and direction of the southern Pacific (Late Eocene) and Antarctic circumpolar (Early Miocene and present day) surface currents. Map background and current position modified from Lawver & Gahagan (2003).

Faunal connections between southern Argentina and New Zealand are also evidenced by the common presence of *Cyathocidaris* and *Diplodetus* in both regions.

Zinsmeister (1979) proposed the existence of the so-called Weddellian faunistic province to account for these faunal similarities along the southern margins of the Pacific at the end of the Cretaceous and early Cenozoic. The Gondwanan break-up and the prevalence of shallow-water seaways between land masses is considered to have promoted the origination and diversification of marine invertebrates in the SW Pacific Ocean. These taxa spread readily to the east in the early Cenozoic (Zinsmeister & Camacho 1980). This hypothesis is mostly supported by biogeographic studies of Struthiolariidae gastropods (Zinsmeister 1979; Zinsmeister & Camacho 1980). Néraudeau *et al.* (2000) highlighted echinoid faunal similarities between Boreal, Tethyan and Austral regions in the Late Cretaceous, contrasting markedly with Zinsmeister's hypothesis of a distinctive Austral province. The high level of echinoid cosmopolitanism at the end of the Cretaceous is documented by the occurrence of the following European genera in Austral areas: *Micraster* (Antarctic Peninsula), *Hemiaster* (Antarctic Peninsula), *Cyathocidaris* (Antarctic Peninsula, southern Argentina and New Zealand), *Echinocorys* (Australia), *Holaster* (southern Argentina) and *Nucleopygus* (southern Argentina) (Néraudeau & Mathey 2000). Unlike Struthiolariidae gastropods (Zinsmeister & Camacho 1980), the sparse echinoid data do not show any evidence of Austral provincialism during the Maastrichtian. A SW Pacific origin of Austral echinoid faunas followed by an eastward dispersal to South America is also not supported.

Late Eocene

BSN results for the Late Eocene show strong connections between southern Argentina and the Antarctic Peninsula on the one hand, and between southeastern and southwestern Australia on the other (Fig. 3). Trans-Pacific connections between southern Argentina and southeastern Australia as well as between southern Argentina and New Zealand are also well supported.

Previous studies of Austral echinoid faunas have suggested the existence of faunal affinities between southern Argentina and Antarctica in the Late Eocene (Bernasconi 1959; Hotchkiss & Fell 1972; Néraudeau *et al.* 2000). This is congruent with Zinsmeister's (1981) study that alluded to the Austral affinity of mollusc, echinoid and brachiopod faunas from southern Argentina. This strong connection between the two areas is typically exemplified by the occurrence of the shallow-water cassiduloid *Stigmatopygus* in both regions. This supports the existence of shallow-water seaways between the two areas during the Late Eocene (Zinsmeister 1981).

Philip & Foster (1971, 1977) and Zinsmeister & Camacho (1980) documented a burst of faunal diversity in Australia and New Zealand in Eocene times. The faunas are interpreted as either of Austral (Philip & Foster 1977) or Indo-Pacific origin (Zinsmeister & Camacho 1980). The diversity and distinctiveness of Australian echinoid faunas are mostly due to cassiduloid and spatangoid irregular echinoids and to the regular family Temnopleuridae, whereas faunal affinities between Chatham Island and New Zealand are the result of holasteroid occurrences (*Giraliaster* and *Corystus*). Foster & Philip (1978) also stressed the existence of recurrent echinoid trans-Tasman dispersals from southeastern Australia to New Zealand throughout the Cenozoic. This is not supported for the Late Eocene in the present analysis (Fig. 3).

The genera *Schizaster*, *Linthia*, *Austrocidaris* and *Prionocidaris* (the Patagonian species '*Cidaris*' *antarctica* Loriol (1902) and '*Cidaris*' *julianensis* Loriol (1902) show strong similarities with species of the genus *Prionocidaris* to which they are herein assigned) have a widespread distribution that supports long-distance connections from Australia to southern Argentina (*Schizaster* and *Linthia*) and to the Antarctic Peninsula (*Austrocidaris*) as well as between New Zealand and southern Argentina (*Prionocidaris*). These southern Trans-Pacific connections between southern Argentina, New Zealand and southeastern Australia are consistent with geographic and oceanographic reconstructions for that time (Lawver & Gahagan 2003) and congruent with other palaeontological studies based on echinoids (Hotchkiss 1982) and molluscs (Beu *et al.* 1997; Del Rio 2002). These connections can be interpreted either as an inheritance of the Zinsmeister's (1979) Weddellian province, which could explain the distribution of the genus *Schizaster*, mentioned in South Argentina in the Late Cretaceous (Parma & Casadio 2005), or as the result of dispersal through the cool surface current that circulated along the Pacific margin of Antarctica (Zinsmeister 1979; Beu *et al.* 1997; Lawver & Gahagan 2003). The genus *Schizaster* has a worldwide distribution and is mainly known as a typical Mediterranean and Tethyan echinoid of the Eocene–Miocene time period. Therefore, the wide austral distribution of the genus cannot be considered an inheritance of the Weddellian province. The same is true for the genus *Linthia*, which is probably of Turonian origin from France (D. Néraudeau pers. comm.). The current that circulated along the Pacific margin of Antarctica is considered to have initiated late in the Eocene (Fig. 3) when the shelf of the South Tasman Rise started to subside (Lawver & Gahagan 2003) and to have favoured the westward dispersal of taxa that are unknown in Austral areas before the Eocene. This could explain the widespread distribution of the genera *Austrocidaris* and *Prionocidaris*.

In spite of the existence of long-distance connections from Australia to the Antarctic Peninsula, most genera are less widely distributed. Within the family Schizasteridae, if genera such as *Schizaster* and *Linthia* are distinguished by a widespread distribution, other genera are restricted to only one region, such as *Abatus* in the Antarctic Peninsula and *Kina* in New Zealand. The origination of the two genera *Abatus* and *Kina* from a schizasterid ancestor along with their endemism could suggest a divergent evolution within the family, an obvious outcome of continental drift. However, phylogenetic relationships within the Schizasteridae require clarification, especially between the genus *Abatus* and its South African relatives. Only marsupiate species with sunken petals were assigned to the genus. They might be closely related to non-Antarctic and non-marsupiate species, and a genus can be represented by marsupiate species in a region of the world ocean, and non-marsupiate species in other areas (Dudicourt *et al.* 2005).

Early Miocene

The most noticeable result of the BSN analysis performed for the Early Miocene is the faunal isolation of southern Argentina from other Austral regions, including a south Pacific disconnection between southern Argentina and New Zealand. Results also show a direct connection between southwestern Australia and New Zealand, and a faunal disconnection between southeastern and southwestern Australia. An unexpected but robust relationship between the Antarctic Peninsula and southwestern Australia is also highlighted (Fig. 3).

The isolation of southern Argentina is partly congruent with Del Rio (2002), who highlighted the Caribbean affinity of part of Lower Miocene Patagonian molluscs. However, a faunal connection between Patagonia and New Zealand was also stressed by Del Rio (2002) and Beu *et al.* (1997). Based on echinoid data, there is no evidence so far for such a connection, despite the establishment of the Antarctic circumpolar current at the Eocene–Oligocene boundary and the possibility of subsequent westward dispersal from New Zealand (Beu *et al.* 1997; Lawver & Gahagan 2003). The fossil record is weak in southern Argentina and Antarctica (only three echinoid genera are known in Argentina and West Antarctica; Table 1), but the faunal distinctiveness between these two areas is well supported by the occurrence of peculiar South American faunas. In southern Argentina, echinoids are typically represented by the clypeasteroid family Monophorasteridae (*Amplaster* and *Monophoraster*) and the genus *Abertella* (Hotchkiss 1982; Mooi *et al.* 2000). These echinoids are unknown from other Austral areas. Hotchkiss & Fell (1972) postulated an Antarctic affinity of southern Argentinian clypeasteroids after the discovery of the genus *Iheringiella* in the Upper Eocene of East Antarctica, but interestingly Mooi *et al.* (2000) excluded the latter genus from the family Monophorasteridae, which is unequivocally restricted to South America. Moreover, the occurrence of the genus *Iheringiella* in the Upper Eocene of Antarctica (Hotchkiss & Fell 1972) is suspect (Table 1) considering that the only known specimen consists of a fragment without diagnostic features and that the genus is recorded from Miocene deposits in South America (R. Mooi pers. comm.). In Australia, the diversification of clypeasteroid echinoids gave rise to the families Fossulasteridae, Scutellinoididae and Arachnoididae. Diversity of Australasian echinoids is also typified by cassiduloids (e.g. *Pisolampas*), spatangoids (e.g. *Eupatagus*) and cidaroids (e.g. *Goniocidaris*). The southwestern Australian genera *Cardabia* and *Tripneustes* are absent in southeastern Australia, while the latter is present in New Zealand and Chatham Island. This supports the direct connection between southwestern Australia and New Zealand, and the faunal disconnection between southeastern and southwestern Australia. However, the genus *Tripneustes* is known from Miocene deposits of Switzerland and France, so that its absence from southeastern Australia could be due to incomplete sampling.

The robust relationship between the Antarctic Peninsula and southwestern Australia is mostly supported by the widespread distribution of the genus *Schizaster* (except in southern Argentina where the genus is not so far recorded in the Early Miocene). This widespread distribution extends back to the Late Eocene, the genus being recorded worldwide (see discussion above). Therefore, the robust relationship between the Antarctic Peninsula and southwestern Australia in the Early Miocene can be interpreted as an inheritance of distribution patterns of the past, and not as the result of contemporary faunal connections. This interpretation is corroborated by the divergent evolution of cidarid echinoids, which gave rise to distinct genera between Antarctica and Australia in the Early Miocene. For example, the genus *Goniocidaris* is first recorded in the Lower Miocene of New Zealand and Australia whereas in Antarctica, cidarids are known by the genus *Notocidaris*. The two genera diversified in different areas during the Neogene, the genus *Notocidaris* being mostly restricted to the Antarctic continental shelf (except for one Recent species in New Zealand), whereas *Goniocidaris* species spread to the Indo-Pacific region (Mortensen 1928; David *et al.* 2005). The geographically limited evolution and diversification of the genus *Notocidaris* in Antarctica supports the relative isolation of the continent from Australasia.

Present day

The comparison of BSN results for present day faunas with those of previously studied time intervals highlights the strong affinity between Australian and New Zealand modern faunas, including those of Chatham Island. The present day configuration also displays a relatively well-supported connection between southern Argentina and the Antarctic Peninsula, and the existence of a weak faunal connection between southern Argentina and Chatham Island (Fig. 3).

The similarity of echinoid faunas between Australia and New Zealand may be explained by the existence of recurrent trans-Tasman faunal exchanges between southeastern Australia and New Zealand throughout the Cenozoic (Foster & Philip 1978). Spatangoid echinoids are highly diversified in Australia and New Zealand (e.g. Maretiidae, Brissidae, among others), but most are restricted to these areas and have no representative in Antarctica or southern Argentina.

The relatively well-supported connection between southern Argentina and the Antarctic Peninsula across the Scotia Arc has been extensively documented in previous studies across various benthic groups, from bivalves and gastropods to polychaetes and isopods (e.g. Hedgpeth 1969; Linse *et al.* 2006; Clarke *et al.* 2007; Brandt *et al.* 2009). Based on echinoids, the connection can be explained by the extended latitudinal range of the schizasterid genera *Abatus* and *Tripylus*, and of the echinid genus *Sterechinus*. The three genera are represented by high Antarctic, Subantarctic and southern Argentinian cold-temperate species (David *et al.* 2005; Diaz *et al.* 2011), and are absent from Australasia. They are only known as fossils in the Upper Eocene (*Abatus*) and Lower Miocene (*Sterechinus*) of the Antarctic Peninsula, which suggests a West Antarctic or Subantarctic (see Diaz *et al.* 2011) origin, then dispersal through the Antarctic circumpolar current downstream of Drake Passage, in the Scotia Arc and southernmost Argentina, as well as in the Weddell Sea (Pearse *et al.* 2009). However, the occurrence of *Abatus* in the Upper Eocene of the Antarctic Peninsula (McKinney *et al.* 1988) might predate the full opening of the Drake Passage, in which case the genus could have migrated to South America before the full establishment of the Antarctic circumpolar current through the shallow waters that separated the two continents.

The existence of a weak faunal connection between southern Argentina and Chatham Island has been demonstrated by previous authors (Beu *et al.* 1997; Del Rio 2002) to have occurred since the Oligocene owing to dispersal through the Antarctic circumpolar current. In the present work, this circum-Antarctic dispersal is only supported by the occurrence of the Subantarctic and cool-temperate genera *Austrocidaris* and *Pseudechinus*, which are present off the coasts of southern Argentina, New Zealand and Chatham Island. *Pseudechinus* first occurs in the Cenozoic of Australia and New Zealand and then in the Pliocene of South America. This suggests the Australasian origin of the genus followed by circum-Antarctic dispersal to the west after the Early Miocene. The genus *Austrocidaris* is known as fossils in the Upper Eocene of West Antarctica, southern Argentina and Australia, and is supposed also to have dispersed from Australia in the Late Eocene (see discussion above).

Conclusion

Our results largely agree with previous biogeographic studies on echinoids and other benthic invertebrates. However, two main points contrast markedly with previous works: (1) there is no evidence of an Austral provincialism of echinoid faunas at the end of the Cretaceous and early Cenozoic – data instead suggest the cosmopolitanism of Austral echinoids that show Boreal and Tethyan affinities; (2) there is no evidence of faunal exchanges between New Zealand and southern Argentina in the Early Miocene – data better support a Caribbean affinity of southern Argentinian faunas as well as the presence of a biogeographic barrier between southern Argentina and other Austral regions. This result remains difficult to elucidate as it postdates the onset of the Antarctic circumpolar current, but it may be related to the transient intense glaciation that took place in Antarctica at the Oligocene-Miocene boundary (Rogers 2007). More fossil data are needed to confirm or invalidate this preliminary result.

In spite of the relative incompleteness of our knowledge of the echinoid fossil record at southern high latitudes, our results show that palaeobiogeographic relationships among echinoid faunas can be studied using a quantitative approach and not just qualitatively discussed. The quantitative analyses performed in this work improved our knowledge of the albeit still unclear biogeographic relationships between the different Austral regions. Moreover, the newly obtained biogeographic maps can be easily and intuitively discussed. More sampling efforts are needed though to improve understanding of the observed palaeobiogeographic relationships. Data from Madagascar, southern Chile and South Africa (Zululand) should be included in the future so as to complete this large-scale palaeobiogeographic analysis of Austral echinoid faunas.

Between the two biogeographic processes invoked to explain the distribution patterns of taxa, our interpretations usually favoured dispersal over vicariance. Two arguments support this lack of balance. First, biogeographic patterns of Australasian and Argentinian faunas have been deeply conditioned by the importance of immigration (Warren *et al.* 2009) from their respective northern areas. This is well typified by the Caribbean affinity of Argentinian echinoids in the Early Miocene or the Indo-Pacific affinity of Australian and New Zealand echinoids in the Late Eocene. Second, demonstration of the divergent evolution of taxa from an ancestral group by vicariance requires reliable phylogenies. The diversity of extant Antarctic echinoids is mostly due to the families Schizasteridae and Cidaridae, which are predominantly composed of endemic species (David *et al.* 2005). However, the origin of this diversity is not fully understood, as phylogenetic relationships with echinoids of other Austral regions remain unclear.

This paper is a contribution of the team BIOME of UMR CNRS 6282 Biogéosciences. The authors are indebted to the staff of the Australian Museum in Sydney, Melbourne Museum, the Museo Argentino de Ciencas Naturales in Buenos Aires, the National Institute of Water and Atmospheric researches in Wellington and the GNS Science in Lower Hutt for access to fossil and present day material. Funding sources included CAML/TOTAL, ANR ANTFLOCKS (no. 07-BLAN-0213-01), BIANZO II, the Région Bourgogne, the Fondation pour la Recherche sur la Biodiversité (FRB) and the Institut national des sciences de l'Univers (INSU) Interrvie. We thank specifically our reviewers, D. Néraudeau and R. Mooi, who contributed to improving the quality of a first draft of the manuscript.

References

Arnaud, H. 1883. Micraster regularis. *In*: Cotteau, G. (ed.) *Echinides jurassiques, crétacés et éocénes du Sud-Ouest de la France*. Annales de la Société des Sciences Naturelles de La Rochelle, 165, pl. XI, figs 1–5.

Batagelj, V. & Mrvar, A. 2005. *PAJEK (Program for Analysis and Visualization of Large Network) version 1.07*. Distributed by the authors.

Bernasconi, I. 1954. Nota sobre una nueva especie de Equinoidea fosil de Tierra del Fuego. *Physis*, **20**, 397–400.

Bernasconi, I. 1959. Equinoideos fosiles de la coleccion del Museo Argentino de Ciencias Naturales. *Physis*, **21**, 137–176.

Beu, A. G. 2009. Before the ice: biogeography of Antarctic Paleogene molluscan faunas. *Palaeogeography, Palaeoclimatology, Palaeoecology*, **284**, 191–226.

Beu, A. G., Griffin, M. & Maxwell, P. A. 1997. Opening of Drake Passage gateway and Late Miocene to Pleistocene cooling reflected in Southern Ocean molluscan dispersal: evidence from New Zealand and Argentina. *Tectonophysics*, **281**, 83–97.

Blake, D. B. & Zinsmeister, W. J. 1991. A new marsupiate cidaroid echinoid from the Maastrichtian of Antarctica. *Palaeontology*, **34**, 629–635.

Brandt, A., Linse, K. & Schüller, M. 2009. Bathymetric distribution patterns of Southern Ocean macrofaunal taxa: Bivalvia, Gastropoda, Isopoda and Polychaeta. *Deep-Sea Research I*, **56**, 2013–2025.

Bray, J. R. & Curtis, J. T. 1957. An ordination of the upland forest communities of southern Wisconsin. *Ecological Monographs*, **27**, 325–349.

Brayard, A., Escarguel, G. & Bucher, H. 2007. The biogeography of Early Triassic ammonoid faunas: clusters, gradients, and networks. *Geobios*, **40**, 749–765.

Clarke, A., Griffiths, H. J., Linse, K., Barnes, D. K. A. & Crame, J. A. 2007. How well do we know the Antarctic marine fauna? A preliminary study of macroecological and biogeographical patterns in Southern gastropod and bivalve molluscs. *Diversity and Distribution*, **13**, 620–632.

Crame, J. A. 1996. Evolution of high-latitude molluscan faunas. *In*: Taylor, J. D. (ed.) *Origin and Evolutionary Radiation of the Mollusca*. Oxford University Press, Oxford, 119–131.

David, B., Choné, T., Mooi, R. & De Ridder, C. 2005. *Antarctic Echinoidea. Synopses of the Antarctic Benthos*. Koeltz Scientific Books, Königstein, **10**.

Del Rio, C. J. 2002. Moluscos del Terciario Marino. *In*: Haller, M. J. (ed.) *Geología y Recursos Naturales de Santa Cruz. XV Congreso Geológico Argentino* (El Calafate, 2002), Relatorio, Buenos Aires, **II-9**, 1–22.

Díaz, A., Féral, J.-P., David, B., Saucède, T. & Poulin, E. 2011. Evolutionary pathways among shallow and deep-sea echinoids of the genus *Sterechinus* in the Southern Ocean. *Deep-Sea Research II*, **58**, 205–211.

Dudicourt, J. C., Néraudeau, D., Nicolleau, P., Ceulemans, L. & Boutin, F. 2005. Les échinides marsupiaux pliocènes de La Motte Foucran à Challans (Vendée, Ouest de la France). *Bulletin de la Société Géologique de France*, **6**, 579–593.

Foster, R. J. & Philip, G. M. 1978. Tertiary holasteroid echinoids from Australia and New Zealand. *Palaeontology*, **21**, 791–822.

Furque, G. & Camacho, H. H. 1949. El Cretácico de la Costa Atlantica de Tierra del Fuego. *Revista de la Sociedad Geológica Argentina*, **4**, 263–98.

Gaździcki, A. J. & Wrona, R. 1982. Late Cretaceous and Tertiary fossils from glacio-marine sediments of Melville Peninsula, King George Island, West Antarctica. *Przeglad Geologiczny*, **8**, 399–404. (In Polish, English summary.)

Hammer, Ø., Harper, D. A. T. & Rya, P. D. 2001. PAST: paleontological statistics software package for education and data analysis. *Palaeontologia Electronica*, **4**, 1–9.

Hedgpeth, J. W. 1969. Introduction to Antarctic zoogeography. Distribution of selected groups of marine invertebrates in water south of 35° S. *In*: Bushnell, V. C. & Hedgpeth, J. W. (eds) *Antarctic Map Folio*

Series, Folio 11. American Geographical Society, New York, 1–29.

Hotchkiss, F. H. C. 1982. Antarctic fossil echinoids: review and current research. *In*: Craddock, C. (ed.) *Antarctic Geoscience*. University of Wisconsin Press, Madison, WI, 679–684.

Hotchkiss, F. H. C. & Fell, H. B. 1972. Zoogeographical implications of a Paleogene echinoid from East Antarctica. *Journal of the Royal Society of New Zealand*, **2**, 369–72.

Jesionek-Szymanska, W. 1987. Echinoid from the Cape Melville Formation, (Lower Miocene) of King George Island, West Antarctica. *Palaeontologia Polonica*, **49**, 163–168.

Lambert, J. 1910. Les échinoides fossiles des îles Snow-Hill et Seymour. *Wissenschaftliche Ergebnisse Schwedischen Südpolar-Expedition, 1901–1903*, **3**, 1–15.

Lawver, L. A. & Gahagan, L. M. 2003. Evolution of Cenozoic seaways in the circum-Antarctic region. *Palaeogeography, Palaeoclimatology, Palaeoecology*, **198**, 11–37.

Lawver, L. A., Gahagan, L. M. & Dalziel, I. W. D. 2011. A different look at gateways: drake passage and Australia/Antarctica. *In*: Anderson, J. B. & Wellner, J. S. (eds) *Tectonic, Climatic, and Cryospheric Evolution of the Antarctic Peninsula*. American Geophysical Union, Washington, DC, 5–33, doi: 10.1029/2010SP001017

Linse, K., Griffiths, H. J., Barnes, D. K. A. & Clarke, A. 2006. Biodiversity and biogeography of Antarctic and sub-Antarctic mollusca. *Deep-Sea Research II*, **53**, 985–1008.

Loriol, P. de 1902. Notes pour servir à l'étude des échinodermes. *Revue Suisse de Zoologie, série 2*, **10**, 1–52.

McKinney, M. L., McNamara, K. J. & Wiedman, L. A. 1988*b*. Echinoids from the La Meseta Formation (Eocene), Seymour Island, Antarctica. *In*: Feldman, R. M. & Woodburne, M. O. (eds) *Geology and Paleontology of Seymour Island, Antarctic Peninsula*. Geological Society of America, Boulder, CO, Memoirs, **169**, 499–503.

Mooi, R., Martinez, S. & Parma, S. G. 2000. Phylogenetic systematics of Tertiary monophorasterid sand dollars (Clypeasteroida: Echinoidea) from South America. *Journal of Paleontology*, **74**, 263–281.

Mortensen, T. 1928. *A Monograph of the Echinoidea. 1, Cidaroidea*. C. A. Reitzel, Copenhagen.

Néraudeau, D. & Mathey, B. 2000. Biogeography and diversity of South Atlantic Cretaceous echinoids: implications for circulation patterns. *Palaeogeography, Palaeoclimatology, Palaeoecology*, **156**, 71–88.

Néraudeau, D., Crame, J. A. & Kooser, M. 2000. Upper Cretaceous echinoids from James Ross Basin, Antarctica. *Geobios*, **33**, 455–466.

Parma, S. G. & Casadio, S. 2005. Upper Cretaceous–Paleocene echinoids from Northern Patagonia, Argentina. *Journal of Paleontology*, **79**, 1072–1087.

Pearse, J. S., Mooi, R., Lockhart, S. J. & Brandt, A. 2009. Brooding and species diversity in the Southern Ocean: selection for brooders or speciation within brooding clades? *In*: Krupnik, I., Lang, M. A. & Miller, S. E. (eds) *Smithsonian at the Poles: Contributions to International Polar Year science*. Smithsonian Institution Press, Washington, DC, 181–196.

Philip, G. M. & Foster, R. J. 1971. Marsupiate Tertiary echinoids from South-Eastern Australia and their zoogeographic significance. *Palaeontology*, **14**, 666–695.

Philip, G. M. & Foster, R. J. 1977. The origins and relationship of the Australasian Cainozoic echinoid fauna. (Abstract.) *Journal of Paleontology*, **51**, 21.

Radwanska, U. 1996. A new echinoid from the Eocene, La Meseta formation of Seymour Island, Antarctic Peninsula. *Palaeontologia Polonica*, **55**, 117–125.

Rogers, A. D. 2007. Evolution and biodiversity of Antarctic organisms: a molecular perspective. *Philosophical Transactions of the Royal Society of London B*, **362**, 2191–2214.

Scher, H. D. & Martin, E. E. 2006. Timing and climatic consequences of the opening of the Drake Passage. *Science*, **312**, 428–430, doi: 10.1126/science.1120044

Stilwell, J. D., Zinsmeister, W. J. & Oleinik, A. E. 2004. Early Paleocene mollusks of Antarctica: systematics, palaeoecology and paleobiogeographic significance. *Bulletins of American Paleontology*, **367**, 1–89.

Warren, B. H., Strasberg, D., Bruggemann, J. H., Prys-Jones, R. P. & Thébaud, C. 2009. Why does the biota of the Madagascar region have such a strong Asiatic flavour? *Cladistics*, **25**, 1–12.

Zinsmeister, W. J. 1979. Biogeographic significance of the late Mesozoic and early Paleogene molluscan faunas of Seymour Island (Antarctic Peninsula) to the final breakup of Gondwanaland. *In*: Gray, J. & Boucot, A. J. (eds) *Historical Biogeography, Plate Tectonics, and the Changing Environment*. 37th Biological Colloquium of the Oregon State University. University of Oregon Press, Corvallis, OR, 347–355.

Zinsmeister, W. J. 1981. Middle to late Eocene Invertebrate fauna from the San Julian Formation at Punta Casamayor, Santa Cruz Province, Southern Argentina. *Journal of Paleontology*, **55**, 1083–1102.

Zinsmeister, W. J. & Camacho, H. H. 1980. Late Eocene Struthiolariidae (Mollusca: Gastropoda) from Seymour Island, Antarctic Peninsula and their significance to the biogeography of early Tertiary shallow-water faunas of the Southern Hemisphere. *Journal of Paleontology*, **54**, 1–14.

Taxonomic diversity of Eocene Antarctic penguins: a changing picture

PIOTR JADWISZCZAK

Institute of Biology, University of Bialystok, Swierkowa 20B, Bialystok 15-950, Poland (e-mail: piotrj@uwb.edu.pl)

Abstract: Eocene Antarctic penguins, at least 10 species in six genera, are known only from the La Meseta Formation, Seymour Island, Antarctic Peninsula. They are most numerous (in terms of individuals, body sizes and taxa) in Late Eocene strata. Specimens from three species and phylogenetic analysis presented in this work shed new light on the systematics and evolution of Antarctic Sphenisciformes. The earliest reported bones of giant penguins from the genus *Anthropornis* set the conservative estimate of its divergence time at *c.* 53 Ma (Early Eocene). They also document the oldest known appearance of quite a high diversity of Sphenisciformes; altogether, three morphotypes (differing in size) have been found within the same sampling locality. A newly described, relatively small and intriguingly elongated, tarsometatarsus from the Late Eocene of the La Meseta Formation, belonging to another genus of large-sized Antarctic penguins (*Palaeeudyptes*), suggests the possible existence of an unnamed species within this long-established genus. The phylogenetic analysis based on tarsometatarsal features shows that the relationship between '*Archaeospheniscus*' *wimani* and three species of *Delphinornis* (all of them co-existed during the Late Eocene time period) does not appear to be close enough to justify merging them into a single genus (as was recently postulated).

Supplementary material: An annotated data matrix used for the phylogenetic analysis is available at www.geolsoc.org.uk/SUP18599

Penguins (Aves: Sphenisciformes) are medium- to large-sized flightless seabirds represented by 17 modern species in six distinct genera (Williams 1995, but see Jadwiszczak 2009, p. 4; Ksepka & Ando 2011). Sphenisciformes originated in (and are still restricted to) the Southern Hemisphere (e.g. Ksepka & Thomas 2012), and appear to have already achieved a circum-Antarctic distribution in the Late Paleocene (Tambussi *et al.* 2005; Slack *et al.* 2006). By the Eocene/Oligocene transition penguins became numerous and diverse (Myrcha *et al.* 2002; Clarke *et al.* 2003; Jadwiszczak 2006*a*, 2008, 2009, 2010*a*; Tambussi *et al.* 2006; Sallaberry *et al.* 2010; Fordyce & Thomas 2011; Ksepka & Ando 2011), their known breeding range extending from the northern Antarctic Peninsula to the present-day Peruvian coast (e.g. Clarke *et al.* 2010; see also Jadwiszczak 2011). Representatives of the crown group penguins (Spheniscidae *sensu* Clarke *et al.* 2003) are not known from the Palaeogene (Ksepka & Thomas 2012, fig. 2).

The most numerous Eocene stem penguins come from the La Meseta Formation (Seymour Island, Antarctic Peninsula; Fig. 1), particularly from its upper part (Late Eocene, all named taxa present; e.g. Jadwiszczak 2009, 2010*a*). The oldest, although not taxonomically identified, bones are known from the two lowermost units of the formation and are Early Eocene in age (Jadwiszczak 2006*b*). It is well agreed upon that Eocene Antarctic sphenisciforms can be assigned to at least 10 species, although their phylogenetic relationships are debatable (Table 1; Jadwiszczak 2006*a*, *b*, 2008; Ksepka & Clarke 2010; Acosta Hospitaleche & Jadwiszczak 2011; Ksepka & Ando 2011). Interestingly, the largest of them, *Anthropornis nordenskjoeldi*, *A. grandis*, *Palaeeudyptes klekowskii* and *P. gunnari*, distinctly exceeded birds from any extant species of Sphenisciformes in size (Livezey 1989; Jadwiszczak 2001; Jadwiszczak & Chapman 2011; Jadwiszczak & Mörs 2011). The purpose of this paper is to report on the earliest record of the *Anthropornis* penguins, present new data on the smallest representatives of the Antarctic *Palaeeudyptes* (in the context of the species-level diversity) and to clarify the phylogenetic relationship between '*Archaeospheniscus*' *wimani* and *Delphinornis* species.

Geology, sedimentology and stratigraphy

Seymour Island is an ice-free island located close to the tip of the Antarctic Peninsula, in the back-arc James Ross Basin (64°17′S, 56°45′W). The Eocene La Meseta Formation (Elliot & Trautman 1982) crops out in the northeastern part of the island (Fig. 1), and is a fossiliferous lens, 6.5 km wide and more than 700 m thick. It comprises deltaic (wave-dominated), estuarine and shallow

From: Hambrey, M. J., Barker, P. F., Barrett, P. J., Bowman, V., Davies, B., Smellie, J. L. & Tranter, M. (eds) 2013. *Antarctic Palaeoenvironments and Earth-Surface Processes*. Geological Society, London, Special Publications, **381**, 129–138. First published online June 24, 2013, http://dx.doi.org/10.1144/SP381.7

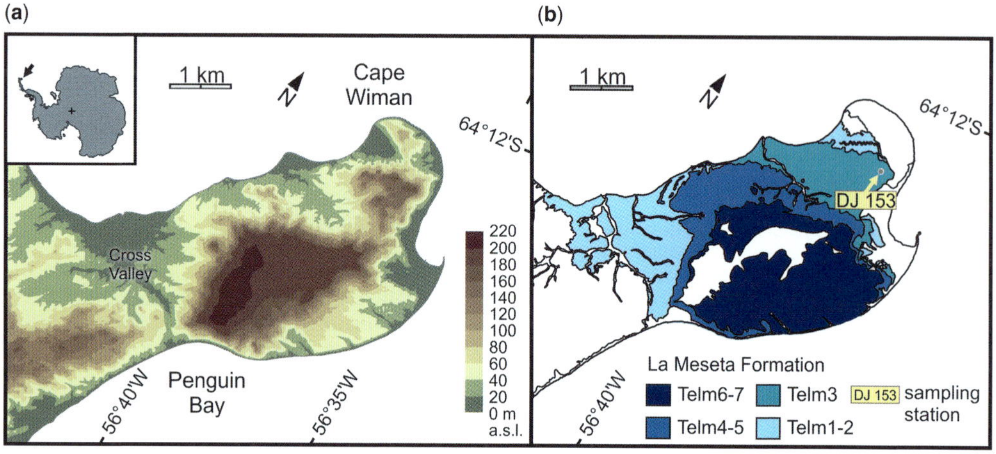

Fig. 1. The location of Seymour Island in Antarctica (inset) and simplified topographical (**a**) and stratigraphic (**b**) map of the Eocene La Meseta Formation (northeastern Seymour Island).

marine, mostly poorly consolidated siliciclastic fine-grained sediments (Porębski 1995, 2000; Marenssi 2006; Tatur *et al.* 2011).

Porębski (2000) proposed that the development of this geometrically compound fill had been governed mainly by local subsidence along fault-controlled shelf valley margins and associated amplified tidal circulation, whereas according to Marenssi (2006), sedimentation had been mainly controlled by eustatic sea-level changes. The Antarctic Peninsula was the most obvious source area and palaeocurrents indicate dispersal to the SE (Marenssi *et al.* 2002; Reguero *et al.* 2002). Ivany *et al.* (2008) suggested that sediments had been transported by longshore drift along the coast combined with mass wasting and localized runoff from valley margins, and diminished the role of an associated river. In any case, the perfect preservation of fragile structures (such as hypotarsal crests) in many bones, the impressive abundance of specimens as well as numerous typically terrestrial (e.g. Myrcha *et al.* 2002) fossils indicate the close proximity of the source area for those remains. Sadler (1988) mapped the La Meseta Formation in terms of seven major lithofacies Telm1 to Telm7 (Fig. 1b), and this system is adopted here; for an alternative scheme, see Marenssi (2006).

Age constraints

Strontium isotope ratios (SIS method) from bivalve shells indicate an Eocene age for the La Meseta Formation (Ivany *et al.* 2008), which is generally in line with results from earlier approaches based primarily on biostratigraphy (Wrenn & Hart 1988; Fordyce 1989; Cocozza & Clarke 1992; Stilwell & Zinsmeister 1992; Askin 1997, and others). Values reported by Marenssi (2006) suggest an Early Eocene age (52.4/54.3 Ma) for the base of Telm2, and those obtained by Dingle & Lavelle (1998) point to the latest Eocene (34.2 Ma) for the uppermost few metres of the formation (Telm7). $^{87}Sr/^{86}Sr$ ratios reported by Dutton *et al.* (2002) suggest an early to earliest Middle Eocene age for Telm2 to Telm5 (50.5 and 47.4–44.5 Ma, respectively), and a Late Eocene age for Telm7 (36.1–34.7 Ma).

The most detailed studies to date on this subject were conducted by Ivany *et al.* (2008). Isotope ratios from Telm2 to Telm5 indicate the Early Eocene (54.0–48.8 Ma), and those from Telm6 and Telm7 suggest late Middle to Late Eocene (41.0–34.0 Ma). These are 'best inferred ages' and age control is better in the upper part of the formation. In fact, the basal part of the La Meseta Formation (lowermost Telm1) may have been deposited close to the Paleocene/Eocene transition (Marenssi 2006), and the uppermost shell-bearing marine bed may be earliest Oligocene in age (Ivany *et al.* 2006). The most significant depositional hiatus may have occurred between Telm5 and Telm6 (48–43 Ma) (Ivany *et al.* 2006).

Material and methods

The limb bones (a humerus and two tarsometatarsi) of *Anthropornis* penguins discussed below were collected at Cape Wiman, Seymour Island (La Meseta Formation, Telm3, BAS (British Antarctic Survey) locality DJ 153; Fig. 1) by J. J. Hooker, a member of the British Antarctic Survey Expedition, in 1989. They are permanently located at the

Table 1. *A brief review of changes in systematics of Eocene Antarctic penguins*

Taxa*	References†									
	W05	M53	B63	S71	M90	M02	T06	J06	K10	J11
Orthopteryx gigas	Y‡	S	as in W05	D		N				S
Anthropornis nordenskjoeldi	Y	Y	Y	Y		Y	Y	Y		Y
Pachypteryx grandis	Y	*Anthropornis*	Y	Y		Y	Y	Y		
Wimanornis seymourensis				Y		N		N		
Eosphaeniscus gunnari	Y	Y	Y	*Palaeeudyptes*		Y	Y	Y	?*Palaeeudyptes*	?SSE
Palaeeudyptes klekowskii					Y	Y	Y	Y	?*Palaeeudyptes*	
Notodyptes wimani		Y	Y	*Archaeospheniscus*		Y		Y	*Delphinornis*	
Delphinornis larseni	Y	Y	Y	Y		Y	Y	Y		
Delphinornis arctowskii						Y	Y	Y		
Delphinornis gracilis						Y		Y		S
Ichtyopteryx gracilis	Y		Y	D		D		N		*Delphinornis*
Mesetaornis polaris						Y		Y		
Marambiornis exilis						Y		Y		
Tonniornis minimum							Y	N	D	
Tonniornis mesetaensis							Y	N	D	

*Graphically documented species only; *Orthopteryx gigas* is based solely on a synsacrum, and *Wimanornis seymourensis*, *Tonniornis mesetaensis* and *T. minimum* on humeri; type specimens of other species are tarsometatarsi.

†*Abbreviated references*: W05, Wiman (1905); M53, Marples (1953); B63, Brodkorb (1963); S71, Simpson (1971); M90, Myrcha *et al.* (1990); M02, Myrcha *et al.* (2002); T06, Tambussi *et al.* (2006); J06, Jadwiszczak (2006*a*, *b*); K10, Ksepka & Clarke (2010); J11, Jadwiszczak & Mörs (2011).

‡*Key to symbols*: 'Y', erected (if first in a row) or clearly confirmed taxon (last generic assignment of the species is valid); 'N', not confirmed taxon; 'D', dubious taxon; 'S', synonymized with a species from the next row; ?SSE stands for 'possible extremely pronounced sexual size dimorphism within a single species of *Palaeeudyptes*'.

Natural History Museum (London) and catalogued under numbers NHMUK BAS A2475, A2478 and A2479. The specimen identified as cf. *Palaeeudyptes gunnari* was collected by A. Tatur and A. Gaździcki within Telm7 in 1988, and is housed at the Institute of Biology, University of Bialystok (Bialystok, Poland; IB/P/B-0967). Comparative material comes from the collections housed at the Natural History Museum (London; a set acquired by the Falkland Island Dependencies Survey in 1946; Marples 1953; Simpson 1971), the Institute of Biology, University of Bialystok (Poland; Myrcha *et al.* 1990, 2002; Jadwiszczak 2006*a*), the Naturhistoriska riksmuseet (Stockholm, Sweden; abbreviated NRM-PZ; Wiman 1905; Simpson 1971) and the Museo de La Plata (La Plata, Argentina; abbreviated MLP).

The vast majority of Antarctic penguin fossils are isolated bones, hence the importance of size categorization during their taxonomic identification (Wiman 1905; Simpson 1971; Myrcha *et al.* 1990, 2002; Jadwiszczak 2006*a*, 2009). As the fossil record consists of thousands of specimens (Jadwiszczak 2009), many of which are well preserved, current diagnoses also incorporate qualitative features (e.g. Myrcha *et al.* 2002; Jadwiszczak 2006*a*). Tarsometatarsi and humeri (the most characteristic elements of the penguin skeleton) have long been used for taxonomic purposes, and in most cases have been designated as holotypes (Jadwiszczak 2009).

The phylogenetic relationships between '*Archaeospheniscus*' *wimani* and three species of *Delphinornis* were inferred from the parsimony analysis of tarsometatarsal morphology. The analysis was conducted using PAUP*4.0b10 (Swofford 2003) with a heuristic search strategy. Starting trees for branch swapping (tree bisection-reconnection algorithm) were obtained via stepwise addition using random addition sequence with 1000 replications; the 'multrees' and 'le50' options were on. The set of equally parsimonious topologies was summarized using a strict consensus tree accompanying by the following measures of topological congruence: component information (consensus fork), Nelson–Platnick term and total information, Mickevich's consensus information and Rohlf's CI. Two data matrices were used for the study. The first of them was extracted from a dataset used by Ksepka & Thomas (2012); the second matrix was adjusted by the author to better reflect differences between morphotypes. It had the same size as the former, but introduced a number of changes to character states and definitions (see Supplementary Material). The Southern Giant Petrel, *Macronectes giganteus*, from Procellariiformes – an order closely related to Sphenisciformes (e.g. Jadwiszczak 2009), was chosen as an outgroup.

The generic name *Palaeeudyptes*, when used alone, encompasses Antarctic species only (see also Ksepka & Clarke 2010). Measurements were taken with digital callipers and rounded to the nearest 0.1 mm.

Systematic palaeontology

Aves Linnaeus 1758
Order Sphenisciformes Sharpe 1891 (also Clarke *et al.* 2003)
Family Spheniscidae Bonaparte 1831

Anthropornis *Wiman 1905*

Type species. Anthropornis nordenskjoeldi Wiman 1905, Seymour Island (Antarctic Peninsula), La Meseta Formation, Eocene.

Other species. Anthropornis grandis (Wiman 1905), Seymour Island (Antarctic Peninsula), La Meseta Formation, Eocene.

Anthropornis *sp.*

Material. Two incomplete right tarsometatarsi (NHMUK BAS A2478 and NHMUK BAS A2479) and a single right distal humerus (NHMUK BAS A2475).

Occurrence. La Meseta Formation (Seymour Island), Telm3 unit (Early Eocene), BAS locality DJ 153 (Cape Wiman; Fig. 1b); outcrop of coquina, surface shelly sand.

Description and comparisons. All specimens, although incomplete, are large bones (within the size range of *Anthropornis* and *Palaeeudyptes*, i.e. larger than their counterparts in any recent species). Tarsometatarsi (Fig. 2a, b, e, f) were assigned to *Anthropornis* based on the following features (Myrcha *et al.* 2002 and the author's personal observation, e.g. NRM-PZ A45 (Fig. 2d) and NHMUK A3358 (Fig. 2c, g)). The distal part of their medial margin is characteristically concave (Fig. 2a, b; black arrows), suggesting a strongly divergent trochlea II (Fig. 2c; Ksepka & Clarke 2010, fig. 30c; moderately divergent in *Palaeeudyptes*; see also Wiman 1905 and Simpson 1971) and the presence of a proximal tuberosity (well developed in the holotype (Fig. 2d); not present in *Palaeeudyptes*). There is a conspicuous medial intermetatarsal foramen (typically small if present in *Palaeeudyptes*) and no sign of its lateral counterpart, the latter obviously resulting from the poor preservation of considered skeletal elements (Fig. 2a, b, e, f). Nevertheless, the lateral foramen appears to be more proximal (dorsal view) relative to the medial one (unlike in

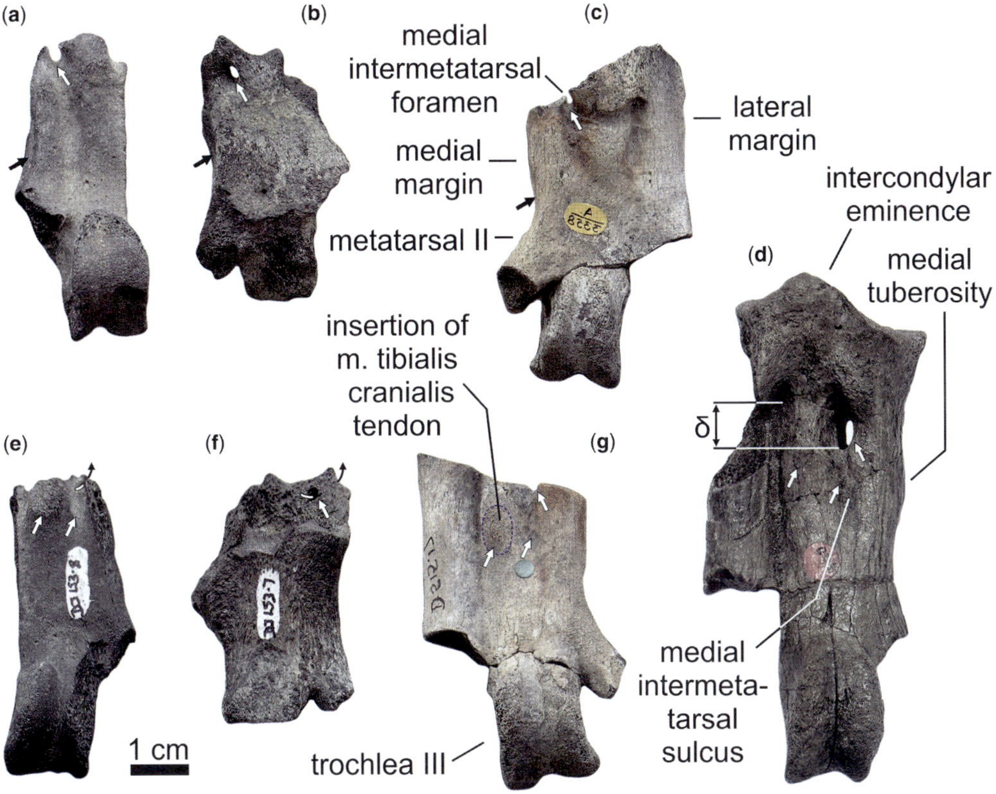

Fig. 2. Tarsometatarsi assigned to *Anthropornis* from the Eocene La Meseta Formation: NHMUK BAS A2479 (**a** and **e**); NHMUK BAS A2478 (**b** and **f**); NHMUK A3358 (**c** and **g**); NRM-PZ A45 (**d**). Plantar (a–c), and dorsal (d–g) views. δ Represents the difference in location (relative to proximal end) of both intermetatarsal foramina. Arrows point to characters discussed in text.

Palaeeudyptes). The medial intermetatarsal sulcus is short (both specimens; Fig. 2e, f), wide and shallow (in NHMUK BAS A2478 this fragment is too damaged (Fig. 2f); typically not present or vestigial in *Palaeeudyptes*). The insertion surface of the *tibialis cranialis* muscle is very wide (NHMUK BAS A2479, Fig. 2e; slightly more slender in most *Palaeeudyptes*).

The humerus considered here (Fig. 3a–c) has the following features found in specimens attributable to *Anthropornis* (e.g. IB/P/B-0307 and NHMUK A3338, Fig. 3d–g) and not present in *Palaeeudyptes* (Jadwiszczak 2006*a* and author's personal observation). Its distal end is massive. There is a row of conspicuous nutritious foramina situated between two bony crests that frame the ventral tendon sulcus (Fig. 3b). The proximal part of the middle crest is clearly visible in ventral view (Fig. 3a). There is a pronounced indentation between the radial condyle and the cranial margin of the bone (Fig. 3a, c).

Measurements (in mm). NHMUK BAS A2475 (humerus): total cranio-caudal width of the distal end, >38.7; dorso-ventral breadth at the level of the ulnar condyle, 17.6. NHMUK BAS A2478 (tarsometatarsus): dorso-ventral thickness measured distally to the *m. tibialis cranialis* insertion surface, 11.1. NHMUK BAS A2479 (tarsometatarsus): dorso-plantar thickness (see above), 11.8. Medio-lateral width of the third trochlea, >14.0. Dorso-plantar thickness of the third trochlea (its central part), 19.3.

Discussion. The three Early Eocene bones described above cannot be reasonably ascribed to a particular species of *Anthropornis*, because of the recently reported (Acosta Hospitaleche & Jadwiszczak 2011) presence of the size-independent morphological heterogeneity within hypotarsi that challenges current specific diagnoses. Still, they are the oldest well-documented specimens assigned to this genus, and together with the tarsometatarsus

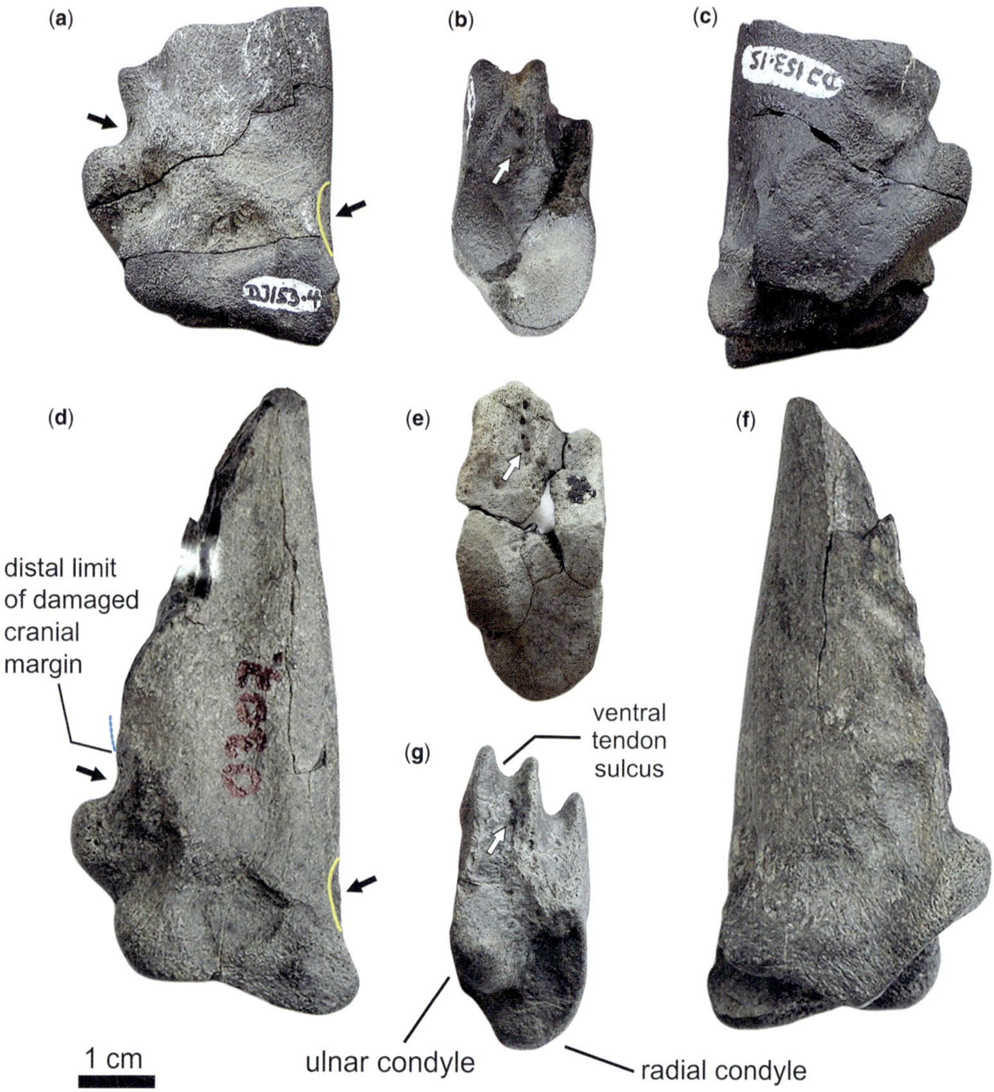

Fig. 3. Humeri assigned to *Anthropornis* from the Eocene La Meseta Formation: NHMUK BAS A2475 (**a**–**c**); IB/P/B-0307 (**d**–**f**); NHMUK A3338 (**e**). Ventral (a, d), distal (b, e, g) and dorsal (c, f), view. Arrows point to characters discussed in text.

MLP 93-X-1-151 of *P. gunnari* from the same unit (Telm3, locality IAA 1/94; Myrcha *et al.* 2002) constitute the earliest record of identified Antarctic penguin taxa. Other and geologically younger bones attributed to *Anthropornis* come from Telm4 and Telm5, but they are most numerous in Telm7 (Myrcha *et al.* 2002; Jadwiszczak 2006*a*). The last mentioned Telm corresponds to the late Middle–Late Eocene (Ivany *et al.* 2008), a time period that witnessed a substantial climate deterioration between the end of the Middle Eocene Climatic Optimum and the onset of the continental glaciation at the Eocene–Oligocene boundary (Ivany *et al.* 2006, 2008; Jadwiszczak 2010*a*).

Interestingly, the sampling station DJ 153 has yielded another important fossil – a distal humerus of a medium-sized sphenisciform (i.e. an individual being within a size range of the vast majority of present-day penguins; Jadwiszczak & Chapman 2011). This means that at least three species of Early Eocene Spheniciformes inhabited the Antarctic Peninsula region just after the Early Eocene Climatic Optimum (between 53.5 and 52.8 Ma; Ivany *et al.* 2008). It does not seem to be a coincidence that observed changes in penguin diversity (in terms of both body size and species richness) took place during cooling periods and associated major reorganizations of a food-web structure (particularly well documented for the upper La Meseta Formation; e.g. Jadwiszczak 2010*a*).

Palaeeudyptes *Huxley 1859*

Type species. Palaeeudyptes antarcticus Huxley 1859; New Zealand (South Island), Ototara Limestone?, latest Eocene or Early Oligocene (Ksepka & Ando 2011).

Other species. Antarctic only; the whole genus may not be monophyletic, see Ksepka & Clarke 2010. *P. gunnari* (Wiman 1905) and *P. klekowskii* Myrcha *et al.* 1990; Seymour Island (Antarctic Peninsula), La Meseta Formation, Eocene.

Cf. Palaeeudyptes gunnari

Material. Incomplete left tarsometatarsus (IB/P/B-0967).

Occurrence. La Meseta Formation (Seymour Island), Telm7 unit (Late Eocene).

Description and comparisons. The tarsometatarsus reported above (Fig. 4a) is (despite abrasion) larger than its counterparts in '*Archaeospheniscus*' *wimani* (at least 30% longer than its holotype; Jadwiszczak unpublished observations), but shorter than most and narrower (at the centre) than all relatively well preserved specimens of those assigned to *Palaeeudyptes* (Myrcha *et al.* 2002, table 1; two bones from this set, MLP 82-IV-23-6 and IB/P/B-0072, are presented in Fig. 4). Its elongation index is uniquely high – an estimated value exceeds 2.56 (Myrcha *et al.* 2002, table 1). There is no sign of intermetatarsal sutures; that is, this bone belonged to an adult bird. It has the following unique combination of features found in specimens attributable to *Palaeeudyptes* (e.g. Fig. 4b, c). The lateral intermetatarsal foramen is well developed, whereas the medial one is vestigial. The lateral intermetatarsal sulcus is deep proximally and its medial counterpart is absent. There is no sign of the distal intermetatarsal foramen. The insertion surface for the *tibialis cranialis* muscle attachment appears to be well developed. The medial margin is smoothly concave.

Measurements (in mm). Length between the proximal intercondylar eminence and the lateral (fourth) trochlea, *c.* 58.0. Medio-lateral width at the centre, 22.7. Dorso-plantar thickness at the centre, *c.* 7.8.

Discussion. The bone described above is not well enough preserved to show specific distinctions that might exist, except for the high elongation index and the overall size, and its general similarity to specimens assigned to *Palaeeudyptes gunnari* appears to be unquestionable. It can be quite safely assumed, however, that if complete it would obviously reduce the negative skewness reported for the length distribution of tarsometatarsi assignable to *P. gunnari* (Jadwiszczak & Mörs 2011; unfortunately, IB/P/B-0967 has broken a trochlea III, so its length along the main axis must be estimated). Taking into account that there are much more fragmentary tarsometatarsi that resemble this relatively non-robust bone in respect to their dimensions and shape (e.g. IB/P/B-0294 (Jadwiszczak unpublished observations) and MLP 83-V-20-41 (Myrcha *et al.* 2002, table 1), the above-mentioned asymmetry could be improved even more. In addition, it becomes conspicuous that the bone length alone is a rather rough estimate of size in

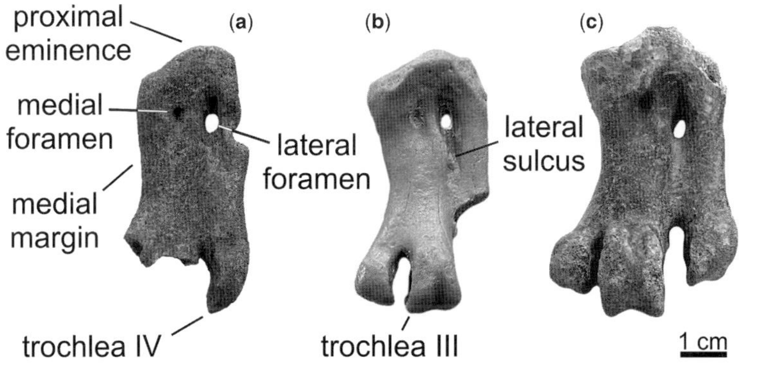

Fig. 4. Tarsometatarsi assigned to *Palaeeudyptes* from the Eocene La Meseta Formation: IB/P/B-0967 (**a**), NHMUK A5575 (a cast of MLP 82-IV-23-6) (**b**) and IB/P/B-0072 (**c**). Dorsal view.

the case of such a morphologically variable genus (Jadwiszczak & Mörs 2011, p. 274). To summarize, this finding demonstrates that small-sized representatives of *Palaeeudyptes* were more numerous (and slender) than previously thought, and warrant further attention. The presence of the pronounced sexual size dimorphism (Jadwiszczak & Mörs 2011), perhaps overlapping the current (primarily size-based) intrageneric systematics, cannot be ruled out, but for the first time the possible existence of an unnamed small-bodied species of Antarctic *Palaeeudyptes* is conceivable.

Phylogenetic position of '*Archaeospheniscus*' *wimani*

Parsimony analysis

The first analysis was based on tarsometatarsal characters of extinct Antarctic penguins extracted from a dataset used by Ksepka & Thomas (2012), and resulted in a tree with '*Archaeospheniscus*' *wimani* included in a clade together with three species from the genus *Delphinornis* (Fig. 5a; see Ksepka & Clarke 2010, fig. 21, and Ksepka & Thomas 2012, fig. S1). The second analysis utilized a corrected data matrix and placed *A. wimani* outside the *Delphinornis* clade (Fig. 5b). The topology of most parsimonius trees in the latter case did not differ substantially; *A. wimani* formed either a trichotomy with both *Delphinornis* and *Mesetaornis* + *Marambiornis* clade or constituted a separate clade with *Mesetaornis* + *Marambiornis*. Importantly, a thorough analysis of the hypotarsus in *A. wimani* (specimens IB/P/B-0284 and NHMUK A3331 (holotype)) revealed, in contrast to previous assessments reported by Myrcha *et al.* (2002) and Ksepka & Clarke (2010), that the intermediate crests (structures important for penguin systematics) were most probably not coalesced (for more details on the corrected dataset, see Material and methods, and Supplementary material).

Discussion

The results presented above demonstrate the need for cautious interpretations of dendrograms resulting from phylogenetic analyses of morphological data (here a case study is restricted to Eocene Antarctic Sphenisciformes). In the case of *A. wimani*, which co-existed with *Delphinornis* species during the Late Eocene, creation of the new combination *Delphinornis wimani* (Ksepka & Clarke 2010, p. 41) appears to be premature. The tarsometatarsal morphology has proved its potential to retain the phylogenetic signal; for example, positions occupied by *Anthropornis* and *Palaeeudyptes* within presented cladograms (Fig. 5), suggesting their close relationship, are in line with intriguing findings (not coded in data matrices) reported by Jadwiszczak (2010*b*) for bones of juvenile individuals from both taxa.

The specimen IB/P/B-0967 (cf. *Palaeeudyptes gunnari*), although very interesting (see above), was not included in analyses. The only feature that distinguishes that bone from other specimens attributable to *Palaeeudyptes* is its (estimated) relative elongation (as defined in Myrcha *et al.* 2002); however, the elongation index *sensu* Ksepka & Thomas (2012) that was used in character matrices cannot be reasonably assessed. Furthermore, for

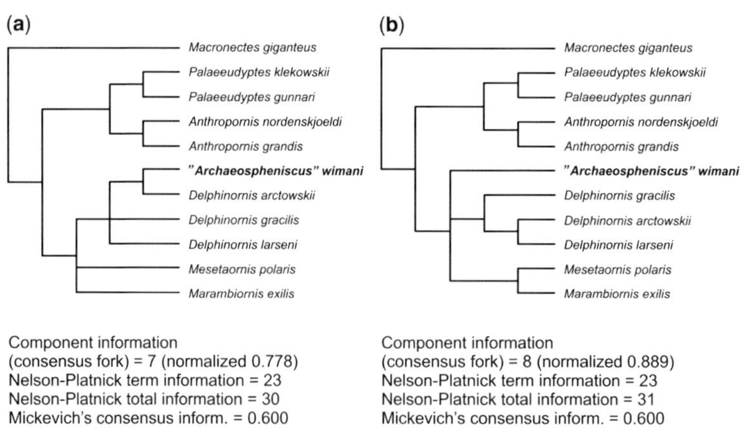

Fig. 5. Results of two phylogenetic analyses of the Eocene Antarctic penguins based on tarsometatarsal characters. A strict consensus of two most parsimonious trees (320 966 rearrangements tried) from the analysis of a dataset extracted from the matrix used by Ksepka & Thomas (2012) (**a**), and a strict consensus of two most parsimonious trees (312 763 rearrangements tried) from the analysis of an adjusted data (this study) (**b**).

technical reasons, some important features (such as a differences in sizes of proximal intermetatarsal foramina in *Delphinornis* and *'Archaeospheniscus'*; small v. large) have not been taken into account. This does not mean, however, that the long-established generic assignment of *A. wimani* (Simpson 1971; Myrcha *et al.* 2002; Jadwiszczak 2006*a*) is correct (Ksepka & Clarke 2010); for example, it may also form a single-species genus (*Notodyptes*; Marples 1953). More complete specimens are required to solve this puzzle.

Conclusions

New data on Early Eocene giant penguins from the genus *Anthropornis* not only extend the stratigraphic range of this taxon, but also demonstrate that the first appearance of the relatively high diversity of Sphenisciformes (at least three differing in size species) is as old as *c.* 53 Ma (Telm3; see also Ivany *et al.* 2008). A newly described small-sized (relative to other giant-penguin bones), although elongated, tarsometatarsus assignable to *Palaeeudyptes* suggests the existence of an unnamed Late Eocene (Telm7) species within this taxon. The relationship between *A. wimani* and three species of *Delphinornis* does not appear to be close enough to justify merging them into a single genus.

I am very grateful to M. Richter (Natural History Museum, London) for the information about the exact localization of the collecting station DJ 153 (same as her DJ 251) as well as to D. Ksepka (North Carolina State University, Raleigh) for sharing his phylogeny data matrix. I would like to thank the editors of this volume for inviting me to write this contribution. I am indebted to Tatsuro Ando (Ashoro Museum of Paleontology, Ashoro, Japan) and an anonymous reviewer for their helpful comments on this work. Last but not least, I appreciate the financial support through SYNTHESYS funding made available by the European Community – Research Infrastructure Action under the FP6 and FP7 'Structuring the European Research Area' Programme, projects SETAF-4399 and GBTAF-987, respectively.

References

Acosta Hospitaleche, C. & Jadwiszczak, P. 2011. Enigmatic morphological disparity in tarsometatarsi of giant penguins from the Eocene of Antarctica. *Polish Polar Research*, **32**, 175–180, doi: 10.2478/v10-183-011-0013-9.

Askin, R. A. 1997. Eocene–?earliest Oligocene terrestrial palynology of Seymour Island, Antarctica. *In*: Ricci, C. A. (ed.) *The Antarctic Region: Geological Evolution and Processes*. Terra Antarctica, Siena, Italy, 993–996.

Brodkorb, P. 1963. Catalogue of fossil birds. *Bulletin of the Florida State Museum*, **7**, 179–293.

Clarke, J. A., Olivero, E. B. & Puerta, P. 2003. Description of the earliest fossil penguin from South America and first Paleogene vertebrate locality of Tierra del Fuego, Argentina. *American Museum Novitates*, **3423**, 1–18, doi: 10.1206/0003-0082(2003)423 <0001:DOTEFP>2.0.CO;2.

Clarke, J. A., Ksepka, D. T. *et al.* 2010. Fossil evidence for evolution of the shape and color of penguin feathers. *Science*, **330**, 954–957, doi: 10.1126/science.1193604.

Cocozza, C. D. & Clarke, C. M. 1992. Eocene microplankton from La Meseta Formation, northern Seymour Island. *Antarctic Science*, **4**, 355–362.

Dingle, R. V. & Lavelle, M. 1998. Antarctic Penisular cryosphere: early Oligocene (*c.* 30 Ma) initiation and a revised glacial chronology. *Journal of the Geological Society, London*, **155**, 433–437.

Dutton, A., Lohmann, K. & Zinsmeister, W. 2002. Stable isotope and minor element proxies for Eocene climate of Seymour Island, Antarctica. *Paleoceanography*, **17**, doi: 10.1029/2000PA000593.

Elliot, D. H. & Trautman, T. A. 1982. Lower Tertiary strata on Seymour Island, Antarctic Peninsula. *In*: Craddock, C. (ed.) *Antarctic Geoscience*. The University of Wisconsin Press, Madison, WI, 287–297.

Fordyce, R. E. 1989. Origins and evolution of Antarctic marine mammals. *In*: Crame, J. A. (ed.) *Origins and Evolution of the Antarctic Biota*. Geological Society, London, Special Publications, **47**, 269–281.

Fordyce, R. E. & Thomas, D. B. 2011. *Kaiika maxwelli*, a new Early Eocene archaic penguin (Sphenisciformes, Aves) from Waihao Valley, South Canterbury, New Zealand. *New Zealand Journal of Geology and Geophysics*, **54**, 43–51, doi: 10.1080/00288306.2011.536521.

Huxley, T. H. 1859. On a fossil bird and a fossil cetacean from New Zealand. *Quarterly Journal of the Geological Society of London*, **15**, 670–677.

Ivany, L. C., Van Simaeys, S., Domack, E. W. & Samson, S. D. 2006. Evidence for an earliest Oligocene ice sheet on the Antarctic Peninsula. *Geology*, **34**, 377–380, doi: 10.1130/G22383.1.

Ivany, L. C., Lohmann, K. C., Hasiuk, F., Blake, D. B., Glass, A., Aronson, R. B. & Moody, R. M. 2008. Eocene climate record of a high latitude continental shelf: Seymour Island, Antarctica. *Geological Society of America Bulletin*, **120**, 659–678, doi: 10.1130/B26269.1.

Jadwiszczak, P. 2001. Body size of Eocene Antarctic penguins. *Polish Polar Research*, **22**, 147–158.

Jadwiszczak, P. 2006*a*. Eocene penguins of Seymour Island, Antarctica: taxonomy. *Polish Polar Research*, **27**, 3–62.

Jadwiszczak, P. 2006*b*. Eocene penguins of Seymour Island, Antarctica: the earliest record, taxonomic problems and some evolutionary considerations. *Polish Polar Research*, **27**, 287–302.

Jadwiszczak, P. 2008. An intriguing penguin bone from the Late Eocene of Seymour Island, Antarctic Peninsula. *Antarctic Science*, **20**, 589–590, doi: 10.1017/S0954102008001405.

Jadwiszczak, P. 2009. Penguin past: the current state of knowledge. *Polish Polar Research*, **30**, 3–28.

Jadwiszczak, P. 2010*a*. Penguin response to the Eocene climate and ecosystem change in the northern

Antarctic Peninsula region. *Polar Science*, **4**, 229–235, doi: 10.1016/j.polar.2010.03.001.

Jadwiszczak, P. 2010*b*. New data on the appendicular skeleton and diversity of Eocene Antarctic penguins. *In*: Nowakowski, D. (ed.) *Morphology and Systematics of Fossil Vertebrates*. DN Publisher, Wroclaw, 45–51.

Jadwiszczak, P. 2011. New data on morphology of late Eocene penguins and implications for their geographic distribution. *Antarctic Science*, **23**, 605–606, doi: 10.117/S0954102011000526.

Jadwiszczak, P. & Chapman, S. D. 2011. The earliest fossil record of a medium-sized penguin. *Polish Polar Research*, **32**, 269–277, doi: 10.2478/v10183-011-0029-x.

Jadwiszczak, P. & Mörs, T. 2011. Aspects of diversity in early Antarctic Penguins. *Acta Palaeontologica Polonica*, **56**, 269–277, doi: 10.4202/app.2009.1107.

Ksepka, D. T. & Ando, T. 2011. Penguins past, present, and future: trends in the evolution of the sphenisciformes. *In*: Dyke, G. & Kaiser, G. (eds) *Living Dinosaurs: The Evolutionary History of Modern Birds*. John Wiley & Sons, Chichester, 155–186.

Ksepka, D. T. & Clarke, J. A. 2010. The basal penguin (Aves: Sphenisciformes) *Perudyptes devriesi* and a phylogenetic evaluation of the penguin fossil record. *Bulletin of the American Museum of Natural History*, **337**, 1–77.

Ksepka, D. T. & Thomas, D. B. 2012. Multiple Cenozoic invasions of Africa by penguins (Aves, Sphenisciformes). *Proceedings of the Royal Society of London B*, **279**, 1027–1032, doi: 10.1098/rspb.2011.1592.

Livezey, B. C. 1989. Morphometric patterns in recent and fossil penguins (Aves, Sphenisciformes). *Journal of the Zoological Society of London*, **219**, 269–307.

Marenssi, S. A. 2006. Eustatically controlled sedimentation recorded by Eocene strata of the James Ross Basin, Antarctica. *In*: Francis, J. E., Pirrie, D. & Crame, J. A. (eds) *Cretaceous–Tertiary High-latitude Palaeoenvironments, James Ross Basin, Antarctica*. Geological Society, London, Special Publications, **258**, 125–133.

Marenssi, S. A., Net, L. I. & Santillana, S. N. 2002. Provenance, depositional, and paleogeographic controls on sandstone composition in an incised valley system: the Eocene La Meseta Formation, Seymour Island, Antarctica. *Sedimentary Geology*, **150**, 301–321.

Marples, B. J. 1953. *Fossil penguins from the mid-Tertiary of Seymour Island*. Falkland Islands Dependencies Survey Scientific Reports, **5**, 1–15.

Myrcha, A., Tatur, A. & Del Valle, R. 1990. A new species of fossil penguin from Seymour Island, West Antarctica. *Alcheringa*, **14**, 195–205.

Myrcha, A., Jadwiszczak, P., Tambussi, C. P., Noriega, J. I., Gaździcki, A., Tatur, A. & Del Valle, R. 2002. Taxonomic revision of Eocene Antarctic penguins based on tarsometatarsal morphology. *Polish Polar Research*, **23**, 5–46.

Porębski, S. J. 1995. Facies architecture in a tectonically-controlled incised-valley estuary: La Meseta Formation (Eocene) of Seymour Island, Antarctic Peninsula. *In*: Birkenmajer, K. (ed.) *Geological Results of the Polish Antarctic Expeditions*. Part XI. Studia Geologica Polonica, Cracow, **107**, 7–97.

Porębski, S. J. 2000. Shelf-valley compound fill produced by fault subsidence and eustatic sea-level changes, Eocene La Meseta Formation, Seymour Island, Antarctica. *Geology*, **28**, 147–150, doi: 10.1130/0091-7613(2000)28<147:SCFPBF>2.0.CO;2.

Reguero, M. A., Marenssi, S. A. & Santillana, S. N. 2002. Antarctic Peninsula and Patagonia Paleogene terrestrial environments: biotic and biogeographic relationships. *Palaeogeography, Palaeoclimatology, Palaeoecology*, **179**, 189–210.

Sadler, P. 1988. Geometry and stratification of uppermost Cretaceous and Paleogene units of Seymour Island, northern Antarctic Peninsula. *In*: Feldmann, R. M. & Woodburne, M. O. (eds) *Geology and Paleontology of Seymour Island, Antarctic Peninsula*. Geological Society of America, Boulder, CO, Memoirs, **169**, 303–320.

Sallaberry, M. A., Yury-Yáñez, R. E., Otero, R. A., Soto-Acuña, S. & Torres, T. 2010. Eocene birds from the western margin of southernmost South America. *Journal of Paleontology*, **84**, 1061–1070, doi: 10.1666/09-157.1.

Simpson, G. G. 1971. Review of fossil penguins from Seymour Island. *Proceedings of the Royal Society of London B*, **178**, 357–387.

Slack, K. E., Jones, C. M., Ando, T., Harrison, G. L., Fordyce, R. E., Arnason, U. & Penny, D. 2006. Early penguin fossils, plus mitochondrial genomes, calibrate avian evolution. *Molecular Biology and Evolution*, **23**, 1144–1155, doi: 10.1093/molbev/msj124.

Stilwell, J. D. & Zinsmeister, W. J. 1992. Molluscan systematics and biostratigraphy: lower Tertiary La Meseta Formation, Seymour Island, Antarctic Peninsula. *American Geophysical Union, Antarctic Research Series*, **55**, 1–192.

Swofford, D. L. 2003. *PAUP*. Phylogenetic Analysis Using Parsimony (*and Other Methods)*. Version 4. Sinauer Associates, Sunderland, MA.

Tambussi, C. P., Reguero, M. A., Marenssi, S. A. & Santillana, S. N. 2005. *Crossvallia unienwillia*, a new Spheniscidae (Sphenisciformes, Aves) from the Late Paleocene of Antarctica. *Geobios*, **38**, 667–675, doi: 10.1016/j.geobios.2004.02.003.

Tambussi, C. P., Acosta Hospitaleche, C., Reguero, M. A. & Marenssi, S. A. 2006. Late Eocene penguins from West Antarctica: systematics and biostratigraphy. *In*: Francis, J. E., Pirrie, D. & Crame, J. A. (eds) *Cretaceous – Tertiary High-latitude Palaeoenvironments, James Ross Basin, Antarctica*. Geological Society, London, Special Publications, **258**, 145–161.

Tatur, A., Krajewski, K. P. & del Valle, R. A. 2011. The facies and biota of the oldest exposed strata of the Eocene La Meseta Formation (Seymour Island, Antarctica). *Geological Quarterly*, **55**, 345–360.

Williams, T. D. 1995. *Bird Families of the World. The Penguins*. Oxford University Press, New York.

Wiman, C. 1905. Über die alttertiären Vertebraten der Seymourinsel. *Wissenschaftliche Ergebnisse der Schwedischen Südpolar-Expedition 1901–1903*, **3**, 1–37.

Wrenn, J. H. & Hart, G. F. 1988. Paleogene dinoflagellate cyst biostratigraphy of Seymour Island, Antarctica. *In*: Feldmann, R. M. & Woodburne, M. O. (eds) *Geology and Paleontology of Seymour Island, Antarctic Peninsula*. Geological Society of America, Boulder, CO, Memoirs, **169**, 321–447.

Heat-flow determinations of basement age in small oceanic basins of the southern central Scotia Sea

P. F. BARKER[1†], L. A. LAWVER[2] & R. D. LARTER[1]*

[1]*British Antarctic Survey, Madingley Rd, Cambridge CB3 0ET, UK*

[2]*University of Texas Institute for Geophysics, 10100 Burnet Rd, R2200, Austin, TX 78758-4445, USA*

**Corresponding author (e-mail: rdla@bas.ac.uk)*

Abstract: Results are reported from seven heat flow stations in small basins of the southern part of the central Scotia Sea (CSS), undertaken in order to determine basement ages. The basins are small, which makes magnetic anomaly-based ages ambiguous and preserves basin subsidence that may have been anomalous as a result of local factors. The fact that these small basins formed in a back-arc setting adds additional uncertainty to depth-based age estimates. The results confirm that basin extension commenced in the Eocene, and indirectly support a relatively young, back-arc origin for the northern CSS, but do not affect previously published suggestions of the age of onset of the Antarctic Circumpolar Current.

The Scotia Sea lies between about 53° and 61°S, 75° and 25°W, and is now and most probably has been for the past 30–40 Ma a back-arc, generally oceanic region. It is surrounded by the mainly continental fragments of the Scotia Arc (see review in Barker *et al.* 1991). The central Scotia Sea (CSS) lies within the Scotia Sea, between Drake Passage to the west, that opened approximately ESE–WNW between *c.* 30 and 6 Ma (Barker & Burrell 1977; LaBrecque 1985; Barker 2001; Livermore *et al.* 2005), and the presently active (from at least 15 Ma ago; Barker 1995; Larter *et al.* 2003) East Scotia Ridge east–west-opening spreading centre in the east. The CSS comprises a relatively large northern area of generally accepted oceanic origin and east–west oceanic magnetic lineations, and a smaller region to the south, of continental fragments and intervening small ocean basins.

The zone of east–west anomalies in the northern CSS extends from 54 to 58°S and from 43°W (Hill & Barker 1980) to at least 38°W (Maldonado *et al.* 2003). A wide range of ages and origins has been proposed for the northern ocean floor of the CSS, ranging from early determinations (Hill & Barker 1980), based on magnetic anomaly identifications, of 21–6 or 32–23.5 Ma (the first being preferred) with an essentially back-arc extensional origin, to a proposal by De Wit (1977) that it was intra-Gondwanaland ocean floor of Late Jurassic–Early Cretaceous age, part of a much earlier back-arc extensional phase. An origin of the Central Scotia Sea as a late Mesozoic part of the northern flank of an intra-Gondwanaland spreading centre (the southern flank being in the Weddell Sea), since captured within the Cenozoic back-arc extensional opening of the Scotia Sea, has been proposed by Eagles (2010*a*, *b*).

The principal cause of the survival of this uncertainty is the relative shortness of the zone of east–west linear magnetic anomalies in the northern CSS, and therefore inevitable non-uniqueness of the anomaly identifications. Hill & Barker (1980) cited the compatibility of the Miocene age with the (unloaded) water depths and sediment cover (with respect to Drake Passage), and the coincidence, at about 16 Ma, of a reduction in spreading rate in both the CSS and western Scotia Sea opening, implying close coupling. Eagles (2010*a*) used the magnetic reversal timescale (MRTS) of Gradstein *et al.* (2004) to argue that a single, sudden spreading rate change was unnecessary (although a reduction of some kind is common to both Drake Passage and the northern CSS), and pointed out that the abandoned spreading centre, and other topographic features expected from the Drake Passage–CSS connection that formed part of Hill & Barker's (1980) preferred model, cannot be identified in gravity data.

In late 1995, 10 heat-flow measurements were made in the CSS, from RRS *James Clark Ross* of

[†]Deceased. Peter Barker died on 25 June 2012 while this paper was in review.

From: HAMBREY, M. J., BARKER, P. F., BARRETT, P. J., BOWMAN, V., DAVIES, B., SMELLIE, J. L. & TRANTER, M. (eds) 2013. *Antarctic Palaeoenvironments and Earth-Surface Processes*. Geological Society, London, Special Publications, **381**, 139–150. First published online April 19, 2013, http://dx.doi.org/10.1144/SP381.3

the British Antarctic Survey (BAS), using the heat flow probe of the University of Texas Institute for Geophysics. The rationale for this enterprise concerned the measurable parameters governing conductive cooling of oceanic lithosphere (Tréhu 1975; Parsons & Sclater 1977). The calculated relationships for ocean floor younger than 80 Ma were: z (water depth, km) $= 2.53 + 0.336t^{\frac{1}{2}}$ (Tréhu 1975), where t is age in millions of years, and h (heat flow, mW m^{-2}) $= 473/t^{\frac{1}{2}}$ (Parsons & Sclater 1977). It is clear that back-arc basin sea floor rarely fits any standard age–depth correlation, which makes the use of depth in such areas as a determinant of age questionable, particularly since crustal thickness may vary over short distances (Martinez & Taylor 2002, 2003). In contrast, geothermal heat flow depends solely on the ocean floor age. In view of the controversy over the age of the northern CSS, and the limited extent of the basins further south, measurement of potentially age-diagnostic oceanic heat flow seemed useful.

The interpretation of heat flow in terms of oceanic basement age is somewhat uncertain. The estimate of Parsons & Sclater (1977) is preferred here to that of Stein & Stein (1992) for young ocean floor, because the former chose sites where sediment cover restricted hydrothermal circulation, causing more evenly distributed heat loss. Hydrothermal circulation in young ocean floor on which sediment cover is thin or absent causes significant spatial variation in heat flow (Davis *et al.* 1996). On the other hand, the estimates of Stein & Stein (1992) of age from unloaded basement depth are preferred for ocean floor older than about 20 Ma ($z = 5.651 - 2.473\exp(-0.0278t)$). These early estimates have persisted because research attention (for both heat flow and water depth) has been focused upon discrimination between models of lithospheric structure, for which older ocean floor is more valuable. Generally, more recent heat flow determinations have not been analysed, but for unloaded depths in the main ocean basins, the relation $a + bt^{\frac{1}{2}}$ for young ocean floor has many published values of b ranging from 0.295 to 0.354 km, with 'young' ranging from 20 to 90 Ma, and ridge-crest depth (a) from 2.420 to 2.890 km (Turcotte & Oxburgh 1967; Davis & Lister 1974; Tréhu 1975; Parsons & Sclater 1977; Heestand & Crough 1981; Schroeder 1984; Phipps Morgan & Smith 1992; Stein & Stein 1992).

The results of three stations within the northern CSS have been published elsewhere (Barker & Lawver 2000). Unfortunately, the thermal gradients there were too low to be primarily the results of vertical conductive cooling of ocean floor of any age, as were the results of two of three earlier measurements (Zlotnicki *et al.* 1980), attributed by those authors to instrumental error (the third was deemed 'consistent with an Oligocene age' for the northern CSS). The anomalously low heat flow values are explained by Barker & Lawver (2000) as the influence of either colder than usual bottom water temperatures for several years prior to the measurements caused by an influx of Antarctic Bottom Water, or lateral pore-water advection in the uppermost crust resulting from the scouring action of the Antarctic Circumpolar Current (ACC). More recently, Eagles (2010*a*) modelled these temperature profiles as curves between Mesozoic basal temperatures and bottom water temperatures that varied in accordance with available observations. Here we report the results of measurements at the remaining seven stations (Fig. 1) from the small oceanic basins further south, which are too far south to have been affected by the Polar Front (see Fig. 1 inset and Naveira Garabato *et al.* 2002), and where present-day bottom-water temperatures are lower, making the effects of a prior episode of even lower temperatures less severe. We then compare them with magnetic anomaly identifications, bearing in mind the essential ambiguity of identifications in small ocean basins, and with estimates from basement depth, corrected for sediment loading.

Method

Barker & Lawver (2000) described the heat flow probe used, the experimental method and attendant uncertainties and assumptions. All stations were chosen on the basis of seismic reflection profiles to be typical of the basins within which they were sited, and were shown from 3.5 kHz profiles obtained on initial crossings to have avoided mass-flow deposits and turbidites that would have invalidated the measurements. Mass-flow deposits are rare in the CSS, although some have been mapped in the southern part of Dove Basin, to the south of sites HF05 and HF06 (Pudsey & Howe 1988). Mass flow deposits in that area have also recently been described from new acoustic sub-bottom profiler data (Ruano *et al.* 2011). Available piston cores from the region provided additional assurance of pelagic or hemipelagic deposition and low sedimentation rates, which has been supported by the finding of Pugh *et al.* (2009) of a close correspondence between magnetic susceptibility logs of CSS cores and EPICA Dome C dust variations.

At each site, the probe was held about 50 m above the sea bed for 6–8 min to ensure stability, then lowered rapidly into the sediments to ensure as great a penetration as possible but also generating an initial frictional component to the temperature rise. After a further 6–8 min in which this

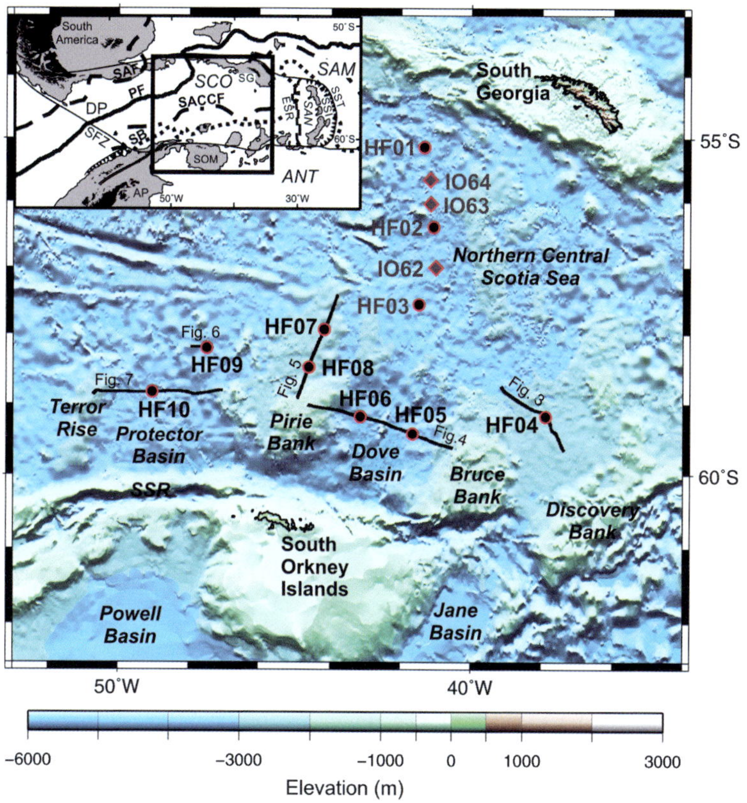

Fig. 1. Locations of CSS heatflow stations HF01–10 by RRS *James Clark Ross* November 1995 (red circles with black fill), IO62–4 by MV *Islas Orcadas* March 1978 (red diamonds with grey fill; Zlotnicki *et al.* 1980). Stations HF01–3, IO62–4 are annotated in grey, HF04–10 described here are annotated in black). Thick black lines are seismic reflection profiles through stations HF04-10 shown in Figures 3–7. Stations are overlaid on elevation data from GEBCO 2003 dataset (IOC, IHO & BODC 2003), displayed with shaded relief illumination from the north. SSR, South Scotia Ridge. Black rectangle on inset shows regional location of main figure, with areas shallower than 2000 m shaded grey and plate boundaries marked as continuous thin lines (single lines indicate boundaries where motion is dominantly strike-slip; double lines indicate extensional boundaries; saw-tooth annotations indicate convergent boundaries). Annotations in large italic lettering identify tectonic plates: ANT, Antarctic; SAM, South America; SAN, Sandwich; SCO, Scotia. Annotations in smaller, upright lettering indentify continental blocks and tectonic features: AP, Antarctic Peninsula; DP, Drake Passage; ESR, East Scotia Ridge; SFZ, Shackleton Fracture Zone; SG, South Georgia; SOM, South Orkney microcontinent; SSI, South Sandwich Islands; SST, South Sandwich Trench. Inset also shows typical locations of the main oceanographic fronts as thick black lines, illustrating how their locations are influenced by tectonically generated topographic features: SAF, Sub-Antarctic Front (dashed line); PF, Polar Front (solid line); SACCF, Southern Antarctic Circumpolar Current Front (dash–dot line); SB, Southern Boundary of the Antarctic Circumpolar Current (dotted line).

component dissipated and true *in situ* temperatures could be determined, a heat pulse was applied so that its decay, over about 15 min, could be used to determine thermal conductivity. A station comprised four to six such sites, about 1 km apart. The ship moved slowly between sites, with the partly recovered probe towed behind.

The 13 temperatures (from 12 thermistors along outrigger bows attached to the probe barrel and an upper thermistor within the weightstand) are shown for each site in Figure 2, after thermistor calibration. Probe penetration was generally to the weightstand base (i.e. top outrigger thermistor at 1.5 m depth with the weightstand thermistor above the sediment), as designed, but was incomplete at Stations 9 and 10, so that the upper thermistors were in the water column, most probably because of the presence of coarse sediments (and, in the case of the former, the deteriorating weather). Coarseness also reduced 3.5 kHz penetration of

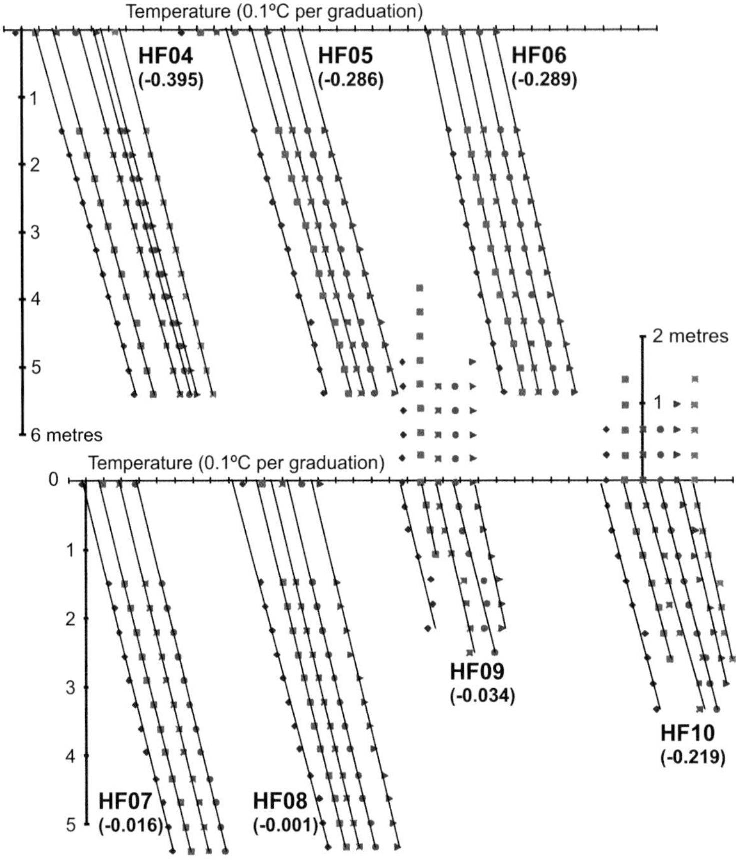

Fig. 2. Temperatures at Stations HF04–10. For all stations, straight lines are best-fitting gradients through the (calibrated) sediment temperatures of an individual penetration. Penetrations at a station are offset by multiples of 0.10 K for clarity. Station locations are shown in Figure 1. Numbers in parentheses beneath each station ID are the mean sea bed temperature at the station in degrees Centigrade.

the sediments, so to some extent incomplete penetration could be anticipated. Of these two stations, temperatures at HF10 showed a constant slope, apart from the third penetration, but HF09 showed signs of variable *in situ* temperature gradients, or (more likely) of sediment disturbance, and was thus of reduced value as a measure of basement age. Note that, even so, the general temperature gradient at HF09 was not greatly reduced, as were those at HF01–3 and two of the three earlier northern CSS stations (Zlotnicki *et al.* 1980; Barker & Lawver 2000). Sea bed temperatures at sites were generally within 0.1°C of the value extrapolated from the best-fitting straight line (Fig. 2), supporting the designed 1.5 m penetration. The offset at HF05 was greater but, since bottom water temperatures at stations HF05 and HF06 were similar, it is possible that the weightstand at the former partially penetrated the sediments.

Results

Thermal conductivities were determined from decay of the applied heat pulse (Hyndman *et al.* 1979; Lister 1979) and of the initial frictional heat component using the program HFRED (Villinger & Davis 1987*a*, *b*), and the thermal gradients of Figure 2 are shown as indications of data quality. The major uncertainties determined by HFRED were the effective origin times of the initial (run-in) frictional heat pulse and of the subsequent applied heat pulse: the thermistor sample interval was 10 s, and the effective origin times lay within this interval. This was done iteratively by HFRED so that, finally, heat flow could be calculated (as the product of temperature gradient and thermal conductivity) and an improved estimate made of basement age. Heat flow and calculated age at each station are shown in Table 1.

Table 1. *Heat flow stations HF04-10 (located in Fig. 1)*

Station	Latitude (south)	Longitude (west)	Number of penetrations	Water depth (m)	Sea bed temperature (°C)	Location	Heat flow (mW m^{-2})	Age (Ma)
4	59°11′	37°54′	5	2918	−0.395	Bruce–Discovery Basin	86.89	30.2
5	59°26′	41°40′	5	3852	−0.286	Dove Basin (Pirie–Bruce)	72.79	42.2
6	59°11′	43°09′	5	3988	−0.289	Dove Basin (Pirie–Bruce)	72.11	43
7	57°56′	44°10′	4	2863	−0.016	N of Pirie Bank	79.96	35
8	58°29′	44°36′	5	2946	0.001	North of Pirie Bank	80.88	34.2
9	58°12′	47°29′	4	3368	−0.034	NE of Pirie Bank	88.84	28.3
10	58°50′	49°02′	6	4045	−0.219	Protector Basin	94.31	25.2

Water depths, sea-bed temperatures and positions are means over a station, and basement ages were calculated using the estimated parameters of Parsons & Sclater (1977).

An additional factor is the possible effect of either of the two mechanisms advanced by Barker & Lawver (2000) for the non-geothermal heat flow values determined in the northern CSS, albeit at a reduced level. The result would be that the heat flow values reported here are minimal values, and therefore that the estimated ages are too old.

Discussion

For the basins considered here, we compared age estimates at each station made from heat flow, from (unloaded) water depth and from magnetic anomalies. Unloaded depth may be obtained from sediment two-way time (TWT; from seismic reflection profiles), converted to thickness (Carlson *et al.* 1986), thence to load assuming compaction of a diatomaceous mud (Sclater & Christie 1980), and for water and mantle densities of 1.03 and 3.3 g cm^{-3}, respectively. For these parameters, there is little difference in the unloading calculation (less than 35 m for up to 2 s TWT of sediment) from the simpler method of Crough (1983), used by some authors. No correction has been made for radiometric heat generation within sediments, or for the reduction of heat flow by sediment 'blanketing' (Hutchison 1985). These last two effects are relatively small for the sediment thicknesses and likely sedimentation rates seen here: they oppose each other (although the latter is usually the larger), need additional information on sediment properties and were not considered by Parsons & Sclater (1977) or Stein & Stein (1992).

The age estimates cover a wide range. Those derived from magnetic anomaly interpretation are inevitably ambiguous, because the basins are small and the anomalies band-limited. Differences in the MRTSs used by different authors are comparatively small, but occasionally important. Ages derived from basement depths also vary: in some cases within the literature, depths are assumed to be the same as those of the same age in the main ocean basins, and in others to have been modified (in one case, profoundly so) by their presence within a back-arc environment. That the back-arc environment may modify depth to some extent is likely, in that the asthenosphere on which the cooling lithosphere floats may not be typically oceanic within the back-arc, and crustal thickness may vary over short distances (Martinez & Taylor 2002, 2003). For example, the present South Sandwich back-arc ridge crest is in places much deeper (Barker 1995; Livermore 2003) than the Parsons & Sclater (1977) estimate. Sclater *et al.* (1980) found a strong similarity of back-arc and main ocean basin depths, although data were sparse and uncertainties large. The data set has since grown, but no more recent review has been made. An additional factor is the 'chaotic' nature of initial opening (Sclater *et al.* 1980), which may involve reduced subsidence that would persist, and dominate in a basin of limited extent.

HF04

Sea ice prevented the siting of this station further south. It was located at the boundary between the Bruce–Discovery Basin (also referred to by some authors as the Discovery Basin and by others as the Scan Basin) and the northern CSS, and gave a heat flow age estimate of 30.2 Ma. Figure 3 shows the seismic reflection profile through the station, which suggests a minor additional basement elevation consistent with the boundary between two oceanic provinces. No magnetic anomaly-based age determinations have been made in the Bruce–Discovery Basin (nor has the direction of opening been established), and the ambiguous magnetic anomaly age of the northern CSS has already been mentioned. The unloaded water depths of the northern CSS (from Hill & Barker 1980, Fig. 3) are equivalent to ages of 7–21 Ma (Parsons & Sclater 1977) or 5–16 Ma (Stein & Stein 1992), and those of the Bruce–Discovery Basin are similar. The unloaded water depth of 3500 m at station HF04 is equivalent to an age of only 6–8 Ma.

Lawver *et al.* (1991) reported six heat flow stations (35 penetrations) in Jane Basin, east of the South Orkney microcontinent and south of the CSS. The basin is assumed to have been back-arc extensional (Barker *et al.* 1984), and its present-day contiguity with the Bruce–Discovery Basin invites consideration of a common origin. However, these two basins could only have opened in approximately their present relative positions if the Dove and Protector basins are older than the Bruce–Discovery Basin, as the Bruce–Discovery Basin would have opened in a position north of the South Orkney microcontinent if it is older than one or both of the other two. Jane Basin is narrow (80–90 km) and water depths are 3200–3600 m. Sediments on the two seismic reflection profiles crossing the basin showed 1.5–1.8 s TWT of sediment. Prior ODP drilling (Leg 113: Shipboard Scientific Party 1988) permitted a relatively precise assessment of sediment type and sedimentation rate, and thus of the effects of sediment blanketing (Hutchison 1985). Lawver *et al.* (1991) also estimated radiogenic heat generation within the sediments and concluded that, combining these two effects, the measured heat flow was 86–89% of the true geothermal value, and that Jane Basin age was 25–32 Ma. This finding differs from that of Bohoyo *et al.* (2002), that magnetic anomalies date Jane Basin opening at 17.6–14.4 Ma, after the cessation of Weddell Sea subduction (Barker *et al.* 1984). However, Jane Basin is narrow and the magnetic anomaly identifications correspondingly ambiguous.

The HF04 age is similar to that of Lawver *et al.* (1991), which might support views of the original unity of the Bruce–Discovery and Jane basins. On the other hand, if station HF04 lies at the southern margin of the northern CSS, then the heat flow determination would support a mid-Cenozoic age of onset for that province.

HF05, 06

The two stations in the Pirie–Bruce (Dove) Basin were symmetrically disposed about the centre (Fig. 4) , and avoided a central bathymetric high that may have been of later origin (Eagles *et al.* 2006). Their similar heat-flow ages (42.2 and 43.0 Ma) strongly support each other. Unloaded water depths at the stations are *c.* 4500 m, equivalent to ages of 34–28 Ma. Magnetic anomalies were interpreted by Eagles *et al.* (2006) as revealing east–west extension between 41 and 35 Ma, the choice being guided by the unloaded depths. The heat flow ages are greater, but the possibility should be borne in mind of true basin age being

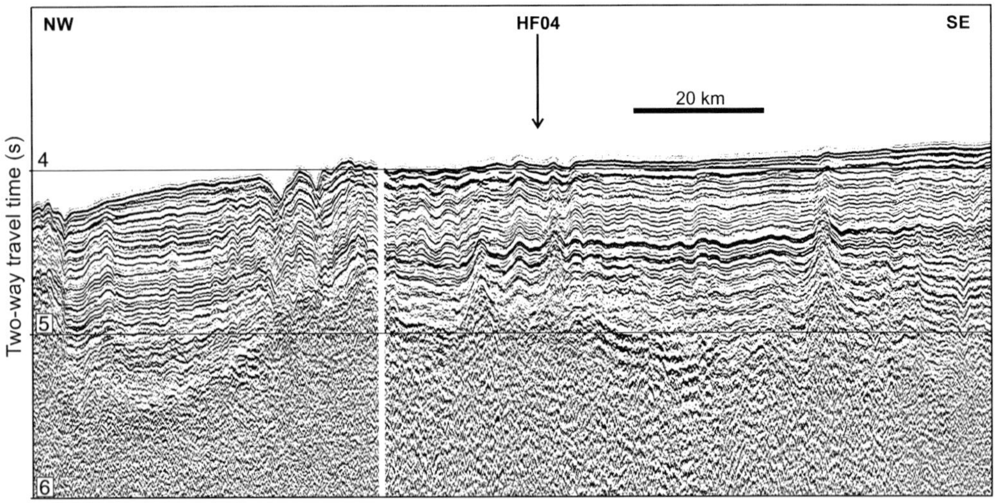

Fig. 3. Station HF04 and the north–south seismic reflection profile through it (RRS *Charles Darwin* 1989: Line BAS889-s08).

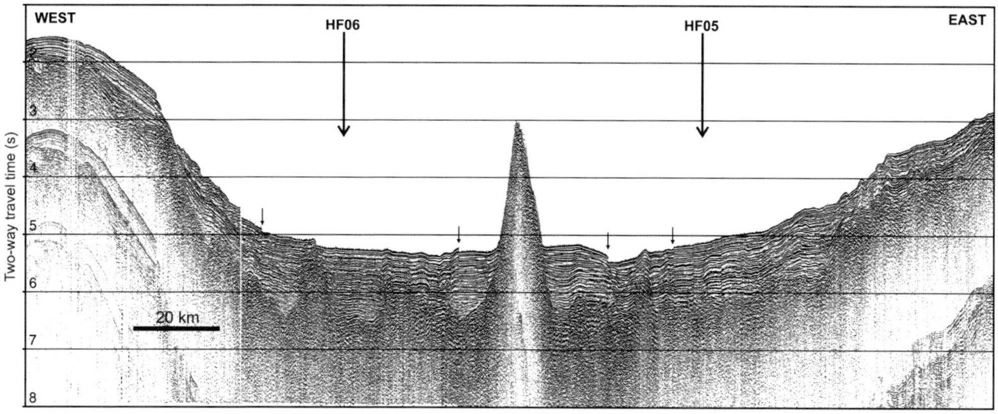

Fig. 4. Locations of stations HF05 and HF06 on the east–west seismic reflection profile through them (RRS *Discovery* Cruise 172, 1988: Line BAS878-02; an interpreted line drawing of this profile is included in Eagles *et al.* 2006). Some apparent small steps in the sea floor (small arrows) on this near trace gather display result from technical problems during data acquisition that caused short breaks in recording.

younger, as these are the closest of any stations to the Weddell Sea, the source of cold bottom water identified by Barker & Lawver (2000).

HF07, 08

Both stations lie in the small basin north of Pirie Bank (Fig. 5), symmetrically disposed about the high in the sub-sedimentary basement. Their similar results (equivalent to ages of 35.0 and 34.2 Ma) are mutually supportive. No magnetic interpretation has been published. The unloaded water depths (*c.* 3300 m) provide a very young age, which is incongruous given the thickness of the sediment cover, and it seems likely that this basin did not subside fully during its initial formation. Such anomalous subsidence could result from a variety of factors, including crustal thickness greater than normal oceanic crust, influence of regional mantle dynamic processes, and flexural support from the surrounding lithosphere. The small size of the basin and the anomalously shallow basement depths raise doubts about whether any of its basement is true oceanic crust. If the basin floor consists of transitional or even continental crust, radiogenic heat production might bias the ages determined from heat flow measurements to younger values, and therefore the ages given above can be considered as minimum ages. The basin shape, particularly its well-defined southern margin and the symmetry observed along a north–south seismic profile (Figs 1 & 5), suggests north–south extension. It lies adjacent to the northern CSS, the east–west magnetic lineations within which show that it also most probably formed by

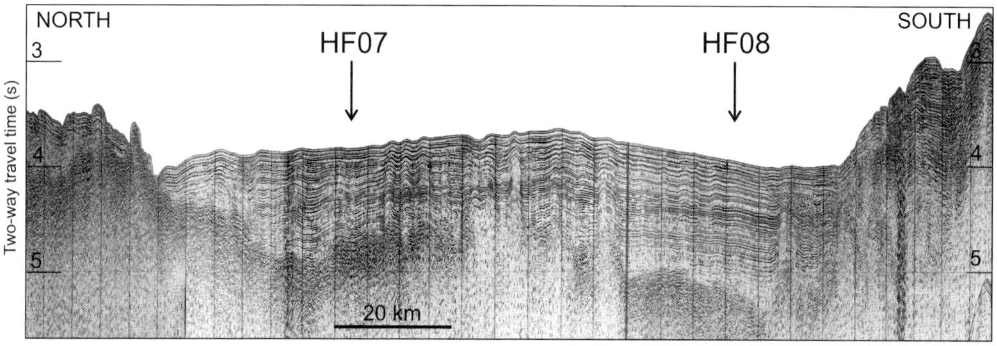

Fig. 5. Stations HF07 and HF08 and the north–south seismic reflection profile through them (RRS *Shackleton*, 1973: U. Birmingham/BAS unpublished data).

north–south extension, so that the small basin sampled could have been an initial, failed part. To this extent, therefore, the heat flow ages support a back-arc origin for the northern CSS, with a mid-Cenozoic onset.

HF09

The tectonic significance of this station is uncertain. It most definitely lies within a small basin that is an extensional oceanic province (Fig. 6), and its estimated age (*c*. 28 Ma) places it clearly within the mid to late Cenozoic back-arc regional framework, but it may lie within a more northeasterly arm of Protector Basin, or at the eastern edge of Drake Passage. Penetration was incomplete, as a result of deteriorating weather and relatively coarse sediments, and this result is less reliable than those at the other stations (see Fig. 2 and Method section).

The unloaded water depth (*c*. 3800 m) gives an unacceptably young age for either of the most likely tectonic provinces, and anomalous initial subsidence seems possible, as the basin is small. Magnetic anomaly identification in Drake Passage is undisputed through many publications, from Barker & Burrell (1977) onwards (e.g. LaBrecque 1985; Eagles *et al.* 2005; Lodolo *et al.* 2006), and the estimated age for HF09 is compatible. Protector Basin is discussed below.

HF10

This station (a single site because time was short) lies clearly within the east–west-opening Protector Basin (Fig. 7), and provides an estimated age of 25.2 Ma. Penetration here was incomplete also, most probably the result of the presence of coarse sediments as at HF09, but the greater depth of penetration means that the result is more reliable (Fig. 2). The unloaded water depth is *c*. 4500 m, equivalent to ages of 34–28 Ma, as at stations HF05 and HF06. Protector Basin is small (although larger than any of the other basins discussed here), and the magnetic anomaly identifications are therefore ambiguous. Hill & Barker (1980) proposed 17–13 Ma, and a similar age range (17.5–14 Ma) was suggested by Galindo-Zaldivar *et al.* (2006). Eagles *et al.* (2006) proposed two alternative opening ages of 34–30 or 49–41 Ma on the basis of marine magnetic anomalies, but preferred the younger of the two estimates as it fitted better with the unloaded water depths.

The heat flow value and estimated age at HF09 are compatible with those at HF10, suggesting that the latter measurement is valid and that the small basin to the NE could be a part of Protector Basin, precursory or late stage as Galindo-Zaldivar *et al.* (2006) suggest.

The coarse sediment at Stations HF09 and HF10 was unexpected, but may have been derived from

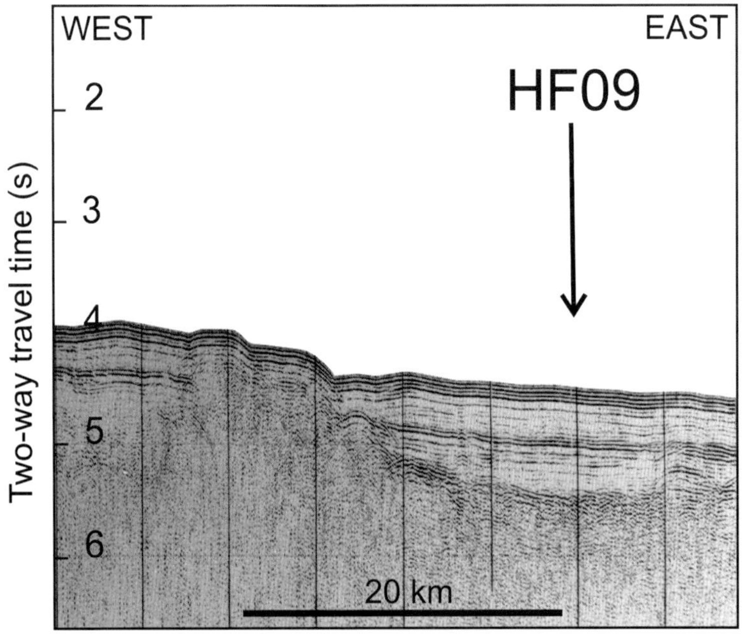

Fig. 6. Station HF09 and the east–west seismic reflection profile through it (RRS *Shackleton* 1976: U. Birmingham/BAS unpublished data). Longer profiles are reported by Galindo-Zaldivar *et al.* (2006).

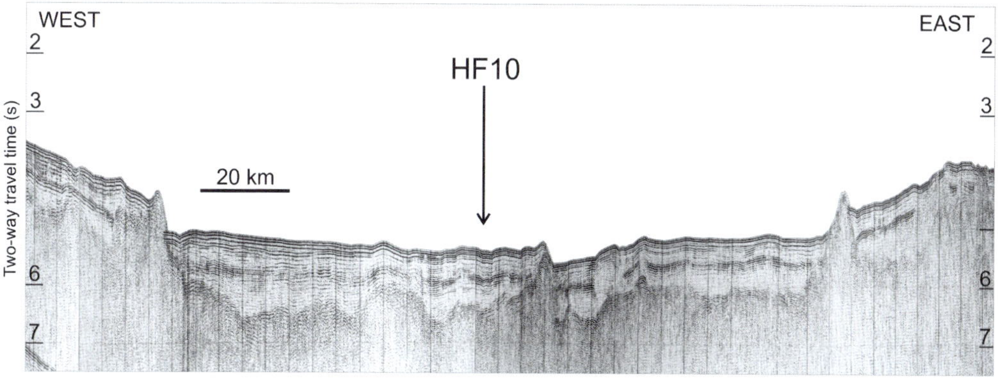

Fig. 7. Station HF10 and the east–west seismic reflection profile through it (RRS *Shackleton* 1976: U. Birmingham/BAS unpublished data; line drawing in Eagles *et al.* 2006).

the nearby South Scotia Ridge, which now shows a fairly flat top in places, and therefore may have been occupied, and eroded, by grounded ice during glacial periods, the resulting coarse debris having been deposited from melting icebergs. This is consistent with the observations of Herron & Anderson (1990), who concluded that the grounding line on the South Orkney Island microcontinent, directly

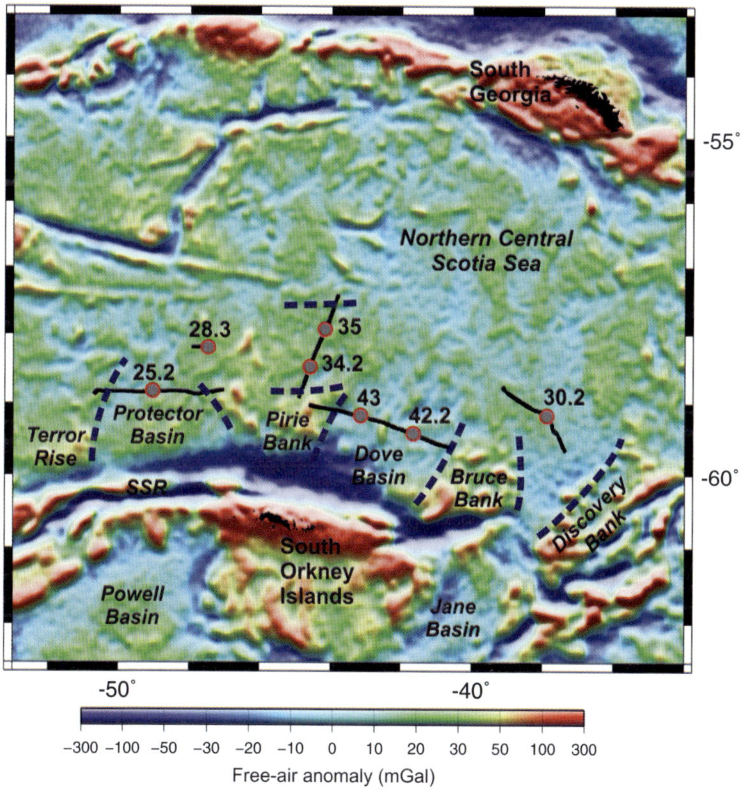

Fig. 8. Ages estimated from heat flow at Stations HF04–10 (numbers in bold type, in millions of years), with outlines of basins within which they lie (blue dashed lines), overlaid on free-air gravity anomaly field derived from satellite altimetry (Sandwell & Smith 2009) displayed with shaded relief illumination from the north. Tectonic environments of HF04 and HF09 are uncertain. SSR, South Scotia Ridge.

to the east of the part of the South Scotia Ridge that is adjacent to Protector Basin, extended to 250 m.

The meaning of the heat flow age for Protector Basin is uncertain. Eagles *et al.* (2006) used long negative magnetic field intervals (*c.* 32 and 45 Ma) within the MRTS of Cande & Kent (1995) in their models to fit the observed magnetic field. There is a long negative interval *c.* 24 Ma, which would permit compatibility with the heat flow age, but it is not long enough to permit an acceptable fit to the observed field except at a very high spreading rate. A possible reason for an abnormally high heat flow is the abundance of radiogenic (south Scotia Ridge continental, glacial) material within the sediments. Another is post-spreading magmatism at the spreading centre (e.g. Haase *et al.* 2011), as HF10 is only *c.* 20 km away.

Conclusions

Heat flow in the five small basins of the southern CSS is compatible with essentially vertical conductive cooling of oceanic basement, and with basin origins by back-arc extension during the mid to late Cenozoic, within the Scotia Sea. Estimated ages range from 43 to 25 Ma, and the effect (at a lower level) of either of the mechanisms proposed by Barker & Lawver (2000) for the non-tectonic nature of northern CSS heat flow measurements is to make these maximum ages. In the majority of cases, these ages are compatible with those estimated from marine magnetic anomalies and sediment-unloaded depth to basement, where those estimates are available. In some (small) basins, initial subsidence may have been anomalous.

Figure 8 shows the heat flow age estimates at each station. The older ages (42.2 and 43.0 Ma within the Dove (Pirie–Bruce) Basin) confirm that back-arc extension at the western end of the SAM–ANT boundary was underway in the Eocene. It is possible that back-arc extension in the Scotia Sea was initiated as a result of a change in the relative motion of the bounding major plates at about 50 Ma (Livermore *et al.* 2005, 2007).

Some ages are older than those assumed in a model that suggested an early Miocene onset of the ACC (Barker 2001). However, none of these estimates modify that suggestion, because the key barrier in the model, until that time, lay along the North Scotia Ridge. In fact, the only change that would modify that time of onset is a different age for the northern CSS, which could have been fully open earlier than assumed. Although different ages have been suggested for the northern CSS (e.g. Eagles 2010*a*, *b*), that issue is not addressed directly here. Indirectly, an old (pre-43 Ma) northern CSS could not have lain in its present position relative to Dove Basin, which must have opened in a position much closer to Tierra del Fuego, and subsequent relative motion would have produced basement topography that has not been detected.

Also not addressed is the issue of an earlier ‘proto-ACC’, flowing over a shallow (continental shelf depth) barrier, sometimes invoked to explain earlier similarities between the Atlantic and Pacific Oceans. Geochemical evidence of such earlier interchange (e.g. neodymium isotopes: Scher & Martin 2004) does not necessarily require circumpolar flow (the ‘gateway’ could have been elsewhere) and may be explained by relatively small fluxes, so there may not be any related deep basins in the Drake Passage region.

The heat-flow equipment was prepared at University of Texas Institute for Geophysics by Keith Najmulski and set up and maintained on-board ship by Vsevelod Afanasyev and Mark Preston. P. F. Barker is grateful to the ship’s company of RRS *James Clark Ross* for advice and willing assistance, and also thanks Arnold Gordon for useful comments on regional oceanography and Bruno della Vedova for sight of unpublished data. We thank Lisa Gahagan for help in producing Figure 8, and we are grateful to Graeme Eagles and Fernando Bohoyo for constructive reviews.

References

BARKER, P. F. 1995. Tectonic framework of the east Scotia Sea. *In*: TAYLOR, B. (ed.) *Back-arc Basins: Tectonics and Magmatism*. Plenum Press, New York, 281–314.

BARKER, P. F. 2001. Scotia Sea regional tectonic evolution: implications for mantle flow and palaeocirculation. *Earth Science Reviews*, **55**, 1–39.

BARKER, P. F. & BURRELL, J. 1977. The opening of Drake Passage. *Marine Geology*, **25**, 15–34.

BARKER, P. F. & LAWVER, L. A. 2000. Anomalous temperatures in central Scotia Sea sediments – bottom water variation, or pore water circulation in old ocean crust. *Geophysical Research Letters*, **27**, 13–16.

BARKER, P. F., BARBER, P. L. & KING, E. C. 1984. An early Miocene ridge crest–trench collision on the South Scotia Ridge. *Tectonophysics*, **102**, 315–332.

BARKER, P. F., DALZIEL, I. W. D. & STOREY, B. C. 1991. Tectonic development of the Scotia Arc region. *In*: TINGEY, R. J. (ed.) *Geology of Antarctica*. Oxford University Press, Oxford, 215–248.

BOHOYO, F., GALINDO-ZALDIVAR, J., MALDONADO, A., SCHREIDER, A. A. & SURINACH, E. 2002. Basin development subsequent to ridge crest-trench collision: the Jane Basin, Antarctica. *Marine Geophysical Researches*, **23**, 413–421.

CANDE, S. C. & KENT, D. V. 1995. Revised calibration of the geomagnetic polarity timescale for the late Cretaceous and Cenozoic. *Journal of Geophysical Research*, **100**, 6093–6095.

CARLSON, R. L., GANGI, R. F. & SNOW, R. F. 1986. Empirical reflection travel time versus depth and velocity versus depth functions for the deep-sea sediment column. *Journal of Geophysical Research*, **91**, 8249–8266.

CROUGH, S. T. 1983. The correction for sediment loading on the seafloor. *Journal of Geophysical Research*, **88**, 6449–6454.

DAVIS, E. E. & LISTER, C. R. B. 1974. Fundamentals of ridge crest topography. *Earth and Planetary Science Letters*, **21**, 405–413.

DAVIS, E. E., CHAPMAN, D. S. & FORSTER, C. B. 1996. Observations concerning the vigor of hydrothermal circulation in young oceanic crust. *Journal of Geophysical Research*, **101**, 2927–2942.

DEWIT, M. J. 1977. The evolution of the Scotia Arc as a key to the reconstruction of southwest Gondwanaland. *Tectonophysics*, **37**, 53–81.

EAGLES, G. 2010*a*. The age and origin of the central Scotia Sea. *Geophysical Journal International*, **183**, 587–600.

EAGLES, G. 2010*b*. South Georgia and Gondwana's Pacific margin: lost in translation? *Journal of South American Earth Sciences*, **30**, 65–70, doi: 10.1016/j.jsames.2010.04.004.

EAGLES, G., LIVERMORE, R. A., FAIRHEAD, J. D. & MORRIS, P. 2005. Tectonic evolution of the west Scotia Sea. *Journal of Geophysical Research*, **110**, B02401, doi: 10.1029/2004JB003154.

EAGLES, G., LIVERMORE, R. & MORRIS, P. 2006. Small basins in the Scotia Sea: the Eocene Drake Passage gateway. *Earth and Planetary Science Letters*, **242**, 343–353.

GALINDO-ZALDIVAR, J., BOHOYO, F., MALDONADO, A., SCHREIDER, A., SURINACH, E. & VAZQUEZ, A. T. 2006. Propagating rift during the opening of a small oceanic basin: the Protector Basin (Scotia Arc, Antarctica). *Earth and Planetary Science Letters*, **241**, 398–412.

GRADSTEIN, F. M., OGG, J. G. & SMITH, A. G. 2004. *A Geologic Time Scale 2004*. Cambridge University Press, Cambridge.

HAASE, K. M., BEIER, C. ET AL. 2011. Magmatic evolution of a dying spreading axis: Evidence for the interaction of tectonics and mantle heterogeneity from the fossil Phoenix Ridge, Drake Passage. *Chemical Geology*, **280**, 115–125, doi: 10.1016/j.chemgeo.2010.11.002.

HEESTAND, R. L. & CROUGH, S. T. 1981. The effect of hot spots on the oceanic age–depth relation. *Journal of Geophysical Research*, **86**, 6107–6114.

HERRON, M. J. & ANDERSON, J. B. 1990. Late Quaternary glacial history of the South Orkney Plateau, Antarctica. *Quaternary Research*, **33**, 265–275.

HILL, I. A. & BARKER, P. F. 1980. Evidence for Miocene back-arc spreading in the central Scotia Sea. *Geophysical Journal of the Royal Astronomical Society*, **63**, 427–440.

HUTCHISON, I. 1985. The effects of sedimentation and compaction on oceanic heat flow. *Geophysical Journal of the Royal Astronomical Society*, **82**, 439–459.

HYNDMAN, R. D., DAVIS, E. E. & WRIGHT, J. A. 1979. The measurement of marine geothermal heat flow by a multipenetration probe with digital acoustic telemetry and insitu thermal conductivity. *Marine Geophysical Researches*, **4**, 185–205.

IOC, IHO AND BODC 2003. *Centenary Edition of the GEBCO Digital Atlas*. Published on CD-ROM on behalf of the Intergovernmental Oceanographic Commission and the International Hydrographic Organization as part of the General Bathymetric Chart of the Oceans. British Oceanographic Data Centre, Liverpool.

LABRECQUE, J. L. (ed.) 1985. *South Atlantic Ocean and Adjacent Antarctic Continental Margin*. Ocean Margin Drilling Program Regional Atlas Series **13**. Marine Science International, Woods Hole, MA.

LARTER, R. D., VANNESTE, L. E., MORRIS, P. & SMYTHE, D. K. 2003. Structure and Tectonic Evolution of the South Sandwich arc. *In*: LARTER, R. D. & LEAT, P. T. (eds) *Intra-Oceanic Subduction Systems: Tectonic and Magmatic Processes*. Geological Society, London, Special Publications, **219**, 255–284.

LAWVER, L. A., DELLA VEDOVA, B. & VON HERZEN, R. P. 1991. Heat flow in Jane Basin, northwest Weddell Sea. *Journal of Geophysical Research*, **96**, 2019–2038.

LISTER, C. R. B. 1979. The pulse-probe method of conductivity measurement. *Geophysical Journal of the Royal Astronomical Society*, **57**, 451–461.

LIVERMORE, R. A. 2003. Back-arc Spreading and Mantle Flow in the East Scotia Sea. *In*: LARTER, R. D. & LEAT, P. T. (eds) *Intra-Oceanic Subduction Systems: Tectonic and Magmatic Processes*. Geological Society, London, Special Publications, **219**, 315–331.

LIVERMORE, R. A., NANKIVELL, A. P., EAGLES, G. & MORRIS, P. 2005. Paleogene opening of Drake Passage. *Earth and Planetary Science Letters*, **236**, 459–470.

LIVERMORE, R. A., HILLENBRAND, C.-D., MEREDITH, M. & EAGLES, G. 2007. Drake Passage and Cenozoic climate: An open and shut case? *Geochemistry, Geophyscis, Geosystems*, **8**, Q01005, doi: 10.1029/2005GC001224

LODOLO, E., DONDA, F. & TASSONE, A. 2006. Western Scotia Sea margins: improved constraints on the opening of the Drake Passage. *Journal of Geophysical Research*, **111**, B06101, doi: 10.1029/2006JB004361.

MALDONADO, A., BARNOLAS, A. ET AL. 2003. Contourite deposits in the central Scotia Sea: the importance of the Antarctic Circumpolar Current and the Weddell Gyre flows. *Palaeoceanography, Palaeoclimatology, Palaeoecology*, **198**, 187–221.

MARTINEZ, F. & TAYLOR, B. 2002. Mantle wedge control on back-arc crustal accretion. *Nature*, **416**, 417–420.

MARTINEZ, F. & TAYLOR, B. 2003. Controls on back-arc crustal accretion: insights from the Lau, Manus and Mariana basins. *In*: LARTER, R. D. & LEAT, P. T. (eds) *Intra-Oceanic Subduction Systems: Tectonic and Magmatic Processes*. Geological Society, London, Special Publications, **219**, 19–54.

NAVEIRA GARABATO, A. C., HEYWOOD, K. J. & STEVENS, D. P. 2002. Modification and pathways of Southern Ocean Deep Waters in the Scotia Sea. *Deep-Sea Research I*, **49**, 681–705.

PARSONS, B. L. & SCLATER, J. G. 1977. An analysis of the variation of ocean floor bathymetry and heat flow with age. *Journal of Geophysical Research*, **82**, 803–827.

PHIPPS MORGAN, J. & SMITH, W. H. F. 1992. Flattening of the sea-floor depth–age curve as a response to asthenospheric flow. *Nature*, **359**, 524–527.

PUDSEY, C. J. & HOWE, J. A. 1998. Quaternary history of the Antarctic Circumpolar Current: evidence from the Scotia Sea. *Marine Geology*, **148**, 83–112.

PUGH, R. S., MCCAVE, I. N., HILLENBRAND, C.-D. & KUHN, G. 2009. Circum-Antarctic age modelling of Quaternary marine cores under the Antarctic Circumpolar Current: ice-core dust–magnetic correlation. *Earth and Planetary Science Letters*, **284**, 113–123.

RUANO, P., LOBO, F. J. ET AL. 2011. Mass-movement deposits and seismic activity in the southern Scotia Sea. *In*: *International Symposium on Antarctic Earth Sciences, Program and Abstracts, John McIntyre Conference Centre*, Edinburgh, 11–15 July 2011, Abstract PS4.3, 73. http://www.isaes2011.org.uk/abstracts_v4_20_07_2011.pdf

SANDWELL, D. T. & SMITH, W. H. F. 2009. Global marine gravity from retracked Geosat and ERS-1 altimetry: Ridge segmentation versus spreading rate. *Journal of Geophysical Research*, **114**, B01411, doi: 10.1029/2008JB006008.

SCHER, H. D. & MARTIN, E. E. 2004. Circulation in the Southern Ocean during the Paleogene inferred from neodymium isotopes. *Earth and Planetary Science Letters*, **228**, 391–405.

SCHROEDER, W. 1984. The empirical age–depth relation and depth anomalies in the Pacific Ocean basin. *Journal of Geophysical Research*, **89**, 9873–9883.

SCLATER, J. G. & CHRISTIE, P. A. F. 1980. Continental stretching: an explanation of the post-mid-Cretaceous subsidence of the Central North Sea basin. *Journal of Geophysical Research*, **85**, 3711–3739.

SCLATER, J. G., JAUPART, C. & GALSON, D. 1980. The heat flow through oceanic and continental crust and the heat loss of the Earth. *Reviews of Geophysics and Space Physics*, **18**, 269–311.

SHIPBOARD SCIENTIFIC PARTY. 1988. *Proceedings of the Ocean Drilling Program, Initial Reports*, **113**. Ocean Drilling Program, Texas A & M University, College Station, TX.

STEIN, C. A. & STEIN, S. 1992. A model for the global variation in oceanic depth and heat flow with lithospheric age. *Nature*, **359**, 123–9.

TRÉHU, A. M. 1975. Depth versus (age)$\frac{1}{2}$: a perspective on mid-ocean rises. *Earth and Planetary Science Letters*, **27**, 287–304.

TURCOTTE, D. L. & OXBURGH, E. R. 1967. Finite amplitude convective cells and continental drift. *Journal of Fluid Mechanics*, **28**, 29–42.

VILLINGER, H. & DAVIS, E. E. 1987*a*. A new reduction algorithm for marine heat flow measurements. *Journal of Geophysical Research*, **92**, 12 846–12 856.

VILLINGER, H. & DAVIS, E. E. 1987*b*. HFRED: a program for the reduction of marine heat-flow data on a microcomputer. *Geological Survey of Canada Open File Reports*, **1627**.

ZLOTNICKI, V., SCLATER, J. G., NORTON, I. O. & VON HERZEN, R. P. 1980. Heat flow through the floor of the Scotia, far South Atlantic and Weddell seas. *Geophysical Research Letters*, **7**, 421–424.

Cenozoic landscape and ice drainage evolution in the Lambert Glacier–Amery Ice Shelf system

DUANNE A. WHITE[1,2]

[1]*Environmental Science, Macquarie University, Sydney, NSW 2109, Australia*

[2]*Institute for Applied Ecology, University of Canberra, Canberra, ACT 2601, Australia (e-mail: duanne.white@canberra.edu.au)*

Abstract: Landforms and sediments in the Prince Charles Mountains record the timing and magnitude of Cenozoic palaeotopographic changes in the Lambert Glacier–Amery Ice Shelf system. A review of geomorphic and sedimentological evidence indicates that considerable (>1–2 km) glacial incision into a pre-glacial palaeosurface occurred along the major outlet glaciers during the Cenozoic. This erosion was in turn the likely driver for uplift that averaged *c.* 50 m/Ma along the flank of the Amery Ice Shelf since at least the mid-Miocene Epoch. The volume of eroded material is an order of magnitude greater than the quantity of sediment presently preserved in Prydz Bay, suggesting considerable export of Cenozoic sediment off the continental shelf. The magnitude of erosion recorded in the Prince Charles Mountains is sufficient to have focussed Cenozoic ice-drainage patterns, but was too slow to have driven Quaternary changes in ice volume.

Sediments preserved in Prydz Bay on the continental shelf (Hambrey *et al.* 1991) and in the trough–mouth fan (Cooper & O'Brien 2004) provide an excellent record with which to interpret the glacial history of the Lambert Glacier–Amery Ice Shelf System during the Cenozoic (Fig. 1). Large-scale topographic changes may also have affected the dynamics of the ice sheet on this timescale (Taylor *et al.* 2004). Thus, landscape evolution in the Prince Charles Mountains must be taken into consideration when interpreting the Cenozoic glacial history of the region.

Cenozoic landscape development in the Prince Charles Mountains has been investigated using many different techniques, including geomorphic analysis of the present landscape (Crohn 1959; Trail 1964; Bardin 1977; Wellman & Tingey 1981; Adamson *et al.* 1997; Hambrey *et al.* 2007), investigation of fjordal marine and terrestrial sediments within the Prince Charles Mountains (McKelvey *et al.* 2001; Whitehead & McKelvey 2002; Whitehead *et al.* 2003), cosmogenic nuclides (Fink *et al.* 2006; Lilly *et al.* 2010; White *et al.* 2011), apatite fission track thermochoronology (Arne 1994; Lisker *et al.* 2003, 2007) and numerical backstacking of the Cenozoic sediments deposited in Prydz Bay (Jamieson *et al.* 2005) and the larger Southern Ocean (Wilson *et al.* 2012) using a numerical model. However, while some correlations between the onshore and offshore sedimentary stratigraphic and biostratigraphic record have been attempted (Whitehead *et al.* 2006; Hambrey *et al.* 2007), a synthesis of the Cenozoic landscape history throughout the Prince Charles Mountains is lacking.

The primary aim of this paper is to constrain the magnitude and timing of relief generation in the Prince Charles Mountains since the onset of glaciation during the Eocene (Ehrmann & Mackensen 1992; Barker *et al.* 2007). To achieve this goal, it will combine results from existing studies, new field observations and a GIS analysis of the BEDMAP data set (Lythe *et al.* 2001). It will initially discuss evidence for the (limited) amount of Cenozoic erosion that has occurred on the presently exposed mountain peaks, and then contrast this with a range of evidence that suggests considerable Cenozoic incision along the present day outlet glaciers. Lastly, it will compare the field results with modelling efforts, and use these data to draw conclusions about how topographic changes may have affected Cenozoic ice volumes and flow patterns.

Methods

Exposures of Cenozoic sediments were identified from 1:50 000 scale colour aerial photographs. Three days of field investigation, stratigraphic logging and sampling of Pagodroma Group-like sediments were completed during the 2002–2003 PCMEGA expedition to the southern Prince Charles Mountains.

Exposed sections were measured with a tape and clinometers, and the sediments described using the facies definitions of Hambrey & McKelvey (2000*a*, *b*). In this classification, sediments with <1% gravel are classified as sand if the predominant grainsize is >63 μm and as mud if it is finer.

From: Hambrey, M. J., Barker, P. F., Barrett, P. J., Bowman, V., Davies, B., Smellie, J. L. & Tranter, M. (eds) 2013. *Antarctic Palaeoenvironments and Earth-Surface Processes*. Geological Society, London, Special Publications, **381**, 151–165. First published online July 1, 2013, http://dx.doi.org/10.1144/SP381.15

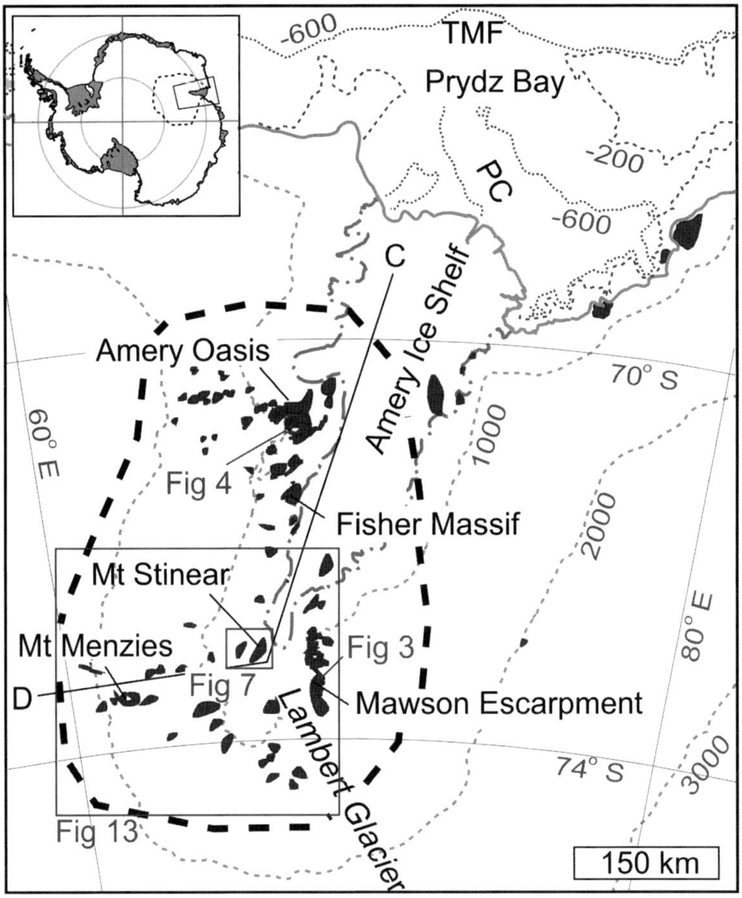

Fig. 1. Geographical features in the lower Lambert Glacier–Amery Ice Shelf system. TMF, the Prydz Bay trough–mouth fan; PC, Prydz Channel. The extent of the Prince Charles Mountains is defined by the thick dashed black line. Inset: rectangle displays location of larger figure, while dotted line indicates the present day Lambert Glacier–Amery Ice Shelf system ice drainage. The black line C–D indicates the location of Figure 10.

Poorly sorted sediments with 1–50% gravel and >10% sand and mud are classified as diamicts, and those with >50% gravel are classified as conglomerate. Prefixes (e.g. bouldery mud) are used to denote the dominant size of the >2 mm sized clasts in the case of diamict.

Selected samples from diamicts interpreted as Pagodroma Group-style deposits were investigated for diatom content. Samples were collected from silt-rich horizons within the diamicts, from which a few grams of the mud (<63 μm) fraction suspended in water were placed on smear slides and investigated on a petrologic microscope at 400× magnification. The qualitative abundance of diatoms was established by systematically scanning each slide 20 times from end to end using the method described by Whitehead *et al.* (2003).

The thickness and distribution of material eroded from the Prince Charles Mountains were calculated by measuring the difference in elevation between the modern fjord and valley floors and a regional pre-Eocene palaeosurface. The elevation of the regional pre-glacial palaeosurface was measured from point elevation data sourced from the Australian Antarctic Division (1993) and USSR (1978*a*–*d*) topographic maps, and field measurements using hand-held GPS. At each nunatak (e.g. Mt Ruker), the palaeosurface undulates with an amplitude of *c.* 200 m, so the map and field data were used to define an average local palaeosurface height (i.e. the dots in Figs 2 & 13a), with a sampling error of *c.* 50 m. Average local palaeosurface heights were then interpolated to a regional surface using the kriging function in ARCGIS 8.2. The

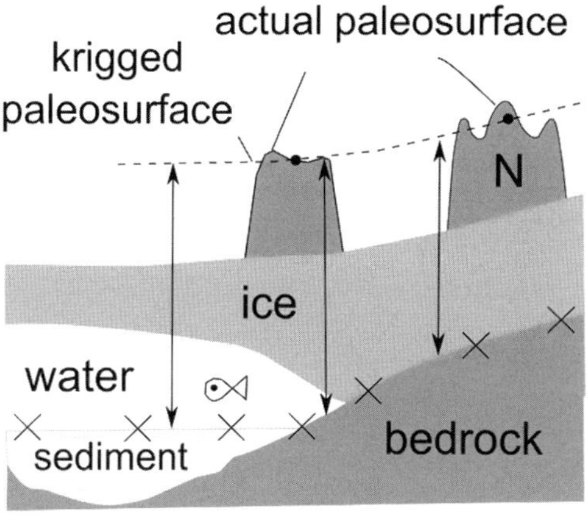

Fig. 2. Surfaces used for calculation of post-Eocene erosion thicknesses. The dots represent the local average height of the palaeosurface at each nunatak (N), used to produce the smoothed (i.e. krigged) height of the palaeosurface across the region. Crosses represent the base of the present day glacial valleys – that is, bedrock or subglacial sediment surface below the ice sheet, and bottom of the water column below the modern ice shelf. Arrows represent the thickness of rock eroded since incision of the palaeosurface began during the Eocene. The left-most arrow terminates at the sediment surface, as most of the sediment deposited by glaciers in this region since the Eocene would have been locally sourced.

volume of material removed since the formation of the palaeosurface was then estimated by subtracting the present height of ice-free outcrop, glacier bed or sub-ice shelf sediment (i.e. the 'valley bottom', derived from the BEDMAP dataset; Lythe *et al.* 2001) from the height of the interpolated palaeosurface (Fig. 2).

Preservation of uplands and summit surfaces

Apatite fission track analyses in both the northern and southern Prince Charles Mountains (Arne 1994; Lisker *et al.* 2003, 2007) indicate that the last major denudational event substantial enough to bring thermally re-set bedrock to the surface occurred during the Cretaceous Period, and that this denudation was likely to have been in response to the break-up of Gondwana. The 100–300 Ma fission track ages recorded from the uplands mean that these areas cannot have undergone more than *c.* 1 km of erosion since the Cretaceous Period, and have feasibly undergone substantially less than this.

The flat-topped nature of the peaks in the Prince Charles Mountains (Fig. 3), particularly in the south, suggests that these were once part of a smooth, undulating land surface with *c.* 200 m of relief (Bardin 1977; Wellman & Tingey 1981). This is hereby termed the Crohn Surface following the first report of the surface by Crohn (1959): 'many of the peaks within the ranges have flat topped summits, often partly covered by moraine deposits, at levels 500 m above the surrounding ice and snow surface'; 'since they truncate all geological structures, they must be assumed to be remnants of an old erosion surface'. The Crohn Surface may pre-date deposition of the Permo-Triassic Amery Group, but if so, had been re-exhumed during the Cretaceous Period (Lisker *et al.* 2007), in time for deposition of a 50 ± 2 or 40 ± 11 Ma BP tephritic phonolite lava flow (Fig. 4) onto the Precambrian basement that underlies the gently rolling uplands at Manning Massif (Wellman & Tingey 1981; Sheraton 1983; Andronikov & Beliatsky 1995).

Cenozoic erosion of the Crohn Surface has occurred largely through alpine or local glacial processes, evidenced by the numerous cirques that incise into it. Fluvial erosion, while not presently dominant, has incised the palaeosurface in some areas (Fig. 3) and fluvial processes remain active in lower altitude areas to the present day (White 2007; Hambrey *et al.* 2007). The magnitude of alpine glacial erosion is greatest in the Aramis, Athos and Porthos ranges in the northern Prince Charles Mountains, where the Crohn Surface has been completely eroded, leaving pointy alpine-style peaks (Trail 1964). Cenozoic erosion magnitudes are lower in the south, where there is preservation of the flat topography of the Crohn Surface itself and the Eocene Lava on Manning Massif.

Cosmogenic nuclide studies confirm that erosion rates were very low in the upland areas during the Quaternary, including summits and the plateau of the Crohn palaeosurface. Cosmogenic nuclide concentrations from both bedrock and glacial sediments from summits around the region consistently provide exposure ages of >1 Ma, and in some areas >2 Ma (Fink *et al.* 2006; White 2007; Lilly *et al.* 2010). The nuclide concentrations indicate maximum Quaternary erosion rates for summit localities of 0.1–1 m/Ma.

Cenozoic incision along major outlet glaciers

The flat-topped peaks in the Prince Charles Mountains are today separated by substantial U-shaped valleys filled by the major outlet glaciers, with relief of up to 3.5 km (Figs 5 & 6; Crohn 1959; Trail 1964; Bardin 1977; Wellman & Tingey 1981). Valley bottoms in the areas occupied by the modern Lambert, Mellor, Fisher, Charybdis and

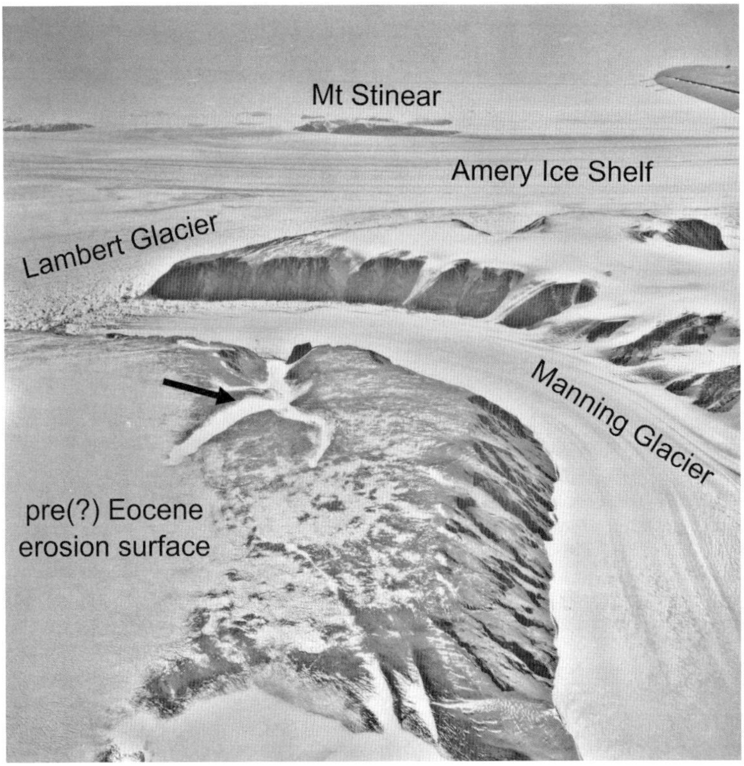

Fig. 3. Pre(?)-Eocene Crohn erosion surface at Mawson Escarpment (73°10′20″S, 68°20′36″E). Both glacial (e.g. Manning Glacier) and fluvial (black arrow) erosion has incised the surface, but it retains a consistent height on either side of the valleys, suggesting limited erosion of this surface since the Eocene. Cliff on far side of Manning Glacier is approximately 1 km high, while the valley at the black arrow is *c*. 300 m wide. Oblique aerial photograph courtesy of the Australian Antarctic Division, series ANT47, run 27, photo L7120, 27 Nov 1956.

many other outlet glaciers are now up to 2 km below sea-level, deep enough to remain substantially below sea-level even if the isostatic depression of the crust owing to the ice sheet is removed (Wellman & Tingey 1981). Prior to the onset of glaciation in Eocene time, large subaerial alluvial plains were present in Prydz Bay (Truswell 1991; Turner & Padley 1991; Truswell *et al.* 1999; O'Brien *et al.* 2004). This implies that the fluvial system was well established and would have filled with sediment any large internal basins at or above the sea-level of the time, which is likely to have been *c*. 100–200 m higher than the present day (Haq *et al.* 1987; Miller *et al.* 1998). Thus, it is clear that substantial valley bottom lowering has occurred along the outlet glaciers since Eocene time.

This valley bottom lowering could have resulted from either tectonic subsidence of the graben or glacial erosion along the base of the major valleys. The former is discounted because of the apparent lack of active Cenozoic tectonics in the region. Prydz Bay hosts the largest accumulation of Neogene sediment in the region, and despite preservation of fault offsets in pre-Cenozoic sediments, the Cenozoic sediments are tectonically unmodified (Cooper *et al.* 1991; Leitchenkov *et al.* 1994; O'Brien & Harris 1996). Further, on the ice-free areas in the Prince Charles Mountains, there is no evidence of fault offsets on the Cenozoic Pagodroma deposits (McKelvey *et al.* 2001). Decametre or greater scale fault offsets are also absent on the planated summit surfaces, despite the low Cenozoic erosion rates in these upland areas. Phillips & Laufer (2009) report small-scale Cenozoic brittle faulting, but relate this to crustal unloading produced by glacial erosion rather than active tectonic processes. Lastly, there is no evidence of post-Eocene volcanic activity in the region, suggesting the region has been volcanically as well as tectonically quiescent during the Cenozoic Era.

In comparison to the absence of evidence for tectonic activity, there is clear evidence for erosion along the valleys occupied by the present-day outlet glaciers, from both provenance studies in

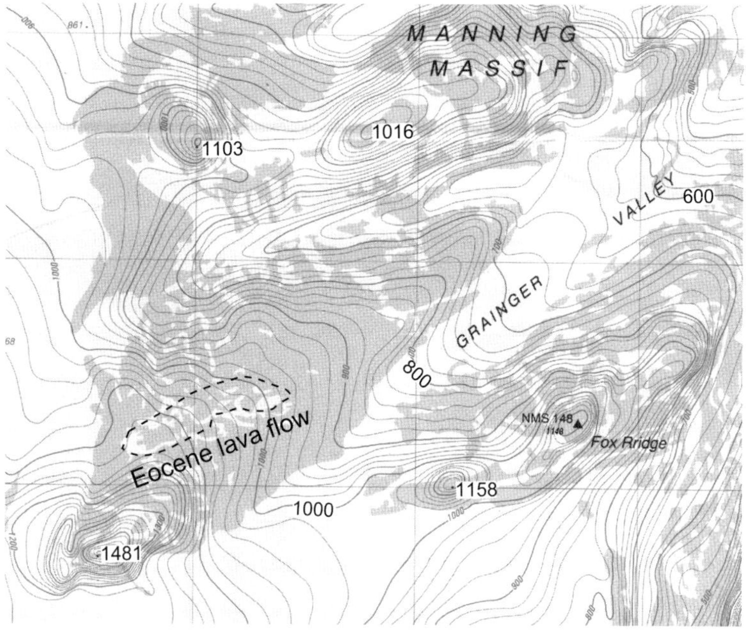

Fig. 4. Location of the Eocene tephritic phonolite lava flow on Manning Massif. The ridge-top position of the lava suggests erosion of the valleys north and south of the lava (Otter (informal name) and Grainger, respectively), implying Cenozoic erosion of the order of a few hundred metres at this site. Base map from the Australian Antarctic Division (1990), with a contour interval of 20 m. Snow and ice cover displayed in white, ice-free sediment and bedrock in grey.

Prydz Bay and geomorphic evidence in the Prince Charles Mountains. Sandstones are the dominant lithology of dropstones in Prydz Bay Cenozoic glacial sediments in Prydz Bay (Cooper & O'Brien 2004). Sandstone outcrop in the Prince Charles Mountains is presently limited to low altitude at Amery Oasis, but sub-ice outcrop is considered to exist along the length of the Lambert Graben (Mishra *et al.* 1999), where geophysical evidence indicates pre-Cenozoic sedimentary fill up to 6 km. Thus, the sandstone clasts exported to Prydz Bay during the Cenozoic Era must have come from glacial erosion of the valleys along the present-day outlet glaciers. Geomorphic evidence indicates that each of the packages of Cenozoic fjordal sediments of the Pagodroma Group in the northern Prince Charles Mountains have subsequently been incised by glacial erosion along the modern glacier margins (Hambrey & McKelvey 2000*a*, *b*; McKelvey *et al.* 2001). Further, the palaeo-fjord walls and floor that underlie Pagodroma Group sediments record the bedrock surface at the time of deposition. At many locations these former fjord surfaces have been truncated by erosion at the massif margin, indicating glacial incision into the palaeo-fjord walls during the late Cenozoic Era (Hambrey & McKelvey 2000*a*, *b*; Whitehead *et al.* 2004). Thus the majority of sediment exported from the northern Lambert Glacier–Amery Ice Shelf system during the Cenozoic must have been excavated from along the valleys occupied by the modern outlet glaciers.

Pagodroma Group equivalent sediments in the central southern Prince Charles Mountains

Thick accumulations Pagodroma Group equivalent sediment were identified on Mt Stinear and Mt Rymill (Fig. 7) by Whitehead *et al.* (2003), and sampled by White & Hermichen (2007). At Mt Stinear, the uppermost unit is informally termed the Edwards Pillar silt, which lies at 1670 masl in the saddle to the south of Edwards Pillar (Fig. 8). It consists of bedded, silt-rich material with rounded dropstones, indicative of a waterlain deposit with ice-rafted material. Because of time and logistical constraints, only the uppermost 10 m were observed in detail. Aerial photographs and helicopter-based observations suggest that the unit is up to 100 m thick and rests directly on bedrock.

A potential correlative of the Edwards Pillar silt drapes the northern end of Mount Rymill at 1120 masl, here termed the Rymill diamict. It consists of a 5–7 m thick sequence of bouldery diamict with weak stratification and gravelly lenses

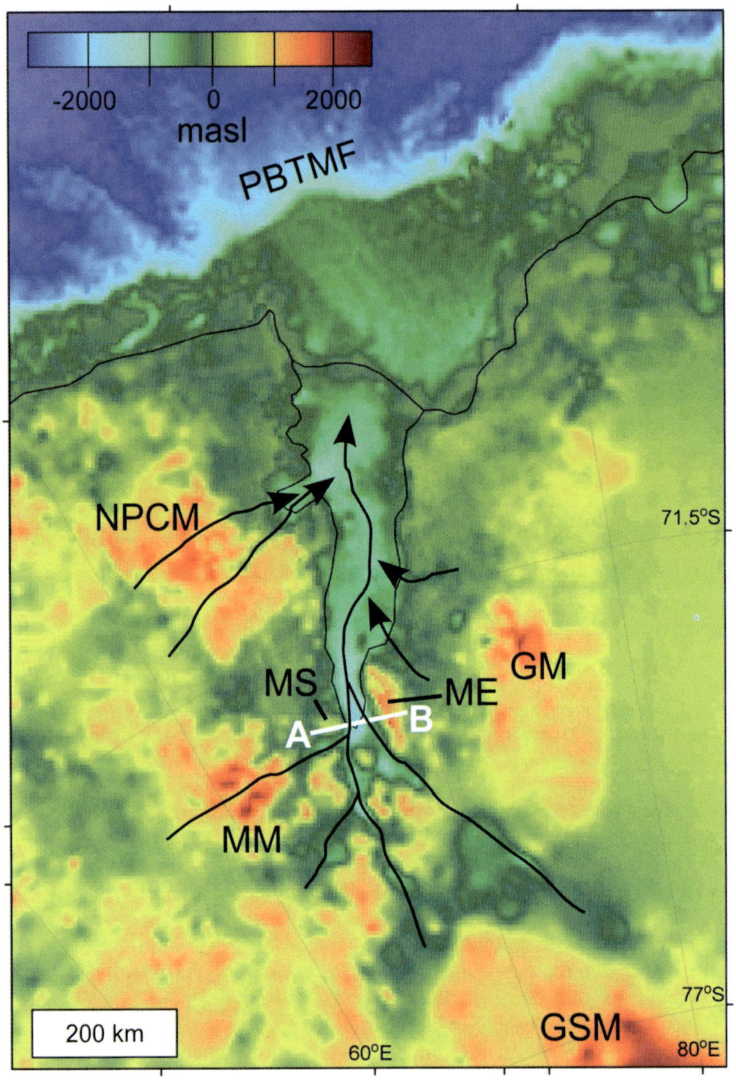

Fig. 5. Modern bedrock elevation in the Lambert Glacier–Amery Ice Shelf system. Data from BEDMAP folio (Lythe *et al.* 2001). Black lines outline the extent of the modern ice sheet and Amery Ice Shelf. White line A–B displays the location of the section in Figure 6. Note the large troughs through which the modern outlet glaciers flow (arrowed) and the heavily dissected nature of the southern Prince Charles Mountains. NPCM, northern Prince Charles Mountains; MS, Mount Stinear; ME, Mawson Escarpment; MM, Mount Menzies; GM, Grove Mountains; GSM, Gamburtsev subglacial mountains. PBTMF, Prydz Bay trough–mouth fan.

(Fig. 8). A >100 m thick unit, dominated by sandy gravel, was exposed in sections along the western margin of the northern bench of Mt Stinear. This unit thins to the south and pinches out to the north of the summit ice cap (Whitehead *et al.* 2000). In the study area, the lower part of the unit lay directly on the mafic bedrock of the summit plateau at 920 masl. The nature of both contacts was difficult to discern because of the poor exposure and lack of stratification in these diamicts. Unlike the northern Prince Charles Mountains Pagodroma Group sediments, no *in-situ* diatoms have been recovered from samples from Edwards Pillar, Mt Rymill or the other southern Prince Charles Mountains Pagodroma Group equivalents (Whitehead *et al.* 2000); therefore, it is possible that the southern Prince Charles Mountains sediments were deposited in a terrestrial rather than a marine setting.

Clay mineral abundances in the Edwards Pillar and Mt Rymill deposits are inconsistent with the

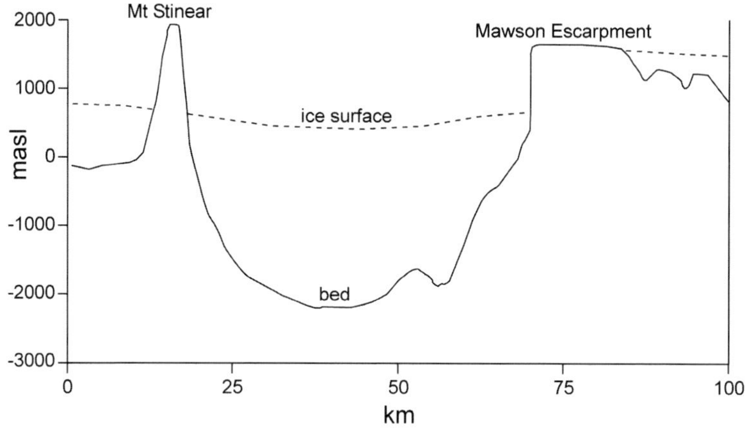

Fig. 6. Profile of U-shaped bed morphology at 73°S based on radio-echo sounding data, modified after Hambrey (1991) and Morgan & Budd (1975). Location of section displayed as the white line A–B in Figure 5.

modern climatic and depositional regime. Both deposits have a clay mineral mix that is similar to Pagodroma Group deposits in the northern Prince Charles Mountains (15–25% kaolin, 0–5% expandable clays, 20–30% chlorite and 50–60% illite; White & Hermichen 2007). Clay abundances in the Pagodroma group equivalent sediments at Mt Stinear and Rymill contrast strongly with Pleistocene deposits at these sites, which lack kaolin and have chlorite–illite ratios of >4:1 (White & Hermichen 2007), similar to the unweathered metamorphic basement of the region (Ehrmann *et al.* 2003). Ehrmann *et al.* (1992, 2003) indicated that the clay mineralogy of the Pagodroma Group

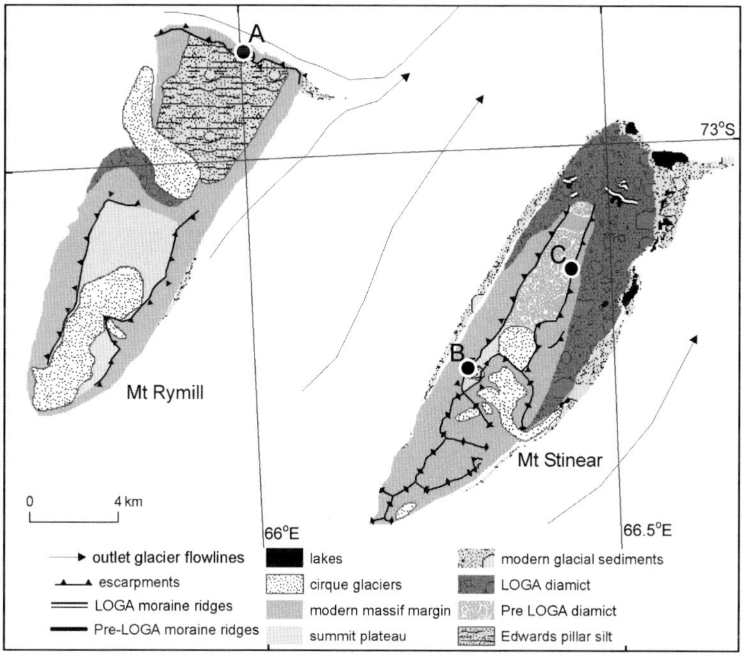

Fig. 7. Geomorphology and glacial sediments at Mt Rymill and Mt Stinear mapped from field reconnaissance and air-photo interpretation. (A–C) Locations of sections investigated on Rymill diamict, Edwards Pillar silt and the northern bench units, respectively. LOGA, Last Outlet Glacier Advance, which at Mt Stinear has been dated using cosmogenic exposure at between 20 and *c.* 9 ka (White *et al.* 2011).

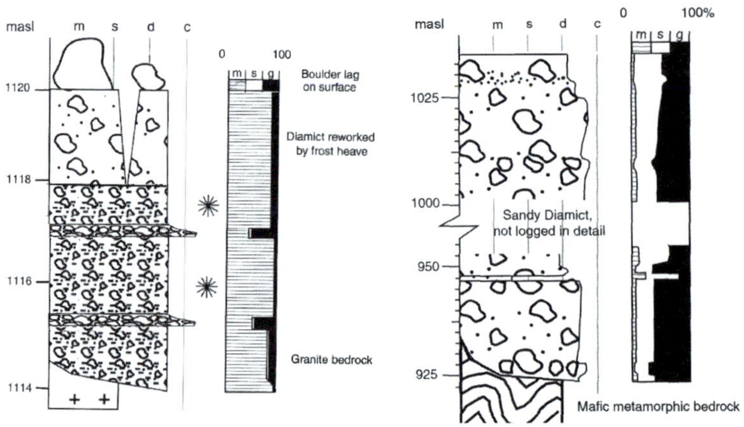

Fig. 8. Stratigraphy of Pagodroma Group-like diamict on Mt Rymill (left panel, location A in Fig. 7) and the northern bench of Mt Stinear (right panel, C, in Fig. 7). Stars indicate locations of sediment samples.

deposits in the northern Prince Charles Mountains reflect the sediment provenance (i.e. erosion from the fluviatile Amery Group) rather than the climatic conditions during deposition. Thus, the difference in clay mineral abundance is attributed to a change in provenance.

Interestingly, there are no kaolin-rich rocks at the bedrock surface (either above or below ice) at the present day, suggesting that the source of kaolin for the Edwards Pillar silt and Mt Rymill diamict has since been removed by erosion. The bedrock presently cropping out in the region surrounding Mt Stinear and Mt Rymill is dominated by high metamorphic-grade gneiss and schist that lacks significant quantities of kaolin, and there are no kaolin-rich Amery Group equivalent sediments exposed on nunataks in the southern Prince Charles Mountains. Further, there are three lines of evidence that fit together to suggest that a subcrop of Amery equivalent sediments is also rare in the area covered by the polar ice sheet: first, while a small component ($<1\%$) of erratics in recent moraine at Mt Stinear contains clasts of low-metamorphic grade sandstone, siltstone, coal and fluvially rounded quartzite cobbles (Fedorov *et al.* 1982; Tingey 1991; White 2007), the subglacial

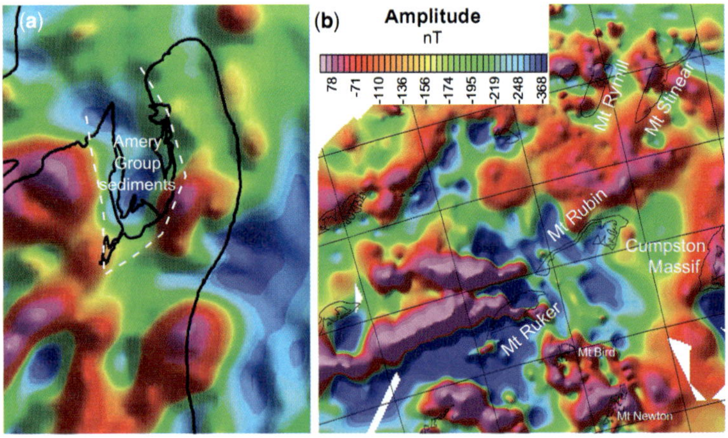

Fig. 9. Aeromagnetic maps of Amery Group sediment exposures and suspected subcrop in the Prince Charles Mountains. (**a**) Beaver Lake (M. McLean, pers. comm. 2005); (**b**) southern Prince Charles Mountains (Damaske & McLean 2005). Magnetic amplitudes are identical for both maps. Note the relatively weak magnetic signature of the Amery Group sediments at Beaver Lake, and the lack of such a signature in the region near Mt Stinear and Mt Rymill. The nearest potential subcrop up-glacier of Mt Stinear and Mt Rymill is the Sodruzhestvo Series, >20 km to the SW of the two massifs.

Table 1. *Cenozoic sediments in the Prince Charles Mountains (after Whitehead* et al. *2006)*

Formation	Location	Environment	Age (Ma)	References
Bardin Bluffs	Amery Oasis	Shallow glaciomarine	3.1–1	1–6, 9–12, 15, 16
Battye Glacier	Amery Oasis	Shallow glaciomarine	10.7–9.0	9, 10, 12, 14
Fisher Bench	Fisher Massif	Shallow glaciomarine	12.1–8.5	7, 9, 10, 12
Mount Johnston	Fisher Massif	Glaciomarine*	35–33*	9, 10, 12
Mt Stinear		Fjordal?	?	8
Pardoe and Trail	Mount Menzies	Partly lacustrine	?	13

*The age and depositional environment of this deposit are based on diatoms that may have been reworked from older deposits into the Mt Johnston Formation by glacial erosion, in which case this would represent a maximum age only.
References: 1, Bardin (1980); 2, Drewry (1981); 3, Bardin (1982); 4, Bardin & Kolosova (1983); 5, Bardin & Belevich (1985); 6, Bardin & Kolosova (1988); 7, Laiba & Pushina (1997); 8, Whitehead *et al.* (2000); 9, Hambrey & McKelvey (2000*a*); 10, Hambrey & McKelvey (2000*b*); 11, Whitehead & McKelvey (2001); 12, McKelvey *et al.* (2001); 13, Whitehead & McKelvey (2002); 14, Whitehead *et al.* (2003); 15, Whitehead *et al.* (2004); 16, Whitehead *et al.* (2006).

source of this material is not large enough to supply modern and Late Glacial Maximum moraines with >5% kaolin. Second, the Amery Group sediments in the northern Prince Charles Mountains have a very weak magnetic signature (Fig. 9), which may be used to locate the subglacial outcrop of these units in the southern Prince Charles Mountains, but there are few areas with a low magnetic signature in the potential source areas to the south, west or east of Mt Stinear and Rymill. Thus, any potential Amery Group subcrop must be significantly smaller than the 5 km aerogeophysical line spacing in this region, and is unlikely to be a significant kaolin source. Third, the other potential source rocks that may contain a high concentration of kaolin up-glacier of Mt Rymill and Mt Stinear are the strongly folded carbonate-rich metasediments of the Sodruzhestvo basin series (Phillips *et al.* 2005). There are no data on the clay content or mineralogy of the Sodruzhestvo series, but the supraglacial outcrop pattern (Phillips *et al.* 2005) and magnetic maps (Damaske & McLean 2005) of the area indicate that this series forms a semicircular band surrounding Mt Stinear and Mt Rymill. Thus, even if this unit was a significant source of kaolin, it is unlikely that a change in glacial transport direction could significantly alter the clay mineralogy of material supplied to Mt Stinear and Mt Rymill. In the absence of other alternatives, it is concluded that that the high kaolin contents from the Pagodroma Group-like deposits on both Mt Stinear and Mt Rymill were supplied by a unit that has been subsequently eroded, and extends the Cenozoic sedimentary evidence for substantial erosion along the outlet glaciers at least as far south as Mt Stinear.

Isostatic uplift

Post-Eocene uplift of the lightly eroded peaks by 0.7–1.5 km concurrent with bedrock erosion owing to erosion and isostatic unloading along the glacial valleys was predicted by Wellman & Tingey (1981). Such uplift has since been documented by the present elevations of glacio-marine fjordal deposits of the Pagodroma Group. At Amery Oasis

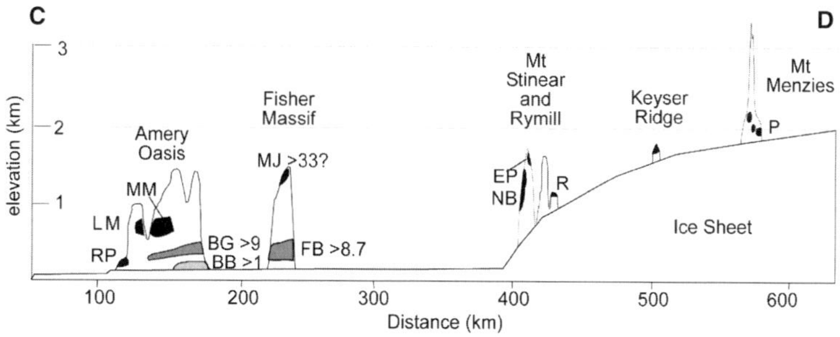

Fig. 10. Cenozoic sediments of the Prince Charles Mountains. Sediments with well-constrained ages (in Ma) are shown in grey. The line of the section from C to D is located in Figure 1. LM, Loewe Massif; RP, Ritchie Point; MM, Manning Massif; BB, Bardin Bluffs Formation; BG, Battye Glacier Formation; FB, Fisher Bench Formation; MJ, Mount Johnston Formation; NB, northern bench diamict*; EP, Edwards Pillar silt*; R, Rymill diamict*; P, Pardoe Formation. *Denotes informal names used for sediments investigated in this study.

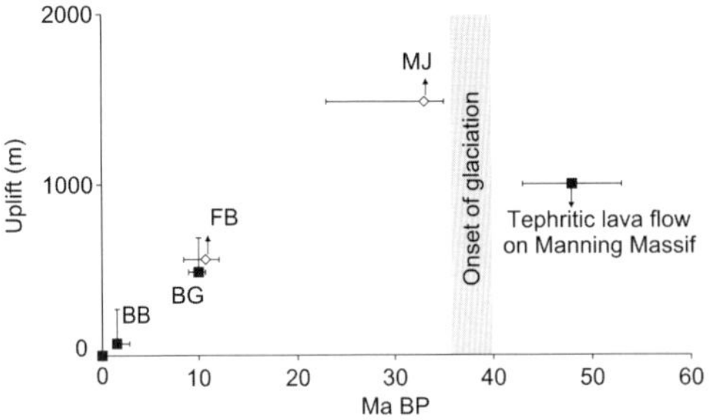

Fig. 11. Cenozoic uplift in the northern Prince Charles Mountains. Solid squares indicate data from the Amery Oasis; open diamonds indicate data from Fisher Massif. BB, BG, FB and MJ indicate uplift indicated by the position of the Bardin Bluffs, Battye Glacier, Fisher Bench and Mount Johnston formations respectively. Horizontal error bars represent uncertainties in the age of the deposits, while vertical error bars represent the scale of variability in global sea-level during these period of deposition (Haq *et al.* 1987). The presence of scoria, and the absence of pillow features associated with the Eocene lava flow on Manning Massif suggest that the 50 ± 2 Ma BP (Sheraton 1983) or 40 ± 11 (Andronikov & Beliatsky 1995) eruption was subaerial (and thus above sea-level), although none of the preserved volcanic features conclusively preclude a subaqueous (or subglacial) eruption.

the presence of diatomaceous units and *in situ* molluscs indicates that the Bardin Bluffs and Battye Glacier Formations of the Pagodroma Group were deposited in a glaciomarine setting (Table 1, Whitehead & McKelvey 2001; Whitehead *et al.* 2003, 2004). Pagodroma Group sediments now lie at altitudes of up to 1500 masl (Fig. 10), so the present elevation of these deposits is significantly greater than the likely magnitude (100–200 m) of sea-level variations since their deposition (Haq *et al.* 1987; Miller *et al.* 1998), indicating substantial uplift following their deposition. The age of these deposits increases with increasing elevation, indicating that this uplift has continued throughout the Cenozoic Era.

As discussed above, there is no evidence for significant Cenozoic tectonic activity, and as such, the uplift recorded by the Pagodroma Group sediments is unlikely to have been tectonically generated. Isostatic uplift owing to a reduction in grounded ice volume is also unlikely, as the sediments themselves indicate that they were deposited during periods where glacial extent was equal to, or even less than, that of today (e.g. Whitehead & McKelvey 2001; Whitehead *et al.* 2003, 2004, 2006). Thus, the uplift of these units is interpreted here as the process of progressive crustal unloading by selective glacial erosion along the existing glacial troughs, similar to that described in the Transantarctic Mountains (Stern *et al.* 2005). The Pagodroma Group sediment elevations suggests a consistent rate of uplift of the peaks of the northern Prince Charles Mountains of *c.* 50 m/Ma (Fig. 11) during the Cenozoic. The rate of uplift is tied to the rate of sediment export from the glacial valleys the Prince Charles Mountains, so the sediment altitudes also suggest that that erosion rates have remained constant on these timescales.

Comparison with landscape evolution models

The overall pattern of incision produced by numerical landscape evolution models that provide detail in the study area (Jamieson *et al.* 2005, 2010; Wilson *et al.* 2012) is similar to that observed through field evidence: that is, limited erosion in the upland areas, with substantial erosion along the adjacent outlet glaciers (Fig. 12).

The plausible magnitude of post-Eocene incision along the major outlet glaciers in the Prince Charles Mountains is constrained by two factors– on the lower side by the amount of post-Eocene isostatic uplift, and on the upper side by the modern day relief between the Crohn palaeosurface and the modern glacial valleys. The 1.5 km of isostatic uplift that has occurred in the northern Prince Charles Mountains, as evidenced by the Pagodroma Group sediments, requires at least the same thickness of bedrock to have been removed to produce such an uplift. If this erosion had occurred on the massifs, and assuming an Airy response, this requires erosion of at least *c.* 1.9 km of bedrock (e.g. Wellman & Tingey 1981). However, as the focus of erosion was along the outlet glaciers, many tens of kilometres distant from the uplifted areas, the thickness of material eroded from the glacial valleys was greater than *c.* 1.9 km. Where the Crohn palaeosurface is preserved, present day relief ranges from 3 to 4 km. Given the small amount of erosion that has occurred on the uplands, post-Eocene incision cannot have been substantially greater than present relief (Fig. 13). Thus, the depth of post-Eocene erosion along the outlet glaciers in the Prince Charles Mountains is considered to be between 2 and 4 km.

The thickness of material removed from the Prince Charles Mountains is consistent with more recent landscape evolution models (Jamieson *et al.* 2010; Wilson *et al.* 2012), which produce modelled erosion thicknesses of 3–4 km in the region presently occupied by the Amery Ice Shelf. It contrasts strongly with the earlier model that was constrained to the volume of the <600 m-thick stack of post-Eocene sediment that is preserved in Prydz Bay (Jamieson *et al.* 2005), and modelled post-Eocene

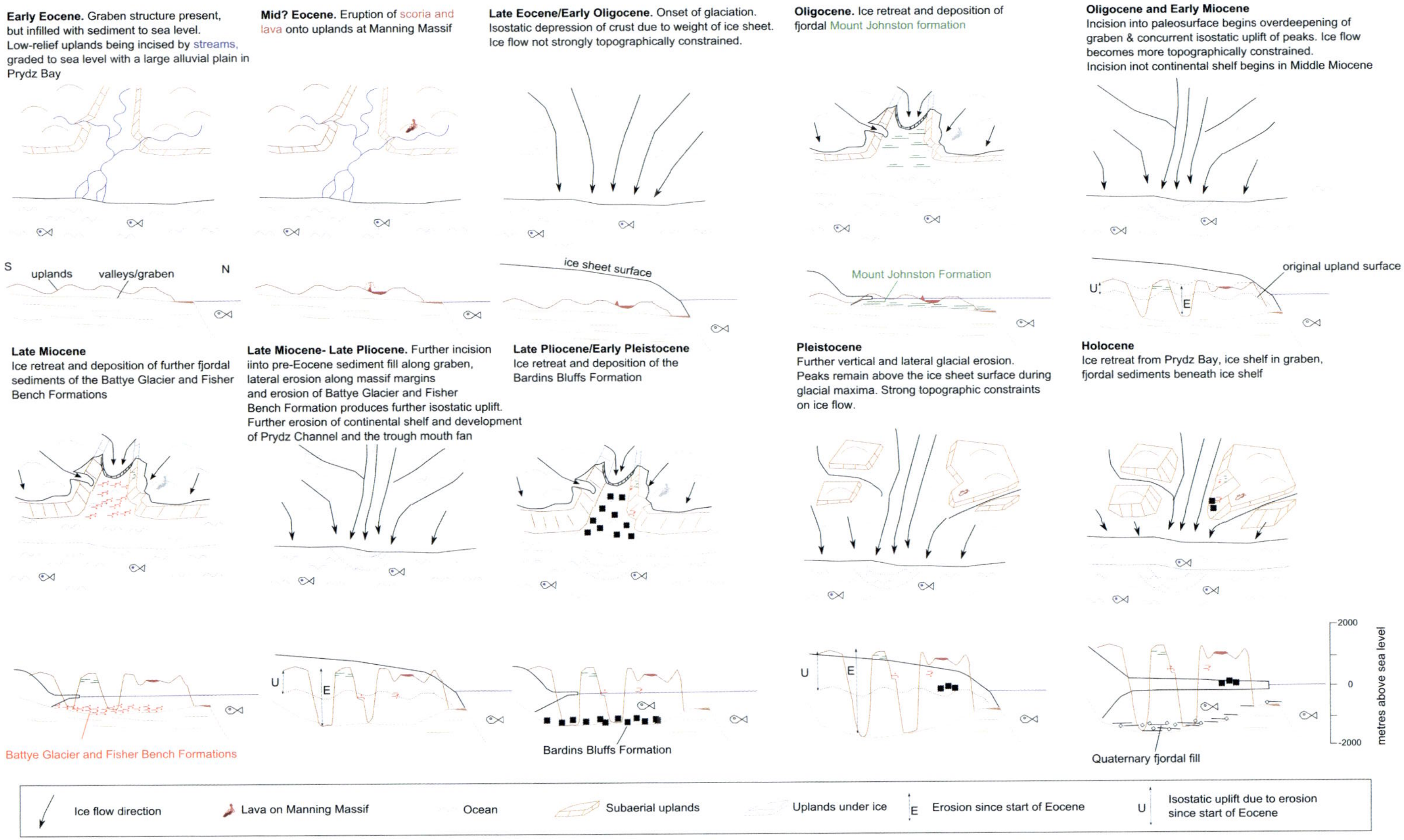

Fig. 12. Cartoon block diagram and south–north cross section of topographic and glacial evolution of the lower Lambert Glacier and Prydz Bay region during the Cenozoic. Selective linear erosion along the troughs occurs during the Cenozoic induces uplift of the more slowly eroding uplands. Fjordal glaciomarine sediments deposited during glacial minima are progressively uplifted above sea-level, producing a sequence that increases in age with increasing elevation. Peaks have been shown to have remained exposed above the ice sheet during the Quaternary, in line with cosmogenic evidence for such. However, the cosmogenic results do not preclude earlier subaerial exposure, so their 'permanent' exposure above the ice sheet may have happened earlier.

erosion thicknesses of only 0.2 km along the major outlet glaciers. This finding lends support to the notion that the vast majority of sediment produced in the Lambert Glacier–Amery Ice Shelf system since the Eocene has been transported to the deep ocean, some of which has been retained on the continental slope and rise (Cooper & O'Brien 2004).

The location of the present Lambert Glacier–Amery Ice Shelf system, and the subsequent valleys that were incised by the outlet glaciers that flow through the Prince Charles Mountains, are likely to have been constrained by the region of relatively poorly consolidated pre-glacial, and perhaps pre-Cretaceous, sediments filling the Lambert Graben (Mishra *et al.* 1999), a tectonic feature considered to have developed prior to deposition of the Amery Group in the Permian, and reactivated during the breakup of Gondwana in the Cretaceous Period (Lisker *et al.* 2003). These weak sediments are ideal for the formation of a major ice stream, as they are more easily eroded than the surrounding basement rocks, leading to enhanced basal deformation (e.g. Bamber *et al.* 2006) and hence an increased rate of ice flow. This produces a positive feedback, with the increased rate of ice flow further enhancing the rate of erosion, leading to the formation of a depression in the area underlain by the weaker sediments, which further focuses ice flow. It follows that over time more and more of the flow in this part of the ice sheet would have been focussed into the Lambert Graben, possibly enhancing the rate of associated glacial erosion. This model is supported by the sediment archives in Prydz Bay, which indicate that the system changed from a largely distributed ice sheet in the early Miocene to the focussed delivery of sediment in the Prydz Bay Trough Mouth Fan by the early Pliocene (O'Brien *et al.* 2004; Fig. 12).

Implications for Cenozoic glacial history

Sediment deposition on the continental shelf and slope slowed during the Cenozoic Era, culminating in a period of very low deposition in the Prydz Bay Trough–Mouth Fan in mid-late Pleistocene time (Passchier *et al.* 2003; Cooper & O'Brien 2004). The limited deposition has been interpreted as a period when ice advances to the continental shelf edge were rare or absent (Cooper & O'Brien 2004). Topographic changes in the basin, particularly the deepening of the glacial valley beneath

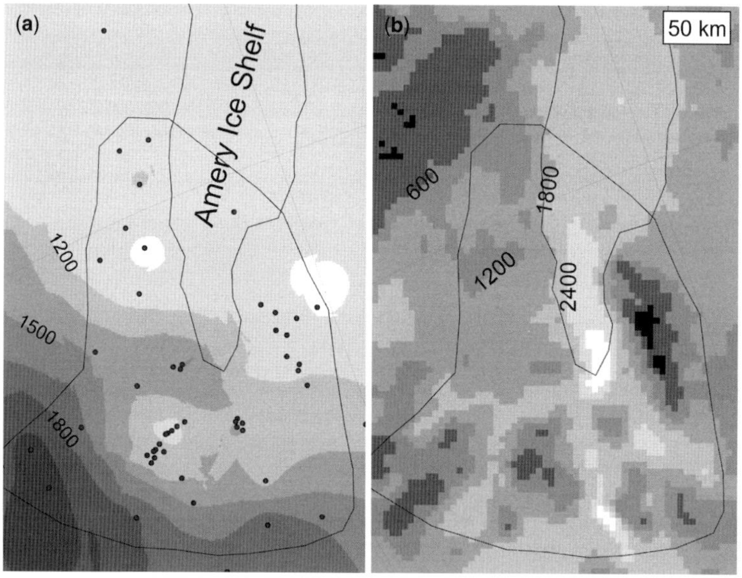

Fig. 13. (**a**) Height of the flat mountain summits in the southern Prince Charles Mountains. (**b**) 'Maximum' thickness of lithosphere (in metres) removed during the Cenozoic era, assuming the flat mountain summits provide a reasonable proxy for pre-Eocene land surface. In (a) dots indicate locations where flat mountain summits occur, while shading and numbers indicate the height of the surface above sea-level. In (b), numbers and shading indicate the depth of erosion that has occurred since incision into this surface began, with dark areas indicating little to no erosion, and light areas indicating substantial erosion. Upper U-shaped line indicates approximate grounding line of the Lambert Glacier–Amery Ice Shelf system Lower, pear-shaped line indicates area where mountain summits are exposed above the ice.

the modern Amery Ice Shelf to greater than 400 m below sea-level, have been shown by a numerical ice model (Taylor *et al.* 2004) to inhibit the expansion of the ice sheet to the continental shelf break during glacial maxima. Both Taylor *et al.* (2004) and O'Brien *et al.* (2007) suggest that ice sheet-topographic evolution crossed a threshold during the Late Pleistocene, resulting in significantly smaller ice advances during the past *c.* 1 Ma.

The history of landscape evolution in the Prince Charles Mountains does not support the idea that topographic changes were a major driver in limiting Pleistocene (i.e. post 1 Ma) ice sheet expansion. While the base of the valley along the Graben in this area below the Amery Ice Shelf has been eroded through the key depth of *c.* 400 m below sea-level during the Cenozoic Era, this probably happened much earlier than the Pleistocene, as follows. The base of the valley began at a few hundred metres above sea-level prior to the onset of glaciation in the Eocene (Wellman & Tingey 1981), and has been eroded to the point that it would remain at least 1000 m below sea-level if the ice sheet were removed (Wilson *et al.* 2012). The relatively constant rates of erosion-induced uplift in the northern Prince Charles Mountains (Fig. 11) suggest that the rate of glacial erosion along the Lambert Graben has remained relatively steady during the Cenozoic Era. Thus, assuming constant erosion rates, the area under the Amery Ice Shelf would have crossed the *c.* 400 m water depth threshold identified by Taylor *et al.* during the middle Miocene. Applying a decreasing erosion rate through time, as suggested by the decreasing rates of clastic sediment deposition on the continental slope and rise (Cooper & O'Brien 2004) would only see the key area cross the 400 m water depth threshold earlier in the Cenozoic. Thus, it is likely that the key region below the Amery Ice Shelf has been eroded by <100 m during the Pleistocene, which would be insufficient to have had a substantial effect on ice sheet extent and height. In the absence of significant Pleistocene erosion, the substantial Pleistocene changes in ice sheet extent in this region, climatic influences, such as climate or sea-level, changes in climate cyclicity during the mid Pleistocene transition (e.g. Zachos *et al.* 2001) or a reduction in precipitation (O'Brien *et al.* 2007) were probably the most important factors in controlling ice sheet size in the Lambert Glacier–Amery Ice Shelf system.

This paper has been possible through logistic support provided by the Australian Antarctic Division (ASAC grants 1071 and 1215). D. Gardner, W.-D. Hermichen and G. Kuhen provided field support. Special thanks go to reviewers of the final (B. McKelvey and P. O'Brien) and previous (D. Gore, M. Hambrey and D. Fabel) versions of the manuscript.

References

ADAMSON, D. A., MABIN, M. C. G. & LULY, J. G. 1997. Holocene isostasy and late Cenozoic development of landforms including Beaver and Radok Lake basins in the Amery Oasis, Prince Charles Mountains, Antarctica. *Antarctic Science*, **9**, 299–306.

ANDRONIKOV, A. V. & BELIATSKY, B. V. 1995. Implication of Sm–Nd isotopic systematics to the events recorded in mantle derived xenoliths (Jetty Peninsula, East Antarctica). *Terra Antartica*, **2**, 103–110.

ARNE, D. A. 1994. Phanerozoic exhumation history of northern Prince Charles Mountains (East Antarctica). *Antarctic Science*, **6**, 69–84.

AUSTRALIAN ANTARCTIC DIVISION 1990. *Beaver Lake, Mac. Robertson Land Topographic Map*. Australian Antarctic Division, Hobart.

AUSTRALIAN ANTARCTIC DIVISION 1993. *Southern Prince Charles Mountains, Mac.Robertson Land.* 1:500,000 satellite image map, SS 40-42b, First edition. Kingston.

BAMBER, J. L., FERRACCIOLI, F., JOUGHIN, I., SHEPHERD, T., RIPPIN, D. M., SIEGERT, M. J. & VAUGHAN, D. G. 2006. East Antarctic ice stream tributary underlain by major sedimentary basin. *Geology*, **34**, 33–36.

BARDIN, V. I. 1977. Glacial geomorphology of the Lambert Glacier and its surrounding mountains. *Polar Geography*, **1**, 251–269.

BARDIN, V. 1980. Moraines of the Antarctic continent. *In*: *International Geological Congress, Abstracts–Congres Geologique Internationale, Resumes. 26th International Geological Congress*, 635.

BARDIN, V. I. 1982. Composition of East Antarctic moraines and some problems of Cenozoic history. *In*: CRADDOCK, C. (ed.) *Antarctic Geoscience*. Wisconsin Press, Madison University, Madison, WI, 1069–1076.

BARDIN, V. I. & BELEVICH, A. M. 1985. On the study of early glacial deposits in the Prince Charles Mountains. *Antarktika, Doklady Komissii*, **24**, 76–81.

BARDIN, V. I. & KOLOSOVA, G. N. 1983. Early glaciation deposits in the Prince Charles Mountains. *Antarktika, Doklady Komissii*, **28**, 35–38.

BARDIN, V. I. & KOLOSOVA, G. N. 1988. Ancient glacial deposits at Radok Lake. *Antarktika, Doklady Komissii*, **27**, 66–75.

BARKER, P. F., DIEKMANN, B. & ESCUTIA, C. 2007. Onset of Cenozoic Antarctic glaciation. *Deep Sea Research Part II: Topical Studies in Oceanography*, **54**, 2293–2307.

COOPER, A. K. & O'BRIEN, P. E. 2004. Leg 188 synthesis: transitions in the glacial history of the Prydz Bay region, East Antarctica, from ODP drilling. *In*: COOPER, A. K., O'BRIEN, P. E. & RICHTER, C. (eds) *Proceedings of the Ocean Drilling Program, Scientific Results*. Ocean Drilling Program. Texas A & M University, College Station, TX, 1–42, http://www-odp.tamu.edu/publications/188_SR/synth/synth.htm (accessed 27 March 2006).

COOPER, A. K., STAGG, H. & GEIST, E. 1991. Seismic stratigraphy and structure of Prydz Bay, Antarctica: implications from Leg 119 drilling. *In*: BARRON, J. A., LARSEN, B. ET AL. (eds) *Proceedings of the Ocean Drilling Program, Scientific Results*. Ocean Drilling Program. Texas A & M University, College

Station, TX, 1–42, 5–25. doi: 10.2973/odp.proc.sr.119.181.1991

CROHN, P. W. 1959. *A Contribution to the Geology and Glaciology of the Western Part of the Australian Antarctic Territory*. Bureau of Mineral Resources, Australia, Canberra, Bulletins, 103.

DAMASKE, D. & MCLEAN, M. 2005. An aerogeophysical survey south of the Prince Charles Mountians, East Antarctica. *Terra Antartica*, **12**, 69–86.

DREWRY, D. J. 1981. The record of late Cenozoic glacial events in East Antarctica (60° – 171°E). *In*: HAMBREY, M. J. & HARLAND, W. B. (eds) *Earth's Pre-Pleistocene Glacial Record*. Cambridge University Press, Cambridge, 212–216.

EHRMANN, W. U. & MACKENSEN, A. 1992. Sedimentological evidence for the formation of an East Antarctic ice sheet in Eocene/Oligocene time. *Palaeogeography, Palaeoclimatology, Palaeoecology*, **93**, 85–112.

EHRMANN, W. U., MELLES, M., KUHN, G. & GROBE, H. 1992. Significance of clay mineral assemblages in the Antarctic Ocean. *Marine Geology*, **107**, 249–273.

EHRMANN, W., BLOEMENDAL, J., HAMBREY, M. J., MCKELVEY, B. & WHITEHEAD, J. 2003. Variations in the composition of the clay fraction of the Cenozoic Pagodroma Group, East Antarctica: implications for determining provenance. *Sedimentary Geology*, **161**, 131–152.

FEDOROV, L. V., GRIKUROV, G. E., KURININ, R. G. & MASOLOV, V. N. 1982. Crustal structure of the Lambert Glacier area from geophysical data. *In*: CRADDOCK, C. (ed.) *Antarctic Geoscience*. University of Wisconsin Press, Madison, WI, 931–936.

FINK, D., MCKELVEY, B., HAMBREY, M. J., FABEL, D. & BROWN, R. 2006. Pleistocene deglaciation chronology of the Amery Oasis and Radok Lake, northern Prince Charles Mountains, Antarctica. *Earth and Planetary Science Letters*, **243**, 229–243.

HAMBREY, M. J. 1991. Structure and dynamics of the Lambert Glacier–Amery Ice Shelf system: implications for the origin of Prydz Bay sediments. *In*: *Proceedings of the Ocean Drilling Program, Scientific Results*, **129**. Ocean Drilling Program, Texas A & M University, College Station, TX, 61–75.

HAMBREY, M. J. & MCKELVEY, B. 2000*a*. Major Neogene fluctuations of the East Antarctic ice sheet; stratigraphic evidence from the Lambert Glacier region. *Geology*, **28**, 887–890.

HAMBREY, M. J. & MCKELVEY, B. 2000*b*. Neogene fjordal sedimentation on the western margin of the Lambert Graben, East Antarctica. *Sedimentology*, **47**, 577–607.

HAMBREY, M. J., EHRMANN, W. U. & LARSEN, B. 1991. Cenozoic glacial record of the Prydz Bay continental shelf, East Antarctica. *In*: *Proceedings of the Ocean Drilling Program, Scientific Results*, **129**, Ocean Drilling Program, Texas A & M University, College Station, TX, 77–132.

HAMBREY, M. J., GLASSER, N. F., MCKELVEY, B. C., SUGDEN, D. E. & FINK, D. 2007. Cenozoic landscape evolution of an East Antarctic oasis (Radok Lake area, northern Prince Charles Mountains), and its implications for the glacial and climatic history of Antarctica. *Quaternary Science Reviews*, **26**, 598–626.

HAQ, B. U., HARDENBOL, J. & VAIL, P. R. 1987. Chronology of fluctuating sea levels since the Triassic. *Science*, **235**, 1156–1166.

JAMIESON, S. S. R., HULTON, N. R. J., SUGDEN, D. E., PAYNE, A. J. & TAYLOR, J. 2005. Cenozoic landscape evolution of the Lambert basin, East Antarctica: the relative role of rivers and ice sheets. *Global and Planetary Change*, **45**, 35–49.

JAMIESON, S. S. R., SUGDEN, D. E. & HULTON, N. R. J. 2010. The evolution of the subglacial landscape of Antarctica. *Earth and Planetary Science Letters*, **293**, 1–27.

LAIBA, A. A. & PUSHINA, Z. V. 1997. Cenozoic glacial-marine sediments from the Fisher Massif (Prince Charles Mountains). *International Symposium on Antarctic Earth Sciences*, **7**, 977–984.

LEITCHENKOV, G., STAGG, H., GANDJUKHIN, V., COOPER, A. K., TANAHASHI, M. & O'BRIEN, P. E. 1994. Cenozoic seismic stratigraphy of Prydz Bay (Antarctica). *Terra Antartica*, **1**, 395–397.

LILLY, K., FINK, D., FABEL, D. & LAMBECK, K. 2010. Pleistocene dynamics of the interior East Antarctic ice sheet. *Geology*, **38**, 703–706.

LISKER, F., BROWN, R. & FABEL, D. 2003. Denudational and thermal history along a transect across the Lambert Graben, northern Prince Charles Mountains, Antarctica, derived from apatite fission track thermochronology. *Tectonics*, **22**, 1055, doi: 1029/2002TC001477.

LISKER, F., GIBSON, H., WILSON, C. J. & LÄUFER, A. 2007. Denudation and uplift of the Mawson Escarpment (eastern Lambert Graben, Antarctica) as indicated by apatite fission track data and geomorphological observation, in Antarctica. *In*: COOPER, A. K. & RAYMOND, C. R. (eds) *A Keystone in a Changing World–Online Proceedings of the 10th ISAES*. USGS Open-File Report **6**.

LYTHE, M. B., VAUGHAN, D. G. ET AL. 2001. BEDMAP; a new ice thickness and subglacial topographic model of Antarctica. *Journal of Geophysical Research, B, Solid Earth and Planets*, **106**, 11 335–11 351.

MCKELVEY, B. C., HAMBREY, M. J., HARWOOD, D. M., MABIN, M. C. G., WEBB, P. N. & WHITEHEAD, J. M. 2001. The Pagodroma Group; a Cenozoic record of the East Antarctic ice sheet in the northern Prince Charles Mountains. *Antarctic Science*, **13**, 455–468.

MILLER, K. G., MOUNTAIN, G. S. ET AL. 1998. Cenozoic global sea level, sequences, and the New Jersey transect: results from coastal plain and continental slope drilling. *Reviews of Geophysics*, **36**, 569–601.

MISHRA, D. C., CHANDRA SEKHAR, D. V., VENKATA RAJU, D. CH. & VIJAYA KUMAR, V. 1999. Crustal structure based on gravity–magnetic modelling constrained from seismic studies under Lambert Rift, Antarctica and Godavari and Mahanadi rifts, India and their interrelationship. *Earth and Planetary Science Letters*, **172**, 287–300.

MORGAN, V. I. & BUDD, W. F. 1975. Radio-echo sounding of the Lambert Glacier basin. *Journal of Glaciology*, **15**, 103–111.

O'BRIEN, P. E. & COOPER, A. K. 2004. Leg 188 synthesis: transitions in the glacial history of the Prydz Bay Region, East Antarctica, from ODP drilling. *In*: COOPER, A. K., RICHTER, C. ET AL. (eds) *Proceedings of the Ocean Drilling Program, Scientific Results*.

Ocean Drilling Program, Texas A & M University, College Station, TX, 1–42, http://www-odp.tamu.edu/publications/188_SR/synth/synth.htm (accessed 11 January 2005).

O'Brien, P. E. & Harris, P. T. 1996. Patterns of glacial erosion and deposition in Prydz Bay and the past behavior of the Lambert Glacier. *Papers and Proceedings of the Royal Society of Tasmania*, **130**, 79–85.

O'Brien, P. E., Goodwin, I., Forseberg, C.-F., Cooper, A. K. & Whitehead, J. 2007. Late Neogene ice drainage changes in Prydz Bay, East Antarctica and the interaction of Antarctic ice sheet evolution and climate. *Palaeogeography, Palaeoclimatology, Palaeoecology*, **245**, 390–410.

Passchier, S., O'Brien, P. E. O., Damuth, J. E., Januszczak, N., Handwerger, D. A. & Whitehead, J. M. 2003. Pliocene–Pleistocene glaciomarine sedimentation in eastern Prydz Bay and development of the Prydz trough–mouth fan, ODP Sites 1166 and 1167, East Antarctica. *Marine Geology*, **199**, 279–305.

Phillips, G. & Laufer, A. L. 2009. Brittle deformation relating to the Carboniferous–Cretaceous evolution of the Lambert Graben, East Antarctica: a precursor for Cenozoic relief development in an intraplate and glaciated region. *Tectonophysics*, **471**, 216–224.

Phillips, G., Wilson, C. J. L. & Fitzsimons, I. C. W. 2005. Stratigraphy and structure of the southern Prince Charles Mountains, East Antarctica. *Terra Antartica*, **12**, 69–86.

Sheraton, J. W. 1983. Geochemistry of mafic igneous rocks of the northern Prince Charles Mountains, Antarctica. *Journal of the Geological Society of Australia*, **30**, 295–304.

Stern, T. A., Baxter, A. K. & Barrett, P. J. 2005. Isostatic rebound due to glacial erosion within the Transantarctic Mountains. *Geology*, **33**, 221–224.

Taylor, J., Siegert, M. J., Payne, A. J., Hambrey, M. J., O'Brien, P. E., Cooper, A. K. & Leitchenkov, G. 2004. Topographic controls on post-Oligocene changes in ice sheet dynamics, Prydz Bay region, East Antarctica. *Geology*, **32**, 197–200.

Tingey, R. J. (ed.) 1991. *The Geology of Antarctica*. Oxford University Press, Oxford.

Trail, D. S. 1964. The glacial geology of the Prince Charles Mountains. *In*: Adie, R. J. (ed.) *Antarctic Geology*. North Holland Publishing Company, Amsterdam, 143–151.

Truswell, E. M. 1991. Palynology of sediments from Leg 119 drill sites in Prydz Bay, East Antarctica. *In*: Barron, J., Larsen, B. *et al.* (ed.) *Proceedings of the Ocean Drilling Program, Scientific Results*. Ocean Drilling Program, Texas A & M University, College Station, TX, **119**, 941–948.

Truswell, E. M., Dettmann, M. E. & O'Brien, P. E. 1999. Mesozoic palynofloras from the Mac.Robertson Shelf, East Antarctica: geological and phytogeographic implications. *Antarctic Science*, **11**, 239–255.

Turner, B. R. & Padley, D. 1991. Lower Cretaceous coal-bearing sediments from Prydz Bay, East Antarctica. *In*: Barron, J., Larsen, B. *et al.* (ed.) *Proceedings of the Ocean Drilling Program, Scientific Results*, **119**, Ocean Drilling Program, Texas A & M University, College Station, TX, 57–60.

USSR. 1978*a*. *Antarktida, G. Udivitel'naya*. Map S-42-XV. Glavnoye upravleniye geodezii i kartografii pri Sovetye Ministrov SSSR.

USSR. 1978*b*. *Antarktida, Gory Prints Charl'z/Gory Sodruzhestvo*. Map S-41-XVII, X VIII. Glavnoye upravleniye geodezii i kartografii pri Sovetye Ministrov SSSR.

USSR. 1978*c*. *Antarktida, Gory Prints Charl'z/Skaly Roborovskogo*. Map S-42-VII, VIII. Glavnoye upravleniye geodezii i kartografii pri Sovetye Ministrov SSSR.

USSR. 1978*d*. *Antarktida, Obryv Mousona (=Mawsona), Khr. Mezhevoy*. Map S-42-IX, X. Glavnoye upravleniye geodezii i kartografii pri Sovetye Ministrov SSSR.

Wellman, P. & Tingey, R. J. 1981. Glaciation, erosion and uplift over part of East Antarctica. *Nature*, **291**, 142–144.

White, D. A. 2007. *Cenozoic Glacial History and Landscape Evolution of Mac.Robertson Land and the Lambert Glacier–Amery Ice Shelf System, East Antarctica*. PhD thesis, Macquarie University, Sydney.

White, D. A. & Hermichen, W.-D. 2007. Glacial and periglacial history of the southern Prince Charles Mountains, East Antarctica. *Terra Antartica*, **14**, 5–12.

White, D. A., Fink, D. & Gore, D. B. 2011. Cosmogenic nuclide evidence for enhanced sensitivity of an East Antarctic ice stream to change during the last deglaciation. *Geology*, **39**, 23–26.

Whitehead, J. M. & McKelvey, B. C. 2001. The stratigraphy of the Pliocene–lower Pleistocene Bardin Bluffs Formation, Amery Oasis, northern Prince Charles Mountains, Antarctica. *Antarctic Science*, **13**, 79–86.

Whitehead, J. M. & McKelvey, B. C. 2002. Cenozoic glacigene sedimentation and erosion at the Menzies Range, southern Prince Charles Mountains, Antarctica. *Journal of Glaciology*, **48**, 226–236.

Whitehead, J. M., McKelvey, B. C. & McMinn, A. 2000. Cenozoic glacial deposits in the southern Prince Charles Mountains of East Antarctica. *Terra Antartica*, **7**, 655–656.

Whitehead, J. M., Harwood, D. M. & McMinn, A. 2003. Ice-distal upper Miocene marine strata from inland Antarctica. *Sedimentology*, **50**, 531–552.

Whitehead, J. M., Harwood, D. M., McKelvey, B., Hambrey, M. J. & McMinn, A. 2004. Diatom biostratigraphy of the Cenozoic fjordal Pagodroma Group, northern Prince Charles Mountains, East Antarctica. *Australian Journal of Earth Sciences*, **51**, 527–547.

Whitehead, J. M., Quilty, P. G., McKelvey, B. & O'Brien, P. E. 2006. A review of the Cenozoic stratigraphy and glacial history of the Lambert Graben–Prydz Bay region, East Antarctica. *Antarctic Science*, **18**, 83–99.

Wilson, D. S., Jamieson, S. S. R., Barrett, P. J., Leitchenkov, G., Gohl, K. & Larter, R. D. 2012. Antarctic topography at the Eocene–Oligocene boundary. *Palaeogeography, Palaeoclimatology, Palaeoecology*, **335–336**, 24–34.

Zachos, J., Pagani, M., Sloan, L., Thomas, E. & Billups, K. 2001. Trends, rhythms, and aberrations in global climate 65 Ma to Present. *Science*, **292**, 686–693.

History of the grounded ice sheet in the Ross Sea sector of Antarctica during the Last Glacial Maximum and the last termination

BRENDA L. HALL[1,2]*, GEORGE H. DENTON[1,2], JOHN O. STONE[3] & HOWARD CONWAY[3]

[1]*Climate Change Institute, University of Maine, Orono, ME, USA*

[2]*School of Earth and Climate Sciences, University of Maine, Orono, ME, USA*

[3]*Department of Earth and Space Sciences, University of Washington, Seattle, WA, USA*

**Corresponding author (e-mail: BrendaH@maine.edu)*

Abstract: Knowledge of variations in the extent and thickness of the Antarctic Ice Sheet is key for understanding the behaviour of Southern Hemisphere glaciers during the last ice age and for addressing issues involving global sea level, ocean circulation and climate change. Insight into past ice-sheet behaviour also will aid predictions of future ice-sheet stability. Here, we review terrestrial evidence for changes in ice geometry that occurred in the Ross Sea sector of Antarctica at the Last Glacial Maximum (LGM) and during subsequent deglaciation. During the LGM, a thick grounded ice sheet extended close to the continental shelf edge in the Ross Embayment. This ice reached surface elevations of more than 1000 m along the coast of the central and southern Transantarctic Mountains and Marie Byrd Land. The local LGM occurred by 18 ka on the coast, but as late as 7–10 ka inland. The first significant thinning took place at roughly 13 ka, with most ice loss happening in the Holocene. This history makes it unlikely that the Ross Sea sector was a major contributor to meltwater pulse 1A (MWP 1A). Resolution of a possible Antarctic origin for MWP 1A awaits detailed reconstructions from all sectors of the ice sheet.

The Antarctic Ice Sheet contains the world's largest reservoir of fresh water and is a major control on Southern Hemisphere climate. A history of changes in ice-sheet extent and thickness is vital for answering questions involving the initiation of Southern Hemisphere glaciation and deglaciation during the last ice age, as well as for addressing issues concerning global sea level (Bassett *et al.* 2005; Clark *et al.* 2009; Deschamps *et al.* 2012), ocean deepwater formation and heat transport (Broecker 1998), origins(s) of abrupt climate change (Blunier & Brook 2001; Weaver *et al.* 2003; Denton *et al.* 2010) and drivers of ice ages (Denton 2000). Moreover, an understanding of ice-sheet behaviour will aid predictions of future ice-sheet stability (Mercer 1978).

One approach toward understanding both past and future ice-sheet behaviour and its relationship to local and global climate is to reconstruct changes in ice-sheet extent and volume. The focus of this paper is the marine-based ice sheet in the Ross Sea sector [the Ross Sea ice sheet; Scott 1905; Stuiver *et al.* 1981], which at the Last Glacial Maximum (LGM) involved an extension of the present-day West Antarctic Ice Sheet combined with an important contribution from the East Antarctic Ice Sheet (Licht *et al.* 2005; Farmer *et al.* 2006). The Ross Sea sector today drains nearly one-quarter of the ice in East and West Antarctica and in the past it underwent one of the largest glacial–interglacial ice-volume changes of any area of the continent (Stuiver *et al.* 1981; Denton *et al.* 1989*a*, *b*; Shipp *et al.* 1999; Denton & Hall 2000; Denton & Hughes 2002; Licht & Andrews 2002; Pollard & DeConto 2009). Data from the Ross Sea sector bear on several key problems. The first concerns the volume and timing of Antarctica's contribution to global sea-level changes at the global LGM (*c.* 19–23 ka; Mix *et al.* 2001) and during the subsequent deglaciation. This information is important not only for a knowledge of how excess ice volume responsible for the change in LGM sea level was divided among the world's ice sheets, but also for determining whether or not enough excess ice existed in Antarctica to produce significant global sea-level changes during the last termination. For example, deglaciation of Antarctica has been implicated in geophysical calculations as the cause of meltwater pulse 1A (MWP 1A), a rise of as much as 20 m in 300–500 years initiated at *c.* 14.6 ka (Hanebuth *et al.* 2000; Kienast *et al.* 2003; Deschamps *et al.* 2009, 2012). An event of this magnitude implies a much larger ice sheet prior to 14.6 ka than currently is envisaged by recent models (Denton & Hughes 2002; Pollard & DeConto 2009; Mackintosh *et al.* 2011; Whitehouse

From: Hambrey, M. J., Barker, P. F., Barrett, P. J., Bowman, V., Davies, B., Smellie, J. L. & Tranter, M. (eds) 2013. *Antarctic Palaeoenvironments and Earth-Surface Processes*. Geological Society, London, Special Publications, **381**, 167–181. First published online April 19, 2013, http://dx.doi.org/10.1144/SP381.5

et al. 2012), and evidence of this larger ice sheet, if it indeed existed, should be found at high elevation in the Transantarctic Mountains bordering the Ross Sea. If MWP 1A could be attributed confidently to collapse of Antarctic ice, it would have important implications for the possibility of polar ice sheets transferring large amounts of ice to the sea on centennial timescales. A second key scientific question involves the long-term future of the marine-based West Antarctic Ice Sheet. This ice sheet, grounded well below sea level, is thought to be inherently unstable (Hughes 1973; Weertman 1976) and susceptible to rapid collapse. Examination of the longer-term evolution of the West Antarctic Ice Sheet in the Ross Sea sector affords insight into the basic mechanisms that control the ice-sheet grounding lines, as well as the sensitivity of the ice sheet to past and future environmental perturbations.

Here, we review terrestrial glacial geological data and glaciological evidence relating to the maximum extent of the Ross Sea ice sheet during the last glacial period and the timing of deglaciation. All dates given here are in calendar years, with radiocarbon dates converted using CALIB 6.0.1 and the INTCAL09 dataset (Reimer *et al.* 2009). Conversions of radiocarbon dates of marine organisms employ the Marine09 dataset (Reimer *et al.* 2009) and the time-dependent delta-*R* calculation of Hall *et al.* (2010*b*) derived from Ross Sea solitary corals that span the Holocene. Conversions may differ from those presented by the original authors because of changes in the calibration datasets. Exposure ages referred to in this paper have been recalculated to account for recent recalibration of the cosmogenic ^{10}Be and ^{26}Al production rates.

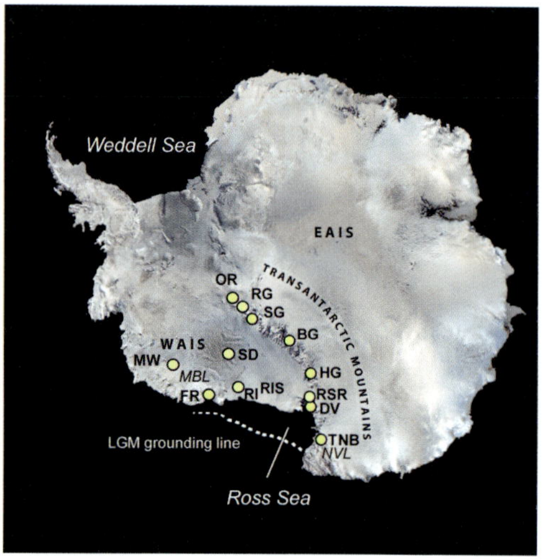

Fig. 1. Index map of Antarctica, showing locations mentioned in the text. MW, Mount Waesche; MBL, Marie Byrd Land; FR, Ford Ranges; RI, Roosevelt Island; SD, Siple Dome; RIS, Ross Ice Shelf; OR, Ohio Range; RG, Reedy Glacier; SG, Scott Glacier; BG, Beardmore Glacier; HG, Darwin/Hatherton Glacier system; RSR, Royal Society Range; DV, Dry Valleys; TNB, Terra Nova Bay; NVL, Northern Victoria Land.

LGM ice configuration

During the LGM, the Ross Embayment filled with a grounded ice sheet (Figs 1 & 2). Scoured channels, grounding-line wedges, megaflutes and widespread drift sheets on the floor of the Ross Sea indicate that the ice sheet extended close to the continental shelf edge (Stuiver *et al.* 1981; Licht *et al.* 1996; Domack *et al.* 1999; Shipp *et al.* 1999; Denton & Hughes 2000). Terrestrial evidence of ice-sheet extent comes from (1) a widespread drift sheet (Ross Sea drift) with far-travelled erratics located on islands and peninsulas in the Ross Embayment, along the southern Scott Coast, and in the mouths of the Dry Valleys and the valleys fronting the Royal Society Range (Stuiver *et al.* 1981; Denton *et al.* 1989*b*; Denton & Marchant 2000; Hall *et al.* 2000); (2) the occurrence of discontinuous, but correlative, little-weathered drift sheets alongside EAIS outlet glaciers in the Transantarctic Mountains from Reedy Glacier to Terra Nova Bay (Mercer 1968; Bockheim *et al.* 1989; Denton *et al.* 1989*a*; Orombelli *et al.* 1990; Denton & Marchant 2000; Bromley *et al.* 2010, 2012; Todd *et al.* 2010); (3) relatively unweathered drift sheets in Marie Byrd Land (Stone *et al.* 2003) and on nunataks in inland West Antarctica (Fig. 1; Ackert *et al.* 1999, 2007); and (4) inferences from glaciological models based on ice-penetrating radar data from ice rises and streams in the central and eastern Ross Embayment (Conway *et al.* 1999; Parizek *et al.* 2003; Parizek & Alley 2004; Waddington *et al.* 2005; Martin *et al.* 2006; Price *et al.* 2007). We discuss each of these data sets below by geographic region.

Southern and Central Transantarctic Mountains

A largely unweathered drift sheet, contrasting sharply with more distal and older, weathered deposits, occurs throughout the Transantarctic Mountains adjacent to many outlet glaciers of the East Antarctic Ice Sheet (Fig. 3). Although widespread, the drift is discontinuous and is absent or consists only of rare scattered erratics in many places. Based on its unweathered appearance, negligible soil development and ice core (in places), this drift sheet has been attributed to the LGM and subsequent deglaciation (Mercer 1968; Bockheim *et al.* 1989;

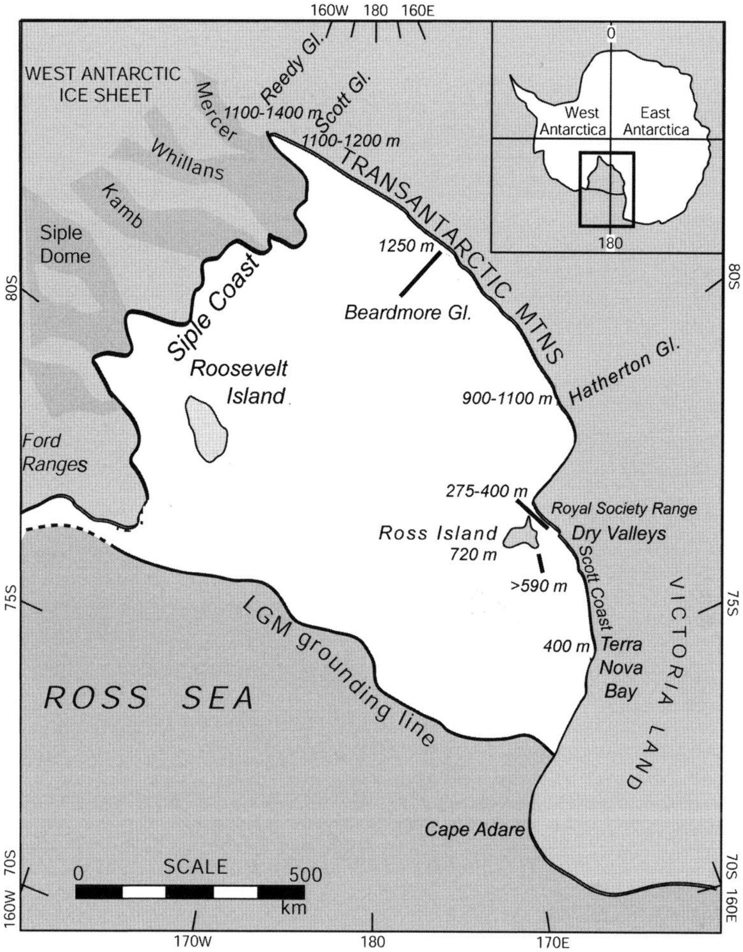

Fig. 2. Map of the Ross Embayment, showing locations mentioned in the text. In addition, the map shows LGM ice-surface elevations reconstructed from glacial geological data along the Transantarctic Mountains coast.

Denton *et al.* 1989*a*; Bromley *et al.* 2010, 2012). Reconstructions of past outlet-glacier extent show progressively less thickening with distance upglacier (Fig. 4). The lower reaches of these glaciers were more than 1000 m thicker than at present at their intersection with the Ross Embayment. In contrast, ice-surface elevations on the East Antarctic plateau remained virtually unchanged (Mercer 1968; Bockheim *et al.* 1989; Denton *et al.* 1989*a*; Bromley *et al.* 2010). The former elevations of these glaciers where they entered the Ross Embayment afford an estimate of past thickness of the WAIS into which they flowed. Results indicate that the elevation at the mouth of Reedy Glacier was 1100–1400 m at the LGM (Bromley *et al.* 2010; Todd *et al.* 2010; note that elevations presented here and elsewhere in the text are uncorrected for isostatic changes). At Scott Glacier (Fig. 2), the ice-sheet elevation exceeded 930 m, but was below 1200 m. Further north, the LGM limit at the mouth of Beardmore Glacier may have reached as much as 1250 m elevation (Denton *et al.* 1989*a*) and at the Darwin/Hatherton Glacier system estimates range from 900–1100 m elevation (Bockheim *et al.* 1989; Anderson *et al.* 2004). All of the data presented above are consistent with a thick grounded ice sheet in the western and southern Ross Embayment during the glacial maximum.

The chronology of the relatively unweathered drift sheet has improved in recent years. Todd *et al.* (2010) produced approximately 80 ^{10}Be exposure ages of erratics along the margin of Reedy Glacier (Fig. 1). The data showed that at the Quartz Hills near the mouth, the glacier was at maximum

Fig. 3. (**a**) Reedy III drift at the Quartz Hills adjacent to Reedy Glacier. The LGM drift limit, delineated by the abrupt transition from unweathered grey drift to brown, heavily stained deposits, occurs *c.* 250 m above the present-day ice surface in this location and has been dated to *c.* 14–17 ka using surface-exposure age dating (Todd *et al.* 2010). (**b**) Drift attributed to the local LGM in the Dominion Range along the uppermost part of Beardmore Glacier. The sharp contact between the little-weathered grey deposits and the highly weathered stained drift marks the former limit at *c.* 30 m above present-day ice level. (**c**) The correlative moraine and drift limit along the headlands of McMurdo Sound in the Royal Society Range. The drift here consists of primarily basalt erratics transported across McMurdo Sound from Ross Island and extends to *c.* 250–320 m elevation. (**d**) Ross Sea drift at Hjorth Hill at the mouth of Taylor Valley. Here, the drift extends from sea level to a composite moraine at *c.* 300–400 m elevation. Portions of the moraine have been dated to 14–18 ka, based on radiocarbon dates of incorporated algae (Hall & Denton 2000).

extent at 14–17 ka. In contrast, further upglacier near the East Antarctic plateau, the glacier reached its maximum at *c.* 7 ka. The thickening was thus time-transgressive, with ages of erratics at the maximum drift limit becoming progressively younger upglacier. Reworked boulders with inherited ^{10}Be were common where the drift was thick and could not be avoided despite careful sampling. This occurrence is thought to reflect the fact that most rocks were transported by cold-based ice that did not remove the effects of earlier exposure. Outliers were excluded from the dataset by taking the youngest (and usually most) significant population of dates as an indication of deposit age. This practice is based on the assumption that the likelihood of rocks with inherited ^{10}Be far outweighs that of post-depositional processes, such as exhumation or erosion, and thus outliers are likely to be older than the true age, rather than younger (Todd *et al.* 2010). Glacier thinning in the Quartz Hills began by *c.* 13 ka and continued into the Holocene (Todd *et al.* 2010). This prolonged deglaciation is confirmed by data from Cohen Nunataks close to the confluence with Mercer Ice Stream, where mountain summits emerged from LGM ice cover around 8 ka and deglaciation continued until 2–3 ka. Recently acquired data from Scott and Beardmore Glaciers (Stone *et al.* 2009) are consistent with maximum glaciation between 15 and 17 ka, followed by rapid thinning in the early Holocene. Ackert & Kurz (2004) dated the correlative, relatively unweathered drift at Beardmore Glacier and obtained exposure ages of *c.* 20 ka for the maximum.

Further north at the Darwin/Hatherton Glacier system, Bockheim *et al.* (1989), on the basis of glacial mapping, soil characterization and correlation with dated deposits, along with an interpretation

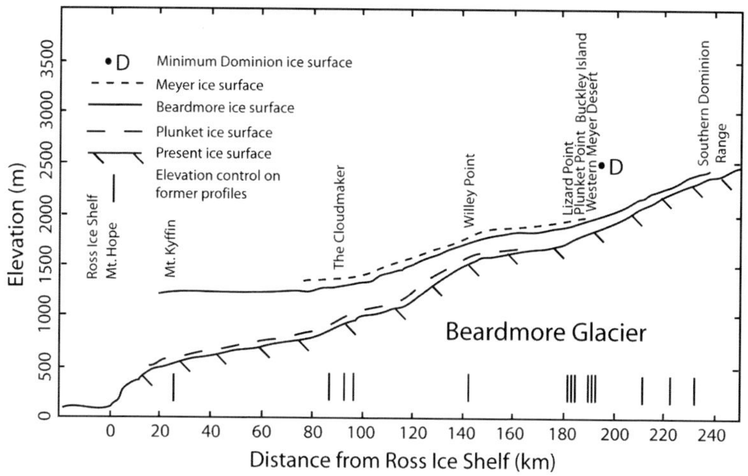

Fig. 4. Longitudinal profile of Beardmore Glacier, reconstructed from glacial deposits (Denton *et al.* 1989*a*). The diagram shows both the modern profile and that reached during former times. The Beardmore ice surface, which was reconstructed from the little-weathered drift limits, represents the local LGM limit. Beardmore Glacier thickened substantially more where it intersected the Ross Embayment than at the East Antarctic plateau. This pattern of thickening is characteristic of all East Antarctic Ice Sheet outlet glaciers that entered the Ross Embayment during the LGM.

of radiocarbon ages of algae deposited in former lateral ice-dammed ponds, suggested significant expansion of the Darwin/Hatherton system at the LGM. Radiocarbon ages of subfossil algae collected in transects from the maximum limit to the present ice level yielded dates ranging from *c.* 13 ka near the fresh drift limit to *c.* 7 ka at the present glacier (Bockheim *et al.* 1989). The simplest explanation for this pattern is that the algae grew in a series of ponds that formed against the glacier margin repeatedly as the ice retreated down the slope towards its present position. If correct, this implies that the maximum ice thickness there was maintained until at least *c.* 13 ka and present-day elevations were reached by *c.* 7 ka. However, this scenario is in contrast with surface-exposure measurements obtained by Storey *et al.* (2010), most of which predate the LGM. On the basis of these measurements, they proposed that ice did not expand in the Darwin/Hatherton system at the LGM. At present, then, there are two different interpretations for the Darwin/Hatherton system that have important ramifications for understanding the glacial history of the Ross Embayment: either the exposure chronology presented for Darwin Glacier is incorrect (probably because the data are compromised by inherited ^{10}Be owing to prior exposure and minimal erosion rates) or the idea of grounded ice in the Ross Embayment at the LGM needs to be revisited. At present, the weight of evidence supports the concept of a widespread grounded ice sheet in the embayment.

Dry Valleys sector

In contrast to areas both to the north and the south, East Antarctic Ice Sheet outlet glaciers in the Dry Valleys region (Fig. 2) from Taylor to Victoria Valleys did not flow into the Ross Embayment at the LGM. Rather, both local alpine glaciers and the small East Antarctic Ice Sheet outlet glaciers in this region retreated (Stuiver *et al.* 1981; Denton *et al.* 1989*b*; Hall *et al.* 2000) and terminated in the valley heads. At the same time, the Ross Sea ice sheet occupied McMurdo Sound and dammed large proglacial lakes that filled many of the valleys (Stuiver *et al.* 1981; Hall *et al.* 2000, 2010*a*). For example, Taylor Glacier receded at least several kilometres at the LGM and is now readvancing over the area of the valley covered by a proglacial lake dammed by the Ross Sea ice sheet between at least *c.* 13–28 ka (Denton *et al.* 1989*b*; Hall *et al.* 2000; Higgins *et al.* 2000). Alpine glaciers both in the Dry Valleys and the nearby Royal Society Range also have overridden lake deposits of the same age, indicating that these glaciers, too, were of lesser extent during the LGM than they are at present. Shrinkage of local ice was probably the result of both reduced precipitation when grounded ice occupied the Ross Sea and high melting ablation in blue-ice zones (Hall *et al.* 2010*a*).

As mentioned above, because of the lack of through-flowing outlet glaciers from the East Antarctic Ice Sheet, the westward-flowing ice from the Ross Embayment filled McMurdo Sound and sent tongues into eastern Taylor Valley and into the valleys fronting the nearby Royal Society Range at the LGM (Péwé 1960; Clayton-Greene *et al.* 1988; Denton & Hughes 2000; Denton & Marchant 2000; Hall *et al.* 2000). The grounded ice sheet deposited well-preserved moraines and kame terraces on the headlands adjacent to McMurdo Sound (Fig. 3), which allow reconstruction of former ice-surface elevations. Elevations of deposits marking the former ice margin range from *c.* 720 m on eastern Ross Island, to >590 m elevation on Cape Bird, to *c.* 350–400 m elevation at the mouth of Taylor Valley, and to *c.* 250–300 m elevation in front of the Royal Society Range (Stuiver *et al.* 1981; Denton & Marchant 2000; Dochat *et al.* 2000; Hall *et al.* 2000). The slope of the former ice surface, as well as the distribution of kenyte erratics that originate from Ross Island, shows that grounded ice flowed westward around both the south and north sides of Ross Island and then into the ice-free valleys during the LGM (Stuiver *et al.* 1981; Denton & Marchant 2000; Hall *et al.* 2000). The moraines produced by this grounded ice descend from the headlands westward into the valleys, where the moraines give way to glaciolacustrine sediments formed in ice-dammed lakes. Lake-ice conveyors (Clayton-Greene *et al.* 1988; Hall *et al.* 2000; Hendy *et al.* 2000) transported drift into the valleys more than 10 km beyond the grounding line.

The chronology of deposits from the Ross Sea ice sheet in the Dry Valleys and adjacent Royal Society Range comes mainly from radiocarbon dates of freshwater algae deposited in lakes and streams adjacent to the former ice body. Moraines at the mouth of Taylor Valley at Hjorth Hill range from *c.* 13 to 18 ka (Hall & Denton 2000), with the maximum being reached by 18 ka. New data from a similar moraine fronting the Royal Society Range appear to confirm this age for the maximum ice extent (Allard *et al.* 2011; Koffman *et al.* 2011). This timing is also consistent with a limited number of ^{3}He surface-exposure age dates from the correlative moraine near Blue Glacier, which place ice at the maximum at *c.* 14–16 ka, although other presumed LGM landforms described in the same study produced mixed ages, probably because of prior exposure of the dated samples (Brook *et al.* 1995). Deltas that formed in the LGM proglacial lakes also afford information on the presence of the Ross Sea ice sheet in McMurdo Sound, because the lakes could not have existed without ice dams at the valley mouths. Radiocarbon dates of deltas range from *c.* 9 to 28 ka, suggesting that grounded Ross Sea ice occupied McMurdo Sound throughout that time span (Stuiver *et al.* 1981; Clayton-Greene *et al.* 1988; Hall & Denton 2000; Hall *et al.* 2010*a*). Evidence of ice-free conditions in McMurdo Sound subsequent to the LGM comes from radiocarbon and uranium–thorium dates of marine organisms found in sediment cores from McMurdo Sound and in debris cones on the McMurdo Ice Shelf, which has incorporated seafloor sediments by basal freezing. These ages are as old as *c.* 7.7 ka (Stuiver *et al.* 1981; Kellogg *et al.* 1990; Licht *et al.* 1996; Hall & Denton 2000; Hall *et al.* 2010*b*), which is similar to that obtained from relative sea-level curves for the adjacent coast that suggest unloading of grounded ice by *c.* 7.8 ka (Hall *et al.* 2004).

Northern Victoria Land

Northern Victoria Land (Fig. 1) displays both large outlet glaciers from the East Antarctic Ice Sheet and local glaciers and ice fields. At the LGM, the outlet glaciers merged with the Ross Sea ice sheet and deposited a widespread drift sheet, Terra Nova drift, which has been correlated to Ross Sea drift on the basis of elevation, morphology, weathering, and limiting ages (Stuiver *et al.* 1981; Orombelli *et al.* 1990). Mapping of glacial geomorphologic features and reconstruction of the former longitudinal profile at Reeves Glacier near Terra Nova Bay indicate that the Ross Sea ice sheet surface where it crossed the present-day coast was at *c.* 400 m elevation (Orombelli *et al.* 1990). Deglaciation of grounded Ross Sea ice at Terra Nova Bay was complete by *c.* 8 ka, based on relative sea-level curves and on the age of abandoned penguin rookeries (Baroni & Hall 2004).

Marie Byrd Land and interior West Antarctica

Although a wealth of information exists for the Transantarctic Mountains, there are relatively few data related to LGM ice expansion in Marie Byrd Land (Fig. 1) and in the interior of the West Antarctic Ice Sheet. From ^{10}Be surface-exposure ages of erratics perched on nunataks, Stone *et al.* (2003) showed that LGM ice thickening exceeded 700 m, compared with today's values, in the coastal Ford Ranges of Marie Byrd Land, consistent with ice-sheet expansion into Sulzberger Bay extending to the continental shelf edge (Wellner *et al.* 2001). The magnitude of ice thickening was greatest at the coast and decreased towards the interior. Because the coastal mountains were overrun, Stone *et al.* (2003) could not determine the LGM ice surface elevation nor the timing of the LGM; however, emergence of the nunataks due to lowering of the ice surface began before 11 ka and continued throughout the Holocene. Further inland, Ackert *et al.* (1999) documented deposits from the last glaciation at as much as 45 m above the present-day ice level at Mt Waesche (Fig. 1). Exposure-age data suggest that this ice level was achieved at *c.* 10 ka. Likewise, ice from the interior Ohio Range (Fig. 1), which was 125 m above present, reached its maximum at approximately the same time (Ackert *et al.* 2007, 2011).

Central Ross Embayment and Roosevelt Island

In contrast to glacial geological evidence from mountain ranges both east and west of the Ross Embayment, glaciological models have suggested only modest ice thickening in the central embayment. For example, Waddington *et al.* (2005) used the observed age–depth relationship at Siple Dome (Fig. 1) to constrain a time-dependent ice-flow model. They concluded that Siple Dome was no more than 200–400 m thicker than at present during the LGM. A low-profile ice sheet in the central embayment during the LGM is also compatible with model results of Price *et al.* (2007). They used borehole temperature measurements as an additional constraint and showed that Siple Dome may have reached its present thickness by 13–15 ka. The low profile in the central embayment was probably maintained by the presence of low-gradient ice streams. Experiments with a thermomechanical model indicated that, although ice streams probably slowed during the LGM, they probably did not stagnate (Parizek *et al.* 2003). To reconstruct grounding-line retreat in the eastern Ross Sea, Conway *et al.* (1999) used the pattern of radar-detected stratigraphy at Roosevelt Island (Fig. 1) to constrain a model of the thinning history of the region. They explored a variety of different accumulation rates and concluded that the ice island first became surrounded by shelf ice at *c.* 3.2 ka. That is, the grounding line of the Ross Sea ice sheet must have passed south of Roosevelt Island at that time. Over the past millennia, evidence from flow features recorded on the Ross Ice Shelf and the modern flow field indicates that Ross Ice Streams have activated and shut down repeatedly (Fahnestock *et al.* 2000; Hulbe & Fahnestock 2004, 2007). Overall, the mass balance of the Ross Sea sector is now positive, primarily because of the shut-down of Kamb Ice Stream (Joughin & Tulaczyk 2002). It is not yet clear if this is a transient effect or whether it signals the end of Holocene grounding-line retreat.

Ice-surface elevations in the Ross Embayment at the LGM

The data mentioned above yield a consistent picture of LGM ice extent and thickness (Figs 2 & 6) in the Ross Embayment. The Ross Sea ice sheet was thick at the mouths of the southern and central Transantarctic Mountains outlet glaciers, reaching as much as 1000 m above present levels. Surface elevations in this region were *c.* 1100–1400 m (Denton *et al.* 1989*a*; Bromley *et al.* 2010, 2012; Todd *et al.* 2010). Glacial geological evidence from Marie Byrd Land on the eastern side of the Ross Embayment suggests similar thickening, with the ice sheet reaching >1165 m elevation. Further north adjacent to the western Ross Embayment, ice-surface elevation decreased, reaching only a little more than 700 m on eastern Ross Island (Fig. 5). From here, elevations decreased both to the west, in the Dry Valleys/Royal Society Range region, where ice terminated at 250–400 m elevation on the headlands, and to the north, where the surface of the Ross Sea ice sheet reached *c.* 400 m at Terra Nova Bay. Evidence from interior nunataks, Marie Byrd Land and East Antarctic Ice Sheet outlet glaciers suggests that the amount of ice thickening tapered off significantly away from the Ross Embayment. Moreover, limited data indicate that the maximum ice elevation was achieved later at inland sites than near the coast. Both of these observations are consistent with major grounding of ice in the Ross Sea as a cause of outlet-glacier thickening. It should be noted that elevations reconstructed from dated glacial limits in the mountains adjacent to the Ross Sea are nearly 150–200 m higher than those calculated from ice models of internal radar layers in Siple Dome (Waddington *et al.* 2005; Price *et al.* 2007), located roughly equidistant between the Transantarctic Mountains and Marie Byrd Land. If correct, this discrepancy suggests that the central portion of the ice sheet fed by ice streams crossing the present-day Siple Coast was lower than the margins near the mountains, requiring streaming, low-gradient ice flow across the Ross Sea floor. Ice must have remained grounded, however, in the central Ross Embayment. We do not favour models that show a central, ice-free embayment at the LGM (i.e. Drewry 1979; Denton *et al.* 1989*b*), because such a configuration would prevent ice from entering McMurdo Sound from the east. So long as ice flowed from the Ross Embayment westward into McMurdo Sound, we suggest that grounded ice occupied the central Ross Sea. Loss of this ice flow is thought to have occurred at some time after *c.* 10 ka (Hall & Denton 2000).

Timing of LGM

As mentioned above, the local LGM ice limit probably was time-transgressive, with interior locations showing a maximum surface elevation that lagged that at the coast by as much as 10 000 years. Coastal sites, such as the Dry Valleys/Royal Society Range region and the mouths of large outlet

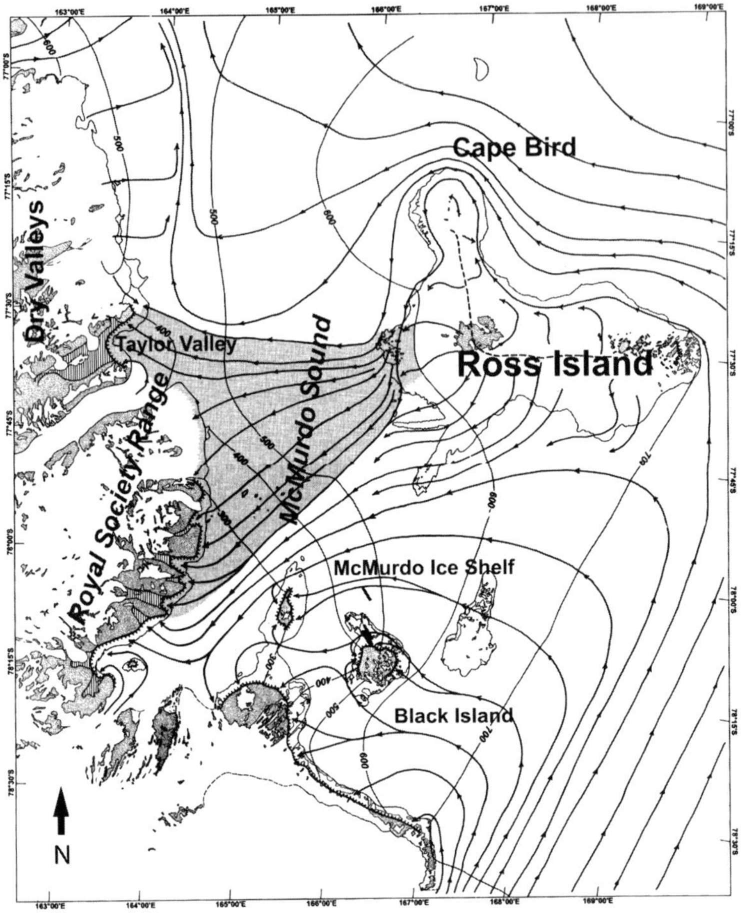

Fig. 5. Reconstruction of ice flow lines in the McMurdo Sound region at the LGM (Denton & Marchant 2000). Ice flowed both north and south around Ross Island from the east toward the west. Ice that passed around northern Ross Island carried kenyte and deposited it in the mouth of Taylor Valley and in the valleys fronting the Royal Society Range (shaded band). Ice-dammed lakes occupied these valleys (diagonal pattern).

glaciers (i.e. Reedy and Scott), typically show maximum ice elevation at *c.* 13–18 ka (Hall & Denton 2000; Todd *et al.* 2010; Allard *et al.* 2011; Koffman *et al.* 2011). In contrast, interior locations, such as Mt Waesche, the Ohio Range, and upper reaches of Reedy Glacier, show a much later maximum, at 7–10 ka (Ackert *et al.* 1999, 2007; Todd *et al.* 2010). This time-transgressive behaviour may reflect the lag time of interior sites to perturbations in the Ross Embayment partially offset by a gradual thickening of the East Antarctic Ice Sheet from increased Holocene precipitation (Ackert *et al.* 2007; Todd *et al.* 2010). Such thickening is consistent with readvance of East Antarctic Ice Sheet outlet glaciers, as well as local alpine glaciers, in the Dry Valleys region in the Holocene (Denton *et al.* 1989*b*).

Thinning history

Sites with available glacial geological data show that ice along the coast of the embayment maintained a position close to its maximum as late as *c.* 13 ka. At 13 ka, the Ross Sea ice sheet began to lower irreversibly at the mouth of Taylor Valley and in front of the Royal Society Range. At the same time, ice levels started to decrease in the Quartz Hills near the mouth of Reedy Glacier and nunataks emerged at the mouth of Scott Glacier (Stone *et al.* 2009; Todd *et al.* 2010). The subsequent rate of thinning at each of these sites varied and was linked to the proximity of the southward-retreating grounding line (Conway *et al.* 1999). In the Dry Valleys/Royal Society Range region, the last evidence of grounded ice comes from

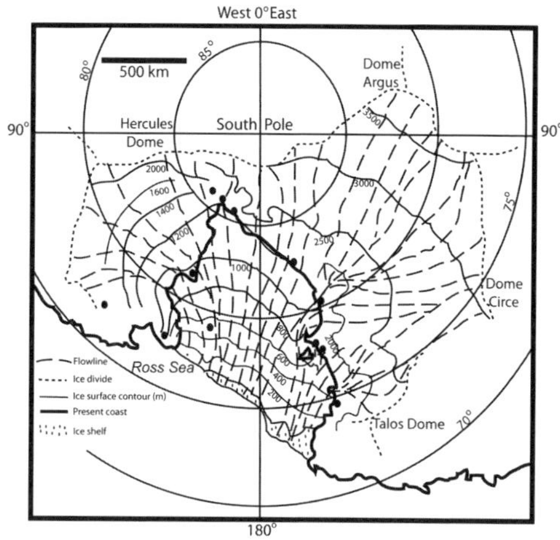

Fig. 6. Example of a reconstruction of ice in the Ross Sea sector during the LGM. This diagram, modified slightly after Denton *et al.* (1989*a*) to account for new data at Reedy and Scott Glaciers (Bromley *et al.* 2010, 2012; Todd *et al.* 2010), shows the major drainage divides, flowlines and ice-surface elevations, based on glacial geological data. Dots show locations (see Fig. 1 for names) where there is information on past ice-surface elevations.

ice-dammed lakes dating to *c.* 9.3 ka (Stuiver *et al.* 1981; Hall & Denton 2000; Hall *et al.* 2010*a*). The presence of extensive ice-cored terrain at low elevations on Hjorth Hill and along the west coast of McMurdo Sound suggests that the last grounded ice stagnated, probably as it was cut off from the main Ross Sea ice sheet by marine incursion.

East Antarctic Ice Sheet outlet glaciers that project into the Ross Embayment have shown a pattern of thinning ice and steepening profiles as the Ross Sea grounding line has retreated south in the Holocene. For example, most thinning at Reedy and Scott Glaciers took place after *c.* 10 ka (Stone *et al.* 2009; Todd *et al.* 2010). Ice did not reach close to present-day levels until *c.* 2 ka (Todd *et al.* 2010). This pattern of Holocene deglaciation matches that in coastal Marie Byrd Land, where the ice sheet dropped 700 m after 11.5 ka (Stone *et al.* 2003). Overall, most of the removal of ice from the Ross Embayment appears to have occurred in the Holocene.

Grounding-line retreat

The timing of initial grounding-line retreat from the outer continental shelf in the Ross Sea is uncertain because of difficulties with obtaining an accurate chronology from marine sediment cores. Dates of total organic carbon (TOC) from marine sediments consistently yield older ages than those of foraminifera (which are rare) from the same levels (Andrews *et al.* 1999). Current practice is to apply a correction for the old carbon within the TOC, based on surface ages, but this method does not take into account down-core variations in the influx of old carbon. For a thorough discussion of deglacial ages from marine cores, see Licht & Andrews (2002), who gave an overview/revision of the earlier papers and commented on the variety of correction methods that have been used to account for problems with dating TOC.

Licht & Andrews (2002) suggested that deglaciation of the Drygalski Trough occurred at *c.* 11 ka, an age younger than that proposed in earlier studies (Licht *et al.* 1996; Domack *et al.* 1999), but older than that from the coast of nearby Terra Nova Bay (Baroni & Hall 2004). Based on a date of an *Adamussium colbecki* shell within a marine sediment core taken near Ross Island, Licht *et al.* (1996) suggested southward retreat of the grounding line into McMurdo Sound by *c.* 7.7 ka, in good agreement with limiting ages from ice-dammed lakes in nearby Taylor Valley (Hall & Denton 2000). Additional ages for retreat of ice from the western Ross Sea come from relative sea-level curves based on the ages of organic material in raised beaches (Baroni & Orombelli 1991, 1994; Hall & Denton 1999; Baroni & Hall 2004; Hall *et al.* 2004). At Terra Nova Bay, radiocarbon dates of shell, seal skin and penguin remains bracket the age of the marine limit and the timing of complete unloading of grounded ice to *c.* 8 ka (Baroni & Hall 2004). Further south along the southern Scott Coast, relative sea-level data indicate grounding-line retreat into McMurdo Sound at *c.* 7.8 ka (Hall *et al.* 2004), in agreement with the date from the marine core mentioned above and with ages of shells and barnacles from the McMurdo Ice Shelf (Kellogg *et al.* 1990; Hall *et al.* 2010*b*).

One reconstruction of Holocene grounding-line recession in the Ross Sea is that of a 'swinging gate' (Stuiver *et al.* 1981; Conway *et al.* 1999; Fig. 7). Conway *et al.* (1999) reconstructed recession southward through the embayment from (1) relative sea-level curves along the Victoria Land coast, (2) longitudinal profiles of Hatherton Glacier during the Holocene, and (3) internal layers within the Roosevelt Island ice dome documented by radar. Although most of their data are from the coast adjacent to the Transantarctic Mountains, the youngest site, Roosevelt Island, is a grounded ice dome in the eastern Ross Sea located just behind the present-day calving front of the Ross Ice Shelf. During the LGM, Roosevelt Island was surrounded by the grounded ice sheet, whereas now it is flanked by the floating ice shelf.

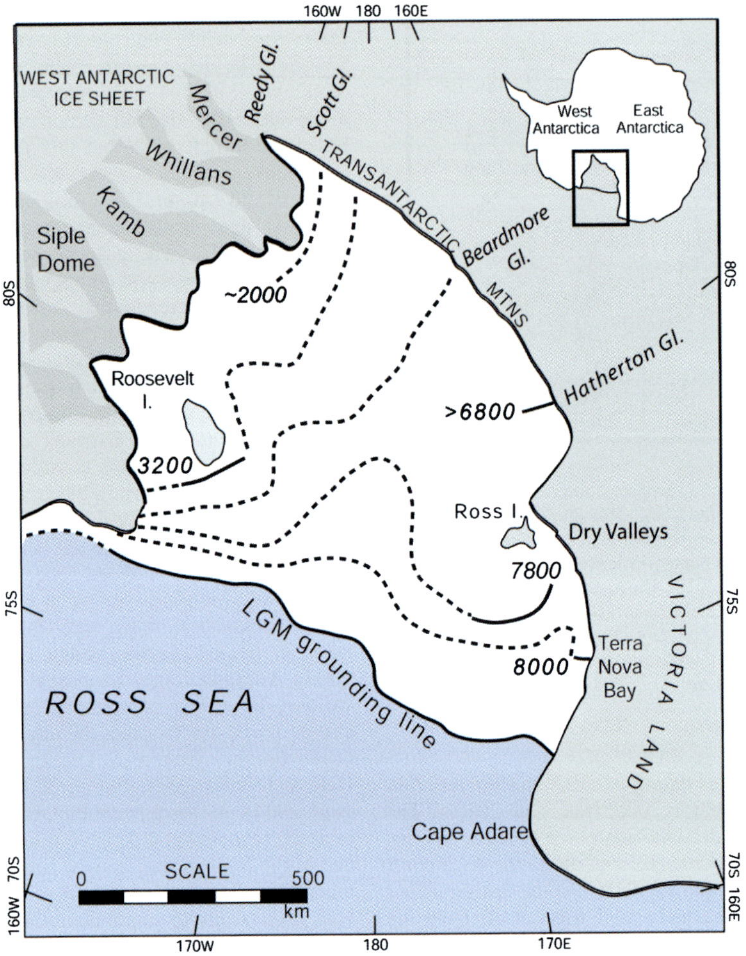

Fig. 7. The Conway *et al.* (1999) model of grounding-line retreat in the Ross Embayment, modified to account for recent data (Baroni & Hall 2004; Stone *et al.* 2009; Bromley *et al.* 2010, 2012; Todd *et al.* 2010). This version shows the development of a central embayment during deglaciation. This embayment is based only on conjecture at present.

Ice-penetrating radar revealed layers that showed the 'Raymond bump', a characteristic feature under ice divides, which results from the rheological properties of ice. Fitting a model to the bump amplitude using a variety of realistic accumulation rates allowed Conway *et al.* (1999) to infer that the grounding line retreated southward past Roosevelt Island about 3.2 ka. New data from Reedy Glacier suggest reduced rates of thinning over the last 2 ka where the glacier merges with the West Antarctic Ice Sheet. Todd *et al.* (2010) suggested that this could indicate relative grounding-line stability in the southern Ross Embayment since that time.

It should be noted that data points constraining the Conway *et al.* (1999) model are sparse and do not exclude the possibility that the central embayment opened early in deglaciation and that subsequent grounding-line recession proceeded both east and west from this embayment towards Marie Byrd Land and the Transantarctic Mountains, rather than in a simple southward direction. However, any such central embayment could not have existed while ice still was flowing westward across McMurdo Sound to the Dry Valleys and Royal Society Range. Such westward flow into McMurdo Sound is not consistent with the existence of an embayment that would have directed ice flow toward the central Ross Sea. On the basis of landforms with kenyte erratics derived from Ross Island, Hall & Denton (2000) suggested that active ice flow across McMurdo Sound was maintained until at least 10 ka.

Behaviour of the Antarctic ice sheet

The cause and mechanisms behind Ross Sea ice sheet advance and recession remain unclear. The traditional paradigm is that the Antarctic ice sheets, particularly the West Antarctic Ice Sheet, are highly sensitive to sea-level forcing (Hollin 1962; Stuiver *et al.* 1981; Denton & Hughes 1986), and thus their fate is tied to the waxing and waning of Northern Hemisphere ice sheets. However, data presented here and elsewhere indicate that most ice recession, particularly grounding-line retreat, in the Ross Embayment was delayed compared with the initiation of eustatic sea-level rise (Fairbanks 1989), making it unlikely that sea-level variations are the sole driver of Ross Sea ice sheet oscillations. There are alternate hypotheses (see a comprehensive review by Joughin & Alley 2011). Some have focused on bed conditions as an explanation for Ross Sea ice sheet retreat. For example, MacAyeal (1992) proposed that the ice sheet could be part of a self-oscillating system that is relatively insensitive to sea-level changes, but rather fluctuates in response to dynamics controlled by basal thermal conditions, which lag changes in air temperature. Likewise, Parizek *et al.* (2003) suggested that significant thinning and increased discharge of ice streams was delayed until warm temperatures reached the bed of the ice sheet. Yet another possibility is that warming ocean temperatures during global deglaciation could have increased melt rates along the grounding lines and caused ice recession. For example, a recent model shows the Ross Sea grounding line to be highly sensitive to ocean temperatures, more so than to sea-level rise (Pollard & DeConto 2009). A modern analogue for this latter process is occurring today in the Amundsen Sea region, where increases in ocean-water temperature and changes in sub-ice geometry have resulted in significant increases in melt rates and rapid glacier drawdown (Jacobs *et al.* 1996, 2011; Thoma *et al.* 2008; Jenkins *et al.* 2010).

The history of the grounded ice sheet in the Ross Sea sector contrasts with recent evidence documenting ice behaviour in the Weddell Sea sector. Early estimates, based on sea-floor sediments and the presence of high-elevation, although undated, trimlines in the Ellsworth Mountains, were that thick grounded ice, similar to that in the Ross Embayment, occupied the Weddell Embayment during the LGM (Elverhøi 1981; Anderson *et al.* 1991; Denton *et al.* 1992). However, new surface-exposure age data from the Ellsworth Mountains suggest that LGM ice levels may have been lower than previously thought (Bentley *et al.* 2010). In addition, surface-exposure age data from the Shackleton Range have been interpreted as indicating little or no LGM ice thickening (Hein *et al.* 2011). If true, then these data limit the contribution of the Weddell Sea sector to sea-level change not only at the LGM, but also to any deglacial event, such as MWP 1A. The reason for the apparent difference in ice behaviour between the Ross and Weddell Embayments is not known at present.

The Ross Sea Sector and MWP 1A

Meltwater pulse 1A is thought to be a *c.* 14–20 m rise in sea level over an interval of 300–500 years beginning at roughly 14.6 ka (Hanebuth *et al.* 2000; Deschamps *et al.* 2012). When calibrated to sea-level data at low latitudes (i.e. Barbados), glacioisostatic adjustment models allow one to ‘fingerprint’ the source of the meltwater (Clark *et al.* 2002). Such fingerprinting suggests that Antarctica was the origin of MWP 1A, either entirely or in part. However, in order for MWP 1A to have come from Antarctica, two requirements must have been met. First, there must have been sufficient ice in Antarctica (in excess of that which exists today) in order to produce the sea-level rise. Second, that ice must have been released to the ocean quickly at 14.6 ka. Because Antarctica lacks widespread surface melting ablation zones, the only mechanism that would release enough ice in a centennial timeframe would be a catastrophic surge of the Antarctic Ice Sheet. If the ice sheet is in fact capable of such a surge, it would be important information for future ice-sheet predictions.

Geological data have not yet emerged to satisfy the two major requirements outlined above. In addition, most ice-sheet models have converged on a relatively modest amount of excess ice in Antarctica at the LGM, well short of that needed to produce MWP 1A. For example, in a recent estimate, Mackintosh *et al.* (2011) calculated that only 10 m of sea-level rise could have come from Antarctica since 14 ka and that this was released gradually to the sea over thousands of years. Records both from the Ross Embayment (described above) and elsewhere (see Mackintosh *et al.* 2011 for a review) suggest that most Antarctic deglaciation postdated *c.* 13 ka. Moreover, the Ross Embayment appears to have experienced a drawn-out grounding-line recession over much of the Holocene, with no evidence yet available of catastrophic ice loss at 14.6 ka. In summary, an Antarctic origin for MWP 1A remains inconclusive; additional data delineating both ice thickness and chronology will help resolve this impasse.

Conclusions

Terrestrial glacial geological data from areas adjacent to the Ross Embayment afford evidence for a thick, grounded ice sheet that extended close to the shelf edge and that reached surface elevations of 1100–1400 m in the central and southern Transantarctic Mountains and Marie Byrd Land. The timing of the initiation of ice expansion is not well known, but the ice sheet was at its local maximum along the coast adjacent to the Transantarctic Mountains by 18 ka. The local LGM was time-transgressive and occurred as late as 7–10 ka in interior locations. Most deglaciation seems to have occurred in the Holocene. The first evidence of significant glacier thinning was at *c.* 13 ka. Reconstruction of thinning history and grounding-line retreat indicates that recession occurred throughout the Holocene and has slowed/stopped only in the last two millennia. This history makes it unlikely that the Ross Sea sector of the Antarctic Ice Sheet was a major contributor to MWP 1A. However, resolution of a possible Antarctic origin of MWP 1A awaits detailed reconstructions from all major sectors of the ice sheet.

The authors thank numerous colleagues and students, past and present, who have worked with them on the glacial history of the Ross Sea region. We also thank logistical support personnel. M. Bentley and D. Sugden provided useful reviews. The work described here was funded largely by the Office of Polar Programs of the National Science Foundation. B. L. Hall and G. H. Denton also acknowledge support from the Italian Antarctic Research Program.

References

ACKERT, R. & KURZ, M. 2004. Age and uplift rates of Sirius Group sediments in the Dominion Range, Antarctica, from surface exposure dating and geomorphology. *Global and Planetary Change*, **42**, 207–225.

ACKERT, R., BARCLAY, D. J. *ET AL.* 1999. Measurement of past ice sheet elevations in interior West Antarctica. *Science*, **286**, 276–280.

ACKERT, R., MUKHOPADHYAY, S., PARIZEK, B. & BORNS, H. W., JR. 2007. Ice elevation near the West Antarctic Ice Sheet divide during the last glaciation. *Geophysical Research Letters*, **34**, L21506, doi: 10.1029/2007GL031412

ACKERT, R., MUKHOPADHYAY, S. *ET AL.* 2011. West Antarctic ice sheet elevations in the Ohio range: geologic constraints and ice sheet modeling prior to the last highstand. *Earth and Planetary Science Letters*, **307**, 83–93.

ALLARD, S., HALL, B. & DENTON, G. 2011. History of the Ross Sea ice sheet based on glcial and lake records from Marshall Valley, Antarctica, *18th Annual West Antarctic Ice Sheet meeting*, Loveland, CO, http://www.waisworkshop.org

ANDERSON, B., HINDMARSH, R. & LAWSON, W. 2004. A modelling study of the response of Hatherton Glacier to Ross Ice Sheet grounding line retreat. *Global and Planetary Change*, **42**, 143–153.

ANDERSON, J., ANDREWS, J. T., BARTEK, L. & TRUSWELL, E. M. 1991. Petrology and palynology of glacial sediments: implications for subglacial geology of the Easter Weddell Sea, Antarctica. *In*: THOMPSON, M. R. A. (ed.) *Geological Evolution of Antarctica*. Cambridge University Press, New York, 231–235.

ANDREWS, J. T., DOMACK, E. W. *ET AL.* 1999. Problems and possible solutions concerning radiocarbon of surface marine sediments, Ross Sea, Antarctica. *Quaternary Research*, **52**, 206–216.

BARONI, C. & HALL, B. 2004. A new Holocene relative sea-level curve for Terra Nova Bay, Victoria Land, Antarctica. *Journal of Quaternary Science*, **19**, 377–396.

BARONI, C. & OROMBELLI, G. 1991. Holocene raised beaches at Terra Nova Bay, Victoria Land, Antarctica. *Quaternary Research*, **36**, 157–177.

BARONI, C. & OROMBELLI, G. 1994. Abandoned penguin rookeries as Holocene paleoclimate indicators in Antarctica. *Geology*, **22**, 23–26.

BASSETT, S., MILNE, G., MITROVICA, J. & CLARK, P. U. 2005. Influences on far-field sea-level histories. *Science*, **309**, 925–928.

BENTLEY, M. J. *ET AL.* 2010. Deglacial history of the West Antarctic ice sheet in the Weddell Sea embayment: constraints on past ice volume change. *Geology*, **38**, 411–414.

BLUNIER, T. & BROOK, E. 2001. Timing of millennial-scale climate change in Antarctica and Greenland during the last glacial period. *Science*, **291**, 109–112.

BOCKHEIM, J., WILSON, S. C., DENTON, G., ANDERSEN, B. G. & STUIVER, M. 1989. Late Quaternary ice-surface fluctuations of Hatherton Glacier, Transantarctic Mountains. *Quaternary Research*, **31**, 229–254.

BROECKER, W. 1998. Paleocean circulation during the last deglaciation: a bipolar seesaw? *Paleoceanography*, **13**, 119–121.

BROMLEY, G., HALL, B., STONE, J. O. H., TODD, C. & CONWAY, H. 2010. Late Cenozoic deposits at Reedy Glacier, Transantarctic Mountains: implications for former thickness of the West Antarctic Ice Sheet. *Quaternary Science Reviews*, **29**, 384–393.

BROMLEY, G., HALL, B., STONE, J. O. H. & CONWAY, H. 2012. Late Pleistocene evolution of Scott Glacier, southern Transantarctic Mountains: implications for the Antarctic contribution to deglacial sea level. *Quaternary Science Reviews*, **50**, 1–13.

BROOK, E., KURZ, M., ACKERT, R., RAISBECK, G. M. & YIOU, F. 1995. Comogenic nuclide exposure ages and glacial history of late Quaternary Ross Sea drift in McMurdo Sound, Antarctica. *Earth and Planetary Science Letters*, **131**, 41–56.

CLARK, P. U., MITROVICA, J., MILNE, G. & TAMISEA, M. E. 2002. Sea-level fingerprinting as a direct test for the source of global meltwater pulse 1a. *Science*, **295**, 2438–2441.

CLARK, P. U., DYKE, A. S. *ET AL.* 2009. The Last Glacial maximum. *Science*, **325**, 710–714.

CLAYTON-GREENE, J., HENDY, C. & HOGG, A. 1988. Chronology of a Wisconsin age proglacial lake in the Miers Valley, Antarctica. *New Zealand Journal of Geology and Geophysics*, **31**, 353–361.

CONWAY, H., HALL, B., DENTON, G., GADES, A. & WADDINGTON, E. 1999. Past and future grounding-line retreat of the West Antarctic Ice Sheet. *Science*, **286**, 280–283.

Denton, G. 2000. Does an asymmetric thermohaline oscillator drive 100 000-yr glacial cycles? *Journal of Quaternary Science*, **15**, 301–318.

Denton, G. & Hall, B. (eds) 2000. Glacial and paleoclimate history of the Ross Ice Drainage System of Antarctica. *Geografiska Annaler*, **82A**, 139–432.

Denton, G. & Hughes, T. 1986. Global ice-sheet system interlocked by sea level. *Quaternary Research*, **26**, 3–26.

Denton, G. & Hughes, T. 2000. Reconstruction of the Ross ice drainage system, Antarctica, at the last glacial maximum. *Geografiska Annaler*, **82A**, 143–166.

Denton, G. & Hughes, T. 2002. Reconstructing the Antarctic Ice Sheet at the last glacial maximum. *Quaternary Science Reviews*, **21**, 193–202.

Denton, G. & Marchant, D. 2000. The geologic basis for a reconstruction of a grounded ice sheet in McMurdo Sound, Antarctica, at the last glacial maximum. *Geografiska Annaler*, **82A**, 167–212.

Denton, G., Bockheim, J., Wilson, S., Leide, J. & Andersen, B. G. 1989*a*. Late Quaternary ice-surface fluctuations of Beardmore Glacier, Transantarctic Mountains. *Quaternary Research*, **31**, 183–209.

Denton, G., Bockheim, J., Wilson, S. & Stuiver, M. 1989*b*. Late Wisconsin and early Holocene glacial history, inner Ross Embayment, Antarctica. *Quaternary Research*, **31**, 151–182.

Denton, G., Bockheim, J., Rutford, R. & Andersen, B. G. 1992. Glacial history of the Ellsworth Mountains, West Antarctica. *Geological Society of America Memoir*, **170**, 403–432.

Denton, G. H., Anderson, R. F. *et al.* 2010. The last glacial termination. *Science*, **326**, 1652–1656.

Deschamps, P., Durand, N. *et al.* 2009. Synchroneity of Meltwater Pulse 1A and the Bolling onset: new evidence from the IODP 'Tahiti Sea-Level' expedition. *Geophysical Research Abstracts*, **11**, EGU2009-10233.

Deschamps, P., Durand, N. *et al.* 2012. Ice-sheet collapse and sea-level rise at the Bolling warming 14 600 years ago. *Nature*, **483**, 559–564.

Dochat, T. M., Marchant, D. & Denton, G. 2000. Glacial geology of Cape Bird, Ross Island, Antarctica. *Geografiska Annaler*, **82A**, 237–247.

Domack, E., Jacobson, E., Shipp, S. & Anderson, J. 1999. Late Pleistocene–Holocene retreat of the West Antarctic Ice Sheet in the Ross Sea: Part 2 – Sedimentologic and stratigraphic signature. *Geological Society of America Bulletin*, **111**, 1517–1536.

Drewry, D. 1979. Late Wisconsin reconstruction for the Ross Sea region, Antarctica. *Journal of Glaciology*, **24**, 230–244.

Elverhøi, A. 1981. Evidence for a late Wisconsin glaciation of the Weddell Sea. *Nature*, **293**, 641–642.

Fahnestock, M., Scambos, T., Bindschadler, R. & Kvaran, G. 2000. A millennium of variable ice flow recorded by the Ross Ice Shelf, Antarctica. *Journal of Glaciology*, **46**, 652–664.

Fairbanks, R. 1989. A 17 000-year glacio-eustatic sea level record: influence of glacial melting rates on the Younger Dryas event and deep-ocean circulation. *Nature*, **342**, 637–642.

Farmer, G. L., Licht, K., Swope, R. J. & Andrews, J. T. 2006. Isotopic constraints on the provenance of fine-grained sediment in LGM tills from the Ross Embayment, Antarctica. *Earth and Planetary Science Letters*, **249**, 90–107.

Hall, B. & Denton, G. 1999. New relative sea-level curves for the southern Scott Coast, Antarctica: evidence for Holocene deglaciation of the western Ross Sea. *Journal of Quaternary Science*, **14**, 641–650.

Hall, B. & Denton, G. 2000. Radiocarbon chronology of Ross Sea drift, eastern Taylor Valley, Antarctica: evidence for a grounded ice sheet in the Ross Sea at the last glacial maximum. *Geografiska Annaler*, **82A**, 305–336.

Hall, B., Denton, G. & Hendy, C. 2000. Evidence from Taylor Valley for a grounded ice sheet in the Ross Sea, Antarctica. *Geografiska Annaler*, **82A**, 275–304.

Hall, B., Denton, G. & Baroni, C. 2004. Holocene relative sea-level history of the Southern Victoria Land Coast, Antarctica. *Global and Planetary Change*, **42**, 241–263.

Hall, B., Denton, G., Fountain, A. G., Hendy, C. & Henderson, G. 2010*a*. Antarctic lakes suggest millennial reorganizations of Southern Hemisphere atmospheric and oceanic circulation. *Proceedings of the National Academy of Sciences*, **107**, 21355–21359.

Hall, B., Henderson, G., Baroni, C. & Kellogg, T. 2010*b*. Constant Holocene Southern-Ocean ^{14}C reservoir ages and ice-shelf flow rates. *Earth and Planetary Science Letters*, **296**, 115–123.

Hanebuth, T., Stattegger, K. & Grootes, P. 2000. Rapid flooding of the Sunda Shelf: a late-glacial sea-level record. *Science*, **288**, 1033–1035.

Hein, A. S., Fogwill, C., Sugden, D. & Xu, S. 2011. Glacial/interglacial ice-stream stability in the Weddell Sea embayment, Antarctica. *Earth and Planetary Science Letters*, **307**, 211–221.

Hendy, C., Sadler, A., Denton, G. & Hall, B. 2000. Proglacial lake-ice conveyors: a new mechanism for the deposition of drift in polar environments. *Geografiska Annaler*, **82A**, 249–270.

Higgins, S., Hendy, C. & Denton, G. 2000. Geochronology of Bonney drift, Taylor Valley, Antarctica: evidence for interglacial expansions of Taylor Glacier. *Geografiska Annaler*, **82A**, 391–409.

Hollin, J. 1962. On the glacial history of Antarctica. *Journal of Glaciology*, **4**, 172–195.

Hughes, T. 1973. Is the West Antarctic Ice Sheet disintegrating? *Journal of Geophysical Research*, **78**, 7884–7910.

Hulbe, C. & Fahnestock, M. 2004. West Antarctic ice-stream discharge variability: mechanism, controls, and pattern of grounding-line retreat. *Journal of Glaciology*, **50**, 471–484.

Hulbe, C. & Fahnestock, M. 2007. Century-scale discharge stagnation and reactivation of the Ross ice streams, West Antarctica. *Journal of Geophysical Research*, **112**, doi: 10.1029/2006JF000603

Jacobs, S., Hellmer, H. & Jenkins, A. 1996. Antarctic ice sheet melting in the southeast Pacific. *Geophysical Research Letters*, **23**, 957–960.

JACOBS, S., JENKINS, A., GIULIVI, C. F. & DUTIRIEUX, P. 2011. Stronger ocean circulation and increased melting under Pine Island Glacier ice shelf. *Nature Geoscience*, **4**, 519–523.

JENKINS, A. P., JACOBS, D. S. S. *ET AL.* 2010. Observations beneath Pine Island Glacier in West Antarctica and implications for its retreat. *Nature Geoscience*, **3**, 468–472.

JOUGHIN, I. & ALLEY, R. B. 2011. Stability of the West Antarctic ice sheet in a warming world. *Nature Geoscience*, **4**, 506–513.

JOUGHIN, I. & TULACZYK, S. 2002. Positive mass balance of the Ross Ice Streams, West Antarctica. *Science*, **295**, 476–480.

KELLOGG, T., KELLOGG, D. & STUIVER, M. 1990. Late Quaternary history of the southwestern Ross Sea: evidence from debris bands on the McMurdo Ice Shelf. *Antarctic Research Series*, **50**, 25–56.

KIENAST, M., HANEBUTH, T., PELEJERO, C. & STEINKE, S. 2003. Synchroneity of meltwater pulse 1a and the Bølling warming: new evidence from the South China Sea. *Geology*, **31**, 67–70.

KOFFMAN, T., HALL, B. & DENTON, G. 2011. New radiocarbon dates from glacial deposits in Miers Valley constrain the past behavior of the Antarctic Ice Sheet. *18th Annual West Antarctic Ice Sheet Meeting*, Loveland, CO, http://www.waisworkshop.org

LICHT, K. & ANDREWS, J. T. 2002. A ^{14}C record of late Pleistocene ice advance and retreat in the central Ross Sea. *Arctic, Antarctic, and Alpine Research*, **34**, 324–333.

LICHT, K., JENNINGS, A. E., ANDREWS, J. T. & WILLIAMS, K. 1996. Chronology of late Wisconsin ice retreat from the western Ross Sea, Antarctica. *Geology*, **24**, 223–226.

LICHT, K. J., LEDERER, J. R. & SWOPE, R. J. 2005. Provenance of LGM glacial till (sand fraction) across the Ross Embayment, Antarctica. *Quaternary Science Reviews*, **24**, 1499–1520.

MACAYEAL, D. 1992. Irregular oscillations of the West Antarctic Ice Sheet. *Nature*, **359**, 29–32.

MACKINTOSH, A., GOLLEDGE, N. *ET AL.* 2011. Retreat of the East Antarctic ice sheet during the last glacial termination. *Nature Geoscience*, doi: 10.1038/NGEOO1061

MARTIN, C., HINDMARSH, R. & NAVARRO, F. 2006. Dating ice flow change near the flow divide at Roosevelt Island using a thermomechanical model to predict radar stratigraphy. *Journal of Geophysical Research*, **111**, doi: 10.1029/2005JF00026.

MERCER, J. 1968. Glacial geology of the Reedy Glacier area, Antarctica. *Geological Society of America Bulletin*, **79**, 471–486.

MERCER, J. 1978. West Antarctic ice sheet and CO_2 greenhouse effect: a threat of disaster. *Nature*, **271**, 321–325.

MIX, A. C., BARD, E. & SCHNEIDER, R. 2001. Environmental Processes of the Ice age: Land, Oceans, Glaciers (EPILOG). *Quaternary Science Reviews*, **20**, 627–657.

OROMBELLI, G., BARONI, C. & DENTON, G. 1990. Late Cenozoic glacial history of the Terra Nova Bay region, Northern Victoria Land, Antarctica. *Geografia Fisica et Dinamica Quaternaria*, **13**, 139–163.

PARIZEK, B. & ALLEY, R. 2004. Ice thickness and isostatic imbalances in the Ross Embayment, West Antarctica: model results. *Global and Planetary Change*, **42**, 265–278.

PARIZEK, R., ALLEY, R. & HULBE, C. 2003. Subglacial thermal balance permits ongoing grounding-line retreat along the Siple Coast of West Antarctica. *Annals of Glaciology*, **36**, 251–256.

PÉWÉ, T. 1960. Multiple glaciations in the McMurdo Sound region, Antarctica – a progress report. *Journal of Geology*, **68**, 498–514.

POLLARD, D. & DECONTO, R. 2009. Modelling West Antarctic ice sheet growth and collapse through the past 5 million years. *Nature*, **458**, 329–333.

PRICE, S. F., CONWAY, H. & WADDINGTON, E. 2007. Evidence for late Pleistocene thinning of Siple Dome, West Antarctica. *Journal of Geophysical Research*, **112**, F03021, doi: 10.1029/2006JF000725

REIMER, P. J., BAILLIE, M. G. L. *ET AL.* 2009. INTCAL09 and Marine09 radiocarbon age calibration curves, 0–50 000 years cal BP. *Radiocarbon*, **51**, 1111–1150.

SCOTT, R. F. 1905. *The Voyage of the Discovery 2*. Macmillan and Company, London.

SHIPP, S., ANDERSON, J. & DOMACK, E. 1999. Late Pleistocene–Holocene retreat of the West Antarctic Ice Sheet system in the Ross Sea: Part 1 – geophysical results. *Geological Society of America Bulletin*, **111**, 1486–1516.

STONE, J. O. *ET AL.* 2003. Holocene deglaciation of Marie Byrd Land, West Antarctica. *Science*, **299**, 99–102.

STONE, J. O., CONWAY, H., COWDERY, S., HALL, B. & BROMLEY, G. 2009. Deglaciation of the Ross Sea: the view from Scott and Beardmore Glaciers. *16th Annual West Antarctic Ice Sheet Meeting*, Eatonville, WA; http://www.waisworkshop.org

STOREY, B. C., FINK, D., HOOD, D., JOY, K., SHULMEISTER, J., RIGER-KUSK, M. I. & STEVENS, M. 2010. Cosmogenic nuclide exposure age constraints on the glacial history of the Lake Wellman area, Darwin Mountains, Antarctica. *Antarctic Science*, **22**, 603–618.

STUIVER, M., DENTON, G., HUGHES, T. & FASTOOK, J. 1981. History of the marine ice sheet in West Antarctica during the last glaciation. *In*: DENTON, G. & HUGHES, T. (eds) *The Last Great Ice Sheets*. Wiley Interscience, New York, 319–436.

THOMA, M., JENKINS, A., HOLLAND, D. & JACOBS, S. 2008. Modelling Circumpolar Deep Water intrusions on the Amundsen Sea continental shelf, Antarctica. *Geophysical Research Letters*, **35**, doi: 10.1029/2008GL034939

TODD, C., STONE, J. O. H., CONWAY, H., HALL, B. & BROMLEY, G. 2010. Late Quaternary evolution of Reedy Glacier, Antarctica. *Quaternary Science Reviews*, **29**, 1328–1341.

WADDINGTON, E. D., CONWAY, H. *ET AL.* 2005. Decoding the dipstick: thickness of Siple Dome, West Antarctica, at the last glacial maximum. *Geology*, **33**, 281–284.

WEAVER, A. J., SAENKO, O. A., CLARK, P. U. & MITROVICA, J. 2003. Meltwater Pulse 1A from Antarctica as a trigger of the Bølling–Allerød warm interval. *Science*, **299**, 1709–1713.

WEERTMAN, J. 1976. Glaciology's grand unsolved problem. *Nature*, **260**, 284–286.

WELLNER, J. S., LOWE, A., SHIPP, S. & ANDERSON, J. 2001. Distribution of glacial geomorphic features on the Antarctic continental shelf and correlation with substrate: implications for ice behavior. *Journal of Glaciology*, **47**, 397–411.

WHITEHOUSE, P. L., BENTLEY, M. J. & LE BROCQ, A. M. 2012. A deglacial model for Antarctica: geoogical constraints and glaciological modelling as a basis for a new model of Antarctic isostatic adjustment. *Quaternary Science Reviews*, **32**, 1–24.

Late Eocene Glaciofluvial to Glaciomarine transition in the Lambert Graben: constraints from lithofacies and mineralogy of ODP Site 1166 sediments, Prydz Bay, Antarctica

K. STRAND[1]*, J. KÖYKKÄ[1] & J. LAMMINEN[2]

[1]*Thule Institute, University of Oulu, PO Box 7300, FIN-90014, Finland*

[2]*Boliden Mineral AB, Klarabergsviadukten 90, PO Box 44, SE-101 20 Stockholm, Sweden*

**Corresponding author (e-mail: kari.strand@oulu.fi)*

Abstract: The transition from a glaciofluvial to a glacial sedimentary environment contains critical information for evaluating the onset of glaciation and early glacial dynamics in Antarctica. This study presents investigations of a sediment core from Ocean Drilling Program Leg 188, Site 1166, on the continental shelf in Prydz Bay, in order to provide evidence for the first occurrence of late Eocene glaciers in the Lambert Graben, East Antarctica. The Lambert Graben is a large fault-bounded structure, at least 700 km long and 100 km wide, and presently hosts the Lambert Glacier–Amery Ice Shelf system. The core consists of middle Eocene glacially influenced alluvial plain sediments, *c.* 110 m in total thickness that are overlain by a *c.* 20 m-thick early Oligocene glaciomarine succession. The upper contact is a major unconformity. Grain-surface microtextures, such as fractured plates and a relatively high content of ilmenite in alluvial outwash-plain sediments, indicate mixed northern and southern Prince Charles Mountains provenance and initiation of glaciers as early as middle Eocene time. Distribution of the major heavy minerals indicates a change to garnet-rich granulite terrain in sediment-supply during later ice-sheet evolution, when a lack of detritus from the southern Prince Charles Mountains becomes apparent. Glacier advance was no longer restricted to the Lambert Graben and the drainage area became wider. Good preservation of these low-stand glaciofluvial outwash-plain deposits was related to regional subsidence that progressed to a marine transgression and glaciomarine deposition when ice was able to advance to the continental shelf edge.

The Antarctic Ice Sheet is the largest continental ice sheet on Earth and plays a fundamental role in the global climate system and sea-level change. There are presently only a few sedimentary records of the earliest evidence of glacial activity in East Antarctica. The Lambert Glacier–Amery Ice Shelf drainage system drains into Prydz Bay from the south, and this system is believed to have the longest-lived record for the Cenozoic Antarctic Ice Sheet (Hambrey 1991). Thus the oldest Prydz Bay sediments, obtained from Site 1166 of the Ocean Drilling Program (ODP) Leg 188, are in a key position for finding evidence of initiation and earliest glacial advances in the region (Fig. 1a). With its long history of sedimentation, Prydz Bay is regarded as one of the best locations to investigate the long-term record of Antarctic glaciation (Hambrey *et al.* 1991; Barker *et al.* 1998), and that is why reconstruction of the onset of the East Antarctic Ice Sheet was one of the key objectives of Leg 188. Leg 119 had previously drilled a transect of holes in Prydz Bay (Barron *et al.* 1991), but these holes did not penetrate fully into and through the early glacial sections to reach preglacial Cenozoic rocks, as Leg 188 did at Site 1166. Leg 119, Site 742 *c.* 40 km NE of Site 1166, however, provided evidence of the existence of late–middle and late Eocene glacial sediments in this region (Barron *et al.* 1991). In that respect, sedimentological and mineralogical study of glacial conditions at Site 1166 from middle Eocene to early Oligocene time (Florindo *et al.* 2003) is an important extension to our understanding of East Antarctic glacial history via the drainage system in the Lambert Graben. Here we present new lithofacies and sediment provenance descriptions and interpretations, which together with previously published sand-grain microtexture information by Strand *et al.* (2003) contribute to our understanding of the arrival of glaciers at the coast via the Lambert Graben.

Geology of the research area

Prydz Bay is in the seaward terminus of the Mesozoic or older graben structure, the Lambert Graben, which hosts the Lambert Glacier–Amery Ice Shelf system, the largest single glacier in Antarctica (Fig. 1a). This system drains presently nearly 20% of the total East Antarctic ice volume (cf.

From: HAMBREY, M. J., BARKER, P. F., BARRETT, P. J., BOWMAN, V., DAVIES, B., SMELLIE, J. L. & TRANTER, M. (eds) 2013. *Antarctic Palaeoenvironments and Earth-Surface Processes*. Geological Society, London, Special Publications, **381**, 183–197. First published online July 30, 2013, http://dx.doi.org/10.1144/SP381.24

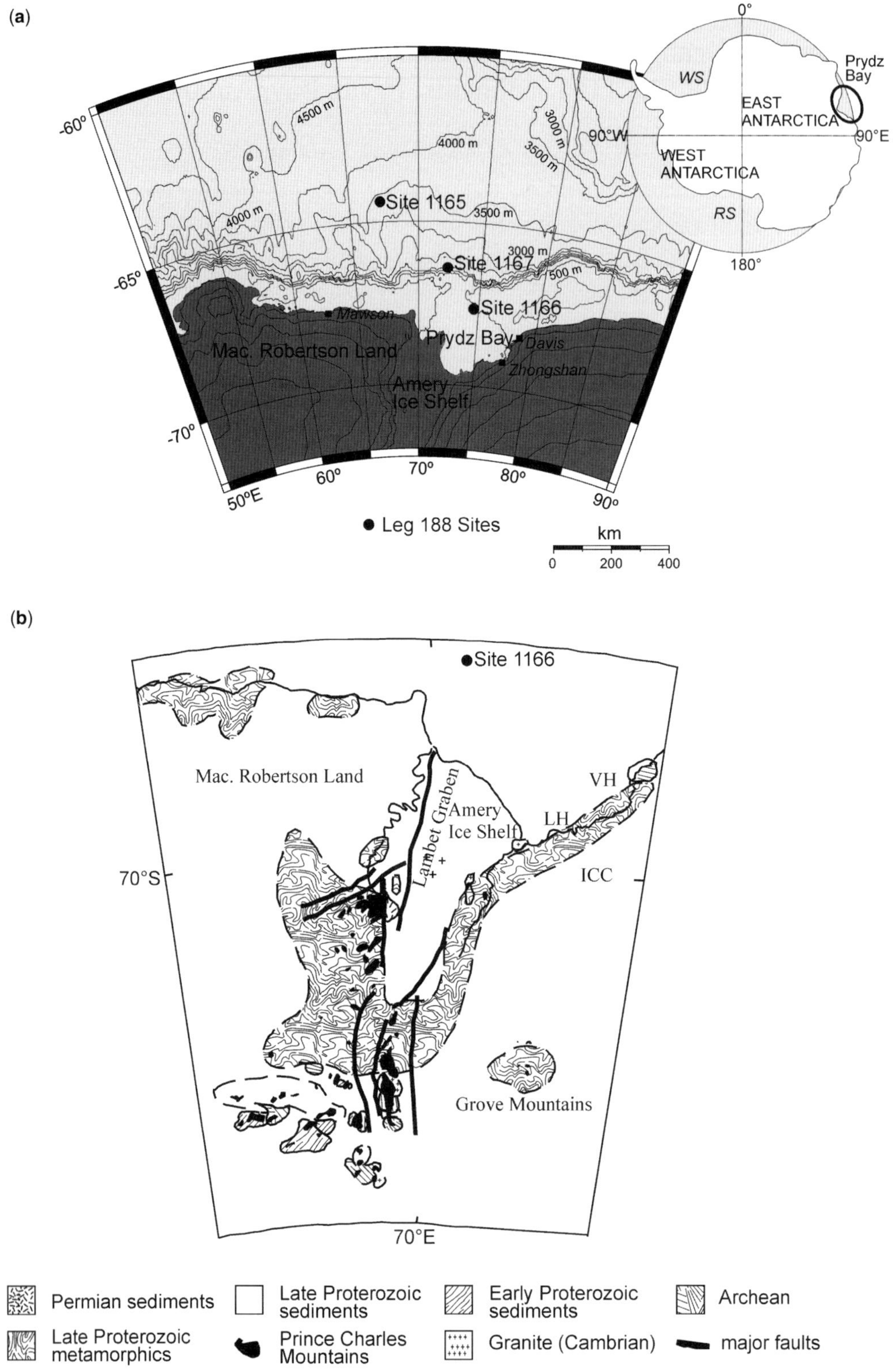
(a)
0°
Prydz Bay
WS
EAST ANTARCTICA
90°W
90°E
WEST ANTARCTICA
RS
180°
-60°
-65°
-70°
4500 m
4000 m
3000 m
3500 m
Site 1165
Site 1167
Site 1166
500 m
Mawson
Mac. Robertson Land
Prydz Bay
Davis
Zhongshan
Amery Ice Shelf
50°E
60°
70°
80°
90°
● Leg 188 Sites
km
0 200 400
(b)
Site 1166
Mac. Robertson Land
Lambert Graben
Amery Ice Shelf
VH
LH
ICC
70°S
Grove Mountains
70°E
Permian sediments
Late Proterozoic sediments
Early Proterozoic sediments
Archean
Late Proterozoic metamorphics
Prince Charles Mountains
Granite (Cambrian)
major faults

Fricker *et al.* 2000; Passchier *et al.* 2003; Forsberg *et al.* 2008). The fault-bounded Lambert Graben is at least 700 km long and 100 km wide and is located between Princess Elizabeth Land and MacRobertson Land. The continental shelf slopes landwards from water depths of 200–400 m near the shelf break to 800–1000 m in the inner-shelf deeps (Passchier *et al.* 2003). The Prydz Bay area is a depositional centre for sediments supplied by the Lambert Glacier–Amery Ice Shelf system. The Prydz Bay sedimentary basin contains as much as 12 km of sediment (Cooper & O'Brien 2004), and is underlain mostly by Precambrian high-grade metamorphic and intrusive rock, as well as Palaeozoic granite (e.g. Mikhalsky *et al.* 2001; Veevers *et al.* 2008; Fig. 1b). Based on U–Pb and Hf analysis of detrital zircons (Veevers *et al.* 2008), the Prince Charles Mountains, Grove Mountains and the entirely subglacial Gamburtsev Mountains form the main geological components for Site 1166 sediment provenance. The northern Prince Charles Mountains comprise primarily Mesoproterozoic to Neoproterozoic granulite and upper amphibolite facies metamorphic rock, intruded by charnockite and granite plutons, and felsic and alkaline dykes of the Beaver-Lambert Terranes (Mikhalsky *et al.* 2001). The down-faulted block of Permo-Triassic sandstone and coal measures form the Amery Group, which in turn are overlain by early Miocene–Quaternary Pagodroma Group comprising glacigenic sediment (Hambrey & McKelvey 2000*a*, *b*). The southern Prince Charles Mountains mainly comprise a variety of Archaean to Palaeoproterozoic metasedimentary and metavolcanic rock of the Ruker Terrane that were metamorphosed under greenschist to amphibolite facies conditions (Mikhalsky *et al.* 2001). A narrow branch of Mesoproterozoic to Neoproteorozoic high-grade gneiss and mafic-ultramafic rock of the Prydz complex reaches the NE from the Mawson Escarpment, east of Amery Ice Shelf all the way to the Larsemann Hills (Veever *et al.* 2008). The Vestfold Hills in the northeastern-most corner consist of late Archaean granulites. The Grove Mountains, which cover an area of *c.* 3200 km^2 in Princess Elizabeth Land, comprise numerous nunataks that are composed mainly of upper amphibolite to granulite facies metamorphic rock, granite and pegmatite (Kong *et al.* 2010). The Gamburtsev Subglacial Mountains form a major mountain range in central Antarctica. Sm–Nd isotope studies show no signs of young volcanism in the region, rendering a hot-spot-related origin for the mountain belt unlikely, although several other tectonic processes of mountain building have been proposed (van de Flierdt *et al.* 2008).

Lithostratigraphy

During Leg 188 three transect sites (1165, 1166 and 1167) on the Prydz Bay continental margin were drilled in order to sample the geological record left by the Lambert Glacier–Amery Ice Shelf drainage system (Fig. 1a). Site 1166 is located on the southwestern flank of Four Ladies Bank, *c.* 150 km east of Prydz Channel in the zone of iceberg-scoured seabed, in a water depth of 475.4 m (Fig. 1a). Hole 1166A (67°41.77′S, 74°47.22′E) penetrated to a maximum depth of 381.3 m below seafloor (mbsf), and consists of a succession of poorly sorted sandstone, mudstone and diamicton of Cretaceous to Holocene age (O'Brien *et al.* 2001; Cooper & O'Brian 2004). The chronology of the succession is based on diatom biostratigraphy and pollen zonations (Macphail & Truswell 2004). Site 1166 sediments are divided into five lithostratigraphic units, numbered from top to bottom (Fig. 2).

Unit I (0.0–135.63 mbsf) is of late Pliocene to Holocene age, and is dominated by dark grey clast-rich structureless diamicton with minor diatom-bearing muddy intervals (O'Brien *et al.* 2001; Passchier *et al.* 2003). A significant unconformity, spanning *c.* 30 Ma, separates the upper Neogene (Unit I) to Holocene sequence from the underlying Cretaceous to lower Oligocene sediment (Units II–V).

Unit II (135.63–156.62 mbsf) is a diatom-bearing claystone with thin interbedded sand, containing lonestones that record ice-rafting of pebbles (O'Brien *et al.* 2001). The bottom of Unit II has rhythmically interbedded centimetre-thick sand and dark claystone. The contact between Units II and III is abrupt, and probably represents an erosion surface.

Unit III (156.62–267.17 mbsf) consists predominantly of massive poorly sorted and deformed sand with a silty clay matrix. A middle–late Eocene (39–34 Ma) age is indicated by pollen of the middle *Nothofagidites asperus* palynozone (Macphail &

Fig. 1. (**a**) Map of East Antarctic coastline between 50°E and 90°E, showing the location of Prydz Bay and Leg 188 drill sites (modified from O'Brien *et al.* 2001; Strand *et al.* 2003). Insert: *RS*, Ross Sea; *WS*, Weddell Sea. (**b**) Pre-Mesozoic geology of the Lambert Glacier drainage basin and Prydz Bay region after Tingey (1982) and Federov *et al.* (1982) (modified from O'Brien *et al.* 2001). The Prydz Bay Basin is underlain mostly by Precambrian high-grade metamorphic and intrusive rock and some Palaeozoic granite. ICC, Ingrid Christensen Coast; LH, Larseman Hills; VH, Vestfjold Hills.

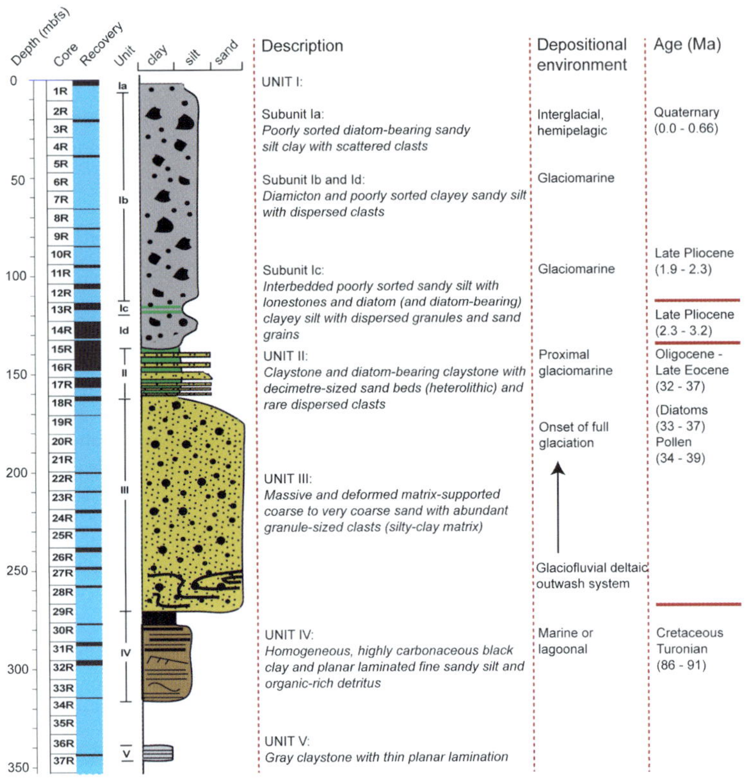

Fig. 2. Lithostratigraphy of Site 1166, showing principal units, description and depositional environments (modified from O'Brien *et al.* 2001); mbsf, metres below seafloor.

Truswell 2004). Evidence for nearby glaciation in the rift-flank mountains to the south is seen in the surface features of grains in Unit III sand (Strand *et al.* 2003). The lower part of Unit III is deformed and folded by soft-sediment deformation of sandy beds. It contains black organic-rich material, including pieces of wood. The contact between Units III and IV is an unconformity (Cooper & O'Brian 2004). Seismic data shows that this unconformity has almost no relief and only occasional displays truncation, but nonetheless it represents a break of at least 47 myr as palynomorphs indicate a Late Cretaceous (Turonian) age (86–91 Ma) for the sediment in Units IV and V (Macphail & Truswell 2004).

Unit IV (276.44–314.91 mbsf) is composed mainly of black carbonaceous clay and fine sandy silt with organic-rich laminae and some bioturbation (O'Brien *et al.* 2001). This unit is interpreted as representing deposition in a restricted marine or lagoonal environment bordered by local austral coniferous woodlands.

Unit V (342.80–342.96 mbsf) consist mainly of finely laminated grey claystone of an alluvial plain system. There is no evidence of Cretaceous glaciation from Leg 188 cores according to Cooper & O'Brian (2004).

Lithofacies characteristics

This study focuses on the late Eocene to Oligocene successions (Units II and III) of ODP Site 1166. In this study lithofacies and their assemblages as well as lithological properties were newly characterized based on visual core description by Shipboard Scientific Party (2001). On the basis of lithology, grain size and sedimentary structures, three major lithofacies are recognized from Unit II and three from Unit III (Table 1). The facies codes, used in this study, are modified from Miall (1977, 1978).

Unit II

Description of lithofacies Sp/Sl (planar and low-angle cross-bedded sandstone). Lithofacies Sp/Sl is common throughout the succession of Unit II

Table 1. *Units II and III main lithofacies classification, associated structures and depositional process interpretation*

Facies abbreviation	Lithofacies	Sedimentary structures	Depositional process/interpretation
Unit II			
Sp/Sl	Fine- to medium-grained sandstone; contains dispersed rock clasts (<8 mm)	Planar and low-angle cross-bedding	Straight- or slightly sinuous-crested low-flow regime sandbars with iceberg-drafted dropstones
Sr	Fine- to medium-grained sandstone	Rippled	Migrating 2-D ripples; waning stage of the current flow or oscillatory waves
Fl/Fr	Mudstone (claystone–siltstone); contains dispersed rock clasts (<8 mm)	Laminated and rippled	Waning stream flow and low-energy suspension deposits with iceberg drafted dropstones; migrating 2-D ripples; possible abandonment and vertical aggradation
Unit III			
Sd	Medium- to coarse-grained sandstone; contains pebbles	Soft sediment deformation	Post-depositional gravitational slumping; glaciotectonic deformation
Sm	Coarse-grained sandstone; contains pebbles	Massive, faint lamination, matrix-rich	Cohensionless grain flows or debris flows; rapid sedimentation
Fl	Mudstone (claystone, siltstone)	Weak lamination	Waning stream flow and low-energy suspension deposits

(Fig. 2). Planar and low-angle cross-bedded sandstone beds generally range from approximately a few millimetres to 2 cm thick. Basal bed-contacts are distinct, and partly erosional with the underlying beds (Fig. 3a, b). The exact nature of the foresets is difficult to establish, but some of them seems to be shallow, dipping <15°. The grain size of the sandstone beds ranges from medium- to fine-grained. The grains are typically subangular to rounded, and they are generally normally graded, although some poorly sorted beds also occur. Beds contain angular to subangular dispersed lithic clasts, which disturb the bedding (Fig. 3b). The clast lithology is variable and includes gneiss, quartz and other dark mafic fragments, and their size varies from a few millimetres up to 8 mm. Colour of the silica-rich matrix varies from pale greyish to dark greyish. Moderate bioturbation in zones a few centimetres thick, rare fibrous organic clasts and horizontal burrow structures also characterize this sand lithofacies.

Interpretation. Similar planar and low-angle cross-bedded sandstone beds are described from many other shallow marine settings (e.g. Plint 1983; Mueller *et al.* 2002; Simpson *et al.* 2002; Köykkä & Lamminen 2011). This lithofacies is interpreted as representing lateral migration of low-flow regime sandbars that are straight- or slightly sinuous-crested. Planar migrating sand-wave foresets were formed when sand was carried up the seaward slope of the ridge by swash, and deposited by avalanching down the steep slope into the standing wave of the rumen. The low-angle features of some of the sandstone foresets indicate a flow regime transition from dune to upper plane-bed stability conditions, or vice versa (e.g. Fielding *et al.* 2006). Alternatively, normally graded bedding could also indicate subaqueous deposition from turbidity currents (Allen 1991). However, occasional burrow structures are likely to indicate stability of the sedimentary environment. Dispersed outsized rock clasts, which disturb the bedding, can be interpreted as iceberg-rafted dropstones, features that have been described from many glaciated settings (e.g. Lønne 1995; Glasser & Hambrey 2002; Jones & Fielding 2008). The process of dropping was related to the melting out of debris from the base of an iceberg, which resulted in the fallout of single or aggregated clasts.

Description of lithofacies Sr (ripple-marked sandstone). Lithofacies Sr is a more common in the upper part of the Unit II (Fig. 2). It occurs as small individual interbeds with the other sandstone and mudstone beds. Ripple-marked sandstone beds are <2.0 cm thick. Basal bed-contacts are mostly gradational with the underlying beds, and grain size varies from medium- to fine-grained. The grains are typically subangular to rounded. In this lithofacies, beds are generally normally graded and they do not contain any visible clasts or pebbles.

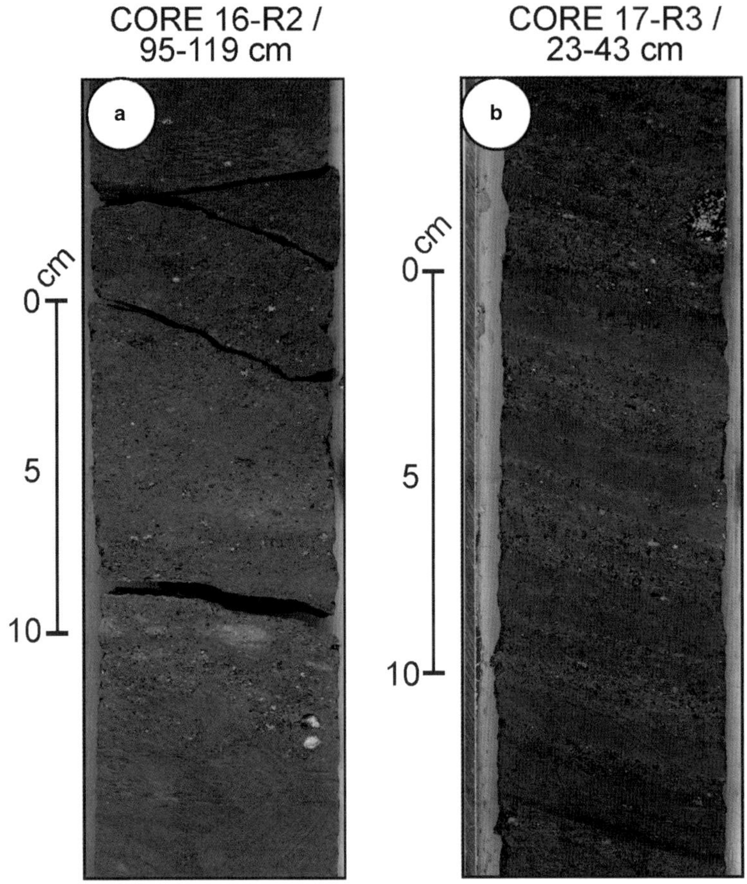

Fig. 3. (**a**) Graded bedded mudstone and sandstone from the upper part of Unit II (Core 16-R2, 95–119 cm) with a few outsized clasts (lithofacies Sp/Sl and Fp/Fr). (**b**) Interbedded mudstone and sandstone with outsized clast (lithofacies Sp/Sl and Fp/Fr) from the lower part of Unit II (core 17-R3, 23–43 cm).

Interpretation. Lithofacies Sr is interpreted as representing a lower flow regime of migrating 2-D ripples, deposited during the waning stage of current flow or by oscillatory waves in shallow water (e.g. Köykkä 2011*a*). The normal grading of the beds is a result of waning flow in a lower flow regime, with sufficiently low concentrations to permit grain-size sorting.

Description of lithofacies Fl/Fr (laminated and rippled mudstone). Lithofacies Fl/Fr is a dominant feature of Unit II (Fig. 2). Planar-laminated and rippled mudstone beds are commonly a few millimetres thick (<2 cm) and contain either silt- or clay-sized detritus. The upper part of Unit II is characterized by dark greenish grey clay-sized detritus with diatoms and some shell fragments, whereas dark grey clay-sized beds become more common in the lower part of the unit. Bed contacts are gradational with the underlying sandstone beds. Beds contain similar angular to subangular dispersed lithic clasts to those observed from lithofacies Sp/Sl. Lithic clasts are similarly gneiss, quartz and other dark mafic fragments, and they disturb the bedding. Grain size of the clasts varies from a few to 8 mm. Moderate bioturbation is also a common feature in this lithofacies.

Interpretation. Lithofacies Fl/Fr is interpreted as representing waning current strength, which was weak enough to deposit mud and silt-sized detritus from the flow (e.g. Bridge & Gabel 1992). Occasional ripple marks represent migrating 2-D ripples deposited from the single current event as a waning stage of the flow within the lower flow regime. Graded bases to this lithofacies suggest a waning of the suspended load followed by more active episodes (Collinson *et al.* 2006; Köykkä &

Lamminen 2011). Based on sedimentary structures and lithofacies assemblages below and above, it is likely that lithofacies Fl/Fr represents abandonment and vertical aggradation of a tidal channel or coastal plain, suggesting low-energy suspension deposits (e.g. Dalrymble 1992). Similarly, outsized rock clasts can interpreted as iceberg-rafted dropstones, as in lithofacies Sp/Sl.

Unit III

Description of lithofacies Sd (deformed sandstone). Lithofacies Sd is a common feature of the lower part of Unit III (Figs 2 & 4). The exact thickness of the individual deformed sand beds cannot be well detected because of intense distortion. The beds range approximately from a few to 10 cm in thickness. Basal bed contacts, however, are distinct, erosional and sometimes curved (Fig. 4). The sand-grain size of the deformed sandstone beds ranges from medium- to coarse-grained. The grains are typically subangular to rounded, and are moderately to poorly sorted. The beds are generally pebbly, containing abundant angular to subrounded gneiss, quartz and other dark mafic fragments in a silica-rich matrix. The size of the pebbles varies from <1 to 5 cm. Diffuse colour-banding is common in all deformed beds, which are formed from greyish to dark greyish matrix and darker organic-rich mudstone layers.

Interpretation. Soft-sediment deformation structures may occur soon after deposition at, or close to, the contemporaneous surface, or associated with later burial and lithification (Collinson *et al.* 2006). These structures are usually related to gravitational instability or collapse as buried ice melts over time in periglacial or glacial settings, in more proximal parts of the basin (e.g. Arnaud 2008, 2011). The lack of any visible water-escape structures and disturbance affecting several beds indicate that this facies is likely to represent larger-scale post-depositional gravitational slumping or deformation as a result of sediment-loading, rather than small-scale syn-sedimentary loss of sediment yield-strength through increased pore-water pressure and liquefaction of sediments. Sandy beds, including some with black organic-rich material and small pieces of wood, remained unconsolidated and thus not so resistant to post-depositional plastic deformation. The diffuse colour-banding indicates former bedding. Rapid deposition could lead to gravitational instability and generation of these structures, which are common on progradational alluvial setting. The contorted and disturbed beds in this lithofacies are consistent with larger-scale deformation with the possibility of sedimentary slump-folding, as indicated by vertical orientation of the bedding (Collinson *et al.* 2006). The occurrence of abundant pebbles, moderate- to poor-sorting, erosional and curved basal bed contacts possibly suggests channel scouring and episodic flow that are likely to have been surge-type sheet-flows (see e.g. Köykkä 2011*a*).

Description of lithofacies Sm (massive sandstone). Lithofacies Sm is a more common in the upper part of the Unit III, although it occurs throughout the succession (Figs 2 & 4). Bed-thickness varies from *c.* 5 to 15 cm. The basal bed-contacts are either amalgamated and gradational, or reveal erosional contacts with the underlying deformed sandstone beds (lithofacies Sd). The grain size varies from medium- to coarse-grained, and the grains are typically subangular to rounded, and are moderately to poorly sorted. The beds are generally pebbly, and contain similar angular to subrounded gneiss, quartz and other dark mafic fragments to those in lithofacies Sd. The size of the pebbles varies from >1 to 5 cm. In some of the beds, pebbles are clearly concentrated in stratified layers within the lower part of the bed (Fig. 4) and show weak normal grading, although most of the beds are ungraded. The colour of the matrix varies from pale grey to dark grey.

Interpretation. Massive sand beds commonly reflect the depositional process itself or are the result of later destruction. Syn-sedimentary massive sandstone beds are thought to have been deposited by subaerial hyperconcentrated flows (Smith 1986) or by subaqueous high-density turbidity currents (Lowe 1982). In both situations, the mechanism is associated with freezing of a sandy flow and collapse fallout from, or continuous aggradation beneath, a high-density turbidity current (Kneller & Buckee 2000). The massive nature, poor sorting, abundant pebbles and mostly ungraded beds indicate that this lithofacies represents cohesionless sediment gravity flows where there is probably rapid fallout of sediment in distributary channels as mass flows entirely consisting of sand (see Björklund-Plink & Steel 2005). Weak normal grading in some of the beds suggests density currents and suspension-settling, or re-working and re-deposition of the previously deposited sands (e.g. Köykkä 2011*a*, *b*).

Description of lithofacies Fl (weakly laminated mudstone). Lithofacies Fl occurs mainly as interbeds within sandstones throughout Unit III, although it becomes more common in the upper part of the succession (Figs 2 & 4). The weakly laminated mudstone beds are only a few centimetres thick, and are characterized by alternating layers of silt- and clay-sized detritus, although

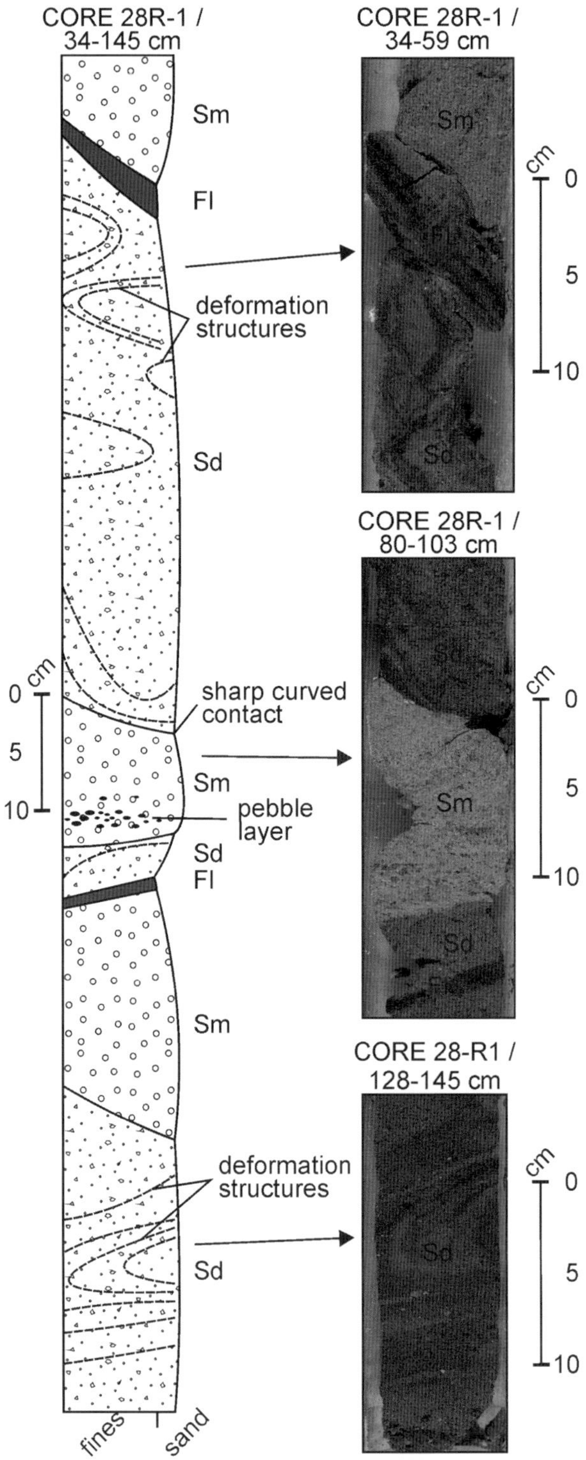

Fig. 4. Detailed lithostratigraphy of the lower part of Unit III (core 28R-1, 34–145 cm), showing the main lithofacies types (see Table 1 for lithofacies codes).

these are not always clearly visible. A darker, clay-sized detritus dominates the mudstone layers, whereas the lighter and coarser layers tend to be thinner. The basal bed contacts with the underlying sand beds generally vary from sharp to gradational. The silt and mud layers are a few millimetres thick. The layers are not arranged in any regular grouping, and they do not show any sign of ripple marks or other sedimentary structures.

Interpretation. The lack of ripple marks, climbing ripples or convolute bedding in the pale silty laminae suggests that these mud deposits do not represent a flood basin, crevasse splay or natural levée. Thus, lithofacies Fl is interpreted to represent waning current strength (e.g. Bridge & Gabel 1992; Köykkä, 2011*a*, *b*). The current was weak enough to deposit mud and silt-sized detritus from the flow. The sharp-based layers suggest relatively sudden events superimposed upon a background of constant sedimentation of fine-grained detritus, whereas the graded bases suggest a waning of the suspended load followed by more active episodes (Collinson *et al.* 2006; Köykkä & Lamminen 2011).

Sediment textural characteristics

Textural characteristics are based on previously published information on sand-grain microtexture by Strand *et al.* (2003); however, several new samples covering Units II and III were selected for updating the existing microtexture database and to determine grain-size distribution by using microphotographs (Fig. 5). Microtextural analysis of quartz sand grains was undertaken within the fraction between 0.250 and 0.6 mm, and from each sample at least 30 grains were analysed using a scanning electron microscope, type Jeol, JSM-6400 SEM with EDS or Jeol, JSM-6300 FESEM at 12 kV at the Centre of Microscopy and Nanotechnology at the University of Oulu. Grains in these samples were counted for frequencies of critical textures.

According to Strand *et al.* (2003), in Upper Pliocene–Holocene glaciomarine Unit I, 50–80% of the grains are angular, compared with only 20–60% of the grains in Units II and III. Grains with rounded outlines and edge- or surface-abrasion are present in the sand of Unit II (Fig. 5a) and Unit III (Fig. 5d, e). However, numerous grains with fractured plates, conchoidal fractures and edge-abrasion (Fig. 5g) and polished surfaces (Fig. 5h) are also present. The highest frequency of grains with rounded edges is present in middle–late Eocene alluvial outwash sands. However, fractured plates with concoidal fractures indicating glacier influence are also present. Rounded outlines and surface abrasion are compatible with an alluvial origin for

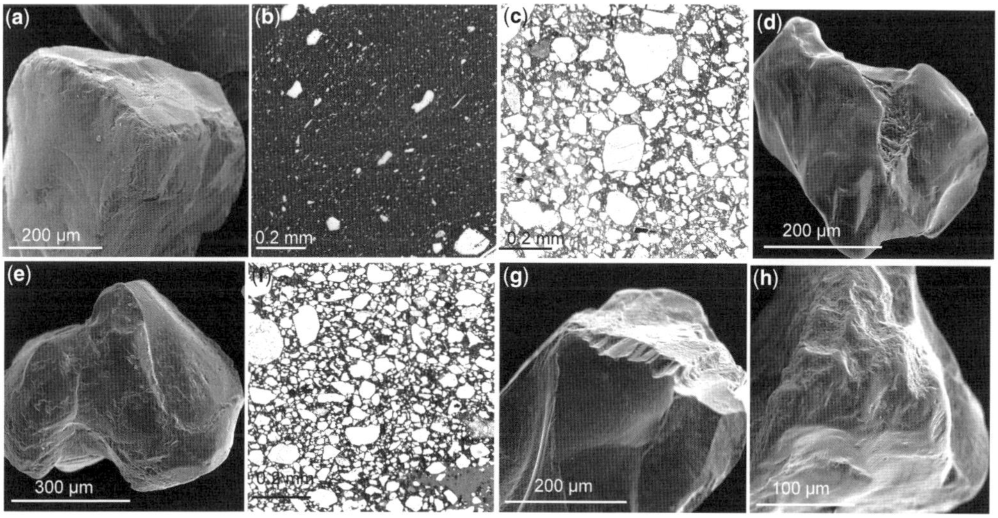

Fig. 5. Selected microphotographs of sediments and SEM micrographs of grains from Site 1166 (Units II and III) showing typical morphological and mechanical textures. (**a**) Rounded quartz grain with edge abrasion (Unit II, sample 1166A-17R-4, 39–42 cm). (**b**) Massive fine-grained sediment with dispersed sand clasts (Unit II, sample 1166A-17R-4, 39–42 cm). (**c**) Bipolar grain distribution in sand (Unit III, sample 1166A-22R, 12–13 cm). (**d**) Rounded outline (Unit III, sample 1166A-22R, 12–13 cm). (**e**) Rounded outline with surface-abrasion (Unit III, sample 1166A-22R, 12–13 cm). (**f**) Bipolar grain distribution in poorly sorted sand (Unit III, sample 1166A-26R-1, 52–54 cm). (**g**) Fractured plates, subparallel linear fractures and edge-abrasion (Unit III, sample 1166A-26R-1, 52–54 cm). (**h**) Rounded outline and polished surfaces (Unit III, sample 1166A-26R, 52–54 cm).

the sediments (Fig. 5h). Grain-size distribution is bimodal (Fig. 5c, f) and sediment was transported by traction currents. The results from microtextural analysis of quartz grains support the transition from a glaciofluvial to a glacimarine environment from middle Eocene to early Oligocene time. The presence of grains with features that were only developed in a subglacial environment suggests that the sedimentary system was also sourced by a glacier. No sequence comparable with the homogeneous coarse sand was recovered at Site 742, suggesting that the deposit is a relatively local phenomenon (cf. Barron *et al.* 1991; Hambrey *et al.* 1991). The coarse sand may record a glaciofluvial outwash plain system or a braided delta in an ice-distal setting. In braided-river deposits, a glacial connection could be demonstrated by identification of glacially generated textures on clasts (cf. Eyles & Eyles 2010). Unit II is a stratified marine sequence that records ice-rafting of pebbles and sand-sized material (Fig. 5b). The high abundance of grains with glacial textures suggests that the depositional environment was proximal to a glacial source. At Site 742 of Leg 119, 40 km eastward, similar or slightly younger sediments were interpreted as proximal glacially influenced proglacial or subglacial deposits (Barron *et al.* 1991), which is in agreement with the interpretation of Unit II presented here.

Heavy mineral distribution

The use of geochemistry of heavy minerals in provenance studies is a well-established technique (e.g. Morton 1991). To obtain the heavy mineral content in each sample, approximately 40–50 randomly selected grains in the 0.125–0.250 mm fraction are typically analysed with an electron probe microanalyser and frequencies counted. In this study, the 0.125 mm fraction was separated in heavy liquid (specific gravity = 2.8 g cm^{-3}). Heavy minerals were cast in epoxy and polished. A WDS-equipped electron microprobe, type Jeol JXA-8200, hosted by the Centre of Microscopy and Nanotechnology at the University of Oulu, was used in analysis. Analytical conditions included an acceleration voltage 15 kV, probe current 3.15 nA and the analyses were ZAF-corrected online. Fifteen elements were analysed. Heavy mineral suites do not always mirror their parent rocks, since they may be influenced by syn-sedimentary weathering and post-depositional alteration. In this study three resistant mineral groups were chosen for close chemical inspection: garnet, Fe-Ti-oxides and amphiboles.

The major unconformity between units I and II (Fig. 6) is seen in the heavy mineral suite. Magnetite and diopside appear while rutile largely disappears. Ilmenite exhibits a drastic drop in abundance while the growing trend of almandine garnet continues. The relatively unstable amphibole occurring even in the lower part of Unit III serves as an indicator of the limited effect of chemical weathering on mineral frequency. Based on garnet data (Fig. 6), the granulitic garnets are prevalent throughout the sedimentary column, but a significant increase in abundance is seen in the younger deposits. This is interpreted as the result of a widening of the drainage area caused by ice-sheet advance. The dominating metamorphic source for the sediment is clear, and the question of provenance is resolved by characterizing the specific metamorphic terrains in the study area and determining their relative contribution to the sedimentary detritus. The lowest grade rocks are those in the southern Prince Charles Mountains and Mawson Escarpment (Mikhalsky *et al.* 2001). These areas are mainly of greenschist to upper amphibolite facies. The area around Site 1166, the Beaver–Lambert Terrane and the Vestfold Hills, comprises felsic and mafic gneiss of granulite facies, with some Neoproteorozoic intrusive. It is reasonable to assume that the drainage patterns in pre-glacial, and even in the glaciofluvial period, were controlled by the Lambert Graben and smaller riftogenic structures. Since the Miocene–Quaternary Pagodroma Group contains abundant clasts derived from the Amery Group (McKelvey & Stephenson 1990), it is obvious that certain recycling of the Amery Group sediments may have occurred also in the glaciofluvial Eocene sequence of Unit III. Relatively stable minerals such as ilmenite and rutile are enriched in Unit III, and may reflect multiple sedimentary cycles.

Reconstruction of the Eocene–Oligocene depositional history in Lambert Graben

Many factors can affect accommodation space and energy regimes in glaciated sedimentary basins (Eyles & Eyles 2010). These factors are a combination of eustatic sea-level changes driven by growth and melting of ice sheets, and localized glacio-isostatic changes driven by ice-loading and unloading of the Earth's crust, and also regional tectonism. Facies analysis combined with sediment-mineralogical analysis may be used for verifying critical periods of cryospheric evolution, such as the transitions from preglacial to glacial conditions. The glacial signature of sand grains in the glaciofluvial outwash plain sands of Unit III indicates that glacier ice was present in the Prydz Bay area already in middle Eocene time. The Lambert Graben was actively subsiding and thus able to aggrade this sandy alluvial plain. Tectonic

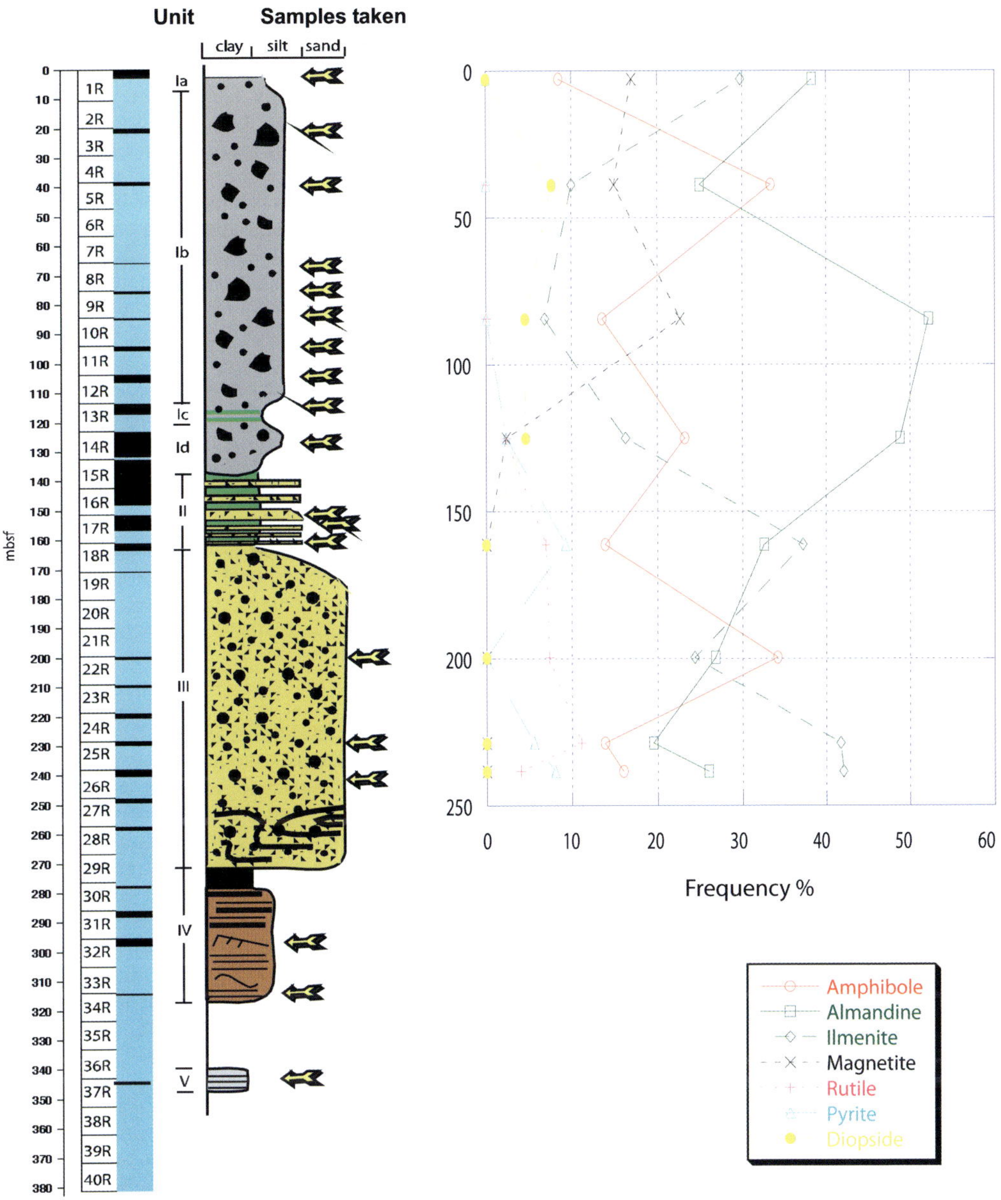

Fig. 6. Dominant heavy minerals and their distribution as frequency (%) at Site 1166 in the size fraction 0.125–0.250 mm.

subsidence was even more active latterly, allowing marine transgression over the alluvial plain and making accommodation space available for the deposition of marine sediment. This marine sediment was influenced by glacial processes, as recorded by dropstones in Unit II. The resulting glaciomarine sequence of Unit II signalled the arrival of ice sheet at sea-level, and the development of the Lambert Glacier–Amery Ice Shelf system, draining the East Antarctic ice-sheet approximately 37–33 Ma ago (e.g. Strand *et al.* 2003).

Sedimentary and textural characteristics of Unit III indicate that during late Eocene time sedimentation in the Lambert Graben was defined by the building of a glaciofluvial outwash-plain system, which was likely to have been characterized by well-developed braided distributary channels (Fig. 7). This is also supported by non-contemporaneous glaciotectonic soft-sediment deformation structures, and grain-flow and suspension deposits. Soft-sediment deformation with associated stratification features are typical of a promixal/mid-part of the outwash-plain system. Fine-grained outwash sediment was carried in a buoyant melt-water plume that rose steeply to the free-water surface and spread fine-grained sediment. Rhythmic vertical lithofacies changes in the textural and structural features of Unit III probably correspond to the characteristic pattern of outwash sedimentation with frequent shifts of both main and secondary channels. Deposition in an unconfined outwash-plain system is accompanied by frequent shifting of sedimentation in the main and secondary braided channels, with clear decrease of flow-energy from the proximal to distal part of the plain. Overall, the whole Unit III succession indicates waning of relatively strong, episodic glaciofluvial currents and resembles the fine-grained outwash-plain system. Based on heavy mineral analysis, this glaciofluvial outwash-plain system stage was characterized by mixed provenance sources from the northern and southern Prince Charles Mountains, and also from the Phanerozoic successions.

Sedimentary and textural features, dropstones and marine dinoflagellates in Unit II (Fig. 7) indicate shallow glaciomarine deposition under conditions of rising then stable sea-level, thus reflecting basin-subsidence during late Oligocene time (cf. Shipboard Scientific Party 2001; Macphail & Truswell 2004). Proximity to an ice margin is evident from the style of glaciomarine sedimentation. Depositional processes were related to a varying supply of sediment, with higher energy sand related to meltwater release. Fine-grained diatom-bearing sediments reflect background sedimentation into which icebergs delivered dropstones. Fine-grained sediments were released from plumes of sediment, probably related to seasonal melt cycles that can produce laminated assemblages (O'Cofaigh & Dowdeswell 2001). Fine-grained sand and mud/silt beds were likely to have been supplied from the buoyant water plume that rises from the submarine outlet of a subglacial or englacial channel to the sea surface (e.g. Powell 1990; Lønne 1995). In some glaciomarine settings, lamination can be related to tidal activity (Eyles & Eyles 2010). Heavy minerals indicate that this glacial stage was characterized by a prominent frozen basement in

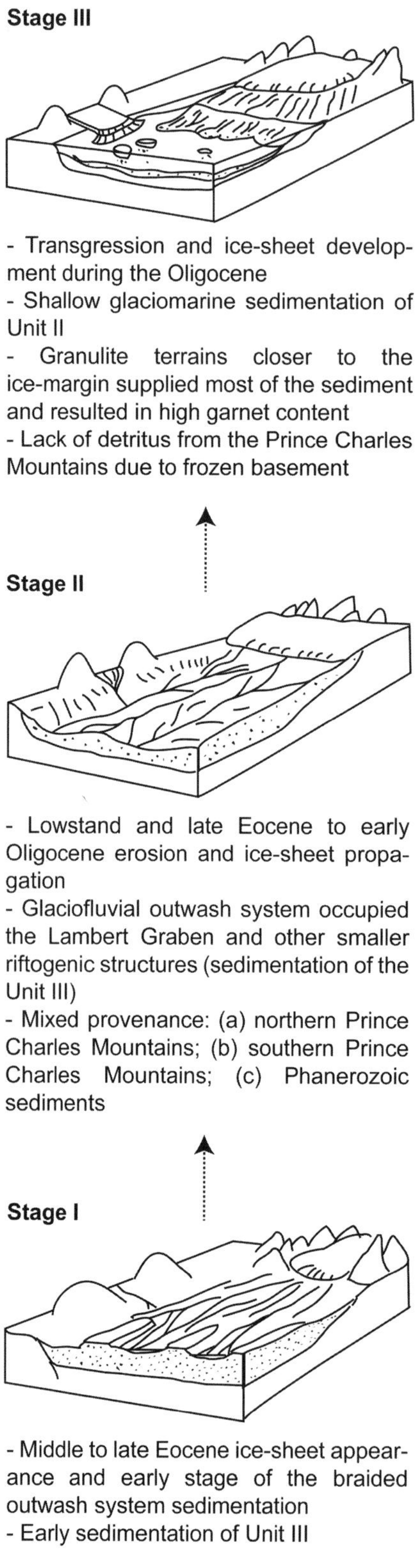

Fig. 7. Reconstruction of the Eocene–Oligocene history in Lambert Graben (modified from O'Brien *et al.* 2001; Strand *et al.* 2003). Sedimentation was characterized mainly by a periglacial outwash plain (Stages I and II), followed by a shallow glaciomarine setting (Stage III).

the southern Prince Charles Mountains, which resulted in an absence of detritus from this area. During that stage, granulite terrain closer to the ice-margin supplied most of the sediment and resulted in the high garnet content. The lack of any visible subaerial shrinkage cracks in the succession supports the concept that the Lambert Graben basin was submerged during entire sedimentation history of Unit II. Tectonic subsidence at this stage allowed the resulting high-stand conditions.

Conclusions

The grains from the middle–upper Eocene alluvial plain sands in the Lambert Graben, East Antarctica, show evidence of a glacier source. The upper Eocene/lower Oligocene diatomaceous claystones with interbedded sand and a few lonestones were also deposited during early glacial conditions, and signal the arrival of the ice sheet at sea-level in Prydz Bay.

The occurrence of a relatively thick unit of glaciofluvial outwash-plain sediment with soft-sediment deformational structures, followed by glaciomarine sediment, indicates that the glacier meltwaters and glaciotectonic basin evolution played an important role in the genesis of the sedimentary sequences in Prydz Bay throughout Palaeogene time. Because of high sediment input during the glaciofluvial stage, the redistribution of sediments caused by changes in base-level was generally low. The well-preserved nature of the low-stand glaciofluvial outwash-plain deposits was related to regional subsidence that progressed into a marine transgression and glaciomarine deposition when ice was able to advance to the continental shelf-edge. A tectonically controlled relative sea-level rise had an important effect on the stability of the ice sheet that occupied the Lambert Graben and the iceberg-calving episodes that occurred.

This paper was financially supported by the Research Council for Natural Sciences and Engineering of the Academy of Finland (project nos 49039 and 129010). We would like to thank the Ocean Drilling Program for providing samples for this study, and especially colleagues from the group of shipboard sedimentologists on Leg 188, as well as co-chief scientists P. O'Brien and A. Cooper for useful discussions regarding the Prydz Bay sediment interpretations. S. Passchier and M. Hambrey provided helpful and constructive reviews.

References

ALLEN, J. R. L. 1991. The Bouma Division A and the possible duration of turbidity currents. *Journal of Sedimentary Petrology*, **61**, 291–295.

ARNAUD, E. 2008. Deformation in the Neoproterozoic Smalfjord Formation, northern Norway: an indicator of glacial depositional conditions? *Sedimentology*, **55**, 335–356.

ARNAUD, E. 2011. The paleoclimatic significance of deformation structures in Neoproterozoic successions. *Sedimentary Geology*, **243–244**, 33–56, doi: 10.1016/j.sedgeo.2011.09.012

BARKER, P. F., BARRETT, P. J. ET AL. 1998. Ice sheet history from Antarctic continental margin sediments: the ANTOSTRAT approach. *Terra Antarctica*, **5**, 737–760.

BARRON, J., LARSEN, B. & BALDAUF, J. G. 1991. Evidence for Late Eocene to Early Oligocene Antarctic glaciation and observations on late Neogene glacial history of Antarctica: results from Leg 119. *In*: BARRON, J., LARSEN, B. ET AL. (eds) *Proceedings of the Ocean Drilling Program, Scientific Results*, **119**. Ocean Drilling Program, Texas A & M University, College Station, TX, 869–891.

BJÖRKLUND-PLINK, P. & STEEL, R. 2005. Deltas on falling-stage and lowstand shelf margins, the Eocene Central Basin of Spitsbergen: importance of sediment supply. *In*: GIOSAN, L. & BHATTACHARYA, J. P. (eds) *River Deltas – Concepts, Models, and Examples*. SEPM, Tulsa, OK, Special Publications, **83**, 179–207.

BRIDGE, J. S. & GABEL, S. L. 1992. Flow and sediment dynamics in a low sinuosity, braided river: Calamus River, Nebraska Sandhills. *Sedimentology*, **39**, 125–142.

COOPER, A. K. & O'BRIEN, P. E. 2004. Leg 188 synthesis: transitions in the glacial history of the Prydz Bay region, East Antarctica, from ODP drilling. *In*: COOPER, A. K., O'BRIEN, P. E. & RICHTER, C. (eds) *Proceedings of the Ocean Drilling Program, Scientific Results*, **188**. Ocean Drilling Program, Texas A & M University, College Station, TX, 1–42, doi: 10.2973/odp.proc.sr.188.001.2004

DALRYMPLE, R. W. 1992. Tidal depositional system. *In*: WALKER, R. G. & JAMES, N. P. (eds) *Facies Models: Response to Sea Level Change*. Geological Association of Canada, Ottawa, 196–218.

EYLES, C. H. & EYLES, N. 2010. Glacial deposits. *In*: JAMES, N. O. & DALRYMPLE, R. W. (eds) *Facies Models 4*. Geological Association of Canada, Ottawa, 73–105.

FEDOROV, L. V., GRIKUROV, G. E., KURININ, R. G. & MASOLOV, V. N. 1982. Crustal structure of the Lambert Glacier Area from geophysical data. *In*: CRADDOCK, C., LOVELESS, J. K., VIERIMA, T. L. & CRAWFORD, K. (eds) *Antarctic Geoscience*. University Wisconsin Press, Madison, 931–936.

FIELDING, C. R., BANN, K. L., MACEACHERN, J. A., TYE, S. C. & JONES, B. G. 2006. Cyclicity in the nearshore marine to coastal, Lower Permian, Pebbly Beach Formation, southern Sydney Basin, Australia: a record of relative sea-level fluctuations at the close the Late Palaeozoic Gondwanan ice age. *Sedimentology*, **53**, 435–463.

FLORINDO, F., BOHATY, S. M., ERWIN, P. S., RICHTER, C., ROBERTS, A. P., WHALEN, P. A. & WHITEHEAD, J. M. 2003. Magnetobiostratigraphic chronology and palaeoenvironmental history of Cenozoic sequences from ODP sites 1165 and 1166, Prydz Bay, Antarctica. *Palaeogeography, Palaeoclimatology, Palaeoecology*, **198**, 69–100.

FORSBERG, C. F., FLORINHO, F., GRUTZNER, J. & SOLHEIM, A. 2008. Sedimentation and aspects of glacial dynamics from physical properties, mineralogy and magnetic properties at ODP Sites 1166 and 1167, Prydz Bay, Antarctica. *Palaeogeography, Palaeoclimatology, Palaeoecology*, **260**, 184–201.

FRICKER, H. A., WARNER, R. C. & ALLISON, J. 2000. Mass balance of the Lambert Glacier-Amery Ice Shelf system, East Antarctica: a comparison of computed balance fluxes and measured fluxes. *Journal of Glaciology*, **46**, 561–570.

GLASSER, N. F. & HAMBREY, M. J. 2002. Sedimentary facies and landform genesis at a temperate outlet glacier: Soler Glacier, North Patagonia Icefield. *Sedimentology*, **49**, 43–64.

HAMBREY, M. J. 1991. Structure and dynamics of the Lambert Glacier-Amery Ice Shelf System: implications for the origin of Prydz Bay sediments. *In*: BARRON, J., LARSEN, B. ET AL. (eds) *Proceedings of the Ocean Drilling Program, Scientific Results*, **119**. Ocean Drilling Program, Texas A & M University, College Station, TX, 61–75.

HAMBREY, M. J. & MCKELVEY, B. 2000*a*. Major Neogene fluctuations of the East Antarctic ice sheet: stratigraphic evidence from the Lambert Glacier region. *Geology*, **28**, 865–960.

HAMBREY, M. J. & MCKELVEY, B. 2000*b*. Neogene sedimentation on the western margin of the Lambert Graben, East Antarctica. *Sedimentology*, **47**, 577–607.

HAMBREY, M. J., EHRMANN, W. U. & LARSEN, B. 1991. Cenozoic glacial record of the Prydz Bay continental shelf, East Antarctica. *In*: BARRON, J., LARSEN, B. ET AL. (eds) *Proceedings of the Ocean Drilling Program, Scientific Results*, **119**. Ocean Drilling Program, Texas A & M University, College Station, TX, 77–132.

JONES, A. T. & FIELDING, C. R. 2008. Sedimentary facies of a glacially influenced continental succession in the Pennsylvanian Jericho Formation, Galilee Basin, Australia. *Sedimentology*, **55**, 531–556.

KNELLER, B. & BUCKEE, C. 2000. The structure and fluid mechanics of turbidity currents: a review of some recent studies and their geological implications. *Sedimentology*, **47**, 62–94.

KONG, P., HUANG, F., FINK, D., DING, L. & LAI, Q. 2010. Late Miocene ice sheet elevation in the Grove Mountains, East Antarctica, inferred from cosmogenic ^{21}Ne–^{10}Be–^{26}Al. *Global and Planetary Change*, **72**, 50–54.

KÖYKKÄ, J. 2011*a*. Precambrian alluvial fan and braidplain sedimentation patterns: example from the Mesoproterozoic Rjukan Rift Basin, southern Norway. *Sedimentary Geology*, **234**, 89–108.

KÖYKKÄ, J. 2011*b*. The sedimentation and paleohydrology of the Mesoproterozoic stream deposits in a strike-slip basin (Svinsaga Formation), Telemark, southern Norway. *Sedimentary Geology*, **236**, 239–255.

KÖYKKÄ, J. & LAMMINEN, J. 2011. Tidally influenced clastic epeiric sea at a Mesoproterozoic continental margin, Rjukan Rift Basin, southern Norway. *Precambrian Research*, **185**, 164–182.

LØNNE, I. 1995. Sedimentary facies and depositional architecture of ice-contact glaciomarine systems. *Sedimentary Geology*, **98**, 13–43.

LOWE, R. D. 1982. Sediment gravity flows. II. Depositional models with special reference to the deposits of high-density turbidity currents. *Journal of Sedimentary Petrology*, **52**, 279–297.

MACPHAIL, M. K. & TRUSWELL, E. M. 2004. Palynology of Site 1166, Prydz Bay, East Antarctica. *In*: COOPER, A. K., O'BRIEN, P. E. & RICHTER, C. (eds) *Proceedings of the Ocean Drilling Program, Scientific Results*, **188**. Ocean Drilling Program, Texas A & M University, College Station, TX, 1–43, doi: 10.2973/odp.proc.sr.188.013.2004

MCKELVEY, B. C. & STEPHENSON, N. C. N. 1990. A geological reconnaissance of the Radok Lake Area, Amery Oasis, Prince Charles Mountains. *Antarctic Science*, **2**, 53–66.

MIALL, A. D. 1977. A review of the braided river depositional environment. *Earth Science Reviews*, **13**, 1–62.

MIALL, A. D. 1978. Lithofacies types and vertical profile models in braided river deposits: a summary. *In*: MIALL, A. D. (ed.) *Fluvial Sedimentology*. Canada Society of Petroleum Geology, Alberta, Memoirs, **5**, 597–604.

MIKHALSKY, E. V., SHERATON, J. W., LAIBA, A. A., TINGEY, R. J., THOST, D. E., KAMENEV, E. N. & FEDOROV, L. V. 2001. *Geology of the Prince Charles Mountains*. AGSO, Geoscience Australia Bulletins, **247**, 1–210.

MORTON, A. C. 1991. Geochemical studies of detrital heavy minerals and their application to provenance research. *In*: MORTON, A. C., TODD, S. P. & HAUGHTON, P. D. W. (eds) *Developments in Sedimentary Provenance Studies*. Geological Society, London, Special Publications, **57**, 31–45.

MUELLER, W. U., CORCORAN, P. L. & DONALDSON, J. A. 2002. Sedimentology of a tide- and wave-influenced high-energy Archean coastline: the Jackson Lake Formation, Slave Province, Canada. *In*: ALTERMANN, W. & CORCORAN, P. L. (eds) *Precambrian Sedimentary Environments – A Modern Approach to Ancient Depositional Systems*. IAS, Special Publications, **3**, 153–182.

O'BRIEN, P. E., COOPER, A. K., RICHTER, C. ET AL. 2001. *Proceedings of the Ocean Drilling Program, Initial Reports*, **188**. Ocean Drilling Program, Texas A&M University, College Station, TX [CD-ROM], http://www-odp.tamu.edu/publications/188_IR/188ir.htm

O'COFAIGH, C. & DOWDESWELL, J. A. 2001. Laminated sediments in glacimarine environments: diagnostic criteria for their interpretation. *Quaternary Science Reviews*, **20**, 1411–1436.

PASSCHIER, S., O'BRIEN, P. E., DAMUTH, J. E., JANUSZCZAK, N., HANDWERGER, D. A. & WHITEHEAD, J. M. 2003. Pliocene-Pleistocene glaciomarine sedimentation in eastern Prydz Bay and development of the Prydz trough-mouth fan, ODP Sites 1166 and 1167, East Antarctica. *Marine Geology*, **199**, 279–305.

PLINT, A. G. 1983. Facies, environments and sedimentary cycles in the Middle Eocene, Bracklesham Formation of the Hampshire Basin: evidence for global sea-level changes? *Sedimentology*, **30**, 625–653.

POWELL, R. D. 1990. Glacimarine processes at grounding-line fans and their growth to ice-contact deltas. *In*: DOWDESWELL, J. A. & SCOURSE, J. D. (eds) *Glacimarine Environments: Processes and Sediments*. Geological Society, London, Special Publications, **53**, 53–73.

SIMPSON, E. L., DILLIARD, K. A., ROWELL, B. F. & HIGGINS, D. 2002. The fluvial-to-marine transition within the post-rift Lower Cambrian Hardyston Formation, Eastern Pennysylvania, USA. *Sedimentary Geology*, **147**, 127–142.

SHIPBOARD SCIENTIFIC PARTY. 2001 Site 1666. *In*: O'BRIEN, P. E., COOPER, A. K., RICHTER, C. ET AL. (eds) 2001. *Proceedings of the Ocean Drilling Program, Initial Reports*, **188**. Ocean Drilling Program, Texas A & M University, College Station, TX, 1–110.

SMITH, G. A. 1986. Coarse-grained nonmarine volcaniclastic sediment: Terminology and depositional processes. *Geological Society of America Bulletin*, **97**, 1–10.

STRAND, K., PASSCHIER, S. & NÄSI, J. 2003. Implications of quartz grain microtextures for onset of Eocene/Oligocene glaciation in Prydz Bay, ODP Site 1166, Antarctica. *Palaeogeography, Palaeoclimatology, Palaeoecology*, **193**, 101–111.

TINGEY, R. J. 1982. The geologic evolution of the Prince Charles Mountains – an Antarctic Archean cratonic block. *In*: CRADDOCK, C. (ed.) *Antarctic Geoscience*. University Wisconsin Press, Madison, 455–464.

VAN DE FLIERDT, T., HEMMING, S. R., GOLDSTEIN, S. L., GEHLERS, G. E. & COX, S. E. 2008. Evidence against a young volcanic origin of the Gamburtsev Subglacial Mountains, Antarctica. *Geophysical Research Letters*, **35**, L21303, doi: 10.1029/2008GL035564

VEEVERS, J. J., SAEED, A. & O'BRIEN, P. E. 2008. Provenance of the Gamburtsev subglacial mountains from U-Pb and Hf analysis of detrital zircons in Cretaceous to Quaternary sediments in Prydz Bay and beneath the Amery Ice Shelf. *Sedimentary Geology*, **211**, 12–32.

The Offshore New Harbour Project: deciphering the Middle Miocene through Late Eocene seismic stratigraphy of Offshore New Harbour, western Ross Sea, Antarctica

STEPHEN F. PEKAR[1,2]*, MARVIN A. SPEECE[3], GARY S. WILSON[4], DAVID S. SUNWALL[3] & KIRSTY J. TINTO[2]

[1]*School of Earth and Environmental Sciences, Queens College, City University of New York, Flushing, NY 11367, USA*

[2]*Lamont Doherty Earth Observatory, Columbia University, Palisades, NY 10964, USA*

[3]*Geophysical Engineering Department, Montana Tech, Butte, MT 59701, USA*

[4]*Department of Marine Science, University of Otago, Dunedin, New Zealand*

**Corresponding author (e-mail: stephen.pekar@qc.cuny.edu)*

Abstract: In the Austral spring of 2008, over 48 km of multi-channel seismic data were collected offshore of New Harbour, which is located in the western most Ross Sea, Antarctica. This project is part of the ANDRILL (Antarctic Geological Drilling) Programme's Offshore New Harbour Project, with its aim to investigate the stratigraphic and tectonic history of the inner shelf of southern McMurdo Sound. Correlating the seismic data to CIROS-1 and ANDRILL AND-2A drill-holes provided age dates ranging from Late Miocene to at least Late Eocene for the reflectors and lithological descriptions for the seismic units. This permitted development of a glacial history of this area, resulting in dividing the seismic data into three units: an Upper Eocene–Lower Oligocene unit; an Upper Oligocene–Lower Miocene unit; and a Middle Miocene and younger unit. The seismic section below 700 mbsf has two possible interpretations: (1) substantial Eocene strata are present downdip of CIROS-1, which would represent a potential future-drilling objective for the ANDRILL Programme; or (2) these reflectors may be equivalent to Devonian strata recovered at the Cape Roberts 3 as the two-way traveltime of Eocene and Devonian strata are nearly identical, making it difficult to discriminate between them in seismic data.

One of the outstanding questions in geology relates to the nature of the evolution of the Antarctic cryosphere, from the mainly ice-free greenhouse world (>34 Ma) to the icehouse world of <34 Ma, in which the Antarctic continent is believed to have been dominated by large continent-sized ice sheets (e.g. Miller *et al.* 1991; Zachos *et al.* 2001; Miller *et al.* 2005). Our understanding of Antarctic cryosphere has been shaped mainly from distal records, such as from deep-sea isotopic and lower latitude shallow marine stratigraphic data (Miller *et al.* 1991, 1996, 1998; Zachos *et al.* 2001; Pekar *et al.* 2002; Lear *et al.* 2004; Pekar & Christie-Blick 2008; Katz *et al.* 2009). In comparison, far fewer records from the Antarctic continent and adjacent shelf exist to provide direct confirmation of ice-volume estimates from these distal records (e.g. Barrett 1989; Thomas *et al.* 1990; Hambrey *et al.* 1991; Cooper *et al.* 1995*a*, *b*; Cape Roberts Project, 1998, 1999, 2000; Naish *et al.* 2007). This has led to numerous conflicting interpretations of the cryospheric evolution in the Cenozoic, such as the timing of the first significant ice sheets on the Antarctic continent, ranging from Late Eocene (*c.* 37 Ma) (e.g. Zachos *et al.* 2001) to late Early Eocene (*c.* 51 Ma) (Miller *et al.* 1998; Pekar *et al.* 2005).

In contrast, proximal records obtained from outcrops, cores and seismic profiles provide a relatively consistent picture of Antarctic climate change, although thus far they are fragmented and sparse, mostly owing to a paucity of data (Barrett 1989; Thomas *et al.* 1990; Barron *et al.* 1991; Hambrey *et al.* 1991; Cooper *et al.* 1995*b*; Bartek *et al.* 1996; Cape Roberts Science Team 1998, 1999, 2000; Francis 2000; Naish *et al.* 2001; Raine & Askin 2001; Thorn 2001; Francis & Poole 2002; Chow & Bart 2003; Prebble *et al.* 2006). These records indicate a gradual cooling of the continent from warm temperate conditions in the Middle Eocene (Barrett 1989; Cape Roberts Science Team 1998, 1999, 2000; Raine & Askin 2001; Thorn 2001; Prebble *et al.* 2006), culminating in tundra-like conditions occurring during glacial minima by the Late Oligocene in the Ross Sea (Naish *et al.* 2001) and grounded ice extending beyond the

From: Hambrey, M. J., Barker, P. F., Barrett, P. J., Bowman, V., Davies, B., Smellie, J. L. & Tranter, M. (eds) 2013. *Antarctic Palaeoenvironments and Earth-Surface Processes*. Geological Society, London, Special Publications, **381**, 199–213. First published online April 19, 2013, http://dx.doi.org/10.1144/SP381.2

present-day coastlines during glacial maxima (Cooper *et al.* 1991; Sorlien *et al.* 2007). These results are consistent with pCO_2 records that indicate decreasing levels throughout the Eocene from over 1000 ppm to near present-day levels by latest Oligocene time (Pagani *et al.* 2005; Pekar & Christie-Blick 2008). Nevertheless, these data are still fragmentary, which is a consequence in part of the difficulty in locating strata to drill, their depth below the sea floor, as well as the extreme environmental conditions present in Antarctica today. Gathering proximal records to resolve these first-order questions provides the ground-truth necessary for global climate modellers to simulate past climate, especially during times of elevated atmospheric pCO_2 and warmer climate than today.

The Victoria Land Basin has been one of the centres for seismic studies (Davy & Alder 1989; Anderson & Bartek 1992; Brancolini *et al.* 1995; Melhuish *et al.* 1995; Bartek *et al.* 1996; Fielding *et al.* 2001, 2006) and deep-drilling projects (e.g. Dry Valley Drilling Project, CIROS-1 and 2, Cape Roberts Project, ANDRILL) for understanding the geological history of Antarctica. These records suggest a history of a rapidly subsiding basin, concomitant with high sedimentation rates during the Oligocene and early part of the Miocene (Early Miocene near Cape Roberts and Middle Miocene in the southern McMurdo Sound area), followed by lower rates during Late Miocene time (Stern *et al.* 1991; Davey & De Santis 2006; Fielding *et al.* 2008*a*). Additionally, Wilson *et al.* (2012) reconstructed the subsidence, depositional and erosional history of the Ross Sea, showing that, during the Oligocene Epoch, the Ross Sea had considerably shallower-water depths and was divided by islands (Wilson & Luyendyk 2009), which was first proposed by De Santis *et al.* (1995, 1999).

The CIROS-1 borehole, located in the southern McMurdo Sound (Fig. 1), recovered Late Eocene to Early Miocene strata (Barrett 1989; Claps *et al.* 1997), with the Late Eocene section being the oldest strata recovered in a borehole this far south (78°S). Since then, strata older than what were recovered at CIROS-1 from the Antarctic are considered to represent one of ‘holy grails’ for palaeoclimatologists studying this area. However, in the southern McMurdo Sound area, Palaeogene strata that are at sufficiently shallow depths to be drilled by current technology (i.e. ANDRILL drilling platform) are exceedingly difficult to locate. Additionally, most seismic studies have had difficulty imaging below the water-bottom multiple in this area. In 2005, over 27 km of multi-channel seismic data were collected using a sea-ice-based platform that imaged Neogene reflectors over 1 km below the sea floor in southern McMurdo Sound (Betterly *et al.* 2007). These results were used for positioning the AND-2A borehole (Harwood *et al.* 2006).

In the austral spring 2008, the Offshore New Harbour Expedition, which is part of the multinational ANDRILL (Antarctic Geological Drilling) Programme, successfully collected over 48 km of multi-channel seismic and gravity data (Wilson *et al.* 2009; Sunwall *et al.* 2012). The programme’s main goal was to investigate the stratigraphic and tectonic history of westernmost southern McMurdo Sound during the greenhouse world (Eocene) and the start of the icehouse world (Oligocene). This survey represents an important step for identifying future drilling targets for ANDRILL, with the aim of recovering stratigraphic intervals to interpret Antarctica’s climate and glacial history over the past 50 Ma. Processing of the seismic data successfully removed the water-bottom multiple, permitting deeper seismic reflectors to be identified.

In this paper, we present the results from the multi-channel seismic survey, which were used to evaluated the seismic stratigraphy of offshore New Harbour. For the first time, these seismic data imaged the major Transantarctic Mountains Border Fault in this area and a nearly 2 km thick sedimentary fill immediately downdip of the fault that extended out into the central part of southern McMurdo Sound. This permitted correlation between the updip CIROS-1 and downdip AND-2A boreholes.

Stratigraphic setting

This seismic study was conducted offshore of New Harbour located in southern McMurdo Sound, Antarctica (Fig. 1), which is part of the Victoria Land Basin (Cooper *et al.* 1987; Davey & Cooper 1987; Davey & Brancolini 1995). Fault and flexure-related subsidence associated with rifting and volcanic loading in Victoria Land Basin has provided the necessary accommodation for thick sedimentary wedges to accumulate (De Santis *et al.* 2001). Low-resolution gravity data have indicated that a complex tectonic structure occurs on the western side of the Victoria Land Basin, with a series of grabens cross-cut by faults (Cooper *et al.* 1991). The Transantarctic Mountains Border Fault is located on the western portion of the Victoria Land Basin and is one of the most tectonically significant structures in the Ross Sea, (e.g. Cooper *et al.* 1991). This fault has been imaged in the Cape Roberts area, revealing a fault system with up to 3 km of vertical separation (Hamilton *et al.* 2001). However, this fault has never been clearly seismically imaged in southern McMurdo Sound. Regional lineaments, outcrops and limited gravity data indicate complex faulting between the

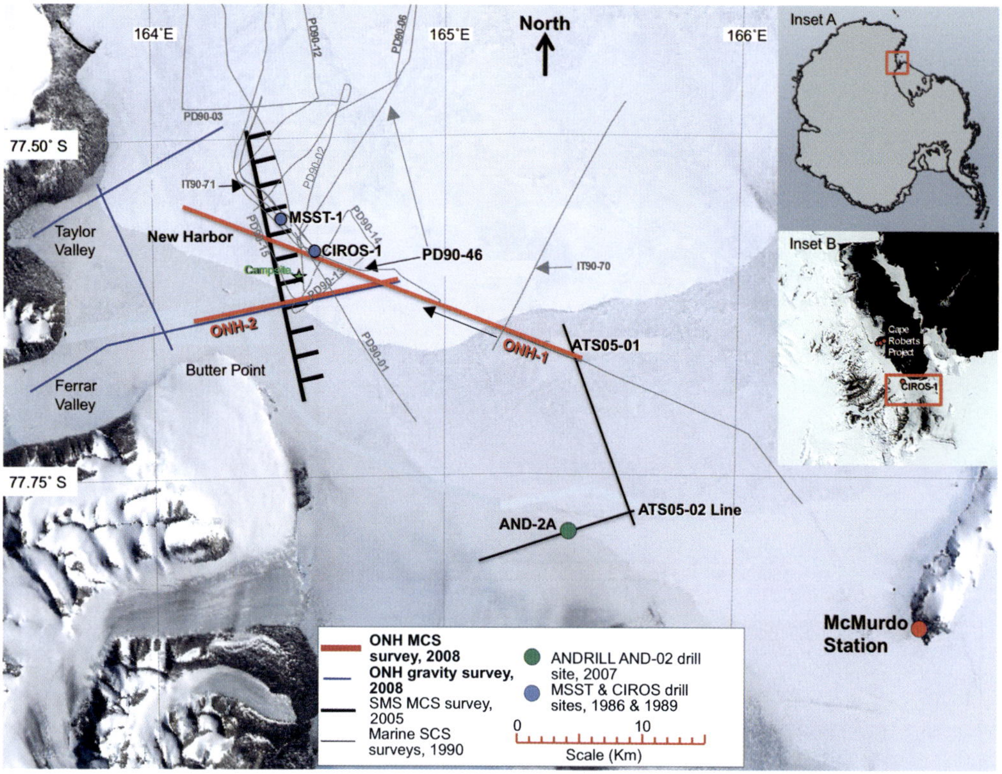

Fig. 1. Base map of southern McMurdo Sound and adjacent Ross Island to the east and East Antarctica to the west, with inset A of the Antarctic continent and inset B of southern McMurdo Sound, which also shows the locations of the Cape Roberts Project and CIROS-1. Boxes show the location of each map. Multi-channel seismic lines, ONH-08-01 and ONH-08-02, are shown in red (Sunwall *et al.* 2010) and gravity data in blue (Wilson *et al.* 2009). Seismic lines from previous surveys are also shown: multi-channel seismic lines ATS-05-01 and ATS-05-02 in black lines (Betterly *et al.* 2007) and single-channel seismic lines in grey (Anderson & Bartek 1992; Cooper *et al.* 1995*a*, *b*; Bartek *et al.* 1996). The CIROS-1 and AND-2A boreholes are shown. The Transantarctic Mountains Border Fault is indicated by a black-hatched line and the location of the expedition field campsite by a green star.

Transantarctic Mountains and Victoria Land Basin (Ben-Avraham *et al.* 1998; Wilson 1999).

Multi-channel seismic data

For decades, marine-based surveys have collected seismic data to develop a better understanding of the geological history in the Victoria Land Basin area (e.g. Cooper *et al.* 1987). Nevertheless, these data were often restricted to single-channel seismic data, which often limited imaging reflectors to those above the water-bottom multiple. In the last seven years, a number of seismic surveys gathered multi-channel seismic data in the Victoria Land Basin using sea-ice and ice-shelf platforms that imaged reflectors over 1 km below the sea floor (McMurdo Ice Shelf, Horgan *et al.* 2005; southern McMurdo South, Betterly *et al.* 2007; Speece *et al.* 2009; McKay Sea Valley, Williams *et al.* 2011).

In 2008, Offshore New Harbour survey collected over 48 km of multi-channel seismic data on two multi-channel seismic lines (ONH-08-01 and ONH-08-02; Sunwall *et al.* 2010, 2012). Line ONH-08-01 extended from central southern McMurdo Sound crossing the multi-channel seismic strike line ATS05-01 and then roughly followed the path of single-channel seismic line PD90-46 trending in a WNW direction, crossing within 100 m of CIROS-1, and then extending 10 km further updip (Fig. 1) for a total of over 32 km. The line is oblique to dip (up to 25°), based on previous work that constrained strike and dip in southern McMurdo Sound (e.g. Betterly *et al.* 2007). Near CIROS-1, ONH-08-01 also crossed a number of single-channel seismic lines (PD90-01, PD90-02, PD90-13, PD90-14, PD90-15; Fig. 1). ONH-08-02 begins

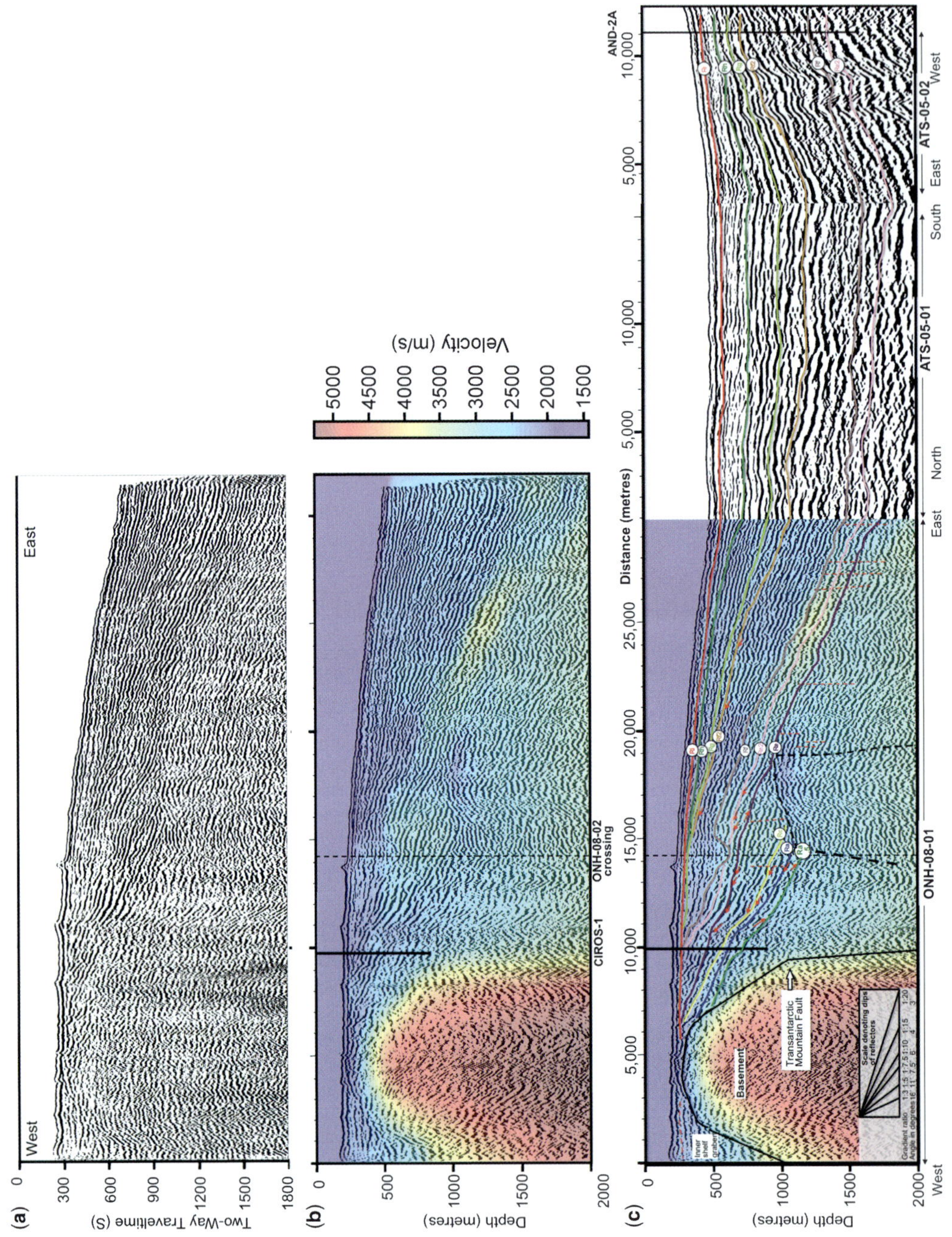
(a)
Two-Way Traveltime (S)
West
East
(b)
Depth (metres)
CIROS-1
ONH-08-02 crossing
Velocity (m/s)
(c)
Depth (metres)
Distance (metres)
Basement
Transantarctic Mountain Fault
ONH-08-01
ATS-05-01
ATS-05-02
AND-2A
North
South
East
West

about 2 km east of km 18 (4 km downdip of CIROS-1) on Line ONH-08-01. It also crosses several other seismic lines (PD90-02, PD90-13). ONH-08-02 trends in a WNW direction, toward Ferrar Valley for a distance of 15.3 km to about 2 km north of Butter Point (Fig. 1). This line was orientated to the dip of the strata in New Harbour based on previous studies (Harwood *et al.* 2006; Betterly *et al.* 2007).

Data acquisition

Multi-channel seismic data acquisition used a sea-ice platform with methods similar to those outlined (Betterly *et al.* 2007). The Offshore New Harbour survey employed a dual-chamber Generator Injector (GI) air gun as the seismic source, which was lowered through 35 cm diameter holes drilled through 2–7 m thick sea ice at 100 m spacing. A 1.5 km long streamer used here contained 60 gimbled geophones installed at 25 m spacing that measured the returning reflected seismic energy, resulting in a nominal 7.5-fold coverage. This allowed the processing of seismic artifacts such as the water-bottom multiple and multiples from other strong reflectors. Two main additions to the data gathering process ultimately improved the seismic survey over the 2005 southern McMurdo Sound (SMS) survey (Betterly *et al.* 2007). These included the use of a larger air compressor, which permitted a more rapid recharge of the air tanks, increasing the number of air storage tanks to six, which permitted more shots to be fired before the compressor needed to recharge the tanks. These improvements also permitted the GI air gun pressure to be increased from 3.4×10^6 to 13.8×10^6 Pa, which resulted in deeper penetration of the seismic waves.

Data processing

Details of the data processing have been discussed in Sunwall *et al.* (2012). Timing errors that were caused by improper calibration between the air-gun control software with the air gun (Williams *et al.* 2011) were corrected using hand statics, which are small timing corrections that are done manually. In addition, static errors were detected on traces that occupied previous shot locations. These static errors were caused by air bubble plumes trapped beneath the ice (Williams *et al.* 2011; Sunwall *et al.* 2012). The bubble plumes were remnants of the released air from the GI air gun from when the source occupied the station. A time correction on the order of 5 ms was applied to the traces that occupied previous shot locations to correct for the timing delays cause by the bubbles.

A frequency–wave number (f-k) filter was used to remove the steeply dipping ice flexural mode from the records. Time-domain gap deconvolution was employed to remove sea-ice-related short-period multiples. Long-period ocean-bottom multiples were removed using parabolic-velocity filtering based on the discrete parabolic radon transform. Otherwise, these data were conventionally processed up to post-stack depth-migrated common-midpoint stacks.

We attempted prestack depth migration on these data but obtained poor results because of the low fold of these data (7.5 nominal fold). Moreover, attempts to improve the prestack migrations results using trace interpolation were unsuccessful. As a consequence, a steep-dip (70°) poststack explicit finite-difference depth migration was used to produce the final depth-migrated sections. Migration velocity analysis was accomplished using a layer-stripping approach (Al-Yahya 1989; Lafond & Levander 1993). Velocities were first determined for the shallow part of the section and then velocity analysis iteratively progressed to deeper parts of the seismic section in a step-wise fashion. During each step, common reflection point gathers were analysed for residual moveout. Residual moveout was picked during each step and used to vertically update the velocity model. The high velocities for the horst block shown in Figures 2 and 3 are poorly constrained because of the lack of useable reflectors within the suspected weak seismically layered crystalline basement of the horst block. Our migration velocities are in reasonable agreement with velocities calculated for the CIROS-1 drillhole

Fig. 2. (**a**) Processed unmigrated multi-channel seismic reflection profile ONH-08-01 with the CIROS-1 borehole shown. (**b**) Migrated multi-channel seismic profile ONH-08-01 converted to depth, with superimposed velocity estimates shown in colour. (**c**) Migrated multi-channel seismic profile ONH-08-01 converted to depth, with superimposed velocity estimates shown in colour and interpreted seismic reflectors. Lines ATS-05-01 and ATS-05-02 are shown with the correlation of the Miocene reflectors to the ANDRILL AND-2A drill site. The CIROS-1 and AND-2A boreholes are shown as thick vertical black lines. Red arrows show truncating, onlapping or downlapping reflectors. The Transantarctic Mountains Border Fault is labelled and is indicated by a thick black line, with crystalline basement rock within the fault and the outline of the horst. Faults are shown as red dashed lines. A gradient scale denoting the dip of the reflectors is shown in the lower left-hand corner. The convex-up shaped feature outlined in a thick black dashed line may represent out-of-plane echoes, an igneous body or horst. Updip is toward the west and the basin is to the east. Unmigrated common-midpoint stacked seismic time sections were migrated and converted to depth using explicit finite-difference post-stack-depth migration.

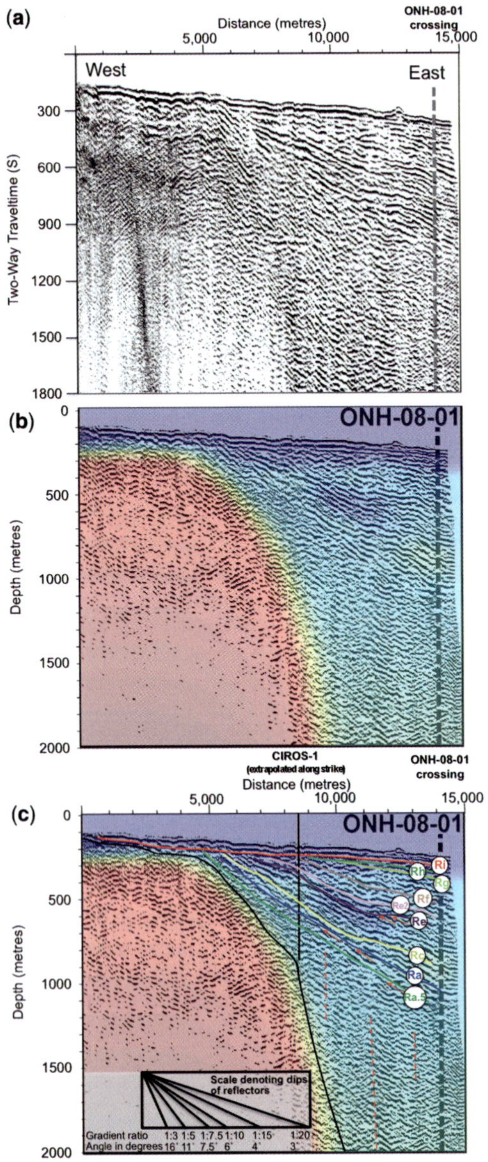

Fig. 3. (**a**) Processed unmigrated multi-channel seismic reflection profile ONH-08-02 with the CIROS-1 borehole shown. (**b**) Migrated multi-channel seismic profile ONH-08-02 converted to depth, with superimposed velocity estimates shown in colour. (**c**) Migrated multi-channel seismic profile ONH-08-02 converted to depth, with superimposed velocity estimates shown in colour and interpreted seismic reflectors. The location of the CIROS-1 borehole was extrapolated along-strike onto the ONH-08-02 line. Red arrows indicate truncation. A gradient scale denoting the dip of the reflectors is shown in the lower left-hand corner. The Transantarctic Mountains Border Fault is labelled and is indicated by a thick black line. Faults are shown as red dashed lines. Updip is toward the west and the basin is to the east. Unmigrated common-midpoint stacked seismic time sections were migrated and converted to depth using explicit finite-difference post-stack-depth migration.

from downhole logging data using porosity–velocity equations (Bucker *et al.* 1998).

Data resolution

We calculated the vertical resolution using the dominant wavelength found in the upper 2000 m of the depth-migrated seismic sections. Using the Rayleigh criterion, the vertical resolution of Offshore New Harbour data is 8.3 m. The vertical resolution of Offshore New Harbour data is 5.7 m better than that of SMS data, which is reported at 14 m (Betterly *et al.* 2007). This increase in Offshore New Harbour data resolution is attributed to bubble-plume static corrections and timing corrections that were not incorporated into the 2005 seismic survey (Betterly *et al.* 2007) processing sequence (Sunwall *et al.* 2012). The overall improvement in data quality from 2005 to 2008 is apparent where the tie between ONH-08-01 and ATS-05-01 provides a good comparison (Fig. 2).

Seismic stratigraphy results

The two multi-channel seismic profiles gathered in this study reveal the offshore stratigraphic architecture for the first time below the water-bottom multiple from near the shore of New Harbour into the central portion of southern McMurdo Sound (Figs 2 & 3). While the upper reflectors have been imaged in the New Harbour area by a number of seismic studies (e.g. Anderson & Bartek 1992; Bartek *et al.* 1996), imaging reflectors below the water-bottom multiple have only been imaged for only a short portion of seismic profile IT90-71 (Claps *et al.* 1997). In this study, we identified and correlated major reflectors to the updip (CIROS-1) and downdip (AND-2A) boreholes located in southern McMurdo Sound, assigning ages for reflectors ranging from the Miocene to Late Eocene, which permitted interpretation of the depositional history for this area.

The transantarctic mountain border fault

A major fault discerned in both lines is located approximately 3 km offshore of Butter Point and is identified as the Transantarctic Mountain Border Fault (Wilson *et al.* 2009). It trends in a NNW direction and is generally parallel to the East Antarctic coastline. Immediately east of the fault, reflectors up to 2 km below the sea floor are indicative of underlying sedimentary strata. The fault separates the deeper basin from a raised horst upon which reflectors onlap.

Description of the seismic stratigraphic architecture

This study used the seismic stratigraphic nomenclature developed by Fielding *et al.* (2006, 2008*a*; Fig. 4). Previous studies in the Victoria Land Basin developed a firm foundation of geological history, dividing the Victoria Land Basin stratigraphy into five phases based on rifting and subsidence history (e.g. Fielding *et al.* 2006, 2008*a*). In this study we identified three seismic units in the offshore New Harbour area (Fig. 5): an Eocene–Lower Oligocene unit; an Upper Oligocene–Lower Miocene unit; and a Middle Miocene and younger unit.

Unit 1: Eocene and Lower Oligocene reflectors. Over 1.5 km of Palaeogene and possibly older strata were imaged in both seismic lines immediately in front of the Transantarctic Mountains Border Fault. Although strong continuous reflectors only extend to about 1200 m below the sea floor, the deepest reflectors have a generally parallel configuration. All of the reflectors in Unit 1 onlap against the Transantarctic Mountains Border Fault on Line ONH-08-01, while the Lower Oligocene and uppermost Eocene reflectors on Line ONH-08-02 truncate against reflector Ri (Figs 2 & 3). These deeper reflectors dip at a relatively high angle (approximate 6° slope, 1:10 gradient) on Line ONH-08-02 (Fig. 3).

Correlation with CIROS-1 provided Upper Eocene ages for reflectors in the upper 700 m. Three Palaeogene reflectors truncate underlying reflectors and were correlated with a sequence boundaries identified in CIROS-1. The lowest of these truncated reflectors, Ra-0.5, contains several downlapping surfaces and intersects CIROS-1 at about 520 mbsf, in which the base of a sandy diamictite occurs at 515 mbsf (Hambrey *et al.* 1989). Two prominent reflectors correlate with the Eocene/Oligocene boundary (404 mbsf) and the '366 mbsf' sequence boundary, which has up to a 9 ma hiatus spanning 33–24 Ma at CIROS-1 (Wilson *et al.* 1998). These reflectors are equivalent to the Ra (blue) and Rc (yellow) reflectors, respectively, which have been described in other parts of the Victoria Land Basin (e.g. Fielding *et al.* 2006, 2008*a*; Fig. 4). Both are strong amplitude reflectors that are associated with underlying truncations (Figs 2 & 3). The Lower Oligocene section is quite thin in both seismic profiles, which is consistent with the thin Lower Oligocene section (*c.* 40 m thick) identified at CIROS-1 (Barrett 1989; Wilson *et al.* 1998). Between *c.* 12 and *c.* 17 km updip of the east end of Line ONH-08-01, deep reflectors appear to reverse dip, resulting in a convex-up configuration. This feature hinders further evaluation of these deeper reflectors downdip along ONH-08-01 (Fig. 2). This feature is not present on line

Brancolini *et al.* (1995)	Bartek *et al.* (1996)	Fielding *et al.* (2008*a*)	This study	Age (Ma, ATS)	Rift phase (Fielding *et al.* 2008*a*)
RSU2	E-F	Ri	Ri	4 Ma*	5
	I/ J	Rh	Rh	14.3 Ma-??*	
	J/ K	Rg	Rg	15.5–14.9 Ma†	
			Rf2	16.1 Ma†	4
RSU4a	K/ L	Rf (Beige)	Rf	17.8–18.7 Ma†	
RSU5	L/ M		Re2	19.4†	
	N/ O	Re	Re	23 Ma‡	3
		Rd			
RSU6	P/ Q	Rc	Rc	25–33 Ma‡	
		Rb			2
	R/ S	Ra	Ra	33.9 Ma‡	
			Ra.5	35 Ma‡	

Fig. 4. Seismic stratigraphic framework for this study, showing correlation to other seismic stratigraphic schemes for the region. See text for explanation of the colour shading. *Age is based on AND-1b borehole data (Naish *et al.* 2007). †Age is based on correlation to AND-02 borehole data (Acton *et al.* 2008). ‡Age is based on correlation to the CIROS-1 borehole (Wilson *et al.* 1998).

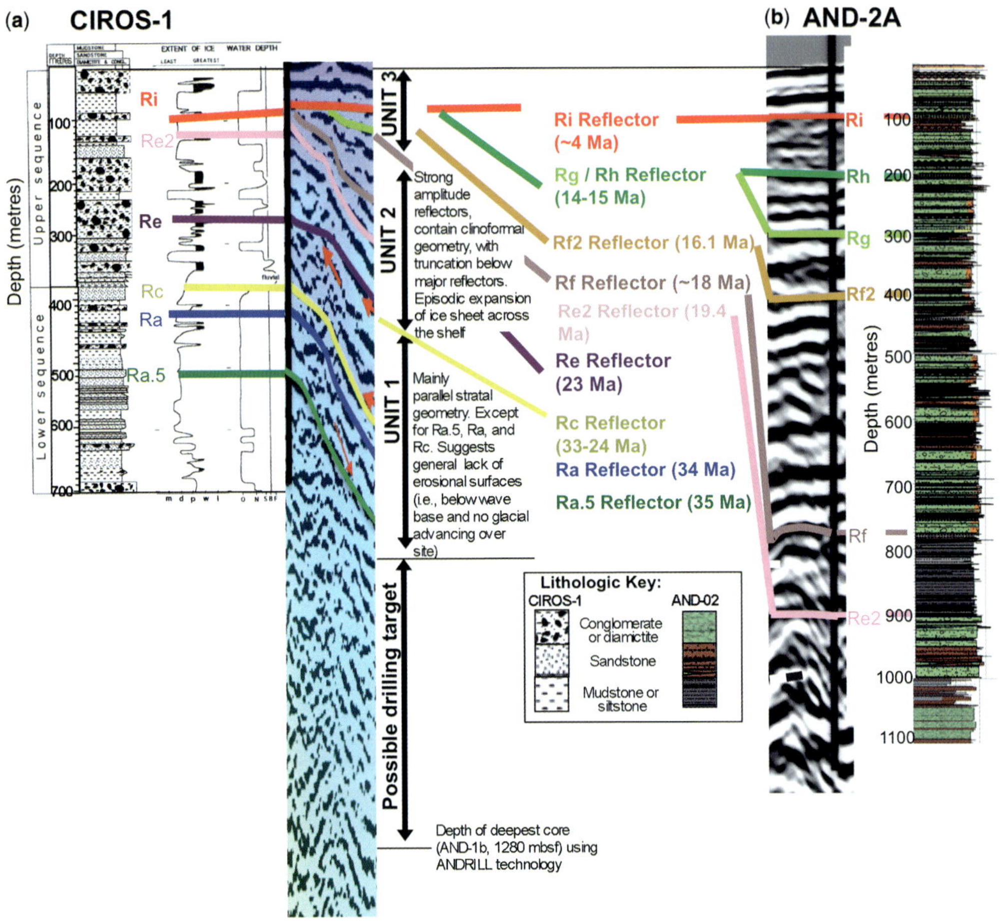

Fig. 5. Summary of lithology of (**a**) CIROS-1 (after Hambrey *et al.* 1989) and (**b**) AND-2A cores (Fielding *et al.* 2008*b*), with seismic profile close-ups of the site and summary of the seismic stratigraphy of ONH-08-01 near the CIROS-1 borehole. Also shown in the seismic section are the potential drilling objectives. The CIROS-1 and AND-2A boreholes are shown on the seismic profiles as thick black lines. Projection of the seismic reflectors onto the cores is shown. Ice extent and water-depth estimates at CIROS-1 site are modified from Hambrey *et al.* (1989). Ice-front position: m, marine; d, distal; p, proximal; w, waterlain till; l, lodgement till. Water depth: O, offshore; N, near shore; F, fluvial, modified from Hambrey *et al.* (1989). The AND-2A borehole lithology is from Fielding *et al.* (2008*b*). Unmigrated common-midpoint stacked seismic time sections were migrated and converted to depth using explicit finite-difference post-stack-depth migration for both core sites.

ONH-08-02, (Fig. 3), although the line extends sufficiently downdip to intersect any along-strike continuation of this feature.

Unit 2: Upper Oligocene and Lower Miocene reflectors. Unit 2 is bounded by the Rc reflector ('mid' Oligocene sequence boundary) and Middle Miocene Rg reflector, and is characterized by a prograding geometry and several reflectors that are truncated (Figs 2 & 3). Four reflectors within Unit 2 truncate underlying reflectors and are identified as sequence boundaries based on correlation to the CIROS-1 and AND-2A coreholes. A prominent reflector that truncates underlying reflectors correlates to a surface in CIROS-1 at 248 mbsf. This surface is dated as the Oligocene/Miocene sequence boundary (Roberts *et al.* 2003), identifying this reflector as the Re (purple) reflector of Fielding *et al.* (2006). A second reflector that truncates underlying reflectors intersects CIROS-1 at 120–130 mbsf and AND-2A at *c.* 900 mbsf. It is not associated with an equivalent reflector from Fielding *et al.* (2008*a*) and is named in this study reflector Re2. A third prominent reflector is probably the Rf beige reflector of Fielding *et al.* (2001). It contains a large scoured-out concave-up feature approximately 1.6 km wide and 130 m deep located *c.* 14 km from the western end of Line ONH-08-01. Similar features have been previously identified in the PD-90 seismic profiles around New Harbour, which have been interpreted to be the result of glacial scour (Brancolini *et al.* 1995; De Santis *et al.* 1995; Bartek *et al.* 1996; Fig. 2). Since ONH-08-01 is oblique to dip, the width of the valley on this line is probably exaggerated. This valley is not seen on ONH-08-02. The Rf reflector is truncated by an overlying reflector immediately downdip of CIROS-1, but ties into AND-2A at 770–780 mbsf (Figs 2 & 5). A sequence boundary at 775 mbsf contains a hiatus of nearly 1 Ma (17.8–18.7 Ma) (Figs 2 & 3). A fourth strong reflector containing underlying truncations is named here as reflector Rf2 (terra cotta) and intersects AND-2A at *c.* 390 mbsf, where it is dated as 16.1 Ma (Fig. 2).

Unit 3: Middle Miocene and younger reflectors. Middle Miocene and younger reflectors have been among the best-studied seismic units in southern McMurdo Sound because they have been imaged and traced throughout the Ross Sea (Davy & Alder 1989; Anderson & Bartek 1992; Brancolini *et al.* 1995; Bartek *et al.* 1996; Fielding *et al.* 2006). While previous studies have identified Middle Miocene and younger reflectors in New Harbour and throughout the Victoria Land Basin (e.g. Bartek *et al.* 1996), this study identified and traced reflectors that were below the bottom water multiple in previous seismic studies from the inner shelf of New Harbour into the deeper waters and then correlated with the AND-2A borehole. Each of the reflectors identified as a sequence boundary, the Rg, Rh, and Ri reflectors, truncate against an overlying sequence boundary and are dated 15.5–14.9, <14.3 and 4 Ma, respectively (Fielding *et al.* 2006; Naish *et al.* 2007; Fig. 4).

Interpretations and discussion

Seismic Unit 1: Eocene and Lower Oligocene seismic stratigraphy of offshore New Harbour

Seismic stratigraphic interpretations of Unit 1. Reflectors in Unit 1 extend down to at least 1000 m below the base of CIROS-1 and may represent a thick Palaeogene succession. However, the age of the strata below CIROS-1 is uncertain. At CRP-3, earliest Oligocene/Late Eocene strata lie unconformably above Devonian sedimentary rocks (Henrys *et al.* 2001) and previous seismic studies suggest that New Harbour may have a similar succession (Cooper *et al.* 1987). One of the main difficulties in resolving this conundrum is that Devonian and Palaeogene strata have almost identical velocity signals in seismic data.

Several pieces of evidence suggest that the reflectors below CIROS-1 are Palaeogene in age. A significant difference between the lowest strata at CRP-3 and CIROS-1 is that palaeo-water-depth estimates at CIROS-1 were interpreted as marine and probably below wave base (Fielding *et al.* 1998), while at CRP-3 they are terrestrial or marginal marine (Cape Roberts Science Team 2000). This suggests that subsidence may have begun earlier in New Harbour than near Granite Harbour. Additionally, gravity data indicate that a thick section of sedimentary rock approximately 4 km thick lies immediately downdip of CIROS-1 (Wilson *et al.* 2009). Since the Devonian Beacon Sandstone has been observed to be no thicker than 2.5 km on land (Barrett 1972, 1991; Bernet & Gaupp 2005), 0.5–1 km of Cenozoic sediments should be preserved below the base of CIROS-1. If the strata below CIROS-1 are Cenozoic in age, then the deepest reflectors may be between 42 and 47 Ma based on sedimentation rates in the lower part of CIROS-1 and would represent potential drilling targets for the ANDRILL Program (Fig. 6).

In contrast, at CRP-3, basement rocks consisting of Devonian Sandstone of the Beacon

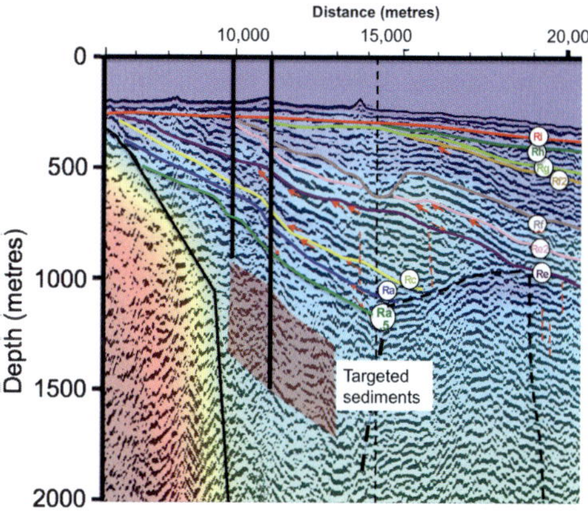

Fig. 6. Close-up of ONH-01 from the Transantarctic Mountains Fault to *c.* 10 km downdip of CIROS-1. Shown in the shaded area are the possible drilling targets assuming that the strata below CIROS-1 are Eocene in age. Red arrows show truncating reflectors. The Transantarctic Mountains Border Fault is labelled and is indicated by a thick black line, with basement rock updip of the Transantarctic Mountains Border Fault and the outline of the horst shown in dashed black line. Faults are shown as red dashed lines. Unmigrated common-midpoint stacked seismic time sections were migrated and converted to depth using explicit finite-difference post-stack-depth migration for CIROS-1.

Supergroup were encountered at 823 mbsf, with this surface interpreted to represent the Ra unconformity (Henrys *et al.* 2001). Down-hole physical properties data at CRP-3 show little difference in the velocity between the Devonian Sandstone and the overlying Palaeogene rocks (both are close to 3.0 km s^{-1}; Henrys *et al.* 2001). In addition, seismic data that cross Cape Roberts contain parallel reflectors including the Ra reflector (Henrys *et al.* 2001) and belong to a single megasequence (V5 of Cooper & Davey 1987). Furthermore, Line 401 ties CRP-3 into the New Harbour area, with Cooper & Davey (1987) interpreting the >1 km reflectors (i.e. sedimentary rocks) beneath CIROS-1 as most likely the Beacon Supergroup.

The question is, how much (if any) Eocene strata are preserved below CIROS-1? Is the sedimentary succession at New Harbour similar to CRP-3 or do significant Palaeogene strata exist below CIROS-1? At this point, both hypotheses have to be considered, with the only way to know for certain being to drill this deeper section.

Reflectors between 400 and 700 mbsf near CIROS-1 have a generally parallel nature, suggesting that limited erosion or by-pass occurred in the offshore New Harbour area during Eocene and Early Oligocene times. This is supported by lithofacies and marine fossils found in Eocene and Lower Oligocene sediments at CIROS-1 that suggest relatively deep water depths (Webb 1989; Claps *et al.* 1997; Galeotti & Coccioni 1998). The three named Palaeogenes, Ra.5, Ra and Rc, truncate underlying reflectors, which could be the result of sea-level fall, more proximal glacial conditions or deep-water processes (i.e. turbidites). The lithological data at CIROS-1 provide evidence that glacier ice reached somewhere along the East Antarctic coastline at both *c.* 35 Ma (Ra.5 reflector) and *c.* 34 Ma (Ra reflector) based on the occurrence of diamictite interpreted to represent waterlain till (Hambrey *et al.* 1989; Claps *et al.* 1997). The lithology above the Rc reflector has been interpreted as fluvial at CIROS-1, suggesting that this sequence boundary may have formed, at least in part, as a result of subaerial exposure (Hambrey *et al.* 1989; Fig. 5).

Unknown structure downdip of the main fault. About 4 km downdip of CIROS-1 on line ONH-08-01, reflectors abruptly change dip for several kilometres before returning to their original dip orientation (Fig. 2). This produces a mounded feature containing strong internal reflectors. This feature could represent an igneous intrusion as volcanic edifices have been observed throughout this region (Anderson & Bartek 1992). However, the internal reflectors within igneous masses on other seismic lines in the New Harbour area are often not well developed, with few to no reflectors in seismically homogeneous strata. Another possibility is that it represents side echoes from steeply dipping faults or intrusive rocks that are out of plane of the seismic line. This is supported by ONH-08-01 being oblique to dip, and the Transantarctic Mountains Border Fault, with its high-angle dip, could have resulted in the production of out-of-plane echoes. Furthermore, this feature is not seen on ONH-08-02, which is orientated close to the dip of the strata. Another possible explanation is that the mounded feature could be the result of faults (Fig. 2) as numerous faults were identified within this section. Finally, this feature could represent a basement horst. Gravity data suggest that a horst-like feature may be buried in front of the main Transantarctic Mountains Border Fault (Wilson *et al.* 2009). If this is a buried horst, then Upper Oligocene strata buried it. Furthermore, the horst would also influence the geometry of the overlying strata.

Seismic Unit 2: Offshore New Harbour Upper Oligocene and Lower Miocene seismic stratigraphy

The strata in Unit 2 were deposited under more ice-proximal conditions than Seismic Unit 1, based on the seismic geometry and correlation to CIROS-1. Glacier ice is interpreted to have overridden CIROS-1 during reflector Re based on the stratal geometry of the underlying truncations and the occurrence of the reflector being within an expanded diamictite section (waterlain till and minor lodgement till; Hambrey *et al.* 1989). The Re2 reflector intersects CIROS-1 at approximately 120–130 mbsf and correlates to a distinct surface at 130 mbsf in CIROS-1. This surface separates fine-grained sandy silts below and a diamcitite interpreted as lodgement till above (Hambrey *et al.* 1989; Fig. 5), suggesting that glacier ice may have advanced over this site. This reflector was also correlated with the AND-2A borehole, which intersects the hole at approximately 890–910 mbsf (Fig. 2). This reflector may correlate with one of two reflectors: a surface at 891.86, which has muddy sandstones with common clasts above and siltstones with dispersed clasts below (Pekar *et al.* 2010); or one at 911.40 mbsf, with clast-rich to clast-poor sandy diamictite above and fine to medium sandstone with dispersed clasts below (Fielding *et al.* 2008*b*, 2011; Passchier *et al.* 2011). The lithologies associated with both surfaces suggest water depths shallower than storm wave base and the presence of icebergs. The age for the reflectors is 19.3–19.4 Ma.

The beige (Rf) reflector correlates with the sequence boundary at 775 mbsf at AND-2A, which contains a hiatus of 17.8–18.7 Ma. This reflector contains a valley trough interpreted as the result of subglacial scour, providing *prima facie* evidence that the ice expanded beyond the coastline during this time. The beige reflector has been dated as less than *c.* 17.5 Ma at CRP-2 (Fielding *et al.* 2001; Henrys *et al.* 2001) and *c.* 20 Ma at CIROS-1, which points to the difficulty in correlating reflectors over long distances in this area or perhaps suggests that this reflector may be time-transgressive. In preparing to drill the AND-2A borehole, the beige reflector was estimated to be located at approximately 1200 mbsf at the site (Harwood *et al.* 2006). However, the base of AND-2A (1134 mbsf) was older than 20 Ma (Acton *et al.* 2008), suggesting either the age of the Beige reflector was highly variable or the reflector had been misidentified.

The correlation of the Rf reflector between CIROS-1 and AND-2A provides the means to resolve the apparent transgressive temporal nature of this reflector. Previous studies have suggested that large uncertainties exist in the age model in the upper 100 m of CIROS-1 (Wilson *et al.* 1998; Florindo *et al.* 2005). In this study, correlation of the Rf reflector from the AND-2A drillhole shows that the Rf reflector truncates against an overlying reflector immediately downdip of reaching CIROS-1. These new direct correlations from CIROS-1 to AND-2A coreholes, coupled with the excellent age dates from AND-2A and the convergence of the reflectors near CIROS-1, suggest that a miscorrelation of a few reflectors near CIROS-1 could result in a difference of several million years in the age of the reflector. Therefore, the Rf (beige) reflector need not be time-transgressive and we speculate that the original seismic correlation from CRP to CIROS-1 may have 'jumped' a few reflectors, which would have resulted in the apparent temporal offset between the two sites.

The Rf2 reflector correlates with a sequence boundary at 389 mbsf in the AND-2A corehole. This surface separates clast-poor sandy diamictite above and fine to very fine sandstone with box-work fractures below, and is interpreted as an advance of glacier ice (Fielding *et al.* 2008*b*, 2011; Passchier *et al.* 2011). Taken together, the lithology and seismic geometry are consistent with glacier ice advancing to the palaeo-coastline and perhaps over the AND-2A site at 16.1 Ma.

Upper and middle Miocene reflectors in western southern McMurdo Sound

The late Neogene reflectors in New Harbour have a similar geometry as elsewhere in the Victoria Land Basin (Davy & Alder 1989; Anderson & Bartek 1992; Brancolini *et al.* 1995; Bartek *et al.* 1996; Fielding *et al.* 2006), in which reflectors are truncated by overlying reflectors (e.g. Fielding *et al.* 2006), suggesting that the ice sheet repeatedly expanded across the shelf in the New Harbour area during these times.

Conclusions

Over 48 km of multi-channel seismic data were collected on two seismic lines offshore from New Harbour, East Antarctica. A fault imaged on both seismic lines is identified as the Transantarctic Mountains Border Fault and projects up to near the ocean bottom several kilometres updip of CIROS-1. Immediately downdip of the fault, multi-channel seismic data imaged up to 1.8 km of sedimentary rocks. The ages of reflectors below CIROS-1 are uncertain, and may be Devonian in age, as was encountered at CRP-3, or may represent a significant thickness of Eocene strata below CIROS-1. The latter is based on gravity data that indicates up to 4 km of sedimentary rock exists. Since the Devonian strata is <2.5 km thick, there could be between 800 and 1300 m of Cenozoic strata preserved below CIROS-1.

Three seismic units were identified: (1) an Eocene–Early Oligocene unit that consists mainly of parallel reflectors and suggests distal glacier ice conditions and depths below the wave base; (2) a Late Oligocene–Early Miocene unit, which contains reflectors interpreted as sequence boundaries and represents increased glacier ice conditions; and (3) a late Early Miocene to Present unit, which contains truncated reflectors and represents repeated advances of glacier ice over the Ross Sea. For the first time, Early Miocene reflectors were traced from AND-2A to CIROS-1, which helped identify two new reflectors, Re2 and Rf2. The Re reflector correlates with a surface at *c.* 130 mbsf at CIROS-1 and to a surface at *c.* 900 mbsf at AND-2A, which is dated at 19.3–19.4 Ma. The Rf2 reflector correlates with a sequence boundary identified in the AND-2A drillhole at 389 mbsf and is dated as 16.1 Ma. The Rf reflector contains a hiatus spanning from 17.8 to 18.7 Ma and is interpreted to be the Beige reflector identified at other locations in the Victoria Land Basin. A valley trough contained within this reflector is interpreted as a glacier-scoured valley, providing *prima facie* evidence that the ice sheet advanced across the palaeo-shelf during Early Miocene time. A new Upper Eocene reflector, Ra.5 (35 Ma), was identified as a sequence boundary in this study, based on its seismic geometry and correlation of the reflector to the CIROS-1.

We wish to thank the editor, Michael Hambrey for his encouragement and excellent feedback as well as P. Barker, S. Henrys and one anonymous reviewer for their extremely insightful and thoughtful comments in reviewing an earlier version of this manuscript, which greatly improved this paper. This research was supported by National Science Foundation grants ANT-0732796 and ANT-0732875 to S. F. Pekar and M. A. Speece, respectively. G. Wilson and K. Tinto were supported by the New Zealand Foundation for Research, Science & Technology Contract CO5X0410. G. Wilson also acknowledges logistical support from Antarctica New Zealand. A special thanks goes to S. Alesandrini and R. Hein of Raytheon Polar Services Company for their help designing and building the Offshore New Harbour air-gun system, and the entire Offshore New Harbour crew and the support personnel at McMurdo Station for field support.

References

ACTON, G., CRAMPTON, J. ET AL. 2008. Preliminary integrated chronostratigraphy of the AND-2A core, ANDRILL southern McMurdo Sound Project Antarctica. *In*: HARWOOD, D. M., FLORINDO, F., TALARICO, F. & LEVY, R. H. (eds) *Studies from the ANDRILL, Southern McMurdo Sound Project, Antarctica. Terra Antartica*, **15**, 211–220.

AL-YAHYA, K. M. 1989. Velocity analysis by iterative profile migration. *Geophysics*, **54**, 718–729.

ANDERSON, J. B. & BARTEK, L. R. 1992. Cenozoic glacial history of the Ross Sea revealed by intermediate resolution seismic reflection data combined with drill site information. *In*: KENNETT, J. P. & WARNKE, D. A. (eds), *The Antarctic Paleoenvironment: a Perspective on Global Change I*. American Geophysical Union, Washington, D. C., Antarctic Research Series, **56**, 231–263.

BARRETT, P. J. 1972. Stratigraphy and palaeogeography of the Beacon Supergroup in the Transantarctic Mountains, Antarctica. *In*: HAUGHTON, S. H. (ed.) *Proceedings and Papers of 2nd Gondwana Symposium, South Africa*. CSIR, Pretoria, 249–256.

BARRETT, P. J. (ed.) 1989. *Antarctic Cenozoic history from the CIROS-1 drillhole, McMurdo Sound*. Science Information Publishing Centre, Wellington, Department of Scientific and Industrial Research Bulletins, **245**.

BARRETT, P. J. 1991. The Devonian to Jurassic Beacon Supergroup of the Transantarctic Mountains and correlatives in other parts of Antarctica. *In*: TINGEY, R. J. (ed.) *The Geology of Antarctica*. Oxford University Press, Oxford, 120–152.

BARRETT, P. J., GRINDLEY, G. W. & WEBB, P. N. 1972. The Beacon Supergroup of East Antarctica. *In*: ADIE, R. (ed.) *Antarctic Geology and Geophysics*. Universitetsforlaget, Oslo, 319–332.

BARRON, J. A., BALDAUF, J. G. ET AL. 1991. Biochronologic and magnetochronologic synthesis of Leg 119 sediments from the Kerguelen Plateau and Prydz Bay, Antarctica. *In*: BARRON, J., LARSEN, B. ET AL. (eds) *Proceedings of the Ocean Drilling Program, Scientific Results*, **119**. Ocean Drilling Program, Texas A&M University, College Station, TX, 813–847.

BARTEK, L. R., HENRYS, S. A., ANDERSON, J. B. & BARRETT, P. J. 1996. Seismic stratigraphy in McMurdo Sound based on Polar Duke-90 data: implications for glacially influenced early Cenozoic eustatic change. *Marine Geology*, **130**, 78–98.

BEN-AVRAHAM, Z., BUSETTI, M. & SPADINI, G. 1998. Transform-normal extension in the Victoria Land Basin (Antarctica), *Rendiconti Lincei*, **9**, 257–269.

BERNET, M. & GAUPP, R. 2005. Diagenetic history of Triassic sandstone from the Beacon Supergroup in central Victoria Land, Antarctica. *New Zealand Journal of Geology and Geophysics*, **48**, 447–458.

BETTERLY, S. J., SPEECE, M. A., LEVY, R. H., HARWOOD, D. M. & HENRYS, S. A. 2007. A Novel Over-Sea-Ice Seismic Reflection Survey in McMurdo Sound, Antarctica. *Terra Antartica*, **14**, 97–106.

BRANCOLINI, G., COOPER, A. K. & COREN, F. 1995. Seismic facies and glacial history in the Western Ross Sea. *In*: COOPER, A. K., BARKER, P. F. & BRANCOLINI, G. (eds) *Seismic Facies and Glacial History in the Western Ross Sea (Antarctica). Geology and Seismic Stratigraphy of the Antarctic Margin*. American Geophysical Union, Washington D.C., Antarctic Research Series, **68**, 209–233.

BUCKER, C. J., HENRYS, S. A. & WONIK, T. 1998. Revision of the Cenozoic seismic velocity structure of the CIROS-1 drillhole, Antarctica, and implication for the further drilling off Cape Roberts. *Terra Antartica*, **5**, 291–289.

CAPE ROBERTS SCIENCE TEAM 1998. Initial Report on CRP-1, Cape Roberts Project, Antarctica. *Terra Antartica*, **5**, 1–713

CAPE ROBERTS SCIENCE TEAM 1999. Studies from the Cape Roberts Project, Ross Sea, Antarctica, Initial Report on CRP-2/2A. *Terra Antartica*, **6**, 1–173

CAPE ROBERTS SCIENCE TEAM 2000. Studies from the Cape Roberts Project, Ross Sea, Antarctica, Initial Report on CRP-3. *Terra Antartica*, **7**, 1–306

CHOW, J. M. & BART, P. J. 2003. West Antarctic Ice Sheet grounding events on the Ross Sea outer continental shelf during the middle Miocene. *Palaeogeography, Palaeoclimatology, Palaeoecology*, **198**, 169–186.

CLAPS, M., COREN, F., DE SANTIS, L., MASETTI, D. & SARTI, M. 1997. Deep-water deposits from CIROS-1 Drillhole, McMurdo Sound, and implications for Ross Sea Late Eocene–Early Oligocene palaeogeography (Western Ross Sea, Antarctica). *Terra Antartica*, **4**, 133–147.

COOPER, A. K. & DAVEY, F. J. 1987. The Antarctic continental margin: geology and geophysics of the western Ross Sea. *Circumpacific Council on Economic and Mineral Resources Earth Science Series*, ***5B***, Houston, 1–253.

COOPER, A. K., DAVEY, F. J. & BEHRENDT, J. C. 1987. Seismic stratigraphy and structure of the Victoria Land Basin, western Ross Sea, Antarctica. *In*: COOPER, A. K. & DAVEY, F. J. (eds) *The Antarctic Continental Margin: Geology and Geophysics of the Western Ross Sea*. Circum-Pacific Council Energy and Mineral Resources, Houston, TX, Earth Science Series, **5B**, 27–65.

COOPER, A. K., BARRETT, P. J., HINZ, K., TRAUBE, V., LEITCHENKOV, G. & STAGG, H. M. J. 1991. Cenozoic prograding sequences of the Antarctic continental margin: a record of glacio-eustatic and tectonic events. *Marine Geology*, **102**, 175–213.

Cooper, A. K., Barker, P. F. & Brancolini, G. (eds) 1995*a*. *Geology and Seismic Stratigraphy of the Antarctic Margin*. Antarctic Research Series. AGU, Washington, DC, **68**.

Cooper, A. K., Barker, P. F. & Brancolini, G. (eds) 1995*b*. *Geology and Seismic Stratigraphy of the Antarctic Margin*. Antarctic Research Series. American Geophysical Union, Washington, DC, **68**, atlas, CD-ROMS.

Davey, F. J. & Brancolini, G. 1995. The Late Mesozoic and Cenozoic structural setting of the Ross Sea region. *In*: Cooper, A. K., Barker, P. F. & Brancolini, G. (eds) *Geology and Seismic Stratigraphy of the Antarctic Margin*. American Geophysical Union, Washington, D. C., Antarctic Research Series, **68**, 167–182.

Davey, F. J. & Cooper, A. K. 1987. Gravity studies of the Victoria Land basin and Iselin Bank. *In*: Cooper, A. K. & Davey, F. J. (eds) *The Antarctic Continental Margin: Geology and Geophysics of the Western Ross Sea*. Circum-Pacific Council for Energy and Mineral Resources, Houston, TX, CPCEMR Earth Science Series, **5B**, 119–138.

Davey, F. J. & De Santis, L. 2006. A multi-phase rifting model for the Victoria Land Basin, western Ross Sea. *In*: Fütterer, D. K., Damaske, D., Kleinschmidt, G., Miller, H. & Tessensohn, F. (eds) *Antarctica, Contributions to Global Earth Sciences*. Springer, Berlin, 303–308.

Davy, B. W. & Alder, G. 1989. Seismic reflection surveys. *In*: Barrett, P. J. (ed.) *Antarctic Cenozoic History from the CIROS-1 Drill Hole, McMurdo Sound*. New Zealand Department of Science and Industrial Research, Wellington, Bulletin in the Miscellaneous Series, **245**, 15–21.

De Santis, L., Anderson, J. B., Brancolini, G. & Zayatz, I. 1995. Seismic record of late Oligocene through Miocene glaciation on the central and eastern continental shelf of the Ross Sea. *Antarctic Research Series*, **68**, 235–260.

De Santis, L., Prato, S., Brancolini, G., Lovo, M. & Torelli, L. 1999. The eastern Ross Sea continental shelf during the Cenozoic: implications for the West Antarctic Ice Sheet development. *In*: Van der Wateren, F. M. & Cloetingh, S. A. P. L. (eds) *Lithosphere Dynamics and Environmental Change of the Cenozoic West Antarctic Rift System. Global and Planetary Change*, Special Issue, **23**, 173–196.

De Santis, L., Davey, F., Prato, S. & Brancolini, G. 2001. Subsidence at the Cape Roberts Project (CRP) drillsites from backstripping techniques, Victoria Land Basin, Antarctica. *Terra Antartica*, **8**, 137–141.

Fielding, C. R., Woolfe, K. J., Howe, J. A. & Lavelle, M. 1998. Sequence stratigraphic analysis of CRP-1, Cape Roberts Project, McMurdo Sound, Antarctica. *Terra Antartica*, **5**, 353–361.

Fielding, C. R., Naish, T. R. & Woolfe, K. J. 2001. Facies architecture of the CRP-3 Drillhole, Victoria Land Basin, Antarctica. *Terra Antartica*, **8**, 217–224.

Fielding, C. R., Henrys, S. A. & Wilson, T. J. 2006. Rift history of the western Victoria Land Basin: a new perspective based on integration of cores with seismic reflection data. *In*: Futterer, D. K., Damaske, D., Kleinschmidt, G., Miller, H. & Tessensohn, F. (eds) *Antarctica: Contributions to Global Earth Sciences*. Springer, Berlin, 309–318.

Fielding, C. R., Whittaker, J., Henrys, S. A., Wilson, T. J. & Naish, T. R. 2008*a*. Seismic facies and stratigraphy of the Cenozoic succession in McMurdo Sound, Antarctica: Implications for tectonic, climatic and glacial history. *Palaeogeography, Palaeoclimatology, Palaeoecology*, **260**, 8–29.

Fielding, C. R., Atkins, C. B. & THE ANDRILL-SMS SCIENCE TEAM 2008*b*. Sedimentology and Stratigraphy of the AND-2A Core, ANDRILL Southern McMurdo Sound Project, Antarctica. *Terra Antartica*, **15**, 77–112.

Fielding, C. R., Browne, G. H. *et al*. 2011. Sequence stratigraphy of the ANDRILL AND-2A drillcore, Antarctica: a long-term, ice-proximal record of Early to Mid-Miocene climate, sea-level and glacial dynamism. *Palaeogeography, Palaeoclimatology, Palaeoecology*, **305**, 337–351.

Florindo, F., Wilson, G. S., Roberts, A. P., Sagnotti, L. & Verosub, K. L. 2005. Magnetostratigraphic chronology of a late Eocene to early Miocene glacimarine succession from the Victoria basin, Ross Sea, Antarctica. *Global and Planetary Change*, **45**, 207–236.

Francis, J. E. 2000. Fossil wood from Eocene high latitude forests McMurdo Sound, Antarctica. *In*: Stilwell, J. & Feldmann, R. (ed.) *Paleobiology and paleoenvironments of Eocene rocks, McMurdo Sound, East Antarctica*. American Geophysical Union, Washington, DC, Antarctic Research Series, **76**, 253–259.

Francis, J. E. & Poole, I. 2002. Cretaceous and early Tertiary climates of Antarctica: evidence from fossil wood. *Palaeogeography, Palaeoclimatology, Palaeoecology*, **182**, 47–64.

Galeotti, S. & Coccioni, R. 1998. Foraminiferal analysis of the Miocene CRP-1 core (Ross Sea, Antarctica). *Terra Antartica*, **5**, 521–526.

Hambrey, M. J., Barrett, P. J. & Robinson, P. J. 1989. Stratigraphy. *In*: Barrett, P. J. (ed.) *Antarctic Cenozoic History from the CIROS-1 Drillhole, McMurdo Sound, Antarctica*. New Zealand Department of Science and Industrial Research, Wellington, **245**, 23–48.

Hambrey, M. J., Ehrmann, W. U. & Larsen, B. 1991. Cenozoic glacial record of the Prydz Bay continental shelf, East Antarctica. *In*: *Proceedings of the Ocean Drilling Program, Scientific Results*, **119**. Ocean Drilling Program, Texas A&M University, College Station, TX, **119**, 77–132.

Hamilton, R. J., Sorlien, C. C., Luyendyk, B. P., Bartek, L. R. & Henrys, S. A. 2001. Cenozoic tectonics of the Cape Roberts Rift Basin and Transantarctic Mountains Front, Southwestern Ross Sea, Antarctica. *Tectonics*, **20**, 325–342.

Harwood, D. M., Florindo, F., Levy, R. H., Fielding, C. R., Pekar, S. F., Speece, M. A. & THE SMS SCIENCE TEAM 2006. *ANDRILL Southern McMurdo Sound Project: Scientific Prospectus*. ANDRILL Science Management Office, University of Nebraska, Lincoln, NE.

Henrys, S. A., Bucker, C. J., Niessen, F. & Bartek, L. R. 2001. Correlation of seismic reflectors with the CRP-3 drillhole, Victoria Land Basin, Antarctica. *Terra Antartica*, **8**, 127–136.

Horgan, H., Naish, T., Bannister, S., Balfour, N. & Wilson, G. 2005. Seismic stratigraphy of the

Plio-Pleistocene Ross Island flexural moat-fill: a prognosis for ANDRILL Program drilling beneath McMurdo-Ross Ice Shelf. *Global and Planetary Change*, **45**, 83–97.

Katz, M. E., Miller, K. G., Wright, J. D., Wade, B. S., Browning, J. V., Cramer, B. S. & Rosenthal, Y. 2009. Stepwise transition from the Eocene greenhouse to the Oligocene icehouse. *Nature Geoscience*, **1**, 329–334.

Lafond, C. F. & Levander, A. R. 1993. Migration moveout analysis and depth focusing. *Geophysics*, **58**, 91–100.

Lear, C. H., Rosenthal, Y., Coxall, H. K. & Wilson, P. A. 2004. Late Eocene to early Miocene ice sheet dynamics and the global carbon cycle. *Paleoceanography*, **19**, PA4015, doi: 10.1029/2004PA001039.

Melhuish, A., Henrys, S. A., Bannister, S. & Davey, F. J. 1995. Seismic profiling adjacent to Ross Island: constraints on Late Cenozoic Stratigraphy and Tectonics. *Terra Antartica*, **2**, 127–136.

Miller, K. G., Wright, J. D. & Fairbanks, R. G. 1991. Unlocking the ice house: Oligocene–Miocene oxygen isotopes, eustasy, and margin erosion. *Journal of Geophysical Research*, **96**, 6,829–6,848.

Miller, K. G., Mountain, G. S. *et al.* 1996. Drilling and dating New Jersey Oligocene–Miocene sequences; ice volume, global sea level, and Exxon records. *Science*, **271**, 1,092–1,095.

Miller, K. G., Mountain, G. S. *et al.* 1998. Cenozoic global sea level, sequences, and the New Jersey Transect: results from coastal plains and continental slope drilling. *Reviews of Geophysics*, **36**, 569–601.

Miller, K. G., Kominz, M. A. *et al.* 2005. The geological record of global-sea level change. *Science*, **310**, 1293–1298.

Naish, T. R., Wolfe, K. J. *et al.* 2001. Orbitally induced oscillations in the East Antarctic ice sheet at the Oligocene/Miocene boundary. *Nature*, **413**, 719–723.

Naish, T. R., Powell, R., Levy, R. THE ANDRILL-MIS SCIENCE TEAM (EDS) 2007. Initial Results of the AND-1B Core, ANDRILL McMurdo Ice Shelf Project, Antarctica. *Terra Antartica*, **14**, 121–130.

Pagani, M., Zachos, J. C., Freeman, K. H., Tipple, B. & Bohaty, S. 2005. Marked decline in atmospheric carbon dioxide concentrations during the Paleogene. *Science*, **309**, 600–603.

Passchier, S., Browne, G., Field, B., Fielding, C. R., Krissek, L. A., Panter, K., Pekar, S. F. & THE ANDRILL SMS SCIENCE TEAM 2011. Early and middle Miocene Antarctic glacial history from the sedimentary facies distribution in the AND-2A drill hole, Ross Sea, Antarctica. *Geological Society of American Bulletin*, **123**, 2352–2365, doi: 10.1130/B30334.1.

Pekar, S. F. & Christie-Blick, N. 2008. Resolving apparent conflicts between oceanographic and Antarctic climate records and evidence for a decrease in $p\mathrm{CO_2}$ during the Oligocene through early Miocene (34–16 Ma). *Palaeogeography, Palaeoclimatology, Palaeoecology*, **260**, 41–49.

Pekar, S. F., Christie-Blick, N., Kominz, M. A. & Miller, K. G. 2002. Calibrating eustasy to oxygen isotopes for the early icehouse world of the Oligocene. *Geology*, **30**, 903–906.

Pekar, S. F., Hucks, A., Fuller, M. & Li, S. 2005. Glacioeustatic changes in the early and middle Eocene (51–42 Ma) greenhouse world based on shallow-water stratigraphy from ODP Leg 189 Site 1171 and oxygen isotope records. *Geological Society of America Bulletin*, **117**, 1081–1093.

Pekar, S. F., Koss, H. C. *et al.* 2010. Glacial extent in the Ross Sea during the Late Early Miocene (17.6–16.2 Ma): results from the ANDRILL AND-2A Drillcore, Southern McMurdo Sound Project, Antarctica. *ANDRILL SMS Project Science Integration Workshop*, Erice, Sicily, Italy.

Prebble, J. G., Raine, J. I., Barrett, P. J. & Hannah, M. J. 2006. Vegetation and climate from two Oligocene glacioeustatic sedimentary cycles (31 and 24 Ma) cored by the Cape Roberts Project, Victoria Land Basin, Antarctica. *Palaeogeography, Palaeoclimatology, Palaeoecology*, **231**, 41–57.

Raine, J. I. & Askin, R. A. 2001. Oligocene and Early Miocene terrestrial palynology of the Cape Roberts Drillhole CRP-2/2A, Victoria Land Basin, Antarctica. *Terra Antartica*, **7**, 389–400.

Roberts, A. P., Wilson, G. S., Harwood, D. M. & Verosub, K. L. 2003. Glaciation across the Oligocene–Miocene boundary in southern McMurdo Sound, Antarctica: new chronology from the CIROS-1 drill hole. *Palaeogeography, Palaeoclimatology, Palaeoecology*, **198**, 113–130.

Sorlien, C. C., Luyendyk, B. P., Wilson, D., Decesari, R., Bartek, L. & Diebold, J. 2007. Oligocene development of the West Antarctic Ice Sheet recorded in eastern Ross Sea strata. *Geology*, **35**, 467–470.

Speece, M. A., Levy, R. H., Harwood, D. M., Pekar, S. F. & Powell, R. D. 2009. New Seismic Methods to Support Sea-Ice Platform Drilling. *Scientific Drilling*, **7**, 40–42.

Stern, T. A., Davey, F. J. & Delisle, G. 1991. Lithospheric flexure induced by the load of the Ross Archipelago, southern Victoria land, Antarctica. *In*: Thomson, M. R. A., Crame, A. & Thomson, J. W. (eds) *Geological Evolution of Antarctica*. Cambridge University Press, Cambridge, 323–328.

Sunwall, D., Speece, M. & Pekar, S. F. 2010. Latest advances in over-sea-ice seismic reflection surveys: offshore New Harbour, Antarctica. *SEG International Exposition and 80th Annual Meeting 2010*, Technical Program Expanded Abstracts, 36–40.

Sunwall, D., Speece, M. & Pekar, S. F. 2012. Advances in on-sea-ice seismic reflection methods using an air gun: McMurdo Sound, Antarctica. *Geophysics*, **77**, 1–12.

Thomas, E., Barrera, E. *et al.* 1990. Upper Cretaceous–Paleogene stratigraphy of Sites 689 and 690, Maud Rise, (Antarctica). *In*: Barker, P. F., Kennett, J. P. *et al.* (eds) *Proceedings of the Ocean Drilling Program, Scientific Results*, **113**. Ocean Drilling Program, Texas A&M University, College Station, TX, 901–914.

Thorn, V. C. 2001. Oligocene and Early Miocene phytoliths from CRP-2/2A and CRP-3, Victoria Land Basin, Antarctica. *Terra Antartica*, **8**, 407–422.

Webb, P.-N. 1989. Benthic foraminifera. *In*: Barrett, P. J. (ed.) *Antarctic Cenozoic History from the CIROS-1 Drillhole, McMurdo Sound*. New Zealand Department of Science and Industrial Research, Wellington, Research Bulletins, **245**, 99–118.

Williams, B. P., Speece, M. A. & Powell, R. D. 2011. A high-resolution, multi-channel, over-sea-ice seismic reflection survey over the Mackay Sea Valley, Granite Harbor, Antarctica. *Marine Geophysical Research*, **32**, 383–395, doi: 10.1007/s11001-011-9115-3.

Wilson, T. J. 1999. Cenozoic structural segmentation of the Transantarctic Mountains rift flank in southern Victoria Land. *Global and Planetary Change*, **23**, 105–127.

Wilson, D. S. & Luyendyk, B. P. 2009. West Antarctic paleotopography estimated at the Eocene–Oligocene climate transition. *Geophysical Research Letters*, **36**, L16302, doi: 10.1029/2009GL039297.

Wilson, G. S., Roberts, A. P., Verosub, K. L., Florindo, F. & Sagnotti, L. 1998. Magnetobiostratigraphic chronology of the Eocene–Oligocene transition in the CIROS-1 core, Victoria Land margin, Antarctica: implications for Antarctic glacial history. *Geological Society of American Bulletin*, **110**, 35–47.

Wilson, G., Tinto, K. *et al.* 2009. Use of gravity data to characterize geologic structures beneath the sea floor, New Harbor, Antarctica. *2009 ACE Symposium Abstracts*.

Wilson, D. S., Jamieson, S. S. R., Barrett, P. J., Leitchenkov, G., Gohl, K. & Larter, R. D. 2012. Antarctic topography at the Eocene–Oligocene boundary. *Palaeogeography, Palaeoclimatology, Palaeoecology*, **335–336**, 24–34.

Zachos, J. C., Pagani, M., Sloan, L., Thomas, E. & Billups, K. 2001. Trends, rhythms and aberrations in global climate 65 Ma to present. *Science*, **292**, 686–693.

Constraints on Antarctic Ice Sheet configuration during and following the Last Glacial Maximum and its episodic contribution to sea-level rise

JOHN B. ANDERSON[1]*, ALEXANDRA E. KIRSHNER[1] & ALEXANDER R. SIMMS[2]

[1]*Rice University, Houston, TX 77005-1892, USA*

[2]*Department of Earth Science, University of California, Santa Barbara, 1006 Webb Hall, Santa Barbara, CA 93106, USA*

**Corresponding author (e-mail: johna@rice.edu)*

Abstract: Marine geological studies provide a record of diachronous expansion and retreat of the Antarctic Peninsula Ice Sheet, West Antarctic Ice Sheet and East Antarctic Ice Sheet during the past *c.* 30 000 cal yr BP. Retreat of these ice sheets and Antarctica's contribution to sea-level rise was largely complete by the early Holocene. Estimates of ice sheet thickness, based on maximum grounding depths, range from 640 to 1640 m on the inner continental shelf. Grounding depths on the outer continental shelf equate to minimum thicknesses of 410–950 m. Geomorphic features indicate that retreat from the continental shelf was mostly step-wise around the continent, a result of the different factors that control ice sheet behaviour and the degree to which these factors vary regionally. Thus, the nature of post-LGM (Last Glacial Maximum) sea-level rise was episodic and believed to have been punctuated by rapid pulses triggered by individual ice stream collapse. Most of these pulses would have been of sub-metre magnitudes and below the resolution of existing sea-level curves, but they would have had significant impact on coastal evolution, especially along low-gradient coasts.

There has been general consensus that Antarctica contributed no more than 30% to the overall 120 ± 15 m rise in sea-level since the Last Glacial Maximum (LGM), with the range of uncertainty ranging from 8 to 30 m (Nakada *et al.* 2000; Denton & Hughes 2002; Huybrechts 2002; Peltier 2002; Ivins & James 2005). More recent estimates are in the 8–10 m range (Whitehouse *et al.* 2012). Peltier & Fairbanks (2006) argue that Antarctica's contribution to the overall sea-level rise came during the late stages of the rise. This is in conflict with arguments that Antarctica may have been an important contributor to Meltwater Pulse 1A (MWP 1A; *c.* 15 000 and *c.* 14 000 cal yr BP), which represents about a 20 m sea-level rise within a few centuries (Bard *et al.* 1990; Clark *et al.* 2002). This disagreement highlights the uncertainty about the magnitude of growth and decay of the Antarctic ice sheets during the most recent glacial/eustatic cycle (Anderson *et al.* 2002; Mackintosh *et al.* 2011; Weber *et al.* 2011; Livingstone *et al.* 2012).

While considerable attention has been given to the timing and magnitude of post-LGM sea-level rise, there is also a need to understand variability in the rate of rise and those factors that influenced ice sheet retreat. This is particularly important for gaining a better perspective on coastal response to sea-level rise and for refining numerical models that relate ice sheet behaviour to various forcing mechanisms.

Anderson *et al.* (2002) summarized the state of knowledge about the configuration of the Antarctic Ice Sheet during and after the LGM. Since that time there has been significant progress in this area, especially as it relates to West Antarctica. These new studies support earlier results that indicated expansion of all three components of the Antarctic cryosphere, the East Antarctic Ice Sheet (EAIS), West Antarctic Ice Sheet (WAIS) and Antarctic Peninsula Ice Sheet (APIS), during the last glacial/eustatic cycle. However, there is now good evidence for the episodic and diachronous nature of ice sheet retreat from the continental shelf. This has important implications for post-LGM sea-level rise. There has also been significant progress in our understanding of the multiple mechanisms that influence ice sheet stability, which provides an explanation for the diachronous and episodic behaviour of the ice sheets.

One of the most important marine geological advances in our understanding of marine ice sheet history has been the dramatic expansion in quality and coverage of multibeam bathymetry data. This has led to the discovery of a myriad of geomorphic features that formed beneath the expanded ice sheets and during their retreat from the continental

From: Hambrey, M. J., Barker, P. F., Barrett, P. J., Bowman, V., Davies, B., Smellie, J. L. & Tranter, M. (eds) 2013. *Antarctic Palaeoenvironments and Earth-Surface Processes*. Geological Society, London, Special Publications, **381**, 215–232. First published online July 1, 2013, http://dx.doi.org/10.1144/SP381.13

shelf. The past decade has also been a time of more systematic marine geological surveys that integrated the results of sub-bottom profiling, multibeam mapping, detailed sedimentological analyses and radiocarbon dating. The result has been improved mapping of palaeo-ice streams and improved understanding of the nature and timing of ice sheet retreat from the continental shelf. This paper summarizes these new results as they relate to Antarctica's contribution to the post-LGM sea-level rise. We benefited from several recent review papers (Wright *et al.* 2008; Bentley *et al.* 2009; Johnson *et al.* 2011; Davies *et al.* 2012; Livingstone *et al.* 2012).

Improvements in marine geological methodology and approach

With the recognition of the important role of ice streams in controlling ice sheet behaviour, recent studies have focused on the glacial troughs that were occupied by ice streams when ice sheets advanced onto the continental shelf. This approach led to the discovery of a progression of geomorphic features from inner to outer shelf that results largely from changes in the geology of the continental shelf; inner shelf areas are typically characterized by rugged, crystalline bedrock and outer shelf areas by seaward dipping sedimentary strata that decrease in age, and therefore consolidation, in an offshore direction (Anderson 1999; Wellner *et al.* 2001, 2006; Larter *et al.* 2009; Graham *et al.* 2009). Sculptured bedrock features, such as drumlins and roches moutonnées, and complex subglacial water drainage networks consisting of basins and interconnecting channels, characterize the more landward portions of troughs. The latter indicate significant subglacial water storage and movement, but there is a lack of information about when these processes were last active (Lowe & Anderson 2003; Domack *et al.* 2006; Anderson & Oakes Fretwell 2008; Smith *et al.* 2009). At the offshore transition from bedrock to sedimentary strata, there are attenuated drumlins and megaflutes, and these extend seaward into mega-scale glacial lineations (MSGL). This geomorphic progression is believed to reflect accelerated ice stream flow associated with the formation of a deforming till layer beneath ice streams as they flowed over more deformable beds (Anderson 1999; Wellner *et al.* 2006). More recent papers have emphasized the complex relationship between geomorphic features, ice–bed interactions and ice stream flow conditions (Larter *et al.* 2009; Graham *et al.* 2009; Reinardy *et al.* 2011).

Multibeam swath bathymetry has also increased our ability to identify and map features formed during ice sheet retreat from the continental shelf. Foremost among these features are grounding-zone wedges (GZW) that record pauses in grounding-line retreat (Shipp & Anderson 1997). With the advent of a newer generation of multibeam instruments came significant improvement in resolution. These new data, coupled with deep-tow side-scan sonar records from earlier cruises, have yielded images of transverse ridges (moraines) that reflect brief pauses in grounding line retreat, and corrugation ridges, which record episodes of ice shelf collapse and associated rapid grounding-line retreat (Shipp *et al.* 1999; Dowdeswell *et al.* 2008; Jakobsson *et al.* 2010, 2012).

The criteria for distinguishing different kinds of subglacial and glacimarine deposits have been the topic of many papers over the past three decades (see Anderson 1999 for review), allowing rigorous use of sedimentary facies to reconstruct the glacial setting during ice sheet retreat from the continental shelf and acquisition of more meaningful age constraints on ice sheet retreat from the continental shelf. In general, proximal glacimarine sediments provide the best timing of grounding-line retreat, whereas distal glacimarine facies, most notably diatomaceous glacimarine sediments, yield ages that reflect the onset of open marine conditions that may lag grounding-line retreat by thousands of years.

Age constraints on ice sheet retreat from the continental shelf are still patchy owing to the scarcity of carbonate material, especially in proximal glacimarine sediments. However, there have been advances in radiocarbon dating methods that have reduced the volume of sample needed to acquire reliable ages and provided better constraints on radiocarbon reservoir effects (Berkman & Forman 1996; Domack *et al.* 2001; Reimer *et al.* 2009; Hall *et al.* 2010). This has led to a significant increase in the number of radiocarbon ages in some areas, but there is a need to re-visit some areas and apply these new techniques. There continues to be strong reliance on acid insoluble organic (AIO) fraction ages in areas where carbonate material is lacking. These ages are strongly influenced by carbon reservoir effects and thus require significant and often poorly constrained corrections (Andrews *et al.* 1999). Progress has also been made on this front (e.g. Rosenheim *et al.* 2008; Ohkouchi & Eglinton 2008), and more recent studies have presented reasonable constraints on the ages of ice-proximal glacimarine sediments (Hillenbrand *et al.* 2010*a*; Smith *et al.* 2011).

We have attempted to estimate minimum ice thicknesses on the continental shelf using water depths at which the ice was grounded on the shelf. This estimate is based on the required buoyancy of a grounded ice shelf from the density differences of liquid to frozen water (equation 1).

$$h = (\rho_i/\rho_w)[d + \Delta d_{LGM}] \quad (1)$$

where h is calculated ice thickness, ρ_i is the approximate density of ice, ρ_w is the approximate density of sea water, d is water depth from bathymetric data and Δd_{LGM} is the sea-level difference at the LGM, which we approximate as 120 m.

The bathymetric data for water depths is from the Marine Geoscience Data System Global Multi-Resolution Topography synthesis (Carbotte *et al.* 2007; Ryan *et al.* 2009). These estimates come from areas where there are subglacial geomorphic features and where the timing of ice sheet grounding has been determined (Table 1). We also provide estimates for other areas where there are subglacial geomorphic features whose ages are unconstrained but noted as such in Table 1. Table 1 includes a quantification of our confidence in each value based on data quality for each measurement. Locations where ice was not grounded during the LGM, based on a literature review (see following sections), are also indicated in Table 1. The purpose of this exercise is to provide the palaeo-ice sheet community estimates of the ice sheet thickness on the shelf that compliment the onshore data of minimum ice thicknesses, data that are needed for ice sheet reconstructions. Additionally, we provide an assessment of the width of the continental shelf around Antarctica. We measure the shelf width from the modern grounding line to the shelf break along major interpreted flow paths or perpendicular to the modern coastline.

Marine record of ice sheet expansion and retreat

Results from marine geological surveys of the past decade have yielded compelling evidence for diachronous behaviour of both the EAIS and WAIS. Likewise, data from the interior of the ice sheet suggests little to no thickening (e.g. Hein *et al.* 2011) to as much as several hundreds of metres at some locations along the coast (Stone *et al.* 2003). Our minimum constraints suggest that the trend of variable thickening of the ice sheets around its perimeter and the present-day coastline continued offshore.

In this paper we summarize the current state of knowledge of ice sheet advance onto and retreat from the Antarctic continental shelf and implications with regard to Antarctica's contribution to post-LGM sea-level rise. Our discussion is divided by region and ice sheet and focuses on those areas where detailed marine geological investigations have been conducted, including multibeam mapping of geomorphic features, sediment facies analysis and radiocarbon dating. We report all ages as calibrated (cal yr BP), using previously reported data where possible and otherwise relying on calibrated ages provided by Livingstone *et al.* (2012).

Antarctic Peninsula ice sheet

It is now well established that an ice sheet covered the Antarctic Peninsula and its continental shelves during the LGM. Several glacial troughs occur on the western side of the peninsula; all of which have been surveyed and found to contain subglacial geomorphic features (Canals *et al.* 2002; Ó Cofaigh *et al.* 2002, 2005*a*; Evans *et al.* 2004; Heroy & Anderson 2005; Willmott *et al.* 2007). The two largest ice streams flowed through Boyd Strait and Marguerite Trough (Figs 1 & 2). Smaller ice streams flowed into Bransfield Strait and onto the Antarctic Peninsula shelf. Ice was also grounded on the South Shetland Plateau and in the bays of the South Shetland Islands (Banfield & Anderson 1995; Simms *et al.* 2011), and an ice shelf covered Bransfield Strait (Heroy *et al.* 2008). Between the South Shetland Islands and Marguerite Bay, there were several small ice streams that extended to the continental shelf break (Heroy & Anderson 2005; Livingstone *et al.* 2012), (Fig. 2). Within each of these troughs there is a transition from drumlins and megaflutes on the inner shelf to MSGL on the outer shelf, and seismic profiles from the area show that this transition corresponds to the contact between crystalline bedrock and sedimentary strata (Heroy & Anderson 2005).

The largest ice stream on the western side of the Antarctic Peninsula had a catchment area in Marguerite Bay and extended to the continental shelf break through Marguerite Trough (Fig. 1). Marguerite Bay is characterized by an elaborate subglacial drainage system comprising interconnected basins and a range of channel morphologies that becomes more organized seaward (Anderson & Oakes Fretwell 2008). This seascape transitions seaward into attenuated drumlins and megaflutes and further seaward into MSGL that extend more or less uninterrupted along the axis of Marguerite Trough (Ó Cofaigh *et al.* 2002, 2005*a*). A separate ice cap occupied Alexander Island with two tidewater outlets that extended to the outer continental shelf (Graham & Smith 2012). Ice flowing from the southern-most peninsula flowed around the south side of Alexander Island and onto the shelf within Latady Trough (Graham *et al.* 2010; Hillenbrand *et al.* 2010*b*; Fig. 2).

On the eastern side of the Antarctic Peninsula there are four small troughs on the inner shelf (Vega Trough, Prince Gustav Trough, Greenpeace

Table 1. *Ice sheet grounding depths and estimated ice sheet thickness. Of note are the minimum, average and maximum approximations listed. These are based on including as many grounded ice geomorphic indicators as possible in each area. Thus we are able to include multiple depths at which ice was grounded. The best estimate of true maximum ice thickness is based on the maximum grounded ice depths in each area and the corresponding calculated ice thicknesses*

Region	Maximum grounded depth inner shelf (m)*†				Minimum ice thickness inner shelf (m)‡			Maximum grounded depth outer shelf trough (m)*†			
	Confidence‖	Minimum	Average	Maximum	Minimum	Average	Maximum	Confidence‖	Minimum	Average	Maximum
Wilkes Land	1	−784	−1102	−1305	738	1091	1316	5	NA	NA	NA
Amery	1	−783	−804	1824	737	760	782	4	666	666	666
Queen Maud Land	1	NG	NG	NG	NG	NG	NG	5	NA	NA	NA
Weddell Sea	3	−600	−620	1700	533	556	644	5	NA	NA	NA
Antarctic Peninsula (AP) average	1	−495	−1054	−1599	416	1038	1643	4	491	680	973
Northern AP (west side)	1	−828	−1063	−1406	787	1048	1429	4	973	973	973
Southern AP-Marguerite Trough (west side)	1	−1599	−1599	−1599	1643	1643	1643	1	577	577	577
East Side AP	1	−495	−906	−1084	416	873	1071	1	491	491	491
Amundsen Sea	1	−942	−1149	−1451	913	1143	1479	2	707	771	836
Ross Sea average	1	−633	−865	−1232	570	828	1236	5	NA	NA	NA
Eastern Ross (Outer Shelf)	1	−633	−715	−789	570	661	753	5	NA	NA	NA
Western Ross (Inner and Outer Shelf)	1	−796	−1014	−1232	751	994	1236	5	NA	NA	NA

*Depths are metres below modern sea-level.
†Bathymetry data from Marine Geoscience Data System (MGDS; www.marine-geo.org), Global Multi-Resolution Topography synthesis (Ryan *et al.* 2009; Carbotte *et al.* 2007).
‡Minimum ice thickness based on ice thickness = (10/9) water depth from the density contrast of frozen vs liquid water and 120 m sea-level lowering at LGM.
§Shelf width measured perpedicular to the coastline from the coastline to the shelf break, or along major flow path.

Trough and Robertson Trough) that contain subglacial geomorphic features, mainly drumlins and megaflutes (Camerlenghi *et al.* 2001; Evans *et al.* 2005). These converge seaward into two cross-shelf troughs with MSGL (Heroy & Anderson 2005; Reinardy *et al.* 2011).

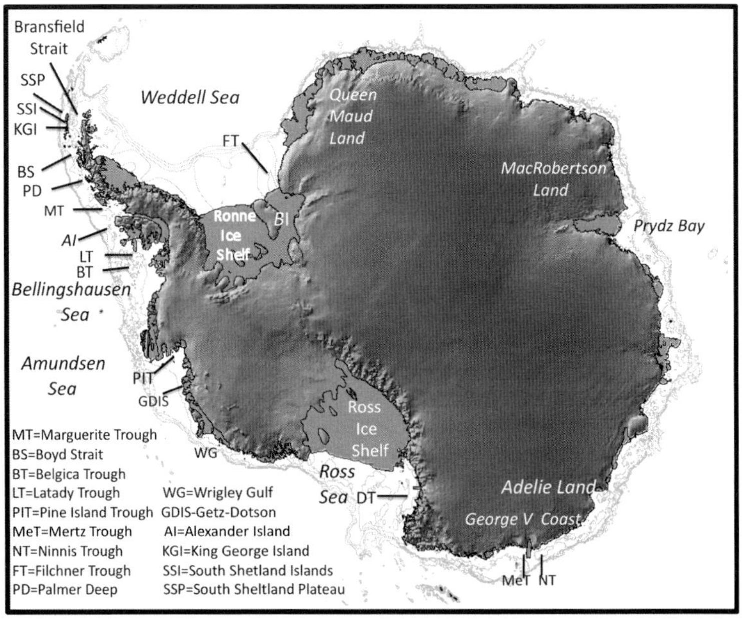

Fig. 1. Geographic locations referred to in text.

Minimum ice thickness outer shelf trough (m)‡				Average depth total continental shelf (m)†			Minimum ice thickness continental shelf (m)‡			Shelf width (km)§		
Minimum	Average	Maximum	Confidence‖	Minimum	Average	Maximum	Minimum	Average	Maximum	Minimum	Average	Maximum
NA	NA	NA	3	321	457	588	223	374	520	63	141	204
606	606	606	3	211	420	518	101	334	442	116	406	776
NG	NG	NG	3	NG	NG	NG	NG	NG	NG	30	131	253
NA	NA	NA	3	325	446	566	228	363	495	692	977	1300
413	623	947	3	135	401	729	17	312	677	145	260	566
947	947	947	3	135	410	729	17	323	677	160	188	202
508	508	508	3	353	413	474	258	326	394	336	451	566
413	413	413	3	279	385	448	177	294	364	145	240	386
652	724	795	3	372	556	619	280	485	555	199	384	608
NA	NA	NA	3	493	566	637	415	496	575	589	931	1199
NA	NA	NA	3	493	557	625	415	485	562	589	839	969
NA	NA	NA	3	552	585	637	479	517	575	1003	1117	1199

‖Confidence based on data quality: 1, high confidence, max depth from mapped streamlined geomorphic features; 2, medium confidence, crystalline geomorphic features, age unconstrained; 3, low confidence, based solely on GMRT data; 4, lower confidence, from only one location; 5, no data.
NA, no data available; NG, not grounded.

The timing of ice sheet retreat from the western Antarctic Peninsula continental shelf is relatively well constrained. Initial retreat from the shelf margin occurred *c.* 18 000 years ago with punctuated landward shifts of the grounding line across the continental shelf at *c.* 14 000 and *c.* 11 000 cal yr BP (Pope & Anderson 1992; Pudsey *et al.* 1994; Ó Cofaigh *et al.* 2005*a*; Heroy & Anderson 2007; Bentley *et al.* 2011). This was followed by diachronous retreat from the inner shelf and inland bays and passages (Evans *et al.* 2004; Heroy & Anderson 2007; Bentley *et al.* 2011; Kilfeather *et al.* 2011; Simms *et al.* 2011). Retreat of grounded ice around the margins of Bransfield Strait occurred initially at *c.* 17 300 cal yr BP, when the grounding line stepped landward from just over 700 m water depth to a shallower mid-shelf location by *c.* 9000 cal yr BP (Banfield & Anderson 1995). On King George Island, ice that was grounded in the deepest part of Maxwell Bay retreated from the bay at *c.* 14 800–14 100 cal yr BP (Milliken *et al.* 2009; Simms *et al.* 2011). Palmer Deep, which is located on the inner shelf, was ice-free by *c.* 13 000 cal yr BP (Domack *et al.* 2001, 2006). Further south, the grounding line within Marguerite Trough was situated near the mouth of Marguerite Bay by *c.* 12 000 cal yr BP. Marguerite Bay, including Neny Fjord in the inner most part of the bay, experienced open marine conditions by *c.* 9600 cal yr BP (Ó Cofaigh *et al.* 2002, 2005*a*; Heroy & Anderson 2007; Allen *et al.* 2010; Bentley *et al.* 2011).

On the eastern side of the Antarctic Peninsula, retreat of the ice sheet from the outer shelf was initiated by *c.* 18 500 cal yr BP (Heroy & Anderson 2007), followed by retreat from the inner shelf between *c.* 13 000 and *c.* 11 000 cal yr BP (Brachfeld *et al.* 2003; Pudsey *et al.* 2006; Evans *et al.* 2005). By *c.* 9400 cal yr BP grounded ice had retreated from the shelf entirely and diatomaceous glacimarine sediments began accumulating in the Firth of Tay, which is situated at the northern tip of the Peninsula (Michalchuk *et al.* 2009).

The timing of ice sheet retreat from the Antarctic Peninsula continental shelf has been related to results from land studies of ice sheet thinning (Bentley & Anderson 1998; Bentley *et al.* 2006, 2011). These land-based studies have demonstrated that lowering of the surface of the ice sheet on the west side of the peninsula was complete by the early Holocene, which is consistent with the marine geological results. The results from the eastern side of the peninsula do not constrain thinning of the ice sheet prior to *c.* 7200 years ago (Bentley *et al.* 2006).

West Antarctic ice sheet

The northern-most ice stream of the WAIS flowed onto the Bellingshausen Sea continental shelf

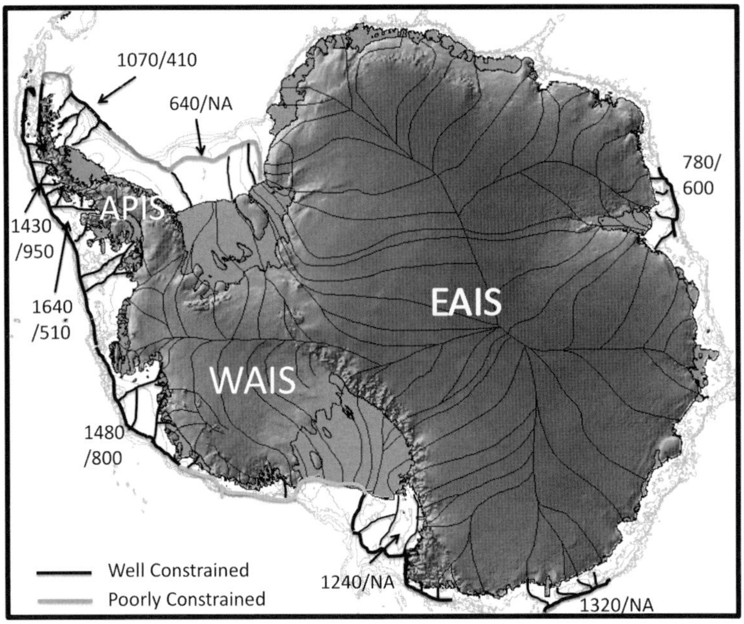

Fig. 2. LGM reconstruction for the Antarctic Peninsula region. Heavy black and grey lines depict maximum grounding line positions. Thin lines show main ice streams. Also shown are maximum ice sheet thickness values for the inner (upper number) and outer shelf (lower number). The maximum thickness values are used as they provide the maximum depth of observed grounded ice based on geomorphic data and are thus the best estimate of true maximum ice sheet thickness.

through Eltanin Bay and across the shelf through Beligica Trough (Fig. 2). Multibeam data collected on the outer and middle shelf reveal a series of back-stepping grounding-zone wedges, suggesting episodic grounding-line retreat (Ó Cofaigh *et al.* 2005*b*). Data from within Eltanin Bay reveal drumlins that merge seaward into MSGL (Wellner *et al.* 2001). Hillenbrand *et al.* (2009, 2010*a*) acquired radiocarbon ages that indicate grounding-line retreat from the outer to middle shelf between *c.* 30 000 cal yr BP and *c.* 24 000 cal yr BP. Deglaciation of the inner shelf occurred by *c.* 14 000 cal yr BP, and by *c.* 4500 cal yr BP the ice stream had retreated into Ronne Entrance.

One of the major marine geophysical accomplishments of the last decade was detailed swath bathymetry mapping of the Amundsen Sea continental shelf (Nitsche *et al.* 2007, 2013; Graham *et al.* 2010; Jakobsson *et al.* 2010, 2012). These new maps reveal large troughs where ice streams once flowed across the continental shelf (Fig. 2). Seismic profiles show a rugged inner shelf with little or no sediment cover and a seaward-dipping wedge of sedimentary strata on the middle and outer shelf (Lowe & Anderson 2003; Graham *et al.* 2009). Geomorphic features on the inner shelf include drumlins, roches moutonnées and a well-organized subglacial drainage network of interconnected basins and channels (Lowe & Anderson 2002; Graham *et al.* 2009). The middle and outer shelf is characterized by MSGL and GZWs with superimposed iceberg furrows and smaller-scale recessional features including transverse ridges and corrugation ridges (Lowe & Anderson 2003; Graham *et al.* 2010; Jakobsson *et al.* 2010, 2012).

Pine Island Trough is the dominant bathymetric feature on the eastern Amundsen Sea shelf (Nitsche *et al.* 2007). Recent marine geological studies have revealed as many as five grounding-zone wedges in the middle and outer part of the trough that record step-wise retreat of the grounding line (Lowe & Anderson 2003; Evans *et al.* 2006; Graham *et al.* 2010; Jakobsson *et al.* 2011, 2012). Kirshner *et al.* (2012) conducted a detailed sediment facies analysis using sediment cores from the bay and provided chronological constraints for grounding-line retreat from the shelf. Their results indicate that initial retreat from the outer shelf occurred shortly after the LGM, with the grounding line shifting from the outer shelf to the central shelf by *c.* 16 400 cal yr BP. They argue that initial retreat from the outer shelf was probably driven by rapid sea-level rise during this time interval. Following this, another landward shift in the grounding line occurred, leaving central Pine Island Bay (PIB) covered by an ice shelf between *c.* 12 300 and *c.* 10 600 cal yr BP. At *c.* 10 600 cal yr BP, rapid grounding-line retreat was prompted by an episode of ice shelf collapse that is believed to have been triggered by warm deep-water incursion onto the shelf (Kirshner *et al.* 2012). This final phase of retreat left PIB ice-free by early Holocene time (Hillenbrand *et al.* 2013). This final phase of grounding-line retreat may have been triggered by the escape of subglacial water from basins on the inner shelf (Lowe & Anderson 2002; Kirshner *et al.* 2012). The evidence for this occurs in the form of a draping silt unit that is devoid of microfossils and ice-rafted debris that is interpreted as a surface plume deposit and a single core from one of the channels that sampled graded gravel and sand.

Detailed marine geological studies have also been conducted in the western Amundsen Sea and these studies yielded a glacial retreat history similar to that of the eastern Amundsen Sea (Graham *et al.* 2009; Hillenbrand *et al.* 2010*a*, 2013; Smith *et al.* 2011). These studies focused on the largest trough in the area, which extends seaward from the Dotson and Getz ice shelves (Fig. 1). The geomorphic features within this trough show the same inner to outer shelf progression as observed in Pine Island Trough. The chronological constraints on grounding-line retreat are based mainly on AIO dates from distal glacimarine sediments. These data indicate that the mid-shelf was ice-free by *c.* 13 800 cal yr BP and the inner shelf by *c.* 12 600–*c.* 10 000 cal yr BP. These data are consistent with the ice sheet retreat history from the eastern Amundsen Sea (Kirshner *et al.* 2012; Hillenbrand *et al.* 2013), and with cosmogenic surface-exposure ages from the onshore eastern Amundsen Sea, which indicate progressive thinning of the WAIS since *c.* 14 500 years ago (Johnson *et al.* 2008).

At least four other troughs occur on the Amundsen Sea continental shelf between 120° and 135°W (Nitsche *et al.* 2007). Multibeam data, a single seismic profile and sediment cores have been collected in only one of these troughs, which is situated in the Wrigley Gulf (Fig. 1), between 127° and 130°W (Wellner *et al.* 2001, 2006; Anderson *et al.* 2002). These data show the same offshore progression of geomorphic features observed elsewhere in the Amundsen Sea, with drumlins and megaflutes on the rugged bedrock of the inner shelf transitioning seaward into MSGL on the mid-shelf (Wellner *et al.* 2001). The outer shelf was not surveyed. Two sediment cores from the mid-shelf that sampled glacimarine sediments resting above till yielded carbonate radiocarbon ages of between *c.* 12 300 and *c.* 12 500 cal yr BP, indicating that the outer and middle shelf was free of grounded ice by this time (Anderson *et al.* 2002).

Eastern Ross Sea

Both the EAIS and WAIS drain into Ross Sea and results from both till provenance (Anderson *et al.* 1992; Licht *et al.* 2005; Farmer *et al.* 2006) and geomorphic evidence (Shipp *et al.* 1999; Mosola & Anderson 2006) can be used to place the drainage divide between ice streams draining the expanded EAIS and WAIS (Fig. 2). However, the interaction between the two ice sheets was complex and appears to have varied through time, especially in the McMurdo Sound area, where westward drainage from the WAIS occurred (Denton & Hughes 2000; Denton & Marchant 2000; Greenwood *et al.* 2013). Greenwood *et al.* (2013) used newly acquired multibeam swath bathymetry data from the south-western Ross Sea to reconstruct the palaeodrainage changes that took place between the EAIS and the WAIS following the LGM.

The eastern Ross Sea continental shelf has three large troughs where drainage from the WAIS was focused (Fig. 2). Swath bathymetry data show MSGLs that extend hundreds of kilometres along the axes of these troughs to the shelf margin (Mosola & Anderson 2006). Seismic profiles and multibeam data collected along the axes of each of the troughs show variable numbers and spacing of GZWs, indicating diachronous retreat of individual WAIS ice streams (Mosola & Anderson 2006). Deep (up to 650 m) iceberg furrows that overprint the MSGL indicate that the overall retreat was punctuated by episodes of rapid break-up, resulting in calving at the ice sheet grounding line. Transverse ridges and corrugation ridges record these episodes of rapid retreat (Mosola & Anderson 2006; Jakobsson *et al.* 2011).

Early attempts to constrain the post-LGM retreat of the WAIS from the eastern Ross Sea were based on 39 AIO ages (Mosola & Anderson 2006) and three carbonate ages (Licht & Andrews 2002). Mosola & Anderson (2006) suggested retreat immediately following, and perhaps even during, the LGM, with ages of the oldest glacimarine sediments ranging back to *c.* 26 700–*c.* 14 000 cal yrs BP. With only two exceptions, the ages are in stratigraphic order. Even with considerable reservoir correction (average surface age of 3990 cal yr BP), these data indicate early retreat of the WAIS from the shelf. These results are supported by three carbonate ages obtained by Licht & Andrews (2002) that, while out of stratigraphic order, indicate glacimarine sedimentation on the shelf as early as 21 000 cal yr BP. Later, Bart & Cone (2011) obtained radiocarbon ages on benthic foraminifera from proximal glacimarine sediments and AIO ages from diatomaceous glacimarine sediments from eastern Ross Sea that support early retreat of the WAIS from the continental shelf. The Bart & Cone (2011) reconstruction differs from that of Mosola & Anderson (2006) in that it does not call for ice having been grounded on the outer continental shelf. In either case, marine evidence calls for significant retreat of the WAIS from the eastern Ross Sea continental shelf prior to the LGM. This is in conflict with Conway *et al.* (1999) and Hall *et al.* (2010) which call for WAIS retreat from the continental shelf during the Holocene.

Southern Weddell Sea

Like the Ross Sea, the Weddell Sea is an outlet for ice discharge from both the EAIS and WAIS. The palaeodrainage divide is generally located within Filchner Trough, with the EAIS drainage to the east and the WAIS drainage to the west (Hillenbrand *et al.* 2012; Stolldorf *et al.* 2012).

Although the Weddell Sea was the first area to yield evidence of ice sheet expansion onto the continental shelf (Anderson 1972), it remains the most understudied marine sector of West Antarctica. This situation is due to severe sea ice conditions, especially over the southern continental shelf where the WAIS would have grounded during the LGM. Bentley *et al.* (2010) provide results of numerical modeling that indicates WAIS expansion across the continental shelf during LGM and glacial geological data that indicates progressive retreat from the sheet since *c.* 15 000. A recent summary of marine geological data from the southern Weddell Sea by Hillenbrand *et al.* (2012) led to the conclusion that the WAIS was grounded on the continental shelf during the LGM, but provided little new chronostratigraphic constraint on the timing of ice sheet retreat from the shelf.

A recent study of multibeam data from the area directly north of the Ronne Ice Shelf by Stolldorf *et al.* (2012) revealed MSGL in three troughs on the southern shelf and provided key information about where ice streams that drained the ice sheet existed (Fig. 2). The pristine nature of the MSGL in these troughs suggests that they were formed during the LGM, but there have been very few long cores collected on this part of the shelf and very few radiocarbon ages exist. Core IWSOE 2-19-1 from the outermost shelf yielded an inverted radiocarbon stratigraphy dating back to *c.* 17 900 cal yr BP. Another core from the western flank of Filchner Trough (IWSOE 69-10) yielded one age (48 212 cal yr BP) that extends beyond the reliable range of radiocarbon dating. The only other sample from Filchner Trough yielded Holocene ages, so they do not help constrain ice sheet retreat history (Crawford *et al.* 1996; Hillenbrand *et al.* 2012; Stolldorf *et al.* 2012).

East Antarctic Ice Sheet

George V and Adelie Land continental shelf. The first marine geological investigation off East Antarctica aimed at reconstructing the LGM extent of the EAIS was conducted in 1979 and yielded evidence, in the form of tills, that the ice sheet had grounded on the continental shelf during the LGM within two troughs, the Mertz and Ninnis troughs (Fig. 2; Anderson *et al.* 1980; Domack 1982). More recently, swath bathymetry data collected within Mertz Trough has revealed spectacular backstepping GZWs and MSGL (McMullen *et al.* 2006). Radiocarbon constraints on the retreat of the ice sheet from the continental shelf indicate that deglaciation of the inner shelf was complete by *c.* 13 600 cal yr BP (Maddison *et al.* 2006; Leventer *et al.* 2006) with open marine conditions supporting diatomaceous sediment accumulation by *c.* 9000 cal yr BP (Domack *et al.* 1989). Results from onshore studies indicate expansion of the ice sheet in coastal areas during the LGM (Goodwin & Zweck 2000).

Western Ross Sea. Swath bathymetry records show that MSGL do not extend to the shelf margin in the western Ross Sea, where the EAIS would have been grounded (Fig. 2). This is in contrast to the eastern sector of the continental shelf where the WAIS advanced to the shelf margin (Mosola & Anderson 2006). In fact, banks on the outer western shelf are covered by bioclastic carbonates, including *in situ* assemblages that have yielded LGM radiocarbon ages, thus indicating the existence of benthic refugia on the outer shelf (Taviani *et al.* 1993).

Retreat of the EAIS from the western Ross Sea was more gradual and lagged the WAIS retreat from eastern Ross Sea, probably owing to the presence of banks that provided penning points for the ice sheet (Shipp *et al.* 1999; Domack *et al.* 1999; Howat & Domack 2003). The grounding line was situated near Drygalski Trough *c.* 11 000 cal yr BP, and by *c.* 7800 cal yr BP had retreated to the vicinity of Ross Island (Licht *et al.* 1996; Domack *et al.* 1999; McKay *et al.* 2008). The grounding line may have stabilized near Ross Island for approximately 2500 years (Licht *et al.* 1996; McKay *et al.* 2008), whereas the onshore record suggests that, by *c.* 6800 cal yr BP, it had shifted approximately 300 km to the south to the location of Hatherton Glacier (Stuiver *et al.* 1981; Hall & Denton 2000*a*, *b*; Dochat *et al.* 2000). In either case, the grounding line shifted landward approximately 1000 km after the LGM. The modern Ross

Ice Shelf covers the entire inner shelf and therefore masks the late stages of retreat of the ice sheet.

MacRobertson Land continental shelf and Prydz Bay. Harris & O'Brien (1998) placed the maximum LGM grounding line of the EAIS on the outer MacRobertson Land continental shelf (Fig. 2) based on the presence of a 'grounding line moraine'. More recently, marine geological investigations of the area focused on two troughs, Nielsen Basin and Iceberg Alley (Leventer *et al.* 2006). Both troughs are characterized by rugged bedrock with drumlins on the inner shelf and both have portions with MSGL and superimposed GZWs. The maximum extent of the ice sheet is marked by GZWs at the shelf margin in both troughs. Harris & O'Brien (1998) acquired a single AIO date from glacimarine sediments of 17 150 cal yr BP. More recently, Mackintosh *et al.* (2011) acquired radiocarbon dates indicating deglaciation of the troughs *c.* 14 000 ^{14}C yr BP, followed by retreat across the inner shelf between *c.* 12 000 and *c.* 7000 cal yr BP. In Prydz Bay (Fig. 1), Svenner Channel on the eastern side of the bay contains likely drumlins and megaflutes, indicating that ice was grounded there at some time in the past (O'Brien & Harris 1996). Sediment cores from Prydz Bay contain diatomaceous glacimarine sediments that date back to *c.* 10 700 ^{14}C yr BP, indicating open marine conditions by this time, with the actual timing of ice sheet retreat from the area being *c.* 12 000 cal yr BP (Domack *et al.* 1998).

Eastern Weddell Sea. The only other location where detailed marine geological investigations have been conducted off East Antarctica is the eastern Weddell Sea, east of Filchner Trough and offshore of Queen Maud Land (Fig. 1). Early studies of the area provided evidence, in the form of tills, of the ice sheet having grounded on the continental shelf, but no radiocarbon ages were obtained from continental shelf sediments (Anderson *et al.* 1980; Elverhøi 1981). The first radiocarbon ages derived from shelf sediments yielded dates in excess of 20 000 cal yr BP, which led to the conclusion that the EAIS did not advance onto the shelf during the LGM (Anderson & Andrews 1999). This argument was countered by a recent study of contourite deposits on the deep-sea floor of the eastern Weddell Sea (Weber *et al.* 2011). Weber *et al.*'s argument is based on deep-sea sedimentary evidence, specifically alternations in laminated v. massive sediments. They argued that these record changes in ice sheet extent that influenced the establishment of coastal polynyas, which in turn led to increased bottom water production and contourite sedimentation. A more direct LGM and post-LGM record for the area is provided by Stolldorf *et al.* (2012). These authors conducted detailed sedimentological analyses of sediment cores from the eastern continental shelf that revealed a retreat stratigraphy with distal glacimarine sediments overlying proximal glacimarine sediments and till. They acquired 21 new radiocarbon ages from foraminifera taken from glacimarine sediments on the shelf. The radiocarbon ages are, with few exceptions, in stratigraphic order and span the last 30 000 years, or to the limits of reliable radiocarbon dating. These authors concluded that the ice sheet was not grounded on the continental shelf during this time interval, a conclusion that is at odds with Hillenbrand *et al.* (2012), who suggested that a low profile ice sheet may have existed in the region during the LGM.

Stolldorf *et al.*'s (2012) results are consistent with cosmogenic nuclide measurements from Hein *et al.* (2011), which show that two East Antarctic glaciers, Slessor and Recovery glaciers, which drain into the southeastern corner of Weddell Sea, were no thicker during the LGM than at present. Further results from the study of an ice core from Berkner Island (Fig. 1) indicate that the island was not completely overridden by ice during the LGM (Mulvaney *et al.* 2007). Finally, these findings are consistent with model results that call for no significant expansion of the EAIS in the region during the LGM (Le Brocq *et al.* 2011).

Discussion

Size and thickness of the ice sheet during the LGM

As estimates of the amount of ice over Antarctica during the LGM continue to decline (recent estimates are around 9.0 ± 1.5 m of eustatic equivalent sea-level, Whitehouse *et al.* 2012), most of that ice mass continues to be placed on the continental shelves. Thus, in addition to constraints on the thickness of the LGM ice sheet from coastal areas through the inversion of sea-level data (e.g. Zwartz *et al.* 1998) and exposure-age dating of trim lines and moraines (e.g. Bentley *et al.* 2006), the most important constraints on the thickness of the LGM ice sheets concerns the thickness of the ice sheets on the shelf. Table 1 provides one of the first syntheses of these minimum-limiting values.

The ice streams were grounded in troughs on the continental shelf in maximum water depths between *c.* 700 and *c.* 1600 m on the inner shelf. These grounding depths translate to a first-order approximation of minimum ice thicknesses between *c.* 780 and *c.* 1640 m (Table 1, Fig. 3). Grounding

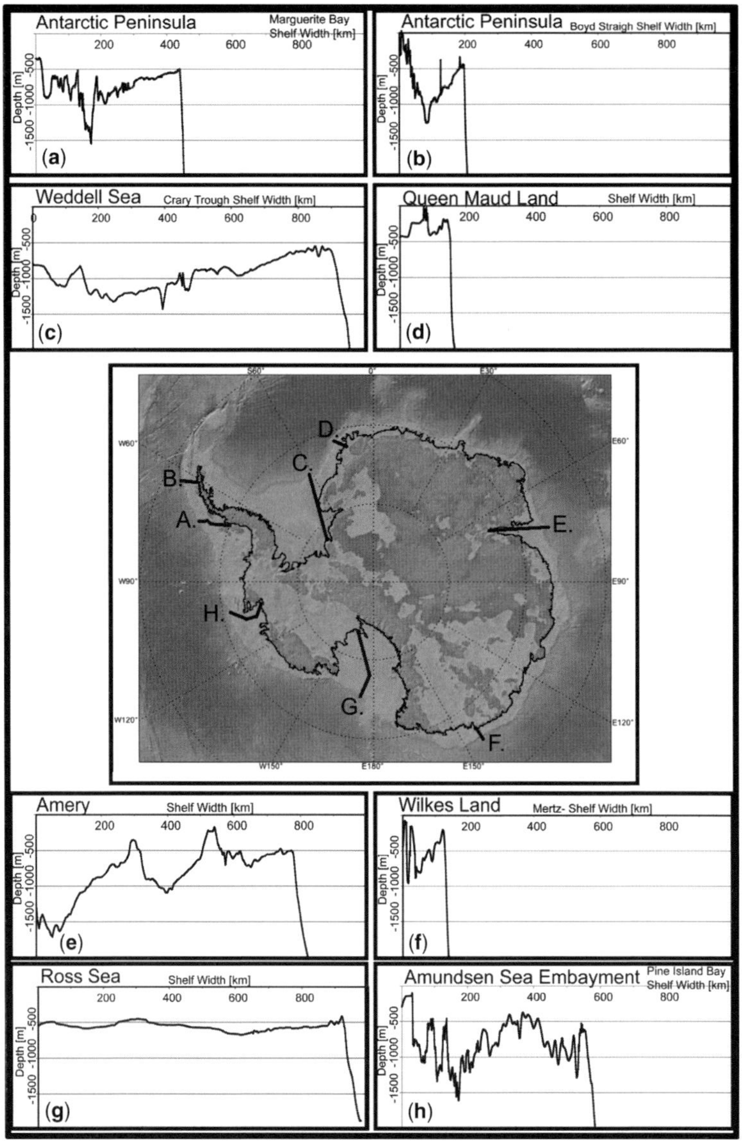

Fig. 3. Representative bathymetric profiles from the Antarctic continental shelf illustrating variations in shelf physiography. Profiles were constructed along major trough axes. Central map of Antarctic displays location of bathymetric profiles. All profiles are oriented from the coastline (left) to offshore (right). (**a**) Antarctic Peninsula region, profile from the Marguerite Bay Trough; (**b**) Antarctic Peninsula region, profile from the Boyd Straight; (**c**) Weddell Sea region, profile from Crary Trough; (**d**) Queen Maud Land; (**e**) Amery (Prydz Bay) region, profile from the Amery Trough; (**f**) Wilkes Land, profile from the Mertz Trough; (**g**) Ross Sea, profile from central Ross Sea Trough; (**h**) Amundsen Sea Region, from Pine Island Trough.

depths on the outer continental shelf equate to minimum thicknesses of *c.* 600–*c.* 950 m. However, the width of the continental shelves around Antarctica varies (Fig. 3), so the profiles of the expanded ice sheets vary widely.

Causes of diachronous and episodic retreat

From the above review of marine ice sheet retreat in Antarctica, the evidence of diachronous retreat is compelling. We argue that this is expected given the number of factors that control marine ice sheet stability and how they would have varied around the continent. These include: (1) the highly variable size of individual drainage basins and volumes of ice flowing toward the coast; (2) differences in the physiography and geology of the continental shelves on which the ice sheets advanced; (3) differences in isostatic depression and rebound; (4) potential for intrusion of warm deep water onto the continental shelf and associated erosion of the ice shelf; (5) variable climatic conditions around the continent; and (6) potential for focused subglacial water outflow leading to underpinning of ice streams.

The modern ice sheets are characterized by considerable variability in the size of glacial drainage basins and the degree to which ice diverges or converges as it flows toward the coast (Fig. 2). The expanded LGM ice sheets were also characterized by spatial and temporal variability in drainage, which was due largely to the fact that they expanded onto continental shelves with highly variable physiography and geology (Fig. 3). The Antarctic Peninsula and West Antarctic continental shelves are generally wider than in East Antarctica (Table 1, Fig. 3). Thus, the APIS and the WAIS were able to advance greater distances across their wide continental shelves (Fig. 3).

There is now reasonably well-constrained palaeodrainage information for the Antarctic Peninsula and West Antarctic continental shelves, based on lineation patterns within troughs (Fig. 2; see Davies *et al.* 2012 and Livingstone *et al.* 2012 for greater detail). Changes in trough orientation and geometry, and in the orientations of lineations within these troughs, indicate that drainage and discharge of individual ice streams changed as the ice sheets retreated from the continental shelf (Mosola & Anderson 2006; Ó Cofaigh *et al.* 2008; Livingstone *et al.* 2012). APIS drainage during the LGM was focused in large ice streams (Fig. 2). As the ice sheet retreated from the shelf, drainage reorganized and became more isolated with flow into smaller ice streams and outlet glaciers (Camerlenghi *et al.* 2001; Heroy & Anderson 2005; Reinardy *et al.* 2011). In West Antarctica, larger more isolated ice streams on the outer shelf diverged into two or more smaller ice streams as the grounding line migrated landward (Shipp *et al.* 1999; Mosola & Anderson 2006; Smith *et al.* 2011; Graham & Smith 2012; Livingstone *et al.* 2012; Jakobsson *et al.* 2012). Most of the East Antarctic continental shelf has not been surveyed using multibeam technology. In general, there are fewer troughs relative to other areas and the ones that exist are associated with modern ice streams. However, the marine geological data that have been collected in East Antarctic troughs reveal that they were occupied by ice streams (Shipp *et al.* 1999; McMullen *et al.* 2006; Leventer *et al.* 2006; Mackintosh *et al.* 2011). Existing bathymetric maps suggest that it is unlikely that additional large troughs are yet to be discovered off East Antarctica (Arndt *et al.* 2013).

Combined multibeam swath bathymetry mapping and seismic profiling have revealed a strong pattern in subglacial bedforms within troughs and geology that is interpreted as reflecting acceleration of ice streams as they flowed from the rugged, nondeformable crystalline bedrock of the inner shelf onto sediments on the outer shelf (Anderson 1999; Wellner *et al.* 2006; Graham *et al.* 2009; Larter *et al.* 2009). There is considerable variability in the location of this geological boundary around the continent (Anderson 1999), so there was considerable spatial variability in geological influence on ice stream behaviour. This effect would have been compounded by variations in trough physiography, as evidenced by the association of GZW's with bottlenecks within some troughs (e.g. Mosola & Anderson 2006; Jakobsson *et al.* 2012).

Eustatic sea-level changes undoubtedly played a key role in the advance of the Antarctic Ice Sheet onto the continental shelf and its early retreat from the shelf, providing a connection between Northern and Southern Hemisphere glaciation (Huybrechts 2002; Pollard & DeConto 2009). However, in areas close to the ice sheet, glacial-isostatic adjustments may overwhelm global changes in ocean volume and hence relative sea-level change (e.g. Lambeck & Chappell 2001). Without knowledge of the ice sheet thickness, it is not possible to calculate isostatic depression of the shelf. Given the variability in grounding depths (Table 1), it is expected that the amount of isostatic depression varied around the continent. Thus, thresholds of ice-sheet stability owing to rates of relative sea-level change would have been crossed at different times in different areas. The isostatic influence on ice sheet stability has only recently been investigated (Gomez *et al.* 2010) and has been largely ignored by those of us working on marine ice sheet histories, but it is indeed a significant factor regulating past marine ice sheet behaviour and the influence of eustatic rise must be counter-balanced with isostatic uplift estimates. Eustatic events like MWP1-A may have been so rapid as to transcend any isostatic effects.

It is also believed that the destabilizing influence of relative sea-level rise may have been augmented by extension and acceleration of ice streams that flowed across deformable sediments of the outer shelf. Without a corresponding increase in precipitation, this could have resulted in 'draw-down' (thinning) of the ice sheet margin that, in turn, could have triggered ice sheet retreat from the outer shelf. Mosola & Anderson (2006) argue that this may have been a key factor contributing to WAIS grounding-line retreat in eastern Ross Sea, where ice streams that extended hundreds of kilometres onto the continental shelf began their retreat from the shelf prior to the LGM. The same was probably true for the ice streams that occupied the extended troughs of the Antarctic Peninsula, such as Marguerite Trough where retreat is argued to have been rapid (Ó Cofaigh *et al.* 2005*a*). When the combined effects of isostatic loading of the continental shelf, rapid sea-level rise in the early post-LGM period from Northern Hemisphere deglaciation and over-extension of ice streams onto the continental shelf and associated thinning of the marine margins are considered, it is not at all surprising that the record of deglaciation of the outer continental shelf occurred during or shortly after the LGM in virtually every area where the timing has been constrained.

As the grounding line migrated landward, there was progressive isolation of glacial drainage into smaller ice streams, outlet glaciers and tidewater glaciers. By early Holocene time, eustatic rise owing to Northern Hemisphere deglaciation had begun to decrease, isostatic rebound of the Antarctic continental shelf was taking place, and the ice was largely grounded on bedrock with highly irregular bathymetry. From this time on, factors such as climate change, warm deep-water intrusion onto the continental shelf, and underpinning of the grounding line by subglacial water probably played more important roles in grounding-line retreat from the continental shelf. A review of the Antarctic and Southern Ocean climate and its relation to the global climate system over the last few millennia has revealed a complex and unstable system characterized by abrupt changes in atmospheric circulation and temperature recorded in ice cores (Mayewski *et al.* 2009). The spatial and temporal variability of the Antarctic climate adds another level of complexity to the ice sheet's behavior through time.

Holocene climate records from the marine sediments in the Antarctic Peninsula region reveal an early Holocene warming event whose onset coincided approximately with the final phase of grounding-line retreat on the eastern side of the peninsula (Brachfeld *et al.* 2003; Evans *et al.* 2005; Heroy & Anderson 2007), with marine incursion into the Firth of Tay at *c.* 9400 cal yr BP (Michalchuk *et al.* 2009), the disappearance of an ice shelf in Bransfield Strait (Heroy *et al.* 2008) and rapid ice shelf and grounding-line retreat in Marguerite Bay at *c.* 9600 cal yr BP (Heroy & Anderson 2007; Allen *et al.* 2010; Bentley *et al.* 2011; Kilfeather *et al.* 2011). Heroy & Anderson (2007) suggested that progressive north to south deglaciation of the Antarctic Peninsula may have been due, at least in part, to warming in the region. Elsewhere in Antarctica, the record of Holocene climate change is not as well known as in the Antarctic Peninsula region. Also, the size of glacial drainage basins and the ice streams and outlet glaciers draining these drainage basins is generally much larger outside the peninsula region. Results of model studies indicate that more extreme climate events are needed to destabilize these larger glacial systems (Pollard & DeConto 2009).

Warm water incursion onto the continental shelf is an important effect today (Jacobs *et al.* 1996, 2011) and there is some evidence, in the form of abundance peaks in planktonic foraminifera in sediments far removed from the shelf break, for past incursions of offshore waters onto the inner shelf and into Pine Island Bay (Kirshner *et al.* 2012) and Marguerite Bay (Kilfeather *et al.* 2011). This is believed to have triggered erosion and eventual collapse of ice shelves, ultimately resulting in rapid grounding-line retreat (Jakobbson *et al.* 2010; Kirshner *et al.* 2012).

Organized subglacial drainage systems characterize the inner parts of both Marguerite Bay and Pine Island Bay (Lowe & Anderson 2002; Anderson & Oakes Fretwell 2008; Nitsche *et al.* 2013) and may have delivered subglacial water to the catchment areas of the ice streams that occupied these bays. The evidence that water flowed through these systems during and after the LGM remains inconclusive, but includes graded gravel and sand in a single core from a channel in Pine Island Bay (Lowe & Anderson 2002) and widespread silt deposits in both Pine Island and Marguerite Bay that are believed to be dispersed by sediment-laden surface water plumes (Kennedy & Anderson 1989; Kirshner *et al.* 2012).

Antarctica's contribution to post LGM sea-level rise

Antarctica's contribution to post LGM sea-level rise, including Meltwater Pulse 1-A, is still unresolved, owing to both poor constraints on ice sheet thickness and the timing of ice sheet retreat from the continental shelf. What is known is that significant retreat occurred all around Antarctica soon after the LGM and the grounding lines of the APIS, WAIS and EAIS had retreated to the inner shelf by the beginning of the Holocene. There were two abrupt steps in the retreat of the APIS from the continental shelf at *c.* 14 000 and *c.* 11 000 cal yr BP (Heroy & Anderson 2007). This indicates that the APIS responded to the rapid rise in sea-level during this time interval and may, in turn, have contributed to MWP1-A, although its contribution would have been minor. Likewise, there was step-wise retreat of the WAIS from the Amundsen Sea continental shelf prior to and during the Holocene, but the exact timing of grounding-line retreat is not as well constrained as in the Antarctic Peninsula area. The same may be true for the Ross Sea, but the WAIS retreat history for the southern Weddell Sea is unconstrained. All that can be said about the EAIS contribution to early post-LGM sea-level rise is that in some areas the continental shelf was largely ice free before the Holocene (Stolldorf *et al.* 2012). This includes the MacRobertson Land shelf, which was ice-free by *c.* 14 000 cal yr BP. Likewise, diatomaceous glacimarine sediments were accumulating on the Wilkes Land and Prydz Bay continental shelves by *c.* 11 000 and 9000 cal yr, respectively.

Episodic sea-level rise

The combined results from marine geological studies around the Antarctic continental shelf indicate that ice sheet retreat from the continental shelf was episodic, which is expected given the multiple factors that regulate marine ice sheet behaviour. Hence, we expect that post-LGM sea-level rise would have been episodic and punctuated by periods of rapid rise caused by the collapse of individual ice streams. This episodic behaviour would have increased in frequency but decreased in magnitude during the Holocene as the ice sheets underwent diachronous retreat from the inner shelf and inland waters. The question is, what evidence is there for this episodic nature to sea-level rise?

Sea-level curves for the post-LGM rise are, with the exception of a few late Holocene records (e.g. Kemp *et al.* 2011), accurate only within a few metres (Shennan 2007). This is not precise enough to show the kinds of changes that would have been caused by collapse of individual ice streams. Pine Island ice stream, which currently shows the most disturbing signs of potential collapse, could potentially contribute several decimetres to sea-level rise (Vaughan 2008). This magnitude of sea-level rise is probably on the high end of what most ice streams would have contributed during observed back-stepping phases. Perhaps more important were the changes in the rate of sea-level rise, which would have been especially influential along low gradient coasts (Anderson *et al.* 2010). The record of estuary evolution for the US Gulf Coast reflects episodic and rapid changes that are equated to sea-level rise by virtue of the fact that some of these flooding events were contemporaneous between different Gulf Coast bays (Anderson *et al.* 2010; Rodriguez *et al.* 2010). The best documented is the *c.* 8200 cal yr BP event, which is argued to have been caused by rapid drainage of the large proglacial lakes of the Laurentide Ice Sheet (Barber *et al.* 1999; Teller *et al.* 2002; Clarke *et al.* 2003; Törnqvist *et al.* 2004; Alley & Ágústsdóttir 2005). However, this was also a time of rapid change in the Antarctic ice sheet. Indeed, there were many such events in Antarctica, so the nature of post-LGM sea-level rise was without question episodic in nature.

Conclusions

Marine geological studies have provided compelling evidence for expansion of the Antarctic Peninsula Ice Sheet, West Antarctic Ice Sheet and East Antarctic Ice Sheet during the past *c.* 30 000 cal yr BP. Estimates of ice sheet thickness, based on maximum grounding depths, range from 620 to 1150 m on the inner continental shelf and from 620 to 950 m on the outer shelf.

The retreat of the marine ice sheet from around Antarctica was diachronous and episodic in nature. This is to be expected based on the number of factors controlling ice sheet stability, including differences in shelf physiography, oceanography, climate history and ice drainage. Deglaciation of the continental shelf was largely complete by the early Holocene. Existing records remain inconclusive with regard to the source of MWP-1A, but the diachronous nature of ice sheet retreat from the continental shelf limits the possible areas where such a significant contribution could have been derived.

Modern eustatic sea-level records do not have the resolution to record the complex nature of sea-level rise since the LGM. However, there is little doubt that the nature of Holocene sea-level rise was episodic in nature and punctuated by rapid, decimetre-scale rises associated with the collapse of individual ice streams. This is in agreement with the record of evolution of low gradient coasts, which is characterized by episodic and rapid change.

Funding for this research was provided by the National Science Foundation Office of Polar Programs grant NSF/ARRA grant ANT-0837925 to J. B. Anderson. This manuscript was greatly improved by reviews from C. D. Hillenbrand and an anonymous reviewer. This paper benefited by discussions at the workshops of PALSEA, a PAGES/INQUA/WUN working group.

References

ALLEN, C. S., OAKES FRETWELL, L., ANDERSON, J. B. & HODGSON, D. A. 2010. A record of Holocene glacial oceanographic variability in Neny Fjord, Antarctic Peninsula. *The Holocene*, **20**, 551–564.

ALLEY, R. B. & ÁGÚSTSDÓTTIR, A. M. 2005. The 8 k event: cause and consequences of a major Holocene abrupt climate change. *Quaternary Science Reviews*, **24**, 1123–1149.

ANDERSON, J. B. 1972. Nearshore glacial-marine deposition from modern sediments of the Weddell Sea. *Nature*, **240**, 189–192.

ANDERSON, J. B. 1999. *Antarctic Marine Geology*. Cambridge University Press, New York.

ANDERSON, J. B. & ANDREWS, J. T. 1999. Radiocarbon constraints on ice sheet advance and retreat in the Weddell Sea, Antarctica. *Geology*, **27**, 179–182.

ANDERSON, J. B. & OAKES FRETWELL, L. 2008. Geomorphology of the onset area of a paleo-ice stream, Marguerite Bay, Antarctic Peninsula. *Earth Surface Processses and Landforms*, **33**, 503–512.

ANDERSON, J. B., KURTZ, D. D., DOMACK, E. W. & BALSHAW, K. M. 1980. Antarctic glacial marine sediments. *Journal of Geology*, **88**, 399–414.

ANDERSON, J. B., SHIPP, S. S., BARTEK, L. R. & REID, D. E. 1992. Evidence for a grounded ice sheet on the Ross Sea continental shelf during the late Pleistocene and preliminary paleodrainage reconstruction. *In*: ELLIOT, D. H. (ed.) *Contributions to Antarctic Research III*. Antarctic Research Series. American Geophysical Union, Washington, DC, **57**, 39–62.

ANDERSON, J. B., SHIPP, S. S., LOWE, A. L., WELLNER, J. S. & MOSOLA, A. B. 2002. The Antarctic ice sheet during the last glacial maximum and its subsequent retreat history: a review. *Quaternary Science Reviews*, **21**, 49–70.

ANDERSON, J. B., RODRIGUEZ, A. B., MILLIKEN, K. T., SIMMS, A. & WALLACE, D. 2010. Is predicted coastal impact from accelerated sea-level rise underestimated? *EOS*, **91**, 205–206.

ANDREWS, J. T., DOMACK, E. W. *ET AL*. 1999. Problems and possible solutions concerning radiocarbon dating of surface marine sediments, Ross Sea, Antarctica. *Quaternary Research*, **52**, 206–216.

ARNDT, J. E. *ET AL*. 2013. The International Bathymetric Chart of the Southern Ocean (IBCSO) Version 1.0. *Geophysical Research Letters*, doi: 10.1029/2013GL055394.

BANFIELD, L. A. & ANDERSON, J. B. 1995. Seismic facies investigation of late quaternary glacial history of Bransfield Basin, Antarctica. *American Geophysical Union Antarctic Research Series*, **68**, 123–140.

BARBER, D. C., DYKE, A. *ET AL*. 1999. Forcing of the cold event of 8200 years ago by catastrophic drainage of Laurentide lakes. *Nature*, **400**, 344–348.

BARD, E., HAMELIN, B. & FAIKRBANKS, R. G. 1990. U–Th ages obtained by mass spectrometry in corals from Barbados: sea level during the past 130,000 years. *Nature*, **346**, 456–458.

BART, P. J. & CONE, A. N. 2011. Early stall of West Antarctic Ice Sheet advance on the eastern Ross Sea middle shelf followed by retreat at 27 500 ^{14}C yr BP. *Palaeogeography, Palaeoclimatology, Palaeoecology*, doi: 10.1016/j.palaeo2011.08.007.

BENTLEY, M. J. & ANDERSON, J. B. 1998. Glacial and marine geological evidence for the extent of grounded ice in the Weddell Sea–Antarctic Peninsula region during the last glacial maximum. *Antarctic Science*, **10**, 307–323.

BENTLEY, M. J., FOGWILL, C. J., KUBIK, P. W. & SUGDEN, D. E. 2006. Geomorphological evidence and cosmogenic Be/Al exposure ages for the last glacial maximum and deglaciation of the Antarctic Peninsula Ice Sheet. *Geological Society of America Bulletin*, **118**, 1149–1159.

BENTLEY, M. J., HODGSON, D. A. *ET AL*. 2009. Mechanisms of Holocene palaeoenvironmental change in the Antarctic Peninsula region. *The Holocene*, **19**, 51–69.

BENTLEY, M. J., FOGWILL, C. J., LE BROCQ, A. M., HUBBARD, A. L., SUGDEN, D. E., DUNAI, T. J. & FREEMAN, S. P. H. T. 2010. Deglacial history of the West Antarctic Ice Sheet in the Weddell Sea embayment: constraints on past ice volume change. *Geology*, **38**, 411–414.

BENTLEY, M. J., JOHNSON, J. S., HODGSON, D. A., DUNAI, T., FREEMAN, S. P. H. T. & Ó COFAIGH, C. 2011. Rapid deglaciation of Marguerite Bay, western Antarctic Peninsula in the early Holocene. *Quaternary Science Reviews*, **30**, 3338–3349.

BERKMAN, P. A. & FORMAN, S. L. 1996. Pre–bomb radiocarbon and the reservoir correction for calcareous marine species in the Southern Ocean. *Geophysical Research Letters*, **23**, 363–366.

BRACHFELD, S., DOMACK, E. W. *ET AL*. 2003. Holocene history of the Larsen-A Ice Shelf con-strained by geomagnetic paleointensity dating. *Geology*, **31**, 749–752.

CAMERLENGHI, A., DOMACK, E. *ET AL*. 2001. Glacial morphology and post-glacial contourites in the north Gustav Channel (NW Weddell Sea, Antarctica). *Marine Geophysical Researches*, **22**, 417–443.

CANALS, M., CASAMOR, J. L. *ET AL*. 2002. Seafloor evidence of a subglacial sedimentary system off the northern Antarctic Peninsula. *Geology*, **30**, 603–606.

CARBOTTE, S. M., RYAN, W. B. F. *ET AL*. 2007. *Antarctic multibeam bathymetry and geophysical data synthesis: an on-line digital data resource for marine geoscience research in the Southern Ocean*. US Geological Survey Open-File Report **2007-1047**, doi: 10.3133/of2007-1047.srp002.

CLARK, P. R., MITROVICA, J. X., MILNE, G. A. & TAMISIEA, M. E. 2002. Sea-level fingerprinting as a direct test for the source of global Meltwater Pulse IA. *Science*, **295**, 2438–2441.

CLARKE, G., LEVERINGTON, D., TELLER, J. & DYKE, A. 2003. Superlakes, megafloods, and abrupt climate change. *Science*, **301**, 922–923.

CONWAY, H., HALL, B. L., DENTON, G. H., GADES, A. M. & WADDINGTON, E. D. 1999. Past and future grounding-line retreat of the West Antarctic Ice Sheet. *Science*, **280**, 283–286.

CRAWFORD, K., KUHN, G. & HAMBREY, M. J. 1996. Changes in the character of glacio- marine sedimentation in the southwestern Weddell Sea, Antarctica: evidence from the core PS1423-2. *Annals of Glaciology*, **22**, 200e204.

DAVIES, B. J., HAMBREY, M. J., SMELLIE, J. L., CRRIVICK, J. L. & GLASSER, N. F. 2012. Antarctic Peninsula Ice Sheet Evolution during the Cenozoic Era. *Quaternary Science Reviews*, **31**, 30–66.

DENTON, G. H. & HUGHES, T. J. 2000. Reconstruction of the Ross ice drainage system, Antarctica at the last glacial maximum. *Geografiska Annaler*, **82A**, 143–166.

DENTON, G. H. & HUGHES, T. J. 2002. Reconstructing the Antarctic Ice Sheet at the last glacial maximum. *Quaternary Science Reviews*, **21**, 193–202.

DENTON, G. H. & MARCHANT, D. R. 2000. The geologic basis for a reconstruction of a grounded ice sheet in

McMurdo Sound, Antarctica, at the last glacial maximum. *Geografiska Annaler*, **82A**, 167–211.

Dochat, T. M., Marchant, D. R. & Denton, G. H. 2000. Glacial geology of Cape Bird, Ross Island, Antarctica. *Geografiska Annaler*, **82A**, 237–247.

Domack, E. W. 1982. Sedimentology of glacial and glacial-marine deposits of the George V–Adelie continental shelf, East Antarctica. *Boreas*, **11**, 79–97.

Domack, E. W., Jull, A. J. T., Anderson, J. B., Linick, T. W. & Williams, C. R. 1989. Application of tandem accelerator mass-spectrometer dating to late Pleistocene–Holocene sediments of the East Antarctic continental shelf. *Quaternary Research*, **31**, 277–287.

Domack, E. W., O'Brien, P., Harris, P., Taylor, F., Quilty, P. G., Santis, L. D. & Raker, B. 1998. Late Quaternary sediment facies in Prydz Bay, East Antarctica and their relationship to glacial advance onto the continental shelf. *Antarctic Science*, **10**, 236–246.

Domack, E. W., Jacobson, E. A., Shipp, S. & Anderson, J. B. 1999. Sedimetologic and stratigraphic signature of the late Pleistocene/Holocene fluctuations of the West Antarctic Ice Sheet in the Ross Sea: a new perspective. Part 2. *Geological Society of America*, **111**, 1516–1536.

Domack, E. W., Leventer, A., Dunbar, R., Taylor, F., Brachfeld, S., Sjunneskog, C. & ODP LEG 178 SCIENTIFIC PART. 2001. Chronology of the palmer deep site, Antarctic Peninsula: a Holocene palaeoenvironmental reference for the circum-Antarctic. *The Holocene*, **11**, 1–9.

Domack, E. W., Amblàs, D. *et al.* 2006. Subglacial morphology and glacial evolution of the Palmer deep outlet system, Antarctic Peninsula. *Geomorphology*, **75**, 125–142.

Dowdeswell, J. A., Ottesen, D., Evans, J., O'Cofaigh, C. O. & Anderson, J. B. 2008. Submarine glacial landforms and rates of ice stream collapse. *Geology*, **36**, 819–822.

Elverhøi, A. 1981. Evidence for a late Wisconsin glaciation of the Weddell Sea. *Nature*, **293**, 641–642.

Evans, J., Dowdeswell, J. & O'Cofaigh, C. 2004. Late Quaternary submarine bedforms and ice-sheet flow in Gerlache Strait and on the adjacent continental shelf, Antarctic Peninsula. *Journal Quaternary Science*, **19**, 397–407.

Evans, J., Pudsey, C. J., ÓCofaigh, C., Morris, P. & Domack, E. W. 2005. Late Quaternary glacial history, flow dynamics and sedimentation along the eastern margin of the Antarctic Peninsula Ice Sheet. *Quaternary Science Reviews*, **24**, 741–774.

Evans, J., Dowdswell, J. A., O'Cofaigh, C., Benham, T. J. & Anderson, J. B. 2006. Extent and dynamics of the West Antarctic Ice Sheet on the outer continental shelf of Pine Island Bay during the last glaciation. *Marine Geology*, **230**, 53–72.

Farmer, G. L., Licht, K., Swope, R. J. & Andrews, J. 2006. Isotopic constraints on the provenance of fine-grained sediment in LGM tills from the Ross Embayment, Antarctica. *Earth and Planetary Science Letters*, **249**, 90–107.

Gomez, N., Mitrovica, J. X., Huybers, P. & Clark, P. U. 2010. Sea level as a stabilizing factor for marine-ice-sheet grounding lines. *Nature Geoscience*, **3**, 850–853.

Goodwin, I. D. & Zweck, C. 2000. Glacio-isostasy and glacial ice load at Law Dome, Wilkes Land. *East Antarctica. Quaternary Research*, **53**, 285–293.

Graham, A. G. C. & Smith, J. A. 2012. Palaeoglaciology of the Alexander Island ice cap, western Antarctic Peninsula, reconstructed from marine geophysical and core data. *Quaternary Science Reviews*, **35**, 63–81.

Graham, A. G. C., Larter, R. D., Gohl, K., Hillenbrand, C.-D., Smith, J. A. & Kuhn, G. 2009. Bedform signature of a West Antarctic palaeo-ice stream reveals a multi- temporal record of flow and substrate control. *Quaternary Science Reviews*, **28**, 2774–2793.

Graham, A. G. C., Larter, R. D. *et al.* 2010. Flow and retreat of the late Quaternary Pine Island-Thwaites palaeo-ice stream, West Antarctica. *Journal of Geophyshysical Research*, **115**, F03025, doi: 10.1029/2009JF001482.

Greenwood, S. L., Gyllencreutz, R., Jakobsson, M. & Anderson, J. B. 2013. Ice flow switching and east/west Antarctic Ice Sheet roles in glaciation of the western Ross Sea. *Geological Society of America Bulletin*, doi: 10.1130/B30643.1.

Hall, B. L. & Denton, G. H. 2000*a*. Radiocarbon chronology of Ross Sea Drift, eastern Taylor Valley, Antarctica: evidence for a grounded ice sheet in the Ross Sea at the Last Glacial maximum. *Geografiska Annaler*, **A82**, 305–336.

Hall, B. L. & Denton, G. H. 2000*b*. Extent and chronology of the Ross Sea ice sheet and the Wilson Piedmont Glacier along the Scott Coast at and since the last glacial maximum. *Geografiska Annaler*, **82A**, 275–303.

Hall, B. L., Henderson, G. M., Baroni, C. & Kellogg, T. B. 2010. Constant Holocene Southern-Ocean ^{14}C reservoir ages and ice-shelf flow rates. *Earth and Planetary Science Letters*, **296**, 115–123.

Harris, P. T. & O'Brien, P. E. 1998. Bottom currents, sedimentation and ice sheet retreat facies successions on the Mac Robertson Land shelf, East Antarctica. *Marine Geology*, **151**, 47–72.

Hein, A. S., Fogwill, C. J., Sugden, D. E. & Xu, S. 2011. Glacial/interglacial ice-stream stability in the Weddell Sea embayment, Antarctica. *Earth and Planetary Science Letters*, **307**, 211–221.

Heroy, D. C. & Anderson, J. B. 2005. Ice sheet extension on the Antarctic Peninsula during the last glacial maximum (LGM)-insights from glacial geomorphology. *Geological Society of America Bulletin*, **117**, 1497–1512.

Heroy, D. C. & Anderson, J. B. 2007. Radiocarbon constraints on Antarctic Peninsula ice sheet retreat following the last glacial maximum. *Quaternary Science Reviews*, **26**, 3286–3297.

Heroy, D. C., Sjunneskog, C. & Anderson, J. B. 2008. Holocene climate change in the Bransfield Basin, Antarctic Peninsula: evidence from sediment and diatom analysis. *Antarctic Science*, **20**, 69–87.

Hillenbrand, C.-D., Ehrmann, W. *et al.* 2009. Clay mineral provenance of sediments in the southern Bellingshausen Sea reveals drainage changes of the West

Antarctic Ice Sheet during the Late Quaternary. *Marine Geology*, **265**, 1–18.

HILLENBRAND, C.-D., LARTER, R. D. *ET AL.* 2010*a*. Age assignment of a diatomaceous ooze deposited in the western Amundsen Sea Embayment after the last glacial maximum. *Journal of Quaternary Science*, **25**, 280–295.

HILLENBRAND, C.-D., LARTER, R. D. *ET AL.* 2010*b*. The sedimentary legacy of a palaeo-ice stream on the shelf of the southern Bellingshausen Sea: Clues to West Antarctic glacial history during the Late Quaternary. *Quaternary Science Reviews*, **29**, 2741–2763.

HILLENBRAND, C.-D., MELLES, M., KUHN, G. & LARTER, R. D. 2012. Marine geological constraints for the grounding-line position of the Antarctic Ice Sheet on the southern Weddell Sea shelf at the last glacial maximum. *Quaternary Science Reviews*, **32**, 25–47.

HILLENBRAND, C.-D., KUHN, G. *ET AL.* 2013. Grounding line retreat of the West Antarctic Ice Sheet from inner Pine Island Bay. *Geology*, **41**, 35–38.

HOWAT, I. M. & DOMACK, E. W. 2003. Reconstructions of western Ross Sea palaeo-ice-stream grounding zones from high-resolution acoustic stratigraphy. *Boreas*, **32**, 56–75.

HUYBRECHTS, P. 2002. Sea-level changes at the LGM from ice-dynamic reconstructions of the Greenland and Antarctic ice sheets during the glacial cycles. *Quaternary Science Reviews*, **21**, 203–231.

IVINS, E. R. & JAMES, T. S. 2005. Antarctic glacial isostatic adjustment: a new assessment. *Antarctic Science*, **17**, 541–553.

JACOBS, S. S., HELLMER, H. H. & JENKINS, A. 1996. Antarctic ice sheet melting in the Southeast Pacific. *Geophysical Research Letters*, **23**, 957–960.

JACOBS, S. S., JENKINS, A., GIULIVI, C. F. & DUTRIEUX, P. 2011. Stronger ocean circulation and increased melting under Pine Island Glacier ice shelf. *Nature Geoscience*, **4**, 519–523.

JAKOBSSON, M., ANDERSON, J. B. *ET AL.* 2011. Geological record of ice shelf break-up and grounding line retreat, Pine Island Bay, West Antarctica. *Geology*, **39**, 691–694.

JAKOBSSON, M., ANDERSON, J. B. *ET AL.* 2012. Ice sheet retreat dynamics inferred from Glacial Morphology of the Central Pine Island Bay Trough, West Antarctica. *Quaternary Science Reviews*, doi: 10.1016/j.quascirev.2011.12.017.

JOHNSON, J. S., BENTLEY, M. J. & GOHL, K. 2008. First exposure ages from the Amundsen Sea embayment, West Antarctica: the late quaternary context for recent thinning of Pine Island, Smith, and Pope Glaciers. *Geology*, **36**, 223–226.

JOHNSON, J. S., BENTLEY, M. J., ROBERTS, S. J., BINNIE, S. A. & FREEMAN, S. P. 2011. Holocene deglacial history of the northeast Antarctic Peninsula – A review and new chronological constraints. *Quaternary Science Reviews*, **30**, 3791–3802.

KEMP, A. C., HORTON, B. P., DONNELLY, J. P., MANN, M. E., VERMEER, M. & RAHMSTORF, S. 2011. Climate related sea-level variations over the past two millennia. *Proceedings of the National Academy of Sciences*, **108**, 11 017–11 022.

KENNEDY, D. S. & ANDERSON, J. B. 1989. Quaternary glacial history of Marguerite Bay, Antarctic Peninsula. *Quaternary Research*, **31**, 255–276.

KILFEATHER, A. A., Ó COFAIGH, C., LLOYD, J. M., DOWDESWELL, J. A., XU, S. & MORETON, S. G. 2011. Ice-stream retreat and ice-shelf history in Marguerite Trough, Antarctic Peninsula: sedimentological and foraminiferal signatures. *Geological Society of America Bulletin*, **123**, 997–1015.

KIRSHNER, A. E., ANDERSON, J. B., JAKOBSSON, M., O'REGAN, M., MAJEWSKI, W. & NITSCHE, F. O. 2012. Post-LGM deglaciation in Pine Island Bay, West Antarctica. *Quaternary Science Reviews*, **38**, 11–26.

LAMBECK, K. & CHAPPELL, J. 2001. Sea level change through the last glacial cycle. *Science*, **292**, 679–686.

LARTER, R. D., GRAHAM, A. G. C. *ET AL.* 2009. Subglacial bedforms reveal complex basal regime in a zone of paoeo-ice stream convergence, Amundsen Sea embayment, West Antarctica. *Geology*, **37**, 411–414.

LE BROCQ, A. M., BENTLEY, M. J., HUBBARD, A. L., FOGWILL, C. J., SUGDEN, D. E. & WHITEHOUSE, P. L. 2011. Reconstructing the last glacial maximum ice sheet in the Weddell Sea embayment, Antarctica, using numerical modeling constrained by field evidence. *Quaternary Science Reviews*, **30**, 2422–2432.

LEVENTER, A., DOMACK, E. *ET AL.* 2006. Marine sediment record from the East Antarctic margin reveals dynamics of ice sheet recession. *GSA Today*, **16**, 4–10.

LICHT, K. J. & ANDREWS, J. T. 2002. The ^{14}C record of Late Pleistocene ice advance and retreat in the central Ross Sea, Antarctica. Arctic. *Antarctic and Alpine Research*, **34**, 324e333.

LICHT, K. J., JENNINGS, A. E., ANDREWS, J. T. & WILLIAMS, K. M. 1996. Chronology of late Wisconsin ice retreat from the western Ross Sea, Antarctica. *Geology*, **24**, 223–226.

LICHT, K. J., LEDERER, J. R. & SWOPE, R. J. 2005. Provenance of LGM glacial till (sand fraction) across the Ross Embayment, Antarctica. *Quaternary Science Reviews*, **24**, 14991520–0, Research Letters, 37 (L13503); doi: 10.1029/2010GL042496.

LIVINGSTONE, S. J., Ó COFAIGH, C., STOKES, C. R., HILLENBRAND, C.-D., VIELI, A. & JAMIESON, S. S. R. 2012. Antarctic paleo-ice streams. *Earth Science Reviews*, **111**, 90–128.

LOWE, A. L. & ANDERSON, J. B. 2002. Late Quaternary advance and retreat of the West Antarctic Ice Sheet in Pine Island Bay, Antarctica. *Quaternary Science Reviews*, **21**, 1879–1897.

LOWE, A. L. & ANDERSON, J. B. 2003. Evidence for abundant subglacial meltwater beneath the paleo-ice sheet in Pine Island Bay, Antarctica. *Journal Glaciology*, **49**, 125–138.

MACKINTOSH, A., GOLLEDGE, N. *ET AL.* 2011. Retreat of the East Antarctic ice sheet during the last glacial termination. *Nature Geoscience*, **4**, 195–202.

MADDISON, E. J., PIKE, J. *ET AL.* 2006. Post-glacial seasonal diatom record of the Mertz Glacier Polynya, East Antarctica. *Marine Micropaleontology*, **60**, 66–88.

MAYEWSKI, P. A., MEREDITH, M. P. *ET AL.* 2009. State of the Antarctic and Southern Ocean climate system. *Reviews of Geophysics*, **47**, RG1003, 1–38.

MCKAY, R. M., DUNBAR, G. B., NAISH, T. R., BARRETT, P. J., CARTER, L. & HARPER, M. 2008. Retreat history of the Ross Ice Sheet (shelf) since the last glacial maximum from deep-basin sediment cores around Ross Island. *Palaeogeography, Palaeoclimatology, Palaeoecology*, **260**, 245–261.

McMullen, K., Domack, E., Leventer, A., Olson, C., Dunbar, R. & Brachfeld, S. 2006. Glacial geomorphology and sediment formation in the Mertz Trough, Antarctica. *Palaeogeography, Palaeoclimatology, Palaeoecology*, **231**, 169–180.

Michalchuk, B. R., Anderson, J. B., Wellner, J. S., Manley, P. L., Majewski, W. & Bohaty, S. 2009. Holocene climate and glacial history of the northeastern Antarctic Peninsula: the marine sedimentary record from a long SHALDRIL core. *Quaternary Science Reviews*, **28**, 3049–3065.

Milliken, K. T., Anderson, J. B., Wellner, J. S., Bohaty, S. M. & Manley, P. L. 2009. High-resolution Holocene climate record from Maxwell Bay, South Shetland Islands Antarctica. *Geological Society of America Bulletin*, **121**, 1711–1725.

Mosola, A. B. & Anderson, J. B. 2006. Expansion and rapid retreat of the West Antarctic Ice Sheet in eastern Ross Sea: possible consequence of over extended ice streams?. *Quaternary Science Reviews*, **25**, 2177–2196.

Mulvaney, R., Arrowsmith, C. *et al.* 2007. A deglaciation climate and ice sheet history of the Weddell Sea region from the Berkner Island ice core. *Quaternary International*, **167**, 294–295.

Nakada, M., Kimura, R., Okuno, J., Moriwaki, K., Miura, H. & Maemoku, H. 2000. Late Pleistocene and Holocene melting history of the Antarctic ice sheet derived from sea-level variations. *Marine Geology*, **167**, 85–103.

Nitsche, F. O., Jacobs, S. S., Larter, R. D. & Gohl, K. 2007. Bathymetry of the Amundsen Sea continental shelf: implications for geology, oceanography, and glaciology. *Geochemistry Geophysics Geosystems*, **8**, 1–10.

Nitsche, F. O., Gohl, K. *et al.* 2013. Paleo ice flow and subglacial meltwater dynamics in Pine Island Bay, West Antarctica. *Cryosphere*, **7**, 249–262.

O'Brien, P. E. & Harris, P. T. 1996. Patterns of glacial erosion and deposition in Prydz Bay and the past behaviour of the Lambert Glacier. *Papers and Proceedings of the Royal Society of Tasmania*, **130**, 79–85.

Ó Cofaigh, C., Pudsey, C. J., Dowdeswell, J. & Morris, P. 2002. Evolution of subglacial bedforms along a paleo-ice stream, Antarctic Peninsula continental shelf. *Geophysical Research Letters*, **29**, 1–4.

Ó Cofaigh, C., Dowdeswell, J. A., Allen, C. S., Hiemstra, J., Pudsey, C. J., Evans, J. & Evans, D. J. A. 2005*a*. Flow dynamics and till genesis associated with a marine-based Antarctic palaeo-ice stream. *Quaternary Science Reviews*, **24**, 709–740.

Ó Cofaigh, C., Larter, R. D., Dowdeswell, J. A., Hillenbrand, C-D., Pudsey, C. J., Evans, J. & Morris, P. 2005*b*. Flow of West Antarctic Ice Sheet on the continental margin of the Bellingshausen Sea at the last glacial maximum. *Journal of Geophysical Research*, **110**, B11103.

Ó Cofaigh, C., Dowdeswell, J. A., Evans, J. & Larter, R. D. 2008. Geological constraints on Antarctic palaeo-ice stream retreat. *Earth Surface Processes and Landforms*, **33**, 513–525.

Ohkouchi, N. & Eglinton, T. I. 2008. Compound-specific radiocarbon dating of Ross Sea sediments: a prospect for constructing chronologies in high-latitude oceanic sediments. *Quaternary Geochronology*, **3**, 235–243.

Peltier, W. R. 2002. On eustatic sea level history, last glacial maximum to Holocene. *Quaternary Science Reviews*, **21**, 377–396.

Peltier, W. R. & Fairbanks, R. G. 2006. Global glacial ice volume and last glacial maximum duration from an extended Barbados sea level record. *Quaternary Science Reviews*, **25**, 3322–3337.

Pollard, D. & DeConto, R. M. 2009. Modelling West Antarctic ice sheet growth and collapse through the past five million years. *Nature*, **458**, 329–332.

Pope, P. G. & Anderson, J. B. 1992. Late Quaternary glacial history of the northern Antarctic Peninsula's western continental shelf: evidence from the marine record. *In*: Elliot, D. H. (ed.) *Contributions to Antarctic Research III, Antarctic Research Series*. American Geophysical Union, Washington, DC, **57**, 63–91.

Pudsey, C. J., Barker, P. F. & Larter, R. D. 1994. Ice sheet retreat form the Antarctic Peninsula shelf. *Continental Shelf Research*, **14**, 1647–1675.

Pudsey, C. J., Murray, J. W., Appleby, P. & Evans, J. 2006. Ice shelf history from petrograph- ic and foraminiferal evidence, northeast Antarctic Peninsula. *Quaternary Science Reviews*, **25**, 2357–2379.

Reimer, P. J., Baillie, M. G. *et al.* 2009. Intcal09 and Marine09 radiocarbon age calibration curves, 0–50 000 years cal Bp. *Radiocarbon*, **51**, 1111–1150.

Reinardy, T. I., Larter, R. D., Hillenbrand, C.-D., Murray, T., Hiemstra, J. F. & Booth, A. D. 2011. Streaming flow of an Antarctic Peninsula palaeo-ice stream, both by basal sliding and deformatin of substrate. *Journal of Glaciology*, **57**, 596–608.

Rodriguez, A. B., Simms, A. R. & Anderson, J. B. 2010. Bayhead deltas across the northern Gulf of Mexico back step in response to the 8.2 ka cooling event. *Quaternary Science Reviews*, **29**, 3983–3993.

Rosenheim, B. E., Day, M. B., Schrum, H., Domack, E. W., Benthien, A. & Hayes, J. M. 2008. Antarctic sediment chronology by programmed temperature pyrolysis: methodology and data treatment. *Geochemistry, Geophysics, Geosystems*, **9**, Q04005, doi: 10.1029/2007GC001816.

Ryan, W. B. F., Carbotte, S. M. *et al.* 2009. Global Multi-Resolution Topography (GMRT) synthesis data set. *Geochem. Geophys. Geosyst.*, **10**, Q03014, doi: 10.1029/2008GC002332; data doi: 10.1594/IEDA.0001000.

Shennan, I. 2007. Overview–Sea-level studies. *In*: Scott, A. E. (ed.) *Encyclopedia of Quaternary Science*. Elsevier, Amsterdam, 2967–2974.

Shipp, S. & Anderson, J. B. 1997. Grounding zone wedges on the Antarctic continental shelf, Ross Sea. *In*: Davies, T. A., Bell, T., Cooper, A. K., Josenhans, H., Polyak, L., Solheim, A., Stoker, M. S. & Stavers, J. A. (eds) *Glaciated Continental Margins: An Atlas of Acoustic Images*. Chapman and Hall, London, 104–105.

SHIPP, S., ANDERSON, J. B. & DOMACK, E. W. 1999. Seismic signature of the late Pleistocene fluctuation of the West Antarctic Ice Sheet System in Ross Sea: a new perspective. Part I. *Geological Society of America Bulletin*, **111**, 1486–1516.

SIMMS, A. R., MILLIKEN, K. T., ANDERSON, J. B. & WELLNER, J. S. 2011. The marine record of deglaciation of the South Shetland Islands, Antarctica since the last glacial maximum. *Quaternary Science Reviews*, **30**, 1583–1601.

SMITH, J. A., HILLENBRAND, C.-D., LARTER, R. D., GRAHAM, A. G. C. & KUHN, G. 2009. The sediment infill of subglacial meltwater channels on the West Antarctic continental shelf. *Quaternary Research*, **71**, 190–200.

SMITH, J. A., HILLENBRAND, C.-D. *ET AL.* 2011. Deglacial history of the West Antarctic Ice Sheet in the western Amundsen Sea Embayment. *Quaternary Science Reviews*, **30**, 488–505.

STOLLDORF, T. D., SCHENKE, H.-W. & ANDERSON, J. B. 2012. Diachronous behavior of the Antarctic Antarctic Ice Sheets: Weddell Sea, Antarctica. *Quaternary Science Reviews*, http://dx.doi.org/10.1016/j.quascirev.2012.05.017

STONE, J. O., BALCO, G. A., SUGDEN, D. E., CAFFEE, M. C., SASS, L. C., COWDERY, S. G. & SIDDOWAY, C. 2003. Holocene deglaciation of Marie Byrd Land, West Antarctica. *Science*, **299**, 99–102.

STUIVER, M., DENTON, G. H., HUGHES, T. J. & FASTOOK, J. L. 1981. History of the marine ice sheet in West Antarctica during the last glaciation: a working hypothesis. *In*: DENTON, G. H. & HUGHES, T. J. (eds) *The Last Great Ice Sheets*. Wiley, New York, 319–436.

TAVIANI, M., REID, D. E. & ANDERSON, J. B. 1993. Skeletal and isotopic composition and paleoclimatic significance of Lat, Pleistocene Carbonates, Ross Sea, Antarctica. *Journal of Sedimentary Petrology*, **63**, 84–90.

TELLER, J. T., LEVERINGTON, D. W. & MANN, J. D. 2002. Freshwater outbursts to the oceans from glacial Lake Agassiz and their role in climate change during the deglaciation. *Quaternary Science Reviews*, **21**, 879–887.

TÖRNQVIST, T. E., BICK, S. J., GONZÁLEZ, J. L., VAN DER BORG, K. & DE JONG, A. F. M. 2004. Tracking the sea-level signature of the 8.2 ka cooling event: new constraints from the Mississippi Delta. *Geophysical Research Letters*, **31**, doi: 10.1029/2004GL021429.

VAUGHAN, D. G. 2008. West Antarctic Ice Sheet collapse – the fall and rise of a paradigm. *Climatic Change*, **91**, 65e79.

WEBER, M. E., CLARK, P. U., RICKEN, W., MITROVICA, J. X., HOSTETLER, S. W. & KUHN, G. 2011. Interhemispheric ice-sheet synchronicity during the last glacial maximum. *Science*, **334**, 1265–1269.

WELLNER, J. S., LOWE, A. L., SHIPP, S. S. & ANDERSON, J. B. 2001. Distribution of glacial geomorphic features on the Antarctic continental shelf and correlation with substrate: implications for ice behavior. *Journal of Glaciology*, **47**, 397–411.

WELLNER, J., HEROY, D. & ANDERSON, J. 2006. The death mask of the Antarctic Ice Sheet: comparison of glacial geomorphic features across the continental shelf. *Geomorphology*, **75**, 157–171.

WHITEHOUSE, P. L., BENTLEY, M. J. & LE BROCQ, A. M. 2012. A deglacial model for Antarctica: geological constraints and glaciological modeling as a basis for a new model of Antarctic glacial isostatic adjustment. *Quaternary Science Reviews*, **32**, 1–24.

WILLMOTT, V., DOMACK, E. W., PADMAN, L. & CANALS, M. 2007. Glaciomarine sediment drifts from Gerlache Strait, Antarctic Peninsula. *In*: GLASSER, N. & HAMBREY, M. J. (eds) *Glacial Sedimentary Processes and Products*. Blackwells, Oxford, IAS Special Publications, 67–84.

WRIGHT, A. P., SIEGERT, M. J., LE BROCQ, A. M. & GORE, D. B. 2008. High sensitivity of subglacial hydrological pathways in Antarctica to small ice-sheet changes. *Geophysical Research Letters*, **35**, L17504.

ZWARTZ, D., BIRD, M., STONE, J. & LAMBECK, K. 1998. Holocene sea-level change and ice-sheet history in the Vestfold Hills, East Antarctica. *Earth and Planetary Science Letters*, **155**, 131–145.

Glaciomarine sedimentation dynamics of the Abbot glacial trough of the Amundsen Sea Embayment shelf, West Antarctica

KATHARINA HOCHMUTH* & KARSTEN GOHL

Department of Geosciences, Alfred Wegener Institute for Polar and Marine Research, Am Alten Hafen 26, 27568 Bremerhaven, Germany

**Corresponding author (e-mail: Katharina.Hochmuth@awi.de)*

Abstract: Sedimentary sequences of the continental shelf of the eastern Amundsen Sea Embayment in West Antarctica represent records of past outlet glaciers and ice streams. The former flow of ice streams was channelled through glacial troughs, which now form large bathymetric depressions. We therefore selected one of the largest troughs, the Abbot glacial trough in the outer shelf, to analyse its glacial depositional and erosional history, based on horizon-stratigraphy derived from seismic data. Several basement highs channellized the delivery of sediment and controlled the grounded ice sheet in early glacial periods. Both pre-glacial and full glacial seismic facies were identified. Glacially transported and deposited sediments extended the shelf break by 75 km from the pre-glacial shelf-edge. The main Abbot glacial trough contains sediment from confluent ice flows of the Pine Island/Thwaites, Cosgrove and Abbot Glacier systems, as well as smaller contributions from local ice streams emanating from Thurston Island. Sherman Island of Peacock Sound played an important role in the dynamics of the Abbot Glacier by dividing the ice flow into two ice streams, which interfered with the main glacial sediment transport paths from the south. This study contributes to an understanding of the formation of the Amundsen Sea shelf and the extent of past ice sheet advances.

The continental shelf of the Amundsen Sea Embayment holds a sedimentary archive that, if deciphered, has the potential to help reconstruct the major glacial periods of the West Antarctic Ice Sheet from early glacial advances to the grounded ice sheet retreat since the Last Glacial Maximum. Unlike the continental shelves of the Ross Sea and Filchne–Ronne embayments, the Amundsen Sea Embayment was fed entirely by the West Antarctic Ice Sheet, since it was unaffected by flow dynamics of the East Antarctic and Antarctic Peninsula ice sheets. Despite this significance for understanding past advance and retreat cycles of West Antarctic Ice Sheet, the shelf of the embayment has rarely been investigated. Most of its sea-floor stratigraphy remains unknown owing to unfavourable sea ice conditions that have prevented systematic seismic survey and drilling programmes. The first seismic profiles and analyses of an improving grid of multibeam bathymetric data (Nitsche *et al.* 1997, 2007; Lowe & Anderson 2002, 2003; Evans *et al.* 2006; Graham *et al.* 2010; Jakobsson *et al.* 2011, 2012) have shown that the ice sheet was grounded on the middle continental shelf, with excursions onto the outer shelf in some glacial periods.

This study focuses on the herein-named Abbot glacial trough, which represents the easternmost glacier outlet of the Amundsen Sea Embayment, incised into the outer continental shelf west and NW of Thurston Island (Fig. 1). Newly acquired seismic-reflection profiles crossing the outer continental shelf allow, in combination with gravimetric measurements, an analysis of the sedimentary strata and their underlying crystalline basement. By introducing a first horizon-stratigraphic model of the area, the glacial and pre-glacial sedimentary history of the Abbot glacial trough can partly be deciphered. In addition, the influence of the crystalline basement on the behaviour of the local ice streams will be discussed. Furthermore, sedimentation in this area, as well as additional sedimentation provided by medium-sized ice streams coming from Peacock Sound (Abbot Ice Shelf) and smaller glaciers originating on Thurston Island, are evaluated.

Geological background

The Amundsen Sea Embayment (Fig. 1) formed as a consequence of the Cretaceous breakup between the former Gondwanian blocks of greater New Zealand (Zealandia) and West Antarctica (Eagles *et al.* 2004; Gohl *et al.* 2007). Later, internal rifting related to the West Antarctic Rift System was active in this region during the Cenozoic Era (Dalziel 2006; Gohl *et al.* 2007; Jordan *et al.* 2009). Therefore, the Amundsen Sea Embayment

From: Hambrey, M. J., Barker, P. F., Barrett, P. J., Bowman, V., Davies, B., Smellie, J. L. & Tranter, M. (eds) 2013. *Antarctic Palaeoenvironments and Earth-Surface Processes*. Geological Society, London, Special Publications, **381**, 233–244. First published online July 24, 2013, http://dx.doi.org/10.1144/SP381.21

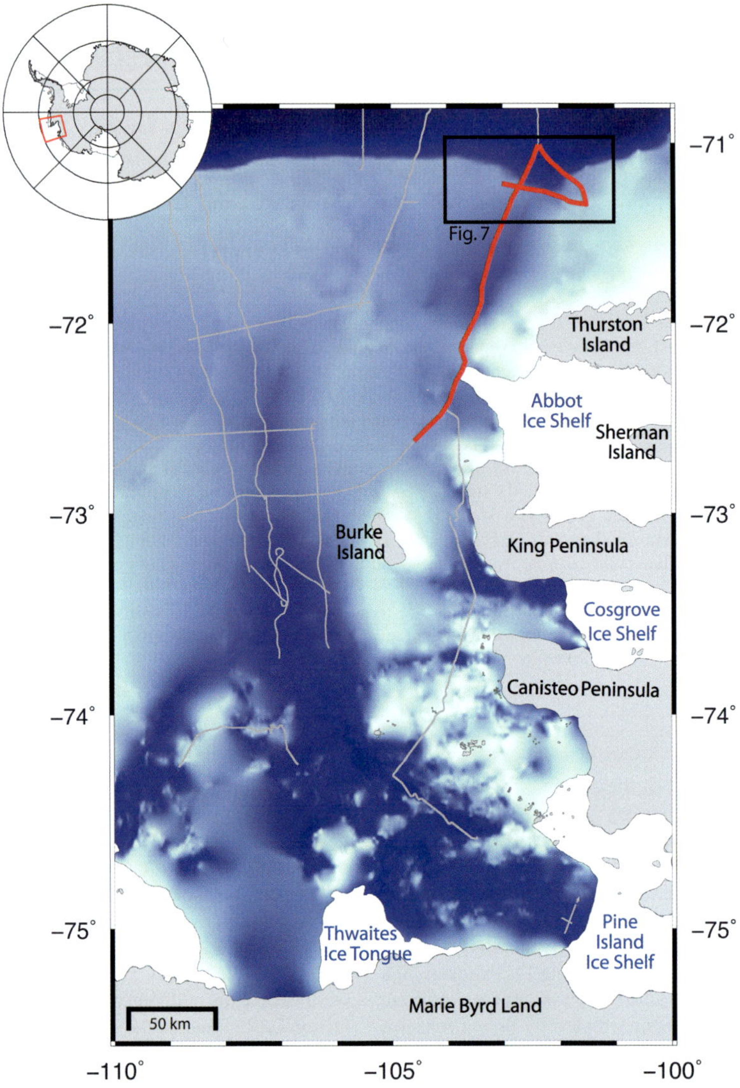

Fig. 1. Overview of the bathymetry of the eastern Amundsen Sea Embayment and Pine Island Bay after *Nitsche et al.* (2007). Red lines mark multichannel seismic profiles discussed in this study. Grey lines mark other existing seismic lines, including single-channel profile NBP-9902 (Lowe & Anderson 2002). Mean water-depths on the continental shelf reach 400–600 m (white to light blue zones) and extend to more than 1200 m (middle to dark blue zones).

was subject to extensional and rifting forces acting in both east–west and north–south directions, which thereby shaped the current structure of the crust and basement (Gohl *et al.* 2013).

Repeated advance and retreat of the West Antarctic Ice Sheet sculpted the present day bathymetry of the continental shelf (Lowe & Anderson 2002; Evans *et al.* 2006; Nitsche *et al.* 2007; Larter *et al.* 2009; Jakobsson *et al.* 2011, 2012). Similar to other continental polar shelves, the Amundsen Sea Embayment shelf deepens inshore and shows a reduced relief on the middle and outer shelf, where the water depth is between 400 and 600 m (Nitsche *et al.* 2007). On the middle shelf, south of Burke Island, the boundary between the outcropping crystalline basement and the oceanward-dipping sedimentary sequences is clearly observed (Lowe & Anderson 2002; 2003; Uenzelmann-Neben *et al.* 2007; Gohl *et al.* 2013).

The large Pine Island/Thwaites glacial trough dominates the bathymetry of the inner and middle continental shelf of the eastern embayment

(Fig. 1). It was carved by the Pine Island and Thwaites ice streams, which converged in the mid-shelf area (Nitsche *et al.* 2007). On the outer shelf, the trough branches into a system of shallow and wide depressions with multiple glacial outlets incising the shelf break (Nitsche *et al.* 2007). Whether these outlets acted simultaneously during an advanced ice sheet or resulted from a change in flow direction of the major ice stream is still being debated. Graham *et al.* (2010) argued that an eastern outlet was established later than the western outlet and therefore the main outlet during later expansions of the West Antarctic Ice Sheet. This concept of flow-switching is supported by a different degree of overprinting of mega-scale glacial lineations in both outlets. Possibly, younger mega-scale glacial lineations are present at the eastern outlet. Little is known about the Abbot glacial trough in the easternmost outer shelf, or its relationship with the ice streams of the main Pine Island/Thwaites trough, the outlet of Cosgrove glacier (Cosgrove Ice Shelf) between Canisteo Peninsula and King Peninsula, and the Abbot glacier (Abbot Ice Shelf) in Peacock Sound. Most of the outer shelf remained outside the range of detailed ship-borne geophysical surveys because of severe sea-ice conditions, until an unusual open-water situation occurred in early 2010, allowing the acquisition of such data.

Data acquisition and processing

As part of the geophysical data-acquisition programme during the R/V *Polarstern* expedition ANT-XXVI/3 in 2010 (Gohl 2010), multichannel seismic reflection profiles were also collected on the outer continental shelf and the Abbot trough in the eastern Amundsen Sea Embayment (Fig. 1). The seismic data discussed here were recorded with a 3000 m-long digital solid streamer, of the type Sercel Sentinel^{TM}, with 240 channels towed at a constant water-depth of 10 m. A cluster of three GI-Guns^{TM}, each with a volume of 0.72 l for the generator and 1.72 l for the injector, was towed at 5 m water-depth. The shot interval was 12 s, and records with 1 ms interval were sampled. Gravimetric measurements were carried out simultaneously with an on-board Bodenseewerke KSS31 gravimeter.

All seismic lines were treated with the same general processing flow. The shot-gathers were

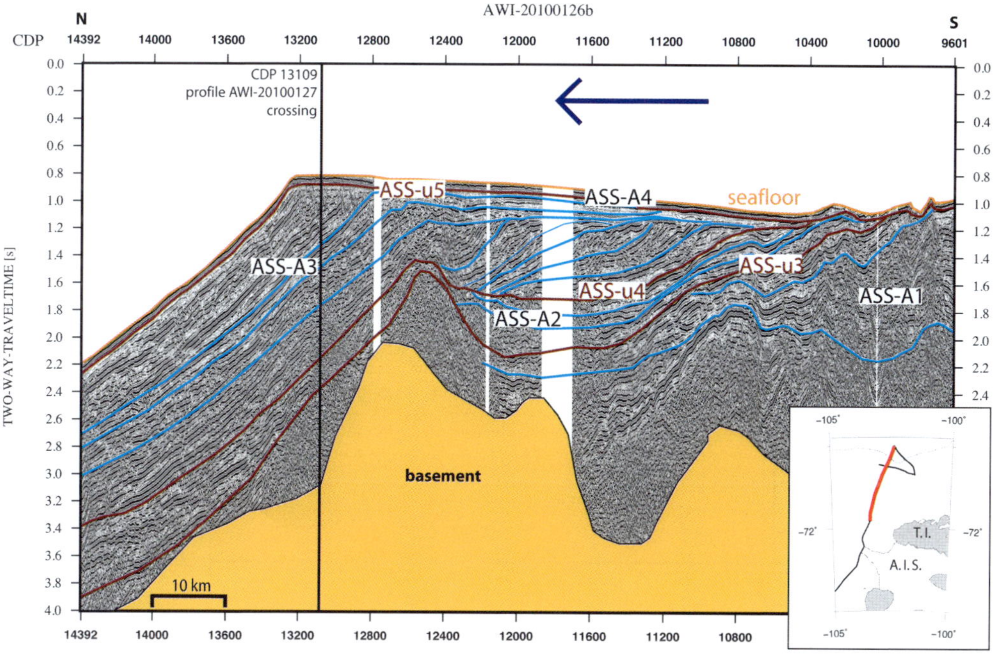

Fig. 2. Northern segment of seismic-reflection profile AWI-20100126b (marked by red line in the inset map; dotted black line shows the shelf edge) with basement depths derived from gravity-anomaly modelling. Unconformities ASS-u3, ASS-u4 and ASS-u5 are marked in red; internal structures of sedimentary units ASS-A1, ASS-A2, ASS-A3 and ASS-A4 are marked in blue. The ice-flow direction is indicated by the dark blue arrow.

sorted into CDP (common depth point) bins of 25 m. General band-pass filtering of 10–120 Hz was applied to the data. A detailed velocity analysis was followed by a parabolic radon transformation to reduce the strong multiple reflections that are typically observed on glaciated continental shelves, as well as to enhance the visibility of underlying reflectors. After stacking, an omega-x finite-difference migration was applied to image the position of dipping reflectors. For further enhancement of the resolution of particular profile sections, multiple filters were applied before final editing.

Even after careful multiple suppression, deeper-lying basement reflections are not visible or remain uncertain in large parts of the profiles, possibly because of the relatively small airgun source and attenuating effect of the multiples. Therefore gravimetric recordings were used and forward density-depth modelling was applied to the free-air anomaly data in order to aid outlining of the sediment-to-basement boundary. A constant crust-mantle depth of 22 km, derived by Gohl *et al.* (2007) for this region of the shelf, was used to constrain the lower model boundaries.

Seismic horizon stratigraphy

Based on the seismic reflection profiles (Figs 2–4 & 6), we derived a basic horizon stratigraphy for the area of the Abbot glacial trough (Fig. 5). Datable sediment samples only exist from shallow gravity-cores and box-samplers that were used to reconstruct the timing of grounding zone retreat since the Last Glacial Maximum (Lowe & Anderson 2002, 2003; Hillenbrand *et al.* 2009; Kirshner *et al.* 2012). Deeply penetrating drill cores from the Amundsen Sea shelf do not exist. Therefore, the seismic horizon stratigraphy that is presented here relies entirely on the observation of seismic reflection characteristics and its comparison with other glaciated continental shelves, such as on dated sequences from the Ross Sea shelf (Anderson & Bartek 1992; Bart & Anderson 2000; Chow & Bart 2003).

Where acoustic basement is relatively shallow, it can be identified mainly in the seismic sections as a reflective band, setting the lower limit of layered sequences. Deeper lying basement of uncertain seismic identification, or lack of it, had to be verified by gravity modelling. Four large sedimentary units

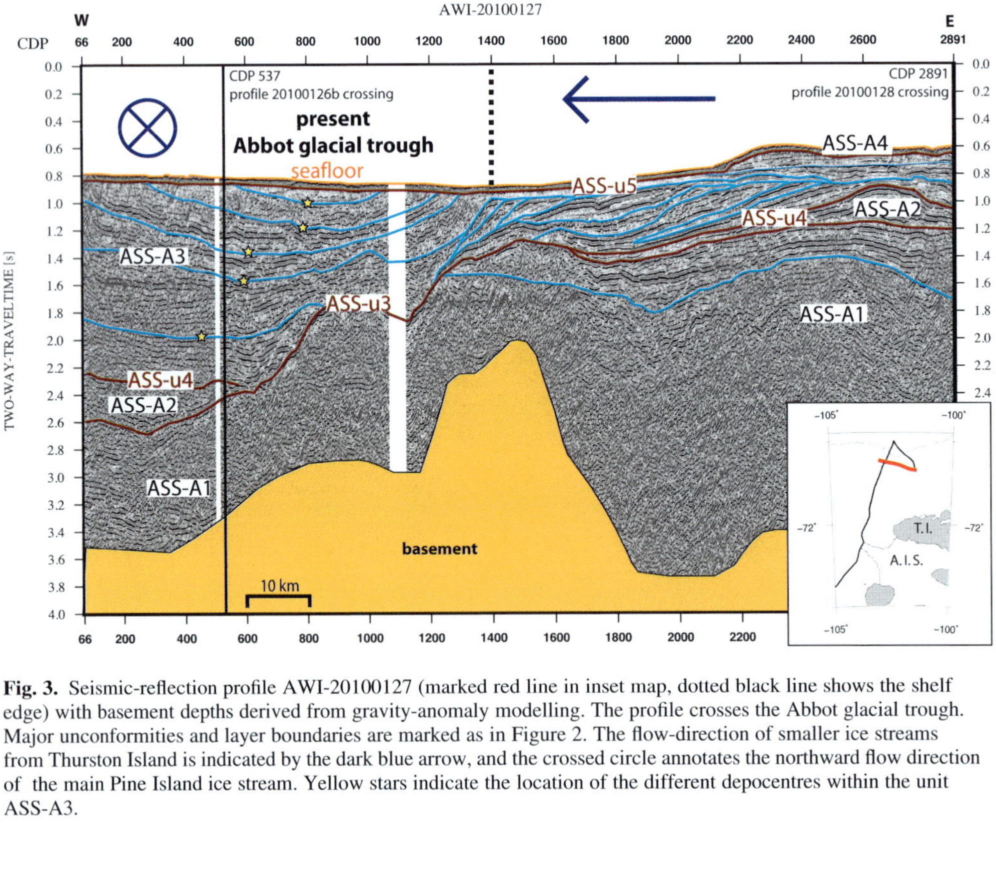

Fig. 3. Seismic-reflection profile AWI-20100127 (marked red line in inset map, dotted black line shows the shelf edge) with basement depths derived from gravity-anomaly modelling. The profile crosses the Abbot glacial trough. Major unconformities and layer boundaries are marked as in Figure 2. The flow-direction of smaller ice streams from Thurston Island is indicated by the dark blue arrow, and the crossed circle annotates the northward flow direction of the main Pine Island ice stream. Yellow stars indicate the location of the different depocentres within the unit ASS-A3.

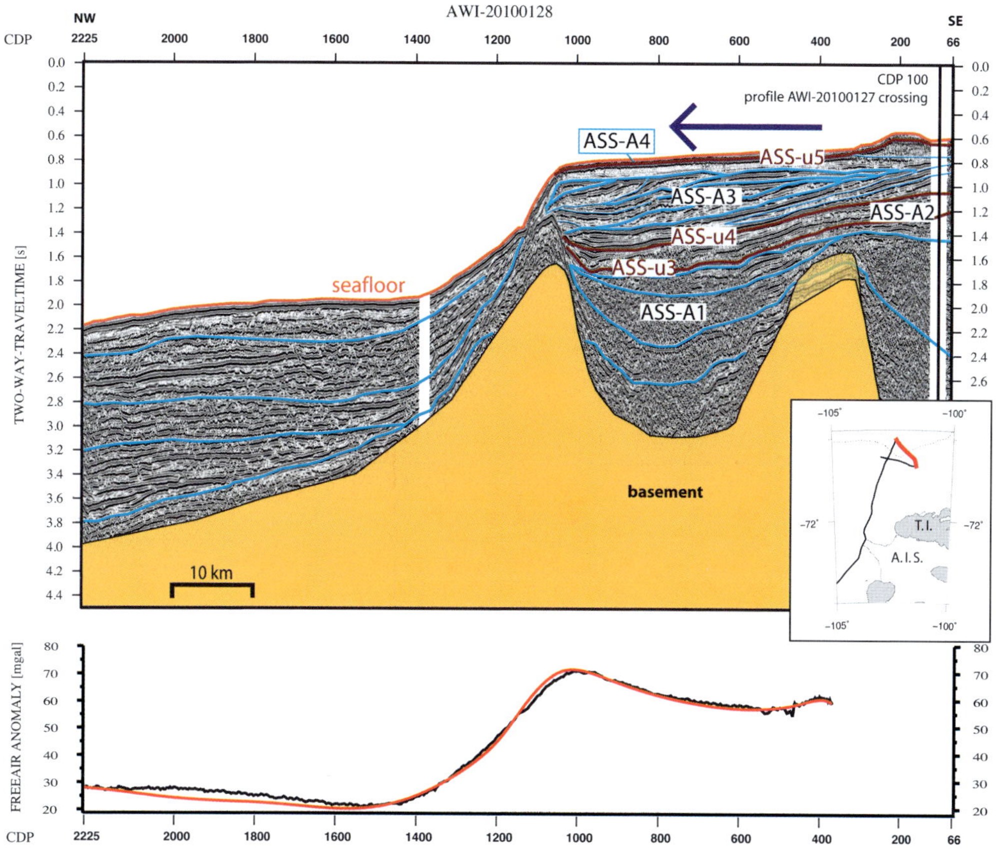

Fig. 4. Seismic-reflection profile AWI-20100128 (marked red line in inset map, dotted black line shows the shelf edge) with basement depths derived from gravity-anomaly modelling. The profile crosses the outer shelf and slope east of the main Abbot glacial trough. Major unconformities and layer boundaries are marked as in Figure 2. The dark blue arrow indicates the flow-direction of smaller ice streams from Thurston Island. The corresponding ship-borne free-air gravity-anomaly profile is shown in the bottom graph, with the black line being the measured data, and the red line being the calculated model response (see text for explanation).

can be distinguished from their seismic facies and from observations of erosional discontinuities. The discontinuities are labelled, from the oldest to the youngest, as ASS-u3 to ASS-u5 (Amundsen Sea Shelf unconformity). The sedimentary units are named ASS-A1 to ASS-A4 (Amundsen Sea Shelf Abbot glacial trough) from oldest to youngest (Fig. 5).

Sedimentary unit ASS-A1 consists of short reflectors, which have mainly low amplitudes and are discontinuously layered (Figs 2–4). This homogeneous unit is interlayered with occasional bands of closely spaced reflectors of high amplitude, which are traceable across the shelf. ASS-A1 is present in all seismic lines as a massive sedimentary package with a thickness of between 3 and 6.5 km that crops out at the sea floor on parts of the middle shelf. On the outer shelf, ASS-A1 thins and is mainly overlain by an erosional unconformity ASS-u3. ASS-u3 is also present on the mid-shelf, cutting into ASS-A1 as a trough-like structure.

Sedimentary unit ASS-A2 occurs on top of unconformity ASS-u3 on the outer shelf (Figs 2–4). This unit shows continuous, strong, slightly inclined reflectors. It varies little in thickness throughout the outer shelf, but could not be identified in the Peacock Sound area.

Above discontinuity ASS-u4, the sedimentary strata of unit ASS-A3 consist of prograding clinoforms and sequences, showing continuous layering between high-and low-amplitude reflections (Figs 2–4). This unit is present mainly on the outer shelf, but also in the trough-like structure on the outer middle shelf.

In all profiles, a third major erosional discontinuity ASS-u5 cuts into the older strata (Figs 2–4).

sedimentary unit	*erosional discontinuity*	*Age (approximated)*	*description of sedimentary unit*	*sedimentary processes & environment*
ASS-A4		Pleistocene	continuous strong reflectors, mimicing the seafloor	retreat of the WAIS after the LGM, glaciomarine deposition
	ASS-u5			
ASS-A3		Pliocene - Pleistocene	prograding clinoforms, continous layering between high and low amplitudes	full glacial environment, glacial-interglacial cyclicity, shelf progadation and aggradation
	ASS-u4			
ASS-A2		late Miocene - early Pliocene	continous, slightly inclined reflectors (not identified in Peacock Sound)	establishment of the WAIS in the Amundsen Sea, partly grounded ice
	ASS-u3			
ASS-A1		Cretaceous (?) - middle/late Miocene (?)	short reflectors, low amplitude, occasional bands of closely spaced reflectors	ice-free shelf with occasional advances of smaller ice streams, aggrational sedimentation on the middle shelf
basement				

Fig. 5. Overview of sedimentary units and erosional discontinuities from the Abbot glacial trough, including description of the unit and sedimentary processes and depositional environment.

Above ASS-u5, a small sedimentary unit, ASS-A4, is present throughout the continental shelf. It mimics the sea floor with continuous and strong reflectors. This sedimentary package varies only slightly in thickness. The sea floor is represented by a very strong reflector, which shows diffraction hyperbolae, as a result of the abundant iceberg-scouring and the resulting roughness of the sea floor.

In order to outline the depth of the crystalline basement more accurately, 2-D forward modelling on ship-borne free-air gravity-anomaly data was performed along all seismic profiles (example in Fig. 4). The depth of the crust-mantle boundary was kept constant at 22 km after Gohl *et al.* (2007), and the crystalline crust subdivided into three layers of typical continental crustal densities of between 2.70 and 2.90 g cm^{-3} from top to bottom. The combination of the gravimetric models and the depth-converted seismic profiles helped identify the weak reflectors that represent the top of the crystalline basement. The resulting basement topography varies strongly towards the continental slope. Here, several areas with elevated basement reach a depth of up to 500 m below the sea floor (Figs 2–4), whereas adjacent basins have a depth of between 3.4 and 4.5 km. Two deeper basins with depths of over 6 km, located north of Thurston Island and at Peacock Sound, were also modelled.

Comparison with horizon stratigraphy of outer Ross Sea shelf

The major sedimentary units and unconformities identified (Figs 2–4) can be related to the drilled (DSDP Leg 28; Hayes *et al.* 1975) and dated strata of the Ross Sea continental shelf by similarities in the seismic facies and erosional pattern. ASS-u3 is a strong erosional discontinuity, representing the change of the depositional environment from aggradational to progradational. On the Ross Sea shelf, this corresponds to RSU-4, marking the interface of the first advance of the West Antarctic Ice Sheet across the continental shelf (Anderson & Bartek 1992). Therefore, the underlying unit ASS-A1 is likely to represent the strata deposited in a pre-glacial environment (possibly Cretaceous to middle/late Miocene; Anderson & Bartek 1992; Chow & Bart 2003). ASS-A2 represents the transition towards a full glacial environment in Antarctica when ice sheets were present on the continent and advanced occasionally onto the shelf (Anderson & Bartek 1992). The discontinuity ASS-u4 marks the changeover between a transitional to a full glacial environment. The prograding sequences of ASS-A3 are comparable with unit RSS-7 of the Ross Sea, which was deposited during the Pliocene and Pleistocene epochs (Anderson & Bartek 1992).

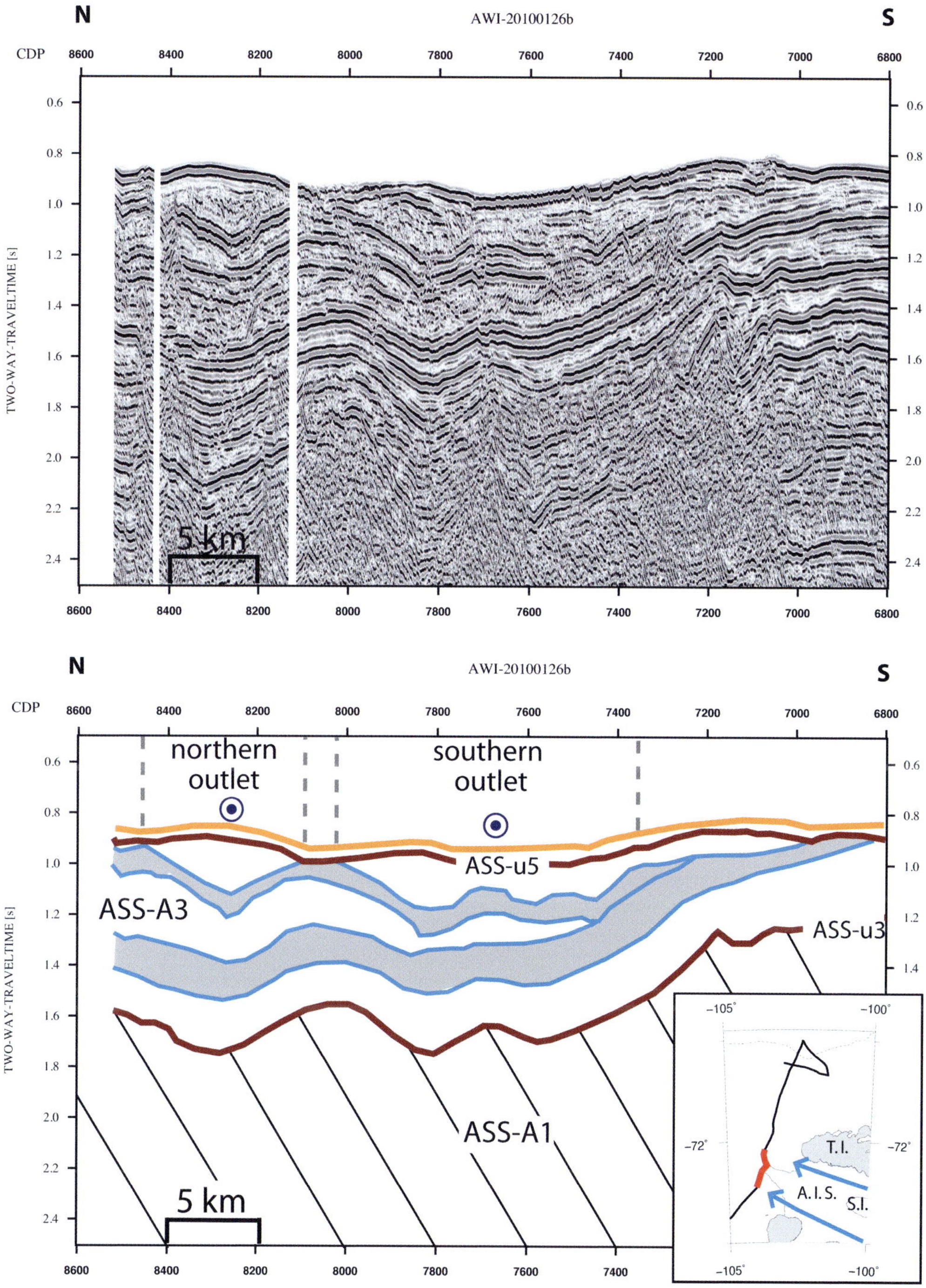

Fig. 6. Southern part of seismic profile AWI-20100126b, which bypasses the Abbot Ice Shelf of Peacock Sound (red line of inset map with ice-flow indicated by blue arrows; dotted black line shows the shelf edge). The interpretation of the bottom profile shows alternating zones of strong (grey) and weak (white) reflectivities in transitional and glacially dominated periods. The pre-glacial strata (ASS-A1) lie beneath unconformity ASSu-1.

ASS-A4 consists of the sediments deposited after the last glacial maximum, and ASS-u5 represents the erosional surface of this last advance of the ice sheet to the outer shelf.

Discussion

Pre-glacial setting of the outer shelf

The outer continental shelf of the eastern Amundsen Sea Embayment is crossed by tectonic lineaments, striking mainly NE–SW as interpreted from gravity and magnetic anomaly data (Gohl 2012; Gohl *et al.* 2013). Our observed and modelled basement highs and lows appear as horst-and-graben structures that correlate with one of the tectonic lineaments, which has been associated with Cretaceous rifting between New Zealand and West Antarctica (Fig. 7; Gohl 2012; Gohl *et al.* 2013).

Despite the lack of any drilling and outcropping sediment samples from the Amundsen Sea Embayment, the observation of seismic-reflection characteristics leads to the inference that sequence unit ASS-A1 represents the building-up of sediments on the continental shelf before the glaciation of West Antarctica took place. Multiple basement highs with adjacent small basins are present on the outer shelf (Fig. 7), channellizing the pre-glacial deposition. At the location where ASS-A1 crops out at the sea floor at its northernmost position (CDP 9700 of Fig. 2), the sediments are highly reworked and show only little lamination. This location represents the pre-glacial shelf break. Therefore, the pre-glacial shelf break in the area of the Abbot trough is 75 km further inshore than the present-day shelf break (Figs 2 & 8). The undulating basement topography of the present outer shelf and slope is covered by a few hundred metres of sediment of pre-glacial origin within the glacial trough (Figs 2 & 3), whereas few to no pre-glacial strata cover the basement ridges outside of the glacial trough off Thurston Island (Fig. 4). A distinct basement-high limits the pre-glacial and glacial shelf extent at this location, highlighting the role of basement tectonics on the outer shelf.

Glacial sedimentation

The Cretaceous tectonic lineaments and basement ridges of the outer shelf appear to have also constrained deposition of glacially eroded and

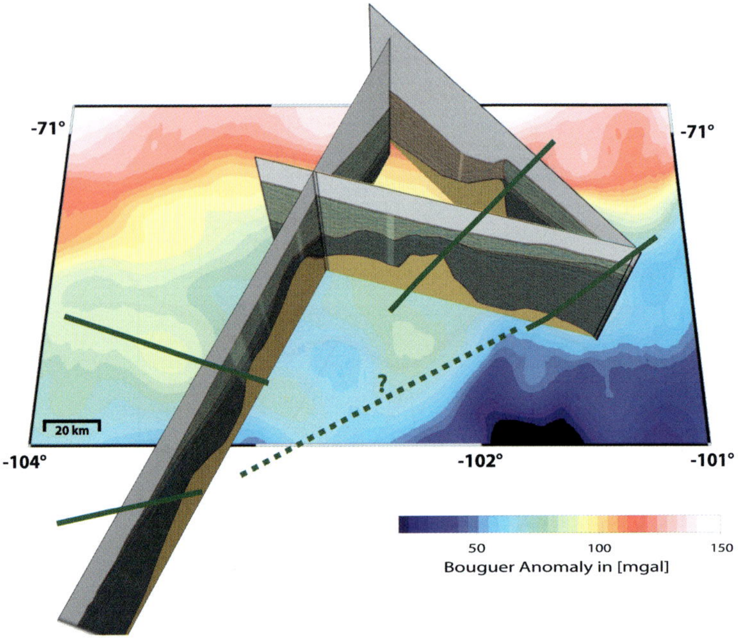

Fig. 7. Bird's-eye view of the three interpreted seismic profiles of the outer Abbot glacial trough, on top of a Bouguer gravity-anomaly map calculated from satellite-derived gravity data of Andersen *et al.* (2010). Solid dark green lines illustrate the locations and directional trends of basement ridges observed from seismic data and gravity modelling, and which correspond to positive Bouguer-anomaly lineaments. The dotted green line symbolizes a possible connecting basement ridge or escarpment between observed ridge segments.

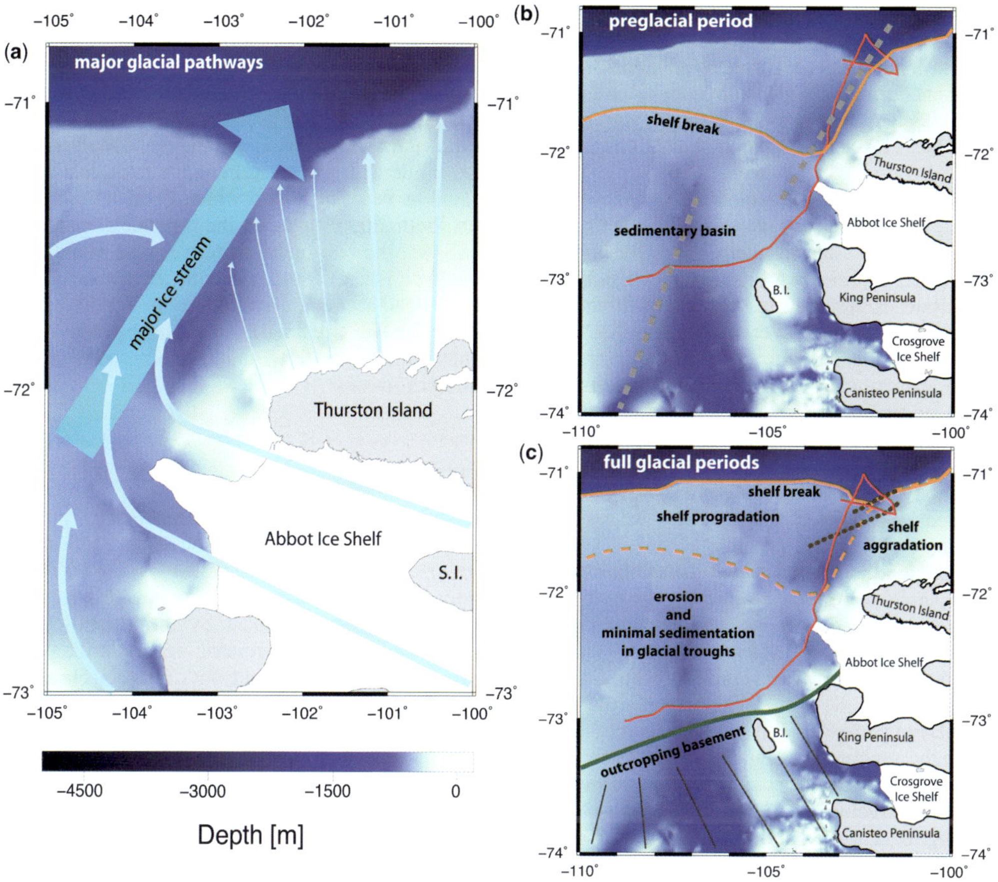

Fig. 8. Bathymetric maps illustrating the ice flow pattern during glacial maxima as interpreted from the seismic data (**a**) and the growth of the outer shelf and outward migration of the shelf break (orange solid and hashed lines) from pre-glacial times (**b**) to full glacial periods (**c**). The solid red line marks the seismic profiles in this paper. The dark green line illustrates the boundary between outcropping inner shelf basement and the shelf sedimentary basin. Basement-highs on the outer shelf are indicated by the brown dotted lines and the grey hashed lines of (b) are projected tectonic lineaments from Gohl (2012). Bathymetry is from Nitsche *et al.* (2007). B.I. is Burke Island. S.I. is Sherman Island.

transported sediment (Figs 2–4). Inside the Abbot trough, the glacial strata (ASS-A3) consists of multiple prograding sequences leading to growth of the outer shelf from the pre-glacial shelf break into the position of the present-day continental slope (Fig. 2). ASS-A3 consists of smaller clinoforms in the south and larger sequences after depositing over the top of the northern basement ridge. Thus, younger strata extend into the former basin, creating a till delta (Anderson & Bartek 1992; Larter & Vanneste 1995; Bart & Anderson 2000; Smith *et al.* 2011). A cross-section of this till delta can be observed in a section across the Abbot trough between CDP 66 and 500 (Fig. 3). Following further examination of the internal structure, the position of the former depositional centre of the glacial trough can be identified (Fig. 3). In the younger strata, an eastward shift of the depocentre is visible. Graham *et al.* (2010) proposed an eastwards trend of the main Pine Island/Thwaites ice stream in younger glacial periods. In earlier glacial cycles, the sedimentation in the Abbot glacial trough was channellized into the basin between the two basement-highs north of the pre-glacial shelf break (Figs 3 & 7). Only after sedimentation had become level with the northern basement-high did progradation proceed northwards down the slope during later glacial cycles, leading to relocation of the depocentre of the till delta to the east. Dowdeswell *et al.* (2006*a*) showed in 3D seismic data from the Norwegian continental margin that the change in ice-stream flow can occur between glacial advances because of the accessibility of new sedimentary basins. Therefore, the Abbot trough may have experienced a higher discharge of ice and sediment during later glacial cycles, since formerly inaccessible accommodation space had become available. This led to continental shelf progradation by the build-up of a till delta.

East of the main Abbot trough, the configuration of the slope differs strongly (Fig. 4). Only thin sediments cover the northern basement-high, and the glacial sedimentary sequences are limited by the basement, resulting in shelf aggradation (Dowdeswell *et al.* 2006*b*). We also observe that the orientation of the clinoforms of ASS-A3 creates a radial pattern (Figs 4 & 7). Therefore, multiple small ice streams originating from Thurston Island seem to have been the source of these sequences (Fig. 8). Such small ice streams had insufficient capacity to overcome the basement-high, and thus enabled significant deposition on the shelf break and slope.

Through the limited supply of sediment to the outer shelf, deposition was restricted to the pre-existing basins between the basement ridges. Thus sedimentation was limited to the outer shelf, leading to aggradation, with minimal to no change in the position of the shelf break between pre-glacial and glacial periods.

In the section crossing the Abbot trough from west to east, a distinct boundary between the glacial infill structure of the till delta and small prograding sequences from the east is exposed (Fig. 3). At this location, the larger, fast-flowing ice stream following the main Abbot trough absorbed the smaller ice streams from the east, merging their ice masses and deflecting sediment transport from a westwards to northwards flow direction towards the shelf slope.

Abbot Ice Shelf

The seismic profile AWI-20100126b bypasses the outlet of the west-flowing Abbot Ice Shelf in Peacock Sound (Fig. 6). The record shows distinct bands of strong reflectors enclosing zones of weak and partially chaotic reflections within the sedimentary unit ASS-A3. A similar record has been interpreted, for the western Amundsen Sea Embayment, as depositional sequences alternating between full glacial and less glacially dominated periods in Miocene times (Weigelt *et al.* 2009). Although the age is speculative, the depositional processes appear to have been similar to those in the eastern embayment. Two small troughs, with a small sedimentary high in between, form a W-shaped reflection pattern between CDPs 7500 and 8500, partly representing local erosional unconformities. It can be inferred from this observation that the ice streams from Peacock Sound carved two individual glacier outlet troughs, a southern and a northern outlet, orthogonally or obliquely into the main Pine Island/Thwaites and Cosgrove glacial trough deposits of the shelf. The dividing sedimentary high lies along the ice flow-direction and inline with the location of Sherman Island, an island in Peacock Sound which divides the Abbot Ice Shelf (Figs 1 & 6). In the pre-glacial sequence ASS-A1, the two-outlet pattern is also visible but less pronounced compared with the glacial strata. This island seems to have played an important role in the ice-sheet dynamics of the Abbot Ice Sheet, channellizing the flow of ice into a southern and northern stream.

Conclusions

Seismic-reflection data, supplemented by gravity modelling results for basement depths, reveal the pre-glacial and glacial sedimentary dynamics of the outer shelf of the eastern Amundsen Sea Embayment, which is dominated by the Abbot glacial trough as a major bathymetric depression. Several basement-highs, inherited from Cretaceous breakup and Cenozoic rifting processes, have constrained sedimentation. These basement ridges played a key role in the development of the Abbot glacial trough. They channellized sediment deposition, and may have controlled the ice-sheet dynamics by acting as pinning points or barriers in early glacial periods.

By establishing a first basic horizon stratigraphy, we identify pre-glacial (ASS-A1), transitional (ASS-A2) and glacial (ASS-A3) seismic facies. The pre-glacial continental shelf-break was located approximately 75 km further inshore than the present-day Abbot glacial trough. When the first ice sheet advances onto the outer shelf took place, the depocentre moved towards the outer shelf and the continental slope. The NE–SW-striking basement-highs restricted deposition to NW of Thurston Island throughout early glacial advances, leading to aggradation of glaciomarine sediment on the continental shelf. Later, this basement barrier was partly overcome, allowing progradational deposition over the shelf break in younger glacial periods.

The main Abbot glacial trough contained merged flows from the Pine Island/Thwaites, Cosgrove and Abbot glacier systems, in addition to small contributions from local ice streams on Thurston Island. Sherman Island of Peacock Sound played an important role in the ice-dynamics of the Abbot glacier, by dividing the northwestwards ice-flow into two ice streams, which interfered with the main glacial sediment transport from the south.

The support of the master and crew of RV *Polarstern* during expedition ANT-XXVI/3 (2010) is gratefully acknowledged. Parts of this paper have resulted from the M.Sc. thesis of K.H., who was co-supervised by H. Keil of University of Bremen. We would also like to thank the joint Volume Editor/Corresponding Editor, M. Hambrey, J. B. Anderson and one anonymous reviewer for their comments and suggestions. Seismic data-processing was performed using the *Paradigm Focus*™ software package. This study has been funded through Work Package 3.2 of the AWI research programme PACES.

References

ANDERSON, J. & BARTEK, L. R. 1992. Cenozoic glacial history of the Ross Sea revealed by intermediate resolution seismic reflection data combined with drill site information. *Antarctic Research Series*, **56**, 231–263.

ANDERSEN, O. B., KNUDSEN, P. & BERRY, P. 2010. The DNSC08GRA global marine gravity field from double retracked satellite altimetry. *Journal of Geodesy*, **84**, 191–199.

BART, P. J. & ANDERSON, J. B. 2000. Relative temporal stability of the Antarctic ice sheets during the late Neogene based on the minimum frequency of outer shelf grounding events. *Earth and Planetary Science Letters*, **182**, 259–272.

CHOW, J. M. & BART, P. J. 2003. West Antarctic Ice Sheet grounding events on the Ross Sea outer continental shelf during the middle Miocene. *Palaeogeography, Palaeoclimatology, Palaeoecology*, **198**, 169–186.

DALZIEL, I. W. D. 2006. On the extent of the active West Antarctic rift system. *In*: SIDDOWAY, C. S. & RICCI, C. A. (eds) *Proceedings of Workshop on Frontiers and Opportunities in Antarctic Geosciences*. Terra Antartica Reports **12**. Terra Antartica, Siena, 193–202.

DOWDESWELL, J. A., OTTENSEN, D. & REISE, L. 2006*a*. Flow switching and large-scale deposition by ice streams draining former ice sheet. *Geology*, **34**, 313–316.

DOWDESWELL, J. A., COFAIGH, C. Ó. & ANDERSON, J. B. 2006*b*. Morphology and sedimentary processes on the continental slope off Pine Island Bay, Amundsen Sea, West Antarctica. *Bulletin of the Geological Society of America*, **118**, 606–619.

EAGLES, G., GOHL, K. & LARTER, R. D. 2004. High-resolution animated tectonic reconstruction of the South Pacific and West Antarctic Margin. *Geochemistry Geophysics Geosystems*, **5**, 1–21, doi: 10.1029/2003GC000657.

EVANS, J., DOWDESWELL, J. A., COFAIGH, C. Ó., BENHAM, T. J. & ANDERSON, J. B. 2006. Extent and dynamics of the West Antarctic Ice Sheet on the outer continental shelf of Pine Island Bay during the last glaciation. *Marine Geology*, **230**, 53–70.

GOHL, K. 2010. *The expedition of the research vessel 'Polarstern' to the Amundsen Sea, Antarctica, in 2010.* Berichte zur Polar- und Meeresforschung/Reports on Polar and Marine Research, **617**, http://epic.awi.de/epic/Main?static=yes&page=abstract&entry_dn=Goh2010f

GOHL, K. 2012. Basement control on past ice sheet dynamics in the Amundsen Sea Embayment, West Antarctica. *Palaeogeography Palaeoclimatology, Palaeoecology*, **335–336**, 35–41, doi: 10.1016/j.palaeo.2011.02.022.

GOHL, K., TETERIN, D. *ET AL*. 2007. Geophysical survey reveals tectonic structures in the Amundsen Sea embayment, West Antarctica. *In*: COOPER, A. K. & RAYMOND, C. R. (eds) *Proceedings of the 10th Int. Symposium of Antarctic Earth Sciences*. USGS Open-File Report 2007–1047. US Geological Survey, Reston, VA, doi: 10.3133/of2007-1047.srp047.

GOHL, K., DENK, A., EAGLES, G. & WOBBE, F. 2012. Deciphering tectonic phases of the Amundsen Sea Embayment shelf, West Antarctica, from a magnetic anomaly grid. *Tectonophysics*, **585**, 113–123, doi: 10.1016/j.tecto.2012.06.036.

GRAHAM, A. G. C., LARTER, R. D. *ET AL*. 2010. Flow and retreat of the Late Quaternary Pine Island-Thwaites paleo-ice stream, West Antarctica. *Journal of Geophysical Research–Earth Surfaces*, **115**, 1–12, doi: 10.1029/2009JF001482.

HAYES, D. E. & FRAKES, L. A. 1975. General synthesis. *Initial Reports of the Deep Sea Drilling Project*, **28**, 919–942.

HILLENBRAND, C.-D., SMITH, J. A. *ET AL*. 2009. Age assignment of a diatomaceous ooze deposited in the western Amundsen Sea Embayment after the Last Glacial Maximum. *Journal of Quaternary Science*, **25**, 280–295, doi: 10.1002/jqs.1308.

JAKOBSSON, M., ANDERSON, J. B. *ET AL*. 2011. Geological record of ice shelf break-up and grounding line retreat, Pine Island Bay, West Antarctica. *Geology*, **39**, 691–694, doi: 10.1130/G32153.1.

JAKOBSSON, M., ANDERSON, J. B. *ET AL*. 2012. Ice sheet retreat dynamics inferred from glacial morphology of the central Pine Island Bay Trough, West Antarctica. *Quaternary Science Reviews*, **38**, 1–10, doi: 10.1016/j.quascirev.2011.12.017.

JORDAN, T. A., FERRACCIOLI, F., VAUGHAN, D. G., HOLT, J. W., CORR, H., BLANKENSHIP, D. D. & DIEHL, T. M. 2009. Aerogravity evidence of a major crustal thinning under the Pine Island Glacier region (West Antarctica). *Bulletin of the Geological Society of America*, **122**, 714–726, doi: 10.1130/B26417.1.

KIRSHNER, A. E., ANDERSON, J. B., JAKOBSSON, M., O'REGAN, M., MAJEWSKI, W. & NITSCHE, F. O. 2012. Post-LGM deglacation in Pine-Island Bay, West Antarctica. *Quaternary Science Reviews*, **38**, 1–16, doi: 10.1016/j.quascirev.2012.01.017.

LARTER, R. D. & VANNESTE, L. E. 1995. Relict subglacial deltas on the Antarctic Peninsula outer shelf. *Geology*, **23**, 33–36.

LARTER, R. D., GRAHAM, A. G. C. *ET AL*. 2009. Subglacial bedforms reveal complex basal regime in a zone of paleo-ice stream convergence, Amundsen Sea Embayment, West Antarctica. *Geology*, **37**, 411–414.

LOWE, A. L. & ANDERSON, J. B. 2002. Reconstruction of the West Antarctic ice sheet in Pine Island Bay during the Last Glacial Maximum and its subsequent retreat history. *Quaternary Science Reviews*, **21**, 1879–1897.

LOWE, A. L. & ANDERSON, J. B. 2003. Evidence for abundant subglacial meltwater beneath the paleo-ice sheet in Pine Island Bay, Antarctica. *Journal of Glaciology*, **46**, 125–138.

Nitsche, F. O., Gohl, K., Vanneste, K. & Miller, H. 1997. Seismic expression of glacially deposited sequences in the Bellinghausen and Amundsen Sea, West Antarctica. *Antarctic Research Series*, **71**, 95–108.

Nitsche, F. O., Jacobs, S. S., Larter, R. D. & Gohl, K. 2007. Bathymetry of the Amundsen Sea continental shelf: implications for geology, oceanography, and glaciology. *Geochemistry Geophysics Geosystems*, **8**, 1–10, doi: 10.1029/2007GC001694.

Smith, J. A., Hillenbrand, C.-D. *et al.* 2011. Deglacial history of the West Antarctic Ice Sheet in the western Amundsen Sea Embayment. *Quaternary Science Reviews*, **30**, 488–505.

Uenzelmann-Neben, G., Gohl, K., Larter, R. D. & Schlüter, P. 2007. Differences in ice retreat across Pine Island Bay, West Antarctica, since the Last Glacial Maximum: indications from multichannel seismic reflection data. *In*: Cooper, A. K. & Raymond, C. R. (eds) *Proceedings of the 10th Int. Symposium of Antarctic Earth Sciences*. USGS Open-File Report **2007-1047**. US Geological Survey, Reston, VA, doi: 10.3133/of2007-1047.srp084.

Weigelt, E., Gohl, K., Uenzelmann-Neben, G. & Larter, R. D. 2009. Late Cenozoic ice sheet cyclicity in the western Amundsen Sea Embayment – evidence from seismic records. *Global and Planetary Change*, **69**, 162–169.

Southern Ocean bioproductivity during the last glacial cycle – new detection method and decadal-scale insight from the Scotia Sea

D. SPRENK[1]*, M. E. WEBER[1], G. KUHN[2], P. ROSÉN[3], M. FRANK[4], M. MOLINA-KESCHER[4], V. LIEBETRAU[4] & H.-G. RÖHLING[5]

[1]*Institute of Geology and Mineralogy, University of Cologne, Zuelpicher Str. 49a, D-50674 Cologne, Germany*

[2]*Alfred-Wegener-Institut Helmholtz-Zentrum für Polar und Meeresforschung, Am Alten Hafen 26, D-27568 Bremerhaven, Germany*

[3]*Climate Impacts Research Centre (CIRC), Umeå University, SE-981 07 Abisko, Sweden*

[4]*GEOMAR Helmholtz Centre for Ocean Research Kiel, Wischhofstr. 1-3, D-24148 Kiel, Germany*

[5]*LBEG Landesamt für Bergbau, Energie und Geologie, Stilleweg 2, D-30655 Hannover, Germany*

**Corresponding author (e-mail: dsprenk@uni-koeln.de)*

Abstract: We present biogenic opal flux records from two deep-sea sites in the Scotia Sea (MD07-3133 and MD07-3134) at decadal-scale resolution, covering the last glacial cycle. In addition to conventional and time-consuming biogenic opal measuring methods, we introduce new biogenic opal estimation methods derived from sediment colour *b**, wet bulk density, Si/Ti-count ratio and Fourier transform infrared spectroscopy (FTIRS). All methods capture the biogenic opal amplitude; however, FTIRS–a novel method for marine sediment – yields the most reliable results. ^{230}Th normalization data show strong differences in sediment focusing with intensified sediment focusing during glacial times. At MD07-3134 ^{230}Th normalized biogenic opal fluxes vary between 0.2 and 2.5 g cm^{-2} kyr^{-1}. Our biogenic opal flux records indicate bioproductivity changes in the Southern Ocean, strongly influenced by sea ice distribution and also summer sea surface temperature changes. South of the Antarctic Polar Front, lowest bioproductivity occurred during the Last Glacial Maximum when upwelling of mid-depth water was reduced and sea ice cover intensified. Around 17 ka, bioproductivity increased abruptly, corresponding to rising atmospheric CO_2 and decreasing seasonal sea ice coverage.

The Southern Ocean plays an important role in transferring carbon dioxide (CO_2) via wind-induced upwelling from the deep sea to the atmosphere (Toggweiler *et al.* 2006; Anderson *et al.* 2009) and is therefore one of the key regions from which to study climate change. Bioproductivity in the Southern Ocean is mainly controlled by the rate of upwelling of cold nutrient- and silica-rich water masses, the extent of sea ice coverage, and the availability of light and micronutrients (Dezileau *et al.* 2003; de Baar *et al.* 2005; Stenni *et al.* 2010). Thus, bioproductivity in the Southern Ocean is mainly coupled to local climate change. Biogenic opal is an important nutrient that fuels biological production in the surface waters around Antarctica. Recent investigations (Pondaven *et al.* 2000; Chase *et al.* 2003; Bradtmiller *et al.* 2007; Anderson *et al.* 2009) revealed that biogenic opal flux in the Southern Ocean primarily reflects variations in bioproductivity rather than changes in preservation, which enables its use for studying palaeoenvironmental changes during the last glacial cycle. Anderson *et al.* (2009) also identified biogenic opal as an upwelling proxy because, south of the Polar Frontal Zone in the Southern Ocean, its production is ultimately limited by the supply of dissolved Si from the deep ocean. Horn *et al.* (2011) argued that biogenic opal flux might not be a direct upwelling proxy, but could also indicate major changes in nutrient demand such as iron. Until now, high-resolution and continuous biogenic opal flux records from the Southern Ocean extending back to the last interglacial are rare (e.g. Anderson *et al.* 2009) and therefore only limited knowledge exists on how Southern Ocean bioproductivity changed in the past.

We studied two deep-sea cores from the Scotia Sea, with linear sedimentation rates of up to 1.2 m

From: Hambrey, M. J., Barker, P. F., Barrett, P. J., Bowman, V., Davies, B., Smellie, J. L. & Tranter, M. (eds) 2013. *Antarctic Palaeoenvironments and Earth-Surface Processes*. Geological Society, London, Special Publications, **381**, 245–261. First published online July 30, 2013, http://dx.doi.org/10.1144/SP381.17

kyr^{-1} (MD07-3134) and 1.8 m kyr^{-1} (MD07-3133). As biogenic opal analysis using the leaching method by Müller & Schneider (1993) is very time-consuming and expensive, we additionally tested several biogenic opal estimation methods, for example Si/Ti-count ratio, wet bulk density, colour b^* and Fourier transform infrared spectroscopy (FTIRS), and compared the results with conventionally measured biogenic opal. FTIRS has been successfully used to determine biogenic opal quantitatively in lacustrine sediment (Vogel *et al.* 2008). We present one of the first studies to assess biogenic opal from FTIRS in marine sediment. Our goal is to provide a high-resolution biogenic opal accumulation rate record, giving the opportunity to study bioproductivity changes at decadal-scale resolution in the Southern Ocean over the last glacial cycle (92.5 ka to present; note that all ages mentioned are calendar ages).

Core material

Sediment cores used in this study originate from the central Scotia Sea (Fig. 1) and were drilled with a Calypso II piston corer during the Marion Dufresne (MD) II cruise 160 in March 2007. Core sites MD07-3133 (57°26′S, 43°27′W, 3101 m water depth, 32.8 m core recovery) and MD07-3134 (59°25′S, 41°28′W, 3663 m water depth, 58.2 m core recovery) are located approximately 450 km apart. Both core sites lie below the carbonate compensation depth. Therefore, the core material is virtually free of biogenic carbonate and consists of varying amounts of lithogenic material and biogenic opal. Homogenous, olive grey to yellow diatomaceous oozes were deposited during warm climatic periods (Marine Isotopic Stages (MIS) 5 and 1), grey to blue-grey diatom-bearing mud during cold periods (MIS 4 and 2), and mostly olive-grey diatomaceous mud during MIS 3.

Chronology

The chronology of sites MD07-3133 and MD07-3134 is detailed in Weber *et al.* (2012). It relies on seven ground-truth data points obtained at site MD07-3134. Accordingly, MIS boundaries 5/4 (71 ka), 4/3 (57 ka), 3/2 (29 ka), and 2/1 (14 ka) comply with 46, 38.5, 16.5 and 11 m core depths, respectively. Relative palaeointensity data indicate a significant drop at approximately 24 m in the core that is considered to represent the Laschamp Event (40.4 ± 1.1 ka; Guillou *et al.* 2004). Furthermore, volcanic ashes also rich in ice-rafted detritus (e.g. site MD07-3134 at 8.9, 13.36, 34.17, and 43.88 m core depth) have been dated and correlated across the Southern Atlantic as layers SA0 (*c.* 14–15 ka) to SA6 (*c.* 55 ka) (Kanfoush *et al.* 2000;

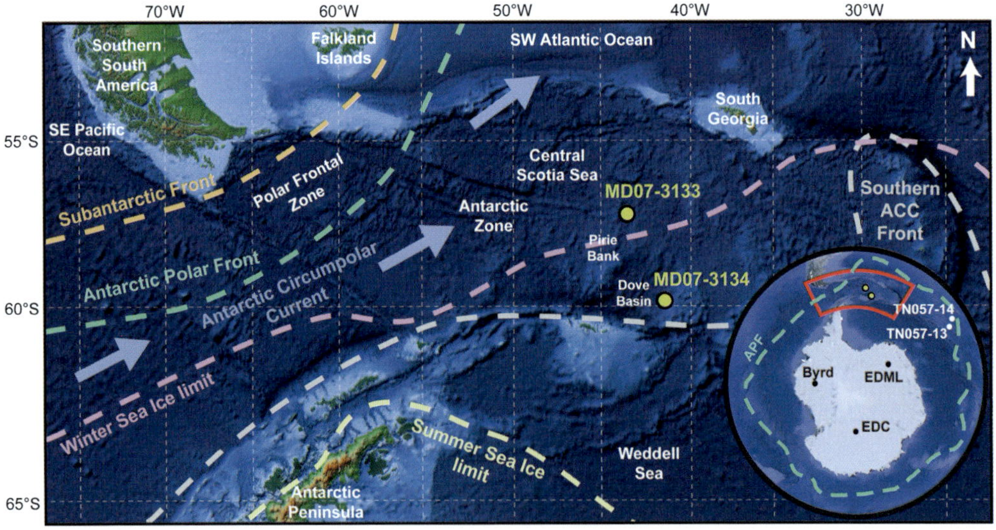

Fig. 1. Location map showing the MD07-3133 and MD07-3134 core sites referred to in this study; Subantarctic Front, Antarctic Polar Front, Southern Antarctic Circumpolar Current Front, limits modified from Gersonde *et al.* (2005). Additionally, the positions of Winter and Summer Sea Ice limits today (Gersonde *et al.* 2005; Allen *et al.* 2011) are shown. Also, the direction of Antarctic Circumpolar Current is highlighted. Underlying map originates from http://maps.ngdc.noaa.gov/viewers/bathymetry/ (access date: 20 October 2011). TN core sites are from Anderson *et al.* (2009); Byrd (Ahn & Brook 2008), EPICA Dronning Maud Land (EDML; EPICA Community Members 2006) and EPICA Dome C (EDC; EPICA Community Members 2004) refer to Antarctic ice cores (for details see text).

Nielsen *et al.* 2007). For site MD07-3134, volcanic layers SA0 and SA6 were found at 13.36 and 34.17 m in the core, respectively. The detection of these two layers further supports the independent MIS ground-truth assignation because SA6 is just slightly younger than MIS boundary 4/3, and SA0 is exactly as old as MIS boundary 2/1.

Weber *et al.* (2012) provided convincing arguments that, from the various supply mechanisms proposed for magnetic susceptibility and Ca^{2+} concentrations, that is icebergs, ocean circulation, atmospheric circulation, volcanic sources and sea ice, only atmospheric circulation remains for the Scotia Sea sites. Hence, they correlated both signals to the non-sea-salt Ca^{2+} flux of the European Project for Ice Coring in Antarctica (EPICA) Dronning Maud Land (EDML) record (Fischer *et al.* 2007*a*) to increase the resolution of their age model for site MD07-3134. They applied three stratigraphic approaches to produce a tuned, high-resolution age model, evaluate its quality and provide error estimates: (i) manual picking of tie points; (ii) an automated pattern-matching algorithm; and (iii) a smoothing spline regression using the manual tie points.

As a result, Weber *et al.* (2012) established 48 confident age control points for the last 92.5 kyr. Average linear sedimentation rates are high, ranging from 0.2 to 1.2 m kyr^{-1} and from 0.3 to 1.8 m kyr^{-1} for sites MD07-3134 and MD07-3133, translating into mass accumulation rates (calculated by multiplying linear sedimentation rate with dry bulk densities) of 10–130 and 9–100 g cm^{-2} kyr^{-1}, and into sample resolution of 50–8 years and 33–5 years (given a sample increment of 1 cm), respectively.

Tuning magnetic susceptibility to the non-sea-salt Ca^{2+} flux has significant implications for the quality and reliability of the age model. The fact that the two signals are more or less phase-locked during this procedure is reasonable because dust transport from Patagonia to Antarctica takes only approximately a week (Li *et al.* 2010). Therefore, no major leads or lags should be involved for atmospheric transport. Accordingly, the chronology of our two sites provides a precise and high-resolution age model for Southern Ocean deep-sea sediment.

Setting

The Scotia Sea is located in the Atlantic Sector of the Southern Ocean where the Antarctic Circumpolar Current dominates oceanic circulation. It is the world's largest current with a transport volume of approximately 134 Sverdrup (1 Sv = 10^6 m^3 s^{-1}; Whitworth & Peterson 1985) flowing eastward around the Antarctic continent. The Antarctic Circumpolar Current includes, in most regions, water masses from the seafloor to the surface. Its velocity generally decreases with depth but is also severely influenced by local seabed topography (Pugh *et al.* 2009). Uniquely connecting the Atlantic, Indian and Pacific, the Antarctic Circumpolar Current allows for heat, nutrient and salt exchange between the oceans (Maldonado *et al.* 2003), accordingly playing an important role in global climate control (Pugh *et al.* 2009). It is mainly wind-driven by the Southern Hemisphere Westerlies between 45 and 55°S. These Westerlies are also responsible for transporting surface waters away from Antarctica through Ekman transport, leading to upwelling of mid-depth water (2–3 km) south of the current, a unique process restricted to the Southern Ocean (Berger & Loutre 1991).

In the Scotia Sea, the Antarctic Circumpolar Current flows between the Subtropical Front and the Southern Antarctic Circumpolar Current Front (Fig. 1). Its flow is concentrated in circumpolar, vertically coherent, seafloor-reaching fronts: the Subantarctic Front, the Antarctic Polar Front and the Southern Antarctic Circumpolar Current Front (Orsi *et al.* 1995). The Antarctic Polar Front is the northern limit of wind-driven, nutrient-rich deep-water upwelling because of northward Ekman transport. It therefore forms an ecological and physical water–mass boundary to colder, more silicate-rich and less saline water masses south of it (Pondaven *et al.* 2000; Dezileau *et al.* 2003; Cassar *et al.* 2007; Diekmann 2007), where both core sites are located (Fig. 1). The area between the Antarctic Polar Front and the northern limit of seasonal sea ice is known as the Circum-Antarctic Opal Belt with high biosiliceous production rates (Geibert *et al.* 2005; Diekmann 2007). The southern boundary of siliceous deposits corresponds well to the Southern Antarctic Circumpolar Current Front in the Atlantic Sector of the Southern Ocean (Geibert *et al.* 2005).

Methods

We used a Minolta spectrophotometer CM-2002 to measure L^*, a^* and b^* colour components at 1 cm intervals (Weber 1998). To obtain information on chemical element distribution (e.g. for Si, Ti), every centimetre of sediment was analysed according to the method of Richter *et al.* (2006) using an Avaatech X-ray Fluorescence Core Scanner (Jansen *et al.* 1998).

Sediment physical properties such as wet bulk density and magnetic susceptibility (κ volume specific) were measured non-destructively every centimetre with a Geotek Multi-Sensor Core Logger (Weber *et al.* 1997). Dry bulk densities, which are

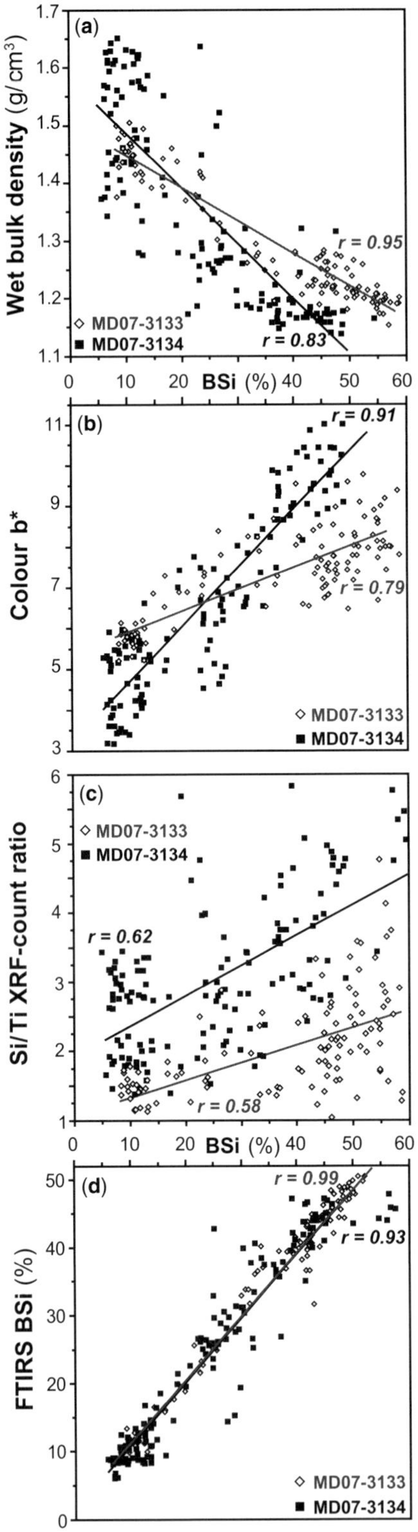

required for the calculation of mass accumulation rates, were calculated from wet bulk densities using an iteration procedure described in Weber *et al.* (1997), by applying specific mass attenuation coefficients to each measurement to account for fluctuating water content and grain density. To check its accuracy we also compared the non-destructive data with dry bulk densities determined on freeze-dried samples.

In addition to measuring biogenic opal on specific samples we implemented several strategies to estimate it at high spatial resolution. We first analysed biogenic opal by leaching the sediment in 1 M NaOH-solution according to the method of Müller & Schneider (1993). The initial values for biogenic silicon were corrected by adding 10 wt% H_2O for the water bound in the amorphous opal skeleton (biogenic opal = biogenic $opal_{M\&S}$ + 10% H_2O). All resulting data needed to be corrected for the sea salt content of the pore fluid (35‰). As this leaching method is relatively expensive and time-consuming, measurements were restricted to 10 cm intervals of specific sections that exhibit large-amplitude variations (MD07-3133: 1655–2785 cm, 112 samples; MD07-3134: 745–1785 cm, 4135–4585 cm, 141 samples).

Biogenic opal-rich, unconsolidated sediments exhibit extremely high porosities because interstitial fluid rests both within the shells (intragranular) and in the pore space (intergranular). The high sediment water contents as well as the lower grain densities of biogenic opal, in contrast to detrital material and biogenic carbonate, cause the extremely low wet and dry bulk densities of biogenic opal-rich sediments. Therefore, wet bulk densities are negatively correlated to biogenic opal (Weber 1998) for diatom-rich deep-sea sediments well below the carbonate compensation depth, and hence biogenic opal can be estimated from wet bulk densities using linear correlation coefficients (Fig. 2a).

Another striking correlation is given for colour *b** and biogenic opal in diatom-rich sediment (Weber 1998), that is, the more yellow the sediment, the higher the amount of biogenic opal, whereas more bluish colours indicate low biogenic opal. Accordingly, we derived linear correlation coefficients to estimate the amount of biogenic

Fig. 2. Scatter plots illustrating linear correlations for conventionally leached biogenic opal (BSi) values (method see Müller & Schneider 1993) and BSi content estimated from wet bulk density (**a**), colour *b** (**b**), XRF-core-scanner measured Si/Ti-count ratio (**c**) and Fourier transform infrared spectroscopy (FTIRS) (**d**; method see Rosén *et al.* 2010). White diamonds and grey lines refer to site MD07-3133; black squares and black lines indicate data from site MD07-3134.

opal from colour b^* for Scotia Sea sites MD07-3133 and MD07-3134 (Fig. 2b).

Also, we used the ratio of Si and Ti XRF-counts to obtain information on biogenic opal changes. Si originates from either detrital siliciclastics or biogenic, mostly diatom sources. Because Ti reflects only clastic input (Murray *et al.* 1993), the XRF Si/Ti ratio should only represent changes in biogenic opal (Francus *et al.* 2009; Balascio *et al.* 2011). Because XRF core-scanner measurements were made at higher resolution than some other proxies (1 cm increments), we implemented a nine-point moving average to smooth the data (using AnalySeries 2.0.4.2.; Paillard *et al.* 1996). Then we used the linear correlation between the Si/Ti XRF-count ratio and leached biogenic opal to calculate biogenic opal values, depending on Si/Ti-count ratios (Fig. 2c).

Furthermore, we implemented FTIRS to obtain high-resolution information on biogenic opal and other biogeochemical sediment components. FTIRS is a fast and relatively inexpensive technique, which requires only 11 mg of sample material (Rosén *et al.* 2011). It was therefore employed at 10 cm intervals over the entire core lengths of sites MD07-3133 and MD07-3134 (316 and 575 samples, respectively) at the Climate Impacts Research Centre in Umeå, Sweden. FTIRS has been successfully used to determine, quantitatively, total inorganic and organic carbon, and biogenic opal in lacustrine sediment (Vogel *et al.* 2008). Here, we present one of the first studies to assess these sediment properties quantitatively for marine sediment. Samples were first freeze-dried and ground to <63 μm using a swing mill. In addition, 500 mg of oven-dried (80 °C) potassium bromide (KBr, Merck), which is transparent in the infrared band, was added to each sample. This is necessary to avoid very high absorbances (>2), which are released by low intensities of infrared light that would otherwise reach the detector and produce higher noise levels in the data and spectral distortions (Griffiths & De Haseth 1986). The samples were measured using an FTIR spectrometer (Bruker Vertex 70) equipped with a diffuse reflectance accessory. Each sample was scanned 128 times and data were collected every 2 cm^{-1} (reciprocal centimetres). The measurement resolution was 4 cm^{-1} for wavenumbers between 3750 and 400 cm^{-1}, which equals wavelengths from 2666 to 25 000 nm, thus yielding 1735 data points per sample. To avoid variations caused by temperature, all samples were placed in the same temperature-controlled laboratory (25 ± 0.2 °C) as the FTIRS device for at least 5 h prior to analysis. Multiple scatter correction and baseline correction were used to linearize spectra and remove variation in spectra caused by noise (Geladi *et al.* 1985; Rosén *et al.* 2010). Baseline correction performs a linear correction of the spectra so that two points (3750 and 2210–2200 cm^{-1}) equal zero. Multiple scatter correction removes spectral variation arising from different effective path lengths and particle sizes (Geladi *et al.* 1985). Partial least squares regression was used to develop quantitative calibration models between FTIR spectra of sediment and conventionally measured sediment properties (Martens & Naes 1989). An internal calibration model based on 253 samples from MD07-3133 and MD07-3134 was regressed on 253 conventionally determined sea salt corrected mass percentage biogenic opal contents using the leaching technique (Müller & Schneider 1993; Vogel *et al.* 2008). Wavenumbers between 450 and 1320 cm^{-1} were used in the partial least squares regression model, that is, a region where different types of silicates absorb. Although different silicates absorb in the FTIRS wavenumber region between 1050 and 1250 cm^{-1}, it is not necessary to subtract spectra as the curve's shape shows characteristic differences for, for example, quartz and biogenic opal. The predictive performance of the partial least squares regression calibration model was assessed by 10% cross-validation. This means that the calibration model was developed using 90% of data of the calibration samples with the remaining 10% used to test the predictions. This process was repeated a total of 10 times as each group, in turn, was set aside. Root mean squared error of cross-validation was used as an estimate of prediction error. All primary sediment properties were square root transformed prior to the analysis. SIMCA-P 11.5 (Umetrics AB, SE-907 19 Umeå, Sweden) was used for all multivariate data analyses.

Correction of sediment focusing using $^{230}Th_{xs}$ normalization

The extremely high sedimentation rates of up to 1.8 m kyr^{-1} clearly indicate that the sediments of this site have been subject to massive sediment focusing induced by bottom currents. This is observed at many locations in the Southern Ocean and results in completely erroneous mass accumulation rates, which can be off by a factor of 12 (Frank *et al.* 1999) or even more (this study). For the calculation of realistic mass accumulation rates (rain rates) of biogenic opal and other components, the effects of sediment focusing have to be quantified and be corrected for applying excess ^{230}Th ($^{230}Th_{xs}$) normalization (Kumar *et al.* 1995; Frank *et al.* 1996, 2000; Francois *et al.* 2004). This method is based on the constant production of ^{230}Th from its global homogenously distributed radioactive parent ^{234}U. In contrast to U, Th is highly particle reactive, resulting in essentially the entire amount of locally produced ^{230}Th being deposited in the sediments below. This means that the mass accumulation rate of ^{230}Th in the sediments should match its production in the overlying water column. Given that ^{230}Th can only be advectively transported in solution over very short distances as a consequence of its particle reactivity, any significant increases over the expected ^{230}Th-mass accumulation rate must be a result of sediment focusing and can be quantitatively corrected for by relating it to the expected flux at any particular location.

For the determination of the activities of ^{230}Th, ^{232}Th, ^{234}U and ^{238}U, between 50 and 90 mg of sediment were weighed and spiked with $^{229}Th/^{233}U/^{236}U$ in the laboratory of GEOMAR in Kiel. The samples were first treated with aqua regia and were then subjected to total dissolution in a mixture of concentrated nitric, hydrofluoric and perchloric acid. Subsequently U and Th were separated from each other and from disturbing matrix elements via ion exchange chromatography using a Uteva resin (Eichrom). The activities of the U and Th isotopes were determined on an Agilent 7500 quadrupole mass spectrometer. The precision of the measurements was better than 1% (2 standard errors) for the respective U isotopes and better than 2% (2 standard errors) for the Th isotopes. For the calculation of $^{230}Th_{xs}$ activities see Frank *et al.* (1996).

Results

Correlation coefficients of $r = 0.95$ and 0.83 indicate that wet bulk density changes at sites MD07-3133 and MD07-3134 are mainly caused by fluctuations in biogenic opal content (Fig. 2a). Also colour b^* changes predominantly reflect fluctuations in biogenic opal content (Fig. 2b), although their relation is, with correlation coefficients of $r = 0.91$ and 0.79, not as striking as for wet bulk density. The generally good correlation coefficients ($\geq$0.79) let us assume that both wet bulk density and colour b^* can be used to estimate realistic biogenic opal contents (within the given errors).

Additionally, XRF-measured Si/Ti-count ratios provide, to some extent, rough information about biogenic opal fluctuations. The correlation coefficient for Si/Ti and biogenic opal is $r = 0.58$ and 0.62 for sites MD07-3133 and MD07-3134 (Fig. 2c). However, its use as a biogenic opal proxy is questionable owing to uncertainties associated with Ti values (see Discussion for further explanation).

The most promising estimation method is FTIRS (Fig. 2d) with correlation coefficients of $r = 0.99$ and $r = 0.93$ for sites MD07-3133 and MD07-3134. The models of leached and FTIRS-estimated biogenic opal show good statistical performance, with cross-validated root mean squared errors of 4.8% (MD07-3133), respectively 8.7% (MD07-3134) of the gradient. FTIRS provides the closest similarities of all estimation methods to the reference method of conventional leaching.

We then calculated biogenic opal contents from the various proxy data sets for sites MD07-3133 and MD07-3134 (Fig. 3), using the linear regression lines discussed above (Fig. 2). In general, the same overall trend is recognizable for all biogenic opal estimation curves; however, the amplitude varies significantly for some parts of the cores. Higher linear sedimentation rates lead to higher-resolution data for site MD07-3133 relative to site MD07-3134, specifically for the Holocene section. However, both cores share the same overall trend and principal features. As expected from the correlation coefficients, FTIRS-estimated and leached biogenic opal share the most commonalities. Biogenic opal contents estimated from colour b^* capture the glacial to interglacial transition; however, colour b^* overestimates biogenic opal contents for the uppermost and lowermost core sections of site MD07-3133. In fact, for the upper 12 m of MD07-3133, estimates from colour b^*, wet bulk density and the Si/Ti-count ratio show substantial discrepancies relative to FTIRS estimates (Fig. 3).

On average, the highest biogenic opal contents occur during interglacial MIS 5 and 1, confirming results of earlier studies on deep-sea sediment in the Atlantic sector (e.g. Charles *et al.* 1991; Mortlock *et al.* 1991; Pudsey & Howe 1998; Diekmann 2007), the Indian Sector (Bareille *et al.* 1998) and the Pacific Sector (Chase *et al.* 2003) of the Southern Ocean, south of the Antarctic Polar Front. Interglacial MIS 3 shows intermediate contents of biogenic opal with an overall decreasing trend towards MIS 2. During glacials MIS 4 and 2, absolute minima in biogenic opal contents are noticeable (Fig. 3). Surprisingly, maximum biogenic opal contents occur during glacial MIS 4 (site MD07-3134). These extraordinary high values seem robust (i.e. not caused by analytical errors), as five data points scatter around 60% biogenic opal in less than 1 m of sediment core. This section also shows low densities and yellower sediment colour, corresponding to substantially elevated biogenic opal contents. FTIRS also indicates high biogenic opal values for this section, whereas the highest FTIRS-estimated biogenic opal values (*c.* 67%) occur at the end of MIS 5. MIS 5 and 4 show high-amplitude and high-frequency changes in biogenic opal, whereas during MIS 2 and 1 biogenic opal contents were relatively stable, showing only low-amplitude fluctuations.

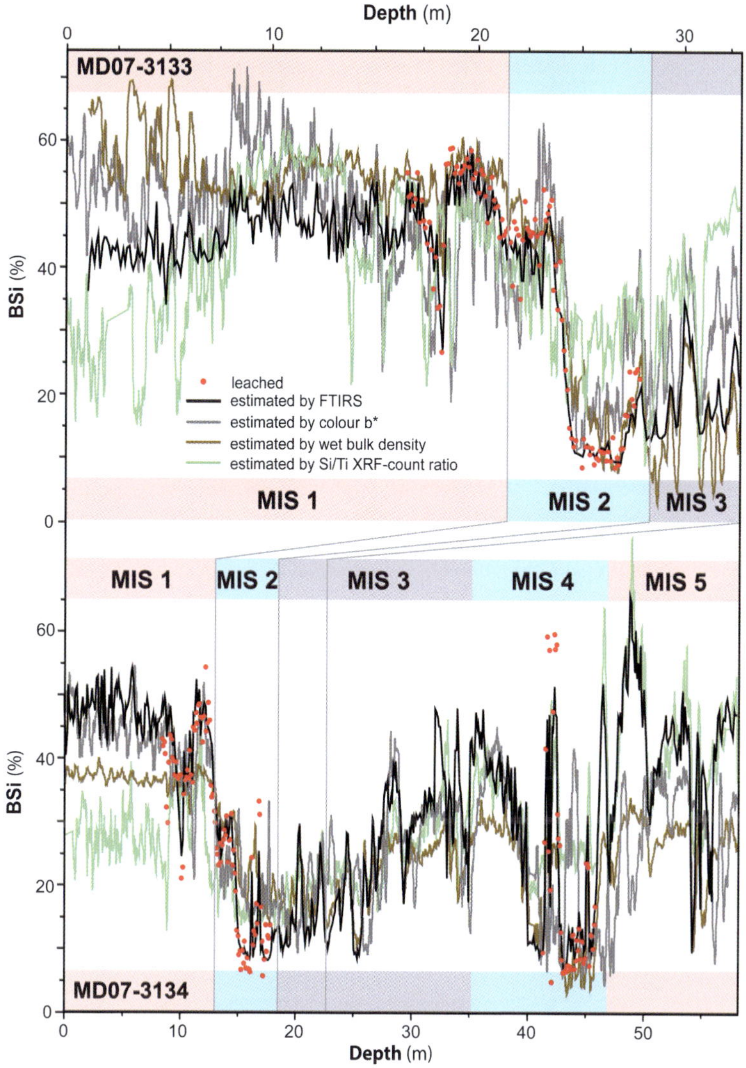

Fig. 3. Comparison of biogenic opal (BSi) contents evaluated with different methods for sites MD07-3133 (top) and MD07-3134 (bottom): conventionally (leached; data points highlighted as red dots; method see Müller & Schneider 1993); and BSi estimated from wet bulk density (brown curve), colour b^* (grey curve), Fourier transform infrared spectroscopy (FTIRS; black curve; method see Rosén *et al.* 2010) and XRF-core-scanner measured Si/Ti-count ratio (green curve; method see Balascio *et al.* 2011). Marine Isotopic Stages (MIS) 1–5 are plotted for reference (Lisiecki & Raymo 2005).

To better understand the biogenic opal sedimentation changes through time, we calculated mass accumulation rates for FTIRS-estimated biogenic opal contents (Fig. 4e), which we consider the most precise estimation method (see Methods; Fig. 2). This was done by multiplying the biogenic opal content as a percentage by the mass accumulation rate, that is, the product of sedimentation rate and dry bulk density (Fig. 4a, b). Instead of linear sedimentation rates, we used sedimentation rates estimated from a cubic smoothing spline age-model (Weber *et al.* 2012; Fig. 4a, grey lines). This overcomes the problems of normal splines, which do not include errors and so might produce outliers and artefacts, hence, representing more realistic sedimentation rates. Both cores exhibit extremely high sedimentation rates in some sections with values up to 1.8 and 1.2 m kyr^{-1} for sites MD07-3133 and MD07-3134. Warmer periods (MIS 3 and 1) show approximately 1.5–2 times

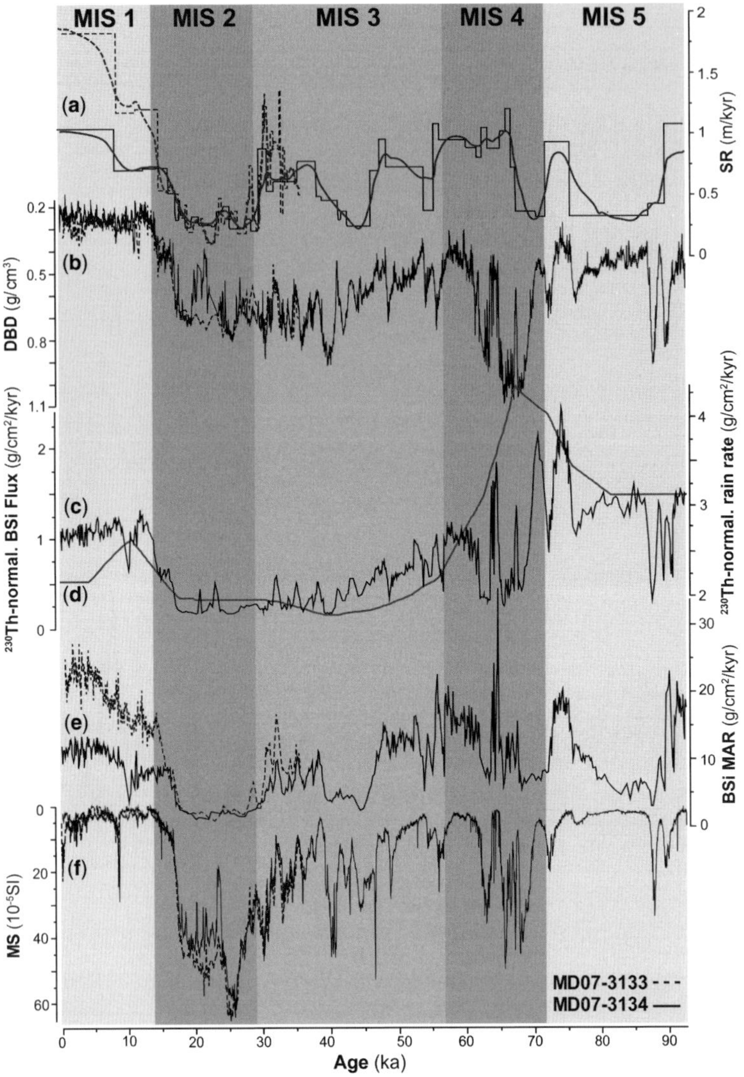

Fig. 4. Sites MD07-3133 (dashed lines) and MD07-3134 (solid lines) records: (**a**) Linear sedimentation rate (LSR) (black lines) and sedimentation rates (SR) estimated from a cubic smoothing spline age-model (grey lines; Weber *et al.* 2012). (**b**) Dry bulk density (DBD). (**c**) ^{230}Th-normalized biogenic opal (BSi) flux evaluated using the FTIRS estimated biogenic opal values (for method see Rosén *et al.* 2010). (**d**) ^{230}Th-normalized average rain rate. (**e**) BSi mass accumulation rate (MAR) evaluated using FTIRS estimated BSi values multiplied by SR from a cubic smoothing spline age-model (a) and DBD (b). (**f**) Magnetic susceptibility (MS) records as dust proxy (Weber *et al.* 2012). MIS 1–5 are plotted for reference (Lisiecki & Raymo 2005).

increased sedimentation rates at sites MD07-3133 relative to site MD07-3134, whereas during MIS 2 nearly identical rates occur at both core sites (Fig. 4a). The sedimentation rates and biogenic opal mass accumulation rates show a relatively complicated glacial-to-interglacial pattern with large-amplitude changes. While the biogenic opal mass accumulation rate reaches its minimum during the Last Glacial Maximum with values of approximately 3 g cm^{-2} kyr^{-1} at both core sites, up to 10 times higher rates are observed during MIS 4 (Fig. 4e).

^{230}Th-normalization data for MD07-3134 show that sediment focusing was very high, ranging from 5 (minimum value at 84.3 ka) to 29 (maximum value at 55.9 ka). During the interglacials the focusing factor ranged between 5 and 13, meaning that 5–13 times more sediment was added laterally to the core site via focusing than vertically through the water column. In the last glacial, in particular MIS 3, the focusing factors were generally even higher and ranged between 11 and 29. Strictly speaking, the ^{230}Th-normalized fluxes can only be applied for the 10 samples on which ^{230}Th was measured because there is no control over changes in sediment focusing between the depths of the measured ^{230}Th-normalized values. We are aware of this limitation and the interpolated ^{230}Th-normalized rain rates are just provided as an approximation for illustrative purposes until we have additional ^{230}Th-normalized data.

The interpolated rain rate for MD07-3134 (Fig. 4d) varies between 2 and 4.3 g cm^{-2} kyr^{-1}, with maximum values around 67 ka resulting from intermediate dry bulk densities (Fig. 4b), a focusing factor of 10 and a relatively high linear sedimentation rate (Fig. 4a). In order to achieve a more realistic record of the biogenic opal flux we multiplied ^{230}Th-normalized rain rates and FTIRS-estimated biogenic opal amount to get ^{230}Th-normalized biogenic opal flux (Fig. 4c). The calculated ^{230}Th-normalized biogenic opal flux varies from 0.2 to 2.5 g cm^{-2} kyr^{-1}. Maxima occur during interglacials, with mean values of 1 (MIS 1) and 1.4 (MIS 5) g cm^{-2} kyr^{-1}, whereas glacial MIS 2 and 3 were generally characterized by lower values near 0.5 g cm^{-2} kyr^{-1}. MIS 4 shows large short-term fluctuations, which, however, would need to be confirmed by additional $^{230}Th_{xs}$ measurements at higher resolution. In general, the ^{230}Th-normalized biogenic opal flux data (Fig. 4c) show a similar pattern over the past 92.5 kyr compared with the uncorrected biogenic opal accumulation rates (Fig. 4d) albeit with realistic numbers when compared with other sites in the Atlantic sector of the Southern Ocean (Kumar *et al.* 1995; Frank *et al.* 2000). As we have no ^{230}Th-normalization data for MD07-3133, we can only suggest that the sediment focusing was analogous and maybe even higher than at the MD07-3134 core site, as the linear sedimentation rate shows a similar trend, but higher values (Fig. 4a).

Discussion

To what extent are biogenic opal estimation methods reliable?

Biogenic opal can be analysed and estimated using various techniques, but no standardized determination methodology exists because each method has advantages and drawbacks. As deduced from Figure 3, the main biogenic opal fluctuations are captured by all estimation methods, so all methods can be used to decipher general biogenic opal trends; however, there are some strong discrepancies. Leaching according to Müller & Schneider (1993) is one of the most commonly used methods because of its simplicity and robustness (Swann 2010). However, samples with $\leq$10% biogenic opal might be affected by the pH value of the leaching solution (Schlüter & Rickert 1998).

XRF Si/Ti ratio is often used to indicate biogenic opal content changes in sediments (e.g. Francus *et al.* 2009; Balascio *et al.* 2011; Johnson *et al.* 2011). Nonetheless, Tjallingii *et al.* (2007) found reduced XRF-element intensities for light elements such as Si and Al of wet material relative to dried material, which they relate to differences in water contents. Accordingly, this can lead to misinterpretations of wet-measured samples. Our Si/Ti-count ratios reflect roughly the main biogenic opal fluctuations, especially at MD07-3134 (Fig. 3), but they should be only seen as rough estimations because of the given measurement uncertainties. Colour b^* can also be a useful parameter to estimate biogenic opal content of carbonate-free sediments that mainly contain varying amounts of biogenic opal, quartz and clay minerals. Our estimation shows that wet bulk density is also useful to roughly estimate biogenic opal content. It would also yield reliable results for sediments containing biogenic carbonate because biogenic opal has significantly lower densities (*c.* 2.2 g cm^{-3}) than biogenic carbonate (*c.* 2.8 g cm^{-3}) or detrital material (*c.* 2.6 g cm^{-3}). However, greater amounts of organic carbon (density of *c.* 1.4 g cm^{-3}) would lead to misinterpretations (Weber 1998).

Estimates from colour b^*, wet bulk density and the Si/Ti-count ratio for the upper 12 m of MD07-3133 show substantial discrepancies relative to FTIRS estimates (Fig. 3). This may be due to a combination of coring effect and sediment facies, that is, the upper parts of core MD07-3133 were partially supersaturated with water. These sections leaked pore water when they were cut into 1 m sections because of missing cohesive forces typical for the fabric of diatomaceous oozes. Accordingly, scanning techniques such as the Multi-Sensor Core Logger and the XRF-scanner measured false values in the uppermost core section, because they rely on completely filled core liners as well as smooth and even surfaces. So these estimations have limitations if sediments are supersaturated with water or show uneven surfaces.

A number of studies (e.g. Vogel *et al.* 2008; Rosén *et al.* 2011) show that FTIRS can be used to gain high-resolution and high-quality geochemical information on lacustrine sediment consisting of varying amounts of inorganic and organic components. Our results (Figs 2d & 3) also indicate that FTIRS is a very promising and useful tool for studying marine sediment composition. Correlation coefficients of $r = 0.99$ (MD07-3133) and $r = 0.93$ (MD07-3134) and an all-gradient accordance (Fig. 2d) underline our suggestion that FTIRS seems to be the most reliable estimation method, and hence can be used successfully for deep-sea sediment analysis. Therefore, we used the FTIRS results for all further discussions and comparisons (Figs 4 & 5).

Biogenic opal flux as a bioproductivity proxy

There is a strong debate on the reliability of biogenic opal flux data for indicating bioproductivity. It is still not entirely clear whether biogenic opal preservation is consistent over longer time periods and how the silicon cycle is coupled with the carbon biogeochemical cycle during interglacials (e.g. Pondaven *et al.* 2000; Ragueneau *et al.* 2000; Dezileau *et al.* 2003). However, in recent years, a number of studies indicate that biogenic opal flux can be used as bioproductivity proxy. Chase *et al.* (2003) and Bradtmiller *et al.* (2007) argued that biogenic opal flux reflects variations in diatom productivity and not changes in biogenic opal preservation. For the Indian sector Pondaven *et al.* (2000) found that local differences in biogenic opal preservation, apparent in the polar frontal zone and south of it, have a modulating but no primary effect on biogenic opal flux fluctuations. The good correlation of site MD07-3133 and MD07-3134 biogenic opal mass accumulation rate records (Fig. 5e), which are approximately 450 km apart (Fig. 1), indicates that this argument is also true for the central Scotia Sea.

Presently, the $^{231}Pa/^{230}Th$ ratio is the most accurate and most commonly used proxy for analysing particle and water mass transport in the Southern Ocean, for example, past opal fluxes (Kretschmer *et al.* 2011). Also Anderson *et al.* (2009) found strong correlation between the biogenic opal flux (i.e. mass accumulation rate) and $^{231}Pa/^{230}Th$ ratios, which are not altered by biogenic material loss during early diagenesis (Chase *et al.* 2003) in Southern Ocean sediment. This led them to the conclusion that biogenic opal flux variations reflect changes in biogenic opal production and silicon supply, rather than changes in preservation.

Estimated mean sedimentation rates are extremely high at both core sites: the mean linear sedimentation rate for MIS 1 is 0.9 (MD07-3134) and 1.6 (MD07-3133) m kyr^{-1}; it is approximately 0.5 m kyr^{-1} (MD07-3133 and MD07-3134) for MIS 2, 0.7 m kyr^{-1} (MD07-3134) for MIS 3 as well as MIS 5, and 0.9 m kyr^{-1} (MD07-3134) for MIS 4. Our values are more than 10–30 times higher than reported in previous studies, where rates of less than 5 cm kyr^{-1} for interglacials, and more than 5 cm kyr^{-1} for glacial times in the Scotia Sea were estimated (Pudsey & Howe 1998; Diekmann *et al.* 2000; Pugh *et al.* 2009). Accordingly, we conclude that both core sites are strongly influenced by sediment focusing, which is also reflected in the ^{230}Th normalized data of MD07-3134 (Fig. 4c, d), showing quite strong variability in sediment focusing with average interglacial values of 8 and average glacial values of 20. To our knowledge, these are some of the highest focusing factors ever determined with this method. The calculated ^{230}Th normalized rain rates (Fig. 4d) now correspond very well to other Southern Ocean sites, for instance, neighbouring core PS2319-1 (Francois *et al.* 2004). However, the ^{230}Th normalized biogenic opal flux overall shows the same trend as our estimated biogenic opal mass accumulation rate (Fig. 4e) for MD07-3134.

High sediment focusing of course is an indicator for bottom currents affecting an area, but these currents obviously did not vary so dynamically that they destroyed the dust-related signal (Pugh *et al.* 2009; Weber *et al.* 2012) within the decadal resolution. In contrast they collected the sediments from an upstream mostly shallower region and focused the material deposition to the core sites, thereby enhancing time resolution of the sedimentary record. By increasing the focusing factor from 8 to 20 during glacials, this sediment advection acted like a slow motion frame. The Antarctic Circumpolar Current has a significant effect on seafloor sedimentation in that area (Pudsey & Howe 1998). As a mainly wind-driven current (Maldonado *et al.* 2003) we assume that it resulted in decadal and centennial variations, which are comparable to dust climate signals, for example, Antarctic EDML non-sea-salt Ca^{2+} flux (Fischer *et al.* 2007*a*).

In regions such as the Antarctic Zone, where nutrient concentration is constantly high, iron availability strongly influences bioproductivity (Martínez-Garcia *et al.* 2011). Different iron fertilization experiments (e.g. Martin *et al.* 1990; Boyd *et al.* 2000) showed that Southern Ocean productivity is limited by Fe-deficiency. Coastal sediments, aerosols, upwelling, ice melting and vertical mixing can supply iron to the Southern Ocean (Cassar *et al.* 2007); however, today the major nutrient and iron supply originates from upwelling water masses or lateral advection (Meskhidze *et al.* 2007; Martínez-Garcia *et al.* 2011). The influence of dust-transported aeolian iron on bioproductivity is still debated. Erickson *et al.* (2003) suggested that bioproductivity in the Antarctic Circumpolar Current region is mainly controlled by atmospheric dust-Fe, mostly from Patagonia, whereas Kaufmann *et al.* (2010) argued that dust flux variations had no major influence on

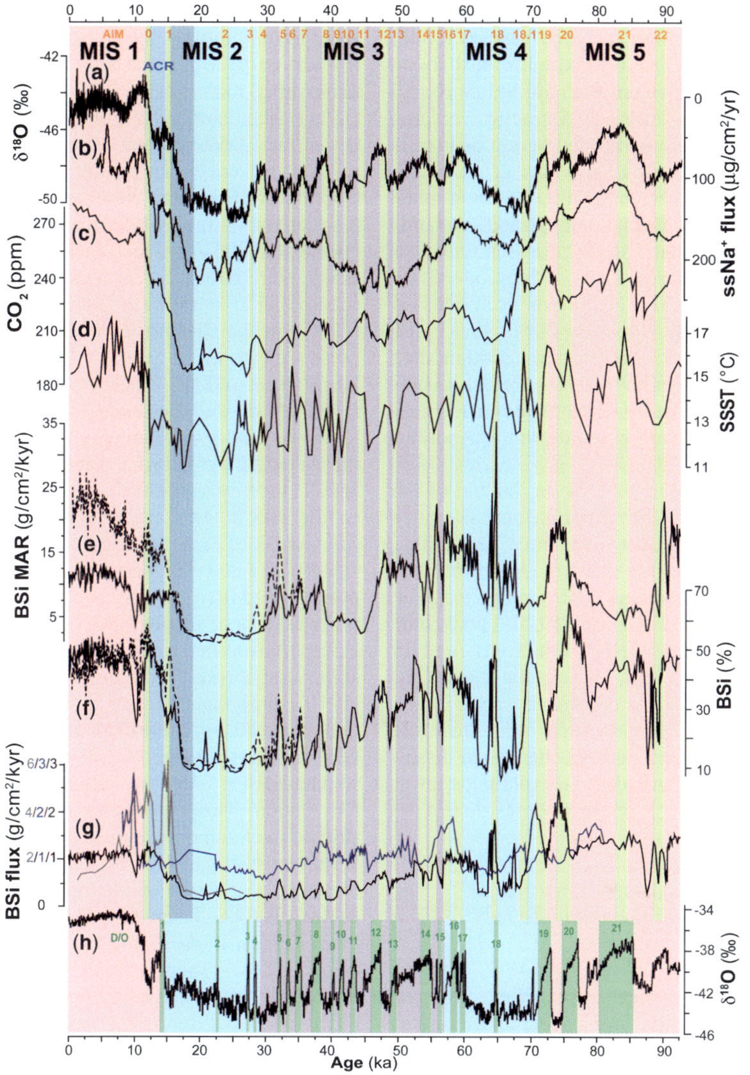

Fig. 5. Palaeoclimate and palaeoproductivity records for the last 92.5 kyr highlighting millennial-scale climate fluctuations during the last glacial: (**a** and **b**) EDML ice core $\delta^{18}O$ (EPICA Community Members 2006) as local ice sheet temperature indicator and sea-salt Na^+ flux (Fischer *et al.* 2007*a*) as sea ice proxy; (**c**) combined record of CO_2 measured at EDC (0–20 ka; Monnin 2006) and Byrd ice core (20–91 ka; Ahn & Brook 2008); (**d**) radiolarian-based summer sea-surface temperature (SSST) reconstruction derived at ODP site 1089 (Cortese & Abelmann 2002) in the Atlantic sector of the Southern Ocean; (**e**) biogenic opal mass accumulation rates (BSi MAR) for MD07-3133 (dashed line) and MD07-3134 (solid line) core sites calculated from multiplying FTIRS-estimated BSi (method see Rosén *et al.* 2010) by dry bulk density and sedimentation rates estimated from a cubic smoothing spline age-model (Weber *et al.* 2012); (**f**) MD07-3133 and MD07-3134 BSi content as percentages estimated with FTIRS; (**g**) ^{230}Th-normalized BSi flux from sediment cores MD07-3134 (black line), TN057-13PC (grey line) and TN057-14PC (blue line; Anderson *et al.* 2009) from the SE Atlantic Sector of the Southern Ocean (Fig. 1); (**h**) is Northern Hemisphere NGRIP ice core $\delta^{18}O$ (NGRIP Members 2004). MIS 1–5 are plotted for reference (Lisiecki & Raymo 2005), as well as Antarctic Isotopic Maxima (EPICA Community Members 2006; Stenni *et al.* 2010), Antarctic Cold Reversal (Rahmstorf 2002) and Daansgard–Oeschger–Cycles (Blunier & Brook 2001).

Southern Ocean productivity over the last 150 kyr. Further, modelling studies suggest that upwelling delivers approximately 99% of micronutrients to the Southern Ocean rather than dust (Lefèvre & Watson 1999). At sites MD07-3133 and MD07-3134 dust-indicator magnetic susceptibility exhibits mostly minima when biogenic opal mass accumulation rate is high and vice versa (Fig. 4e, f). During the Last Glacial Maximum, dust transport was significantly intensified and led to 10–30 times higher concentrations in Antarctic ice cores (Delmonte *et al.* 2004) and in the Scotia Sea (sites MD07-3133 and MD07-3134; Weber *et al.* 2012), whereas biogenic opal mass accumulation rates were extremely low during that period. Accordingly, we believe that iron-fertilization via dust transport is not the main driver of bioproductivity changes in the Scotia Sea. In addition, iron released from melting icebergs (Raiswell 2011), that is, Fe either from terrigenous material or aeolian dust (Lin *et al.* 2011), might have a minor affect on bioproductivity, as both sediment core sites are within the so-called 'iceberg alley' (Anderson & Andrews 1999; Stuart & Long 2011).

Biogenic opal export increase at the end of the last glacial could reflect higher nutrient content in upwelling waters or just intensified upwelling, more fractional utilization of nutrients or, to a lesser extent, better preservation of sinking biogenic opal (Horn *et al.* 2011). Allen *et al.* (2011) conclude that reduced surface water productivity and/or export in the Scotia Sea during the Last Glacial Maximum, caused by enhanced sea ice cover, lower sea surface temperatures and therefore shorter growing seasons, and not changes in dissolution processes, are more likely to explain the lower abundance of diatoms in sediment cores during the Last Glacial Maximum.

Bioproductivity changes since the last interglacial

Given the above arguments, we believe that the Scotia Sea biogenic opal mass accumulation rates at sites MD07-3133 and MD07-3134 mainly provide records of changing bioproductivity since the last interglacial. Interestingly, a relatively complicated glacial-to-interglacial pattern is noticeable (Fig. 5e–g). Large-amplitude millennial-scale fluctuations occur in the Scotia Sea records, which exhibit decadal-scale sample resolution. We estimate the lowest bioproductivity for the Last Glacial Maximum, while Southern Hemisphere Westerlies seem to have been located approximately 7–10° latitude north of their present position (e.g. Moreno *et al.* 1999; Toggweiler *et al.* 2006). During that time upwelling of silica- and CO_2-rich, relatively warm mid-depth water to the sea surface was reduced significantly – approximately 30% relative to today (Horn *et al.* 2011) – which led to low-temperature and low-salinity surface waters that supported sea ice formation in the Southern Ocean (Toggweiler *et al.* 2006). Also, micronutrient availability in the Scotia Sea was at a minimum during the Last Glacial Maximum (Hendry *et al.* 2011). As shown in Figure 5, low biogenic opal mass accumulation rates at the Last Glacial Maximum correspond to (i) low $\delta^{18}O$ values of the EDML ice core (Fig. 5a), which provide a local ice sheet temperature indicator (EPICA Community Members 2006) and (ii) high sea-salt Na^+ fluxes of the EDML ice core (Fig. 5b; Fischer *et al.* 2007*b*), which reflect intense sea ice cover. During MIS 2, decreased biogenic opal mass accumulation rates were estimated by Diekmann (2007) for the southern parts of Antarctic Circumpolar Current in all Southern Ocean sectors. Also Frank *et al.* (2000), Chase *et al.* (2003), Dezileau *et al.* (2003) and Bradtmiller *et al.* (2009) found lower biogenic opal accumulation rates south of the Antarctic Polar Front during MIS 2 relative to MIS 1, thereby suggesting less diatom productivity during MIS 2.

Around 17 ka, atmospheric CO_2 increased rapidly and the Southern Hemisphere Westerlies probably shifted south (Toggweiler 2009). It remains unclear whether the shift in Southern Hemisphere Westerlies caused the rise in CO_2 or if the shift was a result of rising CO_2. Denton *et al.* (2010) show that reorganization of ocean circulation, induced by intensified cooling as a result of large meltwater pulses in the Northern Hemisphere, weakened the Atlantic Meridional Overturning Circulation. This consequently led to a poleward shift of the Southern Hemisphere Westerlies, intensified upwelling of CO_2-rich mid-depth water, and therefore the warming of Antarctica (Toggweiler & Russell 2008; Toggweiler & Lea 2010). Banderas *et al.* (2012), Burke & Robinson (2012) and Kwon *et al.* (2012) support this theory by inferring increased deep-water mixing and less stratification of the water column in the Southern Ocean during the deglacial period. Anderson *et al.* (2009) also interpreted biogenic opal flux as an upwelling proxy because biogenic opal production is ultimately limited by dissolved Si supply, explaining its direct, but not necessarily linear connection. Biogenic opal mass accumulation rate rose abruptly at MD07-3133 and still also the ^{230}Th-normalized biogenic opal flux record of MD07-3134 (Fig. 5e, *g*, grey vertical bar) rose around 17 ka, thus representing probable increases in bioproductivity and upwelling. The rise is followed by a maximum around 15 ka in biogenic opal accumulation rate at MD07-3133, which was also detected at other core sites in the Southern Ocean (Atlantic, Indian, and Pacific sectors; Anderson *et al.* 2009), and coincides with Antarctic Isotopic Maximum 1.

During the Antarctic Cold Reversal (14.54–12.76 ka; Putnam *et al.* 2010), biogenic opal mass accumulation rates at MD07-3133 and

MD07-3134 (Fig. 5e) are only slightly reduced and remain high. Also EDML ice core data show only a minor reduction in $\delta^{18}O$ and rise in sea-salt Na^+ flux (Fig. 5a, b) according to regional differences. This is slightly different from the observation of Anderson *et al.* (2009) of strongly decreased biogenic opal flux (Fig. 5g), indicating reduced Southern Ocean upwelling during that time. The Antarctic Cold Reversal is followed by the Antarctic Isotopic Maximum 0, an abrupt rise in temperature and CO_2, as well as sea ice decrease; also the biogenic opal content rose at both core sites during that time.

Linear sedimentation rate and biogenic opal mass accumulation rate were equivalent for both Scotia Sea sites during the Last Glacial Maximum but increased by a factor of 1.5–2 at site MD07-3133 during MIS 3, the Antarctic Cold Reversal and MIS 1 (Fig. 4a, e). Higher amplitudes further north could be related to an approximately 5° shift in latitude of the Antarctic Circumpolar Current fronts (and associated opal belt) to the north during the Last Glacial Maximum (Gersonde *et al.* 2003). Also, Allen *et al.* (2011) observed a shift of the winter sea ice limit of at least 5° northwards during the Last Glacial Maximum, so that it remained north of the Scotia Sea. Accordingly, at that time, both core sites would have been within the winter sea ice limit, explaining the similarly reduced biogenic opal mass accumulation and linear sedimentation rates. Today, MD07-3133, which is located approximately 450 km NW of MD07-3134 (Fig. 1), lies north of the winter sea ice limit with significantly elevated biogenic opal mass accumulation and linear sedimentation rates, while MD07-3134 is within the winter sea ice limit and exhibits reduced biogenic opal mass accumulation and linear sedimentation rates. Therefore, we suggest that the distribution of sea ice has a strong influence on bioproductivity and could be the reason for pronounced regional differences in the Southern Ocean.

The last glacial period was characterized by abrupt millennial-scale climate fluctuations (e.g. Blunier & Brook 2001; Ahn & Brook 2008), which have been detected worldwide, for example in stalagmites in SE China (Cosford *et al.* 2008), in Antarctic ice cores (e.g. EDML; EPICA Community Members 2006; Fig. 5a, b) and in Greenland ice cores (nGRIP; NGRIP Members 2004; Fig. 5h). Also, in marine sediments (ODP site 1089; Cortese & Abelmann 2002; Fig. 5d) in the Atlantic Sector, the summer sea surface temperature record shows large millennial-scale temperature changes during the last glacial and correlates relatively well with our biogenic opal percentages, and to a lesser extent also with the biogenic opal mass accumulation rate records. During periods with high summer sea surface temperature, maxima in biogenic opal percentages can also be recognized. Around 65 ka in MIS 4 the MD07-3134 data show an absolute maximum in biogenic opal percentages, which is also observed in the ^{230}Th-normalized biogenic opal flux (Fig. 5g). At this time, summer sea surface temperature at ODP site 1089 (Cortese & Abelmann 2002; Fig. 5d), in the Atlantic Sector, also rose about 4 °C.

The precise and detailed chronology of sites MD07-3133 and MD07-3134 in combination with the decadal-scale resolution of biogenic opal determinations provides the opportunity to study regional bioproductivity changes in the Southern Ocean relative to the timing of individual Antarctic Isotopic Maxima. Strong high-frequency fluctuations are also noticeable in the biogenic opal content record (Fig. 5f) and to a lesser extent in biogenic opal mass accumulation rate (Fig. 5e). At site MD07-3134, biogenic opal peaks, representing enhanced bioproductivity, occurred during the relatively warm phases in Antarctica, that is, during Antarctic Isotopic Maxima 5, 7, 8, 10, 12, 13, 14, 16, 17 and 19, while less sea ice occurred (Fig. 5b). Extremely large spikes are documented for Antarctic Isotopic Maxima 15, 18, 20 and 22, when the biogenic opal mass accumulation rate in the central Scotia Sea increases up to more than 20 g cm^{-2} kyr^{-1}, at times when the sea ice cover was only marginally reduced and Antarctic atmospheric temperatures and CO_2 concentration elevated only slightly. These fluxes are, however, clearly biased by sediment focusing as stated by the ^{230}Th normalization results (Fig. 4c, d), but still showing the same incline. In general, biogenic opal content and mass accumulation rates show the same trend as atmospheric CO_2 concentration (Fig. 5c); biogenic opal rises correspond well to increasing CO_2 concentration, while during low CO_2, biogenic opal is also mainly at minima. In addition, Anderson *et al.* (2009) also found an increase in upwelling at each phase of rising CO_2 during MIS 3, illustrated by higher biogenic opal flux in the Southern Ocean (Fig. 5g, blue and grey curve), which corresponds quite well with the ^{230}Th normalized biogenic opal flux from MD07-3134 (Fig. 5g, black curve). Our investigations show that variations in biogenic opal flux reflect bioproductivity changes and strongly correlate to sea ice cover, summer sea surface temperature as well as atmospheric CO_2 variations.

Summary and conclusions

Silica is an important nutrient that fuels biological production in the Southern Ocean. We investigated two long deep-sea sediment cores from the central

Scotia Sea south of the present Antarctic Polar Front, that exhibit exceptionally high sedimentation rates (up to 1.8 and 1.2 m kyr^{-1} for sites MD07-3133 and MD07-3134, respectively) to gain detailed (decadal-scale) insight into past biogenic opal flux variations.

In a first step, we measured biogenic opal conventionally by leaching 253 samples from core sections providing large-amplitude variations according to the method of Müller & Schneider (1993). Then we estimated biogenic opal by testing three methods that rely on *c.* 9100 non-destructive measurements (at 1 cm increments) of sediment colour b^*, wet bulk density and Si/Ti-count ratios. Finally, we determined biogenic opal using FTIRS on 891 samples at 10 cm increments.

One interesting result is that all methods can be used to decipher general biogenic opal trends. However, FTIRS – a novel method for marine sediment – provides the most reliable estimation relative to the conventional leaching method. It is rather inexpensive and requires only a small portion (11 mg) of sample material.

Biogenic opal flux records in the Southern Ocean are only marginally affected by preservation changes and therefore provide valid information on past bioproductivity changes. We provide the first decadal-scale resolution continuous bioproductivity record for the Southern Ocean over the last 92.5 kyr. The biogenic opal mass accumulation rate records of the central Scotia Sea exhibit a relatively complicated glacial-to-interglacial pattern with large-amplitude, millennial-scale fluctuations in bioproductivity.

South of the Antarctic Polar Front, lowest bioproductivity levels deduced from biogenic opal fluxes occurred during the Last Glacial Maximum, when upwelling of mid-depth water was reduced and sea ice cover intensified. ^{230}Th normalized biogenic opal fluxes in core MD07-3134 show a similar pattern albeit at realistic values of between 1 and 1.5 g cm^{-2} kyr^{-1} during interglacials, whereas glacial MIS 2 and 3 were generally characterized by lower values near 0.5 g cm^{-2} kyr^{-1}. These numbers are comparable to other records in the Atlantic sector of the Southern Ocean. Around 17 ka, bioproductivity increased abruptly, corresponding to rising atmospheric CO_2 and decreasing seasonal sea ice coverage. Distribution of sea ice strongly influences bioproductivity and may be the reason for pronounced regional differences in the Southern Ocean. Also summer sea surface temperature changes are strongly correlated to the biogenic opal flux changes in the central Scotia Sea.

We are grateful for financial support from the Deutsche Forschungsgemeinschaft (DFG; grants RI 525/17-1, KU 683/9-1, WE2039/8-1) and the DFG-Priority Programme Antarctic Research 1158. Thanks are also due to C. Olofsson and C. Meyer-Jacob for assisting the FTIRS measurements at Climate Impacts Research Centre in Umeå, Sweden. A. Kolevica is thanked for the U-series isotope measurements on the quadrupole ICPMS at GEOMAR, Kiel. Finally, we thank the reviewers R. Anderson and X. Crosta, who greatly helped to improve the manuscript. Our study was part of the Southern Ocean Initiative of the International Marine Past Global Change Study programme and the AWI research programme 'Polar Regions and Coasts in a changing Earth System', Topic 3, 'Lessons from the Past'. Supplementary data to this paper are available at http://doi.pangaea.de/10.1594/PANGAEA.789348.

References

AHN, J. & BROOK, E. J. 2008. Atmospheric CO_2 and climate on millennial time scales during the last glacial period. *Science*, **322**, 83–85.

ALLEN, C. S., PIKE, J. & PUDSEY, C. J. 2011. Last glacial–interglacial sea-ice cover in the SW Atlantic and its potential role in global deglaciation. *Quaternary Science Reviews*, **30**, 2446–2458.

ANDERSON, J. B. & ANDREWS, J. T. 1999. Radiocarbon constraints on ice sheet advance and retreat in the Weddell Sea, Antarctica. *Geology*, **27**, 179–182.

ANDERSON, R. F., ALI, S., BRADTMILLER, L. I., NIELSEN, S. H. H., FLEISHER, M. Q., ANDERSON, B. E. & BURCKLE, L. H. 2009. Wind-driven upwelling in the Southern Ocean and the deglacial rise in atmospheric CO_2. *Science*, **323**, 1443–1448.

BALASCIO, N. L., ZHANG, Z., BRADLEY, R. S., PERREN, B., DAHL, S. O. & BAKKE, J. 2011. A multi-proxy approach to assessing isolation basin stratigraphy from the Lofoten Islands, Norway. *Quaternary Research*, **75**, 288–300.

BANDERAS, R., ÁLVAREZ-SOLAS, J. & MONTOYA, M. 2012. Role of CO_2 and Southern Ocean winds in glacial abrupt climate change. *Climate of the Past*, **8**, 1011–1021.

BAREILLE, G., LABRACHERIE, M., BERTRAND, P., LABEYRIE, L., LAVAUX, G. & DIGNAN, M. 1998. Glacial–interglacial changes in the accumulation rates of major biogenic components in Southern Indian Ocean sediments. *Journal of Marine Systems*, **17**, 527–539.

BERGER, A. & LOUTRE, M. F. 1991. Insolation values for the climate of the last 10 million years. *Quaternary Science Reviews*, **10**, 297–317.

BLUNIER, T. & BROOK, E. J. 2001. Timing of millennial-scale climate change in Antarctica and Greenland during the last glacial period. *Science*, **291**, 109–112.

BOYD, P. W., WATSON, A. J. *ET AL.* 2000. A mesoscale phytoplankton bloom in the polar Southern Ocean stimulated by iron fertilization. *Nature*, **407**, 695–702.

BRADTMILLER, L. I., ANDERSON, R. F., FLEISHER, M. Q. & BURCKLE, L. H. 2007. Opal burial in the equatorial Atlantic Ocean over the last 30 ka: implications for glacial–interglacial changes in the ocean silicon cycle. *Paleoceanography*, **22**, PA4216, doi: 10.1029/2007PA001443.

BRADTMILLER, L. I., ANDERSON, R. F., FLEISHER, M. Q. & BURCKLE, L. H. 2009. Comparing glacial and

Holocene opal fluxes in the Pacific sector of the Southern Ocean. *Palaeoceanography*, **24**, PA2214, doi: 10.1029/2008PA001693.

Burke, A. & Robinson, L. F. 2012. The Southern Ocean's role in carbon exchange during the last deglaciation. *Science*, **335**, 557–561.

Cassar, N., Bender, M. L., Barnett, B. A., Fan, S., Moxim, W. J., Levy Ii, H. & Tilbrook, B. 2007. The Southern Ocean biological response to aeolian iron deposition. *Science*, **317**, 1067–1070.

Charles, C. D., Froelich, P. N., Zibello, M. A., Mortlock, R. A. & Morley, J. J. 1991. Biogenic opal in southern ocean sediments over the last 450 000 years: implications for surface water chemistry and circulation. *Paleoceanography*, **6**, 697–728, doi: 10.1029/91PA02477.

Chase, Z., Anderson, R. F., Fleisher, M. Q. & Kubik, P. W. 2003. Accumulation of biogenic and lithogenic material in the Pacific sector of the Southern Ocean during the past 40 000 years. *Deep Sea Research II*, **50**, 799–832.

Cortese, G. & Abelmann, A. 2002. Radiolarian-based paleotemperatures during the last 160 kyr at ODP Site 1089 (Southern Ocean, Atlantic Sector). *Palaeogeography, Palaeoclimatology, Palaeoecology*, **182**, 259–286.

Cosford, J., Qing, H., Yuan, D., Zhang, M., Holmden, C., Patterson, W. & Hai, C. 2008. Millennial-scale variability in the Asian monsoon: evidence from oxygen isotope records from stalagmites in southeastern China. *Palaeogeography, Palaeoclimatology, Palaeoecology*, **266**, 3–12.

de Baar, H. J. W., Boyd, P. W. *et al.* 2005. Synthesis of iron fertilization experiments: from the iron age in the age of enlightenment. *Journal of Geophysical Research: Oceans*, **110**, C9, C09S16, doi: 10.1029/2004jc002601

Delmonte, B., Petit, J. R., Andersen, K. K., Basile-Doelsch, I., Maggi, V. & Ya Lipenkov, V. 2004. Dust size evidence for opposite regional atmospheric circulation changes over east Antarctica during the last climatic transition. *Climate Dynamics*, **23**, 427–438.

Denton, G. H., Anderson, R. F., Toggweiler, J. R., Edwards, R. L., Schaefer, J. M. & Putnam, A. E. 2010. The last glacial termination. *Science*, **328**, 1652–1656.

Dezileau, L., Reyss, J. L. & Lemoine, F. 2003. Late Quaternary changes in biogenic opal fluxes in the Southern Indian Ocean. *Marine Geology*, **202**, 143–158.

Diekmann, B. 2007. Sedimentary patterns in the late Quaternary Southern Ocean. *Deep Sea Research II*, **54**, 2350–2366.

Diekmann, B., Kuhn, G. *et al.* 2000. Terrigenous sediment supply in the Scotia Sea (Southern Ocean): response to Late Quaternary ice dynamics in Patagonia and on the Antarctic Peninsula. *Palaeogeography, Palaeoclimatology, Palaeoecology*, **162**, 357–387.

EPICA COMMUNITY MEMBERS 2004. Eight glacial cycles from an Antarctic ice core. *Nature*, **429**, 623–628.

EPICA COMMUNITY MEMBERS 2006. One-to-one coupling of glacial climate variability in Greenland and Antarctica. *Nature*, **444**, 195–198.

Erickson, D. J., Hernandez, J. L., Ginoux, P., Gregg, W. W., McClain, C. & Christian, J. 2003. Atmospheric iron delivery and surface ocean biological activity in the Southern Ocean and Patagonian region. *Geophysical Research Letters*, **30**, 12, 1609, doi: 10.1029/2003GL017241.

Fischer, H., Fundel, F. *et al.* 2007*a*. EPICA EDML chemical concentrations and fluxes. doi: 10.1594/PANGAEA.683642

Fischer, H., Fundel, F. *et al.* 2007*b*. Reconstruction of millennial changes in dust emission, transport and regional sea ice coverage using the deep EPICA ice cores from the Atlantic and Indian Ocean sector of Antarctica. *Earth and Planetary Science Letters*, **260**, 340–354.

Francois, R., Frank, M., Rutgers van der Loeff, M. M. & Bacon, M. P. 2004. ^{230}Th normalization: an essential tool for interpreting sedimentary fluxes during the late Quaternary. *Paleoceanography*, **19**, PA1018, doi: 10.1029/2003PA000939.

Francus, P., Lamb, H., Nakagawa, T., Marshall, M., Brown, E. & Members, S. P. 2009. The potential of high-resolution X-ray fluorescence core scanning: applications in paleolimnology. *PAGES news*, **17**, 93–95.

Frank, M., Gersonde, R., Rutgers van der Loeff, M. M., Kuhn, G. & Mangini, A. 1996. Late Quaternary sediment dating and quantification of lateral sediment redistribution applying $^{230}Th_{ex}$: a study from the eastern Atlantic sector of the Southern Ocean. *Geologische Rundschau*, **85**, 554–566.

Frank, M., Gersonde, R. & Mangini, A. 1999. Sediment redistribution, $^{230}Th_{ex}$-normalization and implications for the reconstruction of particle flux and export paleoproductivity. *In*: Fischer, G. & Wefer, G. (eds) *Use of Proxies in Paleoceanography: Examples from the South Atlantic*. Springer, New York, 409–426.

Frank, M., Gersonde, R. *et al.* 2000. Similar glacial and interglacial export bioproductivity in the Atlantic sector of the Southern Ocean: multiproxy evidence and implications for glacial atmospheric CO_2. *Paleoceanography*, **15**, 642–658, doi: 10.1029/2000PA000497.

Geibert, W., van der Loeff, M. M. R., Usbeck, R., Gersonde, R., Kuhn, G. & Seeberg-Elverfeldt, J. 2005. Quantifying the opal belt in the Atlantic and southeast Pacific sector of the Southern Ocean by means of ^{230}Th normalization. *Global Biogeochemical Cycles*, **19**, GB4001, doi: 10.1029/2005GB002465.

Geladi, P., MacDougall, D. & Martens, H. 1985. Linearization and scatter-correction for near-infrared reflectance spectra of meat. *Applied Spectroscopy*, **39**, 491–500.

Gersonde, R., Abelmann, A. *et al.* 2003. Last glacial sea surface temperatures and sea-ice extent in the Southern Ocean (Atlantic–Indian sector): a multiproxy approach. *Paleoceanography*, **18**(3), 1061, doi: 10.1029/2002PA000809.

Gersonde, R., Crosta, X., Abelmann, A. & Armand, L. 2005. Sea-surface temperature and sea ice distribution of the Southern Ocean at the EPILOG Last Glacial Maximum – a circum-Antarctic view based on siliceous microfossil records. *Quaternary Science Reviews*, **24**, 869–896.

Griffiths, P. R. & De Haseth, J. A. 1986. *Fourier Transform Infrared Spectrometry*. Wiley, New York.

Guillou, H., Singer, B. S., Laj, C., Kissel, C., Scaillet, S. & Jicha, B. R. 2004. On the age of the Laschamp geomagnetic excursion. *Earth and Planetary Science Letters*, **227**, 331–343.

Hendry, K. R., Rickaby, R. E. M. & Allen, C. S. 2011. Changes in micronutrient supply to the surface Southern Ocean (Atlantic sector) across the glacial termination. *Geochemistry Geophysics Geosystems*, **12**, Q09007, doi: 10.1029/2011gc003691.

Horn, M. G., Beucher, C. P., Robinson, R. S. & Brzezinski, M. A. 2011. Southern ocean nitrogen and silicon dynamics during the last deglaciation. *Earth and Planetary Science Letters*, **310**, 334–339.

Jansen, J. H. F., Van der Gaast, S. J., Koster, B. & Vaars, A. J. 1998. CORTEX, a shipboard XRF-scanner for element analyses in split sediment cores. *Marine Geology*, **151**, 143–153.

Johnson, C. M., Brown, E. T. & Shi, J. 2011. Biogenic silica deposition in Lake Malawi, East Africa over the past 150 000 years. *Palaeogeography, Palaeoclimatology, Palaeoecology*, **303**, 103–109.

Kanfoush, S. l., Hodell, D. A., Charles, C. D., Guilderson, T. P., Mortyn, P. G. & Ninnemann, U. S. 2000. Millennial-scale instability of the Antarctic ice sheet during the last glaciation. *Science*, **288**, 1815–1818.

Kaufmann, P., Fundel, F. *et al.* 2010. Ammonium and non-sea salt sulfate in the EPICA ice cores as indicator of biological activity in the Southern Ocean. *Quaternary Science Reviews*, **29**, 313–323.

Kretschmer, S., Geibert, W., Rutgers van der Loeff, M. M., Schnabel, C., Xu, S. & Mollenhauer, G. 2011. Fractionation of ^{230}Th, ^{231}Pa, and ^{10}Be induced by particle size and composition within an opal-rich sediment of the Atlantic Southern Ocean. *Geochimica et Cosmochimica Acta*, **75**, 6971–6987.

Kumar, N., Anderson, R. F., Mortlock, R. A., Froelich, P. N., Kubik, P., Dittrich-Hannen, B. & Suter, M. 1995. Increased biological productivity and export production in the glacial Southern Ocean. *Nature*, **378**, 675–680.

Kwon, E. Y., Hain, M. P., Sigman, D. M., Galbraith, E. D., Sarmiento, J. L. & Toggweiler, J. R. 2012. North Atlantic ventilation of 'southern-sourced' deep water in the glacial ocean. *Paleoceanography*, **27**, PA2208, doi: 10.1029/2011pa002211

Lefèvre, N. & Watson, A. J. 1999. Modeling the geochemical cycle of iron in the oceans and its impact on atmospheric CO_2 concentrations. *Global Biogeochemical Cycles*, **13**, 727–736, doi: 10.1029/1999GB900034.

Li, F., Ginoux, P. & Ramaswamy, V. 2010. Transport of Patagonian dust to Antarctica. *Journal of Geophysical Research*, **115**, D18217, doi: 10.1029/2009jd012356

Lin, H., Rauschenberg, S., Hexel, C. R., Shaw, T. J. & Twining, B. S. 2011. Free-drifting icebergs as sources of iron to the Weddell Sea. *Deep-Sea Research Part II*, **58**, 1392–1406.

Lisiecki, L. E. & Raymo, M. E. 2005. A Pliocene-Pleistocene stack of 57 globally distributed benthic d^{18}O records. *Paleoceanography*, **20**, PA1003, doi: 10.1029/2004pa001071.

Maldonado, A., Barnolas, A. *et al.* 2003. Contourite deposits in the central Scotia Sea: the importance of the Antarctic Circumpolar Current and the Weddell Gyre flows. *Palaeogeography, Palaeoclimatology, Palaeoecology*, **198**, 187–221.

Martens, H. & Naes, T. 1989. *Multivariate Calibration*. John Wiley, New York.

Martin, J. H., Gordon, R. M. & Fitzwater, S. E. 1990. Iron in Antarctic waters. *Nature*, **345**, 156–158.

Martínez-Garcia, A., Rosell-Mele, A., Jaccard, S. L., Geibert, W., Sigman, D. M. & Haug, G. H. 2011. Southern Ocean dust-climate coupling over the past four million years. *Nature*, **476**, 312–315.

Meskhidze, N., Nenes, A., Chameides, W. L., Luo, C. & Mahowald, N. 2007. Atlantic Southern Ocean productivity: fertilization from above or below? *Global Biogeochemical Cycles*, **21**, GB2006, doi: 10.1029/2006GB002711.

Monnin, E. 2006. EPICA Dome C high resolution carbon dioxide concentrations, doi: 10.1594/PANGAEA.472488.

Moreno, P. I., Lowell, T. V., Jacobson, G. L., Jr & Denton, G. H. 1999. Abrupt vegetation and climate changes during the Last glacial Maximum and Last Termination in the Chilean Lake District: a case study from Canal De La Puntilla (41°S). *Geografiska Annaler: Series A, Physical Geography*, **81**, 285–311.

Mortlock, R. A., Charles, C. D., Froelich, P. N., Zibello, M. A., Saltzman, J., Hays, J. D. & Burckle, L. H. 1991. Evidence for lower productivity in the Antarctic Ocean during the last glaciation. *Nature*, **351**, 220–223.

Müller, P. J. & Schneider, R. 1993. An automated leaching method for the determination of opal in sediments and particulate matter. *Deep Sea Research I*, **40**, 425–444.

Murray, R. W., Leinen, M. & Isern, A. R. 1993. Biogenic flux of Al to sediment in the central equatorial Pacific Ocean: evidence for increased productivity during glacial periods. *Paleoceanography*, **8**, 651–670, doi: 10.1029/93PA02195

NGRIP MEMBERS. 2004. High-resolution record of Northern Hemisphere climate extending into the last interglacial period. *Nature*, **431**, 147–151.

Nielsen, S. H. H., Hodell, D. A., Kamenov, G., Guilderson, T. & Perfit, M. R. 2007. Origin and significance of ice-rafted detritus in the Atlantic sector of the Southern Ocean. *Geochemistry, Geophysics, Geosystems*, **8**, Q12005, doi: 10.1029/2007GC001618

Orsi, A. H., Whitworth, T. III & Nowlin, W. D. Jr. 1995. On the meridional extent and fronts of the Antarctic Circumpolar Current. *Deep Sea Research I*, **42**, 641–673.

Paillard, D., Labeyrie, L. & Yiou, P. 1996. Macintosh Program performs time-series analysis. *EOS Transactions*, **77**, 379–379, doi: 10.1029/96eo00259

Pondaven, P., Ragueneau, O., Treguer, P., Hauvespre, A., Dezileau, L. & Reyss, J. L. 2000. Resolving the 'opal paradox' in the Southern Ocean. *Nature*, **405**, 168–172.

Pudsey, C. J. & Howe, J. A. 1998. Quaternary history of the Antarctic Circumpolar Current: evidence from the Scotia Sea. *Marine Geology*, **148**, 83–112.

PUGH, R. S., MCCAVE, I. N., HILLENBRAND, C. D. & KUHN, G. 2009. Circum-Antarctic age modelling of Quaternary marine cores under the Antarctic Circumpolar Current: ice-core dust–magnetic correlation. *Earth and Planetary Science Letters*, **284**, 113–123.

PUTNAM, A. E., DENTON, G. H. *ET AL.* 2010. Glacier advance in southern middle-latitudes during the Antarctic Cold Reversal. *Nature Geoscience*, **3**, 700–704.

RAGUENEAU, O., TREGUER, P. *ET AL.* 2000. A review of the Si cycle in the modern ocean: recent progress and missing gaps in the application of biogenic opal as a paleoproductivity proxy. *Global and Planetary Change*, 317–365.

RAHMSTORF, S. 2002. Ocean circulation and climate during the past 120 000 years. *Nature*, **419**, 207–214.

RAISWELL, R. 2011. Iceberg-hosted nanoparticulate Fe in the Southern Ocean: mineralogy, origin, dissolution kinetics and source of bioavailable Fe. *Deep Sea Research Part II*, **58**, 1364–1375.

RICHTER, T. O., VAN DER GAAST, S. *ET AL.* 2006. The Avaatech XRF Core Scanner: technical description and applications to NE Atlantic sediments. *Geological Society, London, Special Publications*, **267**, 39–50.

ROSÉN, P., VOGEL, H., CUNNINGHAM, L., REUSS, N., CONLEY, D. J. & PERSSON, P. 2010. Fourier transform infrared spectroscopy, a new method for rapid determination of total organic and inorganic carbon and biogenic silica concentration in lake sediments. *Journal of Paleolimnology*, **43**, 247–259.

ROSÉN, P., VOGEL, H. *ET AL.* 2011. Universally applicable model for the quantitative determination of lake sediment composition using Fourier transform infrared spectroscopy. *Environmental Science & Technology*, **45**, 8858–8865.

SCHLÜTER, M. & RICKERT, D. 1998. Effect of pH on the measurement of biogenic silica. *Marine Chemistry*, **63**, 81–93.

STENNI, B., MASSON-DELMOTTE, V. *ET AL.* 2010. The deuterium excess records of EPICA Dome C and Dronning Maud Land ice cores (East Antarctica). *Quaternary Science Reviews*, **29**, 146–159.

STUART, K. M. & LONG, D. G. 2011. Tracking large tabular icebergs using the SeaWinds Ku-band microwave scatterometer. *Deep Sea Research Part II*, **58**, 1285–1300.

SWANN, G. 2010. A comparison of the Si/Al and Si/time wet-alkaline digestion methods for measurement of biogenic silica in lake sediments. *Journal of Paleolimnology*, **44**, 375–385.

TJALLINGII, R., RÖHL, U., KÖLLING, M. & BICKERT, T. 2007. Influence of the water content on X-ray fluorescence core- scanning measurements in soft marine sediments. *Geochemistry, Geophysics, Geosystems*, **8**, Q02004, doi: 10.1029/2006GC001393

TOGGWEILER, J. R. 2009. Shifting westerlies. *Science*, **323**, 1434–1435.

TOGGWEILER, J. R. & LEA, D. W. 2010. Temperature differences between the hemispheres and ice age climate variability. *Paleoceanography*, **25**, PA2212, doi: 10.1029/2009pa001758

TOGGWEILER, J. R. & RUSSELL, J. 2008. Ocean circulation in a warming climate. *Nature*, **451**, 286–288.

TOGGWEILER, J. R., RUSSELL, J. L. & CARSON, S. R. 2006. Midlatitude westerlies, atmospheric CO_2, and climate change during the ice ages. *Paleoceanography*, **21**, 15, doi: 10.1029/2005pa001154

VOGEL, H., ROSÉN, P., WAGNER, B., MELLES, M. & PERSSON, P. 2008. Fourier transform infrared spectroscopy, a new cost-effective tool for quantitative analysis of biogeochemical properties in long sediment records. *Journal of Paleolimnology*, **40**, 689–702.

WEBER, M. E. 1998. Estimation of biogenic carbonate and opal by continuous non-destructive measurements in deep-sea sediments: application to the eastern Equatorial Pacific. *Deep Sea Research I*, **45**, 1955–1975.

WEBER, M. E., NIESSEN, F., KUHN, G. & WIEDICKE, M. 1997. Calibration and application of marine sedimentary physical properties using a multi-sensor core logger. *Marine Geology*, **136**, 151–172.

WEBER, M. E., KUHN, G., SPRENK, D., ROLF, C., OHLWEIN, C. & RICKEN, W. 2012. Dust transport from Patagonia to Antarctica – a new stratigraphic approach from the Scotia Sea and its implications for the last glacial cycle. *Quaternary Science Reviews*, **36**, 177–188.

WHITWORTH, T. & PETERSON, R. G. 1985. Volume transport of the Antarctic Circumpolar Current from bottom pressure measurements. *Journal of Physical Oceanography*, **15**, 810–816.

Palaeoenvironmental records from the West Antarctic Peninsula drift sediments over the last 75 ka

MARYLINE J. VAUTRAVERS[1]*, DAVID A. HODELL[1], JAMES E. T. CHANNELL[2], CLAUS-DIETER HILLENBRAND[3], MIKE HALL[1], JAMES SMITH[3] & ROBERT D. LARTER[3]

[1]*Department of Earth Sciences, Godwin Laboratory for Palaeoclimatic Research, University of Cambridge, Downing Street, Cambridge CB2 3EQ, UK*

[2]*Department of Geological Sciences, University of Florida, PO Box-112120, Gainesville, FL 32611, USA*

[3]*British Antarctic Survey, High Cross, Madingley Road, Cambridge CB3 0ET, UK*

**Corresponding author (e-mail: mv217@cam.ac.uk)*

Abstract: We present results of a multi-proxy study on marine sediment core JR179-PC466 recovered from the crest of a sediment drift off the West Antarctic Peninsula at approximately 2300 m water depth. The 10.45 m-long core consists dominantly of glaciomarine terrigenous sediments, with only traces of calcium carbonate (<1 wt%). Despite the very low abundance of calcareous foraminifera, planktonic shell numbers are sufficient for stable isotope analyses in two-thirds of the samples studied. The core chronology is based on oxygen isotope stratigraphy and correlation of its relative palaeomagnetic intensity (RPI) with a stacked reference curve. According to the age model, core PC466 spans the last 75 ka, with average sedimentation rates of between about 4 and 25 cm ka^{-1}. Planktonic foraminifera abundances fluctuate between 0 and 30 individuals per gram throughout the core, with minima observed during Marine Isotope Stage (MIS) 2 (14–29 ka before present, BP) and MIS4 (57–71 ka BP). Planktonic foraminifera are present in the Holocene but more abundant in sediments deposited during MIS3 (29–57 ka BP), owing to less dilution by terrigenous detritus and/or better carbonate preservation. During MIS3, foraminifera maxima correlate with Antarctic warming events as recorded in the $\delta^{18}O$ signal of the EPICA Dronning Maud Land (EDML) ice core. They indicate higher planktonic foraminifera production and better carbonate preservation west of the Antarctic Peninsula during that time. The abundance of ice-rafted detritus (IRD) in core PC466 increased during the last deglaciation between about 19 and 11 ka BP, when numerous icebergs drifted across the core site, thereby releasing IRD. During this time, sea-level rise destabilized the Antarctic Peninsula (APIS) and West Antarctic (WAIS) ice sheets that had advanced onto the shelf during the sea-level lowstand of the Last Glacial Maximum (LGM; *c.* 19–23 ka BP). Overall, our results demonstrate that it is possible to establish an age model and reconstruct palaeoceanographical and climatic changes at high temporal resolution from sedimentary sequences recovered at 2300 m water depth from a West Antarctic drift.

Mercer (1978) warned of the potential consequences of warming caused by anthropogenic greenhouse gas forcing on the stability of the West Antarctic Ice Sheet. Since the late 1970s, there has been intense scientific and public interest in the potential response of the Antarctic Peninsula Ice Sheet and West Antarctic Ice Sheet to global warming (e.g. Bindschadler 2006; Mayewski *et al.* 2009). Recent observations show that both ice sheets are undergoing rapid changes, including surface warming, ice shelf melting, flow acceleration, outlet glacier thinning and grounding-line retreat, and contribute to present sea-level rise (Vaughan *et al.* 2003; Martinson *et al.* 2008; Pritchard *et al.* 2009, 2012; Joughin & Alley 2011). These recent changes can be understood in a longer geological context by studying continuous, ice-sheet-proximal marine sedimentary records. Although numerous marine sediment cores have been recovered from the West Antarctic continental margin throughout the last 50 years, their predominantly terrigenous composition and lack of foraminiferal carbonate for isotopic analyses has hindered the establishment of precise age models and stymied palaeoenvironmental interpretations.

Here we present results of a multi-proxy study on core PC466 recovered from Sediment Drift 4 on the upper continental rise west of the Antarctic Peninsula (number according to Rebesco *et al.* 2002). The core site lies at a relatively shallow

From: Hambrey, M. J., Barker, P. F., Barrett, P. J., Bowman, V., Davies, B., Smellie, J. L. & Tranter, M. (eds) 2013. *Antarctic Palaeoenvironments and Earth-Surface Processes*. Geological Society, London, Special Publications, **381**, 263–276. First published online July 5, 2013, http://dx.doi.org/10.1144/SP381.12

water depth of approximately 2300 m, where carbonate ion concentrations are high enough that sedimented calcareous particles should be preserved (e.g. Hillenbrand *et al.* 2003) and thus permit stable isotopes analysis of foraminifera. We also used the RPI of core PC466, which should reflect the variations of Earth's magnetic field, as a stratigraphical tool (Sagnotti *et al.* 2001; Macrì *et al.* 2006; Venuti *et al.* 2011).

In this study, we aim to assess the feasibility of reconstructing the Late Quaternary environmental history of the Antarctic Peninsula region by analysing marine sediments from the continental rise that are low in biogenic calcium carbonate. The specific objectives are to: (1) develop an accurate age model for core PC466; (2) infer ice-sheet dynamics (e.g. advance and retreat); and (3) reconstruct changes in surface water conditions, such as sea-surface temperature and sea-ice cover. Although the methods employed are relatively standard at lower latitudes, it is more challenging to apply them in the Antarctic environment. We first evaluate how well each proxy performed and then discuss the inferred palaeoenvironmental history in the final section.

Study area

The surface water current regime on the western Antarctic Peninsula shelf is characterized by a cyclonic gyre. The Antarctic Circumpolar Current (ACC) and the shelf gyre are separated by the southern boundary of the ACC, a front that runs along the shelf break (Orsi *et al.* 1995). As a result of this circulation pattern, some icebergs calved from the Antarctic Peninsula Ice Sheet and the West Antarctic Ice Sheet into the Bellingshausen and Amundsen seas drift clockwise towards the shelf (e.g. Gladstone *et al.* 2001), whereas others may be carried further east through the Drake Passage by the ACC (Young 1998).

The water masses in the study area are from top to bottom: (1) fresh and very cold (near-freezing temperature) Antarctic Surface Water; (2) Upper Circumpolar Deep Water (UCDW), which is characterized by an oxygen minimum and maximum in nutrients and temperature (Ito *et al.* 2010); and (3) relatively warm and salty Lower Circumpolar Deep Water (LCDW), which is a remnant of North Atlantic Deep Water (NADW) input to the ACC, and Weddell Sea Deep Water (WSDW). While most of the LCDW and UCDW are affected by the ACC, the bottom water on the uppermost continental rise, where core PC466 was taken, consists of SW-flowing modified WSDW and LCDW (Withworth *et al.* 1998; Hillenbrand *et al.* 2008*b*).

The annual average sea-surface temperature (SST) near the site of PC466 is about 0 °C (Locarnini *et al.* 2006), and the surface water salinity is lower than 34‰, owing to the winter sea ice maintaining a strong halocline (Antonov *et al.* 2006). The site is located slightly south of the Polar Front (Dong *et al.* 2006) and covered by sea ice for about 4 months each year (Parkinson 2004). The study area has the largest extra-tropical SST response to El Niño Southern Oscillation (ENSO) on Earth (Yuan 2004), with ENSO having a large impact on the annual sea-ice cover (Kwok & Comiso 2002). During the last 60 years, the area has experienced rapid and major amplified anthropogenic warming, with a summer SST increase of up to 1 °C (Meredith & King 2005). The surface water salinity has also increased, while the winter sea-ice concentration has decreased (Vaughan *et al.* 2003).

Biological productivity at the core site occurs mainly during the approximately 8 months of open water in spring and summer, although some primary production (sea-ice diatoms) takes place within the sea ice (Arrigo *et al.* 1998). The planktonic foraminifer species of *Neogloboquadrina pachyderma* sinistral is able to survive in sea-ice cavities and brine channels (Dieckmann *et al.* 1991). Consequently, as elsewhere in the Antarctic marginal ice zone, particle flux in the water column exhibits a high seasonality, with a strong maximum in summer following the sea-ice retreat and the phytoplankton bloom (Wefer *et al.* 1988; Ducklow *et al.* 2008). Species distribution and community composition of the plankton are tightly correlated with water masses, oceanic fronts and the sea-ice edge (Ducklow *et al.* 2007).

The Antarctic Peninsula Ice Sheet is less than 500 m thick (Drewry 1983). Along its western margin, this ice sheet terminates in steep margins and tidewater glaciers. The Antarctic Peninsula Ice Sheet is thicker in the east, where it drains via glaciers and larger ice shelves into the Weddell Sea. During the LGM, the Antarctic Peninsula Ice Sheet and West Antarctic Ice Sheet advanced across the shelf (Anderson *et al.* 2002; Livingstone *et al.* 2012; Davies *et al.* 2012), and grounded ice streams advanced to the shelf break adjacent to Sediment Drift 4 (see references in Livingstone *et al.* 2012) (Fig. 1). The sedimentary changes on drifts west of the Antarctic Peninsula during the Late Quaternary have been described previously by Pudsey & Camerlenghi (1998), Pudsey (2000), Hillenbrand & Ehrmann (2003), Lucchi *et al.* (2002) and Lucchi & Rebesco (2007) but these studies were hindered by the lack of robust age models.

Core material

Piston core PC466 was recovered during expedition JR179 aboard the RRS *James Clark Ross* in 2008.

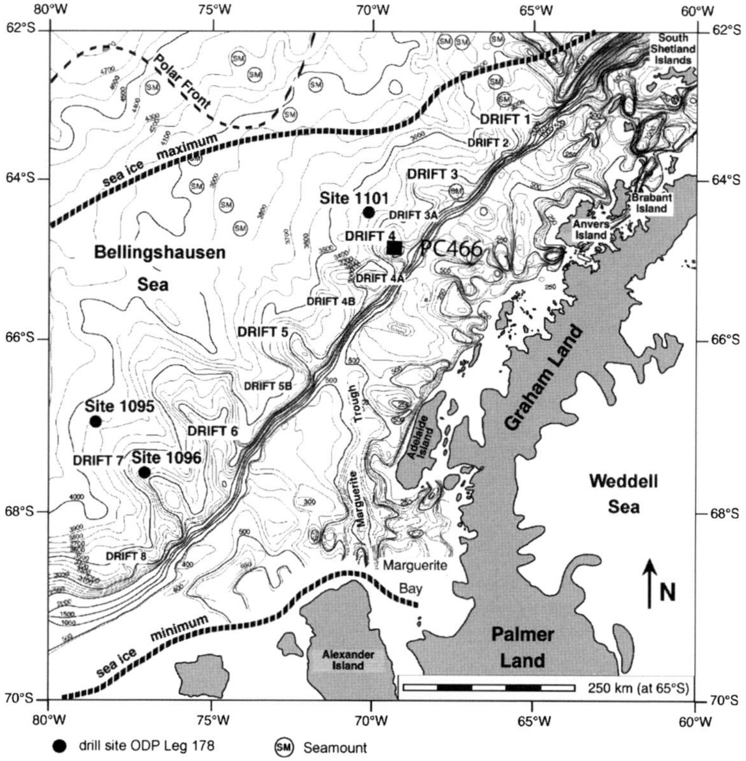

Fig. 1. Regional oceanographical map showing the core site (PC466, black square) on Sediment Drift 4 and previous continental rise drilling sites of Ocean Drilling Program (ODP) Leg 178 (reproduced from Hillenbrand & Ehrmann 2005). Extent of existing ice shelves around the Antarctic Peninsula is not represented on the map.

The 10.45 m-long core is located at 64°54′.60 S and 69°04′.59 W, at a water depth of 2306 m. Core PC466 contains two lithological units: the upper Unit I (0–57 cm) consists of bioturbated olive mud; whereas the lower Unit II (57–1045 cm) consists mainly of homogeneous, occasionally finely laminated, greyish–brownish mud with rare occurrence of thin sand layers and sand lenses. We also investigated a few samples from the 1.12 m-long trigger core TC466 collected at the same location. TC466 recovered grey–olive muds, which are bioturbated (0–47 cm) and homogeneous (47–112 cm), respectively.

Methods

Magnetic sediment properties

The relative palaeo(magnetic) intensity (RPI) (Fig. 2b) was obtained for each u-channel sample, natural remanent magnetization (NRM) components were determined at 1 cm spacing and demagnetized at 5 mT steps in the 10–100 mT peak field range. Anhysteretic remanent magnetization (ARM) was acquired in a peak alternating field of 100 mT and a 50 μT DC bias field. ARM was measured prior to and after demagnetization at 5 mT steps in the 10–60 mT peak field range. The relative strength of the magnetizing field or RPI can be determined by using the intensity of ARM to normalize the NRM intensity for changes in concentration of remanence-carrying grains. The resulting normalized remanence, calculated as the slope of the NRM-intensity v. ARM-intensity in the 20–60 mT peak field range, is a proxy for RPI variations and can be correlated to a calibrated palaeointensity template (reference curve) produced by stacking several independently-dated RPI records (e.g. the 'PISO-1500' stack: Channell *et al.* 2009) to produce the age model. The volume magnetic susceptibility data (Fig. 2b) were measured on u-channels at 1 cm intervals using a custom-built instrument at the University of Florida (Thomas *et al.* 2003). The 3.3 cm square coil of this instrument has a narrow response function (*c.* 3 cm half-peak width) designed to maximize measurement resolution.

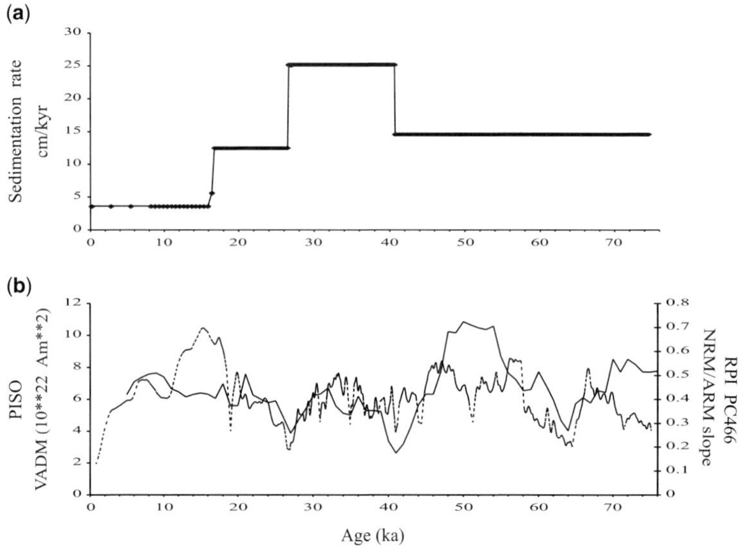

Fig. 2. Stratigraphy and sedimentation rate in core PC466 from relative palaeointensity (RPI) stratigraphic tie-lines (Table 1). (**a**) Sedimentation rate in cm/ka. (**b**) core PC466 RPI record based on the slope of the NRM intensity v. ARM intensity in the 20–60 mT peak field demagnetization range (dash line) compared to the PISO RPI reference curve (Channell *et al.* 2009) scaled to virtual axial dipole moment (VADM) (solid black line).

Sampling and processing

The core was systematically sampled by taking approximately 5 cm^3 of 1 cm-thick slices at 2 cm intervals. Carbonate content was determined on 68 bulk sediment samples. Total inorganic carbon was determined by acidification using an AutoMateFX carbonate preparation system, and evolved CO_2 was measured using a UIC Coulometrics^{TM} 5011 CO_2 coulometer. Weight per cent calcium carbonate ($CaCO_3$, %) was calculated by multiplying the values of total inorganic carbon by 8.33 (the ratio of molecular weight of $CaCO_3$ to C). All samples were dispersed in approximately 50 ml of de-ionized water, then placed for 2 h on a rotating table and, afterwards, were wet sieved through a 63 μm mesh using a gentle spray of de-ionized water using 500 ml of water. The coarse fraction was dried on the sieve in an oven at less than 50 °C and then weighed. The fine fraction was left to settle for up to 2 days and then dried, weighed and retained for grain size analysis.

Coarse fraction examination

Microfossil content

Each >63 μm sample was dry sieved over a 150 μm mesh. The entire >150 μm fraction of each sample was placed on a gridded tray for quantitative examination under a binocular light microscope. All planktonic foraminifera, fragments of planktonic foraminifera, radiolaria, ostracodes and benthic foraminifera (calcareous or agglutinated) were counted in the whole >150 μm fraction. We distinguished and counted two morphotypes of *N. pachyderma* (s) (Fig. 3a, b). The most frequent morphotype A is square shaped with a secondary crust showing various stages of preservation and is often relatively large. Morphotype B has a more lobate shape and thinner shell than morphotype A. In the Arctic, the low proportion of morphotype B in sediments and its high proportion in the water column suggest (Vilkes 1975) that it is the juvenile form of morphotype A, which is consistent with the lack of gametogenic calcite encrustation and the often smaller test size of morphotype B.

Lithogenic content

The >63 μm fraction was dry sieved over a 355 μm sieve, and all terrigeneous particles in the fraction >355 μm were counted as ice-rafted detritus (IRD). Rare large drop-stones (>0.5 cm and >1 cm) were counted separately. Ice-rafted detritus abundance is expressed in grain numbers per gram of dry sediment.

Stable isotopes and trace metal analyses

Because the samples from core PC466 contain only a few calcareous benthic foraminifera tests,

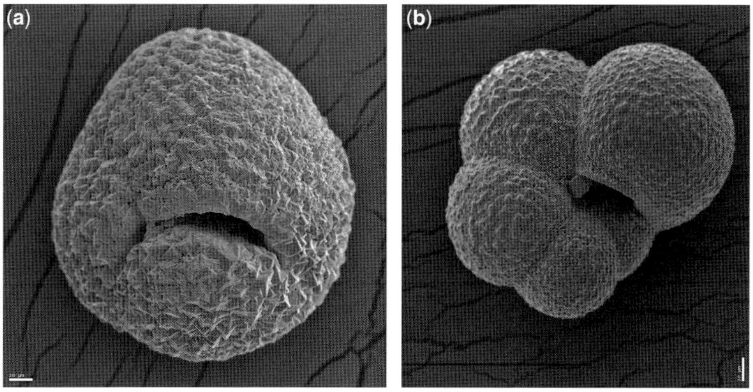

Fig. 3. Morphotypes of *N. pachyderma* (s). (**a**) Adult (encrusted) specimen of *N. Pachyderma* (s) taken from a sample at 1000 cm depth in core PC466. (**b**) Juvenile specimen *N. Pachyderma* (s) taken from a sample at 800 cm depth in core PC466.

geochemical analyses were exclusively performed on *N. pachyderma* (s) morphotype A. Picking was performed once the overall number of shells available per sample was known. Most measurements for stable isotopes and all Mg/Ca analyses were carried out in the Godwin Laboratory at the University of Cambridge. Sample treatment followed standard cleaning procedures. Foraminifera shells were crushed and soaked in a solution of 3% hydrogen peroxide for 30 min in individual vials. Acetone was added and the samples placed in an ultra-sonic bath, after which the liquid was decanted carefully using a tissue. The samples were dried in an oven at 50 °C overnight. Initially, stable isotopes were analysed on 24 samples containing less than 10 individuals of *N. pachyderma* (s) in the 212–250 μm fraction using a VG PRISM mass spectrometer and MultiPrep carbonate device. A second set of 78 samples containing up to 30 specimens was measured using a VG Sira mass spectrometer and MultiPrep device. Analytical error for all these samples was $\pm 0.06‰$ for $\delta^{18}O$ and $\pm 0.08‰$ for $\delta^{13}C$. In addition, 58 samples containing only a few tests or large test fragments were prepared in Cambridge and measured at the University of Florida using a ThermoScientific MAT 252 mass spectrometer coupled to a Kiel carbonate device. Precision of these measurements was $\pm 0.06‰$ for $\delta^{18}O$ and $\pm 0.02‰$ for $\delta^{13}C$.

Twenty-four samples containing ≥ 17 shells in the fraction 212–300 μm were selected for trace metal analyses. Each sample was cleaned following the procedure of Barker *et al.* (2003) and analysed on a Varian Vista ICP-AES (inductively coupled plasma atomic emission spectroscope) using the intensity ratio calibration method of de Villiers *et al.* (2002). We present results as Mg/Ca ratios ($mmol \cdot mol^{-1}$) rather than SSTs, thereby interpreting relative temperature changes only. Additional data obtained from these analyses are individual shell weights, which are relevant to understand down-core changes in carbonate preservation (cf. Moy *et al.* 2009).

Grain size analysis

The grain size distribution within the silt fraction was analysed on 1 g of dry sediment from the >63 μm fraction and expressed as mean sortable silt. We followed the methodology of McCave *et al.* (1995) but without the initial use of hydrogen peroxide and 0.1% sodium hexa-metaphosphate (Calgon). Even though most samples are $CaCO_3$-free, we treated the samples with dilute acetic acid (1 M) to remove all $CaCO_3$ prior to grain size analysis. We also removed biogenic opal with heated sodium carbonate (2 M). Grain sizes were analysed with a Coulter Counter Multisizer 3 after the samples had been suspended in a 0.2% Calgon and 1% NaCl isotonic solution to prevent flocculation. Particle counting was performed using 64 bins of sizes from 10 to 63 μm (the sortable silt fraction). Because PC466 was recovered from a sediment drift, we interpret variations in mean sortable silt as changes in bottom-current speed. Nevertheless, we acknowledge that the interpretation of this proxy can be more complicated in high-latitude continental margins because of the pronounced changes in IRD supply, down-slope transport and the source of the terrigenous detritus (McCave & Hall 2006).

Results

Stratigraphic framework

The age model for core PC466 is based on the integrated lithostratigraphy, oxygen isotopes and

RPI. A common lithostratigraphical succession has been documented previously in numerous cores recovered from the study area (Pudsey & Camerlenghi 1998; Pudsey 2000; Lucchi *et al.* 2002; Hillenbrand *et al.* 2008*a*): (i) MIS1 (0–14 ka BP) and MIS5e are marked by bioturbated muds containing siliceous microfossils and biogenic opal, whilst calcareous microfossils are rare to absent and $CaCO_3$ content is very low; and (ii) MIS2–MIS4 and MIS6 are marked by terrigenous, laminated to homogenous muds largely barren of microfossils. This lithostratigraphical succession was supported by chemostratigraphy (Pudsey & Camerlenghi 1998; Pudsey 2000), biostratigraphy (Villa *et al.* 2003), $^{230}Th_{excess}$ dating (Pudsey 2000), tephrochronology (Hillenbrand *et al.* 2008*a*) and RPI dating (Macrì *et al.* 2006; Venuti *et al.* 2011). Applying this lithostratigraphy to core PC466, Unit I corresponds to MIS1, whereas Unit II corresponds to MIS2–MIS4. The basal stage assignment of the core to MIS4 is supported by the absence of tephra layers in Unit II, which were deposited at nearby core sites during MIS5e (Hillenbrand *et al.* 2008*a*). The $\delta^{18}O$ data on *N. pachyderma* (s) from core PC466 suggest that the boundary between MIS1 and MIS2 (Termination I) lies somewhere between about 30 and 135 cm core depth. Three lows recognized in the RPI record from core PC466 can be correlated to minima in the 'PISO' stack at 25, 40 and 64 ka (Fig. 2b). The age model for core PC466 (Table 1) was constructed and sedimentation rates estimated using the three RPI tie points, and assuming that the top of the core is modern (Fig. 2a).

Stable isotope results

Beyond the general aim of producing a stratigraphy based partly on oxygen isotopes, we also wanted to test whether the sediments at this site contain sufficient foraminifera to generate meaningful palaeoenvironmental records.

In the upper part of core PC466, only three samples from 1 to 30 cm below sea floor (cm b.s.f.) contained enough specimens for isotope analyses, and they all yielded low $\delta^{18}O$ values (Fig. 5c) ranging between 3.3 and 3.4‰, which are consistent with core top values from the region (e.g. Mulitza *et al.* 2004) and comparable to those of Holocene sediments from the Bellingshausen Sea (e.g. Hillenbrand *et al.* 2010). In contrast, $\delta^{18}O$ values in samples below 136 cm b.s.f. to the base of core PC466 vary between approximately 4.7 and 5.2‰, and thus resemble values reported from *N. pachyderma* (s) in Antarctic continental margin sediments deposited during MIS2–MIS 4 (e.g. Mackensen *et al.* 1994; Rickaby *et al.* 2010). The planktonic $\delta^{18}O$ signal of PC466 does not allow a distinction between MIS4, MIS3 and MIS2 by itself. Planktonic $\delta^{13}C$ values (Fig. 6d) decrease slightly from a value of approximately 0.2 to about 0.0‰ from 74 to 64 ka, and increase to approximately 0.4‰ from 58 to 23 ka. After a gap in the record during Termination I, they reach a maximum of about 0.8‰ at 3 ka. Throughout MIS4–MIS3, fluctuations of about 0.4‰ are superimposed on the longer-term changes.

A continuous stable isotope record could not be obtained because foraminifera are very rare during the Holocene (less than 1 per gram), virtually absent during MIS2 and rare during MIS4.

Sedimentological results

Weight per cent $CaCO_3$ ranges from below detection limits to approximately 1 wt%, showing a general decrease up-core (Fig. 5d). Although the carbonate concentrations are close to the analytical error of the measurements (+ to −0.5 wt%), the variations in wt% $CaCO_3$ are supported by abundant variations of planktonic foraminifera.

The coarse fraction (>63 μm) contains, on average, only two foraminifera specimens per g of dry sediment (Fig. 4e). Thus, this fraction is mainly a proxy for the coarse lithogenic material. The fraction >63 μm tends to increase when the mean sortable silt decreases and, therefore, suggests that these variations are primarily related to IRD supply rather than to gravitational down-slope transport. The average percentage of the >63 μm fraction throughout core PC466 is only 2.6 wt%, with a maximum of 20.6% found during the last deglaciation at 11.2 ka. In contrast, values of about 0% occur during the LGM and middle Holocene.

In general, the fluctuations of IRD concentration (Fig. 4d) coincide with variations in the >63 μm fraction. Ice-rafted detritus particles >355 μm are found in 519 of the 538 samples examined, underscoring the fundamental influence of IRD deposition at this site. The average IRD content is approximately 5 grains/g. The maximum is 163 grains/g and found at the top of TC466.

In some intervals, the magnetic susceptibility (MS) is negatively correlated with the content of large lithogenic particles (Fig. 4c, d); for example, the MS minima at the MIS4–MIS3 transition is rich in larger particles. For all the intervals with low amounts of IRD larger than 0.5 cm (for example, during the earlier part of MIS3), a good correspondence exists between the contents of the fraction >63 μm and the MS. This suggests that, for these intervals, the grains responsible for the MS signal are concentrated in the fine sand fraction >63 μm. An examination of large IRD grains deposited during the last deglaciation and the MIS4–MIS3 transition revealed that they are composed of metamorphic rock fragments (gneiss, mica schist, green schist).

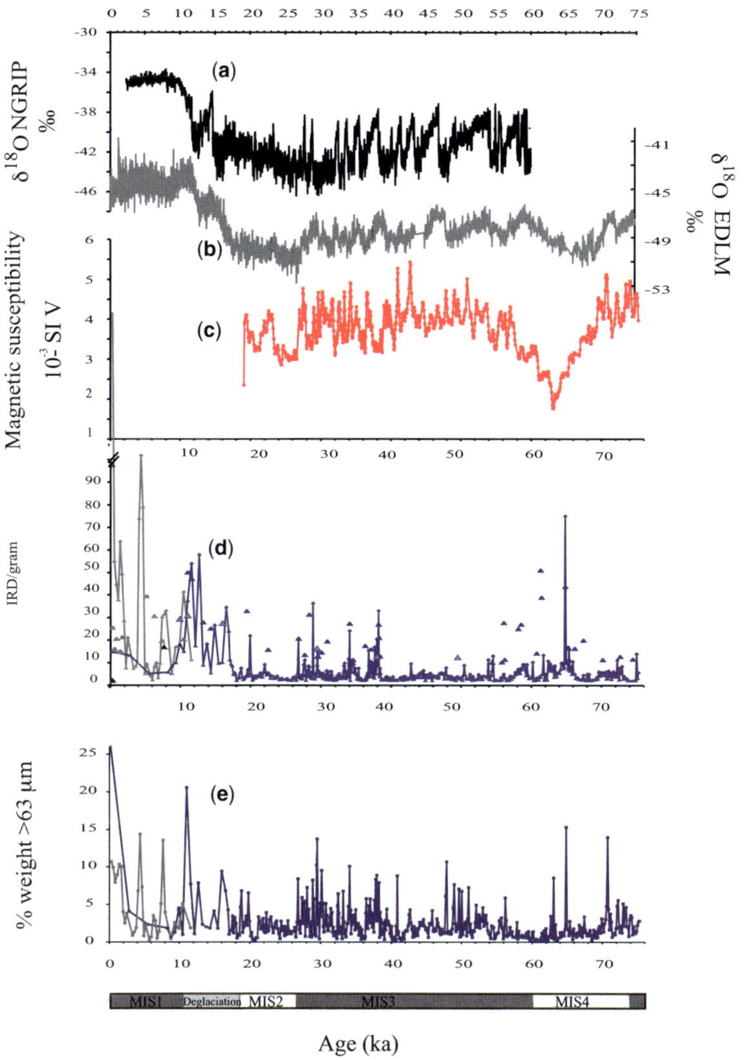

Fig. 4. The lithogenic records of core PC466. (**a**) Ngrip $\delta^{18}O$ (‰) Rasmussen *et al.* 2008. (**b**) EDML $\delta^{18}O$ (‰) (EPICA Community Members 2006). (**c**) Magnetic susceptibility. (**d**) IRD > 335 μm (lines, blue PC466, grey TC466). Medium triangles are for IRD > 0.5 cm and large triangles for IRD > 1 cm. (**e**) Percentage of fraction larger than 63 μm (lines, blue PC466, grey TC466).

Table 1. *Stratigraphical tie points**

Depth in core (cm)	Age BP (ka)
0	0
61	16.8
186	26.8
544	41
880	64

*Depth (cm in core) and ages (ka) used in the correlation between the PC466 RPI and the 'PISO' RPI reference curve (Fig. 2).

The mean sortable silt (Fig. 6c) fluctuates between 15.4 and 26.6 μm (average 18 μm), and shows some positive correlation with the MS (Fig. 2c), which itself is anti-correlated with the concentration of IRD >335 μm (Fig. 4d).

Microfossil content

The planktonic foraminifera content of 405 out of 538 samples is at least of one whole shell of the planktonic foraminifera species *N. pachyderma* (s).

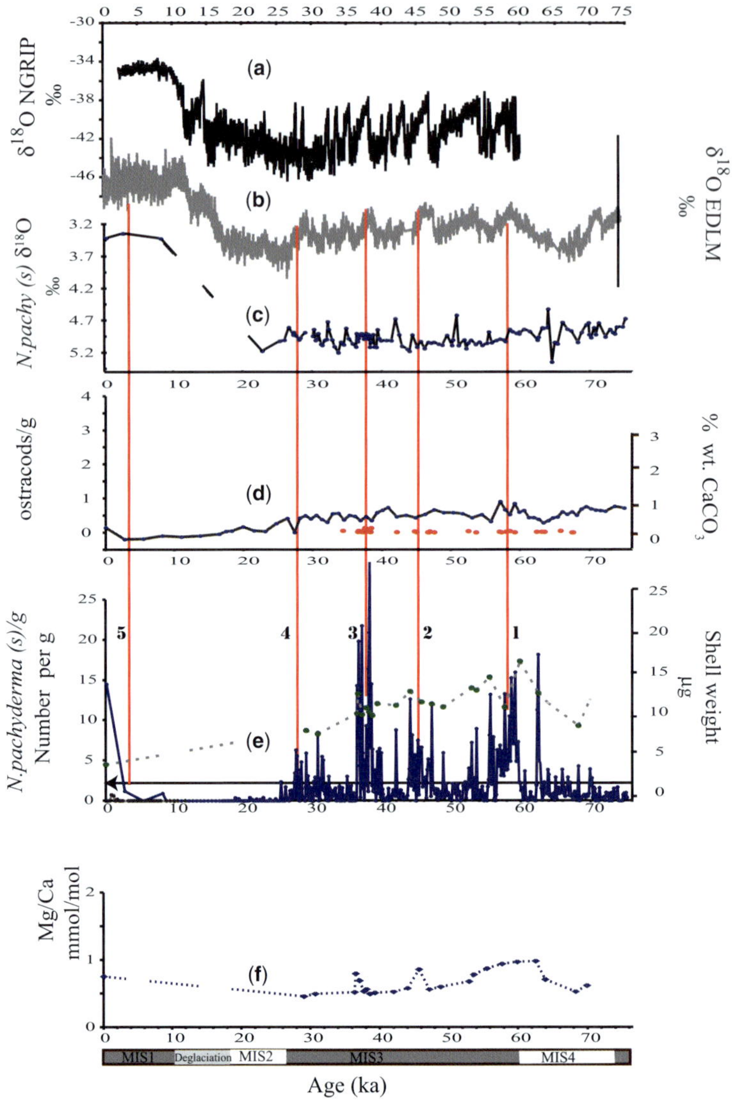

Fig. 5. Coarse fraction biogenic components v. $\delta^{18}O$ (‰). (**a**) Ngrip $\delta^{18}O$ (‰) Rasmussen *et al.* 2008. (**b**) EDML $\delta^{18}O$ (‰) EPICA Community Members, 2006. (**c**) $\delta^{18}O$ (‰) measured on *N. Pachyderma* (s). (**d**) Number of ostracods valves per gram of dry sediment (red dots). % wt $CaCO_3$ line with dots. (**e**) Number of *N. pachyderma* (s) per gram of dry sediments. Horizontal solid line is average *N. Pachyderma* (s) per gram. Dotted line with green dots is for shell weight in µg. (**f**) Mg/Ca measured on *N. Pachyderma* (s).

Maximum abundances are only 30 specimens per g, and occur at about 36 ka (Figs 5e & 6f). The average content for all samples is below two specimens per g of dry sediment. This is a very small amount considering that a more typical range of shells concentrations in deep marine sediments at 2500 m depth (outside polar area) is between a few hundred and a few thousand tests per g of dry sediments. We recognize five intervals with foraminifer concentrations above the average of two specimens per g: between 62 and 56 ka; between 50 and 43 ka: between 40 and 35 ka: between 32 and 27 ka: and from 8 to 0 ka. These intervals appear to correlate with warming events recorded in Antarctic ice cores, such as those expressed in the $\delta^{18}O$ record of the EPICA Dronning Maud Land ice core (EPICA Community Members 2006). Unexpectedly, a few samples

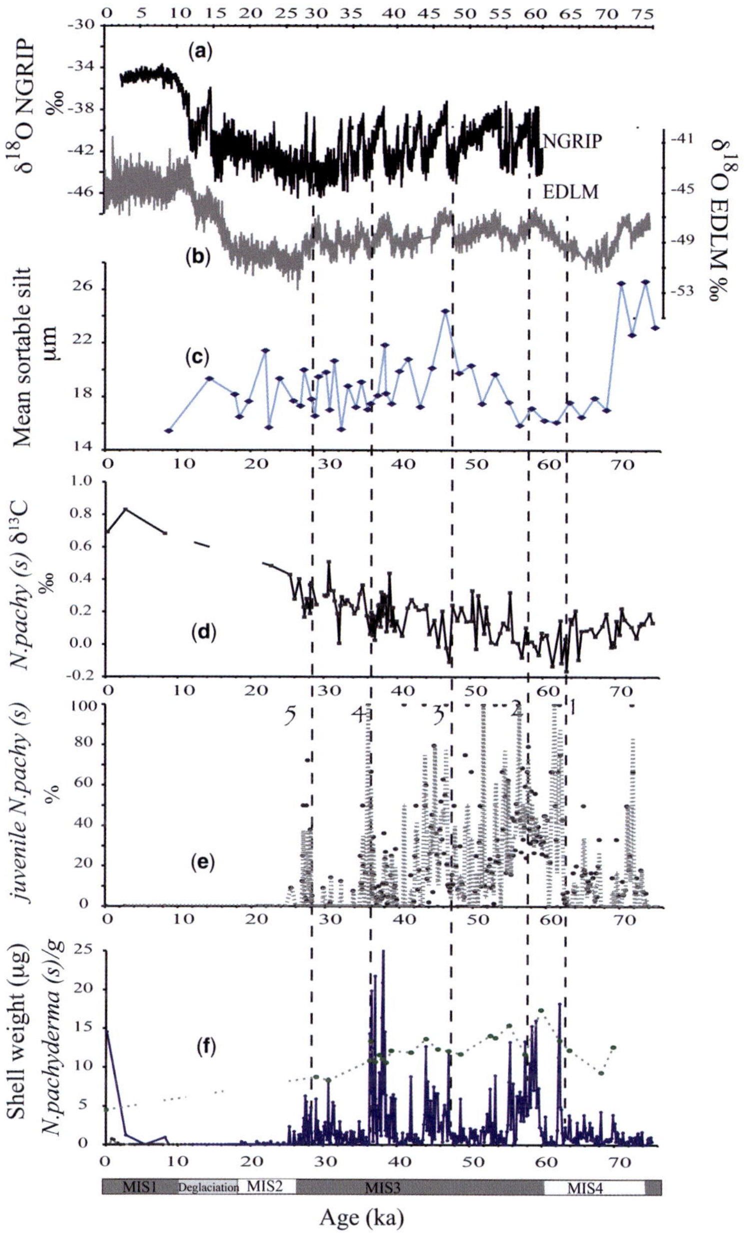

Fig. 6. Sea ice and paleocirculation records. (**a**) Ngrip $\delta^{18}O$ (‰) Rasmussen *et al.* 2008; (**b**) EDML $\delta^{18}O$ (‰) EPICA Community Members, 2006; (**c**) Mean Sortable Silt in mm as a proxy of deep-current speed; (**d**) $\delta^{13}C$ (‰) measured on *N. Pachyderma* (s); (**e**) Percentage of juvenile morphotype (B) of *N. pachyderma* (s) (see Fig. 3) in total *N. pachyderma* (s): dots. The grey line is the smooth curve (2 points moving average). 1 to 5 refer to 5 peaks of representation; (**f**) Reproduced from Figure 5 for visual reference: Number of *N. pachyderma* (s) per gram of dry sediments Horizontal solid line is average *N. Pachyderma* (s) per gram. The dotted line with green dots is for shell weight in µg.

(18 out of 538) contained a few specimens of planktonic foraminifera other than *N. pachyderma* (s), indicating even warmer surface waters.

Trace metals analyses reveal that the highest Mg/Ca ($0.98\ \text{mmol} \cdot \text{mol}^{-1}$) values occurred between 64 and 58 ka and the lowest ratios at

25 ka (0.46 mmol · mol^{-1}) (Fig. 5f). Mg/Ca ratios are 0.76 mmol · mol^{-1} in the core-top sample, which is lower than most of the peak Mg/Ca values observed during MIS3.

The thinned-shell *N. pachyderma* (s) morphoptype B (Fig. 6e), which is abundant between 72 and 26 ka, is less resistant to dissolution than the heavily encrusted morphotype A. We assume that morphotype B represents a juvenile form of morphotype A and its abundance may increase when the sea-ice cover increases. A simple explanation might be that the relative increase of morphotype B is the result of premature death of the few young specimens present at the time of sea-ice formation before they could grow and develop into encrusted adults and reproduce.

For all the samples selected for Mg/Ca (*N. pachyderma* (s), morphotype A) an average shell weight (Fig. 6f) was obtained for samples with at least 10 shells in the 212–300 μm size range. Average shell weight ranges from 4.5 μg during the Holocene to just above 17 μg at 62 ka. In addition, shell weight declines from 62 to 25 ka, in parallel with the $CaCO_3$ content. Decreases in average shell weight are interpreted to represent increasing calcium carbonate dissolution.

Benthic foraminifera in the >150 μm fractions are extremely rare and are present in only 126 samples out of 538 samples. Examination of the fraction below 150 μm for a few samples shows that benthic foraminifera are always extremely small. Agglutinated foraminifera >150 μm (counted as a whole group) are mainly restricted to the Holocene where they dominate the benthic assemblage (between three and 20 specimens per sample).

Finally, we note the rare but remarkable occurrence of a few well-preserved ostracods valves (Fig. 5d) in a few samples clustering around 58 and 38 ka.

Discussion

The development of a coherent age model for core PC466 permits the suite of multi-proxy records to be interpreted in terms of temporal changes in oceanic surface conditions (e.g. temperature, sea-ice cover and influence of icebergs). The results are discussed chronologically in five time intervals: MIS4 (74–60 ka); MIS3 (60–24 ka); Last Glacial Maximum (24–18 ka); the Last Deglaciation (18–11 ka); and MIS1 (11–0 ka) or Holocene. All intervals are shown in Figures 4, 5 and 6.

Carbonate content and foraminiferal abundance was greater during MIS4 than the LGM (MIS2), suggesting that sea-ice cover may have been less extensive during the penultimate glaciation. This interpretation is consistent with evidence for lower sea-salt concentrations in Antarctic ice cores during MIS4 than MIS2, which has been used as a proxy for Southern Ocean sea-ice cover (Wolf *et al.* 2006).

During MIS4 and MIS3, the planktonic $\delta^{18}O$ signal in core PC466 (Fig. 5c) is marked by low variability (*c.* 0.5‰), varying between approximately 4.7 and 5.2‰. In fact, the planktonic $\delta^{18}O$ record from core PC466 displays the subdued characteristics of a benthic $\delta^{18}O$ record during the last glacial interval. The high $\delta^{18}O$ values during MIS3 and MIS4 indicate the persistence of near-freezing temperatures and the absence of input of large amounts of isotopically-depleted meltwater at the depth inhabited by *N. pachyderma* (s) (morphotype A). Isotope analyses were conducted exclusively on morphotype A. of *N. pachyderma* (s). This morphotype has a secondary gameotgenic crust that is added between 50 and 200 m (Kohfeld *et al.* 1996), and maximum abundance of encrusted *N. pachyderma* (s) often occurs at the pycnocline. Thus, $\delta^{18}O$ of *N. pachyderma* does not reflect surface conditions but rather is biased towards the depth interval of the main pycnocline (100–200 m).

Sediments deposited during MIS3 contain the highest carbonate content and the greatest abundances of planktonic foramifera. Enhanced carbonate preservation is supported by the presence of the fragile morphotype B of *N. pachyderma* (s) during MIS3, high shell weight for *N. pachyderma* (s), and the rare occurrence of a few ostracods carapaces close to 58 and 38 ka. According to our age model, four peaks of foraminiferal abundance occurred during MIS3: between 62 and 56 ka; between 50 and 43 ka; between 40 and 35 ka; and between 32 and 27 ka.

Increases in the wt% of carbonate and the planktonic foraminiferal abundances can result from: (i) increased biological carbonate productivity in surface waters; (ii) less carbonate dissolution; and/or (iii) decreased dilution by non-carbonate material (e.g. terrigenous). The four peaks in carbonate and planktonic foraminifera during MIS3 are reminiscent of four warming events (A1–A4) recorded in Antarctic ice cores during MIS3 (EPICA Community Members 2006). Whilst it is tempting to attribute the increased foraminiferal abundances to decreased sea-ice cover during these warming events, four peaks in carbonate preservation have also been identified in the deep Indian (Mleneck-Vautravers 1997) and South Atlantic oceans (Kanfoush *et al.* 2000). These carbonate events occur just after the Antarctic warm events and are associated with increased North Atlantic Deep Water formation (Anderson & Fleisher 2011). The resolution of our age model does not permit us to resolve whether the carbonate events in PC466 are synchronous with the Antarctic warm events or occur just afterwards, and thereby correlate with increases in the deep-sea carbonate preservation.

There is no apparent change in planktonic $\delta^{18}O$ associated with times of increased abundance of *N. pachyderma,* but foraminiferal productivity may be responding to conditions at the surface, whereas $\delta^{18}O$ is biased towards calcite added at greater depth. In contrast, variations in the Mg/Ca of *N. pachyderma* indicate several apparent increases during MIS3 that coincide with increased foraminiferal abundances. Although these could indicate warmings during MIS3, they are difficult to reconcile with the lack of a $\delta^{18}O$ change. Mg/Ca can also be affected by carbonate ion concentration because calcite with higher Mg content preferentially dissolves, thereby lowering Mg/Ca. The peaks in Mg/Ca are associated with better carbonate preservation (less dissolution) that could result in higher Mg concentrations, which would be indistinguishable from increased Mg/Ca related to warming.

Despite clear fluctuations in foraminifera concentrations, which may correspond with interstadials during the early part of MIS3 (prior to 38 ka), these events are not associated with changes in IRD concentrations, suggesting that very few icebergs were being produced and/or melting at the location of PC466 during the first part of MIS3. Generally, IRD concentrations were higher after 40 ka, including the coarsest fraction of detrital grains (larger than 0.5 cm). We speculate that this IRD increase could be related to the size of the nearby APIS, which may have advanced onto the continental shelf when sea level dropped during the later part of MIS3 (e.g. Siddall *et al.* 2008). The relative early expansion of APIS during late MIS3 may have been related to reduced sea-ice cover during MIS3. Seasonally open water would have provided a source of moisture for ice growth.

Palaeoenvironmental interpretations are limited during MIS2 because of the absence of foraminifera. The persistently low concentration of IRD during the LGM may indicate: (i) the APIS was very stable; (ii) the APIS margin terminated in ice shelves; or (iii) permanent sea-ice coverage reduced the drift of icebergs to the site of PC466. The presence of perennial sea ice may also explain the lack of foraminifera, as it would have suppressed biogenic $CaCO_3$ production.

During the last deglaciation, foraminifera were also absent owing to a combination of reduced carbonate production, dissolution and/or dilution by terrigenous sediment. Ice-rafted detritus concentrations reached a maximum during the last deglaciation, beginning at 18 ka BP and continuing until 10 ka BP. This onset of the last deglaciation at the site of PC466 occurred synchronously with the initial APIS retreat from the outer continental shelf (e.g. Livingstone *et al.* 2012), but 3–4 ka earlier than estimated by Clark *et al.* (2009) according to the age model of the core. Once the ice sheets reached their maximum extent prior to the onset of deglaciation, large parts of their margins would have formed marine-based termini. This configuration would have made their grounding lines highly sensitive to sea-level rise and oceanic warming (Pritchard *et al.* 2012), especially after the start of the last deglaciation from 19 ka onwards (Clark *et al.* 2009; Livingstone *et al.* 2012).

Foraminifera in core PC466 are surprisingly rare during the Holocene when temperatures were significantly warmer and sea ice reduced relative to the last glacial period. The low shell weight (5 μg) found for Holocene-aged planktonic foraminifera contrasts with an average of 12 μg during MIS3, suggesting that the main control on carbonate content during the Holocene was dissolution rather than lack of production or dilution. Sediments deposited during the Holocene also contain large amounts of agglutinated benthic foraminifera and only rare benthic calcareous shells. Holocene dissolution would have been driven by lower carbonate ion concentration in the deep water or within sediment pore water, perhaps related to high fluxes of organic carbon from surface-water productivity (e.g. Wollenburg & Kuhnt 2000).

The maximum in IRD abundance for the core is observed in the top few centimetres of the trigger core, indicating either that the IRD delivery was at a maximum during the latest part of the Holocene or that the core-top sediments were affected by strong current winnowing or increased bioturbation during recent times (Singer & Anderson 1984). An ice-core record from James Ross Island reveals rapid warming in the latest part of the Holocene, especially during the past 50 years, that was associated with collapse of ice shelves and accelerating glacier mass loss in West Antarctica (Mulvaney *et al.* 2012). The IRD maximum observed in the top few centimetres of TC466 may record similar recent climate changes in the Antarctic Peninsula region.

Conclusions

One of the challenges of working on marine cores from this area is dating sediment with very low carbonate content. We constructed an age model using relative palaeointensity (RPI) and stable isotope stratigraphy, in concert with complementary lithological, sedimentological (IRD, magnetic susceptibility) and quantitative micropalaeontological data. Because the sediments used in this study contained only a very small coarse fraction no additional subsampling was needed to undertake the point counting (IRD, planktonic foraminifera). Therefore, the record has a remarkably small error in comparison to sites, which present large samples. However, extracting useful data on such tiny samples requires both careful handling, and detailed micropalaeontological and sedimentological analyses of all the samples.

Our results suggest that it is possible to date and extract a long palaeoenvironmental record from shallow water depths (i.e. *c.* 2300 m) from the sediment drifts off West Antarctica with IODP drilling. In glacial-aged sediments from core PC466, foraminifera are rare in MIS4, abundant and well preserved in MIS3, and virtually absent from MIS2. Millennial-scale peaks in planktonic foraminifera abundance (and preservation) observed during MIS3 either coincide with or occur slightly after Antarctic warm events in the EDML ice core, indicating either reduced sea-ice cover and/or increased carbonate preservation at these times. The abundance of IRD increased during the last deglaciation between approximately 19 and 11 ka when sea-level rise destabilized the West Antarctic Ice Sheet and the Antarctic Peninsula Ice Sheet, which had advanced across the shelf during the Last Glacial Maximum. The Holocene is marked by poor carbonate preservation and maximum IRD concentrations in the core-top, which may be related to recent warming of the Antarctic Peninsula region.

This work was supported by NERC grant NE/I006214/1 (to D.A. Hodell and M.J. Vautravers), the BAS Polar Science for Planet Earth programme (to C.-D. Hillenbrand, J. Smith and R.D. Larter) and US National Science Foundation grant OCE-0850413 and OCE-1014506 (to J.E.T. Channell). We are also grateful for technical support from J. Curtis (stable isotopes analysis), M. Greave (trace elements analysis), S. Crowhurst and B. Bowe (sediments preparations and sortable silt analysis).

References

Anderson, R. F. & Fleischer, M. Q. 2011. Deep ocean carbonate chemistry on millennial to Milankovitch time scales. Goldschmidt conference abstract. *Mineralogical Magazine*, **75**, 437.

Anderson, J. B., Shipp, S. S., Lowe, A. L., Wellner, J. S. & Mosola, A. B. 2002. The Antarctic Ice Sheet during the Last Glacial Maximum and its subsequent retreat history: a review. *Quaternary Science Reviews*, **21**, 49–70.

Antonov, J. I., Locarnini, R. A., Boyer, T. P., Mishonov, A. V. & Garcia, H. E. 2006. *In*: Levitus, S. (ed.) *World Ocean Atlas 2005, Volume 2*. NOAA Atlas NESDIS, **62**. US Government Printing Office, Washington, DC.

Arrigo, K., Worthen, D. L., Dixon, P. & Lizotte, M. P. 1998. Primary productivity of near surface communities within Antarctic pack ice. *In*: Lizotte, M. P. & Arrigo, K. (eds) *Antarctic Sea Ice Biological Processes, Interactions and Variability*. American Geophysical Union, Antarctic Research Series, **73**, 23–43.

Barker, S., Greaves, M. & Elderfield, H. 2003. A study of cleaning procedures used for foraminiferal Mg/Ca palaeothermometry. *Geochemistry, Geophysics, Geosystems*, **4**, 9, http://dx.doi.org/10.1029/2003GC000559.

Bindschadler, B. 2006. The environment and evolution of the west Antarctic ice sheet: setting the stage. *Philosophical Transactions of the Royal Society London, Series A*, **364**, 1583–1605.

Channell, J. E. T., Xuan, C. & Hodell, D. A. 2009. Stacking palaeointensity and oxygen isotope data for the last 1.5 Myrs (PISO-1500). *Earth and Planetary Science Letters*, **283**, 14–23.

Clark, P. U., Dyke, A. S. *et al.* 2009. The last glacial maximum. *Science*, **325**, 710–714.

Davies, B. J., Hambrey, M. J., Smellie, J. L., Carrivick, J. L. & Glasser, N. F. 2012. Antarctic Peninsula Ice Sheet evolution during the Cenozoic Era. *Quaternary Science Reviews*, **31**, 30–66.

de Villiers, S., Greaves, M. & Elderfield, H. 2002. An intensity ratio calibration method for the accurate determination of Mg/Ca and Sr/Ca of marine carbonates by ICP-AES. *Geochemistry, Geophysics, Geosystems*, **3**, 1001.

Dieckmann, G. S., Spindler, M., Lange, M. A., Ackley, S. F. & Eicken, H. 1991. Antarctic sea ice: a habitat for the foraminifer *Neogloboquadrina pachyderma*. *Journal of Foraminiferal Research*, **21**, 182–189.

Dong, S., Sprintall, J. & Gille, S. T. 2006. Location of the Antarctic Polar Front from AMSR-E Satellite Sea Surface Temperature Measurements. *Journal of Physical Oceanography*, **36**, 2075–2089.

Ducklow, H. W., Baker, K. *et al.* 2007. Marine pelagic ecosystems: the West Antarctic Peninsula. *Philosophical Transactions of the Royal Society B. Biological Sciences*, **362**, 67–94.

Ducklow, H. W., Erickson, M. *et al.* 2008. Particle export from the upper ocean over the continental shelf of the west Antarctic Peninsula: a long-term record, 1992–2007. *Deep-Sea Research. Part I. Topical Studies in Oceanography*, **55**, 2118–2131.

Drewry, D. J. 1983. *Antarctica: Glaciological and Geophysical Folio.* Scott Polar Research Institute, Cambridge.

EPICA COMMUNITY MEMBERS. 2006. One-to-one coupling of glacial climate variability in Greenland and Antarctica. *Nature*, **444**, 195–198.

Gladstone, R. M., Bigg, G. R. & Nicholls, K. W. 2001. Icebergs trajectory modeling and meltwater injection in the Southern Ocean. *Journal of Geophysical Research*, **106**, 19,903–19,915.

Hillenbrand, C.-D. & Ehrmann, W. 2003. Distribution of clay minerals in drift sediments on the continental rise west of the Antarctic Peninsula, ODP Leg 178, Sites 1095 and 1096. *In*: Barker, P. F., Camerlenghi, A., Acton, G. D. & Ramsay, A. T. S. (eds) *Proceedings of ODP Science Results*, **178**. Ocean Drilling Program, College Station, TX, 1–29.

Hillenbrand, C.-D. & Ehrmann, W. 2005. Late Neogene to Quaternary environmental changes in the Antarctic Peninsula region: evidence from drift sediments. *Global and Planetary Change*, **45**, 165–191.

HILLENBRAND, C.-D., GROBE, H., DIECKMANN, B., KUHN, G. & FÜTTERER, D. K. 2003. Distribution of clay minerals and proxies for productivity in surface sediments of the Bellingshausen and Amundsen seas (West Antarctica): Relation to modern environmental conditions. *Marine Geology*, **193**, 253–271.

HILLENBRAND, C.-D., MORETON, S. G. *ET AL*. 2008*a*. Volcanic time-markers for Marine Isotopic Stages 6 and 5 in Southern Ocean sediments and Antarctic ice cores: implications for tephra correlations between palaeoclimatic records. *Quaternary Science Reviews*, **27**, 518–540.

HILLENBRAND, C.-D., CAMERLENGHI, A., COWAN, E. A., HERNANDEZ-MOLINA, F. J., LUCCHI, R. G., REBESCO, M. & UENZELMANN-NEBEN, G. 2008*b*. The present and past bottom-current flow regime around the sediment drifts on the continental rise west of the Antarctic Peninsula. *Marine Geology*, **255**, 55–63.

HILLENBRAND, C.-D., SMITH, J. A. *ET AL*. 2010. Age Assignment of a diatomaceous ooze deposited in the western Amundsen Sea Embayment after the last Glacial Maximun. *Journal of Quaternary Science*, **25**, 280–295.

ITO, T., WOLOSZYN, M. & MAZLOFF, M. 2010. Anthropogenic carbon dioxide transport in the Southern Ocean driven by Ekman flow. *Nature*, **463**, 80–83, http://dx.doi.org/10.1038/nature08687.

JOUGHIN, I. & ALLEY, R. B. 2011. Stability of the West Antarctic ice sheet in a warming world. *Nature Geoscience*, **4**, 506–513, http://dx.doi.org/10.1038/NGEO1194.

KANFOUSH, S. L., HODELL, D. A., CHARLES, C. D., GUILDERSON, T. P., MORTYN, G. & NINNEMANN, U. S. 2000. Millennial-scale instability of the Antarctic ice sheet during the last glaciation. *Science*, **288**, 1815; doi: 10.1126/science.288.5472.1815.

KOHFELD, K. E., FAIRBANKS, R. G., SMITH, S. L. & WALSH, I. D. , 1996. Neogloboquadrina pachyderma (sinistral coiling) as paleoceanographic tracers in polar oceans: evidence from Northeast Polynya plankton tows, sediment traps, and surface sediments. *Paleoceanography*, **11**, 679–699.

KWOK, R. & COMISO, J. C. 2002. Southern Ocean climate and sea ice anomalies associated with the Southern Oscillation. *Journal of Climate*, **15**, 487–501.

LIVINGSTONE, S. J., Ó COFAIGH, C., STOKES, C. R., HILLENBRAND, C.-D, VIELI, A. & JAMIESON, S. R. 2012. Antarctic palaeo-ice streams. *Earth Science Reviews*, **111**, 90–128, http://dx.doi.org/10.1016/j.earscirev.2011.10.003.

LOCARNINI, R. A., MISHONOV, A. V., ANTONOV, J. I., BOYER, T. P. & GARCIA, H. E. 2006. *In*: LEVITUS, S. (ed.) *World Ocean Atlas 2005, Volume 1*. NOAA Atlas NESDIS, **61**. US Government Printing Office, Washington, DC.

LUCCHI, R. G. & REBESCO, M. 2007. Glacial contourites on the Antarctic Peninsula margin: insight for palaeoenvironmental and palaeoclimatic conditions. *In*: VIANA, A. R. & REBESCO, M. (eds) *Economic and Palaeoceanographic Significance of Contourite Deposits*. Geological Society, London, Special Publications, **276**, 111–127.

LUCCHI, R. G., REBESCO, M., BUSETTI, M., CABURLOTTO, A., COLIZZA, E. & FONTOLAN, G. 2002. Mid-late Pleistocene glacimarine sedimentary processes of a high-latitude, deep- sea sediment drift (Antarctic Peninsula Pacific margin). *Marine Geology*, **189**, 343–370.

MACKENSEN, A., GROBE, H., HUBBERTEN, H-W. & KUHN, G. 1994. Benthic foraminiferal assemblages and the $\delta^{13}C$ signal in the Atlantic sector of the Southern Ocean: glacial to interglacial contrast. *In*: ZAHN, R. (ed.) *Carbon Cycling in the Glacial Ocean: Constraints on the Ocean's Role in Global Change*. NATO ASI Series, **17**, 105–145.

MACRI, P., SAGNOTTI, L. & LUCCHI, R. G. 2006. A stacked record of relative geomagnetic palaeointensity for the past 270 kyr from the western continental rise of the Antarctic Peninsula. *Earth Planetary Science Letter*, **252**, 162–179.

MARTINSON, D. G., STAMMERJOHN, S. E., IANNUZZI, R. A., SMITH, R. C. & VERNET, M. 2008. Western Antarctic Peninsula physical oceanography and spatio-temporal variability. *Deep-Sea Research Part II. Topical Studies in Oceanography*, **55**, 1964–1987.

MAYEWSKI, P. A., MEREDITH, M. P. *ET AL*. 2009. State of the Antarctic and Southern Ocean climate system. *Reviews of Geophysics*, **47**, RG1003, http://dx.doi.org/10.1029/2007RG000231.

MCCAVE, I. N. & HALL, I. R. 2006. Size sorting in marine muds: processes, pitfalls, and prospects for palaeoflow-speed proxies. *Geochemistry, Geophysics, Geosystems*, **7**, 10, http://dx.doi.org/10.1029/2006GC001284.

MCCAVE, I. N., MANIGHETTI, B. & ROBINSON, S. G. 1995. Sortable silt and fine sediment size/composition slicing: parameters for palaeocurrent speed and palaeoceanography. *Palaeoceanography*, **10**, 593–610.

MERCER, J. H. 1978. West Antarctica Ice sheet and CO_2 Greenhouse effect. Threat of disaster. *Nature*, **271**, 321–325.

MEREDITH, M. P. & KING, J. C. 2005. Rapid climate change in the ocean west of the Antarctic Peninsula during the second half of the 20th century. *Geophysical Research Letter*, **32**, L19604, http://dx.doi.org/10.1029/2005GL024042.

MLENECK-VAUTRAVERS, M. J. 1997. *Sédimentation et dissolution des carbonates biogéniques aux moyennes latitudes Nord et Sud*. PhD thesis, University Bordeaux I.

MOY, A. D., HOWARD, W. R., BRAY, S. G. & TRULL, T. W. 2009. Reduced calcification in modern Southern Ocean planktonic foraminifera. *Nature Geoscience*, **2**, 276–280.

MULITZA, S., DONNER, B., FISCHER, G., PAUL, A., PÄTZOLD, J., RÜHLEMANN, C. & SEGL, M. 2004. The South Atlantic oxygen isotope record of planktic foraminifera. *In*: WEFER, G., MULITZA, S. & RATMEYER, V. (eds) *The South Atlantic in the Late Quaternary*. Springer, Berlin, 121–142.

MULVANEY, R., ABRAM, N. J. *ET AL*. 2012. Recent Antarctic Peninsula warming relative to Holocene climate and ice-shelf history. *Nature*, **489**, 141–144, http://dx.doi.org/10.1038/nature11391.

ORSI, A. H., WHITWORTH, T. & NOWLIN, W. D. 1995. On the Meridional extent and fronts of the Antarctic

Circumpolar Current. *Deep-Sea Research Part I. Topical Studies in Oceanography*, **42**, 641–673.

Parkinson, C. L. 2004. Southern Ocean sea ice and its wider linkages; insights revealed from models and observations. *Antarctic Science*, **16**, 387–400.

Pritchard, H. D., Arthern, R. J., Vaughan, D. G. & Edwards, L. A. 2009. Extensive dynamic thinning on the margins of the Greenland and Antarctic ice sheets. *Nature*, **461**, 971–975.

Pritchard, H. D., Ligtenberg, S. R. M., Fricker, H. A., Vaughan, D. G., van den Broeke, M. R. & Padman, L. 2012. Antarctic ice-sheet loss driven by basal melting of ice shelves. *Nature*, **484**, 502–505.

Pudsey, C. J. 2000. Sedimentation on the continental rise west of the Antarctic Peninsula over the last three glacial cycles. *Marine Geology*, **167**, 313–338.

Pudsey, C. J. & Camerlenghi, A. 1998. Glacial-interglacial deposition on a sediment drift on the Pacific margin of the Antarctic Peninsula. *Antarctic Science*, **10**, 286–308.

Rasmussen, S. O., Seierstad, I. K., Andersen, K. K., Bigler, M., Dahl-Jensen, D. & Johnsen, S. J. 2008. Synchronization of the NGRIP, GRIP, and GISP2 ice cores across MIS2 and palaeoclimatic implications. *Quaternary Science Reviews*, **27**, 18–28, http://dx.doi.org/10.1016/j.quascirev.2007.01.016.

Rebesco, M., Pudsey, C. J., Canals, M., Camerlenghi, A., Barker, P., Estrada, F. & Giorgetti, A. 2002. Sediment drift and deep-sea channel systems, Antarctic Peninsula Pacific margin. *In*: Stow, D. A. V., Pudsey, C. J., Howe, J. A., Faugeres, J. C. & Viana, A. R. (eds) *Deep-Water Contourite Systems: Modern Drifts and Ancient Series, Seismic and Sedimentary Characteristics*. Geological Society, London, Memoirs, **22**, 353–371.

Rickaby, R. E. M., Elderfield, H., Roberts, N., Hillenbrand, C. D. & Mackensen, A. 2010. Evidence for elevated alkalinity in the glacial Southern Ocean. *Palaeoceanography*, **25**, PA1209, http://dx.doi.org/10.1029/2004GC000858.

Sagnotti, L., Macrì, P., Camerlenghi, A. & Rebesco, M. 2001. Environmental magnetism of Antarctic late Pleistocene sediments and interhemispheric correlation of climatic events. *Earth and Planetary Sciences Letter*, **192**, 65–80.

Siddall, M., Rohling, E. J., Thompson, W. G. & Waelbroeck, C. 2008. Marine isotopic stage 3 sea level fluctuations: data synthesis and new outlook. *Reviews of Geophysics*, **46**, RG4003, http://dx.doi.org/10.1029/2007RG000226.

Singer, J. K. & Anderson, J. B. 1984. Use of total grain-size distributions to define bed erosion and transport for poorly sorted sediment undergoing simulated bioturbation. *Marine Geology*, **57**, 335–359.

Thomas, R. G., Guyodo, Y. & Channell, J. E. T. 2003. U channel track for susceptibility measurements. *Geochemistry, Geophysics, Geosystems*, **4**, 1050, http://dx.doi.org/10.1029/2002GC000454.

Vaughan, D. G., Marshall, G. J. *et al.* 2003. Recent rapid regional climate warming on the Antarctic Peninsula. *Climatic Change*, **60**, 243–274.

Villa, G., Persico, D., Bonci, M. C., Lucchi, R. G., Morigi, C. & Rebesco, M. 2003. Biostratigraphic characterization and Quaternary microfossil palaeoecology in sediment drifts west of the Antarctic Peninsula – implications for cyclic glacial-interglacial deposition. *Palaeogeography, Palaeoclimatology, Palaeoecology*, **198**, 237–263.

Venuti, A., Florindo, F. *et al.* 2011. Late Quaternary sediments from deep-sea sediment drifts on the Antarctic Peninsula Pacific margin: climatic control on provenance of minerals. *Journal of Geophysical Research, B*, **116**, B06104, http://dx.doi.org/10.1029/2010JB007952.

Vilkes, G. 1975. Comparison of Globorotalia Pachyderma (Ehrenberg) in the water column and sediments of the Canadian Arctic. *Journal of Foraminiferal Research*, **5**, 313–325.

Wefer, G., Fischer, G., Fuetterer, D. & Gersonde, R. 1988. Seasonal particle flux in the Bransfield Strait, Antartica. *Deep Sea Research Part A. Oceanographic Research Papers*, **35**, 891–898.

Withworth, T., III, Orsi, A. H., Kim, S.-J. & Nowlin, W. D., Jr. 1998. Water masses and mixing near the Antarctic Slope Front. *In*: Jacobs, S. & Weiss, R. (eds) *Ocean Ice and Atmosphere Interactions at the Antarctic Continental Margin*. American Geophysical Union, Antarctic Research Series, **75**, 1–27.

Wolff, E. W., Fischer, H. *et al.* 2006. Southern Ocean sea-ice extent, productivity and iron flux over the past eight glacial cycles. *Nature*, **440**, 491–496, http://dx.doi.org/10.1038/nature04614.

Wollenburg, J. E. & Kuhnt, W. 2000. The response of benthic foraminifers to carbon flux and primary production in the Arctic Ocean. *Marine Micropalaeontology*, **40**, 189–231.

Young, N. W. 1998. Antarctic iceberg drift and ocean currents derived from scatterometer image series. *In*: *European Space Agency Work Emerging Scatterometer Applications*. ESA Publications Division, ESA SP-424, 125–132.

Yuan, X. 2004. ENSO-related impacts on Antarctic sea ice: a synthesis of phenomenon and mechanisms. *Antarctic Science*, **16**, 415–425.

Provenance of Pleistocene sediments from Site U1359 of the Wilkes Land IODP Leg 318 – evidence for multiple sourcing from the East Antarctic Craton and Ross Orogen

N. C. PANT[1]*, P. BISWAS[1], PRAKASH K. SHRIVASTAVA[2], S. BHATTACHARYA[2], KAMLESH VERMA[2], MAYURI PANDEY[1] & IODP EXPEDITION 318 SCIENTIFIC PARTY

[1]*Department of Geology, University of Delhi, Delhi 110007, India*

[2]*Geological Survey of India, Faridabad, India*

**Corresponding author (e-mail: pantnc@rediffmail.com)*

Abstract: Site U1359 is located on the eastern levée of the Jussieau submarine channel on the Wilkes Land margin, East Antarctica. The upper approximately 60 m of the sediment core records more than 2.5 Ma of the depositional history. Present work focuses on inferring provenance from the heavy mineral fraction from the Pleistocene sediments. Clay and non-clay fractions were characterized using X-ray diffraction and micro-beam techniques. Metamorphic minerals including orthopyroxene, high-Ca garnet and high-Ti biotite indicate a source in a high-grade metamorphic terrain. Mixing from a low- to medium-grade metamorphic component is also indicated. Several basaltic rock fragments, showing mineralogical affinities to the Ferrar volcanic province in the Ross Sea sector, are present. The metamorphic component is correlatable with the Proterozoic East Antarctic cratonic shield component. Ordovician–Silurian ages for the euhedral xenotime and monazite, coupled with the Ferrar equivalent basalts, indicate an additional sediment source from the Ross Orogen along with that from the craton.

As much as 98% of the Antarctic continent is covered by ice. Although the main elements of Antarctic geology are defined in general terms, a large fraction of the continent's bedrock remains inaccessible to research. The marine sedimentary record of the Antarctic continental margin is a function of Cenozoic–present glacial activity (e.g. Cooper *et al.* 2004), and provides the most accessible information on the palaeoclimatic and tectonothermal evolution of the continent. These glacigenic sediments are particularly useful in documenting the integrated geological characteristics of large source-rock areas in Antarctica because the ice sheet is drained by a vast network of flow units, especially ice streams (Bamber *et al.* 2000), and hence the composition of the sediments is an average of the different lithologies eroded along the length of the flow units (Roy *et al.* 2007).

The continental margin of eastern Wilkes Land formed during the Cretaceous separation of Australia and Antarctica (Cande & Mutter 1982; Veevers 1987; Sayers *et al.* 2001; Colwell *et al.* 2006; Leitchenkov *et al.* 2007; O'Brien & Stagg 2007). The acoustic basement across the margin consists of block-faulted continental crust, thinned and intruded transitional crust, and oceanic crust (Eittreim & Smith 1987; Eittreim 1994). The transition from continental to oceanic crust is generally characterized by deep marginal rift-basins (Eittreim 1994) that have the maximum thickness of sedimentary units (*c.* 8 km). A large impact crater beneath the Wilkes Land part of the East Antarctic Ice Sheet is suggested (e.g. Schmidt 1962; Weihaupt 1976).

Seismic units have been correlated from shelf to rise and abyssal plain, based largely on tracing and projecting of unconformities and seismic units. The units above unconformity WL-U8 downlap and pinch out at the base of the continental slope but deeper units (i.e. between unconformities WL-U8 and WL-U3) continue across the margin (Hampton *et al.* 1987; Eittreim *et al.* 1995; Escutia *et al.* 1997; De Santis *et al.* 2003). Unconformities WL-U3 and WL-U8 are of particular importance as the former marks the first arrival of ice sheet at the coast, and the latter denotes the transition from a dynamic ice sheet to a persistent ice sheet. Unconformity WL-U3 in Wilkes Land can be traced from the shelf, where it marks the onset of progradation on the Wilkes Land margin (Eittreim & Smith 1987), to the rise, where it correlates with an upsection increase in turbidite and contourite deposition (Escutia *et al.* 1997, 2003; Donda *et al.* 2003). On the continental rise, strata above unconformity WL-U3 include six glacially related seismic units (WL-S4–WL-S9) (De Santis *et al.* 2003; Donda *et al.* 2003), of which the two deepest ones, WL-S4 and WL-S5, consist of stratified and continuous reflectors that onlap at the base of the slope.

From: Hambrey, M. J., Barker, P. F., Barrett, P. J., Bowman, V., Davies, B., Smellie, J. L. & Tranter, M. (eds) 2013. *Antarctic Palaeoenvironments and Earth-Surface Processes*. Geological Society, London, Special Publications, **381**, 277–297. First published online July 1, 2013, http://dx.doi.org/10.1144/SP381.11

Site U1359 was drilled off the coast of the eastern margin (south latitude 64°54′ and east longitude 143°57′) of the Wilkes Land sector (110°–160°E longitude) in East Antarctica during Integrated Ocean Drilling Programme (IODP) expedition No. 318 in the period between January and March 2010. The site is located on the continental rise about 270 km from the coast, and the water depth is 4003 m (Fig. 1). The drill-hole is located on the eastern levée of a submarine channel (Jussieu channel: Escutia *et al.* 2000). The sediments represent the upper part of coalescing deep-sea fans (Escutia 2000). Fine-grained components of turbidity flows within the channels and hemipelagic drape are considered the dominant sedimentary processes (Escutia *et al.* 2011). A total of about 600 m of core was raised from this site, representing glaciomarine sediments for about the last 13 Ma. This sequence intersects unconformities WL-U6–WL-U8 (Fig. 1). Eastern Wilkes Land represents the easternmost segment of the Precambrian East Antarctic shield. Site U1359 is located in proximity to the junction of the East Antarctic shield with the northern Victoria Land. The northern Victoria Land is an amalgamation of Cambrian metamorphic belt (Wilson Terrane: Grew & Sandiford 1984; Carmigiani *et al.* 1989), a remnant oceanic volcanic arc of Cambrian age (Bowers Terrane; Weaver *et al.* 1984) and flysch sequence of Cambrian–Lower Ordovician age (Robertson Bay Terrane; Vetter & Tessensohm 1987). These three are separated by terrane boundary faults (Flöttmann & Kleinschmidt 1991). The present work provides inferences on the provenance of the Pleistocene sediments from this crucial site and attempts to comment on the depositional control of these sediments.

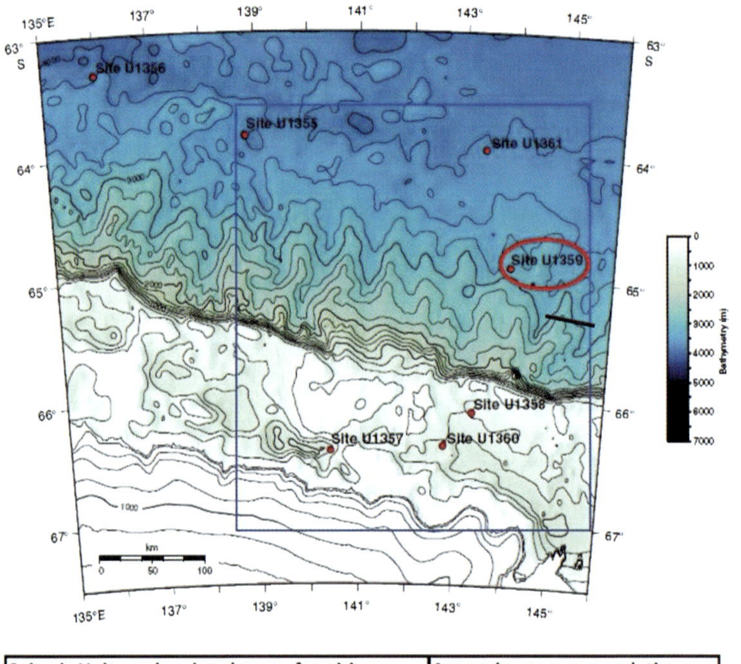

Seismic Units and regional unconformities	Approximate age correlations
WL-S9 and unconformity WL-U8 at its base	Pliocene-Pleistocene
WL-S8 and unconformity WL-U7 at its base	Late Miocene
WL-S7 and unconformity WL-U6 at its base	Middle-Late Miocene
WL-S6 and unconformity WL-U5 at its base	Middle Miocene
WL-S5 and unconformity WL-U4 at its base	Early Miocene
WL-S4 and unconformity WL-U3 at its base	Late Oligocene-Early Miocene

Fig. 1. Bathymetric map with the location of drill-site U1359 and other drill-sites cored during IODP Expedition 318. Note that Site U1359 represents the easternmost drilled location. Summarized seismic stratigraphy of the Wilkes Land margin is also given. Drill-core U1359 intersected unconformities WL-U6, WL-U7 and WL-U8 (modified from Escutia *et al.* 2011).

Table 1. *Laser particle size analysis data of sediments from U1359*

Serial No.	Sample	Depth (m b.s.f.)	Kurtosis	Skewness	Mode
1	B/3H/3W/124-126	19.84	40.27	4.43	16.82
2	B/3H/4W/24-26	20.34	3.91	1.71	19.63
3	C/4H/1W/122-124	30.85	22.9	4.05	12.79
4	C/4H/4W/122-124	35.35	1.76	1.31	19.27
5	C/5H/1W/122-124	40.35	21.62	3.9	14.34
6	B/6H/4W/124-126	49.45	49.07	6.63	8.74
7	C/6H/2W/122-124	51.25	152.7	11.17	8.94
8	C/6H/6W/22-24	56.35	2.43	1.45	11.07
9	C/7H/4W/22-24	63.76	11.72	2.89	5.15
10	B/8H/2W/124-126	67.5	28.19	3.85	9.27

m b.s.f., metres below sea floor.

Sediments of U1359

The sediments recovered from drill-core U1359 have been classified into three units based on the lithology, sedimentary structure and microfossil abundance (Escutia *et al.* 2011). Unit I is present from 0 to 43.5 m b.s.f. (metres below sea floor). It consists of decimetre-scale yellow brown silty clay interbedded with diatom-bearing olive grey silty clays with dispersed clasts (particles >2 mm), and occasional foraminifera-bearing clayey silt and sandy silt. Unit II (43.5–247.1 m b.s.f.) consists of massive to laminated (olive grey–live brown) silty clays interbedded at the millimetre scale with mottled greenish grey diatom-bearing/-rich silty clays. Dispersed clasts occur throughout Unit II. This unit is further subdivided into three subunits (Units IIa, IIb and IIc). Unit II is distinguished from Unit I on the basis of lack of foraminifera and a distinct change in colour. Bioturbated diatom-bearing silty clays interbedded with laminated silty clays constitute Unit III (247.1–596.32 m b.s.f.). The latter clays have sub-millimetre- to millimetre-scale laminations, which distinguishes this unit from Unit II.

The studies mainly pertain to lithostratigraphical Unit I and the upper part of bioturbated lithostratigraphic Unit II of these sediments. The lithology in both is silty clay. Particle-size analysis of 10 samples, representing depths of 19.64–67.5 m b.s.f. (Table 1), shows the variation in mode between 9 and 20 μm, with the lower values associated with deeper sediments. Skewness of these samples is generally between 1 and 4, with one sample at 50.13 m b.s.f. showing significant variation from standard normal distribution with a skewness of more than 15. This sample also has a very high value of kurtosis (*c.* 152), possibly indicating the influence of turbidity currents during its deposition.

Methods

The clay fraction was separated using the method of Jackson (1979). Laser particle-size analysis was carried out on a MALVERN MS 2000 particle-size

Table 2. *Analyses of garnet from the heavy mineral non-clay fraction of U1359 sediments*

Data set/ Point	1 35/1.	2 38/1.	3 39/1.	4 49/1.	5 34/1.
SiO_2	39.02	38.69	38.38	38.92	39.66
TiO_2	0.00	0.14	0.07	0.06	0.01
Al_2O_3	21.41	20.71	20.53	21.93	21.58
Cr_2O_3	0.04	0.07	0.01	0.01	0.06
BaO	0.10	0.00	0.00	0.00	0.00
FeO	24.81	25.18	25.43	24.11	28.59
MnO	0.39	0.91	0.83	0.52	0.77
MgO	10.61	7.68	7.44	10.50	6.75
CaO	3.09	6.31	6.38	3.01	3.33
Na_2O	0.03	0.04	0.00	0.01	0.00
K_2O	0.00	0.03	0.00	0.02	0.00
Total	99.57	99.75	99.07	99.09	100.74
Cations on the basis of 12 oxygen atoms					
Si	3.00	3.01	3.01	2.99	3.06
Ti	0.00	0.01	0.00	0.00	0.00
Al	1.94	1.90	1.90	1.99	1.96
Cr	0.00	0.00	0.00	0.00	0.00
Fe^{2+}	1.59	1.64	1.67	1.55	1.84
Mn	0.03	0.06	0.06	0.03	0.05
Mg	1.22	0.89	0.87	1.20	0.78
Ca	0.25	0.53	0.54	0.25	0.27
Na	0.00	0.01	0.00	0.00	0.00
K	0.00	0.00	0.00	0.00	0.00
Total	8.03	8.04	8.04	8.02	7.96
Almandine	0.52	0.53	0.53	0.51	0.63
Pyrope	0.39	0.29	0.28	0.40	0.26
Spessartine	0.01	0.02	0.02	0.01	0.02
Grossular	0.08	0.17	0.17	0.08	0.09

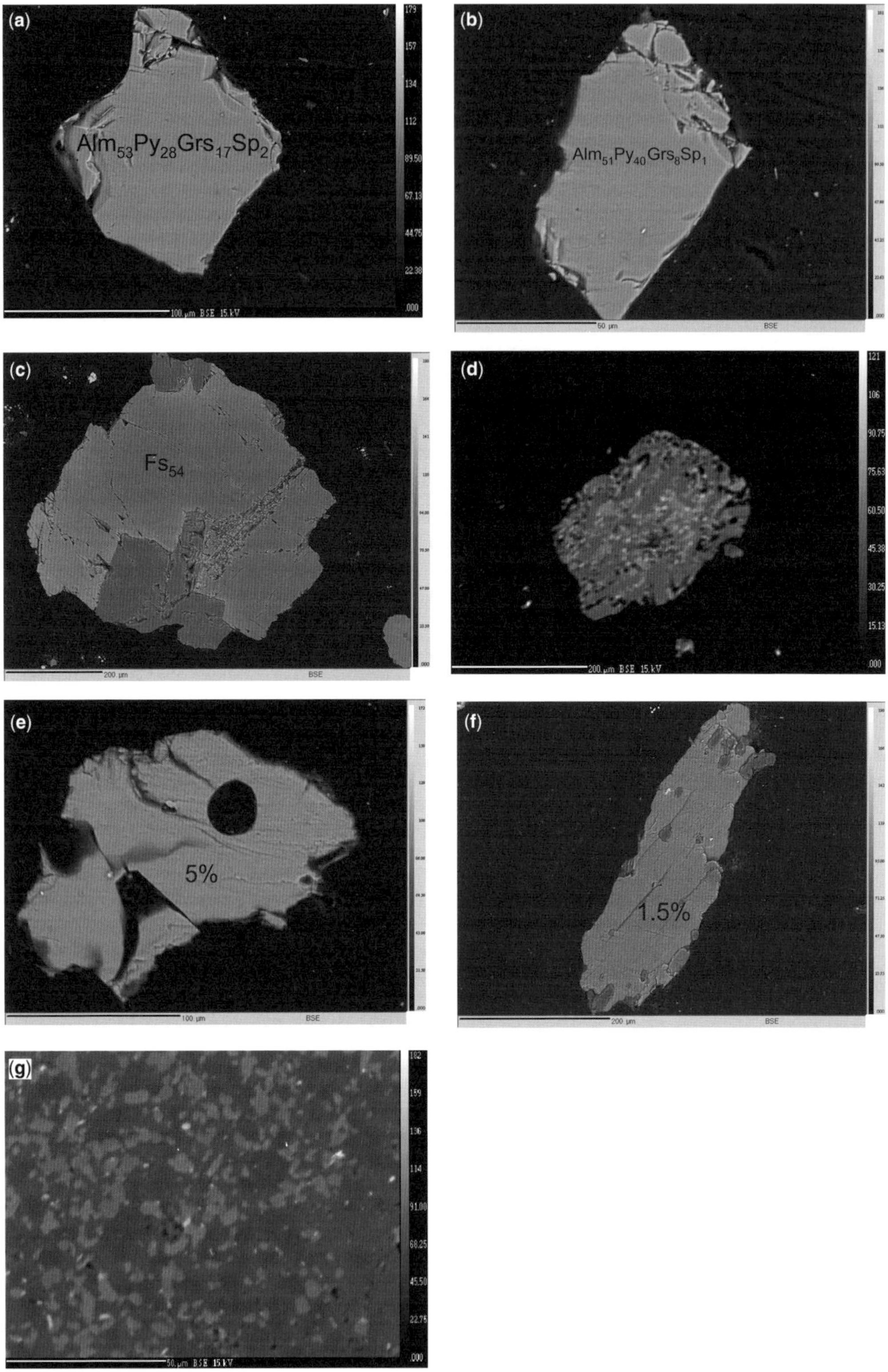
(a)
$Alm_{53}Py_{28}Grs_{17}Sp_2$
179 157 134 112 89.50 67.13 44.75 22.38 .000
100 µm BSE 15.kV
(b)
$Alm_{51}Py_{40}Grs_8Sp_1$
50 µm
BSE
(c)
Fs_{54}
200 µm
BSE
(d)
121 106 90.75 75.63 60.50 45.38 30.25 15.13 .000
200 µm BSE 15.kV
(e)
5%
100 µm
BSE
(f)
1.5%
200 µm
BSE
(g)
182 159 136 114 91.00 68.25 45.50 22.75 .000
50 µm BSE 15.kV

Table 3. *Analyses of orthopyroxene from the heavy mineral non-clay fraction of U1359 sediments*

Data set/point	1 M 30/1.	2 M 32/1.	3 M 32/1.	4 I 28/1.	5 I 35/1.	6 I 70/1.
SiO_2	50.14	50.96	50.96	56.69	56.46	56.46
TiO_2	0.07	0.01	0.01	0.00	0.03	0.08
Al_2O_3	0.97	1.06	1.06	2.43	0.73	1.39
Cr_2O_3	0.02	0.00	0.00	0.64	0.18	0.26
FeO	32.16	31.72	31.72	9.12	10.49	11.09
MnO	0.51	0.40	0.40	0.25	0.25	0.25
MgO	14.94	15.93	15.93	29.41	28.78	28.53
CaO	0.50	0.61	0.61	2.12	2.59	2.76
Na_2O	0.00	0.03	0.03	0.08	0.02	0.03
K_2O	0.00	0.00	0.00	0.00	0.00	0.00
Total	99.32	100.73	100.73	100.73	99.57	100.87
Cations on the basis of 6 oxygen atoms						
Si	1.976	1.971	1.971	1.972	2.003	1.983
Ti	0.002	0.000	0.000	0.000	0.001	0.002
Al	0.045	0.048	0.048	0.100	0.030	0.058
Cr	0.001	0.000	0.000	0.018	0.005	0.007
Fe	1.060	1.026	1.026	0.265	0.311	0.326
Mn	0.017	0.013	0.013	0.007	0.007	0.008
Mg	0.877	0.919	0.919	1.525	1.522	1.494
Ca	0.021	0.025	0.025	0.079	0.098	0.104
Na	0.000	0.002	0.002	0.005	0.001	0.002
K	0.000	0.000	0.000	0.000	0.000	0.000
Total	3.999	4.005	4.005	3.972	3.979	3.983

M, metamorphic; I, igneous.

analyser (range from 20 nm to 2000 μm). X-ray diffraction scans of 174 samples, with one sample from each metre of drilled core for approximately 16–190 m b.s.f., were obtained for orientated as well as glycolated samples mounted on glass slides. Selected samples were heated and analysed by X-ray on a Pan-Analytical instrument (X'Pert PRO) at the Geological Survey of India, Faridabad. For clay-mineral phase identification, the orientated slides of <2 μm clay separates were scanned with Cu Kα radiation (40 kV and 30 mA) from 3° to 30° 2θ at 1.2° 2θ/min using an Ni-filter.

The non-clay fraction of samples, each representing approximately 5 m of sediment core, was subjected to heavy mineral separation using bromoform. The heavy mineral fraction was mounted on glass slides, polished and analysed using an electron microprobe to gather information concerning the provenance. Scanning electron microscope (SEM) analyses were carried out at the Department of Geology, University of Delhi, on a Zeiss SEM fitted with a Cambridge Energy Dispersive Spectrometer (EDS). The electron-probe micro-analysis (EPMA) was carried out at the Geological Survey of India, Faridabad, on a four-spectrometer CAMECA SX100 electron microprobe. The operating conditions for silicate analysis were 15 kV of high voltage and 10 nA beam current, with beam diameter of 1 μm or less. Chemical dating of monazite was carried out at the EPMA Laboratory, Geological Survey of India, Faridabad. Calibrations were carried out at 20 kV and 20 nA, while analyses were obtained at the same high voltage but with a 200 nA current, as it has been shown that

Fig. 2. Back-scattered scanning electron microscopic images of heavy mineral and rock fragments of Pleistocene sediments from U1359. (**a**) Garnet grain inferred from high-grade metamorphic source. The grossular content of this garnet is approximately 10 mol% higher than other grains. (**b**) Garnet grain with lower Ca and higher Mg than the previous grain. Note the absence of any detectable zoning in both grains. (**c**) Rock fragment consisting of orthopyroxene and plagioclase. The grain boundaries of orthopyroxene (bright grey tone) and plagioclases (dark grey tone) are fused, indicating the metamorphic nature of the rock fragment. (**d**) Fine-grained basalt rock fragment. (**e**) Large biotite grain. Note high-Ti content of biotite indicating a granulite-facies affinity. (**f**) Low-Ti biotite showing the variation of Ti content in the biotite and suggestive of multiple protoliths for this mineral. (**g**) Fine-grained rock containing orthopyroxene (bright grey tone), clinopyroxene (similar to orthopyroxene but slightly less bright), plagioclase (grey) and quartz (black), with the mineralogy showing an affinity to magmatic charnockite.

Table 4. *Analyses of clinopyroxene from the heavy mineral non-clay fraction of U1359 sediments*

Data set/point	1 M 14/1.	2 M 16/1.	3 I 17/1.	4 M 28/1.	5 M 10/1.	6 I 43/1.	7 M 43/2.	8 I 55/1.	9 I 25/1.	10 I 50/1.	11 I 11/1.	12 I 17/1.
SiO_2	52.57	51.57	53.55	54.19	53.22	51.07	53.88	52.87	51.19	49.88	53.34	51.44
TiO_2	0.23	0.52	0.26	0.02	0.36	0.20	0.24	0.39	0.20	0.41	0.10	0.39
Al_2O_3	1.43	1.48	0.68	1.57	1.52	1.78	2.14	1.40	0.66	0.92	2.37	1.12
Cr_2O_3	0.04	0.00	0.24	0.32	0.16	0.21	0.37	0.00	0.08	0.06	0.37	0.00
BaO	0.00	0.00	0.00	0.01	0.00	0.00	0.00	0.02	0.00	0.00	0.00	0.00
FeO	10.53	11.98	8.20	6.83	10.51	17.02	8.55	11.60	24.72	24.80	15.78	22.25
MnO	0.32	0.31	0.23	0.22	0.30	0.67	0.16	0.29	0.59	0.44	0.36	0.44
MgO	18.57	14.82	15.36	19.24	16.31	8.08	16.46	15.49	17.59	14.66	22.28	14.47
CaO	15.32	18.55	22.15	18.57	17.90	21.15	19.67	17.88	3.91	7.42	3.58	9.72
Na_2O	0.15	0.12	0.20	0.13	0.11	0.19	0.15	0.16	0.03	0.07	0.06	0.05
K_2O	0.03	0.00	0.00	0.00	0.00	0.03	0.07	0.02	0.02	0.00	0.08	0.01
Total	99.19	99.39	100.86	101.11	100.45	100.40	101.70	100.13	99.00	98.66	98.33	99.90
Cations on the basis of 6 oxygen atoms												
Si	1.955	1.947	1.973	1.957	1.965	1.970	1.954	1.968	1.974	1.955	1.974	1.973
Ti	0.006	0.015	0.007	0.001	0.010	0.006	0.006	0.011	0.006	0.012	0.003	0.011
Al	0.063	0.066	0.029	0.067	0.066	0.081	0.092	0.061	0.030	0.043	0.103	0.051
Cr	0.001	0.000	0.007	0.009	0.005	0.006	0.011	0.000	0.002	0.002	0.011	0.000
Fe^{2+}	0.327	0.378	0.253	0.206	0.325	0.549	0.259	0.361	0.797	0.813	0.489	0.714
Mn	0.010	0.010	0.007	0.007	0.009	0.022	0.005	0.009	0.019	0.015	0.011	0.014
Mg	1.029	0.834	0.844	1.036	0.898	0.465	0.890	0.860	1.012	0.857	1.229	0.827
Ca	0.610	0.750	0.875	0.719	0.708	0.874	0.764	0.713	0.161	0.312	0.142	0.399
Na	0.010	0.009	0.014	0.009	0.008	0.014	0.011	0.012	0.003	0.005	0.004	0.004
K	0.001	0.000	0.000	0.000	0.000	0.001	0.003	0.001	0.001	0.000	0.004	0.000
Total	4.013	4.010	4.009	4.009	3.993	3.988	3.995	3.996	4.006	4.013	3.970	3.993
Fe	0.327	0.378	0.253	0.206	0.325	0.549	0.259	0.361	0.797	0.813	0.489	0.714
Mg	1.029	0.834	0.844	1.036	0.898	0.465	0.890	0.860	1.012	0.857	1.229	0.827
Ca	0.610	0.750	0.875	0.719	0.708	0.874	0.764	0.713	0.161	0.312	0.142	0.399
Total	1.967	1.963	1.971	1.961	1.930	1.888	1.913	1.934	1.970	1.981	1.860	1.940
Cfs	16.64	19.28	12.82	10.52	16.82	29.08	13.55	18.67	40.47	41.03	26.27	36.79
Cen	52.33	42.50	42.81	52.83	46.50	24.62	46.50	44.45	51.34	43.24	66.10	42.63
Wo	31.02	38.23	44.37	36.65	36.68	46.30	39.95	36.88	8.20	15.73	7.63	20.58

Cfs, clinoferrosilite; Cen, clinoenstatite; Wo, wollastonite.

calibrations at 20 nA as well at 200 nA yielded similar results (Pant *et al.* 2009). For monazite Pb Mα, and for xenotime Pb Mβ, was measured on an LPET (large pentaerythyritol) crystal and the peak counting time was 240 s, with the background measured on both sides. For better statistics and counting, subcounting was performed four times with the background measured each time for Pb. For uranium, also measured using an LPET crystal, the U Mβ line was used in order to avoid interference with the Th Mβ line, with a peak counting time of 120 s and background time of 60 s on both sides of the peak. The thorium Mα peak was also counted for 240 s and the background for 120 s on a PET (pentaerythyritol) crystal. Rare earth elements measured and the X-ray lines used included La (Lα), Ce (Lα), Nd (Lα), Pr (Lβ), Sm (Lα), Ho (Lβ), Dy (Lα) and Gd (Lβ). Positions for background measurements were arrived at using simulation software (Virtual WDS (wavelength dispersive spectrometry): Reed & Buckley 1996). Counting time at the peak varied from 40 to 60 s for these elements. The Y Lα line was used for yttrium, and it was counted for 240 s on peak and 120 s for background on both sides. Interference of Y Lλ on Pb Mα and Th Mζ on Pb Mα was corrected after measuring the interfering lines during calibration and thereafter applying the overlap correction.

Standards used for Pb, U and Th were crocite, U-glass and Th-glass. Synthetic silica-aluminium glass containing 4% rare earth elements were used as standards for La, Ce, Nd, Pr, Sm, Ho, Dy and Gd. Phosphorus was calibrated with apatite. For Y, yttrium aluminium garnet (YAG) was used. Al and Fe were calibrated using corundum and hematite; whereas for silica, orthoclase was used. Further details of the analytical procedure are given in Pant *et al.* (2009).

Description of the heavy fraction assemblage

Analysed minerals include minerals with metamorphic as well as magmatic affinities. Metamorphic minerals include garnet, biotite, chlorite, amphibole, plagioclase, muscovite, magnetite, orthopyroxene, xenotime and sphene. Magmatic minerals include clipnopyroxene, pigeonite and plagioclase. Orthopyroxene has also been analysed but its source is difficult to infer. Several rock fragments were also observed.

Heavy minerals

Garnet. Representative analyses of garnet are listed in Table 2. Garnet grains show highly variable composition from $Alm_{53}Py_{28}Grs_{17}Sp_2$ for high-calcium and low-magnesium grains, to $Alm_{51}Py_{40}Grs_8Sp_1$ for high-Mg and low-Ca garnets, within the sediments representing approximately 10 m of sediment thickness. The back-scattered electron (BSE) images of the garnet do not depict any zoning (Fig. 2a, b), indicating simultaneous sediment supply from two different metamorphic provenance sources.

Orthopyroxene. Several orthopyroxene grains, as well as orthopyroxene in rock fragments, have been analysed from the heavy mineral fraction of U1359 sediments. Six representative analyses of these are listed in Table 3. Orthopyroxene in the rock fragment has high Fe, and the fused grain boundaries in associated plagioclase indicate the rock fragment to be a metamorphosed mafic rock (Fig. 2c).

Clinopyroxene. Clinopyroxene analyses show a wide variation in the investigated sediments of U1359 (Table 4). The FeO content varies from approximately 7 to 17% and, generally, the higher Fe also has higher alumina. The presence of basalt rock fragments and their CaO content, being generally higher (CaO > 18%), can be used to consider the relatively higher Ca pyroxenes to be of igneous origin. However, the Fe–Mg content of

Table 5. *Analyses of biotite from the heavy mineral non-clay fraction of U1359 sediments*

Data set/point	1 21/1.	2 27/1.	3 37/1.
SiO_2	35.62	36.24	36.58
TiO_2	4.97	1.57	2.66
Al_2O_3	12.48	16.67	18.22
Cr_2O_3	0.02	0.00	0.12
FeO	25.79	19.98	18.80
MnO	0.25	0.23	0.19
MgO	6.85	10.08	9.71
CaO	0.00	0.00	0.03
Na_2O	0.52	0.05	0.10
K_2O	8.50	9.04	9.84
BaO	0.29	0.04	0.17
Total	95.30	93.91	96.42
Cations on the basis of 11 oxygen atoms			
Si	2.825	2.806	2.751
Ti	0.297	0.092	0.150
Al	1.166	1.521	1.615
Cr	0.001	0.000	0.007
Fe^{2+}	1.710	1.294	1.182
Mn	0.017	0.015	0.012
Mg	0.810	1.163	1.089
Ca	0.000	0.000	0.003
Na	0.080	0.008	0.014
K	0.860	0.893	0.944
Total	7.765	7.792	7.767

Table 6. *Analyses of feldspar from the heavy mineral non-clay fraction of U1359 sediments*

Data set/point	1 31/1.	2 42/1.	3 3/1.	4 78/1.	5 79/1.	6 80/1.	7 5/1.	8 46/1.	9 51/1.	10 74/1.
SiO_2	57.13	49.29	52.91	55.00	61.46	54.60	63.54	53.92	58.01	60.45
TiO_2	0.00	0.12	0.25	0.32	0.20	0.36	0.00	0.33	0.16	0.00
Al_2O_3	25.97	31.10	29.53	27.97	24.33	24.77	23.37	28.24	25.54	21.09
Cr_2O_3	0.00	0.01	0.77	0.02	0.00	0.18	0.00	0.01	0.24	3.39
BaO	0.00	0.00	0.12	0.06	0.22	0.11	0.00	0.00	0.14	0.39
FeO	0.41	0.69	0.62	1.35	0.82	1.55	0.00	0.66	0.50	0.00
MnO	0.00	0.00	0.02	0.01	0.03	0.03	0.00	0.02	0.00	0.04
MgO	0.03	0.14	0.03	0.02	0.26	1.23	0.00	0.00	0.05	0.00
CaO	8.65	15.19	11.98	11.40	6.68	11.24	4.86	12.06	8.25	0.02
Na_2O	6.67	3.09	4.47	4.81	7.81	4.89	9.09	4.81	6.60	0.95
K_2O	0.11	0.21	0.19	0.38	0.89	0.35	0.14	0.21	0.39	12.74
Total	98.95	99.85	100.89	101.45	102.90	99.51	101.08	100.32	99.97	99.07
Cations on the basis of 8 oxygen atoms										
Si	2.592	2.267	2.388	2.468	2.691	2.511	2.785	2.443	2.611	2.821
Ti	0.000	0.004	0.008	0.011	0.007	0.012	0.000	0.011	0.006	0.000
Al	1.388	1.686	1.571	1.479	1.255	1.342	1.207	1.508	1.355	1.160
Cr	0.000	0.000	0.027	0.001	0.000	0.006	0.000	0.000	0.009	0.125
Fe^{2+}	0.015	0.027	0.023	0.051	0.030	0.060	0.000	0.025	0.019	0.000
Mn	0.000	0.000	0.001	0.000	0.001	0.001	0.000	0.001	0.000	0.002
Mg	0.002	0.009	0.002	0.002	0.017	0.084	0.000	0.000	0.003	0.000
Ca	0.420	0.749	0.580	0.548	0.314	0.554	0.228	0.585	0.398	0.001
Na	0.587	0.276	0.391	0.419	0.663	0.436	0.773	0.423	0.576	0.086
K	0.006	0.012	0.011	0.022	0.049	0.021	0.008	0.012	0.022	0.758
Total	5.011	5.030	5.004	5.001	5.030	5.030	5.001	5.009	5.000	4.957
Albite	0.579	0.266	0.398	0.424	0.646	0.431	0.766	0.414	0.578	0.102
Anorthite	0.415	0.722	0.590	0.554	0.306	0.548	0.226	0.574	0.399	0.001
Orthoclase	0.006	0.012	0.011	0.022	0.048	0.021	0.007	0.012	0.023	0.897

Table 7. *Analyses of amphiboles from the heavy mineral non-clay fraction of U1359 sediments*

Data set/point	1 6/1.	2 3/1.	3 6/1.	4 14/1.	5 23/1.	6 47/1.	7 51/1.	8 52/1.
SiO_2	58.14	53.71	54.20	49.07	44.56	48.25	45.53	52.82
TiO_2	0.04	0.16	0.34	0.26	0.92	1.31	1.16	0.39
Al_2O_3	25.64	1.45	3.45	8.22	8.97	0.81	10.22	1.18
Cr_2O_3	0.00	0.21	0.74	0.06	0.13	0.05	0.14	0.03
FeO	0.63	7.17	5.57	11.06	19.07	35.38	15.56	8.34
MnO	0.00	0.17	0.08	0.45	1.29	0.62	0.23	0.17
MgO	0.11	17.59	20.05	14.45	8.46	0.04	10.54	16.41
CaO	9.11	17.11	11.30	11.12	11.06	3.69	11.50	18.34
Na_2O	4.05	0.12	0.56	0.99	1.22	5.89	1.12	0.12
K_2O	0.03	0.00	0.17	0.09	1.08	1.32	0.58	0.00
Total	97.78	97.88	96.71	95.86	97.40	98.98	96.78	97.87
Cations on the basis of 23 oxygen atoms								
Si	7.595	7.659	7.636	7.178	6.837	7.891	6.811	7.607
Ti	0.004	0.018	0.036	0.029	0.106	0.161	0.130	0.042
Al	3.947	0.244	0.573	1.417	1.622	0.157	1.802	0.200
Cr	0.000	0.023	0.082	0.007	0.016	0.007	0.017	0.004
Fe^{2+}	0.069	0.855	0.657	1.353	2.446	4.838	1.947	1.005
Mn	0.000	0.021	0.010	0.056	0.168	0.086	0.029	0.021
Mg	0.021	3.739	4.211	3.151	1.936	0.010	2.351	3.523
Ca	1.276	2.614	1.705	1.743	1.819	0.646	1.843	2.830
Na	1.026	0.034	0.152	0.280	0.364	1.867	0.325	0.034
K	0.005	0.000	0.031	0.016	0.211	0.274	0.111	0.000
Total	13.943	15.207	15.092	15.229	15.525	15.936	15.366	15.266

the clinopyroxenes shows wide variation, possibly on account of the metamorphic effect. High-Fe and -Al, and low-Ca, clinopyroxenes (analyses nos 1, 2, 4, 5 and 7; Table 4), with one exception (analysis no. 6), are inferred to be from metamorphic source (e.g. Binns 1965). The magmatic clinopyroxenes have relatively lower Fe and higher Ca content (Fig. 2d; analyses nos 3 and 8). Pigeonite compositions (analyses nos 9–12) are also inferred to be of magmatic affinity.

Biotite. Representative biotite analyses are listed in Table 5. There are two distinct biotite compositions analysed, with one being high in Ti (TiO *c.* 5%) and high in Fe (FeO *c.* 25%), and the other being lower in Ti and Fe (Fig. 2e, f) and having higher MgO (TiO_2 *c.* 1.5%; FeO *c.*18–20%), indicating two distinct sources for these.

Feldspars. Plagioclases are common feldspar grains in the analysed fractions (Table 6). There are two chemical variants present, with one being high in calcium ($>An_{50}$) and the other of low–intermediate composition (An_{23-42}). Both show considerable variation. High-calcium plagioclase also belongs to two categories, with one representing a narrow range (An_{54-60}) and the other of bytownite ($>An_{70}$) composition. The low–intermediate calcium plagioclase represents at least two metamorphic sources; the narrow range of anorthite content of high Ca represents a basaltic magmatic source, whereas the bytownite composition is from a meteorite fragment (Pant *et al.*, pers. comm.). K-feldspars are relatively rare and contain approximately 7 mol% of albite.

Amphiboles. Amphiboles in the heavy mineral fraction represent a significant proportion, and these are mainly calcic amphiboles (Table 7) with widely varying magnesium (MgO 8–20%) and iron (FeO 7–19%) content. These reflect a medium-grade metamorphic source.

Kyanite. The aluminosilicate phase analysed in the heavy mineral fraction is kyanite, as it shows a low extinction angle, low birefringence and twinning typical of kyanite. It occurs as a nearly pure phase, with Al + Si making up 11.94 atoms per formula unit (a.p.f.u. on 20 O basis) for stoichimetric 12 cations (Table 8).

Oxides. Magnetite, titaniferous magnetite, rutile and ilmenite are the major oxides present in the heavy fraction of the studied sediments (Table 9). A comparison with the analyses of the basalt fragment indicates that the titaniferous magnetite is sourced from this rock, whereas magnetite may have been derived from both magmatic and metamorphic lithologies. Ilmenite shows variation in the picroilmenite component, with MgO varying from 1.3 to >3%.

Table 8. *Analyses of kyanite from the heavy mineral non-clay fraction of U1359 sediments*

Data set/point	1 55/1.	2 56/1.
SiO_2	36.80	37.55
TiO_2	0.07	0.00
Al_2O_3	61.49	62.86
Cr_2O_3	0.00	0.13
FeO	0.42	0.41
MnO	0.01	0.04
MgO	0.00	0.01
CaO	0.03	0.05
Na_2O	0.00	0.02
K_2O	0.00	0.00
Total	99.02	101.19
Cations on the basis of 20 oxygen atoms		
Si	4.026	4.019
Ti	0.006	0.000
Al	7.929	7.930
Cr	0.000	0.011
Fe^{2+}	0.038	0.037
Mn	0.001	0.003
Mg	0.001	0.002
Ca	0.003	0.006
Na	0.000	0.003
K	0.000	0.000
Total	12.004	12.012

Rock fragments

Three major types of rock fragments have been recovered. One is a fine-grained basalt, which has a pyroxene and plagioclase composition similar to that analysed as single mineral fragments, thereby indicating the magmatic nature of calcic pyroxene and plagioclase in the labradorite range. In addition, pigeonite was also analysed from the magmatic rock fragment. Another rock fragment displaying typical magmatic features contains high Ca–Al–Ti clinopyroxenes similar to fassite composition and, in conjunction with an Ir anomaly, has been confirmed as an achondrite meteorite fragment (Pant *et al.*, pers. comm.). The rock fragment containing orthopyroxene–plagioclase has undergone metamorphism, as is evident from fused grain boundaries of plagioclases. A very-fine-grained rock fragment (Fig. 2g) containing orthoyroxene, clinopyroxene, plagioclase and quartz, and showing affinity with magmatic charnockite, is also present in the approximately 48–53 m b.s.f. sample.

Accessory minerals

Besides ilmenite and magnetite, rutile, sphene, zircon and xenotime have also been recorded.

Table 9. *Analyses of accessories from the heavy mineral non-clay fraction of U1359 sediments*

Data set/point	Rutile 43/1.	Rutile 45/1.	Spinel 10/1.	Ilmenite 24/1.	Ilmenite 27/1.
SiO_2	0.11	0.11	0.28	0.06	0.24
TiO_2	97.71	96.29	18.22	46.38	24.81
Al_2O_3	0.02	0.24	2.72	0.09	1.00
Cr_2O_3	0.15	0.20	0.19	0.00	0.21
FeO	0.06	0.09	67.71	46.43	58.62
MnO	0.00	0.04	1.31	0.43	0.69
MgO	0.00	0.02	2.84	1.31	3.04
CaO	0.10	0.05	0.04	0.01	0.05
Na_2O	0.02	0.01	0.03	0.04	0.03
K_2O	0.01	0.00	0.00	0.00	0.03
Total	98.63	97.55	93.52	94.96	88.75
Cations on the basis of 2 oxygen atoms			32(O)	6(O)	6(O)
Si	0.002	0.002	0.093	0.003	0.015
Ti	0.996	0.993	4.616	1.882	1.190
Al	0.000	0.004	1.081	0.005	0.075
Cr	0.002	0.002	0.050	0.000	0.010
Fe^{2+}	0.001	0.001	19.066	2.094	3.127
Mn	0.000	0.000	0.374	0.020	0.037
Mg	0.000	0.000	1.423	0.105	0.289
Ca	0.001	0.001	0.014	0.000	0.004
Na	0.001	0.000	0.017	0.004	0.004
K	0.000	0.000	0.000	0.000	0.002
Total	1.002	1.003	26.734	4.114	4.754

Euhedral zircon, monazite and xenotime are present in the heavy mineral assemblage. Xenotime and monazite were analysed for use in chemical dating.

Inferences on provenance for U1359 Pleistocene sediments

The upper 60 m of the composite thickness of the ocean sediments in U1359 represent Pleistocene sedimentation (Tauxe *et al.* 2012). The heavy mineral fraction, representing the provenance for these sediments deposited off the coast of Wilkes Land in the region of approximately 144°E longitude,

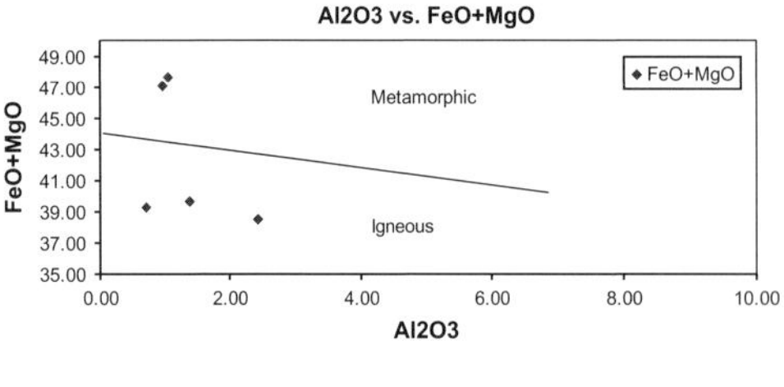

Fig. 3. Plot of Al_2O_3 v. FeO + MgO for distinguishing between igneous and metamorphic orthopyroxens from the heavy fraction of U1359 (field after Bhattacharyya 1971).

was from a composite of magmatic and metamorphic terrain. A total FeO + MgO v. Al_2O_3 plot can be used to distinguish between orthopyroxene from an igneous or metamorphic source (Bhattacharyya 1971), and orthopyroxene from the U1359 sediments indicate the presence of both metamorphic as well as igneous orthopyroxene (Fig. 3). Orthopyroxenes of metamorphic affinity are Fe-rich ($En_{45-47}Fs_{52-54}Wo_1$), while magmatic orthopyroxenes are Mg- and Ca-enriched ($En_{78-82}Fs_{14-17}$ Wo_{4-5}; Table 2).

The metamorphic assemblage indicates upper-amphibolite- to granulite-facies conditions (*c.* 700 °C at intermediate pressure), which are supported by the presence of metamorphic orthopyroxene, plagioclase, high-Ti bioitite and numerous Ca-amphibole fragments. A high content of MgO ilmenite and the presence of kyanite in the heavy mineral assemblages indicate the metamorphism to be intermediate to high pressure. A medium- to low-grade contribution (low-Ti biotite and low-Ca garnet) is also reflected. The provenance has a significant magmatic component, as indicated by clino- as well as orthopyroxenes and calcic plagioclase. It is supported by the euhedral nature of accessory minerals. The rock fragments in the sediments are predominantly represented by basalts and high-grade metamorphic equivalents of mafic and plutonic rocks. Granitic rock fragments are nearly absent.

Table 10. *Xenotime from the 20–26 m b.s.f. sample of U 1359 with chemical geochronology estimated ages*

Data set/point	1/1.	2/1.	3/1.	4/1.	5/1.	6/1.	7/1.	8/1.	9/1.	10/1.	11/1.	12/1.	13/1.	14/1.	15/1.	16/1.	17/1.	18/1.
SiO_2	0.56	0.51	0.43	0.40	0.44	0.35	0.46	0.49	0.43	0.53	0.46	0.50	0.45	0.48	0.84	0.46	0.81	0.54
Al_2O_3	0.00	0.00	0.00	0.00	0.00	0.00	0.00	0.00	0.00	0.00	0.00	0.00	0.00	0.00	0.00	0.00	0.00	0.00
FeO	0.00	0.01	0.01	0.04	0.02	0.02	0.02	0.02	0.03	0.00	0.01	0.03	0.00	0.00	0.00	0.02	0.02	0.01
P_2O_3	34.74	34.59	34.36	34.09	33.91	34.15	34.53	34.59	34.78	35.13	35.26	35.11	35.24	34.75	34.69	35.68	34.36	35.56
CaO	0.14	0.13	0.13	0.12	0.13	0.15	0.14	0.13	0.12	0.10	0.12	0.12	0.10	0.14	0.07	0.09	0.06	0.08
Y_2O_3	41.46	41.09	41.17	41.21	40.75	40.43	41.16	40.84	41.39	42.01	41.58	41.46	42.06	41.50	40.90	42.53	40.79	41.38
La_2O_3	0.00	0.00	0.00	0.01	0.00	0.00	0.00	0.00	0.00	0.02	0.00	0.00	0.00	0.02	0.02	0.00	0.00	0.00
Ce_2O_3	0.11	0.09	0.08	0.08	0.12	0.13	0.10	0.12	0.13	0.10	0.07	0.07	0.14	0.08	0.07	0.05	0.08	0.09
Pr_2O_3	0.00	0.06	0.07	0.00	0.11	0.03	0.02	0.00	0.00	0.03	0.01	0.05	0.14	0.05	0.01	0.00	0.03	0.00
Nd_2O_3	0.41	0.46	0.45	0.46	0.51	0.50	0.47	0.50	0.47	0.44	0.49	0.50	0.43	0.39	0.41	0.30	0.38	0.38
Sm_2O_3	0.56	0.68	0.66	0.66	0.69	0.72	0.67	0.76	0.70	0.57	0.68	0.72	0.62	0.64	0.51	0.59	0.51	0.53
Gd_2O_3	1.78	2.01	1.95	1.93	1.91	2.01	1.96	2.02	1.99	1.81	2.05	2.10	1.94	1.92	1.62	1.89	1.58	1.76
Dy_2O_3	4.57	4.92	4.84	4.85	4.83	4.92	4.80	4.86	4.98	4.62	4.91	4.89	4.77	4.75	4.39	4.73	4.39	4.54
Ho_2O_3	1.16	1.03	1.22	1.01	1.07	1.13	1.10	1.04	1.07	1.11	1.13	1.12	1.10	1.15	1.08	1.17	1.17	1.17
PbO	0.13	0.11	0.10	0.09	0.09	0.09	0.11	0.09	0.10	0.11	0.09	0.08	0.07	0.07	0.14	0.08	0.14	0.10
ThO_2	0.82	0.69	0.48	0.39	0.46	0.39	0.51	0.52	0.54	0.63	0.63	0.63	0.52	0.48	0.96	0.54	0.85	0.57
UO_2	2.17	1.92	1.56	1.55	1.48	1.37	1.87	1.81	1.71	1.96	1.73	1.41	1.38	1.30	2.70	1.52	2.37	2.30
Total	97.74	96.97	96.43	95.82	95.42	95.05	96.70	96.54	97.07	98.55	97.95	97.50	97.86	96.63	98.40	98.94	97.60	99.16
Age (Ma)	380.00	385.00	429.00	381.00	382.00	414.00	403.00	340.00	373.00	385.00	337.00	345.00	345.00	360.00	347.00	355.00	383.00	303.00
Age err	53.00	59.00	74.00	76.00	77.00	86.00	64.00	64.00	68.00	60.00	64.00	76.00	80.00	86.00	44.00	73.00	50.00	51.00
Cations on the basis of 4 oxygen atoms																		
Si	0.020	0.018	0.015	0.014	0.016	0.012	0.016	0.017	0.015	0.019	0.016	0.018	0.016	0.017	0.029	0.016	0.029	0.019
Al	0.000	0.000	0.000	0.000	0.000	0.000	0.000	0.000	0.000	0.000	0.000	0.000	0.000	0.000	0.000	0.000	0.000	0.000
Fe	0.000	0.000	0.000	0.000	0.000	0.000	0.000	0.000	0.000	0.000	0.000	0.000	0.000	0.000	0.000	0.000	0.000	0.000
P	1.034	1.034	1.035	1.034	1.034	1.040	1.036	1.037	1.036	1.036	1.039	1.038	1.039	1.038	1.033	1.041	1.032	1.046
Ca	0.005	0.005	0.005	0.005	0.005	0.006	0.005	0.005	0.004	0.004	0.004	0.005	0.004	0.005	0.003	0.003	0.002	0.003
Y	0.776	0.772	0.780	0.786	0.781	0.774	0.776	0.769	0.775	0.779	0.770	0.771	0.780	0.779	0.766	0.780	0.770	0.765
La	0.000	0.000	0.000	0.000	0.000	0.000	0.000	0.000	0.000	0.000	0.000	0.000	0.000	0.000	0.000	0.000	0.000	0.000
Ce	0.001	0.001	0.001	0.001	0.002	0.002	0.001	0.002	0.002	0.001	0.001	0.001	0.002	0.001	0.001	0.001	0.001	0.001
Pr	0.000	0.001	0.001	0.000	0.001	0.000	0.000	0.000	0.000	0.000	0.000	0.001	0.002	0.001	0.000	0.000	0.000	0.000
Nd	0.005	0.006	0.006	0.006	0.007	0.006	0.006	0.006	0.006	0.006	0.006	0.006	0.005	0.005	0.005	0.004	0.005	0.005
Sm	0.014	0.017	0.017	0.017	0.018	0.019	0.017	0.019	0.018	0.014	0.017	0.018	0.016	0.016	0.013	0.015	0.013	0.013
Gd	0.021	0.024	0.023	0.023	0.023	0.024	0.023	0.024	0.023	0.021	0.024	0.024	0.022	0.022	0.019	0.022	0.019	0.020
Dy	0.052	0.056	0.055	0.056	0.056	0.057	0.055	0.055	0.056	0.052	0.055	0.055	0.053	0.054	0.050	0.052	0.050	0.051
Ho	0.012	0.011	0.013	0.011	0.011	0.012	0.012	0.011	0.011	0.011	0.012	0.012	0.011	0.012	0.011	0.012	0.012	0.012
Pb	0.001	0.001	0.001	0.001	0.001	0.001	0.001	0.001	0.001	0.001	0.001	0.001	0.001	0.001	0.001	0.001	0.001	0.001
Th	0.007	0.006	0.004	0.003	0.004	0.003	0.004	0.004	0.004	0.005	0.005	0.005	0.004	0.004	0.008	0.004	0.007	0.005
U	0.017	0.015	0.012	0.012	0.012	0.011	0.015	0.014	0.013	0.015	0.013	0.011	0.011	0.010	0.021	0.012	0.019	0.018
Total	1.965	1.966	1.968	1.968	1.969	1.966	1.966	1.965	1.966	1.964	1.964	1.965	1.965	1.966	1.960	1.963	1.961	1.957

Table 11. *Monazite from the 48–53 m b.s.f. sample of U 1359 with chemical geochronology estimated ages*

Data set/ point	1/1.	2/1.	3/1.	4/1.	5/1.	6/1.	8/1.	9/1.
SiO_2	0.26	0.12	0.13	0.14	0.13	0.14	0.13	0.15
Y_2O_3	2.22	2.69	2.76	2.74	2.86	2.89	2.74	2.69
P_2O_5	30.59	30.41	30.46	30.76	30.39	30.89	30.42	30.59
La_2O_3	14.49	12.66	12.22	12.51	12.56	12.53	12.52	12.77
Ce_2O_3	28.39	26.71	26.71	26.62	26.57	26.61	26.76	26.45
Pr_2O_3	3.04	3	2.93	2.95	2.92	3.03	2.94	2.96
Nd_2O_3	10.63	10.92	11	10.84	10.84	10.86	11.1	10.82
SmO	1.82	1.92	1.99	1.86	1.79	1.79	1.91	1.82
Gd_2O_3	1.42	1.4	1.48	1.47	1.44	1.34	1.47	1.47
Dy_2O_3	0.63	0.62	0.67	0.75	0.64	0.64	0.63	0.68
ThO_2	3.99	4.06	4.13	4.2	4	3.99	4.09	4.05
K_2O	0	0.01	0	0	0	0	0	0
CaO	2.07	1.46	1.46	1.48	1.57	1.46	1.49	1.49
PbO	0.13	0.14	0.13	0.14	0.15	0.14	0.15	0.14
UO_2	0.91	1.03	1.03	1.07	1.1	1.05	1.07	1.04
Total	100.6	97.15	97.09	97.54	96.96	97.34	97.41	97.12
Age (Ma)	450	437	420	429	457	451	452	442
Age err	40	37	36	36	37	37	37	37
Cations on the basis of 4 oxygen atoms								
Ca	0.09	0.06	0.06	0.06	0.07	0.06	0.06	0.06
Pb	0	0	0	0	0	0	0	0
Y	0.05	0.06	0.06	0.06	0.06	0.06	0.06	0.06
La	0.21	0.18	0.18	0.18	0.18	0.18	0.18	0.18
Ce	0.4	0.38	0.38	0.38	0.38	0.38	0.38	0.38
Pr	0.04	0.04	0.04	0.04	0.04	0.04	0.04	0.04
Nd	0.15	0.15	0.15	0.15	0.15	0.15	0.16	0.15
Sm	0.02	0.03	0.03	0.03	0.02	0.02	0.03	0.02
Gd	0.02	0.02	0.02	0.02	0.02	0.02	0.02	0.02
Dy	0.01	0.01	0.01	0.01	0.01	0.01	0.01	0.01
Si	0.01	0.01	0.01	0.01	0.01	0.01	0.01	0.01
Th	0.04	0.04	0.04	0.04	0.04	0.04	0.04	0.04
U	0.01	0.01	0.01	0.01	0.01	0.01	0.01	0.01
P	0.99	1.01	1.01	1.02	1.01	1.02	1.01	1.01
Total	2.02	2	2	1.99	2	1.99	2	2

Age constraints

Several publications describe the age and the provenance of oceanic sediments off the coast of Wilkes Land (as discussed in the next section). In the present study, a xenotime grain at 20–26 m b.s.f. and a monazite at 48–53 m b.s.f. were used to obtain additional geochronological controls.

Complete analyses of monazite were obtained for chemical dating, and the results with a Si + P + S value of approximately 1.0 (normalization of cations at 4 O basis) were used for further processing. For individual spot ages, the formulation of Montel *et al.* (1996) was followed, whereas the age probability plots and unmixing of ages were obtained using the software Isoplot3 (v.3.71.09.06. 19nx) (Ludwig 2003). Uncertainties in individual analyses in the data table and in weighted mean ages are quoted at the 95% confidence level (2σ). The age estimations were simultaneously checked against that of the Progress Granite from Antarctica (Carson *et al.* 1996: SHRIMP U–Pb age 516 $\pm$ 7 Ma; age obtained during present measurements 520 $\pm$ 13 Ma). The analysis, cations and point ages of xenotime and monazite are given in Tables 10 and 11. Point ages are shown in Figure 4, and the histogram and processed ages are shown in Figure 5. The age obtained from the xenotime grain is 366 $\pm$ 15 Ma, while euhedral monazite is dated as 422 $\pm$ 13 Ma.

Discussion

The Wilkes sub-basin is characterized by relatively thinner crust (30 $\pm$ 2 km) compared with the adjacent Terre Adèlie, where it is approximately 42 km thick (Reading 2004). An impact crater has been proposed based on sub-ice bedrock configuration and gravity, and other geophysical anomalies

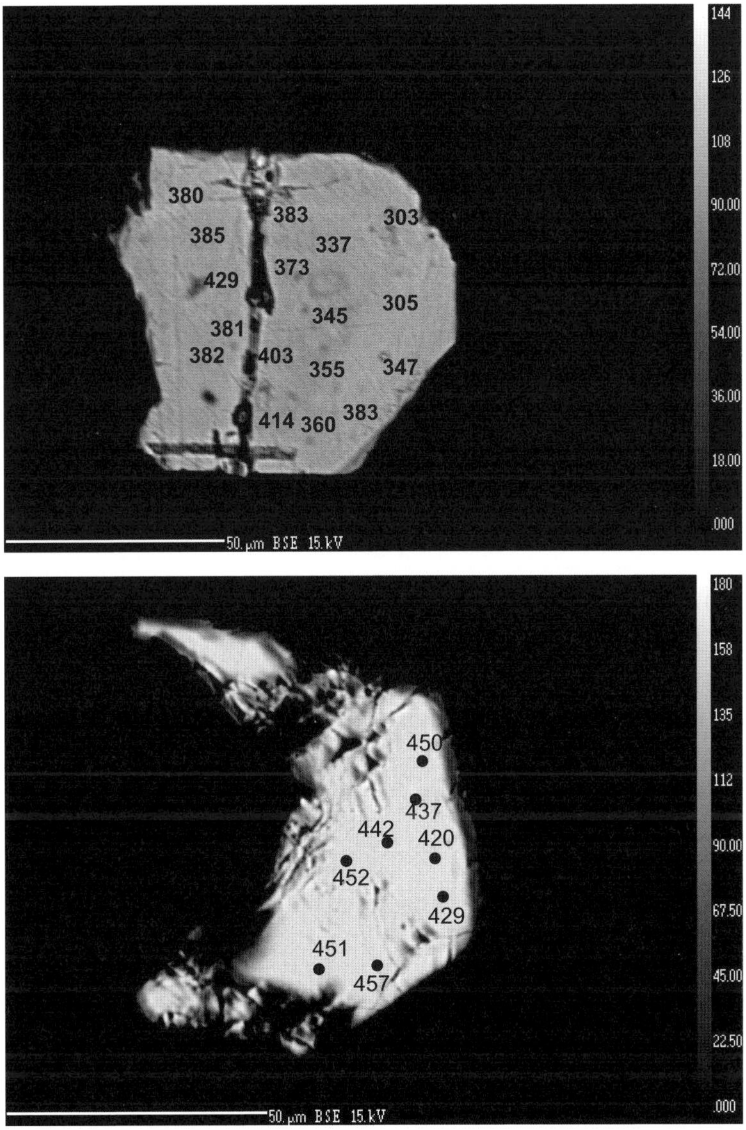

Fig. 4. Back-scattered electron microscopic images of (**a**) xenotime and (**b**) monazite from 20–26 and 48–53 m b.s.f. heavy mineral fractions, respectively, annotated with ages determined using electron microprobe chemical dating.

(Schmidt 1962; Weihaupt 1976; von Frese 2006; von Frese *et al.* 2009). This is significant in the context of supply of sediment from inland sources.

Roy *et al.* (2007) have obtained average Nd isotope crustal residence ages of approximately 2.1 Ga for sediments from the Wilkes Land margin. They also reported a wide spectrum of $^{40}Ar/^{39}Ar$ ages from detrital hornblende, with peaks at 1500–1800, 1100–1300, 400–600 and 0–200 Ma, and two grains of about 2.4 and 3.0 Ga. The paucity of outcrops has led to a near-absence of first-hand geological information on the bedrock geology of the Wilkes Land sector, although Palaeoproterozoic rocks have been inferred (Boger 2011), as supported by the detrital zircon population in the sedimentary rocks of the Transantarctic Mountains that provide ages in the range 2000–1400 Ma (Goodge *et al.* 2002, 2004). The area around the U1359 site lies west of ice drainage exiting at Cap Carr (CC in sector DD'; fig. 1 in Veevers & Saeed 2011), and also towards the east of this area is Victoria Land. Palaeo- to

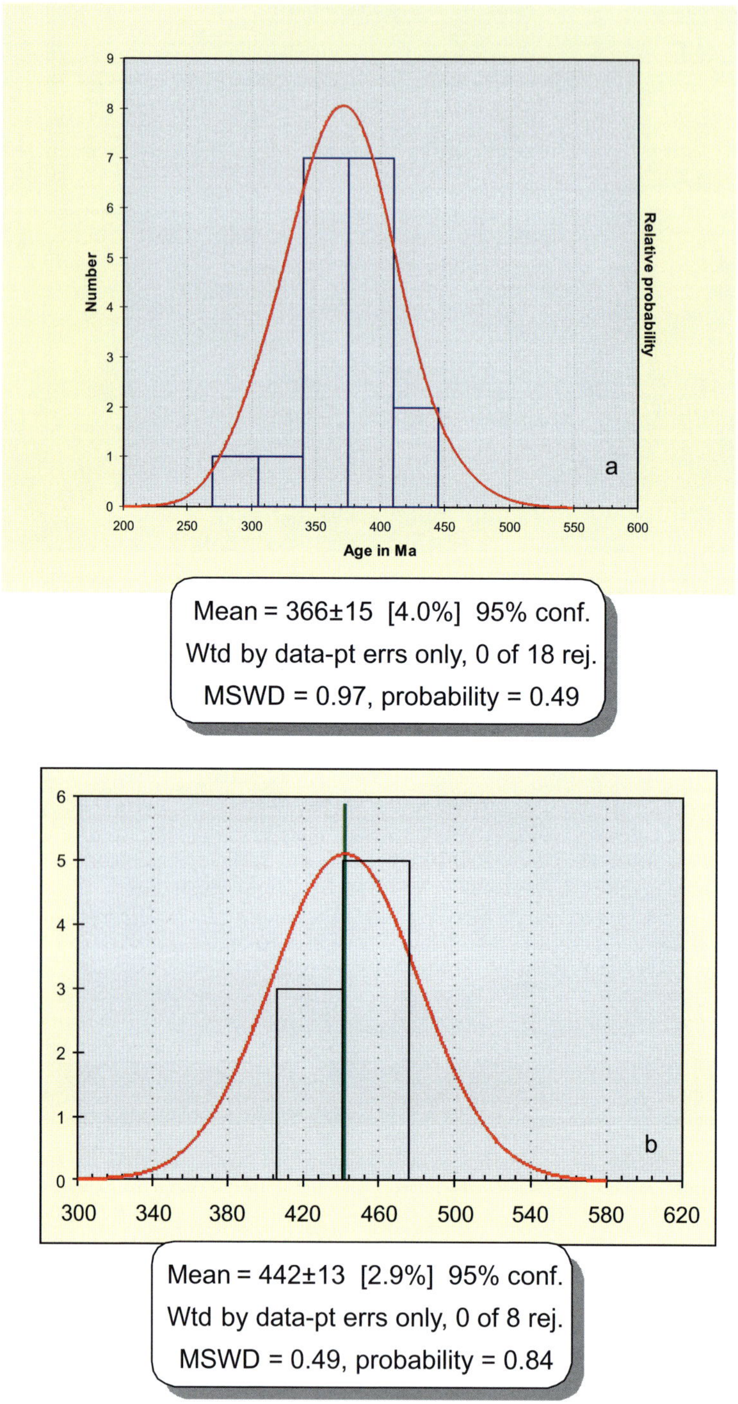

Fig. 5. Frequency diagram and weighted average of the ages obtained from xenotime and monazite.

Mesoproterozoic metamorphic rocks, including granulite-facies ortho- as well as paragneisses, volcanosedimentary rocks and plutonic granitic–granodioritic rocks that have been subsequently

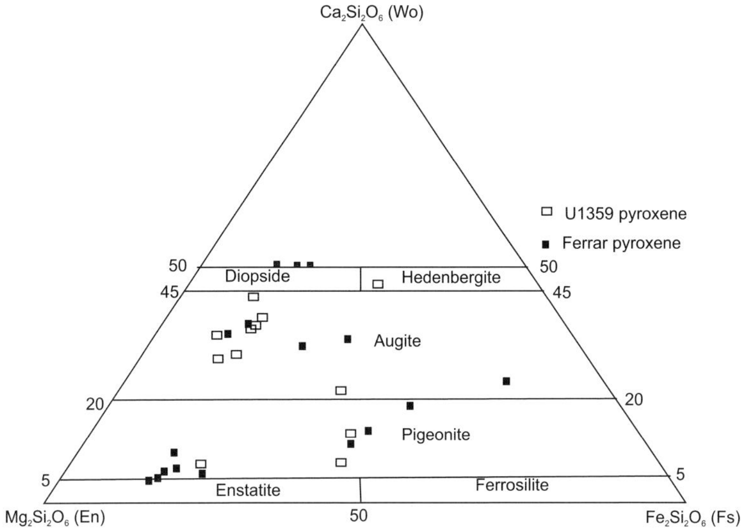

Fig. 6. A plot of the representative pyroxene (open rectangles) from U1359 in the pyroxene classification diagram. Representative compositions of pyroxene from the Ferrar volcanic province (filled squares). Note the overlap of U1359 and Ferrar pyroxene compositions.

metamorphosed, in addition to lithologies from the Ross Orogen, are present in the sector between 120° and 160°E longitude (Fitzsimons 2003; Di Vincenzo *et al.* 2007; Duclaux *et al.* 2008). Based on K–Ar dating, Jurassic dolerite sills have also been reported from this sector (Ravich *et al.* 1968).

The heavy mineral fraction recovered from U1359 shows magmatic as well as metamorphic affinities. The presence of metamorphic orthopyroxene, high-Ti biotite and grossular enriched garnet is suggestive of sediments originating from a high-grade metamorphic province. The presence of amphiboles in significant amounts implies an upper-amphibolite to granulite transition. It is likely that a lower-grade metamorphic province also contributed to these Pleistocene sediments, since lower-Ti biotite and lower-Ca garnet also constitute part of the heavy mineral fraction. The magmatic source includes a major component from mafic volcanic rocks/basalt. Plagioclase constitutes the major feldspar mineral, and the sediment supply from granite appears to have been rather limited. The metamorphic component of the heavy mineral fraction is, as expected, from a Proterozoic-age magmatic–metamorphic provenance. However, the provenance of basaltic fragments and the composition of the igneous-affinity pyroxenes are uncertain. In the marginal zone of Wilkes Land along the Terre Adèlie coast and the George V coast, high-grade gneisses and orthogneisses are reported (Peucat *et al.* 1999; Oliver & Fanning 2002; Ménot *et al.* 2005). In a study made using the dredged glaciomarine sediments of Pleistocene age on the continental margins of the Terre Adèlie and George V coasts, the following lithologies were recovered: muscovite-bearing metapelites and gneisses; muscovite–biotite–chlorite schists; metapelitic granulites; biotite–garnet–orthopyroxene charnockites; garnet–hornblende–clinopyroxene granulites; monzodiorite; microgranite; and feldspathic and arkosic arenites. Of these, charnockite, monzodiorite and garnet–clinopyroxene granulite are of Palaeoproterozoic age, while the granite and the low-grade schists were dated at approximately 500 Ma (Goodge & Fanning 2010). It is apparent that the lithologies inferred from the heavy mineral fraction of the Pleistocene ocean sediments of U1359 show broad similarities. However, differences are reflected in the presence of basalt fragments in U1359 and the relative paucity of granitic fragments. One of the possible sources of basalt in U1359 sediments is the Ferrar volcanic province present in the Transantarctic Mountains. The clinopyroxene composition of the U1359 heavy mineral fraction is plotted in the pyroxene classification diagram along with the representative clinopyroxene from the Ferrar volcanic province (Fig. 6). There is significant overlap, and it is likely that the basaltic component was derived from the extensive Jurassic tholeiites present in the Transantarctic Mountains (e.g. Elliot *et al.* 1986). A large amount of weathering, possibly on account of relatively warmer climatic conditions, can lead to breakdown of K-feldspar, the major mineral of the granite. This process is implied by the paucity of granitic rock fragments. Weathering

Table 12. *Proportion of minerals in the clay fraction of U1359 sediments*

Depth (m b.s.f.)	Smectite (%)	Illite (%)	Chlorite (%)	Kaolinite (%)	Quartz (%)
15.96	17.60	38.77	13.36	15.88	14.39
16.84	14.66	31.86	13.57	28.70	11.21
17.24	14.90	34.42	16.31	23.37	11.00
18.34	15.64	43.46	11.86	17.07	11.98
19.84	16.40	37.83	13.82	24.39	7.56
20.34	20.41	41.51	10.11	12.96	15.01
21.36	13.72	39.85	10.12	20.46	15.85
22.84	23.25	30.30	9.54	19.12	17.79
23.36	21.05	38.75	9.41	15.30	15.47
24.34	18.64	46.99	11.37	7.01	15.99
25.99	12.44	44.55	13.60	14.94	14.46
26.35	13.08	36.70	13.69	22.20	14.33
29.85	32.01	27.92	10.84	11.10	18.12
30.85	22.93	31.48	9.54	18.69	17.36
32.35	29.94	24.50	7.93	18.58	19.04
33.85	11.21	45.59	16.19	14.72	12.29
34.35	15.04	31.65	9.29	22.00	22.01
35.35	7.14	46.11	14.81	14.70	17.24
36.85	9.87	43.79	14.86	14.46	17.01
37.06	9.34	39.89	9.84	40.93	0.00
38.35	12.13	34.82	8.47	25.16	19.43
39.35	14.96	29.37	11.04	26.83	17.80
40.35	10.04	35.70	9.46	26.39	18.41
41.56	7.57	43.91	11.98	21.78	14.76
42.35	10.27	42.78	18.48	16.98	11.50
43.35	11.03	43.33	11.80	20.06	13.78
44.85	7.34	42.05	14.12	23.40	13.09
45.35	6.54	46.54	9.58	18.70	18.64
45.45	10.92	40.84	11.18	23.08	13.99
46.45	20.30	37.12	9.64	18.32	14.62
47.85	15.55	41.93	12.38	10.66	19.47
48.45	6.36	48.32	14.25	17.49	13.58
49.45	17.30	44.47	11.05	9.86	17.32
50.35	11.28	45.89	11.64	19.50	11.69
51.35	23.37	39.67	13.64	10.39	12.93
52.85	15.85	43.80	12.97	10.61	16.78
53.35	9.61	53.76	10.76	9.38	16.48
54.35	14.17	54.97	12.00	0.00	18.86
55.68	10.12	59.79	19.98	0.00	10.10
56.35	20.36	53.02	13.25	0.00	13.37
57.32	14.77	44.23	15.98	0.00	25.02
58.68	8.18	44.16	12.72	7.68	27.26
59.68	15.95	43.21	8.41	9.95	22.48
60.18	9.08	43.68	9.11	11.17	26.96
60.29	23.26	48.01	14.78	0.00	13.95
61.76	10.06	53.10	14.43	8.26	14.16
62.26	8.54	52.70	14.48	11.07	13.21
63.76	4.33	55.95	16.11	10.70	12.91
64.76	7.48	51.83	17.20	10.04	13.44
65.26	7.27	54.25	10.36	10.22	17.90
66.5	8.74	46.01	12.56	11.75	20.94
67.5	8.10	47.62	17.54	11.82	14.92
68	17.46	38.69	12.42	12.47	18.96
69.5	16.33	39.76	12.50	13.91	17.51
70.5	5.08	51.75	15.18	12.35	15.65
71.62	5.96	56.57	18.08	11.00	8.39
72.56	8.10	50.90	11.68	12.08	17.24

(*Continued*)

Table 12. *Continued*

Depth (m b.s.f.)	Smectite (%)	Illite (%)	Chlorite (%)	Kaolinite (%)	Quartz (%)
73.62	4.11	56.78	15.97	12.61	10.53
74.62	6.19	48.33	16.88	9.41	19.19
75.12	4.85	58.41	14.13	11.98	10.62
76.56	13.13	46.09	12.48	0.00	28.31
77.56	7.31	46.90	11.36	16.23	18.19
78.2	5.70	51.23	15.14	14.76	13.17
79.56	9.31	44.25	7.99	12.17	26.28
80.56	8.62	41.22	12.97	10.34	26.85
81.06	9.14	54.06	12.44	0.00	24.36
82.72	20.37	36.83	0.00	11.10	31.70
84.39	6.96	33.61	11.29	18.21	29.93
85.39	8.84	41.52	9.30	14.98	25.36
86.89	12.00	44.14	17.48	8.47	17.92
87.39	13.32	36.94	10.23	10.21	29.30
88.39	7.92	54.37	17.21	0.00	20.49
89.89	7.02	49.30	12.75	12.75	18.18
91.39	12.22	43.86	8.43	10.35	25.14
93.45	9.82	40.21	14.77	12.02	23.17
94.64	7.27	46.98	11.58	12.09	22.09
95.64	6.92	55.23	12.43	10.26	15.16
96.14	9.15	40.15	11.31	10.71	28.68
97.14	24.45	36.94	5.14	9.89	23.58
98.64	10.01	49.10	9.24	9.39	22.27
99.14	15.34	39.21	8.89	8.57	27.99
100.64	10.09	48.50	10.98	7.99	22.44

leads to a higher proportion of kaolinite in the clay fraction. The clay fraction of the U1359 was analysed for clay minerals, and the kaolinite content up to 100 m b.s.f. is plotted against depth (Table 12; Fig. 7). A warm climate at some stage during the Pleistocene Epoch is indicated by the enhanced kaolinite content in these sediments. At times of ice retreat, the Transantarctic Mountains have been suggested as acting as a barrier that diverted ice-flow into the Wilkes Basin (Hambrey & Barrett 1993), and thus a sediment contribution from the Transantarctic Mountains is quite likely.

Further corroborations of this inference come from the chemical geochronological data of xenotime and monazite. Euhedral grains of xenotime and monazite recovered, respectively, from 20–26 and 48–53 m b.s.f. sediments provide ages of 366 $\pm$ 15 and 442 $\pm$ 13 Ma (Figs 3 & 4), indicating Ordovician–Silurian ages for the source rocks. These minerals can come from the Cambrian–Ordovician (490–544 Ma: Allibone & Wysoczanski 2002; Foden *et al.* 2006) Ross Orogen, as well as from intrusive rocks (ages of 360–390 Ma) from the Transantarctic Mountains of

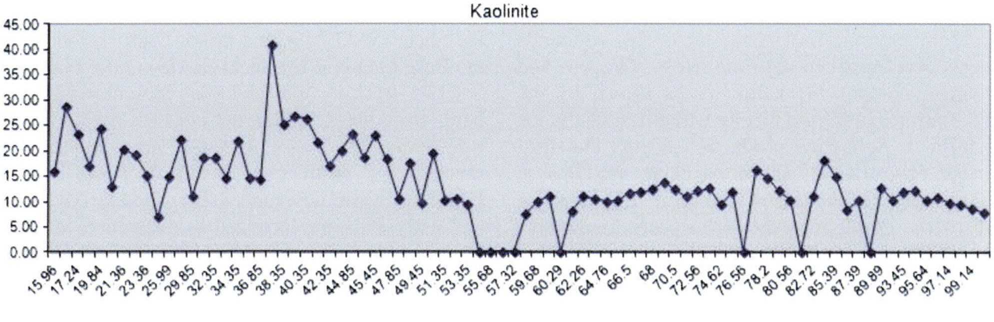

Fig. 7. Plot of kaolinite v. composite depth in U1359. There are significantly high kaolinite levels in the Pleistocene sediments (<55 m b.s.f.) suggesting warmer conditions.

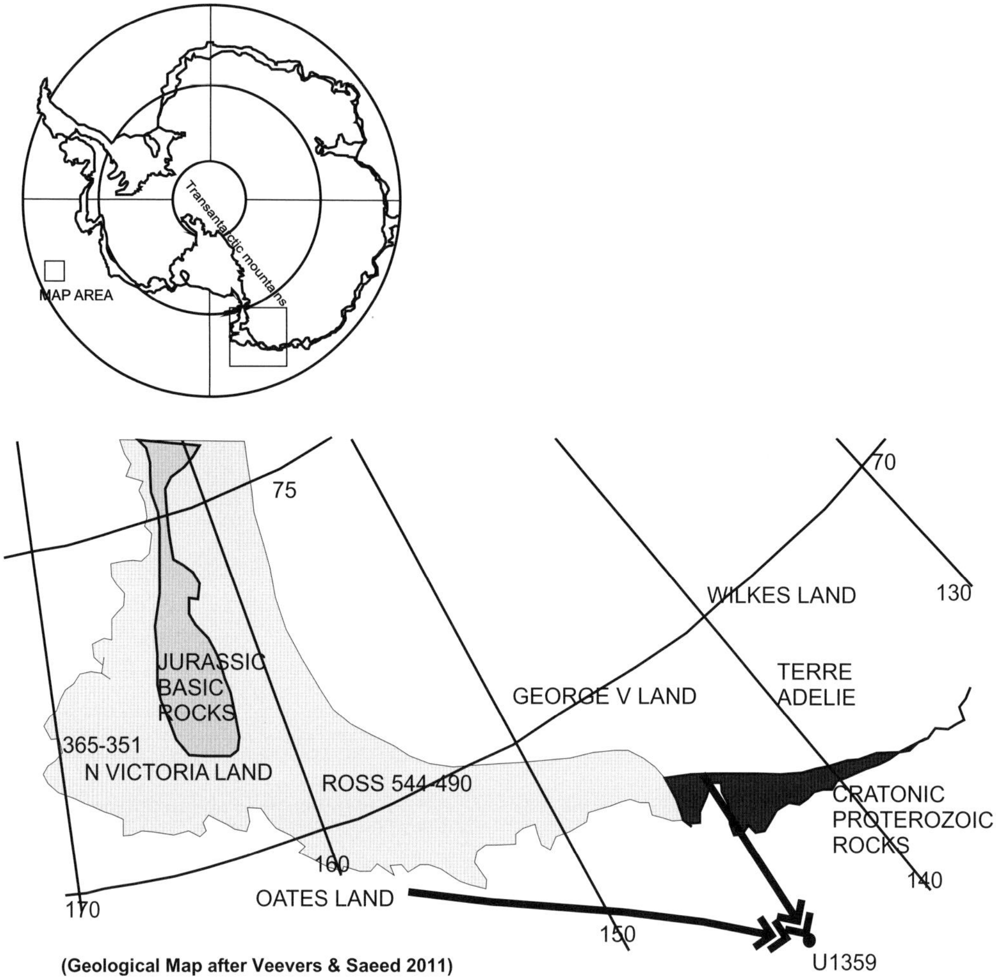

Fig. 8. Composite geological map showing the cratonic East Antarctic shield and components of the Ross Orogen (this figure is drawn from Veevers & Saeed 2011). The thick arrows show the source of the U1359 sediments as inferred from the heavy mineral assemblage.

northern Victoria Land (Fioretti *et al.* 2001; fig. 7 of Veevers & Saeed 2011). Thus, besides sourcing from the cratonic East Antarctic shield, sediments were also supplied by the Ross Orogen and younger components within it, as shown by the thick arrows in Figure 8. The variability of the ice margin in Antarctica produces icebergs that lead to the deposition of ice-rafted debris (IRD) as a component of marine sediments. The presence of clasts throughout the Pleistocene sediments in U1359 indicates ice-rafting as a contributing process to the sedimentation. The provenance of IRD is an indicator of the area in which the ice-sheet became unstable and produced icebergs. The Ross Sea is an embayment, with the Transantarctic Mountains and Victoria Land to the west. The Ross Ice Shelf is fed mainly by the West Antarctic Ice Sheet (WAIS), with a minor contribution from the East Antarctic Ice Sheet (EAIS) outlet glaciers. Satellite tracking indicates that presently the icebergs from the Ross Sea and adjacent region move westwards for a short distance prior to drifting towards the north (e.g. Stuart & Long 2008). However, warming events during the late Miocene and early Pliocene have led to deposition of IRD at distances of more than 1500 km from their sources in Wilkes Land and Adèlie Land (Williams *et al.* 2010), indicating a long westwards drift of icebergs in the past. The presence of IRD, partially sourced from several hundred kilometres east of the depositional site at U1359, indicates large fluctuations in the ice margin in the Ross Sea area during the Pleistocene time.

Conclusions

Some of the important findings of the present study can be summarized as following.

- The predominant source of the Pleistocene U1359 ocean sediments is from magmato-metamorphic provinces.
- The nature of rock fragments and the composition of minerals is suggestive of a poly-metamorphosed provenance, with peak metamorphism at least in the amphibolite–granulite transition. A subordinate low- to medium-grade metamorphic component is also reflected.
- The composition of basalt fragments suggest sediment-sourcing from the Transantarctic Mountains. This implies a change of ocean currents, which is possible in the case of warmer climate and recession of the East Antarctic Ice Sheet. Increased kaolinite content in the clay-mineral fraction supports this inference.
- Chemical geochronology of euhedral accessory minerals, indicating Ordovician–Silurian provenance ages, corroborate sediment sourcing from the Ross Orogen.

This research used samples and data provided by Integrated Ocean Drilling Program (IODP). The National Centre for Antarctic and Ocean Research (NCAOR) funded this project. A. Kundu and S. Joshi, Geological Survey of India assisted in the Microprobe analysis, and N.C. Mehra, University of Delhi in the SEM studies. Laboratory facilities were provided by B. Chattopadhyay of the Geological Survey of India. We gratefully acknowledge reviews of this article by M. Hambrey and S. Hemming.

References

ALLIBONE, A. & WYSOCZANSKI, R. 2002. Initiation of the magmatism during the Cambrian-Ordovician Ross orogeny in southern Victoria Land, Antarctica. *Geological Society of America Bulletin*, **114**, 1007–1018, http://dx.doi.org/10.1130/0016-7606(202)114<1007:IOMDTC>2.0.CO.

BAMBER, J. L., VAUGHAN, D. G. & JOUGHIN, I. 2000. Widespread Complex Flow in the Interior of the Antarctic Ice Sheet. *Science*, **287**, 1248, http://dx.doi.org/10.1126/science.287.5456.1248.

BHATTACHARYYA, A. 1971. An evalution of the chemical distinctions between igneous and metamorphic orthopyroxenes. *The American Mineralogist*, **56**, 499–506.

BINNS, R. A. 1965. The mineralogy of metamorphosed basic rocks from the Willyama complex. Pt. II. Pyroxenes, garnets, plagioclase, and opaque oxides. *Mineralogical Magazine*, **35**, 561–587.

BOGER, S. D. 2011. Antarctica–before and after Gondwana. *Gondwana Research*, **19**, 335–371.

CANDE, S. C. & MUTTER, J. C. 1982. A revised identification of the oldest sea-floor spreading anomalies between Australia and Antarctica. *Earth and Planetary Science Letters*, **58**, 151–160, http://dx.doi.org/10.1016/0012-821X(82)90190-X.

CARMIGIANI, L., GHEZZO, L. *ET AL.* 1989. Geology of the Wilson terrane area between David and Mariner glaciers, Victoria Land (Antarctica). *Società Geologica Italiana, Memorie*, **33**, 77–98.

CARSON, C. J., FANNING, C. M. & WILSON, C. J. L. 1996. Timing of the Progress granite, Larsemann Hills: additional evidence for Early Palaeozoic orogenesis within the East Antarctic Shield and implications for Gondwana assembly. *Australian Journal of Earth Sciences*, **43**, 539–553.

COLWELL, J. B., STAGG, H. M. J. & DIREEN, N. G. 2006. Geology of deep-water margin of East Antarctica between Queen Mary and George V Lands. *In*: FUTTERER, D. K. (eds) *Terra Nostra: Proceedings of the Ninth International Symposium of Antarctic Earth Science.*

COOPER, A. K., O'BRIEN, P. E. & RICHTER, C. (eds). 2004. *Proceedings of the Ocean Drilling Program, Scientific Results*, **188**. Ocean Drilling Program, College Station, TX, http://dx.doi.org/10.2973/0dp.proc.sr.188.2004.

DE SANTIS, L., BRANCOLINI, G. & DONDA, F. 2003. Seismostratigraphic analysis of the Wilkes Land Continental Margin (East Antarctica): influence of glacially driven processes on the Cenozoic deposition. *Deep-Sea Research, Part II*, **50**, 1563–1594, http://dx.doi.org/10.1016/S0967-0645(03)00079-1.

DI VINCENZO, G., TALARICO, F. M. & KLEINSCHMIDT, G. 2007. An ^{40}Ar–^{39}Ar investigation of the Mertz Glacier area (George V Land, Antarctica): implications for the Ross Orogen–East Antarctic Craton relationship and Gondwana reconstruction. *Precambrian Research*, **152**, 93–118.

DONDA, F., BRANCOLINI, G., DE SANTIS, L. & TRINCARDI, F. 2003. Seismic facies and sedimentary processes on the continental rise off Wilkes Land (East Antarctica): evidence of bottom current activity. *Deep-Sea Research, Part II*, **50**, 1509–1527, http://dx.doi.org/10.1016/S0967-0645(03)00075.

DUCLAUX, G., ROLLAND, Y. *ET AL.* 2008. Superimposed Neoarchaean and Paleoproterozoic tectonics in the Terre Adélie Craton (East Antarctica): evidence from Th–U–Pb ages on monazites and ^{40}Ar/^{39}Ar ages. *Precambrian Research*, **167**, 316–338.

EITTREIM, S. L. 1994. Transition from continental to oceanic crust on the Wilkes-Adelie margin of Antarctica. *Journal of Geophysical Research*, **99**, 24 189–24 205, http://dx.doi.org/10.1029/94JB01903.

EITTREIM, S. L. & SMITH, G. L. 1987. Seismic sequences and their distribution on the Wilkes Land margin. *In*: EITTREIM, S. L. & HAMPTON, M. A. (eds) *The Antarctic Continental Margin: Geology and Geophysics of Offshore Wilkes Land.* Circum-Pacific Council for Energy and Mineral Resources, Earth Science Series, **5A**, 15–43.

EITTREIM, S. L., COOPER, A. K. & WANNESSON, J. 1995. Seismic stratigraphic evidence of ice-sheet advances on the Wilkes Land margin of Antarctica. *Sedimentary*

Geology, **96**, 131–156, http://dx.doi.org/10.1016/0037-0738(94)00130-M.

ELLIOT, D. H., SIDERS, M. A. & HABAN, M. A. 1986. Jurassic tholeiites in the region of the upper Rennick Glacier, north Victoria Land. *In*: STUMP, E. (ed.) *Geological Investigations in Northern Victoria Land*. American Geophysical Union, Antarctic Research Series, **46**, 249–265.

ESCUTIA, C., EITTREIM, S. L. & COOPER, A. K. 1997. Cenozoic sedimentation on the Wilkes Land continental rise, Antarctica. *In*: RICCI, C. A. (eds) *The Antarctic Region: Geological Evolution and Processes. Proceedings of 7th International Symposium on Antarctic Earth Science*. Terra Antarctica, Tokyo, 791–795.

ESCUTIA, C., EITTREIM, S. L., COOPER, A. K. & NELSON, C. H. 2000. Morphology and acoustic character of the Antarctic Wilkes Land turbidite systems: ice-sheet-sourced v. river-sourced fans. *Journal of Sedimentary Research*, **70**, 84–93, http://dx.doi.org/10.1306/2DC40900-0E47-11D7-8643000102C1865D.

ESCUTIA, C., WARNKE, D., ACTON, G. D., BARCENA, A., BURCKLE, L., CANALS, M. & FRAZEE, C. S. 2003. Sediment distribution and sedimentary processes across the Antarctic Wilkes Land margin during the Quaternary. *Deep Sea Research, Part II*, **50**, 1481–1508, http://dx.doi.org/10.1016/S0967-0645(03)00073-0.

ESCUTIA, C., BRINKHUIS, H., KLAUS, A. & EXPEDITION 318 SCIENTISTS. 2011. *IODP Expedition 318: Cenozoic East Antractic ice Sheet evolution from Wilkes Land Margin Sediment. Proceedings of the Integrated Ocean Drilling Program*, **318**. Ocean Drilling Program, College Station, TX.

FIORETTI, A. M., BLACK, L. P., VARNE, R. & VISONA, D. 2001. U–Pb zircon SHRIMP dating of Palaeozoic magmatism in northern Victoria Land and the South Tasman Rise; inferences for the correlation with Tasmania. *In*: *The Geological Framework of Tasmania, Abstracts. Volume, 16*. Australian Geological Survey Organisation, Canberra and Mineral Resources Tasmania, Hobart.

FITZSIMONS, I. C. W. 2003. Proterozoic basement provinces of southwestern Australia,and their correlation with Antarctica. *In*: YOSHIDA, Y., WINDLEY, B. F. & DASGUPTA, S. (eds) *Proterozoic East Gondwana: Supercontinent Assembly and Breakup*. Geological Society, London, Special Publications, **206**, 93–130.

FLÖTTMANN, T. & KLEINSCHMIDT, G. 1991. Opposite thrust systems in northern Victoria Land, Antarctica: imprints of Gondwana's Paleozoic accretion. *Geology*, **19**, 45–47, http://dx.doi.org/10.1130/0091-7613(1991)019<0045:OTSINV>2.3.CO;2.

FODEN, J., ELBURG, M. A., DOUGHERTY-PAGE, J. & BURTT, A. 2006. The timing and duration of the Delamerian Orogeny: correlation with the Ross Orogen and implications for Gondwana assembly. *Journal of Geology*, **114**, 189–210.

GOODGE, J. W. & FANNING, C. M. 2010. Composition and age of the East Antarctic Shield in eastern Wilkes Land determined by proxy from Oligocene–Pleistocene glaciomarine sediment and Beacon Supergroup sandstones, Antarctica. *Geological Society of America Bulletin*, **122**, 1135–1159.

GOODGE, J. W., MYROW, P., WILLIAMS, I. S. & BOWRING, S. 2002. Age and provenance of the Beardmore Group, Antarctica: constraints on Rodinia supercontinent breakup. *Journal of Geology*, **110**, 393–406, doi: 10.1086/340629.

GOODGE, J. W., MYROW, P., PHILLIPS, D., FANNING, C. M. & WILLIAMS, I. S. 2004. Siliciclastic record of rapid denudation in response to convergent-margin orogenesis, Ross orogen, Antarctica. *In*: BERNET, M. & SPIEGEL, C. (eds) *Detrital Thermochronology–Provenance Analysis, Exhumation, and Landscape Evolution of Mountain Belts*. Geological Society of America, Special Papers, **378**, 101–122.

GREW, E. S. & SANDIFORD, M. 1984. A staurolite-talc assemblage in tourmaline- phlogopite chlorite schist from northern Victoria Land, Antarctica, and its petrogenetic significance. *Contributions to Mineralogy and Petrology*, **87**, 337–350.

HAMBREY, M. & BARRETT, P. J. 1993. Cenozoic sedimentary and climatic record, Ross Sea region, Antarctica. *In*: KENNETT, J. P. & WARNKE, D. A. (eds) *The Antarctic Environment: A Perspective on Global Change: American Geophysical Union*. Antarctic Research Series, **60**, 91–124.

HAMPTON, M. A., EITTREIM, S. L. & RICHMOND, B. M. 1987. Post-breakup sedimentation on the Wilkes Land margin, Antarctica. *In*: EITTREIM, S. L. & HAMPTON, M. A. (eds) *The Antarctic Continental Margin: Geology and Geophysics of Offshore Wilkes Land*. Circum-Pacific Council for Energy and Mineral Resources, Earth Science Series, **5A**, 75–87.

JACKSON, M. L. 1979. *Soil Chemical Analysis–Advanced Course*. 2nd edn. Published by the Author, Madison, Wisconsin.

LEITCHENKOV, G. L., BELYATSKY, B. V., RODIONOV, N. V. & SERGEEV, S. A. 2007. *Insight into the Geology of the East Antarctic Hinterland: A Study of Sediment Inclusions from Ice Cores of the Lake Vostok Borehole*. Short Research Paper, **014**, http://dx.doi.org/10.3133/of2007-1047.srp014. United States Geological Survey and United States National Academy; **USGS OF-2007-1047**.

LUDWIG, K. R. 2003. *Isoplot 3.00: A Geochronological Toolkit for Microsoft Excel Berkeley, California*. Berkeley Chronology Center, Special Publications, **4**.

MÉNOT, R. P., PÊCHER, A., ROLLAND, Y., PEUCAT, J. J., PELLETIER, A., DUCLAUX, G. & GUILLOT, S. 2005. Structural setting of the Neoarchean terrains in the Commonwealth Bay Area (143–145°E), Terre Adélie Craton, East Antarctica. *Gondwana Research*, **8**, 1–9.

MONTEL, J. M., FOREST, S., VESEHAMBRE, M., CHRISTAN, N. & PROVOST, A. 1996. Electron Microprobe dating of Monazite. *Chemical Geology*, **131**, 37–53.

O'BRIEN, P. E. & STAGG, H. M. J. 2007. Tectonic elements of the continental margin of East Antarctica, 38–164°E. *In*: COOPER, A. K., RAYMOND, C. R. ET AL. (eds) *Online Proceedings of the 10th ISAES X*. United States Geological Survey, Open-File Report, Short Research Paper, **085**, 2007–1047.

OLIVER, R. L. & FANNING, C. M. 2002. Proterozoic geology east and southeast of Commonwealth Bay, George V land, Antarctica, and its relationship to that of adjacent Gondwana terranes. *In*: GAMBLE, J. A., SKINNER, D. N. B. & HENRY, S. (eds) *Antarctic at the Close of a Millennium*. Royal Society of New Zealand Bulletin, **35**, 51–58.

Pant, N. C., Kundu, A., Joshi, S., Dey, A., Bhandari, A. & Joshi, A. 2009. Chemical dating of monazite: testing of an analytical protocol against independently dated standards. *Indian Journal of Geosciences*, **63**, 311–318.

Peucat, J. J., Ménot, R. P., Monnier, O. & Fanning, C. M. 1999. The Terre Adélie basement in the East Antarctica Shield: geological and isotopic evidence for a major 1.7 Ga thermal event; comparison with the Gawler Craton in South Australia. *Precambrian Research*, **94**, 205–224.

Ravich, M. G., Klimov, L. V. & Soloviev, D. S. 1968. *The Pre-Cambrian of East Antarctica*. Israel Program for Scientific Translations, Jerusalem.

Reading, A. M. 2004. The seismic structure of Wilkes Land/Terre Adelie, East Antarctica & comparison with Australia: first steps in reconstructing the Deep Lithosphere of Gondwana. *Gondwana Research*, **7**, 21–30.

Reed, S. J. B. & Buckley, A. 1996. Virtual WDS. *Mikrochimica Acta (Supplement)*, **13**, 479–483.

Roy, M., van de Flierdt, T., Hemming, S. R. & Goldstein, S. L. 2007. 40Ar/39Ar ages of hornblende grains and bulk Sm/Nd isotopes of circum-Antarctic glacio-marine sediments: implications for sediment provenance in the southern ocean. *Chemical Geology*, **244**, 507–519.

Sayers, J., Symonds, P. A., Direen, N. G. & Bernardel, G. 2001. Nature of the continent–ocean transition on the non-volcanic rifted margin of the central Great Australian Bight. *In*: Wilson, R. C. L., Whitmarsh, R. B., Taylor, B. & Froitzheim, N. (eds) *Non-Volcanic Rifting of Continental Margins: A Comparison of Evidence from Land and Sea*. Geological Society, London, Special Publications, **187**, 51–76, http://dx.doi.org/10.1144/GSL.SP.2001.187.01.04.

Schmidt, R. A. 1962. Australites and Antarctica. *Science*, **138**, 443–444, http://dx.doi.org/10.1126/science.138.3538.443.

Stuart, K. M. & Long, D. G. 2008. Analysis of Antarctic iceberg and sea-ice melting patterns using QUIKSCAT. *In*: *Proceedings of IGARSS 2008. International Geoscience and Remote Sensing Symposium: Geoscience and Remote Sensing, the Next Generation, 6–11 July 2008 Boston, MA, Volume 5*. IEEE, Piscataway, NJ, 192–195.

Tauxe, L., Stickley, C. *et al.* 2012. Chronostratigraphic framework for the IODP Expedition 318 cores from the Wilkes Land Margin: Constraints for paleoceanographic reconstruction. *Paleoceanography*, **27**, PA2214, http://dx.doi.org/10.1029/2012PA002308.

Veevers, J. J. 1987. The conjugate continental margins of Antarctica and Australia. *In*: Eittreim, S. L. & Hampton, M. A. (eds) *The Antarctic Continental Margin: Geology and Geophysics of Offshore Wilkes Land*. Circum-Pacific Council for Energy and Mineral Resources, Earth Science Series, **5A**, 45–73.

Veevers, J. J. & Saeed, A. 2011. Age and composition of Antarctic bedrock reflected by detrital zircons, erratics, and recycled microfossils in the Prydz Bay–Wilkes Land–Ross Sea–Marie Byrd Land sector (70°–240°E). *Gondwana Research*, **20**, 710–738.

Vetter, U. & Tessensohn, F. 1987. S- and I-type granitoids of north Victoria Land, Antarctica, and their inferred geotectonic setting. *Geologische Rundschau*, **76**, 233–243.

von Frese, R. R. 2006. Permian–Triassic mascon in Antarctica. *Eos, Transactions of the American Geophysical Union, Joint Assembly Supplement*, **87**(36), Abstract T41A–08

von Frese, R. R., Potts, L., Wells, S., Leftwich, T., Kim, H. *et al.* 2009. GRACE gravity evidence for an impact basin in Wilkes Land, Antarctica. *Geochemistry, Geophysics, and Geosystems*, **10**, 1–14, http://dx.doi.org/10.1029/2008GC002149.

Weaver, S. D., Bradshaw, J. D. & Laird, M. G. 1984. Geochemistry of Cambrian volcanics of the Bowers Supergroup and implications for the early Paleozoic tectonic evolution of northern Victoria Land, Antarctica. *Earth and Planetary Science Letters*, **68**, 128–140.

Weihaupt, J. G. 1976. The Wilkes Land anomaly: evidence for a possible hypervelocity impact crater. *Journal of Geophysical Research*, **81**, 5651–5663, http://dx.doi.org/10.1029/JB081i032p05651.

Williams, T., van de Flierdt, T., Hemming, S. R., Chung, E., Roy, M. & Goldstein, S. L. 2010. Evidence for iceberg armadas from East Antarctica in the Southern Ocean during the late Miocene and early Pliocene. *Earth and Planetary Science Letters*, **290**, 351–361.

Geomorphological evidence of cold-based glacier activity in South Victoria Land, Antarctica

C. B. ATKINS

School of Geography, Environment and Earth Sciences, Victoria University of Wellington, P.O. Box 600, Wellington, New Zealand (e-mail: cliff.atkins@vuw.ac.nz)

Abstract: Cold-based glaciers have long been recognized as capable of covering and protecting landscapes. However, recent studies of modern cold-based glaciers in Antarctica show that, in some situations, erosion, deformation and deposition can occur. Recognizing the dual ability of cold-based glaciers to protect and preserve surfaces on the one hand and erode and modify on the other is important for correctly interpreting the often-subtle imprint of cold-based glaciers on landscapes. A range of geomorphological features related to cold-based glacier activity has now been documented along with an improved understanding of cold-based glacier structure, processes and interaction with various substrates. Collectively, this provides an enhanced ability to understand the impact of cold-based glaciers on landscapes and reappraise the geomorphological record. Such insight allows recognition of previously unknown glacial events and better interpretations of the landscape exposure record. This is particularly important at the margins of the Antarctic Ice Sheets, where past fluctuations in ice sheet volume and its contributions to post glacial sea-level rise are poorly constrained. This paper reviews the known geomorphological evidence associated with cold-based glaciers in the South Victoria Land region of the Transantarctic Mountains. It aims to provide progress towards a set of criteria for recognizing cold-based glacier activity in other regions and to highlight the implications of cold-based glacial activity for surface exposure studies and interpreting glacial history.

Cold-based glaciers are below pressure melting point throughout and are therefore frozen to the substrate and move primarily by internal deformation (Drewry 1986). This led to the long-standing view that cold-based glaciers cannot slide and are thus incapable of erosion and deposition and therefore protect the underlying surface. While numerous studies have shown this to be true in many situations, recent investigations of modern cold-based glaciers have revealed that, in some circumstances, complex processes operate at subfreezing temperatures in the basal zone and that cold-based glaciers can generate a variety of geomorphological features. These studies have forced a reappraisal of the geomorphic effectiveness of cold-based glaciers and have confirmed that they can both protect and modify landscapes (e.g. Fitzsimons 1996; Atkins *et al.* 2002; Fitzsimons 2003; Waller *et al.* 2012). Thus, the debate has moved on and the challenge now is to understand the full range of processes and products associated with cold-based glaciers and the implications for landscape development (Hambrey & Fitzsimons 2010; Waller *et al.* 2012).

Recognizing the often-subtle features indicative of cold-based glacial activity offers the potential to delineate and date previously unknown or poorly constrained glacial events. This is particularly relevant for interpreting glacial history and modelling ice sheet reconstructions. For example, numerous surface exposure dating studies have been carried out on ice marginal landforms with the aim of constraining past variations in the ice sheet volume. However, many of these studies have produced inconsistent, contradictory results, including evidence of complex exposure histories suggesting episodes of burial and reworking by cold-based ice (e.g. Strasky *et al.* 2008; Di Nicola *et al.* 2009). Compared with warm-based glaciers, there are still relatively few studies of cold-based glaciers and significant gaps in our understanding of the processes and products remains. In addition, lingering misconceptions regarding the geomorphic effectiveness of cold-based glaciers and historic use of ambiguous terminology to describe cold-based sediments and landforms mean that the evidence and implications of past cold-based glacial events have yet to be fully appreciated by the wider geomorphological community. This review provides a summary of known geomorphological features associated with cold-based glaciers in the South Victoria Land region of Antarctica. It aims to provide a basis for recognizing similar features elsewhere and attempts to draw attention to the implications of cold-based glacial geomorphology for the correct application of surface exposure dating studies and interpretation of cold-based glacial history.

Cold-based glaciers

Evidence from many locations and on a variety of scales shows that cold-based ice can cover

From: Hambrey, M. J., Barker, P. F., Barrett, P. J., Bowman, V., Davies, B., Smellie, J. L. & Tranter, M. (eds) 2013. *Antarctic Palaeoenvironments and Earth-Surface Processes*. Geological Society, London, Special Publications, **381**, 299–318. First published online July 24, 2013, http://dx.doi.org/10.1144/SP381.18

landscapes and protect them from erosion. For example, there are instances where patterned ground, vegetated surfaces and alluvial fans have survived overriding by cold ice (Goldthwait 1960; Falconer 1966; England 1986). On a larger scale, many studies describe relict landscapes that have escaped erosion despite being in areas known to have been glaciated in the past. For example, landscape protection by cold-based glaciers has been documented in Scandinavia (Kleman 1994; Klemen & Hattestrand 1999; Fabel *et al.* 2002; Hattestrand & Stroven 2002), Scotland (Hall & Phillips 2006), the Canadian Arctic (Dyke 1993; Davis *et al.* 2006; Miller *et al.* 2006) and Antarctica (Naslund 1997; Sugden *et al.* 1991, 2005).

However, empirical evidence exists for subglacial processes remaining active at subfreezing temperatures. For example, Echelmeyer & Zhongxiang (1987) described sliding beneath a cold-based glacier in China and Cuffey *et al.* (1999, 2000) documented sliding and entrainment at −17 °C beneath the Meserve Glacier in Antarctica. Moreover, Fitzsimons (1996) and Fitzsimons *et al.* (1999, 2001) measured shearing in sediment-rich basal ice and ice-rich substrates beneath cold-based glaciers in Antarctica, documenting brittle and ductile deformation and the entrainment of blocks of frozen sand from the substrate. Others have also recognized the complexity of the basal zone of cold-based glaciers that rest on permafrozen ground, describing the deformation that occurs and the difficulty in recognizing the 'effective bed' of such glaciers (e.g. Waller 2001; Waller & Tuckwell 2005; Waller *et al.* 2009). Collectively, these studies showed that the subglacial conditions beneath cold-based glaciers are not as simple as previously thought and that they are in fact capable of limited geomorphic work.

Largely independent from the glaciologically focussed investigations of cold-based glaciers, geomorphologists working in the Dry Valleys region of Antarctica have long recognized small boulder-belt or 'drop' moraines and thin, patchy glacial deposits (often termed 'drifts') deposited by cold-based glaciers, overlying undisturbed desert pavements (e.g. Calkin 1971; Bockheim 1982, 2002; Chinn 1987; Brook *et al.* 1993; Marchant *et al.* 1993, 1994). However, most of these studies concentrated on utilizing the moraines as markers of former ice margins and did not focus on the depositional processes involved. Consequently, the observations did not appear to have much impact on the glaciological debate. The incongruity between the glaciological and geomorphological viewpoints was gradually reduced by several researchers who began linking observed processes operating beneath modern cold-based glaciers resting on unconsolidated, frozen substrates with the ice-contact moraines formed at the margins (e.g. Shaw 1977; Chinn 1989, 1991; Fitzsimons 1996; Humphries & Fitzsimons 1996; Fitzsimons *et al.* 1999, 2001; Fitzsimons 2003). Others reported a wide variety of erosion, deposition and deformation features on bedrock surfaces near the margins of cold-based glaciers (Atkins *et al.* 2002; Lloyd-Davies *et al.* 2009), offering simple conceptual models of cold-based glaciers overriding bedrock promontories, producing minor erosion and deposition in the absence of meltwater or basal sliding.

Further advances in understanding the structure, behaviour and entrainment mechanisms of debris-rich basal ice and frozen substrates were made (e.g. Fitzsimons *et al.* 2001, 2008), along with additional observations of the geomorphology at several modern cold-based glacier margins in Antarctica (Lloyd-Davies & Fitzsimons 2004). More recently, Hambrey & Fitzsimons (2010) used a facies-based approach to describe ice-contact sediments and deformed ice-cored substrate at the margins of cold-based glaciers resting on frozen substrates in the Dry Valleys. This study linked observed processes with the development of ice-marginal landforms, emphasizing the role of glaciotectonic thrusting and folding of frozen sediments. This has further reduced the gap between glaciological and geomorphological viewpoints and provided progress towards understanding sediment–landform associations and development of a 'landsystem' model for cold-based glaciers. Concomitantly, observations of erosion, deformation and deposition features from cold-based glaciers in other Quaternary glacial environments have emerged, for example, bedrock tors modified by cold-based glaciers in the Cairngorm Mountains of Scotland (Hall & Phillips 2006) and small boulder moraines deposited by cold-based glaciers at high elevations in Bolivia (Smith *et al.* 2009). Furthermore, large areas of deformed permafrost have been recognized in both contemporary and Quaternary glacial environments in the Northern Hemisphere, indicating more extensive distribution and influence of past cold-based glaciers (e.g. Waller & Tuckwell 2005; Waller *et al.* 2009; Waller *et al.* 2012).

Collectively, these studies show that cold-based glaciers can cover and protect landscapes but can also modify them in many situations. Although the scale of erosion and deposition is much less than for warm-based glaciers, cold-based glaciers are capable of producing discernible although often subtle traces of their passage over otherwise undisturbed landscapes. Recognizing these traces is essential for correct interpretation of glacial history and landscape development in regions where cold-based glaciers may have existed.

South Victoria Land and the Dry Valleys, Antarctica

In the South Victoria Land region of the Transantarctic Mountains, the East Antarctic Ice Sheet drains through the mountains to the Ross Ice Shelf and the Ross Sea via several outlet glaciers. Numerous nunataks and extensive ice-free glacial valleys occur along the edge of the ice sheet (e.g. McMurdo Dry Valleys). These valleys were probably carved by warm-based glaciers in the Miocene (Denton & Sugden 2005) and since then cold, polar conditions have persisted (e.g. Lewis *et al.* 2007). Mean annual temperatures range from −15 to −30 °C (Doran *et al.* 2002) and the region has a continual moisture deficit with only a small amount of snowfall of up to 10 mm water equivalent per year. Ablation occurs predominantly by sublimation with only 2–3% by melting (Chinn 1985). Temperature and precipitation decrease with increasing distance inland from the coast and with increasing altitude (Marchant & Head 2007). Meltwater from supraglacial and snow melting runs intermittently for a maximum of two months a year at lower elevations (<800 m a.s.l.). These cold, hyper-arid conditions mean that most glaciers in the region are cold-based and have remained so during the Quaternary (e.g. Staiger *et al.* 2006; Lewis *et al.* 2007).

Examples of geomorphological features associated with various types of cold-based glacier presented in this paper come from four main locations in South Victoria Land (Fig. 1). The northernmost site is Allan Hills nunatak approximately 80 km north of the McMurdo Dry Valleys (76°43′S 159°40′E), where a cold-based glacier fed directly by the East Antarctic Ice sheet encroaches onto sedimentary bedrock at high elevation (1700 m a.s.l.). At the northern boundary of the Dry Valleys, Victoria Valley (77°21′S 162°12′E) is a large east–west-trending glacial trough that has small cold-based glaciers descending from ice-fields down the valley walls and also the broad cold-based Victoria Lower Glacier, occupying the

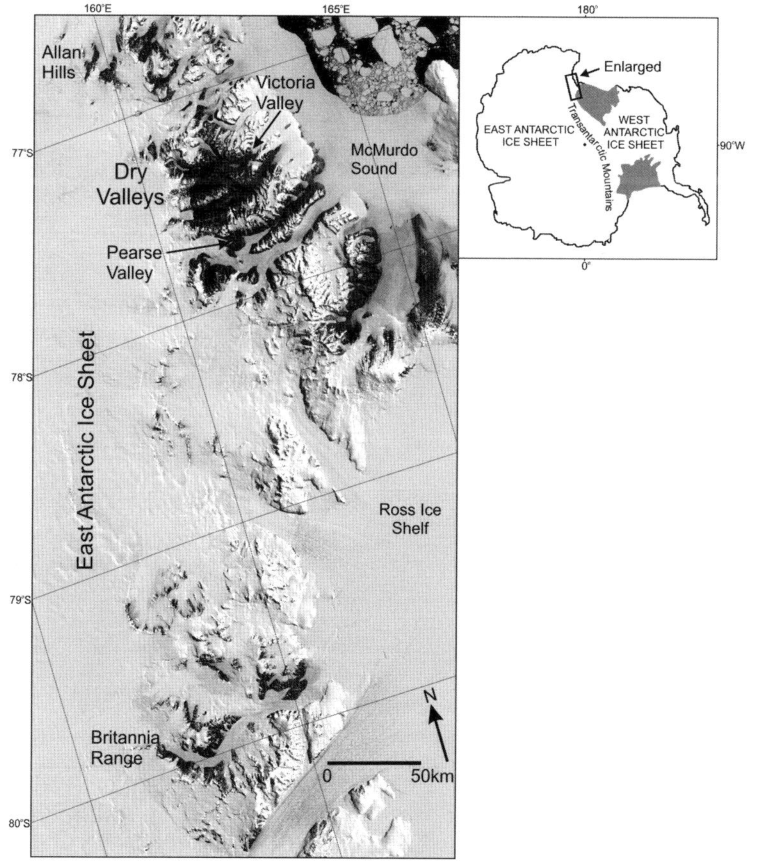

Fig. 1. Satellite image of South Victoria Land region showing the main field locations featured in the text.

lower valley, terminating at about 350 m a.s.l. At the southern limit of the Dry Valleys, Pearce Valley (77°43′S 161°38′E) displays several small cold-based glaciers on the valley walls and a cold-based lobe of the Taylor Glacier occupies the lower valley mouth, terminating at about 400 m a.s.l. Four-hundred kilometres south of the Dry Valleys, in the Britannia Range (80°00′S 155°35′E), cold-based lobes of the Hatherton Glacier extend into the lower reaches of relict glacial valleys about 1300 m a.s.l.

Types of cold-based glacier in South Victoria Land, Antarctica

Although the majority of glaciers in South Victoria Land are cold-based, there is significant variation in the glacier size, catchment area, elevation, debris load, meltwater and substrate conditions. These variables exert a strong control on the characteristics of each glacier, the processes operating and the geomorphological features that occur. Understanding these factors is important for recognizing and interpreting the geomorphological record of cold-based glacial events. Cold-based glaciers in South Victoria Land are here divided into three main types following the classification of Chinn (1991) and Humphries & Fitzsimons (1996).

Type 1: alpine glaciers

Alpine glaciers descend from small cirque basins and ice-fields into ice-free valleys, terminating at altitudes ranging from sea-level to about 900 m a.s.l. The glaciers are commonly less than 5 km long, and narrow (usually a few hundred metres). The terminal faces are typically vertical and up to 20 m high with characteristic ice and debris aprons formed at the base (Fig. 2a). Glaciers that terminate on the valley walls usually rest on bedrock or consolidated glacial deposits (drift) and have very low, mostly supraglacial debris loads and only minor moraines at the margins. Glaciers that reach on the valley floors rest on pre-existing glacial deposits and frozen fluvial or lacustrine sediments and usually have higher basal debris loads, larger debris aprons and ice-cored moraines (Humphries & Fitzsimons 1996). Some glaciers terminating at lower elevations produce supraglacial meltwater during the summer months, creating distinctive meltwater channels along the margins and draining downslope from the glacier.

Type 2: piedmont glaciers

In the Dry Valleys area of South Victoria Land, a broad coastal piedmont glacier has several lobes

Fig. 2. Examples of different types of cold-based glacier in the South Victoria Land region. (**a**) Alpine glacier, Schlatter Glacier, Pearse Valley. Ice cliff is *c.* 15 m high. (**b**) Piedmont glacier, Victoria Lower Glacier, Victoria Valley. Glacier is *c.* 3.5 km wide in mid field of view. (**c**) Cold-based lobe of outlet glacier. Peripheral lobe of Hatherton Glacier in Dubris Valley, Britannia Range. Frozen lake at glacier margin is *c.* 600 m long.

that flow 'inland', terminating at low elevations (<450 m a.s.l.) in the lower reaches of Victoria and Wright valley floors. The distance from piedmont crest to the glacier fronts is 15–20 km. The glacier lobes exhibit broad ice fronts several kilometres wide with terminal faces ranging from vertical (up to 20 m high) to subdued ramps (Fig. 2b). Debris load is variable but usually includes debris layers in the basal zone. Complex ice-cored landforms occur at the margins, developed on a variety of substrates including pre-existing glacial deposits, unconsolidated or frozen lacustrine and fluvial deposits. Supraglacial meltwater channels occur around the margins of the glaciers along with small proglacial lakes and ponds during summer.

Type 3: ice sheet and outlet glaciers

Thin, cold-based peripheral lobes connected directly to the East Antarctic Ice Sheet or to outlet glaciers draining the ice sheet rest on the margins of nunataks or extend into the mouths of relict glacial valleys. The outlet glaciers themselves are polythermal as they have patches of warm-based ice along their trunks owing to pressure melting (e.g. Robinson 1984; Hubbard *et al.* 2004; Golledge & Levy 2011). However, the warm-based ice only occurs in restricted areas where ice thickness exceeds approximately 1000 m (Kavanaugh & Cuffey 2009). Importantly, numerical modelling of the basal thermal regime using mean annual temperatures, glacial profiles and glacier thermodynamics shows that the peripheral lobes are entirely cold-based and must have remained cold-based throughout the Quaternary (Staiger *et al.* 2006; Lewis *et al.* 2007; Marchant *et al.* 1994). The glacier lobes usually terminate at medium to high elevations (700–2000 m a.s.l.) and display fronts ranging from vertical ice cliffs (up to 20 m high) with ice and debris aprons to subdued, actively sublimating ramp margins (Fig. 2c). These cold-based glaciers typically rest on bedrock and consolidated pre-existing drift substrates, and meltwater and channel features are rare. Debris load is highly variable and usually includes basal debris layers and some supraglacial debris. The large ice volume feeding the lobes means that they are likely to advance and retreat over a much wider area than the smaller alpine cold-based glaciers and patches of warm-based ice near the trunks of the outlet glaciers erode and entrain debris that is transported englacially to the glacier margin, where it is released (e.g. Evans 1989*b*). The higher debris load sometimes results in a greater deposition and thus more obvious and persistent geomorphological record at the margins (e.g. Swanger *et al.* 2011). Although the outlet glaciers are polythermal, and some debris is derived subglacially, the extensive lobes described here are entirely cold-based. Therefore, the geomorphological record at the glacier margins reflects cold-based glacier processes (e.g. Marchant *et al.* 1994; Staiger *et al.* 2006).

Geomorphological evidence of cold-based glaciers

The geomorphological evidence of cold-based glaciers presented here is derived from all three types of cold-based glacier in the South Victoria Land region. Some features are common to more than one type of cold-based glacier, whereas others appear to be specific to a particular cold-based glacier type or physiographic setting. The features are here divided into: (1) erosion, (2) deposition, (3) deformation and (4) protection categories. Table 1 displays known geomorphological features related to cold-based glacier type and substrate, together with references to selected published examples.

Erosion features

Broad scrapes and grooves. Abrasion marks consisting of fresh, unweathered broad scrapes and grooves occur at Allan Hills, where the East Antarctic Ice Sheet-fed Manhaul Bay Glacier encroaches onto horizontally stratified sedimentary bedrock and pre-existing glacial deposits (Sirius Till; Atkins *et al.* 2002). The abrasion marks occur sporadically around the glacier margin adjacent to undisturbed weathered rock surfaces. Individual scrapes are variable in shape and size, ranging up to centimetres deep and tens of centimetres wide, although they can extend for over one metre (Fig. 3a). Remnants of the abrading tool are often present and centimetre-scale levees exist around the margins of some abrasions, indicating a complete lack of meltwater. Atkins *et al.* (2002) developed a simple conceptual model of cold-based glacier overriding bedrock promontories. This showed how blocks of plucked bedrock or surface boulders projecting into the overriding and deforming cold-based ice are rotated forward and dragged over the adjacent surface forming crude abrasion marks, as originally envisaged by Drewry (1986). The process of block removal and entrainment is in itself significant but poorly understood. It is probably related to failure along planes of weakness in the sedimentary bedrock, resulting in 'plucking' of bedrock blocks during glacial overriding (Boulton 1974).

Boulder-top abrasion. Irregular scuffs, scrapes and scratches occur on surfaces of exposed boulders embedded in pre-existing glacial deposits at the margins of all three types of cold-based glacier (e.g. Atkins *et al.* 2002; Bockheim 2010). They are particularly common on boulders projecting out of compressed desert pavements (see section on preservation features) around cold-based lobes of outlet glaciers. The abrasion marks consistently occur on the stoss side and highest parts of the

Table 1. *Geomorphological features related to cold-based glacier type and substrate in South Victoria Land, Antarctica with selected published examples*

Geomorphological feature		Cold-based glacier type, substrate and published examples			
		Alpine (*sediment substrate*)	Piedmont (*sediment substrate*)	Ice sheet or outlet glacier lobe	
				(*sediment substrate*)	(*bedrock*)
Erosional	Broad scrapes and grooves	–	–	Allan Hills (Atkins *et al.* 2002)	Allan Hills (Fig. 3a; Atkins *et al.* 2002; Lloyd-Davies *et al.* 2009)
	Boulder top abrasion	Numerous locations, e.g. Meserve Glacier (Holdsworth 1974), Packard Glacier (Atkins & Dickinson 2007)	Victoria Lower Glacier (Atkins & Dickinson 2007)	Britannia Range (Fig. 3b; this paper), Allan Hills (Atkins *et al.* 2002)	–
	Meltwater channels	Numerous locations, e.g. Packard Glacier (Fig. 3c; Atkins & Dickinson 2007), Stocking Glacier (Swanger *et al.* 2010)	Victoria Lower Glacier (Atkins & Dickinson 2007), Wright Lower Glacier (Hambrey & Fitzsimons 2010)	Taylor Glacier lobe in Pearse Valley (Dickinson & Hopkins 2005)	–
Depositional	Thin, patchy drift	Numerous locations, e.g. Bockheim (2002, 2010)	Victoria Valley (Calkin 1971; Chinn 1987) Wright Valley (Hall *et al.* 1993)	Numerous locations, e.g. Britannia Range (Fig. 4a; Bockheim *et al.* 1989, this paper), Dry Valleys area, e.g. (Brook *et al.* 1993; Marchant *et al.* 1993, 1994; Staiger *et al.* 2006; Lewis *et al.* 2007; Swanger *et al.* 2011)	–
	Boulder-belt/drop moraines	Numerous locations, e.g. (Bockheim 1982, 2002; Humphries & Fitzsimons 1996; Swanger *et al.* 2010)	Victoria Valley (Calkin 1971, Chinn 1987), Wright Valley (Hall *et al.* 1993)	Numerous locations, e.g. Britannia Range (Fig. 4b). References as above for 'thin, patchy drift'	–
	Boulder trains and scattered boulders	Numerous locations, e.g. (Bockheim 2002, 2010)	–	Numerous locations, e.g. (Bockheim 2010), Allan Hills (Fig. 4c; Atkins *et al.* 2002; Lloyd-Davies *et al.* 2009) and Britannia Range (this paper)	Allan Hills (Atkins *et al.* 2002; Lloyd-Davies *et al.* 2009)
	Till patches	–	–	Allan Hills (Atkins *et al.* 2002; Lloyd-Davies *et al.* 2009), Britannia Range (Fig. 4d; this paper)	Allan Hills (Atkins *et al.* 2002; Lloyd-Davies *et al.* 2009)

	Ice-cored ridges/ moraines	Numerous locations, e.g. Suess Glacier (Fitzsimons *et al.* 1999), Packard Glacier (Atkins & Dickinson 2007), Stocking Glacier (Swanger *et al.* 2010)	Victoria Lower Glacier (Fig. 4e; Chinn 1987; Atkins & Dickinson 2007), Wright Lower Glacier (Fitzsimons 1996; Hambrey & Fitzsimons 2010)	Britannia Range (this paper; Bockheim *et al.* 1989); Dry Valley (Swanger *et al.* 2011)	–
	Ice cored debris cones	–	–	Allan Hills (Atkins *et al.* 2002, Lloyd-Davies *et al.* 2009), Britannia Range (Fig. 4f; this paper)	Allan Hills (Atkins *et al.* 2002; Lloyd-Davies *et al.* 2009)
	Buried ice	Numerous locations, e.g. Taylor Valley (Swanger *et al.* 2010), Pearse Valley (Fig. 4g; Dickinson & Hopkins 2005)	Victoria Lower Glacier (Chinn 1991), Wright Lower Glacier (Chinn 1991; Hambrey & Fitzsimons 2010)	Beacon Valley (Sugden *et al.* 2005), Pearse Valley (Dickinson & Hopkins 2005)	–
	Glacotectonized bedrock	–	–	–	Allan Hills (Fig. 5a; Atkins *et al.* 2002; Lloyd-Davies *et al.* 2009)
Deformation	Ploughed boulders and overturned boulders	Packard Glacier (Atkins & Dickinson 2007)	Victoria Lower Glacier (Atkins & Dickinson 2007)	Allan Hills (Lloyd-Davies *et al.* 2009), Britannia Range (Fig. 5b; this paper)	–
	Deformed ice-cored substrate/ permafrost	Suess Glacier (Fitzsimons *et al.* 1999)	Victoria Valley (Fig. 5c; Chinn 1987, 1991; Atkins & Dickinson 2007), Wright Lower Glacier (Hambrey & Fitzsimons 2010)	–	–
Protection	Compressed, flattened ground	–	–	Britannia Range (Fig. 6a), Pearse Valley (Atkins *et al.* 2011)	–
	Buried soils/ preserved pavements	Numerous locations (Bockheim 2002, 2010)	Dry Valleys (Bockheim 2002, 2010)	Numerous locations, e.g. Britannia Range (Fig. 6b; this paper; Bockheim *et al.* 1989; Marchant *et al.* 1994; Bockheim 2010)	–

Fig. 3. Examples of erosion features related to cold-based glaciers. (**a**) Broad scrapes and grooves on sandstone bedrock at Allan Hills. The abrasions occur sporadically around the glacier margin adjacent to unmodified surfaces. Ice movement from top to bottom. (**b**) Fresh scuffs and scrapes on the stoss side of a weathered boulder embedded in pre-existing glacial deposits near the margin of Hatherton Glacier, Britannia Range. Ice movement from right to left. (**c**). Curved, nested, supraglacial meltwater channels eroded into pre-existing drift near margin of Packard Glacier, Victoria Valley. The interchannel area comprises either unmodified ground or is covered by an ice-cored moraine. The vertical ice cliff is *c.* 15 m high.

boulders and range from conspicuous but superficial etchings into the 'desert varnish' on boulders, to centimetre-deep scrapes, tens of centimetres long and wide (Fig. 3b). Often the desert pavement surfaces around the boulders appear completely untouched. Atkins *et al.* (2002) suggested that these marks form by the dragging of entrained debris over the upper surface of boulders that protruded up into the overriding and deforming cold-based glacier. This is consistent with observations of 'rock smears' on the top surfaces of boulders embedded in the substrate beneath the Meserve Glacier (Holdsworth 1974).

Meltwater channels. Meltwater channels occur at the margins of all three types of cold-based glaciers in South Victoria Land, but only on sediment substrates and at low elevations (<800 m a.s.l.), where small volumes of supraglacial meltwater are produced during summer months. Atkins & Dickinson (2007) described meltwater channels at the margins of the Packard Glacier (alpine cold-based glacier) and Victoria Lower Glacier (piedmont glacier) in Victoria Valley. At these locations, the hillside topography slopes towards the glacier margin and nested channels up to 3 m deep and 20 m wide occur eroded into pre-existing glacial drift (Fig. 3c). The channels show evidence of fluvial erosion with undercut banks and overturned boulders. Separating the channels are areas of apparently unaltered ground or discontinuous ice-cored ridges up to 2 m high. A simple model proposed by Atkins & Dickinson (2007) describes how supraglacial meltwater is concentrated at the margins of the ice-apron eroding a channel into the hillside. When the glacier retreats, the channel is abandoned and unaltered ground previously protected beneath the glacier or ice-apron is exposed. The discontinuous ice-cored ridges between the channels may represent the remnants of the ice-apron stranded during retreat. Similar lateral channels have been described from the margins of cold, polythermal and, rarely, warm glaciers in the Arctic (Dyke 1993; O'Cofaigh *et al.* 1999; Syverson & Mickleson 2009). However, in the cold, hyper-arid setting of South Victoria Land, these meltwater channels constitute one of the most obvious features of landscape modification associated with cold-based glaciers and highlight the ability of cold-based glaciers to both erode and protect surfaces in close proximity.

Depositional features

Thin, patchy drift. Widespread but thin and irregular deposits of bouldery debris overlying older undisturbed surfaces occur near the margins of all three types of cold-based glacier. A variety of terms, such as 'ground moraine', 'drift sheets' and 'till', have been widely used to describe these deposits. Unfortunately, some of these terms are either outdated or ambiguous, or have been widely used to describe older warm-based glacial deposits as well. Some researchers have used the term 'alpine drift' (e.g. Chinn 1987) to distinguish deposits formed by cold-based glaciers, but this too is problematical as the deposits are not restricted to alpine cold-based glaciers. The descriptive term 'thin, patchy drift' is used here. The deposits are most commonly associated with cold-based lobes of outlet glaciers and are often little more than a patchy, discontinuous scattering of boulders and cobbles superimposed on an older surface (e.g. Hall *et al.* 1993; Marchant *et al.* 1994; Bockheim 2002, 2010; Staiger *et al.* 2006; Lewis *et al.* 2007). However, 'thin, patchy drift' can be up to several metres thick in places, comprising an unconsolidated mix of clasts with minor sand and gravel but with a notable absence of mud-rich matrix. Where the deposits fully cover the underlying surface, buried soils and desert pavements occur (see section on buried soils). Clasts have a conspicuous lack of glacial striations, polish and moulding, and many are 'desert-varnished' clasts reworked from the underlying surface (Marchant *et al.* 1994; Bockheim 2002, 2010). Clear examples of this type of deposition occur in the Britannia Range where various aged 'drifts' are delineated only by subtle colour change, reflecting the deposition of small amounts of fresh, unweathered, light-coloured sandstone debris superimposed over older, dark coloured, weathered drift surfaces (Bockheim *et al.* 1989; Fig. 4a). Although these deposits are widely acknowledged as having been deposited by cold-based glaciers, the processes responsible for the deposition of thin, patchy drift are poorly understood. However, the low volume of debris, inclusion of reworked clasts and lack of deformation of the underlying surface suggest a relatively gentle process of entrainment and deposition. A model originally proposed by Shaw (1977) and modified by Evans (1989*b*) and Chinn (1991) suggests that cold-based glaciers can advance by overriding their frontal aprons, partly entraining ice and supraglacial debris into the basal zone. The debris is then re-deposited at the glacier margin. However, the high percentage of reworked clasts in the drift combined with observations of clasts with well-developed tafoni embedded in the basal ice of modern cold-based glaciers indicate that some clasts must be entrained from the underlying older deposits (e.g. Bockheim 2010). Furthermore, Fitzsimons *et al.* (2008) showed that, where glaciers override unconsolidated or frozen substrates, significant subglacial deformation and entrainment occurs in addition to the apron entrainment.

Boulder-belt/drop moraines. Boulder-belt or drop moraines deposited by cold-based glaciers have been described from numerous locations in the South Victoria Land region (e.g. Calkin 1971; Bockheim 1982, 2002; Bockheim *et al.* 1989; Brook *et al.* 1993; Hall *et al.* 1993; Marchant *et al.* 1993, 1994; Staiger *et al.* 2006; Todd *et al.* 2010; Swanger *et al.* 2011). These moraines are closely associated with 'thin, patchy drift' and occur at margins of all types of cold-based glacier, but are most commonly associated with cold-based lobes of outlet glaciers. The moraines are usually small ridges only a few metres high but can extend for several kilometres (Fig. 4b). They comprise mostly loose collections of clast-supported boulders and cobbles with some gravel and sand matrix. Like the 'thin, patchy drift', clasts lack striations, polish or moulding and often contain a high volume of weathered, reworked clasts and overlie minimally disturbed older drifts and soils (Marchant *et al.* 1994; Bockheim 2002, 2010; Stagier *et al.* 2006). These similarities suggest that the boulder-belt moraines form by the same process of passive, slow release or 'drop' of debris, forming a ridge wherever the glacier remains stationary long enough. Despite the frequent occurrence of these moraines and the widespread use as markers of past glacier limits, the processes responsible for their formation are still poorly known.

Boulder trains and scattered isolated boulders. Atkins *et al.* (2002) and Lloyd-Davies *et al.* (2009) documented boulder trains and scattered isolated boulders at Allan Hills. Trains consisting of cobbles to boulders up to 3 m in diameter extend tens to hundreds of metres on the lee side of glaciotectonized bedrock promontories. These are most clearly visible where freshly broken, light-coloured sandstone boulders derived from the bedrock overlie over weathered desert pavements (Fig. 4c). Lloyd-Davies *et al.* (2009) presented a model describing the formation of boulder trains whereby advancing cold-based ice overrides bedrock promontories, forming a pressure shadow on the lee side, allowing glaciotectonically quarried and entrained debris to accumulate in elongate trains parallel to the direction of flow. Isolated boulders not associated with boulder trains occur scattered over bedrock and older glacial deposits (Sirius Till) at Allan Hills (Atkins *et al.* 2002). The boulders are composed of the local sandstone bedrock but typically have a fresh, unweathered appearance relative to the underlying surface. These are likely to have been 'plucked' from the bedrock. In some cases, there is limited deformation of the sedimentary bedrock underneath the boulders, indicating some loading by overriding ice. Other examples show no abrasion or deformation, suggesting passive lowering

(a)
(b)
(c)
(d)
(e)
(f)
(g)

and deposition of entrained boulders from the ice onto the landscape during cold-based glacier retreat via sublimation (Atkins *et al.* 2002). Occurrences of isolated boulders deposited by cold-based glaciers over minimally disturbed landscapes have been documented in other areas of South Victoria Land (e.g. Bockheim 2010).

Till patches. Localized deposits of unconsolidated diamicton occur around the cold-based margins of ice sheet and outlet glaciers. The deposits are typically 'smeared' onto the stoss side of bedrock promontories or boulders. At Allan Hills, the deposits are thin (typically only tens of centimetres thick) with a very patchy distribution covering areas of only a few metres. The deposits consist of variably brecciated and pulverized sandstone from the underlying substrate and are poorly sorted, unlithified and structureless, but contain sandstone clasts. Thin-section analysis of the deposits shows they are dominated by planar brittle deformation structures (Lloyd-Davies 2004). Lloyd-Davies *et al.* (2009) reasoned that, because the deposits were deposited subglacially by the cold-based glacier, they actually constituted till, using the classification of Dreimanis (1988). Furthermore, they proposed a conceptual model describing how cold-based ice could encroach onto the stoss side of bedrock promontories, creating compressive stress, thereby glaciotectonically crushing the bedrock though simple dry shearing, smearing it onto the adjacent outcrop. Similar deposits occur on the stoss side of boulders at the margin of the Hatherton Glacier in the Britannia Range (Fig. 4d). In this setting, the deposits comprise unconsolidated sand matrix with pebble to cobble sized clasts. Many of the clasts show the same degree of weathering and desert varnish as clasts from the underlying surface.

Ice-cored ridges/moraines. Ice-cored moraines and ridges occur at the margins of all three types of cold-based glacier, particularly where they sit on unconsolidated frozen substrates (e.g. Shaw 1977; Chinn 1987; Fitzsimons 2003). The moraines include a variety of ice-cored ridges including debris-covered ice-aprons in contact with the glacier, moraines parallel to the glacier in the proglacial areas and larger 'thrust-block' moraines up to 20 m high. Considerable research has been carried out on ice-cored moraines and it is now clear that there are many complex processes operating, ranging from simple release of basal debris over the ice-apron to glaciotectonic faulting and folding of the subglacial substrate. For example, tunnels excavated beneath modern cold-based glaciers that rest on thick unconsolidated and frozen sediment (Fitzsimons 1996; Fitzsimons *et al.* 1999, 2001) showed that shear stress propagates down into the substrate causing deformation and failure along planes of weakness, leading to thrusting, detachment and entrainment of intact fragments of the substrate. These 'blocks' of sediment are then transported to the glacier front and stacked into debris-rich moraines in addition to 'thrust-blocks' of frozen proglacial sediments. This shows that the 'apron entrainment' model of Shaw (1977) is inadequate for cold-based glaciers that rest on deformable substrates (Fitzsimons *et al.* 2008). Furthermore, Hambrey & Fitzsimons (2010) emphasized the importance of glaciotectonic folding and thrusting of basal ice and frozen sedimentary substrates at the Wright Lower Glacier and highlighted the complex distribution of reworked glacial, fluvial and lacustrine sediments associated with the ice-cored moraines. Similar complex ice-cored moraines occur at the front of the Victoria Lower Glacier (Fig. 4e).

Fig. 4. Examples of deposition features related to cold-based glaciers. (**a**) Thin, patchy drift comprising boulders and cobbles superimposed over older but undeformed deposits. The thin, patchy drift contains both fresh and weathered (reworked) clasts that lack striae, polish or moulding. Note the boulder-belt moraine and older, darker drift surface beyond at top of image, Britannia Range. Light-coloured boulder in centre foreground is 50 cm diameter. (**b**) Boulder-belt/drop moraine showing stacked, clast-supported boulders and cobbles and lack of fine-grained matrix. Clasts lack striae, moulding and polish. The moraine contains a mix of fresh clasts and weathered, desert-varnished clasts recycled from older deposits. The moraine sits on a weathered, undisturbed, pre-existing drift, visible in the top left of image, Britannia Range. Note person on moraine crest for scale. (**c**) Boulder train composed of light-coloured, unweathered sandstone resting on dark, weathered, but undisturbed surface of older Sirius Till, Allan Hills. Ice movement towards foreground. Large boulder embedded in the Sirius Till has been ploughed into the till creating a small dilated ridge. (**d**) Till patches with reworked ventifcated clasts mixed with sand matrix smeared onto the stoss side of a weathered boulder embedded in pre-existing drift, Britannia Range. Note also the fresh scuff abrasions on the boulder. Ice movement from bottom to top. (**e**) Ice-cored moraines at the ramped margin of the Victoria Lower Glacier showing complex cover of fluvial, aeolian and glacial sediment and debris-rich basal ice. Central moraine is approximately 3 m high. (**f**) Ice-cored debris cones at the margin of the Hatherton Glacier, Britannia Range. Cones range up to 8 m high and debris cover is commonly <30 cm thick. Cones become increasing deflated with less ice away from the glacier margin. Note the englacial boulder disintegrating onto the ice apron showing how debris covered ice forms at the glacier margin. (**g**) Debris-covered buried ice or 'gelifluction' lobes on north wall of Pearce Valley. Bodies of buried ice are widespread in the region and may be relict cold-based glacier ice. Field of view at lake level is *c.* 200 m.

Ice-cored debris cones. Ice-cored debris cones occur rarely near the cold-based margins of ice sheets and outlet glaciers (Campbell & Claridge 1975; Atkins *et al.* 2002). Lloyd-Davies *et al.* (2009) described examples resting on both bedrock and consolidated glacial deposits (Sirius Till) at Allan Hills. The cones are up to 8 m high and 12 m diameter at the base and often occur in lines associated with boulder trains. The cones are typically covered in loose debris ranging from boulders to sand, forming a layer at least 20 cm thick. Similar features occur at the cold-based margin of the Hatherton Glacier in the Britannia Range (Fig. 4f). The cones appear to form as large pockets of englacial debris are released from the glacier front, forming scree cover over parts of the apron ice, protecting it from ablation. These eventually form cone shaped, debris-covered mounds stranded in the proglacial area as the glacier margin retreats. This process is similar to the formation of debris covered aprons and is consistent with the interpretation of similar ice-cored mounds described from many locations in the Arctic (Evans 1989*b*). In the Britannia Range, the cones sometimes occur in lines tens of metres long, perpendicular to the glacier margin, suggesting on-going but intermittent release of englacial sediment from the same point source as the glacier retreats. The cones persist for several kilometres inland from the glacier margin, but become progressively smaller and more deflated, eventually comprising a circular pile of debris with no ice core.

Buried ice. Buried ice is widespread in the South Victoria Land region and associated with all three types of cold-based glacier. Bodies of ice buried beneath a cover of debris have been variously described as ground ice, gelifluction lobes, rock glaciers, debris covered glaciers and viscous flow lobes (e.g. Chinn 1991; Sugden *et al.* 2002; Dickinson & Hopkins 2005; Swanger *et al.* 2010). The origin and age of the ice is poorly known but in some cases has been interpreted as relict cold-based glacier ice protected from sublimation by the debris cover. For example, Swanger *et al.* (2010) described ice-cored viscous flow lobes up to 2 km long and 30 m thick on the sides of central Taylor Valley. Using shallow seismic surveys, displacement surveys and isotopic analyses of the buried ice, they concluded that the lobes represent relict glacier ice derived from former advances of local alpine cold-based glaciers (Fig. 4g). Others have also shown that relict buried ice can extend well beyond the edge of modern cold-based glaciers concealing the former extent of cold-based glacial advances (Fitzsimons 2003, 2011; Hambrey & Fitzsimons 2010). Some buried ice in the Dry Valley area has been interpreted to have existed for millions of years beneath a protective cover of debris (e.g. Sugden *et al.* 2002).

Deformation features

Glaciotectonized bedrock. Atkins *et al.* (2002) recognized brittle deformation on sandstone bedrock at the margin of a cold-based glacier at Allan Hills. The deformation consists of variably fractured and brecciated sandstone found predominantly of the lee side of bedrock promontories and ridges (Fig. 5a). Subsequently, Lloyd-Davies *et al.* (2009) described different scales of glaciotectonic deformation ranging from $<30\ m^2$ to over $100\ m^2$. The deformed rubble contained various sized blocks, some over 3 m in diameter. In some situations, especially where there was a marked topographic step down on the lee side of the ridge, the rubble formed boulder trains tapering away from the ridge. They offered a conceptual model showing how overriding cold-based ice could induce tensional fractures on the top of the ridges, increasing in scale and density towards the lee end until complete brecciation, quarrying and block displacement dominated.

Ploughed and overturned boulders. Boulders ploughed into drift deposits occur at Allan Hills and at the cold-based margin of the Hatherton Glacier in the Britannia Range. In each case, the weathered boulders are embedded in pre-existing drift, but have been ‘ploughed’ further into the enclosing matrix, causing small-scale deformation with dilated ridges of displaced matrix tens of centimetres high on the lee side. The boulders often show scuffs and scratches on the stoss and top sides and are occasionally associated with boulder trains. At Allan Hills, boulders up to 3 m in length appear to have been moved by tens of centimetres into the drift in the direction of cold-based glacier movement (Fig. 4c). Although there has been very little investigation of the processes, Lloyd-Davies *et al.* (2009) suggested that this could occur where the overriding and deforming ice exerts compressive stress onto the stoss side of boulder that projects up into the basal ice. This acts to push the boulder forward further into the substrate. Overturned boulders also occur at the margins of ice sheet or outlet glaciers that rest on pre-existing drifts. Examples have been observed at Allan Hills (Atkins *et al.* 2002), the margin of the Taylor Glacier in Pearse Valley and the margin of the Hatherton Glacier in the Britannia Range. In each case, weathered boulders are found with the desert-varnished side down and unweathered side facing up with the soil burial line exposed (Fig. 5b). The overturning could occur by the same overriding process that produces ploughed boulders, except in

Fig. 5. Examples of deformation features related to cold-based glaciers. (**a**) Small-scale glaciotectonic brecciation on the top and lee side of a sandstone bedrock promontory, Allan Hills. Overriding cold-based ice causes fractures in the bedrock leading to complete brecciation and quarrying. Cold-based glacier in background and ice movement from top left to bottom right. (**b**) Overturned dolerite boulder amongst scattered clasts on pre-existing drift surface. The desert varnished surface is facing down and unweathered (previously buried) side up. Ice movement from bottom right to top left. Geological hammer is 33 cm long. (**c**) Intensely deformed debris-rich buried ice exposed in a stream bank in the proglacial area of Victoria Lower Glacier. Note the discrete blocks of frozen sand with preserved sedimentary structures. Geological hammer is 33 cm long.

this case the forward momentum acts on boulders not fully embedded in the substrate, thereby rolling them over. Alternatively, boulders could be entrained by the cold-based glacier and subsequently lowered onto the surface during glacier retreat in the same manner as envisaged for the deposition of 'scattered, isolated boulders'.

Deformed ice-cored substrate/permafrost. Extensive bodies of buried ice and permafrost near the margins of cold-based glaciers in South Victoria Land commonly show deformation structures that have been interpreted as the result of overriding by cold-based glaciers (e.g. Chinn 1991). As described in the section on ice-cored moraines, several studies have documented the propagation of shear stress down into the substrate, causing subglacial deformation and entrainment of permafrozen sediment (Fitzsimons 1996; Fitzsimons *et al.* 1999, 2001; Hambrey & Fitzsimons 2010). This highlights the difficulty of defining the 'effective bed' of cold-based glaciers overlying permafrost and shows that it is an important mechanism for debris entrainment into the basal zone. Blocks of deformed ice and frozen sediment can be folded, faulted and, in places, stacked into moraines at the margin of glacier. Some of these blocks show a wide range of deformation features ranging from simple en-echelon stacking to complex recumbent folds and thrust faults, suggestive of multiphase deformation by cold-based glaciers. Furthermore, intensively deformed, buried ice extends well beyond the moraines visible on the surface, indicating a greater extent of past advances of cold-based glaciers (e.g. Fitzsimons 2011). Similar intense deformation of buried ice occurs in the lower Victoria Valley in the proglacial area of the Victoria Lower Glacier (Fig. 5c).

Preservation features

Compressed flattened ground. Areas of compressed ground occur at the cold-based margins of outlet glaciers that rest on weathered desert pavements. The compressed ground often ranges over several hundreds of metres and extends beneath the glacier margin. Clear examples occur at the margin of the Taylor Glacier in Pearse Valley and the margin of the Hatherton Glacier in the Britannia Range (Atkins *et al.* 2011). In these cases, the weathered surface of the pre-existing drifts shows typical desert pavement features with common desert-varnished and weathered clasts, but has a distinctive flattened and compressed surface (Fig. 6a). Boulders and cobbles are depressed into the matrix, producing an over-consolidated surface with a markedly reduced roughness compared with the same surface beyond the limit of compressed

Fig. 6. Examples of preservation features related to cold-based glaciers. (**a**) Compressed desert pavement showing compacted sediment around large desert-varnished boulders near the cold-based margin of the Hatherton Glacier, Britannia Range. Flat-topped boulder in foreground is *c.* 1 m diameter. Patches of compressed ground extend away from the glacier margin for several hundred metres. (**b**) Preserved, relict drift surface with only scattered boulders and cobbles superimposed on surface (thin, patchy drift), forming boulder-belt/drop moraines in places. Overriding cold-based ice has not disturbed the pre-existing surface. In places where the drift completely covers the older surface, buried desert pavements and soils occur. Britannia Range. Person for scale (circled).

ground. Furthermore, overturned boulders occur and some boulders that protrude up out of the drift display abrasions on the stoss side and top edge (see section on boulder-top abrasion), consistent with the surface having been overridden and compressed by the weight of cold-based ice but not deformed. In the Britannia Range, common circular mounds of debris from deflated ice-cored cones occur over a wide area. Several examples within the limits of the compressed ground are completely flattened. The mechanism for creating these areas of compressed ground is poorly known. However, the frontal apron overriding (Shaw 1977; Chinn 1991; see 'thin, patchy drift' section) could explain how the desert pavement is overridden and compressed without significant erosion, deformation or deposition. In this situation, dry calved blocks of ice from the glacier create a frontal ice-apron that buries and protects the ice-marginal substrate. As the glacier advances, it overrides and deforms the apron, compressing the substrate beneath, but not eroding it. Only boulders large enough to project into the deforming ice are overturned or abraded, whereas the rest of the surface is held in place by the ice and compressed under the weight of the glacier. Continual apron formation and overriding could act as a 'rolling carpet', allowing the glacier to advance over the landscape, as suggested by Lloyd-Davies *et al.* (2009). Although compressed ground is not a geomorphological landform per se, it nevertheless illustrates the apparent ability of cold-based ice to passively cover and protect subglacial surfaces in some situations.

Buried soils and preserved pavements. Buried soils containing recycled ventifacts are common in areas associated with 'thin, patchy drift' and boulder-belt moraines deposited by cold-based lobes of outlet glaciers (Bockheim 2010). Bockheim (2002, 2010) reviewed a large database of soil profiles from the South Victoria Land region documenting numerous occurrences of buried soils and buried *in situ* desert pavements that have no evidence of deformation. In some cases, the buried pavements show interlocking gravel and ventifacts on soil horizons (Marchant *et al.* 1994). In the Britannia Range, Bockheim *et al.* (1989) recorded several buried soil horizons in a single location suggesting several phases of overriding by cold-based glaciers. With each advance, the underlying deposits and their soil are overridden and buried or partially incorporated into the ice and re-deposited as 'thin, patchy drift' that contains both fresh and reworked clasts. Where the 'thin, patchy drift' is not thick enough to cover the underlying surface, the only evidence of cold-based glacier overriding is a scattering of clasts superimposed on an older, undisturbed relict surface (Fig. 6b).

Discussion and implications

The geomorphological features summarized in this paper show that cold-based glaciers can both protect and modify the landscape. The types of features that form and the extent of landscape protection or modification depend on many factors related to glaciology, debris load, physiographic setting and substrate. Although some geomorphological evidence produced by cold-based glaciers can be large-scale and obvious (e.g. meltwater channels, glaciotectonic deformation and thrust block moraines), in many cases, the evidence is localized, often subtle and challenging to recognize, particularly on bedrock or consolidated substrates (e.g. sparse abrasion marks, overturned boulders, till patches). These subtle features may be the only evidence of otherwise imperceptible cold-based glacier advance and retreat events over the landscape. Even features that are relatively widespread such as 'thin, patchy drift' and boulder-belt moraines can be deceptive as they commonly contain a high percentage of reworked clasts and often do not completely cover or deform the underlying surface, making it difficult to consistently recognize and map the deposits. Furthermore, some evidence of cold-based glacier activity is mostly buried beneath the surface and only recognized in stream banks, drill holes or subsurface imaging, belying the true extent (e.g. buried ice and deformed permafrost). Evidence of landscape protection is even more subtle, comprising areas of compressed desert pavement with clasts impacted into the matrix being the only indication of past passive cover and protection by cold-based ice. The subtleties of the evidence, combined with long-standing misconceptions regarding the geomorphic effectiveness of cold-based glaciers and historic use of some ambiguous terminology, are probably reasons why the influence of cold-based glacier activity has remained poorly recognized for so long.

Recognizing the imprint of cold-based glaciers has profound implications for interpreting polar landscapes and glacial history. The geomorphological record can be extremely variable and potentially complex, consisting of subtle, superimposed, reworked or relict features from which it is challenging to extract the full glacial history. This is clearly illustrated by the difficulty in successfully applying surface exposure dating methods to date cold-based glacier marginal landforms. The technique relies on the ability to measure the accumulation of cosmogenic nuclides produced by cosmic ray spallation reactions in the minerals of rock surfaces (Gosse & Phillips 2001). This provides a measure of how long the rock surface has been exposed to cosmic radiation. However, the accumulation of cosmogenic nuclides stops if the rock surface is buried by ice or debris, and begins again once re-exposed, creating situations where multiple exposures and inherited nuclide concentrations occur. Although these problems can apply to warm-based glacial environments, the problems of multiple burials by ice, relict surfaces and subtle reworking are greatly exacerbated in areas where cold-based glaciers occur. Many surface exposure dating studies in both Antarctica and the Arctic reveal complex distributions of exposure ages, often with contradictory or anomalous dates from the same surface or landform, suggesting one or more episodes of passive burial by cold-based ice (e.g. Bierman *et al.* 1999; Fabel *et al.* 2002; Bentley *et al.* 2006; Miller *et al.* 2006; Strasky *et al.* 2008; Di Nicola *et al.* 2009; Todd *et al.* 2010). Other complex exposure histories arise where previously exposed clasts have been reworked into younger moraines providing anomalously old, 'inherited' ages (Mackintosh *et al.* 2007; Strasky *et al.* 2008; Bockheim 2010), or where a scattering of fresh debris has been passively deposited onto older landscapes, giving an anomalously young age for that surface (e.g. Davis *et al.* 2006). Some of these problems are reasonably well understood and certain episodes of burial and inheritance can be elucidated using multiple nuclides techniques or by sampling bedrock-erratic pairs to identify discontinuous exposure (e.g. Balco & Shuster 2009; Li *et al.* 2008; Storey *et al.* 2010). However, in places where short or multiple burials have occurred or where subtle modification by cold-based glacial events has not yet been recognized, it remains difficult to draw detailed and meaningful conclusions about cold-based glacial history. Improved awareness of the subtleties of cold-based glacial geomorphology may improve the selection of samples for surface exposure studies (Di Nicola *et al.* 2009; Bockheim 2010), thereby ensuring that a sample represents the glacial event that deposited it rather than the age of the underlying relict surface or the age of a previous event.

Understanding the geomorphological record of cold-based glaciers is particularly important at the margins of the East Antarctic Ice Sheet where there are still significant uncertainties regarding the scale of past fluctuations in ice volume (Strasky *et al.* 2008; Di Nicola *et al.* 2009). Cold-based lobes of outlet glaciers advance and retreat into ice-free valleys, directly recording changes in ice sheet volume. Although the lobes have remained cold-based throughout the Quaternary (Staiger *et al.* 2006; Lewis *et al.* 2007), the large volume of ice and high debris load inherited from the outlet glacier feeding them mean that they can advance over wide areas and sometimes deposit a more obvious geomorphological record than other types of cold-based glaciers (e.g. Swanger *et al.* 2011). Correct interpretation and dating of these records is critical for interpreting the past behaviour of the ice sheet. A good example of such a record occurs as 'thin, patchy drifts' and boulder-belt moraines in the Britannia Range alongside the Hatherton Glacier (Fig. 4a, b).

Recognizing the imprint of cold-based glaciers on landscape has other applications. For example, some isolated nunataks and valleys surrounded by cold-based glaciers are thought to have remained ice-free through multiple glacial cycles and have acted as refugia for a variety of terrestrial organisms. However, current reconstructions and models do not have high enough resolution to elucidate the number and past extent of these refugia (Adams *et al.* 2006; Convey *et al.* 2009). Improved understanding of cold-based glacial geomorphology and targeted surface exposure studies may provide important constraints on past cold-based ice coverage and therefore the spatial and temporal distribution of refugia and evolution of terrestrial biota (e.g. Storey *et al.* 2010). Another application is the use of the cold, hyper-arid desert conditions and cold-based glacier geomorphology in the South Victoria Land region as an analogue for interpreting landscapes on Mars. Several comparative studies utilizing high-resolution images of the Martian surface combined with field studies of cold-based glacier geomorphology in Antarctica have provided strong evidence suggesting the occurrence of cold-based glaciers on Mars (Head & Marchant 2003; Marchant & Head 2007). This includes features such as extensive overlapping curved ridges on otherwise undisturbed surfaces interpreted as 'drop' moraines from multiple cold-based glacial events. Other evidence such as lobate debris-aprons and viscous flow features have been interpreted as debris-covered glaciers or ice-cemented rock glaciers (Head *et al.* 2010; Hubbard *et al.* 2011).

Although considerable progress has been made in understanding cold-based glaciers over recent years, there are still significant gaps in knowledge. This reflects the relative paucity of detailed studies combined the wide variation in processes and geomorphological features associated with the different types of cold-based glacier on various substrates. In particular, the underlying substrate conditions strongly influence the processes operating and the geomorphological features produced. For example, cold-based glaciers overlying unconsolidated or permafrozen sediments can deform and entrain the substrate, and develop thick basal debris zones and extensive ice-cored moraines, whereas cold-based glaciers resting on bedrock or consolidated drift substrates often show limited deformation and erosion, and commonly have lower volumes of basal debris and only minor moraine development. Therefore, observations of processes and landforms at one cold-based glacier margin are not necessarily representative of processes and landforms at another (e.g. Fitzsimons *et al.* 1999). Uncertainties remain regarding deformation and entrainment processes within frozen substrates and how these contribute to ice-cored moraine formation. Furthermore, there are large gaps in understanding the processes of erosion, deformation, deposition and protection by cold-based glaciers that rest on consolidated drift or bedrock surfaces. At present, the interpretation of many geomorphological features and processes at these margins is based largely on simple conceptual models developed from sparse field observations.

Conclusions

A range of geomorphological features indicative of erosion, deposition, deformation and protection have been documented at the margins of cold-based glaciers in the South Victoria Land region of Antarctica. This contributes to a growing recognition that cold-based glaciers can both protect and modify landscapes and provides a basis for recognizing similar features elsewhere. The range of features that form is strongly dependent on the type of cold-based glacier and site-specific factors such as elevation, meltwater availability and substrate conditions. The evidence ranges from large-scale features such as meltwater channels and ice-cored moraines to small-scale and subtle features such as minor abrasions on rock surfaces and thin patchy deposits that can be difficult to recognize. In many places, particularly on bedrock and consolidated drift, subtle modification of pre-existing landforms or slight reworking of surface clasts may be the only evidence of past advance and retreat of cold-based glaciers.

Recognizing the often-subtle evidence and the dual ability of cold-based glaciers to both protect and erode landscapes is important for interpreting the geomorphological record and glacial history in polar regions or where cold-based glaciers may have existed in the past. Such insight is directly relevant to the increasing number of surface exposure studies on ice-marginal landforms in polar regions aimed at constraining the timing and scale of past ice sheet fluctuations. Understanding the imprint of cold-based glacier activity can act as a guide to sample selection and interpretation, thereby improving the coherency of exposure age datasets and better constraining cold-based glacial events.

Despite recent advances in understanding cold-based glaciers, significant gaps in knowledge remain. These primarily relate to understanding and interpreting the wide range of processes operating under and around the different types of cold-based glacier resting on different substrates. There are remaining uncertainties regarding deformation and entrainment processes on permafrozen ground and how these generate ice-cored landforms. More significantly, there are clearly large gaps in understanding processes of erosion, deformation, deposition and protection where cold-based glaciers overlie bedrock and consolidated drift surfaces. To reduce these gaps, future investigations focussed on understanding the processes and geomorphological features at all three types of cold-based glaciers and on different substrates are needed. Careful description of the glacier margin structure, substrate properties and geomorphological features is necessary, utilizing a facies-based approach to describe sediments and debris-rich ice. Collectively, this will help identify the full range of geomorphological features and sediments that occur at cold-based glacier margins and provide a more robust connection between processes and products. Such improvement should form a sound basis for interpreting similar features elsewhere, thereby aiding glacial reconstructions and ultimately providing progress towards a 'landsystem' model for cold-based glacial environments.

Fieldwork in the Britannia Range area was carried out in association with K. Joy and B. Storey 'Dynamics of the Darwin-Hatherton Glacial system' (K056), Gateway Antarctica, University of Canterbury and Antarctica New Zealand. Fieldwork in the Pearse Valley was carried out in conjunction with W. Dickinson (K047), Antarctic Research Centre, Victoria University of Wellington and Antarctica New Zealand. I thank Martyn Tranter and Michael Hambrey for helpful editorial assistance. Constructive reviews by Peter Nienow and Sean Fitzsimons helped strengthen this paper significantly.

References

Adams, B. J., Bardgett, R. D. *et al.* 2006. Diversity and distribution of Victoria Land biota. *Soil Biology and Biochemistry*, **38**, 3003–3018.

Atkins, C. B. & Dickinson, W. W. 2007. Landscape modification by meltwater channels at margins of cold-based glaciers, Dry Valleys, Antarctica. *Boreas*, **36**, 47–55, doi: 10.1080/03009480600827306

Atkins, C. B., Barrett, P. J. & Hicock, S. R. 2002. Cold glaciers erode and deposit: evidence from Allan Hills, Antarctica. *Geology*, **30**, 659–662.

Atkins, C. B., Dickinson, W., Joy, K. & Storey, B. 2011. Influence of cold-based glaciers on landscape development. *11th International Symposium on Antarctic Earth Sciences*, Edinburgh, **168**.

Balco, G. & Shuster, D. L. 2009. Production rate of cosmogenic ^{21}Ne in quartz estimated from ^{10}Be, ^{26}Al, and ^{21}Ne concentrations in slowly eroding Antarctic bedrock surfaces. *Earth and Planetary Science Letters*, **281**, 48–58.

Bentley, M. J., Fogwill, C. J., Kubik, P. W. & Sugden, D. E. 2006. Geomorphological evidence and cosmogenic ^{10}Be/^{26}Al exposure ages for the Last Glacial Maximum and deglaciation of the Antarctic Peninsula Ice Sheet. *Geological Society of America Bulletin*, **180**, 1149–1159.

Bierman, P. R., Marsella, K. A., Patterson, C., Davis, P. T. & Caffee, M. 1999. Mid-Pleistocene cosmogenic minimum-age limits for pre-Wisconsinan glacial surfaces in southwestern Minnesota and southern Baffin Island; a multiple nuclide approach. *Geomorphology*, **27**, 25–39, doi: 10.1016/S0169-555X(98)00088-9

BOCKHEIM, J. G. 1982. Properties of a chronosequence of ultraxerous soils in the Trans-Antarctic Mountains. *Geoderma*, **28**, 239–255.

BOCKHEIM, J. G. 2002. Landform and soil development in the McMurdo Dry Valleys, Antarctica: a regional synthesis. *Arctic Antarctic and Alpine Research*, **34**, 308–317, doi: 10.2307/1552489

BOCKHEIM, J. G. 2010. Soil preservation and ventifact recycling from dry-based glaciers in Antarctica. *Antarctic Science*, **22**, 409–417, doi: 10.1017/s0954102010000167

BOCKHEIM, J. G., WILSON, S. C., DENTON, G. H., ANDERSEN, B. G. & STUIVER, M. 1989. Late Quaternary ice-surface fluctuations of Hatherton Glacier, Transantarctic Mountains. *Quaternary Research*, **31**, 229–254, doi: 10.1016/0033-5894(89)90007-0

BOULTON, G. S. 1974. Processes and patterns of glacial erosion. *In*: COATES, D. R. (ed.) *Glacial Geomorphology*. Allen & Unwin, London, 41–87.

BROOK, E. J., KURZ, M. D., ACKERT, R. P., DENTON, G. H., BROWN, E. T., RAISBECK, G. M. & YIOU, F. 1993. Chronology of Taylor Glacier advances in Arena Valley, Antarctica using insitu cosmogenic ^{3}He and ^{10}Be. *Quaternary Research*, **39**, 11–23, doi: 10.1006/qres.1993.1002

CALKIN, P. E. 1971. Glacial geology of the Victoria Valley system, southern Victoria Land, Antarctica. *In*: CRARY, A. P. (eds) *Antactic Snow and Ice Studies II*. American Geophysical Union, Washington, DC, Antarctic Research Series, **16**, 363–341.

CAMPBELL, I. B. & CLARIDGE, G. G. C. 1975. Occurrence of dirt cones in Antarctica. *New Zealand Journal of Geology and Geophysics*, **18**, 349–355.

CHINN, T. J. H. 1985. Structure and Equilibrium of the Dry Valley Glaciers. *New Zealand Antarctic Record*, **6**, 73–88.

CHINN, T. J. H. 1987. Victoria Lower Glacier and Ross Sea Glaciation, Dry Valleys area, South Victoria Land, Antarctica. *Journal of Glaciology*, **33**, 293–299.

CHINN, T. J. H. 1989. Single folds at the margins of dry-based glaciers as indicators of a glacial advance. *Annals of Glaciology*, **12**, 23–30.

CHINN, T. J. H. 1991. Polar glacier margin and debris features. *Memorie della Società Italiana*, **46**, 25–44.

CONVEY, P., STEVENS, M. I. *ET AL*. 2009. Exploring biological constraints on the glacial history of Antarctica. *Quaternary Science Reviews*, **28**, 3035–3048.

CUFFEY, K. M., CONWAY, H., HALLET, B., GADES, A. M. & RAYMOND, C. F. 1999. Interfacial water in polar glaciers and glacier sliding at −17 °C. *Geophysical Research Letters*, **26**, 751–754, doi: 10.1029/1999gl900096

CUFFEY, K. M., CONWAY, H. *ET AL*. 2000. Entrainment at cold glacier beds. *Geology*, **28**, 351–354, doi: 10.1130/0091-7613

DAVIS, P. T., BRINER, J. P., COULTHARD, R. D., FINKEL, R. W. & MILLER, G. H. 2006. Preservation of Artic landscapes overridden by cold-based ice sheets. *Quaternary Research*, **65**, 165–163, doi: 10.1016/j.yqres.2005.08.019

DENTON, G. H. & SUGDEN, D. E. 2005. Meltwater features that suggest Miocene ice-sheet overriding of the Transantarctic Mountains in Victoria Land, Antarctica. *Geografiska Annaler*, **87A**, 67–85.

DICKINSON, W. & HOPKINS, K. 2005. *Preliminary Report on Soil and Permafrost Sampling: Beacon, Arena and Pearse Valleys, Dry Valleys, Antarctica, Nov–Dec, 2001*. Antarctic Data Series, **30**, Victoria University of Wellington, Wellington New Zealand.

DI NICOLA, L., STRASKY, S. *ET AL*. 2009. Multiple cosmogenic nuclides document complex Pleistocene exposure history of glacial drifts in Terra Nova Bay (northern Victoria Land, Antarctica). *Quaternary Research*, **71**, 83–92, doi: 10.1016/j.yqres.2008.07.004

DORAN, P. T., MCKAY, C. P., CLOW, G. D., DANA, G. L., FOUNTAIN, A. G., NYLEN, T. & LYONS, W. B. 2002. Valley floor climate observations from the McMurdo dry valleys, Antarctica, 1986–2000. *Journal of Geophysical Research – Atmospheres*, **107**, 4772, doi: 477210.1029/2001

DREIMANIS, A. 1988. Tills: their genetic terminology and classification. *In*: GOLDTHWAIT, R. P. & MATSCH, C. L. (eds) *Genetic Classification of Glacigenic Deposits*. Balkema, Rotterdam, 17–83.

DREWRY, D. J. 1986. *Glacial Geologic Processes*. Edward Arnold, London.

DYKE, A. S. 1993. Landscapes of cold-centered Late Wisconsinan ice caps, Arctic Canada. *Progress in Physical Geography*, **17**, 223–247, doi: 10.1177/030913339301700208

ECHELMEYER, K. & ZHONGXIANG, W. 1987. Direct observation of basal sliding and deformation at sub-freezing temperatures. *Journal of Glaciology*, **33**, 83–98.

ENGLAND, J. 1986. Glacial erosion of a high Arctic valley. *Journal of Glaciology*, **32**, 60–64.

EVANS, D. J. A. 1989*b*. Apron entrainment at the margins of sub-polar glaciers, northwest Ellesmere Island, Canadian High Arctic. *Journal of Glaciology*, **35**, 317–324.

FABEL, D., STROEVEN, A. P., HARBOR, J., KLEMAN, J., ELMORE, D. & FINK, D. 2002. Landscape preservation under Fennoscandian ice sheets determined from in situ produced Be-10 and Al-26. *Earth and Planetary Science Letters*, **201**, 397–406.

FALCONER, G. 1966. Preservation of vegetation and patterned ground under a thin ice body in northern Baffin Island, N.W.T. *Geographical Bulletin*, **8**, 194–200.

FITZSIMONS, S. J. 1996. Formation of thrust-block moraines at the margins of dry-based glaciers south Victoria Land, Antarctica. In: *Proceedings of the International Symposium on Glacial Erosion and Sedimentation. Annals of Glaciology*, **22**, 68–74.

FITZSIMONS, S. J. 2003. Ice-marginal terrestrial landsystems: polar continental glacier margins. *In*: EVANS, D. J. A. (ed.) *Glacial Landsystems*. Arnold, London, UK, 89–110.

FITZSIMONS, S. J. 2011. Glacial geology and landform evolution beneath and at the margins of cold-based glaciers. *11th International Symposium on Antarctic Earth Sciences*, Edinburgh, 165.

FITZSIMONS, S. J., MCMANUS, K. J. & LORRAIN, R. D. 1999. Structure and strength of basal ice and substrate of a dry-based glacier: evidence for substrate deformation at sub-freezing temperatures. *Annals of Glaciology*, **28**, 236–240.

Fitzsimons, S. J., McManus, K. J., Sirota, P. & Lorrain, R. D. 2001. Direct shear test of materials from a cold glacier: implications for landform development. *Quaternary International*, **86**, 129–137.

Fitzsimons, S. J., Webb, N., Mager, S., MacDonell, S., Lorrain, R. & Samyn, D 2008. Mechanisms of basal ice formation in polar glaciers: an evaluation of the apron entrainment model. *Journal of Geophysical Research*, **113**, F02010, doi: 10.1029/2006JF000698

Goldthwait, R. P. 1960. *Study of Ice Cliff in Nanatarssuag, Greenland.* Technical Report of Snow, Ice and Permafrost Research Establishment, **39**, 1–103.

Golledge, N. R. & Levy, R. H. 2011. Geometry and dynamics of an East Antarctic Ice Sheet outlet glacier, under past and present climates. *Journal of Geophysical Research*, **116**, FO 3025, doi: 10.1029/2011JF002028

Gosse, J. C. & Phillips, F. M. 2001. Terrestrial in situ cosmogenic nuclides: theory and applications. *Quaternary Science Reviews*, **20**, 1475–1560.

Hall, A. M. & Phillips, W. M. 2006. Glacial modification of granite tors in the Cairngorms, Scotland. *Journal of Quaternary Science*, **21**, 811–830, doi: 10.1002/jqs.1003

Hall, B. L., Denton, G. H., Lux, D. R. & Bockheim, J. G. 1993. Late Tertiary Antarctica paleoclimate and ice-sheet dynamics inferred from surficial deposits in Wright Valley. *Geografiska Annaler*, **75A**, 239–267.

Hambrey, M. J. & Fitzsimons, S. J. 2010. Development of sediment-landform associations at cold glacier margins, Dry Valleys, Antarctica. *Sedimentology*, **57**, 857–882, doi: 10.1111/j.1365-3091.2009.01123

Hattestrand, C. & Stroeven, A. J. 2002. A relict landscape in the centre of Fennoscandian glaciation: geomorphological evidence of minimal Quaternary glacial erosion. *Geomorphology*, **44**, 127–143.

Head, J. W. & Marchant, D. R. 2003. Cold-based mountain glaciers on Mars: Western Arsia Mons. *Geology*, **31**, 641–644, doi: 10.1130/0091-7613

Head, J. W., Marchant, D. R., Dickson, J. L., Kress, A. M. & Baker, D. M. 2010. Northern mid-latitude glaciation in the Late Amazonian period of Mars: criteria for the recognition of debris covered glacier and valley glacier landsystem deposits. *Earth and Planetary Science Letters*, **294**, 306–320.

Holdsworth, G. 1974. *Meserve Glacier, Wright Valley, Antarctica: Part 1. Basal Processes.* Institute of Polar Studies Report, Ohio State University, **37**, 104.

Hubbard, A., Lawson, W., Anderson, B., Hubbard, B. & Blatter, H. 2004. Evidence for extensive subglacial ponding across the tongue of Taylor Glacier, Antarctica. *Annals of Glaciology*, **39**, 79–84, doi: 10.3189/172756404781813970

Hubbard, B., Milliken, R. E., Kargel, J. S., Limaye, A. & Souness, C. 2011. Geomorphological characterisation and interpretation of a mid-latitude glacier-like form: Hellas Planitia, Mars. *Icarus*, **211**, 330–346.

Humphries, K. A. & Fitzsimons, S. J. 1996. Landform and sediment associations of dry-based glaciers in arid polar environments. *Zeitschrift fur Geomorphologie, Supplement*, **105**, 21–33.

Kavanaugh, J. L. & Cuffey, K. M. 2009. Dynamics and mass balance of Taylor Glacier, Antarctica: 2. Force balance and longitudinal coupling. *Journal of Geophysical Research*, **114**, F04011, doi: 10.1029/2009JF001329

Kleman, J. 1994. Preservation of landforms under ice sheets and ice caps. *Geomorphology*, **9**, 19–32.

Kleman, J. & Hattestrand, C. 1999. Frozen-bed Fennoscandian and Laurentide ice sheets during the Last Glacial Maximum. *Nature*, **402**, 63–66, doi: 10.1038/47005

Lewis, A. R., Marchant, D. R., Ashworth, A. C., Hemming, S. R. & Machlus, M. L. 2007. Major middle Miocene global climate change: evidence from East Antarctica and the Transantarctic Mountains. *Geological Society of America Bulletin*, **119**, 1449–1461, doi: 10.1130/b26134

Li, Y., Fabel, D., Stroeven, A. P. & Harbor, J. 2008. Unravelling complex exposure-burial histories of bedrock surfaces under ice sheets by integrating cosmogenic nuclide concentrations with climate proxy records. *Geomorphology*, **99**, 139–149.

Lloyd-Davies, M. T. 2004. *A Polar Paradise, the glaciation of South Victoria Land, Antarctica.* PhD thesis, Universiteit van Amsterdam.

Lloyd Davies, M. T. & Fitzsimons, S. J. 2004. Selected case studies of cold-based glaciers, South Victoria Land, Antarctica. *Quaternary Newsletter*, **104**, 30–44.

Lloyd-Davies, M. T., Atkins, C. B., van der Meer, J. J. M., Barrett, P. J. & Hicock, S. R. 2009. Evidence for cold-based glacial activity in the Allan Hills, Antarctica. *Quaternary Science Reviews*, **28**, 3124–3137.

Mackintosh, A., White, D. A., Fink, D., Gore, D. B., Pickard, J. & Fanning, P. C. 2007. Exposure ages from mountain dipsticks in MacRobertson Land, East Antarctica, indicate little change in ice-sheet thickness since the Last Glacial Maximum. *Geology*, **35**, 551–554.

Marchant, D. R. & Head, J. W., III. 2007. Antarctic dry valleys: microclimate zonation, variable geomorphic processes, and implications for assessing climate change on Mars. *Icarus*, **192**, 187–222, doi: 10.1016/j.icarus.2007.06.018

Marchant, D. R., Denton, G. H. & Swisher, C. C., Jr. 1993. Miocene–Pliocene–Pleistocene glacial history of Arena Valley, Quatermain Mountains, Antarctica. *Geografiska Annaler*, **75A**, 269–302.

Marchant, D. R., Denton, G. H., Bockheim, J. G., Wilson, S. C. & Kerr, A. R. 1994. Quaternary changes in level of the upper Taylor Glacier, Antarctica: implications for palaeclimate and East Antarctic Ice Sheet dynamics. *Boreas*, **23**, 29–43.

Miller, G. H., Briner, J. P., Lifton, N. A. & Finkel, R. C. 2006. Limited ice-sheet erosion and complex exposure histories derived from in situ cosmogenic ^{10}Be, ^{26}Al and ^{14}C on Baffin Island, Arctic Canada. *Quaternary Geochronology*, **1**, 74–85.

Naslund, J. O. 1997. Subglacial preservation of valley morphology at Amundsenisen, western Dronning Maud Land, Antarctica. *Earth Surface Processes and Landforms*, **22**, 441–455.

O'Cofaigh, C., Lemmen, D. S., Evans, D. J. A. & Bednarski, J. 1999. Glacial landform-sediment assemblages in the Canadian High Arctic and their implications for late Quaternary glaciation. *Annals of Glaciology*, **28**, 195–201.

Robinson, P. H. 1984. Ice dynamics and thermal regime of Taylor Glacier, south Victoria Land, Antarctica. *Journal of Glaciology*, **30**, 153–160.

Shaw, J. 1977. Tills deposited in arid polar environmnents. *Canadian Journal of Earth Sciences*, **14**, 1239–1245.

Smith, C. A., Lowell, T. V. & Caffee, M. W. 2009. Lateglacial and Holocene cosmogenic surface exposure age glacial chronology and geomorphological evidence for the presence of cold-based glaciers at Nevado Sajama, Bolivia. *Journal of Quaternary Science*, **24**, 360–372, doi: 10.1002/jqs.1239

Staiger, J. W., Marchant, D. R., Schaefer, J. M., Oberholzer, P., Johnson, J. V., Lewis, A. R. & Swanger, K. M. 2006. Plio-Pleistocene history of Ferrar Glacier, Antarctica: implications for climate and ice sheet stability. *Earth and Planetary Science Letters*, **243**, 489–503, doi: 10.1016/j.epsl.2006.01.037

Storey, B. C., Fink, D., Hood, D., Joy, K., Shulmeister, J., Riger-Kusk, M. & Stevens, M. I. 2010. Cosmogenic nuclide exposure age constraints on the glacial history of the Lake Wellman area, Darwin Mountains, Antarctica. *Antarctic Science*, **22**, 603–618.

Strasky, S., Di Nicola, L. *et al.* 2008. Surface exposure ages imply multiple low-amplitude Pleistocene variations in East Antarctic Ice Sheet, Ricker Hills, Victoria Land. *Antarctic Science*, **21**, 59–69, doi: 10.1017/S0954102008001478

Sugden, D. E., Denton, G. H. & Marchant, D. R. 1991. Subglacial meltwater channel systems and ice sheet overriding, Asgard Range, Antarctica. *Geografiska Annaler*, **73A**, 512–527.

Sugden, D. E., Marchant, D. R., Potter, N., Souchez, R. A., Denton, G. H., Swisher, C. C. & Tison, J. L. 2002. Preservation of Miocene glacier ice in East Antarctica. *Nature*, **376**, 412–414, doi: 10.1038/376412a

Sugden, D. E., Balco, G., Cowdery, S. G., Stone, J. O. & Sass, L. C. 2005. Selective glacial erosion and weathering zones in the coastal mountains of Marie Byrd Land, Antarctica. *Geomorphology*, **67**, 317–334, doi: 10.1016/j.geomorph.2004.10.007

Swanger, K. M., Marchant, D. R., Kowalewski, D. E. & Head, J. W., III. 2010. Viscous flow lobes in central Taylor Valley, Antarctica: origin as remnant buried glacial ice. *Geomorphology*, **120**, 174–185.

Swanger, K. M., Marchant, D. R., Schaefer, J. M., Winckler, G. & Head, J. W., III. 2011. Elevated East Antarctic outlet glaciers during warmer-than-present climates in southern Victoria Land. *Global and Planetary Change*, **79**, 61–72.

Syverson, K. M. & Mickelson, D. M. 2009. Origin and significance of lateral meltwater channels formed along a temperate glacier margin, Glacier Bay, Alaska. *Boreas*, **38**, 132–145.

Todd, C., Stone, J., Conway, H., Hall, B. & Bromley, G. 2010. Late Quaternary evolution of Reedy Glacier, Antarctica. *Quaternary Science Reviews*, **29**, 1328–1341, doi: 10.1016/j.quascirev.2010.02.001

Waller, R. I. 2001. The influence of basal processes on the dynamic behaviour of cold-based glaciers. *Quaternary International*, **86**, 117–128, doi: 10.1016/s1040-6182(01)00054-4

Waller, R. I. & Tuckwell, G. W. 2005. Glacier-permafrost interactions and Glaciotectonic Landform generation at the margin of the Leverett Glacier, West Greenland. *In*: Harris, C. & Murton, J. B. (eds) *Cryospheric Systems: Glaciers and Permafrost*. Geological Society, London.

Waller, R. I., Murton, J. B. & Whiteman, C. A. 2009. Geological evidence for subglacial deformation of Pleistocene permafrost. *Proceedings of the Geologists Association*, **120**, 155–162.

Waller, R. I., Murton, J. B. & Kristensen, L. 2012. Glacier-permafrost interactions: processes, products and glaciological implications. *Sedimentary Geology*, **255–256**, 1–28, doi: 10.1016/j.sedgeo.2012.02.005

A review of geomorphic processes and landforms in the Dry Valleys of southern Victoria Land: implications for evaluating climate change and ice-sheet stability

D. R. MARCHANT[1]*, S. L. MACKAY[1], J. L. LAMP[1], A. T. HAYDEN[1] & J. W. HEAD[2]

[1]*Department of Earth and Environment, Boston University, Boston, MA 02215, USA*

[2]*Department of Geological Sciences, Brown University, Providence, RI 02912, USA*

**Corresponding author (e-mail: marchant@bu.edu)*

Abstract: The Dry Valleys are subdivided into three microclimate zones on the basis of summertime measurements of atmospheric temperature, soil moisture, relative humidity and wind-speed/direction. Subtle variations in these climate parameters result in considerable differences in process geomorphology and in the development of unique landforms within each zone. The mapped zones include a coastal thaw zone, an inland mixed zone and a stable upland zone. Landforms within each zone are subdivided into macroscale features (e.g. valleys, slopes and gullies), mesoscale features (e.g. polygons and viscous-flow features) and microscale features (e.g. rock and near-surface soil features, including the effects of salt weathering, wind erosion and pitting). We present a review of landscape development in the Dry Valleys with implications for long-term climate change and ice-sheet stability. Chronological control is afforded by $^{40}Ar/^{39}Ar$ dating of volcanic ash-fall deposits and cosmogenic nuclide analyses of surface boulders. Collectively, the data call for persistent cold and dry conditions in the stable upland zone for approximately the last 14 Ma, although some level of climatic amelioration and landform modification may have occurred within low-lying regions and in the inland mixed zone.

Landscapes represent the physical manifestation of several interrelated processes, including variations in tectonism, lithology and climate. Of these, climate is known to change on timescales relevant to human endeavours and, thus, changing landscapes are commonly viewed as resulting from a change in local climate conditions. Important climate parameters include variations in temperature, rainfall, wind velocity and solar radiation. Notwithstanding the importance of tectonic structure and lithology in shaping landscapes, especially in areas of active tectonics, climate parameters yield diverse geomorphic processes, which in turn produce distinct assemblages of landforms. Among climate parameters, the leading driver of landscape change is the abundance, and state, of water. For example, in comparing arid and humid temperate environments, arid landscapes show far less water-related fluvial activity than humid environments. Climate zones characterized by mean annual temperatures at or below the freezing point of water will also have relatively little influence from fluvial activity. Rather, periglacial and glacial processes will prevail, and the landforms will be formed and modified by freeze–thaw cycles and the movement of ice. These relationships offer a powerful tool for defining climate zones, understanding how they form and evolve, and even for predicting the nature of landforms under different climate conditions. Exploiting such linkages among climate and landscape has led naturally to the notion of morphogenetic zones, with each zone containing a distinctive landform assemblage that corresponds with major climatic belts. In this paper, we use the terms equilibrium landform and equilibrium landform assemblages to describe landforms produced in balance with local microclimate conditions in the Dry Valleys. Although the concept is similar to that of morphogenetic zones, we prefer the phrase equilibrium landform assemblage because it highlights the importance of very small microclimate variations within the Dry Valleys that yield disproportionally large changes in geomorphic process and resulting landscapes. In the absence of buffering vegetation and biological processes, the Dry Valleys landscape is extremely sensitive to minor climate variation. The location of the zero-degree summer isotherm is one of the most critical geomorphic thresholds impacting landscape evolution in the Dry Valleys region, and, indeed, in most high-latitude environments and polar deserts worldwide.

From: Hambrey, M. J., Barker, P. F., Barrett, P. J., Bowman, V., Davies, B., Smellie, J. L. & Tranter, M. (eds) 2013. *Antarctic Palaeoenvironments and Earth-Surface Processes*. Geological Society, London, Special Publications, **381**, 319–352. First published online July 1, 2013, http://dx.doi.org/10.1144/SP381.10

Scientific approach

In this contribution, we examine the nature and evolution of the hyper-arid polar desert of the McMurdo Dry Valleys, defining three microclimate zones. We assess the role of small variations in temperature and precipitation in producing and sustaining different characteristic landforms at a variety of scales. We then explore the application of these concepts towards the assessment of long-term climate change and ice-sheet stability. Throughout the text, we synthesize earlier work from Marchant & Head (2007, 2010), which focused on exploiting geomorphic similarities among landforms in the Dry Valleys and those observed on Mars, with an eye towards understanding climate change and glaciation on Mars. Here, we utilize our geomorphic synthesis to present an assessment of Antarctic climate change and ice-sheet stability, with an eye towards understanding ice dynamics and landscape evolution under warmer-than-present climates (e.g. DeConto *et al.* 2012). We begin with a brief overview of the physical setting of the Dry Valleys, and then highlight several key geomorphic processes endemic to all microclimate zones. We then show how subtle variations in summertime temperature, wind speed, relative humidity and soil moisture across each zone alters these key geomorphic processes and results in the development of an array of different landforms at multiple scales. We introduce new data on microclimate conditions within each zone, discuss the geomorphic evolution of buried-ice deposits as a function of climate zone, provide new age control for landform evolution and assess the implications for ice-sheet dynamics over time.

The McMurdo Dry Valleys

Located in the central Transantarctic Mountains, the McMurdo Dry Valleys, or more commonly the Dry Valleys, lie between the East Antarctic Ice Sheet and seasonally open water of the Ross Sea (Fig. 1). Local relief is approximately 2800 m. Most regions are generally free of surface ice and there are no vascular plants. Taylor, Wright and Victoria are the main east–west-trending valleys that transect the region. These valleys, each about 80 km long and up to 15 km wide, are separated by the Asgard–Olympus ranges and the Quartermain Mountains. Mountainous terrain is highest at the western rim of the Dry Valleys (where peaks commonly exceed 2000 m in elevation), and this bedrock threshold prevents significant influx of East Antarctic ice from flowing into the Dry Valleys. However, two small outlet glaciers, Taylor Glacier and Upper Wright Glacier, just pass over this threshold and terminate on land 50–75 km from the coast; only one outlet glacier reaches the coast, the Ferrar Valley Glacier (Fig. 1). Snowfall is less than 50 mm water equivalent per year, with the highest values nearest the coast and decreasing inland (Fountain *et al.* 2010). Along with wind-blown snow from the polar plateau, this precipitation appears sufficient to nourish alpine glaciers that descend major valley walls and occupy the heads of dozens of tributary valleys.

The local bedrock in the Dry Valleys consists of a basement complex of pre-Cambrian–Palaeozoic age granites, gneisses and marbles, most of which were produced and deformed during the Cambrian–Ordovician Ross Orogeny. Typically, these rocks crop out below approximately 800 m in elevation near the coast, as well as in low-lying portions of the central Dry Valleys region. Further inland, and at higher elevations, the exposed bedrock consists of relatively flat-lying sedimentary rocks of the Beacon Supergroup (Devonian–Triassic-age sandstones, siltstones, orthoquartzites and conglomerates) and 200–300 m thick sills of Ferrar Dolerite (Jurassic-age intrusives representing the break-up of Gondwana: Elliot & Fleming 2004). The Kukri Peneplain separates the basement complex from overlying sedimentary formations and Jurassic-age intrusives. Offshore, large stratovolcanoes occur within 100 km of the Dry Valleys region. Volcanic detritus recovered in drill cores in the western Ross Embayment indicates that regional volcanism extends back to at least late Eocene time (Kyle 1990).

Early geomorphic studies in the Dry Valleys

The earliest investigations in the Dry Valleys set the stage for modern geomorphic analyses. Based on logistical considerations, the earliest investigations were largely centred near the coast, and the implicit assumption was that observed geomorphic processes operating in low-lying, relatively warm and humid coastal sites could be applied uniformly across the entire Dry Valleys region. Ultimately, this approach perpetuated the notion of a single, hyper-arid cold-polar desert with spatially uniform geomorphic processes. Paradoxically, early soil scientists in the region exploited the influence of local microclimate conditions on soil development (e.g. Campbell & Claridge 1969; see also Barrett *et al.* 2006) but only recently have geomorphologists begun to appreciate and utilize linkages among microclimate and geomorphic process to both constrain past climate variation and predict future variations with global change (Swanger *et al.* 2010). Of paramount importance in this endeavour has been the deployment of weather-sensitive

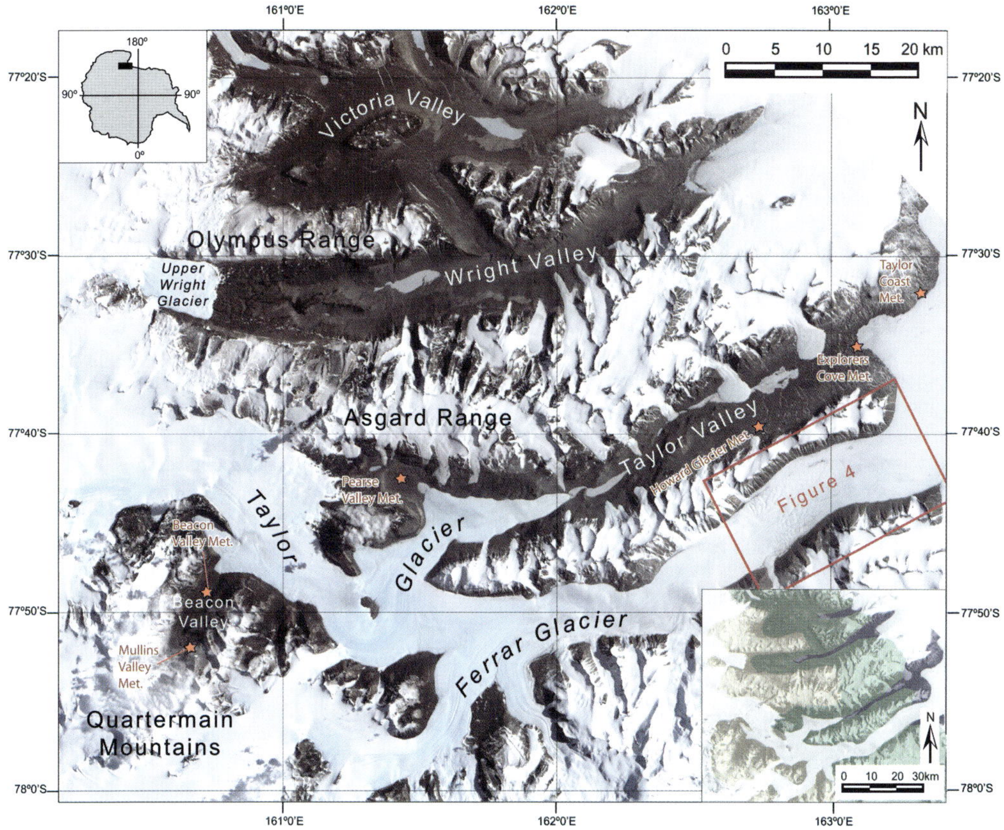

Fig. 1. Location map for the Dry Valleys region showing major valleys and topographical features, as well as locations for automatic weather stations in the costal thaw zone (Taylor Coast Met. and Explorers Cove Met.), inland mixed zone (Howard Glacier Met. and Pearse Valley Met.) and stable upland zone (Beacon Valley Met. and Mullins Valley Met.). Inset shows the distribution of microclimate zones: coastal thaw zone in blue, inland mixed zone in green and stable upland zone in tan (adapted from Marchant & Head 2007).

data loggers across the region (Doran *et al.* 2002; see also the next section).

Fundamental geomorphic processes

Several fundamental geomorphic processes impact landscape evolution across the Dry Valleys. These include aeolian erosion from katabatic winds, active-layer cryoturbation (including thermal expansion/contraction), cold-based glaciation, fluvial activity, salt weathering and mass wasting. These processes vary in intensity and rate across each microclimate zone.

Katabatic winds

Gravity-driven katabatic winds flow off the East Antarctic Ice Sheet and gather speed as they pass through the Dry Valleys, commonly exceeding 50 km h^{-1} (Schwerdtfeger 1984; Marshall & Turner 1997; Nylen *et al.* 2004). Katabatic winds warm adiabatically as they descend ice-sheet margins, resulting in elevated air temperatures and excursions that are particularly apparent during winter months (Fountain *et al.* 1999; Doran *et al.* 2002; Nylen *et al.* 2004); adiabatic warming can increase local air temperatures by 30 °C in just a few hours (Fountain *et al.* 1999). Alternatively, some authors have suggested that a significant portion of these wind events should be reinterpreted as foëhn winds, resulting from topographical modification of flow in the lee of mountain barriers (Speirs *et al.* 2010). However, regardless of the exact origin of the winds, they increase air temperatures and transport significant snow from the Polar Plateau, some of which is deposited in the lee of topographical obstacles throughout the Dry Valleys. The fate of this snow – to melt or sublimate – is a critical factor in the development of most unconsolidated landforms in the Dry Valleys. In addition to changing surface albedo, this snow influences the areal distribution and depth to ice-cemented ground (Bockheim 2002; McKay 2009), the longevity of buried-ice deposits (Kowalewski *et al.* 2006, 2011, 2012; Hagedorn *et al.* 2007, 2010), the distribution of near-surface salt accumulations (Campbell & Claridge 1987; Bockheim 1997; Bao *et al.* 2008), and, more generally, the potential for liquid water (Gooseff *et al.* 2003*b*; Hall *et al.* 2010) and runoff (Head *et al.* 2007*a*, *b*; Levy *et al.* 2011). Moreover, high-velocity winds entrain sand grains that scour the ground, producing cavernous weathering and ventifacts (Malin 1984, 1987; Selby 1977; Lancaster 2002, 2004; Gillies *et al.* 2009). In places, rates of erosion from wind action have been estimated to be approximately 0.1 $\mu m\ a^{-1}$ (Malin 1984, 1987), although the magnitude varies greatly over the region (Malin 1991). The largest concentration of ventifacts occurs on ground surfaces that have been exposed for more than 1 Ma; the presence of these wind-polished clasts has been used to infer the relative age and stability of surfaces in the Dry Valleys (Bockheim 2002).

Active-layer cryoturbation and thermal contraction/expansion

Permafrost is any material that is at or below 0 °C for at least 2 consecutive years. It may be classified as dry, in which it lacks sufficient interstitial water to be cohesive (Bockheim 2002), or wet. In either case, most permafrost displays a near-surface layer (active layer) that experiences seasonal temperatures both above and below 0 °C (Yershov 1998; Davis 2001). The presence and thickness of this active layer depends primarily on atmospheric temperature, and, secondarily, on substrate heat conduction; the latter being strongly modulated by water content. Given the importance of active-layer cryoturbation on landform development, we distinguish a 'wet' active layer from a 'dry' active layer. A wet active layer contains visible ice or liquid water; changes in the state of water to ice, and back again, result in considerable variations in landscape morphology and soil geochemistry (Barrett *et al.* 2009). In contrast, dry active layers contain minimal soil moisture, generally less than 5% gravimetric water content; given their lack of water, temperature oscillations above and below 0 °C result in minimal geomorphic change. In interior regions of the Dry Valleys, thin, dry active layers (<15 cm thick) are the norm, whereas relatively thick and wet active layers (25+ cm in thickness) are common in coastal settings (Campbell *et al.* 1997*a*, *b*; Bockheim 2002; Campbell & Claridge 2006; Bockheim *et al.* 2007). The greater thickness of active layers in coastal settings is due both to elevated atmospheric temperatures and to high soil-moisture concentrations (Ikard *et al.* 2009). Soil moisture increases apparent bulk soil thermal diffusivity, which in turn leads to deeper summertime thaw and deeper active-layer cryoturbation (Ikard *et al.* 2009).

Strong seasonal changes in temperature not only produce active layers but may also induce volumetric changes in ice-cemented/cohesive ground, resulting in the development of thermal contraction-crack polygons (Lachenbruch 1962). The polygons form by thermal cracking of ice-rich permafrost and subsequent infilling of cracks with a variety of materials. In plan view, the cracks intersect to form a network of polygonal patterns and shapes. Thermal cracking is favoured in regions that experience abrupt seasonal cooling (Berg & Black 1966; Black 1973; Hallett *et al.* 2011) and where near-surface materials are cohesive; cracks may be absent in dry permafrost, even though atmospheric conditions are suitable for polygon formation. Different classes of polygons arise from microclimate variation: ice-wedge polygons occur near the coast; sand-wedge polygons dominate inland regions; and sublimation polygons, a special type of sand-wedge polygon that forms in areas with dry active layers underlain by massive ice (Marchant *et al.* 2002), occupy portions of the coldest and driest areas (see details below).

Cold-based glaciers

Alpine glaciers in the Dry Valleys are cold-based. In contrast to wet-based glaciers, which move through a combination of internal creep, basal sliding, regelation and subglacial sediment deformation, cold-based glaciers move only through creep deformation. Rates of bedrock erosion beneath cold-based glaciers are extremely low, of the order of about $10^{-7}\ m\ a^{-1}$ (Cuffey *et al.* 2000). This rate is several orders of magnitude less than that of typical wet-based alpine glaciers (Hallet *et al.* 1996; Spotila *et al.* 2004; Brook *et al.* 2006; Koppes & Hallet 2006), and even below estimates for wind erosion in the region (Malin 1984, 1987). There is a large body of evidence that suggests that cold-based glaciers tend to preserve landscapes, rather than erode them (e.g. Borgstrom 1999; Kleman & Hatterstrand 1999; Fabel *et al.* 2002; Briner *et al.* 2006; Swanger *et al.* 2010). However, in suitable areas, the passage of cold-based alpine glaciers has resulted in observable landscape change, including minor scrapes and gouges on bedrock, rotated blocks, and patches of isolated, coarse-grained till (e.g. Fitzsimons *et al.* 2001; Atkins *et al.* 2002; Davies *et al.* 2009). Further, well-documented evidence exists for distinct landform assemblages for cold-based glaciers (Fitzsimons 2005). At glacier snouts, for example, debris entrainment and modification may occur through localized detachment of frozen sediment and glaciotectonic thrusting (Hambrey & Fitzsimons 2010). Thus, it is clear that some erosion is possible beneath cold-based ice but it is also clear that in places cold-based glaciers have advanced across unconsolidated landforms without modification of the underlying substrate (Marchant *et al.* 1993*a*; fig. 2 in Swanger & Marchant 2007).

Notwithstanding the above, the lack of appreciable regelation beneath cold-based glaciers (Cuffey *et al.* 2000) limits entrainment, and direct rockfall onto glacier ice is the primary mechanism of debris entrainment. However, because rates of rockfall are low throughout the region (Summerfield *et al.* 1999), most alpine glaciers are relatively free of debris. Moraines, if present, may be little more than a single row of aligned cobbles. Furthermore, given the stochastic nature of rockfall, moraines are commonly discontinuous over relatively short distances (measured over 10s of metres) and most occur only where glacier ice has passed alongside steep bedrock cliffs.

Fluvial activity

Water produced from melting along the surface of outlet, alpine and piedmont glaciers may generate considerable gullying and erosion of unconsolidated deposits (Chinn 1987; Atkins *et al.* 2002; Atkins & Dickinson 2007). This type of intermittent fluvial activity is most pronounced in the coastal thaw zone but surface melting during extreme warm events may also promote gullying at higher elevations (Doran *et al.* 2008).

Salt weathering

Almost all geomorphic processes in the Dry Valleys region are modified by variations in local soil salinity (e.g. Claridge & Campbell 1968, 1977, 2005; Bockheim 1997, 2002; Mahaney *et al.* 2001; Campbell & Claridge 2006). Salts can occur as coatings and efflorescences at the ground surface (Campbell & Claridge 1987; Bockheim 2002), and as discrete salt-cemented horizons, sometimes up to 15 cm in thickness, in the subsurface (e.g. Bockheim 1997; Campbell & Claridge 2006). Local areas of relatively low surface albedo may induce snowmelt on solar-heated rocks. This snowmelt, over time, may generate brines that migrate laterally, and evaporate beneath rocks to produce salt coatings, efflorescences and indurated pans. These local concentrations of salts can induce variations in soil moisture by affecting the state of water. Surface water has been observed at temperatures of less than −4 °C in small damp, salty hollows (e.g. Campbell & Claridge 2006), and has even been observed to remain unfrozen at temperatures below −40 to −50 °C in upper Wright Valley at Don Juan Pond (e.g. Marion 1997; Takamatsu *et al.* 1998; Healy *et al.* 2006). Water cycled laterally through hyporheic zones (regions beneath and lateral to stream beds, where mixing of shallow groundwater and surface water occurs) can mobilize salts, forming saline solutions, brines and seeps, and can locally redeposit salts in soil horizons at and just below the surface (e.g. McKnight *et al.* 1999; Lyons *et al.* 2005; Harris *et al.* 2007). Vertical migration of water through soils can also result in mobilization and concentration of salts (e.g. Gibson *et al.* 1983; Dickinson & Rosen 2003; Wentworth *et al.* 2005). These variations in vertical concentrations result in salt aggregations and layers that can be characterized by significant locally elevated water contents; such variations can produce local complexities in the geochemistry and thermal states of soils, and can influence geomorphic processes and resultant surface landforms. Salt types include a very wide variety of crystalline phases of sodium, potassium and magnesium chlorides, nitrates, and sulphates. Although salts increase in importance and abundance with increasing soil age (e.g. Claridge & Campbell 1977; Mahaney *et al.* 2001; Wentworth *et al.* 2005; Bao & Marchant 2006; Campbell & Claridge 2006), local changes in rock-surface albedo, which induce or retard surface snowmelt, play significant roles in salt occurrence and distribution.

Viscous flow

There are several classes of viscous flow in the Dry Valleys. For micro- and mesoscale flow features (<100 m long), the terms solifluction and gelifluction are sometimes used interchangeably (e.g. Matsuoka 2001), and for large-scale viscous flow features the terms rock-glacier and debris-covered glacier are often interchanged (e.g. Giardino *et al.* 1987; Whalley & Palmer 1998). Misconceptions in terminology arise because the features all form by a continuum of processes that are related to the slow movement of hill slope materials. We define solifluction as the slow flow of saturated materials, with or without the presence of nearby ice (e.g. as originally implied in Andersson 1906). Our usage of the term gelifluction is different in that in addition to saturated flow, gelifluction lobes advance by internal deformation of (and, perhaps, slippage along) buried-ice lenses or pore ice. Although rock-glaciers and debris-covered glaciers both flow by creep deformation of internal ice, we distinguish rock-glaciers as having ice of secondary origin (i.e. not glacial), including ice that forms by the freezing of pore water. Debris-covered glaciers are those viscous-flow features that have a demonstrable core of buried glacier ice.

Division of microclimate zones

Observed and measured weather conditions over a 10 year time span show that the Dry Valleys region is best divided into a series of microclimate zones (e.g. Marchant & Denton 1996; Marchant & Head 2007; see long-term climate records available at http://www.mcmlter.org/). Slight changes in summertime atmospheric temperature and soil moisture, particularly along microclimate boundaries, are sufficient to produce major changes in geomorphic processes and surface topography. As is the case for most desert regions, the geomorphic impacts of minor changes in soil moisture are disproportionately large when compared to similar shifts in humid temperate latitudes due to the crossing of geomorphic thresholds (Langbein & Schumm 1958; Schumm 1965; Schumm & Lichty 1965). On the basis of measured summertime climate conditions (Table 1), we distinguish a coastal thaw zone, an inland mixed zone and a stable upland zone (Fig. 1, inset). Alternative subdivisions are possible (e.g. Campbell & Claridge 1969, 1987, 2006; Bockheim 2002, 2003) but this three-fold classification captures the major trends in geomorphic processes and equilibrium landforms across the region. We focus on summertime climate because of the prevalence of liquid water during this time and because climate variation across the Dry Valleys during the winter months is relatively minor (Doran *et al.* 2002). Figures 2 and 3, as well as Table 1, show a subset of meteorological data collected by the authors. Although relatively short in duration, the records highlight typical variability in summertime climate parameters across the Dry Valleys region; see also additional long-term records available at http://www.mcmlter.org/.

Coastal thaw zone

Modern summertime air temperatures in the coastal thaw zone show a mean season-long temperature of around −4 °C and a mean daily maximum of approximately −1.5 °C. Snowfall in the coastal thaw zone reaches 50 mm of water equivalent per year (Schwerdtfeger 1984; Fountain *et al.* 2010), and a large fraction melts and infiltrates near-surface soils. Owing to large variations in surface albedo (which arise from the presence of numerous tills capped by different lithologies in the coastal thaw zone: Hall *et al.* 1993; Denton & Marchant 2000), surface and subsurface ground temperatures vary considerably over relatively short horizontal distances of about 10 m. In all places measured,

Table 1. *Summer (annual) climate data for three microclimate zones in the Dry Valleys*

		Air temperature (°C)			Soil temperature (°C)		Relative humidity (%)
		Mean	Mean daily (max)	Mean daily (min)	Mean at 0 cm depth	Mean at 10 cm depth	Mean
CTZ	Explorer's Cove*,† (26 m a.s.l, 4 km from the coast)	−5 (−20)	−2 (−15)	−7 (−24)	2 (−18)	1 (−17)	64 (63)
	Taylor Coast‡ (4 m a.s.l, 0.1 km from the coast)	−3 (−17)	−1 (−13)	−7 (−21)	4 (−16)	− (−)	59 (61)
IMZ	Howard Glacier* (472 m a.s.l., 16 km from the coast)	−7 (−18)	−4 (−14)	−10 (−21)	− (−)	− (−)	67 (55)
	Pearse Valley‡ (451 m a.s.l, 58 km from the coast)	−4 (−18)	−2 (−13)	−7 (−22)	2 (−19)	−1 (−18)	40 (46)
SUZ	Beacon Valley* (1176 m a.s.l., 74 km from the coast)	−10 (−22)	−8 (−18)	−12 (−25)	−5 (−22)	−6 (−21)	41 (43)
	Mullins Valley‡ (1401 m a.s.l, 81 km from the coast)	−11 (−23)	−7 (−18)	−15 (−27)	−8 (−23)	−10 (−23)	47 (50)

CTZ, coastal thaw zone; IMZ, inland mixed zone; SUZ, stable upland zone; m a.s.l, metres above sea level.
Summer defined as December, January and February; annual conditions shown in parenthesis.
*Data from McMurdo LTER weather stations. Available at http://www.mcmlter.org/data_home.htm.
†Soil data from location near Lake Bonney (not Explorers Cove) at 60 m a.s.l, 25 km from the coast.
‡Data from weather stations installed by the authors (HOBO micro-station data loggers and sensors; Onset Computer Corporation). Time period: 2007–present. For details on sensor specifications and general data acquisition procedures see Kowalewski *et al.* (2006) .

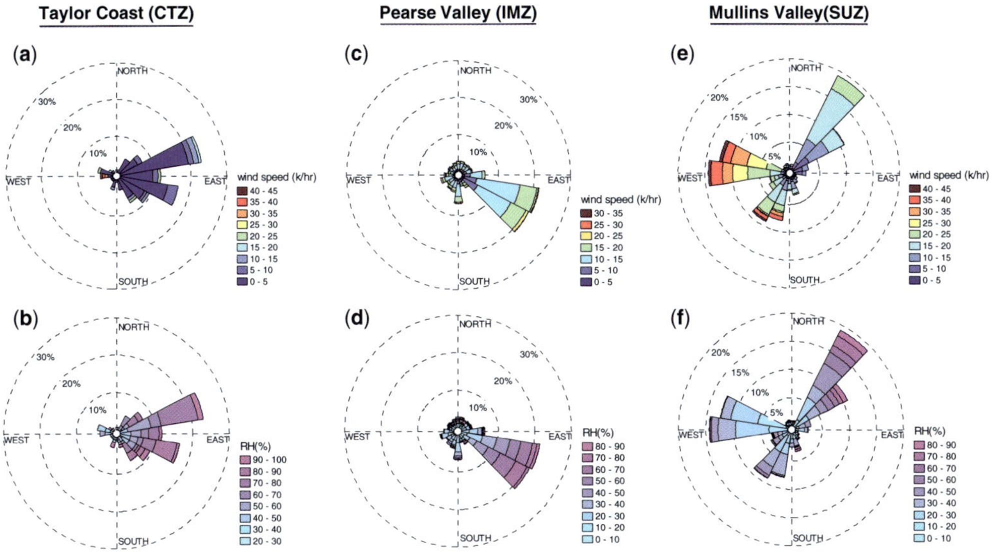

Fig. 2. Wind-rose diagrams showing values for relative humidity, wind-speed and wind direction as measured at multiple automatic weather stations (AWS) in the Dry Valleys region. Data in (**a**) and (**b**) come from lower Taylor Valley, Taylor Coast Met. (coastal thaw zone); data from (**c**) and (**d**) are from Pearse Valley (inland mixed zone); and data from (**e**) and (**f**) are from lower Mullins Valley (stable upland zone) (see Fig. 1 for specific locations of the AWS). Data shown are for the months of December 2007 and January 2008. In all panels, wind directions, divided into 20 bins from 0° to 360°, are represented radially, with longer slices indicating greater winds from these directions during the study interval. The proportion and magnitude of both wind velocity and relative humidity are represented by colour. Although local valley topography influences prevailing wind directions at each location, the general trends call for less frequent, stronger and dryer winds from the west/SW and more frequent, weaker and wetter winds from the east /NE. Note also the general increase in the proportion of westerly winds in the stable upland zone relative to the inland mixed zone and coastal thaw zone.

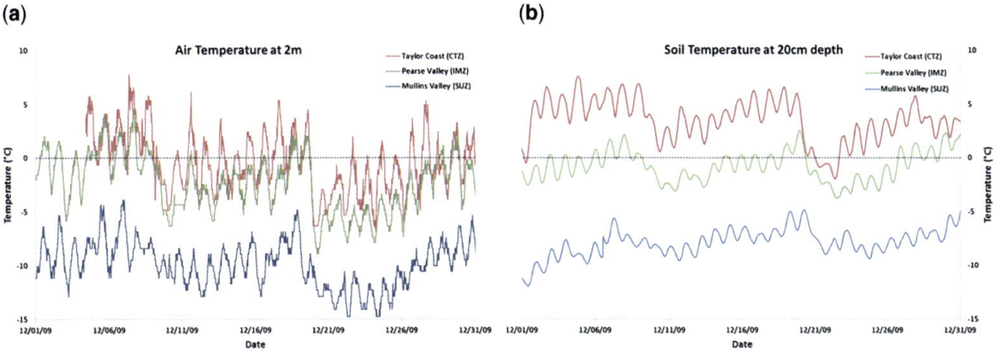

Fig. 3. (**a**) Atmospheric temperatures (measured 2 m above the ground surface) and (**b**) coeval soil temperatures (measured at a 20 cm depth) for weather stations in Taylor Valley (coastal thaw zone, red), Pearse Valley (inland mixed zone, green) and Mullins Valley (stable upland zone, blue); time interval is December 2009. See Figure 1 for the locations of the weather stations. Atmospheric temperatures in the coastal thaw zone and inland mixed zone oscillate above and below 0 °C (dashed horizontal line), but remain less than 0 °C in the stable upland zone. Likewise, soil temperatures remain largely above 0 °C in the coastal thaw zone (red), oscillate above and below 0 °C in the inland mixed zone, and remain well below 0 °C in the stable upland zone. Dates reported in month/day/year format.

however, subsurface soil temperatures in the coastal thaw zone rise well above 0 °C (Table 1, Figs 2 & 3). Most soils in the coastal thaw zone have a gravimetric water content in excess of 30% (Campbell *et al.* 1997*a*, b) and display relatively thick (>20 cm) wet active layers. As seen from data plotted in Figure 2 and Table 1, the winds passing across the coastal thaw zone show an average summertime relative humidity of approximately 62%, reflecting the predominance of easterly winds passing over the Ross Sea (Table 1, Fig. 2). Figure 2 also shows that very strong, but infrequent, westerly winds impact the region. These westerly winds possess low relative humidity values, averaging about 40%. Gooseff *et al.* (2006) showed that, at the valley scale, surface snow, streams and lakes show a decreasing amount of deuterium (δD) with respect to hydrogen with increasing distance from McMurdo Sound.

Inland mixed zone

Alternating winds passing across the inland mixed zone (westerly katabatic and easterly from the Ross Sea) produce variable relative humidity in the inland mixed zone but the mean summertime relative humidity is close to that of the coast, around 54%. Summertime air temperatures show a mean of approximately −5.5°C, with a mean daily maximum of about −3°C (Table 1). Snowfall is most probably less than that of the coastal thaw zone, although uncertainty arises from the unknown quantity of wind-blown snow from the Polar Plateau and the unconstrained, and most probably variable, rate of ablation via sublimation and melting (Fountain *et al.* 2010). Subsurface soil temperatures (Fig. 3) rise above 0°C for most of the summer months and, apart from regions alongside ephemeral streams and isolated snow patches, near-surface soils in the inland mixed zone contain less than 30% gravimetric water content (Campbell *et al.* 1997*a*, 1997b). Such conditions foster widespread, dry active layers.

Stable upland zone

The low values for summertime relative humidity in the stable upland zone, approximately 44% (Table 1), reflect the passage of dry winds originating over the East Antarctic Ice Sheet. Summertime air temperatures show a mean of about −10.5°C, with a mean daily maximum of around −7.5°C (Table 1, Fig. 3). Precipitation is limited but snow blown off the Polar Plateau accumulates on small glaciers and feeds perennial snow banks. Glaciers and snow banks lose mass by sublimation, and the upper horizons of most soils contain less than 5% gravimetric water content (Campbell *et al.* 1997*a*). Shallow, buried-ice deposits and ice-cemented ground do not undergo melting in this zone. Instead, ice loss is almost entirely by sublimation, with unique, diffusion-dominated sublimation profiles preserved in the top few centimetres of remaining ice (Lacelle *et al.* 2011). These conditions foster the development of widespread, dry active layers.

Landform analysis

We now outline the specific geomorphic processes that give rise to various equilibrium landforms in the three microclimate zones. In doing so, we document the spatial variability in landform assemblage across each zone, and assess surface ages and rates of geomorphic change on the basis of dated ash-fall deposits and cosmogenic nuclide analyses of surface boulders. Ultimately, we consider the implications of these results in terms of assessing the stability of both the East and West Antarctic ice sheets over time.

Our analysis is organized by the size of mapped geomorphic features and by microclimate zone. Subdivisions include: (1) macroscale landforms (>250 m in areal extent or those features incised in bedrock); (2) mesoscale landforms (unconsolidated deposits from 1 to *c.* 250 m in areal extent); and (3) microscale landforms (features <1 m in size). One benefit of this scale-dependent evaluation is that similar analyses can be performed for other regions of the Transantarctic Mountains on the basis of high-resolution satellite imagery coupled with digital elevation models (DEMs).

Macroscale landforms: gullies and valley-side asymmetry

Coastal thaw zone. Slopes in the coastal thaw zone display a systematic asymmetry. Although some variability exists, north-facing slopes that receive relatively high levels of incident solar radiation (e.g. Dana *et al.* 1998) are shallower (averaging *c.* 20°) than south-facing slopes (*c.* 25°). The size and spacing of gullies on valley walls in the coastal thaw zone also display local variation with aspect. In lower Ferrar Valley, for example, gullies on the north-facing slopes appear deeper and spaced further apart than those on south-facing slopes (Fig. 4). Melting occurs preferentially on north-facing slopes, with meltwater commonly percolating centimetres to tens of centimetres into soils. In places, this meltwater elevates soil pore pressures sufficiently to induce downslope movement via solifluction; that is, the slow flow of saturated materials. Elsewhere, evaporation of meltwater produces visible salts that coat rock surfaces and

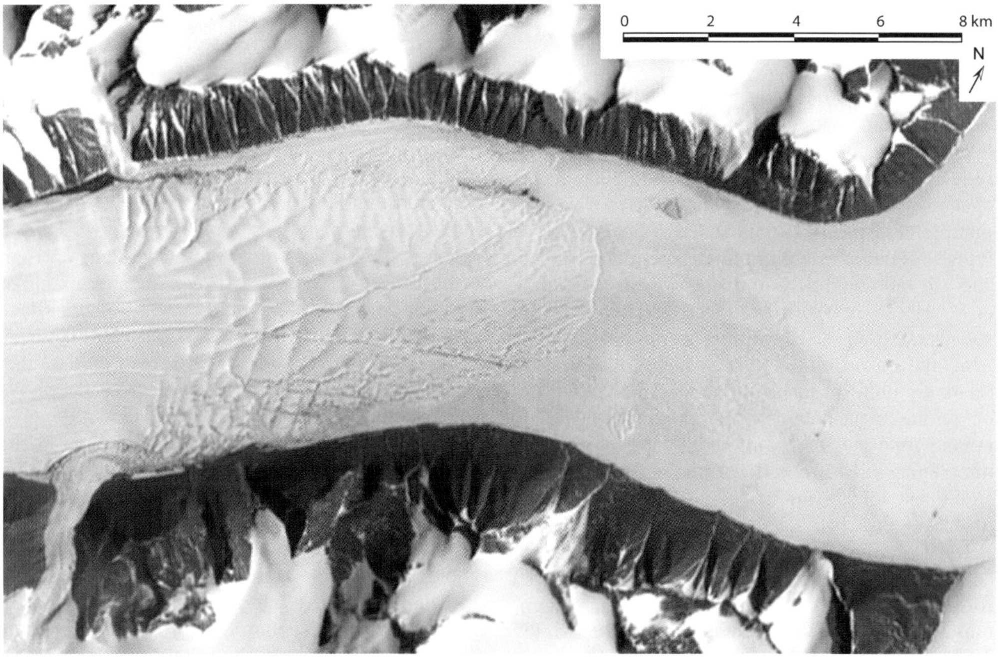

Fig. 4. Landsat satellite image of lower Ferrar Valley (coastal thaw zone) illustrating gully and slope asymmetry the on north-facing (bottom of image) and south-facing (top of image) slopes. The south-facing slope is steep, 30°, and contains numerous shallow and closely spaced gullies; the north-facing slope is gentle, 25°, and possesses fewer, although far-better developed, gullies (adapted from Marchant & Head 2007). Horizontal field of view is 23 km.

intervening soil. Variations in the size of these salt crystals, arising from expansion and contraction upon hydration–dehydration, permits rock breakdown, as salts pry away loosely bound crystals (Huinink *et al.* 2004). In the coastal thaw zone, the most common salt species are chlorides and sulphates, reflecting proximity to the Ross Sea (Claridge & Campbell 1977; Bao *et al.* 2000). Overall, the melting of snow and glaciers on north-facing slopes in the coastal thaw zone is a rather minor process of ice loss, only accounting for about 10% of the total ablation, the majority of the loss takes place by sublimation (Chinn 1980, 1981, 1987; Frezotti 1997; Lewis *et al.* 1998; Fountain *et al.* 1999). Nonetheless, the meltwater produced is sufficient in places to cut channels 3–5 m deep in loose debris. A saturated hyporheic zone, 1–2 m wide (Gooseff *et al.* 2003*a*), commonly fringes these channels during summer months and helps sustain a unique biota of several varieties of cold-adapted nematodes (Gooseff *et al.* 2003*b*; Nkem *et al.* 2006); this glacial meltwater may also lead to the formation of ephemeral ponds (Lyons *et al.* 2012). Snow that falls on south-facing slopes in the coastal thaw zone tends to lose mass via sublimation and, in a relative sense, produces far less geomorphic impact.

Inland mixed zone. The inland mixed zone also shows considerable valley-side asymmetry. Although the average air temperature in the inland mixed zone is less than that of the coastal thaw zone (Table 1, Figs 2 & 3), minor snowmelt fringes most snow banks because snow melts alongside rocks that are heated by solar radiation to temperatures above 0°C. In addition, snow may be blown from local snowbanks onto the surface of solar-heated rocks. In this way the distribution of meltwater is increased well beyond the immediate margins of snowbanks, although it is localized in the downwind direction. The total meltwater contributed by these processes is relatively minor (and does not commonly support solifluction) but appears sufficient to maintain a shallow, discontinuous ice-cemented layer within the upper few centimetres of debris, and may also contribute to shallow subsurface meltwater flow in saline soils (e.g. Lyons *et al.* 2005; Harris *et al.* 2007). Perennial snowbanks and seasonal wind-blown snow trapped in lows (alcoves, channels, polygon troughs) may form significant sources for meltwater that can lead to gully formation in the inland mixed zone (Head *et al.* 2007*a*, *b*; Levy *et al.* 2007). However, gullies in the inland mixed zone are typically relatively shallow (in comparison with those of the coastal thaw zone), appear closely spaced and show sharp, knife-like interfluves.

Stable upland zone. Given that the current climate conditions in the stable upland zone prohibit significant meltwater (Table 1, Figs 2 & 3), there is little evidence for ongoing, macroscale geomorphic change. Gullies, where present in this zone, are interpreted to be relict and inactive because overlying colluvium is commonly interbedded with *in situ* ash fall dated to as much as 11 Ma (Marchant *et al.* 1993*a*, *b*, 1996; Marchant & Denton 1996; see also below). Unlike slopes in the coastal thaw zone and inland mixed zone, which are incised in relatively uniform igneous and metamorphic rocks of the basement complex, the slopes of the stable upland zone are incised in sedimentary rocks and intervening sills of dolerite. Lithology thus plays a significant role in the macroscale geomorphology of the region: cliffs are formed in resistant sills of Ferrar Dolerite and Beacon Heights Orthoquartzite, whereas smooth, rectilinear slopes (*c.* 25°–30°) form on weakly cemented sandstones (Altar Mountain Sandstone, New Mountain Sandstone) (Selby 1971*a*, 1974; Augustinus & Selby 1990; Sugden *et al.* 1995*a*; Prentice *et al.* 1998). Glaciers in the stable upland zone lose mass almost entirely by sublimation. The margins of some glaciers and perennial snow banks may experience melting where intense re-radiation off nearby surfaces elevates local temperatures (e.g. Dana *et al.* 1998; Fountain *et al.* 1999) or where a thin cover of wind-blown sands lowers the surface albedo and induces melting of underlying ice and snow (this phenomenon occurs in all microclimate zones but it is the only mechanism for relatively significant melting in the stable upland zone). In all cases, however, the geomorphic impact of this meltwater in the stable upland zone is negligible. Although meltwater may moisten the upper 5–15 cm of soil, it does not induce solifluction, and evaporation removes most of this water within hours (Kowalewski *et al.* 2006, 2011, 2012).

Synthesis. Macroscale landforms: gullies and valley-side asymmetry

The observed measures for gullies and valley-side asymmetry in the Dry Valleys are plausibly related to spatial variations in the melting of snow and ice (and, where appropriate, the distribution of resistant bedrock lithologies). The mature gully system of the coastal thaw zone most probably reflects: (1) the preponderance of rock breakdown associated with snowmelt, freeze–thaw and salt weathering; and (2) the downslope transport of these weathering products by water, solifluction and wind. Given the direct correlations between the magnitude of observed surface melting and the size and spacing of gullies in the inland mixed zone and coastal thaw zone, we suggest that gully maturation may be analogous to the developmental sequence of rills to master rills observed in cohesive sands in humid temperate regions (Bocco 1991). If so, mature gullies form at the expense of immature gullies by progressive capture, a process that requires significant meltwater and some component of lateral flow. As gullies grow, they are capable of trapping increasing amounts of wind-blown snow, which on melting enables cross-grading (i.e. lateral flow on walls of newly formed gullies) that promotes the wide spacing of gullies. If the above sequence is correct, then the relatively high density and narrow morphology of gullies in the inland mixed zone could reflect stagnation within the maturation sequence due to insufficient meltwater. Among additional parameters that could be responsible for the observed variation in gully morphology is the *duration* of gully incision. We contend that time is not *the* most critical factor in differentiating gully morphology because, as noted below, many slopes in the inland mixed zone (as well as in the stable upland zone) have been dated on the basis of $^{40}Ar/^{39}Ar$ analyses of overlying ash fall to more than 7 Ma (Marchant & Denton 1996, see below). Given the time available for slope evolution in the inland mixed zone, we postulate that gullies would have achieved mature forms if sufficient meltwater had been available (i.e. comparable to that now found in the modern coastal thaw zone). The extremely low levels of meltwater produced in the stable upland zone lead to a preponderance of inherited slopes that probably formed under wetter, and most likely warmer, climate conditions before the onset of cold-polar desert conditions (e.g. Sugden *et al.* 1995*a*, *b*; Lewis *et al.* 2007, 2008).

Mesoscale landforms: contraction-crack polygons and viscous-flow features

Coastal thaw zone. Locally, ice-wedge polygons are the most common form of contraction-crack polygon in the coastal thaw zone (Fig. 5). The seasonal influx of liquid water from wet active layers into open thermal-contraction cracks, along with subsequent growth of ice on freezing, leads to the development of downward-tapering ice wedges outlining raised rim polygons (Berg & Black 1966). Polygons in the coastal thaw zone average about 10–20 m in diameter. The maximum width of most ice wedges is around 2 m, considerably less than that for comparable ice-wedge polygons of Arctic regions (e.g. Mackay 1977). The smaller size of

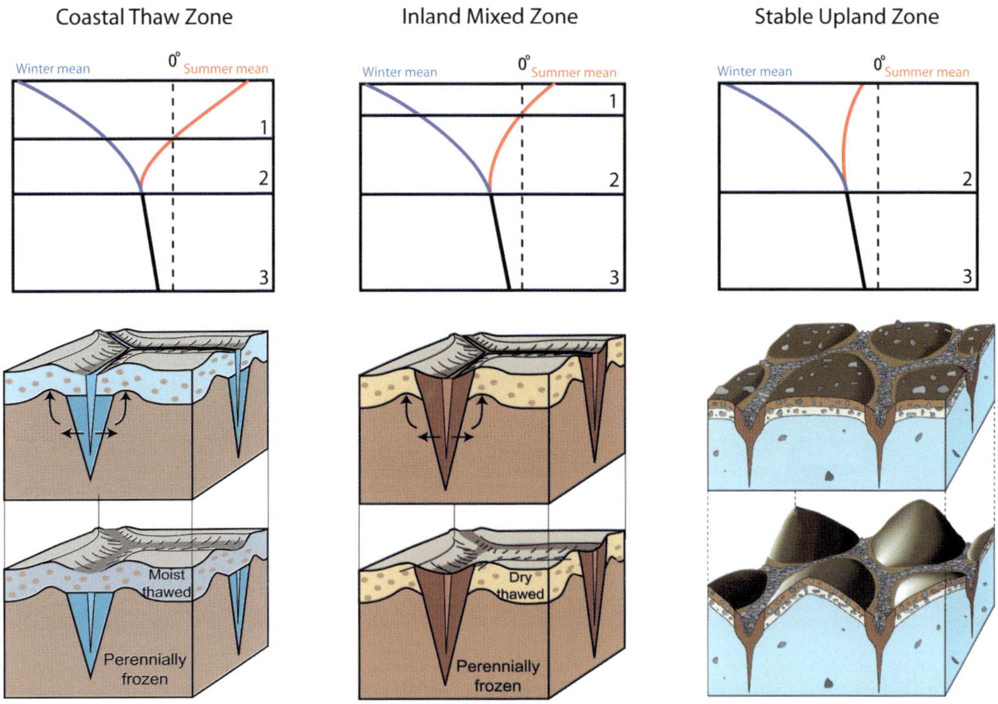

Fig. 5. Schematic soil-temperature profiles and block diagrams for polygon subtypes as a function of microclimate zone. For all soil profiles, layer 1 (if present) is the active layer; layer 2 is the zone of thermal expansion and contraction (with the base being the level of zero annual amplitude); and layer 3 is the remaining permafrost. The top left (coastal thaw zone) soil profile shows a relatively thick active layer (layer 1) with warm mean summer soil temperatures. These conditions foster the development of saturated active layers and ice-wedge polygons, with downward-tapering wedges of ice forming as meltwater percolates into open thermal cracks and subsequently refreezes. The addition of ice to the subsurface in this manner creates compressional stresses (straight arrows) that induce upwards soil movement (curved arrows) yielding raised-rim polygons; blue colours indicate significant soil moisture/ice content. In contrast, mean summer temperatures in the inland mixed zone (centre column) are lower than those of the coastal thaw zone, and these conditions foster relatively shallow active layers. The increased aridity in the inland mixed zone means that most of these active layers are dry (signified by the light tan colour in the polygon schematic) and that sand-wedge polygons, rather than ice-wedge polygons, are the norm. If sufficient sand is available, raised rim, sand-wedge polygons may be produced (Berg & Black 1966). Finally, in the stable upland zone (far-right column), mean summer soil temperatures are below zero, and many regions lack active layers (if active layers exist, they are relatively thin and dry). In the special case where thermal contraction occurs in regions with excess subsurface ice (e.g. ice exceeding pore space), differential sublimation at crack sites promotes the formation of deep troughs, and high-centred sublimation polygons are produced (Marchant *et al.* 2002). Importantly, there is relatively little, if any, net motion at depth in mature sublimation polygons because cohesive plugs, which form at the tops of cracks, typically prevent sediment infill; if some minor sediment is able to penetrate mature cracks, however, ice sublimation along crack margins typically generates sufficient accommodation space to prevent compressive stress and ice deformation (figure adapted from Marchant & Head 2007).

ice wedges in the coastal thaw zone, relative to those of the Arctic, probably reflects the relatively limited availability of liquid water in the coastal thaw zone, rather than the duration of ice-wedge growth.

Solifluction is the dominant form of viscous flow in the coastal thaw zone (Fig. 6). Solifluction lobes are commonly 20–30 cm thick and occur en echelon on slopes as low as 5° (Nichols 1968; Selby 1971*b*). In places, soil pore pressures, and thus rates of solifluction, are elevated where meltwater is unable to penetrate impermeable subsurface horizons, such as a local ice table. As the summer season progresses, shallow ice commonly melts and gives rise to typical, although relatively small, thermokarst features (shallow depressions, planar slides: Swanger *et al.* 2010). The wetting of near-surface debris also commonly produces saline brines (Campbell & Claridge 1987; Bockheim 1997; Campbell *et al.* 1998; Lyons *et al.* 2005). These brines remain in a liquid state, facilitating solifluction even as temperatures drop below the freezing point for pure water.

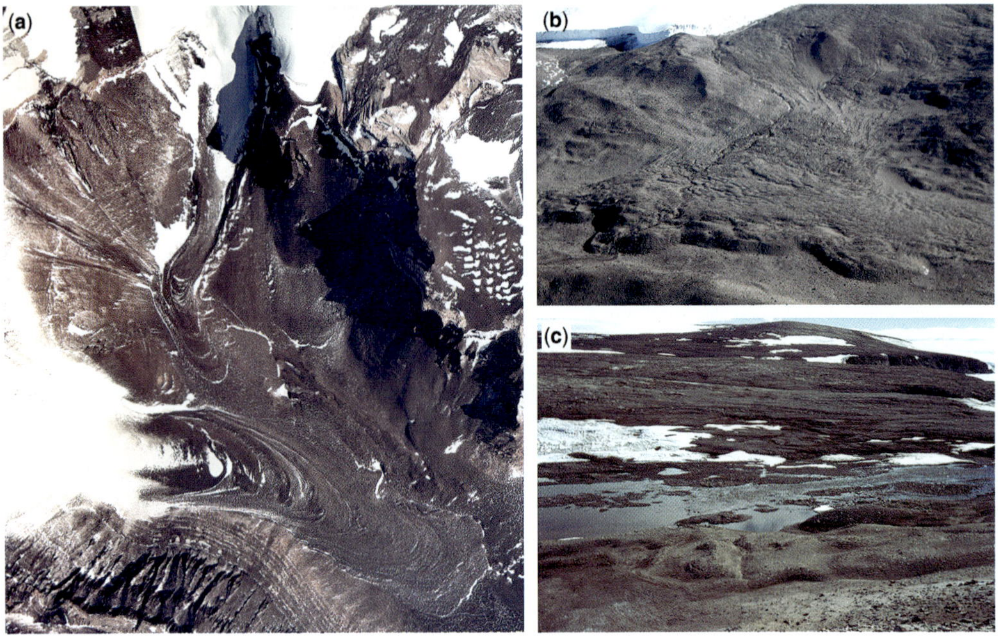

Fig. 6. Examples of viscous-flow features in the Dry Valleys. (**a**) Debris-covered glaciers in upper Beacon Valley, stable upland zone (image from USGS aerial photograph, TMA 3080, frame 227; 11-21-98; horizontal field of view is 3.9 km). The debris is sourced from rockfall at valley heads. Rocks that come to rest on ice accumulation zones are probably buried beneath subsequent snow and ice, facilitating englacial transport before ultimately returning to the ice surface as overlying ice sublimes; rocks that fall onto ice ablation zones, however, travel exclusively as supraglacial debris. The percentage of debris that has been transported solely as supraglacial debris, or that has included a component of englacial flow, is unknown. However, the ratio has probably varied over time as a function of changes in the rate of rockfall and snow accumulation, as well changes in the size and geometry of the ice accumulation zone and nearby bedrock slopes (e.g. Ackert 1998; Kowalewski *et al.* 2011). (**b**) Deformed, ice-cored moraines in central Taylor Valley, inland mixed zone; horizontal field of view is 1 km (Swanger *et al.* 2010). (c) Solifluction lobes in lower Taylor Valley, coastal thaw zone; horizontal field of view is 4 m.

Inland mixed zone. Sand-wedge polygons and composite polygons are the dominant forms of contraction-crack polygons in the inland mixed zone (Fig. 5). Sand-wedge polygons develop in a manner analogous to ice-wedge polygons except that contraction cracks fill with sand, rather than with ice (Péwé 1959; Murton *et al.* 2000). Composite polygons show wedges that exhibit alternating lenses of ice and sand (Ghysels & Heyse 2006). The retention of open contraction cracks at the ground surface, and the availability of sands to fill these cracks are two key factors that determine the growth of sand-wedge polygons (Berg & Black 1966; e.g. Marchant *et al.* 2002). Because cracks tend to be wider and remain open longer in soils with cohesive ice or salt-cemented horizons, sand-wedge polygons are most active near the margins (and downwind) of perennial snow banks that experience minor snowmelt. To a first order, growth rates of sand wedges appear correlated with the abundance of available sediment and with the presence of shallow ice and salt-cemented soil horizons (e.g. Marchant *et al.* 2002). In areas with dry active layers, polygons may be absent even though atmospheric conditions are suitable for thermal expansion and contraction.

Lobate viscous flow features are common along the north wall of central and upper Taylor Valley (Fig. 6). Recent work by Swanger *et al.* (2010) demonstrates that widespread flow features on the north wall of central Taylor Valley probably originated as debris-covered glacier ice (initially, widespread, ice-cored, lateral moraines) about 130 ka ago. The front of the largest lobe (Fig. 6b) is currently advancing downslope at a maximum rate of approximately 7 mm a^{-1}. Interestingly, this creep deformation has transported lobe fronts (with ice up to 30 m thick) well beyond the maximum extent of alpine advance at 130 ka. Overall, this finding highlights the importance of direct burial of ice and snow in the formation of viscous lobes, and is in keeping with recent modelling studies that suggest that soil-moisture conditions in the central and western Dry Valleys region (e.g. areas inland of the coastal thaw zone) appear too dry to permit formation of excess ice (e.g. ice exceeding pore space, including segregation ice), as is observed in these viscous flow features (e.g. Kowalewski *et al.* 2006; see also Heldmann *et al.* 2012).

Stable upland zone. As for the case in the inland mixed zone, sand-wedge polygons are common in the stable upland zone. However, a special type of sand-wedge polygon, a sublimation polygon, forms where sediment overlies massive ice in the stable upland zone (Marchant *et al.* 2002; Marchant & Head 2007) (Fig. 5). The formation and evolution of sublimation-type polygons is tied to the thermal cracking and sublimation of underlying ice (Fig. 5). As cracks form in buried, massive ice, the finest fraction of overlying, dry debris (<2 cm) percolates down into the cracks. Material that is too large (>2 cm) collects at the buried-ice surface. This process of passive sifting and removal of fine-grained material creates a zone of relatively coarse-grained debris at polygon margins (i.e. above contraction cracks). Owing to the relatively high porosity and permeability of this debris, sublimation of the underlying ice is locally enhanced (Marchant *et al.* 2002). Ultimately, elevated rates of ice sublimation at polygon margins leads to the development of deep troughs, more than 2–3 m deep, that outline conical, sediment-covered mounds of buried ice (Fig. 5). Several negative feedbacks prevent runaway ice loss. First, as troughs deepen, they cool relative to polygon centres, as their bases are in shadow for a greater proportion of the day; the lower temperatures reduce rates of ice sublimation (Kowalewski *et al.* 2012). Second, deep troughs become preferred sites for collection of wind-blown snow (and volcanic ash fall, see the section on ‘Chronological control: ash-fall deposits and cosmogenic nuclide dating’). The downwards flux of vapour or minor melt from the base of these snowbanks creates a thin layer of superposed, pore ice (and salt-cemented horizons) that effectively seals the remaining ice from sublimation (Marchant *et al.* 2002; Kowalewski *et al.* 2006, 2011, 2012; Levy *et al.* 2006). Ultimately, cohesive plugs of ice- and salt-cemented sands form at the top of contraction cracks. These plugs prevent further infiltration of overlying sediment, and reduce the potential for compressive stress and ice deformation at depth (e.g. Marchant *et al.* 2002).

Sublimation polygons, particularly those found in central Beacon Valley, were first studied in detail because of an unusual set of seemingly contradictory observations. At first glance, the high-centred morphology of sublimation polygons implies widespread cryoturbation, with near-surface sediment cycling towards troughs and subsurface debris moving towards polygon centres (Sletten *et al.* 2003). However, the presence of ancient, *in situ* ash fall in drift cut by sublimation polygons (Sugden *et al.* 1995*b*; see also below) implies very limited cryoturbation (Marchant *et al.* 1996, 2002; Kowalewski *et al.* 2011). The conceptual model presented in Marchant *et al.* (2002), and later in Levy *et al.* (2006) and Kowalewski *et al.* (2011, 2012), which highlights the effects of localized ice sublimation at polygons troughs (outlined above) appears to resolve this conundrum. The model highlights the potential for equifinality in geomorphology, which postulates that similar-appearing landforms may arise from a number of different processes or from the same process acting over different time spans (Nicholas & Quine 2010). Interestingly, the initial report that called for rapid growth and recycling of sediment at polygon troughs and centres (Sletten *et al.* 2003), did not include data on sublimation polygons. Instead, measures of cryoturbation and recycling were based on analyses of active sand-wedge polygons in lower Beacon Valley (in a region with especially abundant sand supply) and on a variety of other sand-wedge polygons located elsewhere in the Dry Valleys, including coastal sites. Thus, the relatively rapid recycling times reported in Sletten *et al.* (2003) are not necessarily applicable to the sublimation polygons with ancient ash fall in central Beacon Valley, as initially implied. Overall, this discussion highlights the importance of recognizing local variations in process geomorphology, and cautions against the broad application of assigning rates of change across various polygon subtypes and in areas with differing substrates, environmental conditions (sand supply, salt concentrations, etc.) and microclimates.

In terms of viscous-flow features, debris-covered glaciers are the most common form of flow in the stable upland zone. Most debris-covered glaciers occur downwind from dolerite-capped cliffs and originate through the accumulation of rockfall on extant alpine glaciers (Fig. 6a). Consequently, debris-covered glaciers are especially common in the Quartermain Mountains and in the western Asgard Range, where small alpine glaciers typically form in the lee of cliffed-bedrock headwalls (Rignot *et al.* 2002; Shean *et al.* 2007; Shean & Marchant 2010). Rocks and debris that fall onto ice-accumulation zones may travel englacially before rising to the surface as overlying ice sublimes. The debris-covered glaciers in the stable upland zone have extremely low accumulation area ratios (AAR), defined as the ratio of accumulation area to total area (accumulation + ablation). AAR is of the order of about 0.11–0.15 for debris-covered glaciers in Beacon Valley, compared to approximately 0.65 typically assumed for traditional alpine glaciers (Meierding 1982; see also Hambrey *et al.* 2008).

The stratigraphical contact between underlying glacier ice and overlying debris is smooth and dry, and the metre-scale geometry of this surface mimics the modern ground topography (Marchant *et al.* 2002). As dirty ice sublimes, the thickness of the surficial debris layer increases (Kowalewski *et al.* 2011). The rate of ice sublimation is dependent on the thickness, porosity and permeability of this overlying layer (e.g. sublimation till) (Schaefer *et al.* 2000; Marchant *et al.* 2002; Kowalewski *et al.* 2006, 2011, 2012; Levy *et al.* 2006). Recent modelling studies show that, for central Beacon Valley, debris that asymptotically thickens to about 0.5–0.7 m effectively stops sublimation. Sugden *et al.* (1995*b*) argued that remnant debris-covered ice from Taylor Glacier in central Beacon Valley has survived for least 8.1 Ma beneath a thin, approximately 50 cm-thick, dry layer of sublimation till (see also Kowalewski *et al.* 2006). Marchant *et al.* (2007) and Kowalewski *et al.* (2011) have suggested that equally old ice exists in Mullins Glacier (an alpine glacier occupying Mullins Valley and terminating in central Beacon Valley alongside ancient buried ice from Taylor Glacier). Although initial estimates for sublimation of buried ice in the Dry Valleys were extremely high, and at odds with long-term ice preservation (Hindmarsh *et al.* 1998), more recent modelling using local meteorological data show that ice loss is probably an order of magnitude lower than initial estimates, and in central Beacon Valley values for ice loss via sublimation approach 0.022 mm a^{-1} (Kowalewski *et al.* 2012). Further, stable conditions (e.g. no net ice loss or gain) may be achieved with relatively minor deviations from modern summertime climate conditions (Shorghofer 2005, 2007; Kowalewski *et al.* 2006, 2011, 2012); for example, a stable ice surface for ancient Taylor Glacier ice (e.g. the ice described in Sugden *et al.* 1995*a*, *b*) in central Beacon Valley is achieved with a 1.9°C decrease in mean annual atmospheric temperature or a 12% increase in mean annual relative humidity (Kowalewski *et al.* 2012).

Lastly, as a result of the characteristic increase in till thickness as a function of distance down-valley (e.g. Kowalewski *et al.* 2011), ablation rates are highest near valley headwalls and decrease to almost zero near thickly covered glacier termini. This pattern, which is opposite that of clean ice glaciers, leaves uppermost ablation areas vulnerable to climate fluctuations (Whalley 1979; Ackert 1998; Konrad *et al.* 1999). Beheaded debris-covered glaciers (those where topographical hollows exist in former snow-and-ice accumulation areas and in the upper ablation areas of debris-covered glaciers) occur in several places in the stable upland zone, especially in Beacon and adjacent valleys and in the western Asgard Range (Fig. 7). These hollows may reflect a change in climate leading to a marked reduction in ice accumulation or, alternatively, may reflect a reduction in rockfall debris, leading to greater expanses of exposed glacier ice, rapid sublimation and ice-surface lowering (e.g. Marchant & Head 2007).

Synthesis. Mesoscale landforms: contraction-crack polygons and viscous-flow features

Contraction-crack polygons of different types are strong morphological indicators of climate conditions prevailing in a particular zone. As shown in Figure 5, morphological variation arises from the abundance of liquid water in near-surface soils. Ice-wedge polygons of the coastal thaw zone require wet active layers. Sand-wedge polygons of the inland mixed zone signal soil moisture/cohesion sufficient to maintain thermal cracks but insufficient to fill cracks with ice. Sublimation polygons of the stable upland zone indicate the location of near-surface buried ice in regions without wet active layers (Fig. 5). All three types of contraction-crack polygons provide information on the distribution and concentration of subsurface ice. For ice-wedge polygons, the highest concentration of ice occurs in polygon troughs and underlying wedges. For sand-wedge polygons, the ice commonly occurs distributed evenly as pore ice in near-surface sediment (generally $<30\%$ ice by volume). For sublimation polygons, excess ice (ice exceeding available pore space and with values $\gg 30\%$ ice by volume) generally occurs less than 1 m from the ground surface (Marchant *et al.* 2002). In sharp contrast with ice-wedge polygons, troughs at the boundary of sublimation polygons contain the lowest concentration of ice.

The style of viscous-flow developed in each microclimate zone depends on the availability and state of water. If water saturates near-surface soils, as it does in the coastal thaw zone (aided in places by the freezing point depression associated with brines), then flow is best accommodated by solifluction. If water freezes at depth to form excess pore ice, then gelifluction lobes may dominate, although at present the soil-moisture conditions of the inland mixed zone are too dry to permit widespread development of gelifluction lobes (Swanger *et al.* 2010); rather viscous lobes in the inland mixed zone appear related to burial of pre-existing surface-ice deposits, either glaciers or snow banks (Swanger *et al.* 2010; see also Hassinger &

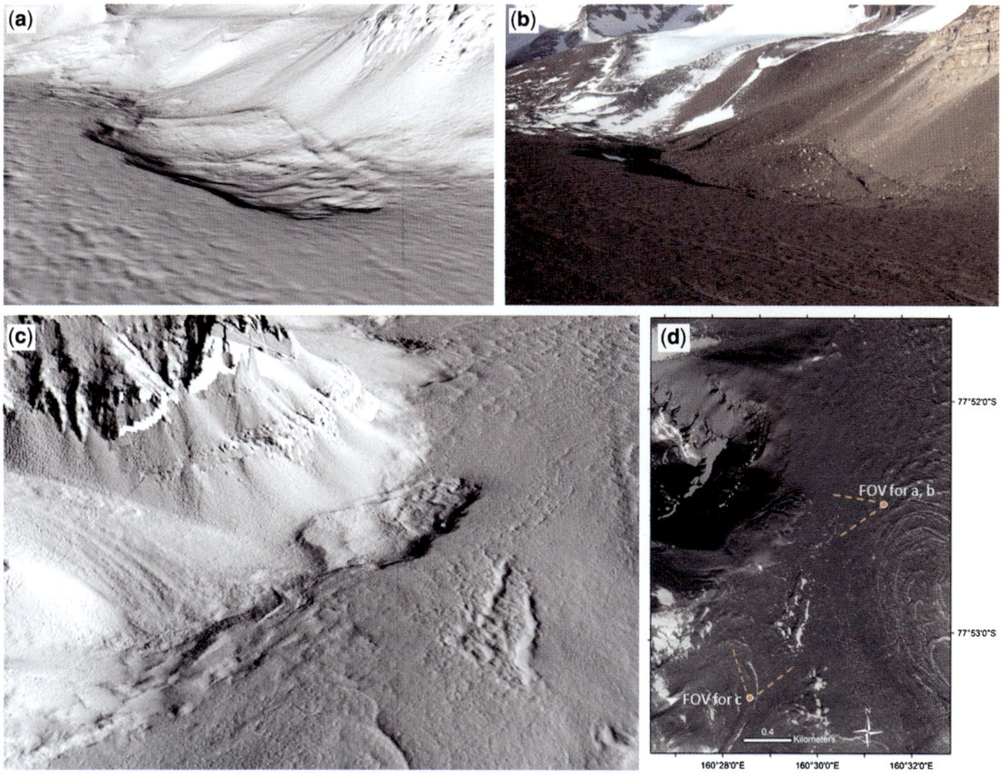

Fig. 7. Multiple images of a beheaded debris-covered glacier in central Beacon Valley, stable upland zone. Beheaded glaciers are those in which topographical hollows exist in former snow and ice accumulation areas or in upper ablation areas (Marchant & Head 2007). Topographical hollows form by down wasting of glacier ice (via sublimation) in areas with minimal supraglacial debris (see the text for details). The hollows reflect either a change in climate, leading to a marked reduction in ice accumulation, or alternatively, a reduction in rockfall debris, leading to greater expanses of exposed glacier ice. Imagery in (**a**), (**c**) and (**d**) show various views of a LiDAR DEM (LIght Detection and Ranging digital elevation model) (*c.* 2 m resolution) with the hillshade applied at 90° azimuth and an angle of 30° off the horizontal. In (a) the view is to the south, (e.g. up glacier) and in (c) the view is down glacier; debris-covered ice exists beyond the hollow. (**b**) Oblique aerial view showing a similar perspective as in (a). (d) Vertical perspective showing field of view (FOV) for imagery in panels (a)–(c); note the scale and north arrow. LiDAR imagery comes from a joint NSF/NASA/USGS effort, with basic processing from the Byrd Polar Research Center.

Mayewski 1983). In the stable upland zone, where there is insufficient meltwater to promote solifluction or to produce excess subsurface ice, debris-covered glaciers dominate. As a check on alternate source mechanisms for subsurface ice in the stable upland zone, Kowalewski *et al.* (2006) showed that the downwards flux of vapour into soils is currently incapable of producing requisite pore-ice volumes that would sustain viscous flow (however, see also Lacelle 2011).

The manner in which buried ice degrades varies as a function of microclimate zone (Swanger *et al.* 2010) (Fig. 8). In the coastal thaw zone, buried ice associated with the late Wisconsinan Ross Sea glaciation (Denton & Hughes 1981; Denton & Marchant 2000; Hall & Denton 2000; Pollard *et al.* 2002) shows extensive physical evidence for melting and topographical degradation by thermokarst processes and wet active-layer cryoturbation (Fig. 8) (Denton & Marchant 2000; Swanger & Marchant 2007; see also Glasser *et al.* 2006 for ice ablation along the nearby McMurdo Ice Shelf). In the inland mixed zone, the top surfaces of buried-ice deposits experience some melting but the evidence for this comes from isotopic variation, rather than from physical changes in the morphology of the buried-ice surface. Meltwater may rework overlying supraglaical debris but the ice surface itself lacks the characteristic hollows and undulating topography typical of thermokarst at the coast. In the stable upland zone, multimillion-year-old buried glacier ice retains its original isotopic signature and

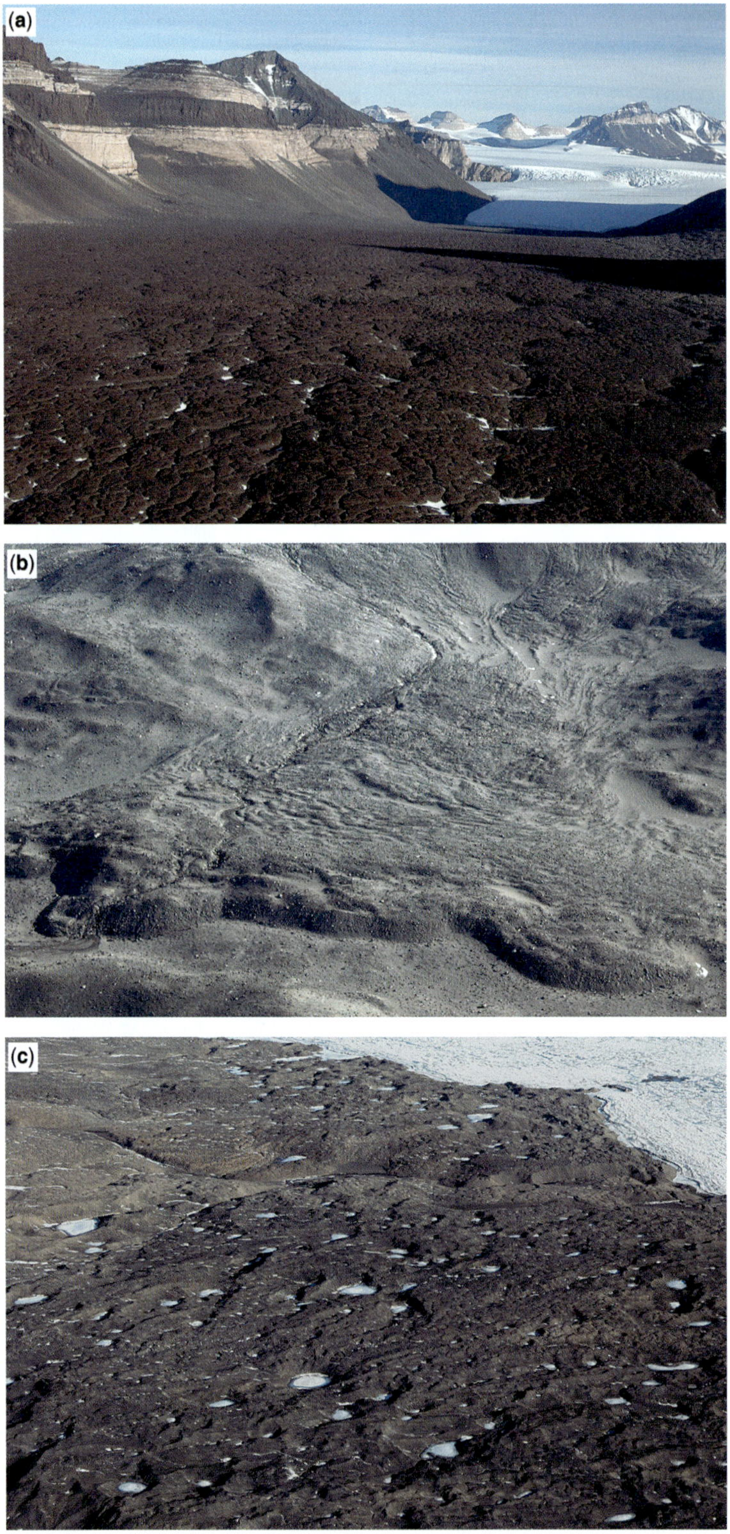
(a)
(b)
(c)

overlying debris shows no evidence for transport by meltwater (Marchant *et al.* 2002; Souchez & Lorrain 2006). Recently, however, Lacelle *et al.* (2011) reported minor, diffusion-driven isotopic variation within the upper 7 cm of some buried-ice deposits in the stable upland zone. At a minimum, the finding confirms that ablation in this region is entirely by sublimation. The important point that follows from this comparison of buried-ice deposits is that, in polar deserts, minor variations in atmospheric temperature play major roles controlling the rate and style of ablation of shallow ice deposits and resultant surface morphologies (Marchant & Head 2007) (Fig. 8). Along microclimate boundaries, as occur between the inland mixed zone and stable upland zones, the impact of slope aspect and variable surface topography may blur the distinction between ice loss wholly by sublimation and ice loss through a combination of widespread sublimation and partial melting on equator-facing slopes (Kowalewski *et al.* 2011, 2012).

Recent numerical-modelling studies of slope stability/regolith transport in the stable upland zone show that values for topographical diffusivity are the lowest of any region yet measured on Earth (the topographical diffusivity relates the rate of sediment transport per unit slope width to the local slope angle, and is measured in $m^2\ a^{-1}$) (Putkonen *et al.* (2008*a*, *b*, 2012). Putkonen *et al.* (2012) calculated topographical diffusivities using a standard regolith-transport model with input from direct measurements of topographical bulges and depressions alongside breaking blocks on 10 different slopes in the stable upland zone. Results show that topographical diffusivities are as low as $10^{-6}\ m^2\ a^{-1}$ which is one or two orders of magnitude lower than for any other reported topographical diffusivity on Earth, and nearly four orders of magnitude lower than most short slopes outside Antarctica (Hanks *et al.* 1984; Fernandes & Dietrich 1997). Importantly, Putkonen *et al.* (2012) noted a wide range of modelled diffusivities across the region, including variable diffusivity along a single slope, and that most of the approximately 1000 boulders measured had not yet achieved steady state conditions (with respect to topographical bulge and down-slope trough formation). For example, of the 997 boulders examined, they found measurable cavity–bulge pairs on 190; and for these 190 boulders, significant variation in calculated topographical diffusivity is implied. Although the results were initially reported as being at odds with long-term preservation of ancient ash fall in the Dry Valleys (Putkonen *et al.* 2012; see also below), the main conclusions appear to be that topographical diffusivities in the stable upland zone are among the lowest on Earth (by several orders of magnitude) and that calculated values are highly variable and site specific. Indeed, some slopes/portions of slopes appear to show no morphological evidence for the development of breaking boulders and downslope movement. Given this, the most parsimonious interpretation of the data is that the ash deposits preserved for millions of years occur on slope segments with very low topographical diffusivities (see discussion below).

Microscale landforms: tafoni, weathering pits, desert pavements and duricrusts

Coastal thaw zone. Most rocks at the surface in the coastal thaw zone show numerous salt encrustations (Nichols 1968; Gibson *et al.* 1983; Campbell & Claridge 1987; Hall 1991) (Fig. 9). These salts are produced by evaporation of near-surface brines and saline meltwater. The growth and expansion of salts pry loose mineral grains, particularly on coarse-grained rocks, and lead to the development

Fig. 8. Variation in the degradation of buried ice as a function of microclimate zone. (**a**) Debris-covered glacier ice in the stable upland zone (distal Mullins Glacier in the foreground, and buried Miocene-age ice in background – the latter is described in Sugden *et al.* 1995*a*, *b* and Marchant *et al.* 2002) about 80 km from the coast and approximately 1300 m elevation. The surface debris is approximately 50–70 cm thick, dry and marked by sublimation polygons (see Fig. 6). Buried ice below this till retains its primary (meteoric) isotopic composition (e.g. Sugden *et al.* 1995*a*, *b*; Souchez & Lorrain 2006; unpublished data), indicating that it has not been modified by extensive melting and refreezing (however, see also Lacelle *et al.* 2011); the field of view in foreground is about 600 m. (**b**) Deformed, ice-cored moraine in central Taylor Valley, inland mixed zone, about 40 km from the coast and around 600 m elevation. Stable isotopic data from the buried-ice surface at this location indicate periods of melting and refreezing, coupled with evaporation and sublimation (Swanger *et al.* 2010). Morphological change associated with widespread thermokarst is lacking, and active meltwater is not pervasive but is, instead, localized alongside melting snow banks and within stream channels; field of view in the foreground is about 400 m. (**c**) Widespread thermokarst degradation of late Wisconsinan-age buried glacier ice in the coastal thaw zone, approximately 1 km from the Ross Sea coast and about 120 m elevation. Circular depressions are filled with ponds that thaw during peak summer conditions (Denton & Marchant 2000); field of view in the foreground is about 700 m (see also Glasser *et al.* 2006 for ablation and melting characteristics on the surface of nearby ice shelves). Altogether, the data show a progressive increase in the degradation of buried-ice deposits from the high-elevation stable upland zone, to the inland mixed zone and, finally, to the coastal thaw zone (figure adapted from Swanger *et al.* 2010).

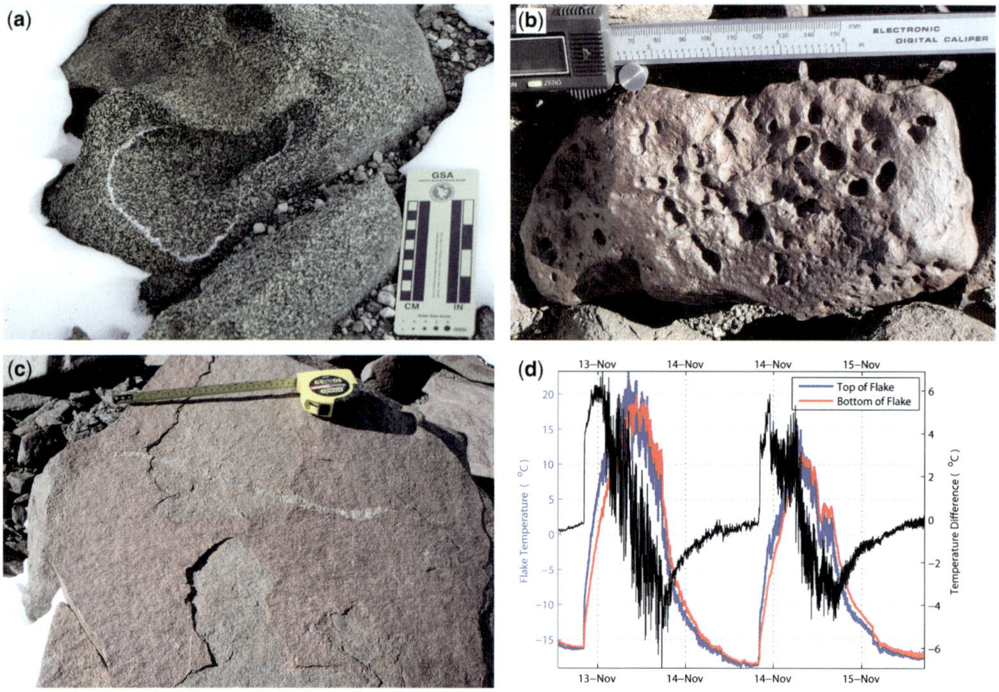

Fig. 9. Microscale rock weathering features in the Dry Valleys. (**a**) Salt weathering on a coarse-grained dolerite clast in the stable upland zone; note the salt ring left after the melting and evaporation of snow from the surface hollow. (**b**) Well-formed pits at the surface of a fine-grained dolerite cobble in the stable upland zone; the top scale on the digital calipers show mm demarcations (see the text for details). (**c**) Thin, mm-scale weathering rinds exfoliating from the surface of a dolerite cobble in the stable upland zone. One possible explanation for these features is compressional and extensional stresses associated with thermal fatigue. (**d**) Field measurements show that rock-surface temperatures vary considerably over short timescales, and that the temperature difference between the top and base of the measured flakes can be up to 8°C, greatly exceeding the rate shown in laboratory experiments to initiate and propagate microfractures (Richter & Simmons 1974); black lines highlight the temperature difference between the top and bottom of flake. Stresses alternate between tension and compression, which may lead to the fracture and detachment of the rinds, as seen in (c). Temperature data were acquired using a type E fine wire thermocouple (FW3, 0.003 inch wire diameter) and a CR850 datalogger, both from Campbell Scientific.

of widespread grus (loose collections of mineral grains). Some salt-coated rocks are weathered flat to the soil surface (Beyer *et al.* 1999); traces of these former upstanding boulders appear as tails of grus, elongated in the downwind direction. Salt weathering in the coastal thaw zone also leads to the development of tafoni through cavernous weathering of coarse-grained rocks (Conca & Astor 1987). Typical salt species in the coastal thaw zone include gypsum ($CaSO_4 \cdot 2H_2O$), halite (NaCl), mirabilite ($Na_2SO_4 \cdot 10H_2O$) and nitratine ($NaNO_3$). As measured over 10 m baselines, the long-term effect of salt weathering in the coastal thaw zone is to smooth bedrock slopes. Observations of slopes at sub-metre scales, however, show considerable relief (Conca & Astor 1987). In addition to salt weathering, surfaces in the coastal thaw zone display evidence for alteration associated with freeze–thaw processes or seasonal meltwater (Kelly & Zumberge 1961; Campbell & Claridge 1987; Hall 1993).

Inland mixed zone. The climate conditions of the inland mixed zone foster the development of widespread desert pavements, with wind-faceted cobbles (ventifacts) and intervening lags of coarse-grained sand and gravel. In regions with fine-grained extrusive rocks (best suited for developing wind-polished facets: Selby 1977), the spacing of ventifacts varies approximately with soil age. Surfaces with interlocking ventifacts and little intervening sand are generally older than are sandy surfaces with widely spaced ventifacts. As in the coastal thaw zone, surface clasts within the inland mixed zone are commonly coated with salts and efflorescences. In areas unaffected by surface snow banks, the current land surface in the inland mixed zone shows little evidence for wet active-layer cryoturbation (Schiller *et al.* 2009).

Stable upland zone. Because rocks in the stable upland zone are not subjected to episodes of repeated burial and exposure, as commonly occurs in areas with traditional, wet active layers (e.g. Hallet & Waddington 1991), rock-surface textures are almost exclusively a function of minor chemical weathering, wind erosion, thermal fracture and salt weathering. Given the very dry conditions of the stable upland zone, salt weathering proceeds at an extremely slow pace. Rates of surface erosion in the stable upland zone that are based on analyses of *in situ*-produced cosmogenic nuclides in boulders are as low as about 6 cm Ma^{-1}, the lowest measured on Earth (Brook *et al.* 1995; Summerfield *et al.* 1998, 1999; Margerison *et al.* 2005; Staiger *et al.* 2006). Micro-relief on rock surfaces is initiated by minor snowmelt that forms on solar-heated rocks. This meltwater tends to occupy shallow surface depressions, which originally may have formed by wind scour on exposed rock surfaces (Selby 1977). Repeated cycles of melting and concentration of ions through water evaporation in depressions are sufficient to produce visible salt encrustations. Microfracturing and undercutting of rock minerals by the growth and hydration of these salts creates pits that, through positive feedback, attract more snow, producing more salts and ever deeper and wider pits (Allen & Conca 1991; Parsons *et al.* 2005; Staiger *et al.* 2006) (Fig. 9). According to Staiger *et al.* (2006), the average width and depth of weathering pits increases linearly with exposure age, with pits at one location in the stable upland zone increasing in width and depth by approximately 10 and 6.7 mm Ma^{-1}, respectively. If inclined rock surfaces are included, this process can also produce a network of pits and micro-rills (Marchant & Head 2007). Micro-rills form where rock surfaces are inclined, such that saline meltwater spills out from pits and flows down the rock surface. The salts in the stable upland zone are most commonly enriched in nitrates, reflecting derivation from snow blown off the ice-sheet margin (Claridge & Campbell 1977; Bao & Marchant 2006). Puzzle rocks, those rocks that are fringed with thermally cleaved and spalled fragments, are also common in the stable upland zone (see also Marchant & Head 2007; Levy *et al.* 2009). For rocks composed of low albedo Ferrar Dolerite (*c.* 0.07: Campbell *et al.* 1997*b*), the transition from solid boulders to partially disintegrated puzzle rocks in the stable upland zone probably requires millions of years (Marchant *et al.* 1993*a*; Staiger *et al.* 2006). Infiltration of snowmelt into the upper few centimetres of silty soils in the stable upland zone helps create a fragile, salt-cemented duricrust. This fragile crust, held together by crystals that bind detrital grains, is susceptible to breakdown upon wetting. Wetted crusts may yield saline solutions that locally enhance downslope movement of near-surface soil–water mixtures (mm- to cm-scale thickness). However, in the absence of surface water, as is commonly the case in the stable upland zone, the salts are enduring and tend to retard downslope movement and aeolian deflation. In addition, owing to their chemical properties and physical ability to reduce near-surface porosity and permeability, salts in duricrusts may slow the flux of water vapour into and out of underlying soil (Marchant *et al.* 2002; Kowalewski *et al.* 2006, 2012).

Recent studies have additionally focused on the role of thermal stress fatigue in the breakdown of rocks in dry, polar regions (Hall 1999, 2003, 2006; Hall & Andre 2001, 2003; Hall *et al.* 2002, 2008*a*, *b*; Elliot 2008; McKay 2009) (Fig. 9c, d). The expansion and contraction of mineral grains from solar heating and cooling can lead to the propagation of microcracks (Yatsu 1988), which may be responsible for the breakdown of coarse-grained granites (Gomez-Heras *et al.* 2006; Hall *et al.* 2008*b*), as well as finer-grained igneous rocks (McKay 2009). By monitoring the *in situ* surface and internal temperatures of exposed clasts, studies have shown that the rate of surface temperature change in the Dry Valleys exceeds that determined by experimental studies (e.g. Richter & Simmons 1974) to create and extend microcracks (Hall & Andre 2003; McKay 2009). Albedo-induced heating of rock surfaces, coupled with convective cooling from gusting winds and cloud-induced shadows, allows for rapid temperature fluctuations (Hall *et al.* 2005; McKay 2009). Although this process is likely to be active in all zones of the Dry Valleys, the lack of liquid water in the stable upland zone implies that thermal fatigue weathering may play a relatively larger role in rock breakdown in this area (Fig. 9).

Synthesis. Microscale topography and rock weathering: tafoni, weathering pits, desert pavements and duricrusts

The relatively wet climate conditions of the coastal thaw zone foster the development of widespread near-surface brines. Successive hydration and dehydration cycles lead to the formation of multiple salt species and relatively rapid rates of salt weathering. In the coastal thaw zone, these rates of rock disintegration appear to outpace the development of well-formed ventifacts and small, rock-weathering pits on most rock surfaces. The rate of degradation via salt weathering appears to decrease inland, with measured rates for bedrock erosion in the stable upland zone being an order of magnitude lower than that measured near the coast (Summerfield *et al.* 1999). The effects of salt weathering are, however, visible in the stable upland zone as thin duricrusts and pitted-rock surfaces; well-formed pits occur only on rocks in the stable upland zone that have been exposed at the ground surface for more than 1–2 Ma (Parsons *et al.* 2005; Staiger *et al.* 2006). Duricrusts tend to be most extensive where wind-blown snow commonly melts on the surface of adjacent, solar-heated rocks. The meltwater may infiltrate the upper few centimetres of soil, evaporate and, over time, leave behind appreciable salts (e.g. Claridge & Campbell 1977; Bockheim 1997; Bao & Marchant 2006). In binding adjacent sand grains, the salt mixture creates a thin duricrust from 1 to 2 cm thick. A simple but important point to emphasize is that surface and near-surface brines in the Dry Valleys may remain in a liquid state at temperatures below 0°C. In extreme situations, such as the case for Don Juan Pond, upper Wright Valley, surface water remains unfrozen even as temperatures drop below −40 to −50°C (e.g. Marion 1997; Takamatsu *et al.* 1998; Healy *et al.* 2006). In addition to effecting geomorphic change, such a situation could impact microbial activity (Siegel *et al.* 1979; Torii *et al.* 1989). Finally, in the coldest and driest regions of the stable upland zone, rock breakdown via thermal fatigue may be an important process and, on a relative scale, of greater importance than heretofore imagined.

Chronological control: ash-fall deposits and cosmogenic nuclide dating

Given the widespread and abundant meltwater in the coastal thaw zone, it is likely that all deposits and landforms in this zone are seasonally active. Indeed, most gullies, solifluction lobes, stream channels, thermokarst depressions, fans, ice-wedge polygons, buried-ice deposits and desert pavements are unequivocally younger than 12 000 years old because they occur superimposed on late Wisconsin Ross Sea drift (Denton *et al.* 1989; Denton & Marchant 2000). Pliocene-age deposits also occur in the coastal thaw zone (Hall *et al.* 1993; Hall & Denton 2000), distal to the late Wisconsin limit, and these appear similarly modified. However, it is not immediately clear whether landforms in other microclimate zones are equally active, occasionally active or even relict (Schiller *et al.* 2009). Unknown, for example, is whether some landforms in the inland mixed zone are active only during extreme climate events (e.g. dramatic warmth following a high snowfall event: Chinn 1993; Barrett *et al.* 2008; Doran *et al.* 2008) or whether they are seasonally active. Likewise, it also is unclear to what extent several of the landforms in the stable upland zone are relict, produced only in the distant past when climate conditions were warmer and wetter than today. To address these questions, and to determine the activity and age of landforms in the inland mixed and stable upland zones, researchers have employed a combination of cosmogenic nuclide analyses of surface boulders, laser-fusion $^{40}Ar/^{39}Ar$ analyses of *in situ* ash fall in unconsolidated deposits (Marchant *et al.* 1993*a–c*, 1996, 2002; Schaefer *et al.* 1999, 2000; Summerfield *et al.* 1999; Margerison *et al.* 2005; Balco *et al.* 2008; Putkonen *et al.* 2008*a*; Balco & Shuster 2009; Morgan *et al.* 2010*a*, *b*) and measures of atmospheric ^{10}Be in Antarctic soils (Schiller *et al.* 2009). Table 2 lists the isotopic age and depositional setting of several ash-fall deposits in the region.

$^{40}Ar/^{39}Ar$ analyses of volcanic ash fall

The benefits of dating landforms with volcanic ash fall are threefold. First, isotopically dated ash-fall deposits afford minimum ages for the surfaces on which they rest. Second, ash-fall deposits can also be used to resolve local climate conditions (temperature and soil moisture) at the time of volcanic deposition. For example, sand-wedge polygons form only in cold and dry microclimates with low relative humidity, such as occur today in the inland mixed zone and stable upland zone. Hence, *in situ* ash-fall deposits that occur in relict sand-wedge troughs in the western Asgard Range, Quatermain Mountains and Olympus Range indicate cold and dry climatic conditions at 15.15 ± 0.02, 13.94 ± 0.75, 13.65 ± 0.06, 12.56 ± 0.1, 12.07 ± 0.03, 10.12 ± 0.03, 10.071 ± 0.989, 8.98 ± 0.25, 8.07 ± 0.06, 7.87 ± 0.43, 7.69 ± 0.10, 7.39 ± 0.08, 7.10 ± 0.11, 4.00 ± 0.47, 3.89 ± 0.5 and 3.73 ± 0.14 Ma (Table 2, Fig. 10). Third, the preservation of *in situ* ash fall for millions of years in near-surface settings provides key data on palaeoclimate change *since* ash-fall deposition. For example, the lack of evidence for subsequent reworking by wet active-layer cryoturbation (e.g. involutions, dissemination of ash in host matrix and waterlain reworking) places important constraints on maximum soil temperatures and soil moisture conditions (see also Schiller *et al.* 2009; Morgan *et al.* 2010*a*). To a first order, the absence of wet active-layer cryoturbation and concomitant deformation of ash-fall deposits, including negligible geochemical alteration of glass shards, implies that the relatively warm and wet conditions of the present-day coastal thaw zone have not advanced inland to the areas that feature widespread

Table 2. *$^{40}Ar/^{39}Ar$ analyses of in situ ash-fall deposits from the Dry Valleys region*

Sample	Crystal age	*N**	Glass age	Latitude (°)	Longitude (°)	Valley/ location[†]	References
Colluvium							
DMS87-113	11.28 ± 0.05	3	12.9 ± 0.06	−77.824	161.011	Lower Arena	Marchant *et al.* (1993*a*)
NPS93-08	10.52 ± 0.67	7		−77.871	160.611	Upper Beacon	Unpublished data
DMS86-110C			8.50 ± 0.33	−77.828	160.973	Lower Arena	Marchant *et al.* (1993*a*)
DMS86-131	7.42 ± 0.04	5		−77.873	160.819	Upper Arena	Marchant *et al.* (1993*a*)
DMF86-141	6.37 ± 0.16	3	7.38 ± 0.14	−77.875	160.836	Upper Arena	Marchant *et al.* (1993*a*)
Lacustrine deposits							
DMS89-211B	14.55 ± 0.06	3	15.08 ± 0.03	−77.600	161.159	W. Asgard	Marchant *et al.* (1993*b*)
ALS05-55B	14.07 ± 0.05	24		−77.459	161.418	W. Olympus	Lewis *et al.* (2008)
DMS96-25D	13.664 ± 0.096	23		−77.596	161.146	W. Asgard	Unpublished data
DMS-91-60	13.262 ± 0.285	15		−77.595	160.804	W. Asgard	Unpublished data
DMS97-1B	10.15 ± 0.11	7		−77.703	162.149	Central Taylor	Unpublished data
DMS97-01	10.09 ± 0.16	7		−77.702	162.145	Central Taylor	Unpublished data
TWS87-142C	7.11 ± 0.12	8		−77.698	162.298	Central Taylor	Marchant & Denton (1996)
Moraines							
ALS04-31k	13.62 ± 0.10	12		−77.473	160.809	W. Olympus	Lewis *et al.* (2007)
ALS00-57			13.36 ± 0.03	−77.508	160.936	W. Olympus	Unpublished data
ALS04-32	12.44 ± 0.16	33		−77.481	160.763	W. Olympus	Lewis *et al.* (2007)
DMS87-27	11.59 ± 0.07	9		−77.856	160.974	Upper Arena	Marchant *et al.* (1993*a*)
DMS94-10	11.401 ± 0.564	12		−78.147	163.829	Royal Society	Unpublished data
DMS87-26	11.27 ± 0.12	5		−77.855	160.964	Upper Arena	Marchant *et al.* (1993*a*)
DMS97-2B	8.10 ± 0.10	7		−77.702	162.122	Central Taylor	Unpublished data
DMS86-100B			1.4 ± 0.2	−77.834	160.990	Lower Arena	Unpublished data
Pavements							
DMS94-004	16.959 ± 0.169	11		−78.087	164.283	Royal Society	Unpublished data
DMS90-124B	14.84 ± 0.28	4	15.24 ± 0.08	−77.603	161.334	W. Asgard	Marchant *et al.* (1993*b*)
ALS04-31d	13.85 ± 0.06	13		−77.474	160.817	W. Olympus	Lewis *et al.* (2007)
ALS02-38c	13.72 ± 0.03	23		−77.474	160.801	W. Olympus	Lewis *et al.* (2007)
ALS00-102	12.37 ± 0.66	6		−77.517	160.896	Upper Wright	Lewis *et al.* (2006)
DMS-86-86B	4.343 ± 0.108	18		−77.862	160.867	Central Arena	Marchant *et al.* (1993*c*)
Hart Ash			3.9 ± 0.3	−77.494	162.393	Lower Wright	Hall *et al.* (1993)
Sand wedges							
DMS91-22	15.15 ± 0.02	10	14.24 ± 0.05	−77.616	161.312	W. Asgard	Marchant *et al.* (1993*b*)
ALS00-30	13.94 ± 0.75			−77.461	160.832	W. Olympus	Lewis *et al.* (2007)
DME91-41	13.65 ± 0.06	5	14.75 ± 0.03	−77.597	160.805	W. Asgard	Marchant *et al.* (1993*b*)
DMS89-143B			13.56 ± 0.69	−77.661	161.014	W. Asgard	Marchant *et al.* (1993*a*)

(Continued)

Table 2. *Continued*

Sample	Crystal age	*N**	Glass age	Latitude (°)	Longitude (°)	Valley/ location[†]	References
DMS91-46	12.56 ± 0.11	2	13.56 ± 0.69	−77.601	160.770	W. Asgard	Marchant *et al.* (1993*b*)
DMS90-38B	12.07 ± 0.03	5		−77.591	161.349	W. Asgard	Marchant *et al.* (1993*b*)
NPS84-327			10.66 ± 0.29	−77.848	160.628	Upper Beacon	Marchant *et al.* (1996)
DMS89-132B	10.12 ± 0.03	5		−77.600	160.933	W. Asgard	Marchant *et al.* (1993*b*)
DMS90-36B			10.08 ± 0.17	−77.590	161.345	W. Asgard	Marchant *et al.* (1993*b*)
DMS00-6-A	10.071 ± 0.989	4		−77.517	160.896	W. Olympus	Unpublished data
NPS93-10			9.95 ± 1.45	−77.855	160.652	Upper Beacon	Unpublished data
DMS97-13B	8.98 ± 0.25	7		−77.841	160.607	Central Beacon	Unpublished data
DMS91-107	8.07 ± 0.06	7		−77.844	160.601	Central Beacon	Sugden et al. (1995*a*, *b*)
NPS88-2	7.87 ± 0.43	7		−77.842	160.602	Central Beacon	Sugden *et al.* (1995*a*, *b*)
EMS98-13A	7.69 ± 0.10	6		−77.836	160.604	Central Beacon	Unpublished data
EMS98-4B	7.39 ± 0.08	3		−77.839	160.588	Central Beacon	Unpublished data
DMS-98-19D	7.10 ± 0.11	3		−77.841	160.590	Central Beacon	Unpublished data
NPS93-02	4.00 ± 0.47	4		−77.872	160.561	Upper Beacon	Unpublished data
NPS93-19	3.89 ± 0.5	4		−77.870	160.551	Upper Beacon	Unpublished data
NPS93-20	3.73 ± 0.14	4		−77.871	160.546	Upper Beacon	Unpublished data

*Number of analyses; one–three volcanic crystals (typically anorthoclase or sanidine) were analysed during each analysis.
[†]W. Asgard , western Asgard Range; W. Olympus, western Olympus Range.
$^{40}Ar/^{39}Ar$ laser total-fusion of volcanic crystals and incremental-heating of volcanic glasses follow previously outlined methods and procedures (Swisher *et al.* 1992; Marchant *et al.* 1993*a*–*c*; Lewis *et al.* 2006, 2007, 2008). Where both crystal and glass fractions were analysed, none of the glasses yielded significantly younger ages than the youngest crystal from the same sample. We consider these findings to verify that the crystal and glass fractions were most probably co-magmatic, and pertain to the same volcanic event. In some cases, the age obtained for the glass was older than the age of crystals extracted from the same sample. In these cases, we attribute the older glass age to the mobility and subsequent uniform loss of potassium during low-temperature hydration (Cerling *et al.* 1985). We therefore consider the youngest crystal age to represent the best age of the deposit (adapted from Marchant *et al.* 1993*b*).

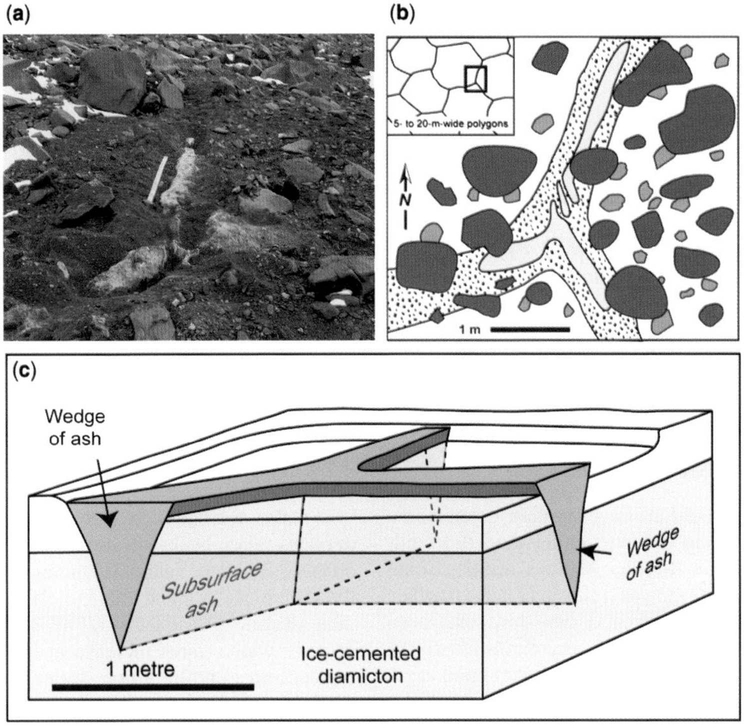

Fig. 10. *In situ* ash-fall deposit within a relict trough from an inactive sand-wedge polygon, western Olympus Range, stable upland zone. (**a**) Oblique view of a shallow soil pit excavated across the intersection of three, formerly buried wedges of concentrated volcanic ash (white) (sample ALS 00-30, dated to 13.94 $\pm$ 0.75 Ma; Table 2) (Lewis *et al.* 2007). Near-surface sand- and-gravel deposits, 5–10 cm thick (e.g. slumped material along the walls of former polygon-troughs: e.g. Kowalewski *et al.* 2012), were removed to expose the underlying ash. The white tape measure in (a) is 1 m long. The ash deposit indicates minimal post-depositional reworking, such as might occur with wet active-layer cryoturbation and development of saturated active layers, for the last 13.94 Ma (the isotopic age of the ash-fall deposit; Table 2: see the text for details). (**b**) Interpretive sketch of ash deposit (vertical view); inset shows the location of excavation with respect to inactive sand-wedge polygons. Based its morphology, the ash was most probably collected in a Y-shaped intersection at the juncture of three former polygon troughs (e.g. Marchant *et al.* 1996) (inset in b). (**c**) Interpretive sketch, showing that the wedge-shaped ash deposit extends more than about 50 cm into the subsurface (adapted from Lewis *et al.* 2007).

in situ ash fall (Marchant & Denton 1996; Marchant *et al.* 1996). Indeed, the preservation of pristine ash fall within ancient sand-wedge troughs, or superposed on ancient ventifact surfaces (Hall *et al.* 1993; Marchant *et al.* 1993*c*, 1996), not only implies cold and dry conditions at the time of ash fall but also cold and dry conditions *at all subsequent times up to the present* (Fig. 10) (Schiller *et al.* 2009; Morgan *et al.* 2010*b*).

Table 2 provides a compendium for dated *in situ* ash-fall deposits in the inland mixed and stable upland zones of the Dry Valleys region (*in situ* and unweathered ash does not occur in the coastal thaw zone). The dated ashes represent direct fallout during volcanic eruptions (Marchant *et al.* 1996): most contain very little detrital contamination (basal sections typically show less than 5% non-volcanic contamination, for example: Marchant *et al.* 1993*c*, 1996); glass shards exhibit delicate spires and intact bubble vesicles (Lewis *et al.* 2007); ashes of different isotopic ages and geochemical compositions are not mixed together (Staiger *et al.* 2006); and ash deposits typically are poorly sorted and show characteristic bi-modal grain size distributions. Such evidence calls for limited post-depositional reworking, and argues against aeolian erosion and re-deposition. We recognize that some ash deposits, particularly those that fell into troughs bounding active sand-wedge polygons (Marchant *et al.* 2002; Kowalewski *et al.* 2012), may have been displaced by several centimetres or more by continued ground contraction (or sublimation of subjacent ice in the case of sublimation polygons: Kowalewski *et al.*

2011, 2012). However, the ashes have not been re-transported since initial deposition in that they still rest within their original morphological traps (Marchant *et al.* 1996).

It is important to note that ash-fall deposits have not been observed in the coastal thaw zone. However, halloysite, a common alteration product produced by the chemical breakdown of altered volcanic glass, is found admixed in sediment along the floor of Wright Valley (Jones *et al.* 1973) and elsewhere in the coastal thaw zone (see also Claridge & Campbell 1974, 2007). The occurrence of halloysite in surficial deposits of the coastal thaw zone is consistent with extensive meltwater, wet active-layer cryoturbation and relatively high rates of chemical weathering.

Cosmogenic nuclide analyses

Numerous investigators have applied cosmogenic nuclide analyses in order to understand the landscape evolution in the Dry Valleys region. Some of the major findings are as follows: (1) near-surface deposits in the stable upland zone have not been reworked by wet active-layer cryoturbation for at least the last several million years (Putkonen *et al.* 2008*a*; Morgan *et al.* 2010*a*, *b*; see also Schiller *et al.* 2009); (2) the topographical diffusivity in the stable upland zone is the lowest of any region measured on Earth (Putkonen *et al.* 2012); (3) erosion rates for exposed bedrock (Schaefer *et al.* 1999; Summerfield *et al.* 1999; Ackert & Kurz 2004) and regolith (Morgan *et al.* 2010*a*) are among the lowest on Earth , reaching approximately 5–20 cm Ma^{-1} in the stable upland zone (and *c.* 1–2 m Ma^{-1} in the coastal thaw zone); and (4) the oldest surfaces at high elevations in the stable upland zone are at least about 12–14 Ma (Schaefer *et al.* 1999, 2000; Margerison *et al.* 2005).

Discussion

Climate and landscape stability

Changes in the spatial distribution of equilibrium landforms over time, or the presence of relict equilibrium landforms, provide a morphological basis for interpreting past and on-going climate change. The preservation of *in situ* Miocene and Pliocene ash-fall deposits in the far-western Dry Valleys region indicates that the relatively warm and wet microclimate of the coastal thaw zone, which permits extensive downslope movement and saturated active-layer cryoturbation, did not advance westwards into the Asgard–Olympus ranges, nor into the Quartermain Mountains (e.g. stable upland zone), during Pliocene time (nor during other post-Miocene warm-Earth intervals: DeConto *et al.* 2012). If atmospheric temperature, soil moisture and relative humidity had risen to levels now experienced at the coast during the summer, then ash-fall deposits and unconsolidated diamictons in the stable upland zone would show evidence of solifluction, erosion by liquid water and widespread cryoturbation from wet active layers, which is not the case. One cannot argue that such features formed during Pliocene time but have since been eroded because the preservation of older *in situ* Miocene and Pliocene ash-fall deposits, sand wedges, gullies and desert pavements indicates that the surface morphology of the stable upland zone is ancient, and that it was not modified extensively during late Miocene, Pliocene or Quaternary time. Hence, the volcanic data of the stable upland zone strongly suggest persistent hyper-arid, cold desert conditions since at least mid-Miocene time (see also Table 2).

Finally, the chemical stability of *in situ* ash-fall deposits also bears on the question of climate change. Volcanic glass is chemically unstable at the ground surface and alters to clay at a rate dependent on atmospheric temperature and the abundance of pore water (rates increase at high atmospheric temperatures and high pore-water pressures: Lowe & Nelson 1983; Lowe 1986). In all but one case, ash deposits in the Dry Valleys contain less than 5% clay, and associated volcanic crystals lack evidence of chemical etching (Marchant *et al.* 1993*c*, 1996). The one exception is a highly altered and weathered ash dated at 16.9 Ma, which occurs along the coast in the foothills of the Royal Society Range (*c.* 250 m elevation). This ash is the oldest yet dated from the McMurdo region, and represents a particularly interesting finding in that it was deposited during the relatively warm interval known as the Mid Miocene climate optimum, a time when atmospheric temperatures in the Dry Valleys were substantially above modern values (Warny *et al.* 2009). In fact, all ash deposits that post-date the subsequent atmospheric cooling that occurred during the Mid Miocene Climate Transition (*c.* 13.8 Ma: Lewis *et al.* 2008) are comprised of fresh-appearing glass shards and little-weathered volcanic crystals. On this basis, we argue that the relatively warm and wet climate conditions of the Mid Miocene Climate Optimum did not return to the stable upland zone during approximately the last 13.8 Ma. Indeed, evidence from fossil tundra and glaciological modelling indicates that the Mid Miocene Climate Transition registered a permanent jump towards colder conditions in the stable upland zone, equivalent to a reduction in summertime atmospheric temperatures of at least 8°C (Lewis *et al.* 2008). This extreme shift, and subsequent climate and geomorphic stability of the stable upland zone, is in stark contrast with measured, large-scale changes in climate and landscape evolution that have occurred in the high Arctic regions throughout the Pliocene and Quaternary Periods (for example, Baffin Island: Kleman *et al.* 2001; Miller *et al.* 2005). In fact, unlike the case for the stable upland zone, most macroscale (and, perhaps, mesoscale) landforms in the Arctic reflect the culmination of alternating geomorphic processes operating under a variety of climates over the last several million years (Bradley 1999). This last point emphasizes the unique aspect of the stable upland zone, and makes it an ideal terrestrial analogue for understanding the origin and evolution of cold-desert landforms on Mars (Head & Marchant 2003; Head *et al.* 2003*b*, 2006, 2011; Marchant & Head 2007).

If correct, and climate conditions in the stable upland zone have remained relatively stable for over 10 Ma, could the same be true of the inland-mixed and coastal thaw zones? Unfortunately, because the vast majority of ash-fall deposits occurs within the stable upland zone, not within the inland mixed zone or coastal thaw zone, the data are inconclusive. However, we find no substantial evidence for dramatic changes in the style of cryoturbation in the inland mixed zone during proposed warm-Earth intervals. Localized warming at low elevations in the Dry Valleys would be registered as a landwards shift of the coastal thaw zone at the expense of the inland mixed zone (e.g. Marchant & Denton 1996). This would pobably increase the areal extent of wet active-layer cryoturbation, promote solifluction and salt weathering, and possibly lead to the development of wide, low-density gullies with rounded interfluves such as those that today occur in the coastal thaw zone. Although this does not appear to be the case, the available geomorphic data do not preclude modest warming of about 3°C at inland sites (Marchant *et al.* 1993*c*; Denton *et al.* 1993; Marchant & Denton 1996) and up to around 8°C at coastal sites (Hall *et al.* 1993). However, detailed geomorphic inspection of the inland mixed zone reveals geomorphic evidence for what may be very slight ongoing climatic warming. Although the majority of landforms in the inland mixed zone appear in equilibrium with local environmental conditions, some ice-cored, viscous flow lobes show evidence of stream dissection, with channels bisecting lobes and producing locally extensive fans of stratified debris (Swanger *et al.* 2010). We suggest that these stream-dissected lobes, the core of which probably formed during the last interglacial (*c.* 130 ka: Swanger *et al.* 2010), may no longer be in equilibrium with current microclimate conditions, and that they may be a harbinger of ongoing climate warming. Regardless, the ice-cored lobes and superposed channels do appear to reflect a palimpsest landscape, hinting at the local migration of dominant geomorphic processes of the coastal thaw zone into the inland mixed zone. If correct, this assertion appears consistent with observations that call for rising lake levels from minor atmospheric warming in the Dry Valleys (Chinn 1993; Priscu 1998; see also Lyons *et al.* 1997, 1998). However, as shown by Hall *et al.* (2010), lake levels in the Dry Valleys show a complex response to varied atmospheric conditions, and rising lake levels are not necessarily tied to regional-scale atmospheric warming.

Implications for Antarctic ice-sheet dynamics

The geomorphic stability of the stable upland zone probably arises from the lasting presence of a robust polar East Antarctic Ice Sheet (e.g. one without extensive surface melting zones: Marchant & Denton 1996). Although east–west thermal gradients across the Dry Valleys may have steepened slightly during the middle Pliocene, climate conditions in the stable upland zone and, hence, regions proximal to the East Antarctic Ice Sheet most probably remained little modified over approximately the last 14 Ma; for example, it is unlikely that the margins of the East Antarctic Ice Sheet in this sector of Antarctica experienced significant surface melting (see also DeConto *et al.* 2012).

Is the record from the Dry Valleys unique? Can it be applied to other regions of the Transantarctic Mountains, or to offshore regions in the Ross Sea? Recent results from the ANDRILL consortium have documented significant, post-Miocene-age changes in the areal extent of grounded ice in the Ross Embayment, with strong implications for dynamic oscillations of the West Antarctic Ice Sheet. Although these documented fluctuations of grounded ice and inferred atmospheric warming in the Ross Embayment (Scherer *et al.* 2008; Naish *et al.* 2009) appear in sync with Milankovitch forcing (Naish *et al.* 2009; Pollard & DeConto 2009), it is unclear what mechanisms drove ice advance and retreat. Rising and falling atmospheric temperatures have been implicated, as have changes in sea level and ocean temperature, but the geomorphic data from the Dry Valleys suggest that inland regions remained relatively cold and dry throughout periods of inferred ice advance and retreat. As such, one likely possibility is that ice-volume fluctuations in the Ross Embayment were most probably related to changes in ocean temperature/circulation, with interior regions of the Dry Valleys and, by extension, ice further inland remaining largely unaffected. If so, the magnitude of ocean warming sufficient to drive retreat of grounded, marine-based ice could have been relatively modest, although at the present time it is unclear how much ice from West Antarctica was lost during recent warm-Earth intervals (Fogwill *et al.* 2012). Indeed, results from numerical modelling using a coupled three-dimensional (3D) ice shelf–ice sheet model show that during Marine Isotope Stage 31 (a warm-Earth interval from *c.* 1.08 to 1.06 Ma), and perhaps during other warm-Earth intervals, ice-volume fluctuations in the West Antarctic Ice Sheet were driven by variations in sub-ice-shelf oceanic melting, with 'insignificant surface melting of East Antarctic ice', even during times of peak warmth (DeConto *et al.* 2012). If correct, this underscores the sensitivity of marine-based ice sheets to subtle changes in ocean circulation and ocean temperature, and also highlights the great stability of local climate conditions in the stable upland zone. On this basis, ongoing research in the Dry Valleys appears focused on determining the magnitude of potential warming in Antarctica during recent warm-Earth intervals. At the present time, for example, there is no clear consensus on the magnitude of warming during the Pliocene Epoch (Lewis *et al.* 2007, 2008; Naish *et al.* 2009; Fielding *et al.* 2012). However, it appears that the initial debate that focused on Pliocene instability of the *terrestrial* East Antarctic Ice Sheet (e.g. Webb *et al.* 1984) has given way to assessing the potential instability of the *marine-based* West Antarctic Ice Sheet.

Summary

- On the basis of measured atmospheric temperature, soil moisture, relative humidity, and prominent wind-speed and direction, we divide the Dry Valleys region into three microclimate zones: a coastal thaw zone; an inland mixed zone; and a stable upland zone. Each zone fosters a unique suite of geomorphic processes and resultant landforms.
- The coastal thaw zone shows numerous active solifluction lobes, streams, gullies and debris flows. The inland-mixed zone features late Miocene- and Pliocene-age landforms, as well as some modern viscous-flow lobes and gullies that occur in protected areas with relatively high moisture content. The stable upland zone exhibits an ancient microtopography; there are no solifluction lobes, active gullies or widespread regions with wet active-layer cryoturbation. Instead, the stable upland zone shows Miocene-age landforms that appear little modified by subsequent cryoturbation or erosion by liquid meltwater.
- The absence of solifluction lobes and stream channels on *in situ* Miocene- and Pliocene-age landforms in the stable upland zone indicates that the relatively mild climates of the coastal thaw zone and inland mixed zone (with mean summer temperature and relative humidity of about $-4^{\circ}C/57\%$ and $-5^{\circ}C/52\%$, respectively) have not advanced westwards into the stable upland zone during at least approximately the last 13.8 Ma (see also Lewis *et al.* 2008).
- Although minor warming may have occurred in the Dry Valleys region during Pliocene time, maximum warming in the stable upland zone was probaby less than 3°C (e.g. Marchant *et al.* 1993*c*) and perhaps up to 8°C in the coastal thaw zone (Hall *et al.* 1993). The implication is that the warm climate conditions necessary for the development of extensive surface-melting ablation zones, ice-margin retreat (Huybrechts 1993; Marchant *et al.* 1993*c*) and growth of vascular vegetation in the Dry Valleys (e.g. Webb & Harwood 1993; Fielding *et al.* 2012) could not have occurred during Pliocene time. In addition, the preservation of *in situ* ash-fall deposits and delicate desert pavements, all of Miocene or Pliocene age, strongly suggests that wet-based erosive glaciers also failed to advance into the stable upland zone (above 1200 m elevation) during Pliocene time.
- Overall, the geomorphic record from the Dry Valleys implies an enduring polar East Antarctic Ice Sheet since Middle Miocene time, and makes it difficult to ascribe large-scale Pliocene sea-level fluctuations (Miller *et al.* 2012) to ice-volume variations on the East Antarctic Craton (e.g. Dowsett & Cronin 1990; Krantz 1991; DeConto *et al.* 2012).

Funding for this research was provided by NSF Polar Programs grants NSF ANT-0944702 and NSF ANT-0739700 to D.R. Marchant. Logistical support for this project in Antarctica was provided by the U.S. National Science Foundation through the U.S. Antarctic Program. We thank A. Lewis, K. Swanger and D. Kowalewski for excellent assistance in the field and many excellent discussions. The manuscript benefited greatly from excellent reviews and commentary from W. Dickinson and N. Glasser.

References

Ackert, R. P., jr. 1998. A rock glacier/debris-covered glacier system at Galena Creek, Absaroka Mountains, Wyoming. *Geografiska Annaler, Series A. Physical Geography*, **80**, 267–276.

Ackert, R. P., jr & Kurz, M. D. 2004. Age and uplift of Sirius Group sediments in the Dominion Range, Antarctic, from surface exposure dating and geomorphology. *Global and Planetary Change*, **42**, 207–225.

Allen, C. C. & Conca, J. L. 1991. Weathering of basaltic rocks under cold, arid conditions. *Antarctica and Mars, Proceedings of Lunar Science Conference*, **21**, 711–717.

ANDERSSON, J. G. 1906. Solifluction, a component of subaerial denudation. *Journal of Geology*, **14**, 91–112.

ATKINS, C. B. & DICKINSON, W. W. 2007. Landscape modification by meltwater channels at margins of cold-based glaciers, Dry Valleys, Antarctica. *Boreas*, **36**, 47–55.

ATKINS, C. B., BARRETT, P. J. & HICOCK, S. R. 2002. Cold glaciers erode and deposit: evidence from Allan Hills, Antarctica. *Geology*, **30**, 659–662.

AUGUSTINUS, P. C. & SELBY, M. J. 1990. Rock slope development in McMurdo Oasis, Antarctica, and implications for interpretations of glacial history. *Geografiska Annaler*, **72**, 55–62.

BALCO, G. & SHUSTER, D. L. 2009. Production rate of cosmogenic Ne-21 in quartz estimated from Be-10, Al-26 and Ne-21 concentrations in slowly eroding Antarctic bedrock surfaces. *Earth and Planetary Science Letters*, **281**, 48–58.

BALCO, G., STONE, J., LIFTON, N. & DUNAI, T. 2008. A simple, internally consistent, and easily accessible means of calculating surface exposure ages and erosion rates from Be-10 and Al-26 measurements. *Quaternary Geochronology*, **3**, 174–195.

BAO, H. & MARCHANT, D. R. 2006. Quantifying sulfate components and their variations in soils of the McMurdo Dry Valleys, Antarctica. *Journal of Geophysical Research*, **111**, D16301, http://dx.doi.org/10.1029/2005JD006669

BAO, H., CAMPBELL, D. A., BOCKHEIM, J. G. & THIEMENS, M. H. 2000. Origin of sulfate in Antarctic Dry Valley soils as deduced from anomalous ^{17}O compositions. *Nature*, **407**, 499–502.

BAO, H., BARNES, J. D., SHARP, Z. D. & MARCHANT, D. R. 2008. Two chloride sources in soils of the McMurdo Dry Valleys, Antarctica. *Journal of Geophysical Research*, **113**, D03301, http://dx.doi.org/10.1029/2007JD008703

BARRETT, J. E., VIRGINIA, R. A. *ET AL.* 2006. Terrestrial ecosystem processes of Victoria Land, Antarctica. *Soil Biology and Biochemistry*, **38**, 301–303.

BARRETT, J. E., VIGINIA, R. A., WALL, D. H., DORAN, P. T., FOUNTAIN, A. G., WELCH, K. A. & LYONS, W. B. 2008. Persistent effects of a discrete warming event on a polar desert ecosystem. *Global Change Biology*, **14**, 1–13.

BARRETT, J. E., GOOSEFF, M. N. & TAKACS-VESBACH, C. 2009. Spatial variation in active-layer geochemistry across hydrologic margins in a polar desert ecosystems. *Hydrology and Earth System Science*, **13**, 2349–2358.

BERG, T. E. & BLACK, R. F. 1966. Preliminary measurements of growth of non-sorted polygons, Victoria Land, Antarctica. *In*: TEDROW, J. C. F. (ed.) *Antarctic Soils and Soil Forming Processes*. American Geophysical Union, Antarctic Research Series, **8**, 61–108.

BEYER, L., BOCKHEIM, J. G., CAMPBELL, I. B. & CLARIDGE, G. G. C. 1999. Review: genesis, properties, and sensitivity of Antarctic gelisols. *Antarctic Science*, **11**, 387–398.

BLACK, R. F. 1973. Growth of patterned ground in Victoria Land, Antarctica. *In*: *Permafrost: North American Contribution to the Second International Permafrost Conference, Yakutsk, Siberia*. National Academies of Science, Washington, DC, 193–203.

BOCCO, G. 1991. Gully erosion; process and models. *Progress in Physical Geography*, **15**, 392–406.

BOCKHEIM, J. G. 1997. Properties and classification of cold desert soils from Antarctica. *Soil Society of America Journal*, **61**, 224–231.

BOCKHEIM, J. G. 2002. Landform and soil development in the McMurdo Dry Valleys: a regional synthesis. *Arctic, Antarctic & Alpine Research*, **34**, 308–317.

BOCKHEIM, J. G. 2003. *University of Wisconsin Antarctic Soils Database*. National Snow and Ice Data Center/World Data Center for Glaciology, Boulder, CO.

BOCKHEIM, J. G., CAMPBELL, I. B. & MCLEOD, M. 2007. Permafrost distribution and active-layer depths in the McMurdo Dry Valleys, Antarctica. *Permafrost and Periglacial Processes*, **18**, 217–227.

BORGSTROM, I. 1999. Basal ice temperatures during late Weichselian deglaciation; comparison of landform assemblages in west-central Sweden. *Annals of Glaciology*, **28**, 9–15.

BRADLEY, R. S. 1999. *Palaeoclimatology: Reconstructing Climates of the Quaternary*. 2nd edn. International Geophysics Series, **68**. Academic Press, London.

BRINER, J. P., MILLER, G. H., DAVIS, P. T. & FINKEL, R. C. 2006. Cosmogenic radionuclides from fiord landscapes support differential erosion by overriding ice sheets. *Geological Society of America Bulletin*, **118**, 406–420.

BROOK, E. J., BROWN, E. T., KURZ, M. D., ACKERT, R. P., RAISBECK, G. M. & YIOU, F. 1995. Constraints on age, erosion, and uplift of Neogene glacial deposits in the Transantarctic Mountains determined from in situ cosmogenic ^{10}Berillium and ^{26}Aluminum. *Geology*, **23**, 1063–1066.

BROOK, M. S., KIRKBRIDE, M. P. & BOCK, B. W. 2006. Quantified time scale for glacial valley cross-profile evolution in alpine mountains. *Geology*, **34**, 637–640.

CAMPBELL, I. B. & CLARIDGE, G. G. C. 1969. A classification of frigic soils- the zonal soils of the Antarctic continent. *Soil Science*, **107**, 75–85.

CAMPBELL, I. B. & CLARIDGE, G. G. C. 1987. *Antarctica: Soils, Weathering Processes, and Environment*. Developments in Soil Science, **16**. Elsevier, New York.

CAMPBELL, I. B. & CLARIDGE, G. G. C. 2006. Permafrost properties, patterns and processes in the Transantarctic Mountains region. *Permafrost and Periglacial Processes*, **17**, 215–232.

CAMPBELL, I. B., CLARIDGE, G. G. C., BALKS, M. R. & CAMPBELL, D. I. 1997*a*. Moisture content in soils of the McMurdo Sound and Dry Valley region of Antarctica. *In*: LYONS, W. B., HOWARD-WILLIAMS, C. & HAWES, I. (eds.) *Ecosystem Processes in Antarctic Ice Free Landscapes*. A.A. Balkema, Rotterdam, 61–76.

CAMPBELL, D. I., MACCULLOCH, R. J. L. & CAMPBELL, I. B. 1997*b*. Thermal regimes of some soils in the McMurdo Sound and Dry Valley region. *In*: LYONS, W. B., HOWARD-WILLIAMS, C. & HAWES, I. (eds) *Ecosystem Processes in Antarctic Ice Free Landscapes*. A.A. Balkema, Rotterdam, 45–56.

CAMPBELL, I. B., CLARIDGE, G. G. C., CAMPBELL, D. I. & BALKS, M. R. 1998. The Soil Environment of the McMurdo Dry Valleys, Antarctica. *In*: PRISCU, J. C. (ed.) *Ecosystem Dynamics in a Polar Desert: The McMurdo Dry Valleys, Antarctica*. American Geophysical Union, Antarctic Research Series, **72**, 297–322.

Cerling, T. E., Brown, E. H. & Bowman, J. R. 1985. Low-temperature alteration of volcanic glass: hydration, Na, K, 0 and Ar mobility. *Chemical Geology: Isotope Geoscience Section*, **52**, 281–293.

Chinn, T. J. 1980. Glacier balances in the Dry valleys area, Victoria Land, Antarctica. *In*: *Proceedings of the Riederlap Workshop*. IAHS-AISH Publication, **126**, 237–247.

Chinn, T. J. 1981. Hydrology and climate in the Ross Sea area. *Journal of the Royal Society of New Zealand*, **11**, 373–386.

Chinn, T. J. H. 1987. Accelerated ablation at a glacier ice-cliff margin, Dry Valleys, Antarctica. *Arctic and Alpine Research*, **19**, 71–80.

Chinn, T. J. 1993. Physical hydrology of the Dry Valley lakes. *In*: Green, W. J. & Friedman, E. I. (eds) *Physical and Biogeochemical Processes in Antarctic Lakes*. American Geophysical Union, Antarctic Research Series, **59**, 1–51.

Claridge, G. G. C. & Campbell, I. B. 1968. Soils of the Shackleton Glacier, Queen Maude Range, Antarctica. *New Zealand Journal of Science*, **11**, 171–218.

Claridge, G. G. C. & Campbell, I. B. 1974. Presence of Halloysite as an indication of volcanic ash in the McMurdo Sound Region. *New Zealand Journal of Geology and Geophysics*, **17**, 679–585.

Claridge, G. G. C. & Campbell, I. B. 1977. The salts in Antarctic soils, their distribution and relationship to soil processes. *Soil Science*, **28**, 377–384.

Claridge, G. G. C. & Campbell, I. B. 2005. Weathering processes in arid cryosols. *In*: Kimble, J. (ed.) *Cryosols Permafrost Affected Soils*. Springer, Berlin, 447–458.

Claridge, G. C. G. & Campbell, I. B. 2007. The 'Golden Shale': an indicator of coastal stability for Marble Point, McMurdo Sound over the last four million years. *In*: Cooper, A. K. & Raymond, C. R. (eds) *Antarctica: A Keystone in a Changing World – Online Proceedings of the 10th ISAES*, extended abstract 087, http://dx.doi.org/10.3133/of2007-1047.srp087. United States Geological Survey, Boulder, CO.

Conca, J. L. & Astor, A. M. 1987. Capillary moisture flow and the origin of cavernous weathering in dolerites of Bull Pass, Antarctica. *Geology*, **15**, 151–154.

Cuffey, K. M., Conway, H. *et al.* 2000. Entrainment at cold glacier beds. *Geology*, **28**, 351–354.

Dana, G. L., Wharton, R. A. & Dubayah, R. 1998. Solar radiation in the McMurdo dry valleys, Antarctica. *In*: Priscu, J. C. (ed.) *Ecosystem Dynamics in a Polar Desert: The McMurdo Dry Valleys, Antarctica*. American Geophysical Union, Antarctic Research Series, **72**, 39–64.

Davies, M. T. L., Atkins, C. B., van der Meer, J. J. M., Barrett, P. J. & Hicock, S. R. 2009. Evidnce for cold-based glacial activity in the Allan Hills, Antarctica. *Quaternary Science Reviews*, **28**, 3124–3137.

Davis, N. 2001. *Permafrost: A Guide to Frozen Ground in Transition*. University of Alaska Press, Fairbanks, AK.

DeConto, R. M., Pollard, D. & Kowalewski, D. E. 2012. Modelling Antarctic ice sheet and climate variations during marine isotope stage 31. *Global and Planetary Change*, **88–89**, 45–52.

Denton, G. H. & Hughes, T. J. 1981. *The Last Great Ice Sheets*. Wiley, Chichester.

Denton, G. H. & Marchant, D. R. 2000. The geologic basis for a reconstruction of a grounded ice sheet in McMurdo Sound, Antarctica, at the last glacial maximum. *Geografiska Annaler*, **82**, 167–211.

Denton, G. H., Bockheim, J. G., Wilson, S. C. & Stuiver, M. 1989. Late Wisconsin and early Holocene glacial history, inner Ross Embayment, Antarctica. *Quaternary Research*, **31**, 151–182.

Denton, G. H., Sugden, D. E., Marchant, D. R., Hall, B. L. & Wilch, T. I. 1993. East Antarctic Ice Sheet sensitivity to Pliocene climatic change from a Dry Valleys perspective. *Geografiska Annaler*, **75A**, 155–204.

Dickinson, W. W. & Rosen, M. R. 2003. Antarctic permafrost: an analogue for water and diagenetic minerals on Mars. *Geology*, **31**, 199–202, http://dx.doi.org/10.1130/0091-7613

Doran, P. T., McKay, C. P., Clow, G. D., Dana, G. L., Fountain, A. G., Nylen, T. & Lyons, W. B. 2002. Valley floor climate observations from the McMurdo dry valleys, Antarctica, 1986–2000. *Journal of Geophysical Research*, **107**, 4772, http://dx.doi.org/10.1029/2001JD002045

Doran, P. T., McKay, C. P., Fountain, A. G., Nylen, T., McKnight, D. M., Jaros, C. & Barrett, J. E. 2008. Hydrologic response to extreme warm and cold summers in the McMurdo Dry Valleys, East Antarctica. *Antarctic Science*, **20**, 499–509, http://dx.doi.org/10.1017/S0954102008001272

Dowsett, H. J. & Cronin, T. M. 1990. High eustatic sea level during the middle Pliocene: Evidence from the southeastern U.S. Atlantic coastal plain. *Geology*, **18**, 435–438.

Elliot, C. 2008. Influence of temperature and moisture availability on physical rock weathering along the Victoria Land coast, Antarctica. *Antarctic Science*, **20**, 61–67.

Elliot, D. E. & Fleming, T. H. 2004. Occurrence and dispersal of magmas in the Jurrasic Ferrar large igneous province, Antarctica. *Gondwana Research*, **7**, 223–237.

Fabel, D., Stroeven, A. P., Harbor, J., Kleman, J., Elmore, D. & Fink, D. 2002. Landscape preservation under Fennoscandian ice sheets determined from in situ produced ^{10}Be and ^{26}Al. *Earth and Planetary Science Letters*, **201**, 397–406.

Fernandes, N. F. & Dietrich, W. E. 1997. Hillslope evolution by diffusive processes: the timescale for equilibrium adjustments. *Water Resources Research*, **33**, 1307–1318.

Fielding, C. R., Harwood, D. M., Winter, D. M. & Francis, J. E. 2012. Neogene stratigraphy of Taylor Valley, Transantarctic Mountains, Antarctica: evidence for climate dynamism and a vegetated early Pliocene coastline of McMurdo Sound. *Global and Planetary Change*, **96–97**, 97–104, http://dx.doi.org/10.1016/j.gloplacha.2010.09.003

Fitzsimons, S. J. 2005. Ice-marginal terrestrial landsystems: polar–continental glacier margins. *In*: Evans, D. (ed.) *Glacial Landsystems*. Hodder Education, London.

Fitzsimons, S. J., McManus, K. J., Sirota, P. & Lorrain, R. D. 2001. Direct shear tests of materials from a cold glacier: implications for landform development. *Quaternary International*, **86**, 129–137.

FOGWILL, C. J., HEIN, A. S., BENTLEY, M. J. & SUGDEN, D. E. 2012. Do blue ice moraines in the Heritage Range show the West Antarctic Ice Sheet survived the last interglacial? *Palaeogeography, Palaeoclimatology, and Palaeocology*, **335–336**, 61–70, http://dx.doi.org/10.1016/j.palaeo.2011.01.027

FOUNTAIN, A. G., LEWIS, K. J. & DORAN, P. T. 1999. Spatial climatic variation and its control on glacier equilibrium line altitude in Taylor Valley, Antarctica. *Global and Planetary Change*, **22**, 1–10.

FOUNTAIN, A. G., NYLEN, T. H., MONAGHAN, A., BASAGIC, H. J. & BROMWICH, D. 2010. Snow in the McMurdo Dry Valleys, Antarctica. *International Journal of Climatology*, **30**, 633–642, http://dx.doi.org/10.1002/joc.1933

FREZOTTI, M. 1997. Ice front fluctuation, iceberg calving flux and mass balance of Victoria Land glaciers. *Antarctic Science*, **9**, 61–73.

GHYSELS, G. & HEYSE, I. 2006. Composite-wedge pseudomorphs in Flanders, Belgium. *Permafrost and Periglacial Processes*, **17**, 145–161.

GIARDINO, J. R., SHRODER, J. F. & VITEK, J. D. 1987. *Rock Glaciers*. Allen and Unwin, London.

GIBSON, E. K., WENTWORTH, S. T. & MCKAY, D. S. 1983. Chemical weathering and diagenesis of a cold desert soil from Wright Valley, Antarctica: an analog for Martian weathering processes. *Journal of Geophysical Research*, **88**, (suppl.), A812–A918.

GILLIES, J. A., NICKLING, W. G. & TILSON, M. 2009. Ventifacts and wind-abraded rock features in the Taylor Valley, Antarctica. *Geomorphology*, **107**, 149–160.

GLASSER, N. F., GOODSELL, B., COPLAND, L. & LAWSON, W. 2006. Debris characteristics and ice-shelf dynamics in the ablation region of the McMurdo Ice Shelf, Antarctica. *Journal of Glaciology*, **52**, 223–234.

GOMEZ-HERAS, M., SMITH, B. J. & FORT, R. 2006. Surface temperature differences between minerals in crystalline rocks: implications for granular disaggregation of granites through thermal fatigue. *Geomorphology*, **78**, 236–249.

GOOSEFF, M. N., MCKNIGHT, D. M., RUNKEL, R. L. & VAUGHN, B. H. 2003*a*. Determining long time-scale hyporheic zone flow paths in Antarctic streams. *Hydrological Processes*, **17**, 1691–1710.

GOOSEFF, M. N., BARRETT, J. E. ET AL. 2003*b*. Snow-patch influence on soil biogeochemical proceses and ibertebrate distribution in the McMurdo Dry Valleys, Antarctica. *Arctic, Antarctic, and Alpine Research*, **35**, 92–100.

GOOSEFF, M. N., LYONS, W. B., MCKNIGHT, D. M., VAUGHN, B. H., FOUNTAIN, A. G. & DOWLING, C. 2006. A stable isotopic investigation of a polar desert hydrologic system, McMurdo Dry Valleys, Antarctica. *Arctic, Antarctic, and Alpine Research*, **38**, 60–71.

HAGEDORN, B., SLETTEN, R. S. & HALLET, B. 2007. Sublimation and ice condensation in hyperarid soils: modelling results using field data from Victoria Valley, Antarctica. *Journal of Geophysical Research, Earth Surface*, **112**, F03017, http://dx.doi.org/10.1029/2006JF000580F03017

HAGEDORN, B., SLETTTEN, R. S., HALLETT, B., MCTIGUE, D. F. & STEIG, E. J. 2010. Ground ice recharge via brine transport in frozen soils of Victoria Valley, Antarctica: insights from modelling 18 O and D profiles. *Geochimica et Cosmochimica Acta*, **74**, 435–448.

HALL, K. J. 1991. Mechanical weathering in the Antarctic: a maritime perspective. *In*: DIXON, J. C. & ABRAHAMS, A. D. (eds) *Periglacial Geomorphology*. John Wiley & Sons, Chichester, 103–123.

HALL, K. 1993. Rock moisture data from Livingston Island (maritime Antarctic) and implications for weathering processes. *Permafrost and Periglacial Processes*, **4**, 245–253.

HALL, K. 1999. The role of thermal stress fatigue in the breakdown of rock in cold regions. *Geomorphology*, **31**, 47–63.

HALL, K. 2003. Micro-transducers and high-frequency rock temperature data: changing our perspectives on rock weathering in cold regions. *In*: PHILLIPS, M., SPRINGER, S. M. & ARENSON, L. U. (eds) *Permafrost*, **Volume 1**. A.A. Balkema, Lisse, 349–354.

HALL, K. 2006. Perceptions of rock weathering in cold regions: a discussion on space and time attributes of scale. *Geomorphologie-Relief Processus Environnement*, 187–196.

HALL, K. & ANDRE, M. F. 2001. New insights into rock weathering from high-frequency rock temperature data: an Antarctic study of weathering by thermal stress. *Geomorphology*, **41**, 23–35.

HALL, K. & ANDRE, M. F. 2003. Rock thermal data at the grain scale: applicability to granular disintegration in cold environments. *Earth Surface Processes and Landforms*, **28**, 823–836.

HALL, B. L. & DENTON, G. H. 2000. Radiocarbon chronology of Ross Sea drift, eastern Taylor Valley, Antarctica; evidence for a grounded ice sheet in the Ross Sea at the last glacial maximum. *Geografiska Annaler*, **82A**, 305–336.

HALL, B. L., DENTON, G. H., LUX, D. R. & BOCKHEIM, J. G. 1993. Late Tertiary Antarctic palaeoclimate and ice-sheet dynamics inferred from surficial deposits in Wright Valley. *Geografiska Annaler*, **75**, 239–268.

HALL, K., GUGLIELMIN, M. & STRINI, A. 2008*a*. Weathering of granite in Antarctica: I. Light penetration into rock and implications for rock weathering and endolithic communities. *Earth Surface Processes and Landforms*, **33**, 295–307.

HALL, K., GUGLIELMIN, M. & STRINI, A. 2008*b*. Weathering of granite in Antarctica: II. Thermal stress at the grain scale. *Earth Surface Processes and Landforms*, **33**, 475–493.

HALL, K., THORN, C. E., MATSUOKA, N. & PRICK, A. 2002. Weathering in cold regions: some thoughts and perspectives. *Progress in Physical Geography*, **26**, 577–603.

HALL, K., LINDGREN, B. S. & JACKSON, P. 2005. Rock albedo and monitoring of thermal conditions in respect of weathering: some expected and some unexpected results. *Earth Surface Processes and Landforms*, **30**, 801–811.

HALL, B. L., DENTON, G. H., FOUNTAIN, A. G., HENDY, C. H. & HENDERSON, G. M. 2010. Antarctic lakes suggest millennial reorganizations of Southern Hemisphere atmospheric and oceanic circulation. *Proceedings of the National Academy of Sciences*, **107**, 21,355.

HALLET, B. & WADDINGTON, D. E. 1991. Buoyancy forces induced by freeze-thaw in the active layer:

implications for diapirism and soil circulation. *In*: DIXON, J. C. & ABRAHAMS, A. D. (eds) *Periglacial Geomorphology*. John Wiley & Sons, Chichester, 251–279.

HALLET, B., HUNTER, L. & BOGEN, J. 1996. Rates of erosion and sediment evacuation by glaciers: a review of field data and their implications. *Global and Planetary Change*, **12**, 213–235.

HALLET, B., SLETTEN, R. & WHILDEN, K. 2011. Micro-relief development in polygons patterned ground in the Dry Valleys of Antarctica. *Quaternary Research*, **75**, 347–355.

HAMBREY, M. J. & FITZSIMONS, S. J. 2010. Development of sediment-landform associations at cold glacier margins, Dry Valleys, Antarctica. *Sedimentology*, **57**, 857–882.

HAMBREY, M. J., QUINCEY, D. J., GLASSER, N. F., REYNOLDS, J. M., RICHARDSON, S. J. & CLEMMENS, S. 2008. Sedimentological, geomorphological and dynamic context of debris-mantled glaciers, Mount Everest (Sagarmatha) region, Nepal. *Quaternary Science Reviews*, **27**, 2361–2389.

HANKS, T. C., BUCKNAM, R. C., LAJOIE, K. R. & WALLACE, R. E. 1984. Modification of wave-cut and faulting-controlled landforms. *Journal of Geophysical Research*, **89**, 5771–5790.

HARRIS, K. J., CAREY, A. E., LYONS, W. B., WELCH, K. A. & FOUNTAIN, A. G. 2007. Solute and isotope geochemistry of subsurface ice melt seeps in Taylor Valley, Antarctica. *Geological Society of America Bulletin*, **119**, 548–555.

HASSINGER, J. M. & MAYEWSKI, P. A. 1983. Morphology and dynamics of rock glaciers in southern Victoria Land, Antarctica. *Arctic and Alpine Research*, **15**, 351–368.

HEAD, J. W. & MARCHANT, D. R. 2003. Cold-based mountain glaciers on Mars: Western Arsia Mons. *Geology*, **31**, 641–644.

HEAD, J. W., MARCHANT, D. R., AGNEW, M. C., FASSETT, C. I. & KRESLAVSKY, M. A. 2006. Extensive valley glacier deposits in the northern mid-latitudes of Mars: evidence for Late Amazonian obliquity-driven climate change. *Earth and Planetary Science Letters*, **241**, 663–671.

HEAD, J. W., MARCHANT, D. R., DICKSON, J. L., LEVY, J. S. & MORGAN, G. A. 2007*a*. Transient streams and gullies in the Antarctic Dry Valleys: geological setting, processes and analogs to Mars. *In*: COOPER, A. K. & RAYMOND, C. R. (eds) *Antarctica: A Keystone in a Changing World*. Online Proceedings of the 10th ISAES, extended abstract 151. United States Geological Survey, Boulder, CO.

HEAD, J. W., MARCHANT, D. R., DICKSON, J. L., LEVY, J. S. & MORGAN, G. A. 2007*b*. Slope streaks in the Antarctic Dry Valleys: characteristics, candidate formation mechanisms, and implications for slope streak formation in the Martian environment. *In*: COOPER, A. K. & RAYMOND, C. R. (eds) *Antarctica: A Keystone in a Changing World*. Online Proceedings of the 10th ISAES, extended abstract 177. United States Geological Survey, Boulder, CO.

HEAD, J. W., KRESLAVSKY, M. A. & MARCHANT, D. R. 2011. Pitted rock surfaces on Mars: a mechanism of formation by transient melting of snow and ice. *Journal of Geophysical Research*, **116**, E09007, http://dx.doi.org/10.1029/2011JE003826

HEALY, M., WEBSTER-BROWN, J. G., BROWN, K. L. & LANE, V. 2006. Chemistry and stratification of Antarctic meltwater ponds II: inland ponds in the McMurdo Dry Valleys, Victoria Land. *Antarctic Science*, **18**, 525–533.

HELDMANN, J. L., MARINOVA, M. ET AL. 2012. Formation and evolution of buried snowpack deposits in Pearse Valley, Antarctica, and implications for Mars. *Antarctic Science*, **24**, 299–316, http://dx.doi.org/10.1017/S0954102011000903

HINDMARSH, R. C. A., VAN DER WATEREN, F. M. & VERBERS, A. 1998. Sublimation of ice through sediment in Beacon Valley, Antarctica. *Geografiska Annaler Series A, Physical Geography*, **80A**, 209–219.

HUININK, H. P., PEL, L. & KOPINGA, K. 2004. Stimulating the growth of Tafoni. *Earth Surface Process Landforms*, **29**, 1225–1233.

HUYBRECHTS, P. 1993. Glaciological modeling of the late Cenozoic East Antarctic Ice Sheet: stability or dynamism? *Geografiska Annaler*, **75A**, 221–238.

IKARD, S. J., GOOSEFF, M. N., BARRETT, J. E. & TAKACS-VESBACH, C. 2009. Thermal characterization of active layer across a soil moisture gradient in the McMurdo Dry Valleys, Antarctica. *Permafrost and Periglacial Processes*, **20**, 27–39.

JONES, L. M., WHITNEY, J. A. & STORMER, J. C., JR. 1973. A volcanic ash deposit, Wright Valley. *Antarctic Journal of the United States*, **8**, 270–272.

KELLY, W. C. & ZUMBERGE, J. H. 1961. Weathering of a Quartz Diorite at Marble Point, McMurdo Sound, Antarctica. *Journal of Geology*, **69**, 433–446.

KLEMAN, J. & HATTESTRAND, C. 1999. Frozen-bed Fennoscandian and Laurentide ice sheets during the last glacial maximum. *Nature*, **402**, 63–66.

KLEMAN, J., MARCHANT, D. R. & BORGSTROM, I. 2001. Late glacial ice dynamics on southern Baffin Island and in Hudson Strait. *Arctic Antarctic and Alpine Research*, **33**, 249–257.

KONRAD, S., HUMPHREY, N., STEIG, E., CLARK, D., POTTER, N. & PFEFFER, W. 1999. Rock glacier dynamics and palaeoclimatic implications. *Geology*, **27**, 1131–1134.

KOPPES, M. & HALLET, B. 2006. Erosion rates during rapid deglaciation in Icy Bay, Alaska. *Journal of Geophysical Research*, **111**, F02023, http://dx.doi.org/10.1029/2005JF000349

KOWALEWSKI, D. E., MARCHANT, D. R., LEVY, J. S. & HEAD, J. W., III. 2006. Quantifying low rates of summertime sublimation for buried glacier ice in Beacon Valley, Antarctica. *Antarctic Science*, **18**, 421–428.

KOWALEWSKI, D. E., MARCHANT, D. R., SWANGER, K. M. & HEAD, J. W. 2011. Modelling vapor diffusion within cold and dry supraglacial tills of Antarctica: implications for the preservation of ancient ice. *Geomorphology*, **126**, 159–173.

KOWALEWSKI, D. E., MARCHANT, D. R., HEAD, J. W. & JACKSON, D. W. 2012. A 2D model for characterizing first-order variability in sublimation of buried glacier ice, Antarctica: assessing the influence of polygon troughs, desert pavements, and shallow-subsurface salts. *Permafrost and Periglacial Processes*, **23**, 1–14, http://dx.doi.org/10.1002/ppp.731

KRANTZ, D. 1991. A chronology of Pliocene sea-level fluctuations: The U.S. middle Atlantic Coastal Plain record. *Quaternary Science Reviews*, **10**, 163–174.

KYLE, P. R. 1990. The McMurdo Volcanic Group–Western Ross Embayment. *In*: LE- MASURIER, W. E. & THOMSON, J. W. (eds) *Volcanoes of the Antarctic Plate and Southern Oceans*. American Geophysical Union, Antarctic Research Series, **48**, 19–134.

LACELLE, D., DAVILA, A. F., POLLARD, W. H., ANDERSEN, D., HELDMANN, J., MARINOVA, M. & MCKAY, C. P. 2011. Stability of massive ground ice bodies in University Valley, McMurdo Dry Valleys of Antarctica: using stable O-H isotope as tracers of sublimation in hyper-arid regions. *Earth and Planetary Science Letters*, **301**, 403–411, http://dx.doi.org/10.1016/j.epsl.2010.11.028

LACHENBRUCH, A. H. 1962. *Mechanics of Thermal Contraction Cracks and Ice-Wedge Polygons in Permafrost*. Geological Society of America, Special Papers, **70**.

LANCASTER, N. 2002. Flux of Eolian sediment in the McMurdo Dry Valleys, Antarctica: a preliminary assessment. *Arctic, Antarctic, and Alpine Research*, **34**, 318–323.

LANCASTER, N. 2004. Relations between aerodynamic and surface roughness in a hyper-arid cold desert: McMurdo Dry Valleys, Antarctica. *Earth Surface Processes and Landforms*, **29**, 853–867.

LANGBEIN, W. B. & SCHUMM, S. A. 1958. Yield of sediment in relation to mean annual precipitation. *American Geophysical Union Transactions*, **39**, 1023–1036.

LEVY, J. S., MARCHANT, D. R. & HEAD, J. W. 2006. Distribution and origin of patterned ground on Mullins Valley debris-covered glacier, Antarctica: the roles of ice flow and sublimation. *Antarctic Science*, **18**, 385–397.

LEVY, J., HEAD, J. W., MARCHANT, D. R., MORGAN, G. A. & DICKSON, J. L. 2007. Gully surface and shallow subsurface structure in the South Fork of Wright Valley, Antarctic Dry Valleys: implications for gully activity on Mars. *Lunar and Planetary Science*, **38**, Abstract 1728.

LEVY, J. S., HEAD, J. W. & MARCHANT, D. R. 2009. Cold and Dry Processes in the Martian Arctic: Geomorphic Observations at the Phoenix Landing Site and Comparisons with Terrestrial Cold Desert Landforms. *Geophysical Research Letters*, **36**, L21203, doi: 10.1029/2009GL040634.

LEVY, J. S., FOUNTAIN, A. G., GOOSEFF, M. N., WELCH, K. A. & LYONS, W. B. 2011. Water tracks and permafrost in Taylor Valley, Antarctica: extensive and shallow groundwater connectivity in a cold desert ecosystem. *Geological Society of America Bulletin*, **123**, 2295–2311, http://dx.doi.org/10.1130/B30436.1

LEWIS, K. J., FOUNTAIN, A. G. & DANA, G. L. 1998. Surface energy balance and meltwater production for a dry valleys Glacier, Taylor Valley, Antarctica. *Annals of Glaciology*, **27**, 603–609.

LEWIS, A. R., MARCHANT, D. R., BALDWIN, S. L. & WEBB, L. E. 2006. The age and origin of the Labyrinth, western Dry Valleys, Antarctica: evidence for extensive middle Miocene subglacial floods and freshwater discharge to the Southern Ocean. *Geology*, **34**, 513–516.

LEWIS, A. R., MARCHANT, D. R., ASHWORTH, A. C., HEMMING, S. R. & MACHLUS, M. L. 2007. Major middle Miocene global climate change: evidence from East Antarctica and the Transantarctic Mountains. *Geological Society of America Bulletin*, **119**, 1449–1461, http://dx.doi.org/10.1130B26134.1

LEWIS, A. R., MARCHANT, D. R. ET AL. 2008. Mid-Miocene cooling and the extinction of tundra in continental Antarctica. *Proceedings of the National Academy of Sciences*, **105**, 10,676–10,689, http://dx.doi.org/10.1073pnas.0802501105

LOWE, D. J. 1986. Controls on the rates of weathering and clay mineral genesis in airfall tephras: a review and New Zealand case study. *In*: COLMAN, S. M. & DETHIER, D. P. (eds) *Rates of Chemical Weathering of rocks and Minerals*. Academic Press, New York, 265–330.

LOWE, D. J. & NELSON, C. S. 1983. *Guide to the Nature and Methods of Analysis of the cla Fraction of Tephras from the South Auckland Region*. Occasional Report, **11**. University of Waikato, Department of Earth Sciences, Hamilton, New Zealand.

LYONS, W. B., HOWARD-WILLIAMS, C. & HAWES, I. (eds) 1997. *Ecosystem Processes in Antarctic Ice-Free Landscapes*. A.A. Balkema, Rotterdam.

LYONS, W. B., TYLER, S. W., WHARTON, R. A., JR., MCKNIGHT, D. M. & VAUGH, B. H. 1998. A late Holocene desiccation of Lake Hoare and Lake Fryxell, McMurdo Dry Valleys, Antarctica. *Antarctic Science*, **10**, 247–256.

LYONS, W. B., WELCH, K. A. ET AL. 2005. Groundwater seeps in Taylor Valley, Antarctica: an example of a subsurface melt event. *Annals of Glaciology*, **40**, 200–206.

LYONS, W. B., WELCH, K. E., GARDNER, C. B., JAROS, C., MOORHEAD, D. L., KNOEPFLE, J. L. & DORAN, P. T. 2012. The geochemistry of upland ponds, Taylor Valley, Antarctica. *Antarctic Science*, **24**, 3–14.

MACKAY, J. R. 1977. *The Widths of Ice Wedges*. Geological Survey of Canada, Professional Papers, **77–1A**, 43–44.

MAHANEY, W. C., DOHM, J. M. ET AL. 2001. Morphogenesis of Antarctic Paleosols: Martian analogue. *Icarus*, **154**, 113–130.

MALIN, M. C. 1984. Abrasion rate observations in Victoria Valley, Antarctica: 340-day experiment. *Antarctic Journal of the United States*, **19**, 14–16.

MALIN, M. C. 1987. Abrasion in ice-free areas of southern Victoria Land, Antarctica. *Antarctic Journal of the United States*, **21**, 38–39.

MALIN, M. C. 1991. Short-term variations in the rate of eolian processes, southern Victoria Land, Antarctica. *Antarctic Journal of the United States*, **26**, 27–29.

MARCHANT, D. R. & DENTON, G. H. 1996. Miocene and Pliocene paleoclimate of the Dry Valleys region, southern Victoria Land. A geomorphological approach. *Marine Micropaleontology*, **27**, 253–271.

MARCHANT, D. R. & HEAD, J. W. 2007. Antarctic Dry Valleys: Microclimate zonation, variable geomorphic processes, and implications for assessing climate change on Mars. *Icarus*, **192**, 187–222, http://dx.doi.org/10.1016/j.icarus.2007.06.018

MARCHANT, D. R. & HEAD, J. W. 2010. Geologic analogies between the surface of Mars and the

McMurdo Dry Valleys: microclimate-related geomorphic features and evidence for climate change. *In*: Doran, P. T., Lyons, W. B. & McKnight, D. M. (eds) *Life in Antarctic Deserts and Other Cold Dry Environments: Astrobiological Analogues*. Astrobiology Series. Cambridge University Press, Cambridge, 9–78.

Marchant, D. R., Denton, G. H. & Swisher, C. C., III. 1993*a*. Miocene-Pliocene-Pleistocene Glacial History of Arena Valley, Quartermain Mountains, Antarctica. *Geografiska Annaler*, **75A**, 269–302.

Marchant, D. R., Denton, G. H. & Sugden, D. E. 1993*b*. Miocene glacial stratigraphy and landscape evolution of the western Asgard Range, Antarctica. *Geografiska Annaler*, **75A**, 303–330.

Marchant, D. R., Swisher, C. C., III, Lux, D. R., West, D. P., jr. & Denton, G. H. 1993*c*. Pliocene paleoclimate and East Antarctic ice-sheet history from surficial ash deposits. *Science*, **260**, 667–670.

Marchant, D. R., Denton, G. H., Swisher, C. C., III & Potter, N., jr. 1996. Late Cenozoic Antarctic paleoclimate reconstructed from volcanic ashes in the Dry Valleys region, south Victoria Land. *Geological Society of America Bulletin*, **108**, 181–194.

Marchant, D. R., Lewis, A., Phillips, W. C., Moore, E. J., Souchez, R. & Landis, G. P. 2002. Formation of patterned-ground and sublimation till over Miocene glacier ice in Beacon Valley, Antarctica. *Geological Society of America Bulletin*, **114**, 718–730.

Marchant, D. R., Phillips, W. M. et al. 2007. Establishing a chronology for the world's oldest glacier ice. *In*: Cooper, A. K. & Raymond, C. R. (eds) *International Symposium on Antarctic Earth Sciences, Santa Barbara, CA*. US Geological Survey and The National Academies, Santa Barbara, CA.

Margerison, H. R., Phillips, W. M., Stuart, F. M. & Sugden, D. E. 2005. Cosmogenic He-3 concentrations in ancient flood deposits from the Coombs Hills, northern Dry Valleys, East Antarctica: interpreting exposure ages and erosion rates. *Earth and Planetary Science Letters*, **230**, 163–175.

Marion, G. M. 1997. A theoretical evaluation of mineral stability in Don Juan Pond, Wright Valley, Victoria Land, Antarctica. *Antarctic Science*, **9**, 92–99.

Marshall, G. J. & Turner, J. 1997. Katabatic wind propoagation over the western Ross Sea observed using ERS-1 scatterometer data. *Antarctic Science*, **9**, 221–226.

Matsuoka, N. 2001. Solifluction rates, processes, and landforms: a global review. *Earth-Science Reviews*, **44**, 107–134.

McKay, C. P. 2009. Snow recurrence sets the depth of dry permafrost at high elevation in the McMurdo Dry Valleys of Antarctica. *Antarctic Science*, **21**, 89–94.

McKay, C. P., Molaro, J. L. & Marinova, M. M. 2009. High-frequency rock temperature data from hyper-arid desert environments in the Atacama and the Antarctic Dry Valleys and implications for rock weathering. *Geomorphology*, **110**, 182–187.

McKnight, D. M., Niyogi, D. K., Alger, A. S., Bomblies, A., Cenovitz, P. A. & Tate, C. M. 1999. Dry Valley streams in Antarctica: ecosystems waiting for water. *Bioscience*, **49**, 985–995, http://dx.doi.org/10.2307/1313732

Meierding, T. C. 1982. Late Pleistocene glacial equilibrium-line altitudes in the Colorado Front Range: a comparison of methods. *Quaternary Research*, **18**, 289–310.

Miller, G. H., Wolfe, A. P., Briner, J. P., Sauer, P. E. & Nesje, A. 2005. Holocene glaciation and climate evolution of Baffin Island, Arctic Canada. *Quaternary Science Reviews*, **24**, 1703–1721.

Miller, K. G., Wright, J. D., Browning, J. V., Kulpecz, A., Kominz, M., Naish, T. R., Cramer, B. S., Rosenthal, Y., Peltier, W. R. & Sosdian, S. 2012. High tide of the warm Pliocene: Implications of global sea level for Antarctic deglaciation. *Geology*, First published online March 19, 2012, doi: 10.1130/G32869.1

Morgan, D., Putkonen, J., Balco, G. & Stone, J. 2010*a*. Quantifying regolith erosion rates with cosmogenic nuclides 10Be and 26Al in the McMurdo Dry Valleys, Antarctica. *Journal of Geophysical Research*, **115**, F03037, http://dx.doi.org/10.1029/2009JF001443

Morgan, D., Putkonen, J., Balco, G. & Stone, J. 2010*b*. Degradation of glacial deposits quantified with cosmogenic nuclides, Quartermain Mountains, Antarctica. *Earth Surface Processes and Landforms*, **36**, 217–228.

Murton, J. B., Worsley, P. & Gozdzik, J. 2000. Sand veins and wedges in cold Aeolian environments. *Quarterly Science Review*, **19**, 899–922.

Naish, T., Powell, R. et al. 2009. Obliquity-paced Pliocene West Antarctic ice sheet oscillations. *Nature*, **458**, 322–329.

Nichols, R. L. 1968. Coastal Geomorphology, McMurdo Sound. *Antarctica Journal of Glaciology*, **7**, 449–478.

Nicholas, A. P. & Quine, T. A. 2010. Quantitative assessment of landform equifinality and palaeoenvironmental reconstruction using geomorphic models. *Geomorphology*, **121**, 167–183.

Nkem, J. N., Virginia, R. A., Barrett, J. E., Wall, D. H. & Li, G. 2006. Salt tolerance and survival thresholds for two species of Antarctic soil nematodes. *Polar Biology*, **28**, 643–651.

Nylen, T., Fountain, A. G. & Doran, P. 2004. Climatology of katabatic winds in the McMurdo Dry Valleys, southern Victoria land, Antarctica. *Journal of Geophysical Research*, **109**, D03114, http://dx.doi.org/10.10292-2003JD003937

Parsons, R., Head, J. W. & Marchant, D. R. 2005. Weathering pits in the Antarctic Dry Valleys: insolation-induced heating and melting and applications to Mars. *Lunar and Planetary Science Conference*, **36**, Abstract 1138.

Péwé, T. L. 1959. Sand-wedge polygons (tesselations) in the McMurdo Sound region, Antarctica–a progress report. *American Journal of Science*, **257**, 545–552.

Pollard, D. & DeConto, R. M. 2009. Modelling West Antarctic ice sheet growth and collapse through the past five million years. *Nature*, **458**, 329–332, http://dx.doi.org/10.1038/nature07809

Pollard, W., Doran, P. & Wharton, R. 2002. The nature and significance of massive ground ice in Ross

Sea Drift, Garwood Valley, McMurdo Sound. *Royal Society of New Zealand Bulletin*, **35** (Antarctica at the Close of the Millennium), 397–404.

PRENTICE, M. L., KLEMAN, J. K. & STROEVEN, A. P. 1998. The composite glacial erosional landscape of the northern McMurdo Dry Valleys: implications for Antarctic Tertiary glacial history. *In*: PRISCU, J. C. (ed.) *The McMurdo Dry Valleys: A Cold-Desert Ecosystem.* American Geophysical Union, Antarctic Research Series, **72**, 1–38.

PRISCU, J. C. (ed.) 1998. *The McMurdo Dry Valleys: A Cold-Desert Ecosystem.* American Geophysical Union, Antarctic Research Series, **72**.

PUTKONEN, J., BALCO, G. & MORGAN, D. 2008*a*. Slow regolith degradation without creep determined by cosmogenic nuclide measurements in Arena Valley, Antarctica. *Quaternary Research*, **69**, 242–249.

PUTKONEN, J., ROSALES, M., TURPEN, N., MORGAN, D., BALCO, G. & DONALDSON, M. 2008*b*. Regolith Transport in the Dry Valleys of Antarctica. *In*: COOPER, A. K., BARRETT, P. J., STAGG, H., COOPER, A. K., STOREY, B., STUMP, E. & WISE, W. (eds) *Antarctica: A Keystone in Changing World. Proceedings of the 10th International Symposium on Antarctic Earth Sciences.* The National Academies Press, Santa Barbara, CA, http://dx.doi.org/10.3133/of2007-1047.srp103

PUTKONEN, J., MORGAN, D. J. & BALCO, G. 2012. Regolith transport quantified by braking block, McMurdo Dry Valleys, Antarctica. *Geomorphology*, **155–156**, 80–87.

RICHTER, D. & SIMMONS, G. 1974. Thermal-expansion behavior of igneous rocks. *International Journal of Rock Mechanics and Mining Sciences*, **11**, 403–411.

RIGNOT, E., HALLET, B. & FOUNTAIN, A. 2002. Rock glacier surface motion in Beacon Valley, Antarctica, from synthetic-aperture radar interferometry. *Geophysical Research Letters*, **29**, 1607, http://dx.doi.org/10.1029/2001GL013494

SCHAEFER, J. M., IVY-OCHS, S., WIELER, R., LEYA, I., BAUR, H., DENTON, G. H. & SCHLUCTHER, C. 1999. Cosmogenic noble gas studies in the oldest landscape on Earth: surface exposure ages of the Dry Valleys, Antarctica. *Earth and Planetary Science Letters*, **167**, 215–226.

SCHAEFER, J. M., BAUER, H., DENTON, G. H., IVY-OCHS, S., MARCHANT, D. R., SCHLÜCHTER, C. & WIELER, R. 2000. The oldest ice on Earth in Beacon Valley, Antarctica: new evidence from surface exposure dating. *Earth and Planetary Science Letters*, **179**, 91–99.

SCHERER, R. P., BOHATY, S. M. & EARTH AND PLANETARY SCIENCE LETTERS *ET AL*. 2008. Antarctic record of precession-paced insolation-driven warming during early Pleistocene Marin Isotope Stage 31. *Geophysical Research Letters*, **35**, L03505, http://dx.doi.org/10.1029/2007GL032254

SCHILLER, M., DICKINSON, W., DITCHBURN, R. G., GRAHAM, I. J. & ZONDERVAN, A. 2009. Atmospheric 10Be in an Antarctic soil: implications for climate change. *Journal of Geophysical Research*, **114**, F01033, http://dx.doi.org/10.1029/2008JF001052

SCHUMM, S. A. 1965. Quaternary paleohydrology. *In*: WRIGHT, H. E. & FREY, D. G. (eds) *The Quaternary of the United States*. Princeton University Press, Princeton, NJ.

SCHUMM, S. A. & LICHTY, R. W. 1965. Time, space and causality in geomorphology. *American Journal of Science*, **263**, 110–119.

SCHWERDTFEGER, W. 1984. *Weather and Climate of the Antarctic*. Developments in Atmospheric Science, **15**. Elsevier, Amsterdam.

SELBY, M. J. 1971*a*. Slopes and their development in an ice-free, arid area of Antarctica. *Geografiska Annaler*, **53**, 235–245.

SELBY, M. J. 1971*b*. Some solifluction surfaces and terraces in the ice-free valleys of Victoria Land, Antarctica. *New Zealand Journal of Geology and Geophysics*, **14**, 469–476.

SELBY, M. J. 1974. Slope Evolution in an Antarctic Oasis. *New Zealand Geographer*, **30**, 18–34.

SELBY, M. J. 1977. Transverse erosional marks on ventifacts from Antarctica. *New Zealand Journal of Geology and Geophysics*, **20**, 949–969.

SHEAN, D. E. & MARCHANT, D. R. 2010. Seismic and GPR surveys of Mullins Glacier, McMurdo Dry Valleys, Antarctica: ice thickness, internal structure and implications for surface ridge formation. *Journal of Glaciology*, **56**, 48.

SHEAN, D. E., HEAD, J. W. & MARCHANT, D. R. 2007. Shallow seismic surveys and ice thickness estimates of the Mullins Valley debris-covered glacier, McMurdo Dry Valleys, Antarctica. *Antarctic Science*, **19**, 485–496.

SCHORGHOFER, N. 2005. A physical mechanism for long-term survival of ground ice in Beacon Valley, Antarctica. *Geophysical Research Letters*, **32**, L19503, http://dx.doi.org/10.1029/2005GL023881

SCHORGHOFER, N. 2007. Theory of ground ice stability in sublimation environments. *Physical Review E*, **75**, 041201, http://dx.doi.org/10.1103/PhysRevE.75.041201

SIEGEL, B. Z., MCMURTY, G., SIEGEL, S. M., CHEN, J. & LAROCK, P. 1979. Life in the calcium chloride environment of Don Juan Pond, Antarctica. *Nature*, **280**, 828–829.

SLETTEN, R. S., HALLET, B. & FLETCHER, R. C. 2003. Resurfacing time of terrestrial surfaces by the formation and maturation of polygonally patterned ground. *Journal of Geophysical Research*, **108**, 8044, http://dx.doi.org/10.1029/2002JE001914

SOUCHEZ, R. & LORRAIN, R. 2006. The environmental significance of deuterium excess in meteoric and non-meteoric Antarctic ice. *In*: KNIGHT, P. (ed.) *Glacier Science and Environmental Change.* Blackwell, Oxford, 179–184.

SPEIRS, J. C., STEINHOFF, D. F., MCGOWAN, H. A., BROMWICH, D. H. & MONAGHAN, A. J. 2010. Foehn winds in the McMurdo Dry Valleys, Antarctica: the origin of extreme warming events. *Journal of Climate*, **23**, 3577–3598.

SPOTILA, J. A., BUSCHER, J. T., MEIGS, A. J. & REINERS, P. W. 2004. Long-term glacial erosion of active mountain belts: example of the Chugach-St. Elias Range, Alaska. *Geology*, **32**, 501–504.

STAIGER, J. W., MARCHANT, D. R., SCHAEFER, J. M., OBERHOLZER, P., JOHNSON, J. V., LEWIS, A. R. & SWANGER, K. M. 2006. Plio-Pleistocene history of

Ferrar Glacier, Antarctica: implications for climate and ice sheet stability. *Earth and Planetary Science Letters*, **243**, 489–503.

SUGDEN, D. E., DENTON, G. H. & MARCHANT, D. R. 1995*a*. Landscape evolution of the Dry Valleys, Transantarctic Mountains: tectonic implications. *Journal of Geophysical Research*, **100**, 9949–9967.

SUGDEN, D. E., MARCHANT, D. R., POTTER, N., JR., ROLAND SOUCHEZ, DENTON, G. H., SWISHER, C. C. & TISON, J.-L. 1995*b*. Miocene glacier ice in Beacon Valley, Antarctica. *Nature*, **376**, 412–416.

SUMMERFIELD, M. A., STUART, F. M., COCKBURN, H. A. P., SUGDEN, D. E., DENTON, G. H., DUNAI, T. & MARCHANT, D. R. 1998. Long-term rates of denudation in the Dry Valleys region of the Transantarctic Mountains, southern Victoria Land based on *in situ* produced cosmogenic Ne-21. *Geomorphology*, **27**, 113–129.

SUMMERFIELD, M. A., SUGDEN, D. E., DENTON, G. H., MARCHANT, D. R., COCKBURN, H. A. P. & STUART, F. M. 1999. Cosmogenic isotope data support previous evidence of extremely low rates of denudation in the Dry Valleys region, southern Victoria Land, Antarctica. *In*: SMITH, B. J., WHALLEY, W. B. & WARKE, P. A. (eds) *Uplift, Erosion and Stability: Perspectives on Long-term Landscape Development*. Geological Society, London, Special Publications, **162**, 255–267.

SWANGER, K. M. & MARCHANT, D. R. 2007. Sensitivity of ice-cemented Antarctic soils to greenhouse-induced thawing: are terrestrial archives at risk? *Earth and Planetary Science Letters*, **259**, 347–359.

SWANGER, K. M., MARCHANT, D. R., KOWALEWSKI, D. E. & HEAD, J. W. 2010. Viscous flow lobes in central Taylor Valley, Antarctica: origin as remnant buried glacial ice. *Geomorphology*, **120**, 174–185, http://dx.doi.org/10.1016/j.geomorph.2010.03.024

SWISHER, C. C., GRAJALES-NISHIMURA, J. M. ET AL. 1992. Chicxulub crater melt- rock and K-T boundary tektites from Mexico and Haiti yield coeval Ar/39Ar ages of 65 Ma. *Science*, **257**, 954–958.

TAKAMATSU, N., KATO, N. & MATSUMOTO, G. I. 1998. The origin of salts in water bodies of the McMurdo Dry Valleys. *Antarctic Science*, **10**, 439–448.

TORII, T., NAKAYA, S., MATSUBAYA, O., MATSUMOTO, G. I., MASUDA, N., KAWANO, T. & MURAYAMA, H. 1989. Chemical characteristics of pond waters in the Labyrinth of southern Victoria Land, Antarctica. *Hydrobiologia*, **172**, 255–264.

WARNY, S., ASKIN, R., HANNAH, M., MOHR, B., RAINE, I., HARWOOD, D. M., FLORINDO, F. & THE SMS SCIENCE TEAM. 2009. Palynomorphs recovered from sediment core reveal a remarkably warm Antarctica during the Mid Miocene. *Geology*, **37**, 955–958, http://dx.doi.org/10.1130/G30139A.1

WEBB, P. N. & HARWOOD, D. M. 1993. Pliocene fossil Norhofagus (Southern Beech) from Antarctica: phytogeographyy, dispersal strategies, and survival in high lattitude glacial and deglacial environments. *In*: ALDEN, J., MASTRANTONIO, J. L. & ØDUM, S. (eds) *Forest Development in Cold Regions*. NATO ASI, Series A: Life Sciences, **244**. Plenum, New York, 135–165.

WEBB, P. N., HARWOOD, D. M., MCKELVEY, B. C., MERCER, J. H. & STOTT, L. D. 1984. Cenozoic marine sedimentation and ice-volume variation on the East Antarctic craton. *Geology*, **12**, 287–291.

WENTWORTH, S. K., GIBSON, E. K., VELBEL, M. A. & MCKAY, D. S. 2005. Antarctic Dry Valleys and indigenous weathering in Mars meteorites: implications for water and life on Mars. *Icarus*, **174**, 383–395.

WHALLEY, W. 1979. The relationship of glacier ice and rock glacier at Grubengletscher, Kanton Wallis, Switzerland. *Geografiska Annaler. Series A, Physical Geography*, **61**, 49–61.

WHALLEY, W. B. & PALMER, C. F. 1998. A glacial interpretation for the origin and formation of the Marinet Rock Glacier, Alpes Maritimes, France. *Geografiska Annaler*, **80**, 221–236.

YATSU, E. 1988. *The Nature of Weathering: An Introduction*. Sozosha, Tokyo.

YERSHOV, E. D. 1998. *General Geocryology, Studies in Polar Research*. Cambridge University Press, Cambridge.

Landscape evolution and ice-sheet behaviour in a semi-arid polar environment: James Ross Island, NE Antarctic Peninsula

BETHAN J. DAVIES[1]*, NEIL F. GLASSER[1], JONATHAN L. CARRIVICK[2], MICHAEL J. HAMBREY[1], JOHN L. SMELLIE[3] & DANIEL NÝVLT[4]

[1]*Institute of Geography and Earth Sciences, Aberystwyth University, Aberystwyth SY23 3DB, UK*

[2]*School of Geography, University of Leeds, Leeds LS2 9JT, UK*

[3]*Department of Geology, University of Leicester, University Road, Leicester LE1 7RH, UK*

[4]*Czech Geological Survey, Brno Branch, Leitnerova 22, 658 69 Brno, Czechia*

**Corresponding author (e-mail: bdd@aber.ac.uk)*

Abstract: This study of landscape evolution presents both new modern and palaeo process-landform data, and analyses the behaviour of the Antarctic Peninsula Ice Sheet through the Last Glacial Maximum (LGM), the Holocene and to the present day. Six sediment-landform assemblages are described and interpreted for Ulu Peninsula, James Ross Island, NE Antarctic Peninsula: (1) the Glacier Ice and Snow Assemblage; (2) the Glacigenic Assemblage, which relates to LGM sediments and comprises both erratic-poor and erratic-rich drift, deposited by cold-based and wet-based ice and ice streams respectively; (3) the Boulder Train Assemblage, deposited during a Mid-Holocene glacier readvance; (4) the Ice-cored Moraine Assemblage, found in front of small cirque glaciers; (5) the Paraglacial Assemblage including scree, pebble-boulder lags, and littoral and fluvial processes; and (6) the Periglacial Assemblage including rock glaciers, protalus ramparts, blockfields, solifluction lobes and extensive patterned ground. The interplay between glacial, paraglacial and periglacial processes in this semi-arid polar environment is important in understanding polygenetic landforms. Crucially, cold-based ice was capable of sediment and landform genesis and modification. This landsystem model can aid the interpretation of past environments, but also provides new data to aid the reconstruction of the last ice sheet to overrun James Ross Island.

Polar deserts were widespread during the Pleistocene (in high-latitude continental interiors) and are common today, in both the Arctic (e.g. Svalbard, Canadian High Arctic) and the Antarctic (e.g. Vestfold Hills, Dry Valleys; Alexander Island). However, geomorphological processes are poorly understood within polar deserts despite their widespread occurrence. In particular, the interrelationship of geomorphological (such as cold-based and warm-based glacial, periglacial and paraglacial) processes in polar deserts has been the subject of relatively few papers (Fitzsimons 2003; Haeberli 2005; Waller *et al.* 2009). This paucity of data and information is surprising because geomorphological processes within polar deserts provide information on past and present ice sheet and glacier dynamics and stability, mass and contribution to sea-level rise (Waller 2001; Kleman & Glasser 2007).

Landsystems use modern analogues to understand geomorphological processes and process-form relationships through the characterization of terrain with repeated patterns in surface form and sediments (e.g. Evans & Twigg 2002; Evans 2003*b*; Benn & Lukas 2006; Golledge 2007; Hambrey & Fitzsimons 2010). When constrained by detailed modern analogues, large-scale geomorphological mapping and a landsystem approach can provide proxy data for palaeo-glaciological models and palaeo-climatic reconstructions (Evans 2011). Landsystem models have been presented, among others, for cold-based glaciers in East Antarctica (French & Guglielmin 1999; Fitzsimons 2003; Hall & Denton 2005), glaciated Alpine valleys (Spedding & Evans 2002), active temperate ice sheets (Evans & Twigg 2002; Evans 2003*a*) and polythermal glaciers (Glasser & Hambrey 2003), but an original model has yet to be developed for a semi-arid polar desert.

Palaeo-ice sheet reconstructions require a thorough understanding of current and past processes and subglacial conditions. Existing reconstructions of Antarctic Peninsula Ice Sheet extent, character and behaviour rely on poorly dated and sparse marine-dominated geological evidence (Heroy & Anderson 2005, 2007; Davies *et al.* 2012*a*), and there have been few detailed terrestrial studies (e.g. Ingólfsson *et al.* 1992; Hjort *et al.* 1997, 2003; Johnson *et al.* 2011). The relatively accessible

From: Hambrey, M. J., Barker, P. F., Barrett, P. J., Bowman, V., Davies, B., Smellie, J. L. & Tranter, M. (eds) 2013. *Antarctic Palaeoenvironments and Earth-Surface Processes*. Geological Society, London, Special Publications, **381**, 353–395. First published online April 19, 2013, http://dx.doi.org/10.1144/SP381.1

Ulu Peninsula on James Ross Island (Fig. 1) is one of the largest ice-free regions in the Antarctic Peninsula, and therefore provides an ideal opportunity to investigate palaeo- and modern geomorphological and sedimentary processes. Previous studies on James Ross Island have focused on only small individual landforms (e.g. Chinn & Dillon 1987; Rabassa 1987; Lundqvist *et al.* 1995; Strelin & Sone 1998; Fukui *et al.* 2007, 2008), and there is a need for a larger-scale, holistic study. Furthermore, the higher precipitation rate and relatively northern latitude of James Ross Island means that meltwater is more readily available than, for example, the Dry Valleys of East Antarctica, making this a different landsystem that requires a new conceptual model. The aims of this paper are to provide a modern analogue to aid the interpretation and understanding of ancient polar deserts, and through interrogation of this, to present new data regarding the character and behaviour of the Antarctic Peninsula Ice Sheet through the Last Glacial Maximum (LGM) and Holocene Epoch on James Ross Island.

Study area

Location and climate

James Ross Island (northeastern tip of the Antarctic Peninsula; Fig. 1) has a cold, polar-continental climate (Martin & Peel 1978), influenced by the orographic barrier of the Trinity Peninsula mountains (Domack *et al.* 2003; King *et al.* 2003; Davies *et al.* 2012*b*). It is separated from Trinity Peninsula by the 4–8 km wide and 450–1200 m deep Prince Gustav Channel (British Antarctic Survey 2010). Mean annual air temperatures on Ulu Peninsula are around −7 °C (Láska *et al.* 2010, 2011) and can reach up to +8 °C in January, with more than 200 positive degree days and 100 freeze–thaw days per year, which are highly variable year-on-year (NASA 2004; Láska *et al.* 2011). Estimates of precipitation range from 200 (Strelin & Sone 1998) to 500 mm a^{-1} w.e. (water equivalent; van Lipzig *et al.* 2004), which falls into the range of a 'semi-arid' climate (cf. Laity 2008). However, prevailing strong south to southwesterly winds on James Ross Island are effective in encouraging ablation through snow drift. The combination of relatively dry air and high wind speeds exacerbates evaporation and sublimation. Ice and snow melt and ground ice thaw are restricted to the short summer season and vary strongly diurnally in response to cooler night time temperatures and a low angle of incidence of the sun.

Place names on Ulu Peninsula in this study follow the Czech Geological Survey (2009), the British Antarctic Survey (2010) and the Antarctic Digital Database (www.add.scar.org), with the following important exception: the informal name 'Whisky Glacier' for the small land-terminating glacier extending north from Lookalike Peaks (Chinn & Dillon 1987; Láska *et al.* 2010, 2011; Engel *et al.* 2012) is not used (Fig. 2). 'Whisky Glacier' has never been formally recognized. It is also known as 'IJR-45' in the glacier classification by Rabassa *et al.* (1982) (also Hjort *et al.* 1997; Davies *et al.* 2012*b*). Subsequently, the much larger tidewater glacier draining Dobson Dome into Whisky Bay was officially designated as Whisky Glacier (British Antarctic Survey 2010). Accordingly, we use 'IJR-45' when referring to the land-terminating glacier described by Chinn & Dillon (1987). Elsewhere, informal names have been allocated where no formal names exist and are indicated by the use of parentheses (e.g. 'Unnamed Glacier'). Finally, in this paper we refer only to that part of Ulu Peninsula between Whisky Bay, Cape Lachman and Shrove Cove (Fig. 2).

Geological and glaciological history

The Permo-Triassic metamorphic Trinity Peninsula Group crops out widely on Trinity Peninsula (Fig. 1; Barbeau *et al.* 2010). It was intruded by Jurassic and Cretaceous felsic plutons, including granite, granodiorite, quartz monazite, tonalite and gabbro (Bibby 1966; Riley *et al.* 2011; Žák *et al.* 2012). James Ross Island comprises Late Cretaceous sedimentary strata with locally fossiliferous marine sandstone, shale and conglomerate, overlain by the Neogene James Ross Island Volcanic Group (Fig. 1; Bibby 1966; Nelson 1975; Ineson *et al.* 1986; Košler *et al.* 2009; Riley *et al.* 2011). The latter includes tuff, hyaloclastite, basaltic lava and glacigenic diamictite (Pirrie *et al.* 1997; Jonkers *et al.* 2002; Smellie *et al.* 2006, 2008; Nelson *et al.* 2009) and marine laminites (Nývlt *et al.* 2011). The tillites and glaciomarine diamictites contain pelitic metamorphic, altered volcanic and granitic erratics derived from Trinity Peninsula, indicating that the area was inundated on multiple occasions by ice caps during the Neogene (Pirrie *et al.* 1997; Hambrey & Smellie 2006; Smellie *et al.* 2006; Hambrey *et al.* 2008; Nývlt *et al.* 2011). The basalt lavas form resistant caps on the Cretaceous strata, resulting in flat-topped mountains (mesas) surrounded by scree.

During Quaternary glaciations, James Ross Island was overwhelmed by ice sheets from Trinity Peninsula, which deposited numerous metamorphic and granitic erratics across the island (Bibby 1966; Johnson *et al.* 2011). Glacial drift, drumlins and moraines have been described from the interior of Ulu Peninsula (Rabassa 1983, 1987; Hjort *et al.* 1997; Strelin *et al.* 2006). Prince Gustav Ice Stream developed during the LGM,

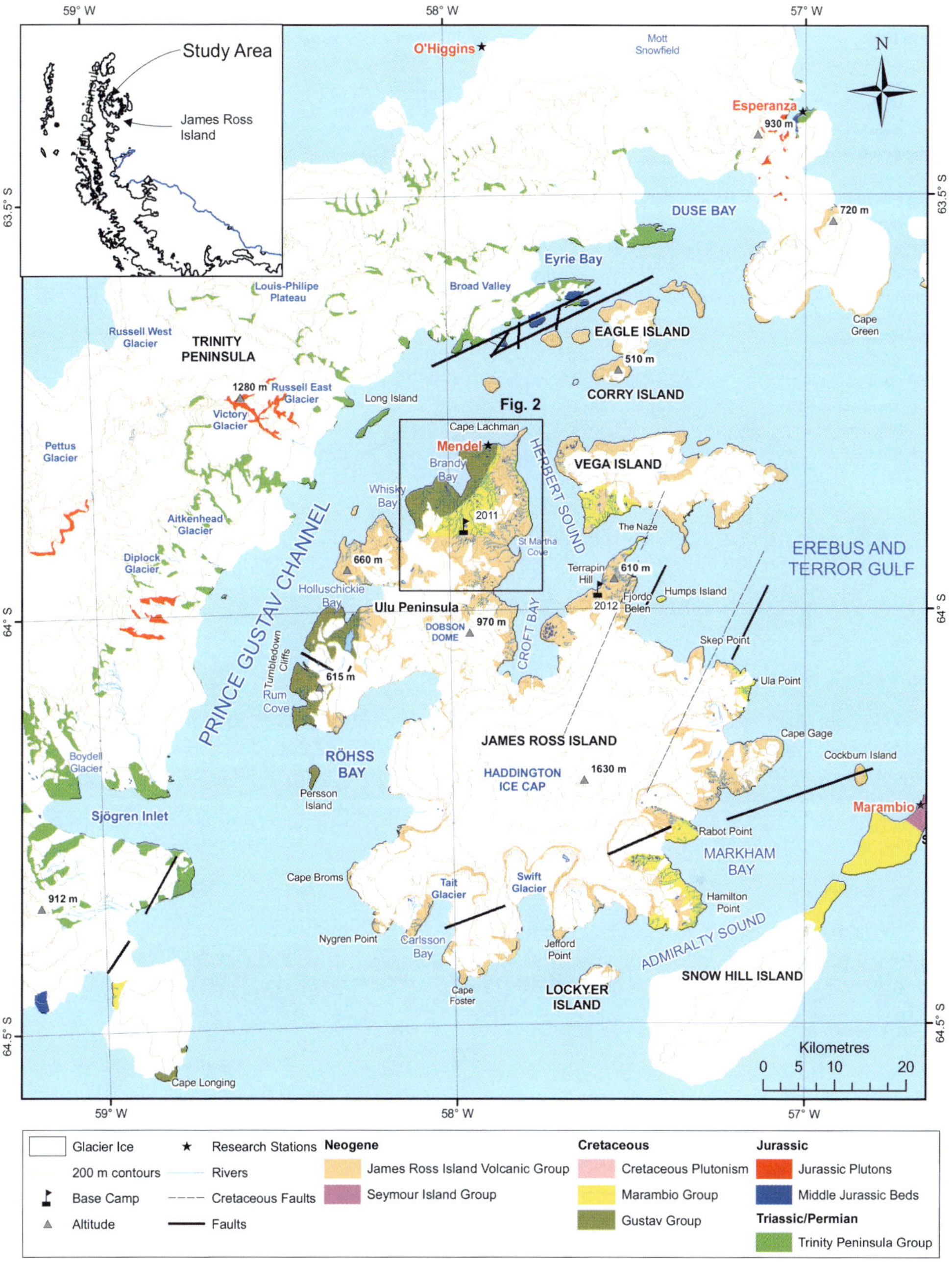

Fig. 1. James Ross Island and Ulu Peninsula, NE Antarctic Peninsula. Study region is in boxed area on main map. Inset shows the wider regional context of James Ross Island. Main map shows the location of Cretaceous, James Ross Island Volcanic Group, Trinity Peninsula Group and granite lithologies (Bibby 1966; Nelson 1975; British Antarctic Survey 2010; Riley *et al.* 2011). James Ross Island Volcanic Group: Miocene to Recent basaltic lava, tuff, hyaloclastite and breccia. Seymour Island Group: Palaeogene–Eocene richly fossiliferous, shallow water, fine-grained volcaniclastic sedimentary rocks. Marambio Group: Santonian–Paleocene fossiliferous, shallow water fine-grained volcaniclastic sedimentary rocks. Gustav Group: Barremian–Albian coarse-grained volcaniclastic sedimentary rocks; deeper water environment. Cretaceous plutons: typically granodiorite. Jurassic plutons: granite–tonalite–quartz mononzite. Mapple, Mt Flora and Kenney Glacier formations and Botany Bay Group: mainly rhyolitic ignimbrite. Trinity Peninsula Group: Permo-Triassic siliclastic turbidite succession with interbedded basic volcanic rocks.

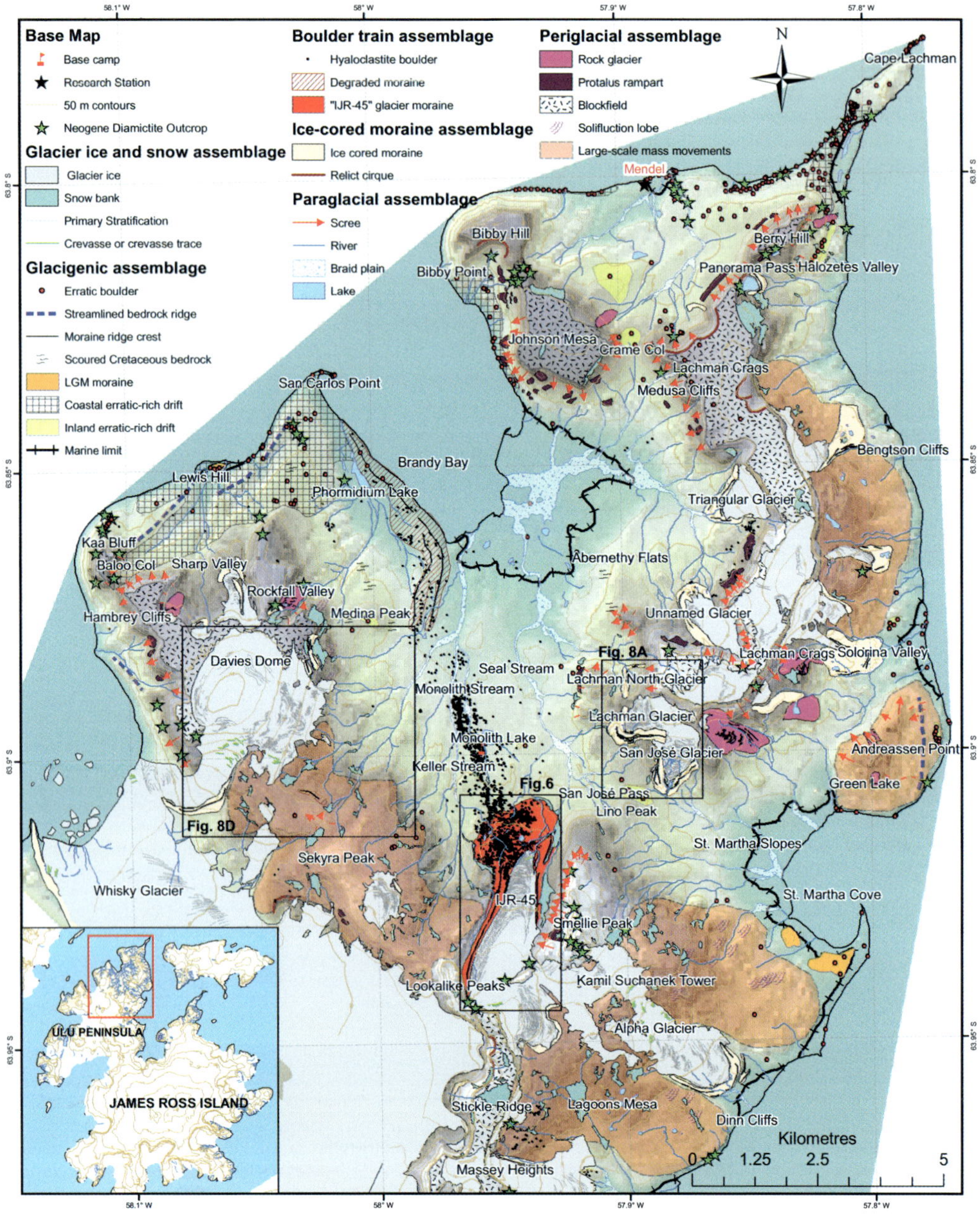

Fig. 2. Geomorphological map of northern Ulu Peninsula, showing the main features from the six sediment-landform assemblages. Overlain on the 2006 digital elevation model (DEM) and using ice, perennial snow, braid plains and rivers mapped by the Czech Geological Survey (2009). Glacier structures are mapped from aerial photographs from the year 2006.

diverting ice flow north and south around the island (Gilbert *et al.* 2003; Evans *et al.* 2005). This ice stream receded from the shelf edge after 18 ka BP (Heroy & Anderson 2007; Davies *et al.* 2012*a*), with retreat to the inner shelf occuring between 17.5 and 9.1 ka (Johnson *et al.* 2011). Mean exposure ages on granite erratics at Cape Lachman are dated to 8.0 ± 0.8 ka, suggesting recession of the ice stream from northern James Ross Island by that time (Johnson *et al.* 2011). Radiocarbon dates from glaciomarine sediments on The Naze, James Ross Island (Fig. 1) indicate that Herbert Sound became ice free by 6.3 ± 0.3 to 7.8 ± 0.2 cal. ka BP (Hjort *et al.* 1997; calibrated and corrected by Davies *et al.* 2012*a*), and this is supported by recent cosmogenic nuclide exposure ages on erratic boulders from 6.8 ± 0.7 ka from Terrapin Hill. Johnson Mesa maintained a small independent ice dome until 6.6 ± 0.7 ka (Johnson *et al.* 2011).

Several Mid-Holocene glacial advances have been documented on James Ross Island. There was an advance of Croft Bay glaciers to Cape Lachman from 7.3 to 7.0 cal. ka BP (Strelin *et al.* 2006; Johnson *et al.* 2011). A later Mid-Holocene readvance of 'IJR-45' deposited the very prominent line of hyaloclastite and tillite boulders from the existing margin of 'IJR-45' northwards to the southern coastline of Brandy Bay. The large lateral moraine related to this advance and flanking Brandy Bay was radiocarbon dated to between 5.4 and 4.7 cal. ka BP (Hjort *et al.* 1997; recalibrated by Johnson *et al.* 2011), suggesting that this glacier readvance may have coincided with the Mid-Holocene collapse of Prince Gustav Ice Shelf (absent from 6 to 2 ka; cf. Pudsey & Evans 2001). Lake cores indicate that the climate was warmer at this time (Björck *et al.* 1996). Prominent ice-cored moraines in front of cirque glaciers also suggest a Late Holocene or Little Ice Age readvance (Carrivick *et al.* 2012).

James Ross Island is now 75% glacierised, and is dominated by the Mount Haddington Ice Cap. The disintegration of Prince Gustav Ice Shelf in 1995 was associated with rapid recession of tributary glaciers (Glasser *et al.* 2011; Davies *et al.* 2012*b*). Periglacial landforms are today well developed on James Ross Island and on surrounding islands (Lundqvist *et al.* 1995; Ermolin *et al.* 2002, 2004; Strelin *et al.* 2006; Fukui *et al.* 2008).

Methods

Geomorphological mapping was achieved through visual interpretation of remotely sensed images (Table 1) and through an intensive field campaign that covered 200 km^2 of Ulu Peninsula from January to March 2011. Identification of glacially associated landforms from remote images is well established (Glasser & Jansson 2005; Glasser *et al.* 2005, 2009*a*, 2009*b*). Remotely sensed structural glaciological mapping followed standard procedures (Hambrey & Dowdeswell 1994, 2000; Glasser *et al.* 1998, 2011; Hambrey *et al.* 2005). Landforms and sediments and were mapped and logged in the field following procedures outlined in Gale & Hoare (1991), Jones *et al.* (1999) and Evans & Benn (2004). Sediments were characterized and quantitatively differentiated through the collection of 101 samples each of 50 clasts (with an *a*-axis from 4 to 32 mm) from a 50 × 50 cm area. Stone lithologies were identified according to the methods of Bridgland (1986), Gale & Hoare (1991) and Walden (2004). A minimum of 30 basalt pebbles were measured to the nearest 0.5 cm for shape-roundness data, following Benn & Ballantyne (1994). Shape-roundness analysis included calculation of the co-variance of the C_{40}

Table 1. *Sources used for remote mapping of landforms*

Data source	Resolution/scale	Date captured	Notes
BAS/JR/2/79	1:95 000	February 1979	BAS and Royal Navy aerial photographs taken over Brandy Bay, Vega Island, Rum Cove and Terrapin Hill. Digitally scanned and georeferenced in ArcGIS to SPOT-5 satellite images (WGS84 UTM 21 S). Photographs copyright British Antarctic Survey and Crown copyright Controller of Her Majestey's Stationary Office and the UK Hydrographic Office. The 2006 aerial photographs were processed, georeferenced and orthorectified by Nývlt and Šerák (Czech Geological Survey 2009).
BAS/JR/1/79	1:95 000	January 1979	
BAS/JR/2/80	1:95 000	February 1980	
RN/6/89	1:16 000	January 1989	
BAS/RN/06	1:25 000 (<1 m)	2006 (various dates)	
Digital elevation model produced by Nývlt and Šerák (Czech Geological Survey 2009)	<5 m	Created from BAS 1970/80 and 2006 aerial photographs	James Ross Island – Northern Part. Heights: geoid EGM96. Map projection: WGS84 UTM 21S.

RMS, Root mean square error.

and RA indices (cf. Adam & Knight 2003). The C_{40} index is the percentage of clasts with a c/a ratio of ≤ 0.4. The RA index is the percentage of very angular and angular clasts within the sample, using the Powers (1953) roundness chart. Only basalt clasts were used to avoid introducing bias into the dataset through the use of different lithologies. These analyses have resulted in the identification of six genetically grouped sediment–landform associations (Fig. 2; Table 2), which are described and interpreted sequentially in the next section.

Glacier ice and snow assemblage

Description. James Ross Island has numerous small cirque, valley, tidewater and plateau dome glaciers (Table 2). Most of these glaciers are receding (Table 3), but the land-terminating cirque glaciers on Ulu Peninsula (Fig. 3) have only small rates of recession or their margins are currently stationary. However, 100–200 m of recession has occurred since the most recent glacial advance, when prominent ice-cored moraines were formed (Carrivick *et al.* 2012). Volume loss has been observed on Davies Dome and 'IJR-45', of 29.7 and 10.9%, respectively, from 1979 to 2006 (Engel *et al.* 2012).

Glacier surfaces are characterized by continuous layering (*sensu* Hambrey *et al.* 2005), that is clearly visible both in the field and in aerial photographs (e.g. Figs 2 & 3). On 'IJR-45', there is well-developed continuous layering down glacier (Fig. 2) and supraglacial debris on either side of the ice divide. 'IJR-45' is 150 m thick, with a uniform bed topography (Engel *et al.* 2012) and a large debris-covered terminus ('IJR-45 Glacier Moraine'). No crevasses or crevasse traces were visible on any glaciers on the aerial photographs although crevasse traces were noted on the surface of 'Unnamed Glacier' and 'IJR-45' during fieldwork. The

Table 2. *Summary of the six sediment-landform assemblages on Ulu Peninsula, James Ross Island*

Assemblage	Principle land elements	Age of formation	Interpretation
Glacier snow and ice assemblage	Glacier ice (cirque, tidewater, valley, dome glaciers) Perennial snow-banks	Holocene to present day	Contemporary glaciation. Small cold-based cirque and plateau glaciers dominate on Ulu Peninsula
Glacigenic assemblage	Erratic-poor drift Cretaceous bedrock with scattered basalts Erratic-rich drift (in coastal regions) Erratic-rich drift (in cols and passes) Moraine fragments	LGM	James Ross Island overwhelmed by the Antarctic Peninsula Ice Sheet. Development of Prince Gustav Ice Stream in Prince Gustav Channel
Boulder train assemblage	Boulder train Erratic-poor drift sheet 'Brandy Bay Moraine' (with some reworked granite erratics) 'IJR-45 Glacier Moraine'	Mid-Holocene	Mid-Holocene readvance of 'IJR-45' to form a marine-terminating glacier in Brandy Bay
Ice-cored moraine assemblage	Ice-cored moraines in front of cirques (both with and without modern glaciers)	Late-Holocene	Late-Holocene readvance of small polythermal glaciers, possibly during a southern Hemisphere equivalent of the 'Little Ice Age'
Paraglacial assemblage	Scree slopes Boulder lags on beaches Ventifacts; cobble–boulder pavements Spits and beaches Marine terraces Rivers and streams	Holocene to present day	Paraglacial reworking of glacigenic sediments and landforms
	Large-scale mass movements	Neogene to present day	Paraglacial relaxation of basalt deltas
Periglacial assemblage	Rock glaciers Nivation hollows Protalus and pronival ramparts Freeze–thaw shattering of boulders Solifluction	Holocene to present day	Periglaciation of James Ross Island, with deep permafrost and a seasonal active layer
	Mesas and blockfields	(Neogene to present day)	Development of blockfields shortly after their genesis in the Neogene. Occasionally glaciated by small, cold-based plateau glaciers

Table 3. *Principal characteristics of the land-terminating glaciers on Ulu Peninsula (after Davies* et al. *2012*b*)*

Glacier	Type	Length	Width	Area	Mean slope	**ELA (m a.s.l.)	Recession rate 1988–2009
'IJR-45'*	Valley glacier	3.26 km	0.8 km	2.36 km^2	5°	270 m	0.144 km^2 a^{-1}
'Unnamed Glacier'	Cirque glacier	2.53 km	0.5 km	3.61 km^2	12°	350 m	0.017 km^2 a^{-1}
Triangular Glacier	Cirque glacier	0.9 km	0.8 km	0.62 km^2	17°	180 m	0.004 km^2 a^{-1}
Lachman and San José Glaciers	Cirque glacier	1.44 km	1.5 km	0.62 km^2	12°	230 m	0.000 km^2 a^{-1}
Alpha Glacier	Valley glacier	3.94 km	1.0 km	3.19 km^2	8°	230 m	0.016 km^2 a^{-1}
Davies Dome	Dome (plateau) glacier	3.46 km	2.7 km	6.46 km^2	9°	300 m	0.041 km^2 a^{-1}

San José and Lachman glaciers are treated as one glacier. Width was measured at the widest point on the glacier. ELA is calculated from topographical data (refer to Davies *et al.* 2012*b* for more information). Refer to Figure 2 for glacier locations.
*Informally named 'Whisky Glacier' by Chinn & Dillon (1987); name not used here as it clashes with officially adopted name for the glacier (Whisky Glacier) between Dobson Dome and Whisky Bay (British Antarctic Survey 2010). 'IJR-45' from Rabassa *et al.* (1982).
**, Equilibrium Line Altitude.

continuous layering in 'Unnamed Glacier' is folded about the flow-parallel axis. Supraglacial scree that falls onto the surface of this glacier at the headwall apparently emerges near the snout (Fig. 3b). Glaciers on the eastern side of Lachman Crags (Fig. 3c) are characteristically small and detached from the plateau ice dome and thus presumably from their accumulation areas. They also have prominent ice-cored moraines. Davies Dome flows radially from the top of a mesa. Its lightly folded continuous layering follows the lobate ice margin. Whisky Glacier is characterized by sparse crevasses, crevasse traces and continuous layering (Fig. 2). Perennial snow banks lie in the lee of hill slopes (Fig. 2), feeding ephemeral streams that flow on positive degree days. These snow banks are sometimes associated with protalus ramparts and nivation hollows.

Fig. 3. Principal characteristics of Glacier Ice and Snow Assemblage on Ulu Peninsula. (**a**) Lachman (left) and San José (right) glaciers, with stratification. Note large ice-cored moraines. (**b**) 'Unnamed Glacier', with supraglacial debris falling onto the glacier head from scree (arrowed, 'b') and emerging near the snout (arrowed, 'a'). (**c**) Stagnant and down-wasting ice mass with prominent ice-cored moraines. View downwards from the top of Lachman Crags, looking east. Arrow indicates an ice scar with stratified ice. (**d**) Stratification in San José glacier (arrowed).

Interpretation. The continuous layering (e.g. Fig. 3) is interpreted as primary stratification formed through firnification, superimposed ice formation, and melting and refreezing of snow. Folding of primary stratification in 'Unnamed Glacier' occurred during down-glacier transport about the flow-parallel axis in a compressive regime (cf. Wadham & Nuttall 2003; Hambrey *et al.* 2005). However, the continuous stratification in marginal areas (and lack of foliation) suggests that ductile deformation and shear are not prevalent on these small cirque glaciers. The lobate planform pattern of stratification on Davies Dome indicates that the ice is flowing radially out of the dome. Crevasse traces on 'Unnamed Glacier' and 'IJR-45' suggest that these glaciers were more active in the past, and that these are now relict features (cf. Hambrey & Lawson 2000).

The supraglacial debris on the surface of some of these glaciers (e.g. 'Unnamed' and Triangular glaciers; Fig. 3b) indicates limited subglacial and supraglacial transport, probably through small-scale creep (Hambrey & Lawson 2000). The small glaciers on the eastern side of Lachman Crags (Fig. 3c) are stagnating and down-wasting *in situ*, presumably because the prevalent southwesterly winds starve them of precipitation. In this location, the abundant scree provided by the basalt and hyaloclastite cliffs provides ample debris, creating ice-cored moraines or rock glaciers (e.g. Strelin *et al.* 2007).

There is little evidence of contemporary thrusting and active moraine formation on these small cirque glaciers today; they are now passively down-wasting and slowly receding, with a negative mass balance (Carrivick *et al.* 2012). The separation of some of these glaciers from their plateau accumulation areas has further encouraged ice stagnation. Despite a climate similar to that in Svalbard, none of these glaciers have structures indicative of past surge behaviour, such as looped moraines, pot holes or distorted longitudinal foliation (cf. Sharp *et al.* 1994; Bennett *et al.* 2000*b*; Hambrey & Lawson 2000; Murray *et al.* 2000, 2002; Hambrey *et al.* 2005; Kjær *et al.* 2008; Grant *et al.* 2009).

Glacigenic assemblage

Ulu Peninsula is covered with a superficial drift sheet dominated by basalt pebbles and occasional erratics, which together form an armoured surface that overlies sand (Table 4). In some coastal areas the superficial drift is overprinted by marine terraces.

Erratic-poor drift

Description. A basalt pebble–cobble gravel is widespread across Ulu Peninsula and comprises unlithified subangular basalt pebbles and cobbles forming a lag on the surface, with frequent basalt boulders and rare granite boulders (Table 4). Samples from Solorina Valley (Fig. 2) and the foreground of 'IJR-45' show that this erratic-poor 'drift sheet' is typically >90% basalt, and clasts on the surface are angular to subangular. In each case, there are few Trinity Peninsula erratics. These basalt and erratic boulders can be either well-embedded in the surface or rest on top of it (compare Fig. 4a, b), and are generally faceted and occasionally striated. Stone stripes and polygons are typically well developed on this surface. This 'drift sheet' is prevalent on topographically flat and smooth, relatively featureless parts of the landscape.

The erratic-poor drift comprises two further subelements, the first of which is an unlithified, poorly compacted, broken regolith of local Cretaceous siltstone, sandstone and flaggy sandstone slabs, which is frequently scattered with subrounded to subangular erratic basalt cobbles (Table 4). Patterned ground is often well developed, as illustrated in Figure 4c. Large isolated basalt and occasionally granite cobbles and boulders are present on the otherwise topographically smooth surfaces (Fig. 4d). The second subelement comprises exposed Cretaceous bedrock with few or no erratics and little surficial regolith, and clearly visible bedding planes (e.g. Fig. 4e).

Interpretation. Drift sheets comprising an unconsolidated sandy boulder gravel with a pebble–cobble armour are regularly observed in deglaciated Antarctic regions, particularly in the Dry Valleys of East Antarctica (Higgins *et al.* 2000; Augustinus 2002; Hall & Denton 2005; Bockheim 2010), where they are inferred to have been deposited by cold-based ice. The climate of the Dry Valleys today (which is colder and drier than that of modern James Ross Island; Fitzsimons 2003) is probably comparable with the LGM climate on James Ross Island. Sediment–landform assemblages observed in the forefields of cold-based glaciers in the Dry Valleys, Antarctica, include striated boulders, sandstone and siltstone breccia, isolated boulders on drift sheets, boulder trains and ice-cored debris cones. Diamicton, silt and clay are typically absent (Waller 2001; Atkins *et al.* 2002; Lloyd Davies *et al.* 2009; Hambrey & Fitzsimons 2010). These drift sheets are distinctly different from subglacial till, which is formed by wet-based ice at the ice-bed interface by a combination of sliding, shearing, lodgement and deformation. Subglacial till is characteristically a diamicton with a high bulk density and shear strength, an unlithified, over-consolidated, heterogeneous and unsorted matrix, and normally bearing polished,

Table 4. *Glacigenic assemblages: LGM sediments and landforms of James Ross Island*

Land element	Sediments	Landforms	Processes and interpretation	Age
Erratic-poor drift; basalt pebble–cobble gravel	Subangular to subrounded clasts, numerous basalt and rare granite boulders either well-embedded in sediments or perched on surface. Pebble lags, ventifacts and desert varnish are common	Smooth, flat surfaces. Widespread across the interior of Ulu Peninsula	Deposition beneath a slow-moving cold-based ice sheet	LGM
Erratic-poor drift; sandstone and siltstone breccia	Unlithified sandstone/ siltstone regolith, poorly compacted, occasional isolated granite or basalt boulders; scattered basalt cobbles across the surface	Smoothed surfaces, occasionally bedding planes are visible	Scouring and deposition beneath a cold-based ice sheet	LGM
Erratic-rich coastal 'drift sheet' with abundant TPG and granite erratics	Found along the western coast of Ulu Peninsula. Subangular to subrounded clasts, numerous boulders and pebbles of Trinity Peninsula origin	Constructional ridges, moraine fragments, smooth slopes. Focused in coastal areas and where ice flow is concentrated	Deposition beneath fast-flowing, warm-based ice, adjacent to Prince Gustav Ice Stream. Lateral moraines were formed by Prince Gustav Ice Stream	LGM
Erratic-rich patchy 'drift sheet' with TPG and granite erratics	Found in cols and passes. Subangular to subrounded clasts, numerous boulders and pebbles of Trinity Peninsula origin	Occasional moraine fragments; streamlined bedrock (hyaloclastite) ridges; smoothed cols and passes	Deposition beneath wet-based ice; sheet flow. Moraines and streamlined features were formed by inland wet-based sheet flow	LGM

TPG, Trinity Peninsula Group metamorphic rocks (cf. Fig. 1).

striated and faceted stones; the diamicton may be massive, fissile and jointed, and may contain numerous inclusions and other sedimentary structures (cf. Evans *et al.* 2006; Benn & Evans 2010).

The cold-based sediment–landform assemblages noted in other parts of Antarctica (e.g. Atkins *et al.* 2002; Lloyd Davies *et al.* 2009; Hambrey & Fitzsimons 2010) are very similar to the erratic-poor drifts on James Ross Island (cf. Table 4). Key similarities include the lack of till, the pebble–cobble armour with occasional striated boulders and the lack of constructional landforms. The smoothed and sculpted surfaces evolved over millennia with repeated warm- and cold-based ice sheets throughout the Cenozoic (Davies *et al.* 2012*a*), and with only very minor modification during LGM cold-based glaciation. This erratic-poor drift is therefore interpreted as having been deposited by a cold-based ice sheet. Basal sliding is likely to be limited to bedrock regions (Cuffey *et al.* 1999; Waller 2001). The relatively soft, poorly consolidated Cretaceous sedimentary strata that underlay the last ice sheet on James Ross Island were likely to have been deformed, encouraging sheet-flow in much the same way as under warm-based ice sheets above the pressure melting point, but at lower speeds (cf. Fitzsimons *et al.* 1999; Cuffey *et al.* 2000; Fitzsimons *et al.* 2008). Based on the extent of these drift sheets and their rare granite erratics, the ice sheet that overwhelmed the island during the LGM was sourced from Trinity Peninsula, during large-scale continental-wide glaciation that extended to the shelf edge (Heroy & Anderson 2007). The minimal meltwater produced at the base of cold-based ice minimized entrainment, erosion, transport and deposition, and reduced the ability of the ice sheet to construct moraines and landforms; this resulted in the smooth land surfaces with minimal glacigenic landforms observed on Ulu Peninsula.

The principal surficial sediment lithofacies are differentiated as erratic-rich or erratic-poor on the

Fig. 4. Glacial drifts on Ulu Peninsula. Drift sheets in photographs A–G were emplaced during the LGM. (**a**) Erratic-poor drift. Lone striated, faceted granite erratic with red staining in a region otherwise devoid of erratics in Solarina Valley. The basalt-rich drift is a sandy gravel with scattered cobbles with an *a*-axis of up to 30 cm. The gravel is almost 100% angular to subangular basalt. (**b**) Erratic-poor drift. Lone granite erratic in the forefield of 'IJR-45', where it rests as an isolated boulder on an otherwise entirely basaltic surficial drift. Basalt boulders can be seen protruding through the glacial drift surface. The boulder appears to be a ventifact. 'IJR-45' moraine can be seen in the distance. (**c**) Erratic-poor drift with broken and brecciated Cretaceous bedrock with stone stripes emanating from basalt cobbles.

presence or absence of Trinity Peninsula Group erratics (Fig. 5a). The pebble shapes in most samples show considerable modification from scree, which is presumed to supply the majority of the debris. Material is inferred to have been transported via supraglacial and subglacial pathways. Two samples from the erratic-poor drift from Solarina Valley and St Martha Cove (Fig. 1) have particularly angular clasts, which could reflect intense post-depositional frost shattering.

Erratic-rich drift

Description. In some coastal regions adjacent to Prince Gustav Channel, such as Lewis Hill, San Carlos Point and Cape Lachman (Figs 1 & 2), there are drift sheets comprising poorly compacted, unsorted sandy boulder gravel with high percentages of Trinity Peninsula erratics from pebble to boulder size (Table 4). At Cape Lachman, the central neck of the promontory is significantly enriched with erratics (see Figs 2 & 4f). Where the ground rises, there is a sharp contrast with the erratic-poor drift that is almost entirely composed of the James Ross Island Volcanic Group. There are large numbers of granite boulders in the topographically lower neck of the promontory, but none above the lateral margins of the coastal erratic-rich drift. These erratic-rich sediments are associated with constructional ridges and moraine fragments, such as Kaa Bluff (Fig. 4g). Similar erratic-rich drifts also occur in isolated patches in cols and passes of Ulu Peninsula (cf. Fig. 2), such as Baloo Col, San José Pass, Crame Col, Andreassen Point and in St Martha Cove, where there is a number of small unconsolidated mounds. North of St Martha Cove, hyaloclastite mounds are streamlined north–south.

Interpretation. During the LGM, Prince Gustav Ice Stream flowed northwards along the northwestern coast of James Ross Island (Camerlenghi *et al.* 2001; Gilbert *et al.* 2003; Evans *et al.* 2005) and impinged upon its coastal regions, resulting in lateral moraines and the erratic-rich drift. The sharp boundary between erratic-poor and erratic-rich sandy boulder gravel on Cape Lachman results from an arm of the ice stream flowing over the lower col, perhaps with cold-based ice on the higher parts of the promontory.

Patches of erratic-rich drift also occur in cols and passes (Table 4). However, there is no evidence of ice streaming over these cols and passes (cf. Stokes & Clark 1999, 2001). Rather, the large accumulations of erratics indicate enhanced deposition and therefore wet-based subglacial conditions. As ice flowed over the cols it became focused and compressional stresses increased. Pressure melting point was reached, allowing subglacial deposition and resulting in the erratic-rich sandy boulder gravels and further erosion of the cols (Lloyd Davies *et al.* 2009). Both ice deformation and frictional sliding can increase basal temperatures (Waller 2001). Furthermore, overriding escarpments can lead to ice-bed separation, the formation of lee-side cavities, and further entrainment and deposition. These patches of erratic-rich drift are therefore interpreted to be a result of changes in the subglacial thermal regime during LGM glaciation, which can create mosaics of selective erosion and deposition (Hall & Glasser 2003; Kleman & Glasser 2007).

The accumulation of Trinity Peninsula boulders and erratics on San José Pass is therefore a result of wet-based ice flowing over the col, before flowing out of St Martha Cove and north up Herbert Sound (cf. Fig. 1), depositing rich accumulations of erratics near Green Lake and ultimately on the eastern slopes of Berry Hill (Fig. 2). Fragments of moraine are associated with this land element at St Martha Cove, further supporting the suggestion that the erratic-rich drift in this location was deposited by wet-based ice, capable of entraining, modifying and depositing subglacial debris. The ice cap that overwhelmed Ulu Peninsula during the LGM was therefore polythermal, with regions of both warm and cold-based ice. These patches of erratic-rich drift are at odds with the traditional view of ice sheet flow, which envisaged 'islands' of cold-based ice, surrounded and

Fig. 4. (*Continued*) (**d**) Erratic-poor drift on broken and brecciated Cretaceous bedrock with scattered basalt erratics on the surface. (**e**) Erratic-poor drift on scoured Cretaceous bedrock with few erratics or cobbles on the surface. The different beds are clearly visible. Note the basaltic mesas in the distance (Johnson Mesa (left) and Lachman Crags). (**f**) Coastal erratic-rich drift with numerous white granite boulders protruding through the drift at Cape Lachman. People are encircled for scale. Note the contrast in the colour of the drift with the brown surfaces with no Trinity Peninsula erratics in the distance. Numerous granite erratics are well embedded in the surficial sediments at this location. Note the abundant granite pebbles, giving the surface a speckled white appearance. (**g**) Erratic-rich drift on fragment of moraine behind Kaa Bluff on the edge of Prince Gustav Channel (cf. Fig. 2). Note the difference in surface texture to the smooth slopes with stone runs on the steep slopes of Kaa Bluff in the background. Note also the presence of several large granite boulders perched on the surface, whose white colour contrasts with the brown basalts that make up the majority of the surface. No stone stripes or polygons are evident on the moraine fragment. (**h**) Boulder Train Assemblage, near Monolith Lake. Note the large hyaloclastite monoliths. Prince Gustav Channel and Trinity Peninsula in the distance.

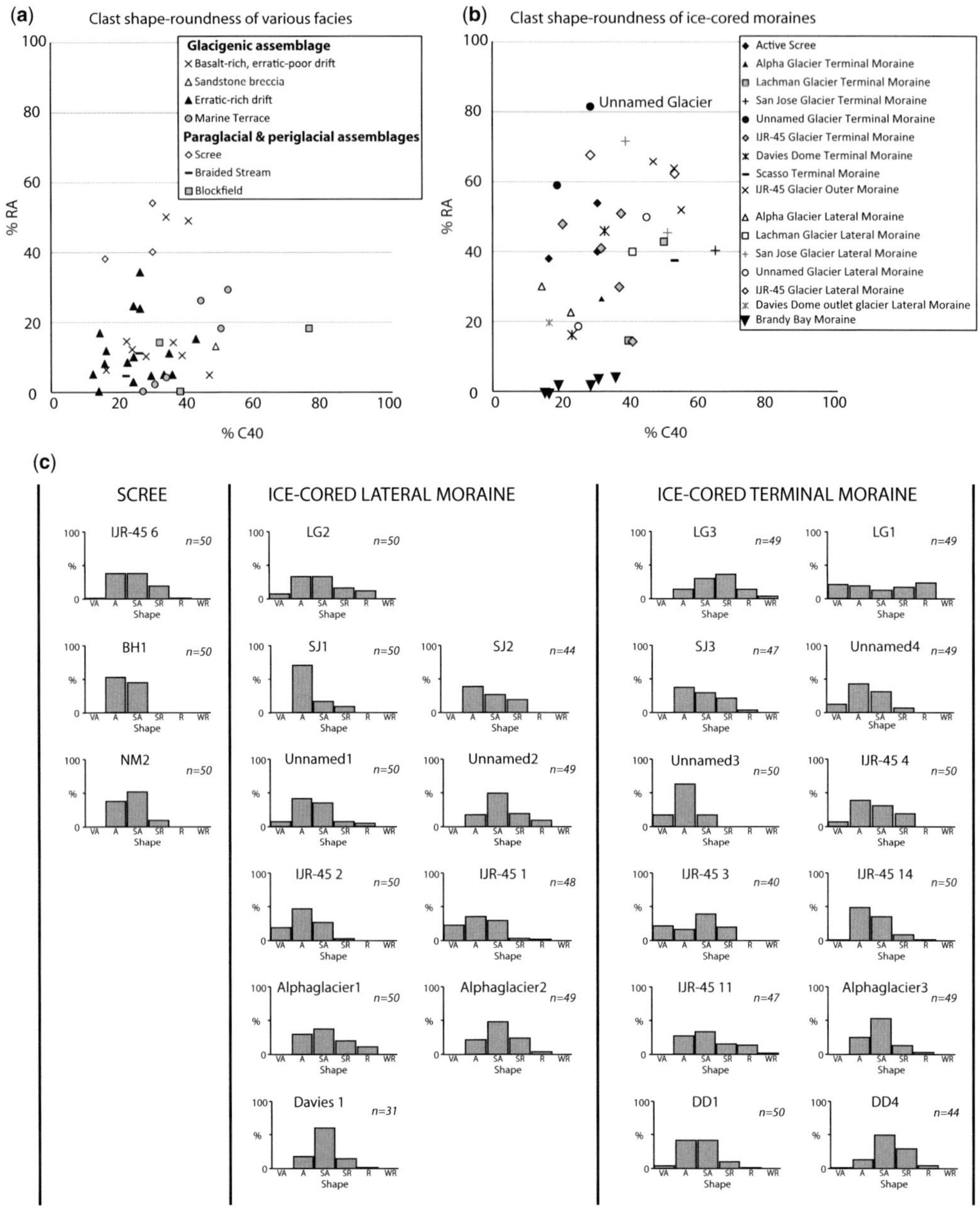

Fig. 5. C_{40}/RA plots and clast-roundness histograms of the different land elements of the sediment-landform assemblages: (**a**) glacial drifts; (**b**) ice-cored moraines. All shape-roundness measurements were carried out on basalt clasts. All samples show considerable modification from the original source material (scree) but there is considerable overlap between those samples with Trinity Peninsula erratics and those without, perhaps highlighting the similar transport distances for all samples; all the basalts are locally derived and so have similar transport histories. (**c**) Angularity histograms for ice-cored moraine samples.

enclosed by faster-flowing wet-based ice (Hughes 1981, 1995).

Care must be taken when interpreting numerous erratics as evidence of a warm-based ice sheet, as the Neogene diamictites on James Ross Island contain Trinity Peninsula erratics (cf. Nelson *et al.* 2009; Nývlt *et al.* 2011). For example, at the northern end of Lachman Crags, east of Crame

Col, there is an abandoned cirque. Beneath this is a vast accumulation of granite boulders and numerous Trinity Peninsula-derived pebbles and cobbles. However, in the cliffs bounding the cirque there is a thickness of Neogene tillite (Fig. 2), and these erratics are most likely reworked from Neogene sedimentary strata. The same is true for erratic boulders on the coast between Brandy Bay and Cape Lachman.

Boulder train assemblage

Boulder train and glacial drift

Description. This assemblage comprises 'IJR-45 Glacier Moraine', 'Brandy Bay Moraine' and a boulder train and erratic-poor drift between the two (Table 2). A train of large (7–10 m *b*-axis), freestanding hyaloclastite and diamictite boulders (Figs 2 & 4h) stretches from the western side of 'IJR-45 Moraine' to a low ridge flanking western Brandy Bay. The surficial sediments associated with the hyaloclastite boulder train are a sandy boulder gravel, characterized by smooth surfaces and subangular to subrounded basalt pebbles, with rare granite or basaltic boulders embedded within the sediments or perched on the surface. The surficial sediments are dissected by rivers and lack clear boundaries.

Interpretation. The surficial sediments are lithologically similar to the LGM erratic-poor drift (Tables 2 & 4), and are stratigraphically younger than the coastal erratic-rich drift. The intact large hyaloclastite boulders are likely to have been transported supraglacially by a readvance of 'IJR-45' during the Mid-Holocene (cf. Björck *et al.* 1996; Hjort *et al.* 1997). The hyaloclastite boulders are perched on, rather than lodged in, the surficial sediments. In addition, these boulders are friable and delicate, and unlikely to survive subglacial transport. Recession was rapid with little reworking of these boulders, and no recessional moraines are visible in the landscape. This boulder train was therefore most probably formed by the marginal dumping of supraglacial debris. Rapid (and possibly cold-based) recession may have protected these friable boulders.

'Brandy Bay Moraine'

Description. The ridge bounding the southwestern coastline of Brandy Bay (Fig. 2) is approximately 30 m high with a rounded, undulating crest, 3.5 km long, and up to 0.7 km wide. There is a sharp break in slope at its base, and it declines in elevation from east to west (seawards). The surficial sediments are a basalt pebble–cobble lag with a fine silt matrix buried beneath the cobble armour. There are rare granite boulders, increasing in number seawards. The clast shape-roundness data shows considerably rounder, blockier pebbles than the ice-cored moraines (Fig. 5b). The percentage of subrounded to rounded pebbles increases northwards along the crest of the moraine. Hyaloclastite boulders decrease in size and degradation and weathering increase southeastwards down the crest of the moraine. Adjacent to the ridge near Phormidium Lake, the glacial drift has abundant Trinity Peninsula erratics, contrasting with the surficial sediments on the moraine (see Fig. 2).

On the crest of the ridge are numerous low surface depressions with arcuate scars, 0.5–1 m high. Slump scars, debris flows and stone polygons characterize the surface of the moraine. For example, a slump in the slope facing Brandy Bay is more than 7 m wide with a 40 m run-out, and has an arcuate headwall scar. This slump has abrupt lateral margins, visible failure planes and desiccated edges. Slumping at the base of the seaward terminus of the moraine resulted in a natural exposure. At the base of the exposed section, 0.6 m of stratified dark blue ice with large coarse crystals is present, and contains occasional coarse gravel clasts. It is overlain by 0.5 m of silty clast-rich diamicton that grades into 1.4 m of crudely bedded sand. From 1.9 to 3.3 m in the section, there is a diamicton with a gradually increasing gravel content, overlain by crudely bedded sand.

Interpretation. The ridge in Brandy Bay has previously been interpreted as a moraine (Hjort *et al.* 1997), and in our study it is called 'Brandy Bay Moraine'. The moraine and associated 'Bahiá Bonita Drift' were deposited by a readvance of 'IJR-45' (Rabassa 1983, 1987; Ingólfsson & Hjort 2002). The readvance has been dated to the Mid-Holocene by radiocarbon dates from marine molluscs in overridden glaciomarine sediments at the base of the moraine, bordering Prince Gustav Channel (Hjort *et al.* 1997), and radiocarbon dates from lake sediments in the boulder train (Björck *et al.* 1996). Drumlins have been reported in association with the moraine and boulder train (Rabassa 1987; Hjort *et al.* 1997). However, our fieldwork indicates that these 'drumlins' are simply remnants of a thicker Mid-Holocene drift sheet, subsequently dissected by ephemeral streams and reduced by periglacial slope processes.

The basaltic debris on the moraine was derived from reworking of basalt scree and hyaloclastite boulders from Lookalike Peaks and Smellie Peak, which were also the source for the diamictite and hyaloclastite boulders in the boulder train. The percentage of Trinity Peninsula erratics and numbers of granite boulders increases northwestwards along

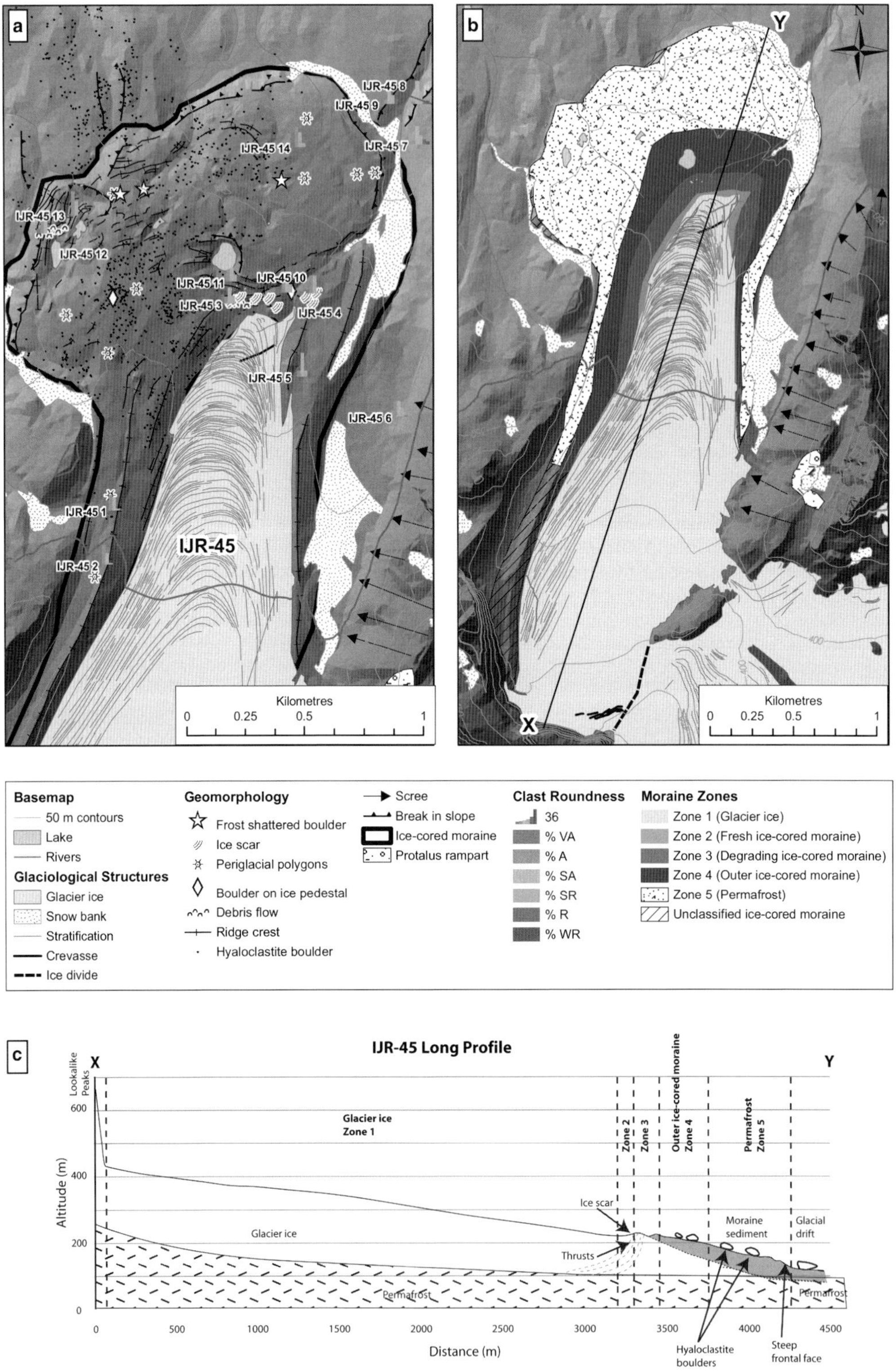

a
IJR-45 8
IJR-45 9
IJR-45 14
IJR-45 7
IJR-45 13
IJR-45 12
IJR-45 10
IJR-45 11
IJR-45 3
IJR-45 4
IJR-45 5
IJR-45 6
IJR-45 1
IJR-45 2
IJR-45
Kilometres
0 0.25 0.5 1
b
Y
N
X
400
Kilometres
0 0.25 0.5 1
Basemap
50 m contours
Lake
Rivers
Glaciological Structures
Glacier ice
Snow bank
Stratification
Crevasse
Ice divide
Geomorphology
Frost shattered boulder
Ice scar
Periglacial polygons
Boulder on ice pedestal
Debris flow
Ridge crest
Hyaloclastite boulder
Scree
Break in slope
Ice-cored moraine
Protalus rampart
Clast Roundness
36
% VA
% A
% SA
% SR
% R
% WR
Moraine Zones
Zone 1 (Glacier ice)
Zone 2 (Fresh ice-cored moraine)
Zone 3 (Degrading ice-cored moraine)
Zone 4 (Outer ice-cored moraine)
Zone 5 (Permafrost)
Unclassified ice-cored moraine
c
IJR-45 Long Profile
X
Y
Lookalike Peaks
Glacier ice
Zone 1
Zone 2
Zone 3
Outer ice-cored moraine
Zone 4
Permafrost
Zone 5
Ice scar
Moraine sediment
Glacial drift
Glacier ice
Thrusts
Permafrost
Permafrost
Altitude (m)
0
200
400
600
0 500 1000 1500 2000 2500 3000 3500 4000 4500
Distance (m)
Hyaloclastite boulders
Steep frontal face

the crest of the moraine, indicating the reworking by wet-based ice of the pre-existing and underlying coastal erratic-rich drift. Conversely, the hyaloclastite boulders decrease in number and size northwards along the crest of the moraine, as a result of increasing degradation and *in situ* weathering.

Subglacial transport is suggested by the clast shape-roundness data (Fig. 5), with considerably less angular material in 'Brandy Bay Moraine' than the ice-cored moraines and the basalt-rich drift. This glacier advance was synchronous with the recession of ice from Johnson Mesa and the collapse of Prince Gustav Ice Shelf (Pudsey & Evans 2001; Johnson *et al.* 2011), suggesting a warmer climate. This is supported by evidence from lake cores (Björck *et al.* 1996). However, constraints on radiocarbon dating limit our ability to refine the timing of these events. The moraine remains ice-cored at depth because of its large size and thick debris cover (which is considerably thicker than the active permafrost layer of 1 m and thus protects the ice core from further melting), and this ice is genetically related to the abundant permafrost features found on its crest. The large thickness of sediment cover is sufficient to protect the ice core from further ablation.

'IJR-45' currently has only a small accumulation area. For the creation of such a large moraine, coalescent glaciers with accumulation areas on Lachman Crags, Davies Dome and Johnson Mesa must be invoked. In addition, shifting ice divides may have favoured 'IJR-45' over Alpha Glacier, resulting in glacier advance. The survival of delicate hyaloclastite boulders in the boulder train suggests that detachment from its accumulation area during the Mid-Holocene resulted in rapid recession.

'IJR-45 Glacier Moraine'

Description. 'IJR-45 Glacier Moraine' is up to 1 km wide, with low slope angles until the edge of the moraine, where it drops off sharply (Fig. 6a). The glacier trunk comprises clean ice and ice with a thin debris cover. The lateral-frontal complex comprises a chaotic assemblage of small, sharp-crested ridges 1–3 m high and up to 1 m wide and lines of boulders. Stratified ice is frequently exposed in back scars in this location. However, the region of intensely ice-cored moraine only extends for 50 m from the glacier snout. In this frontal region, the surficial sediments are highly variable, with localized patches of fossiliferous sandstone, mudstone and diamicton amongst the largely hyaloclastite-derived basalt sandy boulder gravel. In the immediate ice frontal region, the abundant meltwater mixes with fines to produces muddy, clast-rich diamictons. Glacier ice is inferred to continue westwards beneath the high frontal-lateral moraines (Engel *et al.* 2012).

The next 50–160 m from the 'IJR-45' snout are characterized by increasingly degraded down-wasting back scars with exposed stratified blue and white ice, small sharp-crested ridges 1–3 m high, and numerous small perched lakes. The ridges follow the arcuate form of the primary stratification in the glacier ice, and are often asymmetric, with steep stoss faces and gentle lee faces. There is abundant meltwater dissecting the moraine-mound complex. The ice exposed in a back scar close to the snout ('IJR-45 4'; see Fig. 6a for location) has 5 m of bubble-poor, large, dark blue and well-defined ice crystals 0.05–0.10 m long, overlain by 0.50 m of sediment.

From 160–400 m back from the glacier snout, there are subdued crescent-shaped scars, circular niches and ridges with no ice visible, and numerous large hyaloclastite boulders, down-wasting and weathering *in situ*. The moraine surface is characterized by a pebble–boulder lag of subangular to angular basalts. There is a large frozen lake (>100 m diameter) and numerous small perched lakes, as well as several circular depressions from drained lakes.

From 400–1000 m back from the glacier snout, on the glacier true left-lateral moraine, the ridges widen and flatten down-slope into a 100 m wide ridge, sometimes with small 5 m high subsidiary ridges and isolated mounds. The ridges become increasingly subdued with distance from the ice margin. In this lateral position, stone stripes and patterned ground are clearly developing. There are no ice scars, although there are numerous ridges topped with lines of hyaloclastite boulders that reflect the orientation of the primary stratification in the glacier ice (Fig. 6a). These hyaloclastite boulders are up to 7 m in diameter. The moraine surface texture varies spatially, with areas of striated, faceted boulders, and areas with subangular to angular basalts showing little sign of glacial transport. There are rare small lakes. The true right-lateral moraine is similar. For example, 'IJR-45 5' is characterized by a flattened ridge, 100 m wide, scattered boulders and isolated mounds and periglacial polygons. In regions more distal to the terminus, there is a basalt-rich boulder gravel with a

Fig. 6. Boulder train assemblage. (**a**) Geomorphological map of 'IJR-45 Glacier Moraine', showing degradation of ice-cored moraine on 'IJR-45'. Refer to Figure 2 for regional situation. (**b**) Landform zonation on 'IJR-45 Glacier Moraine', from fresh ice-cored moraine to an outer zone of permafrost. Refer to Table 5. (**c**) Annotated long profile of 'IJR-45', along line XY marked on (b). Maps are overlain on the 2006 DEM (Czech Geological Survey 2009).

Table 5. *Description and characteristics of the principle zones associated with the large moraine in front of 'xIJR-45'*

Zone	Sediments	Landforms	Interpretation
Zone 1	Thin drape of fine pebble-gravel at the margin. Largely clean glacier ice	Stratified white and blue ice, supraglacial streams	Glacier ice; minor debris cover
Zone 2	Surface varies from silt-rich diamicton to angular sandy boulder gravel, angular to striated boulders and cobbles. Occasional striated boulders	Series of small, thin, sharp-crested ridges, numerous small lakes (<10 m diameter), scars with exposed stratified ice (sediment drape is 20–40 cm thick), conical mounds of sediment, large boulders resting on pedestals of ice. Complex and chaotic topography. Lines of boulders on ridges. Small debris flows	Fresh ice-cored moraine
Zone 3	Sandy boulder gravel surface with matrix composed of weathered hyaloclastite. Occasional fines in association with numerous sandstone boulders	Moraine widens and flattens; 5 m high sharp-crested to more subtle ridges. Isolated mounds. Ice collapse pits. Occasional small ice scars and buried ice under ridges. Lines of hyaloclastite boulders on ridges. Perched boulders. Crescent-shaped scars. Perched lakes. Ridges increasingly subdued with distance from the ice margin	Degrading ice-cored moraine
Zone 4	Lag of pebbles and boulders on surface. Sandy boulder-gravel. Frost-shattered basalt	Subsiding and subdued ridge crests, debris flows, perched and frozen lakes, no visible ice, sandy boulder gravel surface. Basalt and hyaloclastite boulders wasting down *in situ*. Incipient stone stripes and polygon nets	Dead-ice moraine
Zone 5	Fines are absent from the surface, leaving a cobble-pebble lag. Basalt and weathered hyaloclastite boulders. Well-developed stone stripes and polygon nets. Frost-shattered basalt	Low-relief and subtle ridge crests, no visible ice, sandy boulder gravel surface. Small incised and snow-patch fed streams. Steep frontal edge off moraine. Solifluction lobes move down-slope on steep frontal face and where streams have incised. Evidence of former ice-core, including debris flows, subsiding ridge crests, weathering hyaloclastite boulders. Drained lakes leaving depressions	Dead-ice moraine; permafrost

coarse orange sand matrix, derived from the *in situ* weathering of massive hyaloclastite boulders.

From 1000 m from the glacier terminus to the upslope limit of the visible moraine, the moraine is characterized by smooth slopes with loose scree, frost-shattered basalt boulders, well-developed periglacial stone nets and stripes, drained lakes and subdued ridges. There are large numbers of hyaloclastite boulders, particularly to the true left of the moraine, and subdued arcuate ridges. The outer face of the moraine is steep, lobate, and contrasts sharply with the glacial drift in the forefield (the photograph Figure 4b was taken in front of 'IJR-45', which is visible in the distance). On the western flank of Smellie Peak, adjacent to the moraine terminus, a stream has incised a steep 10 m deep channel. A series of three small, dry tributaries join this stream, running consecutively across the steep slope and at an angle to the fall line.

Interpretation. 'IJR-45 Glacier Moraine' is characterized by five different zones (Table 5 and Fig. 6b, c). Fresh, actively back-wasting ice scars in the frontal-lateral complex grade outwards to increasingly smoother slopes, more uniform and compacted sediments, and increasing weathering and disintegration of hyaloclastite boulders. 'IJR-45 Glacier Moraine' is separated from the ice-cored moraines of the Late Holocene Assemblage because of its older age, deduced on morphostratigraphical grounds, such as advanced periglaciation and ice-core degradation.

Zone 1 comprises clean glacier ice and glacier ice with a thin debris cover. The moraine near the snout of 'IJR-45' (zones 2–3) has many features typical of ice-cored moraines formed through the thrusting of polythermal ice at the glacier margin (summarized in Table 5), including stratified, inclined glacier ice, ice scars, perched ponds, steep slopes, ridges and scarps (e.g. Østrem 1964; Knight *et al.* 2000; Schomacker & Kjær 2007). Sediment accumulates through the meltout of thrusted subglacial sediments, englacial debris bands and channel-fill materials. Supraglacial material is also a significant component of the debris mantling the glacier surface. De-icing processes include backwasting and downwasting (cf. Krüger *et al.* 2010). This causes surface lowering and flattening of ice-cored slopes. Collapse results in the breaking of sediment cover and the exposure of new ice walls and new cycles of ice degradation. Resedimentation occurs through slumping, sliding and sediment flows, and reworking from fluvial, aeolian and periglacial (freeze–thaw) action. The occasional striated boulders indicate active subglacial transport. This zone may have been contemporaneous with the Late Holocene assemblage.

Zones 3–5 are typical of dead-ice (stagnant glacier ice) environments (cf. Krüger *et al.* 2010), where melting of buried ice results in resedimentation processes (Bennett *et al.* 2000*a*). This is dead-ice-cored moraine, with large volumes of supraglacial, englacial and subglacially transported debris (cf. Klint *et al.* 2011). Water allows liquefaction and reduces the shear strength of the sediment, encouraging further mass movement. The large depth of sediment cover in the dead-ice-cored and permafrost zones protects the ice from further ablation. Indeed, once the accumulated sediment on the moraine becomes thicker than the active layer, estimated to be 1 m (Ermolin *et al.* 2002, 2004), the ice core is preserved. There is little sign of active ice melt in zones 4 and 5, illustrating the nature of dead-ice terrain (Klint *et al.* 2011). Landforms of this type are typical in polar arid environments in East Antarctica (Fitzsimons 2003), Svalbard (Lønne & Lyså 2005; Lukas *et al.* 2007; Jacobs *et al.* 2011) and Iceland (Kjær & Krüger 2001; Krüger *et al.* 2010).

'IJR-45 Glacier Moraine' is interpreted as an ice-cored moraine rather than a rock glacier (cf. Chinn & Dillon 1987) because of limited evidence for down-slope movement, the lack of a continuous supply of talus, and the evidence of thrusting (such as long, arcuate, asymmetric ridges, often with exposed scars of glacial ice, and linear trains of bouldery debris; cf. Hambrey & Huddart 1995; Hambrey *et al.* 1997) from polythermal glacial processes. The ridges on the outer moraine show former ice positions, and are not furrows and ridges related to down-slope movement. The distribution of large hyaloclastite boulders across the moraine reflects ice flow, as they are derived from Lookalike Peaks to the SW of the glacier headwall, and are distributed across the true glacier left. Rather, 'IJR-45 Glacier Moraine' is interpreted as a coherent body of stagnant glacier ice, downwasting *in situ* and overlain by debris.

The landform succession reflects the age and increasing degradation of the moraine, with three stages of landform succession: an initial, dynamic, young phase, with thrusting and upwards ice movement transporting debris to the surface (zones 1 and 2), a mature phase of ice-disintegration and sediment reworking (zones 3 and 4) and a final or old phase of partially ice-cored terrain (zone 5). The presence of well-developed periglacial features on 'IJR-45 Glacier Moraine' suggests that zones 4 and 5 formed some time ago, and that this dead-ice feature has been slowly degrading for centuries and possibly millennia. The ridges and lobate margin of 'IJR-45 Glacier Moraine' (Fig. 6a) suggest that, until 'IJR-45' began to stagnate and downwaste *in situ*, it was an active piedmont glacier, expanding out of the valley.

The clast roundness on 'IJR-45 Glacier Moraine' is difficult to interpret, as the majority of the basalt clasts are derived from *in situ* weathering of large hyaloclastite boulders. However, they are largely more angular than those clasts on 'Brandy Bay Moraine'. The tuff from these boulders also provides the majority of the sand fraction on the moraine. Clasts on the lateral moraine are more angular than those on the terminal moraine. The amount of frost-shattering towards the distal margins of the moraine increases the tendency towards angularity (Fig. 5b). Large basalt screes on Smellie Peak abut the true right moraine of the glacier, and probably represent a major source of clasts. Cretaceous sandstones occur in limited locations in the moraine, reflecting localized bedrock outcrops and low transport distances. The presence of more fines (from Cretaceous sandstone and siltstone) exerts a strong control on moraine character, with patchy outcrops of diamicton and deformed muds in 'IJR-45 Glacier Moraine'. Dry channels in front of the moraine may be palaeo-meltwater channels, recording the recession of the glacier up the valley.

Ice-cored moraine assemblage

Description. Ice-cored moraines in front of cirque glaciers and abandoned cirques are amongst the youngest landforms on Ulu Peninsula (Rabassa 1987), and date from a readvance of the small cirque glaciers that was most likely in the Late

(a)
(b)
(c)
(d)
(e)
(f)
(g)
DD1
(h)
DD2
(i)

Holocene (Carrivick *et al.* 2012). The cirques are characterized by steep bedrock headwall cliffs and a rounded, over-deepened basin, and they are occupied by small glaciers or occasionally lakes (e.g. by Davies Dome and the western face of Lachman Crags or the northeastern slopes of Johnson Mesa). The ice-cored moraines (e.g. Fig. 3) typically have multiple sharp-crested ridges and numerous small lakes and ponds (Fig. 7a–c). The surfaces of the moraines range from sandy boulder gravel, through unsorted openwork basalt boulders to diamicton. Pebbles and cobbles are subangular to subrounded and show increasing roundness of shape with proximity to the glacier snout. Exposures reveal stratified glacier ice (layered blue or white ice, bubble poor or bubble rich) that is exposed in ice scars (Fig. 7d, e), and superficial slumps and debris flows. The exposed ice has a fine crystal structure and a low bubble content, and contains debris ranging from sand and fine gravel to small boulders. In San José Glacier moraine, an exposure of ice with stratification dipping at 40° up ice reveals clean blue bubble-free ice, interbedded with white bubble-rich ice, diamicton, fine to boulder-sized clasts and muddy laminations (Carrivick *et al.* 2012). On the terminal moraines there are occasional striated boulders (e.g. Fig. 7i).

Clast lithological data gathered from San José Glacier moraine shows that the predominant lithology is the James Ross Island Volcanic Group (Fig. 8a). These basaltic rocks are well exposed in the bedrock cliffs and on the mesa above the glacier. Transport distances, especially for the friable and non-durable Cretaceous sandstone, are very low, with Cretaceous rocks occurring only in specific locations close to their outcrops. Unlike the smoothed glacial surfaces outside the moraines, no Antarctic Peninsula erratics were observed. The presence of Cretaceous lithologies, especially mudstone, significantly alters the composition of the moraine, with far higher percentages of fines and muds observed. This results in patches of diamicton amongst the largely sandy boulder-gravel of the moraines.

Clast-roundness histograms from gravel-sized basalt clasts vary with sample location. Generally, clasts in terminal moraines exhibit more rounding than those in high lateral moines. For example, in Lachman and San José glaciers, there is a trend towards increasing angularity of gravel clasts towards the headwalls of the glacier (Fig. 8b). However, this trend is disrupted on the outermost edges of Lachman Glacier moraine, where there are increasing percentages of very angular clasts (Fig. 5). However, there is considerable inter-moraine variability. In 'Unnamed Glacier' moraine, sample Unnamed 3 has particularly high percentages of angular clasts despite being located in a terminal moraine position. In general, the ice-cored moraine clasts show more angularity than the older glacial drifts (Fig. 5a, b).

The moraines around Davies Dome (Fig. 7g, h) are texturally and morphologically highly variable. For example, on top of the mesa, where the ice margin is rounded in both planform and surface slope, there is no evidence of landform construction. In contrast, the lobed southeastern margin comprises regions of angular unsorted gravel, diamicton ridges and ridges of angular, frost-shattered basalt. At sample location DD2 (234 m a.s.l.; Fig. 8d), there are small, sharp-crested ridges up to 20 m high and 1 m wide, transverse to ice flow. There are back scars with exposed stratified ice. The proportion of sandstone in the clast count and matrix is highly variable. Mud content varies considerably over small distances, clearly dependent on local bedrock geology, that is, the presence of Cretaceous deposits. Adjacent to these moraines, ice-contact scree is formed below an ice-cliff, forming angular open-work basalt cobbles and boulders, such as Sample DD1 (217 m a.s.l.), which was taken in front of an ice cliff (Fig. 8). The stratified and dipping ice has a structural unconformity with a layer of angular basalt boulders and cobbles.

Lateral moraines on the outlet glacier, which coalesces with Whisky Glacier in Whisky Bay, are also typical ice-cored moraines. Sample Davies 1 (Fig. 8d) is characterized by a series of small, sharp-

Fig. 7. Late-Holocene moraines on James Ross Island. (**a**) Inner moraines on Lachman Glacier. (**b**) Kettle lakes on Lachman Glacier moraine. (**c**) Thrusts (red dashed lines) on the inner face of Lachman Glacier latero-terminal moraine. (**d**) Ice scars (arrowed) exposed in the true right latero-terminal moraine on Lachman Glacier. (**e**). Section 2; exposure of stratified ice in the lateral moraine on San José Glacier. (**f**) The ice-cored moraine of Triangular Glacier. (**i**) Striated sandstone boulder at the proximal terminal moraine of 'IJR-45'. (**g**) Sample location DD1; margin of Davies Dome. Refer to Figure 2 for regional situation. Samples DD1 and DD2 are in close proximity but are strikingly different. Sample DD1 is in front of an ice cliff, with ice-contact scree forming and flowing down-slope under gravity. The surface is openwork gravel and boulders. Note the unconformity within the stratified ice, which contains boulders of basalt. The stone lithologies are all angular to subangular basalt. (**h**) Sample DD2 is characterized by chaotic mounds and ridge crests on the ice-cored moraine in front of the lower lobed margin of Davies Dome. Refer to Figure 5c for angularity histograms. (**i**) Striated Cretaceous sandstone boulder in the proximal terminal moraine for 'IJR-45'. Camera lens cap for scale.

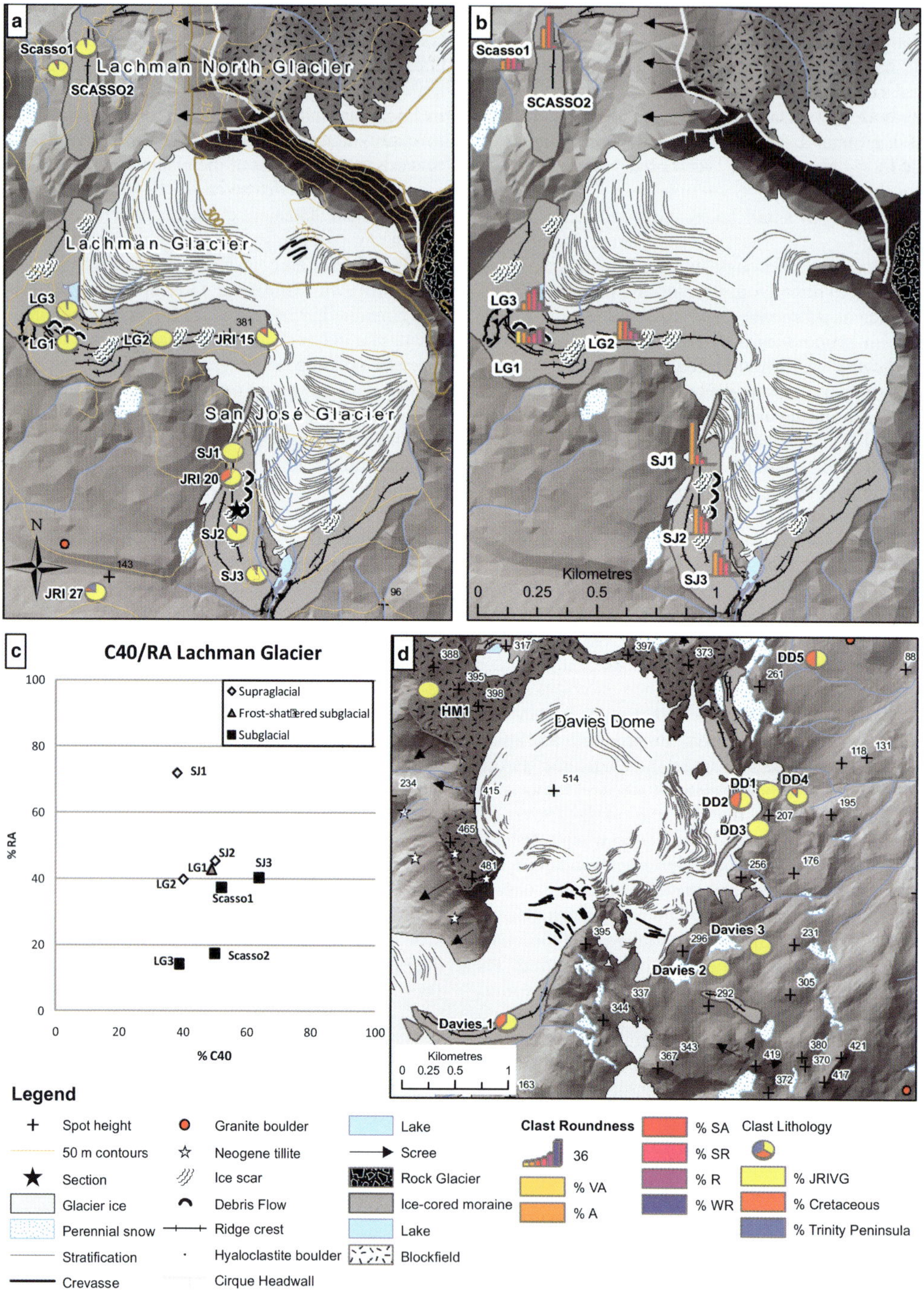

Fig. 8. Ice-cored moraine assemblage. (**a** and **b**) Glaciological and geomorphological map of San José and Lachman glaciers and their associated moraines. Refer to Figure 2. for regional situation. Note how the moraines mirror the primary stratification. Overlain on the 2006 DEM (Czech Geological Survey 2009). The lithological composition of moraine surface samples ($n = 50$) is shown. (b) Map showing roundness data of surface pebble samples ($n = 50$). (**c**) C_{40}/RA plot of shape-roundness data of all samples. (**d**) Map of Davies Dome Glacier, showing sample locations DD1 and DD2.

crested ridges, 2–20 m high, with basalt boulders perched on their crests. Ice scars expose stratified ice, dipping at 80–90°, with 5–30 cm thick layers of dark blue, bubble-poor and white, bubble-rich ice, both with occasional cobbles; 10 cm thick layers of grey mud; and thin mud layers with occasional fine to medium gravel. The ice has a strong unconformity with the 60 cm thick sediment drape above it.

Interpretation. These hummocky ice-cored moraines are assumed to have formed during the last 200 years, when the Antarctic Peninsula experienced a pronounced cooling event, approximately synchronous with the Northern Hemisphere Little Ice Age (Bertler *et al.* 2011). The moraines are principally composed of stratified ice. Ice with dispersed debris is interpreted as ice from above the basal zone. Ice with laminated debris is formed through the attenuation of debris through creep. The ice with stratified debris is interpreted as being derived from basal adfreezing (Hubbard & Sharp 1995; Knight *et al.* 2000; Waller *et al.* 2000; Hubbard *et al.* 2009).

The debris on the surface of the moraines is composed almost entirely of basalt (cf. Fig. 8a, c), originating principally from rockfall from the headwalls onto the glacier surface. However, sand and mud derived from subglacial reworking of Cretaceous sandstone is evident in the fine-grained matrix in particular areas of the lateral moraines. As transport distances of Cretaceous lithologies are low, sandstone gravel forms only a minor component of the moraine lithologies. Subglacially transported debris (such as striated boulders) is apparent, particularly in the terminal moraines. Moraine surface character and matrix content are therefore closely controlled by the bedrock geology, with patches of purely basaltic boulder–cobble gravel and diamicton occurring in close proximity to one another.

Ice-cored moraines are formed by the release of material from proglacial or englacial thrusts, meltout and supraglacial debris (Glasser & Hambrey 2003; Schomacker 2008; Schomacker & Kjær 2008). Subglacial thrusting entrains sediment, with moraines forming through the thrusting upwards of basal glacial ice at the ice margin, induced through marginal shear. This behaviour is typical of small polythermal glaciers (Hambrey *et al.* 1999; Glasser & Hambrey 2003). Folding, thrusting and stacking of basal ice because of longitudinally compressive flow in polythermal glaciers (where the margin is frozen to the bed) results in the development of moraine ridges with intercalated debris-rich and debris-poor stratification (Boulton 1972; Evans 2009). The angle of thrusting controls not only the angle of moraine slope, but also moraine height, width and character. These are *controlled moraines* (Ó Cofaigh *et al.* 2003; Evans 2009), where the linearity of the moraine is controlled by primary stratification in the glacier ice (e.g. Figs 3, 6 & 8). These moraines indicate that the glaciers were previously more active in the past, with at least a temperate ice regime in the snout, which is a common phenomenon in small modern glaciers in the High Arctic (e.g. Hambrey *et al.* 2005; Rippin *et al.* 2011).

The complexity and variation in moraine around Davies Dome (ice-cored moraines composed of sandy boulder gravel and diamicton; ice-contact scree; flat and featureless on the mesa plateau; Fig. 8d) is likely to reflect variations in thermal regime. The lobes and outlet glaciers are likely to be polythermal, with associated ice-cored moraines. Ice-contact scree forms on steep slopes below ice cliffs, where meltout and sublimation result in the deposition of angular englacial and supraglacial debris. Ice-contact scree suggests that the ice margin is stable; this is a common landform in cold-based glaciers (Fitzsimons 2003). The smooth, flat margin on top of the mesa suggests that this part of the dome is fully cold-based, with little or no movement or modification of the landscape.

The shape-roundness data show some differences between samples from the high lateral moraines and samples from the distal terminal moraines within the same moraine (Fig. 5). Samples from the terminal moraines clearly have blockier shapes and edge-rounded corners, which is indicative of subglacial transport. Samples from the lateral moraines are more angular, with less blocky clast shapes. This is indicative of shorter transport distances, and a higher input of supraglacially derived material. This trend reflects the influence of transport distance on the shape of gravel clasts, and that these small cold- to wet-based glaciers are locally capable of modifying their bedload. Angularity again increases towards the older, outer portions of the moraine, as post-depositional frost-shattering fragments basalt clasts. This difference is clearly shown in the C_{40}/RA plot in Figure 8c, with two distinct populations. However, there is considerable variation between the moraines; this is likely to be as a result of variations in the input material (scree), which is dependent on its origin (refer to Paraglacial assemblage).

Part of the complexity in clast shape is highlighted in 'Unnamed Glacier' (Figs 2 & 3c). Sample Unnamed 3 (Fig. 5b) is directly below steep cliffs from which scree is constantly falling onto the glacier surface. This scree is lightly buried by snow, and then observed melting out near the terminus; sample Unnamed 3, although from the terminal moraine, is therefore largely derived from supraglacial material that has undergone little modification. This contrasts with Unnamed 4, also from a terminal position on the moraine (Fig. 5c), but from a position away from supraglacially transported material. The edge-rounding of clasts in this location is indicative of subglacial transport.

Post-depositional processes play a key part in moraine development, with aeolian deflation rapidly removing fines, resulting in a lag of cobbles and pebbles on the surface of the moraine. Melting of the ice-core provides abundant meltwater, and numerous debris flows, especially in regions rich in matrix material. Preferential melting of ice core also results in the formation of ice-collapse pits, which subsequently become perched ponds (kettles).

With only a thin sedimentary covering, these moraines are particularly prominent only because of the presence of this ice. After ice-core melt-out, there will be little geomorphological expression, which explains the lack of other prominent moraines from previous readvances on Ulu Peninsula. Following the terminology of Clayton (1964) and Kjær Krüger (2001), these are mature, fully ice-cored moraines, characterized by mass movement and resedimentation processes. They can be compared with 'IJR-45 Glacier Moraine' and 'Brandy Bay Moraine', which are older and are classified as 'partially ice-cored terrain'.

Paraglacial assemblage

The 'paraglacial period' refers to the period of readjustment from glacial to nonglacial conditions, with the reworking and relaxation of glacial landforms and sediments, such as steep debris-mantled slopes, cliffs and large fluvial systems (Fitzsimons 1996; Ballantyne 2002). Mass-wasting of rock walls, movement on debris-mantled slopes, aeolian processes and limited fluvial transportation dominate sediment transfers, which are limited to the short summer season on James Ross Island. On Ulu Peninsula, paraglacial sediments and landforms overprint and modify landforms and transport material.

Marine terraces and raised beaches

Description. Below 30 m on the Abernethy Flats there are smooth, flat slopes, an absence of large boulders, and rounded pebbles. Shape-roundness sample Aber 3, at 30 m a.s.l. on the Abernethy Flats (Fig. 9a), illustrates the high degree of rounding on the clasts, with spherical shapes (Fig. 5a). Sample Aber 1 was taken further away from the coastline but still at 30 m a.s.l. It has 18% angular pebbles, thereby possibly reflecting frost shattering. In this aspect sample Aber 1 bears a closer resemblance to sample Aber 2, taken from the basalt cobble–gravel glacial drift. The surface near sample Aber 3 is flat and smooth with a deflated pebble–cobble gravel. There are sparsely distributed hyaloclastite boulders.

A series of flat terraces with rounded pebbles were observed near the sea shore in front of Alpha Glacier (Fig. 9b, c). The uppermost bench, at 7 m a.s.l., has relatively few rounded pebbles. The surface is composed of a sorted basalt pebble-gravel, with no large boulders and a sandy silt matrix. The middle bench (2 m a.s.l.) is separated from the others by steep slopes with active solifluction lobes. This bench dips gently seaward. Frost shattering of pebbles is evident.

Interpretation. Associated with and overprinting the glacial drifts in coastal areas on James Ross Island are a series of marine terraces, formed during and after deglaciation. Marine terraces on James Ross Island occur up to a height of 90 m a.s.l., and date from the Pliocene to the Holocene (Rabassa 1983; Strelin & Malagnino 1992; Roberts *et al.* 2011). Hjort *et al.* (1997) recorded a series of strandlines and beach deposits near San Carlos Point, including beach gravels at 16, 30 and 90 m a.s.l. Hjort *et al.* suggested that the 30 m a.s.l. marine limit observed primarily on the Abernethy Flats was associated with deglaciation of Prince Gustav Channel. A 16 m a.s.l. terrace is a Mid-Holocene marine level following further terrestrial deglaciation (Ingólfsson *et al.* 1992; Hjort *et al.* 1997).

Spits and modern beaches

Description. Northern James Ross Island is fringed by beaches, occasionally with sandy spits. Spits occur at San Carlos Point (500 m long, 80 m wide), Cape Lachman and St Martha Cove (900 m long, 300 m wide). Furthermore, over a distance of 1.8 km on the beach west of Mendel Station (Fig. 2), 79 granite, 19 gneiss and 22 pelitic boulders with a *b*-axis >1 m were observed. Boulders were also noted on the beach east of Mendel station to the tip of Cape Lachman.

Interpretation. The isolated sand and gravel spits formed through the reworking of glacial and fluvial sediments in the littoral zone (Ballantyne 2002). Littoral longshore currents transported formerly deposited glacial and fluvial material alongside the beach. This coastal reworking has also resulted in a large number of boulders, especially of Trinity Peninsula origin, forming a boulder lag on the beaches, which is typical of paraglacial environments (Ballantyne 2002). Unstable steep slopes behind beaches are subject to solifluction

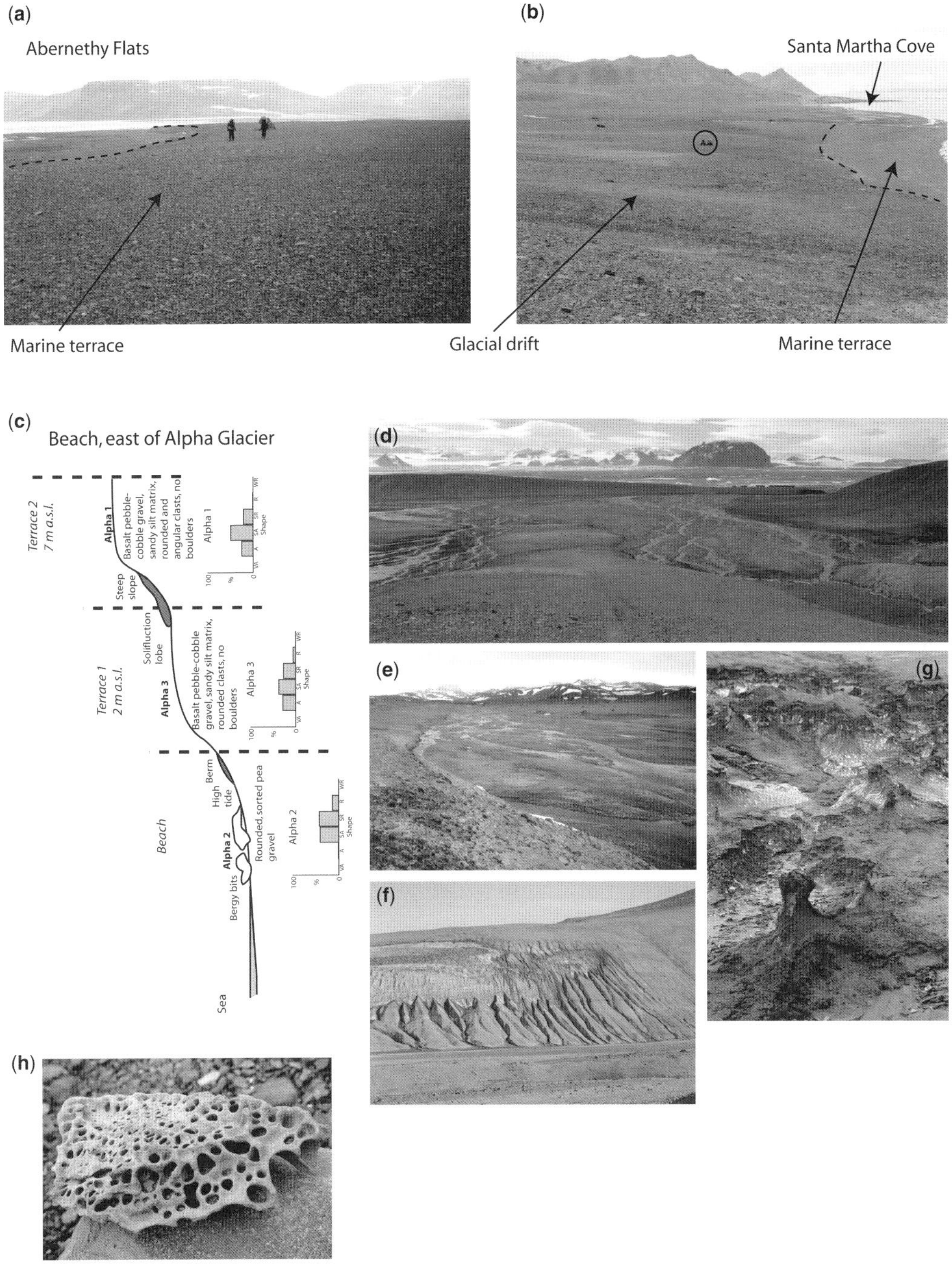

Fig. 9. Paraglacial assemblage. (**a**) Marine terrace at site Aber 3 site, Abernethy Flats, facing Brandy Bay. Surface is flat and featureless, with few boulders. Clasts are 100% locally derived basalts and hyaloclastites. Clast shape shows a high degree of clast rounding. Dashed line indicates margin of modern beach. (**b**) Marine terrace in front of Alpha Glacier. People circled for scale Dashed line indicates margin of marine terrace. (**c**) Idealized sketch of the marine terraces in front of Alpha Glacier. (**d**) Bohemian Stream, draining towards Mendel Station. (**e**) Monolith Stream, Brandy Bay. Braided stream incised through the marine terrace on Abernethy Flats, draining into Brandy Bay. Note the point bars, multiple channels and low-volume discharge. (**f**) Wind-blown sand on snow patch below Crame Col. (**g**) Detail of windblown sand on snow patch below Crame Col. (**h**) Aeroxysts on a basalt boulder.

and over-steepening by wave action, further encouraging accumulations of boulders on the beach.

Rivers and streams

Description. Ephemeral incised streams and braided streams on Ulu Peninsula typically have multiple channels with an active river width of up to 100 m, incised stream cuts, small islands, point-bars and longitudinal mid-channel bars (cf. Miall 1977), which shift rapidly in response to widely variable flows (Figs 2 & 9d, e). The braided river draining Alpha Glacier is characterized by rounded, poorly sorted and clast-supported gravel on point bars (cf. Fig. 5a). The braided channel is incised 1–2 m, with numerous small channels actively downcutting. Seal Stream and Monolith Stream (Fig. 2) are associated with the seaward progradation of the outwash plain in Brandy Bay.

Interpretation. Ephemeral rivers and streams on James Ross Island are fed by perennial snowfields and melting glaciers (Fig. 2) and their discharge varies considerably on diurnal, weekly and monthly timescales. On warm days, these ephemeral rivers and streams are capable of winnowing glacial sediments and incising Cretaceous bedrock. Two contrasting fluvial styles occur: the low-discharge, multi-channel, high-sinuosity, braided stream stage, and the high-discharge low-sinuosity stage with one principal channel. In this stage, the whole of the channel belt is flooded.

Aeolian sediments and landforms

Description. Glacial drifts on Ulu Peninsula are frequently covered by a basalt pebble–cobble armour, commonly only one or two clasts thick, with sand beneath. Accumulations of wind-blown sand are frequently found on the surfaces of snowfields, where it can produce aeolian accumulations several metres thick after snowfield decay (Fig. 9e, f). Many of the basalt and granite boulders associated with the erratic-poor LGM drift (cf. Table 4) have smooth plano-concave and plano-convex faces (e.g. Fig. 4b; Knight 2008). Red staining is common on granite boulders in the LGM drifts on Ulu Peninsula (e.g. Fig. 4a).

Interpretation. Aeolian processes are strong and pervasive on Ulu Peninsula. Winnowing of fine particles from surficial deposits is pervasive on the flatter regions such as the Abernethy Flats. This winnowing has resulted in a pebble–boulder lag on the surface of glacial drifts. The dry, unvegetated climate of James Ross Island makes it particularly susceptible to aeolian deflation, and strong katabatic winds exacerbate the process. Similar processes have been observed in East Antarctica (French & Guglielmin 1999; Bockheim 2010) and in Iceland, where, over four years, a recently deglaciated surface went from being 30–40% clast-covered to 90% clast-covered (Boulton & Dent 1974; Ballantyne 2002). Most of the wind-transported material terminates in the sea or is caught on snow banks. Basalt ventifacts with smooth plano-convex faces are common, which are typical of recently deglaciated, periglacial terrain, because of the lack of vegetation cover, the presence of large unburied boulders, the availability of sand through glacial abrasion and mechanical frost weathering, and the strong katabatic winds (Knight 2008). In addition to ventifacts, aeroxysts are present on the surface of basalt and tuff boulders, along with other types of aeolian erosional micro-forms (e.g. tafoni; Fig. 10h).

The red staining on numerous granite boulders is a red desert varnish, enriched in iron, rather than the manganese frequently associated with hot desert environments. Iron oxides are leached out from the internal rock and are deposited as varnish. Granite is particularly prone to the development of a red desert varnish because it is resistant to physical breakdown and contains iron-bearing biotite (Matsuoka 1995). Well-developed desert varnish forms in semi-arid, sheltered areas, away from wind abrasion. Ventifacts and red desert varnishes are both a common feature of ice-free regions of Antarctica (Matsuoka 1995; French & Guglielmin 1999; Hall & Denton 2005; McLeod *et al.* 2008; Bockheim 2010).

Scree slopes

Description. Scree slopes are common beneath the steep Neogene basalt and hyaloclastite cliffs on James Ross Island, and are an important input into moraines, rock glaciers and protalus ramparts. Three shape-roundness samples were taken from modern and active scree slopes (Fig. 5a, b), where they show considerably more angularity than the Glacigenic assemblage drifts but are similar to the lateral moraines in small cirque glaciers. Scree derived from hyaloclastite (NM2) is more rounded, whilst scree from flood basalts (IJR-45 6; BH1) is more angular (Fig. 5c).

Interpretation. The steep cliffs of the James Ross Island Volcanic Group, with vertically jointed hyaloclastite deltaic deposits, are particularly susceptible to rock weathering and scree slope formation. When deposited, the hyaloclastite lacks a fine binding matrix so it is often only weakly bound by palagonite clay from alteration of original glass (away from any pore-filling zeolite sediment; Johnson & Smellie 2007). After recession of glacier ice after

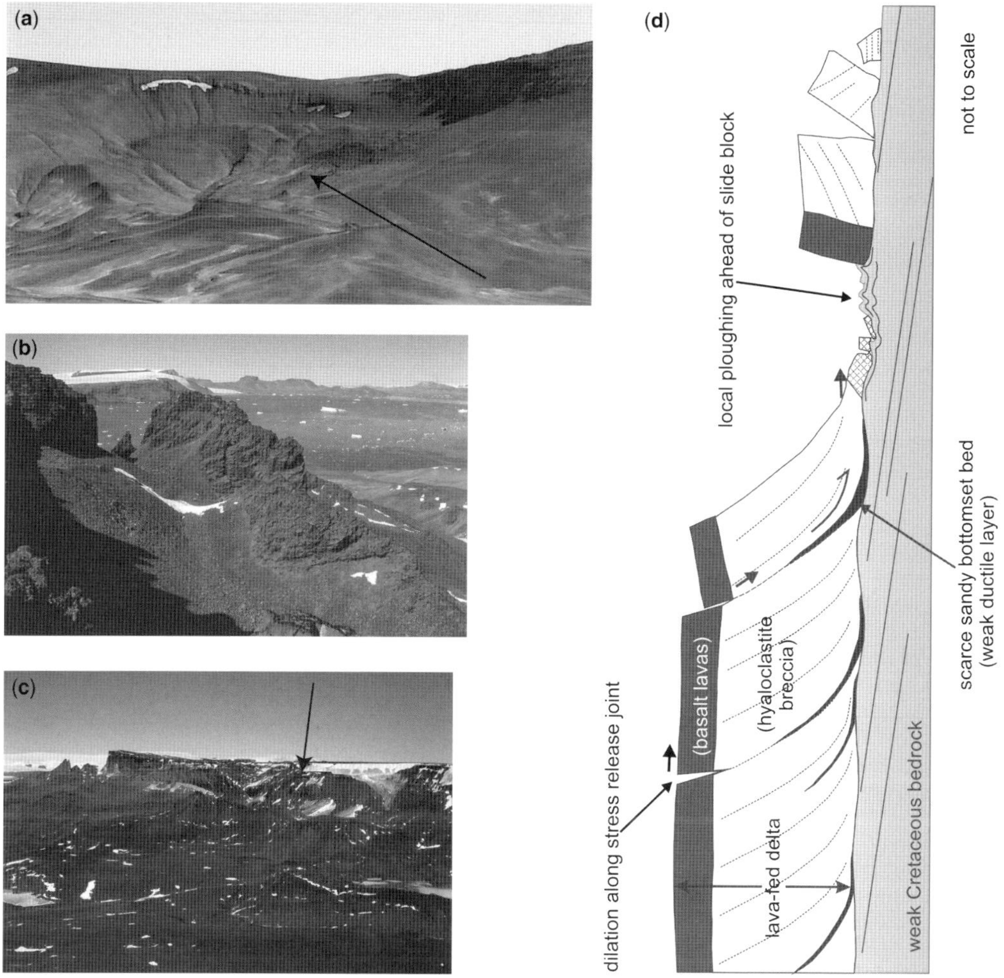

Fig. 10. Paraglacial landforms: large scale mass movements. (**a**) Landslide below the col between Johnson Mesa and Bibby Hill. (**b**) South side of Berry Hill. There is a large block of hyaloclastite with subaerial lava cap that has detached, tilted outwards (to right hand side) slightly and started to fall away from Berry Hill (subaerial lavas still *in situ* on Berry Hill top partly seen at left). A few smaller degraded hyaloclastite blocks can be seen in the background right hand side at a lower level. (**c**) Berry Hill looking down the E side of Lachman Crags. The middle distance and far left are hummocky terrain underlain by multiple large mass movement blocks derived from Lachman Crags. Also visible (arrowed) is a block that has dropped a few tens of metres but is still upright – that is, frozen in action as it detaches and begins its down-slope journey. (**d**) Schematic diagram illustrating processes involved in the emplacement of the large-scale mass movement of sections of the lava-fed deltas on James Ross Island. Instability near the margins of the delta following the removal of supporting ice or sea water is envisaged causing stress-release fracturing, with fractures potentially propagating down to intersect any fine-grained (sandy) bottomset layers in the predominantly very coarse hyaloclastite breccias. Note that, although the sandy layers probably acted as weak ductile surfaces of décollement, they are comparatively uncommon and delta flank collapses also occurred in sections lacking sandy interbeds. However, their presence and greater structural weakness relative to the enclosing hyaloclastite breccias ensures that, if present, they probably act preferentially as décollement surfaces. For clarity, the diagram does not illustrate the presence and potential importance of high porewater pressures beneath a subglacial and/or proglacial permafrost layer, which would reduce the effective stress of the substrate and promote lateral instability and collapse.

the LGM, stress-release fracturing and subaerial exposure of the cliffs occurred, making them susceptible to freeze–thaw activity in a periglacial climate (Harris 2007). This has resulted in rapid readjustment and the formation of active scree slopes (cf. Ballantyne 2002), which are a key input for rock glaciers and protalus ramparts, and which are continuing to form actively today.

Large-scale mass movements

Description. Large-scale mass-movement is evident on Ulu Peninsula where the James Ross Island Volcanic Group rests upon gently inclined poorly consolidated Cretaceous mudstone (Fig. 2). There is a large landslide on the northern foothill just below the col between Johnson Mesa and Bibby Hill (Fig. 10a–c). There are many smaller rockfalls and debris flows at the Andreasson Point massif (Fig. 2). In areas affected by large-scale mass movement, volcanic blocks many tens of metres high (representing the full thickness of the local volcanic sequence) and a few hundred metres long, along with smaller fragments, form enormous jumbled heaps, for example, forming the hilly high ground backing onto Andreasson Point. These blocks have detached from the main ice-capped volcanic outcrops nearby (e.g. the Lachman Crags lava-fed delta is the source for the blocks at Andreasson Point). Elsewhere, similar chaotic blocky terrain formed by mass movement sourced in lava-fed deltas is present on the east side of Lachman Crags (Fig. 10c) to the north of Andreasson Point, and forming the hummocky ground between Davies Dome and Lookalike Peaks (surrounding Sekyra Peak) and between Smellie Peak and Dinn Cliffs.

Interpretation. The large-scale mass-movements are probably controlled by gravity acting on steep-fronted brittle rock masses (delta margins) resting on soft ductile substrate (Cretaceous sediments). They are clearly geologically controlled, the upper surface of the Cretaceous mud representing a décollment surface on which the volcanic blocks can slide. Plough structures suggest that the process is still active, as does the presence of a slab of intact delta several hundred metres along the NE side of Lachman Crags that is down-dropped tens of metres and separated from the main mesa by a snow-filled gap (Fig. 10d). Rates of movement involved in the formation and advection of slabs are not known. Moreover, the formation of the slabs involved in mass movement has not been investigated in detail. However, instability probably occurred after removal of any surrounding ice, or when the deltas were uplifted above sea-level, thus removing the hydraulic support from steep delta fronts; this suggests that formation of the large-scale mass movements began soon after uplift and continued during each interglacial, including the Holocene. Sliding may involve initiation along finer-grained (sandy) 'bottom set' layers in the lava-fed deltas, or a deep décollment surface lubricated by water in a permafrost environment. It is also conceivable that the underlying weak Cretaceous mudstone (and probably weakly lithified sandstone) can deform under the weight of the deltas, which are individually a few hundred metres thick. The blocks then move down-slope with the active layer. Similar landslides have been observed in Scotland, particularly on the Isle of Skye (Ballantyne 1991; Ballantyne *et al.* 1998). However, they are unusual, because these landscape features rely on the uncommon juxtaposition of thick brittle basalt deltas on weak and deformable Cretaceous sedimentary strata. They should therefore not be regarded as 'typical' of semi-arid polar landsystems, unless the appropriate lithologies are present.

Periglacial assemblage

Periglacial and paraglacial processes are closely intertwined on James Ross Island. Massive ground ice and glacier ice underlie many of the landforms on James Ross Island, resulting in abundant periglacial processes. There have been several studies investigating permafrost on James Ross Island and surrounding islands. Electrical resistivity studies have suggested that the active layer here is approximately 1 m deep (Fukuda *et al.* 1992; Ermolin *et al.* 2002, 2004). Slightly lower measurements for the active laver (22–93 cm) and important inter-annual variability (2009–2010) were found by using mechanical vertical probing on a transect from the eastern coast of Brandy Bay, over Johnson Mesa and up to Mendel Station, with variation being controlled partly by altitude, substrate and aspect (Engel *et al.* 2010).

Rock glaciers

Description. Near the southeastern corner of Lachman Crags, there are six rock glaciers (Figs 2 & 11a, b). Lachman II Rock Glacier is located at the end of an ice-cored moraine, in front of stagnating glacier ice (Strelin & Sone 1998; Strelin *et al.* 2006, 2007). Where the sediment cover in the moraine in front of the glacier reaches 0.60 m deep, it is interpreted as a rock glacier, with ridges and furrows delimiting flow-like features. The rock glacier extends 700 m down-valley and is approximately 500 m wide, with a surface dip to 4–5° in the flow direction. It ends in a steep talus apron of 24–42°. The central part is incised by an ephemeral stream and conical holes, occasionally occupied by meltwater, occur on the surface of the rock glacier. Exposures of glacier ice are visible in the central channel, in the conical holes, and in the marginal talus (Strelin & Sone 1998). Lachman II Rock Glacier has a mean annual flow velocity of 0.2 m per year with a travel distance of 400 m from the

Fig. 11. Periglacial phenomena. (**a**) Talus-derived rock glaciers on Andreasson Point (arrowed). (**b**) Talus-derived rock glacier below Berry Hill. (**c**) Protalus rampart on the eastern face of Lachman Crags (arrowed). (**d**) Protalus ramparts below Davies Dome mesa. (**e**) Solifluction lobes (arrowed) mantling the slopes of a hill made of hyaloclastite. Person for scale (circled).

initiation of the rock glacier, suggesting it is approximately 2000 years old. Lakes on this rock glacier may have previously been the source for glacier lake outburst floods (Sone *et al.* 2007).

The rock glacier beneath Berry Hill (Fig. 11b) is formed below steep hyaloclastite cliffs. It is at a low angle with a steep frontal face (60° slope), and is detached from the cliff face. There are several small arcuate ridges, 1–2 m high, which are moderately sharp-crested, with numerous circular depressions. There are several flat benches. The surface is 100% basalt and hyaloclastite, with an orange coarse sandy matrix composed of weathered tuff from the hyaloclastite. There are large blocks of hyaloclastite moving down-slope.

Interpretation. Rock glaciers are lobate or tongue-shaped landforms comprising a mixture of rock debris and ice, typically with a furrowed form, ridges, ponds and a steep terminus and sides. They can be effective transporters of material from cliffs. Rock glaciers can be talus-derived (with debris burying snow and forming interstitial ice) or glacier-derived (Humlum 2000; Degenhardt 2009). Glacier-derived rock glaciers form part of a continuum with, and can evolve from, ice-cored moraines (Evans 1993; Ó Cofaigh *et al.* 2003; Table 6). Distinguishing between ice-cored moraines and glacier-derived rock glaciers can be challenging; however, rock glaciers must move down-slope to be classified as such (similar to glaciers; Østrem 1971), and would therefore normally occur on slopes, where they advance through the internal deformation of ice (see comparison in Table 6). Lachman II Rock Glacier is a glacier-derived rock glacier. Examples of talus-derived rock glaciers can be found west of Andreassen Point (Fig. 11a), below Berry Hill (Fig. 13b) and in Rockfall Valley (Fig. 2).

Protalus ramparts

Protalus ramparts are curved, flat features, found in association with scree and perennial snow banks. They are bounded on their down-slope side by a sharp break in slope, where the talus rests at the angle of repose. Numerous protalus ramparts are formed along Johnson Mesa, the western slopes of Lachman Crags (Fig. 11c) and below Davies Dome mesa (Fig. 11d). They form through the rolling of clasts down perennial snow banks, and indicate the presence of periglacial conditions.

Slope processes

Description. Solifluction lobes are apparent on many moderate- to low-gradient debris-mantled slopes on the island (e.g. Figs 2 & 11e), with many boulders on these slopes ‘ploughing’ into the sediment, with a keel at the down-slope edge. Solifluction processes (French 1988) are also active in modifying ice-cored moraines and degraded ice-cored moraines on James Ross Island. This is apparent, for example, on the outer portions of ‘IJR-45 Glacier Moraine’, and has, in conjunction with slumping and slopewash, reduced slope gradients and subdued ridge crests.

Alluvial fans and valley-fills are also important on James Ross Island, with streams, snow avalanches, debris flows and solifluction resulting in gentle fan-shaped sediment accumulations in many valleys. Some of the valley fill has subsequently been dissected by rivers and streams, with sorted material being deposited downstream. In the valleys near Alpha Glacier and St Martha Cove, the region is dominated by a large hyaloclastite hill mantled with extensive scree. The valley-fill deposits are bounded by a sharp break in slope. The ground is moist, with strongly developed patterned ground and limited incision by small snow bank-fed ephemeral streams. These valley-fill deposits are smooth and flat, and dip down-slope at very low angles. Rock streams (*sensu* Wilson 2007) were also observed in small valleys, with coarse rock debris and snow accumulating in a linear deposit with a down-slope alignment. They typically have a single thread down the valley axis and extend for hundreds of metres, but are typically only a few metres wide.

Interpretation. Periglacial slope processes are controlled by freeze–thaw activity, rock weathering, frost heave and thaw consolidation. The permafrost table inhibits the downward percolation of water, and the melt of segregated ice lenses provides water, which in turn reduces the internal friction and cohesion in the regolith. Frost creep is one of the main components of solifluction, with slow down-slope gravitational deformation of slope materials through freeze–thaw activity (Harris 2007). These unvegetated slopes with thick glacial drift are susceptible to erosion by slope failure, debris flows, tributary streams and surface wash, resulting in gullying, slope-foot debris cones and valley floor deposits (Ballantyne 2002). These processes were observed widely on degraded ice-cored moraines, at the base of scree slopes, on slopes mantling hills (e.g. Fig. 11e), and on slopes covered with Cretaceous regolith.

Freeze–thaw sediments and landforms

Description. Evidence for modification of surface sediments by freeze–thaw activity on James Ross Island includes frost-shattered boulders (Fig. 12a), nivation hollows and processes (Fig. 12b), sorted stone polygons and stripes, sometimes vegetated by moss and lichens (Fig. 12c, d), and surface cracks. The sorted polygons comprise cells of sand and fine to coarse gravel, surrounded by angular coarse gravel and cobbles. Pattern widths vary between 0.5 and 4 m. Sorted polygons (*sensu* Ballantyne 2007) are particularly prevalent in areas with a large supply of water, such as near streams and snow banks, but also occur on the summits of mesas and on degraded moraines. Stone stripes with alternating coarse and fine sediment are well developed on some slopes. Nivation hollows were frequently observed on Ulu Peninsula, related to small late-lying snow patches, and are usually a few metres square in size. Weathering and down-slope transportation of basalt cobbles on Cretaceous sandstones result in striking black stone stripes (Fig. 4c). Patterned ground overprints all of the other land elements on James Ross Island.

Interpretation. Shattered boulders occur through mechanical weathering induced through freezing

Table 6. *Continuum of ice-cored features in a glacial–periglacial landscape, and criteria for recognition in the landform record*

Landform	Criteria for identification	Processes of formation	References and examples	Geomorphological significance
Debris covered glacier	Uniformly thin debris layer on glacier surface.	Thrusting and upwards movement of ice transport subglacial debris to surface. Meltout of englacial debris bands. Ridges mimic the surface of the glacier. Thin debris cover	Lukas *et al.* (2007); Jacobs *et al.* (2011)	Stagnation of glacier snout; melting and down-wasting *in situ*
Ice-cored moraine	• Thrusts • Exposed stratified and debris-rich ice • Perched ponds and boulders • Debris flows and slumps • Possibly striated clasts/boulders	Thrusts are caused by compressive stresses at frozen glacier margins in small polythermal glaciers. Debris entrained in basal ice is thrust towards the surface. Melting and back wasting of ice results in release of debris to the surface, where it is redistributed and resedimented by meltwater, debris flows and slumps. Debris may be supplemented by supraglacial material	Østrem (1964), (1971); Hambrey *et al.* (1999); Glasser & Hambrey (2003); Schomacker & Kjær (2007); Schomacker (2008)	Polythermal glacial conditions. Crest denotes the maximum extent of the glacier
Dead-ice moraine	• Hummocky landscape • Subdued ridges • Perched boulders • No visible ice scars	Down-wasting and back-wasting of ice scars and increased burial by talus. Top and bottom melt. Lowering of moraine surface. Extensive sediment redistribution. Further ice melting is limited by the insulating effect of the debris mantle. Thick debris cover. Ice stagnation and little active melting	Kjær & Krüger (2001); Schomacker & Kjær (2007), (2008); Krüger *et al.* (2010); Klint *et al.* (2011)	Periglacial environment. May indicate the extent of a former advance
Glacier-derived rock glacier	• Core of glacial ice (with stratification) • Down-slope movement • Lobate shape. Furrows, ridges and mounds on surface • Continuous, thick debris cover • Very small accumulation area • Steep frontal and side slopes	Downslope movement of ice/debris mixtures by creep. Talus falls onto glacier surface, resulting in a thick debris cover, accumulating on low-angle slopes. This insulates the ice and limits ablation. Surface debris rolling down the stoss side is overridden, forming a subrock glacier debris layer. Strongly developed surface relief can indicate compressive stress	Hamilton & Whalley (1995); Konrad *et al.* (1999)	May indicate location of palaeo-ice margin. Periglacial conditions. Forms a continuum with ice-cored moraine. Indicative of limited precipitation
Talus-derived rock glacier (also known as protalus lobes)	• Downslope movement • Large talus supply (e.g. scree) • Form below cliffs or steep terrain • Distinct ridge and furrow pattern • Steep front and side slopes • Ponds and glacial karst landscape	Accumulation of ice and debris in the upper part of the rock glacier. The ice–rock mixture flows down-slope, with ice ablating slowly at the base of the active layer	Hamilton & Whalley (1995); Serrano & López-Martínez (2000); Degenhardt (2009)	Periglacial; frequent freeze–thaw cycles, subzero ground temperatures, abundant talus supply, limited precipitation

Fig. 12. Periglacial phenomena. (**a**) Frost-shattered basalt boulder near DD2, with the detritus moving down-slope under periglacial processes. (**b**) Nivation processes south of the neck of Cape Lachman promontory. (**c**) Stone polygon nets, near Green Lake. (**d**) Stone stripes with coarser basalt cobbles concentrated in the centre of the stripes, northern Brandy Bay area. (**e**) Blockfield of basalt fragments on top of Lachman Crags mesa. (**f**) Basalt boulder on top of Lachman Crags mesa, turned subvertical under freeze–thaw processes.

and thawing under a periglacial environment (Murton 2007). Polygons on the ground surface may form through the development of vertical ice wedges in the ground (Harry 1988). Sorted stone polygons and stone stripes both form through diurnal needle-ice growth, with the former on flat and the latter on inclined surfaces. The diurnal needle-ice freezing and thawing produces creep on inclined surfaces (Ballantyne 2007). Stone stripes may form in combination with seasonal gelifluction and creep through solifluction, resulting in lobes. The presence of patterned ground indicates that the glacigenic deposits on James Ross Island have undergone significant modification since their initial deposition. Networks of cracks have been interpreted as being the product of the fissuring of seasonally frozen ground (Ballantyne 2002).

Late-lying snow is common on James Ross Island, and snow falls throughout the summer season, leading to numerous snow banks and small perennial snow patches. These snow banks feed many of the ephemeral streams and form nivation hollows (Thorn 1988). The close association of patterned ground with streams and snow banks suggests that water may be a limiting factor in the freeze–thaw cycles that form these features. More liquid water is present down-slope of snow banks and in moist areas, encouraging ice wedge growth and resulting in more well-developed polygons and stripes.

Mesas and blockfields

Description. The surface of Lachman Crags mesa (400 m a.s.l.) is smooth and flat, but stone stripes and sorted polygon nets occur in most regions. The surficial sediments are almost entirely basalt clasts, with rare, well-embedded, rounded granite boulders. The pebbles are generally edge-rounded with few angular pebbles (Fig. 5a). At the highest point (above 380 m), the sediments form a blockfield of large, angular basalt boulders (Fig. 12e). These boulders are not faceted or striated, and many are subvertical as a result of frost upheaval (Fig. 12f).

Johnson Mesa (320 m a.s.l.) is characterized by stone stripes and polygonal nets, with very rare cobbles and boulders of Trinity Peninsula origin. In the highest parts, the surface comprises a blockfield with some boulders turned subvertical through frost action. In a col between Bibby Hill and Johnson Mesa, Trinity Peninsula erratics make up 20% of the clast lithologies. This col has scattered Trinity Peninsula granites and metamorphic boulders, with an *a*-axis of up to 60 cm.

On Davies Dome mesa (370 m a.s.l.), the topography is smooth with gentle undulation. The margins of Davies Dome plateau glacier have little geomorphological expression and moraines are absent. The surface is a sandy boulder gravel with well-developed stone polygons, and is 100% basalt. Within the polygons, there is coarse, poorly sorted sand. No Trinity Peninsula erratics were observed on this mesa. Again, above approximately 380 m, the surface comprises a basalt-derived blockfield, with angular, non-faceted and non-striated basalt boulders.

Interpretation. Blockfields are assumed to develop under periglacial conditions on levelled or gently undulating relief, with autochthonous blockfields developing from the *in situ* weathering of bedrock with limited down-slope displacement of blocks (Rea 2007; Goodfellow *et al.* 2008; Ballantyne 2010). The blockfields on James Ross Island have developed on the Neogene basalt plateaux (Fig. 2). Vertical frost sorting and patterned ground with sorted polygon nets and stone stripes indicates periglacial frost action. Some authors argue that initial crack formation in blockfields occurred during the warmer climates of the Neogene (cf. Ballantyne 2010). These cracks were then exploited by ice, resulting in frost wedging of bedrock and granular disaggregation under a periglacial climate. However, recent studies have found no evidence of chemical weathering (Goodfellow *et al.* 2008). The Ulu Peninsula blockfields may have been actively forming through frost action throughout the Quaternary (and possibly Neogene), protected by cold-based plateau ice caps during glacials or remaining unglaciated as nunataks above the height of the ice sheet (cf. Goodfellow *et al.* 2008).

The clast size on the blockfields is highly variable. On Lachman Crags mesa, clast size ranges from boulder-size, open-work blockfields, with boulder size being controlled by joint spacing (e.g. Fig. 12e), to sandy boulder-gravels with well-developed stone circles and fine matrix in the interior of the polygons. The pebble fraction has rounded shapes (Fig. 5a), which is not unusual in long-established blockfields (Ballantyne 2010). Edge rounding through granular disaggregation may imply long-term stability of the surface. There is a clear altitudinal control, with coarser, bouldery blockfields only occurring above 380 m.

The presence of granite and Trinity Peninsula cobbles and boulders on the surfaces of some mesas suggests that they might have previously been overridden by glacier ice from Trinity Peninsula. The absence of erratics on Davies Dome mesa, in comparison with Johnson Mesa or Lachman Crags mesa, could be a result of an expansion of Davies Dome, a cold-based ice cap that preserved and protected the blockfield below (cf. Kleman & Glasser 2007) and deflected Antarctic Peninsula ice during the LGM and previous glaciations. An alternative explanation is that the erratics were deposited subaqueously from icebergs derived from Neogene ice sheets under glaciomarine conditions, and that the plateaux were subsequently uplifted above the height of the Quaternary ice sheets. There is observational support for the latter suggestion. The Lachman Crags lava-fed delta was emplaced in a marine setting whereas that at Davies Dome was englacial (Smellie *et al.* 2008). The erratics are found only on the Lachman Crags delta, on areas of the mesa top underlain by hyaloclastite, that is, where the overlying lava 'topsets' have been stripped off. The hyaloclastite surface is flat and at a uniform elevation below the adjacent lava outcrop (except where the lavas are downfaulted, as on Johnson Mesa), suggesting that the lava cover was removed by marine erosion during the years and decades following emplacement of the delta in the sea (i.e. contemporaneous erosion). If ice was still present in the Peninsula (cf. Smellie *et al.* 2009), icebergs would be able to drift over the stripped-off areas of Lachman Crags delta and deposit erratics. Significantly, the Davies Dome

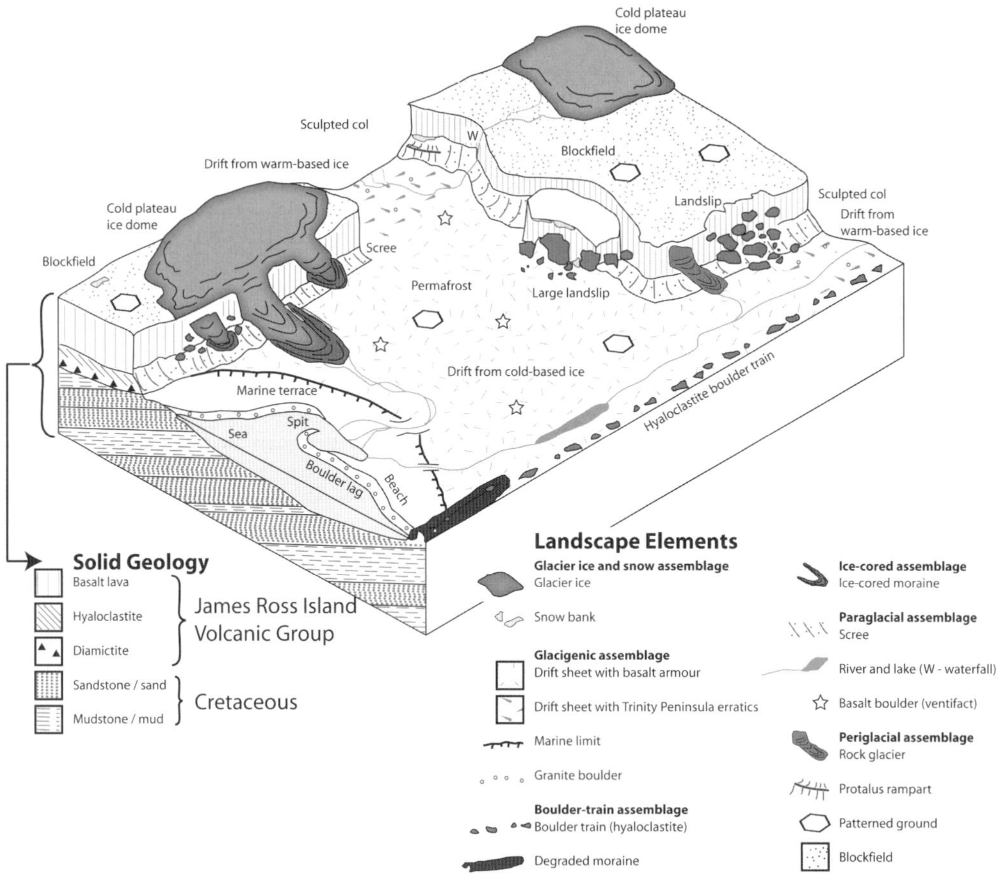

Fig. 13. Conceptual model illustrating the principal processes and sediment-landform assemblages in a semi-arid polar desert, based on Ulu Peninsula of James Ross Island, showing simplified geology and surficial sediments.

delta has not suffered similar erosion, nor does it have any erratics. The lower col on Johnson Mesa, however, is assumed to be a flow pathway for Quaternary ice sheets, with concentrated flow encouraging deposition of numbers of Trinity Peninsula erratics.

Discussion

Processes of landscape evolution in a semi-arid polar environment

Using the geological, geomorphological and glaciological studies undertaken on James Ross Island, we present a new landsystems model for landscape evolution on James Ross Island (Fig. 13), which will aid the interpretation of past, present and future glacierized and glaciated environments. A key feature of this new model is the multi-temporal approach. Six sediment-landform assemblages were described and interpreted in the above sections (Table 2), and they mark a change from landscape dominance by large-scale glaciation during the LGM, Mid- and Late-Holocene glacial readvances, and paraglacial and periglacial processes throughout the Holocene and into the present day. The polar-desert landsystem on Ulu Peninsula is strongly controlled by the geology; basalt mesas dominate the landscape, while being underlain by softer Cretaceous marine deposits, a basalt pebble–cobble gravel covers almost all surfaces, and differences between the six identified sediment-landform assemblages are subtle. The large landslides (Fig. 10) are predicated by the presence of basalt deltas, likely to be an uncommon feature of other polar desert environments. However, similar features have been noted elsewhere (e.g. the Isle of Skye, where basalt columns overlie Jurassic mudstones; Ballantyne 1991). The landscape was initially overprinted by Antarctic Peninsula ice during the LGM. This ice sheet, and previous incarnations, has sculpted and moulded Ulu Peninsula. Variations in the thermal regime and ice stream activity have resulted in differences in the glacial drift. Following deglaciation, paraglacial and periglacial processes immediately began to modify the landscape (summarized in Fig. 13 and Table 2).

Our landsystem model (Fig. 13; Table 2) is a modern analogue to aid the interpretation of other high-latitude semi-arid environments during the Late Glacial in the Northern Hemisphere. Similar semi-arid environments are recognized to have existed during the Late Pleistocene in the Canadian High Arctic (Ó Cofaigh *et al.* 1999) and in Svalbard (Glasser & Hambrey 2003), and may be expected in high-latitude continental interiors glaciated during the LGM, such as Siberia and European Russia.

Glacial–paraglacial–periglacial interactions

A latitudinal transect from northern Chile through to the Dry Valleys highlights the changing dominance of different processes, driven primarily by the availability of meltwater (Fig. 14). It is clear that meltwater is increasingly important at more northerly latitudes. The fast-flowing temperate glaciers of the Northern Patagonian Icefield (47°S) produce large volumes of meltwater, which rapidly rework and remove fines generated subglacially. Subglacial landforms, such as drumlins, flutes, eskers or crevasse-squeeze rides, are rare (Glasser *et al.* 2009*a*). The large volume of meltwater results in numerous glaciofluvial ice-contact landforms, with abundant kame terraces, glacial lakes and raised lake shorelines, and rapid reworking of subglacially deposited sediments and landforms. The meltwater also facilitates basal sliding and enables glaciers to flow quickly (Glasser *et al.* 2009*a*), resulting in abundant ice-scoured bedrock. In contrast, the sediment-landform assemblage around Tierra del Fuego, southernmost Chile (53°S), is dominated by abundant subglacial till, flutes, drumlins and moraines, deposited by temperate glaciers but with reduced proglacial meltwater. Smaller sandur and ice-contact lakes, however, remain important (Benn & Clapperton 2000; Bentley *et al.* 2005). At these latitudes, there is only discontinuous permafrost at high altitudes (Benn & Clapperton 2000; Trombotto 2002), resulting in fewer rock glaciers, protalus ramparts, frost-shattered boulders or stone polygons than on James Ross Island. Aeolian processes are also comparatively less important; ventifacts and pebble–cobble lags are more numerous in East Antarctica than in Chile because of the comparatively faster, sand-bearing winds, less vegetation and less surficial sediments than in temperate regions.

On James Ross Island, freeze–thaw and fluvial activity have far greater relative importance when compared with the Dry Valleys (cf. French & Guglielmin 1999; Sugden *et al.* 1999; Hall & Denton 2005; McLeod *et al.* 2008), which significantly aids the development of patterned ground and solifluction landforms (Fig. 14). Indeed, the prevalence of moisture-driven periglacial processes on Ulu Peninsula contrasts sharply with their absence in more southerly, colder parts of Antarctica. In contrast, the Glacigenic Assemblage on James Ross Island resembles closely the cold-based glacial sediment-landform assemblages recognized in East Antarctica (Fig. 14; Atkins *et al.* 2002; Lloyd Davies *et al.* 2009; Hambrey & Fitzsimons 2010), perhaps indicating similarities between the climate during the LGM on James Ross Island and the climate in the Dry Valleys and East Antarctica today (cf. Fitzsimons 2003). Some paraglacial processes, such as glacio-isostatic adjustment, relaxation of rock walls, littoral reworking (González Bonorino *et al.* 1999) and fluvial reworking (albeit limited in the Dry Valleys) operate throughout the transect, although glacio-isostatic uplift, marine terraces and raised beaches are more prevalent at James Ross Island (cf. Roberts *et al.* 2011) and Tierra del Fuego, as a result of their proximity to the ocean.

The scope for landscape modification at the intersection of a paraglacial–periglacial–glacial landsystem is of great importance. Even in more temperate latitudes, the interplay between periglacial, paraglacial and glacial processes has been recognized as important (Coleman & Carr 2008). Ground ice is prevalent on James Ross Island today, in the form of permafrost, rock-glaciers, buried glacier ice and ice-cored moraines. Buried glacier ice can survive almost indefinitely in the permafrost environment (e.g. 8 Ma old ice in the Dry Valleys; Sugden *et al.* 1995; Schafer *et al.* 2000). However, despite the wealth of research on glacial and periglacial processes, there are relatively few papers that discuss the actions of one set of processes upon the other (Harris & Murton 2005).

The presence of permafrost beneath cold-based glaciers on James Ross Island during the LGM may have facilitated forward movement, as high porewater pressures beneath subglacial permafrost can reduce the shear strength of unfrozen substrate. This may also be a factor in initiating large-scale mass movements. Further, proglacial permafrost can reduce effective stress by generating overpressuring in groundwater (Mathews & Mackay 1960; Boulton *et al.* 1995). Entrainment of debris (as observed in debris-rich basal ice exposed in ice-cored moraines) occurs through the transmission of basal shear stress from the glacier bed into the frozen subglacial sediment (Waller *et al.* 2009).

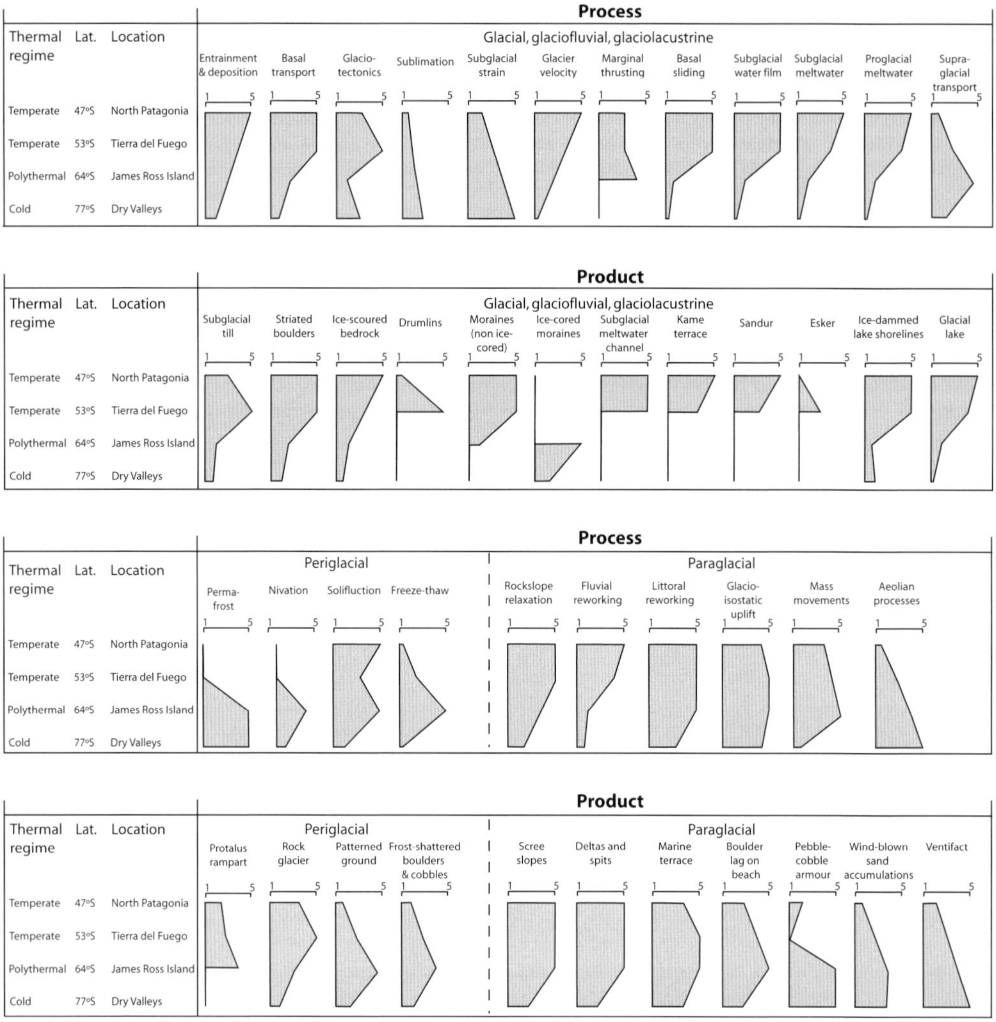

Fig. 14. Schematic cartoon illustrating the relative importance of various modern processes and the abundance of different products across a latitudinal transect from Northern Patagonia (northern Chile), through Tierra del Fuego (southern Chile) and James Ross Island to the Dry Valleys (Northern Victoria Land, East Antarctica). Thermal regimes change from temperate to cold-based; glacier velocities decrease southwards; meltwater is progressively abundant northwards. The shaded shapes illustrate the relative importance or abundance of a process or product, with 1 being low importance and 5 being high importance or very abundant. Sources include Trombotto (2002); Bentley *et al.* (2005); Glasser *et al.* (2009*a*); Hambrey & Fitzsimons (2010).

The glacier couples with the permafrost (e.g. Cuffey *et al.* 2000), and the plane of movement (décollment plane) moves downwards to a plane of weakness within the underlying substrate, itself composed of relatively soft, easily deformable Cretaceous sediments. In this instance (cold-based glaciers overlying frozen ground), the base of the glacier is a heterogeneous zone, with the lower boundary of glacier flow varying spatially and temporally (Fitzsimons 2006; Waller *et al.* 2009).

This landsystem model includes many ice-cored landforms, such as ice-cored moraines, debris-covered glacier snouts, glacier-derived rock glaciers and talus-derived rock glaciers. The continuum and classification of ice-cored landforms has generated much debate (e.g. Hamilton & Whalley 1995; Humlum 1998; Konrad *et al.* 1999; Humlum 2000; Serrano & López-Martínez 2000; Lønne & Lyså 2005; Lukas *et al.* 2007; Degenhardt 2009; Jacobs *et al.* 2011). The morphologies of landforms generated at the termini of polythermal and cold-based glaciers overlap significantly. For example, the processes governing the genesis of ice-cored moraine, debris-covered glacier termini, dead-ice moraine and glacier-derived rock glaciers may be very similar. However, these morphological terms describe different end members of the continuum, and their different processes, sediments and landforms are summarized in Table 6.

In the contemporary permafrost environment on James Ross Island, the seasonal melting of glacier ice and snow dominates sedimentary processes and products. However, this is modulated and modified by the melting and mobilization of the permafrost active layer.

In this semi-arid polar environment, with predominantly cold-based glacial processes, periglaciation and paraglaciation dominate sediment transport. Sediment transfers occur through fluvial erosion, rock glaciers, slope instability and solifluction (cf. Harris & Murton 2005), and result in chaotic dead-ice terrain with abundant re-sedimentation processes (Etzelmüller & Hagen 2005). When analysing geomorphological processes in cold environments, glacial, periglacial and paraglacial processes must be considered as an intrinsically coupled system.

Character and behaviour of the LGM ice sheet on Ulu Peninsula

The basal thermal regime of ice sheets represents important empirical data required for numerical ice sheet models (Kleman *et al.* 1999; Siegert & Dowdeswell 2004; Siegert 2009). The subglacial thermal regime on Ulu Peninsula comprised four principal types: a wet-based ice stream, warm-based sheet flow, cold-based sheet flow and cold-based plateau ice caps. Each leaves a distinctive sediment-landform assemblage behind in the geological record (Table 7). The frozen-bed patches identified by Kleman & Glasser (2007) are more complex and variable than previously assumed. Ice streaming and rapid basal sliding may have been encouraged by slippery marine sediments in Prince Gustav Channel. Marine geological evidence has been taken to suggest that cold-based ice developed on the Antarctic Peninsula mountains (Ó Cofaigh *et al.* 2001; Reinardy *et al.* 2009). Unglacierized mesas are assumed to have been above the height of the LGM ice sheet, but they could have harboured cold-based plateau ice caps (Johnson *et al.* 2009). If the mesas were above the LGM ice sheet, the summit plateau elevations provide an estimate for maximum ice thicknesses at the LGM. However, an ice surface elevation of *c.* 750 m a.s.l. has been demonstrated for Dobson Dome, just south of our field area (Fig. 1), with an age of <80 ka (with large errors; Smellie *et al.* 2008).

Sharp boundaries in thermal regime and styles of erosion and deposition have long been recognized in the geological record at the shear margins of ice streams (Stokes & Clark 1999, 2001). Less is known about flow boundary patterns in sheet-flow areas. However, on Ulu Peninsula, the mosaic in the thermal regime of the cold-based ice sheet operating under sheet-flow was subtle, and the boundaries more dispersed than the shear margins of ice streams.

The glacial sandy boulder gravel that characterizes James Ross Island today is a significant contrast to the Neogene diamictite logged in many locations across the island (Hambrey & Smellie 2006; Hambrey *et al.* 2008; Smellie *et al.* 2009). Similar contrasts have been noted in the Dry Valleys, where pebble gravels deposited by cold-based

Table 7. *Principal land elements of the Glacigenic Assemblage and inferred subglacial thermal regime*

Glacial drift	Characteristics	Inferred thermal regime
Erratic-poor glacial drift	Sandy boulder gravel found widely across the island; smooth slopes. Basalt and Cretaceous lithologies only	Cold-based sheet flow
Erratic-poor glacial drift	On mesas and around Davies Dome. Smooth and flat; no landforms	Cold-based plateau ice cap
Erratic-rich drift (in cols and passes)	Sandy boulder gravel at higher elevations in cols and passes. Up to 30% Trinity Peninsula erratics. Smoothed and sculpted cols	Warm-based sheet flow
Erratic-rich drift (in coastal regions)	Sandy boulder gravel with up to 50% erratics from Trinity Peninsula. Associated with moraine fragments	Ice stream

ice differ from Neogene the Sirius Group diamictons (Hambrey & Fitzsimons 2010). This is indicative of changing climatic regimes, with diamicton, silt and clay forming only under the warmer climatic conditions of the Neogene (cf. DeConto & Pollard 2003; Mayewski *et al.* 2009). Warmer atmospheric temperatures encouraged subglacial sliding, erosion, transportation and deposition under a wet-based ice sheet, with abundant production of fines and therefore diamictites. Abrasion under cold-based glaciers is insufficient to generate the large quantities of fines required for subglacial till formation, which is associated with warm-based glaciers above the pressure melting point. Geothermal heating may also have aided the production of basal meltwater and subglacial till formation during the Neogene Period on James Ross Island (Hambrey *et al.* 2008). Effects of volcanically enhanced geothermal gradients on meltwater availability during the Neogene are difficult to determine. However, those effects were probably localized mainly to within several kilometres of the erupting vents (at Mount Haddington, *c.* 40 km distant), with only a very limited influence likely in the more distal regions where our study was focused. Quaternary volcanic activity was somewhat closer, situated on western Vega Island, Terrapin Hill and Dobson Dome (990 ka, *c.* 660 ka and <80 ka, respectively, the latter two with large errors; Smellie *et al.* 2008), situated *c.* 10, 20 and 10 km, respectively, from our field area. The Quaternary centres are therefore probably too old and too distant to have had a major effect in the Brandy Bay area.

This study has shown that cold-based ice sheets are capable of entraining and moving subglacial sediments and boulders, and, in conjunction with numerous recent studies that illustrate the erosive and depositional capabilities of ice sheets that reach $-17\,^{\circ}$C at their bed (e.g. Phillips *et al.* 2006; Hambrey & Fitzsimons 2010), supports a new and emerging paradigm of sheet flow in high latitude regions. The old paradigm suggested that ice sheets comprised a patchwork of cold-based ice on mountain tops, with very little erosion or deposition, surrounded by highly active warm-based ice that typically occupied valleys (e.g. Hättestrand & Stroeven 2002; Stroeven *et al.* 2002; Briner *et al.* 2003; Davis *et al.* 2006). Delicate structures and blockfields are preserved at high altitudes (Kleman *et al.* 1999; Johnson *et al.* 2009). New and emerging research (for examples, see Kleman & Glasser 2007) suggests that this is an over-simplification, and that sheet flow beneath high-latitude ice sheets is more complex. Although cold-based glaciers are weak erosional and depositional agents compared with warm-based glaciers, the interaction between glaciers, their thermal regime and their beds is more subtle than is usually assumed (Cuffey *et al.* 2000; Lloyd Davies *et al.* 2009). The ice sheet that overwhelmed James Ross Island during the LGM was a composite of ice streaming in the trough of Prince Gustav Channel, which impinged on the coastal margins of Ulu Peninsula, sheet flow on the main part of Ulu Peninsula with both warm- and cold-based ice flowing across the island, and cold-based and stationary plateau ice domes on high plateau blockfields (Table 7). The lateral margins of Prince Gustav Ice Stream were likely to have been dominated by warm-based sheet flow; the evidence for this includes moraines and enhanced deposition of erratics. The key point of this new paradigm is that the cold-based ice was capable of movement, erosion, transportation of material and deposition, with enhanced deposition occurring in small areas as a result of changes in the thermal regime of the ice sheet, and that sheet flow and cold-based ice are more complex than previously thought.

Conclusions

This holistic and systematic study of sediment-landform assemblages from Ulu Peninsula on James Ross Island (Table 2) has used detailed sedimentary descriptions, clast lithology, shape-roundness counts and geomorphology mapped both in the field and from remotely sensed images (Table 1) to discriminate six sediment-landform assemblages and thus to present the first landsystem model from a semi-arid subpolar desert from the Antarctic Peninsula. The conceptual model in Figure 13 emphasizes the interrelationship and importance of not only glacial, but also periglacial and paraglacial processes. The six sediment-landform assemblages identified were: (1) the Glacier Snow and Ice Assemblage; (2) the Glacigenic Assemblage; (3) the Boulder Train Assemblage; (4) the Ice-Cored Moraine Assemblage; (5) the Paraglacial Assemblage; and (6) the Periglacial Assemblage. Analysis of these assemblages provides a detailed understanding of landscape evolution on James Ross Island. Sediments and landforms were deposited during LGM glaciation and from Mid to Late Holocene glacier readvances. These were subsequently reworked and redeposited by periglacial and paraglacial processes throughout the Holocene and into the present day. Crucially, when compared with the few other polar landsystem models, we find that the availability of meltwater encourages strong landform modification by periglacial processes. These processes would have been similarly important in the Northern Hemisphere during the Late Pleistocene epoch. Therefore the landsystem model presented here is a modern analogue to be used in the interpretation of past glaciated environments.

This paper presented new information regarding the thermal regime of the Antarctic Peninsula Ice Sheet during LGM glaciation (Table 7). The data and model for the interplay between cold-based, warm-based and streaming ice challenges the theory that cold-based glaciers do not erode or deposit. We have presented geomorphological and sedimentological data to help interpret the presence of cold-based sheet flow in other Quaternary glaciated environments. Finally, this paper presented important new data regarding the thermal regime and character and behaviour of the LGM ice sheet on James Ross Island, which will aid reconstructions of the LGM ice sheet in northern Antarctic Peninsula.

The authors gratefully acknowledge funding from the Natural Environment Research Council grant through the Antarctic Funding Initiative (grant AFI 9-01; NE/F012942/1). Fieldwork was logistically supported by the British Antarctic Survey, and the authors thank Alan Hill for assistance in the field and the captains and crew of the *RRV Ernest Shackleton* and the *RRV James Clark Ross* for their cooperation and support. Daniel Nývlt's field research and digital data preparation was funded from the R & D project VaV SP II 1a9/23/07, financed by the Ministry of the Environments of the Czech Republic. We thank the crew of Mendel Station for logistic support in the field. The authors acknowledge Etienne Berthier and the Centre National d'Etudes Spatiales for the SPOT-5 images. The Czech Geological Survey are thanked for providing the orthophoto mosaic from 2006 aerial photographs, digital elevation model and GIS vector files from their map of Ulu Peninsula, published in 2009. The Antarctic Digital Database is acknowledged for providing formal glacier names.

References

Adam, W. G. & Knight, P. G. 2003. Identification of basal layer debris in ice-marginal moraines, Russell Glacier, West Greenland. *Quaternary Science Reviews*, **22**, 1407–1414.

Atkins, C. B., Barrett, P. J. & Hicock, S. R. 2002. Cold glaciers erode and deposit: evidence from Allan Hills, Antarctica. *Geology*, **30**, 659–662.

Augustinus, P. C. 2002. Weathering characteristics of the glacial drifts, Bunger Hills, East Antarctica. *Arctic, Antarctic, and Alpine Research*, **34**, 65–75.

Ballantyne, C. K. 1991. Scottish landform examples 2 – the landslides of Trotternish, Isle of Skye. *Scottish Geographical Magazine*, **107**, 130–135.

Ballantyne, C. K. 2002. Paraglacial geomorphology. *Quaternary Science Reviews*, **21**, 1935–2017.

Ballantyne, C. K. 2007. Periglacial landforms: patterned ground. *In*: Scott, A. E. (ed.) *Encyclopedia of Quaternary Science*. Elsevier, Oxford, 2182–2191.

Ballantyne, C. K. 2010. A general model of autochthonous blockfield evolution. *Permafrost and Periglacial Processes*, **21**, 289–300.

Ballantyne, C. K., Stone, J. O. & Fifield, L. K. 1998. Cosmogenic Cl-36 dating of postglacial landsliding at The Storr, Isle of Skye, Scotland. *The Holocene*, **8**, 347–351.

Barbeau, D. L., Davis, J. T., Murray, K. E., Valencia, V., Gehrels, G. E., Zahid, K. M. & Gonbosi, D. J. 2010. Detrital-zircon geochronology of the metasedimentary rocks of north-western Graham Land. *Antarctic Science*, **22**, 65–78.

Benn, D. I. & Ballantyne, C. K. 1994. Reconstructing the transport history of glacigenic sediments: a new approach based on the co-variance of clast form indices. *Sedimentary Geology*, **91**, 215–227.

Benn, D. I. & Clapperton, C. M. 2000. Pleistocene glaciotectonic landforms and sediments around central Magellan Strait, southernmost Chile: evidence for fast outlet glaciers with cold-based margins. *Quaternary Science Reviews*, **19**, 591–612.

Benn, D. I. & Evans, D. J. A. 2010. *Glaciers & Glaciation*. Hodder Education, London.

Benn, D. I. & Lukas, S. 2006. Younger Dryas glacial landsystems in North West Scotland: an assessment of modern analogues and palaeoclimatic implications. *Quaternary Science Reviews*, **25**, 2390–2408.

Bennett, M. R., Huddart, D., Glasser, N. F. & Hambrey, M. J. 2000*a*. Resedimentation of debris on an ice-cored lateral moraine in the high-Arctic (Kongsvegen, Svalbard). *Geomorphology*, **35**, 21–40.

Bennett, M. R., Huddart, D. & Waller, R. I. 2000*b*. Glaciofluvial crevasse and conduit fills as indicators of supraglacial dewatering during a surge, Skeiðarárjökull, Iceland. *Journal of Glaciology*, **46**, 25–34.

Bentley, M. J., Sugden, D. E., Hurlton, N. R. J. & McCulloch, R. D. 2005. The landforms and pattern of deglaciation in the Strait of Magellan and Bahía Inútil, southernmost America. *Geografiska Annaler*, **87A**, 313–333.

Bertler, N. A. N., Mayewski, P. A. & Carter, L. 2011. Cold conditions in Antarctica during the Little Ice Age – implications for abrupt climate change mechanisms. *Earth and Planetary Science Letters*, **308**, 41–51.

Bibby, J. S. 1966. *The Stratigraphy of Part of North-East Graham Land and the James Ross Island Group*. British Antarctic Survey, London.

Björck, S., Olsson, S., Ellis-Evans, C., Håkansson, H., Humlum, O. & de Lirio, J. M. 1996. Late Holocene palaeoclimatic records from lake sediments on James Ross Island, Antarctica. *Palaeogeography, Palaeoclimatology, Palaeoecology*, **121**, 195–220.

Bockheim, J. G. 2010. Evolution of desert pavements and the vesicular layer in soils of the Transantarctic Mountains. *Geomorphology*, **118**, 433–443.

Boulton, G. S. 1972. Modern Arctic glaciers as depositional models for former ice sheets. *Journal of the Geological Society, London*, **128**, 361–393.

Boulton, G. S. & Dent, D. L. 1974. The nature and rate of post-depositional changes in recently deposited till from south-east Iceland. *Geografiska Annaler*, **56A**, 121–134.

Boulton, G. S., Caban, P. E. & van Gijssel, K. 1995. Groundwater flow beneath ice sheets: Part II – its impact on glacier tectonic structures and moraine formation. *Quaternary Science Reviews*, **14**, 563–587.

Bridgland, D. R. 1986. *Clast Lithological Analysis. Technical Guide No. 3*. Quaternary Research Association, Cambridge.

Briner, J. P., Miller, G. H., Davis, P. T., Bierman, P. R. & Caffee, M. 2003. Last Glacial Maximum ice sheet dynamics in Arctic Canada inferred from young erratics perched on ancient tors. *Quaternary Science Reviews*, **22**, 437–444.

BRITISH ANTARCTIC SURVEY. 2010. *Antarctic Sound and James Ross Island, Northern Antarctic Peninsula*. Series BAS (UKAHT) Sheets 3A and 3B, 1:2 50 000, Cambridge.

Camerlenghi, A., Domack, E. *et al.* 2001. Glacial morphology and post-glacial contourites in northern Prince Gustav Channel (NW Weddell Sea, Antarctica). *Marine Geophysical Researches*, **22**, 417–443.

Carrivick, J. L., Davies, B. J., Glasser, N. F. & Nývlt, D. 2012. Late Holocene changes in character and behaviour of land-terminating glaciers on James Ross Island, Antarctica. *Journal of Glaciology*, **58**, 1176–1190.

Chinn, T. J. H. & Dillon, A. 1987. Observations on a debris-covered polar glacier 'Whisky Glacier', James Ross Island, Antarctic Peninsula, Antarctica. *Journal of Glaciology*, **33**, 300–310.

Clayton, L. 1964. Karst topography on stagnant glaciers. *Journal of Glaciology*, **5**, 107–112.

Coleman, C. G. & Carr, S. J. 2008. Complex relationships between Younger Dryas glacial, periglacial and paraglacial landforms, Brecon Beacons, South Wales. *Proceedings of the Geologists' Association*, **119**, 259–276.

Cuffey, K. M., Conway, H., Hallet, B., Gades, A. M. & Raymond, C. F. 1999. Interfacial water in polar glaciers and glacier sliding at $-17\ ^{\circ}C$. *Geophysical Research Letters*, **26**, 751–754.

Cuffey, K. M., Conway, H. *et al.* 2000. Entrainment at cold glacier beds. *Geology*, **28**, 351–354.

CZECH GEOLOGICAL SURVEY. 2009. *James Ross Island—Northern Part. Topographic Map 1:25 000*. Czech Geological Survey, Praha.

Davies, B. J., Hambrey, M. J., Smellie, J. L., Carrivick, J. L. & Glasser, N. F. 2012*a*. Antarctic Peninsula Ice Sheet evolution during the Cenozoic Era. *Quaternary Science Reviews*, **31**, 30–66.

Davies, B. J., Carrivick, J. L., Glasser, N. F., Hambrey, M. J. & Smellie, J. L. 2012*b*. Variable glacier response to atmospheric warming, northern Antarctic peninsula, 1988–2009. *The Cryosphere Discussions*, **6**, 1031–1048.

Davis, P. T., Briner, J. P., Coulthard, R. D., Finkel, R. W. & Miller, G. H. 2006. Preservation of Arctic landscapes overridden by cold-based ice sheets. *Quaternary Research*, **65**, 156–163.

DeConto, R. M. & Pollard, D. 2003. Rapid Cenozoic glaciation of Antarctica induced by declining atmospheric CO_2. *Nature*, **421**, 245–249.

Degenhardt, J. J. Jr. 2009. Development of tongue-shaped and multilobate rock glaciers in alpine environments – interpretations from ground penetrating radar surveys. *Geomorphology*, **109**, 94–107.

Domack, E. W., Burnett, A. & Leventer, A. 2003. Environmental setting of the Antarctic Peninsula. *In*: Domack, E., Leventer, A., Burnett, A., Bindshadler, R., Convey, P. & Kirby, M. (eds) *Antarctic Peninsula Climate Variability: Historical and Paleoenvironmental Perspectives*. Antarctic Research Series. American Geophysical Union, Washington, DC, **79**, 1–13.

Engel, Z., Láska, K., Franta, T., Máčka, Z. & Marvánek, O. 2010. Recent changes of permafrost active layer on the James Ross Island, Maritime Antarctic. *In*: *Abstracts from the Third European Conference on Permafrost*, 13 June 2010, Svalbard, 129.

Engel, Z., Nývlt, D. & Láska, K. 2012. Ice thickness, bed topography and glacier volume changes on James Ross Island, Antarctic Peninsula. *Journal of Glaciology*, **58**, 904–914.

Ermolin, E., De Angelis, H. & Skvarca, P. 2002. Mapping of permafrost on Vega Island, Antarctic Peninsula, using satellite images and aerial photography. *Annals of Glaciology*, **34**, 184–188.

Ermolin, E., de Angelis, H., Skvarca, P. & Rau, F. 2004. Ground ice in permafrost on Seymour (Marambio) and Vega Islands, Antarctic Peninsula. *Annals of Glaciology*, **39**, 373–378.

Etzelmüller, B. & Hagen, J. O. 2005. Glacier-permafrost interaction in Arctic and alpine mountain environments with examples from Norway and Svalbard. *In*: Harris, C. & Murton, J. B. (eds) *Cryospheric Systems: Glaciers and Permafrost*. Geological Society, London, Special Publications, **242**, 11–27.

Evans, D. J. A. 1993. High-latitude rock glaciers: a case study of forms and processes in the Canadian arctic. *Permafrost and Periglacial Processes*, **4**, 17–35.

Evans, D. J. A. 2003*a*. Ice-marginal terrestrial landsystems: active temperate glacier margins. *In*: Evans, D. J. A. (ed.) *Glacial Landsystems*. Arnold, London, 12–43.

Evans, D. J. A. 2003*b*. Introduction to glacial landsystems. *In*: Evans, D. J. A. (ed.) *Glacial Landsystems*. Arnold, London, 1–11.

Evans, D. J. A. 2009. Controlled moraines: origins, characteristics and palaeoglaciological implications. *Quaternary Science Reviews*, **28**, 183–208.

Evans, D. J. A. 2011. Glacial landsystems of Satujökull, Iceland: a modern analogue for glacial landsystem overprinting by mountain icecaps. *Geomorphology*, **129**, 225–237.

Evans, D. J. A. & Benn, D. I. 2004. Facies description and the logging of sedimentary exposures. *In*: Evans, D. J. A. & Benn, D. I. (eds) *A Practical Guide to the Study of Glacial Sediments*. Arnold, London, 11–50.

Evans, D. J. A. & Twigg, D. R. 2002. The active temperate glacial landsystem: a model based on Breiethamerkurjokull and Fjallsjokull, Iceland. *Quaternary Science Reviews*, **21**, 2143–2177.

Evans, J., Pudsey, C. J., Ó Cofaigh, C., Morris, P. & Domack, E. 2005. Late Quaternary glacial history, flow dynamics and sedimentation along the eastern margin of the Antarctic Peninsula Ice Sheet. *Quaternary Science Reviews*, **24**, 741–774.

Evans, D. J. A., Phillips, E. R., Hiemstra, J. F. & Auton, C. A. 2006. Subglacial till: formation, sedimentary characteristics and classification. *Earth-Science Reviews*, **78**, 115–176.

FITZSIMONS, S. J. 1996. Paraglacial redistribution of glacial sediments in the Vestfold Hills, East Antarctica. *Geomorphology*, **15**, 93–108.

FITZSIMONS, S. J. 2003. Ice-marginal terrestrial landsystems: polar continental glacier margins. *In*: EVANS, D. J. A. (ed.) *Glacier Landsystems*. Hodder, London, 89–110.

FITZSIMONS, S. 2006. Mechanical behaviour and structure of the debris-rich basal ice layer. *In*: KNIGHT, P. G. (ed.) *Glacier Science & Environmental Change*. Blackwell, Oxford, 329–335.

FITZSIMONS, S. J., MCMANUS, K. J. & LORRAIN, R. D. 1999. Structure and strength of basal ice and substrate of a dry-based glacier: evidence for substrate deformation at sub-freezing temperatures. *Annals of Glaciology*, **28**, 236–240.

FITZSIMONS, S., WEBB, N., MAGER, S., MACDONELL, S., LORRAIN, R. & SAMYN, D. 2008. Mechanisms of basal ice formation in polar glaciers: an evaluation of the apron entrainment model. *Journal of Geophysical Research*, **113**, F02010.

FRENCH, H. M. 1988. Active layer processes. *In*: CLARK, M. J. (ed.) *Advances in Periglacial Geomorphology*. Wiley-Interscience, Chichester, 151–178.

FRENCH, H. M. & GUGLIELMIN, M. 1999. Observations on the ice-marginal, periglacial geomorphology of Terra Nova Bay, Northern Victoria Land, Antarctica. *Permafrost and Periglacial Processes*, **10**, 331–347.

FUKUDA, M., STRELIN, J. A., SHIMOKAWA, K., TAKAHASHI, N., SONE, T. & TROMBOTT, D. 1992. Permafrost occurrence of Seymour Island and James Ross Island, Antarctic Peninsula region. *In*: YOSHIDA, N., KAMINUMA, K. & SHIRAISHI, K. (eds) *Recent Progress in Antarctic Earth Sciences*. Terra Scientific Publishing Company (TERRAPUB), Tokyo, 745–750.

FUKUI, K., SONE, T., STRELIN, J. A., TORIELLI, C. A. & MORI, J. 2007. Ground penetrating radar sounding on an active rock glacier on James Ross Island, Antarctic Peninsula region. *Polish Polar Research*, **28**, 13–22.

FUKUI, K., SONE, T., STRELIN, J. A., TORIELLI, C. A., MORI, J. & FUJII, Y. 2008. Dynamics and GPR stratigraphy of a polar rock glacier on James Ross Island, Antarctic Peninsula. *Journal of Glaciology*, **54**, 445–451.

GALE, S. J. & HOARE, P. G. 1991. *Quaternary Sediments: Petrographic Methods for the Study of Unlithified Rocks*. John Wiley & Sons, New York.

GILBERT, R., DOMACK, E. W. & CAMERLENGHI, A. 2003. Deglacial history of the Greenpeace trough: ice sheet to ice shelf transition in the Northwestern Weddell Sea. *In*: DOMACK, E., LEVENTER, A., BURNETT, A., BINDSHADLER, R., CONVEY, P. & KIRBY, M. (eds) *Antarctic Peninsula Climate Variability: Historical and Paleoenvironmental Perspectives*. Antarctic Research Series. American Geophysical Union, Washington, DC, **79**, 195–204.

GLASSER, N. F. & HAMBREY, M. J. 2003. Ice-marginal terrestrial landsystems: Svalbard polythermal glaciers. *In*: EVANS, D. J. A. (ed.) *Glacier Landsystems*. Hodder, London, 65–87.

GLASSER, N. F. & JANSSON, K. N. 2005. Fast-flowing outlet glaciers of the Last Glacial Maximum Patagonian Icefield. *Quaternary Research*, **63**, 206–211.

GLASSER, N. F., HAMBREY, M. J., CRAWFORD, K. R., BENNETT, M. R. & HUDDART, D. 1998. The structural glaciology of Kongsvegen, Svalbard, and its role in landform genesis. *Journal of Glaciology*, **44**, 136–148.

GLASSER, N. F., JANSSON, K. N., HARRISON, S. & RIVERA, A. 2005. Geomorphological evidence for variations of the North Patagonian Icefield during the Holocene. *Geomorphology*, **71**, 263–277.

GLASSER, N. F., HARRISON, S. & JANSSON, K. N. 2009*a*. Topographic controls on glacier sediment-landform associations around the temperate North Patagonian Icefield. *Quaternary Science Reviews*, **28**, 2817–2832.

GLASSER, N. F., KULESSA, B. *ET AL.* 2009*b*. Surface structure and stability of the Larsen C Ice Shelf, Antarctic Peninsula. *Journal of Glaciology*, **55**, 400–410.

GLASSER, N. F., SCAMBOS, T. A., BOHLANDER, J. A., TRUFFER, M., PETTIT, E. C. & DAVIES, B. J. 2011. From ice-shelf tributary to tidewater glacier: continued rapid glacier recession, acceleration and thinning of Röhss Glacier following the 1995 collapse of the Prince Gustav Ice Shelf on the Antarctic Peninsula. *Journal of Glaciology*, **57**, 397–406.

GOLLEDGE, N. R. 2007. An ice cap landsystem for palaeoglaciological reconstructions: characterizing the Younger Dryas in western Scotland. *Quaternary Science Reviews*, **26**, 213–229.

GONZÁLEZ BONORINO, G., BUJALESKY, G., COLOMBO, F. & FERRERO, M. 1999. Holocene coastal palaeoenvironments in Atlantic Patagonia, Argentina. *Journal of South American Earth Sciences*, **12**, 325–331.

GOODFELLOW, B. W., FREDIN, O., DERRON, M.-H. & STROEVEN, A. P. 2008. Weathering processes and Quaternary origin of an alpine blockfield in Arctic Sweden. *Boreas*, **38**, 379–398.

GRANT, K. L., STOKES, C. R. & EVANS, I. S. 2009. Identification and characteristics of surge-type glaciers on Novaya Zemlya, Russian Arctic. *Journal of Glaciology*, **55**, 960–972.

HAEBERLI, W. 2005. Investigating glacier-permafrost relationships in high-mountain areas: historical background, selected examples and research needs. *In*: HARRIS, C. & MURTON, J. B. (eds) *Cryospheric Systems: Glaciers and Permafrost*. Geological Society, London, Special Publications, **242**, 29–37.

HALL, A. M. & GLASSER, N. F. 2003. Reconstructing the basal thermal regime of an ice stream in a landscape of selective linear erosion: Glen Avon, Cairngorm Mountains, Scotland. *Boreas*, **32**, 191–207.

HALL, B. L. & DENTON, G. H. 2005. Surficial geology and geomorphology of eastern and central Wright Valley, Antarctica. *Geomorphology*, **64**, 25–65.

HAMBREY, M. J. & DOWDESWELL, J. A. 1994. Flow regime of the Lambert Glacier-Amery Ice Shelf system, Antarctica: structural evidence from LANDSAT imagery. *Journal of Glaciology*, **20**, 401–406.

HAMBREY, M. J. & FITZSIMONS, S. J. 2010. Development of sediment-landform associations at cold glacier margins, Dry Valleys, Antarctica. *Sedimentology*, **57**, 857–882.

HAMBREY, M. J. & HUDDART, D. 1995. Englacial and proglacial tectonic processes at the snout of a thermally complex glacier in Svalbard. *Journal of Quaternary Science*, **10**, 313–326.

HAMBREY, M. J. & LAWSON, W. 2000. Structural styles and deformation fields in glaciers: a review. *In*: MALTMAN, A. J., HUBBARD, B. & HAMBREY, M. J. (eds) *Deformation of Glacial Materials*. Geological Society, London, Special Publications, **176**, 59–83.

HAMBREY, M. J. & SMELLIE, J. L. 2006. Distribution, lithofacies and environmental context of Neogene glacial sequences on James Ross and Vega Islands, Antarctic Peninsula. *In*: FRANCIS, E. A., PIRRIE, D. & CRAME, J. A. (eds) *Cretaceous–Tertiary High-Latitude Palaeoenvironments, James Ross Basin, Antarctica*. Geological Society, London, Special Publications, **258**, 187–200.

HAMBREY, M. J., HUDDART, D., BENNETT, M. R. & GLASSER, N. F. 1997. Genesis of 'hummocky moraines' by thrusting in glacier ice: evidence from Svalbard and Britain. *Journal of the Geological Society, London*, **154**, 623–632.

HAMBREY, M. J., BENNETT, M. R., DOWDESWELL, J. A., GLASSER, N. F. & HUDDART, D. 1999. Debris entrainment and transfer in polythermal valley glaciers. *Journal of Glaciology*, **45**, 69–86.

HAMBREY, M. J., MURRAY, T. *ET AL.* 2005. Structure and changing dynamics of a polythermal glacier on a centennial timescale: Midre Lovénbreen, Svalbard. *Journal of Geophysical Research*, **110**, 1–19.

HAMBREY, M. J., SMELLIE, J. L., NELSON, A. E. & JOHNSON, J. S. 2008. Late Cenozoic glacier–volcano interaction on James Ross Island and adjacent areas, Antarctic Peninsula region. *Geological Society of America Bulletin*, **120**, 709–731.

HAMILTON, S. J. & WHALLEY, B. W. 1995. Rock glacier nomenclature: a re-assessment. *Geomorphology*, **14**, 73–80.

HARRIS, C. 2007. Periglacial landforms: slope deposits and forms. *In*: SCOTT, A. E. (ed.) *Encyclopedia of Quaternary Science*. Elsevier, Oxford, 2207–2217.

HARRIS, C. & MURTON, J. B. 2005. Interactions between glaciers and permafrost: an introduction. *In*: HARRIS, C. & MURTON, J. B. (eds) *Cryospheric Systems: Glaciers and Permafrost*. Geological Society, London, Special Publications, **242**, 1–9.

HARRY, D. G. 1988. Ground ice and permafrost. *In*: CLARK, M. J. (ed.) *Advances in Periglacial Geomorphology*. Wiley-Interscience, Chichester, 113–150.

HÄTTESTRAND, C. & STROEVEN, A. P. 2002. A relict landscape in the centre of Fennoscandian glaciation: geomorphological evidence of minimal Quaternary glacial erosion. *Geomorphology*, **44**, 127–143.

HEROY, D. C. & ANDERSON, J. B. 2005. Ice-sheet extent of the Antarctic Peninsula region during the Last Glacial Maximum (LGM) – insights from glacial geomorphology. *GSA Bulletin*, **117**, 1497–1512.

HEROY, D. C. & ANDERSON, J. B. 2007. Radiocarbon constraints on Antarctic Peninsula Ice Sheet retreat following the Last Glacial Maximum (LGM). *Quaternary Science Reviews*, **26**, 3286–3297.

HIGGINS, S. M., DENTON, G. H. & HENDY, C. H. 2000. Glacial geomorphology of Bonney Drift, Taylor Valley, Antarctica. *Geografiska Annaler. Series A, Physical Geography*, **82**, 365–389.

HJORT, C., INGÓLFSSON, Ó., MÖLLER, P. & LIRIO, J. M. 1997. Holocene glacial history and sea-level changes on James Ross Island, Antarctic Peninsula. *Journal of Quaternary Science*, **12**, 259–273.

HJORT, C., INGÓLFSSON, Ó., BENTLEY, M. J. & BJÖRCK, S. 2003. The Late Pleistocene and Holocene glacial and climate history of the Antarctic Peninsula region as documented by the land and sediment records – a review. *In*: DOMACK, E. W., LEVENTER, A., BURNETT, A., BINDSCHADLER, R., CONVEY, P. & KIRBY, M. (eds) *Antarctic Peninsula Climate Variability: Historical and Palaeoenvironmental Perspectives*. Antarctic Research Series. American Geophysical Union, Washington, DC, **79**, 95–102.

HUBBARD, B. & SHARP, M. 1995. Basal ice facies and their formation in the Western Alps. *Arctic and Alpine Research*, **27**, 301–310.

HUBBARD, B., COOK, S. & COULSON, H. 2009. Basal ice facies: a review and unifying approach. *Quaternary Science Reviews*, **28**, 1956–1969.

HUGHES, T. J. 1981. Numerical reconstruction of palaeo-ice sheets. *In*: DENTON, G. H. & HUGHES, T. J. (eds) *The Last Great Ice Sheets*. John Wiley & Sons, New York, 221–261.

HUGHES, T. J. 1995. Ice sheet modelling and the reconstructions of former ice sheets from Geo(morpho)logical field data. *In*: MENZIES, J. (ed.) *Modern Glacial Environments – Processes, Dynamics and Sediments*. Butterworth-Heinemann, Oxford, 77–99.

HUMLUM, O. 1998. The climatic significance of rock glaciers. *Permafrost and Periglacial Processes*, **9**, 375–395.

HUMLUM, O. 2000. The geomorphic significance of rock glaciers: estimates of rock glacier debris volumes and headwall recession rates in West Greenland. *Geomorphology*, **35**, 41–67.

INESON, J. R., CRAME, J. A. & THOMSON, M. R. A. 1986. Lithostratigraphy of the Cretaceous Strata of West James Ross Island, Antarctica. *Cretaceous Research*, **7**, 141–159.

INGÓLFSSON, Ó. & HJORT, C. 2002. Glacial history of the Antarctic Peninsula since the Last Glacial Maximum – a synthesis. *Polar Research*, **21**, 227–234.

INGÓLFSSON, Ó., HJORT, C., BJÖRCK, S. & SMITH, L. R. I. 1992. Late Pleistocene and Holocene glacial history of James Ross Island, Antarctic Peninsula. *Boreas*, **21**, 209–222.

JACOBS, S. S., JENKINS, A., GIULIVI, C. F. & DUTRIEUX, P. 2011. Stronger ocean circulation and increased melting under Pine Island Glacier ice shelf. *Nature Geoscience*, **4**, 519–523.

JOHNSON, J. S. & SMELLIE, J. L. 2007. Zeolite compositions as proxies for eruptive palaeoenvironment. *Geochemistry, Geophysics, Geosystems*, **8**, doi: 10.1029/2006GC001450, 1–19.

JOHNSON, J. S., SMELLIE, J. L., NELSON, A. E. & STUART, F. M. 2009. History of the Antarctic Peninsula Ice Sheet since the early Pliocene – evidence from cosmogenic dating of Pliocene lavas on James Ross Island, Antarctica. *Global and Planetary Change*, **69**, 205–213.

JOHNSON, J. S., BENTLEY, M. J., ROBERTS, S. J., BINNEY, S. A. & FREEMAN, S. P. H. T. 2011. Holocene deglacial history of the north east Antarctic Peninsula – a review and new chronological constraints. *Quaternary Science Reviews*, **30**, 27–28.

JONES, A. P., TUCKER, M. E. & HART, J. K. 1999. *The Description and Analysis of Quaternary Stratigraphic*

Field Sections. Technical Guide No. 7. Quaternary Research Association, Cambridge.

Jonkers, H. A., Lirio, J. M., Del Valle, R. A. & Kelly, S. P. 2002. Age and environment of Miocene–Pliocene glaciomarine deposits, James Ross Island, Antarctica. *Geological Magazine*, **139**, 577–594.

King, J. C., Turner, J., Marshall, G. J., Connelly, W. M. & Lachlan-Cope, T. A. 2003. Antarctic Peninsula climate variability and its causes as revealed by analysis of instrumental records. *In*: Domack, E. W., Leventer, A., Burnett, A., Bindschadler, R., Convey, P. & Kirby, M. (eds) *Antarctic Peninsula Climate Variability: Historical and Palaeoenvironmental Perspectives*. Antarctic Research Series. American Geophysical Union, Washington, DC, **79**, 17–30.

Kjær, K. H. & Krüger, J. 2001. The final phase of dead-ice moraine development: processes and sediment architecture, Kötlujökull, Iceland. *Sedimentology*, **48**, 935–952.

Kjær, K. H., Korsgaard, N. J. & Schomacker, A. 2008. Impact of multiple glacier surges – a geomorphological map from Brúarjökull, East Iceland. *Journal of Maps*, **2008**, 5–20.

Kleman, J. & Glasser, N. F. 2007. The subglacial thermal organisation (STO) of ice sheets. *Quaternary Science Reviews*, **26**, 585–597.

Kleman, J., Hättestrand, C. & Clarhäll, A. 1999. Zooming in on frozen-bed patches: scale-dependent controls on Fennoscandian ice sheet basal thermal zonation. *Annals of Glaciology*, **28**, 189–194.

Klint, K. E. S., Richardt, N. & Krüger, J. 2011. 9 Evidence for subglacial deformation and deposition during a complete advance-stagnation cycle of Kötlujökull, Iceland – a case study. *In*: Schomacker, A., Krüger, J. & Kjær, K. H. (eds) *The Mýrdalsjökull Ice Cap, Iceland*. Elsevier, Amsterdam, 145–158.

Knight, J. 2008. The environmental significance of ventifacts: a critical review. *Earth-Science Reviews*, **86**, 89–105.

Knight, P. G., Patterson, C. J., Waller, R. I., Jones, A. P. & Robinson, Z. P. 2000. Preservation of basal-ice sediment texture in ice-sheet moraines. *Quaternary Science Reviews*, **19**, 1255–1258.

Konrad, S. K., Humphrey, N. F., Steig, E. J., Clark, D. H., Potter, N. & Pfeffer, W. T. 1999. Rock glacier dynamics and paleoclimatic implications. *Geology*, **27**, 1131–1134.

Košler, J., Magna, T., Mlčoch, B., Mixa, P., Nývlt, D. & Holub, F. V. 2009. Combined Sr, Nd, Pb and Li isotope geochemistry of alkaline lavas from northern James Ross Island (Antarctic Peninsula) and implications for back-arc magma formation. *Chemical Geology*, **258**, 207–218.

Krüger, J., Kjær, K. H. & Schomacker, A. 2010. Dead-Ice environments: a landsystems model for a debris-charged, stagnant Lowland Glacier Margin, Kötlujökull. *In*: Schomacker, A., Krüger, J. & Kjær, K. H. (eds) *Developments in Quaternary Science*. Elsevier, Amsterdam, 105–126.

Laity, J. 2008. *Deserts and Desert Environments*. Wiley-Blackwell, Chichester.

Láska, K., Prosek, P. & Budik, L. 2010. Seasonal variation of air temperature at the Mendel Station, James Ross Island, in the period of 2006–2009. *Geophysical Research Abstracts*, **12**, EGU2010–3880.

Láska, K., Nývlt, D., Engel, Z. & Kopačková, V. 2011. Meteorological data and mass balance measurements on Davies Dome and Whisky Glacier in 2006–2010, James Ross Island, Antarctica. *Geophysical Research Abstracts*, **13**, EGU2011–4858.

Lloyd Davies, M. T., Atkins, C. B., van der Meer, J. J. M., Barrett, P. J. & Hicock, S. R. 2009. Evidence for cold-based glacial activity in the Allan Hills, Antarctica. *Quaternary Science Reviews*, **28**, 3124–3137.

Lønne, I. & Lyså, A. 2005. Deglaciation dynamics following the Little Ice Age on Svalbard: implications for shaping of landscapes at high latitudes. *Geomorphology*, **72**, 300–319.

Lukas, S., Nicholson, L. I. & Humlum, O. 2007. Comment on Lønne and Lyså (2005): 'Deglaciation dynamics following the Little Ice Age on Svalbard: implications for shaping of landscapes at high latitudes', *Geomorphology* 72, 300–319. *Geomorphology*, **84**, 145–149.

Lundqvist, J., Lillieskold, M. & Ostmark, K. 1995. Glacial and periglacial deposits of the Tumbledown Cliffs area, James Ross Island, West Antarctica. *Geomorphology*, **11**, 205–214.

Martin, P. J. & Peel, D. A. 1978. The spatial distribution of 10 m temperatures in the Antarctic Peninsula. *Journal of Glaciology*, **20**, 311–317.

Matsuoka, N. 1995. Rock weathering processes and landform development in the Sør Rondane Mountains, Antarctica. *Geomorphology*, **12**, 323–339.

Mathews, W. H. & Mackay, J. R. 1960. Deformation of soils by glacier ice and the influence of pore pressure and permafrost. *Royal Society of Canada Transactions*, **LIX**, Series III, Section 4, 27–36.

Mayewski, P. A., Meredith, M. P. *et al.* 2009. State of the Antarctic and Southern Ocean climate system. *Reviews of Geophysics*, **47**, 1–38.

McLeod, M., Bockheim, J. G. & Balks, M. R. 2008. Glacial geomorphology, soil development and permafrost features in central-upper Wright Valley, Antarctica. *Geoderma*, **144**, 93–103.

Miall, A. D. 1977. A review of the braided river depositional environment. *Earth Science Reviews*, **13**, 1–62.

Murray, T., Stuart, G. W., Miller, P. J., Woodward, J., Smith, A. M., Porter, P. R. & Jiskoot, H. 2000. Glacier surge propagation by thermal evolution at the bed. *Journal of Geophysical Research*, **105**, 13 491–13 507.

Murray, T., Strozzi, T., Luckman, A., Pritchard, H. & Jiskoot, H. 2002. Ice dynamics during a surge of Sortebrae, East Greenland. *Annals of Glaciology*, **34**, 323–329.

Murton, J. 2007. Periglacial landforms, rock forms: rock weathering. *In*: Scott, A. E. (ed.) *Encyclopedia of Quaternary Science*. Elsevier, Oxford, 2249–2256.

NASA. 2004. Air temperature conditions and climatic-geomorphological characteristics of James Ross Island, Antarctic Peninsula Entry ID: CDA_AR_GEO_J.ROSS_CLIMATOLOGY. Data Centre, Centro de Datos Antarcticos, Argentina: Antarctic Master Directory, http://gcmd.nasa.gov

NELSON, P. H. H. 1975. The James Ross Island Volcanic Group of north-east Graham Land. *British Antarctic Survey Scientific Reports*, **54**, 1–62.

NELSON, A. E., SMELLIE, J. L. *ET AL.* 2009. Neogene glacigenic debris flows on James Ross Island, northern Antarctic Peninsula, and their implications for regional climate history. *Quaternary Science Reviews*, **28**, 3138–3160.

NÝVLT, D., KOŠLER, J., MLČOCH, B., MIXA, P., LISÁ, L., BUBÎK, M. & HENDRIKS, B. W. H. 2011. The Mendel Formation: evidence for Late Miocene climatic cyclicity at the northern tip of the Antarctic Peninsula. *Palaeogeography Palaeoclimatology Palaeoecology*, **299**, 363–384.

Ó COFAIGH, C., LEMMEN, D. S., EVANS, D. J. A. & BEDNARSKI, J. 1999. Glacial landform-sediment assemblages in the Canadian High Arctic and their implications for late Quaternary glaciation. *Annals of Glaciology*, **28**, 195–201.

Ó COFAIGH, C., DOWDESWELL, J. A. & PUDSEY, C. J. 2001. Late Quaternary Iceberg Rafting along the Antarctic Peninsula Continental Rise and in the Weddell and Scotia Seas. *Quaternary Research*, **56**, 308–321.

Ó COFAIGH, C., EVANS, D. J. A. & ENGLAND, J. H. 2003. Ice-marginal terrestrial landsystems: sub-polar glacier margins of the Canadian and Greenland High Arctic. *In*: EVANS, D. J. A. (ed.) *Glacier Landsystems*. Arnold, London, 44–64.

ØSTREM, G. 1964. Ice-cored moraines in Scandinavia. *Geografiska Annaler*, **46**, 282–337.

ØSTREM, G. 1971. Rock glaciers and ice-cored moraines, a reply to D. Barsch. *Geografiska Annaler. Series A, Physical Geography*, **53**, 207–213.

PHILLIPS, W. M., HALL, A. M., MOTTRAM, R., FIFIELD, L. K. & SUGDEN, D. E. 2006. Cosmogenic ^{10}Be and ^{26}Al exposure ages of tors and erratics, Cairngorm Mountains, Scotland: timescales for the development of a classic landscape of selective linear glacial erosion. *Geomorphology*, **73**, 222–245.

PIRRIE, D., CRAME, J. A., RIDING, J. B., BUTCHER, A. R. & TAYLOR, P. D. 1997. Miocene glaciomarine sedimentation in the northern Antarctic Peninsula region: the stratigraphy and sedimentology of the Hobbs Glacier Formation, James Ross Island. *Geological Magazine*, **134**, 745–762.

POWERS, M. C. 1953. A new roundness scale for sedimentary particles. *Journal of Sedimentary Petrology*, **23**, 117–119.

PUDSEY, C. J. & EVANS, J. 2001. First survey of Antarctic sub-ice shelf sediments reveals Mid-Holocene ice shelf retreat. *Geology*, **29**, 787–790.

RABASSA, J. 1983. Stratigraphy of the glacigenic deposits in Northern James Ross Island, Antarctic Peninsula. *INQUA Symposium on the Genesis and Lithology of Quaternary Deposits*, 329–339.

RABASSA, J. 1987. Drumlins and Drumlinoid forms in northern James Ross Island, Antarctic Peninsula. *In*: MENZIES, J. & ROSE, J. (eds) *Drumlin Symposium*. Balkema, Rotterdam, 267–288.

RABASSA, J., SKVARCA, P., BERTANI, L. & MAZZONI, E. 1982. Glacier inventory of James Ross and Vega Islands, Antarctic Peninsula. *Annals of Glaciology*, **3**, 260–264.

REA, B. 2007. Periglacial Landforms, Rockforms: Blockfields (Felsenmeer). *In*: SCOTT, A. E. (ed.) *Encyclopedia of Quaternary Science*. Elsevier, Oxford, 2225–2236.

REINARDY, B. T. I., PUDSEY, C. J., HILLENBRAND, C.-D., MURRAY, T. & EVANS, J. 2009. Contrasting sources for glacial and interglacial shelf sediments used to interpret changing ice flow directions in the Larsen Basin, Northern Antarctic Peninsula. *Marine Geology*, **266**, 156–171.

RILEY, T. R., FLOWERDEW, M. J. & HASELWIMMER, C. E. 2011. *Geological Map of Eastern Graham Land, Antarctic Peninsula (1:6 25 000 Scale)*. BAS *GEOMAP 2 Series, Sheet 1*. British Antarctic Survey, Cambridge.

RIPPIN, D. M., CARRIVICK, J. L. & WILLIAMS, C. 2011. Evidence towards a thermal lag in the response of Kårsglaciären, northern Sweden, to climate change. *Journal of Glaciology*, **57**, 895–903.

ROBERTS, S. J., HODGSON, D. A. *ET AL.* 2011. Geological constraints on glacio-isostatic adjustment models of relative sea-level change during deglaciation of Prince Gustav Channel, Antarctic Peninsula. *Quaternary Science Reviews*, **30**, 3603–3617.

SCHAFER, J. M., BAUR, H., DENTON, G. H., IVY-OCHS, S., MARCHANT, D. R., SCHLUCHTER, C. & WIELER, R. 2000. The oldest ice on Earth in Beacon Valley, Antarctica: new evidence from surface exposure dating. *Earth and Planetary Science Letters*, **179**, 91–99.

SCHOMACKER, A. 2008. What controls dead-ice melting under different climate conditions? A discussion. *Earth-Science Reviews*, **90**, 103–113.

SCHOMACKER, A. & KJÆR, K. H. 2007. Origin and de-icing of multiple generations of ice-cored moraines at Brúarjökull, Iceland. *Boreas*, **36**, 411–425.

SCHOMACKER, A. & KJÆR, K. H. 2008. Quantification of dead-ice melting in ice-cored moraines at the high-Arctic glacier Holmströmbreen, Svalbard. *Boreas*, **37**, 211–225.

SERRANO, E. & LÓPEZ-MARTÍNEZ, J. 2000. Rock glaciers in the South Shetland Islands, Western Antarctica. *Geomorphology*, **35**, 145–162.

SHARP, M. J., JOUZEL, J., HUBBARD, B. & LAWSON, W. 1994. The character, structure and origin of the basal ice layer of a surge-type glacier. *Journal of Glaciology*, **40**, 327–340.

SIEGERT, M. J. 2009. Numerical modelling of the Antarctic Ice Sheet. *In*: SIEGERT, M. J. & FLORINDO, F. (eds) *Antarctic Climate Evolution*. Elsevier, Rotterdam, 235–256.

SIEGERT, M. J. & DOWDESWELL, J. A. 2004. Numerical reconstructions of the Eurasian Ice Sheet and climate during the Late Weichselian. *Quaternary Science Reviews*, **23**, 1273–1283.

SMELLIE, J. L., MCARTHUR, J. M., MCINTOSH, W. C. & ESSER, R. 2006. Late Neogene interglacial events in the James Ross Island region, northern Antarctic Peninsula, dated by Ar/Ar and Sr-isotope stratigraphy. *Palaeogeography, Palaeoclimatology, Palaeoecology*, **242**, 169–187.

SMELLIE, J. L., JOHNSON, J. S., MCINTOSH, W. C., ESSER, R., GUDMUNDSSON, M. T., HAMBREY, M. J. & VAN WYK DE VRIES, B. 2008. Six million years of glacial history recorded in volcanic lithofacies of the James

Ross Island Volcanic Group, Antarctic Peninsula. *Palaeogeography, Palaeoclimatology, Palaeoecology*, **260**, 122–148.

SMELLIE, J. L., HAYWOOD, A. M., HILLENBRAND, C.-D., LUNT, D. J. & VALDES, P. J. 2009. Nature of the Antarctic Peninsula Ice Sheet during the Pliocene: geological evidence and modelling results compared. *Earth-Science Reviews*, **94**, 79–94.

SONE, T., FUKUI, K., STRELIN, J. A., TORIELLI, C. A. & MORI, J. 2007. Glacier lake outburst flood on James Ross Island, Antarctic Peninsula region. *Polish Polar Research*, **28**, 3–12.

SPEDDING, N. & EVANS, D. J. A. 2002. Sediments and landforms at Kviarjokull, southeast Iceland: a reappraisal of the glaciated valley landsystem. *Sedimentary Geology*, **149**, 21–42.

STOKES, C. R. & CLARK, C. D. 1999. Geomorphological criteria for identifying Pleistocene ice streams. *Annals of Glaciology*, **28**, 67–74.

STOKES, C. R. & CLARK, C. D. 2001. Palaeo-ice streams. *Quaternary Science Reviews*, **20**, 1437–1457.

STRELIN, J. A. & MALAGNINO, E. C. 1992. *Geomorfologìa de la Isla James Ross. Geologia de la Isla James Ross*. Instituto Antàrctico Argentino, Buenos Aires, 7–36.

STRELIN, J. A. & SONE, T. 1998. Rock glaciers on James Ross Island, Antarctica. *Permafrost – Seventh International Conference (Proceedings)*, **55**, 1027–1033.

STRELIN, J. A., SONE, T., MORI, J., TORIELLI, C. A. & NAKAMURA, E. 2006. New data related to Holocene landform development and climatic change from James Ross Island, Antarctic Peninsula. *In*: FÜTTERER, D. K., DAMASKE, D., KLEINSCHMIDT, G., MILLER, H. & TESSENSOHN, F. (eds) *Antarctica: Contributions to Global Earth Sciences. Proceedings of the IX International Symposium of Antarctic Earth Sciences, Potsdam, 2003*. Springer, New York, 455–460.

STRELIN, J. A., TORIELLI, C. A., SONE, T., FUKUI, K. & MORI, J. 2007. Particularidades geneticas de glaciares de roca la Isla James Ross, Peninsula Antartica. *Revista de la Asociacion Argentina*, **62**, 627–634.

STROEVEN, A. P., FABEL, D., HÄTTESTRAND, C. & HARBOR, J. 2002. A relict landscape in the centre of Fennoscandian glaciation: cosmogenic radionuclide evidence of tors preserved through multiple glacial cycles. *Geomorphology*, **44**, 145–154.

SUGDEN, D. E., MARCHANT, D. R., POTTER, N., SOUCHEZ, R. A., DENTON, G. H., SWISHER, C. C. & TISON, J.-L. 1995. Preservation of Miocene glacier ice in East Antarctica. *Nature*, **376**, 412–414.

SUGDEN, D. E., SUMMERFIELD, M. A., DENTON, G. H., WILCH, T. I., MCINTOSH, W. C., MARCHANT, D. R. & RUTFORD, R. H. 1999. Landscape development in the Royal Society Range, southern Victoria Land, Antarctica: stability since the mid-Miocene. *Geomorphology*, **28**, 181–200.

THORN, C. E. 1988. Nivation: a geomorphic chimera. *In*: CLARK, M. J. (ed.) *Advances in Periglacial Geomorphology*. Wiley-Interscience, Chichester, 3–32.

TROMBOTTO, D. 2002. Inventory of fossil cryogenic forms and structures in Patagonia and the mountains of Argentina beyond the Andes. *South African Journal of Science*, **98**, 171–180.

VAN LIPZIG, N. P. M., KING, J. C., LACHLAN-COPE, T. A. & VAN DEN BROEKE, M. R. 2004. Precipitation, sublimation and snow drift in the Antarctic Peninsula region from a regional atmospheric model. *Journal of Geophysical Research*, **109**, D24106.

WADHAM, J. L. & NUTTALL, A.-M. 2003. Multiphase formation of superimposed ice during a mass-balance year at a maritime high-Arctic glacier. *Journal of Glaciology*, **48**, 545–551.

WALDEN, J. S. 2004. Particle Lithology (or mineral and geochemical analysis). *In*: EVANS, D. J. A. & BENN, D. I. (eds) *A Practical Guide to the Study of Glacial Sediments*. Arnold, London, 145–180.

WALLER, R. I. 2001. The influence of basal processes on the dynamic behaviour of cold-based glaciers. *Quaternary International*, **86**, 117–128.

WALLER, R. I., HART, J. K. & KNIGHT, P. G. 2000. The influence of tectonic deformation on facies variability in stratified debris-rich basal ice. *Quaternary Science Reviews*, **19**, 775–786.

WALLER, R. I., MURTON, J. B. & KNIGHT, P. G. 2009. Basal glacier ice and massive ground ice: different scientists, same science? *In*: KNIGHT, J. & HARRISON, S. (eds) *Periglacial and Paraglacial Processes and Environments*. Geological Society, London, Special Publications, **320**, 57–69.

WILSON, P. 2007. Periglacial landforms: rock forms. *In*: SCOTT, A. E. (ed.) *Encyclopedia of Quaternary Science*. Elsevier, Oxford, 2217–2225.

ŽÁK, J., SOEJONO, I., JANOUŠEK, V. & VENERA, Z. 2012. Magnetic fabric and tectonic setting of the Early to Middle Jurassic felsic dykes at Pitt Point and Mount Reece, eastern Graham Land, Antarctica. *Antarctic Science*, **24**, 45–58.

Windows on Antarctic soil–landscape relationships: comparison across selected regions of Antarctica

MEGAN R. BALKS[1]*, JERÓNIMO LÓPEZ-MARTÍNEZ[2], SERGEY V. GORYACHKIN[3], NIKITA S. MERGELOV[3], CARLOS E. G. R. SCHAEFER[4], FELIPE N. B. SIMAS[4], PETER C. ALMOND[5], GRAEME G. C. CLARIDGE[6], MALCOLM McLEOD[7] & JOSHUA SCARROW[1]

[1]*Earth and Ocean Sciences, University of Waikato, Private Bag 3105, Hamilton 3240, New Zealand*

[2]*Dept. Geologia y Geoquimica, Facultad de Ciencias, Universidad Autonoma de Madrid, 28049, Spain*

[3]*Institute of Geography, RAS, Staromonetny 29, Moscow, 119017, Russia*

[4]*Universidade Federal de Viçosa, Departamento de Solos. Av P.H. Rolfs sn Centro 36571-000 – Vicosa, MG, Brazil*

[5]*Department of Soil and Physical Sciences, Lincoln University, Christchurch, New Zealand*

[6]*19 Campbell Terrace, Petone 5012, New Zealand*

[7]*Landcare Research, Private Bag 3127, Hamilton 3240, New Zealand*

**Corresponding author (e-mail: m.balks@waikato.ac.nz)*

Abstract: This paper brings together topographic cross-section 'windows' from across Antarctica to illustrate soil–landscapes from the margins of the polar plateau in the Transantarctic Mountains and McMurdo Dry Valleys, through East Antarctic coastal areas, to the northern Antarctic Peninsula Region. Soils identified range from Gelisols in the Ross Sea Region, through Gelisols and Entisols in coastal East Antarctica, to a mixture of Gelisols, Entisols, Spodosols and Inceptisols in the northern Antarctic Peninsula Region where permafrost is not ubiquitous. The relative impacts of the soil-forming factors are considered. At a continental scale climate is the main driver of the differences observed between soils in different areas. At local scales strong soil–topographic relationships are observed. Organisms, time and parent material are dominant influences on soil properties only in relatively localized situations. Organisms dominate in areas of organic matter or guano accumulation and time is a dominant influence on exceptionally old upland surfaces in the McMurdo Dry Valleys. The US Department of Agriculture's Soil Taxonomy gives a useful overall appraisal of Antarctic soils; however, for detailed work, there is a need to introduce some new categories at subgroup level to better capture the range of soils described.

The objectives of this paper are: to investigate Antarctic soil–landscape relationships as captured by example 'windows' by bringing together information from selected regions across the Antarctic (Fig. 1); to compare and contrast the range of soils found in contrasting regions; to assess soil development in relation to the soil-forming factors (Jenny 1941), and to explore soil classification issues common to workers across the Antarctic continent.

The Antarctic continent comprises 13.9×10^6 km^2 (including ice shelves) of which 44 890 km^2 (0.32%) is ice-free (Fox & Cooper 1994; British Antarctic Survey 2005) with potential for soil development. Ice-free areas are mainly confined to the Antarctic Peninsula and surrounding islands, a few places around the perimeter of the continent and along the Transantarctic Mountains. Some of the ice-free areas are extensive, the largest, the McMurdo Dry Valleys, being approximately 5000 km^2. A wide range of soils occur in the ice-free areas (Campbell & Claridge 1987).

The first Antarctic soil map (derived from fieldwork undertaken by Claridge and McCraw in the Taylor Valley, in the 1959–1960 summer; McCraw 1967) gave an initial insight into Antarctic soil extent, variety and distribution in the dry valleys in the Ross Sea Region. Investigations of soil and weathering processes were made in the rare coastal ice-free areas by Glazovskaya (1958). The first soil studies in the Antarctic Peninsula region,

From: HAMBREY, M. J., BARKER, P. F., BARRETT, P. J., BOWMAN, V., DAVIES, B., SMELLIE, J. L. & TRANTER, M. (eds) 2013. *Antarctic Palaeoenvironments and Earth-Surface Processes*. Geological Society, London, Special Publications, **381**, 397–410. First published online May 16, 2013, http://dx.doi.org/10.1144/SP381.9

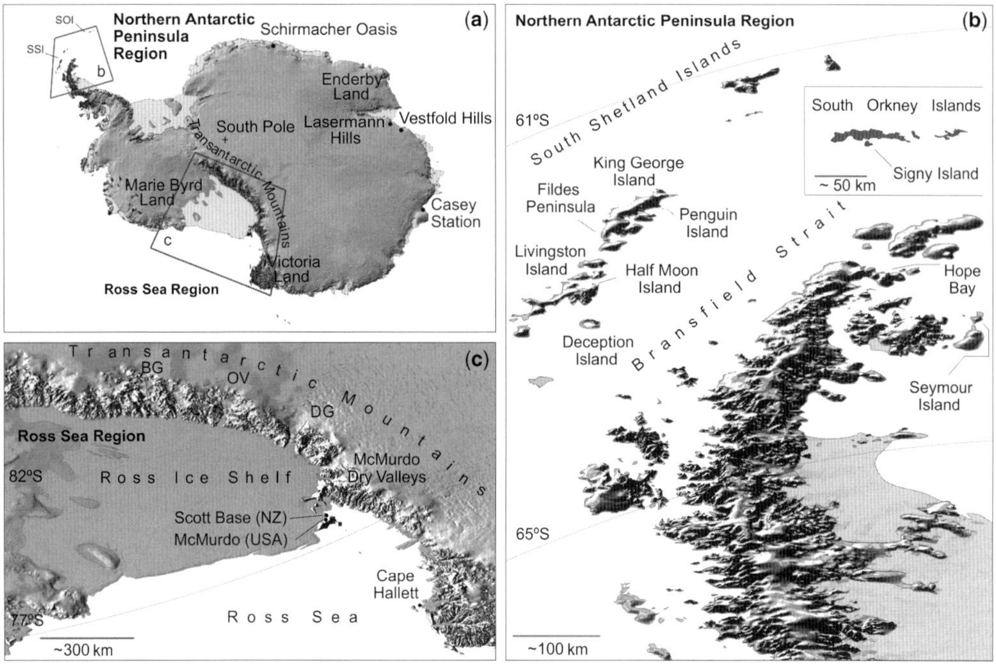

Fig. 1. Location of the studied sites and places mentioned in the text. (**a**) Antarctica: SSI, South Shetland Islands; SOI, South Orkney Islands. (**b**) Northern Antarctic Peninsula Region. (**c**) Ross Sea Region: BG, Beardmore Glacier; OV, Ong Valley; DG, Darwin Glacier. The Taylor Valley, Wright Valley and Onyx River are within the McMurdo Dry Valleys.

for example, those of Half Moon Island and other places (Corte & Somoza 1954), Signy Island (Holdgate *et al.* 1967; Allen & Heal 1970), and Livingston Island and other places (Everett 1976), reported that soils generally show little profile development because of their comparative youth and extensive cryoturbation. More recent soil studies in the Antarctic Peninsula and surrounding islands have revealed a diversity of soil types, in some cases with well-developed profiles, pronounced chemical weathering and organic matter accumulation (Blume *et al.* 2001; Navas *et al.* 2005, 2008; Michel *et al.* 2006; Simas *et al.* 2006, 2007, 2008; Schaefer *et al.* 2008; Francelino *et al.* 2011; Moura *et al.* 2012).

From 1960 to 2000 Antarctic soil research generally focused on describing the soils and understanding their age and genesis, without much attempt to capture spatial distribution. Since about 2000, with more geological and geomorphological maps available, the arrival of GIS, high-resolution satellite data, detailed digital elevation models and the need for spatial databases to underpin decisions on environmental management, opportunities have arisen to develop soil maps. Detailed soil investigations are starting to elucidate the soil–landscape relationships at a range of scales. The approach taken by most workers has been to develop relatively detailed maps that capture the soil–landscape pattern in discrete areas, then the data can be extrapolated to cover the wider region.

Recently soil map 'windows' have been developed on a latitudinal gradient (from 72 to 85°S) in the Ross Sea Region of Antarctica, including the Cape Hallett (Hofstee *et al.* 2006), Wright Valley (McLeod *et al.* 2008, 2009; McLeod 2012), Darwin Glacier (Aislabie *et al.* 2012) and the Beardmore Glacier areas. Soil data are being captured and developed for sites around the continent by the Russian programme, with some soil mapping in the East Antarctic Oases (Gilichinsky *et al.* 2010). In coastal East Antarctica, near Casey Station, a small area has been mapped in detail (Blume & Bölter 1992). Around the Antarctic Peninsula, some of the main ice-free areas have been mapped from 62 to 64°S, comprising the South Shetlands archipelago (King George, Deception and Livingston islands), the northernmost tip of the Peninsula (Esperanza/Hope Bay) and Marambio/Seymour Island, marking a sharp transition from wetter areas west of the peninsula to drier coastal areas, east of the peninsula (Michel *et al.* 2006; Simas *et al.* 2008; Francelino *et al.* 2011; Moura *et al.* 2012).

Within this paper soil names and classifications used are those from the Soil Taxonomy (Soil Survey Staff 2010) unless otherwise stated.

Influences on soil formation

The climate of Antarctica ranges from relatively warm (mean annual temperature about 0 °C) and wet in the marine-influenced Antarctic Peninsula region to the extreme cold (mean annual temperature about −20 °C) and dry environment at higher altitudes in inland regions of the Transantarctic Mountains (Table 1). A diverse suite of mineral parent materials is found across the ice-free regions of Antarctica, often comprising basement granites, granodiorites and/or marbles overlain by basalt and sandstones and intruded by dolerite. Many soils are formed on mixed tills with some directly formed on the bedrock. The influence of organic materials varies with climate and proximity to the coast. Organic matter is minimal (0.1%) in drier and colder inland areas; however, in areas where free moisture occurs, mosses and, on the Antarctic Peninsula, higher plants grow and accumulate to form soil parent materials. Accumulations of penguin guano and other ornithogenic materials are locally dominant in soil parent materials at many coastal locations.

Surface ages in coastal and Antarctic Peninsula regions tend to be predominantly Holocene, exposed by the retreat of the Last Glacial Maximum ice. In contrast, at higher elevations in the McMurdo Dry Valleys, surfaces as old as Mid-Miocene (14 Ma) have been reported (Sugden *et al.* 2006), indicating low erosion rates under a stable polar desert climate. Many regions of Antarctica contain strong topographic drivers for soil/landscape variability in terms of slope, altitude and aspect.

Soil landscape relationships and soil properties

Soils of the Ross Sea Region

Within the Ross Sea Region all soils are recognized as Gelisols (after Soil Survey Staff 2010) or Cryosols (International Union of Soil Sciences

Table 1. *Summary of soil-forming factors in the Ross Sea Region, Antarctic Peninsula Region, and East Antarctic Coastal ice-free areas*

Soil-forming factor	Ross Sea Region	East Antarctic Coastal Oases	Antarctic Peninsula Region
Climate	Cold desert/moist coastal	Moist coastal	Maritime Antarctic
Mean annual temperature (°C)	−15 to −20	−9 to −12	0 to −12
Precipitation (mm water equivalent)	Minimal inland, moist at coast	200–500, mainly as snow	200–1000
Parent material	Bedrock and glacial tills predominantly derived from dolerites, sandstones, and granodiorites	Bedrock and glacial tills predominantly derived from granitic and metamorphic rocks. Small areas of penguin guano and organic matter from mosses	Bedrock and glacial tills derived from igneous, metamorphic and sedimentary rocks. Small areas influenced by penguins and plant materials
Topography/geomorphology	Large ice-carved inland U-shaped valleys and rolling to hilly coastal slopes, terraces and plains	Rolling and hilly topography exposed by retreat of inland ice cap	Mountainous with small ice caps and till-covered slopes
Organisms	Minimal, occasional mosses where moisture is available, penguins on isolated, discrete coastal sites. Soil microbes are now widely described and recognized but have lower numbers in older, saltier, soils. Colembola and Mites occur but are not common	Patchy distribution of mosses, lichens and algae, with some microfauna. Soil microbes are now widely described and recognized	Almost continuous cover of mosses in poorly drained areas, epilithic lichens on hyperskeletal soils and communities of mosses, lichens and some vascular plants in well-drained areas with some ornithogenic influence. Colembola, and mites present
Time	Holocene to mid-Miocene	Holocene to late Pleistocene	Holocene surfaces exposed by retreat of Last Glacial Maximum ice

2006) with a wide range of soil properties that reflect the range of climate and parent materials. Within the coastal regions there is an increase in the depth of the active layer from about 30 cm at Scott Base (near the edge of the Ross ice shelf; Adlam *et al.* 2010), to over 80 cm at Cape Hallett (Hofstee *et al.* 2006). Further inland soil moisture contents are lower (Seybold *et al.* 2010) and many land surfaces are much older, leading to salt accumulation and 'dry permafrost' in soils that have insufficient water to form ice-cement (Bockheim 1978, 2002; Campbell & Claridge 1987; McLeod 2012).

Soil 'windows' illustrate typical soil–landscape relationships within the Ross Sea Region from Cape Hallett at about 72°S (Fig. 2) and an inland cross-section of the Wright Valley (77°S, Fig. 3), to a long-section of the Ong Valley floor near the edge of the polar plateau (83°S, Fig. 4).

At Cape Hallett, at the northern end of the Ross Sea, the soils impacted by penguin activity ('Ornithic' Haplorthels, Fig. 2) are localized but important. They comprise up to 80 cm of very sticky dark brown or reddish organic material derived from penguin guano and remains intermixed with 2–5 cm diameter stones that were deposited by penguins for nest construction, over un-weathered basaltic gravelly sand beach deposits. A more common soil of the area is the Typic Haploturbel that occurs on patterned ground on the flat to gently sloping areas at the foot of the hills with Typic Aquorthels, where a perched water table forms above the ice-cemented permafrost during the summer season. Typic Haplorthels form in angular scree material on the steep hillslope.

Within the McMurdo Dry Valleys (at 77°S) a typical cross-section of the Wright Valley (Fig. 3)

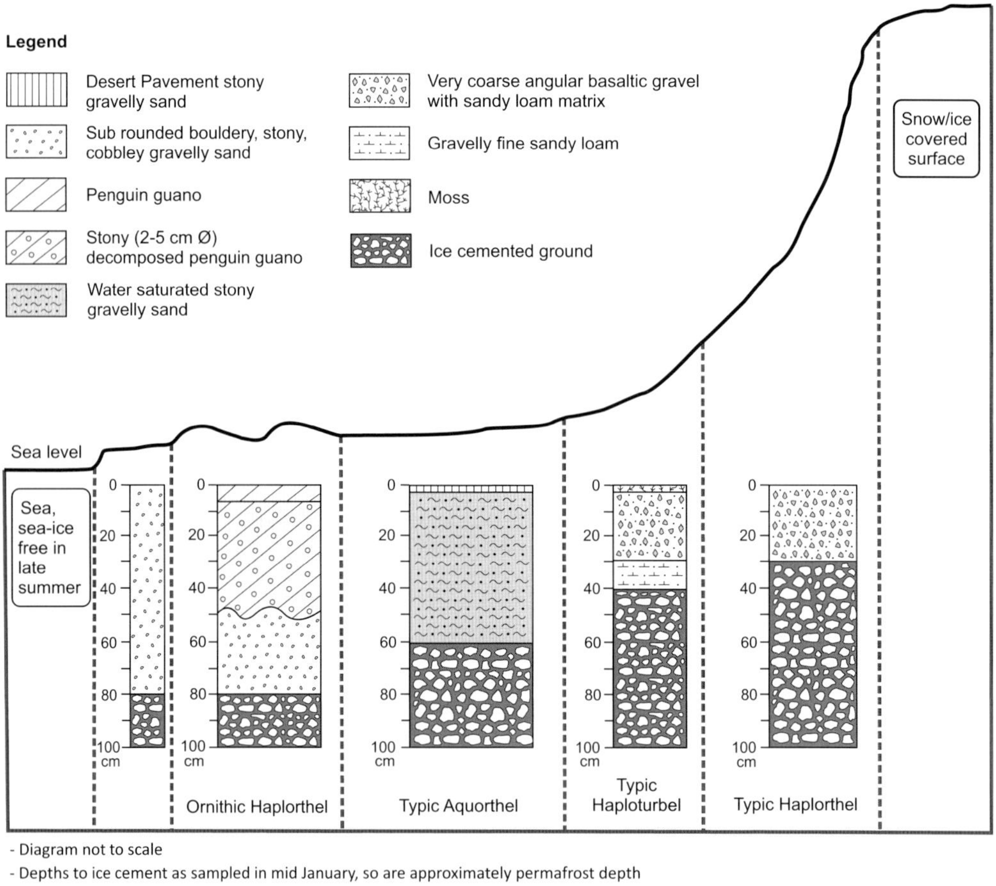

Fig. 2. Generalized cross-section showing soil–landscape relationships at Cape Hallett, a coastal area, at 72°S. Horizontal axis represents approximately 1 km, vertical axis extends from sea level to about 500 m above sea level. The profile adjacent to the sea is the active beach deposit (Typic Haplorthel).

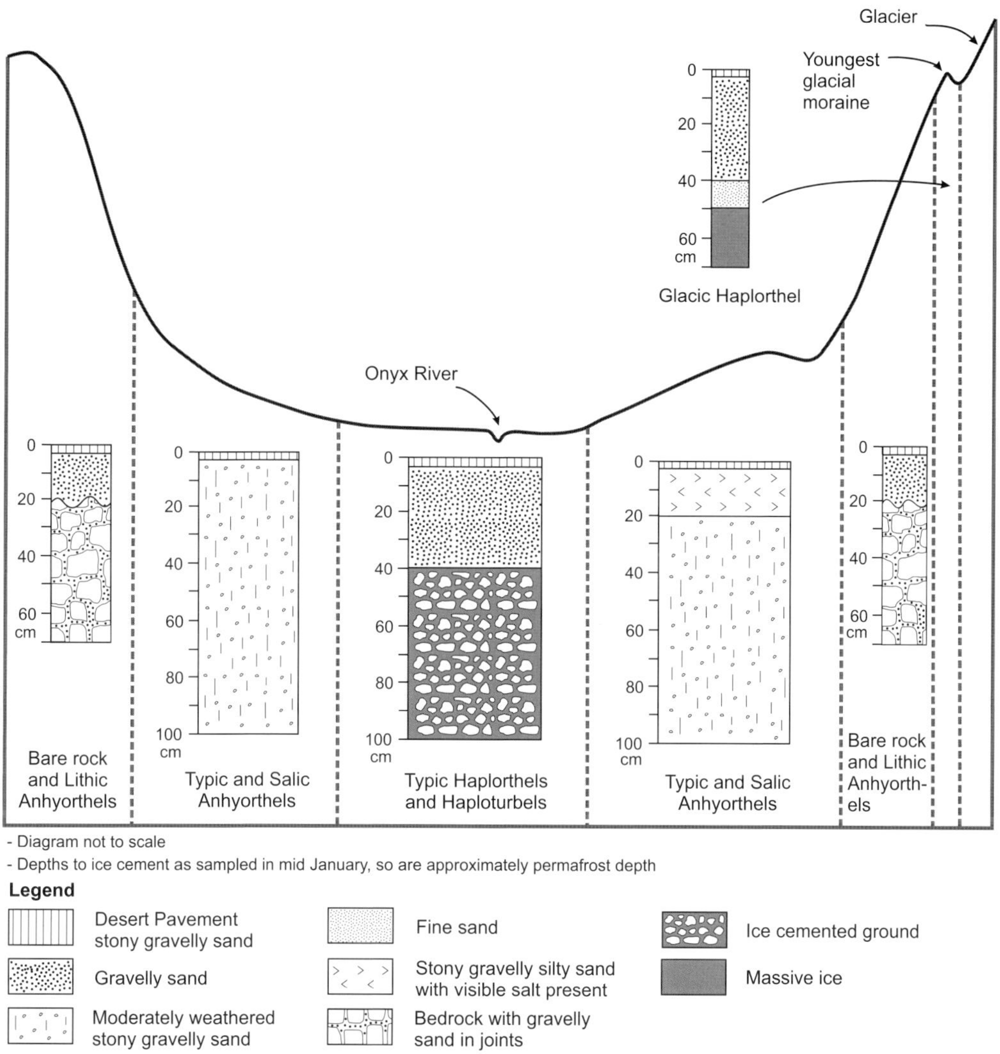

Fig. 3. Generalized cross-section illustrating soil–landscape relationships in Central Wright Valley at 77°S, about 50 km from the coast. Horizontal axis represents approximately 5 km, vertical axis extends from 150 to 1000 m above sea level.

shows Glacic Haploturbels on the young moraines adjacent to present-day alpine glaciers with rock outcrops on steep valley walls. Typic or Salic Anhyorthels occur on dry valley walls and lower slopes where there is insufficient moisture to form ice-cement in the permafrost, and Haplorthels and Haploturbels form on valley floors where water seeps from the adjacent Onyx River or where seepage, from ephemeral side streams, provides a source of soil moisture.

The soils in the Ong Valley, at about 83°S, are mainly formed on relatively young ablation till exposed as the glacier has retreated down-valley, and ablated over the Holocene (Fig. 4). The pattern is one of relatively unweathered gravelly sand over stagnant glacier ice forming Glacic Haploturbels with patterned ground development increasing with surface age. The depth to ice increases from about 15 cm at the youngest end adjacent to the present-day glacier to about 60 cm towards the terminal moraine towards the head of the valley. The terminal moraine marks the highest point that the glacier extended up the valley. Towards the head of the valley (Fig. 4), a Typic Anhyorthel is formed on weathered colluvium derived from the adjacent valley walls.

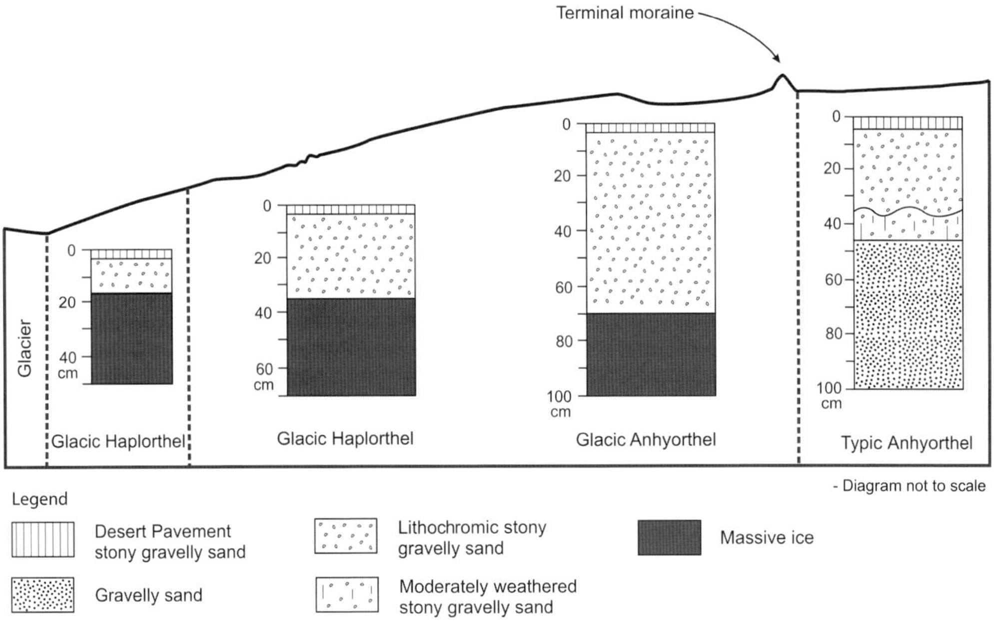

Fig. 4. Generalized long-section illustrating soil–landscape relationships in the Ong Valley floor, 83°S, Central Transantarctic Mountains. Horizontal axis represents approximately 7 km, vertical axis extends from 1500 to 1700 m altitude.

Soils from East Antarctic coastal oases

Ice-free areas in East Antarctica are located mostly close to the continental coast at latitudes of 66–71.5°S. Entisols and Gelisols dominate. The active layer depth is often more than the thickness of loose materials, and is sometimes more than 1 m in sandy and stony materials (MacNamara 1969*a*; Gilichinsky *et al.* 2010).

The soils of coastal Enderby Land generally have well-developed horizonation, bright colours and visible structural elements, and are deep and moist. The average soil moisture contents throughout the profile are three to four times greater than those reported from the Dry Valleys of the Ross Sea Region. There is a net downward migration of soluble materials associated with the movement of moisture through coastal Enderby Land soils (MacNamara 1969*a*). No horizons with indurated salts are found in coastal Enderby Land soils. Subsurface zones of maximum weathering with finer texture, colour development and structure development form incipient B or B/C horizons. Enderby Land soils have acid soil environments with relatively high cation exchange capacities, low quantities of water-soluble salts and varying ratios of exchangeable cations (MacNamara 1969*b*). Lithic Gelorthents and Lithic and Typic Haplorthels and Haploturbels are all characteristic soils for Enderby Land.

In the Larsemann Hills a major geomorphic unit includes the inter-hill valleys which cover at least one-quarter of this 50 km^2 coastal oasis. The 250 mm/year water equivalent precipitation falls mainly as snow; however melting, rather than sublimation, dominates, resulting in intensive summer snowmelt releasing water into the soil. The valley floors receive most of the melt-water impact since the larger snow drifts are concentrated in the valley floors, which also receive runoff from upper slopes. Most of the ground moisture is concentrated at a depth of 20–30 cm from the surface. The valley floors have the greatest abundance of biota and thus the highest biodiversity and soil variety owing to the presence of abundant soil moisture. Moss, lichen and algae ground covers occur, as well as algal–bacterial mats and micromycetes-rich soil horizons. The organisms influence soil development, in the sandy 'granitoid' sediments, to depths ranging from 10 to 80 cm with horizon sequences such as D (gravel pavement)–O–Ah–BC occurring (Fig. 5).

In the Larsemann Hills the availability of surface, and soil, moisture accounts for high soil variety across a humidity gradient ranging from profiles with pronounced humic horizons (up to 10% C) to ahumic soils and barren rock with endolithic soil-like bodies (Fig. 5, Gilichinsky *et al.* 2010; Mergelov *et al.* 2012). All of these features of the

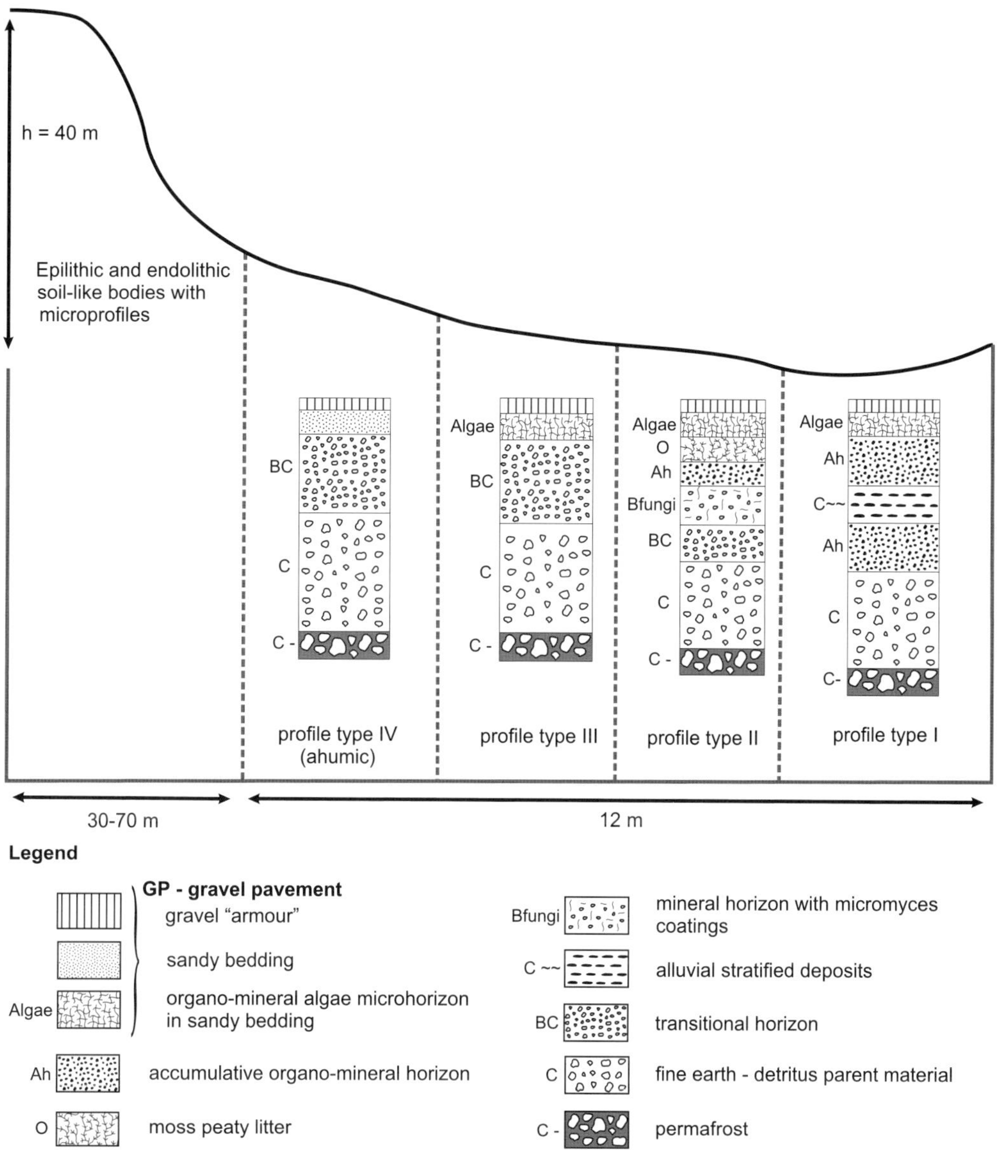

Fig. 5. Generalized cross-section showing soils of a 'wet valley' (Larsemann Hills). Profile type I, Fluventic Haploturbels; type II, 'Oxyaquic' Haploturbels; type III, Typic Haploturbels; type IV, Typic Anhyturbel (ignoring criterium for <50 mm of mean annual precipitation).

valley landscapes suggest a 'wet valleys' denomination (Mergelov 2011), thus underlining the difference from the McMurdo Dry Valleys. Similar patterns occur in the valleys of the Vestfold Hills, Schirmacher Oasis and other areas with comparable landscapes and climates. Fluventic Haploturbels form on the valley floors with typical oxyaquic features because of snow melt-water. Typic Anhyturbels (in relatively snowless, dry, positions on valley walls with northerly aspects), satisfy all the criteria for Anhyturbels, except that of <50 mm of mean annual precipitation.

The soils around Casey Station, in the Windmill Islands off Wilkes Land, have been studied in detail by Blume & Bölter (1992), Beyer & Bölter (1999) and Beyer *et al.* (2000). The area is somewhat subdued in topography and became ice-free following the retreat of the East Antarctic Ice Sheet about

5000 years ago, and consequently the soils are all young in comparison with soils in other regions of Antarctica. The soils are formed on a mixture of solifluction debris and glacial tills, from gneisses and schists, and are strongly cryoturbated. There is little evidence of chemical weathering. The soils have a discontinuous ground cover of lichens and mosses, covering as much as 50% of the ground surface. There is slight staining, resulting from the release and precipitation of iron. Clay contents are very low, although traces of illite and smectite have been detected. The soils are acid, with pH values as low as 4 in places. Because of the low temperatures there is little decomposition of organic matter, and humic materials with narrow C/N ratios accumulate. Organic carbon contents of the surface horizons range from 0.3 to 8 kg m^{-2}, while nitrogen contents range from 21–76 g m^{-2}. The soil organic matter consists of carbohydrates and lipids derived mainly from the lichen ground cover and the mosses, together with traces of degradation products, leading to the formation of true humic material (Beyer 2000).

All of the East Antarctic coastal areas should be recognized as a part of a 'Mid-Antarctic snow-patch barrens' (Goryachkin *et al.* 2004) – landscapes, that lack higher plants but have contrasting hydrological regimes where strong paludification (bog formation), as a result of moisture inputs from thawing snow patches, occurs in combination with soil dryness in adjacent 'snowless' areas.

Different groups have used different criteria for classifying the shallower soils. Beyer & Bölter (2000) recognized Gelisols everywhere in the Grearson Hills area even though the soils they observed often had depths of up to only 30–40 cm over bedrock. At the Grearson Hills, the soils described should be classified (after Soil Survey Staff 2010) as Lithic Gelorthents, Lithic Haplogelods, Lithic Hemistels, Lithic and Typic Haplorthels, Lithic and Typic Haploturbels, and Spodic Haplorthels with Lithic and Typic Aquiturbels under penguin rookeries. All these soils have inclusions of areas of barren rocks (Beyer *et al.* 1999; Beyer & Bölter 2000).

Soils of the Antarctic Peninsula Region

In the South Shetlands archipelago at the northern end of the Antarctic Peninsula the relatively high water availability and mean air temperatures impact soil characteristics in a number of ways. Cryoclastic weathering and cryoturbation are active with a high number of daily freeze-and-thaw cycles. The maritime climate also favours vegetation development, which affects soil genesis by way of organic C inputs and biogeochemical nutrient cycling. The summer relative abundance of water in the terrestrial ecosystem favours chemical weathering and percolation of solutes in the soil (Blume *et al.* 2001; Michel *et al.* 2006; Schaefer *et al.* 2008; Simas *et al.* 2008). As a result soils in the South Shetland Islands are cryoturbated and generally have higher organic C contents than in other parts of Antarctica (Table 2).

In the maritime Antarctic, for example at Signy Island (Holdgate *et al.* 1967; Allen & Heal 1970), soils generally show little profile development because of their comparative youth and intense cryoturbation as a result of relatively frequent freeze-and-thaw events and high water contents. Where mosses and lichens are abundant, the decomposition of organic matter leads to extensive peat formation and development of Spodosols. However, for much of the area raw mineral soils predominate.

Soil classification and mapping in the South Shetland archipelago is a challenge owing to the variable permafrost spatial distribution and active layer depth. There is a general topographic effect on the presence and continuity of the ice-cemented permafrost and the resulting soils (Fig. 6). Geomorphological studies and periglacial landform mapping have contributed to providing information about the presence and distribution of permafrost-related processes and landforms (e.g. López-Martínez *et al.* 1996, 2000, 2002, 2012; López-Martínez & Serrano, 2002; Serrano *et al.* 2008). It is generally accepted that permafrost is absent or sporadic in raised beaches and marine terraces below 25–30 m a.s.l., discontinuous at intermediate altitudes and continuous at altitudes higher than 100 m a.s.l. (Bockheim *et al.* 2008; Serrano *et al.* 2008; Francelino *et al.* 2011; Moura *et al.* 2012).

Field observations and temperature monitoring sites illustrate the changes with altitude, indicating that the active layer is between 20 cm and about 1 m deep at altitudes close to 100 m a.s.l. (Serrano *et al.* 1996, 2008; Vieira *et al.* 2010). Therefore, most soils above 100 m altitude key out as Typic Haploturbels (Soil Survey Staff 2010), corresponding to Turbic Cryosols (International Union of Soil Sciences 2006; Fig. 6).

At midslope, as permafrost is assumed to be discontinuous or sporadic, Entisols and Inceptisols (Soil Survey Staff 2010) form soil complexes with Gelisols (Francelino *et al.* 2011; Moura *et al.* 2012). Below 25–30 m a.s.l., permafrost is assumed to be absent or it is sporadic and therefore Gelisols normally do not occur. On low-lying terraces and raised beaches, Fluvents are common, as well as Orthents and Psamments (Francelino *et al.* 2011; Moura *et al.* 2012). Deception and Penguin islands are exceptions because soils are formed from much younger, Quaternary pyroclastic materials. At Deception and Penguin Islands

Table 2. *Properties of Antarctic soils*

Soil property	Ross Sea Region	East Antarctic Coastal Areas	Antarctic Peninsula Region
USDA Classification	Gelisols ranging from Typic Haplorthels to Salic Anhyorthels	Entisols, Spodosols, and Gelisols	Gelisols, Entisols, Spodosols, and Inceptisols
Texture	Predominantly gravelly sands, rare silt-dominated soils occur	Gravelly sands to sandy loams	Gravelly sands to Sandy clay loams and raw peats
Structure	Generally loose, single grained. Some weakly developed platy structures occur where soil materials are weakly cemented	Skeletal, single-grained	Inorganic coarse-textured soils single-grained, medium and small granules on loamy soils, crumb structures on horizons with high organic C content
Colours	Often reflect parent materials (lithochromic). Some redder colours on older soils owing to oxidation of iron minerals	Greyish lithochromic or brown to red dependent on organic matter and plant cover distribution	Greyish brown derived from the lithochromic gravel matrix with traces of peaty organic matter
pH	Predominantly alkaline	Slightly acid to neutral (5.5–7.0). Ornithogenic soils are acid	Strongly acid on peats, acid sulphate and ornithogenic soils. Alkaline on soils from igneuos and sedimentary rocks with no ornithogenic influence
Nutrient availability	Water, rather than nutrients, is the main limitation to microbial and plant growth	Spatially variable C and N availability	High Ca, Mg, K and P on igneous rocks. Anomolously high P and N on ornithogenic soils. High organic C on vegetated sites, especially when vascular plants are present
Presence of ice-cemented permafrost	Occurs in coastal soils, younger ablation tills and areas with a moisture source, absent in older soils in inland dry valley regions where moisture is scarce	Present at depths >0.7 m	Probably present at depths from 1 to 2 m. Absent or sporadic at altitudes below 30 m a.s.l., discontinuous at altitudes between *c.* 30 and 80 m.a.s.l., continuous above 100 m a.s.l.
Depth of active layer	Varies with altitude, ranging from 0 to about 80 cm, predominantly 20–50 cm	0.7 to >1 m	Over 1 m. Permafrost not always present

permafrost occurs at lower altitudes, even close to the sea level, and the active layer varies from 40 cm to about 1 m in depth (López-Martínez & Serrano 2002; Resck 2011), but there is no systematic trend. Entisols occur in areas without permafrost. Soils are incipient with low contrast between horizons. Although some soils have vitric character, poorly ordered minerals, such as allophane, are not present in soils studied in Deception Island but did occur at Penguin Island.

Soils from some of the main, and most representative, ice-free areas in the Antarctic Peninsula Region have been studied and mapped on a gradient from 62° to 64°S, comprising the South Shetlands archipelago (King George, Deception and Livingston Islands among others), the northernmost tip of the Peninsula (Esperanza/Hope Bay) and Marambio/Seymour Island in the Weddell Sea. There is a sharp transition from warmer and wetter areas west of the peninsula dominated by Haploturbels in higher altitudes, Inceptisols and Entisols at lower areas to drier and colder coastal areas, east of the Peninsula dominated by Haplorthels and Anhyorthels.

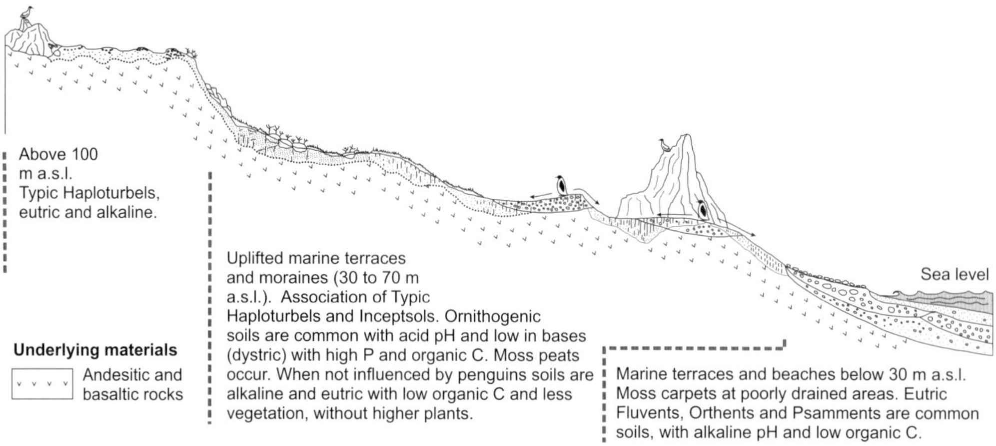

Fig. 6. Generalized cross-section showing the typical soil sequence observed on andesites and basalts in the western coast of Admiralty Bay, King George Island, where penguin activity is intense and ornithogenic soils occur.

Acid sulphate soils are common in King George Island, occurring in Keller (Simas *et al.* 2006, 2008), Barton Peninsula (Jeong & Yoon 2001) and Fildes Peninsula. Such soils have jarosite and ferrihidrite in the clay fraction and are invariably acid, loamy, with yellowish hues with much lower exchangeable Ca and Mg than soils on basalts and andesites not affected by sulphides (Simas *et al.* 2008).

Acid sulphate soils also occupy a large area of Seymour Island, in the Weddel Sea. The oxidation of sulphides present in the sediments results in the formation of acid, jarositic horizons. Differently from acid sulphate soils found in the South Shetland Archipelago, these are not turbic. Most pedons have a yellowish, oxidized, strongly acidic horizon in the upper part of the profile with a sharp transition to the underlying greyish, alkaline, horizon, which usually coincides with the dry permafrost.

Discussion

Soil properties: a comparison between the three regions studied

The overall summary of key soil properties (Table 2) illustrates the pattern across the three regions with an increase in soil development and thus the variety of Soil Orders occurring, from Gelisols in the Ross Sea Region to some Spodosols and Inceptisols in warmer areas of the Antarctic Peninsula Region.

Antarctic soils, though dispersed over a vast continent, have much in common. Gravelly sand soil textures and single grain structures predominate in all areas (Table 2), reflecting limited soil weathering, even in the relatively warm Northern Antarctic Peninsula Region. Likewise, the lack of weathering is reflected in generally lithochromic soil colours in mineral soils, except in the extremely old soil surfaces of the McMurdo Dry Valleys, where oxidation of mineral material is evident in redder soil colours, and in acid sulfate soils where the presence of jarosite and poorly crystalline iron minerals result in yellowish soils.

There is an overlap in the soils that occur in the three regions considered in this paper, with soils similar to those described in the East Antarctic Coastal areas occurring in the warmer more northerly parts of the Ross Sea Region and in the cooler areas in the Antarctic Peninsula Region. The concept of 'wet valleys', introduced in the description of the East Antarctic coastal areas, applies also to the similar environments in the Ross Sea Region and on the Antarctic Peninsula Region, and provides an interesting contrast with the well-known and described 'dry valleys' of the McMurdo Sound region.

Soil pH (Table 2) varies from predominantly alkaline in the Ross Sea Region through to strongly acid in peat, acid sulphate and ornithogenic soils in the Antarctic Peninsula region. Owing to the coarse nature of many soil materials, the total nutrient levels are not high, although nutrients are not generally limiting to plant growth, which is often more strongly restricted by moisture availability and temperature. In the small areas of ornithogenic soils, exceptionally high P levels occur, along with high levels of a range of elements such as arsenic, copper and zinc (Hofstee *et al.* 2006; Simas *et al.* 2008).

Permafrost distribution ranges from continuous at higher latitudes and altitudes through to absent or sporadic near sea level in northern parts of the

Antarctic Peninsula Region, with corresponding changes in active layer depth.

Relationship between soil-forming factors and properties reported

At the continent-wide scale, captured in the examples in this paper, out of all of the soil-forming factors (Jenny 1941), climate has the greatest impact with regard to driving the differences between soils. There is an overall reduction in cryoturbation, active-layer depth, chemical weathering, ice content in permafrost, vegetation development and soil organic C content as the climate becomes drier and colder at higher altitudes and higher latitudes, and in areas with precipitation gradients such as the colder, drier, eastern side of the Antarctic Peninsula and areas inland from the coast in the Ross Sea Region. There is also an interaction between climate and other soil-forming factors, particularly organisms and topography. Where climate warms and becomes wet enough to support vegetation, the accumulation of organic matter can dominate soil properties. Within similar climatic contexts, over smaller distances, other soil-forming factors such as topography, the influence of organisms, particularly birds, and parent material determine the differences in soil types.

The cold Antarctic environment means that weathering processes are slow and in the coastal and Antarctic Peninsula regions most surfaces are relatively young, thus time is not a strong factor in influencing differences between soils. However, the impact of time is important in the McMurdo Dry Valleys where some extremely (millions of years) old surfaces exhibit increased weathering and salt accumulation in spite of the exceptionally cold, dry, environment that slows weathering processes. The effects of time are evident in small areas of particularly young soils, with very little soil development, in the sequence illustrated from the Ong Valley (Fig. 4) and on active surfaces in coastal and Antarctic Peninsula regions.

The influence of organisms is limited to a small portion of the overall area. However, where bird colonies provide accumulation of guano, or where warmer, wet environments are conducive to accumulation of organic matter, the impacts of organisms dominate the soil parent material and soil properties. However mineralogical differences in parent materials tend to be subordinate to the other drivers of soil processes, and generally do not impact on the soil classifications at the Group or Order levels, although they do influence soil colours as a result of the lithochromic nature of most Antarctic soils. Soil development is relatively greater on the more easily weathered till and colluvial materials compared with soils formed from bedrock which often remain within the 'Lithic' subgroups.

Topography is regionally and locally important, as exemplified by Figures 2–6. Altitude, slope and aspect all have a strong influence on the soil microclimate and snow accumulation, and thus soil moisture availability, which drives many other soil processes.

Overall the soil-forming factor model of Jenny (1941) is a useful explanatory tool for understanding the relationships between the varying soils of the relatively uniformly extreme Antarctic environment, at scales ranging from continental to local. However, the *predictive* ability of the soil-forming factor approach has yet to be quantified in Antarctica. Soils, ultimately, are the result of pedogenic processes, not factors. In a maturing understanding of Antarctic soil morphology, patterns and properties, the value of theory such as soil evolution, progressive and regressive pedogenesis (Johnson & Watson-Stegner 1987), upbuilding pedogenesis (Nikiforoff 1949), pedogenic thresholds (Muhs 1984), and inheritance and contingency (Phillips 2001) needs to be evaluated. The complexities about which these theoretical paradigms have arisen are not captured by analysis of soil-forming factors at any scale; nor are they adequately captured in soil classification systems. While mapping soil subgroups of Soil Taxonomy (Soil Survey Staff 2010) achieves one expression of the soil pattern, much of fundamental and ecological significance remains to be understood about the variability of soils in Antarctica.

Soil classification using the USDA Soil Taxonomy

The Gelisol Order of Soil Taxonomy (Soil Survey Staff 2010) is proving useful in its overall concepts, and is generally clear and practical to apply. As the soils are studied, and the classification is applied, in more detail, some shortcomings have become apparent.

There are a number of instances where soils of widely differing characteristics get grouped together under 'Typic' subgroups owing to a lack of suitable categories to capture their often unique properties. There is a need to add an 'Oxyaquic' subgroup to Turbels and Orthels to account for soils that have running water passing over and through them for parts of the year at the foot of slopes and on valley floors. Both Hofstee (2006) and Simas *et al.* (2008) have proposed inclusion of an Ornithic subgroup to highlight the distinctive and unique properties of soils formed under penguin colonies. Guano and other organic materials such as dead birds and egg shell, along with the small stones that penguins use for nest building, often accumulate to depths of at least 0.5 m and dominate the soil properties. Within the Anhyorthels both Bockheim (1997) and McLeod (2012) proposed the inclusion of a Petrosalic subgroup to capture soils with salt layers that are so strongly developed as to be cemented.

In the current USDA Soil Taxonomy soil classification system, all of the acid sulphate soils key out as Typic Haploturbels. Therefore, we suggest the inclusion of a Sulfuric Haploturbel subgroup in Soil Taxonomy and a Thionic prefix qualifier for the Cryosols group in the World Reference Base system to properly classify these special soils.

Owing to the dry and cold climate, most acid sulphate soils on Seymore Island do not show evidence of cryoturbation, being keyed out in one of the following subgroups, according to the Soil Taxonomy (SSS 2010): Salic Aquorthels, Salic Psammorthel, Typic Psammorthel and Salic Anhyorthel. These correspond to the Oxyaquic–Salic Cryosol, Arenic–Salic Cryosol and Aridic–Salic Cryosols. Here the inclusion of the Sulphuric (or Thionic) and the Ornithic or Ornithogenic qualifiers in the Soil Taxonomy and World Reference Base classification systems would enable a better separation of the soils found in Seymour Island.

According to the 2010 version of Soil Taxonomy 'in areas where soil has thin cemented horizons that are impermeable to roots, the soil [that we classify] extends as deep as the deepest cemented horizon, but not below 200 cm' and 'soils in [Gelic] temperature regime have a mean annual soil temperature at or below 0 °C (in Gelic suborders and Gelic great groups) or 1 °C or lower (in Gelisols) either at a depth of 50 cm below the soil surface or at a densic, lithic, or paralithic contact, whichever is shallower' (Soil Survey Staff 2010). Therefore, according to these criteria, soils with mean annual temperature <0 °C, but without permafrost within the soil profile (depth <2 m or on lithic (or paralithic) contact) should not be classified as Gelisols, but as soils of different orders with a Gelic suborder or Gelic great group. Thus many shallow soils that were previously classified as 'Gelisols' do not fit the current definition and thus key out as Entisols. In an environment that generally supports Gelisols it seems somewhat arbitrary for shallow soils to be separated out as Entisols, as in all other ways they are essentially similar to their Gelisol neighbours. There is a need for further consideration of the effects of this arbitrary distinction between Gelisols and Entisols within Soil Taxonomy.

This research has been carried out thanks to research projects funded by Antarctic Programmes from Brazil, New Zealand, Spain and Russia. The paper has been undertaken within the ANTPAS project supported by the Scientific Committee on Antarctic Research and the International Permafrost Association.

References

ADLAM, L. S., BALKS, M. R., SEYBOLD, C. A. & CAMPBELL, D. I. 2010. Temporal and spatial variation in active layer depth in the McMurdo Sound Region, Antarctica. *Antarctic Science*, **22**, 45–52.

AISLABIE, J., BOCKHEIM, J., MCLEOD, M., HUNTER, D., STEVENSON, B. & BARKER, G. M. 2012. Microbial biomass and community structure changes along a soil development chronosequence near Lake Wellman, southern Victoria Land. *Antarctic Science*, **24**, 154–164, doi: 10.1017/S0954102011000873

ALLEN, S. E. & HEAL, O. W. 1970. Soils of the maritime Antarctic zone. *In*: HOLDGATE, M. W. (ed.) *Antarctic Ecology*, **2**. Academic Press, London, 693–696.

BEYER, L. 2000. Properties, formation and geological significance of organic soils in the coastal regions of East Antarctica (Wilkes Land). *Catena*, **39**, 79–93.

BEYER, L. & BÖLTER, M. 1999. Formation, ecology and geography of cryosols of an ice-free oasis in coastal East Antarctica near Casey station (Wilkes Land). *Australian Journal of Soils Research*, **39**, 209–244.

BEYER, L. & BÖLTER, M. 2000. Chemical and biological properties, formation, occurrence and classification of Spodic Cryosols in a terrestrial ecosystem of East Antarctica, Wilkes Land. *Catena*, **39**, 95–119.

BEYER, L., BOCKHEIM, J. G., CAMPBELL, I. B. & CLARIDGE, G. G. C. 1999. Review. Genesis, properties and sensitivity of Antarctic Gelisols. *Antarctic Science*, **11**, 387–398.

BEYER, L., PINGPANK, K., WREIDEL, G. & BÖLTER, M. 2000. Soil formation in coastal continental Antarctica (Wilkes Land). *Geoderma*, **95**, 283–304.

BLUME, H. P., BÖLTER, M. 1992. Soils of Casey Station, Wilkes Land, Antarctica. *Proceedings of the 1st International Conference on Cryopedology Russian Academy of Sciences (Puschino)*, 96–103.

BLUME, H. P., KUHN, D. & BÖLTER, M. 2001. Soils and soilscapes. *In*: BEYER, L. & BÖLTER, M. (eds) *Geoecology of Antarctic Ice-Free Coastal Landscapes*. Springer, Berlin, 92–113.

BOCKHEIM, J. G. 1978. Relative age and origin of soils of the eastern Wright Valley, Antarctica. *Soil Science*, **128**, 142–152.

BOCKHEIM, J. G. 1997. Properties and classification of cold desert soils from Antarctica. *Soil Science Society of America Journal*, **61**, 224–231.

BOCKHEIM, J. G. 2002. Landform and soil development in the McMurdo Dry Valleys, Antarctica: a regional synthesis. *Arctic, Antarctic and Alpine Research*, **34**, 308–317.

BOCKHEIM, J. G., CAMPBELL, I. B., GUGLIELMIN, M. & LÓPEZ-MARTÍNEZ, J. 2008. Distribution of permafrost and buried ice in ice-free areas of Antarctica. *In*: KANE, D. L. & HINDEL, K. M. (eds) *Ninth International Conference on Permafrost*. Fairbanks, AK, Institute of Northern Engineering, University of Alaska, Fairbanks, **1**, 125–130.

BRITISH ANTARCTIC SURVEY 2005. *Antarctic Factsheet Geographical Statistics*. British Antarctic Survey, Cambridge, http://www.antarctica.ac.uk/about_antarctica/teacher_resources/resources/fact sheets/factsheet_geostats_screen.pdf

CAMPBELL, I. B. & CLARIDGE, G. G. C. 1987. *Antarctica: Soils, Weathering Processes and Environment*. Elsevier Science, New York.

CORTE, A. & SOMOZA, A. L. 1954. *Algunas observaciones geológicas y criopedológicas en la Antártida*. Instituto Antártico Argentino, Buenos Aires, **4**.

EVERETT, K. R. 1976. *A survey of the soils in the region of the South Shetland Islands and adjacent parts of the Antarctic Peninsula*. Institute of Polar Studies, Report 58. Ohio State University, Columbus, OH.

FOX, A. J. & COOPER, P. R. 1994. Measured properties of the Antarctic Ice Sheet derived from the SCAR digital database. *Polar Record*, **30**, 201.

FRANCELINO, M. R., SCHAEFER, C. E. G. R., SIMAS, F. N. B. E. I., FERNANDES, F., DE SOUZA, J. J. L. & DA COSTA, L. L. M. 2011. Geomorphology and soils distribution under paraglacial conditions in an ice-free area of Admiralty Bay, King George Island, Antarctica. *Catena*, **85**, 194–204.

GILICHINSKY, D., ABAKUMOV, E. ET AL. 2010. Soils of Mid and Low Antarctic: diversity, geography, temperature regime. *In*: GILKES, R. J. & PRAKONGKEP, N. (eds) *Proceedings of the 19th World Congress of Soil Science; Soil Solutions for a Changing World*. 1–6 August 2010, Brisbane. IUSS, published on DVD; Symposium WG 1.4; Cold soils in a changing world, 32–35, http://www.iuss.org

GLAZOVSKAYA, M. A. 1958. Vyvetrivanie i pervichnoe pochvoobrazovanie v Antarktide (Weathering and primary soil formation in Antarctica). *Nauchnie Dokladi Vishei Shkoli, Geologo-Geograficheskie Nauki, Moskovskii Universitet*, **1**, 63–76 (in Russian).

GORYACHKIN, S. V., BLUME, H. P. ET AL. 2004. Similarities and Differences in Arctic and Antarctic Soil Zones. *In*: KIMBLE, J. (ed.) *Cryosols. Permafrost-Affected Soils*. Springer, Berlin, 49–70.

HOFSTEE, E. H. 2006. *Soils and Hydrology of Seabee Hook, Cape Hallett, Antarctica*. MSc thesis, University of Waikato, New Zealand.

HOFSTEE, E. H., BALKS, M. R., PETCHEY, F. & CAMPBELL, D. I. 2006. Soils of Seabee Hook, Cape Hallett, Northern Victoria Land, Antarctica. *Antarctic Science*, **18**, 473–486.

HOLDGATE, M. W., ALLEN, S. E. & CHAMBERS, M. J. G. 1967. A preliminary investigation of the soils of Signy Island, South Orkney. *Islands British Antarctic Survey Bulletin*, **12**, 53–71.

INTERNATIONAL UNION OF SOIL SCIENCES 2006. *World Reference Base for Soil Resources 2006*. World Soil Resources Report No. 103. FAO, Rome.

JENNY, H. 1941. *Factors of Soil Formation. A System of Quantitative Pedology*. McGraw-Hill, New York.

JEONG, G. Y. & YOON, H. I. 2001. The origin of clay minerals in soils of King George Island, South Shetlands Islands, West Antarctica and its implication for the clay-mineral composition of marine sediments. *Journal of Sedimentary Research*, **71**, 833–842.

JOHNSON, D. L. & WATSON-STEGNER, D. 1987. Evolution model of pedogenesis. *Soil Science*, **143**, 349–366.

LÓPEZ-MARTÍNEZ, J. & SERRANO, E. 2002. Geomorphology of Deception Island. *In*: LÓPEZ-MARTÍNEZ, J., SMELLIE, J. L., THOMSON, J. W. & THOMSON, M. R. A. (eds) *Geology and Geomorphology of Deception Island*. BAS Geomap Series, Sheets 6-A and 6-B. British Antarctic Survey, Cambridge, 31–39.

LÓPEZ-MARTÍNEZ, J., MARTÍNEZ DE PISÓN, E., SERRANO, E. & ARCHE, A. 1996. *Geomorphological Map of Byers Peninsula, Livingston Island*. BAS Geomap Series, Sheet 5-A, 1:25 000. British Antarctic Survey, Cambridge.

LÓPEZ-MARTÍNEZ, J., SERRANO, E., REY, J. & SMELLIE, J. L. 2000. *Geomorphological Map of Deception Island*. BAS Geomap Series, Sheet 6-B, 1:25 000. British Antarctic Survey, Cambridge.

LÓPEZ-MARTÍNEZ, J., SERRANO, E. & LEE, J. I. 2002. *Geomorphological Map of Barton and Weaver Peninsulas. King George Island, Antarctica (1:10 000)*. Polar Science Laboratory, Korean Ocean Research and Development Institute KARP, Seoul.

LÓPEZ-MARTÍNEZ, J., SERRANO, E., SCHMID, T., MINK, S. & LINÉS, C. 2012. Periglacial processes and landforms in the South Shetland Islands (Northern Antarctic Peninsula region). *Geomorphology*, **155–156**, 62–79. doi: 10.1016/j.geomorph.2011.12.018

MACNAMARA, E. E. 1969*a*. Active layer development and soil moisture dynamics in Enderby Land, Antarctica. *Soil Science*, **108**, 345–349.

MACNAMARA, E. E. 1969*b*. Soils and geomorphic surfaces in Antarctica. *Biuletyn Peryglacialny*, **200**, 299–320.

MCCRAW, J. D. 1967. Soils of the Taylor Dry Valley, Victoria Land, Antarctica, with notes on soils from other localities in Victoria Land. *New Zealand Journal of Geology and Geophysics*, **10**, 498–539.

MCLEOD, M. 2012. *Soil and Permafrost Distribution, Soil Characterisation, and Soil Vulnerability to Human Foot Trampling, Wright Valley, Antarctica*. PhD thesis, University of Waikato, New Zealand.

MCLEOD, M., BOCKHEIM, J. G. & BALKS, M. R. 2008. Glacial geomorphology, soil development and permafrost features in central-upper Wright Valley, Antarctica. *Geoderma*, **144**, 93–103.

MCLEOD, M., BOCKHEIM, J., BALKS, M. R. & AISLABIE, J. 2009. Soils of Western Wright Valley, Antarctica. *Antarctic Science*, **21**, 355–365.

MERGELOV, N. 2011. Soils and soil-like bodies of the wet valleys in East Antarctica (Larsemann Hills Oasis). *11th International Symposium on Antarctic Earth Sciences*, 10–15 July 2011, Edinburgh, Abstracts, 381.

MERGELOV, N. S., GORYACHKIN, S. V., SHORKUNOV, I. G., ZAZOVSKAYA, E. P. & CHERKINSKY, A. E. 2012. Endolitic pedogenesis and rock varnish on massive crystalline rocks in East Antarctica. *Eurasian Soil Science*, **45**, 901–917.

MICHEL, R. F. M., SCHAEFER, C. E. G. R., DIAS, L. E., SIMAS, F. N. B. & SÁ MENDONÇA, E. 2006. Ornithogenic Gelisols (Cryosols) from maritime Antarctica: pedogenesis, vegetation, and carbon studies. *Soil Science Society of America Journal*, **70**, 1370–1376.

MOURA, P. A., FRANCELINO, M. R., SCHAEFER, C. E. G. R., SIMAS, F. N. B. & DE MENDONCA, B. A. F. 2012. Distribution and characterization of soils and landform relationship in Byers Peninsula, Livingston Island,

Maritime Antarctica. *Geomorphology*, **155**, 45–54, doi: 10.1016/j.geomorph.2011.12.011

MUHS, D. R. 1984. Intrinsic thresholds in soil systems. *Physical Geography*, **5**, 99–110.

NAVAS, A., SOTO, J. & LÓPEZ-MARTÍNEZ, J. 2005. Radionucleids in soils of Byers Peninsula Byers Peninsula, South Shetland Islands, Western Antarctica. *Applied Radiation and Isotopes*, **62**, 809–816.

NAVAS, A., LÓPEZ-MARTÍNEZ, J. *ET AL.* 2008. Soil characteristics on varying lithological substrates in the South Shetland Islands, maritime Antarctica. *Geoderma*, **144**, 123–139.

NIKIFOROFF, C. C. 1949. Weathering and soil evolution. *Soil Science*, **67**, 219–230.

PHILLIPS, J. D. 2001. Contingency and generalization in pedology, as exemplified by texture-contrast soils. *Geoderma*, **102**, 347–370.

RESCK, B. 2011. *Chemistry and mineralogy of volcanic soils from Decepcion and Penguin Islands, Maritime Antarctica*. MSc thesis, Universidade Federal de Viçosa, in Portuguese.

SCHAEFER, C. E. G. R., SIMAS, F. N. B., GILKES, R. J., MATHISON, C., COSTA, L. M. & ALBUQUERQUE, M. A. 2008. Micromorphology and microchemistry of selected Cryosols from Maritime Antarctica. *Geoderma*, **144**, 104–115.

SERRANO, E., MARTÍNEZ DE PISÓN, E., & LÓPEZ-MARTÍNEZ, J. 1996. Periglacial and nival landforms and deposits. *In*: LÓPEZ-MARTÍNEZ, J., THOMSON, M. R. A. & THOMSON, J. W. (eds) *Geomorphological Map of Byers Peninsula, Livingston Island*. BAS GEOMAP Series, Sheet 5-A, 1:25 000, with Supplementary Text. British Antarctic Survey, Cambridge, 28–34.

SERRANO, E., LÓPEZ-MARTÍNEZ, J., CUCHÍ, J. A., DURÁN, J. J., MINK, S. & NAVAS, A. 2008. Permafrost in the South Shetland Islands (Maritime Antarctica): spatial distribution pattern. *In*: KANE, D. L. & HINDEL, K. M. (eds) *Ninth International Conference on Permafrost, Fairbanks, AK*. Institute of Northern Engineering, University of Alaska, Fairbanks, 1621–1625.

SEYBOLD, C. A., BALKS, M. R. & HARMS, D. S. 2010. Characterization of active layer water contents in the McMurdo Sound region, Antarctica. *Antarctic Science*, **22**, 633–645. doi: 10.1017/S0954102010000696

SIMAS, F. N. B., SCHAEFER, C. E. G. R., DE MELO, V. F., GUERRA, M. B. B., SAUNDERS, M. & GILKES, R. J. 2006. Clay-sized minerals in permafrost-affected soils (Cryosols) from King George Island, Antarctica. *Clays and Clay Minerals*, **54**, 721–736.

SIMAS, F. N. B., SCHAEFER, C. E. G. R. *ET AL.* 2007. Ornithogenic Cryosols from Maritime Antarctica: phosphatization as a soil forming process. *Geoderma*, **38**, 191–203.

SIMAS, F. N. B., SCHAEFER, C. E. G. R., ALBUQUERQUE FILHO, M. R., FRANCELINO, M. R., FERNANDES FILHO, E. I. & COSTA, L. M. 2008. Genesis, properties and classification of Cryosols from Admiralty Bay, maritime Antarctica. *Geoderma*, **144**, 116–122.

SOIL SURVEY STAFF 2010. *Keys to Soil Taxonomy*, 11th edn. USDA–Natural Resources Conservation Service, Washington, DC.

SUGDEN, D. E., BENTLEY, M. J. & COFAIGH, C. O. 2006. Geological and geomorphological insights into Antarctic ice sheet evolution. *Philosophical Transactions of the Royal Society*, **A364**, 1607–1625. doi: 10.1098/rsta.2006.1791

VIEIRA, G., BOCKHEIM, J. *ET AL.*. 2010. Thermal state of permafrost and active-layer monitoring in the Antarctic: advances during the International Polar Year 2007–09. *Permafrost and Periglacial Processes*, **21**, 182–197.

Sedimentary processes in two different polar periglacial environments: Examples from Schirmacher Oasis and Larsemann Hills, East Antarctica

RAJESH ASTHANA[1]*, PRAKASH K. SHRIVASTAVA[1], M. JAVED BEG[2], ASHIT K. SWAIN[1], AMIT DHARWADKAR[1], SANDIP K. ROY[1] & HARI B. SRIVASTAVA[3]

[1]*Antarctica Division, Geological Survey of India, Faridabad-121001, India*

[2]*National Centre for Antarctic and Ocean Research, Goa-403804, India*

[3]*Department of Geology, Banaras Hindu University, Varanasi-221005, India*

**Corresponding author (e-mail: rajeshasthana.antarctica@gmail.com)*

Abstract: Schirmacher Oasis and Bharati Promontory in Larsemann Hills of East Antarctica are currently ice-free coastal areas exposed in physiographically different polar periglacial environments. Schirmacher Oasis is bound by the presence of a vast stretch of ice shelf in the north and the polar ice sheet in the south. It exhibits well-developed patterned ground, abundant till deposition, block-fields, episodic development of curvilinear morainic ridges, extensive outwash plains and erratics. In contrast, the occurrence of such depositional features is sparse on Bharati Promontory, where landmass is directly in contact with the ocean. The sedimentary processes vary in their magnitude owing to different physiographic settings in these two geographically separated locations. Scanning electron microscopy of quartz grains shows subsequent reworking under glaciofluvial environment and final deposition of material in glacially scoured basins. The fluvial action is more pronounced in the Schirmacher Oasis than on Bharati Promontory. The transport of sediment by polar ice is mainly through englacial pathways with a minor contribution from the supraglacial component. Discharge of all sizes of sediments ranging from large boulders to glacially abraded rock-flour in varying proportions takes place at the ice–bedrock interface. Granulometric analysis also shows turbulence of transporting media and reworking of sediments before final deposition.

Sedimentological parameters, such as grain-size distribution and surface textures, are important tools to decipher the processes involved in various depositional elements of periglacial environments. As such, the Antarctic continent is regarded as a cold desert (Heywood 1984) with about 98% of the area covered by a continental ice sheet (up to 4000 m thick) and the remaining area comprising interior mountains and coastal landmasses. There are relatively few deglaciated regions in Antarctica like Schirmacher Oasis (Fig. 1a) and Bharati Promontory of Larsemann Hills (Fig. 1b), which host a repository of sedimentary depositional features of agencies like glaciers, seasonally active streams and wind under periglacial environments that have been active in the geological past. The glaciation and deglaciation of Antarctica play an important role in global climatic history as well as leaving behind signatures of erosional and depositional processes that have been operational in periglacial environments. A considerable amount of work on glacial sediments has been undertaken in the Vestfold Hills and East Antarctic coastal oases on paraglacial sedimentary processes, as well as to develop models for moraine formation (Fitzsimons 1996, 1997). Different aspects of sediment characteristics from Schirmacher Oasis have been described by Asthana & Chaturvedi (1998), who worked on grain-size behaviour and morphoscopy of modern supraglacial sediments. Shrivastava *et al.* (2009) have worked on climatic fluctuation imprinted on quartz grains from lake sediments in both regions. Asthana *et al.* (2009) highlighted surface microtextures of quartz grains from glaciolacustrine sediment core from Schirmacher Oasis.

Bharati Promontory enjoys a polar lowland periglacial climate with relatively lower amounts of precipitation than Schirmacher Oasis, restricted period of snow melting, and low rates of ice melt, sediment transport and accumulation. Meltwater from ice and snow forms the primary control on the type and rate of depositional processes. Unconsolidated glacial deposits are negligible and most of the surficial deposits occur as thin mantles, erratic boulders, isolated patches of glacial deposits and valley-fills. Debris-mantles, as surficial sediment cover on the relatively stable positions of flatter slopes and valley portions, are the only signatures

From: Hambrey, M. J., Barker, P. F., Barrett, P. J., Bowman, V., Davies, B., Smellie, J. L. & Tranter, M. (eds) 2013. *Antarctic Palaeoenvironments and Earth-Surface Processes*. Geological Society, London, Special Publications, **381**, 411–427. First published online July 30, 2013, http://dx.doi.org/10.1144/SP381.22

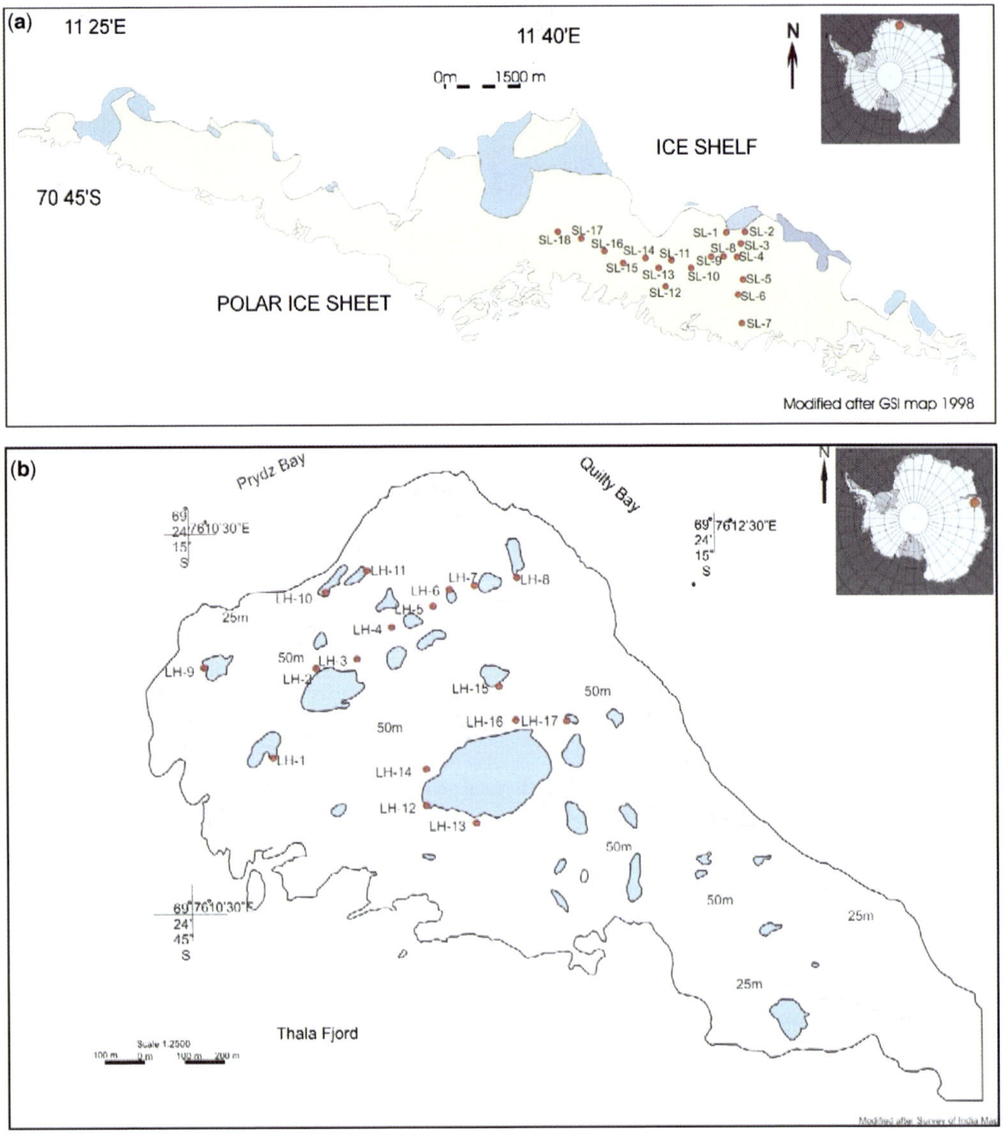

Fig. 1. (**a**) Sample locations in Schirmacher Oasis, East Antarctica. (**b**) Sample locations in Schirmacher Oasis, East Antarctica.

of glacially deposited debris. Sinuous bands of debris adjacent to the margin of the icesheet are also found to occur in a few places. Reworked and wind-blown finer sediments are ultimately deposited within the low-lying lake basins under lacustrine conditions by seasonal glacial meltwater. Multiproxy palynological, palaeontological, geochemical and sedimentological data obtained from lake-sediment cores are important tools for climate change studies as they contain a reliable organic chronometer as well as potentially useful luminescence dating materials (Krause *et al.* 1997). Gillieson (1991) and Hodgson *et al.* (2005) carried out environmental studies with the help of freshwater lake sediments. Verleyen *et al.* (2004) constructed the deglaciation and climate history for late Quaternary time.

Sedimentary processes in perennially frozen ground in the periglacial zone are influenced by the -thaw oscillations, and give rise to Felsenmeer (block fields) and patterned ground. The scarcity of porous glacial detritus at Larsemann Hills

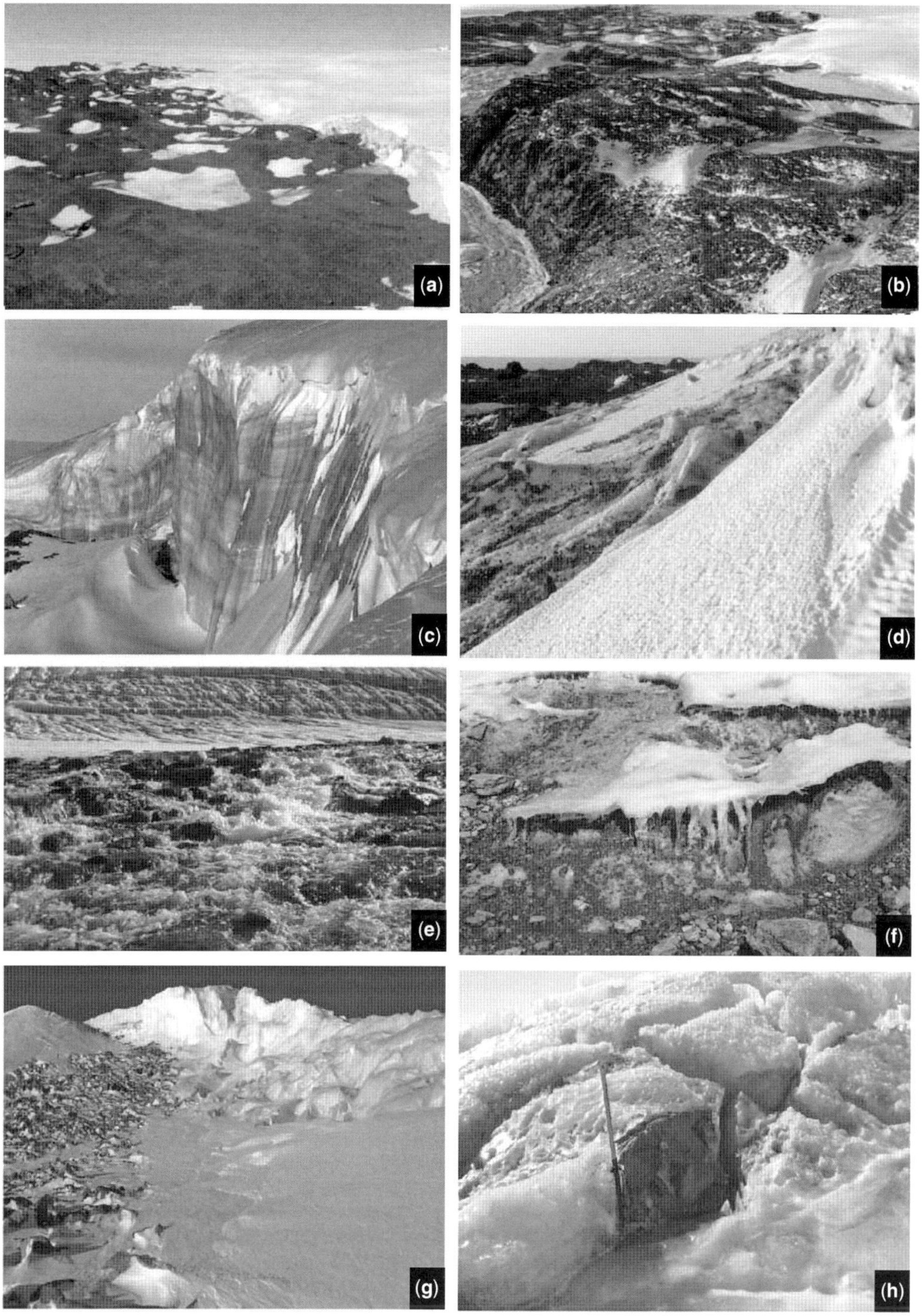

Fig. 2. (**a**) The central part of Schirmacher Oasis exhibiting 'roche moutonnée' landscape with overriding polar ice sheet; (**b**) the northern escarpment; (**c**) fine-grained entrainment of glacial sediments within the polar ice; (**d**, **f** and **g**) sediment discharge of englacial material from polar ice margin; (**e**) meltwater discharge carrying fine-grained sediments; (**h**) lensoid body of thinly laminated sandy silt at the polar ice margin.

Fig. 3. (**a**) Glacial terrace comprising heterogeneous moraines with sediment particle-size ranging from boulders to silt; (**b**) curvilinear recessional moraine ridge; (**c**) recessional lake margins marked by sand accumulation in the vicinity of blockfields (Felsenmeer); (**d**) patterned ground; (**e**) pressure ridges and push-moraines at the northern margins of

has resulted in poor development of such features. Till-covered areas and moraines, as well as irregular surfaces, signify accumulation of glacial debris from the ice sheet following retreat. These features are abundant in Schirmacher Oasis, whereas they are very sparse in Larsemann Hills. The complexity of genesis, sediment-entrainment and quantitative modes of sediment transport is still poorly documented in these two areas. Qualitative field observations in these two Antarctic Oases are made here, with a view to gaining insight into various sedimentary processes that involve glacial, glaciofluvial and aeolian processes.

Geomorphology and geology of field areas

Schirmacher Oasis, covering approximately 35 km^2, occurs in central Dronning Maud Land, East Antarctica (70°5′S; 11°40′E) and represents an exhumed topography dominated by decametric-scale undulating ridges. The elevation ranges from local sea-level in its northern periphery to a maximum of 212 m above mean sea-level in the central part, with an average elevation of about 90 m. The southern part is characterized by a morainic ridge and overriding polar ice sheet (Fig. 2a). The northern margin of Schirmacher Oasis usually forms a steep escarpment followed by epishelf lakes with immature sediments (Shrivastava *et al.* 2012) and ice shelf extending for about 80 km to the open sea (Fig. 2b). The entrainment of fine-grained glacial sediments within the polar ice has been a common feature (Fig. 2c). The englacial material comprises fine-grained sediments, discharged from the polar ice margin (Fig. 2d, e–g) by meltwater. Lensoid bodies of thinly laminated sandy silt commonly occur at the polar ice margin (Fig. 2h). The Oasis consists of more than 100 lakes of varying dimensions that owe their origin to the recession of inland ice. These lakes are recharged from ephemeral meltwater streams originating from the continental ice margin (Ravindra *et al.* 2002). Extensively developed outwash plains, thinly laminated lensoid bodies of fine sands (Fig. 2h), occur in Schirmacher Oasis. The geomorphology of Schirmacher Oasis is further characterized by structurally controlled hills, a glacial terrace (Fig. 3a), curvilinear ridges (Fig. 3b), recessional lakes (Fig. 3c) and patterned ground (Fig. 3d). Pressure ridges and push moraines (Fig. 3e) are some other depositional features observed in the Schirmacher region. Well-rounded, huge erratic boulders some more than 3 m in diameter to finer-grained sediments commonly occur on the ridge-tops throughout the area (Fig. 3f). Pingo (Fig. 3g), deglaciated broad U-shaped hanging and tunnel valleys, glacially polished and striated bedrock hummocks (roches moutonnées) and scoured surfaces, as well as glacial terraces, and polished, striated and grooved surfaces are some other features observed in this region. The pockets of layered clay exhibit penecontemporaneous deformation (Fig. 3h). The glacial deposits and recessional morainic ridges do not occur as laterally persistent and complete sequences, but as moraine dumps, or in the form of linear or curvilinear ridges with heights of 5–8 m and lengths of 20 to over 100 m. Although morainic material is strewn all over the Oasis, the occurrence of prominent morainic ridges is sparse. Such ridges occur in the central and western parts adjacent to epishelf lakes. These ridges are devoid of any internal stratification, and consist of randomly oriented clasts embedded in finer matrix. Ice-cored moraines occur as small ridges or small mounds having sérac-faces and kettle lakes, as well as pressure ridges in the northern parts. In Schirmacher Oasis, the most extensive basement lithology is an amphibolite-facies biotite-, hornblende- and garnet-bearing quartzofeldspathic gneiss (Ravikant & Kundu 1998).

The Larsemann Hills area is a transitional zone between marine and glacial ecosystems, and attempts to reconstruct the palaeoenvironment reveal a heterogeneous and complex history (Burgess *et al.* 1994, 1997; Gillieson 1991; Verleyen *et al.* 2004). This area is an ice-free polar oasis on the Ingrid Christensen Coast, Princess Elizabeth Land and is located approximately midway between the eastern extremity of the Amery Ice Shelf and the southern boundary of the Vestfold Hills. These reconstructions vary with regard to the evolutionary sequence of Larsemann Hills during the Last Glacial Maximum. The area was either wholly or partly ice-free during the Last Glacial Maximum (Burgess *et al.* 1994, 1997) or completely covered by 200–500 m-thick continental ice (Gillieson 1991). The available ages for deglaciation also vary from 9.4 ka BP to as much as 25 ka BP. Proxy records from some of the lakes indicate that biogenic sedimentation set in shortly after deglaciation and has remained unaffected by glacier advances since then (Verleyen *et al.* 2004). Lakes in the Larsemann Hills became seasonally ice-free after 10 500 cal BP (Hodgson *et al.* 2005). The Larsemann Hills area consists of gently

Fig. 3. (*Continued*) Schirmacher Oasis; (**f**) well-rounded boulders, cobbles and pebbles; (**g**) a pingo or hydrolaccolith (mound of earth-covered ice); (**h**) penecontemporaneously deformed sediments occurring near an outwash plain in northern part of Schirmacher Oasis.

Fig. 4. (**a–c**) Glacially eroded ridges and valleys producing undulatory topography showing polar ice sheet, intervening fjords and structurally controlled lineaments; (**d**) rounded, glacially polished granitic outcrops with a thin veneer of

rolling hills and broad valleys (Fig. 4a, b), and an open ocean in the north (Fig. 4c–h). Many valley floors are filled with multilayer ice formed from precipitation, but those that are ice-free expose various landforms such as glacial polish and striae (Fig. 4d). Glacially carved ridges and depressions with negligible amount of glacial drift but redistribution of available sediments by glaciofluvial action can be noticed (Fig. 4e, f). Exposed ridges (Fig. 4g), glacially abraded flat-topped hills (Fig. 4h), sparse morainic sediments and erratics (Fig. 5a), and glaciofluvial material (Fig. 5b, c) depict a complex geomorphological evolution. Wind-blown fine sand forming supraglacial sediment pockets (Fig. 5d) is a characteristic sediment feature in this region. Redistribution of locally available surficial sediments within steep-sided valleys which harbour snow-fed shallow lakes (Fig. 5e–h), and small snow-fed land-locked ponds confined within bare bedrock with no sediment supply except by wind (Fig. 5g) indicate mixed action of snow and wind in sculpting the land.

Lithological and structural elements in bedrocks, including joints and lineaments, have a strong control on the geomorphology. The Bharati Promontory has been selected as a representative area in the Larsemann Hills region. The area is characterized by gently undulating topography, comprising low rounded hills, interspersed by lakes of varying dimensions and U-shaped glaciated valleys. Larsemann Hills is the second largest of only four major ice-free oases of Antarctica, having an area of approximately 50 km^2. This oasis consists of a number of scattered offshore islands having a maximum elevation of 180 m above sea-level. These islands are in direct contact with the ocean. Over 150 freshwater lakes occur in the Larsemann Hills (Gillieson 1991), ranging from small ephemeral ponds to large water bodies. The bedrock geology of Larsemann Hills consists mainly of medium- to coarse-grained granitoids with alkali feldspar as porphyroclasts, which are exposed on the southeastern part of the promontory. These granitoid rocks are traversed by thin aplitic veins. Garnetiferous granite gneisses are predominantly pink and are exposed in the eastern part of the area, with biotites defining the foliation planes sympathetic to the general trend of NE–SW. The quartz–magnetite–biotite-bearing gneisses occupy higher grounds, which are characterized by evenly distributed small magnetite crystals and absence of garnet. However, in places small garnets are found in pockets. These gneisses are traversed by pegmatitic and aplitic veins of granitic composition that have huge, but partly digested, enclaves of older gneisses. Garnet-bearing quartz–feldspar–biotite gneiss is the predominant lithological unit exposed in the northwestern part.

Methods

Qualitative field observations in these two periglacial environments have been made that have a bearing on past sedimentary processes. The sediment samples were collected during Austral summers of 2004–2005 and 2006–2007. All particles larger than 3 mm were removed for further treatment. After coning and quartering, about 100 g of representative samples were taken for mechanical sieving. These samples were treated with acetic acid and H_2O_2 to remove carbonate and organic matter, respectively. After treatment, sieve analysis of these sediments was undertaken using a Retsch AS-200 Automated Sieve Shaker, using different sieves with a 0.5 Φ interval to examine the grain-size distribution pattern. A total of 30 sand-sized grains of quartz were selected for scanning electron microscopy (SEM) from the sediments of Schirmacher Oasis and Larsemann Hills. The quartz grains were first mounted on specially designed aluminium stubs and then coated with 150 Å gold palladium film. They were then studied under an SEM LEO 440 with 94 energy dispersive X-ray analyser.

Results

The depositional landforms are the product of erosional and entrainment processes that deposit material at various favourable locations and evolve over time. The polar ice sheet acts like a conveyor belt carrying sediment loads in supraglacial, englacial (Fig. 2c, d, f, g) and subglacial environments and ultimately in the periglacial zone. The temperature gradient within the ice mass also plays an important role in sedimentation. The temperature is colder at the surface and warmest at the base, where it reaches melting point (Chaturvedi & Singh 1999), generating meltwater and discharging its sediment load (Fig. 2e, f). The sediments were collected from two different parts of East Antarctica to decipher the sedimentary processes, with the help of field observations, grain-size distribution analysis and scanning electron microscopy. A complex interplay of the erosive action of

Fig. 4. (*Continued*) surficial sediment; (**e**, **f**) glacially carved ridges and depressions with negligible amount of glacial drift, but redistribution of available sediments by glaciofluvial action can be noticed; (**g**) exposed ridges devoid of any sediment cover; (**h**) flat-topped glacially abraded floor having inconsistently distributed fine sediments.

Fig. 5. (**a**) Erratics strewn over the ridge-tops; (**b**, **c**) glaciofluvial materials; (**d**) seasonally snow-filled small valley having wind-blown fine-grained sediment cover; (**e**, **f**, **h**) redistribution of locally available surficial sediments within the steep-sided valleys harbouring snow-fed shallow lakes; (**g**) small snow-fed land-locked ponds confined within bare bedrock with no sediment supply except by wind.

ice-mass, generating glacial detritus ranging from huge boulders to silt-clay grade rock-flour, distance of transportation, glacier as well as basin morphology and its hydrological characteristics all determine the final depositional product. Recessional moraines are composed of unconsolidated debris of -garnet gneissic rock fragments as a series of transverse ridges of varying dimensions (Fig. 3b) and glacial terraces (Fig. 3a).

Schirmacher Oasis

Recessional moraines forming small ridges occur at different topographic levels. The morainic ridges are made up of heterogeneous sediments with a range of particle sizes varying from boulders (up to approximately 70 cm in diameter) to clay. Sometimes the finer sediments are partly indurated as compact silty clay with <5% rock fragments and granules. However, inconsistently developed laminae within the finer sediments in confined places (Fig. 2h) and a section of partly indurated layered clay, showing penecontemporaneous deformation (Fig. 3h), have also been observed. Many lakes exhibit distinct recessional levels characterized by silty clay or sandy patches. Ratios of clasts to matrix vary from 30:70 to 20:80. Glacial boulders as well as some smaller clasts are well rounded (Fig. 3f), indicating longer distance of transport and deposition under a glaciofluvial regime. Periodic thawing of ice during the austral summer and accumulation of meltwater in topographic depressions has resulted in the formation of numerous lakes in Schirmacher Oasis. Some of these lakes often overflow over small distances to give rise to local lacustrine pockets. Strong polar winds also contribute a small amount of sedimentation within the lake and other shadow zones. Algal mat overlying sediment generally covers the lake bottoms. Generally no benthic fauna is found, but thecamoebians have been reported (Mathur *et al.* 2006). The mean size-distribution of 18 sediment samples from Schirmacher Oasis is given in Table 1. The distribution pattern shows a polymodal pattern (Fig. 6a). The mean size of the sediments ranges from 0.95 to 2.34 Φ, that is from coarse to fine sand. About 75% of the sediments belong to the medium sand grade, followed by 25% occurring in the coarse sand category. The sorting (standard deviation σ) varies from 0.97 to 1.53. The samples are finely to coarsely skewed. Kurtosis ranges from 0.75 to 1.6 (Table 2). The central tendencies of sediments from Schirmacher Oasis do not show any correlation with each other, which is characteristic of glacial deposits. The majority of sediments show transportation by saltation (Fig. 6b). Kuenen & Perdok (1962), Mahaney & Kalm (1995) and Mahaney *et al.* (1996) have worked extensively on SEM data of the quartz grains from glaciated terrain. Similar studies on the quartz grains from glacial sediments show a preponderance of features such as moderate to high angularity, moderate relief, edge and surface relief and collision pits (Table 3). The majority of the quartz grains (*c.* 75%) show moderate to high angularity (Fig. 7a–c, g, h). Most of the quartz grains have moderate relief. High- and low-relief features are less abundant. Almost all the quartz grains have undergone edge and surface abrasion. All of these features indicate the importance of fluvial action on these sediment grains. Along with these features (angularity, relief, edge abrasion and meandering ridges),

Table 1. *Mean grain-size distribution of sediments samples from Schirmacher Oasis and Bharati Promontory, Larsemann Hills*

Class interval (μm)	Schirmacher Oasis			Larsemann Hills		
	Weight (g)	Weight percentage	Cumulative weight percentage	Weight (g)	Weight percentage	Cumulative weight percentage
0.02–2	0.2	0.149768	100.0018	0.3	0.168256	99.99628
2.00–4.00	1.3	0.973491	99.85207	4.9	2.748177	99.82803
4.00–8.00	3.7	2.770705	98.87858	13.7	7.683679	97.07985
8.00–16.00	13.7	10.2591	96.10787	22.5	12.61918	89.39617
16.00–32.00	14.2	10.63352	85.84878	12.1	6.786315	76.77699
32.00–64.00	11.5	8.611652	75.21526	10.8	6.057207	69.99067
64.00–125.00	14.5	10.85817	66.60361	12.5	7.010656	63.93347
125.00–250.00	10.54	7.892766	55.74544	23.3	13.06786	56.92281
250.00–500.00	12.5	9.360491	47.85267	18.5	10.37577	43.85495
500.00–1000.00	13.3	9.959563	38.49218	18.1	10.15143	33.47918
1000.00–2000.00	27.9	20.89262	28.53262	30.2	16.93775	23.32775
>2000	10.2	7.638161	7.64	11.4	6.393718	6.39
	133.54			178.3		

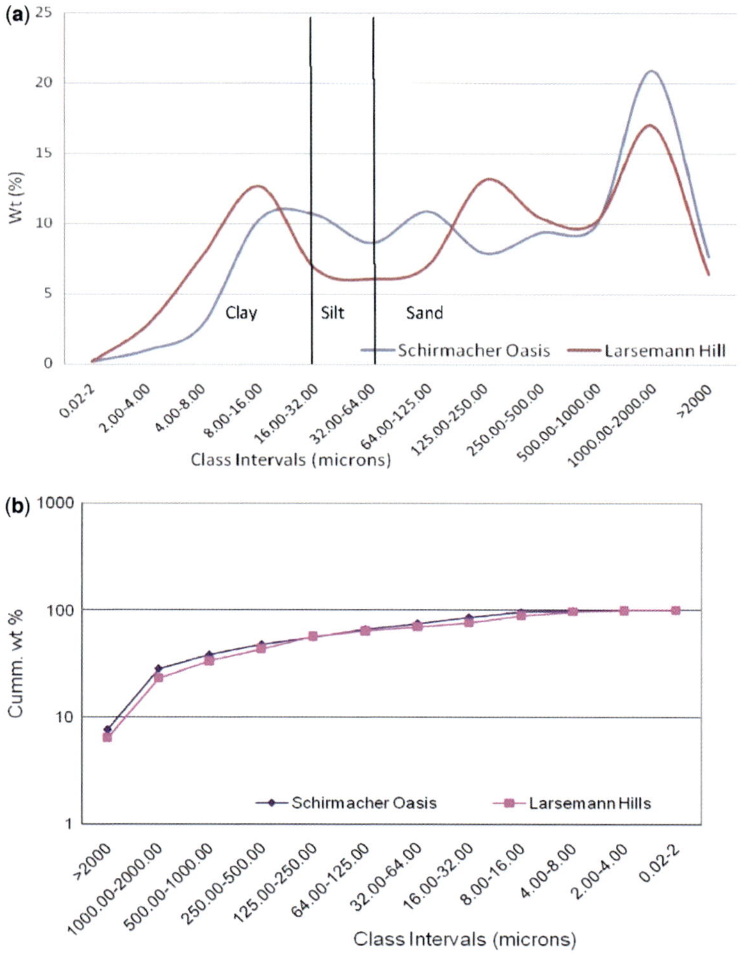

Fig. 6. (**a**) Average weight percentage of different grain size fractions from Schirmacher Oasis and Bharati Promontary, Larsemann Hills. (**b**) Average cumulative weight percentage (logarithmic scale) of different grain size fractions from Schirmacher oasis and Bharati Promontory, Larsemann Hills (after Visher 1969).

arcuate and straight steps (Fig. 7e, f), collision pits (Fig. 7a, b, d) and curved grooves (Fig. 7e, f) also constitute important features. Weathered surfaces, and small discontinuous and random precipitation features (Fig. 7d, e, h), cover about half of the total surface area in some quartz grains. Adhering particles are also noticed on many particles (Fig. 7a–e, h).

Larsemann Hills

The mean size-distribution of 17 sediments samples collected from the Larsemann Hills area is given in Table 1. A polymodal distribution pattern is revealed (Fig. 6a) with modes of 8–16 and 100–200 μm. Again in this region, the majority of sediment grains show transportation by saltation (Fig. 6b). Mean particle size range from 1.05 to 2.72 Φ, that is, from medium sand (dominant) to fine sand. The sorting coefficient (standard deviation σ) varies from 1.059 to 2.015. The sediments are finely to coarsely skewed. Kurtosis ranges from 0.75 to 1.6 (Table 4). SEM studies of quartz grains collected from the Larsemann Hills area show a slightly different character (Table 5). Here most of the quartz grains (>85%) have high angularity (Fig. 8a–f) and conchoidal fractures (Fig. 8a, b, d). All of these features indicate the dominance of glacial over fluvial action. Almost all quartz grains show sharp edges (Fig. 8a–f), angular outline (Fig. 8a–f), conchoidal fractures (Fig. 8a–f), curved grooves (Fig. 8b, d) and breakage blocks (Fig. 8a–d). Few quartz grains have developed deep troughs (Fig. 8b, c) and polished

Table 2. *Mean* (Mz), *Standard Deviation (SD), Skewness* (Sk) *and Kurtosis* (Ks) *of lake sediments from Schirmacher Oasis (after Folk & Ward 1957)*

Sample no.	Mean (*Mz*)	Standard deviation (SD)	Skewness (*Sk*)	Kurtosis (*Ks*)	Sediment environment
Schirmacher Oasis					
SL-1	1.617	1.029	0.097	1.39	Glaciofluvial
SL-2	1.927	1.254	0.089	1.333	Glacial
SL-3	1.617	1.194	0.078	1.607	Glacial
SL-4	1.733	1.177	0.072	1.242	Glacial
SL-5	1.907	1.342	−0.036	0.75	Glacial
SL-6	2.133	1.013	−0.076	1.548	Glacial
SL-7	2.343	1.024	−0.142	1.189	Glaciofluvial
SL-8	2.133	1.013	−0.076	1.548	Glaciofluvial
SL-9	2.08	1.23	−0.031	1.363	Glacial
SL-10	0.95	0.97	0.24	1.18	Glaciofluvial
SL-11	1.17	1.03	0.27	1.14	Glaciofluvial
SL-12	1.81	1.53	0.26	1.11	Glacial
SL-13	1.15	1.23	0.095	0.91	Glacial
SL-14	1.2	1.14	0.112	1.12	Glacial
SL-15	1.11	1.054	−0.035	1.1	Glaciofluvial
SL-16	2.01	0.99	0.099	0.99	Glaciofluvial
SL-17	0.99	1.21	0.089	1.31	Glacial
SL-18	1.52	1.1	0.115	1.23	Glacial

surfaces (Fig. 8c, e). Along with this, there are many quartz grains that show some percentage of collision pits (Fig. 8a–f). Some quartz grains show small-scale precipitation features (Fig. 8e, f) and adhering particles (Fig. 8b, e, f). Lake systems in Larsemann Hills are typically closed and confined to the valleys that dissect the area. These valleys provide a conduit for summer meltwater streams carrying and depositing the surficial sediments (Fig. 5b, c, f, h).

Discussion

A range of different types of mechanical microtextures, including conchoidal fractures, straight

Table 3. *SEM properties of quartz grains (in percentages) from sediment of Schirmacher Oasis*

Sample number	a	b	C	d	e	f	g	h
Morphological textures								
Angular outline	85	70	100	55	50	85	90	60
Low relief	55	10	10	10	10	30	70	15
Medium relief	25	20	20	30	70	50	20	60
High relief	20	70	70	60	20	20	10	25
Mechanical textures								
Large conchoidal fractures	10		80		30	30	15	5
Small conchoidal fractures	25	10	20	25	15	10	10	20
Arcuate steps	15			15	5	5	5	
Straight steps	15	10				5	10	
Breakage blocks	10		15	5	5			
Curved grooves					5	5		
Straight grooves/striations					10	10		
V-shaped percussion cracks								
Deep troughs	5		25			10		
Pitted surface/collision pits	5	10		20	5	10	5	5
Chemical textures								
Precipitation features				20				20
Adhering particles	15	5	5	10	5			5

Fig. 7. Scanning electron microscopy of glacial and glaciofluvial quartz grains from Schirmacher Oasis. (**a**) Medium relief, conchoidal surface, adhered particles; (**b**) medium relief, polished surface, conchoidal surface; (**c**) angular, sharp edges, medium to high relief, breakage block, conchoidal fracture planes; (**d**) medium relief, subround, arcuate steps; (**e**) low relief, subangular, polished surface, large groves; (**f**) low relief, angular, straight ridges; (**g**) angular, low relief, straight ridges, polished surface, sharp edges; and (**h**) medium to low relief, large precipitation features.

Table 4. Mz, *SD*, Sk *and* Ks *of sediments from Bharati Promontory, Larsemann Hills area (after Folk & Ward 1957)*

Sample no.	Mean (*Mz*)	Standard deviation (SD)	Skewness (*Sk*)	Kurtosis (*Ks*)	Sediment environment
Larsemann Hills					
LH-1	2.48	1.355	−0.114	0.868	Glacial
LH-2	2.71	1.256	−0.189	0.911	Glacial
LH-3	2.717	1.218	−0.253	0.969	Glacial
LH-4	2.727	1.144	−0.069	0.921	Glaciofluvial
LH-5	1.943	1.059	0.036	1.506	Glaciofluvial
LH-6	1.928	1.088	0.19	1.44	Glaciofluvial
LH-7	2.005	1.095	0.21	1.343	Glaciofluvial
LH-8	2.011	2.015	0.215	0.998	Glacial
LH-9	1.859	2.005	0.098	1.421	Glacial
LH-10	1.975	1.099	0.109	1.321	Glaciofluvial
LH-11	2.05	1.098	−0.029	1.105	Glaciofluvial
LH-12	2.14	1.086	0.113	0.98	Glaciofluvial
LH-13	1.88	1.132	0.106	0.896	Glaciofluvial
LH-14	1.99	1.121	0.109	1.112	Glaciofluvial
LH-15	2.135	1.113	0.096	1.056	Glaciofluvial
LH-16	2.125	1.243	−0.04	1.124	Glacial
LH-17	1.948	1.096	−0.077	1.212	Glaciofluvial

and curved grooves and troughs, are evident in the sand fraction, and are formed as a result of continental-scale glacial transport. Quartz grains with sharp edges, parallel striations, conchoidal fractures, high relief and little V-shaped percussion marks suggest a glacial, rather than fluvial, origin. A range of different types of mechanical micro-textures (including angular to rounded outlines, shape, surface relief, conchoidal fractures, crescentic gouges, edge and surface abrasion features) were identified from SEM in the work carried out by Mahaney & Kalm (1995) and Mahaney *et al.* (1996). Edge sharpness is more pronounced in quartz grains from sediment samples from Bharati Promontory, Larsemann Hills than those from Schirmacher Oasis. Hence, it can be inferred that fluvial processes are more effective in the later area. The conchoidal fractures and arcuate steps may have originated during the removal of the grains from the crystalline source rocks (Krinsley & Margolis 1969; Krinsley & Smith 1981). The different sizes of pits and cavities with secondary precipitates indicate solution processes at a later stage, depending on the environment of deposition (Armstrong-Altrin *et al.* 2005). The presence of angular to subangular quartz grains and preservation of arcuate steps suggests that these grains have undergone short transportation and rapid deposition. Parallel steps and subrounded edges, indicate strong glaciofluvial action. Strong glacier action is indicated by deep striations and polished surfaces on the quartz grains (Krinsley & Margolis 1969; Krinsley & Smith 1981). The most common glacial features are sharp edges, angular outlines, large-scale conchoidal fractures, arcuate steps, parallel striations, uneven surfaces and irregular small-scale indentations. Most of the angular grains of Larsemann Hills area are freshly derived having composite surface features. The glaciofluvial

Table 5. *SEM properties of quartz grains (in percentages) from sediment Bharati Promontory, Larsemann Hills area*

Sample number	a	b	c	d	e,f
Morphological textures					
Angular outline	100	100	85	95	90
Low relief	5	10	10	10	10
Medium relief	10	10	10	10	10
High relief	85	80	80	80	80
Mechanical textures					
Large conchoidal fractures	20	10		30	10
Small conchoidal fractures	80	80	10	60	10
Arcuate steps					
Straight steps					
Large breakage blocks		40	45		
Curved grooves		10		5	
Straight grooves/striations			10	5	10
V-shaped percussion cracks					
Deep troughs		40	40		
Pitted surface/collision pits	10	10	20	15	20
Chemical textures					
Large precipitation features	26	45	10	20	40
Adhering particles		5			5

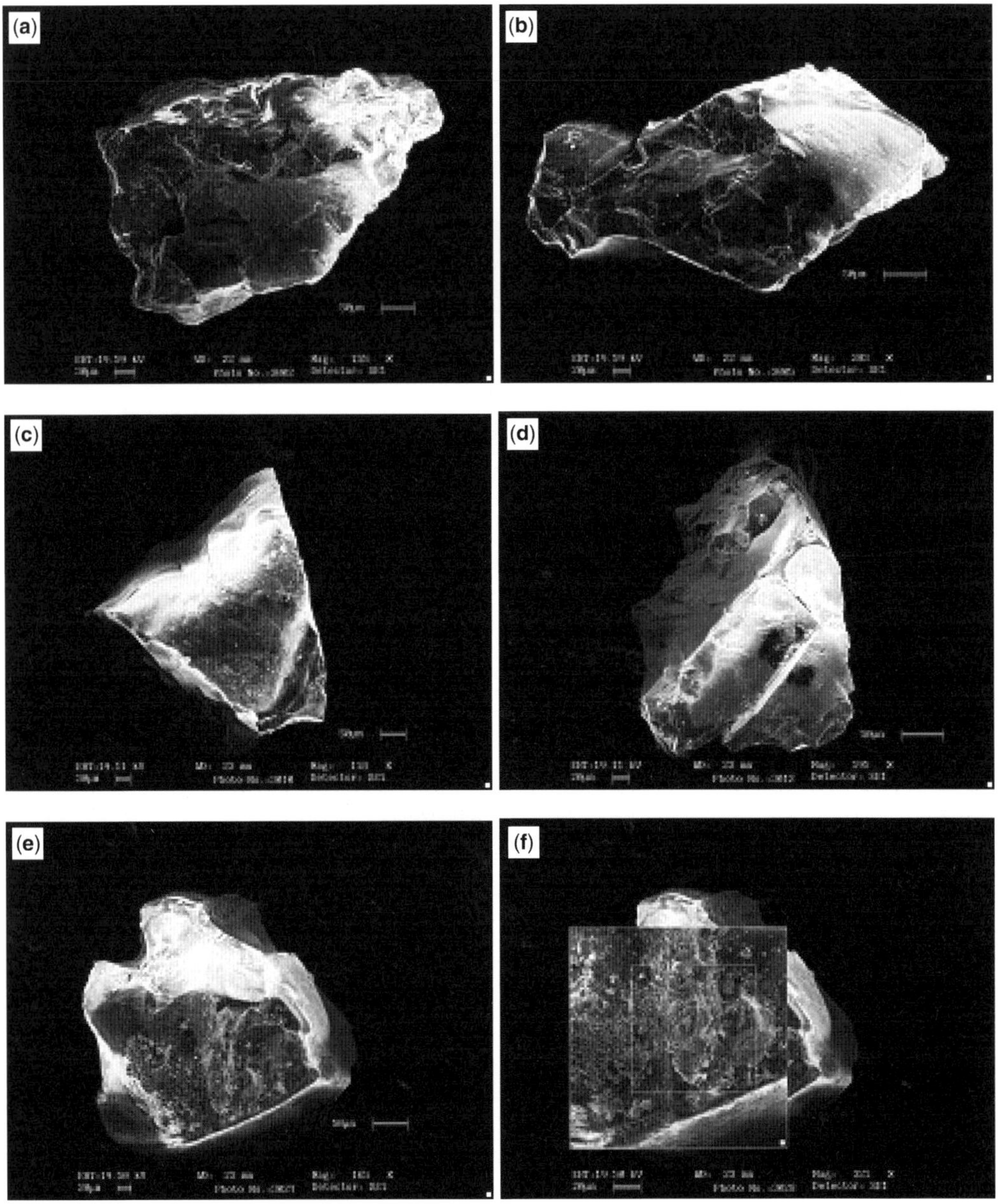

Fig. 8. Scanning electron microscopy of quartz grains from glacial and glaciofluvial sediments from Bharti promontory and Mcleod Island of Larsemann Hills area of East Antarctica. (**a**) Angular, conchoidal surfaces, few arcuate steps; (**b**) angular, large breakage block, straight to arcuate steps, pitted surface; (**c**) angular, straight to curved grooves and ridges, polished surface; (**d**) angular, large breakage blocks, conchoidal surfaces, sharp edges; (**e**, **f**) angular to subangular, large precipitation features, polished surfaces.

action led to the modification of earlier glacial features of reworked quartz grains (Kuenen & Perdok 1962). Chemical activity has been also imprinted on some grains and is represented by dissolution pits, silica precipitation and surface etching.

The mean frequency-distribution curves of glaciofluvial sediments (Fig. 6a) from the Schirmacher and Larsemann Hills area show a polymodal nature, for which various possible explanations exist. It is possible that a subdued class size has

Table 6. *The similarity and differences of SEM and granulometric analyses from Schirmacher Oasis and Bharati Promontory, Larsemann Hills*

	Schirmacher Oasis	Bharati Promontory, Larsemann Hills
Similarities		
Sediment Distribution	Polymodal	Polymodal
Mode of fluvial transport	Suspension and saltation with little traction	Suspension and saltation with little traction
Sorting	Poorly sorted	Poorly sorted
Skewness	Slightly positive to slightly negative	Slightly positive to slightly negative
Differences		
Sharpness of edges	Less	More
Fluvial effect	More	Less
Glacial effect	Less	More
Aeolian effect	Present	Absent
Mean grain size	0.95–2.344 Φ	1.859–2.727 Φ
Maturity	Immature	More immature

not been generated because of specific denudational processes. The instability of certain minerals or reworking of sediments may also cause a scarcity of certain classes of sediments (Pettijohn 1984). The bimodal and polymodal distribution of sediments from both places indicates a change in energy conditions and the dominance of more than one process during deposition. The plot of glaciofluvial sediments on a log–log scale (Visher 1969), between average cumulative weight percentage and particle sizes, shows suspension and saltation as the dominant modes of grain transport (Fig. 6b). The effect of traction is evident from the presence of pebbles. On the basis of the standard deviations and grain size distribution patterns, sediments from both oases were subjected to turbulence within transporting media, followed by reworking. Pronounced skewness and kurtosis values suggest multiple sources for these sediments (Folk & Ward 1957, Mason & Folk 1958, Asthana & Chaturvedi 1998). In the case of Schirmacher Oasis, wind has imprinted some features on quartz grains (like frosted surface), but has failed to do so on the quartz grains of Larsemann Hills. Large sediment grains are distributed irregularly in a fine-grained matrix. A complex interplay of the erosive action of the ice-mass, which generates glacial detritus ranging from boulders to silt-clay-sized rock flour, distance of transportation, glacier as well as basin morphology and its hydrological characteristics determines the final depositional product. The presence of numerous labile minerals like feldspar, ferromagnesian minerals as angular and unaltered sediment grains marks an important feature of glacial sediments (Reineck & Singh 1980) in both places. The mean grain size of sediments in Schirmacher Oasis varies from 0.95 to 2.343 Φ (Table 2), whereas from Larsemann Hills the range is from 1.859 to 2.727 Φ (Table 4). Sediments are poorly sorted and skewness values range from slightly positive to slightly negative. Negative skewness values from both areas may be correlated with a high-energy environment (Friedman 1961, 1962, 1967), whereas the slight positive value of skewness results from the supply of fine material. In general, the silt + clay weight percentage is much less, indicating its removal by localized melting and fluvial action. The sorting of the glacial deposits is poor. The similarity and differences of SEM and granulometric analyses from both areas are summarized in Table 6.

Conclusions

The periglacial environment of Schirmacher Oasis and Bharati Promontory, Larsemann Hills is characterized by a wide range of sedimentary and glacial processes, resulting in different combinations of erosional and depositional features. Field observations, granulometric analyses of sediment and SEM of selected quartz grains in both areas show similar results, except for the extent and strength of glacial and glaciofluvial processes. The field observations and SEM features as observed on quartz grains show more glacial effect in the Bharati Promontory area of Larsemann Hills, whereas Schirmacher Oasis is marked by greater fluvial influence. The grain size distribution data indicate that the sediments in both areas are poorly sorted, comprising pebble, sand, silt and minor clay fractions. The silt + clay weight percentage is much less than that of sand in both areas, indicating its removal by localized melting and fluvial action. A change in energy conditions is evident from the bimodal and polymodal distribution of sediments from both places.

The authors wish to express their gratitude to the Director General, Geological Survey of India and Director, National Centre for Antarctic and Ocean Research for facilitating their participation in East Antarctica. One of the authors, R. Asthana, thanks the Ministry of Earth Sciences and Ministry of Mines for allowing his participation in the XI International Symposium on Antarctic Earth Sciences at Edinburgh under the aegis of SCAR in 2011, where this work was presented. Critical and constructive suggestions by Dr M. J. Hambrey and the anonymous reviewer helped in improving the manuscript and they are gratefully acknowledged.

References

ARMSTRONG-ALTRIN, J. S., MADHAVARAJU, J., RAMASAMY, S. & GLADWIN GANA ASIR, N. 2005. Provenance and depositional history of sandstones from upper Miocene Kudamkulam Formation, Tamilnadu. *Journal of Geological Society of India*, **66**, 59–65.

ASTHANA, R. & CHATURVEDI, A. 1998. The grain size behavior and morphoscopy of supraglacial sediments, South of Schirmacher Oasis, E Antarctica. *Journal of Geological Society of India*, **52**, 557–568.

ASTHANA, R., SHRIVASTAVA, P. K., BEG, M. J., SHOME, S. & KACHROO, K. 2009. Surface microtextures of quartz grains from glaciolacustrine sediment core from Priyadarshini lake, Schirmacher Oasis, East Antarctica as revealed under scanning electron microscope. *Indian Journal of Geosciences*, **63**, 205–214.

BURGESS, J. S., SPATE, A. P. & SHEVLIN, J. 1994. The onset of deglaciation in the Larsemann Hills, Eastern Antarctica. *Antarctic Science*, **6**, 491–495.

BURGESS, J., CARSON, C., HEAD, J. & SPATE, A. 1997. Larsemann Hills: not heavily glaciated during the Last Glacial Maximum. *In*: RICCI, C. A. (ed.) *The Antarctic Region: Geological Evolution and Processes*. Universtà degli Studi di Siena, Siena, 841–843.

CHATURVEDI, A. & SINGH, A. 1999. *Thermal variations withion a glacier from the surface to the bedrock, Fifteenth Indian Expedition to Antarctica*. Scientific Report, DOD, Technical Publications, **13**, 303–312.

FITZSIMONS, S. J. 1996. Paraglacial redistribution of glacial sediments in the Vestfold Hills, East Antarctica. *Geomorphology*, **15**, 93–108.

FITZSIMONS, S. J. 1997. Depositional models for moraine formation in East Antarctic coastal oases. *Journal of Glaciology*, **43**, 256–264.

FOLK, R. L. & WARD, W. C. 1957. Brazos river bar, a study in the significance of grain size parameters. *Journal of Sedimentary Petrology*, **27**, 3–26.

FRIEDMAN, G. M. 1961. Distinction between dune, beach and river sands from their textural characteristics. *Journal of Sedimentary Petrology*, **31**, 514–529.

FRIEDMAN, G. M. 1962. On sorting, sorting coefficients and the lognormality of the grain size distribution of sandstones. *Journal of Geology*, **70**, 737–756.

FRIEDMAN, G. M. 1967. Dynamic processes and statistical parameters compared for size frequency distribution of beach and river sands. *Journal of Sedimentary Petrology*, **37**, 327–354.

GILLIESON, D. S. 1991. An environmental history of two freshwater lakes in the Larsemann Hills, Antarctica. *Hydrobiologia*, **214**, 327–331.

HEYWOOD, R. B. 1984. Antarctic inland waters. *In*: LAWS, R. M. (ed.) *Antarctic Ecology*. Academic Press, London, **1**, 279–344.

HODGSON, D. A., VERLEYEN, E. *ET AL*. 2005. Late quaternary Climate driven environmental change in the Larsemann Hills, East Antarctica, Multi Proxy evidence from lake sediment core. *Quaternary Research*, **64**, 83–99.

KRAUSE, W. E., KRBETSCHEK, M. R. & STOLZ, W. 1997. Dating of quaternary lake sediments from the Schirmacher Oasis (East Antarctica) by infra-red stimulated luminescence (Irsl) detected at the wavelength of 560 nm. *Quaternary Science Reviews (Quaternary Geochronology)*, **16**, 387–392.

KRINSLEY, D. H. & MARGOLIS, S. 1969. A study of quartz sand surface texture with the scanning electron microscope. *Transactions of the New York Academy of Science, Series II*, **31**, 457–477.

KRINSLEY, D. H. & SMITH, D. B. 1981. A selective SEM study of grains from Permian yellow sands of northeast England. *Proceedings of the Geologists Association England*, **92**, 242–247.

KUENEN, P. H. & PERDOK, W. G. 1962. Experimental abrasion 5. Frosting and defrosting of quartz grains. *Journal of Geology*, **70**, 648–658.

MAHANEY, W. C. & KALM, V. 1995. Scanning electron microscopy of Pleistocene tills in Estonia. *Boreas*, **24**, 13–29.

MAHANEY, W. C., CLARIDGE, G. C. & CAMPBELL, I. 1996. Microtextures on quartz grains in tills from Antarctica. *Palaeogeography, Palaeoclimatology, Palaeoecology*, **121**, 89–103.

MASON, C. C. & FOLK, R. L. 1958. Differentiation of beach, dune and Aeolian flat environments by size analysis, Mustang Island, Texas. *Journal of Sedimentary Petrology*, **28**, 211–226.

MATHUR, A. K., ASTHANA, R. & RAVINDRA, R. 2006. Arcellaceans (thecamoebians) from core sediments of Priyadarshini Lake, Schirmacher Oasis, Eastern Antarctica. *Current Science*, **90**, 1603–1605.

PETTIJOHN, F. J. 1984. *Sedimentary Rock, The Texture of Sediments*, 3rd Indian edn. CBS, New Delhi, 31–50.

RAVIKANT, V. & KUNDU, A. 1998. Reaction texture of retrograde pressure temperature deformations in relation to metamorphism – migmatite events in the Schirmacher Hills, East Antarctica. *Journal of Geological Society of India*, **32**, 295–315.

RAVINDRA, R., CHATURVEDI, A. & BEG, M. J. 2002. Melt water lakes of Schirmacher Oasis – their genetic aspects and classification. *In*: SAHOO, D. B. & PANDEY, P. C. (eds) *Advances in Marine and Antarctic Science*. A.P.H. Publishing Corporation, New Delhi, 301–303.

REINECK, H. E. & SINGH, I. B. 1980. Depositional sedimentary environments. *Sedimentary Geology*, **32**, 153, doi: 10.1016/0037-0738(82)90019-7

SHRIVASTAVA, P. K., ASTHANA, R., BEG, M. J. & SINGH, J. 2009. Climatic fluctuation imprinted in quartz grains of lake sediments from Schirmacher Oasis and Larsemann Hills area, East Antarctica. *Indian Journal of Geosciences*, **63**, 87–96.

SHRIVASTAVA, P. K., ASTHANA, R., ROY, S. K., SWAIN, A. K. & DHARWADKAR, A. 2012. Provenance and depositional environment of epi-shelf lake sediment from Schirmacher Oasis, East Antarctica, vis-a-vis scanning electron microscopy of quartz grain, size distribution and chemical parameters. *PolarScience*, **6**, 165–182.

VERLEYEN, E., HODGSON, D. A., SABBE, K. & VYVERMAN, W. 2004. Late Quaternary deglaciation and climate history of the Larsemann Hills (East Antarctica). *Journal of Quaternary Science*, **19**, 361–375.

VISHER, G. S. 1969. Grain size distribution and depositional processes. *Journal of Sedimentary Petrology*, **39**, 1074–1106.

Periglacial processes and landforms of the Antarctic: a review of recent studies and directions

KEVIN HALL

Geography Programme, University of Northern British Columbia, 3333 University Way, Prince George, BC, Canada V2N 4Z9 (e-mail: hall@unbc.ca)

Abstract: The wide range of periglacial environments of the Antarctic, from the wet, mild oceanic sub-Antarctic through to the cold, dry continent, provides not only an extensive modern laboratory, but also one that offers insights into conditions in the Northern Hemisphere at the height of the last glacial, is an analogue for periglacial conditions on other planets, and can be used for monitoring climatic change. Almost the whole known suite of periglacial landforms is present. Recent research directions show strong linkages between the biotic components and the abiotic responses, offering new insights into periglacial synergies and hence landform development. Other new directions are those of using the Antarctic as an analogue for periglacial conditions on Mars, and multinational undertakings monitoring permafrost and active layer changes. During the past three decades there have been a number of reviews of periglacial landforms and processes for both the area as a whole as well as for specific locations or regions. Here information regarding material post-dating the most recent reviews is provided and an attempt is made to highlight new directions and findings. This broad-based review provides a foundation for more detailed accounts on some of the periglacial attributes provided in other papers within these volumes.

The Antarctic periglacial environment offers a range of processes and landforms from those of the mild, wet, permafrost-free sub-Antarctic with diurnal frosts (e.g. Nougier 1964; Hall 1984; Boelhouwers *et al.* 2003) through to the cold, dry continent with extensive permafrost and thermal contraction polygons (e.g. Péwé 1959; Levy *et al.* 2006); the continent also provides a periglacial analogue for other planets (e.g. Anderson *et al.* 1972; Parsons *et al.* 2005; Levy *et al.* 2011). Block (1994) provided a most useful broad, simple description of the range of terrestrial conditions for this whole region. The maritime Antarctic (including the Antarctic Peninsula) offers a wide range environments that fully bridge the gap between those of the sub-Antarctic and the continent proper: from the extensive, continuous permafrost of James Ross Island (Fukuda *et al.* 1992) to the milder, wetter, discontinuous permafrost environment of Livingston Island (e.g. Ramos & Vieira 2003; Hauck *et al.* 2007) or King George Island (Bockheim & Hall 2002). While almost the whole gamut of cryogenic landforms is present within the region, pingos have yet to be identified (Fitzsimons 1989), despite the initial suggestions of Pickard (1983). Conversely, on Alexander Island, active cryoplanation (e.g. Hall 1997; Hall & André 2010) offers process insights not currently available in the Arctic. Thus, the Antarctic is an extremely important region offering not just a very wide range of cryogenic processes and landforms but also a number of analogues for past climates, future climatic scenarios (e.g. Hall & Walton 1992), and other planets. It is also a region, partly owing to its diversity, that offers a significant opportunity for present and future climatic monitoring by means of various larger-scale undertakings across the whole region, that is, the CALM (Circum-Antarctic Active Layer Monitoring) protocol (e.g. Cannone *et al.* 2006; Cannone & Guglielmin 2010), IPY/ANTPAS undertakings (the International Polar Year/Antarctic Permafrost and Soils; Vieira *et al.* 2010) and the LGM (Latitudinal Gradient Project) consideration of ecosystems, that include within them the role of weathering and permafrost (Howard-Williams *et al.* 2006; Adlam *et al.* 2010). Past climates, and their interaction with cryogenic processes, have been investigated by means of such proxy evidence as weathering rinds on glacially deposited clasts (e.g. Hall & Bühmann 1989; Hall *et al.* 2001) and cryogenic carbonates (e.g. Lacelle *et al.* 2006).

In respect of climatic conditions, volumes by van Rooy (1957), Venter & Burdecka (1966), and Weyant (1967) provide extensive information regarding Antarctic meteorology and climates, while there are many journal articles that deal with the climate of specific areas (e.g. Le Roux 2008; Le Roux & McGeoch 2008). There are also a number of works that provide some region-specific (e.g. Hall 1987; Boelhouwers *et al.* 2003), national territory-specific (e.g. Knight 1987) or location-specific (e.g. Löffler *et al.* 1983; Bormann & Fritzsche 1995, pp. 171–206) syntheses of periglacial information. Importantly, Walton (1980) and Hall & Boelhouwers (2003) offer an extensive

From: HAMBREY, M. J., BARKER, P. F., BARRETT, P. J., BOWMAN, V., DAVIES, B., SMELLIE, J. L. & TRANTER, M. (eds) 2013. *Antarctic Palaeoenvironments and Earth-Surface Processes*. Geological Society, London, Special Publications, **381**, 429–453. First published online July 1, 2013, http://dx.doi.org/10.1144/SP381.16

bibliography of material pertaining to permafrost and periglacial landforms and processes for the whole Antarctic region, while Bockheim (1995) and Bockheim & Hall (2002) have dealt in detail with permafrost distribution. At the time of writing a series of papers hosted by Guglielmin (2012) provide a review of much new periglacial work in respect to the Antarctic; some papers are cited, but this volume offers a further update (and reference resource). Despite the extensive work on both glacial and periglacial attributes, little has been done regarding the linkages, in terms of sediments and landforms, between the two, the work of Gordon & Birnie (1986) being both an exception and a model for future undertakings.

Given the extensive material regarding climate and several existing reviews regarding periglacial processes and landforms as well as permafrost, the aim here is to add to them rather than to repeat what is already available. Some reference to these earlier works will be noted where applicable to the context being considered, but not the details. The aim is to attempt to build upon that already existing and to showcase the newer material and provide a wider foundation for the more detailed papers within this volume. In some instances periglacial work now occurs within the cross-over to biological studies (much more so than with the more obvious glacial studies), and so some works may appear more botanical in nature than periglacial, but the connection is decidedly there.

The periglacial environments

For the present discussion there are a number of ways that the Antarctic region could be divided (e.g. based on geomorphic processes or on water availability; Guglielmin, pers. comm.), but each approach has its own benefits and problems. Here (Fig. 1), to be in line with earlier reviews, the Antarctic region is divided into three main regions: (1) the sub-Antarctic (or 'periantarctic'); (2) the Maritime Antarctic and Peninsula region; and (3) the continent (Block 1994; Bölter *et al.* 2002). The sub-Antarctic region lies to the north of the Antarctic Convergence while the Maritime region includes those islands and a part of the Antarctic Peninsula to the south of the Convergence (see Block 1994, Fig. 1). Although clearly a diverse periglacial environment, none of it is characterized by 'intense freeze–thaw activity of water (as) the basis of periglacial processes and landforms' (Kalvoda 2005, p. 79), because it is water, rather than temperature, that is the limiting factor in process activity for much of the Antarctic (see Balke *et al.* 1991). Indeed, given that (potentially) freezing temperatures (although to varying degrees and for varying extents) occur throughout the Antarctic region, it is water that is commonly the controlling factor in determining processes, and hence many of the landforms. Even areas such as Marion Island, which would be considered as 'wet' (based on an annual precipitation of 1975 mm; Le Roux 2008), fail to exhibit frost-based landforms in the higher regions as a result of the highly porous scoria (Fig. 2) that covers much of the area such that, despite the precipitation, it is essentially a 'polar desert' (Gremmen & Smith 2008, p. 227). Moving south from the sub-Antarctic through to the continent is to pass along a moisture gradient wherein liquid water availability becomes more and more limited in terms of amounts, both spatially and temporally. Indeed, as Balke *et al.* (1991) and Hall & André (2010) have shown, ground and rock temperatures are, during the summer, more than adequate for chemical weathering, but this is constrained as a result of moisture limitations. Thus, conceptually (see final discussion), the Antarctic is an important region for our understanding of cold-based landform processes, and for encouraging a move away from the all-too-common (see Hall & Thorn 2011) general assumption of water-based periglacial processes.

Periglacial assemblages of the different regions

The point needs be made that, although a tripartite periglacial division has been employed, there are really a number of 'overlaps' between the regions such that, for example, parts of Kerguelen (generally located within the sub-Antarctic) may have attributes more akin to the Maritime Antarctic, while the >90% ice-covered (see Engelskjøn 1986, Fig. 1) island of Bouvetøya in the Maritime Antarctic is not greatly dissimilar to the continent, with blockfields located only 250 m above sea-level (Jørgensen 1986, p. 26). Other such examples are the large (>2 m width) sorted stripes observed (Fig. 3) at altitude near the Cook ice cap on Kerguelen (see Hall 1983), which are similar to those observed in the Maritime Antarctic (Chambers 1966) with its deep freezes and permafrost. As a corollary, the extensive small sorted stripes observed on Livingston Island (see Hall 1994) in the Maritime Antarctic are closer to those seen in the sub-Antarctic (e.g. Hall 1983; Holness 2004) with its large number of shallow-penetration, diurnal freeze cycles. As with any attempt at simplifying the complexities of reality, some places are not a 'perfect fit'. This is important to note for two reasons. First, the reader may themselves question the location of some places within this framework based upon the nature of the periglacial activity/environment

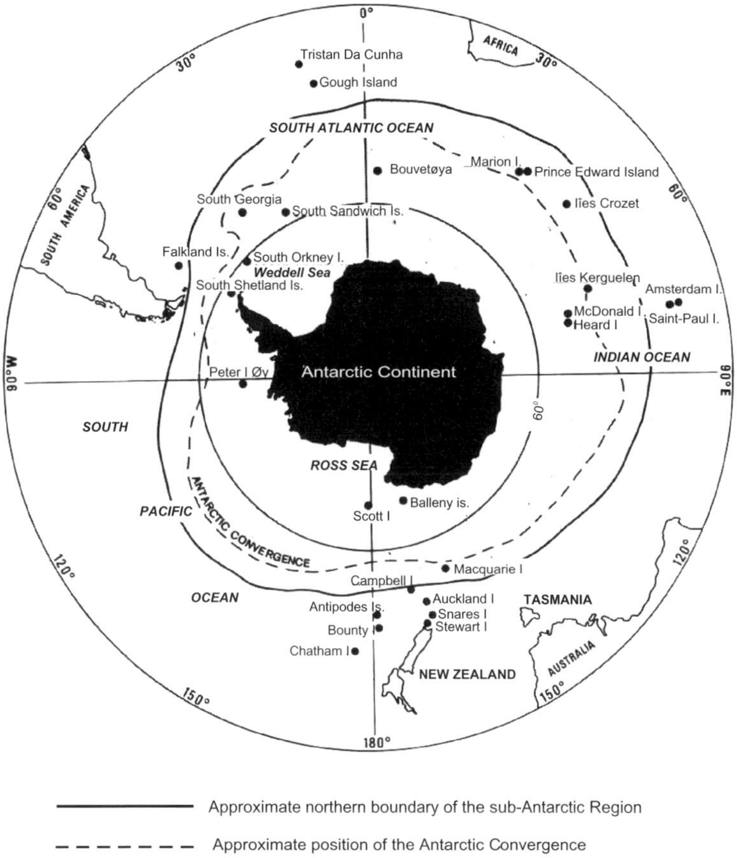

Fig. 1. Map to show the different periglacial regions.

observed. Second, this division has itself led to some oversimplification and 'blurring' of conditions in regard to the nature of the environment (e.g. French 2007, p. 75). Also, recent undertakings in this region are somewhat spatially limited, with much work on Marion Island and a new review of Heard Island, but very little from Kerguelen, Macquarie or the Crozet islands.

The sub-Antarctic

The sub-Antarctic is somewhat of an enigmatic periglacial environment as it is influenced by wet, mild conditions as well as snowy, freezing periods; often vertical zonation plays a significant role in process determination for the same geographic location. Further, conditions during the LGM in much of the sub-Antarctic were significantly more severe than at present, such that there are landforms (e.g. the large stone-banked lobes of Marion Island; Holness 2003*a*) that are no longer active but subject to lower-intensity frost-based processes that keep them from (yet) appearing as fossils. In addition, the sub-Antarctic islands are all volcanic in origin (Hall 2004), often with (to varying degrees) still active volcanism (e.g. Verwoerd *et al.* 1981; Frenot *et al.* 2005) such that there is uncertainty regarding both the spatial extent of the island and the ice distribution during the last glacial (e.g. Hall *et al.* 2011; Hall & Meiklejohn 2011). In addition, volcanic ash has also impacted landforms and vegetation on the island but, in so doing, has also offered dateable events in the sedimentary sequences (e.g. Yeloff *et al.* 2007). Thus, Holocene and earlier periglacial activity can be difficult to ascertain with any certainty as records are both spatially poor and frequently lacking meaningful dates to help determine age. Further, owing to the available materials for glacial and/or periglacial activity being exclusively volcanic in origin (scoria and lavas), identification of both glacial and/or periglacial sediments and landforms can be extremely complicated (e.g. Boelhouwers *et al.* 2008, p. 69), especially for distinguishing them in

Fig. 2. Scoria cones with *Azorella* cushions in the 'polar desert' of the higher elevations on sub-Antarctic Marion Island.

vertical sequences from sediments of purely volcanic origin. Thus, glacial reworking of periglacial sediments or periglacial reworking of glacial sediments, where the sole origin is recent volcanics, can be particularly difficult to unravel (e.g. Boelhouwers *et al.* 2008, p. 69).

Boelhouwers *et al.* (2003, p. 52) have argued that the sub-Antarctic is a 'distinct periglacial environment' characterized by 'high frequency, shallow frost cycles', the influence of wind on sorted forms, as well as being an area within which there is particularly effective surface sediment transport and patterned ground development. This is also a region said to be 'highly sensitive to climatic change' (Boelhouwers *et al.* 2003, p. 53) such that (for Marion Island) a 1 °C decrease in temperature would cause frost to occur to at least 80 cm depth, over a period of 4 months, compared with the

Fig. 3. Large-scale sorted stripes/garlands near the Cook Ice Cap on sub-Antarctic Kerguelen Island.

present penetration depth of 20 cm. Conversely, a warming of 1 °C would reduce frost frequency by 25% and decrease current frost penetration from 20 to 10 cm depth. Ultimately, Boelhouwers *et al.* (2003, p. 53) argued that 'While a grouping with other climates of low annual temperature range is appropriate as far as it identifies highly active diurnal frost environments, a fundamental difference exists owing to contrasting wind conditions and diurnal temperature ranges' such that, coupled with the high sensitivity to climatic change, this should be considered a distinct periglacial environment and not part of a larger grouping. Certainly Selkirk & Saffigna (1999, p. 419) clearly demonstrated the geomorphic importance of wind for sub-Antarctic Macquarie Island, effects being both erosional (360 mm loss in 3.3 yr for a bare surface) and depositional (40 mm gain in 3.3 yr for a bare surface). Kiernan & McConnell (2008, p. 9) also identified the role of the wind on Heard Island (see below).

With the same understanding of current sub-Antarctic environmental conditions as outlined by Boelhouwers *et al.* (2003), Holness (2003*b*) investigated the relict periglacial features of Marion Island and how these provide a record of Holocene environmental change. Holness (2003*b*, p. 73) suggested that the nature of relict forms argues strongly for 'deep seasonal freezing and possibly permafrost' and that temperatures were 'depressed 4 °C to 5.5 °C below present during the period 15–17 k BP'. However, Sumner & Meiklejohn (2004, p. 130) contested that, despite the colder Holocene environmental conditions, and any possible influence these may have had on patterned ground, mechanical weathering was such that only 'a single layer of blocks and plates on previously glaciated surfaces' was able to be produced and that the 'Holocene has provided insufficient time for mechanical bedrock weathering required for the formation of blockfields.' Sumner (2004, p. 126), in respect of weathering, made the point that current lava weathering rates (in the order of 0.1% yr^{-1} at 730 m a.s.l.) are comparable to those found for the maritime Antarctic (e.g. Hall 1990). However, during past, colder conditions (Holness & Boelhouwers 1998), weathering may have been faster, while the current rates found at 730 m would have occurred at lower elevations (Sumner 2004). Conversely, Nel *et al.* (2003) have shown that at higher elevations there are large periglacial landforms which required long periods of exposure to cold for their development. Following Holness & Boelhouwers (1998), Nel *et al.* (2003, p. 180) argued that 'given the current inactivity of debris sources ... the major component of debris production must have been during a different climatic regime'. This is quite important for, as recently revised ice cover reconstructions suggest (Hall *et al.* 2011), it implies that many areas previously thought to be ice-covered were actually ice-free and subject to colder-than-present conditions. Hedding *et al.* (2007) record the occurrence of a pronival rampart, resulting from the transport of weathered rock debris over a snow surface, at an elevation of 900 m a.s.l. on Marion Island and argue this is the first such active form documented for the sub-Antarctic.

An interesting new approach regarding the periglacial of Marion Island has been in regard to the recognition of the interaction between biotic and abiotic attributes. Le Roux & McGeoch (2004), following the work of Frenot *et al.* (1993) on Kerguelen, attempted to use the size of the cushion plant *Azorella selago* to measure age and, through this, to reconstruct past environmental conditions and thus help date glacial and periglacial activity. They argued (le Roux & McGeoch 2004, p. 509) that 'the species could potentially be used to estimate the age of landforms and contribute to the understanding of geomorphological processes across the region'. This work was extended (Le Roux & McGeoch 2008) to directly investigate the complex responses of biotic elements to climate change and hence whether they can be used (and, if so, in what manner) as a proxy for understanding environmental change and hence as indicators of past, present or future processes. Following from this work, and extending it directly into biotic interactions with periglacial processes, Haussmann *et al.* (2008, 2009) discussed the interaction of the cushion plant *Azorella selago* with both sediment transport and soil frost dynamics on Marion Island (Fig. 2). Haussmann *et al.* (2008) suggested that, owing to the prevalence of *Azorella* over such a large portion of the island, it interacts with sediment transport and can be a major factor in the development of vegetation-banked terraces with (p. 146) 'coarser material accumulated on the upslope cushion sides'. Apropos the comments above by Boelhouwers *et al.* (2003) regarding the importance of wind, Haussmann *et al.* (2008, p. 146) also noted the interaction of wind direction and exposure with *Azorella* growth patterns – something also noted earlier for sub-Antarctic Macquarie Island by Taylor (1955). The downslope movement of particles on Marion Island has been ascribed by a number of authors (e.g. Hall 1979; Boelhouwers *et al.* 2003; Holness 2003*a*) to needle ice (Fig. 4), as well frost heave and frost creep, and thus there is a direct interaction between abiotic slope processes with biotic factors which, combined, leads to the resulting landforms. Indeed, Haussmann *et al.* (2009, p. 257) went on to show that the plants themselves can have an impact on 'soil thermal regimes, soil moisture and weathering

Fig. 4. Needle ice typical of one night's growth on sub-Antarctic Marion Island.

and erosion'. *Azorella* cushions were observed (p. 266) to influence winter soil temperatures such that they 'were found to be lower and less variable on the eastern sides' as compared with the western sides of cushions. Although the understanding of the biotic–abiotic interactions and feedbacks in the periglacial of the sub-Antarctic (see similar biotic–abiotic relationships for the other regions outlined below) are still being investigated (e.g. Yeloff *et al.* 2007), work to date does indicate that there is a significant relationship that can have an impact upon both periglacial processes and landforms.

The most recent review of information relating to the periglacial landforms and processes of Marion Island can be found in Boelhouwers *et al.* (2008, pp. 75–82). The main points emerging from this review are that, while there is extensive small-scale sorting and solifluction on most unvegetated surfaces, including scoria slopes, the larger forms are relict. Apropos the observation of Boelhouwers *et al.* (2003) regarding the importance of wind for this region, Holness (2003*a*), following the original observations of Hall (1979) for Marion Island and for Kerguelen Island (Hall 1983), showed that sorted stripes exhibit an alignment with the dominant wind direction and this he ascribed to surface cooling by the wind linked to needle ice growth. Overall, Boelhouwers *et al.* (2008, p. 76) suggested that 'These observations suggest that wind may play an important role in enhancing diurnal frost effectiveness and its spatial distribution in the sub-Antarctic.' It is also commented (p. 76) that stone- and vegetation-banked lobes currently forming are the product of diurnal frost creep, but that their dimensions may not reflect the actual 'severity of the ground frost climate'; see also Berg (2009, pp. 15–19). Sediment movement rates were said (Berg 2009, p. 78) to compare well with elsewhere in the sub-Antarctic as well as with alpine diurnal frost environments, but more significantly, they 'are among the highest in any periglacial environment'. Thus, these descriptions for Marion Island echo that found for most of the rest of the sub-Antarctic and recently observed for Heard Island (Kiernan & McConnell 2008).

Kiernan & McConnell (2008) have provided much needed information regarding the periglacial of Heard Island. These authors placed Heard Island within the sub-Antarctic (p. 10), but the current degree of ice cover, and only the recent exposure of deglaciated land, 'confuses' the understanding of ground temperatures and permafrost and hence in which region the island should be placed. Ice has only recently retreated sufficiently to expose (currently above sea-level) ground such that any previous ice-free ground, that experienced much colder conditions, is now below the present sea-level. This same recent exposure of ice-free land also explains the observed limited development of periglacial landforms (Kiernan & McConnell 2008, p. 9). In many ways, the positioning of Heard Island within the adopted tripartite system exemplifies the problematic issues alluded to earlier (*and* the use of any other divisional criteria). However, following the presentation of Kiernan & McConnell (2008), Heard Island is here considered within the sub-Antarctic.

Although the presence of permafrost or seasonal ice in the ground is unresolved (Kiernan & McConnell 2008, p. 9), and may possibly exist (p. 10), a host of periglacial features and processes have been identified for Heard Island: periglacial aeolian forms, sorted polygons and stripes, stone-banked and vegetation-banked terraces, cryoturbation phenomena, and both mechanical and chemical weathering processes. Although they observed that periglacial landform development appeared to be less than that on other comparable islands, this could, in part, reflect (p. 10) the duration of ground exposure resulting from (only recent) ice retreat such that the recognized landforms show similarities (p. 9) with those on sub-Antarctic Marion, Crozet and Kerguelen islands. Although they stated (Kiernan & McConnell 2008, p. 9) that no thermokarst features or 'Landforms indicative of permafrost, such as ice wedge polygons, ice-cored mounds and pingos, and palsas or peat plateaus, were (not) recognized' this is but a reflection of the recentness of exposed ground; such forms may well exist on land that extended beyond the ice cover but which is now below sea-level. In part, this issue reflects a complexity that is applicable to the sub-Antarctic region, where extensive volcanism has confounded accurate delineation of previous coastlines, ice extent and ice-free areas (see Hall & Meiklejohn 2011). Equally, the absence of permafrost on Heard Island may be but a reflection of the extensive ice cover at high elevations (where it would otherwise have occurred); just as Boelhouwers (2003, p. 71) has suggested that permafrost may occur on Marion Island above 1000 m in an area recently deglaciated (see Sumner & Meiklejohn 2004).

The most problematic observations of Kiernan & McConnell, (2008) relate to weathering. The observations are interesting in regard to how they differ from those of Holness & Boelhouwers (1998) and Sumner (2004), and problematic insofar as they lack supportive data and are not well founded within more recent theory (see Hall & Thorn 2011 for a review). The presumption of freeze–thaw weathering (Kiernan & McConnell 2008, p. 3) remains in spite of the many questions raised over the years in regard to this process (e.g. Hall *et al.* 2002; Hall 2006; Hall & Thorn 2011), although Kiernan & McConnell did acknowledge that 'relatively limited frost shattering of recent glacial boulders is to be anticipated'. The main problem with the allocation of this process is not temperatures (usually discussed at length in most texts), but rather moisture availability (which is commonly, as here, not even mentioned). More importantly, given the mélange of available weathering processes, the comments that no evidence could be found of salt weathering (p. 3) and that (p. 4) sub-surface chemical weathering was found as well as the development of minor karren in limestone are intriguing. Given the relatively short period of rock exposure, this degree of weathering was considered surprising by the authors (Kiernan & McConnell 2008, p. 4). Further, given the oceanic situation with strong winds and high seas, it is surprising to find no evidence of salt weathering.

In the Falkland Islands, which can be considered to be a northern extremity of the sub-Antarctic, a recent presentation (André *et al.* 2008) has provided evidence to suggest that the stone runs (Fig. 5), long considered the archetype site for solifluction (e.g. Andersson 1906), are possibly of tropical (rather than the generally assumed cold-region) origin. André *et al.* (2008) have provided a review of the literature pertaining to the stone runs and their periglacial associations, and while a periglacial 'association' at some time in their history is not questioned, it is their (Cenozoic) tropical origin that is somewhat revolutionary – and is perhaps a finding that may have applicability to many Antarctic continent forms (see below). This is a major transition as the stone runs were largely thought to be the result of (Baker 1924 cited in André *et al.* 2008, p. 529) 'frost action on nunataks', a concept that, through many iterations has been maintained until today, for example, Clapperton (1975, p. 216), 'the stone runs were derived by frost-shattering', or Clark *et al.* (1995, p. 218), the stone runs 'are part of a suite of periglacial weathering, transport and deposition'. However, recent work by Wilson *et al.* (2008), using cosmogenic isotope dating, and Hansom *et al.* (2008), using optically stimulated luminescence, has pushed back the age of these features – although, not yet, to the degree suggested by André *et al.* (2008). Wilson *et al.* (2008, p. 471) suggested that the 'stone runs are composite landforms that developed over several cold stages' and are (at least) as old as 700–800 kyr – and may be older.

Clark & Wilson (1992, p. 35) offered support to the foundational argument of Boelhouwers *et al.* (2003) regarding the role of wind within the sub-Antarctic, citing the occurrence of ventifacts developed prior to the restoration of vegetation. In the same vein, Wilson & Clark (2006) reported the development of miniature sorted stripes following recent soil (and vegetation) erosion. Both of these papers express the delicate balance between periglacial activity and plant colonization for this region. Wilson & Edwards (2004) extend both these earlier papers with further information pertaining to both ventifacts and miniature patterned ground – the key attribute of which is the identification of some further forms whose origin(s) appear related to (in the case of stone polygons on peat) the role of wind, and of stone polygons on

Fig. 5. Stone runs on the Falkland Islands.

clay loam for which a *non*-frost-related origin is proposed. This may be important insofar as, once a periglacial environment is identified, polygons are commonly assumed to have a frost origin.

The Maritime Antarctic and Antarctic Peninsula

This is an extensive, climatically diverse region that encompasses (almost) sub-Antarctic cloudy, wetter environments such as King George Island or Livingston Island through to locations just to the north of Alexander Island that, in terms of temperatures and snowfall, are almost continental in character. Despite this climatic diversity there is often regional oversimplification (e.g. French 2007, p. 66) that can obscure the cryogenic environment for a particular site, and this has been exacerbated by the large amount of research undertaken by many national programmes in the more northerly region of the Peninsula and on the westerly islands. Graphically, Hall & Walton (1992, figure 1) showed a (simplified) north–south and east–west permafrost and cold asymmetry for the Antarctic Peninsula and surrounding islands. The east–west asymmetry is significant, with Livingston Island to the west being extremely wet, with a deep active layer and permafrost not extending throughout the island (e.g. Hauck *et al.* 2007; Vieira *et al.* 2008), while to the east, at a comparable latitude, James Ross Island has extensive permafrost, a much more shallow active layer and experiences much less rain than Livingston (e.g. Strelin & Malagnino 1992; Strelin *et al.* 2006). Moving South, to Alexander Island, on the western side of the Antarctic Peninsula, which is largely ice-covered, has a shallow active layer, continuous permafrost and little precipitation, with none in the form of rain (e.g. Hall & Walton 1992); it can thus be better considered (as shown by Block 1994) to be an extension of the continent. Equally, islands such as Bouvetøya, although north of 60°S, are so extensively ice-covered that they are effectively more like the continent and have so little exposed rock there is almost no periglacial activity; climatically they are, however, considered part of the Maritime Antarctic (Engelskjøn & Jørgensen 1986, p. 77; Block 1994, p. 294).

On South Georgia, where the linkage between rock debris production, debris transfer mechanisms and resulting landforms was so eloquently synthesized by Gordon & Birnie (1986), there is a range of landforms which ‘bridge the gap’ between the sub-Antarctic and the Maritime Antarctic, a point indirectly identified by Birnie & Thom (1982, p. 377) when they noted South Georgia as being a ‘maritime sub-Antarctic island’. Birnie & Thom (1982) located South Georgia as south of the Antarctic Convergence (whereas most sub-Antarctic islands occur to its north), being some 58% ice-covered, and with elevational factors associated with the mountainous spine of the island as having a significant climatic influence.

Active rock glaciers, potential identifiers of permafrost conditions, have been found on South Georgia (e.g. Birnie & Thom 1982), James Ross Island (e.g. Lundqvist *et al.* 1995) and for the South Shetland Islands (see Serrano & López-Martínez (2000) for a review of locations). However, some forms thought to be rock glaciers (e.g. Zhu *et al.* 1996) are now thought to be block streams (Serrano & López-Martínez 2000) which may not require the presence of permafrost. Serrano & López-Martínez (2000, p. 147) suggested that rock glaciers are 'represented more in the periphery of Antarctica than in the interior ice-free land of the continent, being especially common in the Antarctic Peninsula region'. Of great interest is the argument of Serrano & López-Martínez (2000, p. 158) that the age of rock glaciers on the South Shetlands is 'only a few hundred years' and that (p. 157) they are the product of 'high frequency, low intensity processes'. This rapid development, also suggested for South Georgia (Birnie & Thom 1982, 17 000–10 000 yr for one group but between 10 000 and 200 yr for another), contrasts strongly with Antarctica proper where ages of 100 000 yr BP have been suggested (e.g. Mjagkov, in Barsch 1996). Serrano & López-Martínez (2000, pp. 158–159) outlined and discussed the veracity and meaning of the various rock glacier ages, concluding that all of those in the South Shetlands are located in recently deglaciated areas (p. 160), that tongue-shaped forms are more common than lobate-shaped, that they rarely extend more than a few hundred metres from their source areas, that they are all active, and that they have formed 'as a morphogenetic response to recent glacial retreat and short-period climate changes. They are landforms and processes associated with deglaciation and the transition from a glacial environment to a periglacial one in an oceanic climate initially not favourable for rock glacier development.'

There is extensive research regarding permafrost and active layer thicknesses in the South Shetlands (e.g. Ramos & Vieira 2003; Ramos *et al.* 2007; Hauck *et al.* 2007; Ramos *et al.* 2008; Schwamborn *et al.* 2008). Ramos *et al.* (2008, p. 1463) argued that Livingston Island, in the South Shetlands, was 'one of the most sensitive regions of Earth to climate change' and had experienced 'a warming trend over the past 50 years'. It is also (p. 1463) a region located close to the climatic limit of permafrost 'since Mean Annual Air Temperature (MAAT) at sea-level is close to $-2\,^{\circ}$C'. The warming, and thickening, of the active layer, with commensurate thinning of the permafrost, links directly to resulting geomorphological processes and landforms (Vieira & Ramos 2003; Vieira *et al.* 2008). Vieira & Ramos (2003, figure 2) provided a series of geomorphological maps identifying eight landforms and/or processes while López-Martínez & Serrano-Cañadas (2005) recognized 14 types of periglacial landforms for Livingston Island in the South Shetland Islands. Key amongst the identified periglacial forms are stone-banked solifluction lobes, rock glaciers and non-sorted circles. López-Martínez *et al.* (in press) provided a recent synthesis of periglacial processes and landforms for the South Shetland Islands.

Further to the early attempts to elucidate weathering processes on Livingston Island (Hall 1993) that identified the action of wetting and drying, subsequent observations (López-Martínez & Serrano-Cañadas 2005) pertaining to 'frost-shattered debris' (Fig. 6) suffer the same lack of data and conceptual failings as the work cited earlier in relation to Heard Island and so should be viewed with caution. Equally, the comments by Navas *et al.* (2008, p. 137) regarding the action of freeze–thaw weathering in soil development on Livingston Island suffer these same weaknesses: 'the main influences on soil development and evolution are physical weathering processes related to cryogenic processes resulting from freeze–thaw cycles. The cryogenic processes lead to rock disintegration'. Vieira *et al.* (2008) also identified, as a result of the permafrost degradation, the occurrence of thermokarst, debris flows and active-layer detachment slides on both Livingston Island and Deception Island. As climate continues to warm, so the South Shetlands, already at the northern boundary for permafrost, can expect to experience on-going geomorphic processes, and resulting landforms, associated with ground-ice and permafrost loss (Vieira *et al.* 2008, p. 1843).

Cannone *et al.* (2006) detailed the relationship and interactions between climate, vegetation and the active layer for both King George Island and Signy Island (in the South Orkney Islands). This is important work for, like the plant–periglacial (biotic–abiotic) relationships noted above in the sub-Antarctic, here too there is a strong influence on the ground thermal regime by the vegetation such that there can be an effect on the active layer thickness. Guglielmin *et al.* (2008) extended this work on Signy Island and showed that, as climate change impacts both biotic and abiotic components of our environment, so too does the biotic (the vegetation) affect the abiotic (the active layer). Interestingly they found that mosses (p. 73) can 'play an important role in cooling the ground'. Unfortunately, subsequent to the extensive work on patterned ground by such as Chambers (1966) and rock weathering (e.g. Hall 1990, 1992), there has been little research aimed directly at periglacial landforms and/or processes. Guglielmin *et al.* (2012) provided some recent data regarding macrolichens and rock weathering for Adelaide Island and Anchorage Island; they also used the data to help date Holocene deglaciation for this region. On King George Island (also in the South Shetlands but at the northern end of the chain) there has been extensive work on sorted patterned ground (e.g. Zhu *et al.* 1996; Jeong 2006), but little since the

Fig. 6. Weathered bedrock on Livingston Island (Maritime Antarctic), typically described as 'frost shattered'.

earlier reviews noted in the Introduction. Liu *et al.* (2004, p. 972), in the foreland of the Nelson ice cap on King George Island, found that amongst the glacigenic sediments may be material resulting from 'weathering due to the frequent diurnal freeze–thaw cycles in summer'. More important than the assumption of freeze–thaw weathering is the recognition of chemical weathering (Liu *et al.* 2004, p. 976) as indicated by the geochemical analysis of proglacial sediments that show, for instance (p. 979), that proglacial sediments have 'been depleted of most elements, as compared with the local bedrocks, except for a slight enrichment of the element K. Of the depleted elements Ca has the largest loss.' The loss of Ca is thought to result from chemical weathering of Ca-bearing plagioclase while the K enrichment (p. 979) may be explained by 'the fact that K released from a small quantity of feldspar is almost retained by the clay minerals'. The wider significance is that of identifying chemical weathering in this cold environment, a process facilitated by the presence of adequate moisture fed by meltwater from ice and snow. It is important to recognize that such weathering is *not* limited by temperature but by moisture availability (see Balke *et al.* 1991 regarding chemical weathering in Antarctica).

On James Ross Island several authors (e.g. Pirrie *et al.* 1997) ascribed off-shore sediments to a periglacial origin. Rock glaciers occur within the region of continuous permafrost on James Ross Island (e.g. Strelin & Malagnino 1992; Strelin & Sone 1998; Fukui *et al.* 2007) and various studies have identified both lobate ice-cemented forms and ice-cored types. Fukui *et al.* (2007, p. 20), based on ground penetrating radar surveys, found that the rock glacier 'has an internal structure similar to that of the "nested spoons" structure common to valley glaciers. This result indicates that the rock glacier was formed through the deformation of a glacier ice core and a thick and continuous debris mantle.' Once more, findings such as this indicate the need for a closer relationship between glacial undertakings and periglacial studies. Further to the glacial–periglacial relationships, Strelin *et al.* (2006) provided some new data related to Holocene landform development and climate change for James Ross Island. They dealt with an ice-free area deglaciated since 10 000 ^{14}C yrs BP and showed that large gelifluction lobes and stone-banked terraces are now inactive. These forms were active *c.* 3900–3000 BP when there was water made available from the melting glaciers. Now, in the absence of sufficient moisture, they are inactive. In a somewhat similar fashion, but not, in this case, linked to glaciers, the protalus ramparts and protalus lobes are still, just, active but their coming demise is suggested based on less snow being available for their operation as the climate warms.

The Antarctic continent

With most of the continent ice-covered, studies are limited to nunataks (e.g. Matsuoka *et al.* 2006), mountain ranges (e.g. Viera *et al.* in press), coastal exposures of rock (e.g. Beyer & Bölter 2002;

Fig. 7. Circum-Antarctic Active Layer Monitoring site near the Italian Station at Terra Nova Bay, Continental Antarctica.

Arocena & Hall 2003) and ice-free oases such as the McMurdo Dry Valleys (e.g. Bockheim *et al.* 2007) or the Schirmacher Oasis (e.g. Bormann & Fritzsche 1995). Experiencing extremes of temperature, dryness and wind (Block 1994), the continent provides a valuable analogue both for other planets and also for parts of the Northern Hemisphere during the last glacial. Here, again following Block (1994, Fig. 1), the islands (e.g. Alexander Island) at the southern end of the Antarctic Peninsula as well as those around the margins of the continent (e.g. the Balleny Islands) will all be considered as part of the continent from the perspective of periglacial activity and landforms. Nichols (1966) provided a valuable review of the geomorphology of Antarctica within which he exemplified many periglacial landforms and provided detail on some of the early work on permafrost. Vieira *et al.* (in press) provided a synthesis of periglacial features for the Patriot Hills, Ellsworth Mountains, mostly linked with debris transfer processes on slopes. Putkonen *et al.* (in press), referring to the McMurdo Dry Valleys, also dealt with regolith transport and how 'braking blocks' can cause cavities below and accumulation above where such blocks occur. Šabacká *et al.* (in press) dealt with aeolian transport in the Taylor Valley, especially in terms of organic matter transport. Meiklejohn & Hall (1997) provided some preliminary information relating to chemical weathering, an attribute limited more by water availability than temperature.

Much of the recent work is related to active layer–permafrost investigations and with associated non-sorted polygonal ground. Details of recent permafrost monitoring boreholes (Fig. 7) are provided in Guglielmin *et al.* (2011*a*) for southern Victoria Land, and by Vieira *et al.* (2010) as part of the CALM-S initiative (Circumpolar Active Layer Monitoring Network–Southern Hemisphere). Active layer thickness typically ranges from 0.2 to 0.7 m, being deeper (>0.9 m) near the coast and shallower (<0.1 m) at higher elevations (Vieira *et al.* 2010, p. 194). Ultimately, as the number of boreholes increases and there is a more extensive database, so it is hoped to develop a detailed model of active layer variation around Antarctica and how this changes with time. As an 'off-shoot' of the extensive permafrost/active layer studies on the continent have come new investigations examining the relationship between periglacial features and vegetation (e.g. Cannone & Guglielmin 2010). While such relationships are still poorly understood for periglacial regions in general, this is more so in Antarctica where there have been few investigations. While vegetation can influence process rates and resultant landforms, the landforms can also have an impact on the vegetation, thereby creating an interactive linkage that is particularly significant in permafrost regions (Cannone & Guglielmin 2009). Cannone & Guglielmin (2010, p. 711) suggested that 'the feature type is more effective than the site location in influencing the patterns of vegetation floristic composition' and that there is 'a possible control of the feature type and ground texture on vegetation cover as well as its floristic composition, independent of the slope'. Further, the micro-relief resulting from periglacial forms (p. 711) 'provides several micro-niches ... relating to wind exposure, snow accumulation and persistence, and thus water availability' that influence vegetation type and cover. Thus, as was reported above for the more vegetation-rich sub-Antarctic (e.g. Haussmann *et al.* 2008, 2009), so even in the vegetation-sparse continent there is a relationship between periglacial processes, landforms and vegetation, an attribute that Cannone & Guglielmin (2010) argued is in need of greater study for this region. Indeed, there are even spatial variations (both horizontally and vertically) within the sorted polygons of Coal nunatak on Alexander Island of cyanobacteria communities (Brinkmann *et al.* 2007, p. 1506) that, as pioneer colonizers, mirror 'the development stage achieved by terrestrial Antarctic ecosystems shortly after the commencement of deglaciation'. Further, such organisms (microbial autotrophs) can play as significant role in 'rock weathering and the release of nutrients that are a prerequisite for successful colonization' (Brinkmann *et al.* 2007, p. 1506); Vincent (1988) provided an extensive bibliography of earlier Antarctic material relating to Antarctic micro-organism habitats, controls and resulting weathering.

A number of studies have focused on rock weathering and resulting landforms, for example, André & Hall (2005), Hall & André (2006) and Strini *et al.* (2008) on tafoni (Fig. 8), Parsons *et al.* (2005) on weathering pits, Hall & André (2010) on cryoplanation, Hall *et al.* (2008*a*, *b*) relating to weathering of granite, and French & Guglielmin (2000), Guglielmin *et al.* (2005) and Matsuoka *et al.* (2006) on granitic landforms, while Elliott (2003) provided a valuable synthesis of information pertaining to rock weathering studies for Antarctica as a whole. Matsuoka *et al.* (2006) attempted to integrate rock weathering with soil formation and erosion, and hence to landscape evolution in the Sør Rondane Mountains, within which they also referred to the formation of tafoni. In respect of mechanical rock weathering processes, Hall *et al.* (2008*a*, *b*) provided new insights both into the role of translucent minerals on the surface energy regime and into mineral-to-mineral temperature differences and micro-scale thermal conditions across a coarse-grained granite rock. Using high-frequency data acquisition (<1 min) and very small thermocouples ($\Phi = <0.15$ mm), as well as small-area (<1 mm Φ) thermal infra-red monitoring, it was possible to monitor temperatures of different individual minerals as well as temperatures at depths of 0.5, 1 and 2 mm below both quartz and feldspar crystals. These studies provided new information relating to granular disintegration in cold, dry climates, while Hall & André (2010) applied such findings to the development of cryoplanation terraces (Fig. 9). In respect of tafoni (Fig. 8), Strini *et al.* (2008, p. 1516) discussed the importance of the thermal regime as a driver for the weathering processes but emphasised that 'it is fracture occurrence that seems to favour tafoni development' as shown by the 'high percentage of tafoni orientated along fracture planes'. Strini *et al.* (2008), as well as Hall & André (2006), presented detail regarding thermal conditions in and around tafoni, while André & Hall (2005) linked process activity with changes in climate; these three papers also provided a

Fig. 8. A number of large tafoni near Terra Nova Bay, Continental Antarctica.

Fig. 9. Cryoplanation terrace near Terra Nova Bay, Continental Antarctica.

good bibliographic background to Antarctic tafoni research.

Arocena & Hall (2003) dealt with biotic linkages to weathering and rock coverings in the Yalour Islands (Fig. 10), while Chen & Blume (2002) dealt with recent studies of rock weathering by lichens and provided a valuable review of available literature for the whole of the Antarctic region. Interestingly they (Chen & Blume 2002, p. 394), like Arocena & Hall (2003) and Guglielmin *et al.* (2011*b*), also identified the possible *protective* role that lichens and biotic coatings may have in respect to weathering. Within the framework of Antarctic debris production processes and landscape development, this opens a further avenue for research: to what extent, and when, might biotic attributes inhibit, rather than enhance, weathering and debris production? Other landforms resulting from rock weathering (e.g. pronival ramparts and rock glaciers) have also been identified for the

Fig. 10. Organic coatings on bedrock in the Yalour Islands (Antarctica).

Antarctic continent. Hedding *et al.* (2010) described the formation of a pronival rampart in Western Dronning Maud Land while rock glaciers have been recorded for the Transantarctic Mountains (e.g. Bockheim 1995) and in Victoria Land (e.g. Baroni & Orombelli 1991; Guglielmin *et al.* 1997); Strelin & Sone (1994) provided a review of rock glaciers in the Antarctic. Hedding *et al.* (2010, p. 359) argued that the constructive rampart feature they observed is neither of glacial origin nor related to a protalus rock glacier, but is solely related to the accumulation of clasts derived from over-snow transport of weathered debris. They do argue (p. 360) that, while a pronival rampart may well be a stage within the development of a rock glacier, this is not the case here as 'the very cold climate of the site' means 'the permafrost (that occurs within the ridge) is very cold' and so unlikely able to facilitate movement. However, in the absence of any supportive data from this site and considering the occurrence elsewhere of rock glaciers, this argument may not be particularly strong.

Indirectly linked with aspects of rock weathering is the cracking of the permafrost ground (Fig. 11) to produce non-sorted polygons (e.g. Berg & Black 1966); the fracturing of the ground, and the resulting patterns, are driven by the same forces that occur in rock under thermal shock (Fig. 12). With regard to this thermal contraction cracking in permafrost, there is some discussion regarding the resulting form, with arguments regarding rectilinear and/or polygonal forms and what the meaning of these may be (e.g. van Gesselt *et al.* 2005; Chan *et al.* 2008; Ulrich *et al.* 2011). The main difference between polygons in rock and most of those in permafrost is that the latter are normally in unconsolidated sediments. However, van Gesselt *et al.* (2005) did find, at least for Mars, a coincidence of polygons on sunlit slopes, and that pattern development is linked to illumination conditions. Although van Gasselt *et al.* (2005, p. 12) discussed such features on Mars (which could be used as an analogue for Antarctica – see below), they did suggest that the 'long sinuous cracks predate the perpendicular and the fractures that make up the orthogonal polygons are caused by a single cracking', and that 'during further development, patterned ground matures and the polygons become smaller and more regular'. Levy *et al.* (2011, Fig. 1) provided a plot of mean annual temperature against mean annual precipitation and showed, for both Earth and Mars, the ideal conditions for polygon formation. However, as Ulrich *et al.* (2011, p. 214) pointed out, there is great 'difficulty (in) addressing the question of what factors govern the size and shape of polygons'. Ultimately, the Antarctic provides an ideal site, with both ice wedge and sand wedge polygons, for such studies, and as Ulrich *et al.* (2011, p. 214) concluded, 'Field surveys … in terrestrial permafrost environments are the necessary complement and also provide the ground truth for remote sensing data analyses' – the field attribute being that which characterizes the nature of much periglacial science in the Antarctic.

Marchant & Head (2004, 2007) detailed three microclimate zones for the Antarctic Dry Valley area: a coastal thaw zone, an inland mixed zone,

Fig. 11. Non-sorted contraction crack polygons in permafrost (Antarctica).

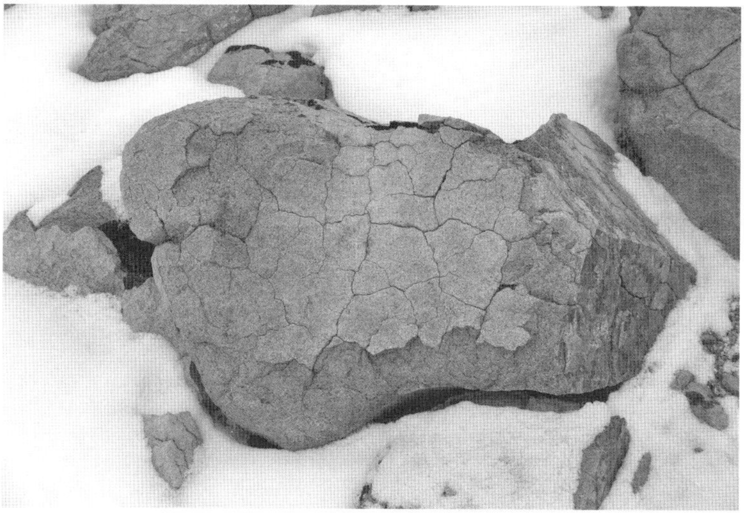

Fig. 12. Polygonal fracturing (owing to thermal shock?) in rock (Antarctica).

and a stable upland zone. They found that these different zones exhibited distinct differences in landforms and landform morphology. For example, they identified (Marchant & Head 2007, p. 192) that slopes in the coastal zone exhibited asymmetry owing to the north-facing slopes experiencing greater radiation inputs and hence warmer ground temperatures that, in turn, caused more melting and deeper meltwater penetration of the ground and hence more solifluction; drainage basins are also found to exhibit the same asymmetry. Here there is also salt action on rocks owing to the evaporation of (relatively abundant) water, and the meltwater also helps in the development of gullies (that may be as deep as 5 m when in loose debris). The inland mixed zone shows less valley asymmetry, but blown snow could redistribute water sources (when it melts) and this might, particularly in conjunction with polygons, begin to form gullies. The cold, stable upland zone inhibits significant meltwater and any gullies found here are thought to be inactive. Extending the applicability of this three Antarctic Dry Valley microclimate zones and the landforms described above, Levy *et al.* (2009*a*, p. 114) considered the geomorphological effects of 'water moving through temperature-dependent phase transitions (freezing, melting, sublimation, evaporation)' in respect of the development of extensive gully–polygon systems. Levy *et al.* (2009*a*, p. 114) observed that, in areas of 'extensive pore ice, the ground commonly showed well-developed thermal contraction crack polygons' and that 'active and recently active gullies are typically present in association with contraction-crack polygons'. Ultimately Levy *et al.* (2009*a*, p. 122) discerned that there was 'evidence of polygon-influenced water ice accumulation, polygonal patterns of elongate alcoves, annexation of polygon troughs by gully channels, and dissection of fans by underlying polygons'. Kowalewski *et al.* (2012) further investigated the Antarctic polygons using a 2D numerical model to evaluate vapour diffusion through porous media and the sublimation rates for buried glacier ice. It was found that 'the development of thermal contraction-crack polygons modulates subsurface ice loss', but that, during the initial stages of polygon formation, 'locally enhanced sublimation beneath immature sand wedges leads to the development of deep, marginal troughs (and high polygon centres).'

As a useful segue to the next section, the early work by Hall (1989), which suggested wind-blown sand as a weathering (rather than just erosion) agent, a process suggested as potentially highly active on the inner planets of our solar system, has been recently expanded by Várkonyi & Laity (2012). Together with Seppälä (2004) and Laity & Bridges (2009), these papers provided a valuable bibliography of ventifact studies relating to both Earth (especially the Antarctic) and Mars. Hall (1989), from work in the McMurdo Dry Valleys, suggested that wind-blown particles may 'wedge' open previously existing cracks and thereby act as a weathering agent; Figure 13 shows a recent photo of such action from the Terra Nova Bay region with quartz particles wedged into a dolerite. Várkonyi & Laity (2012), the work of whom is equally applicable to the other Antarctic regions where the role of wind was previously noted,

Fig. 13. Sand grains wedged into rock by wind (Antarctica).

observed that (in particular) sand abrasion does not ultimately destroy the features it creates, but rather maintains them and that features are initiated by small initial surface irregularities. In Antarctica, the movement of sand grains by the wind also produces a lag gravel (Seppälä 2004) – an 'armoured surface' or 'desert pavement' (Seppälä 2004), as seen in the McMurdo Dry valleys. Further, Kowalewski *et al.* (2012) investigated the role this rocky pavement might play on subsurface ice sublimation, especially when salts are present, and found that it can decrease ice sublimation by as much as 60%. Overall, it is clear that these new avenues of investigation are important but as yet understudied, and that there are important ramifications for the understanding of processes on Mars. As a meaningful analogue on Earth, the Antarctic is the key location for such studies (see comments within Conclusions below).

From Antarctica to other planets

An interesting component of the continental permafrost and periglacial milieu is its capacity as an analogue for nearby planets (e.g. Parsons *et al.* 2005) and asteroids (e.g. Dombard *et al.* 2010). The continental Antarctic's cold, windy, dry environment makes for the best available Earth-bound analogue for the Martian surface (Berkley & Drake 1981; Laity & Bridges 2009, Mellon *et al.* 2008, 2009). Although this comparability has been known for some time (e.g. Lucchitta 1981; Berkley & Drake 1981), there has been, in response to the ever-greater data and visual information from Mars, a significant increase in such comparative undertakings in recent years. For example, Marchant & Head (2004), extended in Marchant & Head (2007), have looked at microclimatic zonation in the Antarctic Dry Valleys and used this as a foundation for evaluating landscape evolution and climatic change on Mars. The Dry Valley microclimatic zonation is seen to have comparability with Mars, and this could (Marchant & Head 2007, p. 218) offer an 'important insight into the current climate on Mars, and the geomorphological evidence for climatic change in the past history of Mars'.

Levy *et al.* (2009*b*) investigated the cold, dry processes of the Martian Arctic and found imagery indicative of features thought to be due to both physical weathering and periglacial processes, comparable to those seen in the McMurdo Dry Valleys. Certainly, one of the most commonly identified forms is that of thermal contraction cracks in the regolith that are 'strikingly similar' (Levy *et al.* 2009*b*, p. L210203) to 'stable (non-churning) permafrost processes' seen in Antarctica; Helfenstein (1981), Fisher (2005) and Gasselt *et al.* (2005) all discussed such thermal cracking (up to tens of metres in size) relating to non-sorted polygonal ground. Others who see similarities between Mars and Antarctica include Chan *et al.* (2008), Fisher (2005), Helfenstein (1981), Levy *et al.* (2009*b*, 2011) and Hallet *et al.* (2011). Indeed, Hallet *et al.* (2011) extended the work on polygonal patterned ground to include discussion regarding the associated micro-relief and its linkage with movement of loose granular material. Mahaney *et al.* (2009) have also identified stone-banked lobes on the Asteroid 25143-Itokawa, the finding of which argued in favour of water being available at some time, possibly related to melting of interstitial ice. Pain *et al.* (2007), in a discussion regarding relief inversion on Mars, used a number of identified process mechanisms from the Antarctic as potential explanations, while Morgan *et al.* (2010) argued that gully formation on Mars, associated with fluvial action, has similarities (p. 658) to gullies cut by 'seasonal and perennial snowmelt in the Antarctic Dry Valleys'. Importantly, as extraterrestrial periglacial studies increase, so they provide another way of looking at our Earth-based concepts and perhaps may offer new, different, insights into how we evaluate both processes and landforms. With the Antarctic as the key Martian analogue, it may offer new approaches and concepts that can be applied to the past and present situation here on Earth.

Rock weathering has also elicited much interest, with Leask & Wilson (2003), Jagoutz (2006), Levy *et al.* (2009*b*), Chan *et al.* (2008) and Head & Marchant (2005) discussing this for Mars, while Mahaney *et al.* (2009) and Dombard *et al.* (2010) consider weathering on asteroids. Not surprisingly, the dominant theme is mechanical weathering, related primarily to thermal stresses (e.g. Dombard *et al.* 2010), but also considering salt weathering. However, Wyatt *et al.* (2004, p. 647) suggested that the Dry Valleys are also a 'terrestrial analogue for characterizing chemical weathering on Mars because of the cold hyperarid environment, stable permafrost, and ground ice'. In respect of the mechanical weathering, Jagoutz (2006) noted that the salt weathering, resulting from progressive salt concentration owing to the evaporative environment, as observed in the Dry Valleys, offers some insights into that taking place on Mars. Jagoutz (2006, p. 697) further suggested that the possibility of salt-induced fragmentation of rock on Mars is intimated because the 'situation in some of these locations (i.e. the Antarctic Dry Valleys) is very similar to the landing sites on Mars'. However, Levy *et al.* (2009*b*, p. 4), based on information from the Phoenix landing site, suggested that there are some distinct geomorphic *dis*similarities between that site and the Antarctic Dry Valleys: 'two notable geomorphic elements typical of the Beacon Valley are not observed at Phoenix: salt accumulations beneath rocks and soil weathering horizons'. Importantly, in contrast to Jagoutz (2006), Levy *et al.* (2009*a*, *b*) noted that analysis of sediment beneath rocks on Mars did not identify increased salt concentrations and that this (p. 4) suggests 'a lack of recently flowing, wicking, or percolating brines'.

In many ways the potential for the Antarctic as an analogue for Mars and other planetary bodies is only just beginning. As more and more data and imagery become available, the ability to test ideas and hypotheses will rely more and more on the terrestrial analogues and, hopefully, in so doing we will begin to understand more, and better, the Antarctic processes and geomorphic history.

Discussion

The periglacial milieu of the Antarctic has long remained somewhat under-studied, rarely featuring large within most national programmes. This has begun to change in recent years as reflected by the ever-increasing number of publications, aided in part by implementation of the multinational 'Circum-Antarctic Active Layer Monitoring' and 'Latitudinal Gradient Project' undertakings, the linkage of permafrost to preservation of early bacteria (e.g. Gilichinsky *et al.* 2007), and the extension of this to the role of permafrost in respect to both life and landforms on other planets.

Where all of this becomes particularly significant is in regard to our certainty regarding the climatic implications of the Antarctic landforms such that they can be used as analogues for past (or future) conditions, or as Martian counterparts. While some aspects such as contraction cracking in permafrost appear to be solidly founded on extensive research (e.g. Hallet *et al.* 2011), other aspects, particularly rock weathering, are much more contentious (e.g. 'frost riving' as suggested by Mahaney *et al.* 2009, p. 50: see Hall & Thorn 2011). The understanding of weathering and its linkage with landforms (Fig. 14), particularly within cold regions (encompassing all of the Antarctic environments), has remained somewhat insular and stilted. Although Gordon & Birnie (1986) attempted to link glacial deposits to weathering production of debris for South Georgia, few others have attempted to integrate the different regimes in this manner. Equally, much of the periglacial weathering concepts have remained fixated on freeze–thaw weathering (e.g. Matsuoka 1995), despite the absence of data regarding water availability or evidence that water ever actually froze within the rock (see Hall & Thorn 2011). However, in recent years there has been an increase in the recognition of the role of salts (e.g. Matsuoka *et al.* 1996), thermal stresses (e.g. Hall *et al.* 2008*b*) and biological processes (e.g. Friedmann 1982; Guglielmin *et al.* 2011*b*) as weathering agents, particularly within the Antarctic environment. In respect to landforms, recent studies have suggested that the 'classic' solifluction forms, the stone runs, of the Falkland Islands, may have a tropical origin (André *et al.* 2008) and process studies on active cryoplanation terraces on Alexander Island failed to show any of the freeze–thaw activity (Hall & André 2010) that is normally

Fig. 14. Enigmatic weathering form in coarse granite, Antarctica. Many questions (not least, the rounded nature of the weathering form) arise in trying to explain the origin of features such as this.

cited as their causative mechanism (e.g. Reger & Péwé 1976; Czudek 1995). Even the 'rock furrows' of North Victoria Land (Fig. 15), said to be a result of 'cryogenic weathering' by French & Guglielmin (2000, 2002), may have an origin more closely aligned with solution weathering (e.g. Wray 1997) and the development of rinnenkarren (e.g. Veress 2010) or röhrenkarren (e.g. Simms 2002) by flowing water under a warmer climate; the recent study by Yokoyama & Matsukura (2006) regarding the weathering of granodiorite under both dry and wet conditions lends some credence to this idea. Indeed Guglielmin *et al.* (2005, p. 83) alluded to this possible 'hot and humid tropics' origin for the North Victoria Land rock furrows but did not develop the idea.

With water being the key limiting factor for weathering in the Antarctic today, so studies there provide a venue for (perhaps) a more realistic evaluation of weathering during and at the end of the

Fig. 15. Furrows and 'tubes' developed in granite, Antarctica. The question arises as to whether these are the product of cold climate processes or simply being 'preserved' (or little altered) under the present cold, dry climate.

last glacial in the Northern Hemisphere when conditions were substantially drier than present. Present day investigations of weathering in much of Europe may well not be suitable analogues for past, glacial-climate conditions. Equally, with such an approach, it may be possible to begin to discern landforms that, while they are *found* in present cold (Antarctic or other) environments, the nature of cold-region weathering indicates owe their *origin* to warmer wetter climates and are merely being preserved or modified under the present cold conditions. For example, thoughts regarding blockfields (e.g. Whalley *et al.* 2004) are moving away from the earlier, simplistic cold-based (periglacial) origins (e.g. Ballantyne & Harris 1994) to more 'complex process interactions and climatic histories that defy straightforward palaeoclimatic interpretations' (Boelhouwers 2004, p. 133); see also André (2003) for a wider discussion on the *non*-periglacial origin of 'periglacial' landscapes. Such ideas might have a significant impact on the viability of landforms currently used as palaeoenvironmental indicators (e.g. the common use of cryoplanation as an indicator of permafrost; Reger & Péwé 1976; Péwé *et al.* 1982; Priesnitz 1988) and hence their application to reconstructions on Mars.

Finally, it is worth noting the importance of the development, across the whole of the Antarctic region, of studies dealing with biotic–abiotic interactions within the periglacial domain. This is a very important direction, for often those studying biotic attributes ignore the abiotic interactions, and vice versa, as Butler (1995) discussed. Further, the recognition of the biotic–abiotic synergies is of particular importance at the micro-scale when considering rock weathering (see Hall & André 2006; Hall *et al.* 2008*a*). As the studies to date clearly show, there are important biotic–abiotic synergies that we need to take in to account when considering long-term landform development in cold regions. Not only do these studies help us understand present day processes and permafrost responses, but they also provide a valuable analogue for the conditions in the Northern Hemisphere as the great ice sheets withdrew and plants slowly moved in to colonize. Such studies helps provide a better understanding of permafrost/active layer changes and the factors that influenced these, plus the often ignored biotic influence on landform development.

Conclusions

In some ways, the significance and importance, as well as the reality, of Antarctic periglacial studies is perhaps best summed up by Elliott (2003, p. 58) in her remarks relating to rock weathering: 'Unfortunately there appears little evidence that research in the Northern hemisphere recognizes the contribution that the Antarctic studies make to the better understanding of rock weathering in cold climates, and yet future advances in this understanding may well occur in Antarctica where the emphasis is on field work rather than laboratory studies.' In truth, many Antarctic periglacial studies are dealing with processes and/or landforms for which there is no active component within the Northern Hemisphere (e.g. cryoplanation terraces), while at the same time being our best analogue for other planets and asteroids. Further, following the observation by Elliott (2003), it is the reality of the situation, at least in the periglacial domain, that almost *all* the work is done in the field. While field work alone has both pros and cons, the reality cannot be ignored that the veracity of a laboratory study, following Warren (1914, p. 413), is such that it would be unsound to 'assume that the result of a certain experiment must also be produced by natural agencies, without *evidence* that similar conditions exist in Nature to those employed in the experiment' (my italics). Thus, as valuable as laboratory studies must be, they may be redundant and even misleading if not founded upon conditions *known* to occur in the field, and certainly *cannot* be applied back to the field unless those conditions are known to occur: 'laboratory experiments are ultimately only worthwhile if it can be shown that they mimic natural conditions successfully' (Thorn 1988, p. 13) and, further, 'the general validity of laboratory experiments is open to question because of the scale of those experiments' (see also comments on scale in Hall 2006). Thus, Antarctica does indeed provide the ideal 'laboratory' for periglacial studies and the very nature of the undertakings there has been one that continues to produce valuable baseline data for the underpinning of meaningful experimentation in laboratory simulations of periglacial processes.

Ultimately, the Antarctic still offers a remarkable laboratory for periglacial studies on a host of topics and provides a number of landforms that have a complex climatic history that may confound many classic periglacial assumptions. The biggest constraints are the combination of distance/costs and of their inclusion within national programmes. To date, periglacial attributes have experienced a very low priority within many programmes, but this is certainly changing. A further driver to this change is the more recent recognition regarding the connectivity between biotic and abiotic attributes, and how these may adjust under a changing climate – and how some of these may be excellent indicators of climatic change. Glacial–periglacial links still remain, sadly, highly tenuous, but even here there is a slow change and it is certain that the Antarctic, with its range of environments, is an ideal region for such studies. Linkages to near-solar system studies also provide a strong impetus for better, longer and more far-reaching Antarctic studies. On a final, personal, note, I believe that we have only just begun in respect of the vast Antarctic potential for periglacial studies, a potential that crosses all three zones (and indeed has strong temporal/climatic linkages across them), and that future findings will confound many of our long-cherished concepts; I certainly hope so!

I would like to thank the people who have, over the past 35 years, provided me with the opportunity to follow periglacial studies in the Antarctic: E. M. van Zinderen Bakker (sadly now deceased), who gave me my first opportunity with the South African National Programme, D. Walton of the British Antarctic Survey and M. Guglielmin of the Italian National Programme deserve special mention. D. Walton was also a superb mentor and challenger of academic boundaries whose support has meant much during my career. I would also like to thank M. Hambrey for offering me this opportunity to update the review of Antarctic periglacial work. There are too many others to name whom I have worked with or been in contact with during my career, but they have spent much time educating, correcting, 'misdirecting' and helping me; I thank them all.

References

ADLAM, L. S., BALKS, M. R., SEYBOLD, C. A. & CAMPBELL, D. J. 2010. Temporal and spatial variation in active layer depth in the McMurdo Sound region, Antarctica. *Antarctic Science*, **22**, 45–52.

ANDERSSON, J. G. 1906. Solifluction: a component of subaerial denudation. *Journal of Geology*, **14**, 91–112.

ANDERSON, D. M., GATTO, L. W. & UGOLINI, F. C. 1972. An Antarctic analog of Martian permafrost terrain. *Antarctic Journal of the United States*, **7**, 114–116.

ANDRÉ, M.-F. 2003. Do periglacial landscapes evolve under periglacial conditions? *Geomorphology*, **52**, 149–164.

ANDRÉ, M.-F. & HALL, K. 2005. Honeycomb development on Alexander Island, glacial history of George VI sound and palaeoclimatic implications (two steps Cliffs/Mars Oasis, W. Antarctica). *Geomorphology*, **65**, 117–138.

ANDRÉ, M.-F., HALL, K., BERTRAN, P. & AROCENA, J. 2008. Stone runs in the Falklands: periglacial or tropical? *Geomorphology*, **95**, 524–543.

AROCENA, L. & HALL, K. 2003. Calcium phosphate accumulation on the Yalour Islands (Antarctica): surface coatings and geomorphic implications. *Arctic, Antarctic and Alpine Research*, **35**, 233–241.

BALKE, J., HAENDAL, D. & KRÜGER, W. 1991. Contribution to the weathering-controlled removal of chemical elements from the active debris layer of the Schirmacher Oasis, East Antarctica. *Zeitschrift für Geologische Wissenschaften*, **19**, 153–158.

BALLANTYNE, C. K. & HARRIS, C. 1994. *The Periglacial of Great Britain*. Cambridge University Press, Cambridge.

BARONI, C. & OROMBELLI, G. 1991. Holocene raised beaches at Terra Nova Bay, Victoria Land, Antarctica. *Quaternary Research*, **36**, 157–177.

BARSCH, D. 1996. *Rock Glaciers*. Springer, Berlin.

BERG, O. 2009. *Small scale spatial and temporal variability of micro-climate in a fellfield landscape, Marion Island*. Arbetsrapporter Kulturgeografiska institutionen, Uppsala Universitet, **690**, 46.

BERG, T. E. & BLACK, R. F. 1966. Preliminary measurements of growth of nonsorted polygons, Victoria Land, Antarctica. *In*: TEDROW, J. C. F. (ed.) *Antarctic Soils and Soil Forming Processes*. American Geophysical Union, Washington, DC, Publication **1418**, 61–108.

BERKLEY, J. L. & DRAKE, M. J. 1981. Weathering of Mars: Antarctic analog studies. *Icarus*, **45**, 231–249.

BEYER, L. & BÖLTER, M. (eds) 2002. *Geoecology of Antarctic Ice-Free Coastal Landscapes*. Springer, Berlin.

BIRNIE, R. V. & THOM, G. 1982. Preliminary observations on two rock glaciers in South Georgia, Falkland Island Dependencies. *Journal of Glaciology*, **28**, 377–386.

BLOCK, W. 1994. Terrestrial ecosystems: Antarctica. *Polar Biology*, **14**, 293–300.

BOCKHEIM, J. G. 1995. Permafrost distribution in the southern circumpolar region and its relation to the environment: a review and recommendations for further research. *Permafrost and Periglacial Processes*, **6**, 27–45.

BOCKHEIM, J. & HALL, K. 2002. Periglacial processes and landforms of the Antarctic continent: a review. *South African Journal of Science*, **98**, 82–101.

BOCKHEIM, J. G., CAMPBELL, I. B. & MCLEOD, M. 2007. Permafrost distribution and active-layer depths in the McMurdo Dry Valleys, Antarctica. *Permafrost and Periglacial Processes*, **18**, 217–227.

BOELHOUWERS, J. 2003. Sensitivity and responses to climate change in the Subantarctic periglacial environment. *In*: PHILIPS, M., SPRINGMAN, S. M. & ARENSON, L. U. (eds) *Permafrost*. A.A. Balkema, Lisse, 67–72.

BOELHOUWERS, J. 2004. New perspectives on autochthonous blockfield development. *Polar Geography*, **28**, 133–146.

BOELHOUWERS, J., HOLNESS, S. & SUMNER, P. 2003. The maritime Subantarctic: a distinct periglacial environment. *Geomorphology*, **52**, 39–55.

BOELHOUWERS, J. C., MEIKLEJOHN, K. I., HOLNESS, S. D. & HEDDING, D. W. 2008. Geology, geomorphology and Climate. *In*: CHOWN, S. L. & FRONEMAN, P. W. (eds) *The Prince Edward Islands: Land–Sea Interactions in a Changing Ecosystem*. Sun Press, Stellenbosch, 65–96.

BÖLTER, M., BEYER, L. & STONEHOUSE, B. 2002. Antarctic coastal landscape: characteristics, ecology and research. *Ecological Studies*, **154**, 5–15.

BORMANN, P. & FRITZSCHE, D. (eds) 1995. *The Schirmacher Oasis, Queen Maud Land, East Antarctica, and its Surroundings*. Justus Perthes Verlag Gotha, Darmstadt.

BRINKMANN, M., PEARCE, D. A., CONVET, P. & OTT, S. 2007. The cyanobacterial community of polygon soils at an inland Antarctic nunatak. *Polar Biology*, **30**, 1505–1511.

Butler, D. R. 1995. *Zoogeomorphology: Animals as Geomorphic Agents*. Cambridge University Press, Cambridge.

Cannone, N. & Guglielmin, M. 2009. Influence of vegetation on the ground thermal regime in continental Antarctica. *Geoderma*, **151**, 215–223.

Cannone, N. & Guglielmin, M. 2010. Relationships between periglacial features and vegetation development in Victoria Land, continental Antarctica. *Antarctic Science*, **22**, 703–713.

Cannone, N., Ellis Evans, J. C., Strachan, R. & Guglielmin, M. 2006. Intercations between climate, vegetation and the active layer in soils at two Maritime Antarctic sites. *Antarctic Science*, **18**, 323–333.

Chambers, M. J. G. 1966. Investigations of patterned ground at Signy Island, South Orkney Islands: I, Interpretation and mechanical analyses. *British Antarctic Survey Bulletin*, **9**, 21–40.

Chan, M. A., Yonkee, W. A., Netoff, D. I., Seiler, W. M. & Ford, R. L. 2008. Polygonal cracks in bedrock on Earth and Mars: implications for weathering. *Icarus*, **194**, 65–71.

Chen, J. & Blume, H.-P. 2002. Rock-weathering by lichens in Antarctic: patterns and mechanisms. *Journal of Geographical Sciences*, **12**, 387–396.

Clapperton, C. M. 1975. Further observations on the stone runs of the Falkland Islands. *Biuletyn Peryglacjalny*, **24**, 211–217.

Clark, R. & Wilson, P. 1992. Occurrence and significance of ventifacts in the Falkland Islands, South Atlantic. *Geografiska Annaler*, **74A**, 3546.

Clark, R., Edwards, E., Luxton, S., Shipp, T. & Wilson, P. 1995. Geology in the Falkland Islands. *Geology Today*, **11**, 217–223.

Czudek, T. 1995. Cryoplanation terraces – A brief review and some remarks. *Geografiska Annaler, Series A, Physical Geography*, **77**, 95–105.

Dombard, A. J., Barnouin, O. S., Prockter, L. M. & Thomas, P. C. 2010. Boulders and ponds on the Asteroid 433 Eros. *Icarus*, **210**, 713–721.

Elliott, C. 2003. Rock weathering processes in Antarctica: a comparison of some recent studies with those from the Northern hemisphere. *New Zealand Geographer*, **59**, 50–60.

Engelskjøn, T. 1986. General outline of the botanical investigations on Bouvetøya. *Norsk Polarinstitutt, Skrifter*, **185**, 5–9.

Engelskjøn, T. & Jørgensen, P. M. 1986. Phytogeographical relations of the cryptogamic flora of Bouvetøya. *Norsk Polarinstitutt, Skrifter*, **185**, 71–79.

Fisher, D. A. 2005. A process to make massive ice in the martian regolith using long-term diffusion and thermal cracking. *Icarus*, **179**, 387–397.

Fitzsimons, S. J. 1989. Reinterpretation of pingos in Antarctica. *Quaternary Research*, **32**, 114–116.

French, H. M. 2007. *The Periglacial Environment*. Wiley, Chichester.

French, H. M. & Guglielmin, M. 2000. Cryogenic weathering of granite, Northern Victoria Land, Antarctica. *Permafrost and Periglacial Processes*, **11**, 305–314.

French, H. M. & Guglielmin, M. 2002. Observations on granite weathering phenomena, Mount Keinath, Nortehrn Victoria Land, Antarctica. *Permafrost and Periglacial Processes*, **13**, 231–236.

Frenot, Y., Gloaguen, J. C., Picot, G., Bougère, J. & Benjamin, D. 1993. *Azorella selago* Hook, used to estimate glacier fluctuations and climatic history in the Kerguelen Islands over the last two centuries. *Oecologia*, **95**, 140–144.

Frenot, Y., Bergstrom, D. M., Gloaguen, J. C., Tavenard, R. & Strullu, D. G. 2005. The first record of mycorrhizae on sub-Antarctic Heard Island: a preliminary examination. *Antarctic Science*, **17**, 205–210.

Friedmann, E. I. 1982. Endolithic microorganisms in the Antarctic cold desert. *Science*, **215**, 1045–1053.

Fukuda, M., Shimokawa, K., Takahashi, N. & Sone, T. 1992. Permafrost in Seymour Island and James Ross Island, Antarctic Peninsula region. *Geographical Review of Japan, Series A*, **65**, 124–131.

Fukui, K., Sone, T., Strelin, J., Torielli, C. & Mori, J. 2007. Ground penetrating radar sounding on an active rock glacier on James Ross Island, Antarctic Peninsula Region. *Polish Polar Research*, **28**, 13–22.

Gasselt, van S., Reiss, D., Thorp, A. K. & Neukum, G. 2005. Seasonal variations of polygonal contraction crack patterns in a south polar trough, Mars. *Journal of Geophysical Research*, **110**, E08002, 15.

Gilichinsky, D. A., Wilson, G. S. *et al.* 2007. Microbial populations in Antarctic permafrost: biodiversity, state, age, and implications for astrobiology. *Astrobiology*, **7**, 275–311.

Gordon, J. E. & Birnie, R. V. 1986. Production and transfer of subaerially generated rock debris and resulting landforms on South Georgia: an introductory perspective. *British Antarctic Survey Bulletin*, **72**, 25–46.

Gremmen, N. J. M. & Smith, V. R. 2008. Terrestrial vegetation and dynamics. *In*: Chown, S. L. & Froneman, P. W. (eds) *The Prince Edward Islands: Land–Sea Interactions in a Changing Ecosystem*. Sun Press, Stellenbosch, 215–244.

Guglielmin, M. 2012. Advances in permafrost and periglacial research in Antarctica: a review. *Geomorphology*, **155-156**, 1–6.

Guglielmin, M., Biasini, A. & Smiraglia, C. 1997. The contribution of geoelectrical investigations in the analysis of periglacial and glacial landforms in ice free areas of the Northern Foothills (Northern Victoria Land, Antarctica). *Geografiska Annaler*, **79A**, 17–24.

Guglielmin, M., Cannone, N., Strini, A. & Lewkowicz, A. G. 2005. Biotic and abiotic processes on granite weathering landforms in a cryotic environment, Northern Victoria Land, Antarctica. *Permafrost and Periglacial Processes*, **16**, 69–85.

Guglielmin, M., Ellis Evan, C. J. & Cannone, N. 2008. Active layer thermal regime under different vegetation conditions in permafrost areas. A case study at Signy Island (Maritime Antarctica). *Geoderma*, **144**, 73–85.

Guglielmin, M., Balks, M. R., Adlam, L. S. & Baio, F. 2011*a*. Permafrost thermal regime from two 30 m deep boreholes in southern Victoria Land, Antarctica. *Permafrost and Periglacial Processes*, **22**, 129–139.

Guglielmin, M., Favero-Longo, S. E., Cannone, N., Piervittori, R. & Strini, A. 2011*b*. Role of lichens in cold and aride environments of continental Antarctica. *In*: Martini, I. P., French, H. M. & Perez Alberti, A. (eds) *Ice-Marginal and Periglacial*

Processes and Sediments. Geological Society, London, Special Publications, **354**, 195–203.

Guglielmin, M., Worland, R., Convey, P. & Cannone, N. 2012. Schmidt hammer studies in the maritime Antarctic: application to dating Holocene deglaciation and estimating the effects of macrolichens on rock weathering. *Geomorphology*, **155-156**, 34–44.

Hall, K. 1979. Sorted stripes orientated by wind action: some observations from sub-Antarctic Marion Island. *Earth Surface Processes*, **4**, 281–289.

Hall, K. 1983. Sorted stripes on sub-Antarctic Kerguelen Island. *Earth Surface Processes and Landforms*, **8**, 115–124.

Hall, K. 1984. Observations of some periglacial features and their palaeoenvironmental implications on sub-Antarctic islands Marion and Kerguelen. *South African Journal of Antarctic Research*, **13**, 35–40.

Hall, K. 1987. Periglacial landforms and processes of the subantarctic and Antarctic islands. *Palaeogeography of Africa, Antarctica and the Surrounding Islands*, **18**, 383–392.

Hall, K. 1989. Wind blown particles as weathering agents? An Antarctic example. *Geomorphology*, **2**, 405–410.

Hall, K. 1990. Mechanical weathering rates on Signy Island, maritime Antarctic. *Permafrost and Periglacial Processes*, **1**, 61–67.

Hall, K. 1992. Mechanical weathering in the Antarctic: a maritime perspective. *In*: Dixon, J. & Abrahams, A. (eds) *Periglacial Geomorphology*. Binghampton Series in Geomorphology, John Wiley, Chichester, 103–123.

Hall, K. 1993. Rock moisture data from Livingston Island, (Maritime Antarctic) and implications for weathering studies. *Permafrost and Periglacial Processes*, **4**, 245–253.

Hall, K. 1994. Some observations regarding sorted patterned ground on Livingston Island, Antarctica. *Permafrost and Periglacial Processes*, **5**, 119–126.

Hall, K. 1997. Observations regarding 'cryoplanation' benches in Antarctica. *Antarctic Science*, **9**, 181–187.

Hall, K. 2004. Quaternary glaciations of the sub-Antarctic islands. *In*: Ehlers, J. & Gibbard, P. C. (eds) *Asia, Africa, Latin America, Australia and Antarctica*. INQUA Commission on Glaciation, Work Group: Extent and Chronology of Glaciations, **3**, 339–346.

Hall, K. 2006. Perceptions of rock weathering: some thoughts on attributes of scale. *Géomorphologie*, **3**, 187–196.

Hall, K. & André, M.-F. 2006. Temperature observations in Antarctic tafoni: implications for weathering, biological colonization, and tafoni formation. *Antarctic Science*, **18**, 377–384.

Hall, K. & André, M.-F. 2010. Cryoplanation terraces on Alexander Island: weathering data and observations regarding formation and development. *Antarctic Science*, **22**, 175–183.

Hall, K. & Boelhouwers, J. 2003. Bibliography on Southern hemisphere permafrost and periglacial landforms and processes. *Circumpolar Active-Layer Permafrost System* (National Snow and Ice Data Center CD-ROM), **3**, 97.

Hall, K. & Bühmann, D. 1989. Palaeoenvironmental reconstruction from redeposited weathered clasts in the CIROS-1 drill core. *Antarctic Science*, **1**, 235–238.

Hall, K. & Meiklejohn, K. I. 2011. Glaciation in southern Africa and in the sub-Antarctic. *In*: Ehlers, J., Gibbard, P. C. & Hughes, P. D. (eds) *Quaternary Glaciations – Extent and Chronology: A Closer Look*. Elsevier, Amsterdam, 1081–1085.

Hall, K. & Thorn, C. 2011. The historical legacy of spatial scales in cold region weathering: misrepresentation and resulting misdirection. *Geomorphology*, **130**, 83–90.

Hall, K. & Walton, D. W. H. 1992. Rock weathering, soil development and colonisation under a changing climate. *Philosophical Transactions of the Royal Society*, **B338**, 269–277.

Hall, K., Arocena, J. & Smellie, J. 2001. Weathering rinds as palaeoenvironmental indicators: evidence from the Cape Roberts drill core (CRP-3) Victoria Land Basin, Antarctica. *Terra Antartica*, **8**, 275–280.

Hall, K., Thorn, C., Matsuoka, N. & Prick, A. 2002. Weathering in cold regions: some thoughts and perspectives. *Progress in Physical Geography*, **4**, 576–602.

Hall, K., Guglielmin, M. & Strini, A. 2008*a*. Weathering of granite in Antarctica I: light penetration in to rock and implications for rock weathering and endolithic communities. *Earth Surface Processes and Landforms*, **33**, 295–307.

Hall, K., Guglielmin, M. & Strini, A. 2008*b*. Weathering of granite in Antarctica II: thermal data at the grain scale. *Earth Surface Processes and Landforms*, **33**, 475–493.

Hall, K., Meiklejohn, I. & Bumby, A. 2011. Marion Island (sub-Antarctic) volcanism and glaciation new findings and reconstructions. *Antarctic Science*, **23**, 155–163.

Hallet, B., Sletten, R. & Whilden, K. 2011. Microrelief development in polygonal patterned ground in the Dry Valleys of Antarctica. *Quaternary Research*, **75**, 347–355.

Hansom, J. D., Evans, D. J. A., Sanderson, D. C. W., Bingham, R. G. & Bentley, M. J. 2008. Constraining the age and formation of stone runs in the Falkland Islands using Optically Stimulated Luminescence. *Geomorphology*, **94**, 117–130.

Hauck, C., Vieira, G., Gruber, S., Blanco, J. & Ramos, M. 2007. Geophysical identification of permafrost in Livingston Island, maritime Antarctica. *Journal of Geophysical Research*, **112**, F02S19, 1–19.

Haussmann, N. S., McGeoch, M. A. & Boelhouwers, J. C. 2008. Interactions between a cushion plant (*Azorella selago*) and surface sediment transport on sub-Antarctic Marion Island. *Geomorphology*, **107**, 139–148.

Haussmann, N. S., Boelhouwers, J. C. & McGeoch, M. A. 2009. Fine scale variability in soil frost dynamics surrounding cushions of the dominant vascular plant species (*Azorella selago*) on sub-Antarctic Marion Island. *Swedish Society for Anthropology and Geography*, **2009**, 257–268.

Head, J. W. & Marchant, D. R. 2005. Weathering pits in the Antarctic dry valleys: insolation-induced heating and melting, and applications to Mars. *36th Annual Lunar and Planetary Science Conference*, League City, TX, Abstract 1138.

HEDDING, D. W., SUMNER, P. D., HOLNESS, S. D. & MEIKLEJOHN, K. I. 2007. Formation of a pronival rampart on sub-Antarctic Marion Island. *Antarctic Science*, **19**, 443–450.

HEDDING, D. W., MEIKLEJOHN, K. I., LE ROUX, J. J., LOUBSER, M. & DAVIS, J. K. 2010. Some observations on the formation of an active pronival rampart at Grunehogna Peaks, Western Dronning Maud Land, Antarctica. *Permafrost and Periglacial Processes*, **21**, 355–361.

HELFENSTEIN, P. 1981. Martian fractured terrain: coalescing polygons. *Lunar and Planetary Sciences*, **12**, 433–435.

HOLNESS, S. D. 2003*a*. Sorted circles in the maritime Subantarctic, Marion Island. *Earth Surface Processes and Landforms*, **28**, 337–347.

HOLNESS, S. D. 2003*b*. The periglacial record of Holocene environmental change, Subantarctic Marion Island. *Permafrost and Periglacial Processes*, **14**, 69–74.

HOLNESS, S. D. 2004. Sediment movement rates and processes on cinder cones in the maritime Subantarctic (Marion Island). *Earth Surface Processes and Landforms*, **29**, 91–103.

HOLNESS, S. D. & BOELHOUWERS, J. 1998. Some observations on Holocene changes in periglacial activity at Long Ridge, Marion Island. *South African Journal of Science*, **94**, 399–403.

HOWARD-WILLIAMS, C., PETERSON, D., LYONS, W. B., CATTANEO-VIETTI, R. & GORDON, S. 2006. Measuring ecosystem response in a rapidly changing environment: the Latitude Gradient project. *Antarctic Science*, **18**, 465–471.

JAGOUTZ, E. 2006. Salt-induced rock fragmentation on Mars: the role of salt in the weathering of Martian rocks. *Advances in Space Research*, **38**, 696–700.

JEONG, G. Y. 2006. Radiocarbon ages of sorted circles on King George Island, South Shetland Islands, West Antarctica. *Antarctic Science*, **18**, 265–270.

JØRGENSEN, P. M. 1986. Macrolichens of Bouvetøya. *Norsk Polarinstitutt, Skrifter*, **185**, 23–34.

KALVODA, J. 2005. Dynamics of glacial and periglacial processes as evidence of global change. *Acta Universitatis Carolinae, Geographica*, **1–2**, 69–88.

KIERNAN, K. & MCCONNELL, A. 2008. Periglacial processes on Heard Island, southern Indian Ocean. *Papers and Proceedings of the Royal Society of Tasmania*, **142**, 1–12.

KNIGHT, R. W. 1987. *Australian Antarctic Bibliography*. Institute of Antarctic and Southern Ocean Studies, University of Tasmania, Hobart.

KOWALEWSKI, D. E., MARCHANT, D. R., HEAD, J. W. & JACKSON, D. W. 2012. A 2D model for characterising first-order variability in sublimation of buried glacier ice, Antarctica: assessing the influence of polygon troughs, desert pavements and shallow subsurface salts. *Permafrost and Periglacial Processes*, **23**, 1–14.

LACELLE, D., LAURIOL, B. & CLARK, I. D. 2006. Effect of chemical composition of water on the oxygen-18 and carbon-13 signature preserved in cryogenic carbonates, Arctic Canada: implications for palaeoclimatic studies. *Chemical Geology*, **234**, 1–16.

LAITY, J. E. & BRIDGES, N. T. 2009. Ventifacts on Earth and Mars: analytical, field, and laboratory studies supporting sand abrasion and windward feature development. *Geomorphology*, **105**, 202–217.

LEASK, H. J. & WILSON, L. 2003. Heating and cooling of rocks on Mars: consequences for weathering. *34th Annual Lunar and Planetary Science Conference*, League City, TX, Abstract 1320.

LE ROUX, P. C. 2008. Climate and climate change. *In*: CHOWN, S. L. & FRONEMAN, P. W. (eds) *The Prince Edward Islands: Land–Sea Interactions in a Changing Ecosystem*. Sun Press, Stellenbosch, 39–64.

LE ROUX, P. C. & MCGEOCH, M. A. 2004. The use of size as an estimator of age in the subantarctic cushion plant *Azorella selago* (Apiaceae). *Arctic, Antarctic and Alpine Research*, **36**, 509–517.

LE ROUX, P. C. & MCGEOCH, M. A. 2008. Changes in climate extremes, variability and signature on sub-Antarctic Marion Island. *Climate Change*, **86**, 309–329.

LEVY, J. S., MARCHANT, D. R. & HEAD, J. W. 2006. Distribution and origin of patterned ground on Mullins Valley debris-covered glacier, Antarctica: the roles of ice flow and sublimation. *Antarctic Science*, **18**, 385–397.

LEVY, J. S., HEAD, J. W., MARCHANT, D. R., DICKSON, J. L. & MORGAN, G. A. 2009*a*. Geologically recent gully–polygon relationships on Mars: insights from the Antarctic Dry Valeys on the roles of permafrost, microclimates, and water sources for surface flow. *Icarus*, **201**, 113–126.

LEVY, J. S., HEAD, W. & MARCHANT, D. R. 2009*b*. Cold and dry processes in the Martian Arctic: Geomorphic observations at the Phoenix landing site and comparisons with terrestrial cold desert landforms. *Geophysical Research Letters*, **36**, L21203, 5.

LEVY, J. S., HEAD, W. & MARCHANT, D. R. 2011. Gullies, polygons and mantles in Martian permafrost environments: cold desert landforms and sedimentary processes during recent Martian geological history. *Geological Society, London, Special Publication*, **354**, 167–182.

LIU, X., SUN, L. & YIN, X. 2004. Textural and geochemical characteristics of proglacial sediments: A case study in the foreland of the Nelson Ice Cap, Antarctica. *Acta Geologica Sinica*, **78**, 970–981.

LÖFFLER, E., SULLIVAN, M. E. & GILLISON, A. N. 1983. Periglacial landforms on Macquarie Island, subAntarctic. *Zeitschrift für Geomophologie*, **27**, 223–236.

LÓPEZ-MARTÍNEZ, J. & SERRANO-CAÑADAS, E. 2005. Permafrost and periglacial processes in Hurd Peninsula, Livingston *Island. Proceedings of the 2nd European Conference on Permafrost*, Potsdam, 71.

LÓPEZ-MARTÍNEZ, J., SERRANO, E., SCHMID, T., MINK, S. & LINÉS, C. 2012. Periglacial processes and landforms in the Shetland Islands (northern Antarctic Peninsula region). *Geomorphology*, **155-156**, 62–79.

LUCCHITTA, B. K. 1981. Mars and earth: Comparison of cold-climate features. *Icarus*, **45**, 264–303.

LUNDQVIST, J., LILLIESKOLD, M. & OSTMARK, K. 1995. Glacial and periglacial deposits of the Tumbledown Cliffs area, James Ross Island, West Antarctica. *Geomorphology*, **11**, 205–214.

MAHANEY, W. C., KALM, V., KAPRAN, B. & HEWITT, K. 2009. Clast fabric and mass wasting on Asteroid 25143-Itokawa: Correlation with talus and other periglacial features on Earth. *Sedimentary Geology*, **219**, 44–57.

MARCHANT, D. R. & HEAD, J. W. 2004. Microclimate zones in the Dry Valleys of Antarctica: implications for landscape evolution and climate change on Mars. *Lunar and Planetary Science*, **35**, Abstract 1405, 2.

MARCHANT, D. R. & HEAD, J. W. 2007. Antarctic dry valleys: microclimate zonation, variable geomorphic processes, and implications for assessing climate change on Mars. *Icarus*, **192**, 187–222.

MATSUOKA, N. 1995. Rock weathering processes and landform development in the Sør Rondane Mountains, Antarctica. *Geomorphology*, **12**, 323–339.

MATSUOKA, N., MORIKAWA, K. & HIRAKAWA, K. 1996. Field experiments on physical weathering and wind erosion in an Antarctic cold desert. *Earth Surface Processes and Landforms*, **21**, 687–699.

MATSUOKA, N., THOMACHOT, C. E., OGUCHI, C. T., HATTA, T., ABE, M. & MATSUZAKI, H. 2006. Quaternary bedrock erosion and landscape evolution in the Sør Rondane Mountains, East Antarctica: reevaluating rates and processes. *Geomorphology*, **81**, 408–420.

MEIKLEJOHN, I. & HALL, K. 1997. Chemical weathering in the Antarctic: some data from eastern Alexander Island. *Polar Geography*, **2**, 101–112.

MELLON, M. T., ARVIDSON, R. E., MARLOW, J. J., PHILLIPS, R. J. & ASPHAUG, E. 2008. Periglacial landforms at the Phoenix landing site and the northern plains of Mars. *Journal of Geophysical Research*, **113**, E00A23, 15.

MELLON, M. T., MALIN, M. C. ET AL. 2009. The periglacial landscape at the Phoenix landing site. *Journal of Geophysical Research*, **114**, E00E06, 13.

MORGAN, G. A., HEAD, J. W., FORGET, F., MADELEINE, J.-P. & SPIGA, A. 2010. Gully formation on Mars: two recent phases of formation suggested by links between morphology, slope orientation and insolation history. *Icarus*, **208**, 658–666.

NAVAS, A., LÓPEZ- MARTÍNEZ, J. ET AL. 2008. Soil characteristics on varying lithological substrates in the South Shetland Islands, maritime Antarctica. *Geoderma*, **144**, 123–139.

NEL, W., HOLNESS, S. & MEIKLEJOHN, K. I. 2003. Observations on rapid mass movement on screes on Sub-Antarctic Marion Island. *South African Journal of Science*, **99**, 177–181.

NICHOLS, R. L. 1966. Geomorphology of Antarctica. *In*: TEDROW, J. C. F. (ed.) *Antarctic Soils and Soil Forming Processes*. American Geophysical Union, Washington, DC, Publication 1418, 1–46.

NOUGIER, J. 1964. Observations périglaciaires dans l'Archipel de Kerguelen, Terres Australes et Antarctiques Français. *Zeitschrift für Geomorphologie*, **8**, 143–149.

PAIN, C. F., CLARKE, J. D. A. & THOMAS, M. 2007. Inversion of relief on Mars. *Icarus*, **190**, 478–491.

PARSONS, R. L., HEAD, J. W. & MARCHANT, D. R. 2005. Weathering pits in the Antarctic Dry Valleys: insolation-induced heating and melting, and application to Mars. *Lunar and Planetary Science*, **36**, Abstract 138.

PÉWÉ, T. L. 1959. Sand-wedge polygons (tessellations) in the McMurdo Sound region, Antarctica. *American Journal of Science*, **257**, 545–552.

PÉWÉ, T. L., ROWAN, D. E., PÉWÉ, R. H. & STUCKENRATH, R. 1982. Glacial and periglacial geology of northwest Bloesletta Peninsula, Spitsbergen, Svalbard. *Norsk Polarinstitutt, Skrifter*, **177**, 32.

PICKARD, J. 1983. Pingos in Antarctica. *Quaternary Research*, **20**, 105–109.

PIRRIE, D., CRAME, J. A., RIDING, J. B., BUTCHER, A. R. & TAYLOR, P. D. 1997. Miocene glaciomarine sedimentation in the northern Antarctic Peninsula region: the stratigraphy and sedimentology of the Hobbs Glacier Formation, James Ross Island. *Geological Magazine*, **136**, 745–762.

PRIESNITZ, K. 1988. Cryoplanation. *In*: CLARK, M. J. (ed.) *Advances in Periglacial Geomorphology*. John Wiley, Chichester, 49–67.

PUTKONEN, J., MORGAN, D. J. & BALCO, G. 2012. Regolith transport quantified by braking block, McMurdo Dry Valleys, Antarctica. *Geomorphology*, **155-156**, 80–87.

RAMOS, M. & VIEIRA, G. 2003. Active layer and permafrost monitoring in Livingston Island, Antarctic. First results from 2000 to 2001. *In*: PHILIPS, M., SPRINGMAN, S. M. & ARENSON, L. U. (eds) *Permafrost*. A.A. Balkema, Lisse, 929–933.

RAMOS, M., VIEIRA, G. ET AL. 2007. *Permafrost and active layer monitoring in the maritime Antarctic: preliminary results from CALM sites on Livingston and Deception Islands*. US Geological Survey and The National Academies, Reston, VA, Short Research Paper, **070**, 1–5.

RAMOS, M., VIEIRA, G., BLANCO, J. J., GRUBER, S., HAUCK, C., HIDALGO, M. A. & TOMÉ, D. 2008. Active layer temperature monitoring in two boreholes in Livingston Island, Maritime Antarctic: first results for 2000–2006. *Proceedings of the 9th International Conference on Permafrost*, Fairbanks, AK, 1463–1467.

REGER, R. D. & PÉWÉ, T. L. 1976. Cryoplanation terraces: indicators of a permafrost environment. *Quaternary Research*, **6**, 99–109.

ŠABACKÁ, M., PRISCU, J. C. ET AL. 2012. Aeolian flux of biotic and abiotic material in Taylor Valley, Antarctica. *Geomorphology*, **155-156**, 102–111.

SCHWAMBORN, G., WAGNER, D. & HUBBERTEN, H.-W. 2008. The use of GPR to detect active layer in young periglacial terrain of Livingston Island, Maritime Antarctica. *New Surface Geophysics*, **6**, 327–332.

SELKIRK, J. M. & SAFFIGNA, L. J. 1999. Wind and water erosion of a peat and sand area on subantarctic Macquarie Island. *Arctic, Antarctic and Alpine Research*, **31**, 412–420.

SEPPÄLÄ, M. 2004. *Wind as a Geomorphic Agent in Cold Climates*. Studies in Polar Research. Cambridge University Press, Cambridge.

SERRANO, E. & LÓPEZ- MARTÍNEZ, J. 2000. Rock glaciers in the South Shetland Islands, West Antarctica. *Geomorphology*, **35**, 145–162.

SIMMS, M. J. 2002. The origin of enigmatic tubular, lake-shore karren: a mechanism for rapid dissolution of limestone in carbonate-saturated waters. *Physical Geography*, **23**, 1–20.

STRELIN, J. A. & MALAGNINO, E. C. 1992. Geomorfología de la isla James Ross. *In*: RINALDI, C. A. (ed.) *Geologica de la Isla James, Antártida*. Instituto Antártico Argentine, Buenos Aires, 7–36.

STRELIN, J. A. & SONE, T. 1994. *Rock glaciers in Antarctica. The Termination of the Pleistocene in South America*. IGCP Project 235, Abstract. CADIC, Ushuaia, 36–38.

STRELIN, J. A. & SONE, T. 1998. Rock glaciers on James Ross Island, Antarctica. *Permafrost – Seventh International Conference (Proceedings), Collection Nordicana*, **55**, 1027–1033.

STRELIN, J. A., SONE, T., MORI, J., TORIELLI, C. A. & NAKAMURA, T. 2006. New data related to Holocene landform development and climatic change from James Ross Island, Antarctic Peninsula. *In*: FÜTTERER, D., DAMASKE, D., KLEINSCHMIDT, G., MILLER, H. & TESSENSOHN, F. (eds) *Antarctica: Contributions to Global Earth Sciences*. Springer, Berlin, 455–600.

STRINI, A., GUGLIELMIN, M. & HALL, K. 2008. Tafoni development in a cryotic environment: an example from Northern Victoria Land. *Earth Surface Processes and Landforms*, **33**, 1502–1519.

SUMNER, P. D. 2004. Rock weathering rates on Subantarctic Marion Island. *Arctic, Antarctic and Alpine Research*, **36**, 123–127.

SUMNER, P. D. & MEIKLEJOHN, K. I. 2004. On the development of autochthonous blockfields in the grey basalts of sub-Antarctic Marion Island. *Polar Geography*, **28**, 120–132.

TAYLOR, B. W. 1955. Terrace formation on Macquarie Island. *Journal of Ecology*, **43**, 133–137.

THORN, C. 1988. Nivation: a geomorphic chimera. *In*: CLARK, M. J. (ed.) *Advances in Periglacial Geomorphology*. John Wiley, London, 3–31.

ULRICH, M., HAUBER, E., HERZSHUH, U., HÄRTEL, S. & SCHIRRMEISTER, L. 2011. Polygon pattern morphology on Svalbard (Norway) and western Utopia Planita (Mars) using high-resolution stereo remote-sensing data. *Geomorphology*, **134**, 197–216.

VAN GESSELT, S., REISS, D., THORPE, A. K. & NEUKUM, G. 2005. Seasonal variations of polygonal thermal contraction crack patterns in a south polar trough, Mars. *Journal of Geophysical Research*, **110**, E08002, 15.

VAN ROOY, M. P. 1957. *Meteorology of the Antarctic*. Weather Bureau, Department of Transport, Pretoria.

VÁRKONYI, P. L. & LAITY, J. E. 2012. Formation of surface features on ventifacts: modeling the role of sand grains rebounding within cavities. *Geomorphology*, **130–140**, 220–229.

VENTER, R. J. & BURDECKA, J. M. 1966. *Bibliography of Regional Meteorological Literature, Volume III, the Antarctic 1739–1957*. The Weather Bureau, Department of Transport, Pretoria.

VERESS, M. 2010. *Karst Environments*. Springer, Berlin.

VERWOERD, W. J., RUSSELL, S. & BERUTTI, A. 1981. 1980 volcanic eruption reported on Marion Island. *Earth and Planetary Science Letters*, **54**, 153–156.

VIEIRA, G. & RAMOS, M. 2003. Geographic factors and geocryological activity in Livingston Island, Antarctic. Preliminary results. *In*: PHILIPS, M., SPRINGMAN, S. M. & ARENSON, L. U. (eds) *Permafrost*. A.A. Balkema, Lisse, 1183–1188.

VIEIRA, G., LÓPEZ- MARTÍNEZ, J., SERRANO, E., RAMOS, M., GRUBER, S., HAUCK, C. & BLANCO, J. J. 2008. Geomorphological observations of permafrost and ground-ice degradation on Deception and Livingston Islands, Maritime Antarctica. *Proceedings of the 9th International Conference on Permafrost*, Fairbanks, AK, 1839–1844.

VIEIRA, G., BOCKHEIM, J. *ET AL*. 2010. Thermal state of permafrost and active-layer monitoring in the Antarctic: advances during the International Polar Year 2007–2009. *Permafrost and Periglacial Processes*, **21**, 182–197.

VIEIRA, R., HINATA, S., KELLEN DE ROSA, K., ZILBERSTEIN, S. & SIMEOS, J. C. 2012. Periglacial features in Patriot Hills, Ellsworth Mountains, Antarctica. *Geomorphology*, **155-156**, 96–101.

VINCENT, W. F. 1988. *The Microbial Ecosystems of Antarctica*. Studies in Polar Research. University of Cambridge, Cambridge.

WALTON, D. W. H. 1980. An annotated bibliography of Antarctic and sub-Antarctic pedology and periglacial processes. *British Antarctic Survey Data*, **5**, 75.

WARREN, S. H. 1914. The experimental investigation of flint fracture and its application to problems of human implements. *Journal of the Royal Anthropological Institute*, **44**, 412–450.

WEYANT, W. S. 1967. *The Antarctic Atmosphere: Climatology of the Surface Environment*. American Geographical Society, Folios, New York, **8**.

WHALLEY, W. B., REA, B. R. & RAINEY, M. M. 2004. Weathering, blockfields, and fracture systems and the implications for long-term landscape formation: some evidence from Lyngen and Øksfordjøkelen areas in north Norway. *Polar Geography*, **28**, 93–119.

WILSON, P. & CLARK, R. 2006. Development of miniature sorted patterned ground following soil erosion in East Falkland, South Atlantic. *Earth Surface Processes and Landforms*, **16**, 369–376.

WILSON, P. & EDWARDS, E. J. 2004. Further examples of ventifacts and unusual patterned ground from the Falkland Islands, South Atlantic. *Geografiska Annaler*, **86A**, 107–115.

WILSON, P., BENTLEY, M. J., SCHNABEL, C., CLARK, R. & XU, S. 2008. Stone run (block stream) formation in the Falkland Islands over several cold stages, deduced from cosmogenic isotope (^{10}Be and ^{26}Al) surface exposure dating. *Journal of Quaternary Science*, **23**, 461–473.

WRAY, R. A. L. 1997. Quartzite dissolution: karst or pseudokarst? *Cave and Karst Science, Transactions of the British Cave Research Association*, **24**, 81–86.

WYATT, M. B., MCSWEEN, H. Y., TANAKA, K. L. & HEAD, J. W. 2004. Global geologic context for rock types and surface alteration on Mars. *Geology*, **32**, 645–648.

YELOFF, D., MAUQUOY, D., BARBER, K., WAY, S., VAN GEEL, B. & TURNEY, C. S. M. 2007. Volcanic ash deposition and long-term vegetation change on Subantartic Marion Island. *Arctic, Antarctic and Alpine Research*, **39**, 500–511.

YOKOYAMA, T. & MATSUKURA, Y. 2006. Field and laboratory experiments on weathering rates of granodiorite: separation of chemical and physical processes. *Geology*, **34**, 809–812.

ZHU, C., CUI, Z. & ZHANG, J. 1996. Relationship between the distribution of periglacial landforms and glaciation history, Fildes Peninsula, King George Island. *Permafrost and Periglacial Processes*, **7**, 95–100.

Palaeoshoreline records of glacial isostatic adjustment in the Dry Valleys region, Antarctica

STEPHANIE A. KONFAL[1]*, T. J. WILSON[1] & B. L. HALL[2]

[1]*The Ohio State University, 275 Mendenhall Laboratory, 125 S. Oval Mall, Columbus, OH 43210, USA*

[2]*University of Maine, 303 Bryand Global Sciences Center, Orono, ME 04469, USA*

**Corresponding author (e-mail: skonfal@hotmail.com)*

Abstract: We present a new record of crustal deformation for the Dry Valleys and surrounding region of Antarctica. Values of crustal tilt resulting from the differential uplift of lacustrine strandlines are derived and linked with age data to provide a history of solid earth deformation since deglaciation. We present tilt directions and gradients for 13 strandlines formed *c.* 18 100–2100 cal yr BP. Derived gradient magnitudes increase exponentially with age and indicate an ongoing response to deglaciation since the Last Glacial Maximum. Azimuths of crustal tilting are consistently down to the SE towards West Antarctica. This tilt pattern is opposite to that predicted by models of glacial isostatic adjustment for Antarctica. Tilt magnitudes are significantly larger than tilted strandlines documented elsewhere in the world, suggesting an influence from thin crust and weak mantle underlying the region. This study presents the first use of lacustrine strandline tilts to document crustal deformation due to glacial unloading in Antarctica and provides an important new datum for constraining glacial isostatic adjustment models.

For more than a century, observations of uplifted and tilted marine and lacustrine strandlines have been used to understand glacial isostatic adjustment (GIA) from retreating ice sheets (e.g. Spencer 1891; Goldthwait 1908). As the Earth's crust rebounds in response to ice mass loss, ice-proximal shorelines are raised and tilted away from ice load centres. These geological shoreline records can be used in different ways depending on whether original elevation and/or age of the features are known.

When both the original elevation and age are known, uplift through time can be reconstructed. The most common application of this case is analysis of marine raised beaches and shorelines where past elevation is determined through comparison to present-day sea-level. Radiocarbon dates from organic material extracted from shoreline deposits are correlated with raised beach elevations to produce a relative sea-level curve (e.g. Hall & Denton 1999; Lambeck *et al.* 2002*b*; Baroni & Hall 2004). When corrected for eustatic sea-level change, relative sea-level curves reflect the influence of solid earth and geoid variations resulting from deglaciation, providing essential constraints for models of glacial isostatic adjustment that solve the sea-level equation (Farrell & Clark 1976; Peltier *et al.* 1978; Lambeck & Chappell 2001; Mitrovica & Milne 2003; Peltier 2004).

For lake palaeoshorelines, the original shoreline elevation is unknown, but differential uplift since the time of deposition results in shoreline tilt. Analysis of strandline tilt through time provides a record of solid earth deformation since deglaciation. For the case of glacial isostatic adjustment, the total amount of uplift is a function of shoreline position relative to former ice mass centres. Greatest uplift occurs closer to the centre of ice mass loss and decreases away from the load; subsidence occurs in the forebulge region (e.g. Clark *et al.* 1978). Shorelines located between the ice load and forebulge tilt away from the centre of ice loss, whereas shorelines outboard of the forebulge tilt towards the load centre. In addition to tilt direction, both temporal and spatial variances in tilt magnitude manifest ice history and geometry. Older deglacial shorelines have a greater magnitude of tilt than younger shorelines from the same location, and shorelines closer to the centre of ice mass loss have a greater magnitude of tilt than shorelines of the same age located farther away from the load. Given these relationships, analysis of the direction, magnitude and rate of tilting provides valuable information regarding ice history. Intersections of projected strandline tilt directions have been used to interpret the location of past ice mass centres, improving reconstructions of former ice geometry (Barnett & Peterson 1964; Andrews & Barnett 1972). Tilted palaeoshorelines linked with age data provide a relaxation curve for the Earth's crust since deglaciation, enabling studies of mantle rheology and lithospheric thickness, parameters crucial for understanding the geodynamic processes involved in

From: Hambrey, M. J., Barker, P. F., Barrett, P. J., Bowman, V., Davies, B., Smellie, J. L. & Tranter, M. (eds) 2013. *Antarctic Palaeoenvironments and Earth-Surface Processes*. Geological Society, London, Special Publications, **381**, 455–467. First published online July 30, 2013, http://dx.doi.org/10.1144/SP381.26

glacial isostatic adjustment (e.g. James *et al.* 2000; Lambeck *et al.* 2002*a*).

In this study, we present a new set of crustal deformation data derived from uplifted and tilted palaeoshorelines of ice-marginal lakes in the Dry Valleys and surrounding region of Antarctica (Fig. 1). We examine a sector of raised marine beaches along the adjacent Ross Sea coastline, previously studied from a relative sea-level perspective (Butler 1999; Hall & Denton 1999; Baroni & Hall 2004), to test for a differential uplift record. Patterns of tilting are assessed in the context of current models for ice reconstruction at the Last Glacial Maximum (LGM) and models of glacial isostatic adjustment. Where available, radiocarbon age data are linked with tilt values to produce a gradient curve of tilt v. time for the region, providing a record of crustal relaxation since deglaciation.

Strandline records in southern Victoria Land

Lacustrine records

Ice-free terrain in Antarctica is scarce, yielding relatively few preserved strandline records. The Dry Valleys and surrounding region of Antarctica comprise an exceptionally large ice-free region, hosting one of the largest assemblages of lakes on the continent (Fig. 1). Wright and Taylor valleys trend WSW to ENE and represent two of the largest valleys within the region. To the south, several smaller valleys, including Miers Valley, trend WNW to ESE. Wright, Taylor, and Miers valleys all remain predominantly free of ice throughout the year and are host to enclosed drainage basins with present-day lakes (e.g. Hendy 2000). Deltas,

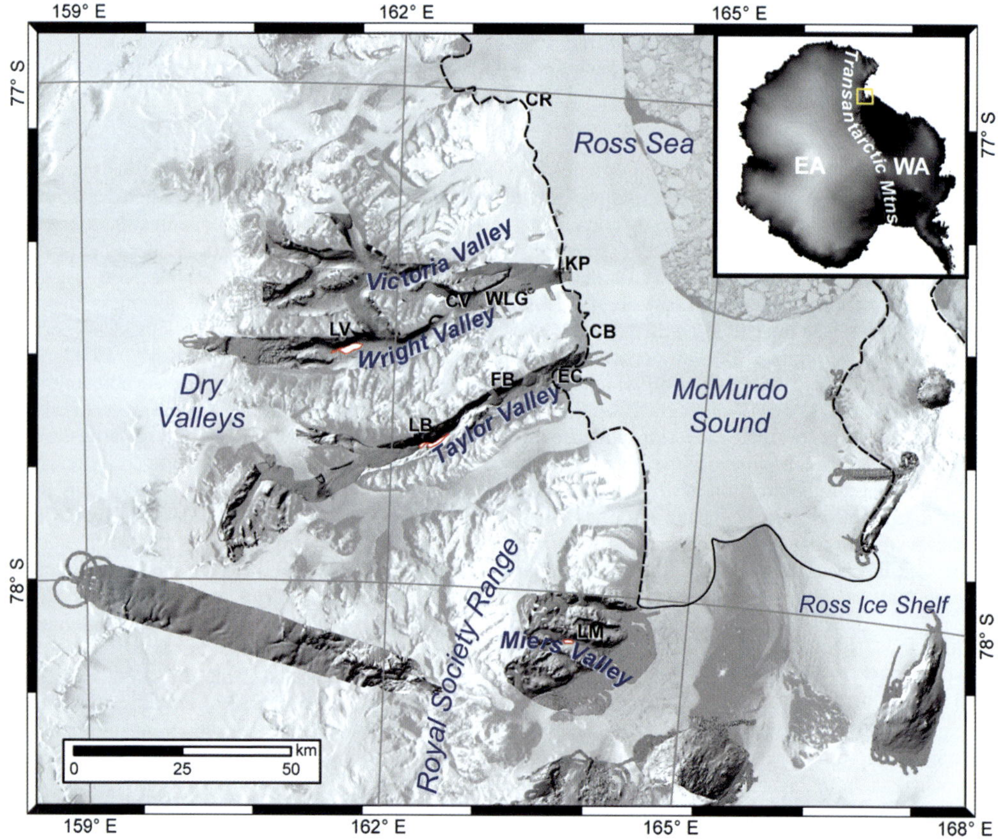

Fig. 1. Lakes and geography of the Dry Valleys and surrounding region, including Cape Roberts (CR), Kolich Point (KP), Wright Lower Glacier (WLG), Clark Valley (CV), Lake Vanda (LV), Cape Bernacchi (CB), Explorers Cover (EC), Fryxell Basin (FB), Lake Bonney and Basin (LB), and Lake Miers (LM). Darker regions are shaded relief representations of LiDAR DEM data. Inset map shows location of study region within the Transantarctic Mountains between East Antarctica (EA) and West Antarctica (WA).

shorelines and glaciolacustrine deposits from within these valleys suggest significantly larger lake levels at the LGM (Clayton-Greene *et al.* 1988; Denton *et al.* 1989; Hall *et al.* 2001). Since the LGM, changes in lake size have resulted in preservation of morphological strandlines associated with former water levels. In addition to the preserved lacustrine records, raised marine strandlines exist adjacent to the Dry Valleys along the Ross Sea coastline (e.g. Colhoun *et al.* 1992; Berkman *et al.* 1998; Hall & Denton 1999; Baroni & Hall 2004; Hall *et al.* 2004; Gardner *et al.* 2006). Strandline records preserved within Wright, Taylor and Miers valleys and a portion of the western coast of McMurdo Sound are the focus of this study.

Wright Valley. Wright Valley drains internally, with the valley floor progressively decreasing in elevation from Wright Lower Glacier to Lake Vanda, which occupies the lowest part of the basin (e.g. Hendy *et al.* 2000). Well-preserved raised palaeoshorelines are visible in Wright Valley on the northern and southern flanks of Lake Vanda (Fig. 2) and on the western flank of Clark Valley. A series of prominent shorelines located directly above Lake Vanda up to an elevation of 60 m represents one of the most complete and well-preserved sets of raised shorelines in the Dry Valleys and surrounding region. These shorelines extend laterally for more than 7 km and date between *c.* 3700 and 2100 ^{14}C yr BP (Hall, unpublished data). East of Lake Vanda, a single, prominent shoreline located on the western flank of Clark Valley is located at 566 m and extends for approximately 1.5 km. This shoreline dates between 15 500 and 14 500 ^{14}C yr BP and is representative of Glacial Lake Wright, the high-level lake that filled Wright Valley at the LGM (Hall *et al.* 2001).

Taylor Valley. Taylor Valley is composed, from west to east, of the Bonney, Fryxell and Explorers Cove basins. Bonney and Fryxell basins, host to Lake Bonney and Lake Fryxell, have internal drainage, whereas Explorers Cove drains eastward to the Ross Sea. A threshold at 118 m elevation separates Bonney and Fryxell basins and a threshold at 78 m elevation separates Fryxell and Explorers Cove basins (Hall & Denton 2000). Glacial Lake Washburn, the high-level lake that occupied Taylor Valley up to an elevation of 350 m at the LGM, was the result of the westward-advancing Ross Ice Sheet plugging the valley mouth, enabling water levels to exceed valley thresholds (Hall & Denton 2000). Radiocarbon dating of perched deltas suggests that water levels fell below the Bonney–Fryxell threshold approximately 11 500 ^{14}C yr BP (Hall & Denton 2000). Preserved deltas, moat deposits and shoreline features exist in Taylor Valley up to the former level of Glacial Lake Washburn.

Miers Valley. Miers Valley is host to Lake Miers, which, unlike the lakes of Wright and Taylor valleys, does not have internal drainage but rather drains to the SE into the Ross Sea. Glacial Lake Trowbridge, the high-level lake that occupied Miers Valley at the LGM, drained from the eastern portion of the valley by approximately 10 000

Fig. 2. Preserved palaeoshorelines visible above present-day Lake Vanda, looking westward along Wright Valley (image source USGS aerial photograph). Length of exposed shorelines between arrows is approximately 5 km.

^{14}C yr BP; present-day Lake Miers in the western valley is a remnant of the former lake (Clayton-Greene *et al.* 1988; Hendy 2000). Highest recorded lake levels date between 19 000 and 18 000 yr BP (Hendy 2000), and, while not dated, lower, well-preserved strandlines visible up to approximately 25 m above present water levels are continuous on both the northern and southern valley walls around Lake Miers.

Marine records

Raised marine deposits are preserved along the Ross Sea coast from Terra Nova Bay to McMurdo Sound, with the densest collection of raised beaches extending along less than 40 km of coastline from Cape Roberts to Explorers Cove in McMurdo Sound. No known raised beaches or marine deposits exist south of Explorers Cove (Stuiver *et al.* 1981) and the oldest marine strandlines, dating no older than 8000 yr BP (Baroni & Hall 2004), are exposed along the coast near Terra Nova Bay. Individual marine shoreline segments range in length up to several kilometres and have an aggregate lateral spatial extent of more than 300 km.

Methods

Elevation data

During the 2001–2002 austral summer, through a collaborative project between the USGS, NSF and NASA, light detection and ranging (LiDAR) data were acquired by NASA's Airborne Topographic Mapper system, yielding 2 m resolution digital elevation models (DEMs) for the Dry Valleys and surrounding region (Csatho *et al.* 2005). These LiDAR DEM data allow sampling of strandline elevations every 2 m along their trace. Whereas traditional methods rely on surveying individual, widely spaced points along strandlines of lakes with large lateral extents (May *et al.* 1991; Fjeldskaar 1994, 1997; Tackman *et al.* 1998; Clague & James 2002), the high-resolution LiDAR data enable quantification of the rate of tilting over the small spatial extent of lake shorelines preserved in the Antarctic ice-free valleys.

Shoreline depositional assumptions

In the simplest case, water level in a closed basin represents an equipotential surface where all points are at the same elevation owing to gravity, resulting in horizontal deposition of strandlines. However, the character of shoreline deposition is also dependent upon sediment character, wave energy, sea-level or lake-level change, and basement topography, with wave action accepted as the dominant influence in most situations (e.g. Orford 1977; Möller *et al.* 2002). In addition to these factors, the presence of ice has a strong influence on shoreline formation in polar environments (e.g. Butler 1999; St-Hilaire-Gravel *et al.* 2010).

All lakes utilized in this study have a floating ice cover (Hendy 2000). The depositional regime of shorelines for lakes with a floating ice cover is dominated by wind-blown sediment accumulation in open water moats around the perimeter of the lake ice cover and minor slumping of water-saturated sediments at the shore (Hendy *et al.* 2000). Moat size remains small enough for ice-covered Antarctic lakes that wave action is insignificant (Hendy *et al.* 2000) and the assumption of horizontal deposition of lake shoreline features is justified.

Fluctuations in lake level are largely controlled by meltwater input and ablation (Hall *et al.* 2010*a*). Therefore, preserved palaeoshorelines are not necessarily in stratigraphic order with an age progression from highest to lowest. When considering strandlines with relative age constraints, this relationship is of particular significance and is therefore addressed case-by-case in this study.

Raised marine shorelines in McMurdo Sound were formed during open-water storm conditions, resulting in strandlines with distinct, high-relief (up to 4 m) morphology (Butler 1999; Hall & Denton 2000; Baroni & Hall 2004; Hall *et al.* 2004). Records from Kolich Point within McMurdo Sound are utilized in this study, and uncertainties in depositional elevation for these strandlines are addressed in the Discussion section.

Procedure

Precise shoreline identification and correlation are essential for accurate crustal tilt derivation. Utilizing published descriptions of the location, elevation and signature of preserved palaeoshorelines as a starting point, imagery derived from LiDAR coverages around known modern and ancient lake margins were scrutinized to visually identify preserved strandlines. In most cases, published literature provided the specific location of individual shorelines and provided evidence for a shoreline origin not easily attainable through remote-sensing techniques, including the presence of lacustrine algae and internal lithology (Clayton-Greene *et al.* 1988; Butler 1999; Hall & Denton 1999, 2000; Hendy 2000; Higgins *et al.* 2000; Hall *et al.* 2001, 2004; Baroni & Hall 2004). Palaeoshorelines in the Dry Valleys region are characterized by 'step-and-tread' cross-sectional profiles (Hendy 2000), distinct from the morphology of other glacial features. Laterally, palaeoshorelines are associated with deltas and, where preserved, may be traced around the full extent of modern lake shorelines. Also, while the shorelines in this study have been tilted, the degree of elevation change is on the order of a few metres per kilometre in nearly all cases. Therefore, the near-horizontality of linear features indicates a shoreline provenance rather than glacial or structural origin. In some cases, contrasts in grain size above and below shorelines manifest visible colour changes, and were also used to aid in the shoreline mapping process.

Individual shorelines were mapped by manually digitizing their traces using the ArcGIS software suite. The best resource for lateral shoreline correlation and mapping in this study was LiDAR DEM-derived hillshades, where features perpendicular to the illumination azimuth were enhanced. Manual digitization permitted careful examination of individual shorelines on the hillshades as well as comparison with previously published descriptions. However, challenges in shoreline correlation and mapping remained owing to lateral discontinuity, post-depositional erosion or sediment accumulation and LiDAR elevation data artefacts. Shorelines mapped from LiDAR using published descriptions were further validated by examination of air photos, LiDAR DEM-derived cross-sectional profiles, and sub-metre resolution QuickBird (2) and GeoEye imagery.

Elevation values were extracted every 2 m from the LiDAR DEMs along the trace of digitized shorelines. Two-dimensional gradient plots of elevation v. distance along the profile were generated for each digitized feature and analysed for anomalous elevation values. Regions of slump (artificial elevation lows) or post-formation deposition (artificial elevation highs), appear as negative or positive spikes, respectively. Each elevation spike was investigated individually using high-resolution imagery to confirm the hypothesis of a non-shoreline origin, and removed accordingly. After shoreline profiles were cleaned by this procedure, linear, polynomial order 1 regression was performed to fit a 3D least squares surface to each set of shoreline elevation data, resulting in a best-fit tilted plane. Anomalously high- and low-elevation values with respect to the fitted plane were examined as in the 2D case, and features of non-shoreline origin were removed. This multi-step process of data editing maximizes the ability to accurately constrain tilt for shorelines with small spatial extents by ensuring that only elevation values representative of the original shoreline are utilized.

Shoreline ages were converted from radiocarbon dates to calendar years using the CALIB program (Stuiver *et al.* 2004). Midpoints representative of shoreline age in calendar years BP were calculated from probability density function curves. To account for the higher regional radiocarbon content of the Southern Ocean when compared with the atmosphere, marine samples associated with shorelines were corrected using a reservoir effect average delta-*R* value of 791 $\pm$ 121 for the Holocene, calculated by Hall *et al.* (2010*b*) for the Southern Ocean. All other radiocarbon records utilized in this study are from algae of lacustrine origin, and do not require a reservoir correction given the minimal influence associated with algae in shallow-water environments in the Dry Valleys (Doran *et al.* 1994; Hall & Denton 2000; Hendy & Hall 2006). Where multiple ages for a single shoreline were available, cumulative calibration curves representative of all samples were generated. For lacustrine shorelines lacking direct radiocarbon sampling, relative age assessments based on geological and glaciological constraints were used to estimate shoreline ages in calendar years BP using the CALIB program (Stuiver *et al.* 2004). Specifically, cumulative calibration curves were generated using estimated date inputs of radiocarbon shoreline age. Owing to the full possible age range of these shorelines, standard uncertainty associated with calculated calendar year BP values for these records are significantly larger than for directly dated shorelines (Table 1). The geological and glaciological constraints used to estimate depositional age for undated shorelines are discussed in detail in the following section.

Results

In total, 13 strandline gradients are derived from the LiDAR data within the Dry Valleys and surrounding region. Of these gradients, 10 are determined from four different lakes within Wright, Taylor and Miers valleys, and three are derived from raised beaches along the adjacent marine coastline at Kolich Point. Figures 3 and 4 show the mapped strandlines on LiDAR-derived hillshade images for each location within this study, along with the correlation of preserved shoreline segments. Absolute ages determined from radiocarbon dates are available for 10 shorelines (Hall & Denton 2000; Hall *et al.* 2001, 2002; Hall, unpublished data) and relative age constraints are available for all other strandlines (Clayton-Greene *et al.* 1988; Hendy 2000; Hall & Denton, 2000; Hall *et al.* 2002). Figure 5 presents tilt direction vectors for tilt planes fitted to the strandlines evaluated in this study, and Figure 6 shows a corresponding gradient plot of tilt magnitude v. age.

Wright Valley

A set of numerous preserved shorelines is visible in the LiDAR data directly above modern Lake Vanda

Table 1. *Shoreline tilts and ages*

Shoreline	Gradient (m/km)	Downhill direction (deg)	RMS height residual (m)	age (cal yr BP)
Wright Valley				
Lake Vanda				
1	0.4640 ± 0.0181	103.39 ± 0.05	0.702	4296 ± 2402
2	0.5061 ± 0.0114	99.68 ± 0.03	0.454	**2861 ± 94**
3	0.5093 ± 0.0093	85.02 ± 0.04	0.518	**2854 ± 106**
4	0.6482 ± 0.0241	124.48 ± 0.02	0.658	**2959 ± 116**
5	0.7015 ± 0.0551	135.51 ± 0.04	0.570	**3148 ± 226**
6	0.3756 ± 0.0168	106.20 ± 0.04	0.475	**2125 ± 86**
Clark Valley				
7	27.513 ± 0.4568	143.21 ± 0.02	0.486	**18116 ± 719**
Taylor Valley				
Lake Bonney				
8	6.154 ± 0.1748	151.80 ± 0.00	1.261	15789 ± 4386
Miers Valley				
Lake Miers				
9	4.079 ± 0.0630	169.38 ± 0.01	0.723	9227 ± 3761
10	1.815 ± 0.1101	166.76 ± 0.05	1.068	7316 ± 4453
Marine records				
Kolich Point				
11	2.292 ± 0.3665	128.13 ± 0.05	0.404	6352 ± 526
12	2.305 ± 0.1943	154.30 ± 0.02	0.233	**6266 ± 324**
13	1.767 ± 0.2328	149.68 ± 0.03	0.325	**6276 ± 361**

Shoreline tilts and ages in calendar years BP based on calibrated and calculated radiocarbon dates. Ages with large standard deviations reflect shorelines lacking absolute age data and represent the full possible age brackets derived from geological data within the region, as discussed in the text. Shorelines with absolute age constraints are in bold.

(Figs 2 & 3). Radiocarbon samples of algae and carbonate intertwined with algae date the shorelines and associated deltas between *c.* 3400 and 2100 ^{14}C yr BP (Hall, unpublished data), resulting in calibrated ages ranging from 3150 to 2100 cal yr BP. Assuming normal ablation rates and no meltwater input, significant lake level changes are possible within the time period of radiocarbon

Fig. 3. Lake Vanda mapped strandlines overlaid on LiDAR-derived hillshade. Shoreline numbers correspond to Table 1.

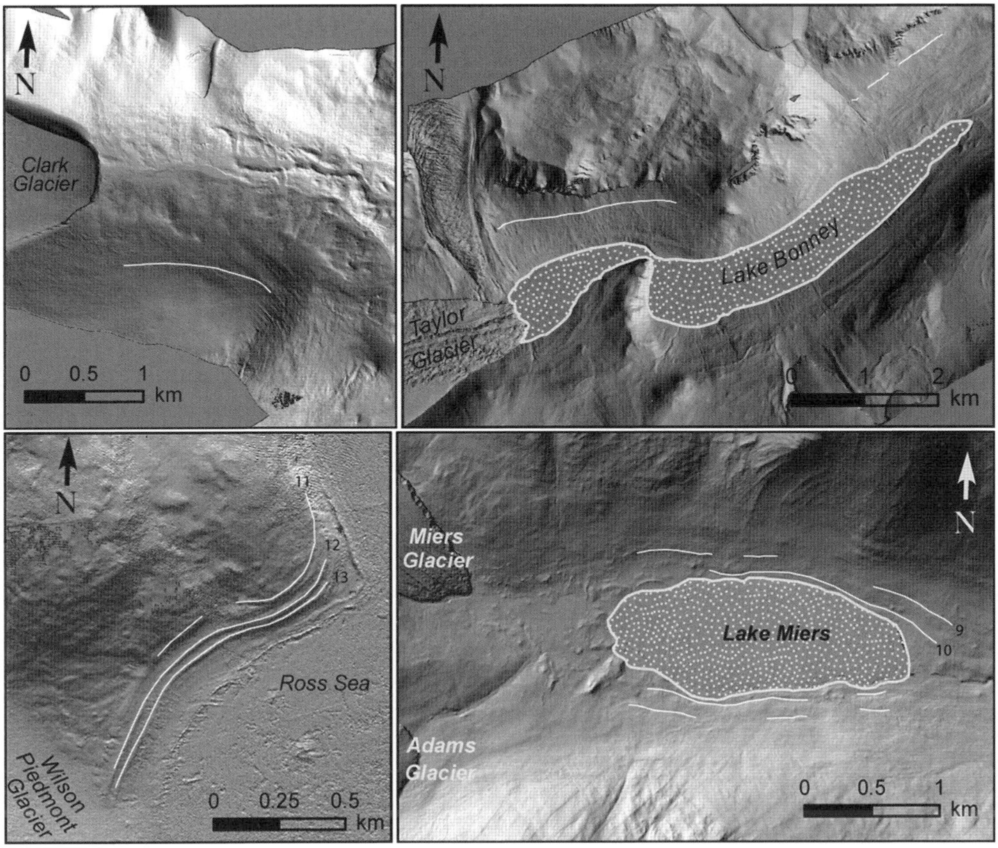

Fig. 4. Mapped strandlines overlaid on LiDAR-derived hillshades for Clark Valley in eastern Wright Valley (upper left), Lake Bonney in Taylor Valley (upper right), marine records at Kolich Point (lower left) and Lake Miers in Miers Valley (lower right). Shoreline numbers correspond to Table 1.

dating error. Therefore, linking shoreline segments on the basis of ^{14}C age was not a viable approach for this site. Instead, lateral and cross-sectional morphology were used to correlate individual shoreline segments. In total, six strandlines were identified with sufficient continuity to determine statistically significant tilt directions and magnitudes. Of these, five records are associated with radiocarbon ages. Owing to uncertainties in lake level fluctuations and evidence for a 6000 ^{14}C yr BP highstand within the shoreline set (Hall, unpublished data), the tilt derived from the remaining shoreline is assigned an error bracket representative of the full possible age range of 6000–2000 ^{14}C yr BP, resulting in a calculated age of 4296 $\pm$ 2402 cal yr BP. Tilt directions are down to the east and SE towards directions ranging between 85 and 136°, and gradient values range from *c.* 0.4 to *c.* 0.7 m/km (Table 1). These strandlines represent the youngest records obtained in our analysis and are associated with the smallest derived tilt magnitudes.

In eastern Wright Valley, an isolated, higher, laterally continuous shoreline located on the southwestern face of Clark Valley is clearly distinguishable in the LiDAR data (Fig. 4). The morphology, lateral continuity, sediment character and presence of layered lacustrine algae within the feature strongly support a shoreline origin (Hall *et al.* 2001). Radiocarbon samples from this shoreline date to 15 500 and 14 500 ^{14}C yr BP (Hall *et al.* 2001), resulting in a calibrated age of 18 116 $\pm$ 719 cal yr BP. The tilt direction is down to the SE towards 143° with a magnitude of 27.5 m/km. This record is the oldest strandline included in our analysis and is associated with the largest derived tilt magnitude.

Taylor Valley

Numerous linear, lateral ridges are visible in the LiDAR data above present-day Lake Fryxell and range in age from approximately 20 000 to

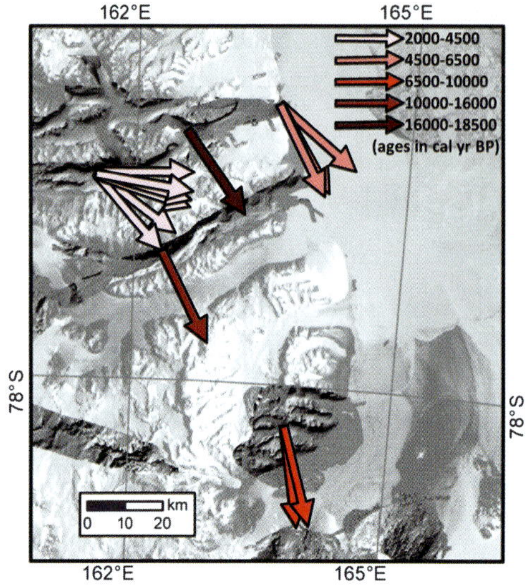

Fig. 5. Gradient direction vectors for shorelines in the Dry Valleys and surrounding region. Colour of vectors denotes age of strandline records. All ages are shown in cal yr BP.

8000 ^{14}C yr BP (Hall & Denton 2000). While some of these features resemble shorelines, studies of the depositional mechanisms for enclosed Dry Valleys proglacial lakes suggest that they are probably moat ridges resulting from debris deposited by lake ice conveyors (Hendy *et al.* 2000). Owing to the non-horizontal depositional processes associated with lake ice conveyor deposits, Lake Fryxell is excluded from our analysis.

In central Taylor Valley, an abundance of linear features, including moat deposits, deltas and shorelines, are visible above the present-day water level of Lake Bonney (e.g. Hall & Denton 2000; Higgins

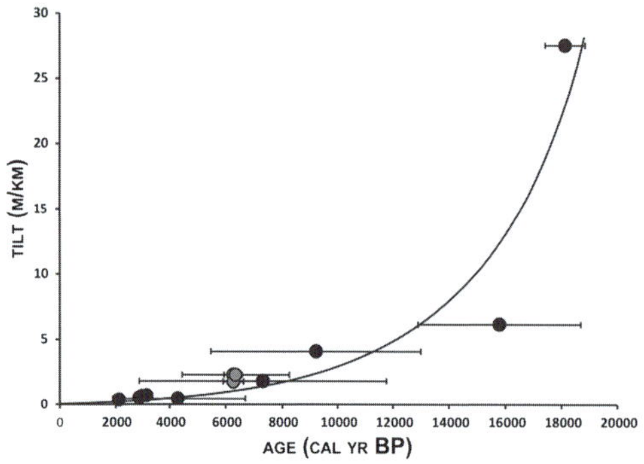

Fig. 6. Gradient curve of shoreline tilt magnitude v. age. Marine shorelines are shown in grey. See text and Table 1 for discussion of age assignments and uncertainties.

et al. 2000). While numerous lacustrine strandlines are preserved, identifying and correlating individual shoreline features in the LiDAR data is difficult owing to overprinting of features. One prominent, laterally continuous strandline visible within the LiDAR data is clearly discernable from other linear features (Fig. 4). However, both shorelines and near-horizontal moraines are present in this area, precluding the definitive interpretation of this feature as a shoreline. Examined in the field at a single location, this feature has been found to be composed of till (Hall, unpublished data), arguing for a glacial origin. The possibility that this feature is an erosional shoreline cut into till cannot be ruled out. The existence of deltas at the same elevation indicates that water reached the level of this feature in the past, and supports a shoreline origin. Based on the lateral continuity, association with deltas and a derived tilt in agreement with other records in the region, this strandline is included in our analysis. The resulting tilt direction is down to the SE towards 151.8° and the tilt magnitude is 6.2 m/km. Because this strandline is not dated directly, we rely on relative age constraints provided by radiocarbon dating of deltas that indicate Glacial Lake Washburn existed in central Taylor Valley from 23 800 to 10 000 ^{14}C yr BP (Higgins *et al.* 2000; Hall *et al.* 2010*a*). Hall & Denton (2000) report a peak in the number of deltas dated within Taylor Valley at 11 000–15 000 ^{14}C. The 13 000 ^{14}C yr BP midpoint from this range also corresponds to the midpoint for deltas at the elevation of the proposed Bonney shoreline included in our analysis (above 300 m). We therefore choose to represent the prominent Bonney shoreline with an age of 13 000 ^{14}C yr BP, but account for the full 23 800–10 000 ^{14}C BP possible age bracket, resulting in a calculated age of 15 789 ± 4386 cal yr BP. This age is broadly consistent with the tilt v. age results from the region (Fig. 5). The age bracket for deposition qualifies this record, at a minimum, as the second oldest strandline utilized in our analysis.

Miers Valley

Two prominent strandlines directly above present-day Lake Miers are visible in the LiDAR data (Fig. 4). These features show lateral and cross-sectional morphologies characteristic of shorelines deposited at the margins of lakes with floating ice cover and are interpreted as shorelines in this study. Resulting tilts are *c.* 1.81 m/km at 167° for the lower strandline, and *c.* 4.1 m/km at 169° for the upper strandline.

Miers Valley is divided into eastern and western basins. Both observed strandlines included in our analysis are associated with the western basin, where a lake has existed since the formation of Glacial Lake Trowbridge (e.g. Clayton-Greene *et al.* 1988). Radiocarbon dating of lacustrine deposits indicate that the east and west basins became isolated from each other after 13 000 yr BP (Koffman and Hall, unpublished data). The morphology of the strandlines included in this study indicates isolation within the western basin, suggesting that their origin post-dates 13 000 yr BP. Assuming tilting owing to glacial unloading, the larger gradient of *c.* 4.1 m/km derived from the upper strandline suggests an older age of deposition than the lower strandline associated with a derived gradient of *c.* 1.81 m/km. Assuming these age relations, we represent the upper Miers strandline with an age of 10 000 ^{14}C yr BP (chosen to correspond with the time when lake levels in eastern Miers Valley are thought to have been below the threshold between the basins; Hendy 2000) and the lower Miers strandline with an arbitrary age of 7000 ^{14}C yr BP (chosen to conform to the assumption that the observed tilt is due to glacial unloading), resulting in calculated ages of 9227 $\pm$ 3761 cal yr BP and 7316 $\pm$ 4453 cal yr BP, respectively. While our justifications for assigning relative age constraints for the upper and lower Miers shorelines utilized in this study seem logical, limiting ages for the separation between the eastern and western basins remain the only absolute geological evidence of shoreline age. Therefore, we must assign exceptionally large error bars to these shorelines to represent the full 13 000 yr BP to present possible age bracket.

While not dated conclusively, a favourable sampling geometry resulting from shoreline preservation around nearly the entire extent of the current lake yields a statistically strong set of tilt values for the two Miers strandlines (Table 1) and the southern location of this site increases the overall spatial coverage of the study region.

Marine shorelines

A prominent set of marine raised beaches is clearly visible within the LiDAR data at Kolich Point (Fig. 4). While the lateral extent of these strandlines is limited (less than 1 km), the high resolution of the LiDAR data combined with the preservation style of these strandlines (ridge crests up to 4 m) yields statistically strong gradient values for three individual strandlines (Table 1). Resulting tilts range from *c.* 1.8 to 2.3 m/km towards 128–154° (Table 1). For two of these strandlines, associated radiocarbon samples provide maximum age constraints between *c.* 6800 and 6500 ^{14}C yr BP (prior to a reservoir effect correction; Hall *et al.* 2004), resulting in calibrated ages of *c.* 6300 cal yr BP. These maximum-limiting ages come from shells in a sheet of marine deposits that underlies all the beaches at this site. For the uppermost strandline, radiocarbon samples cannot be conclusively associated; however, based on the pattern of relative sea-level change, it is reasonable to assume that this higher shoreline is older in age than those that exist below. In addition, the same sheet of marine deposits extends beneath this beach, indicating that it, too, is younger than *c.* 6300 cal yr BP. Therefore, we assign a relative age constraint representative of the uppermost Kolich Point samples and the assumption that the higher shoreline is older, resulting in a calculated age of 6352 $\pm$ 526 cal yr BP.

Discussion

Ten lacustrine and three marine strandlines dip towards the east to SE (Fig. 5) at an average azimuth of 132° and with gradients that increase exponentially with age (Fig. 6). Across the spatial extent of the study region, all shorelines tilt towards West Antarctica. Temporally, there is no consistent change in shoreline tilt azimuth (Fig. 5 & Table 1). The youngest set of shorelines, at Lake Vanda, vary in tilt direction from SE to east but, again, show no systematic change in tilt azimuth with time. The more easterly trend of some younger shoreline records may suggest an alternate or additional mechanism of deformation; however, southeastward shoreline tilts occur at each palaeolake locality and the absence of systematic temporal change in tilt direction argues for a spatially coherent tilt pattern. The agreement in tilt direction and the consistent age–gradient relationship, with older strandlines yielding a greater tilt than younger strandlines, show that the lacustrine shorelines in the Dry Valleys and surrounding region constitute a robust record of solid earth deformation.

Many of the tilts derived from the Dry Valleys region shorelines have significantly larger gradients than tilted strandlines documented elsewhere in the world. In regions rebounding from the Northern Hemisphere ice sheets, shoreline tilt gradients are typically in the range of *c.* 0.2–1.58 m/km (Andrews *et al.* 1970; Andrews & Barnett 1972). Only the youngest Dry Valleys strandlines, approximately half of the records in our analysis, fall within this range; higher gradients are between 2 and 4 m/km, one is *c.* 6 m/km and a particularly large gradient of *c.* 27 m/km was derived for the oldest shoreline. Some strandline studies have yielded large tilt gradients, commonly associated with tilt azimuths that diverge from regional trends, and these have been explained by deformation owing to local loads, faults, intrusive structures or hydrothermal features (Meyer & Locke 1986; McCalpin *et al.* 1994; Pierce & Geological Survey (US) 2007). All of the Dry Valleys shoreline locations share southeastward tilt azimuths towards West Antarctica. Given the commonality in shoreline tilt direction in both time and space, it is unreasonable to invoke local phenomena that would systematically increase tilt, but maintain constant tilt trend.

The age range of the tilted lacustrine shorelines we have analysed is between *c.* 15 500 and *c.* 2000 ^{14}C yr BP, providing an extended record of solid earth response that coincides with deglaciation of Antarctica since the LGM. In the western Ross Sea, retreat of the grounding line from a point south of the continental shelf edge commenced at *c.* 11 000 ^{14}C yr BP (Domack *et al.* 1999), and migrated southward past the coastline adjacent to the Dry Valleys region at about 6600 ^{14}C yr BP (Hall *et al.* 2004). The smooth, exponential form of the tilt–age curve is indicative of an ongoing response to deglaciation since the LGM without significant influence of recent ice mass change. The consistent direction of shoreline tilt azimuths towards West Antarctica and the exponential form of shoreline tilt magnitude through time, typical of the relaxation of the earth in response to glacial unloading, support the interpretation that lacustrine shoreline tilts in this study record long-term crustal deformation owing to GIA.

The elevations along individual marine beach ridges formed at the Ross Sea coast vary considerably owing to depositional processes (e.g. Baroni & Hall 2004). Just to the south of Kolich Point, at Cape Bernacchi, Butler (1999) attributes a southeastward decrease in elevation of a shoreline of approximately 2 m/km to changes in wave energy resulting from variances in coastline aspect. Therefore, the tilt of shorelines measured at Kolich Point can be explained by the original depositional geometry of the beach ridges. It is intriguing, however, that these marine shorelines have a tilt direction parallel to tilt azimuths of the lacustrine shorelines of similar age (Fig. 5) and the magnitude of tilt fits the exponential curve of tilt magnitude v. time defined by the lacustrine shorelines (Fig. 6). Both of these attributes are consistent with tilting of the marine shorelines by solid earth deformation owing to GIA. This hypothesis could be tested by acquisition and analysis of high-resolution digital elevation data along the *c.* 300 km length of the Ross Sea coast where marine shorelines are preserved, to identify any systematic tilt patterns.

The southeastward tilt direction of the lacustrine shorelines towards West Antarctica is surprising, because the dominant centre of ice loss since the LGM is estimated to be in the southern Ross Embayment (Fig. 7). Large ice mass change occurred to the east of the Dry Valleys region as the Ross Ice Sheet retreated towards interior West Antarctica (Denton & Hughes 2002). Predicted uplift owing to GIA produces tilts to the NW, away from West Antarctica. Primary influences on the response of the crust to GIA are (1) earth properties, including lithospheric thickness and mantle viscosity, and (2) ice history, or rather the spatial and temporal pattern of ice mass change following the LGM (e.g. Peltier 2004). If the observed shoreline tilts resulted from GIA, then the discrepancy between observed and predicted tilt directions must be explained by differences in earth properties, ice history or some combination thereof.

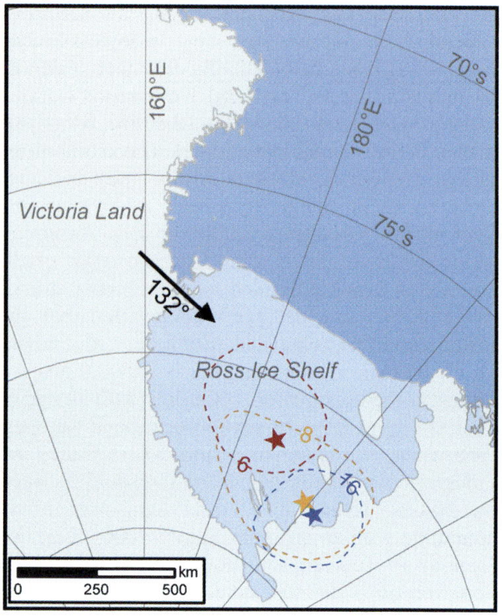

Fig. 7. Average shoreline tilt azimuth for the Dry Valleys compared with present-day uplift rates. Uplift contours in millimetres are shown for predicted ice mass centres from the IJ05 with ocean loading (in red; Ivins & James 2005; Simon *et al.* 2010), ICE-5G (in blue; Peltier 2004) and W12a (in orange; Whitehouse *et al.* 2012*a*) GIA models.

Ice reconstructions show the Dry Valleys region at the interface between the East Antarctic Ice Sheet and the Ross Sea ice sheet at the LGM (e.g. Denton & Hughes 2002). Thus, ice mass changes in both the West Antarctic and East Antarctic ice sheets may play a role in driving deformation of the crust in the Dry Valleys and surrounding region. The SE-directed shoreline tilts documented here indicate maximum ice mass loss to the NW, within East Antarctica. However, geological and glaciological data suggest little change in East Antarctic ice volume since the LGM (e.g. Whitehouse *et al.* 2012*b* and references therein). Current ice histories represented in GIA models for Antarctica place maximum LGM ice volume loss in West Antarctica, and therefore predict tilting for the region to be down towards the NW (Huybrechts 2002; Peltier 2004; Ivins & James 2005; Simon *et al.* 2010; Whitehouse *et al.* 2012*a*), opposite to our derived shoreline tilts.

Sensitivity studies have demonstrated a relation between larger load-induced tilt gradients and earth models characterized by thin lithosphere and low mantle viscosity, particularly at the edge of the load (James *et al.* 2000). As evidenced by seismic studies, thin crust and presumed warm mantle extend from West Antarctica, underlain by a rift system, beneath the Transantarctic Mountains for *c.* 100 km (Lawrence *et al.* 2006*a*, 2006*b*), underlying the entire area of this study. The uniformly high tilt gradients observed in the Dry Valleys region could possibly be attributed to increased crustal tilting associated with this weak rheology. Further, the weak earth properties beneath the study region probably affect the time scale and wavelength of earth response to both local and regional changes in ice mass. Typically, a weak rheological profile results in a shorter-wavelength response, localized around the ice mass centre, and crustal motions of higher amplitude (e.g. Peltier 2004).

At present, it is unclear what combination of ice history and earth property variations can explain the discrepancy between shoreline tilts observed in this study and GIA-predicted crustal motions. A solution must be sought through new GIA modelling and through comparison with modern vertical and horizontal crustal motions measured by GPS (Wilson *et al.* 2011).

This study presents the first case of using lacustrine strandline tilts to investigate crustal deformation owing to glacial unloading in Antarctica. The availability of high-resolution DEMs, such as the LiDAR data used in this study, makes shoreline studies from small lakes in the limited ice-free terrain of Antarctica possible for the first time. We note that even a single new datum, such as the shoreline tilts presented here, provides significant constraints for GIA models. As models of GIA for Antarctica move towards incorporating lateral variation in earth structure, geological constraints spanning the rheological boundary between East and West Antarctica, such as those that we present here, will be of particular importance. Future efforts to acquire additional LiDAR and chronological data for strandlines elsewhere along the Transantarctic Mountains, or in other sectors of Antarctica, would enable a more robust analysis of crustal deformation owing to glacial unloading across the continent. Integrated with modelling studies, these results could provide important new constraints on the volume of ice contributed by Antarctic ice sheets to the global ocean.

Conclusions

Crustal deformation in the Dry Valleys region is documented by lacustrine shoreline tilts showing increasing tilt magnitudes with shoreline age, typical of earth response to glacial unloading. Assuming deformation owing to the removal of an LGM ice mass load, the strandline tilts presented in this study suggest a maximum ice load at the LGM located to the NW of the Dry Valleys and surrounding region. This pattern of ice mass change and tilting is opposite to current ice histories and crustal motions predicted by GIA models, suggesting influences from earth properties and/or ice history currently not defined in model inputs. Lateral variations in earth properties, a factor not yet represented in Antarctic models of GIA, are present in the region and probably play a significant role in the observed signal.

Our results provide an important new geological field constraint for models of GIA. The disparity between GIA-predicted motions and those observed in this study highlight the need for incorporation of lateral variability into models of GIA for Antarctica, as well as increased geological observations of ice history along the Transantarctic Mountains and around the East Antarctic margin. Analysis of additional lacustrine strandlines within the region, as well as analysis of modern vertical and horizontal crustal motions measured by GPS, can provide further insight into the pattern of tilting observed in this study and the causal ice and earth property influences.

LiDAR data acquisition was supported jointly by NSF Office of Polar Programs, the NASA ICESat programme, and the US Geological Survey. Generation of DEMs from the LiDAR data was led by B. Csatho and supported by NSF Office of Polar Programs and the NASA ICESat programme. S. Konfal was supported under the NASA Earth and Space Science Fellowship under grant no. NNX07A036H. M. Bevis provided valuable help in data editing and evaluation of uncertainties for tilt gradients. We thank the editors J. Smellie and M. Hambrey for their assistance, and C. Baroni and D. Zwartz for their constructive reviews of this manuscript.

References

Andrews, J. T. & Barnett, D. M. 1972. Analysis of strandline tilt directions in relation to ice centers and postglacial crustal deformation, Laurentide Ice Sheet. *Geografiska Annaler: Series A, Physical Geography*, **54**, 1–11.

Andrews, J. T., Buckley, J. T. & England, J. H. 1970. Late-Glacial Chronology and Glacio-Isostatic Recovery, Home Bay, East Baffin Island, Canada. *Geological Society of America Bulletin*, **81**, 1123–1148.

Barnett, D. M. & Peterson, J. A. 1964. The significance of Glacial Lake Naskaupi 2 in the deglaciation of

Labrador–Ungava. *Canadian Geographer/Le Géographe canadien*, **8**, 173–181.

BARONI, C. & HALL, B. L. 2004. A new Holocene relative sea-level curve for Terra Nova Bay, Victoria Land, Antarctica. *Journal of Quaternary Science*, **19**, 377–396.

BERKMAN, P. A., ANDREWS, J. T. *ET AL.* 1998. Circum-Antarctic coastal environmental shifts during the Late Quaternary reflected by emerged marine deposits. *Antarctic Science*, **10**, 345–362.

BUTLER, E. R. T. 1999. Process environments on modern and raised beaches in McMurdo Sound, Antarctica. *Marine Geology*, **162**, 105–120.

CLAGUE, J. J. & JAMES, T. S. 2002. History and isostatic effects of the last ice sheet in southern British Columbia. *Quaternary Science Reviews*, **21**, 71–87.

CLARK, J. A., FARRELL, W. E. & PELTIER, W. R. 1978. Global changes in postglacial sea level: a numerical calculation. *Quaternary Research*, **9**, 265–287.

CLAYTON-GREENE, J. M., HENDY, C. H. & HOGG, A. G. 1988. Chronology of a Wisconsin age proglacial lake in Miers Valley, Antarctica. *New Zealand Journal of Geology and Geophysics*, **31**, 353–361.

COLHOUN, E. A., MABIN, M. C. G., ADAMSON, D. A. & KIRK, R. M. 1992. Antarctic ice volume and contribution to sea-level fall at 20,000 yr BP from raised beaches. *Nature*, **358**, 316–319.

CSATHO, B., SCHENK, T. *ET AL.* 2005. Airborne laser scanning for high-resolution mapping of Antarctica. *American Geophysical Union EOS Transactions*, **86**, 237–238.

DENTON, G. H. & HUGHES, T. J. 2002. Reconstructing the Antarctic Ice Sheet at the Last Glacial Maximum. *Quaternary Science Reviews*, **21**, 193–202.

DENTON, G. H., BOCKHEIM, J. G., WILSON, S. C. & STUIVER, M. 1989. Late Wisconsin and early Holocene glacial history, inner Ross Embayment, Antarctica. *Quaternary Research*, **31**, 151–182.

DOMACK, E. W., JACOBSON, E. A., SHIPP, S. & ANDERSON, J. B. 1999. Late Pleistocene-Holocene retreat of the West Antarctic Ice-Sheet system in the Ross Sea: Part 2 – sedimentologic and stratigraphic signature. *Geological Society of America Bulletin*, **111**, 1517–1536.

DORAN, P. T., WHARTON, R. A., JR. & LYONS, W. B. 1994. Paleolimnology of the McMurdo Dry Valleys, Antarctica. *Journal of Paleolimnology*, **10**, 85–114.

FARRELL, W. E. & CLARK, J. A. 1976. On postglacial sea level. *Geophysical Journal of the Royal Astronomical Society*, **46**, 647–667.

FJELDSKAAR, W. 1994. Viscosity and thickness of the asthenosphere detected from the Fennoscandian uplift. *Earth and Planetary Science Letters*, **126**, 399–410.

FJELDSKAAR, W. 1997. Flexural rigidity of Fennoscandia inferred from postglacial uplift. *Tectonics*, **16**, 596–608.

GARDNER, N., HALL, B. & WEHMILLER, J. 2006. Pre-Holocene raised beaches at Cape Ross, Southern Victoria Land, Antarctica. *Marine Geology*, **229**, 273–284.

GOLDTHWAIT, J. W. 1908. A reconstruction of water planes of the extinct Glacial Lakes in the Lake Michigan Basin. *The Journal of Geology*, **16**, 459–476.

HALL, B. L. & DENTON, G. H. 1999. New relative sea-level curves for the southern Scott Coast, Antarctica: evidence for Holocene deglaciation of the western Ross Sea. *Journal of Quaternary Science*, **14**, 641–650.

HALL, B. L. & DENTON, G. H. 2000. Radiocarbon chronology of Ross Sea Drift, Eastern Taylor Valley, Antarctica: evidence for a grounded ice sheet in the Ross Sea at the Last Glacial Maximum. *Geografiska Annaler: Series A, Physical Geography*, **82**, 305–336.

HALL, B. L., DENTON, G. H. & OVERTURF, B. 2001. Glacial Lake Wright, a high-level Antarctic lake during the LGM and early Holocene. *Antarctic Science*, **13**, 53–60.

HALL, B. L., DENTON, G. H., OVERTURF, B. & HENDY, C. H. 2002. Glacial Lake Victoria, a high-level Antarctic Lake inferred from lacustrine deposits in Victoria Valley. *Journal of Quaternary Science*, **17**, 697–706.

HALL, B. L., BARONI, C. & DENTON, G. H. 2004. Holocene relative sea-level history of the Southern Victoria Land Coast, Antarctica. *Global and Planetary Change*, **42**, 241–263.

HALL, B. L., DENTON, G. H., FOUNTAIN, A. G., HENDY, C. H. & HENDERSON, G. M. 2010*a*. Antarctic lakes suggest millennial reorganizations of Southern Hemisphere atmospheric and oceanic circulation. *Proceedings of the National Academies*, **107**, 21355–21359.

HALL, B. L., HENDERSON, G. H., BARONI, C. & KELLOGG, T. B. 2010*b*. Constant Holocene Southern-Ocean ^{14}C reservoir ages and ice-shelf flow rates. *Earth and Planetary Science Letters*, **296**, 115–123.

HENDY, C. H. 2000. Late Quaternary Lakes in the McMurdo Sound Region of Antarctica. *Geografiska Annaler: Series A, Physical Geography*, **82**, 411–432.

HENDY, C. H. & HALL, B. L. 2006. The radiocarbon reservoir effect in proglacial lakes: examples from Antarctica. *Earth and Planetary Science Letters*, **241**, 413–421.

HENDY, C. H., SADLER, A. J., DENTON, G. H. & HALL, B. L. 2000. Proglacial Lake-ice Conveyors: a new mechanism for deposition of drift in polar environments. *Geografiska Annaler: Series A, Physical Geography*, **82**, 249–270.

HIGGINS, S. M., HENDY, C. H. & DENTON, G. H. 2000. Geochronology of Bonney Drift, Taylor Valley, Antarctica: evidence for Interglacial Expansions of Taylor Glacier. *Geografiska Annaler: Series A, Physical Geography*, **82**, 391–409.

HUYBRECHTS, P. 2002. Sea-level change at the LGM from ice-dynamic reconstructions of the Greenland and Antarctic ice sheets during the glacial cycles. *Quaternary Science Reviews*, **21**, 203–231.

IVINS, E. R. & JAMES, T. S. 2005. Antarctic glacial isostatic adjustment: a new assessment. *Antarctic Science*, **17**, 541–553.

JAMES, T. S., CLAGUE, J. J., WANG, K. & HUTCHINSON, I. 2000. Postglacial rebound at the northern Cascadia subduction zone. *Quaternary Science Reviews*, **19**, 1527–1541.

LAMBECK, K. & CHAPPELL, J. 2001. Sea level change through the last glacial cycle. *Science*, **292**, 679–686.

LAMBECK, K., SMITHER, C. & EKMAN, M. 2002*a*. Tests of glacial rebound models for Fennoscandinavia based on instrumented sea- and lake-level records. *Geophysical Journal International*, **135**, 375–387.

Lambeck, K., Smither, C. & Johnston, P. 2002*b*. Sea-level change, glacial rebound and mantle viscosity for northern Europe. *Geophysical Journal International*, **134**, 102–144.

Lawrence, J. F., Wiens, D. A., Nyblade, A. A., Anandakrishnan, S., Shore, P. J. & Voigt, D. 2006*a*. Crust and upper mantle structure of the Transantarctic Mountains and surrounding regions from receiver functions, surface waves, and gravity: implications for uplift models. *Geochemistry, Geophysics, and Geosystems*, **7**, 1–23.

Lawrence, J. F., Wiens, D. A., Nyblade, A. A., Anandakrishnan, S., Shore, P. J. & Voigt, D. 2006*b*. Rayleigh wave phase velocity analysis of the Ross Sea, Transantarctic Mountains, and East Antarctica from a temporary seismograph array. *Journal of Geophysical Research*, **111**, 1–15.

May, G. M., Bills, B. G. & Hodge, D. S. 1991. Far-field flexural response of Lake Bonneville from paleopluvial lake elevations. *Physics of the Earth and Planetary Interiors*, **68**, 274–284.

McCalpin, J. & UTAH GEOLOGICAL SURVEY 1994. *Neotectonic Deformation Along the East Cache fault zone, Cache County, Utah*. Utah Geological Survey, Salt Lake City, UT.

Meyer, G. A. & Locke, W. W. 1986. Origin and deformation of Holocene shoreline terraces, Yellowstone Lake, Wyoming. *Geology*, **14**, 699–702.

Mitrovica, J. X. & Milne, G. A. 2003. On post-glacial sea level: I. general theory. *Geophysical Journal International*, **154**, 253–267.

Møller, J. J., Yevzerov, V. Y. & Corner, G. D. 2002. Holocene raised-beach ridges and sea-ice-pushed boulders on the Kola Peninsula, northwest Russia: indicators of climate change. *The Holocene*, **12**, 169–176.

Orford, J. D. 1977. A proposed mechanism for storm beach sedimentation. *Earth Surface Processes*, **2**, 381–400.

Peltier, W. R. 2004. Global glacial isostasy and the surface of the ice-age Earth – the ICE-5G (VM2) model and GRACE. *Annual Review of Earth and Planetary Sciences*, **32**, 111–149.

Peltier, W. R., Farrell, W. E. & Clark, J. A. 1978. Glacial isostasy and relative sea level: a global finite element model. *Tectonophysics*, **50**, 81–110.

Pierce, K. L. & GEOLOGICAL SURVEY (US) 2007. *Post-glacial Inflation–Deflation Cycles, Tilting, and Faulting in the Yellowstone Caldera Based on Yellowstone Lake Shorelines*. US Geological Survey, Reston, VA, Professional Papers, **1717**, 1–41.

Simon, K. M., James, T. S. & Ivins, E. R. 2010. Ocean loading effects on the prediction of Antarctic glacial isostatic uplift and gravity rates. *Journal of Geodesy*, **84**, 305–317.

Spencer, J. W. 1891. Deformation of the Algonquin beach and birth of Lake Huron. *American Journal of Science Third Series*, **41**, 12–21.

St-Hilaire-Gravel, D., Bell, D. & Forbes, D. 2010. Raised gravel beaches as proxy indicators of past sea-ice and wave conditions, Lowther Island, Canadian Arctic Archipelago. *Arctic*, **63**, 213–226.

Stuiver, M., Denton, G. H., Hughes, T. J. & Fastook, J. L. 1981. History of the marine ice sheet in West Antarctica during the last glaciation: a working hypothesis. *In*: Denton, G. H. & Hughes, T. J. (eds) *The Last Great Ice Sheets*. Wiley, New York, 319–436.

Stuiver, M., Reimer, P. J. & Reimer, R. 2004. *CALIB radiocarbon calibration*, http://radiocarbon.pa.qub.ac.uk/calib

Tackman, G. E., Currey, D. R., Bills, B. G. & James, T. 1998. Paleoshoreline evidence for postglacial tilting in Southern Manitoba. *Journal of Paleolimnology*, **19**, 343–363.

Whitehouse, P. L., Bentley, M. J., Milne, G. A., King, M. A. & Thomas, I. D. 2012*a*. A new glacial isostatic adjustment mmodel for Antarctica: calibrated and tested using observations of relative sea-level change and present-day uplift rates. *Geophysical Journal International*, **190**, 1464–1482.

Whitehouse, P. L., Bentley, M. J. & Le Broq, A. M. 2012*b*. A deglacial model for Antarctica: geological constraints and glaciological modeling as a basis for a new model of Antarctic glacial isostatic adjustment. *Quaternary Science Reviews*, **32**, 1–24.

Wilson, T. & THE POLENET GROUP 2011. The Antarctic-POLENET (ANET) GPS Network in West Antarctica. *11th International Symposium on Antarctic Earth Sciences*, Edinburgh, 10–16 July, 189, http://www.isaes2011.org.uk/abstracts_v4_20_07_2011.pdf

Characteristic atmosphere–ocean–solid earth interactions in the Antarctic coastal and marine environment inferred from seismic and infrasound recording at Syowa Station, East Antarctica

MASAKI KANAO[1]*, ALESSIA MAGGI[2], YOSHIAKI ISHIHARA[3], ELEONORE STUTZMANN[4], MASA-YUKI YAMAMOTO[5] & GENTI TOYOKUNI[6]

[1]*National Institute of Polar Research, Research Organization of Information and Systems, 10-3, Midori-cho, Tachikawa-shi, Tokyo 190-8518, Japan*

[2]*Institut de Physique du Globe de Strasbourg, CNRS and University of Strasbourg, 5 rue Rene Descartes, Strasbourg Cedex F-67084, France*

[3]*National Astronomical Observatory, National Institutes of Natural Sciences, 2-12 Hoshigaoka-cho, Mizusawa-ku, Oshu, Iwate 023-0861, Japan*

[4]*Institut de Physique du Globe de Paris, Bureau 304, 1 rue Jussieu, Paris Cedex 05 75238, France*

[5]*Kochi University of Technology, 185 Miyanokuchi, Tosayamada-cho, Kami-shi, Kochi 782-8502, Japan*

[6]*Research Center for Prediction of Earthquakes and Volcanic Eruptions, Graduate School of Science, Tohoku University, 6-6 Aza-Aoba, Aramaki, Aoba-ku, Sendai 980-8578, Japan*

**Corresponding author (e-mail: kanao@nipr.ac.jp)*

Abstract: Several characteristic waves detected by seismographs in Antarctic stations have been recognized as originating from the physical interaction between the solid earth and the atmosphere–ocean–cryosphere system surrounding the Antarctic and may be used as a proxy for characterizing ocean wave climate. A Chaparral-type infrasound sensor was installed at Syowa Station (SYO; 39.6E, 69.0S), East Antarctica, in April 2008 during the International Polar Year (IPY2007–2008). Matching data are also available for this time period from the existing broadband seismic recorder located close by. Continuous infrasound data for 2008–2009 include background signals (microbaroms) with a broad peak in the wave period between the values of 4 and 10 s. Signals with the same period are recorded by the broadband seismograph at SYO (microseisms). This period band is identified as double-frequency microseisms/baroms (DFM). The DFM have relatively lower amplitudes during winter. We suggest that this is due to the sea-ice extent around the coast causing a decreased ocean loading effect. In contrast, the single frequency microseisms/baroms with a peak in period between 12 and 30 s are observed under storm conditions, particularly in winter. On the infrasound data, stationary signals are identified with harmonic overtones at a few Hertz to lowermost human audible band, which we suggest is due to local effects such as sea-ice cracking and vibration. Microseism measurements are a useful proxy for characterizing ocean wave climate, complementing other oceanographic and geophysical data. At SYO, continuous monitoring by both broadband seismograph and infrasound contributes to the Federation of Digital Seismographic Networks, the Comprehensive Nuclear-Test-Ban Treaty in the high southern latitudes and the Pan-Antarctic Observations System under the Scientific Committee on Antarctic Research.

In the half-century after the International Geophysical Year (IGY; 1957–1958), the number of global seismic network and other seismic stations has increased in Antarctica. The Japanese Syowa Station (SYO; 69S, 39E), East Antarctica, has contributed data since 1959 (Eto 1962). A three-component broadband seismometer (STS-1; Wielandt & Steim 1986) was installed in 1989, as SYO joined the Federation of Digital broadband Seismographic Network stations (Butler & Anderson 2008; Fig. 1). The density of both permanent stations belonging to the global network and temporary stations deployed at specified locations has improved over time, and has recently permitted studies of local

From: HAMBREY, M. J., BARKER, P. F., BARRETT, P. J., BOWMAN, V., DAVIES, B., SMELLIE, J. L. & TRANTER, M. (eds) 2013. *Antarctic Palaeoenvironments and Earth-Surface Processes*. Geological Society, London, Special Publications, **381**, 469–480. First published online May 16, 2013, http://dx.doi.org/10.1144/SP381.8

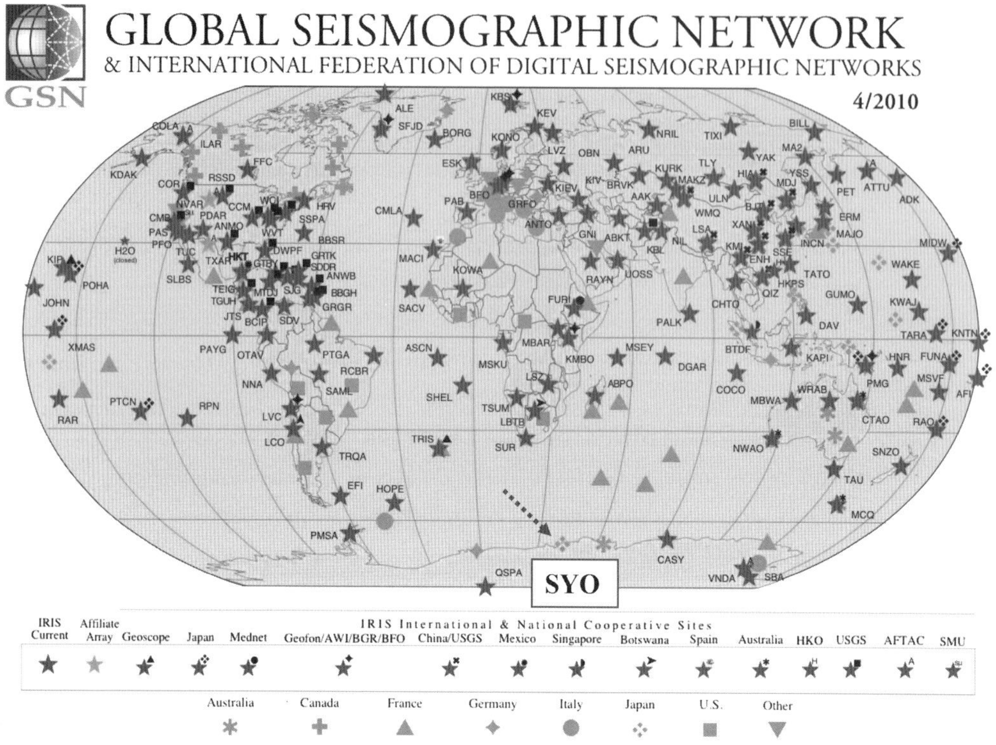

Fig. 1. Global map showing the Federation of Digital Seismological Network (modified after Butler & Anderson 2008). Several regional networks developed by involved nations are indicated; such as IRIS, Geoscope, Geophone, Japan. Syowa Station (SYO) in East Antarctica is pointed out by a broken arrow.

seismicity in the Antarctic (Kanao *et al.* 2006; Reading 2006; Reading 2007). Most recently, the 'Polar Earth Observing Network' (POLENET; http://www.polenet.org) has made a significant contribution of the International Polar Year (IPY2007–2008) towards improving broadband seismic recording in Antarctica.

Several kinds of environmental signals associated with the atmosphere, notably the ocean–solid earth system, have been detected in Antarctica and the surrounding regions. Seismic motions for small-magnitude ice-related events are generally named 'ice-quakes' (or 'ice-shocks') and can be generated by glacier dynamic activity (Anandakrishnan *et al.* 2003; Kanao & Kaminuma 2006). Such cryoseismic sources include the movements of ice sheets, glaciers and sea-ice, oceanic tide-cracks, oceanic gravity waves, icebergs and the calving process. Cryoseismic sources are likely to be influenced by environmental conditions, and the study of their temporal variation may provide indirect evidence of climate change.

Almost all of the seismic stations deployed on the Earth's surface record ubiquitous signals at periods of between 4 and 25 s, commonly referred to as 'microseisms'. In the absence of large earthquakes, microseismic waves are the strongest-amplitude signals worldwide, including the polar regions. Microseisms are considered to be dominated by Rayleigh waves, which arise from gravity waves in the ocean that are forced by surface winds. The period ranges of microseisms are dictated by the physics of gravity-wave generation, and are constrained by the speed and extent of Earth's surface winds (Aster *et al.* 2008; Bromirski 2009; Ardhuin *et al.* 2011). On a global scale, microseism amplitudes are generally highest during the local winter, because nearby oceans are stormier in winter than in summer (Stutzmann *et al.* 2009). Within the polar regions, particularly in Antarctica, there is no clear picture of the behaviour of microseisms in space and time. Recently the seasonal trend has been found to be reversed, most probably because of the effect of winter sea-ice (Stutzmann *et al.* 2009; Grob *et al.* 2011).

Alongside global seismic stations, there is a smaller but increasing number of infrasound stations. Infrasound is defined as sub-audible sound, that is, pressure waves with frequencies ranging from the cut-off frequency of sound (3.21 mHz, for a 15 °C isothermal atmosphere) to the lowest frequency of the human audible band (20 Hz). This frequency range is a new horizon for the remote sensing of the Earth's atmospheric physical environment, and shares the same frequency range as the broadband seismometer. It is known that Earth's free vibrations can excite similar modes in the upper atmosphere. The great Sumatra–Andaman earthquake of 26 December 2004 not only produced the seismic waves that were recorded as far away as Antarctic stations (Nawa *et al.* 2007), but also produced infrasound waves that were recorded after propagating through the atmosphere (Iyemori *et al.* 2005). Another example of simultaneous observation of infrasound and seismic signals concerns the shock waves generated by a meteorite that passed over Japanese Island (Ishihara *et al.* 2004). Continuous infrasound observation was started at SYO as one of the projects of IPY2007–2008, and in addition to the four existing Test-Ban Treaty network certified Antarctic infrasound stations at Davis (Australia), Windless Bight (near McMurdo, USA), Palmer (USA) and Neumayer (Germany).

In this paper, several features of seismic and infrasound wave-propagation in the continental margin of East Antarctica are illustrated by using the data at SYO (Fig. 2). Background infrasound signals (microbaroms) with peak amplitudes at periods of 4 and 10 s are analysed, and compared with those recorded by broadband seismographs (microseisms). Loading effects from the Southern Ocean are investigated through analysis of time-dependent variations of amplitudes and frequency content of power spectral densities (PSD). The oceanic effects on seismic and infrasound data are modulated by the presence of sea-ice, and can therefore be related to the polar atmosphere–ocean–cryosphere system. Measurements of microseisms/microbaroms are a useful proxy for characterizing ocean-wave climate and global storm intensity, and will help to reveal the physical interactions between these multi-spheres. Moreover, they provide an indicator of the extent of very recent environmental change, as inferred from Antarctica.

Microseisms recorded at Syowa Station

In general, the microseism spectrum has a bimodal composition, caused by the existence of two distinct physical mechanisms that transfer ocean-wave energy to seismic waves in the geosphere (Fig. 2). The first spectral peak between approximately 12 and 30 s, commonly called 'primary' or 'single-frequency microseism' (SFM), arises from the transfer of ocean-gravity wave (swell) energy to seismic waves as oceanic waves shoal and break in the shallow waters. The highest-amplitude and longest-period swells are created by large and intense storms that generate strong sustained winds over a large area. Swell propagates dispersively across ocean basins, which results in a longer-period swell arriving at the coast before the shorter-period swell. This period-dependent delay is readily measured in data recorded by seismic stations, ocean buoys and seismographs, even those deployed recently on a giant Antarctic iceberg (MacAyeal *et al.* 2009).

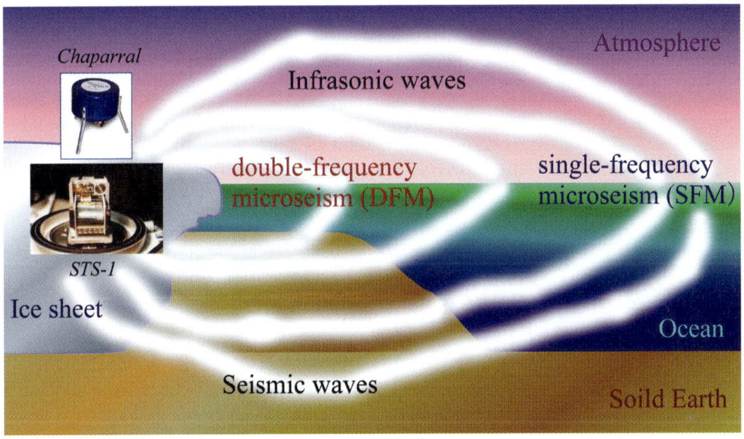

Fig. 2. Schematic illustration of atmosphere–ocean–solid earth environments in Antarctic margins and Southern Ocean. Infrasound and seismic waves propagate from the source regions in Southern Ocean (SFM) and continental shelf area (DFM) to the margins of Antarctic continent; and recorded by Chaparral infrasound sensor as well as broadband seismometer (STS-1) at SYO.

The second, more prominent, microseism peak between approximately 4 and 10 s, commonly called 'secondary' or 'double-frequency microseism' (DFM), arises from nonlinear interaction of interfering ocean-wave components that produce a pressure pulse at double their frequency. This pressure pulse propagates with little or no attenuation to the sea floor where it generates seismic waves. The DFM is thought to be generated both near the coasts, where coastal swell reflection can provide the requisite opposing wave components, and to a lesser extent also in the deep ocean (Fig. 2).

On a global scale, microseism amplitudes are generally highest during the local winter, because nearby oceans are stormier in winter than in summer (Stutzmann *et al.* 2009). In Antarctica, the opposite observation is made: microseism amplitude is attenuated during the local winter for both primary and secondary microseisms (Hatherton 1960; Grob *et al.* 2011). Figure 3 represents examples of short-period seismograms (Hagiwara 1958) recorded at SYO, during both the austral summer (Fig. 3d) and winter (Fig. 3e) of 1997. Microseisms with their dominant frequency of a few seconds are recognized clearly throughout the austral summer recording, and are missing on the winter recording. These observations are most readily explained by winter sea-ice impeding both the direct ocean-to-continent coupling that generates the SFM, and the coastal reflection which is an important component in the generation of the DFM (Grob *et al.* 2011).

In order to illustrate the variability of microseismic amplitude over time, we analyse the PSD for data from the STS-1 seismometer at SYO. During the time period investigated in this paper, the data were continuously sampled at 20 Hz and transmitted from SYO to the National Institute of Polar Research by an Intersat telecommunication system (Aoyama & Kanao 2010). Figure 4 represents the PSD of the vertical component of ground motion during a typical austral summer day in November 2010. In spite of the presence of large teleseismic signals towards the end of the day, the DFM is clearly visible with sufficient amplitude through the whole time period. Figure 5 represents similar PSDs for the whole year of 2004 over the period band 0.1–80 s (modified after Grob *et al.* 2011; Kanao *et al.* 2012). The DFM is in general more strongly excited when storms or blizzards visit the station, and to a large degree is probably generated

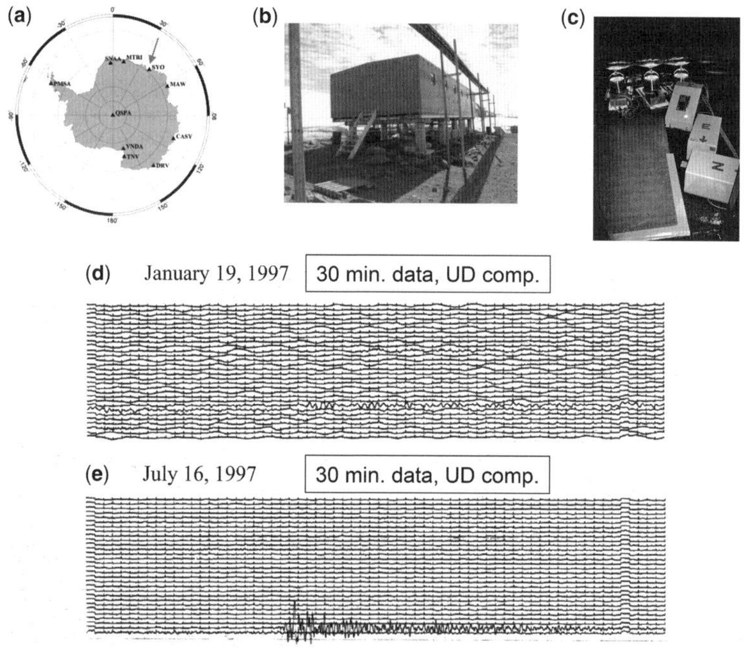

Fig. 3. (**a**) Map of global seismic network stations in Antarctica, including SYO. (**b**) Earth science laboratory at SYO, where continuous monitoring takes place. (**c**) A three-component broadband seismometer (STS-1; behind) and short period seismometers (HES; front) in the seismographic hut at SYO. (**d**) Examples of short-period analogue records during austral summer (19 January 1997; 30 min data; vertial component HES seismograph). (**e**) Examples of analogue records during austral winter (16 July 1997; 30 min data; vertical component HES seismograph). Microseisms are clearly visible in the austral summer data.

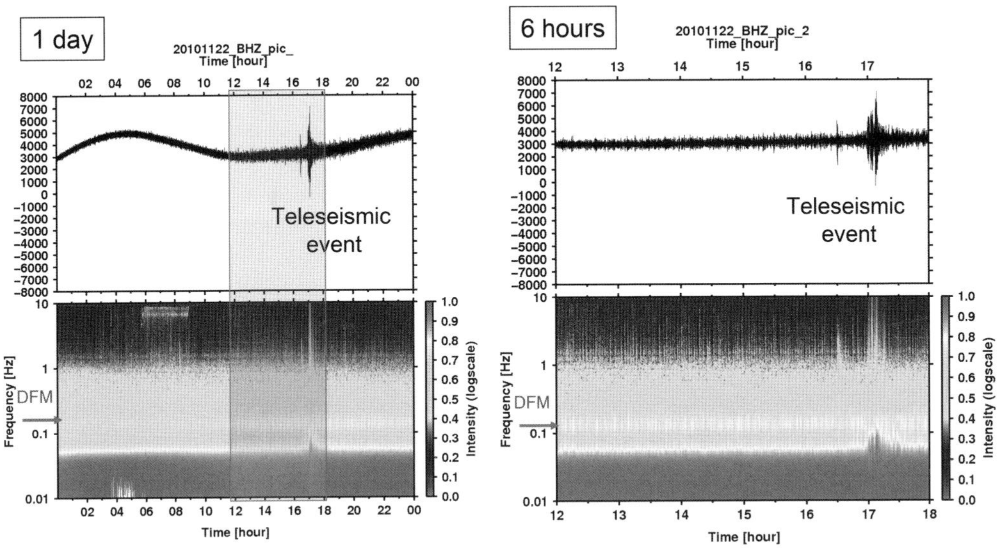

Fig. 4. The PSD of the vertical broadband seismograph (STS-1V) at SYO for 22 November 2010. (Left) one day of data; (right) 6 h of data, corresponding to the shadowed area of the left figure. A teleseismic event (Northern NZ, Mw = 5.9) is clearly recorded on the seismographs. The frequency range of DFM is indicated by the arrow.

in the vicinity of the Lützow–Holm Bay. The DFM can be identified all year round, although with distinctly lower amplitude during the local winter (April–October). The relatively high degree of inter-annual variability presumably reflects the large influence of extratropical cyclonic storms that commonly affect both the northern and southern oceans. On the contrary, the SFM is observed only under excellent storm conditions during the austral local winter. The strengths of both DFM and SFM are strongly related to the seasons, but presumably also to the local sea-ice conditions in the vicinity of the station. For example, summers with lower-amplitude microseisms at SYO seem to correspond to the presence of residual sea-ice near the Enderby Land coast (see Grob *et al.* 2011).

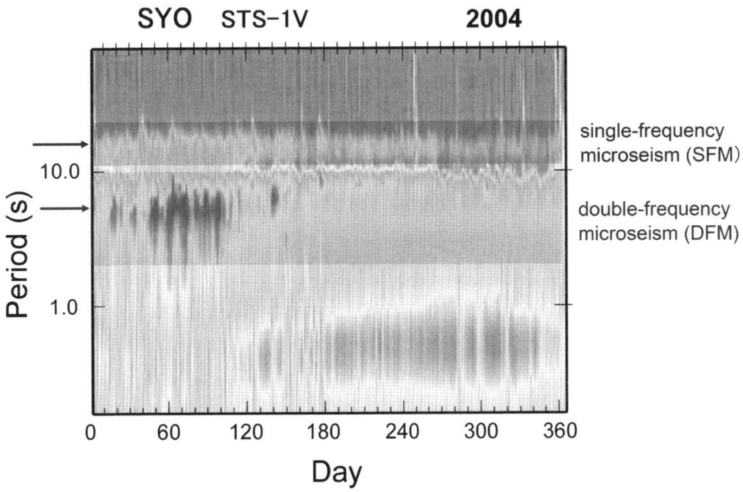

Fig. 5. Power spectral densities of the broadband seismograph (STS-1V) at SYO, for a whole season in 2004. Predominant frequencies corresponding to SFM and DFM are indicated by two arrows, respectively (modified after Grob *et al.* 2011; Kanao *et al.* 2012).

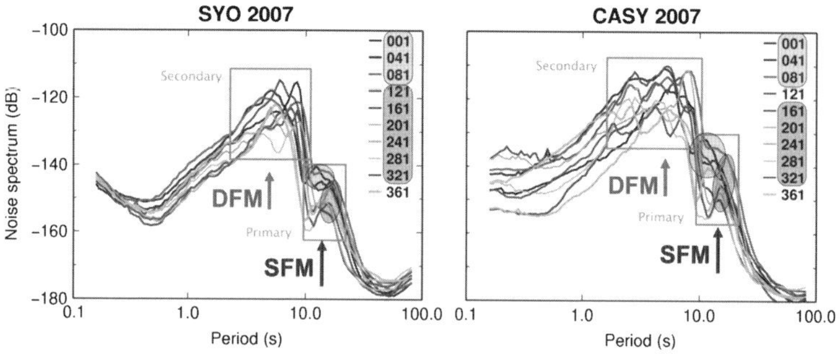

Fig. 6. Microseismic spectra over a year in 2007 for stations SYO and CASY. Each curve represents the seismic noise spectra for one day, coded by day of the year (see legend). Primary (SFM) and secondary (DFM) microseisms are outlined by boxes (modified after Grob *et al.* 2011).

Figure 6 shows in more detail the microseismic spectra over a year in 2007 for both SYO and Casey stations in East Antarctica (modified after Grob *et al.* 2011). Each curve represents the seismic noise spectrum for one day, and is colour-coded by day of the year. Primary (SFM) and secondary (DFM) microseisms are outlined by boxes. The peak frequencies of the SFM are highlighted in red and blue, for local summer and winter respectively. At both stations, it can be clearly seen that the spectral amplitudes for the SFM decrease during local winter, and migrate to longer periods.

An important parameter monitored by station operators is the teleseismic detectability, that is, the capability of a station to detect a seismic event occurring at teleseismic distances (over 30° away). This parameter is strongly correlated to the noise level at the station. Temporal variations in teleseismic detectability at SYO were investigated for the period from 1987 to 2007 (Kanao 2010). During the austral summer, the station shows less teleseismic detectability (i.e. the station detects fewer low-magnitude events) than during the austral winter, because of high noise level in the local summer arising from both natural factors and human activity close to the station. These variations in teleseismic detectability also indicate a strong relationship with the surrounding environment, such as meteorological events, sea-ice thickness and its spreading area (Ushio 2003), and more particularly the amplitude of microseisms, which is strongest during the austral summers.

Microbaroms observed during the IPY

Over the last few decades, in order to monitor nuclear tests, a global infrasound network has been constructed by the Comprehensive Nuclear-Test-Ban Treaty Organization (Fig. 7, top; Butler & Anderson 2008). One objective of the Treaty is to estimate the detection and location capabilities of this network at regional and global distances. Another is to explore ways to improve these capabilities and enhance the understanding of wave propagation through the atmosphere of the observed events. At this time, the Test-Ban Treaty network has 60 infrasound stations, each containing at least four sensors (array stations), that can detect a several-kilotonne TNT-level explosion at a range of *c.* 1000 km. Although the full capability of the global infrasonic network is yet to be established, it is fully adequate for monitoring nuclear tests, but too sparse for analysing natural infrasound phenomena in detail.

In April 2008, a Chaparral-type sensor was installed on a rock outcrop at SYO (Fig. 7a, b) as a project of the International Polar Year of 2007–2008 (Ishihara *et al.* 2009), making Syowa the fifth permanent infrasound station in Antarctica. From the analysis of 100 Hz trial data recorded at SYO during two winter seasons in 2008–2009, we found continuous background infrasound noise (Figs 8–10) that appear to correspond to co-oscillation with the DFM and SFM as observed by seismographs at SYO. Although the DFM is widely recorded, the SFM arising from the direct transfer of ocean-gravity wave (swell) energy to infrasonic waves, has not been recorded to date. Theory predicts radiation of DFM into the atmosphere from the microbarom source process, but theory does not predict for the SFM (Waxler & Gilbert 2006). Time variations similar to those observed on seismic spectra are also observed in infrasound data, and also seem to correlate with storms that occur with intervals of a few days (as indicated by the strong energy arrivals in Figs 8 & 9). These observations suggest the presence of physical interaction between the atmosphere–ocean system and the solid earth in the microbarom

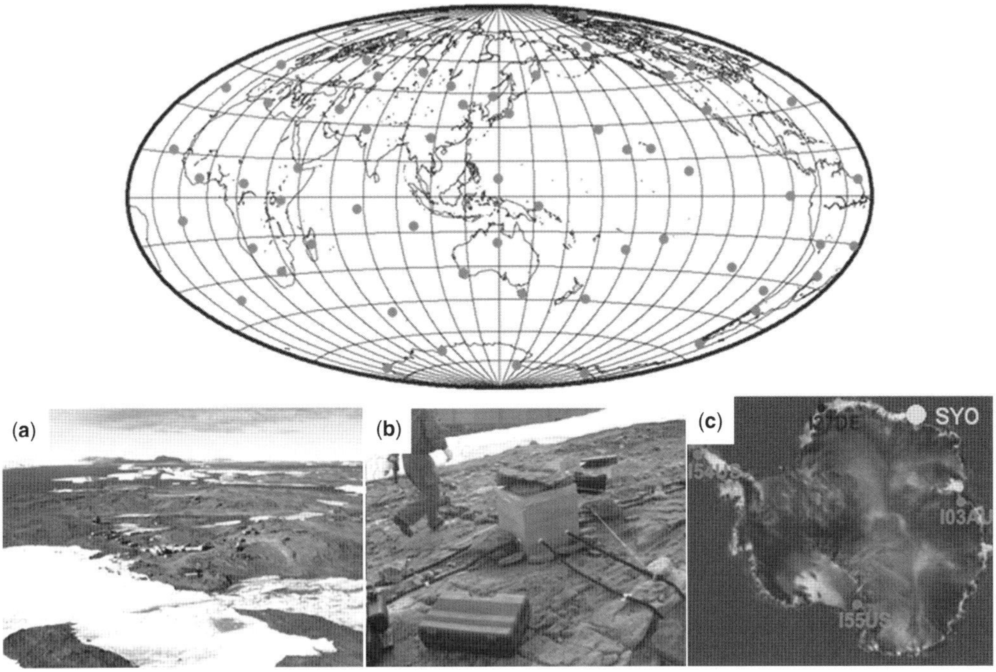

Fig. 7. Top: a global distribution map of the Comprehensive Nuclear Test-Ban Treaty Organization infrasound network. Bottom: (**a**) photograph of Syowa Station (SYO) in East Ongul Island, in Lützow–Holm Bay; (**b**) the infrasound sensor inside the wood box installed at SYO; (**c**) infrasound station distribution in Antarctica.

(microseism) frequency ranges. In addition to the microseisms, characteristic repeating signals over 10 Hz with harmonic overtones are occasionally recorded, as indicated by broken green squares in these figures.

Figure 8 represents the PSD of infrasound signals at SYO for one day during austral summer in January 2010. The DFM frequency is indicated by a red arrow, and varies significantly in both amplitude and frequency content over the day. These variations may correspond to atmospheric variations related to changes in weather conditions in the surrounding Lützow–Holm Bay. High-frequency repeating signals above 10 Hz with harmonic overtones are once more identified, and indicated by a broken green square. The harmonic overtone signals may be associated with local phenomena, such as katabatic winds, or with ice dynamics caused by various environmental changes near the station (i.e. sea-ice movement, tide-crack opening shocks, iceberg tremors, basal sliding of the ice-sheet and calving of glaciers, etc.). The other explanation for signal of 'high-frequency harmonic signals' is resonance effects in infrasound pipes used as a wind filter at the station. For example, a resonance effect in pipes is discussed in Hedlin & Alcoverro (2005).

Loading effects from the Southern Ocean are clearly recorded by both broadband seismographs and infrasound sensors deployed at SYO. The oceanic effects on infrasound will be modulated in the same way as for the seismic data by the presence of sea-ice, and are even more directly related to the atmosphere–ocean–cryosphere system. The measurement of 'microbaroms' may therefore be a useful parameter for characterizing ocean-wave climate and global storm intensity, complementing other estimates from oceanic or satellite measurements. The infrasound data at SYO, moreover, contains signals with harmonic overtones of uncertain origin, but which may be related to the ice dynamics in the Antarctic. Monitoring the infrasound waves could therefore be a new way to record regional environmental changes in Antarctica.

For further investigation, a theoretical modelling approach is required to determine the actual sources of several kinds of infrasound signals. It would be also useful to compare these signals in more detail with other data, such as broadband seismograms that share sensitivity over part of the infrasound frequency range. The array alignment of the infrasound stations, moreover, could firmly provide robust information about the arrival direction and epicentral distance from the infrasound sources.

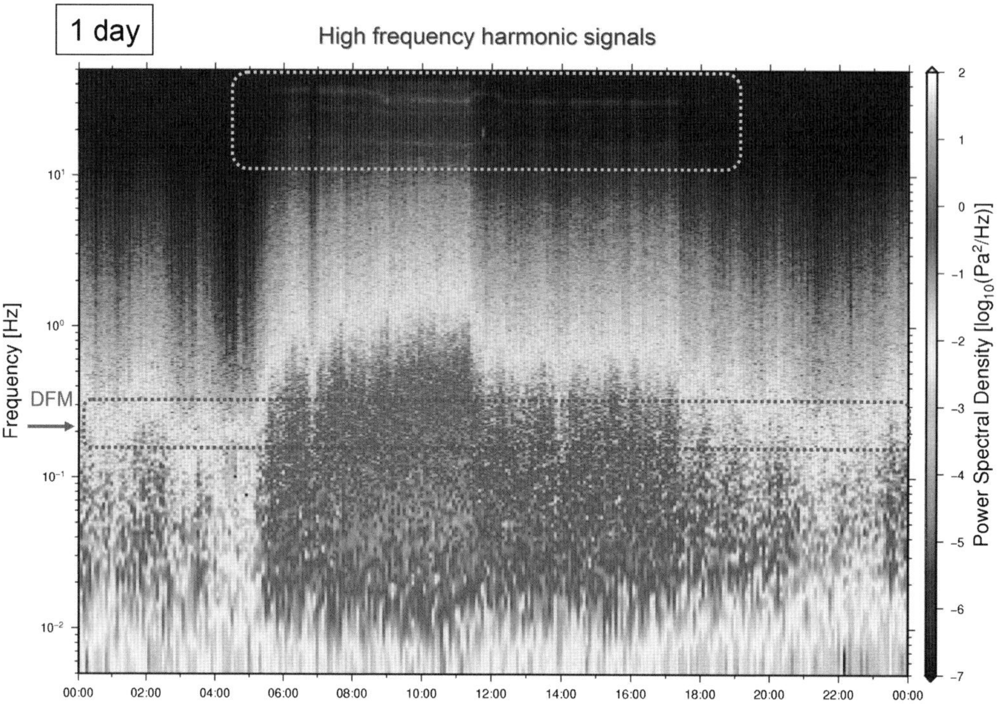

Fig. 8. One-day PSD of infrasound data at SYO on 2 January 2010. The DFM is indicated by an arrow. High-frequency signals above 10 Hz are identified clearly, labelled by the dotted green square.

Additional observations with an array station near the current SYO station would be helpful.

Cryosphere wave signals

Over the past few decades, more seismic observations in the polar regions by both temporary seismic networks and permanent stations have detected local seismicity. Seismic signals involving ice-related phenomena are called 'ice-quakes' (ice-shocks for smaller ones), and are most frequently reported in association with glacier-related mass movements of ice sheets, or with sea-ice, tide-cracks and icebergs (Anandakrishnan & Alley 1997; Kanao & Kaminuma 2006; Wiens 2007; Nettles & Ekström 2010). The so-called 'ice-micity' detected around the Bransfield Strait and Drake Passage by a local network of hydrophone arrays in 2006–2007 illustrated the dynamic behaviour of sea-ice in the Antarctic Peninsula (Dziak *et al.* 2009). Local seismicity around the Lützow–Holm Bay was reported by Kanao & Kaminuma (2006), who detected 17 events using a local network deployed around the Bay. Most of the hypocentres were located along the coast, apart from a few on the northern edge of the continental shelf. Several of these events may have been large ice-quakes associated with the sea-ice around the Lützow–Holm Bay or in the Southern Ocean.

Sea-ice dynamics and icebergs clearly affect seismic signals. A large volume of sea-ice was discharged from Lützow–Holm Bay during the 1997 austral winter, and was clearly imaged by the NOAA satellite (Ushio 2003). The broadband seismographs at SYO recorded distinct waveforms associated with the discharge events. The long-duration sea-ice tremors had very distinct spectral characteristics that distinguished them clearly from ordinary teleseismic events, together with local tectonic events. Several sequences of signals with strong harmonic overtones, presumably associated with the merging of multiple ice volumes, appeared on the PSDs. The PSDs also showed surge events that seemed more closely related to the break-up process of the sea-ice mass. Both kinds of cryoseismic waves occurred continuously for a few hours, and repeated themselves several times within a few days during late July 1997.

Similar cryoseismic phenomena have also been reported around the Ross Sea region (MacAyeal *et al.* 2009), the marginal sea of the Antarctic Peninsula (Bohnenstiehl *et al.* 2005; Dziak *et al.*

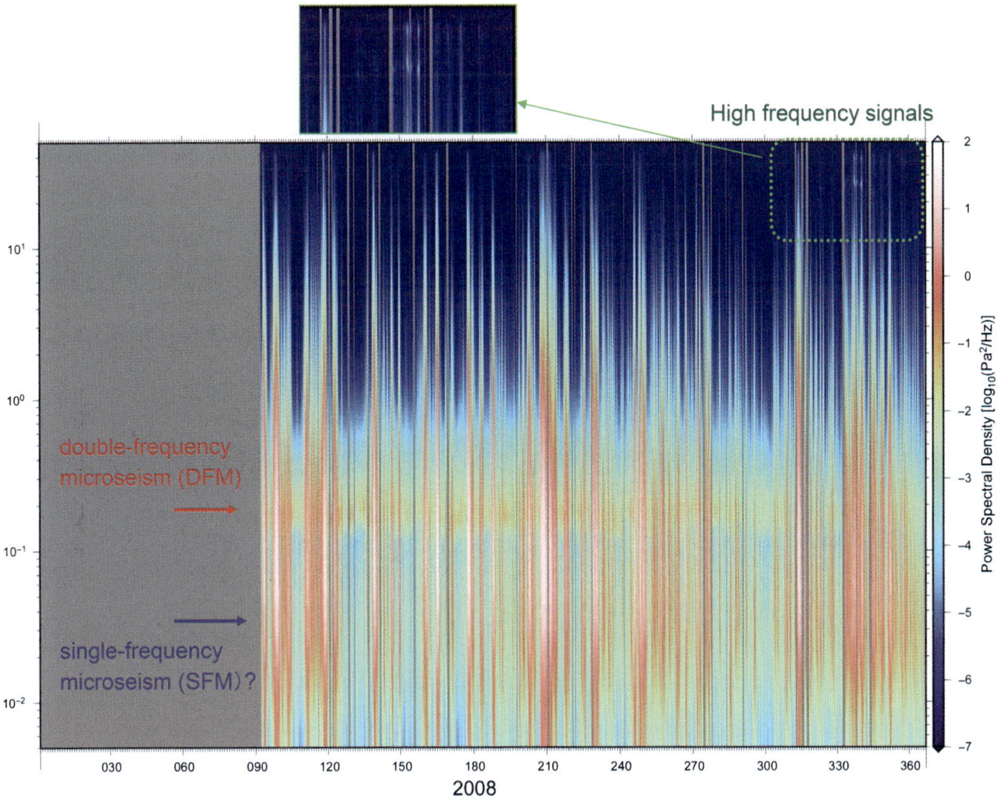

Fig. 9. Power spectral densities of the infrasound signals in April–December 2008, from the start of pilot observation at SYO. The grey-coloured time zones imply lack of data, otherwise any kinds of errors during the PSD processing. Predominant frequencies corresponding to SFM and DFM are indicated by blue and red arrows, respectively. Characteristic harmonic signals >10 Hz are represented by the dotted green square.

2009), as well as the continental margin of Dronning Maud Land (Muller & Eckstaller 2003). In particular, iceberg-originated harmonic tremor emanating from tabular icebergs has been observed by both seismo-acoustic and local broadband seismic signals (MacAyeal *et al.* 2009). The tremor signals consisted of extended episodes of stick-slip ice-quakes generated when the ice-cliff edges of two tabular icebergs rubbed together during a glancing, 'strike-slip' collision. Source mechanisms of such harmonic tremors might provide useful information for the study of iceberg behaviour, and a possible method for remotely monitoring iceberg activity.

However, it should also be pointed out that, in addition to these cryoseismic phenomena, several small- to medium-magnitude local/regional natural seismic events have been recorded. Unfortunately, many of these could not be located accurately, since they have unclear arrivals when recorded by the present global network, which is particularly sparse near Antarctica.

Interaction of the atmosphere–ocean–solid earth system

Most of the community agrees that the Polar Regions play a critical role in the Earth's system. The Southern Ocean and the Antarctic ice sheet and their response to climate change potentially have a great impact upon humanity, through both long-term sea-level rise and modulation of freshwater input to the oceans. Monitoring the dynamic response of the polar ice sheets between the ocean and atmosphere, as well as the underlying solid earth, should be important components in a long-term effort to observe climate change on a global scale. Future directions and targets in the global monitoring effort will probably emerge from multidisciplinary projects combining data from several global networks. For example, it would be useful to compare infrasound signals with other data, such as broadband seismograms and hydro-acoustic waves that share sensitivity over part of the infrasound frequency range.

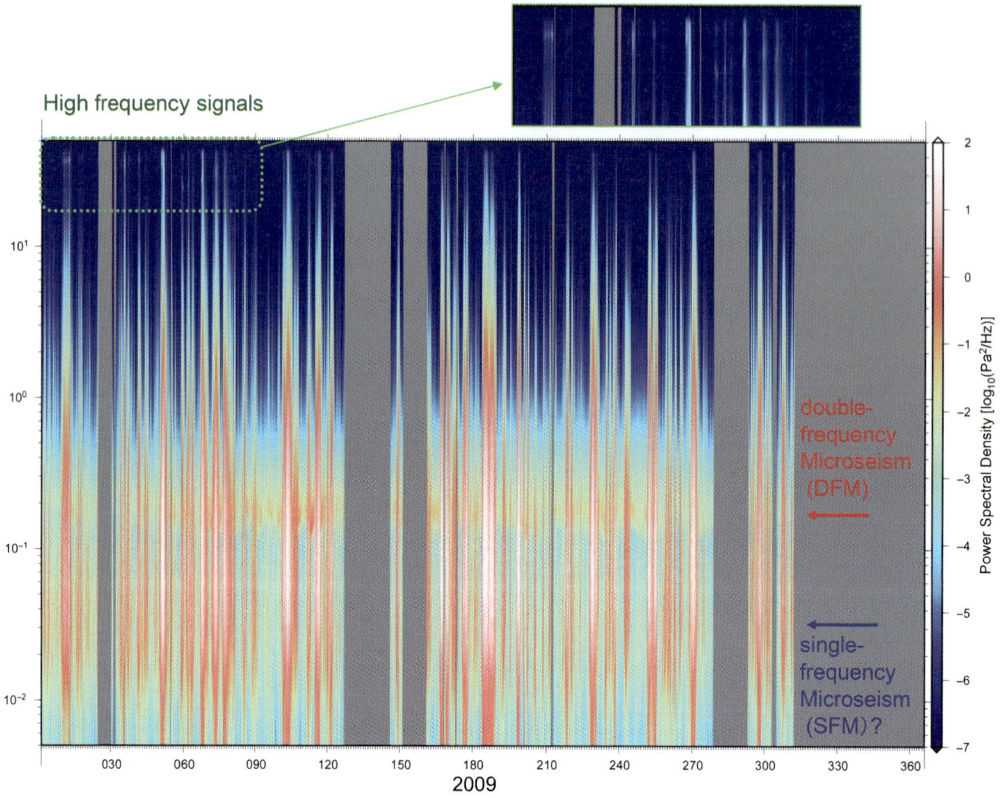

Fig. 10. Power spectral densities of infrasound signals in 2009. The grey-coloured time zones imply lack of data, or the errors during PSD processing. Predominant frequencies corresponding to SFM and DFM are indicated by blue and red arrows, respectively. Characteristic harmonic signals >10 Hz are identified by the dotted green square.

There is still much to be learned about the physical mechanisms of interaction between the atmosphere–ocean–cryosphere–solid earth system in Antarctica. Continuous observation by a sufficiently large number of high-quality monitoring stations, as well as theoretical work, will probably be necessary to make progress in this field. Given the high cost and technical difficulties of continuous observation in the Antarctic, such work would require strong international collaboration beyond the end of the International Polar Year.

The place of Antarctica in the global network

In previous sections, we discus the information that seismic data contain regarding wave activity, ice-dynamics and weather-related phenomena (e.g. storms). Continuous digital records from the global seismographic network (Fig. 1), and their precursor networks extend back more than 40 years, and hence open up the possibility of using seismic data from Antarctica to investigate climate change and palaeoenvironments.

A large seismology research programme consisting of several field-campaigns was held during the IPY2007–2008 in order to complement the network of permanent stations at the high latitudes in the Polar Regions. In Antarctica, the most ambitious seismological field campaign conducted for the IPY was the 'Gamburtsev Mountain SEISmic experiment' or GAMSEIS, an internationally co-ordinated deployment of more than 50 broadband seismographs over the crest of the Gamburtsev Subglacial Mountains in East Antarctica (Wiens 2007; Hansen *et al.* 2010). GAMSEIS and many other seismological deployments, including a French deployment between Concordia and Vostok Lake (CASE-IPY), and a Japanese deployment around SYO (GARNET), were co-ordinated under the larger programme called POLENET. Some of the stations installed for these campaigns are destined to remain deployed well into the future, and this network will be one of the International Polar Year's most significant legacies.

These seismic networks, together with the infrasound stations in the Antarctic, strongly contribute to both the Test-Ban Treaty network (Fig. 7) and the Pan-Antarctic Observations System under the Scientific Committee on Antarctic Research. Hydroacoustic observations contribute to the understanding of wave-propagation inside the ocean, and have already been exploited to investigate seismicity and sea-ice dynamics around the Bransfield Strait, Antarctic Peninsula (Dziak *et al.* 2009). The combination of seismic, infrasound and hydroacoustic observations is most likely required to understand in more detail the atmosphere–ocean–cryosphere–solid earth system and its variations with time. We are hopeful that the large quantity of data of these three types accumulated over the past decades by the Test-Ban Treaty network will soon be distributed to the scientific community in order to advance research in this domain.

Conclusions

Several characteristic features of seismic and infrasound waves propagating in the coastal region of East Antarctica, and how they are related to the atmosphere–ocean–solid earth system, are treated in this paper. Microseisms and microbaroms originating from the Southern Ocean are clearly recorded by both the broadband seismograph and infrasound sensor deployed at SYO, in the Lüztow–Holm Bay, East Antarctica.

The double frequency microseism/microbarom (with peaks at 4–10 s period) reflects the large influence of extratropical cyclonic storms that commonly affect the southern oceans. It can be identified all year round, although with distinctly lower amplitude during austral winter, presumably because of the large sea-ice extent and the consequent reduction in oceanic loading. The single frequency microseism/microbarom (with peaks at 12–30 s) is observed only under excellent storm conditions during the austral local winter. The strengths of both DFM and SFM are strongly related to the seasons, but presumably also to the local sea-ice conditions in vicinity of the station.

Microseism/microbaroms are a useful proxy for characterizing ocean wave climate and global storm intensity, complementing other measurements such as ocean surface and satellite images. Continuous monitoring by seismograph and infrasound networks in Antarctica contribute strongly to the global seismic network and the Test-Ban Treaty network in the high southern latitudes, and to the Pan-Antarctic Observations System.

We would like to express our sincere appreciation for many collaborators to the observations at SYO. We also thank all the members of the Antarctic programmes of the International Polar Year, such as POLENET (T. Wilson of the Ohio State University, and other members), as well as the Japanese Antarctic Research Expeditions (JARE; K. Shiraishi, Director-General of the National Institute of Polar Research, and many other members). Infrasound observation at SYO was partially supported by the Ministry of Education, Science, Sports and Culture, Grant-in-Aid for Young Scientists (B) 19740265, 2007 (P.I. for Y. Ishihara). The authors would like to express sincere appreciation to M. Grob of the University of Alberta, and A. Reading of the University of Tasmania, for their critical reading and giving many useful comments. We would also like to express sincere appreciation to M. J. Hambrey of Aberystwyth University, for his efforts relating to the editorial matters for the Special Publication of the Geological Society, *Antarctic Earth Surface Processes and Paleoenvironments*.

References

Anandakrishnan, S. & Alley, R. B. 1997. Tidal forcing of basal seismicity of ice stream C, West Antarctica, observed far inland. *Journal of Geophysical Research*, **102**, 15 183–15 196.

Anandakrishnan, S., Voigt, D. E., Alley, R. B. & King, M. A. 2003. Ice stream D flow speed is strongly modulated by the tide beneath the Ross Ice Shelf. *Geophysical Research Letters*, **30**, doi: 10.1029/2002GL016329.

Aoyama, Y. & Kanao, M. 2010. *Seismological Bulletin of Syowa Station, Antarctica, 2008*. JARE Data Report, **317** (Seismology 44), 1–85.

Ardhuin, F., stutzmann, E., Schimmel, M. & Mangeney, A. 2011. Revealing ocean wave sources of seismic noise. *Journal of Geophysical Research*, **116**, C09004, doi: 10.1029/2011JC006952.

Aster, R., McNamara, D. & Bromirski, P. 2008. Multidecadal climate-induced variability in microseisms. *Seismological Research Letters*, **79**, 194–202, doi: 10.1785/gssrl.79.2.194.

Bohnenstiehl, D. R., Dziak, R. P., Parlk, M. & Matsumoto, H. 2005. Seismicity of the polar seas: the potential for hydroacoustic monitoring of tectonic and volcanic processes. *The 12th Seoul International Symposium on Polar Science*, 11–14 May 17–19, Ansan, Korea.

Bromirski, P. D. 2009. Earth vibrations. *Science*, **324**, 1026–1027.

Butler, R. & Anderson, K. 2008. *Global Seismographic Network (GSN)*. IRIS Report, 6–7.

Dziak, R. P., Parlk, M., Lee, W. S., Matsumoto, H., Bohnenstiehl, D. R. & Haxel, J. H. 2009. Tectonomagmatic activity and ice dynamics in the Bransfield Strait back-arc basin, Antarctica. *The 16th International Symposium on Polar Science*, 10–12 June, Incheon, Korea, 59–68.

Eto, T. 1962. On the electromagnetic seismographs at Syowa Base, Antarctica. *Antarctic Record*, **14**, 48–50.

Grob, M., Maggi, A. & Stutzmann, E. 2011. Observations of the seasonality of the Antarctic microseismic signal, and its association to sea ice variability. *Geophysical Research Letters*, **38**, L11302, doi: 10.1029/2011GL047525.

HAGIWARA, T. 1958. A note on the theory of the electromagnetic seismograph. *Bulletin of the Earthquake Research Institute*, **36**, 139–164.

HANSEN, S. E., NYBLADE, A. A., HEESZEL, D. S., WIENS, D. A., SHORE, P. & KANAO, M. 2010. Crustal Structure of the Gamburtsev Mountains, East Antarctica, from S-wave Receiver Functions and Rayleigh Wave Phase Velocities. *Earth and Planetary Science Letters*, **300**, 395–401, doi: 10.1016/j.epsl.2010.10.022.

HATHERTON, T. 1960. Microseisms at Scott Base. *Geophysical Journal*, **3**, 381–405.

HEDLIN, M. & ALCOVERRO, B. 2005. The use of impedance matching capillaries for reducing resonance in rosette infrasonic spatial filters. *Journal of the Acoustical Society of America*, **117**, 1880–1888.

ISHIHARA, Y., FURUMOTO, M., SAKAI, S. & TSUKUDA, S. 2004. The 2003 Kanto large bolide's trajectory determined from shockwaves recorded by a seismic network and images taken by a video camera. *Geophysical Research Letters*, **31**, L14702.

ISHIHARA, Y., YAMAMOTO, M. & KANAO, M. 2009. *Current status of infrasound pilot observation at Japanese Islands and SYOWA Antarctica, and development of new infrasound sensor using optical sensing method.* AGU Fall 2009 Meeting, San Francisco, CA, A13D-0244.

IYEMORI, T., NOSE, M. ET AL. 2005. Geomagnetic pulsations caused by the Sumatra earthquake on December 26, 2004. *Geophysical Research Letters*, **32**, L20807, doi: 10.1029/2005GL024083.

KANAO, M. 2010. Detection capability of teleseismic events recorded at Syowa Station, Antarctica – 1987–2007. *Antarctic Record*, **54**, 11–31.

KANAO, M. & KAMINUMA, K. 2006. Seismic activity associated with surface environmental changes of the Earth system, East Antarctica. *In*: FUTTERER, D. K., DAMASKE, D., KLEINSCHMIDT, G., MILLER, H. & TESSENSOHN, F. (eds) *Antarctica: Contributions to Global Earth Sciences*. Springer, Berlin, 361–368.

KANAO, M., NOGI, Y. & TSUBOI, S. 2006. Spacial distribution and time variation in seismicity around Antarctic Plate – Indian Ocean. *Polar Geoscience*, **19**, 202–223.

KANAO, M., MAGGI, A. ET AL. 2012. Interaction on seismic waves between atmosphere–ocean–cryosphere and geosphere in Polar Region. *In*: KANAO, M. (ed.) *Seismic Waves – Research and Analysis*. InTech Publishers, Rijeka, 1–20.

MACAYEAL, D., OKAL, E., ASTER, R. & BASSIS, J. 2009. Seismic observations of glaciogenic ocean waves on icebergs and ice shelves. *Journal of Glaciology*, **55**, 193–206.

MULLER, C. & ECKSTALLER, A. 2003. Local seismicity detected by the Neumayer seismological network, Dronning Maud Land, Antarctica: tectonic earthquakes and ice-related phenomena. *IX International Symposium on Antarctic Earth Science Programme and Abstracts*, 236.

NAWA, K., SUDA, N., SATAKE, K., SATO, T., DOI, K., KANAO, M. & SHIBUYA, K. 2007. Loading and gravitational effects of the 2004 Indian Ocean tsunami at Syowa Station, Antarctica. *Bulletin of the Seismological Society of America*, **97**, S271–278, doi:10.1785/0120050625.

NETTLES, M. & EKSTRÖM, G. 2010. Glacial earthquakes in Greenland and Antarctica. *Annual Reviews of Earth and Planetary Sciences.*, **38**, 467–491.

READING, A. M. 2006. On seismic Strain-Release within the Antarctic Plate. *In*: FUTTERER, D. K., DAMASKE, D., KLEINSCHMIDT, G., MILLER, H. & TESSENSOHN, F. (eds) *Antarctica: Contributions to Global Earth Sciences*. Springer, Berlin, 351–356.

READING, A. M. 2007. The seismicity of the Antarctic Plate. *In*: STEIN, S. & MAZZOTTI, S. (eds) *Continental Intraplate Earthquakes: Science, Hazard and Policy Issues*. Geological Society of America, Boulder, CO, Special Papers, **425**, 285–298.

STUTZMANN, E., SCHIMMEL, M., PATAU, G. & MAGGI, A. 2009. Global climate imprint on seismic noise. *Geochemistry, Geophysics, Geosystems*, **10**, Q11004, doi: 10.1029/2009GC002619.

USHIO, S. 2003. Frequent sea-ice breakup in the Lützow–Holmbukta, Antarctica, based on analysis of sea ice condition from 1980 to 2003. *Antarctic Record*, **47**, 338–348.

WAXLER, R. & GILBERT, K. E. 2006. The radiation of atmospheric microbaroms by ocean waves. *Journal of the Acoustical Society of America*, **119**, 2651–2664.

WIELANDT, E. & STEIM, J. M. 1986. A digital very-broadband seismograph. *Annales Geophysicae*, **4**, 227–232.

WIENS, D. A. 2007. Broadband seismology in Antarctica: recent progress and plans for the International Polar Year. *Proceedings of the International Symposium–Asian Collaboration in IPY 2007–2008*, Tokyo, 21–24.

Evaluation of Envisat ASAR IMP imagery for snow mapping at varying spatial resolution (Deception Island, South Shetlands – Antarctica)

CARLA MORA[1]*, GONÇALO VIEIRA[1] & MIGUEL RAMOS[2]

[1]*Centre for Geographical Studies, IGOT, University of Lisbon, Lisbon, Portugal*

[2]*Department of Physics, University of Alcalá de Henares, Alcalá de Henares, Spain*

**Corresponding author (e-mail: carlamora@campus.ul.pt)*

Abstract: Advanced synthetic aperture radar image mode precision (ASAR IMP) scenes of Deception Island from December 2008 to September 2010 have been analysed to assess its potential for snow cover classification. Backscattering was checked against ground truth. Despite the good spatial resolution of the ASAR, its applicability for detecting snow cover, and especially wet snow, was only possible at much lower resolutions, since noise was found to be very high. Scenes with bare ground or with dry snow cover showed highest backscattering, with averages from −10 to −12 dB. Wet snow showed a shift towards lower values, peaking at −15 dB. A threshold of −13 to −14 dB was identified between dry/bare ground and wet snow scenes at Crater Lake. The backscatter difference to a reference snow-free scene usually provided better classification results, and a threshold ranging from −2 to −3 dB was found. Results show that, despite the relative ease of use of C-band ASAR, special care is necessary since the results show significant noise, and it should only be applied to large areas. Large seasonal patterns of snow melt were identified on Deception Island.

Seasonal snow cover is one of the most important components of Earth's cryosphere and is a fundamental factor in the global climate system, especially owing to its influence on planetary albedo (Slaymaker & Kelly 2007). Snow cover is also a prominent factor controlling the thermal regime of permafrost, an essential climate variable as defined by the Global Climate Observing System and a key element of the Earth System (Smith & Brown 2009). The properties of snow cover are of major significance for the ground energy balance and therefore for the ground thermal regime, since thick snow provides excellent insulation. Permafrost is therefore influenced by snow cover properties, spatial distribution and seasonal regime (Williams & Smith 1989). In order to accurately model permafrost distribution, one of the main objectives of the research community is to have accurate data on snow cover. In the South Shetlands archipelago permafrost is widespread, but its temperatures are high and close to thawing (Vieira *et al.* 2010). A good assessment of snow cover regime and distribution is key for mapping and modelling permafrost, and this paper aims to contribute to the effort towards snow mapping by means of remote sensing, using a Global Terrestrial Network for Permafrost site at Deception Island for validation.

Optical remote sensing systems are limited to daylight and cloud-free conditions, and therefore, microwave imagery analysis is the optimum technique to be used in high-cloudiness environments such as the Maritime Antarctic. Radar imagery has been widely used for detection of snow cover (see Shi & Dozier 1995; Baghdadi *et al.* 1999; Bernier *et al.* 1999; Rees & Steel 2001; Magagi & Bernier 2003). To our knowledge, its potential for the analysis of snow in the ice-free areas of the Maritime Antarctic has not been addressed. Larger radar frequencies show a better ability to discriminate snow cover from snow-free terrain and the best accuracy is obtained through the X-band (Rees 2006). It has, however, been shown that more accessible C-band imagery is also adequate for discriminating wet snow (Shi & Dozier 1995; Rees & Steel 2001; Valenti *et al.* 2008). This paper presents the results of testing the Advanced Synthetic Aperture Radar (ASAR) C-band sensors in Image Mode Precision (IMP) for detecting snow cover in Deception Island. Validation was performed using ground truth data obtained from a time-lapse camera and air-temperature and snow-temperature mini-loggers installed in a test site at Crater Lake.

Study area

Deception Island is located in the South Shetlands archipelago, *c.* 100 km NW of the Antarctic

From: Hambrey, M. J., Barker, P. F., Barrett, P. J., Bowman, V., Davies, B., Smellie, J. L. & Tranter, M. (eds) 2013. *Antarctic Palaeoenvironments and Earth-Surface Processes*. Geological Society, London, Special Publications, **381**, 481–493. First published online July 24, 2013, http://dx.doi.org/10.1144/SP381.19

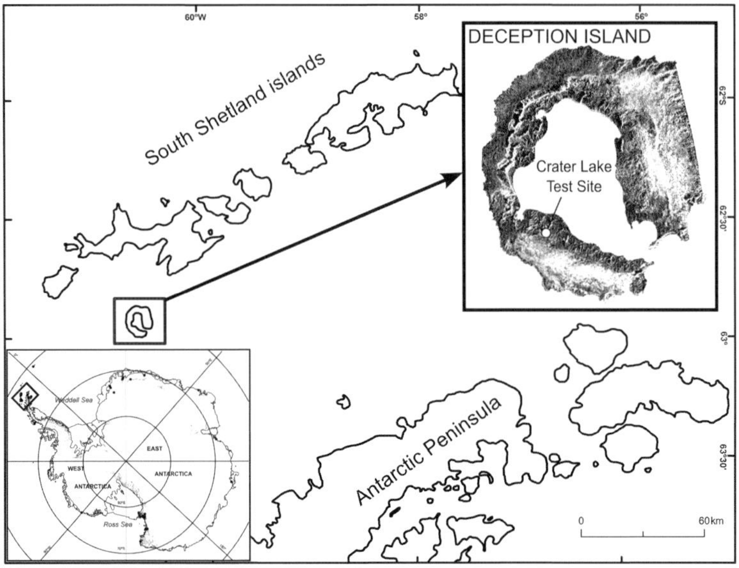

Fig. 1. Location of the study area.

Peninsula (Fig. 1). The island is an active volcano about 15 km in diameter and with a horseshoe shape owing to a large collapsed caldera (Port Foster). The ring has its maximum altitude at Mount Pond (539 m) and some 57% of the island is covered by glaciers (Smellie & López-Martínez 2002). The geology is composed of volcano-sedimentary deposits, including pyroclastic flows and deposits, strombolian scoriae and lava, volcanic and hypovolcanic indurated ashes and phreatomagmatic deposits.

The climate is Maritime Antarctic with average air temperatures *c.* −2 °C at sea-level. The annual temperature range is small, with January and February showing a large number of days above 0 °C. Accurate precipitation data do not exist, but snowfalls are frequent year-round, with abundant summer rainfall events (Vieira *et al.* 2010; Bockheim *et al.* 2013).

The island has been a focus of interest for the permafrost research community, mainly owing to the interactions with volcanic activity since several geothermal anomalies are present. Warm permafrost is widespread down to sea-level and is lacking in narrow areas along geothermally anomalous fault lines. Near Crater Lake a permafrost monitoring site of the Global Monitoring Network for Permafrost (GTN-P) has been installed by our group, as well as a Circumpolar Active Layer Monitoring site (CALM). Other active layer monitoring sites are installed at Collado Vapor (Irizar), Chilean refuge, and by the Argentinean station Decepción.

Methods

Analysis of snow cover was conducted by means of both satellite imagery and ground truthing data for assessment of classification accuracy. Owing to the persistent and dense cloud coverage typical of Deception Island, Advanced Synthetic Aperture Radar imagery was used. Twelve scenes from a single orbit (467) of Envisat ASAR (C-Band) using Image Mode Precision, IS2 Beam and HH polarization from December 2008 to March 2010 were acquired. A single orbit was chosen to minimize angular influences on backscatter and to allow accurate image-to-image co-registration.

Processing was conducted using NEST 4B-1.0 (ESA SAR Toolbox) algorithms. The process includes the transformation of DN into power values, geometric and radiometric correction, image filtering and computation of the backscattering coefficient for each pixel (Fig. 2).

The radiometric correction was performed by compensating for effects of local illuminated area and incident angle on the local backscattering, using the absolute calibration algorithm of Rosich & Meadows (2004). The speckle filtering was carried out using the frost filter, which is considered well suited for snow mapping (Nagler & Rott 2000). For the geocoded terrain image, geometrical correction was performed using the Range Doppler orthorectification method, as in Small & Schubert (2008). The terrain correction operator was used with a digital elevation model with 10 m pixel resolution derived from the Spanish Topographical

Fig. 2. Analytical procedures for the evaluation of ASAR IMP imagery for snow cover mapping on Deception Island.

Map, 1:25 000. Gamma nough backscattering from the topography-corrected imagery was used, while layover areas were masked and excluded from the analysis (Lewis & Henderson 1998; Small 2011).

Ground reference data was obtained from the Crater Lake GTN-P site, located in a wide and flat interfluve at *c.* 80 m a.s.l., north of the Spanish Station Gabriel de Castilla (Figs 1 & 3). The ground surface is a flat and smooth cover of pyroclastic pumice granules without any prominent features. These conditions mitigate topographical as well as varying ground effects on the radar signal. They are representative of wind-protected sites on the island, since on ridges the ground may remain free of snow even during winter periods. Permafrost with mean annual temperatures close to 0 °C occurs at *c.* 50 cm depth (Ramos *et al.* 2007).

Snow extent data were collected by analysing pictures from a Campbell CC640 time-lapse camera making three images per day around local solar noon (11:00, 12:00 and 13:00 LST). This allowed comparison at the time of satellite overpass, that is, 12:30, and the use of daylight conditions in winter and in the case of fog, improving the chances of having at least one good image per day. Several instrument masts installed in the test site served as markers in the pictures for identifying snow extent and were used to classify snow conditions for comparison with backscatter values of the satellite images.

Air temperatures were measured at 1.2 m with a single-channel TinyTag2 data logger, with an accuracy of 0.3 °C. Snow temperatures were measured at 4 h intervals using poles with ibutton mini-loggers (accuracy of 0.5 °C) at 2, 5, 10, 20, 40, 80 and 160 cm. The comparison of temperatures at different heights on the poles allowed estimation of snow thickness (data not shown), and the near-surface snow temperature was obtained from the buried mini-logger lying closer to the snow surface at each measurement. Near-surface snow temperatures and the analysis of the snow cover extent from the time-lapse photos allowed the identification of dry vs wet snow and continuous, discontinuous or snow-free conditions.

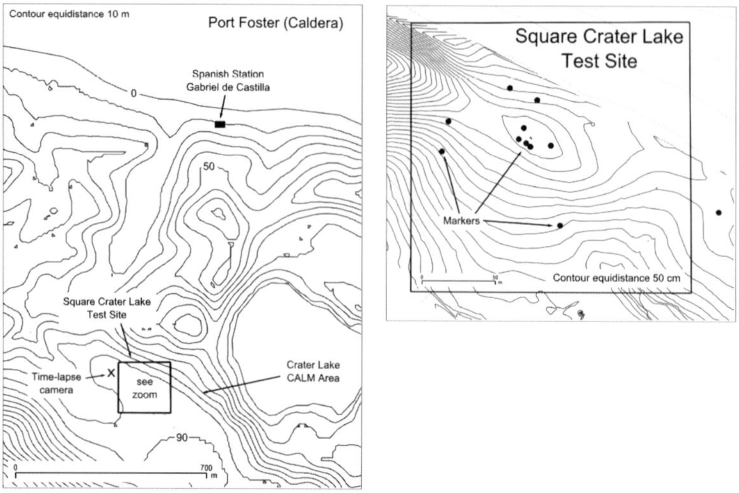

Fig. 3. Location and topography of the Square Crater Lake (SqCL) detailed study site.

Radar backscattering values of the 12 ASAR scenes were compared with the observations from the Crater Lake study site on snow conditions and temperature. The objective was identifying thresholds of backscattering, enabling classification and mapping of wet snow, dry snow and snow-free surfaces.

We also used a multitemporal approach by comparing backscattering of all scenes with backscattering from a snow-free scene, which is a well-known approach to derive indices for wet snow classification (Baghdadi *et al.* 1999; Bernier *et al.* 1999; Rees & Steel 2001; Magagi & Bernier 2003). Malnes & Guneriussen (2002) and Valenti *et al.* (2008) have shown that the C-band backscattering for images at 23° incidence angle is 2–3 dB lower for wet snow than for snow-free terrain. Nagler & Rott (1997) and Malnes & Guneriussen (2003) indicate that the backscattering from wet snow is *c.* 3 dB lower than for dry snow before melting or for bare surfaces. However, Baghdadi *et al.* (1997) and Koskinen & Pulliainen (1997) have shown that it may not be possible to distinguish wet snow and snow-free terrain using C-band imagery, owing to the type of surface or the local incidence angle. We aimed at assessing the applicability of such methodology at Deception Island and identifying its limitations.

Results

Snow conditions at Crater Lake in the period of study

Time-lapse photos of the Crater Lake monitoring site from December 2008 to January 2010 show significant interannual differences in the snow cover. Figure 4 shows the daily classification of snow cover into totally snow-free, partially

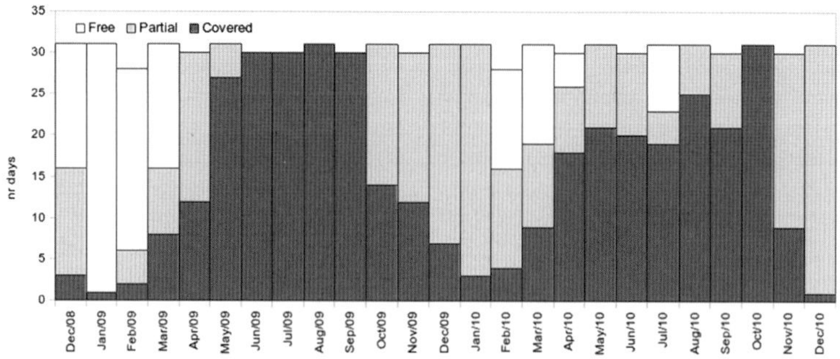

Fig. 4. Evolution of snow cover at the Crater Lake test site. Data derived from time-lapse photographs.

snow-covered and totally snow-covered. This simple classification was done by visual inspection of daily pictures. Snow was present in most of the period and was continuously absent from the plain at Crater Lake only for about 22 days in February and March 2009. All of the other snow-free periods lasted for only a few days. Periods with snow partially covering the monitoring site were frequent, mainly from October to April. These periods are generally associated with the snow melt period during spring or with snow melt during snow settling in autumn, but also with wind drift of snow during drier episodes. They also occurred during winter in 2010. Continuous snow cover occurred on 50% of the days, discontinuous in 34% and snow-free in 16% (Fig. 4). The period from December 2008 to March 2009 showed 68% snow-free days, 21% partially snow-covered days and 11% snow-covered days. The same period in 2009–2010 showed 20, 61 and 19%, respectively, which indicates significant differences in snow cover conditions, mainly between the two summers.

Daily average air temperatures were highly variable and irregular in winter, with average minima close to −20 °C and frequent daily average maxima close to −4 °C (Fig. 5). The winter of 2010 was significantly warmer than that of 2009, with the occurrence of days with average air temperatures above −1 °C in July and August. From December to February, average daily air temperatures were generally over 0 °C, with a much smaller range than in winter. The spring–summer of 2008–2009 was warmer than that of 2009–2010, especially with warm temperatures in December that should have been the cause of the faster snow melt in that year.

Near-surface snow temperatures, presented in Figure 4 at the dates of the satellite scenes, followed closely air temperatures, with less extreme values. They were typically *c.* −4 °C in winter and *c.* −1 °C in summer for the dates with satellite imagery.

The time-lapse photos of the Crater Lake plain for the days of ASAR scenes showed snow-covered conditions on most days (Fig. 6). The only snow-free scene was from 20 December 2008 and this was chosen as the reference image for the calculation of backscattering differences aimed at detecting the presence of wet snow. The day 5 December 2009 shows snow-free conditions on the hill slope close to the camera, but the test site is snow-covered. Air temperatures were at 1 °C, with near-surface snow temperatures close to 0 °C. The ground shows a very dark surface and the window of the camera box shows water droplets, indicating melting conditions and suggesting the presence of wet snow at the test site. All other scenes show the test site covered by a continuous snow mantle.

Point analysis of radar backscattering at Crater Lake

Instrumentation installed at the Crater Lake CALM site to monitor permafrost and active layer served as markers and allowed the precise identification of snow cover by analysing the time-lapse photos. The objective was to evaluate the potential for discrimination of detailed-scale spatial distribution of snow cover of the ASAR sensor by comparing the backscattering at each marker with the snow cover. The original satellite scenes show a pixel size of 12.5 m, but in order to filter interpixel noise, the backscatter at each marker was calculated by averaging (to the unit) the values of the four nearest-neighbour pixels.

Backscattering for different dates was compared with the snow conditions at each marker (presence/absence and snow temperature), but results were inconclusive and no systematic differences allowing for classification were found. This shows that single-pixel backscatter cannot be compared with individual terrain features and that it is too complex to georeference the scenes at

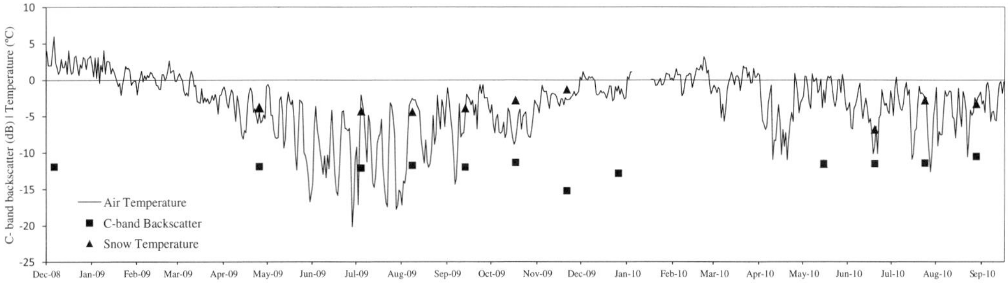

Fig. 5. Mean daily air temperatures from December 2008 to September 2010, mean C-band backscatter (average of four pixels neighbouring the temperature sensors) and near-surface snow temperatures at the Crater Lake test site at dates of ASAR scenes.

Fig. 6. Time-lapse photos of the Crater Lake test site showing three days with different snow conditions: 20 December 2008 – snow-free and dry ground, used as reference scene in the multitemporal approach; 5 December 2009 – snow-covered at the test site, with wet snow (note also droplets on the screen shield); 22 August 2009 – snow covered with dry snow. The yellow circles show the location of georeferenced markers. The outer frame and lines drawn over the relief were used for reference of relief features in the image, since owing to wind the camera shifted slightly during the study period.

very detailed scales. It is also possible that noise remained after speckle filtering.

Figure 7 shows backscattering at each study date v. the snow-free reference scene (20 December 2008). The values which are plotted were obtained for each marker by averaging the values of the closest four pixels. No clear threshold between markers with differing snow conditions is observable. Dry snow scenes generally show higher backscatter, but several scenes with dry snow show lower backscatter than the reference scene. This fact should be related to signal penetration into the snow surface and to sub-nivean roughness effects. Despite the lack of a clear threshold, scenes with wet snow show lower backscatter than the reference snow-free scene, which indicates the potential of the comparison for mapping wet snow cover.

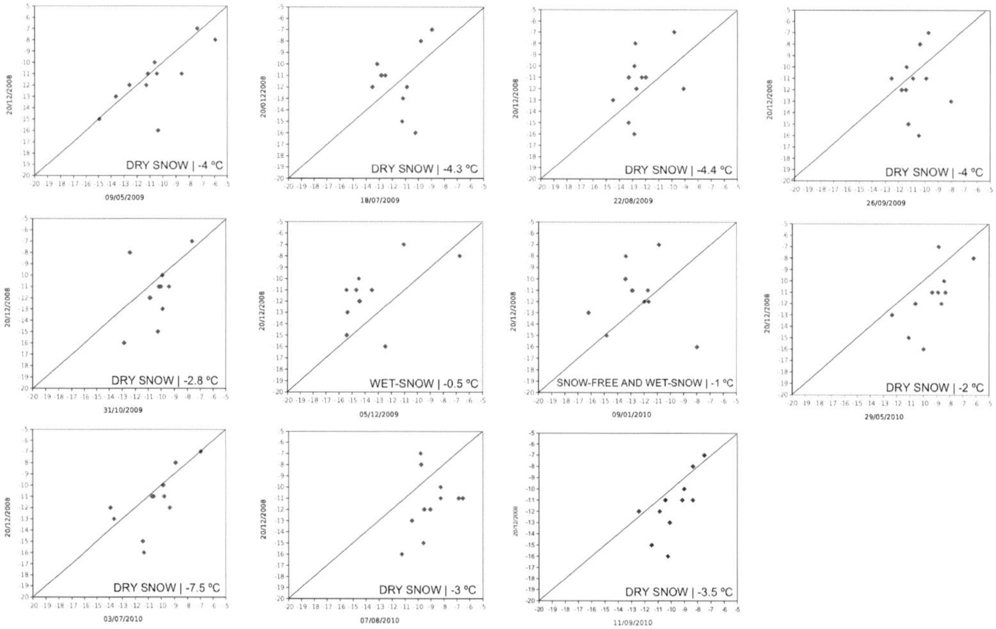

Fig. 7. Scatter plot of backscattering obtained at markers from individual scenes v. reference snow-free scene of 20 December 2008 in the Crater Lake area. Wet-snow scenes show a shift towards higher backscatter in the reference image. Dry snow scenes tend to show lower backscatter, but there are several exceptions.

Table 1. *Envisat ASAR IMP radar backscatter statistics at the Crater Lake Square site*

	Snow conditions	Backscatter (dB)	STD (dB)	Maximum (dB)	Minimum (dB)	Range (dB)	Air temperature (°C)	Snow temperature (°C)	Accuracy −2 dB (%)	Accuracy −3 dB (%)
20 December 2008	Free	−11.89	2.22	−6.4	−17.74	11.34	8.5	–	–	–
9 May 2009	Dry	−11.92	2.22	−5.74	−17.85	12.11	−4.5	−3.8	83	90
18 July 2009	Dry	−12.12	2.07	−5.07	−116.59	11.52	−1.8	−4.3	74	82
22 August 2009	Dry	−12.73	2.03	−7.66	−17.56	9.90	−1.3	−4.4	72	79
27 September 2009	Dry	−11.97	1.91	−7.29	−18.32	11.03	−6.7	−3.9	75	89
31 October 2009	Dry	−11.34	1.86	−4.76	−15.9	11.14	−5.4	−2.8	83	91
5 December 2009	Wet	−15.21	3.03	−6.25	−24.06	17.81	−1.8	−1.3	65	49
9 January 2010	Free/wet	−12.82	2.05	−8.04	−18.00	9.96	−0.8	–	–	–
29 May 2010	Dry	−11.50	2.3	−5.83	−18.08	12.25	−0.4	−11.5	79	89
3 July 2010	Dry	−11.49	2.23	−5.90	−17.62	11.72	−8.8	−6.8	84	91
7 August 2010	Dry	−11.42	2.1	−6.48	−16.53	10.05	−2.3	−2.8	81	88
11 September 2010	Dry	−10.51	2.11	−5.60	−16.54	10.94	−1.6	−3.3	94	97

Areal analysis of radar backscattering at Crater Lake

The poor results of the marker approach led us to enlarge the test area to a 190 × 190 m square around the Crater Lake CALM site (SqCL) and to analyse the spatial variability of the backscattering and its ability to pick the terrain characteristics signal at a coarser spatial scale. The mostly flat characteristics of the central part of the area avoid problems of layover and foreshortening (Fig. 3). While above we aimed at a resolution of 10–20 m, here we evaluated the capacity for discriminating at a coarser resolution, accounting for the variability of radar backscattering inside SqCL, but we still analysed the full-image resolution. The backscattering values were obtained from the individual ASAR scenes following the procedure described in the methodology. SqCL was imaged using time-lapse photos, allowing for identification of the snow-covered v. snow-free surface.

Radar backscatter values showed a high spatial variability in all scenes inside the SqCL site (Table 1). The standard deviation ranged from 1.9 to 3.0 dB at individual scenes, with absolute values of intrascene backscattering ranging from 9.9 to 17.8 dB. The range was similar in most scenes (10–12 dB), but 5 December 2009 showed much larger figures (17.8 dB).

Despite the relatively small number of scenes from December to April, the two scenes for this period in 2009 showed decreases in backscattering to average values of −15.2 dB (5 December 2009) and −12.8 dB (9 January 2010; Table 1). From May to November and also on 20 December 2008 the average backscatter was −10.5 to −12.7 dB. A drop in late spring-summer is therefore noticeable, with a clear shift towards lower values (Fig. 8).

The spatial analysis of the backscattering inside SqCL shows that the standard deviation tends to increase irregularly from SE to NW (Fig. 9a, see arrow). This indicates that the variability is lowest in the SE part SqCL, which relates to more stable snow cover. The arithmetic mean and maximum values show highest values in a patch in the SW part of the site (Fig. 9b, d). The same patch is visible for the minimum, but it is diffuse (Fig. 9c).

Comparison of radar backscattering with a snow-free scene

In order to evaluate on the possibility of applying a multitemporal approach for snow mapping, we used the snow-free scene of 20 December 2008 as reference and used the SqCL area for spatial analysis. The relatively small area, small altitudinal range and gentle slopes of SqCL allow us to assume that the snow wetness conditions should not vary significantly within each scene.

Figure 10 shows a plot of the pixel-by-pixel differences in backscattering between individual scenes and the reference scene, using the percentage of area of the SqCL as a measure on the *y*-axis. Backscattering differences to the reference scene peak near 0 dB in most scenes, with nearly symmetrical distributions (Fig. 10). The exceptions are 5 December 2009, with a distribution shifted towards negative values, peaking at −3 dB and with a negative tailing, and 9 January 2010, which peaks at −2 dB but with a more symmetrical distribution.

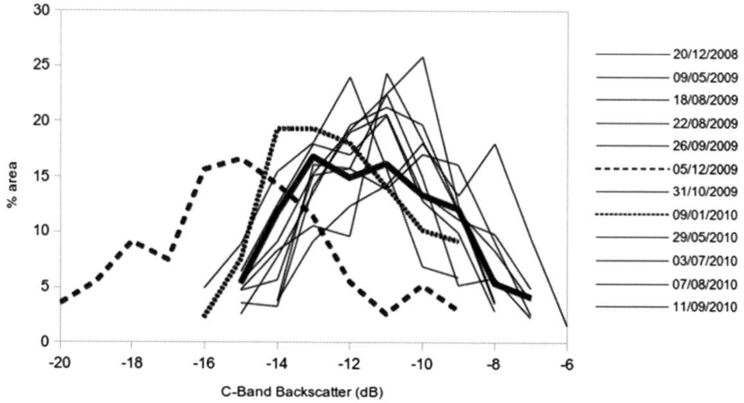

Fig. 8. Backscatter at Crater Lake Square test site: 5 December 2009 is a wet snow scene; 9 January 2010 is a scene with a mix of bare ground and wet snow; 20 December 2008 is a scene with bare ground. All other dates show dry snow.

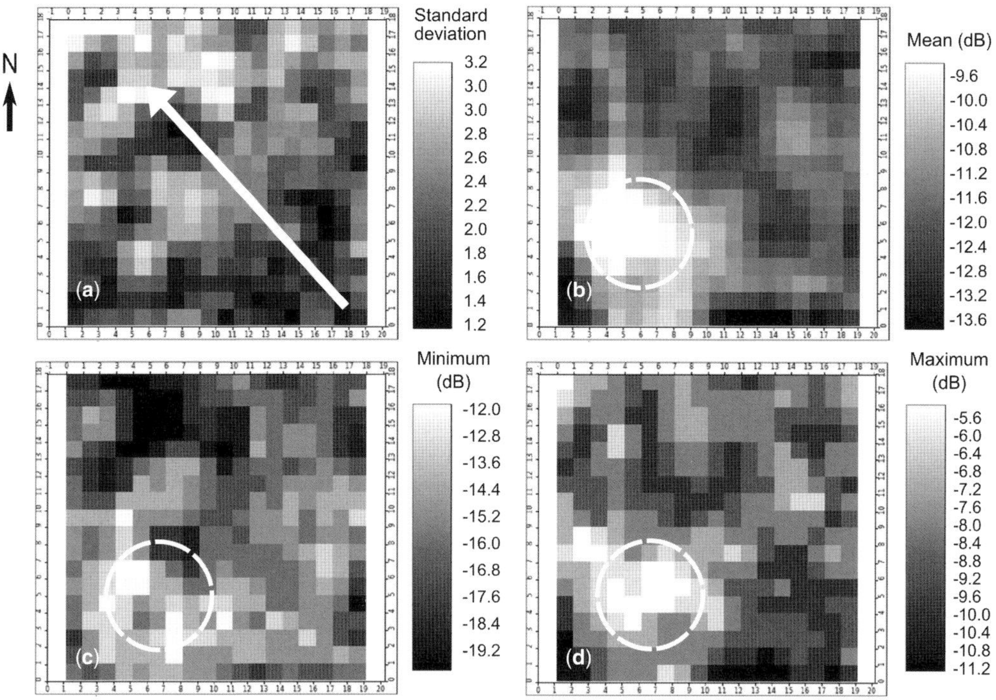

Fig. 9. Spatial variability of backscattering in Crater Lake Square test site.

Discussion

Radar backscattering and snow conditions

The detailed-scale (using markers) evaluation of the accuracy of the C-band scenes for detecting snow cover variability and wet snow shows poor results. Averaging the backscatter of four adjacent pixels showed no clear relation to the snow conditions at the markers, suggesting that the between-pixel noise disturbs the classification potential.

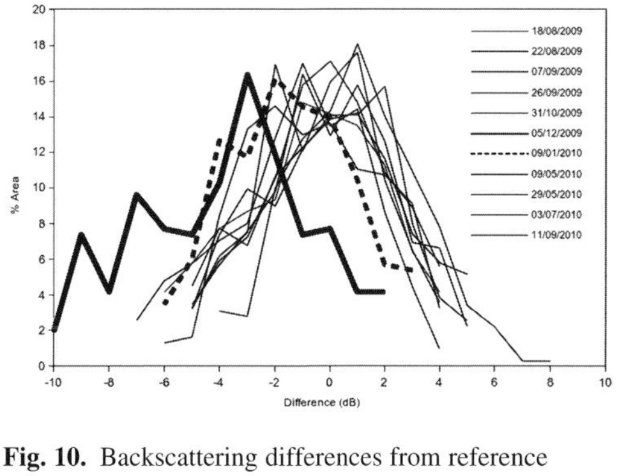

Fig. 10. Backscattering differences from reference snow-free scene at Crater Lake.

Given the very homogeneous topography and ground material at the test site, local roughness effects should be small and should not dominate in the signal.

The results from the analysis of SqCL at a coarser scale show that the backscatter in the snow-free scene is similar to the values of most dry snow-covered scenes. This suggests that the radar signal crosses the dry snow cover. The shift towards lower backscattering values on 5 December 2009 reflects the wet snow conditions. Intermediate values, between those of the wet snow and the dry snow scenes on 9 January 2010, reflect snow melt conditions, already with a significant area of bare ground present. This is confirmed by the time-lapse photos of SqCL that show the melting taking place. This dependence of the radar signal on the melting snow is also noticeable when plotting it against near-surface snow temperature, with a sharp decrease in dB when temperatures rises to -1 °C (± 0.5 °C owing to sensor accuracy), reflecting melting at the snow surface (Fig. 11). The near-surface snow temperature was obtained from the nearest ibutton below the snow surface, which could be buried up to 10 cm below the surface.

The analysis of the frequency distribution curves of backscattering at the SqCL test site suggests a threshold value at values close to -13 to

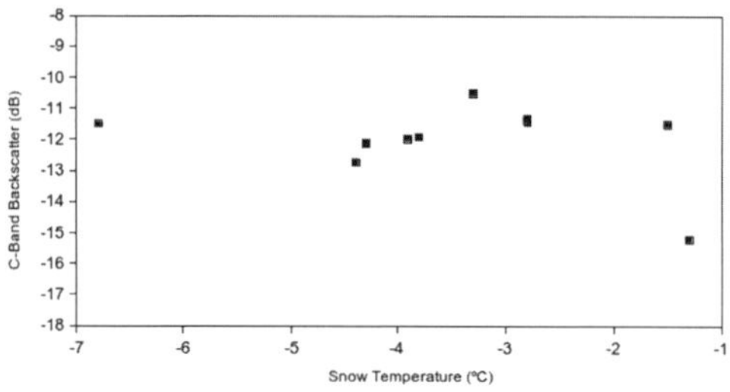

Fig. 11. Near-surface snow temperature v. average C-band backscatter at the Crater Lake test site.

−14 dB, enabling the detection of wet snow (Fig. 8). Application of these thresholds to the scenes of 5 December 2009, 20 December 2008, 9 January 2010 and 29 May 2010 (a seasonal sequence, but from different years) shows the altitudinal migration of the wet snow zone, starting from low altitude (5 December 2009) and at the highest altitude in the scene of 20 December 2008 (Fig. 12a). It is not yet possible to evaluate the overall accuracy of the application of the classification to the whole island; however, such an insight would be useful to determine differences in snow conditions between scenes and should be very useful for intermonthly and interannual comparisons of differences in the snow melt regime. The scenes shown as examples indicate an earlier melt in 2008–2009 than in 2009–2010, which is in agreement with the snow data shown in Figure 4. The scene of 29 May 2010 shows a relatively high area of wet snow pixels, which should be an overestimation.

Difference between snow-free scene and snow conditions

The last two columns of Table 1 show the results of a pixel-by-pixel assessment of the accuracy of the thresholds of −2 and −3 dB when compared with the reference snow-free scene inside the SqCL test area. The percentages account for the number of correctly classified pixels and allow for the estimation of accuracy of the thresholds at an upscaling to the regional scale. For example, if a majority filter of 190 × 190 m is applied to the original 12.5 m resolution scene, for the whole of Deception Island, percentages of accuracy above 50% show a well-classified pixel at the regional scale. This is based on the assumption that SqCL is representative of the snow conditions of Deception Island.

The results suggest that −2 dB is a valid threshold for such a regional scale approach. The −2 dB threshold in the dry snow scenes resulted in a wrong classification of 6–28% of pixels as wet snow, while a threshold of −3 dB improved the result to 3–21% (Table 1).

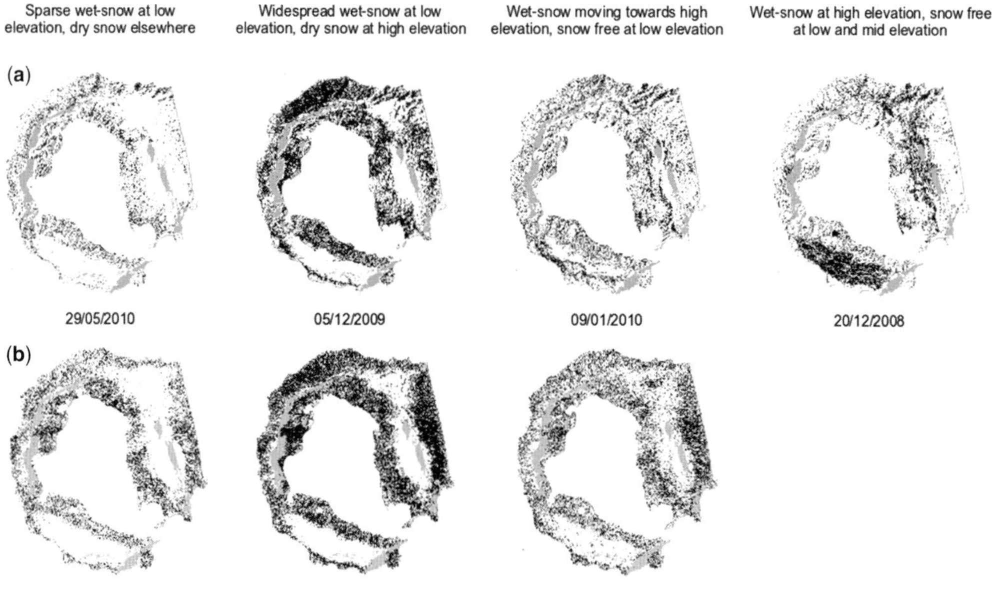

Fig. 12. Distribution of wet snow (black) at Deception Island on different dates using the backscattering threshold of −14 dB (**a**) and a −2 dB threshold in comparison with the snow-free scene of 20 December 2008 (**b**). Grey area indicates the layover not visible to the radar.

For the wet snow scene of 5 December 2009, a −2 dB threshold resulted in 65% of the area being classified as wet snow, while with −3 dB, only 49% was accurately classified. It is noteworthy, however, that we have no data indicating that all pixels in this scene show wet snow, but the small altitude range allows us to assume this. For regional applications to the whole island, if used together with an air temperature model dependent on the DEM, such a threshold might mitigate the poor classification in the winter scenes (Braun 2001).

Figure 13 shows the snow cover at SqCL derived from the time-lapse photo of 5 December 2009 and the wet snow pixels classified using the −2 dB threshold. The results show the spatial limitations of the ASAR image for high-resolution snow mapping. Even if the overall statistics of SqCL show good results, detecting the snow patterns in a pixel-by-pixel analysis is not possible and only a very rough correspondence is identifiable between the patterns. The snow-free area in the NW corner agrees with differences greater than −2 dB and most of the snow-covered area agrees with the wet snow classification. It is possible that part of the area with snow cover is dry snow, which would explain its classification as greater than −2 dB. The snow-free fringe on the northern part of the test-site was not detected in the satellite scene and shows a mixed classification.

This data indicates that the Envisat ASAR Image Precision mode imagery will only show acceptable results for kilometre-scale analysis at the landscape level. Details below at least *c.* 190 × 190 m, the approximate size of SqCL, will not be accurately detected.

The application of the −2 dB threshold to scenes of Deception Island (Fig. 12b) shows a similar general pattern to the non-transformed backscattering presented above (Fig. 12a). SqCL showed very significant noise in the imagery, but a general overview of the wet snow distribution becomes apparent at the regional scale. Altitudinal control on snow melt is shown to be the main factor and the imagery can provide a useful insight into the snow-melt regime and may even provide a basis for interannual comparison by analysing a larger Envisat dataset. The same limitations indicated for the backscattering scenes occur in these scenes, but possibly the overestimation of wet snow is larger, since in the 29 May 2010 scene the wet snow area is increased.

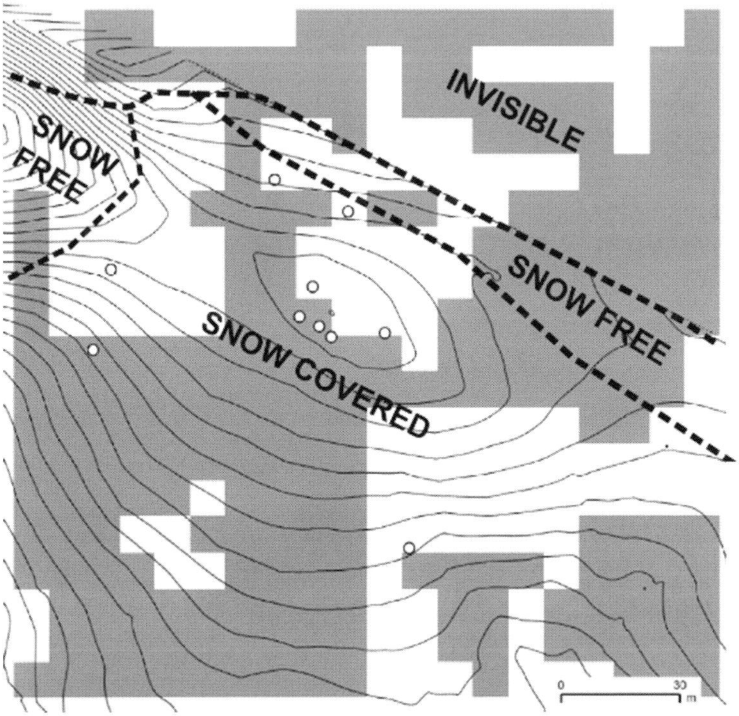

Fig. 13. Classification of snow cover using a threshold of −2 dB below snow-free scene and observed snow cover at the Crater Lake test site.

Conclusions

Radar backscattering in the C-Band has been used by several authors for snow cover mapping, but different regional thresholds have been identified. The method presented here was applied for the first time to the ice-free areas of a volcanic island in the Maritime Antarctic Region. The analysis of 12 Envisat ASAR Image Precision Mode scenes of Deception Island from December 2008 to September 2010 and comparison with ground truth data from a test site at Crater Lake allowed for insight into the quality of the product.

Despite the good spatial resolution of the scenes at 12.5 m, its applicability for detecting wet snow was shown only to be possible at much lower resolutions. Noise, but possibly also the influence of subnivean features on the signal, was found to be very high, resulting in a heterogeneous signal that varied significantly in short distances. Analysis of time-lapse photos with precisely georeferenced markers in the terrain showed poor results with a lack of relationship between backscattering and snow conditions. It was not possible to accurately assess if this was due to poor georeferencing of the image or to image noise or subnivean effects. Lakes in the SAR scenes showed a good visual fit with the DEM used for geocoding, which reduced the possibility for georeferencing error.

The analysis of backscatter descriptive statistics in a larger area of *c.* 190 × 190 m at the Crater Lake test site and comparison with time-lapse photos, as well as with near-surface snow temperatures, allowed for better results. Images with bare ground or with dry snow cover showed the highest backscattering values, with average values close to −10.5 to −12.8 dB. Wet snow showed a clear shift towards lower values, peaking at −15 dB. A threshold of −13 to −14 dB was identified between dry snow/bare ground and wet snow scenes at Crater Lake. As suggested by several authors, using the backscatter difference from a reference snow-free scene usually provides better classification results. A good threshold for the Crater Lake study site ranges from −2 to −3 dB, but should only be applied for the analysis of snow cover in kilometre-size areas.

The analysis of scenes for the whole Deception Island, which has a diameter of *c.* 15 km, for different dates in the melting season shows promising results, allowing for a future interannual assessment of the snow melt regime. The control of altitude in the development of melt is quite clear and the timing of melt is also different between years, 2008–2009 showing an earlier snow melt than 2009–2010.

This approach using ground truthing in a remote area shows that, despite the relatively easy use of ASAR imagery, it has to be used with care, since the results show significant heterogeneity. However, the backscattering thresholds that we have found are within the range of values found by other authors. This type of product should only be used for a kilometre-scale analysis of the spatial patterns of snow melt.

This work was supported by the Portuguese Science Foundation-funded project Permafrost and Climate Change in the Maritime Antarctic (PERMANTAR-2 PTDC/AAC-CLI/098885/2008) and by the Spanish Antarctic Program project PERMAPLANET. Envisat/ASAR data were obtained through the ESA Category-1 project: Snow Cover Characteristics and Regime in the South Shetlands (Maritime Antarctic)–SnowAntar. The work was possible owing to the collaboration of several colleagues who helped with field data collection and instrument maintenance. We greatly appreciate the support of D. Tomé, J. J. Blanco, M. Neves, M. A. de Pablo, M. A. Hidalgo and V. Batista. We warmly thank the Portuguese Polar Program, the Spanish Antarctic Program, the personnel at the Gabriel de Castilla Station and the Spanish Navy ship *Las Palmas*. One anonymous referee and the two editors are thanked for the significant comments that contributed to improving the final version of the manuscript.

References

Baghdadi, N., Gauthier, Y. & Bernier, M. 1997. Capability of multitemporal ERS-1 SAR data for wet-snow Mapping. *Remote Sensing of Environment*, **60**, 174–186.

Baghdadi, N., Fortin, J.-P. & Bernier, M. 1999. Accuracy of wet snow mapping using simulated Radarsat backscattering coefficients from observed snow cover characteristics. *International Journal of Remote Sensing*, **20**, 2049–2068.

Bernier, M., Fortin, J.-P., Gauthier, Y., Gauthier, R., Roy, R. & Vicent, P. 1999. Determination of snow water equivalent using RADARSAT SAR data in eastern Canada. *Hydrological Process*, **13**, 3041–3051.

Bockheim, J., Vieira, G. *et al.* 2013. Climate warming and permafrost dynamics in the Antarctic Peninsula region. *Global and Planetary Change*, **100**, 215–223.

Braun, M. 2001. *Ablation on the ice cap of King George Island (Antarctica)*. Doctoral thesis at the Faculty of Earth Sciences Albert-Ludwigs, Universität Freiburg, 165.

Koskinen, J. T. & Pulliainen, J. T. 1997. The use of ERS-1 SAR data in snow melt monitoring. *IEEE Transactions on Geoscience and Remote Sensing*, **35**, 601–610.

Lewis, A. J. & Henderson, F. M. 1998. Radar fundamentals: the geosciences perspective. *In*: Lewis, A. & Henderson, J. (eds) *Principles and Applications of Imaging Radar*. 3rd edn. Manual of Remote Sensing, **2**, 131–176, John Wiley & Sons Inc., New York.

Magagi, R. & Bernier, M. 2003. Optimal conditions for wet snow detection using RADARSAT SAR data. *Remote Sensing of Environment*, **84**, 221–233.

Malnes, E. & Guneriussen, T. 2002. Mapping of snow covered area with Radarsat in Norway. *Proceedings to IGARSS*, Toronto.

Malnes, E. & Guneriussen, T. 2003. Comparison of satellite SAR based Snow Cover Area (SCA) algorithms. *EARSel Workshop*, Bern.

Nagler, T. & Rott, H. 1997. The application of ERS-1 SAR for snowmelt runoff modeling area. *Remote Sensing and Geographic Information Systems for Design and Operation of Water Resources Systems. Proceedings of Rabat Symposium*, IAHS 242.

Nagler, T. & Rott, H. 2000. Retrieval of wet snow by means of multitemporal SAR data. *IEEE Transactions on Geoscience and Remote Sensing*, **38**, 754–765.

Ramos, M., Vieira, G. et al. 2007. *Permafrost and Active Layer Monitoring in the Maritime Antarctic: Preliminary Results from CALM sites on Livingston and Deception Islands*. US Geological Survey and the National Academies, Reston, VA, USGS OF-2007-1047, Short Research Paper **070**, http://dx.doi.org/10.3133/of2007-1047.srp070

Rees, W. G. 2006. *Remote Sensing of Snow and Ice*. Taylor & Francis, London.

Rees, W. S. & Steel, M. 2001. Radar backscatter coefficients and snow detectability for upland terrain in Scotland. *International Journal of Remote Sensing*, **22**, 3015–3026.

Rosich, B. & Meadows, P. 2004. *Absolute calibration of ASAR Level 1 products*. ESA/ESRIN, ENVI-CLVL-EOPG-TN-03-0010, Issue 1, Revision.

Shi, J. & Dozier, J. 1995. Inferring snow wetness using C-band data from SIR-C's polarimetric synthetic aperture radar. *IEEE Transactions on Geoscience and Remote Sensing*, **33**, 905–914.

Slaymaker, O. & Kelly, R. E. J. 2007. *The Cryosphere and Global Environmental Change*. Blackwell, Oxford.

Small, D. 2011. Flattening Gamma: radiometric terrain correction for SAR imagery. *IEEE Transactions on Geoscience and Remote Sensing*, **49**, 3081–3093.

Small, D. & Schubert, A. 2008. *Guide to ASAR Geocoding*. RSL-ASAR-GC-AD, Issue 1.0.

Smellie, J. L. & López-Martínez, J. 2002. Introdution to the geology and geomorphology of deception island. *In*: López Martínez, J., Smellie, J. L., Thomson, J. W. & Thomson, M. R. A. (eds) *Geology and Geomorphology of Deception Island*. British Antarctic Survey, Cambridge, BAS Geomap Series, 1–6.

Smith, S. & Brown, J. 2009. *Permafrost and Seasonally Frozen Ground–T7. Assessment of the Status of the Development of the Standards for the Terrestrial Essential Climate Variables. GTOS 62*. FAO, Rome.

Valenti, L., Small, D. & Meier, E. 2008. Snow cover monitoring using multi-temporal Envisat/Asar data. *Proceedings of 5th EARSeL LISSIG (Land, Ice, Snow) Workshop*, Bern.

Vieira, G., Bockheim, J. et al. 2010. Thermal state of permafrost and active-layer monitoring in the Antarctic: advances during the International Polar year 2007–09. *Permafrost and Periglacial Processes*, **21**, 182–197.

Williams, P. J. & Smith, M. W. 1989. *The Frozen Earth: Fundamentals of Geocryology*. Cambridge University Press, Cambridge.

Index

Page numbers in *italic* denote figures. Page numbers in **bold** denote tables.